Consumer Information

Education

Environment

Finance

General Interest

(continued on inside back cover)

A SURVEY OF MATHEMATICS WITH APPLICATIONS

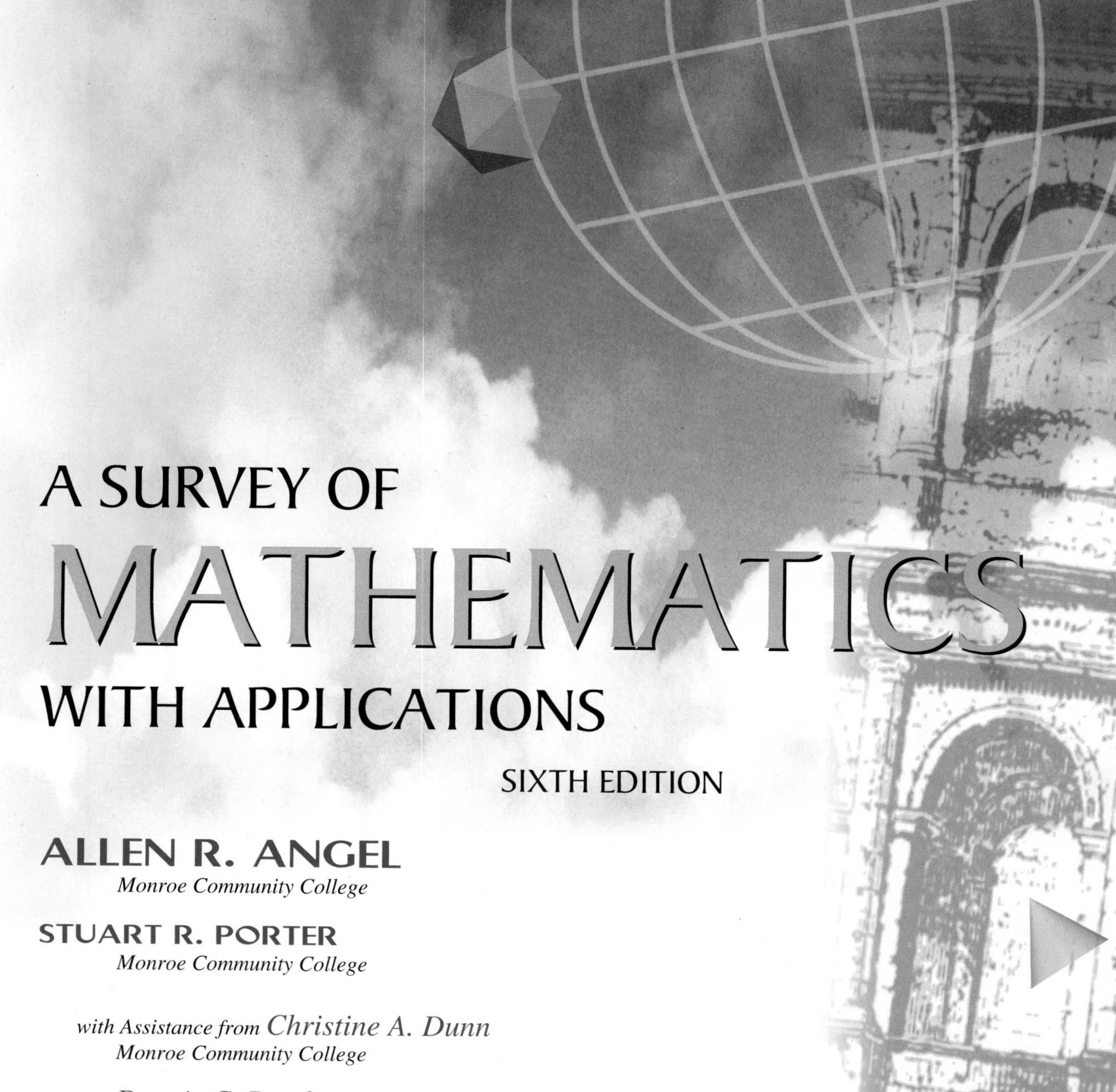

A SURVEY OF MATHEMATICS WITH APPLICATIONS

SIXTH EDITION

ALLEN R. ANGEL
Monroe Community College

STUART R. PORTER
Monroe Community College

with Assistance from Christine A. Dunn
Monroe Community College

and Dennis C. Runde
Manatee Community College

Addison Wesley

Boston San Francisco New York
London Toronto Sydney Tokyo Singapore Madrid
Mexico City Munich Paris Cape Town Hong Kong Montreal

Senior Editor: Bill Poole
Editorial Project Manager: Rachel Reeve
Production Services: UG / GGS Information Services, Inc.
Cover and Text Designer: Barbara T. Atkinson
Senior Marketing Manager: Carter Fenton
Manufacturing Buyer: Evelyn Beaton
Senior Prepress Supervisor: Caroline Fell
Composition and Prepress Services: UG / GGS Information Services, Inc.
Illustrations: Tech Graphics, James A. Bryant
Cover Photo: Tony Stone Images

For permission to use copyrighted material, grateful acknowledgment is made to the copyright holders in the Credits section beginning on page C-1, which is hereby made part of this copyright page.

Library of Congress Cataloging-in-Publication Data

Angel, Allen R., 1942–
A survey of mathematics with applications / Allen Angel, Stuart Porter.—6th ed.
p. cm.
ISBN 0-201-38407-8
1. Mathematics. I. Porter, Stuart R., 1932– II. Title.
QA39.2 .A54 2001
510—dc21 99-056911

Reprinted with corrections.

2 3 4 5 6 7 8 9 10—DOW—0 2 0 1

To my wife, Kathy Angel (photo on page 73)

A.R.A.

CONTENTS

MATH: IT'S ALL AROUND US!

We present *A Survey of Mathematics with Applications*, sixth edition, with that vision in mind. Our primary goal in writing this book was to give students a text that they can read, understand, and enjoy while learning how mathematics affects the world around them. Numerous applied examples motivate topics. A variety of interesting applied exercises demonstrate the real-life nature of mathematics and its importance in the students' lives.

The text is intended for students who require a broad-based general overview of mathematics, especially those majoring in the liberal arts, elementary education, the social sciences, business, nursing, and allied health fields. It is particularly suitable for those courses that satisfy the minimum competency requirement in mathematics for graduation or transfer.

New and Expanded Features

In this edition we made several important improvements in presentation.

- The interior design has been modified and many new photographs were added to make the book more inviting and motivational.
- The Problem Solving/Group Activity Exercises from the Fifth Edition are now titled Challenge Problem/Group Activity Exercises, and their number has been increased.
- The exercise sets have been redesigned and reclassified to include: Concept/Writing Exercises; Practice the Skills Exercises; Problem Solving Exercises; Challenge Problem/Group Activity Exercises; and Research Activities.
- Sources have been added, up-to-date tables, graphs and charts make the material more relevant and encourage students to read graphs and analyze data.
- Mathematical Modeling is now introduced where appropriate.
- Approximately 40% of the exercises are new.
- The number of writing exercises was increased significantly.
- The number of Research Activities was increased.
- The number of examples was increased throughout the text to promote student understanding.

CONTENT REVISION

In addition we revised and expanded certain topics to introduce new material and to increase understanding.

CHAPTER 1, Critical Thinking Skills, was updated with exciting and current examples and exercises.

CHAPTER 3, Logic, now includes examples and exercises that make use of circle and bar graphs. Material on the negation of the conditional statement also has been added.

CHAPTER 5, Number Theory and the Real Number System, was expanded to include coverage of improper fractions and mixed numbers. The most current number theory information (largest prime number, most accurate value of pi, etc) has been included. The use of the scientific calculator has been expanded.

CHAPTER 6, Algebra, Graphs and Functions, has expanded coverage of exponential functions, including additional examples of exponential growth and decay. Section 6.10, Functions and Their Graphs, has been reorganized for greater clarity. Also, material on using a graphing calculator has been included.

CHAPTER 7, Systems of Linear Equations and Inequalities, covers linear programming in more detail. Coverage of using a graphing calculator to find the solution to a system of linear equations is now included.

CHAPTER 8, The Metric System, now includes many examples and interesting photographs of real life (metric) situations taken from around the world.

CHAPTER 9, Geometry, includes expanded coverage of the Klein bottle, the Jordan curves, the 4 color map theorem, and fractals.

CHAPTER 11, Consumer Mathematics, includes current interest rates and updated information on items which may be of interest to students including updated material on sources of credit, and mutual funds. There is also a greater variety of examples and exercises.

CHAPTER 12, Probability, has a greater variety of examples and exercises. More examples and exercises have been added that deal with real life situations.

CHAPTER 13, Statistics, now includes stem-and-leaf displays in the section on statistical graphs.

CONTINUING FEATURES

Several features appear throughout the book, adding interest and provoking thought.

- **Problem Solving** Beginning in Chapter 1, students are introduced to problem solving and critical thinking. The theme of problem solving is then continued throughout the text, and special problem-solving exercises are presented in the exercise sets.
- **Critical Thinking Skills** In addition to a focus on *Problem Solving*, the book also features sections on *Inductive Reasoning* and the important skills of *Estimation* and *Dimensional Analysis*.
- **Profiles in Mathematics** Brief historical sketches and vignettes present the stories of people who have advanced the discipline of mathematics.
- **Chapter Openers** Interesting and motivating photo essays introduce each chapter and illustrate the real-world nature of the chapter topics.
- **Did You Know . . .** These colorful, engaging, and lively boxed features highlight the connections of mathematics to history, to the arts and sciences, to technology, and to a broad variety of disciplines.

INSTRUCTOR'S SUPPLEMENTS

Instructor's Solutions Manual

ISBN 0-201-61324-7

This manual contains detailed, worked-out solutions to all the exercises in the text, and answers to Group Projects.

Instructor's Edition

ISBN 0-201-61326-3

This version of the text includes answers to all of the problems in addition to all of the material found in the student edition.

Instructor's Testing Manual

ISBN 0-201-61325-5

The testing manual includes three alternative tests per chapter. These items may be used as actual tests or as references for creating tests with or without a computer.

TestGen-EQ with QuizMaster-EQ

ISBN 0-201-61330-1

TestGen-EQ is a computerized test generator with algorithmically defined problems organized specifically for this textbook. Its user-friendly graphical interface enables instructors to select, view edit and add test items, then print tests in a variety of fonts and forms. A built-in question editor gives the user the power to create graphs, import graphics, insert mathematical symbols and templates, and insert variable numbers or text. An "Export to HTML" feature lets instructors create practice tests that can be posted to a Web site. Tests created with TestGen-EQ can be used with QuizMaster-EQ, which enables students to take exams on a computer network. QuizMaster-EQ automatically grades the exams, stores results on disk, and allows the instructor to view or print a variety of reports for individual students, classes or courses. Contact your Addison Wesley Longman Sales Consultant.

InterAct Math Plus Software

ISBN 0-201-72140-6

InterActMath Plus combines course management and on-line testing with the features of the basic InterAct Math Tutorial Software to create an invaluable teaching resource. Contact your Addison Wesley Longman Sales Consultant.

Videotapes

ISBN 0-201-61328-X

Videotapes, which correlate to each important topic in the book, are available to departments. Contact your Addison Wesley Longman Sales Consultant.

STUDENT'S SUPPLEMENTS

Student's Solutions Manual

ISBN 0-201-61323-9

This manual contains detailed worked-out solutions to all the odd-numbered section exercises and to all Review and Chapter Test exercises. Students will find this manual very helpful.

Web site

http://www.awl.com/angel

The Web site contains additional resources for instructors and students.

Guide to CLAST Mathematical Competency (State of Florida)

ISBN 0-201-61327-1

This guide provides all the necessary material to help students prepare for the computational portion of the CLAST test. It includes worked-out examples and practice for CLAST skills, as well as a practice test. Optional topics in trigonometry are provided for those who wish to brush up in this area as well.

InterAct Math Tutorial Software

ISBN 0-201-61332-8

InterAct Math Tutorial Software has been developed and designed by professional software engineers working closely with a team of experienced math educators. InterAct math Tutorial Software includes exercises that are lined with every objective in the textbook and require the same computational and problem-solving skills as their companion exercises in the text. Each exercise has an example and an interactive guided solution that are designed to involve students in the solution process and to help them identify precisely where they are having trouble. The software recognizes common student errors and provides students with appropriate customized feedback. With its sophisticated answer recognition capabilities, InterAct Math Tutorial Software recognizes appropriate forms of the same answer for any kind of input. It also tracks student activity and scores for each section, which can then be printed out. The software is free to qualifying adopters or can be bundled with books for sale to students.

ACKNOWLEDGMENTS

I would like to thank my wife, Kathy, and my children Bob and Steve. Kathy helped with the project in many ways, including typing parts of the manuscript. She was very supportive and was always willing to lend a hand when I needed it.

Stuart Porter did not participate in this revision, but I would like to thank him for his contributions to the past editions of this book. For this edition, I received excellent help from Christine Dunn of Monroe Community College and Dennis Runde of Manatee Community College. Chris and Dennis contributed in a great many ways, and I appreciate their conscientious efforts in behalf of the project.

I would also like to thank Lauri Semarne and Alexis T. Mogill for accuracy checking of the text and answers.

There are many people at Addison Wesley Longman who deserve thanks. I would like to thank all those listed on the Library of Congress categorizing page. In particular, I would like to thank Bill Poole, Senior Acquisitions Editor; Rachel Reeve, Project Manager; Peggy McMahon, Senior Production Supervisor; Carter Fenton, Marketing Manager; Barbara Atkinson, Senior Designer; Bobbie Lewis, Developmental Editor; and Karen Guardino, Managing Editor. I would also like to thank Danielle Meckley and Terri O'Prey of UG / GGS Information Services, Inc., for their assistance as Production Editors for this project.

Gary Egan and Aimee Calhoun of Monroe Community College also deserve my thanks for the excellent work they did on the *Student's Solution Manual* and the *Instructor's Solution Manual*.

Finally, I would like to thank the reviewers from all editions of the book, and all the students who have offered suggestions for improving the book. A list of reviewers for all editions of this book follows. Thanks to all of you for helping to make *A Survey of Mathematics with Applications* the most successful Liberal Arts book in the country.

Allen R. Angel

Reviewers for This and Previous Editions

Frank Asta, *College of DuPage, IL*

Hughette Bach, *California State University–Sacramento*

Madeline Bates, *Bronx Community College, NY*

Rebecca Baum, *Lincoln Land Community College, IL*

Vivian Baxter, *Fort Hayes State University, KS*

Una Bray, *Skidmore College, NY*

David H. Buckley, *Polk Community College, FL*

Robert C. Bueker, *Western Kentucky University*

Carl Carlson, *Moorhead State University, MN*

Kent Carlson, *St. Cloud State University, MN*

Donald Catheart, *Salisbury State College, MD*

*Joseph Cleary, *Massasoit Community College, MA*
Donald Cohen, *SUNY Ag & Tech College at Cobleskill, NY*
David Dean, *Santa Fe Community College, FL*
Charles Downey, *University of Nebraska*
Ruth Ediden, *Morgan State University, MD*
*Lee Erker, *Tri-County Community College, NC*
Karen Estes, *St. Petersburg Junior College, FL*
Kurtis Fink, *Northwest Missouri State University*
Raymond Flagg, *McPherson College, KS*
Penelope Fowler, *Tennessee Wesleyan College*
Gilberto Garza, *El Paso Community College, TX*
Judith L. Gersting, *Indiana University–Purdue University at Indianapolis*
Lucille Groenke, *Mesa Community College, AZ*
John Hornsby, *University of New Orleans, LA*
Nancy Johnson, *Broward Community College, FL*
Daniel Kimborowicz, *Massasoit Community College, MA*
Mary Lois King, *Tallahassee Community College, FL*
David Lehmann, *Southwest Missouri State University*
Peter Lindstrom, *North Lake College, TX*
James Magliano, *Union College, NJ*
*Yash Manchanda, *East Los Angeles College & Fullerton College, CA*
Don Marsian, *Hillsborough Community College, FL*
Marilyn Mays, *North Lake College, TX*
Robert McGuigan, *Westfield State College, MA*
Maurice Monahan, *South Dakota State University*
Julie Monte, *Daytona Beach Community College, FL*
*Karen Mosely, *Alabama Southern Community College, AL*
Edwin Owens, *Pennsylvania College of Technology*
Wing Park, *College of Lake County, IL*
Bettye Parnham, *Daytona Beach Community College, FL*
Joanne Peeples, *El Paso Community College, TX*
Nelson Rich, *Nazareth College, NY*
*Kenneth Ross, *University of Oregon, OR*
Ronald Ruemmler, *Middlesex County College, NJ*
Rosa Rusinek, *Queensborough Community College, NY*
Len Ruth, *Sinclair Community College, OH*
John Samoylo, *Delaware County Community College, PA*
Sandra Savage, *Orange Coast College, CA*
Gerald Schultz, *Southern Connecticut State University*

Richard Schwartz, *College of Staten Island, NY*
*Kara Shavo, *Mercer County Community College, NJ*
Minnie Shuler, *Chipola Junior College, FL*
*Kristin Stoley, *Blinn College–Bryan, TX*
Steve Sworder, *Saddleback College, CA*
Shirley Thompson, *Moorhead College, GA*
Alvin D. Tinsley, *Central Missouri State University*
William Trotter, *University of South Carolina*
Sandra Welch, *Stephen F. Austin State University, TX*
Joyce Wellington, *Southeastern Community College, NC*
Sue Welsch, *Sierra Nevada College*
Robert F. Wheeler, *Northern Illinois University*
Susan Wirth, *Indian River Community College, FL*
James Wooland, *Florida State University*
*Jean Woody, *Tulsa Community College, OK*
Michael A. Zwick, *Monroe Community College, NY*

*Denotes reviewers for the sixth edition.

TO THE STUDENT

Mathematics is an exciting, living study. It has applications that shape the world around you and influence your everyday life. I hope that as you read through this book you will realize just how important mathematics is and gain an appreciation of both its usefulness and its beauty. I also hope to teach you some practical mathematics that you can use in your everyday life and that will prepare you for further courses in mathematics.

My primary purpose in writing this text was to provide material that you could read, understand, and enjoy. To this end I have used straightforward language and tried to relate mathematical concepts to everyday experiences. I have also provided many detailed examples for you to follow.

The concepts, definitions, and formulas that deserve special attention have been either boxed or set in boldface type. Within each category the exercises are graded so that the more difficult problems appear at the end. The problems with exercise numbers set in color are writing exercises. At the end of most exercise sets are Challenge Problem/Group Activity exercises that contain challenging or exploratory exercises. At the end of each chapter are Group Projects which reinforce the material learned or provide related material.

Each chapter has a summary, review exercises, and a chapter test. When studying for a test, be sure to read the chapter summary, work the review exercises, and take the chapter test. The answers to the odd-numbered exercises, all review exercises, and all chapter test exercises appear in the Answer section in the back of the text. However, you should use the answers only to check your work.

It is difficult to learn mathematics without becoming involved. To be successful, I suggest you read the text carefully *and work each exercise in each assignment in detail*. Check with your instructor to determine which supplements are available for your use.

I welcome your suggestions and your comments. You may contact me at:

Allen Angel
c/o Marketing
Mathematics & Statistics
Addison Wesley Longman
One Jacob Way
Reading, MA 01867-3999

or by email at:

math@awl.com
Subject: for Allen Angel

Good luck in your adventure in mathematics!

Allen R. Angel

A SURVEY OF MATHEMATICS WITH APPLICATIONS

CHAPTER 1

Critical Thinking Skills

Life constantly presents new problems. The great inventors, scientists, scholars, politicians, and artists make their contributions to civilization by confronting and solving problems. To learn the techniques for solving problems requires practice and patience, but once the basic principles are understood, they can be applied to each new challenge.

The goal of this chapter is to help you master the skills of reasoning, estimating, and problem solving. These skills will aid you in solving the problems in the remainder of this book as well as problems that you will encounter in everyday life.

Every day, you make decisions that require you to use critical thinking skills. For example, in a drugstore, you need to choose between two bottles of shampoo. One is 6 ounces and costs $2.95, the other is 8 ounces and costs $3.50. Which bottle should you buy? Why? Sometimes solving a problem may require you to make a reasonable estimate, to look for clues, or to experiment with several possible solutions before choosing the best solution. Often, the most important part of solving a problem is just understanding what question must be answered. ■

Scientists and engineers who designed the space shuttle had to solve many problems, such as the trajectory, the correct fuel supply, surviving the heat of re-entry, and staying within budget to name just a few. If you saw the movie *Apollo 13,* you saw teams of engineers and scientists working together to solve a multitude of problems.

Designing the multibillion dollar space station, involving many countries, involves even more problem solving. Scientists from around the world are working together to solve the many problems they face.

1.1 INDUCTIVE REASONING

The goal of this chapter is to help you improve your reasoning and problem-solving skills. This section introduces inductive and deductive reasoning, which are used in problem solving. The next section introduces the concept of estimation. Estimation is a technique that can be used to determine if an answer obtained for a problem or from a calculation is "reasonable." Section 1.3 introduces and applies problem-solving techniques.

Before looking at some examples of inductive reasoning and problem solving, let us first review a few facts about certain numbers. The **natural numbers** or **counting numbers** are the numbers 1, 2, 3, 4, 5, 6, 7, 8, ... The three dots, called an **ellipsis**, mean that 8 is not the last number but that the numbers continue in the same manner. A word that we sometimes use is "divisible." If $a \div b$ has a remainder of zero, then *a is divisible by b.* The counting numbers that are divisible by 2 are 2, 4, 6, 8, ... These are called the *even counting numbers.* The numbers that are not divisible by 2 are 1, 3, 5, 7, 9, ... These are the *odd counting numbers.* When we refer to *odd numbers* or *even numbers,* we mean odd or even counting numbers.

Recognizing patterns is sometimes helpful in solving problems, as Examples 1 and 2 illustrate.

EXAMPLE 1 ***The Product of Two Odd Numbers***

If two odd numbers are multiplied together, will the product always be an odd number?

SOLUTION To answer this question, we will examine the products of several pairs of odd numbers to see if there is a pattern.

$$1 \times 3 = 3 \qquad 3 \times 5 = 15 \qquad 5 \times 7 = 35$$
$$1 \times 5 = 5 \qquad 3 \times 7 = 21 \qquad 5 \times 9 = 45$$
$$1 \times 7 = 7 \qquad 3 \times 9 = 27 \qquad 5 \times 11 = 55$$
$$1 \times 9 = 9 \qquad 3 \times 11 = 33 \qquad 5 \times 13 = 65$$

None of the products is divisible by 2. Thus, we might predict from these examples that the product of any two odd numbers is an odd number. ▲

EXAMPLE 2 ***The Sum of an Odd Number and an Even Number***

If an odd and an even number are added, will the sum be an odd or an even number?

SOLUTION Let's look at a few examples where one number is odd and the other number is even.

$$1 + 2 = 3 \qquad 9 + 6 = 15 \qquad 23 + 18 = 41$$
$$3 + 12 = 15 \qquad 5 + 14 = 19 \qquad 81 + 32 = 113$$

None of these sums is divisible by 2. Therefore, we might predict that the sum of an odd and an even number is an odd number. ▲

In Examples 1 and 2, we cannot conclude that the results are true for all counting numbers. From the patterns developed, however, we can make predictions. This type of reasoning process, arriving at a general conclusion from specific observations or examples, is called **inductive reasoning**, or **induction**.

> **Inductive reasoning** is the process of reasoning to a general conclusion through observations of specific cases.

Induction often involves observing a pattern and from that pattern predicting a conclusion. Imagine an endless row of dominoes. You knock down the first, which knocks down the second, which knocks down the third, and so on. Assuming the pattern will continue uninterrupted, you conclude that eventually all the dominoes will fall, even though you may not witness the event.

Inductive reasoning is often used by mathematicians and scientists to predict answers to complicated problems. For this reason, inductive reasoning is part of the **scientific method**. When a scientist or mathematician makes a prediction based on specific observations, it is called a **hypothesis** or **conjecture**. After looking at the products in Example 1, we might conjecture that the product of two odd numbers will be an odd number.

Examples 3 and 4 illustrate how we arrive at a conclusion using inductive reasoning.

EXAMPLE 3 *Fingerprints and DNA*

What reasoning process has led to the conclusion that no two people have the same fingerprints or DNA? This conclusion has resulted in fingerprints and DNA being used in courts of law as evidence to convict persons of crimes.

SOLUTION In millions of tests, no two people have been found to have the same fingerprints or DNA. By induction, then, we believe that fingerprints and DNA provide a unique identification and can therefore be used in a court of law as evidence. Is it possible that, sometime in the future, two people will be found who do have exactly the same fingerprints or DNA? ▲

EXAMPLE 4 *A Divisibility Conjecture*

Consider the conjecture "If the sum of the digits of a number is divisible by 3, then the number is divisible by 3." Test several numbers to determine whether the conjecture appears true or false.

SOLUTION Let's look at some numbers, the sum of whose digits are divisible by 3.

Number	Sum of the digits	Sum of the digits divided by 3	Number divided by 3
114	$1 + 1 + 4 = 6$	$6 \div 3 = 2$	$114 \div 3 = 38$
234	$2 + 3 + 4 = 9$	$9 \div 3 = 3$	$234 \div 3 = 78$
7020	$7 + 0 + 2 + 0 = 9$	$9 \div 3 = 3$	$7020 \div 3 = 2340$
2943	$2 + 9 + 4 + 3 = 18$	$18 \div 3 = 6$	$2943 \div 3 = 981$
9873	$9 + 8 + 7 + 3 = 27$	$27 \div 3 = 9$	$9873 \div 3 = 3291$

In each of the examples, we find that the sum of the digits is divisible by 3 and the number itself is divisible by 3. From these specific examples, we might be tempted to generalize that the conjecture "If the sum of the digits of a number is divisible by 3, then the number is divisible by 3" is true. ▲

DID YOU KNOW

The Eyes Tell It All

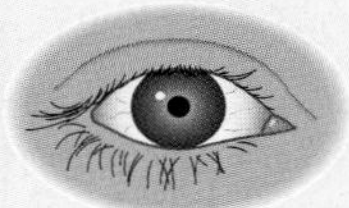

Most of us realize that thus far, no two people have been found to have the same fingerprints, so fingerprints have become a form of identification. As technology improves, so do identification techniques. Today, when you purchase a season pass at Disney World, you place your forefinger and middle finger in a device that measures and records your finger size, shape and density. Then you are given a seasonal pass. Each time you enter Disney, you put those two fingers in the device, which indicates whether you are allowed or not allowed entry under the seasonal pass that you show.

Automated teller machines (ATMs) are now experimenting with determining identity using a person's eye. When you open an account, a photo of your eye is taken and entered into a computer. When you use an ATM, a powerful camera automatically (without you even knowing it) checks the veins in your eye against the computer's files. Shortly, we will no longer need personal identification numbers (PINs) to withdraw money from our accounts. All these techniques are based on inductive reasoning. Based on past experiences, no two people have been found to have the same fingerprints; finger size, shape, and density; or eyes.

EXAMPLE 5 *Pick a Number, Any Number*

Pick any number, multiply the number by 4, add 6 to the product, divide the sum by 2, and subtract 3 from the quotient. Repeat this procedure for several different numbers and then make a conjecture about the relationship between the original number and the final number.

SOLUTION Let's go through this one together.

Pick a number:	say, 5
Multiply the number by 4:	$4 \times 5 = 20$
Add 6 to the product:	$20 + 6 = 26$
Divide the sum by 2:	$26 \div 2 = 13$
Subtract 3 from the quotient:	$13 - 3 = 10$

Note that we started with the number 5 and finished with the number 10. If you start with the number 2, you will end with the number 4. Starting with 3 would result in a final number of 6, 4 would result in 8, and so on. On the basis of these few examples, many of you would conjecture that when you follow the given procedure, the number you end with will always be twice the original number. ▲

DID YOU KNOW

An Experiment Revisited

Apollo 15 astronaut David Scott used the moon as his laboratory to show that a heavy object (a hammer) does indeed fall at the same rate as a light object (a feather). Had Galileo dropped a hammer and feather from the Tower of Pisa, the hammer would have fallen more quickly to the ground and he still would have concluded that a heavy object falls faster than a lighter one. If it is not the object's mass that is affecting the outcome, then what is it? The answer is air resistance or friction: Earth has an atmosphere, the moon does not.

The result reached by inductive reasoning is often correct for the specific cases studied but not correct for all cases. History has shown that not all conclusions arrived at by inductive reasoning are correct. For example, Aristotle (384–322 B.C.) reasoned inductively that heavy objects fall at a faster rate than light objects. About 2000 years later, Galileo (1564–1642) dropped two pieces of metal—one 10 times heavier than the other—from the Leaning Tower of Pisa in Italy. He found that both hit the ground at exactly the same moment, so they must have traveled at the same rate.

When forming a general conclusion using inductive reasoning, you should test it with several special cases to see whether the conclusion appears correct. If a special case is found that satisfies the conditions of the conjecture but produces a different result, such a case is called a **counterexample**. A counterexample proves that the conjecture is false because only one exception is needed to show that a conclusion is not valid. Galileo's counterexample disproved Aristotle's conjecture. If a counterexample cannot be found, the conjecture is neither proven nor disproven.

A second type of reasoning process is called **deductive reasoning**, or **deduction**. Mathematicians use deductive reasoning to *prove* conjectures true or false.

Deductive reasoning is the process of reasoning to a specific conclusion from a general statement.

EXAMPLE 6 *Pick a Number, n*

Prove, using deductive reasoning, that the procedure in Example 5 will always result in twice the original number selected.

SOLUTION To use deductive reasoning, we begin with the *general* case rather than specific examples. In Example 5, specific cases were used. Let's select the letter n to represent *any number.*

Pick any number: n

Multiply the number by 4: $4n$ ($4n$ means 4 times n)

Add 6 to the product: $4n + 6$

Divide the sum by 2: $\frac{4n + 6}{2} = \frac{\overset{2}{\cancel{4}}n}{\underset{1}{\cancel{2}}} + \frac{\overset{3}{\cancel{6}}}{\underset{1}{\cancel{2}}} = 2n + 3$

Subtract 3 from the quotient: $2n + 3 - 3 = 2n$

Note that, for any number n selected, the result is $2n$, or twice the original number selected. ▲

In Example 5, you may have conjectured, using specific examples and inductive reasoning, that the result would be twice the original number selected. In Example 6, we proved, using deductive reasoning, that the result will always be twice the original number selected.

SECTION 1.1 EXERCISES

CONCEPT/WRITING EXERCISES

1. **a)** List the natural numbers.
   b) What is another name for the natural numbers?

2. **a)** What does it mean to say, "a is divisible by b," where a and b represent natural numbers?
   b) List three natural numbers that are divisible by 4.
   c) List three natural numbers that are divisible by 9.

In Exercises 3–6, explain your answer in one or two sentences.

3. What is a conjecture?

4. What is inductive reasoning?

5. What is deductive reasoning?

6. What is a counterexample?

7. You try to log on a computer account using both your Social Security number and your password. You type in your Social Security number and what you believe is your password. The computer indicates that a mistake has been made and asks you to try again. You retype your Social Security number and the same password, and again you get the same message from the computer. You decide not to try again, reasoning you will get the same message from the computer. What type of reasoning are you using? Explain.

8. In the 1950s, doctors noticed that many of their lung cancer patients were also cigarette smokers. Doctors reasoned that cigarette smoking increased a person's chance of getting lung cancer. What type of reasoning did the doctors use? Explain.

PRACTICE THE SKILLS

In Exercises 9–12, use inductive reasoning to predict the next line in the pattern.

9.

1
1 1
1 2 1
1 3 3 1
↘↙↘↙↘↙
1 4 6 4 1

10. $1 \times 9 = 9$
$2 \times 9 = 18$
$3 \times 9 = 27$
$4 \times 9 = 36$

11. $1 = 1$
$1 + 2 = 3$
$1 + 2 + 3 = 6$
$1 + 2 + 3 + 4 = 10$
$1 + 2 + 3 + 4 + 5 = 15$

12. $11 \times 11 = 121$
$11 \times 12 = 132$
$11 \times 13 = 143$

In Exercises 13–16, draw the next figure in the pattern (or sequence).

13. 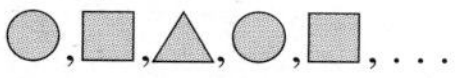 ○, □, △, ○, □, . . .

14. 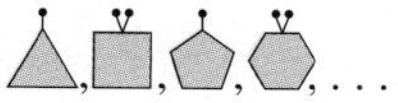 [figures], . . .

15. 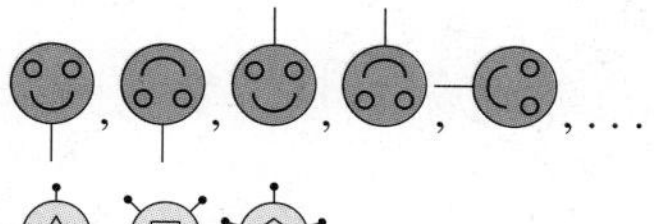 [figures], . . .

16. 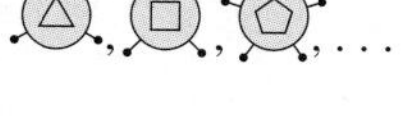 [figures], . . .

In Exercises 17–26, use inductive reasoning to predict the next three numbers in the pattern (or sequence).

17. 5, 7, 9, 11, . . .

18. 13, 9, 5, 1, . . .

19. 1, −1, 1, −1, 1, . . .

20. 5, 3, 1, −1, −3, . . .

21. 1, 1/2, 1/4, 1/8, . . .

22. 2, −6, 18, −54, . . .

23. 1, 4, 9, 16, 25, . . .

24. 0, 3, 8, 15, 24, 35, . . .

25. 1, 1, 2, 3, 5, 8, 13, 21, . . .

26. $5, -\frac{10}{3}, \frac{20}{9}, -\frac{40}{27}, \ldots$

PROBLEM SOLVING

27. a) Select a variety of one- and two-digit numbers between 1 and 99 and multiply each by 9. Record your results.

b) Find the sum of the digits in each of your products in part (a). If the sum is not a one-digit number, find the sum of the digits again until you obtain a one-digit number.

c) Make a conjecture about the sum of the digits when a one- or two-digit number is multiplied by 9.

28. Find the letter that is the 118th entry in the following sequence. Explain how you determined your answer.

Y, R, R, Y, R, R, Y, R, R, Y, R, R, Y, R, R, . . .

29. *A Square Pattern* The ancient Greeks labeled certain numbers as **square numbers**. The numbers 1, 4, 9, 16, 25, and so on are square numbers.

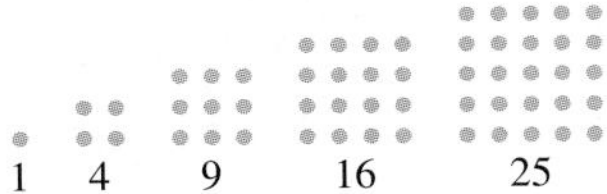

a) Determine the next three square numbers.

b) Describe a procedure to determine the next five square numbers without drawing the figures.

c) Is 72 a square number? Explain how you determined your answer.

30. *A Triangular Pattern* The ancient Greeks labeled certain numbers as **triangular numbers**. The numbers 1, 3, 6, 10, 15, 21, and so on are trianglular numbers.

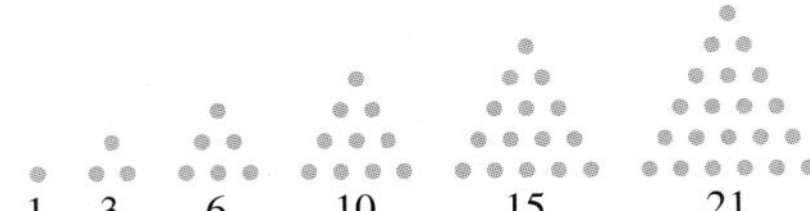

a) Can you determine the next two triangular numbers?

b) Describe a procedure to determine the next five triangular numbers without drawing the figures.

c) Is 72 a triangular number? Explain how you determined your answer.

31. *Quilt Design* The pattern shown here is taken from a quilt design known as a triple Irish chain. Complete the color pattern by indicating the color assigned to each square.

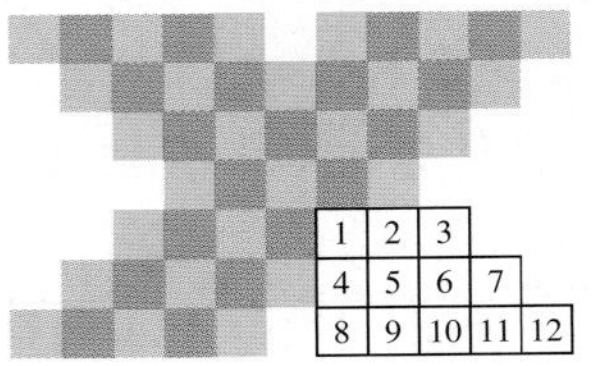

32. *Triangles in a Triangle* Four rows of a triangular figure are shown.

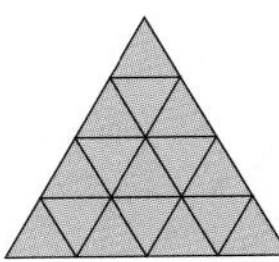

a) If you added six additional rows to the bottom of this triangle, using the same pattern displayed, how many triangles would appear in the 10th row?

b) If the triangles in all 10 rows were added, how many triangles would appear in the entire figure?

33. *Health Care Costs*

a) The graph shows health care inflation. If you had to make a prediction of the annual percent change in health maintenance organization (HMO) premiums in the year 2000, what would you predict?

b) Explain how you are using inductive reasoning in determining your answer.

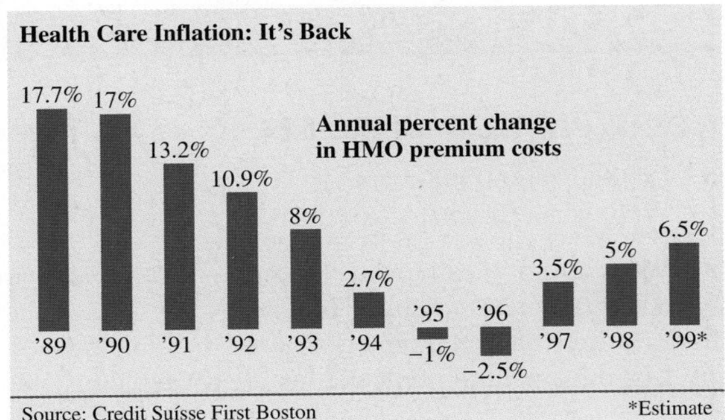

34. *Car Prices* The graph shows the average (median) family income and purchase price for new cars from 1959 through 1999.

a) Using the graph, make a prediction of the average family income in 2009.

b) Using the graph, make a prediction of the average purchase price for new cars in 2009.

c) Explain how you are using inductive reasoning in determining your answer.

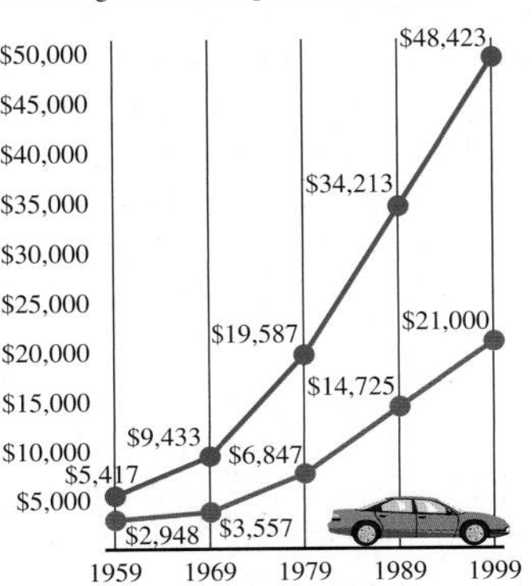

In Exercises 35 and 36, draw the next diagram in the pattern (or sequence).

35.

36.

37. Pick a number, multiply the number by 3, add 6 to the product, divide the sum by 3, and subtract 2 from the quotient. See Example 5.
   a) What is the relationship between the number you started with and the final number?
   b) Arbitrarily select some different numbers and repeat the process, recording the original number and the result.
   c) Can you make a conjecture about the relationship between the original number and the final number?
   d) Try to prove, using deductive reasoning, the conjecture you made in part (c). See Example 6.

38. Pick any number and multiply the number by 6. Add 3 to the product. Divide the sum by 3 and subtract 1 from the quotient.
   a) What is the relationship between the number you started with and the final answer?
   b) Arbitrarily select some different numbers and repeat the process, recording the original number and the results.
   c) Can you make a conjecture about the relationship between the original number and the final number?
   d) Try to prove, using deductive reasoning, the conjecture you made in part (c).

39. Pick any number and add 1 to it. Find the sum of the new number and the original number. Add 9 to the sum. Divide the sum by 2 and subtract the original number from the quotient.
   a) What is the final number?
   b) Arbitrarily select some different numbers and repeat the process. Record the results.
   c) Can you make a conjecture about the final number?
   d) Try to prove, using deductive reasoning, the conjecture you made in part (c).

40. Pick a number and add 10 to the number. Divide the sum by 5. Multiply this quotient by 5. Subtract 10 from the product. Then subtract your original number.
   a) What is the result?
   b) Arbitrarily select some different numbers and repeat the process, recording the original number and the result.
   c) Can you make a conjecture regarding the result when this process is followed?
   d) Try to prove, using deductive reasoning, the conjecture you made in part (c).

In Exercises 41–46, find a counterexample to show that each of the statements is incorrect.

41. The product of 2 two-digit numbers is a three-digit number.

42. The sum of 3 two-digit numbers is a three-digit number.

43. When a counting number is added to 3 and the sum is divided by 2, the quotient will be an even number.

44. The product of any two counting numbers is divisible by 2.

45. The product of a number multiplied by itself is even.

46. The sum of any two odd numbers is divisible by 4.

47. **a)** Construct a triangle and measure the three interior angles with a protractor. What is the sum of the measures?
   b) Construct three other triangles, measure the angles, and record the sums. Are your answers the same?
   c) Make a conjecture about the sum of the measures of the three interior angles of a triangle.

48. **a)** Construct a quadrilateral (a four-sided figure) and measure the four interior angles with a protractor. What is the sum of their angle measures?
   b) Construct three other quadrilaterals, measure the angles, and record the sums. Are your answers the same?
   c) Make a conjecture about the sum of the measures of the four interior angles of a quadrilateral.

CHALLENGE PROBLEMS/GROUP ACTIVITIES

49. Complete the following square of numbers. Explain how you determined your answer.

1	2	3	4
2	5	10	17
3	10	25	52
4	17	52	?

50. Find the next three numbers in the sequence.

1, 8, 11, 88, 101, 111, 181, 1001, 1111, . . .

RESEARCH ACTIVITIES

51. **a)** Using newspapers, magazines, and other sources, find examples of conclusions arrived at by inductive reasoning.
   b) Explain how inductive reasoning was used in arriving at the conclusion.

52. When a jury decides the guilt or innocence of a defendant, do the jurors collectively use primarily inductive reasoning, deductive reasoning, or an equal amount of each? Write a brief report supporting your answer.

1.2 ESTIMATION

An important step in solving mathematical problems—or, in fact, *any* problem—is to make sure that the answer you've arrived at makes sense. One technique for determining whether an answer is reasonable is to estimate. **Estimation** is the process of arriving at an approximate answer to a question. This section demonstrates several estimation methods.

To estimate, or approximate, an answer, we often round numbers as illustrated in the following examples. The symbol ≈ means *is approximately equal to.*

EXAMPLE 1 *Estimate the Cost of Wood Chips*

Estimate the cost of 18 bags of wood chips at $0.92 each.

SOLUTION We may round the amounts as follows to obtain an estimate.

$$\begin{array}{rcr} 18 & \rightarrow & 20 \\ \times\ 0.92 & \rightarrow & \times\ 0.90 \\ \hline & & 18.00 \end{array}$$

Thus, the 18 bags of wood chips will cost approximately $18.00, written ≈ $18. ▲

In Example 1, we could have rounded the 18 to 20 and the $0.92 to $1.00, which would result in an estimate of $20. The true cost is $0.92 × 18, or $16.56. *Estimates are not meant to give exact values for answers but are a means of determining whether your answer is reasonable.* If you calculated an answer of $16.56 and then did a quick estimate to check it, you would know that the answer is reasonable because it is close to your estimated answer.

EXAMPLE 2 *Two Ways to Estimate*

At a local Publix supermarket, Kaitlyn purchased milk for $2.39, lettuce for $1.29, bread for $1.75, hot dogs for $3.29, ground beef for $2.89, bananas for $0.64, and a green onion for $0.83. The total bill was $19.08. Use estimation to determine whether this amount is reasonable.

SOLUTION The most expensive item is $3.29, and the least expensive is $0.64. How should we estimate? We will estimate two different ways. First, we will round the cost of each item to the nearest 10 cents. Then we will round the cost to the nearest dollar. Rounding to the nearest 10 cents is more accurate. To determine whether the total bill is reasonable, however, we may need to round only to the nearest dollar.

	Rounding to the nearest 10 cents		**Rounding to the nearest dollar**	
Milk	$2.39 →	$2.40	$2.39 →	$2.00
Lettuce	1.29 →	1.30	1.29 →	1.00
Bread	1.75 →	1.80	1.75 →	2.00
Hot dogs	3.29 →	3.30	3.29 →	3.00
Ground beef	2.89 →	2.90	2.89 →	3.00
Bananas	0.64 →	0.60	0.64 →	1.00
Onion	0.83 →	0.80	0.83 →	1.00
		$13.10		$13.00

Using either estimate, we find that the bill of $19.08 is quite high. Therefore, Kaitlyn should check the bill carefully before paying it. Adding the prices of all seven items gives the true cost of $13.08. ▲

EXAMPLE 3 *Select the Best Estimate*

The number of bushels of grapes produced at a vineyard are 62,408 Cabernet Sauvignon, 118,916 French Colombard, 106,490 Chenin Blanc, 5960 Charbono, and 12,104 Chardonnay. Select the best estimates of the total number of bushels produced by the vineyard.

a) 500,000 b) 30,000 c) 300,000 d) 5,000,000

SOLUTION Following are suggested roundings. On the left, the numbers are rounded to thousands. For a less close estimate, round to ten thousands, as illustrated on the right.

Round to the nearest thousand		Round to the nearest ten thousand	
62,408 →	62,000	62,408 →	60,000
118,916 →	119,000	118,916 →	120,000
106,490 →	106,000	106,490 →	110,000
5,960 →	6,000	5,960 →	10,000
12,104 →	12,000	12,104 →	10,000
	305,000		310,000

Either rounding procedure indicates that the best estimate is (c), or 300,000. ▲

Little Mistake, Big Discovery

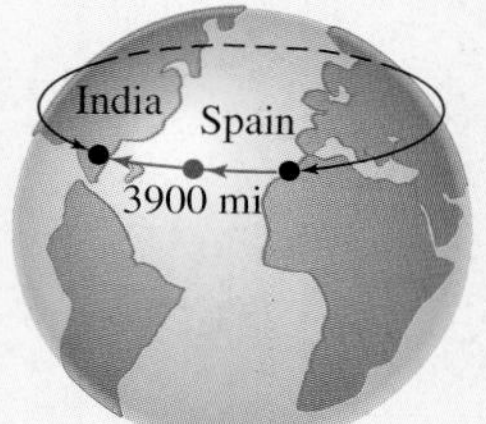

The World, 1492

The red dot represents the area near America, the Canary Islands, where Columbus landed.

It was a mistake in estimation that brought Christopher Columbus to our part of the world in 1492. He believed the world to be round and thought of distance from one place to another in terms of degrees, with 360° longitude representing a "round trip." He estimated 56.6 miles to a degree instead of the approximately 61.6 miles that it should have been. This resulted in a mile equaling about 4848 ft compared with our 5280 ft per mile. The outcome of his miscalculations put India about 3900 miles west of Spain, more or less where the Americas happen to be. Columbus died in 1506 still unaware that he had discovered a new continent.

EXAMPLE 4 *Using Estimation in Calculations*

The odometer of an automobile reads 52,367.2 miles.

a) If the automobile averaged 21.4 miles per gallon (mpg) for that mileage, estimate the number of gallons of gasoline used.
b) If the cost of the gasoline averaged $1.19 per gallon, estimate the total cost of the gasoline.

SOLUTION

a) To estimate the number of gallons, divide the mileage by the number of miles per gallon.

$$\frac{52,367.2}{21.4}$$

Round these numbers to obtain an estimate.

$$\frac{50,000}{20} = 2500$$

Therefore, the car used approximately 2500 gallons (gal) of gasoline.

b) Rounding the price of the gasoline to $1.20 per gallon gives the cost of the gasoline as 2500 × $1.20, or $3000. ▲

Now let's look at some different types of estimation problems.

EXAMPLE 5 *Using Estimation in Determining Distances*

The June 21, 1999, issue of *U.S. News and World Report* carried an article about our national parks. The article contained the map of the Grand Canyon shown below.

a) Estimate the distance, in miles, of the part of Route 64 shown in the map.
b) Estimate the total distance of Center Road.
c) If Faith Sherlock walked the distance determined in part (a) in 3.2 hours, estimate her average walking speed.

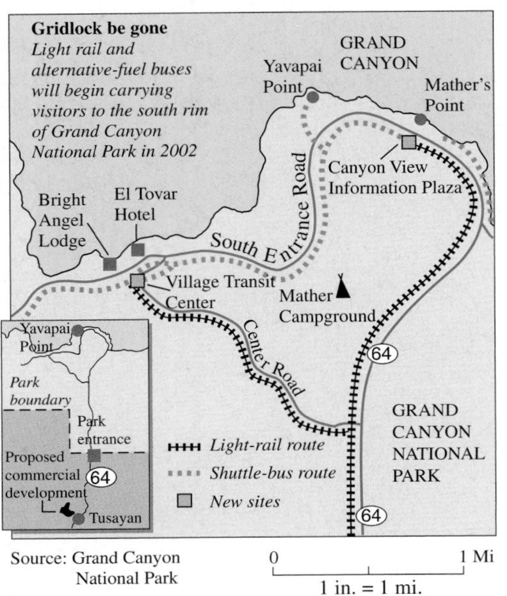

Source: Grand Canyon National Park

SOLUTION

a) We are told that the scale is 1 inch equals 1 mile. We need to estimate the length of Route 64. One way to do this is to estimate and mark off 1 inch intervals. You may obtain a value of about $6\frac{1}{4}$ in. Thus, the distance shown is about $6\frac{1}{4}$ mi. Sometimes in a map like this one, it may be difficult to get an accurate estimate because of the curves in the road. To get a more accurate estimate, you may want to use a piece of string. Place the beginning of the string at the beginning of the road and, using tape or pins, align the string with the road. Indicate on the string where the road ends. Then remove the string and measure the length of string you have marked off. If, for example, your string over this distance measures about $6\frac{3}{8}$ in., then the distance is about $6\frac{3}{8}$ mi.
b) Using a ruler we find that Center Road is about $1\frac{1}{2}$ mi long.
c) The average speed is found by dividing the distance by the time. In part (a), we found the distance to be $6\frac{1}{4}$ or 6.25 mi. Thus, Faith's average speed is

$$\text{Average speed} = \frac{\text{distance}}{\text{time}} = \frac{6.25}{3.2} \approx 1.95$$

Therefore, she averaged about 1.95 miles per hour. ▲

DID YOU KNOW

Estimating Techniques in Medicine

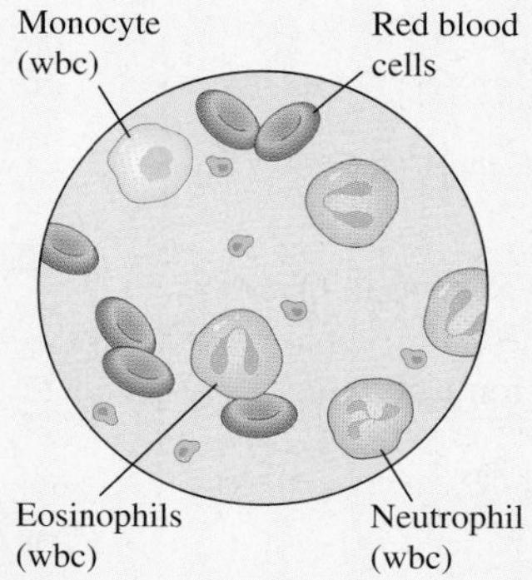

A high count of esinophils (a particular kind of white blood cell, or wbc) can be an indicator of an allergic reaction.

Estimating is one of the diagnostic tools used by the medical profession. Physicians take a small sample of blood, tissue, or body fluids to be representative of the body as a whole. Human blood contains different types of white blood cells that fight infection. When a bacterium or virus gets into the blood, the body responds by producing more of the type of white blood cell whose job it is to destroy that particular invader. Thus, an increased level of white blood cells in a sample of blood not only indicates the presence of an infection but also helps identify its type. A trained medical technician estimates the relative number of each kind of white blood cell found in a count of 100 white blood cells. An increase in any one kind indicates the type of infection present. The accuracy of this diagnostic tool is impressive when you consider that there are normally 5000 to 9000 white blood cells in a dropful of blood (1 cubic millimeter).

Most analysis of blood samples done today is performed using computers.

EXAMPLE 6 ***Estimated Energy Use***

Utility bills sometimes contain graphs illustrating the amount of electricity and gas used. The following graphs show gas and electric use at a specific residence for a period of 1 year, starting in April 1998, as well as the current month's (April, 1999) use by the average residential customer. Using these graphs, can you answer the questions?

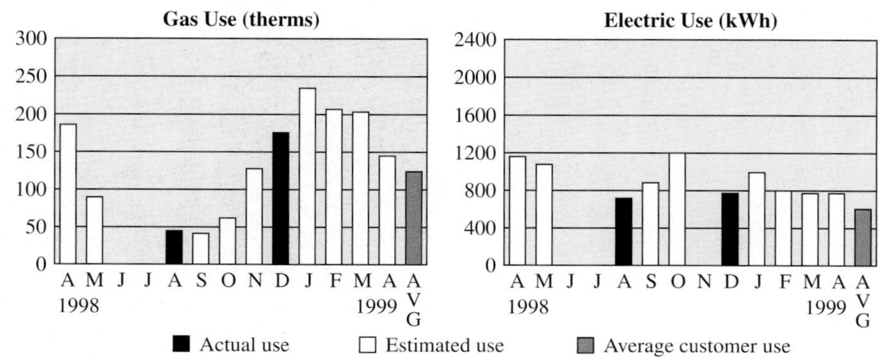

a) How often was an actual gas and electric reading made?
b) Estimate the number of therms of gas used by the average residential customer in April 1999.
c) Estimate the amount of gas used by the resident in April 1999.
d) If the cost of gas is 69.3672 cents per therm, estimate the gas bill in November.
e) In which month was the most electricity used? How many kilowatt hours (kWh) were used in this month?
f) If the cost of electricity is 9.0983 cents per kilowatt hour, estimate the cost of electricity in February.

DID YOU KNOW

Early Estimation

The ancient Egyptians and Greeks used sticks to estimate heights. To measure height, a person held a stick in his or her outstretched arm until it just covered the height of the object. The person then turned the stick through a right angle. Measuring the distance the stick appeared to cover on the ground provided an estimate of the object's height.

SOLUTION

a) Actual readings were made in only two months, August and December.
b) Approximately 125 therms were used, as shown by the height of the red bar.
c) Approximately 145 therms were used (slightly less than 150).
d) In November, about 125 therms were used. The rate, 69.3672 cents per therm, is the same as \$0.693672 per therm. To get a rough approximation, round the rate to \$0.70 per therm.

$$0.70 \times 125 = 87.50^*$$

Thus, the cost of gas used was about \$88.

e) The most electricity (approximately 1200 kWh) was used in October.
f) In February, about 800 kWh were used. Write 9.0983 cents as \$0.090983. Rounding the rate to \$0.09 per kilowatt hour and multiplying by 800 yields an estimate of \$72.

$$0.09 \times 800 = \$72$$

EXAMPLE 7 ***Estimating the Number of Birds in a Photo***

Scientists who are concerned about dwindling animal populations often use aerial photography to make estimates. Estimate the number of birds in the accompanying photograph.

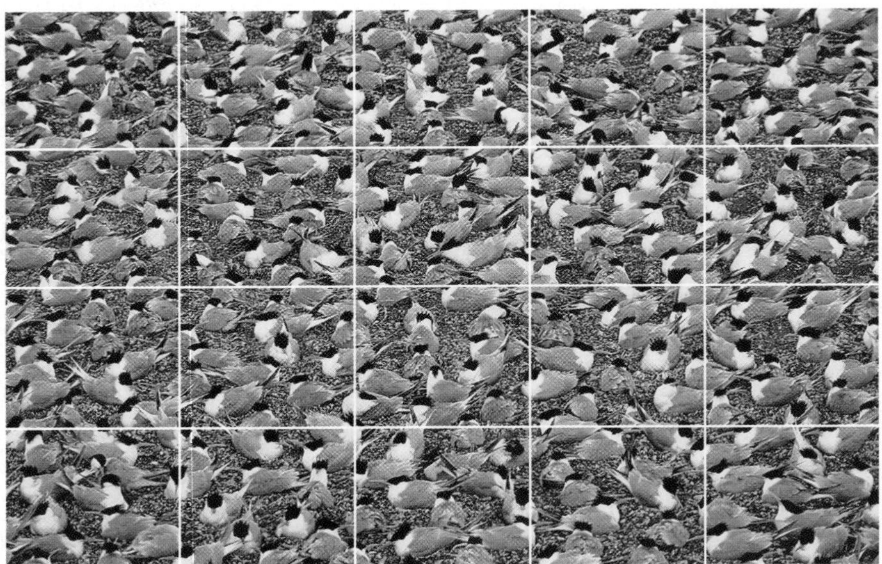

SOLUTION To estimate the number of birds, we can divide the photograph into rectangles with equal areas, then select one area that appears to be representative of all the areas. Estimate (or count) the number of birds in this single area, and then multiply this number by the number of equal areas.

*The amounts here and in part (f) do not include the basic monthly charge, fuel adjustment, taxes, and other extra charges often included on utility bills.

Let's divide the photo into 20 approximately equal areas. We will select the middle region in the bottom row as the representative region. We enlarge this region and count the birds in it. If half a bird is in the region, we count it (see enlargement). There are 13 birds in this region. Multiplying by 20 gives $13 \times 20 = 260$. Thus, there are about 260 birds in the photo.

In problems similar to that in Example 7, the number of regions or areas into which you choose to divide the total area is arbitrary. Generally, the more regions, the better the approximation, as long as the region selected is representative of the other regions in the map, diagram, or photo.

When you estimate an answer, the amount that your approximation differs from the actual answer will depend on how you round the numbers. Thus, in estimating the product of $196{,}000 \times 0.02520$, using the rounded values $195{,}000 \times 0.025$ would yield an estimate much closer to the true answer than using the rounded values $200{,}000 \times 0.03$. Without a calculator, however, the product of $195{,}000 \times 0.025$ might be more difficult to find than $200{,}000 \times 0.03$. When estimating, you need to determine the accuracy desired in your estimate and round the numbers accordingly.

SECTION 1.2 EXERCISES

PRACTICE THE SKILLS

In Exercises 1–24, your answers may vary from the answers given in the back of the text, depending on how you round your answers.

In Exercises 1–12, estimate the answer.

1. $333 + 296.4 + 93.5 + 20.4 + 315.9$

2. $2.53 + 202.6 + 156.9 + 189 + 0.23 + 416$

3. $297{,}700 \times 4087$

4. 1854×0.0096

5. $\dfrac{405}{0.049}$

6. $196.43 - 85.964$

7. 0.048×1964

8. 9% of 2164

9. $31{,}640 \times 79{,}264$

10. $\dfrac{0.0498}{0.00052}$

11. $592 \times 2070 \times 992.62$

12. $296.3 \div 0.0096$

PROBLEM SOLVING

In Exercises 13–24, estimate the answer.

13. The annual distance Jacob walks if he walks about 12.1 mi each week for a year.

14. The cost of twenty-eight 33-cent postage stamps.

15. The average cost per plant if 29 plants cost \$68.90.

16. The number of miles Donna drives in a year if she averages 1690 mi each month.

17. Your salary if you work for 42.8 hr at \$7.95 per hour.

18. A 7% sales tax on a refrigerator that costs \$789.

19. The total weight of the three people on a small plane if their weights are 167, 203, and 137 pounds.

20. A 3.84 lb package of ground beef divided into five approximately equal parts.

21. One fourth of a profit of \$102,272.

22. The difference in the load of two 18-wheeler trucks. One truck has a load of 18,928 lb, and the other truck has a load of 32,090 lb.

23. The weight of each of eight identical cars on a transport truck if the total weight of the eight cars is 24,300 lb.

24. The weight of 18 shovels of dirt, if one shovel of dirt weighs 3.92 lb.

25. *Frequent Flier Miles* Phil Adelfea travels round trip by air from Raleigh, North Carolina, to Denver, Colorado, twice a month for work. Estimate the frequent flier miles he will accumulate in a year if the one-way distance between Raleigh and Denver is 1690 mi.

26. *Estimating the Tip* Ed and Dorothy go out for dinner and spend $38.60 for their meal. If they want to leave a 15% tip, estimate the amount that they should leave.

27. *Estimating Weights* In a tug of war, the weight of the members of the two three-person teams is given below. Estimate the difference in the weights of the teams.

Team A	Team B
189	183
172	229
191	167

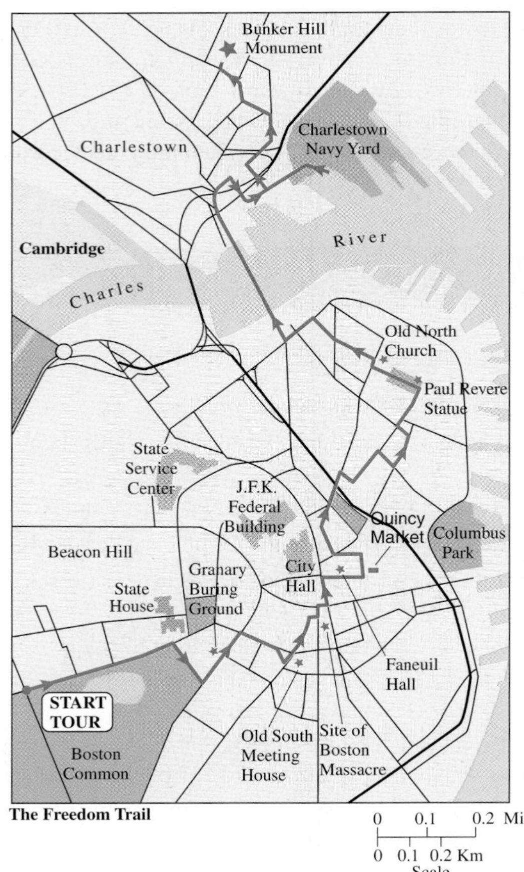

The Freedom Trail

28. *Currency* Estimate the difference in the value of 80 Mexican pesos and 50 American dollars. Assume that one Mexican peso is about 0.302 U.S. dollars.

29. *Estimating Area* Mrs. Sanchez determines that her lawn contains an average of 3.8 grubs per square foot (ft^2). If her rectangular lawn measures 60 ft by 80.2 ft, estimate the total number of grubs in her lawn.

30. *The Cost of a Vacation* The Elways are planning a vacation in Yosemite National Park. Their round-trip airfare from Louisville, Kentucky, to Santa Barbara, California, totals $892. Car rental is $42 per day, lodging is a total of $97 per day, and they estimate a total of $90 per day for food, gas, and other miscellaneous items. If they are planning to stay six full days and nights, estimate their total expenses.

31. *The Freedom Trail* The map at the bottom left shows the Freedom Trail walk in Boston, Massachusetts. Using the scale on the map, estimate the distance via the route indicated in color

a) in miles.

b) in kilometers.

32. *A Scenic Route* The map below shows the famous scenic drive that goes past Pebble Beach, California. Using the scale on the map, estimate the distance via the route marked in color

a) in miles.

b) in kilometers.

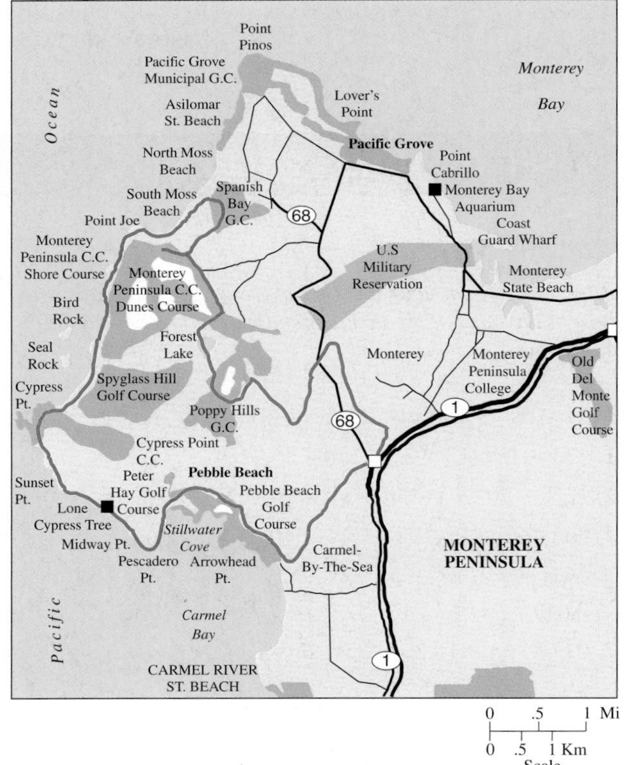

33. *Retirement* The circle graphs show sources of retirement income for retirees with at least $20,000 in annual income for the years 1995 and 1999. Assume Sheila Abbruzzo had a retirement income of $41,105 from the various sources shown in the figures. Estimate her income from investment savings in

a) 1995.
b) 1999.

Source of Retirement Income 1995

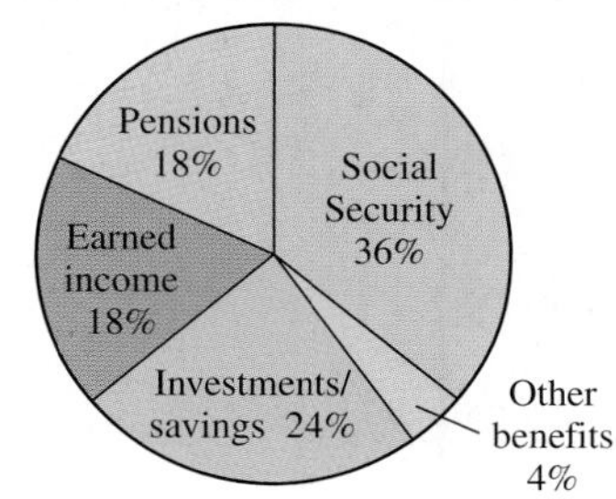

Source of Retirement Income 1999

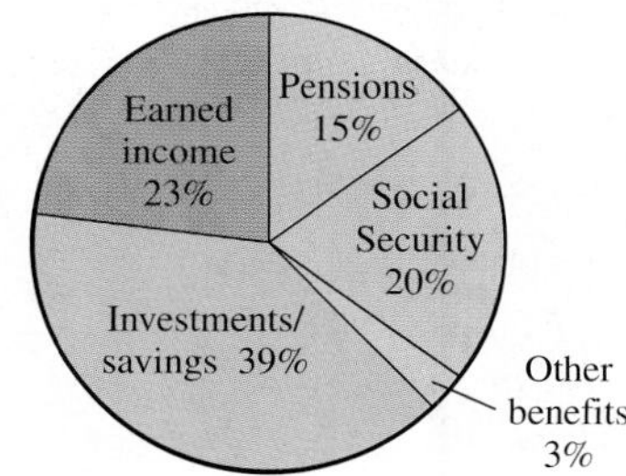

Source: Social Security Administration

34. *A Gold Necklace* The circle graph shows what you are paying for, on average in December 1999, when you buy a gold necklace. Suppose you purchase a gold necklace for $599. Estimate, in dollars

a) the price of the gold in the necklace.
b) the retailer's profit on the necklace.
c) the manufacturer's profit.

What's in a Necklace?
Here's what you're paying for when you buy one.

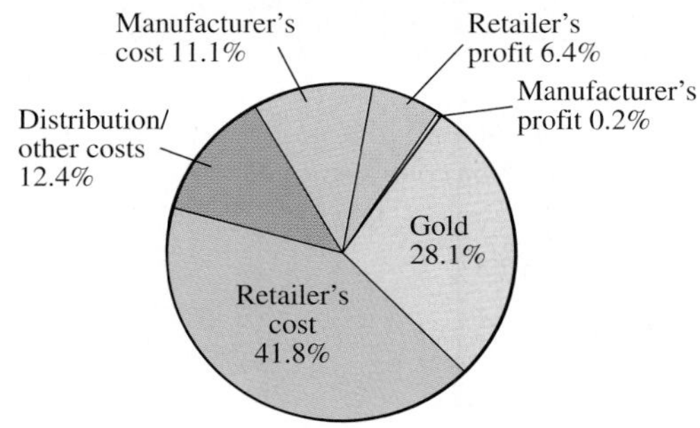

Source: Jewelry Information Center

35. *An Aging Population* The bar graph shows population figures for 1900 and expected population figures for 2000 and 2050.

a) Estimate the number of people 65 and over in 1900.
b) Estimate the expected number of people 65 and older in 2050.
c) Estimate the increase in the expected growth in the number of people 65 and older from 2000 to 2050.
d) Estimate the total U.S. population in 2000 by adding the five categories.

Population by Age

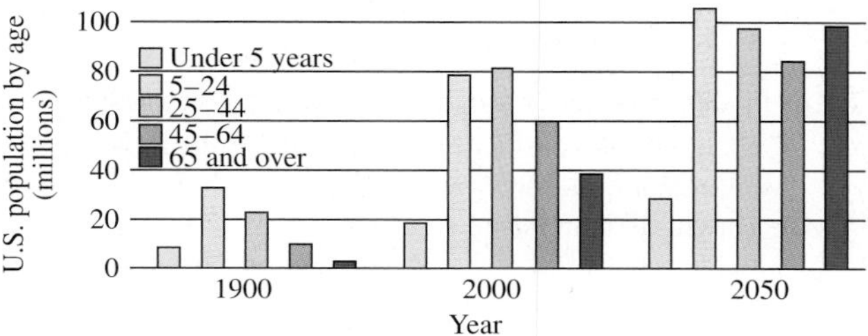

Source: *Newsweek*

36. *Gaining Weight* As the graph shows, as a society we tend to get heavier as we get older. Also, with age, the amount of muscle tends to drop, and fat accounts for a greater percentage of weight.

a) Estimate the average percent of body fat for a male, age 18 to 25.
b) Estimate the average percent of body fat for a female, age 56+.
c) Greg, an average 40-year-old, weighs 179 lb. Estimate the number of pounds of body fat he has.

Getting Older Usually Means Getting Fatter

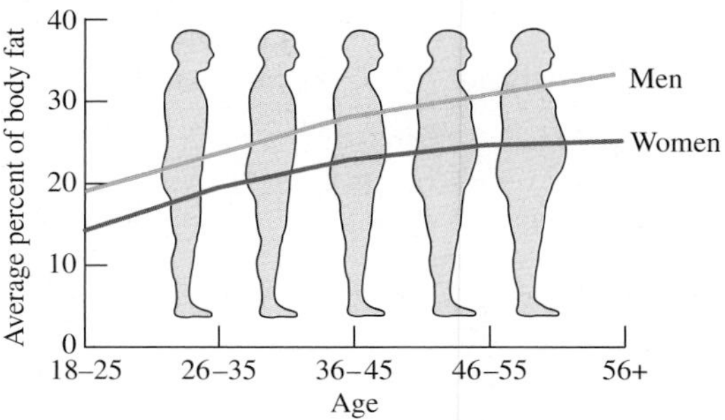

Source: Mayo Clinic Newsletter

37. *Auto Depreciation* The bar graph on page 16 shows the average depreciation of certain types of vehicles 3 years after being purchased.

a) Which type of vehicles will depreciate by more than 40% after 3 years?

Average Depreciation After 3 Years, By Type of Vehicle

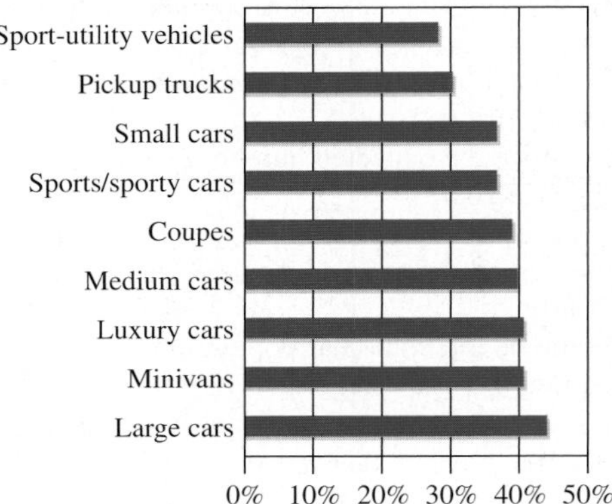

Source: *Consumer Reports*

b) Todd just purchased a pickup truck for $18,209. Estimate the amount of its value it will lose because of depreciation after 3 years.

c) Dawn just purchased a large car for $27,937. Estimate the amount of its value it will lose because of depreciation after 3 years.

38. *Calories and Exercise* The chart shows the calories burned per hour for an average person who weighs 150 lb.

a) Estimate the number of calories Mary burns in a week if she stair-climbs for 2 hours each week and jogs at 5 miles per hour (mph) 4 hours each week.

b) Estimate the difference in the calories Mary will burn each week if she runs for 4 hours at 8 miles per hour rather than does casual bike riding for 4 hours.

c) Assume Mary jogs at 5 miles per hour for 3 hours and bicycles at 13 miles per hour for 3 hours each week. Estimate the number of calories she will burn in a year from these exercises.

Activity	Calories* per hour
Running, 8 mph	920
Bicycling, 13 mph	545
Jogging, 5 mph	545
Air-walking	480
Stair-climbing	410
Weight-lifting	410
Walking, 4 mph	330
Casual bike riding	300

*For a 150 lb person.

In Exercises 39 and 40, estimate the maximum number of smaller figures (at left) that can be placed in the larger figure (at right) without the small figures overlapping.

39.

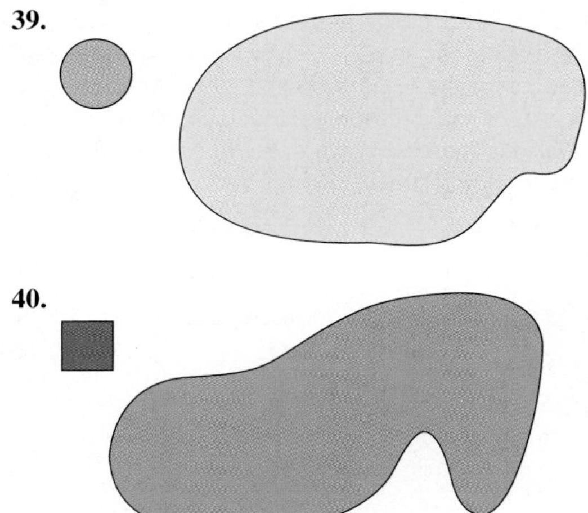

40.

41. Estimate the number of marbles shown in the photo.

42. Estimate the number of ladybugs shown in the photo.

In Exercises 43 and 44, estimate, in degrees, the measure of the angles depicted. For comparison purposes a right angle, , measures 90°.

43. **44.**

In Exercises 45 and 46, estimate the percent of area that is shaded in the following figures.

45.

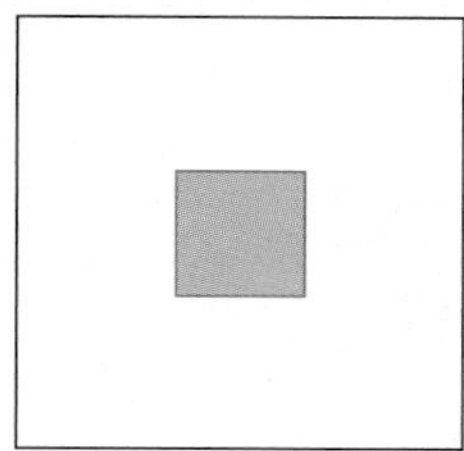

46.

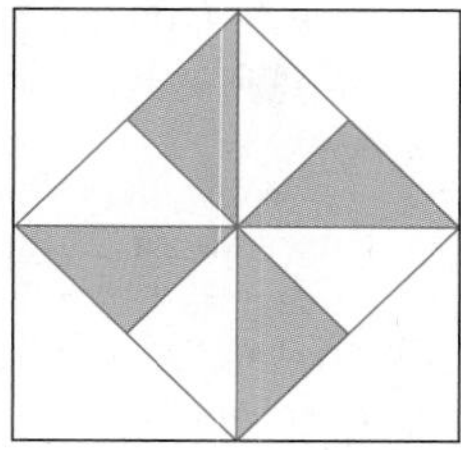

In Exercises 47 and 48, if each square represents one square unit, estimate the area of the shaded figure in square units.

47.

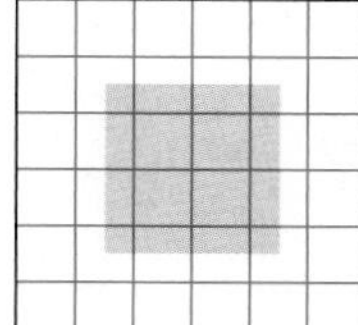

48.

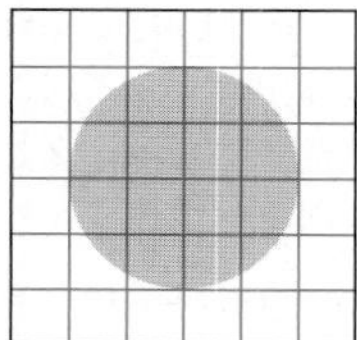

49. *Mt. Rushmore* If the length of the nose on George Washington's carving at Mt. Rushmore is 15 ft, estimate the total height of the carving of Washington's face.

50. *Estimating Heights* If the front legs of the giraffe shown are 5 ft long, estimate the height of the giraffe if he is standing up straight.

51. Estimate, without a ruler, a distance of 12 in. Measure the distance. How good was your estimate?

52. In a bag place objects that you feel have a total weight of 10 lb. Weigh the bag to determine the accuracy of your estimate.

53. Estimate the number of times the phone will ring in 1 minute if unanswered. Have a classmate phone you so that you can count the rings and thus test your estimate.

54. Fill a glass with water and estimate the water's temperature. Then use a thermometer to measure the temperature and check your estimate.

55. Estimate the ratio of your height to your neck size. Then have a friend measure you to determine this ratio and check the accuracy of your estimate.

56. Estimate the number of pennies that will fill a 3-ounce (oz) paper cup. Then actually fill a 3 oz paper cup with pennies, counting them to determine the accuracy of your estimate.

57. Estimate how fast you can walk 60 ft. Then mark off a distance of 60 ft and use a watch with a second hand to time yourself walking it. Determine the accuracy of your estimate.

CHALLENGE PROBLEMS/GROUP ACTIVITIES

58. *Shopping* Make a shopping list of 20 items you use regularly that can be purchased at a supermarket. Beside each item write down what you estimate to be its price. Add these price guesses to estimate the total cost of the 20 items. Next, make a trip to your local supermarket and record the actual price of each item. Add these prices to determine the actual total cost. How close was your estimate? (Don't forget to add tax on the taxable items.)

59. *A Ski Vacation* Two friends, Tiffany Connolly and Ana Pott, are planning a skiing vacation in the Rockies. They plan to purchase round-trip airline tickets from Atlanta, Georgia, to Denver, Colorado. They will fly into Denver on a Friday morning, rent a midsize car, and drive to Aspen that same day. They will stay at the Holiday Inn in Aspen. They will begin skiing at the Buttermilk Ski Area on Saturday, ski up to and including Wednesday, drive back to Denver on Thursday, and fly out of Denver Thursday evening.

a) Estimate the total cost of the vacation for the two friends. Do not forget items such as food, tips, gas, and other incidentals.

b) Using informational sources, including the Internet, determine the airfare cost, hotel cost, cost of ski tickets, cost of a car rental, and so forth. You will need to make an estimate for food and other incidentals.

c) How close was your estimate in part (a) to the amount you found in part (b)? Was your estimate in part (a) lower or higher than the amount obtained in part (b)?

RESEARCH ACTIVITIES

60. *Water Usage*

a) About how much water does your household use per day? Use the following data to estimate your household's daily water usage.

How much water do you use?

Activity	Typical use
Running clothes washer	40 gal
Bath	35 gal
5-minute shower	25 gal
Doing dishes in sink, water running	20 gal
Running dishwasher	11 gal
Flushing toilet	4 gal
Brushing teeth, water running	2 gal

Source: U.S. Environmental Protection Agency

b) Determine from your water department (or company) your household's average daily usage by obtaining the total number of gallons used per year and dividing that amount by 365. How close was your estimate in part (a)?

c) Current records indicate that the average household uses about 300 gal of water per day (the average daily usage is 110 gal per person). Based on the number of people in your household, do you feel your household uses more or less than the average amount of water? Explain your answer.

61. Develop a monthly budget by estimating your monthly income and your monthly expenditures. Your monthly income should equal your monthly expenditures.

62. Identify three ways that you use estimation in your daily life. Discuss each of them briefly and give examples.

1.3 PROBLEM SOLVING

Solving mathematical puzzles and real-life mathematical problems can be enjoyable. You should work as many exercises in this section as possible. By doing so, you will sample a variety of problem-solving techniques.

You can approach any problem by using a general procedure developed by George Polya. Before learning Polya's problem-solving procedure, let's consider an example.

EXAMPLE 1 *Saving Money When Purchasing Video Tapes*

Businesses, to maximize their profits, try to keep their expenses down. We, as individuals, also try to keep our expenses down, and we often look for "bargains" or the "best deal."

A video store owned by Yesha Brill plans to purchase a large number of blank videotapes. One supplier, the Johnny Melton Company, is selling boxes of 20 tapes for \$48 and boxes of 12 tapes for \$30. Only complete boxes of tapes are sold.

a) Find the maximum number of tapes that can be purchased for \$280 or less. Indicate how many boxes of 20 and how many boxes of 12 will be purchased.

b) If the maximum number of tapes determined in part (a) is purchased in the most economical way, how much will the tapes cost?

SOLUTION

a) The first thing to do is to read the problem carefully. Read it at least twice and be sure you understand the facts given and what you are being asked to find. Next, make a list of the given facts and determine which are relevant to answering the question asked.

Given information

Store owner: Yesha Brill

Supplier: the Johnny Melton Company

A box of 20 tapes costs \$48.
A box of 12 tapes costs \$30.
Only complete boxes of tapes can be purchased.

We need to determine the maximum number of tapes that the video store can purchase for \$280 or less. To determine this, we need to know the number of tapes in each of the boxes and the cost of the boxes. We also need to know that only complete boxes of tapes may be purchased.

Relevant information

A box of 20 tapes costs \$48.
A box of 12 tapes costs \$30.
Only complete boxes of tapes may be purchased.

The next step is to determine the answer to the question. That is, we need to determine the maximum number of tapes that can be purchased for \$280 or less.

We now need a plan for solving the problem. One method is to set up a table or chart to compare costs of different combinations of boxes of tapes. Start by using the maximum number of boxes of 20 tapes. Then reduce the number of boxes of 20 tapes, and add more boxes of 12 tapes. In each case, we need to keep the cost at \$280 or less.

Since 1 box of 20 tapes costs \$48, we can determine the number of boxes of 20 tapes that can be purchased by dividing 280 by 48. Since the quotient is about 5.83, and since only whole boxes of tapes may be purchased, only 5 boxes of 20 may be purchased. Five boxes would cost $5 \times 48 = 240$. The remaining \$40 from the \$280 could be used to purchase boxes of 12 tapes. Since each box of 12 tapes costs \$30, only one box of 12 tapes could be purchased. Thus, for \$280 or less, one option is 5 boxes of 20 tapes and 1 box of 12 tapes. This option is indicated in the first row of the table. Also given in the table is the cost of this option, which is \$270. We complete the other rows of the table in a similar manner.

Boxes of 20 and boxes of 12 tapes	Number of tapes	Cost
5 boxes of 20 and 1 box of 12	$(5 \times 20) + (1 \times 12) = 112$	\$270
4 boxes of 20 and 2 boxes of 12	$(4 \times 20) + (2 \times 12) = 104$	\$252
3 boxes of 20 and 4 boxes of 12	$(3 \times 20) + (4 \times 12) = 108$	\$264
2 boxes of 20 and 6 boxes of 12	$(2 \times 20) + (6 \times 12) = 112$	\$276
1 box of 20 and 7 boxes of 12	$(1 \times 20) + (7 \times 12) = 104$	\$258
0 boxes of 20 and 9 boxes of 12	$(0 \times 20) + (9 \times 12) = 108$	\$270

The question asks us to find the maximum number of tapes that can be purchased for \$280 or less. From the second column of the table, we see that the answer is 112 tapes. This result can be done in two different ways: either 5 boxes of 20 and 1 box of 12 tapes, or 2 boxes of 20 and 6 boxes of 12 tapes.

b) When comparing the two possibilities for purchasing the 112 tapes discussed in part (a), we see that the most economical way to purchase the tapes is to purchase 5 boxes of 20 and 1 box of 12 tapes. The cost is \$270. ▲

PROFILE IN MATHEMATICS

George Polya

George Polya (1877–1985) was educated in Europe and taught at Stanford University. In his book *How to Solve It*, Polya outlines four steps in problem solving. We will use Polya's four steps as guidelines for problem solving.

Following is a general procedure for problem solving as given by George Polya. Note that Example 1 demonstrates many of these guidelines.

Guidelines for Problem Solving

1. *Understand the problem.*
   - Read the problem *carefully* at least twice. In the first reading, get a general overview of the problem. In the second reading, determine (a) exactly what you are being asked to find, and (b) what information the problem provides.
   - Try to make a sketch to illustrate the problem. Label the information given.
   - Make a list of the given facts. Are they all pertinent to the problem?
   - Determine if the information you are given is sufficient to solve the problem.
2. *Devise a plan to solve the problem.*
   - Have you seen the problem or a similar problem before? Are the procedures you used to solve the similar problem applicable to the new problem?
   - Can you express the problem in terms of an algebraic equation? (We explain how to write algebraic equations in Chapter 6.)
   - Look for patterns or relationships in the problem that may help in solving it.
   - Can you express the problem more simply?
   - Can you substitute smaller or simpler numbers to make the problem more understandable?
   - Will listing the information in a table help in solving the problem?
   - Can you make an educated guess at the solution? Sometimes if you know an approximate solution, you can work backward and eventually determine the correct procedure to solve the problem.
3. *Carry out the plan.*
   Use the plan you devised in step 2 to solve the problem.
4. *Check the results.*
   - Ask yourself, "Does the answer make sense?" and "Is the answer reasonable?" If the answer is not reasonable, recheck your method for solving the problem and your calculations.
   - Can you check the solution using the original statement?
   - Is there an alternative method to arrive at the same conclusion?
   - Can the results of this problem be used to solve other problems?

The following examples show how to apply the guidelines for problem solving.

EXAMPLE 2 *Bus Revenue*

Many buses, provided by Carey Transportation, operate between John Fitzgerald Kennedy airport and downtown Manhattan, 25 miles away. One particular bus makes 10 round trips per day carrying an average of 42 passengers per trip. The fare each way (as of August 1999) is \$13.00. What are the receipts from one day's operation for this particular bus?

SOLUTION A careful reading of the problem shows that the task is to find the total receipts from one day's operation. Make a list of all the information given and determine whether it is all pertinent to the problem. The facts given are:

Distance from airport to city = 25 mi
Number of round trips daily = 10
Average number of passengers per trip = 42
Fare each way = \$13.00

DID YOU KNOW

Two Medieval Riddles

A boatman must transport a wolf, a goat, and a bundle of herbs across a river. If the boat is so small he can carry only one of them at a time, how can he make the crossing so that the wolf will not devour the goat and the goat will not eat the herbs?

(Clue: The boatman must make multiple round trips.)

As I was going to St. Ives,
I met a man with seven wives;
Every wife had seven sacks,
Every sack had seven cats,
Every cat had seven kits.
Kits, cats, sacks, and wives,
How many were going to
St. Ives?

What information do you need to determine the total receipts for the day? Is all the information given needed in solving the problem? Some thought should reveal that the distance between the airport and the city is not needed to determine the answer. You should realize that the total receipts depend on (a) the number of one-way trips per day, (b) the average number of passengers per trip, and (c) the cost per passenger each way. The product of these three numbers will yield the total daily receipts. For the 10 round trips daily, there are 2×10 or 20 one-way trips daily.

$$\begin{pmatrix}\text{Receipts}\\\text{for one}\\\text{day}\end{pmatrix} = \begin{pmatrix}\text{number of}\\\text{one-way}\\\text{trips per day}\end{pmatrix} \times \begin{pmatrix}\text{number of}\\\text{passengers}\\\text{per trip}\end{pmatrix} \times \begin{pmatrix}\text{cost per}\\\text{passenger}\\\text{each way}\end{pmatrix}$$
$$= 20 \times 42 \times \$13.00 = \$10{,}920$$

In Example 2, we could have used 10 round trips at a fare of \$26.00 per person to obtain the answer. Why?

Is the answer obtained in Example 2 reasonable for the information given? A quick estimate may be obtained as follows:

Cost of one round trip for 1 person	\$26
Cost of one round trip for 40 people	$\$26 \times 40 = \$1040$
Cost of 10 round trips for 40 people	$\$1040 \times 10 = \$10{,}400$

With an estimate of \$10,400, the answer \$10,920 seems reasonable.

EXAMPLE 3 *Determining a Tip*

The cost of Sahar's meal before tax is \$23.60.

a) If a $7\frac{1}{2}\%$ sales tax is added to her bill, determine the total cost of the meal including tax.
b) If Sahar wants to leave a 10% tip on the *pretax* cost of the meal, how much should she leave?
c) If she wants to leave a 15% tip on the *pretax* cost of the meal, how much should she leave?

SOLUTION

a) The sales tax is $7\frac{1}{2}\%$ of \$23.60. To determine the sales tax, first change the $7\frac{1}{2}\%$ to a decimal number. $7\frac{1}{2}\%$ when written as a decimal number is 0.075 (if you have forgotten how to change a percent to a decimal number, read Section 11.1). Next, multiply the decimal number, 0.075, by the amount, \$23.60.

$$\begin{aligned}\text{Sales tax} &= 7\tfrac{1}{2}\% \text{ of } \$23.60\\ &= 0.075(23.60) = 1.77\end{aligned}$$

The sales tax is \$1.77. The total bill is the cost of the meal plus the sales tax.

$$\begin{aligned}\text{Total bill} &= \text{cost of meal} + \text{sales tax}\\ &= 23.60 + 1.77 = 25.37\end{aligned}$$

Thus, the bill, including sales tax, is \$25.37.

b) To find 10% of any number, we can multiply the number by 0.10.

$$\begin{aligned} 10\% \text{ of pretax cost} &= 0.10(23.60) \\ &= 2.36 \end{aligned}$$

Thus, a 10% tip is $2.36.

A simple way to find 10% of any number is to simply move the decimal point in the number one place to the left. Moving the decimal point in $23.60 one place to the left gives $2.36, the same answer we obtained by our calculations.

c) To find 15% of $23.60, multiply as follows.

$$15\% \text{ of } 23.60 = 0.15(23.60) = 3.54$$

Thus, 15% of $23.60 is $3.54. A second method to find a 15% tip is to find 10% of the cost, as in part (b), then add half that amount. Following this procedure we get

$$\$2.36 + \frac{\$2.36}{2} = \$2.36 + \$1.18 = \$3.54$$

In most cases, tips are rounded. If the service is excellent, some people leave a 20% tip. Can you give two methods for determining a 20% tip on $23.60? Determine the 20% tip now. ▲

EXAMPLE 4 *A Recipe for 6*

The following chart shows the amount of each ingredient recommended to make 2, 4, and 8 servings of Betty Crocker Potato Buds®. Determine the amount of each ingredient necessary to make 6 servings of Potato Buds by using the following procedures.

a) Multiply the amount for 2 servings by 3.*
b) Add the amounts for 2 servings to the amounts for 4 servings.
c) Find the average of the amounts for 4 servings and for 8 servings.
d) Subtract the amounts for 2 servings from the amounts for 8 servings.
e) Compare the answers for parts (a) through (d). Are they the same? If not, explain why not.
f) Which is the correct procedure for obtaining 6 servings?

Servings	**2**	**4**	**8**
Water	$\frac{2}{3}$ cup	$1\frac{1}{3}$ cups	$2\frac{2}{3}$ cups
Milk	2 tbsp	$\frac{1}{3}$ cup	$\frac{2}{3}$ cup
Butter or margarine	1 tbsp	2 tbsp	4 tbsp
Salt†	$\frac{1}{4}$ tsp	$\frac{1}{2}$ tsp	1 tsp
Potato Buds®	$\frac{2}{3}$ cup	$1\frac{1}{3}$ cups	$2\frac{2}{3}$ cups

†*Less salt can be used if desired.*

*Addition, subtraction, multiplication, and division of fractions are discussed in detail in Section 5.3.

DID YOU KNOW

"The Adventure of the Creeping Man" by Sir Arthur Conan Doyle

July 2, 11, and 20 and then on August 25 and September 3: On these days, Professor Presbury's assistant Bennett had seen the well-bred, elderly professor behaving very strangely, creeping about like a monkey on his hands and feet. Bennett consulted the famous detective Sherlock Holmes. Holmes couldn't explain the behavior, but he knew when to expect the next episode. Do you?

Holmes had discovered a pattern involving nine-day intervals. On September 12, Holmes and Dr. Watson were on hand to witness Presbury bounding about the grounds. It seems that the professor, recently engaged to a girl half his age, was taking periodic injections of a serum derived from certain glands of a monkey, hoping to recapture his virility. The professor had indeed made a monkey of himself.

SOLUTION

a) We multiply the amounts for 2 servings by 3.

Water: $3(\frac{2}{3}) = 2$ cups
Milk: $3(2) = 6$ tablespoons (tbsp)
Butter or margarine: $3(1) = 3$ tbsp
Salt: $3(\frac{1}{4}) = \frac{3}{4}$ teaspoon (tsp)
Potato Buds®: $3(\frac{2}{3}) = 2$ cups

b) We find the amount of each ingredient by adding the amount for 2 and 4 servings.

Water: $\frac{2}{3}$ cup + $1\frac{1}{3}$ cup = 2 cups
Milk: 2 tbsp + $\frac{1}{3}$ cup   oh oh!

To add these two amounts, we must convert one of them so that they have the same units. By looking in a cookbook or a book of conversion factors, we see that 16 tbsp = 1 cup. The milk in part (a) was given in tablespoons, so we convert $\frac{1}{3}$ cup to tablespoons to compare answers. One third cup equals $\frac{1}{3}(16) = \frac{16}{3}$ or $5\frac{1}{3}$ tbsp. Therefore,

Milk: 2 tbsp + $5\frac{1}{3}$ tbsp = $7\frac{1}{3}$ tbsp

Let's continue with the rest of the ingredients:

Butter: 1 tbsp + 2 tbsp = 3 tbsp
Salt: $\frac{1}{4}$ tsp + $\frac{1}{2}$ tsp = $\frac{3}{4}$ tsp
Potato Buds®: $\frac{2}{3}$ cup + $1\frac{1}{3}$ cups = 2 cups

c) We compute the amounts of the ingredients by finding the average of the amounts for 4 and 8 servings. We do so by adding the amounts for each ingredient and dividing the sum by 2.

$$\text{Water: } \frac{1\frac{1}{3}\text{ cups} + 2\frac{2}{3}\text{ cups}}{2} = \frac{4\text{ cups}}{2} = 2\text{ cups}$$

$$\text{Milk: } \frac{\frac{1}{3}\text{ cup} + \frac{2}{3}\text{ cup}}{2} = \frac{1\text{ cup}}{2} = \frac{1}{2}\text{ cup (or 8 tbsp)}$$

$$\text{Butter: } \frac{2\text{ tbsp} + 4\text{ tbsp}}{2} = \frac{6\text{ tbsp}}{2} = 3\text{ tbsp}$$

$$\text{Salt: } \frac{\frac{1}{2}\text{ tsp} + 1\text{ tsp}}{2} = \frac{\frac{3}{2}\text{ tsp}}{2} = \frac{3}{4}\text{ tsp}$$

$$\text{Potato Buds®: } \frac{1\frac{1}{3}\text{ cups} + 2\frac{2}{3}\text{ cups}}{2} = \frac{4\text{ cups}}{2} = 2\text{ cups}$$

d) We obtain the amounts of ingredients by subtracting the amounts for 2 servings from the amounts for 8 servings.

Water: $2\frac{2}{3}$ cups − $\frac{2}{3}$ cup = 2 cups

$$\begin{aligned}\text{Milk: } \tfrac{2}{3}\text{ cup} - 2\text{ tbsp} &= \tfrac{2}{3}(16)\text{ tbsp} - 2\text{ tbsp}\\ &= \tfrac{32}{3}\text{ tbsp} - \tfrac{6}{3}\text{ tbsp}\\ &= \tfrac{26}{3}\text{ tbsp, or } 5\tfrac{1}{3}\text{ tbsp}\end{aligned}$$

Butter: 4 tbsp − 1 tbsp = 3 tbsp
Salt: 1 tsp − $\frac{1}{4}$ tsp = $\frac{3}{4}$ tsp
Potato Buds®: $2\frac{2}{3}$ cups − $\frac{2}{3}$ cup = 2 cups

e) Comparing the answers in parts (a) through (d), we find that the amounts of all ingredients, except milk, are the same. For milk, we get the following results.

Part (a): Milk = 6 tbsp
Part (b): Milk = $7\frac{1}{3}$ tbsp
Part (c): Milk = 8 tbsp
Part (d): Milk = $5\frac{1}{3}$ tbsp

Why are all these answers different? After rechecking, we find that all our calculations are correct, so we must look deeper. Note that milk is the only ingredient that has different units for 2 servings and 4 servings. Let's check the relationship between 2 tbsp and $\frac{1}{3}$ cup. In going from 2 servings to 4 servings, we would expect that $\frac{1}{3}$ cup should be twice 2 tbsp. We know that 1 cup = 16 tbsp, so

$$\tfrac{1}{3}\text{ cup} = \tfrac{1}{3}(16) = \tfrac{16}{3} = 5\tfrac{1}{3}\text{ tbsp}$$

Therefore, instead of the 4 tbsp of milk we expected for 4 servings, we get $5\frac{1}{3}$ tbsp. This change causes all our calculations for milk to be different.

f) Which is the correct answer? Because all our calculations for milk are correct, there is no single correct answer. All our answers are correct. Using 8 tbsp instead of $5\frac{1}{3}$ tbsp might make the Potato Buds® a little thinner. When we cook, we generally do not add the *exact* amount recommended. We rely on experience to alter the recommended amounts according to individual taste. ▲

Many real-life problems, such as the one in Example 5, can be solved by using proportions.* A proportion is a statement of equality between two ratios (or fractions).

EXAMPLE 5 *Preparing Garden Sprays*

The instructions on a bottle of insecticide indicate that 1.5 oz of insecticide should be mixed with 5 gal of water. Dave Noyes wishes to spray his garden. How much insecticide should he mix with 8 gal of water to get the proper strength solution?

SOLUTION Use the fact that 1.5 oz of insecticide is to be mixed with 5 gal of water to set up a proportion.

$$\text{Given ratio}\left\{\frac{1.5\text{ oz}}{5\text{ gal water}}\right. = \frac{?\text{ oz}}{8\text{ gal}} \quad \begin{array}{l}\leftarrow \text{Item to be found}\\ \leftarrow \text{Other information given}\end{array}$$

Note in the proportion that ounces and gallons are placed in the same relative positions. Often the unknown quantity is replaced with an x. The proportion may be written as follows and solved using cross multiplication.

$$\frac{1.5}{5} = \frac{x}{8}$$
$$1.5(8) = 5x$$
$$12.0 = 5x$$
$$\frac{12.0}{5} = \frac{\cancel{5}x}{\cancel{5}} \quad \text{Divide both sides of the equation by 5 to solve for } x.$$
$$2.4 = x$$

Thus, Dave must mix 2.4 oz insecticide with 8 gal water to get the proper strength solution. ▲

*Proportions are discussed in greater detail in Section 6.2.

DID YOU KNOW

East Meets West: Magic Squares

A Chinese myth says that in about 2200 B.C., a divine tortoise emerged from the Yellow River. On his back was a special diagram of numbers from which all mathematics was derived. The Chinese called this diagram Lo Shu. The Lo Shu diagram is the first known magic square.

Arab traders brought the Chinese magic square to Europe during the Middle Ages, when the plague was killing millions of people. Magic squares were considered strong talismans against evil, and possession of a magic square was thought to ensure health and wealth.

Most of the problems solved so far have been practical ones. Many people, however, enjoy solving brainteasers. One example of such a puzzle follows.

EXAMPLE 6 ***Magic Squares***

A magic square is a square array of numbers such that the numbers in all rows, columns, and diagonals have the same sum. Use the digits 1, 2, 3, 4, 5, 6, 7, 8, and 9 to construct a magic square.

SOLUTION The first step is to create a figure with nine cells as in Fig. 1.1(a). We must place the nine numbers in the cells so that the same sum is obtained in each row, column, and diagonal. Common sense tells us that 7, 8, and 9 cannot be in the same row, column, or diagonal. We need some small and large numbers in the same row, column, and diagonal. To see a relationship, we list the numbers in order:

1, 2, 3, 4, 5, 6, 7, 8, 9

Note that the middle number is 5 and the smallest and largest numbers are 1 and 9, respectively. The sum of 1, 5, and 9 is 15. If the sum of 2 and 8 is added to 5, the sum is 15. Likewise 3, 5, 7, and 4, 5, 6 have sums of 15. We see that in each group of three numbers the sum is 15 and 5 is a member of the group.

9	5	1

(a)

		8
9	5	1
2		

(b)

4		8
9	5	1
2		6

(c)

(d)

Figure 1.1

Because 5 is the middle number in the list of numbers, place 5 in the center square. Place 9 and 1 to the left and right of 5 as in Fig. 1.1(a). Now we place the 2 and the 8. The 8 cannot be placed next to 9 because 8 + 9 = 17, which is greater that 15. Place the smaller number 2 next to the larger number 9. We elected to place the 2 in the lower left-hand cell and the 8 in the upper right-hand cell as in Fig. 1.1(b). The sum of 8 and 1 is 9. To arrive at a sum of 15, we place 6 in the lower right-hand cell as in Fig. 1.1(c). The sum of 9 and 2 is 11. To arrive at a sum of 15, we place 4 in the upper left-hand cell as in Fig. 1.1(c). Now the diagonals 2, 5, 8 and 4, 5, 6 have sums of 15. The numbers that remain to be placed in the empty cells are 3 and 7. Using arithmetic, we can see that 3 goes in the top middle cell and 7 goes in the bottom middle cell as in Fig. 1.1(d). A check shows that the sum in all the rows, columns, and diagonals is 15. ▲

The solution to Example 6 is not unique. Other arrangements of the nine numbers in the cells will produce a magic square. Also, other techniques of arriving at a solution for a magic square may be used. In fact, the process described will not work if the number of squares is even, for example, 16 instead of 9. Magic squares are not limited to the operation of addition or to the set of counting numbers.

SECTION 1.3 EXERCISES

PRACTICE THE SKILLS/PROBLEM SOLVING

1. *Patio Plans* Ann Kuick is making plans to add a patio to the back of her house. On her drawing, she is using a scale of 1 in. = 4.5 ft. If the length of the patio on the scale drawing is 7.5 in., what will be the actual length of the patio?

2. *Blueprints* Chalon Bridges, an architect, is designing a shopping mall. The scale of her plan is 1 in. = 12 ft. If one store in the mall is to have a frontage of 82 ft, how long will the line representing that store's frontage be on the blueprint?

3. *Height of a Tree* At a given time of day, the ratio of the height of an object to the length of its shadow is the same for all objects. If a 3-ft stick in the ground casts a shadow of 1.2 ft, find the length of the shadow of a 48.4 ft tree?

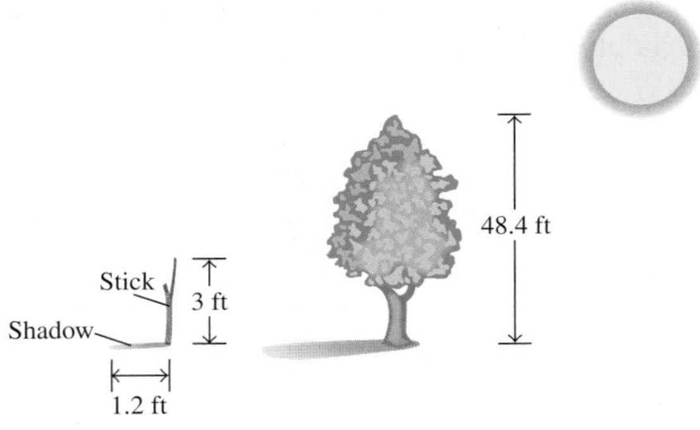

4. *Use of Insecticide* A 40 lb bag of insecticide covers an area of 6000 ft^2 of lawn. How much insecticide is needed to cover 22,000 ft^2 of lawn?

5. *Cab Cost* A taxicab charges \$1.70 for the first $\frac{1}{8}$ mi and 15 cents for each additional $\frac{1}{8}$ mi. Determine the cost of a 5 mi trip.

6. *Cleaning Mixture* The instructions on a bottle of concentrated liquid cleaner reads, "Mix 3 ounces with a gallon of water." If a building custodian wants to mix the solution in a $2\frac{1}{2}$ gal bucket of water, how much concentrate should he use to obtain the proper strength mixture?

7. *The Price of Pigs* The graph shows that the price farmers received for their pigs plunged to just 10 cents a pound in December 1998. (Although the price of pork the farmers were receiving had plunged, the price of pork at grocery stores had not dropped significantly.)
   a) In 1998, what was the farmers' break-even price per pound for pig?
   b) If a farmer delivered 22,000 pounds of pig in 1996, estimate his or her profit.
   c) If a farmer delivered 22,000 pounds of pig in 1998, estimate his or her loss.

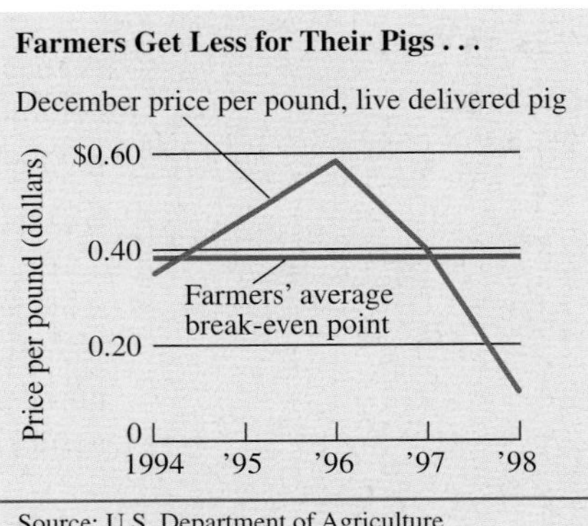

8. *Spending on Cosmetics* Use the graph to answer the questions.
   a) Estimate the total amount spent in the United States in 2000 on cosmetics (facial makeup, eye makeup, lip products, and nail products).
   b) The U.S. population on January 1, 2000, was about 273,300,000. Use this figure to estimate the average amount spent per person on cosmetics in the United States in 2000.

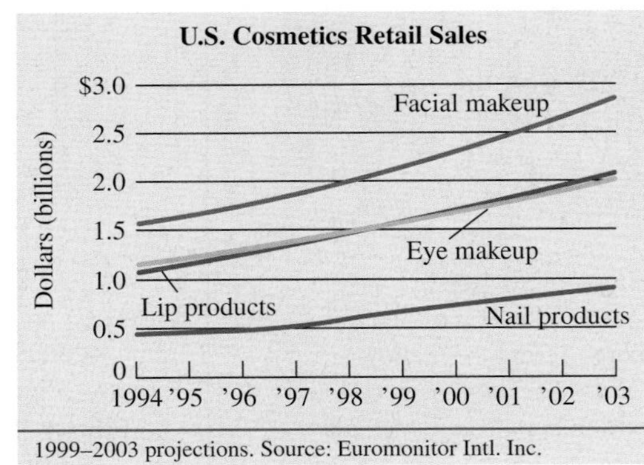

9. *Speed Estimate* The fastest U.S. aircraft, the Lockheed SR-71 Blackbird, travels at about Mach 3 (Mach 1 is the speed of sound, and Mach 3 is 3 times the speed of sound). If the speed of this aircraft is about 2310 mph, estimate Mach 1.

10. *Parking Costs* The Main Street Garage charges \$2.50 for the first hour of parking and \$1.00 for each additional hour or part thereof. Denise Tomey parks her car in the garage from 9 A.M. to 5 P.M., 5 days a week. How much money does she save by paying a weekly parking rate of \$35.00?

11. *Airport Parking* The chart at the top of the next page shows parking rates at Logan Airport in Boston, Massachusetts (as of August 1, 1999).
    a) Mike Drago is going out of town for 5 full days. He plans to park in Terminal B. How much will he save

Logan Airport parking rates

Location	First hour	Daily rate	Weekly rate	Additional day 0–6 hours	Additional day 6–24 hours
Terminal B	\$4.00	\$18.00	\$70.00	\$9.00	\$18.00
Central parking	\$4.00	\$18.00	\$70.00	\$9.00	\$18.00
Satellite economy lots	\$4.00	\$12.00	\$50.00	\$6.00	\$12.00
Short-term rates	**First hour**	**1–1½ hours**	**1½–2 hours**	**Each additional hour**	**First day maximum**
Terminal E west	\$4.00	\$6.00	\$8.00	\$2.00	\$48.00

Source: Logan Airport

by purchasing the weekly rate rather than paying the daily rate?

b) If he plans to use the satellite economy lots, would it be cheaper for him to pay the daily or weekly rate? How much is the difference?

c) What is the cost of short-term parking in Terminal E west if a person were to park there for 7 hours?

12. *Buying a Computer* Emily Falcon wants to purchase a computer that sells for \$1250. She can either pay the total amount at the time of purchase, or she can agree to pay the store \$120 down and \$80 a month for 15 months. How much money can she save by paying the total amount at the time of purchase?

13. *Calling Person to Person* Andrea Sheehan makes a person-to-person call from Houston, Texas, to Seattle, Washington. The call costs \$3.75 for the first 3 minutes and \$0.50 for each additional minute. How much did Andrea's 21 minute call cost?

14. *Getting an 80 Average* On four exams, Bill Leonard's grades were 77, 93, 90 and 76. What grade must he obtain on his fifth exam to have an 80 average?

15. *World Cup Soccer* An article in the July 19,1999, *Newsweek* stated that 650,000 tickets were purchased for the 32 different 1999 women's world cup soccer matches, which resulted in an income of about \$23,000,000. (Attendance at the final game in the Rose Bowl, in which the United States won the World Cup, was 90,185 fans.) What was the average income per game?

16. *Playing a Lottery* In one state lottery game, you must select a four-digit number (digits may be repeated). If your number matches exactly the four-digit number selected by the lottery commission, you win.

a) How many different numbers may be chosen?

b) If you purchase one lottery ticket, what is your chance of winning?

17. *Energy Value and Energy Consumption* The table gives the approximate energy values of some foods, in kilojoules (kJ), and the energy requirements of some activities. How soon would you use up the energy from

a) a fried egg by swimming?

b) a hamburger by walking?

c) a piece of strawberry shortcake by cycling?

d) a hamburger and a chocolate milkshake by walking?

Food	Energy value (kJ)	Activity	Energy consumption (kJ/min)
Chocolate milkshake	2200	Walking	25
Fried egg	460	Cycling	35
Hamburger	1550	Swimming	50
Strawberry shortcake	1400	Running	80
Glass of skim milk	350		

18. *Gas Mileage* Wendy Weisner fills her gas tank completely and makes a note that the odometer reads 38,451.4 mi. The next time she stops to put gas in her car, filling the tank takes 12.6 gal, and the odometer reads 38,687.0 mi. Determine the number of miles per gallon that Wendy's car gets.

19. *Diversity of Doctors* In 2000, the U.S. population was about 273,300,000 and the number of U.S. doctors was about 970,000. Use the graph to determine the number of

a) Hispanics in the United States.

b) Asian doctors in the United States.

c) African-American doctors in the United States.

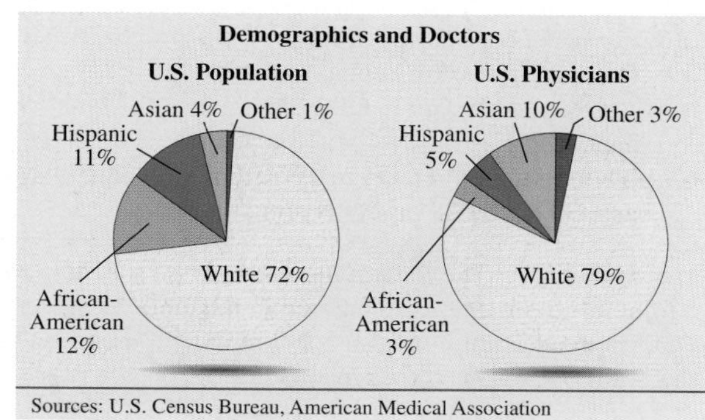

20. *Saving for a Boat*

a) Quinton works 20 hours per week and makes $5.40 per hour. How much money can he expect to earn in 1 year (52 weeks)?

b) If he saves all the money that he earns, how long will he have to work to save for a boat that costs $750?

21. *Mail Order Purchase* Mary purchased 4 tires by mail order. She paid $52.80 per tire plus $5.60 per tire for shipping and handling. There is no sales tax on this purchase because they were purchased out of state. She also had to pay $8.56 per tire for mounting and balancing. At a local tire store, her total for the 4 tires with mounting and balancing would be $324 plus an 8% tax. How much did Mary save by purchasing the tires through the mail?

22. *Sealing a Gym Floor* A gymnasium floor has an area of 2400 square yards (yd^2). Each gallon of floor sealant covers an area of 350 ft^2. How many gallons of sealant are needed to cover the gymnasium floor?

23. *Profit Margins* The following chart shows retail stores' average percent profit margin on certain items.

Product category	Average profit margin (%)
Video equipment	12%
Audio components	14
Stereo speakers	20–25
Extended warranties	50–60

Source: *Consumer Reports*

a) Determine the average profit of a store that has the list price on a camcorder of $620.

b) Determine the average profit of a store that has a list price on a pair of speakers for $1200 (use a 22% profit margin).

c) If you negotiate with the salesperson and get him or her to sell the speakers for $1000, find the store's profit.

24. *Leaking Faucet* A faucet is leaking at a rate of one drop of water per second. If the volume of one drop of water is 0.1 cubic centimeter (0.1 cm^3), find

a) the volume of water in cubic centimeters lost in a year.

b) how long it would take, in days, to fill a rectangular basin 30 cm by 20 cm by 20 cm.

25. *Income Taxes* The federal income tax rate schedule for a joint return in 1998 is illustrated in the table. If Steve and Maureen Tomlin paid $10,200 in federal taxes, find the family's adjusted gross income.

Adjusted gross income	Taxes
$0–$42,350	15% of income
$42,350–$102,300	$6,352.50 + 28% in excess of $42,350
$102,300–$155,950	$23,138.50 + 31% in excess of $102,300
$155,950–$278,450	$39,770 + 36% in excess of $155,950
$278,450 and up	$83,870 + 39.6% in excess of $278,450

26. *Wasted Water* A faucet leaks 1 oz of water per minute.

a) How many gallons of water are wasted in a year? (A gallon contains 128 oz.)

b) If water costs $5.20 per 1000 gal, how much additional money is being spent on the water bill?

27. *Shipping a Package* The chart shows what you would pay to ship a 3 lb, shoebox-sized parcel 2544 miles from Portland, Oregon, to Portland, Maine. If you sent the package discussed once a week for a year, determine how much you would save in a year by sending it

a) by Federal Express 2-day service rather than by Federal Express priority overnight.

b) by Post Office express mail rather than by United Parcel Service, next day air.

Service	Cost	Claimed delivery time
Federal Express		
SameDay	$159.00	Later that day
First Overnight	52.75	1 day by 8 A.M.
Priority Overnight	28.00	1 day by 10:30 A.M.
Standard Overnight	24.25	1 day by 4:30 P.M.
2 Day	11.75	2 days by 7 P.M.
Express Saver	10.70	3 days
Post Office		
Express Mail	17.25	1 day
Priority Mail	4.00	2–3 days
Parcel Post	3.95	8 days
United Parcel Service		
SonicAir Service	159.00	Later that day
Next Day Air	25.75	1 day by 10:30 A.M.
Next Day Air Saver	23.75	1 day by 3 P.M.
2nd Day Air	12.25	2 days by end of day
3 Day Select	9.40	3 days by end of day
Ground Residential	6.48	6 business days

Source: *Consumer Reports*

28. *Tire Pressure* When a car's tire pressure is 28 pounds per square inch (psi), it averages 15.8 mpg of gasoline. If the tire pressure is increased to 30 psi, the car averages 16.4 mpg of gasoline.

a) If Mr. Lee drives an average of 15,000 mi per year, how many gallons of gasoline will he save in a year by increasing his tire pressure from 28 to 30 psi?

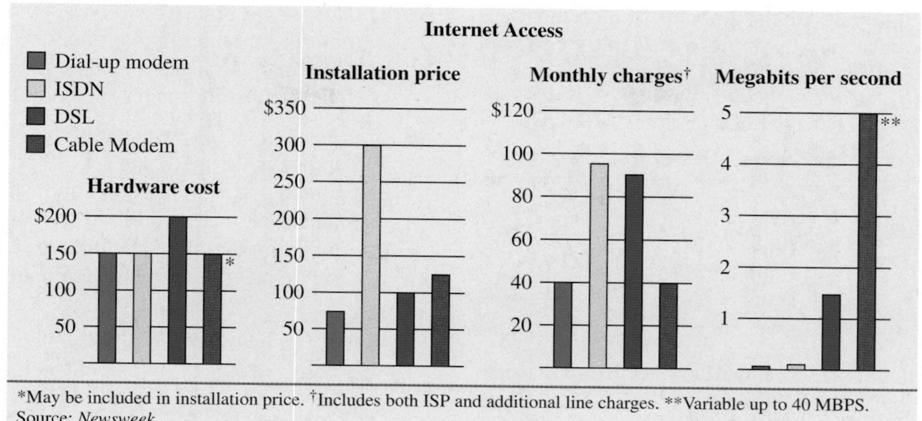

*May be included in installation price. †Includes both ISP and additional line charges. **Variable up to 40 MBPS.
Source: *Newsweek*

b) If gasoline costs $1.20 per gallon, how much will he save in a year?
c) If we assume that there are about 140 million cars in the United States and that these changes are typical of each car, how many gallons of gasoline would be saved if all drivers increased their cars' tire pressure?

29. *Internet Costs* Each year, the number of Americans using the Internet grows dramatically. The graphs above provide information regarding Internet access using four different means.
a) Estimate the total cost for a year, including the hardware, the installation charge, and 12 months' use, using a dial-up modem.
b) Repeat part (a) for a cable modem.
c) Using the chart for megabits per second, estimate the number of times faster a cable modem is than a dial-up modem.

30. *Car Prices* The cost of a car increases by 20% and then decreases by 20%. Is the resulting car price greater than, less than, or equal to the original price of the car?

31. *Adjusting for Inflation* Assume the rate of inflation is 6% for the next 2 years. What will be the cost of goods 2 years from now, adjusted for inflation, if the goods cost $450.00 today?

32. *Buying a Futon* Mary Mahan wants to purchase a twin-size and a full-size futon and 2 matching pillows for each futon. The twin-size futon regularly costs $498, the full-size futon regularly costs $598, and each matching pillow regularly costs $15. In a package deal, Mary was able to obtain the two futons and four pillows for $772.
a) How much did she save off the regular price?
b) What percent of the regular price did she pay?

33. *A Photo Safari* Deirdre McGill is planning a trip to Africa where she will participate in a photo safari. She is planning on bringing a great deal of film. A photography store is selling 4 packs of film for $17 and 10 packs of the same film for $41.
a) If she wishes to purchase only the 4 packs and 10 packs and to spend a maximum of $200 on film, what is the maximum number of rolls of film she can purchase?
b) What will be the cost?

34. *Buying Film* Erika Gutierrez is planning a vacation to Australia and wishes to bring a large supply of film. At Wal-Mart, 4 packs of 24 exposure film costs $4.08 and 4 packs of the same film with 36 exposures costs $5.76.
a) If she wishes to spend a maximum of $50 on film and get the most exposures, how many 4 packs of 24 exposures and how many 4 packs of 36 exposures should she purchase?
b) How many exposures will she get?
c) What will be the cost? If there is more than one choice in part (a), give the minimum cost.

35. *Making Cream of Wheat* The following amounts of ingredients are recommended to make various servings of Nabisco Instant Cream of Wheat. *Note:* 16 tbsp = 1 cup.

Ingredient	1 Serving	2 Servings	4 Servings
Mix water or milk	1 cup	2 cups	$3\frac{3}{4}$ cups
With salt (optional)	$\frac{1}{8}$ tsp	$\frac{1}{4}$ tsp	$\frac{1}{2}$ tsp
Add Cream of Wheat	3 tbsp	$\frac{1}{2}$ cup	$\frac{3}{4}$ cup

Determine the amount of each ingredient needed to make 3 servings using the following procedures.
a) Multiply the amounts for 1 serving by 3.
b) Find the average of the amounts for 2 and 4 servings.
c) Subtract the amounts for 1 serving from the amounts for 4 servings.
d) Compare the answers obtained in parts (a) through (c) and explain any differences.

36. *Making Rice* Following are the amounts of ingredients recommended to make various servings of Uncle Ben's Original Converted Rice. *Note:* 1 tbsp = 3 tsp.

Ingredient	2 Servings	4 Servings	6 Servings	12 Servings
Rice (cups)	$\frac{1}{2}$	1	$1\frac{1}{2}$	3
Water (cups)	$1\frac{1}{3}$	$2\frac{1}{4}$	$3\frac{1}{3}$	6
Salt (teaspoons)	$\frac{1}{4}$	$\frac{1}{2}$	$\frac{3}{4}$	$1\frac{1}{2}$
Butter or margarine	1 tsp	2 tsp	1 tbsp	2 tbsp

Determine the amount of each ingredient needed to make 8 servings using the following procedures.

a) Multiply the amount for 2 servings by 4.
b) Multiply the amount for 4 servings by 2.
c) Add the amounts for 2 and 6 servings.
d) Compare the answers obtained in parts (a) through (c) and explain any differences.

Solve the following problems.

37. How many square inches, 1 in. by 1 in., fit in an area of 1 square foot, 1 ft by 1 ft?

38. How many cubic inches fit in 1 cubic foot?

39. If the length and width of a rectangle each double, what happens to the area of the rectangle?

40. If the length, width, and height of a cube all double, what happens to the volume of the cube?

41. Fill in the three boxes using the symbols $+$, $-$, $\times$, and $\div$ to make a true statement of equality:

$$7 \square 7 \square (7 \square 7) = 13$$

42. A 24-ft-by-24-ft carpet is partitioned into 4-ft-by-4-ft squares. How many squares will there be?

43. *Zebras and Cranes* While on a safari in Africa, I saw a small herd of zebras and cranes wandering over the veldt. Counting heads, I got 18. Counting feet, I got 60. How many zebras and how many cranes were on the veldt?

44. *Palindromes* A *palindrome* is a number that reads the same forward and backwards. The numbers 1991 and 43234 are examples of palindromes. How many palindromes are there between the numbers 2000 and 3000? List them.

45. *Supermarket Display* The figure shows oranges in a supermarket display stacked in a *square pyramid* (the base is a square).

a) How many oranges are in the pyramid shown if the base is 4 oranges by 4 oranges?
b) How many oranges would be in a square pyramid if the base was 7 oranges by 7 oranges?

46. *Financial Loss* A woman purchased a dress that cost \$45 and gave the merchant a \$100 bill. After the woman had gone home with her dress and her change, the merchant took the \$100 bill to the bank. The bank clerk informed him that the \$100 bill was counterfeit. What was the financial loss to the merchant?

47. *Balancing a Scale* If you have a balance scale and only the four weights 1 gram (g), 3 g, 9 g, and 27 g, explain how you could show that an object had the following weights.

a) 5 g **b)** 16 g

(*Hint:* Weights must be added to both sides of the balance scale.)

48. Create a magic square by using the numbers 2, 4, 6, 8, 10, 12, 14, 16, and 18. The sum of the numbers in every column, row, and diagonal must be 30.

49. Create a magic square by using the numbers 1, 3, 5, 7, 9, 11, 13, 15, and 17. The sum of the numbers in every column, row, and diagonal must be 27.

In Exercises 50–52, use the three magic squares illustrated to obtain the answers.

6	5	10
11	7	3
4	9	8

3	2	7
8	4	0
1	6	5

10	9	14
15	11	7
8	13	12

50. Examine the 3 by 3 magic squares and find the sum of the four corner entries of each magic square. How can you determine the sum by using a key number in the magic square?

51. For a 3 by 3 magic square, how can you determine the sum of the numbers in any particular row, column, or diagonal by using a key value in the magic square?

52. For a 3 by 3 magic square, how can you determine the sum of all the numbers in the square by using a key value in the magic square?

53. *Dominos* Consider a domino with six dots, as shown. Two ways of connecting the three dots on the left with the three dots on the right as illustrated. In how many ways can the three dots on the left be connected with the three dots on the right?

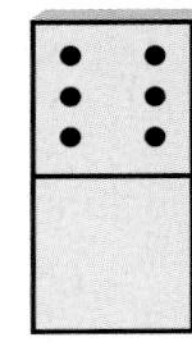
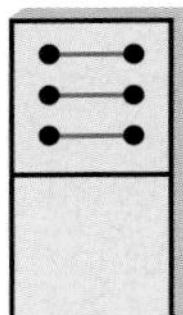
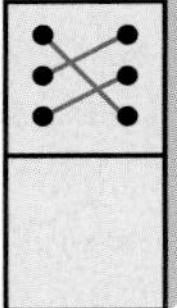

54. *Stack of Blocks* A solid pyramid-like stack is made from a number of identical blocks. A side and top view of the stack follow. How many blocks were used for the stack?

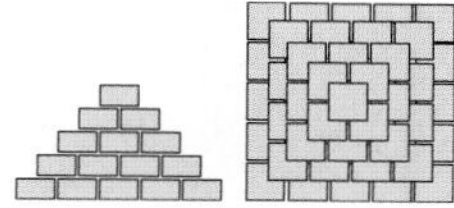

55. *Stack of Cubes* Identical cubes are stacked in the corner of a room, as shown. How many of the cubes are not visible?

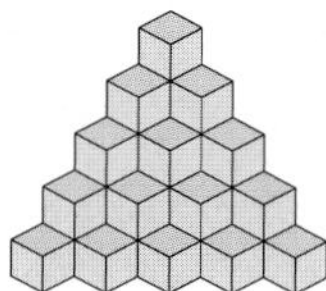

56. *Handshakes All Around* Five salespeople gather for a sales meeting. How many handshakes will each person make if each must shake hands with each of the four others?

57. *Consecutive Digits* Place the digits 1 through 8 in the eight boxes so that each digit is used exactly once and no two consecutive digits touch horizontally, vertically, or diagonally.

58. *A Digital Clock* Digital clocks display numerals by lighting some or all of the seven parts of the pattern shown. If each digit 0 through 9 is displayed once, which part is used least often? Which part is used most often?

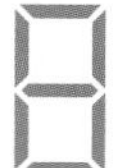

59. *A Grid* Place five 1's, five 2's, five 3's, five 4's, and five 5's in a 5×5 grid so that each digit—that is, 1, 2, 3, 4, 5—appears exactly once in each row and exactly once in each column.

PROBLEM SOLVING/GROUP ACTIVITIES

60. *Insurance Policies* Ray Kelley owns two cars (a Ford Mustang and a Ford Escort), a house, and a rental apartment. He has auto insurance for both cars, a homeowner's policy, and a policy for the rental property. The costs of the policies are

Mustang: $695 per year
Escort: $650 per year
Homeowner's: $412 per year
Rental property: $597 per year

Ray is considering taking out a $1 million personal umbrella liability policy. The annual cost of the umbrella policy would be $392. If he has the umbrella policy, he can lower the limits on parts of his auto policies and still have equal or better protection. If Ray purchases the umbrella policy, he can reduce his premium on the Mustang by $60 per year and his premium on the Escort by 36%. If he purchases the umbrella policy and reduces the amount he pays for auto insurance, what is the net amount he is actually paying for the umbrella policy?

61. *A Sports Puzzle* Peter, Paul, and Mary are three sports professionals. One is a tennis player, one is a golfer, and one is a skier. They live in three adjacent houses on City View Drive. From the following information determine which is the professional skier. (*Hint:* A table may be helpful.)

Mary does not play tennis.
Peter skis and plays tennis, but does not golf.
The golfer and the skier live next to each other.
Three years ago, Paul broke his leg skiing and has not tried it since.
Mary lives in the last house.
The golfer and the tennis player share a common backyard swimming pool.

62. *Counting Triangles* How many triangles are in the figure?

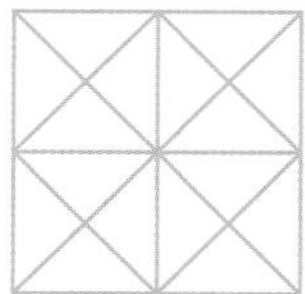

63. *Finding the Area* Rectangle ABCD is made up entirely of squares. The black square has a side of 1 unit. Find the area of ABCD.

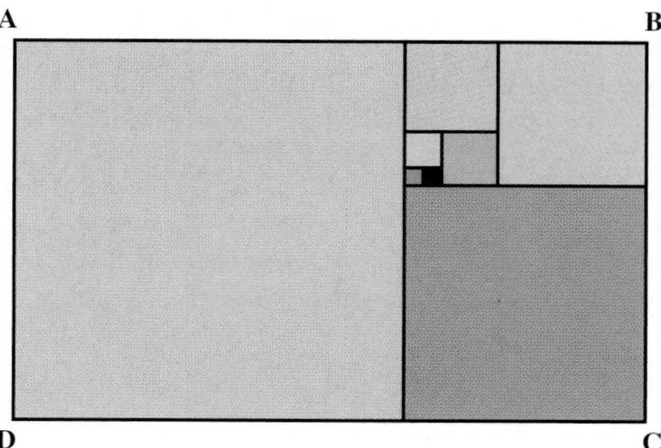

CHAPTER 1 SUMMARY

IMPORTANT FACTS

The **natural numbers** or **counting numbers** are 1, 2, 3, 4, ...

A **conjecture** is a prediction based on specific observations.

A **counterexample** is a special case that satisfies all the conditions of a conjecture, but proves the conjecture false.

Inductive reasoning is the process of reasoning to a general conclusion through observations of specific cases.

Deductive reasoning is the process of reasoning to a specific conclusion from a general statement.

GUIDELINES FOR PROBLEM SOLVING

1. Understand the problem.
2. Devise a plan to solve the problem.
3. Carry out the plan.
4. Check the results.

CHAPTER 1 REVIEW EXERCISES

1.1*

In Exercises 1–8, use inductive reasoning to predict the next three numbers or figures in the pattern.

1. 3, 8, 13, 18, . . .

2. 1, 4, 9, 16, . . .

3. 4, −8, 16, −32, . . .

4. 5, 7, 10, 14, 19, . . .

5. 25, 24, 22, 19, 15, . . .

6. 6, 3, $\frac{3}{2}$, $\frac{3}{4}$, . . .

7. 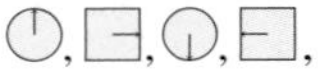 , , , , . . .

8. 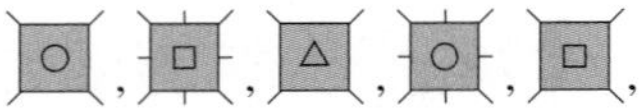 , , , , , . . .

9. Pick any number and multiply the number by 2. Add 10 to the product. Divide the sum by 2. Subtract 5 from the quotient.

- **a)** What is the relationship between the number you started with and the final number?
- **b)** Arbitrarily select some different numbers and repeat the process, recording the original number and the results.
- **c)** Make a conjecture about the original number and the final number.
- **d)** Prove, using deductive reasoning, the conjecture you made in part (c).

10. Choose a number between 1 and 20. Add 5 to the number. Multiply the sum by 6. Subtract 12 from the product. Divide the difference by 2. Divide the quotient by 3. Subtract the number you started with from the quotient. What is your answer? Try this process with a different number. Make a conjecture as to what your final answer will always be.

11. Find a counterexample to the statement "The difference between two squares is an odd number."

1.2

In Exercises 12–20, estimate the answer. Your answers may vary from those given in the back of the book, depending on how you round to arrive at the answer.

12. 204,600 × 1963

13. $\dfrac{19{,}254.5}{524.3}$

14. 346.2 + 96.402 + 1.04 + 897 + 821

15. 21% of 1012

16. Estimate the distance from your wrist to your elbow and estimate the length of your foot. Which do you think is greater? With the help of a friend, measure both lengths to determine which is longer.

17. Estimate the cost of eight compact discs if each disc costs $12.99.

18. Estimate the amount of a 6% sales tax on a coat that costs $202.

---

*The number in color indicates the section in which the material is covered.

19. Estimate your average walking speed in miles per hour if you walked 1.1 mi. in 22 min.

20. Estimate the total cost of six grocery items that cost \$2.49, \$0.79, \$1.89, \$0.10, \$2.19, \$6.75.

21. *A Walking Path* The scale of the map is $\frac{1}{4}$ in. = 0.1 mi. Estimate the distance of the walking path indicated in red.

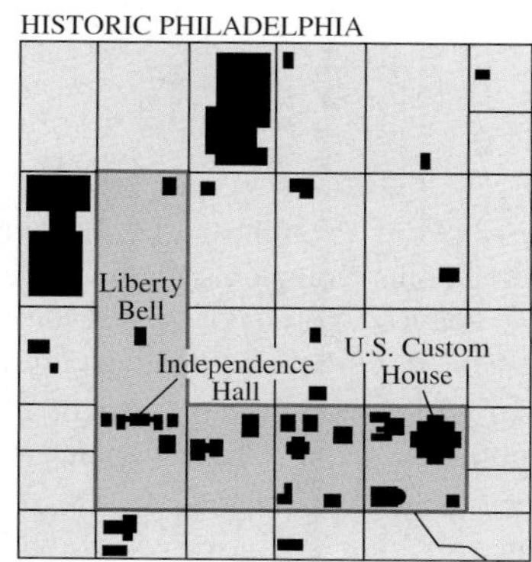

In Exercises 22 and 23, refer to the following graph.

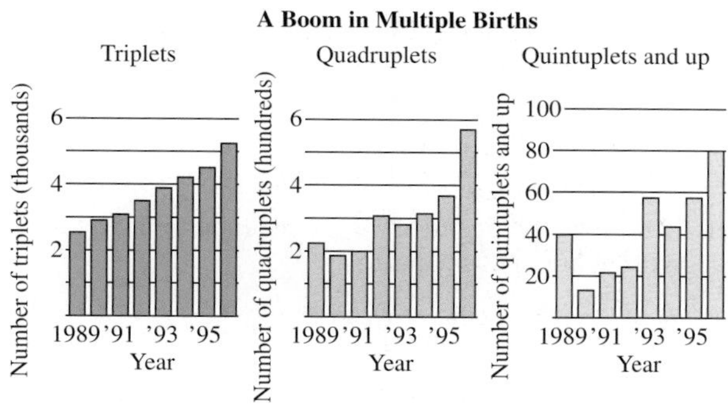

22. Estimate the difference in the number of births of quintuplets and up between 1990 and 1996.

23. Estimate the total number of multiple births (triplets and up) in 1996.

24. *Estimating an Area* If each square represents one square unit, estimate the size of the shaded area.

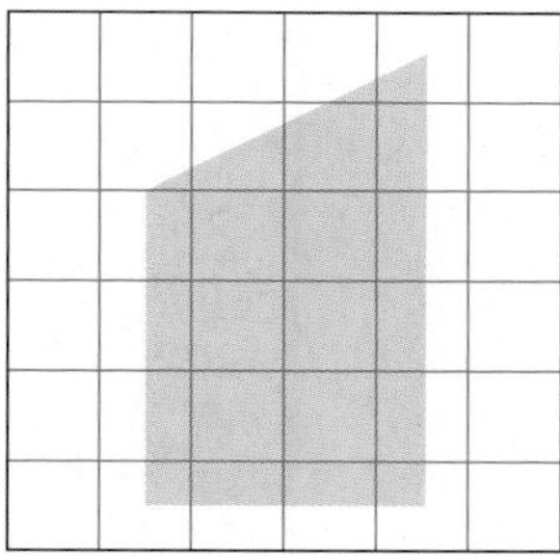

25. *Railroad Car Estimation* The scale of a model railroad is 1 in. = 12.5 ft. Estimate the size of an actual box car if this drawing is the same size as the model box car.

1.3

Solve the following problems.

26. *Dialing Direct* Dorothy Moore made an operator-assisted telephone call to another state. The call costs \$7.50 for 15 min. The same call dialed directly costs \$1.20 for the first minute and \$0.30 for each additional minute. How much money would Dorothy have saved by dialing direct?

27. *Change from a Ten* Jeff Howard parked his car in a lot that charged \$2.00 for the first hour and \$0.90 for each additional hour. He left the car in the lot for 8 hr. How much change did he receive from a \$10 bill?

28. *Buying in Quantity* A six-pack of cola costs \$3.45. A carton of 4 six-packs costs \$12.60. How much will be saved by purchasing the carton rather than 4 individual six-packs?

29. *Jet Ski Rental* The rental cost of a jet ski from Josh Eurich's Ski Rental is \$10 per 15 min, and the cost from Kirsten Starr's Ski Rental is \$25 per half hour. If you plan to rent the jet ski for 2 hr, which is the better deal, and by how much?

30. *Cab Fare* A taxicab charges \$1.35 for the first 1/5 mi and 20 cents for each additional 1/5 mi. Determine the cost of a 10-mi trip.

31. *Auto Insurance* Most insurance companies reduce premiums by 10% until age 25 for people who successfully pass a driver education course. A particular driver education course costs \$60. Richard Semmler, who just turned 18, has auto insurance that costs \$530 per year. By taking the driver education course, how much would he save in auto insurance premiums from the age of 18 until the age of 25?

32. *Pediatric Dosage* If 1.5 milligrams (mg) of a medicine is to be given for 10 lb of body weight, how many milligrams should be given to a child who weighs 47 lb?

33. *Qualifying for a Mortgage* Banks will grant an applicant a mortgage if the monthly payments are not greater than 25% of the person's take-home pay. What is the maximum monthly mortgage payment you can make if your gross salary is \$3800 a month and your payroll deductions are 30% of your gross salary?

34. *Flying West* New York City is on eastern standard time, St. Louis is on central standard time (1 hr earlier than eastern standard time), and Las Vegas is on Pacific standard time (3 hr earlier than eastern standard time). A flight leaves New York City at 9 A.M. eastern standard time, stops for 50 min in St. Louis, and arrives in Las Vegas at 1:35 P.M. Pacific time. How long is the plane actually flying?

35. *Crossing Time Zones* The international date line is an imaginary line of longitude (from the North Pole to the South Pole) on Earth's surface between Japan and Hawaii in the Pacific Ocean. Crossing the line east to west adds a day to the present date. Crossing the line west to east subtracts a day. At 3:00 P.M. on July 25 in Hawaii, what is the time and date in Tokyo, Japan, which is four time zones to the west?

36. 1 in. = 2.54 cm.

a) How many square centimeters are in a square inch?
b) How many cubic centimeters are in a cubic inch?
c) How long is a centimeter in terms of inches?

37. If the following pattern is continued, how many dots will be in the hundredth figure?

38. Complete the magic square by using the numbers 6 through 21 exactly once.

21	7		18
10		15	
14	12	11	17
9	19		

39. Create a magic square by using the numbers 13, 15, 17, 19, 21, 23, 25, 27, and 29.

40. *Microbes in a Jar* A colony of microbes doubles in number every second. A single microbe is placed in a jar, and in an hour the jar is full. When was the jar half full?

41. *Brothers and Sisters* Jim has four more brothers than sisters. How many more brothers than sisters does his sister Mary have?

42. *A Missing Dollar* Three friends check into a single room in a motel and pay \$10 apiece. The room costs \$25 instead of \$30, so a clerk is sent to the room to give \$5 back. The friends each take back \$1, and the clerk is given \$2 for his trouble. Now each of the friends paid \$9, a total of \$27, and the clerk received \$2. What happened to the missing dollar?

43. *The Average Weight* Four women in a room have an average weight of 130 lb. A fifth woman who weighs 180 lb enters the room. Find the average weight of all five women.

44. *Change for a Dollar* Could a person have \$1.15 worth of change in his pocket and still not be able to give someone change for a dollar bill? If so, what coins might he have?

45. *Volume of a Cube* Here is a flat pattern for a cube to be formed by folding. The sides of each square are 6 cm. Find the volume of the cube.

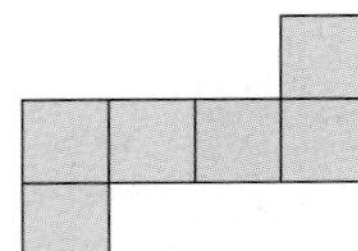

46. *The Heavier Coin* You have 13 coins, which all look alike. Twelve coins weigh exactly the same, but the other one is heavier. You have a pan balance. Tell how to find the heavier coin in just three weighings.

47. *The Sum of Numbers* Find the sum of the first 500 counting numbers. (*Hint:* Group in pairs.)

48. *Balancing a Scale* On a balance scale, three green balls balance six blue balls, two yellow balls balance five blue balls, and six blue balls balance four white balls. How many blue balls are needed to balance four green, two yellow, and two white balls?

49. How many three-digit numbers greater than 100 are palindromes?

50. Describe the fifth figure.

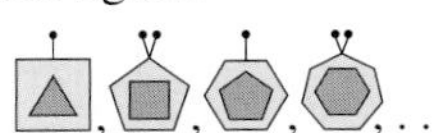

51. How many orange tiles will be required to build the sixth figure in this pattern?

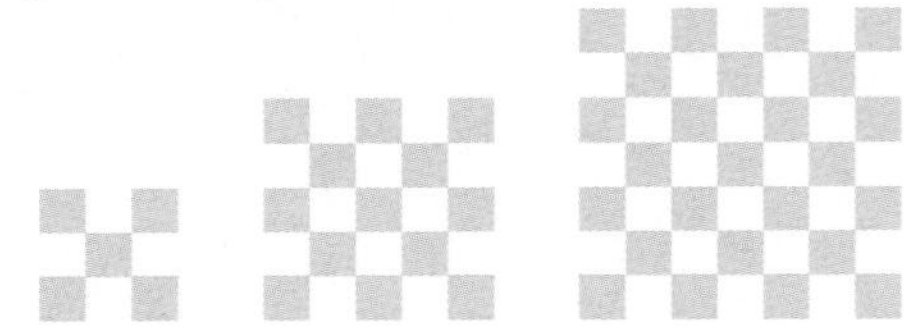

52. Place the numbers 1 through 12 in the 12 circles so that the sum in each of the six rows and the sum of the six points is 26. Use each number from 1 through 12 exactly once.

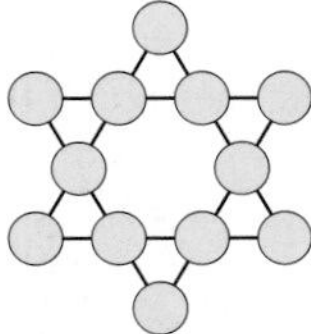

53. In how many ways can

a) two people stand in a line?
b) three people stand in a line?
c) four people stand in a line?
d) five people stand in a line?
e) Using the results from parts (a) through (d), make a conjecture about the number of ways in which n people can stand in a line.

CHAPTER 1 TEST

In Exercises 1 and 2, use inductive reasoning to determine the next three numbers in the pattern.

1. 6, 9, 12, 15, . . .

2. $1, \frac{1}{3}, \frac{1}{9}, \frac{1}{27}, \ldots$

3. Pick any number, multiply the number by 5, and add 10 to the number. Divide the sum by 5. Subtract 1 from the quotient.

a) What is the relationship between the number you started with and the final answer?

b) Arbitrarily select some different numbers and repeat the process. Record the original number and the results.

c) Make a conjecture about the relationship between the original number and the final answer.

d) Prove, using deductive reasoning, the conjecture made in part (c).

In Exercises 4 and 5, estimate the answers.

4. $0.00417 \times 990{,}000$

5. $\dfrac{91{,}000}{0.00302}$

6. If each square represents one square unit, estimate the area of the shaded figure.

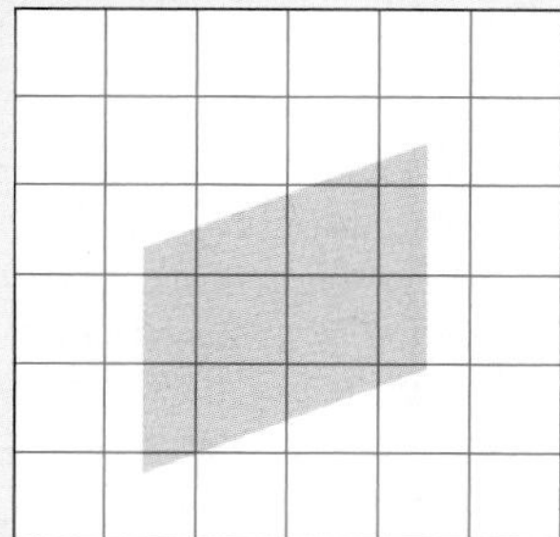

7. *Whose Hurting?* The following graph shows the percent of men and women who get migraine headaches in various age groups.

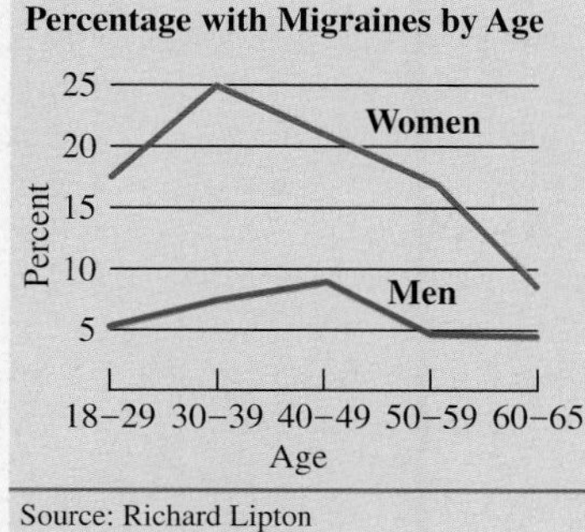

a) Estimate the difference in the percent of men and women who get migraine headaches in the 30–39 age group.

b) Estimate the difference in the percent of women who get migraines in the 30–39 age group than women in the 60–65 age group.

8. *Gas Usage* A gas company charges $7.42 for the basic monthly fee, which includes the first 3 therms of gas used. It charges 62 cents for each additional therm used. If the Smiths' gas bill for December was $100.42, how many therms of gas did they use during that month?

9. *Cans of Soda* At a local store a six pack of soda costs $3.60 and individual cans cost $0.90. What is the maximum number of cans of soda that can be purchased for $15?

10. *Cutting Wood* How much time does it take Carla Knab, a carpenter, to cut a 10 ft length of wood into four equal pieces, if each cut takes $2\frac{1}{2}$ min?

11. *Determining Size* In this photo of the *Mona Lisa,* 1506, by Leonardo da Vinci, 1 in. equals 12 in. on the actual painting. Find the dimensions of the actual painting.

12. *Payment Shortfall* Monica Wilson gets $12.75 per hour with time and a half for any time over 40 hours per week. If she works a 50-hr week and gets paid $652.25, by how much was she underpaid?

13. Create a magic square by using the numbers 5, 10, 15, 20, 25, 30, 35, 40, and 45. The sum of the numbers in every row, column, and diagonal must be 75.

14. *A Drive to the Beach* Christine Ngau Tibbits drove from her home to the beach that is 30 mi from her house. The first 15 mi she drove at 60 mph, and the next 15 mi she drove at 30 mph. Would the trip take more, less, or the same time if she traveled the entire 30 mi at a steady 45 mph?

15. From the six numbers 2, 6, 8, 9, 11, and 13, pick five that, when multiplied, give 11,232.

16. *Jelly Bean Guess* One guess is off by 9, another guess is off by 17, and yet another guess is off by 31. How many beans are in the jar?

My guess is 260
HOW MANY BEANS
My guess is 274
My guess is 234

17. *Buying Plants* David Mackin wants to purchase nine herb plants. Countryside Nursery has herbs that are on sale at three for \$3.99. David has a coupon for 25% off an unlimited number of herb plants at the original price of \$1.75 per plant.

a) Determine the cost of purchasing nine plants at the sale price.

b) Determine the cost of purchasing nine plants if the coupon is used.

c) Which is the least expensive way to purchase the nine plants, and by how much?

18. *Setting the Switches* In how many different ways can a panel of four on–off switches be set if no two adjacent switches may be off?

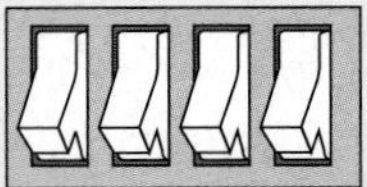

GROUP PROJECTS

HOLIDAY SHOPPING

1. It is December 1, and John needs to begin his holiday shopping. He intends to purchase gifts for his girlfriend Melissa, his mother Ruth, and his father Don. He doesn't want to spend more than a total of \$325, including the 7% sales tax.

a) If John were to spend the \$325 equally among the three people, approximate the amount that would be spent on each person.

b) If John were to spend the \$325 equally among the three people, determine the maximum amount, *before tax,* that he could spend on each person and not exceed the maximum of \$325, including tax.

c) John decides to get a new set of wrenches for his father. He sees the specific set he wants on sale at Sears. He calls four Sears stores to see if they have the set of wrenches in stock. They all reply that the set is out of stock. He decides that calling additional Sears stores is useless, for he believes that they will also tell him that the set of wrenches is out of stock. What type of reasoning did John use in arriving at his conclusion? Explain.

d) John finds an equivalent set of wrenches at a True Value hardware store. The set he is considering is a combination set that contains both standard U.S. size and metric size wrenches. Its regular price before tax is \$62, but it is selling for 10% off its regular price. He can also purchase the same wrenches by purchasing two separate sets, one for standard U.S. size wrenches and the other for metric sizes. Each of these sets has a regular price, before tax, of \$36, but both are on sale for 20% off their regular prices. Can John purchase the combination set or the two individual sets less expensively?

e) How much will John save, *after tax,* by using the less expensive method?

GOING ON VACATION

2. Bill and Kristen Url and their 4-year-old daughter Betty decide to go on a vacation. They live in San Francisco, California, and plan to drive to New Orleans, Louisiana.

a) Obtain a map that shows routes that they may take from San Francisco to New Orleans. Write directions for them from San Francisco to New Orleans via the shortest distance. Use major highways whenever possible.

b) Use the scale on the map to estimate the one-way distance to New Orleans.

c) If the Urls estimate that they will average 50 mph (including comfort stops), estimate the travel time, in hours, to New Orleans.

d) If the Urls want to travel about 400 miles per day, locate a town in the vicinity of where they will stop each evening.

e) If they begin each segment of the trip each day at 9 A.M., about what time will they look for a hotel each evening?

f) Use the information provided in parts (a) through (e) to estimate the time of day they will arrive in New Orleans.

g) Estimate the mileage of a typical mid-sized car and the cost per gallon of a gallon of regular unleaded gasoline. Then estimate the cost of gasoline for the Urls's trip.

h) Estimate the cost of a typical breakfast, a typical lunch, and a typical dinner for two adults and a 4-year-old child, and the cost of a typical motel room. Then estimate the total cost, including meals, gas, and lodging, for the Urls's trip from San Francisco to New Orleans (one way).

PROBLEM SOLVING

3. Four acrobats who bill themselves as the "Tumbling Tumbleweeds" finish up their act with the amazing "Human Pillar," in which the acrobats form a tower, each one standing on the shoulders of the one below. Each acrobat (Ernie, Jed, Tex, and Zeke Tumbleweed) wears a different distinctive item of western garb (chaps, holster, Stetson hat, or leather vest) in the act. Can you identify the members of the "Human Pillar," from top to bottom, by name and apparel?

a) Jed Tumbleweed is not on top, but he is somewhere above the man in the Stetson.

b) Zeke Tumbleweed does not wear the holster.

c) The man in the vest is not on top.

d) The man in the chaps is somewhere above Tex but somewhere below Zeke.

Order	Name	Apparel
______	______	______
______	______	______
______	______	______
______	______	______

CHAPTER 2

Sets

One of the most basic human impulses is to sort and classify things. Consider yourself, for example. How many different sets are you a member of? You might start with some simple categories, such as whether you are male or female, your age group, and state you live in. Then you might think about your family's ethnic group, socioeconomic group, and nationality. These are but some of the many ways you could describe yourself to other people.

Of what use is this activity of categorization? As you will see in this chapter, putting elements into sets helps you order and arrange your world. It allows you to deal with large quantities of information. Set building is a learning tool that helps answer the question, "What are the characteristics of this group?"

Sets underlie other mathematical topics, such as logic and abstract algebra. In fact, the book *Eléments de Mathématique,* written by a group of French mathematicians under the pseudonym Nicolas Bourbaki, states, "Nowadays it is possible, logically speaking, to derive the whole of known mathematics from a single source, the theory of sets." ■

Set building is a fundamental learning tool for even the smallest children. As babies, they learn to distinguish "me" from "mom" and "dad." As toddlers, they learn to distinguish and categorize objects as members of a set according to size, color, or shape. The TV show "Sesame Street" teaches children set building in the game "One of these things is not like the other."

2.1 SET CONCEPTS

We encounter sets in many different ways every day of our lives. A **set** is a collection of objects, which are called **elements** or **members** of the set. For example, the United States is a collection or set of 50 states. The 50 individual states are the members or elements of the set that is called the United States.

A set is **well defined** if its contents can be clearly determined. The set of justices presently serving on the U.S. Supreme Court is a well-defined set because its contents, the justices, can be named. The set of the three best cars is not a well-defined set because the word *best* is interpreted differently by different people. In this text, we use only well-defined sets.

Three methods are commonly used to indicate a set: (1) description, (2) roster form, and (3) set-builder notation.

The method of indicating a set by **description** is illustrated in Example 1.

PROFILE IN MATHEMATICS

Georg Cantor

Georg Cantor (1845–1918), born in St. Petersburg, Russia, is recognized as the founder of set theory. Cantor's creative work in mathematics was nearly lost when his father insisted that he become an engineer rather than a mathematician. His two major books on set theory, *Foundations of General Theory of Aggregates* and *Contributions to the Founding of the Theory of Transfinite Numbers,* were published in 1883 and 1895, respectively.

EXAMPLE 1 *Description of Sets*

Write a description of the set containing the elements Monday, Tuesday, Wednesday, Thursday, Friday, Saturday, Sunday.

SOLUTION The set is the days of the week. ▲

Listing the elements of a set inside a pair of **braces,** { }, is called **roster form.** The braces are an essential part of the notation because they identify the contents as a set. For example, $\{1, 2, 3\}$ is notation for the set whose elements are 1, 2, and 3, but $(1, 2, 3)$ and $[1, 2, 3]$ are not sets because parentheses and brackets do not indicate a set. For a set written in roster form, commas separate the elements of the set. The order in which the elements are listed is not important.

Sets are generally named with capital letters. For example, the name commonly selected for the set of **natural numbers** or **counting numbers** is N.

Natural Numbers

$$N = \{1, 2, 3, 4, 5, \ldots\}$$

The three dots after the 5, called an *ellipsis,* indicate that the elements in the set continue in the same manner. An ellipsis followed by a last element indicates that the elements continue in the same manner up to and including the last element. This notation is illustrated in Example 2(b).

EXAMPLE 2 *Roster Form of Sets*

Express the following in roster form.

a) Set A is the set of natural numbers less than 4.
b) Set B is the set of natural numbers less than or equal to 70.
c) Set P is the set of planets in Earth's solar system.

SOLUTION

a) The natural numbers less than 4 are 1, 2, and 3. Thus, set A in roster form is $A = \{1, 2, 3\}$.

The planets of Earth's solar system.

b) $B = \{1, 2, 3, 4, \ldots, 70\}$. The 70 after the ellipsis indicates that the elements continue in the same manner through the number 70.
c) P = {Mercury, Venus, Earth, Mars, Jupiter, Saturn, Uranus, Neptune, Pluto} ▲

EXAMPLE 3 *The Word* Inclusive

Express the following in roster form.

a) The set of natural numbers between 5 and 8.
b) The set of natural numbers between 5 and 8, inclusive.

SOLUTION
a) $A = \{6, 7\}$
b) $B = \{5, 6, 7, 8\}$. Note that the word *inclusive* indicates that the values of 5 and 8 are included in the set. ▲

The symbol $\in$, read, is an element of, is used to indicate membership in a set. In Example 3, since 6 is an element of set A, we write $6 \in A$. This may also be written $6 \in \{6, 7\}$. We may also write $8 \notin A$, meaning that 8 is not an element of set A.

Set-builder notation (sometimes called *set-generator notation*) may be used to symbolize a set. Set-builder notation is frequently used in algebra. The following example illustrates its form.

D	$=$	$\{$	x	$\vert$	Conditions(s) $\}$
↑	↑	↑	↑	↑	↑
Set D	is	the set of	all elements x	such that	the condition(s) x must meet in order to be a member of the set.

Consider $E = \{x \mid x \in N \text{ and } x > 10\}$. The statement is read: "Set E is the set of all the elements x such that x is a natural number and x is greater than 10." The conditions that x must meet to be a member of the set are $x \in N$, which means that x must be a natural number, and $x > 10$, which means that x must be greater than 10. The numbers that meet both conditions are the set of natural numbers greater than 10. The set in roster form is

$$E = \{11, 12, 13, 14, \ldots\}$$

EXAMPLE 4 *Using Set-Builder Notation*

a) Write set $B = \{1, 2, 3, 4, 5\}$ in set-builder notation.
b) Write, in words, how you would read set B in set-builder notation.

SOLUTION
a) Since set B consists of the natural numbers less than 6, we write

$$B = \{x \mid x \in N \text{ and } x < 6\}$$

Another acceptable answer is $B = \{x \mid x \in N \text{ and } x \leq 5\}$.

b) Set B is the set of all elements x such that x is a natural number and x is less than 6. ▲

EXAMPLE 5 ***Roster Form to Set-Builder Notation***

a) Write set S = {Maine, Maryland, Massachusetts, Michigan, Minnesota, Mississippi, Missouri, Montana} in set-builder notation.
b) Write in words how you would read set S in set-builder notation.

SOLUTION

a) $S = \{x \mid x$ is a state in the United States whose name begins with the letter M$\}$.
b) Set S is the set of all elements x such that x is a state in the United States whose name begins with the letter M. ▲

EXAMPLE 6 ***Set-Builder Notation to Roster Form***

Write set $A = \{x \mid x \in N \text{ and } 2 \leq x < 8\}$ in roster form.

SOLUTION $A = \{2, 3, 4, 5, 6, 7\}$ ▲

EXAMPLE 7 ***Fastest-Growing Cities***

The chart shows the 10 fastest-growing U.S. counties (in terms of percent growth) with populations of 10,000 or more for 1998–1999. Also given is a map that shows Atlanta, Georgia, and its surrounding counties. Let G be the set of counties that surround Atlanta that are among the 10 fastest-growing counties in the United States. Write set G in roster form.

Ten fastest-growing U.S. counties with populations of 10,000 or more:
Increase, 1997–1998

Forsyth, GA	**13.0%**
Douglas, CO	11.2%
Loudoun, VA	8.2%
Henry, GA	**7.2%**
Collin, TX	6.9%
Paulding, GA	**6.9%**
Dawson, GA	**6.5%**
Elbert, CO	6.5%
Williamson, TX	6.2%
Nye, NV	6.1%

Source: U.S. Census Bureau

Atlanta and surrounding counties

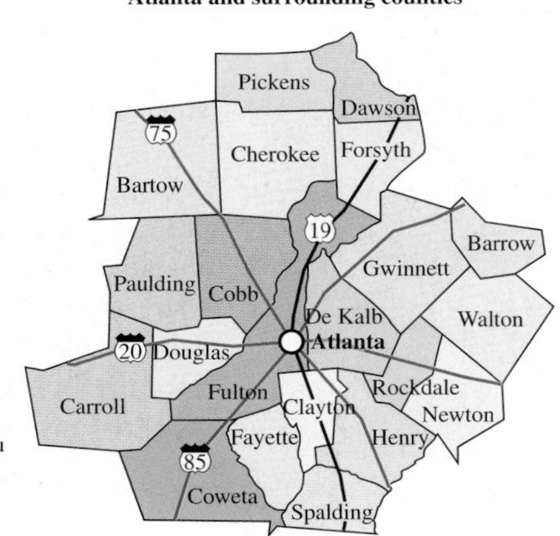

SOLUTION By examining the map and the chart we find that G = {Forsyth County, Henry County, Paulding County, Dawson County}. ▲

A set is said to be **finite** if it either contains no elements or the number of elements in the set is a natural number. The set $B = \{2, 4, 6, 8, 10\}$ is a finite set because the number of elements in the set is 5, and 5 is a natural number. A set that is not finite is said to be **infinite**. The set of counting numbers is one example of an infinite set. Infinite sets are discussed in more detail in Section 2.6.

DID YOU KNOW

Creature Cards

We learn to group objects according to what we see as the relevant distinguishing characteristics. One way used by educators to measure this ability is through visual cues. An example can be seen in this test, called "Creature Cards," offered by the Education Development Center. How would you describe membership in the set of Jexums?

Another important concept is equality of sets.

> Set A is **equal** to set B, symbolized by $A = B$, if and only if set A and set B contain exactly the same elements.

For example, if set $A = \{1, 2, 3\}$ and set $B = \{3, 1, 2\}$, then $A = B$ because they contain exactly the same elements. The order of the elements in the set is not important. If two sets are equal, both must contain the same number of elements. The number of elements in a set is called its *cardinal number*.

> The **cardinal number** of set A, symbolized by $n(A)$, is the number of elements in set A.

Both set $A = \{1, 2, 3\}$ and set $B = \{\text{England, France, Japan}\}$ have a cardinal number of 3; that is, $n(A) = 3$, and $n(B) = 3$. We can say that set A and set B both have a cardinality of 3.

Two sets are said to be *equivalent* if they contain the same number of elements.

> Set A is **equivalent** to set B if and only if $n(A) = n(B)$.

Any sets that are equal must also be equivalent. Not all sets that are equivalent are equal, however. The sets $D = \{a, b, c\}$ and $E = \{\text{apple, orange, pear}\}$ are equivalent, since both have the same cardinal number, 3. Because the elements differ, however, the sets are not equal.

Two sets that are equivalent or have the same cardinality can be placed in **one-to-one correspondence**. Set A and set B can be placed in one-to-one correspondence if every element of set A can be matched with exactly one element of set B and every element of set B can be matched with exactly one element of set A. For example, there is a one-to-one correspondence between the student names on a class list and the student identification numbers because we can match each name with a student identification number.

Consider set B, product brand name, and set D, drinks.

$$B = \{\text{Sealtest, A \& W, Lipton, Folgers}\}$$
$$D = \{\text{ice tea, milk, coffee, root beer}\}$$

Two different one-to-one correspondences for sets B and D follow.

$$B = \{\text{Sealtest, A \& W, Lipton, Folgers}\}$$
$$\downarrow \quad \swarrow \quad \swarrow \quad \swarrow$$
$$D = \{\text{ice tea, milk, coffee, root beer}\}$$

$$B = \{\text{Sealtest, A \& W, Lipton, Folgers}\}$$
$$D = \{\text{ice tea, milk, coffee, root beer}\}$$

Other one-to-one correspondences between sets B and D are possible. Do you know which drink goes with which product brand name?

NULL OR EMPTY SET

Some sets do not contain any elements, such as the set of zebras that are in this room.

> The set that contains no elements is called the **empty set** or **null set** and is symbolized by { } or ∅.

Note that {∅} is not the empty set. This set contains the element ∅ and has a cardinality of 1. The set {0} is also not the empty set because it contains the element 0. It also has a cardinality of 1.

EXAMPLE 8 ***Natural Number Solutions***

Indicate the set of natural numbers that satisfies the equation $x + 2 = 0$.

SOLUTION The values that satisfy the equation are those that make the equation a true statement. Only the number -2 satisfies this equation. Because -2 is not a natural number, the solution set of this equation is { } or ∅. ▲

UNIVERSAL SET

Another important set is a **universal set**.

> A **universal set,** symbolized by U, is a set that contains all the elements for any specific discussion.

When a universal set is given, only the elements in the universal set may be considered when working the problem. If, for example, the universal set for a particular problem is defined as $U = \{1, 2, 3, 4, \ldots, 10\}$, then only the natural numbers 1 through 10 may be used in that problem.

SECTION 2.1 EXERCISES

CONCEPT/WRITING EXERCISES

In Exercises 1–12, answer each question with a complete sentence.

1. What is a set?
2. What is an ellipsis, and how is it used?
3. What are the three ways that a set can be written? Give an example of each.
4. What is an infinite set?
5. What is a finite set?
6. What are equal sets?
7. What are equivalent sets?
8. What is the cardinal number of a set?
9. Write the set of counting numbers in roster form.
10. What does a one-to-one correspondence of two sets mean?
11. What is a universal set?
12. What is the empty set?

PRACTICE THE SKILLS

In Exercises 13–18, determine whether each set is well defined.

13. The set of the best Internet web sites
14. The set of people who own large dogs
15. The set of the four states in the United States having the largest areas
16. The set of states that have a common border with Tennessee

17. The set of students in this class who were born in the United States

18. The set of the most interesting students in this class

In Exercises 19–24, determine whether each set is finite or infinite.

19. $\{1, 3, 5, 7, \ldots\}$

20. The set of even numbers greater than 15

21. The set of multiples of 6 between 0 and 60

22. The set of fractions between 1 and 2

23. The set of odd numbers greater than 15

24. The set of crickets chirping in the town park on a warm July 4 at 10:00 P.M.

In Exercises 25–34, express each set in roster form. You may need to use a world almanac or some other reference source.

25. The set of states in the United States whose name begins with the letter N

26. The set of continents in the world

27. The set of natural numbers between 10 and 178

28. $B = \{x \mid x \in N \text{ and } x \text{ is even}\}$

29. $C = \{x \mid x + 6 = 10\}$

30. The set of states west of the Mississippi that have a common border with the state of Florida

31. The set of football players over the age of 70 who are still playing in the National Football League

32. The set of states in the United States that have no common border with any other state

33. $E = \{x \mid x \in N \text{ and } 6 \leq x < 72\}$

34. The set of professional baseball players in the major leagues who have hit at least 70 home runs in a season prior to 2000

In Exercises 35–42, express each set in set-builder notation.

35. $A = \{1, 2, 3, 4, 5, 6, 7, 8, 9\}$

36. $B = \{4, 5, 6, 7, 8\}$

37. $C = \{3, 6, 9, 12, \ldots\}$

38. $D = \{5, 10, 15, 20, \ldots\}$

39. E is the set of odd natural numbers

40. A is the set of national holidays in the United States in September

41. C is the set of the three manufacturers of calculators with the greatest sales in the United States

42. $F = \{15, 16, 17, \ldots, 100\}$

In Exercises 43–50, write a description of each set.

43. $A = \{1, 2, 3, 4, 5, 6, 7\}$

44. $D = \{4, 8, 12, 16, 20, \ldots\}$

45. L = {Superior, Michigan, Huron, Erie, Ontario}

46. S = {Bashful, Doc, Dopey, Grumpy, Happy, Sleepy, Sneezy}

47. B = {Sears Tower, World Trade Center, Empire State, Amoco, Chrysler}

48. C = {IBM, Apple, Dell, Compaq, Hewlett-Packard, Gateway, . . .}

49. $E = \{x \mid x \in N \text{ and } 5 < x \leq 12\}$

50. T = {Jose Carreras, Placido Domingo, Luciano Pavarotti}

In Exercises 51–58, state whether each statement is true or false. If false, give the reason.

51. $\{b\} \in \{a, b, c, d, e, f\}$

52. $b \in \{a, b, c, d, e, f\}$

53. $h \in \{a, b, c, d, e, f\}$

54. Mickey Mouse ∈ {cartoon characters created by Disney studios}

55. $3 \notin \{x \mid x \in N \text{ and } x \text{ is odd}\}$

56. Maui ∈ {capital cities in the United States}

57. *Titanic* ∈ {top 10 motion pictures with the greatest revenues}

58. $2 \in \{x \mid x \text{ is an odd natural number}\}$

In Exercises 59–62, for the sets $A = \{2, 4, 6, 8\}$, $B = \{1, 3, 7, 9, 13, 21\}$, $C = \{\ \}$, *and* $D = \{\#, \&, \%, \square, *\}$, *determine*

59. $n(A)$.

60. $n(B)$.

61. $n(C)$.

62. $n(D)$.

In Exercises 63–68, determine whether the pairs of sets are equal, equivalent, both, or neither.

63. $A = \{x, y, z\}$, $B = \{z, x, y\}$

64. $A = \{7, 9, 10\}$, $B = \{a, b, c\}$

65. $A = \{\text{red, green, blue}\}$
$B = \{\text{red, green, blue, yellow}\}$

66. A is the set of collies.
B is the set of dogs.

67. A is the set of letters in the word *tap*.
B is the set of letters in the word *ant*.

68. A is the set of states.
B is the set of state capitals.

PROBLEM SOLVING

69. Set-builder notation is often more versatile and efficient than listing a set in roster form. This versatility is illustrated with the two sets.

$$A = \{x \mid x \in N \text{ and } x > 2\}$$
$$B = \{x \mid x > 2\}$$

a) Write a description of set A and set B.
b) Explain the difference between set A and set B. (*Hint:* Is $4\frac{1}{2} \in A$? Is $4\frac{1}{2} \in B$?}
c) Write set A in roster form.
d) Can set B be written in roster form? Explain your answer.

70. Start with sets

$$A = \{x \mid 2 < x \leq 5 \text{ and } x \in N\}$$

and

$$B = \{x \mid 2 < x \leq 5\}$$

a) Write a description of set A and set B.
b) Explain the difference between set A and set B.
c) Write set A in roster form.
d) Can set B be written in roster form? Explain your answer.

A cardinal number answers the question "How many?" An ***ordinal number*** *describes the relative position that an element occupies. For example, Molly's desk is the third desk from the aisle.*

In Exercises 71–74, determine whether the number used is a cardinal number or an ordinal number.

71. Mariah Carey had 19 hit singles.

72. Study the chart on page 25 in the book.

73. Lincoln was the sixteenth president of the United States.

74. Emily paid \$35 for her new blouse.

75. Describe three sets of which you are a member.

76. Describe three sets that have no members.

77. Write a short paragraph explaining why the universal set and the empty set are necessary in the study of sets.

CHALLENGE EXERCISE/GROUP ACTIVITY

78. a) In a given exercise, a universal set is not specified, but we know that actor Brad Pitt is a member of the universal set. Describe five different possible universal sets of which Brad Pitt is a member.
b) Write a description of one set that includes all the universal sets in part (a).

RESEARCH ACTIVITY

79. Georg Cantor is recognized as the founder and a leader in the development of set theory. Do research and write a paper on his life and his contributions to set theory and the field of mathematics. References include history of mathematics books, encyclopedias, and the Internet.

2.2 SUBSETS

In our complex world, we often break larger sets into smaller more manageable sets, called *subsets*. For example, consider the set of people in your class. Suppose we categorize the set of people in your class according to the first letter of their last name (the A's, B's, C's, etc.). When we do this, each of these sets may be considered a subset of the original set. Each of these subsets can be separated further. For example, the set of people whose last name begins with the letter A can be categorized as either male or female or by their age. Each of these collections of people are also subsets. A given set may have many different subsets.

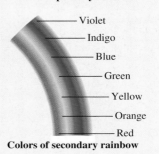

> Set A is a **subset** of set B, symbolized by $A \subseteq B$, if and only if all the elements of set A are also elements of set B.

The symbol $A \subseteq B$ indicates that "set A is a subset of set B." The symbol $\not\subseteq$ is used to indicate "is not a subset." Thus, $A \not\subseteq B$ indicates that set A is not a subset of set B. *To show that set A is not a subset of set B, we must find at least one element of set A that is not an element of set B.*

EXAMPLE 1 *A Subset?*

Determine whether set A is a subset of set B.

a) $A = \{$lettuce, tomato, cucumber$\}$
   $B = \{$tomato, cheese, lettuce, cucumber$\}$
b) $A = \{2, 3, 4, 5\}$ $\quad B = \{2, 3\}$
c) $A = \{x \mid x$ is a yellow fruit$\}$
   $B = \{x \mid x$ is a red fruit$\}$
d) $A = \{$Butterfinger, Baby Ruth, Power House$\}$
   $B = \{$Power House, Butterfinger, Baby Ruth$\}$

SOLUTION

a) All the elements of set A are contained in set B, so $A \subseteq B$.
b) The elements 4 and 5 are in set A but not in set B, so $A \not\subseteq B$ (A is not a subset of B). In this example, however, all the elements of set B are contained in set A; therefore, $B \subseteq A$.
c) There are fruits, such as bananas, that are in set A that are not in set B, so $A \not\subseteq$ B.
d) All the elements of set A are contained in set B, so $A \subseteq B$. Note that set $A =$ set B. ▲

PROPER SUBSETS

> Set A is a **proper subset** of set B, symbolized by $A \subset B$, if and only if all the elements of set A are elements of set B and set $A \neq$ set B (that is, set B must contain at least one element not in set A).

Consider the sets $A = \{$red, blue, yellow$\}$ and $B = \{$red, orange, yellow, green, blue, violet$\}$. Set A is a *subset* of set B, $A \subseteq B$, because every element of set A is also an element of set B. Set A is also a *proper subset* of set B, $A \subset B$, because sets A and B are not equal. Now consider $C = \{$car, bus, train$\}$ and $D = \{$train, car, bus$\}$. Set C is a subset of set D, $C \subseteq D$, because every element of set C is also an element of set D. Set C, however, is not a proper subset of set D, $C \not\subset D$, because set C and set D are equal sets.

DID YOU KNOW

Rainbows

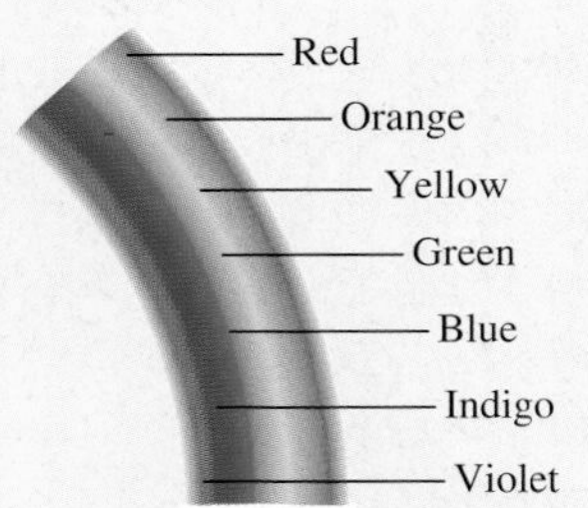

Colors of primary rainbow

Violet
Indigo
Blue
Green
Yellow
Orange
Red

Colors of secondary rainbow

Most rainbows we see are primary rainbows, but there are rare moments when a second, fainter rainbow can be seen behind the first. In this secondary rainbow, the light pattern has been reversed. Both rainbows contain the same set of colors, so each set of colors is a subset of the other.

EXAMPLE 2 ***A Proper Subset?***

Determine whether set A is a proper subset of set B.

a) $A = \{$Corvette, Viper, Trans Am$\}$
   $B = \{$Viper, Trans Am, Camero, Corvette, Mustang$\}$
b) $A = \{a, b, c, d\}$ $\quad B = \{a, c, b, d\}$

SOLUTION

a) All elements of set A are contained in set B, and sets A and B are not equal; thus, $A \subset B$.
b) Set A = set B, so $A \not\subset B$. (However, $A \subseteq B$.)

Every set is a subset of itself, but no set is a proper subset of itself. For all sets A, $A \subseteq A$, but $A \not\subset A$. For example, if $A = \{1, 2, 3\}$, then $A \subseteq A$ because every element of set A is contained in set A, but $A \not\subset A$ because set A = set A.

Let $A = \{\ \}$ and $B = \{1, 2, 3, 4\}$. Is $A \subseteq B$? To show $A \not\subseteq B$, you must find at least one element of set A that is not an element of set B. As this cannot be done, $A \subseteq B$ must be true. Using the same reasoning, we can show that *the empty set is a subset of every set, including itself.*

EXAMPLE 3 ***Element or Subset?***

Determine whether the following are true or false.

a) $3 \in \{3, 4, 5\}$
b) $\{3\} \in \{3, 4, 5\}$
c) $\{3\} \in \{\{3\}, \{4\}, \{5\}\}$
d) $\{3\} \subseteq \{3, 4, 5\}$
e) $3 \subseteq \{3, 4, 5\}$
f) $\{\ \} \subseteq \{3, 4, 5\}$

SOLUTION

a) $3 \in \{3, 4, 5\}$ is a true statement because 3 is a member of the set $\{3, 4, 5\}$.
b) $\{3\} \in \{3, 4, 5\}$ is a false statement because $\{3\}$ is a set, and the set $\{3\}$ is not an element of the set $\{3, 4, 5\}$.
c) $\{3\} \in \{\{3\}, \{4\}, \{5\}\}$ is a true statement because $\{3\}$ is an element in the set. The elements of the set $\{\{3\}, \{4\}, \{5\}\}$ are themselves sets.
d) $\{3\} \subseteq \{3, 4, 5\}$ is a true statement because every element of the first set is an element of the second set.
e) $3 \subseteq \{3, 4, 5\}$ is a false statement because the 3 is not in braces, so it is not a set and thus cannot be a subset. The 3 is an element of the set as indicated in part (a).
f) $\{\ \} \subseteq \{3, 4, 5\}$ is a true statement because the empty set is a subset of every set.

NUMBER OF SUBSETS

How many distinct subsets can be made from a given set? The empty set has no elements and has exactly one subset, the empty set. A set with one element has two subsets. A set with two elements has four subsets. A set with three elements has eight subsets. This information is illustrated in Table 2.1 on page 48. How many subsets will a set with four elements contain?

Table 2.1 Number of Subsets

Set	Subsets	Number of subsets
{ }	{ }	$1 = 2^0$
$\{a\}$	$\{a\}$ { }	$2 = 2^1$
$\{a, b\}$	$\{a, b\}$ $\{a\}, \{b\}$ { }	$4 = 2 \times 2 = 2^2$
$\{a, b, c\}$	$\{a, b, c\}$ $\{a, b\}, \{a, c\}, \{b, c\}$ $\{a\}, \{b\}, \{c\}$ { }	$8 = 2 \times 2 \times 2 = 2^3$

By continuing this table with larger and larger sets, we can develop a general formula for finding the number of distinct subsets that can be made from any given set.

The **number of distinct subsets** of a finite set A is 2^n, where n is the number of elements in set A.

With seven distinct Scrabble tiles, there are 5040 different ways the tiles can be arranged. The four letters S, L, E, D can be arranged in 24 distinct ways.

EXAMPLE 4 *Distinct Subsets*

a) Determine the number of distinct subsets for the set {S, L, E, D}.
b) List all the distinct subsets for the set {S, L, E, D}.
c) How many of the distinct subsets are proper subsets?

SOLUTION

a) Since the number of elements in the set is 4, the number of distinct subsets is $2^4 = 2 \times 2 \times 2 \times 2 = 16$.

b)

{S, L, E, D}	{S, L, E}	{S, L}	{S}	{ }
	{S, L, D}	{S, E}	{L}	
	{S, E, D}	{S, D}	{E}	
	{L, E, D}	{L, E}	{D}	
		{L, D}		
		{E, D}		

c) There are 15 proper subsets. Every subset except {S, L, E, D} is a proper subset. ▲

EXAMPLE 5 *Variations of a Hamburger*

Shawn is going to purchase a hamburger at Wendy's. He can add any of the following items: ketchup, cheese, lettuce, mayonnaise, mustard, onions, pickles, tomatoes. How many different variations of the hamburger can be made?

SOLUTION He can order the hamburger with no extra items, any one item, any two items, any three items, and so on, up to all eight items. One technique used in problem solving is to consider similar problems that you have solved previously. If you think about this problem, you will realize that this problem is the same as, "How many distinct subsets can be made from a set with eight elements?" The number of different variations of the hamburger is the same as the number of possible subsets of a set that has 8 elements. There are 2^8 or 256 possible subsets of a set with 8 elements, so there are 256 possible variations of the hamburger. ▲

DID YOU KNOW

The Ladder of Life

Scientists use sets to classify and categorize knowledge. In biology, the science of classifying all living things is called *taxonomy* and was probably practiced by the earliest cave-dwellers. Over 2000 years ago, Aristotle formalized animal classification with his "ladder of life": higher animals, lower animals, higher plants, lower plants.

Contemporary biologists use a system of classification called the Linnaean system, named after Swedish biologist Carolus Linnaeas (1707–1778). The Linnaean system starts with the smallest unit (member) and assigns it to a specific genus (set) and species (subset).

Even more general groupings of living things are made according to shared characteristics. The groupings, from most general to most specific are: kingdom, phylum, class, order, family, genus, and species.

A zebra, *Equus burchelli,* is a member of the genus, *Equus,* as is the horse, *Equus caballus.* Both the zebra and the horse are members of the universal set called the kingdom of animals and the same family, Equidae; they are members of different species (*E. burchelli* and *E. caballus*), however.

SECTION 2.2 EXERCISES

CONCEPT/WRITING EXERCISES

In Exercises 1–6, answer each question with a complete sentence.

1. What is a subset?
2. What is a proper subset?
3. Explain the difference between a subset and a proper subset.
4. What is the formula for determining the number of distinct subsets for a set with n distinct elements?
5. What is the formula for determining the number of distinct proper subsets for a set with n distinct elements?
6. Can any set be a proper subset of itself? Explain.

PRACTICE THE SKILLS

In Exercises 7–24, answer true or false. If false, give the reason.

7. English $\subseteq$ {French, German, Spanish, English}
8. { } $\in$ {1, 2, 3, 4}
9. { } $\subseteq$ {Curly, Larry, Moe}
10. red $\subset$ {red, green, blue}
11. $5 \notin \{2, 4, 6\}$
12. {Pete, Mike, Amy} $\subseteq$ {Amy, Kaitlyn, Brianna}
13. { } = {$\varnothing$}
14. $\{1\} \subseteq \{1, 5, 9\}$
15. $\varnothing$ = { }
16. 0 = { }
17. $\{0\} = \varnothing$
18. $\{1, 5, 9\} \subseteq \{1, 9, 5\}$
19. $\{5\} \in \{5, 6, 7\}$
20. $\{3, 5, 9\} \not\subset \{3, 9, 5\}$
21. { } $\subset$ { }
22. $\{1\} \in \{\{1\}, \{2\}, \{3\}\}$
23. {apple, orange, plum} $\subseteq$ {plum, orange, apple}
24. $\{b, a, t\} \subseteq \{t, a, b\}$

In Exercises 25–32, determine whether $A = B$, $A \subseteq B$, $B \subseteq A$, $A \subset B$, $B \subset A$, or none of these. (There may be more than one answer.)

25. $A = \{b, c, e, f\}$
$B = \{c, f\}$

26. $A = \{x \mid x \in N \text{ and } x < 6\}$
$B = \{x \mid x \in N \text{ and } 1 \leq x \leq 5\}$

27. Set A is the set of states east of the Mississippi River. Set B is the set of states east of the Pacific Ocean.

28. $A = \{1, 3, 5, 7, 9\}$
$B = \{3, 9, 5, 7, 6\}$

29. $A = \{x \mid x$ is a brand of ice cream$\}$
$B = \{$Breyers, Ben & Jerry's, Sealtest$\}$

30. $A = \{x \mid x$ is a sport that uses a ball$\}$
$B = \{$basketball, soccer, tennis$\}$

31. Set A is the set of natural numbers between 2 and 7. Set B is the set of natural numbers greater than 2 and less than 7.

32. Set A is the set of white keys on a piano.
Set B is the set of keys on a piano.

In Exercises 33–38, list all the subsets of the sets given.

33. $D = \emptyset$

34. $A = \{\bigcirc\}$

35. $B = \{$car, boat$\}$

36. $C = \{$apple, peach, banana$\}$

PROBLEM SOLVING

37. For set $A = \{a, b, c, d\}$,
a) list all the subsets of set A.
b) state which of the subsets in part (a) are not proper subsets of set A.

38. A set contains eight elements.
a) How many subsets does it contain?
b) How many proper subsets does it contain?

In Exercises 39–50, if the statement is true for all sets A and B, write "true." If it is not true for all sets A and B, write "false." Assume that $A \neq \emptyset$, $U \neq \emptyset$, and $A \subset U$.

39. If $A \subset B$, then $A \subseteq B$.

40. If $A \subseteq B$, then $A \subset B$.

41. $A \subset A$

42. $A \subseteq A$

43. $\emptyset \subset A$

44. $\emptyset \subseteq A$

45. $A \subseteq U$

46. $\emptyset \subset \emptyset$

47. $\emptyset \subset U$

48. $U \subseteq \emptyset$

49. $\emptyset \subseteq \emptyset$

50. $U \subset \emptyset$

51. *Computer Upgrade* Jason Jackson is considering having his computer upgraded. He can leave the computer as it is, or he can upgrade any of the following set of items:

{RAM, modem, video card, hard drive, processor, sound card}

How many possible options for upgrading does Jason have?

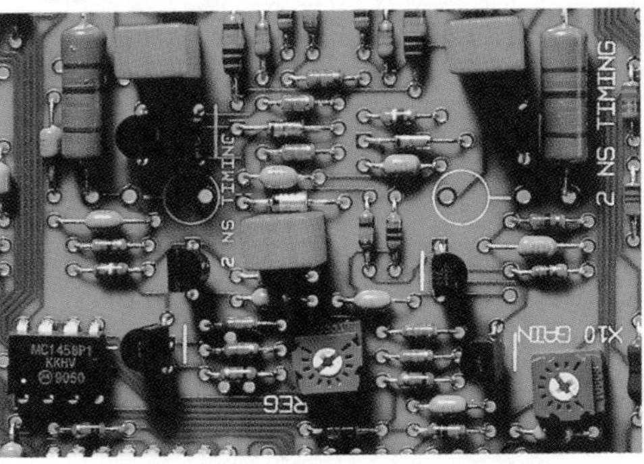

52. *Car Options* A person can order a new car with some, all, or none of the following set of options: {air conditioning, power windows, CD player, leather interior, alarm system, sun roof}. How many different variations of the set of options are possible?

53. *Pizza Toppings* At Pizzeria Uno, a pizza can be ordered with some, all, or none of the following set of toppings: {cheese, pepperoni, peppers, mushrooms, anchovies, olives, sausage}. How many different variations are there for ordering a pizza?

54. *Hamburger Variations* Customers ordering hamburgers at Vic and Irv's Hamburger stand are always asked, "What do you want on it?" The choices are members of the set {ketchup, mustard, relish, hot sauce, onions, lettuce, tomato}. How many different variations are there for ordering a hamburger?

55. If $E \subseteq F$ and $F \subseteq E$, what other relationship exists between E and F? Explain.

56. How can you determine whether the set of boys is equivalent to the set of girls at a roller skating rink?

57. For the set $D = \{a, b, c\}$
a) is a an element of set D? Explain.
b) is c a subset of set D? Explain.
c) is $\{a, b\}$ a subset of set D? Explain.

CHALLENGE EXERCISE/GROUP ACTIVITY

58. *Stock Purchase* An investment club has four members: Allen, Brenda, Chuck and Donna.
a) When they vote on the purchase of a particular stock, how many different ways can the members vote (abstentions are not allowed)? For example, Allen—yes, Brenda—no, Chuck—no, Donna—yes, is one of the many possibilities.
b) Make a listing of all the possible outcomes of the vote. For example, the vote described in part (a) could be represented as (Y, N, N, Y).
c) How many of the outcomes given in part (b) would result in a majority supporting the purchase of the stock? That is, how many of the outcomes have three or more Y's?

2.3 VENN DIAGRAMS AND SET OPERATIONS

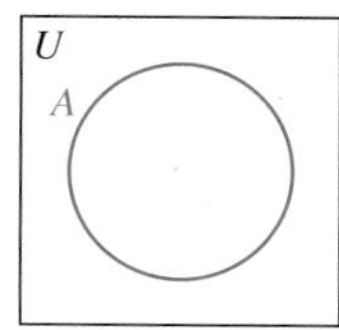

Figure 2.1

A useful technique for picturing set relationships is the Venn diagram, named for the English mathematician John Venn (1834–1923). Venn invented the diagrams and used them to illustrate ideas in his text on symbolic logic, published in 1881.

In a Venn diagram, a rectangle usually represents the universal set, U. The items inside the rectangle may be divided into subsets of the universal set. The subsets are usually represented by circles. In Fig. 2.1, the circle labeled A represents set A, which is a subset of the universal set.

Two sets may be represented in a Venn diagram in any of four different ways (see Fig. 2.2). Two sets A and B are **disjoint** when they have no elements in common. Two disjoint sets A and B are illustrated in Fig. 2.2(a). If set A is a proper subset of set B, $A \subset B$, the two sets may be illustrated as in Fig. 2.2(b). If set A contains exactly the same elements as set B, that is, $A = B$, the two sets may be illustrated as in Fig. 2.2(c). Two sets A and B with some elements in common are shown in Fig. 2.2(d), which is regarded as the most general form of a Venn diagram.

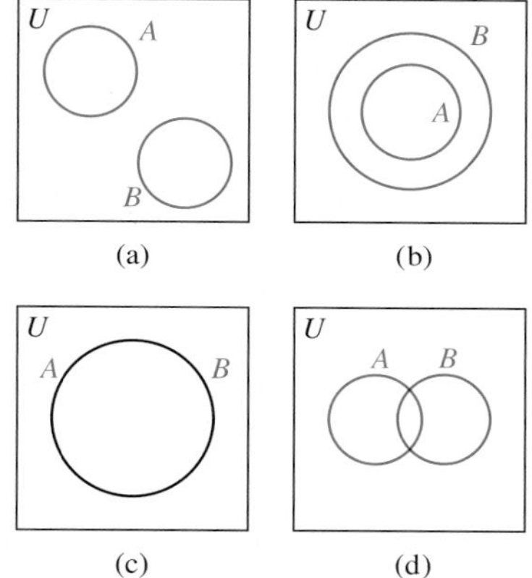

Figure 2.2

If we label the regions of the diagram in Fig. 2.2(d) using I, II, III, and IV, we can illustrate the four possible cases with this one diagram, Fig. 2.3.

CASE 1: DISJOINT SETS When sets A and B are disjoint, they have no elements in common. Therefore, region II of Fig. 2.3 is empty.

CASE 2: SUBSETS When $A \subseteq B$, every element of set A is also an element of set B. Thus, there can be no elements in region I of Fig. 2.3. If $B \subseteq A$, however, then region III of Fig. 2.3 is empty.

CASE 3: EQUAL SETS When set A = set B, all the elements of set A are elements of set B and all the elements of set B are elements of set A. Thus, regions I and III of Fig. 2.3 are empty.

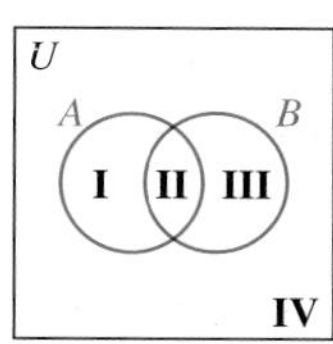

Figure 2.3

CASE 4: OVERLAPPING SETS When sets A and B have elements in common, those elements are in region II of Fig. 2.3. The elements that belong to set A but not to set B are in region I. The elements that belong to set B but not to set A are in region III.

In each of the four cases, any element not belonging to set A or set B is placed in region IV.

Venn diagrams will be helpful in understanding set operations. The operations of arithmetic are $+$, $-$, $\times$, and $\div$. When we see these symbols, we know what procedure to follow to determine the answer. Some of the operations in set theory are $'$, $\cup$, and $\cap$. They represent complement, union, and intersection, respectively.

COMPLEMENT

The **complement** of set A, symbolized by A', is the set of all the elements in the universal set that are not in set A.

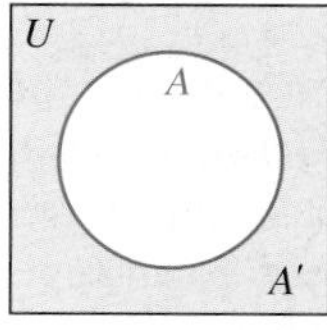

Figure 2.4

In Fig. 2.4, the shaded region outside of set A within the universal set represents the complement of set A, or A'.

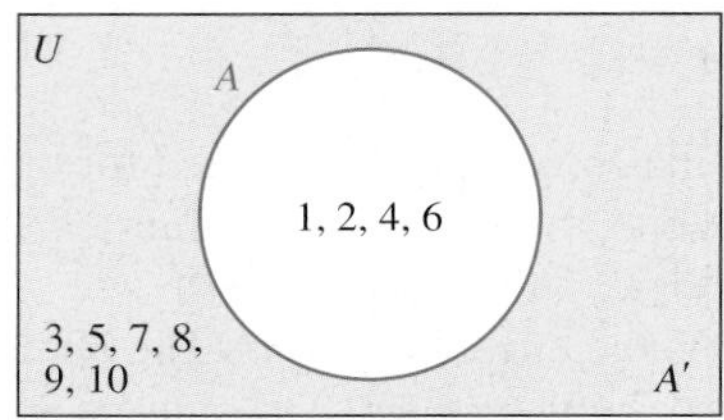

Figure 2.5

EXAMPLE 1 *A Set and Its Complement*

Given

$$U = \{1, 2, 3, 4, 5, 6, 7, 8, 9, 10\} \quad \text{and} \quad A = \{1, 2, 4, 6\}$$

find A' and illustrate the relationship among sets U, A, and A' in a Venn diagram.

SOLUTION The elements in U that are not in set A are 3, 5, 7, 8, 9, 10. Thus, $A' = \{3, 5, 7, 8, 9, 10\}$. The Venn diagram is illustrated in Fig. 2.5. ▲

INTERSECTION

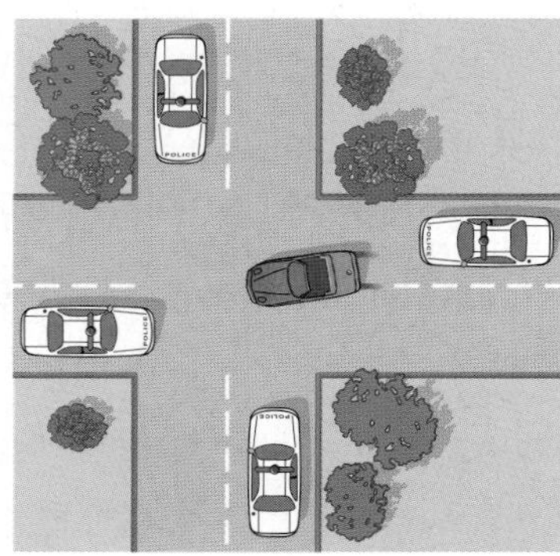

The word *intersection* brings to mind the area common to two crossing streets. The red car in the figure is in the intersection of the two streets. The set operation is defined as follows.

The **intersection** of sets A and B, symbolized by $A \cap B$, is the set containing all the elements that are common to both set A and set B.

The shaded region, region II, in Fig. 2.6 represents the intersection of sets A and B.

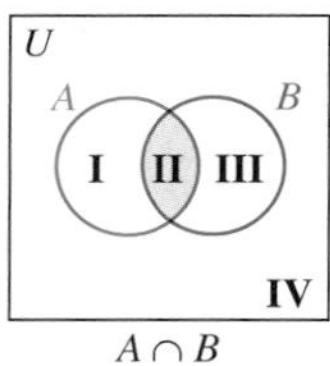

Figure 2.6

EXAMPLE 2 *Sets with Overlapping Regions*

Let the universal set, U, represent the 50 states in the United States. Let A represent the set of states that have at least one building with a height of 1000 feet or taller (see the chart). Let B represent the set of states with a population over 10 million as of January 1998 (see the chart). Draw a Venn diagram illustrating the sets.

States with at least one building 1000 feet tall or taller	States with a population over 10 million
Illinois (Chicago)	California
New York (New York City)	Texas
Georgia (Atlanta)	New York
California (Los Angeles)	Florida
Texas (Houston)	Pennsylvania
	Illinois
	Ohio

SOLUTION First determine the intersection of sets A and B. Illinois, New York, California, and Texas are common to both sets, so

$$A \cap B = \{\text{IL, NY, CA, TX}\}$$

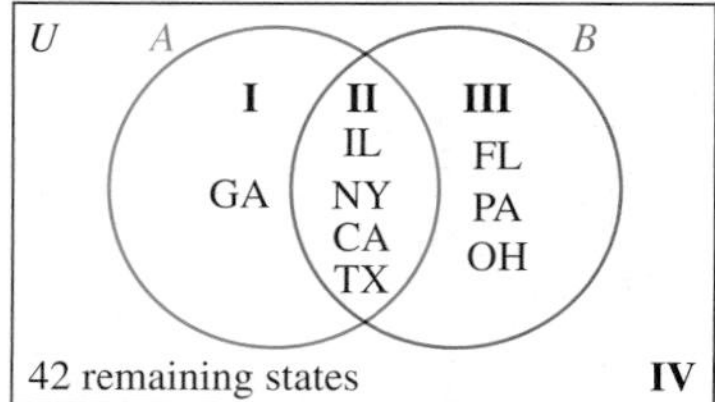

Figure 2.7

Place these elements in region II of Fig. 2.7. Now place, in region I, the elements in set A that have not been placed in region II. Therefore, Georgia goes in region I. Complete region III by determining the elements in set B that have not been placed in region II. Thus, Florida, Pennsylvania, and Ohio go in region III. Finally, place those elements in U that are not in either set outside both circles. This group includes the remaining 42 states, which go in region IV. ▲

EXAMPLE 3 ***The Intersection of Sets***

Given

$$U = \{1, 2, 3, 4, 5, 6, 7, 8, 9, 10\}$$
$$A = \{1, 2, 4, 6\}$$
$$B = \{1, 3, 6, 7, 9\}$$
$$C = \{\ \}$$

find

a) $A \cap B$. b) $A \cap C$. c) $A' \cap B$. d) $(A \cap B)'$.

SOLUTION

a) $A \cap B = \{1, 2, 4, 6\} \cap \{1, 3, 6, 7, 9\} = \{1, 6\}$. The elements common to both set A and set B are 1 and 6.

b) $A \cap C = \{1, 2, 4, 6\} \cap \{\ \} = \{\ \}$. There are no elements common to both set A and set C.

c) $A' = \{3, 5, 7, 8, 9, 10\}$
$A' \cap B = \{3, 5, 7, 8, 9, 10\} \cap \{1, 3, 6, 7, 9\}$
$= \{3, 7, 9\}$

d) To find $(A \cap B)'$, first determine $A \cap B$.

$$A \cap B = \{1, 6\} \text{ from part (a)}$$
$$(A \cap B)' = \{1, 6\}' = \{2, 3, 4, 5, 7, 8, 9, 10\}$$

UNION

The word *union* means to unite or join together, as in marriage, and that is exactly what is done when we perform the operation of union.

The **union** of sets A and B, symbolized by $A \cup B$, is the set containing all the elements that are members of set A or of set B (or of both sets).

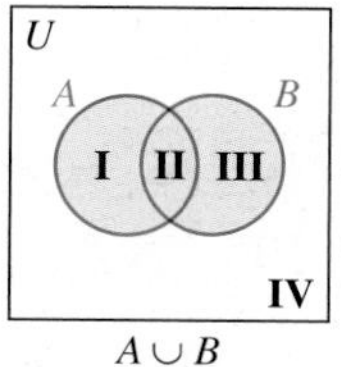

Figure 2.8

The three shaded regions of Fig. 2.8, regions I, II, and III, together represent the union of sets A and B. If an element is common to both sets, it is listed only once in the union of the sets.

EXAMPLE 4 ***Determining Sets from a Venn Diagram***

Use the Venn diagram in Fig. 2.9 to determine the following sets.

a) U b) A c) B' d) $A \cap B$
e) $A \cup B$ f) $(A \cup B)'$ g) $n(A \cup B)$

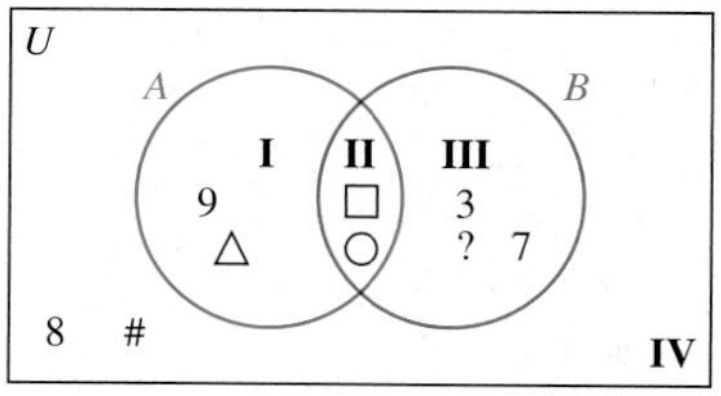

Figure 2.9

SOLUTION

a) The universal set consists of all the elements within the rectangle. Thus, $U = \{9, \triangle, \square, \bigcirc, 3, 7, ?, \#, 8\}$.

b) Set A consists of the elements in regions I and II. Thus, $A = \{9, \triangle, \square, \bigcirc\}$.

c) B' consists of the elements outside set B, or the elements in regions I and IV. Thus, $B' = \{9, \triangle, \#, 8\}$.

d) $A \cap B$ consists of the elements that belong to both set A and set B (region II). Thus, $A \cap B = \{\square, \bigcirc\}$.
e) $A \cup B$ consists of the elements that belong to set A or set B (regions I, II, or III). Thus, $A \cup B = \{9, \triangle, \square, \bigcirc, 3, 7, ?\}$.
f) $(A \cup B)'$ consists of the elements in U that are not in $A \cup B$. Thus, $(A \cup B)' = \{\#, 8\}$.
g) $n(A \cup B)$ represents the *number of elements* in the union of sets A and B. Thus, $n(A \cup B) = 7$, as there are seven elements in the union of sets A and B. ▲

EXAMPLE 5 *The Union of Sets*

Given

$$\begin{aligned} U &= \{1, 2, 3, 4, 5, 6, 7, 8, 9, 10\} \\ A &= \{1, 2, 4, 6\} \\ B &= \{1, 3, 6, 7, 9\} \\ C &= \{\ \} \end{aligned}$$

find

a) $A \cup B$. b) $A \cup C$. c) $A' \cup B$. d) $(A \cup B)'$.

SOLUTION

a) $A \cup B = \{1, 2, 4, 6\} \cup \{1, 3, 6, 7, 9\} = \{1, 2, 3, 4, 6, 7, 9\}$
b) $A \cup C = \{1, 2, 4, 6\} \cup \{\ \} = \{1, 2, 4, 6\}$. Note that $A \cup C = A$.
c) To determine $A' \cup B$, we must determine A'.

$$\begin{aligned} A' &= \{3, 5, 7, 8, 9, 10\} \\ A' \cup B &= \{3, 5, 7, 8, 9, 10\} \cup \{1, 3, 6, 7, 9\} \\ &= \{1, 3, 5, 6, 7, 8, 9, 10\} \end{aligned}$$

d) Find $(A \cup B)'$ by first determining $A \cup B$, and then find the complement of $A \cup B$.

$$\begin{aligned} A \cup B &= \{1, 2, 3, 4, 6, 7, 9\} \text{ from part (a)} \\ (A \cup B)' &= \{1, 2, 3, 4, 6, 7, 9\}' = \{5, 8, 10\} \end{aligned}$$

▲

EXAMPLE 6 *Union and Intersection*

Given

$$\begin{aligned} U &= \{a, b, c, d, e, f, g\} \\ A &= \{a, b, e, g\} \\ B &= \{a, c, d, e\} \\ C &= \{b, e, f\} \end{aligned}$$

find

a) $(A \cup B) \cap (A \cup C)$. b) $(A \cup B) \cap C'$. c) $A' \cap B'$.

SOLUTION

a) $(A \cup B) \cap (A \cup C) = \{a, b, c, d, e, g\} \cap \{a, b, e, f, g\}$
$= \{a, b, e, g\}$

b) $(A \cup B) \cap C' = \{a, b, c, d, e, g\} \cap \{a, c, d, g\}$
$= \{a, c, d, g\}$

c) $A' \cap B' = \{c, d, f\} \cap \{b, f, g\}$
$= \{f\}$

THE MEANING OF *AND* AND *OR*

The words *and* and *or* are very important in many areas of mathematics. We use these words in several chapters in this book, including the probability chapter. The word **or** is generally interpreted to mean **union**, whereas **and** is generally interpreted to mean **intersection**. Suppose $A = \{1, 2, 3, 5, 6, 8\}$ and $B = \{1, 3, 4, 7, 9, 10\}$. Then the elements that belong to set A *or* set B are 1, 2, 3, 4, 5, 6, 7, 8, 9, and 10. These are the elements in the union of the sets. The elements that belong to set A *and* set B are 1 and 3. These are the elements in the intersection of the sets.

THE RELATIONSHIP BETWEEN $n(A \cup B)$, $n(A)$, $n(B)$, AND $n(A \cap B)$

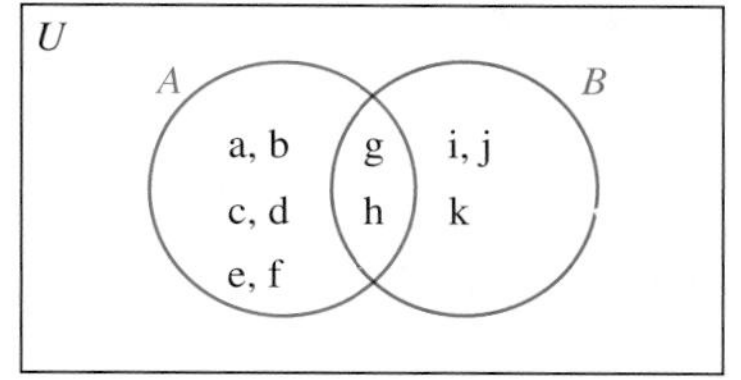

Figure 2.10

Having looked at unions and intersections, we can now determine a relationship between $n(A \cup B)$, $n(A)$, $n(B)$, and $n(A \cap B)$. Suppose set A has eight elements, set B has five elements, and $A \cap B$ has two elements. How many elements are in $A \cup B$? Let's make up some arbitrary sets that meet the criteria specified and draw a Venn diagram. If we let $A = \{a, b, c, d, e, f, g, h\}$, then set B must contain five elements, two of which are also in set A. Let $B = \{g, h, i, j, k\}$. We construct a Venn diagram by filling in the intersection first, as shown in Fig. 2.10. The number of elements in $A \cup B$ is 11. The elements g and h are in both sets, and if we add $n(A) + n(B)$, we are counting these elements twice.

To find the number of elements in the union of sets A and B, we can add the number of elements in sets A and B and then subtract the number of elements common to both sets.

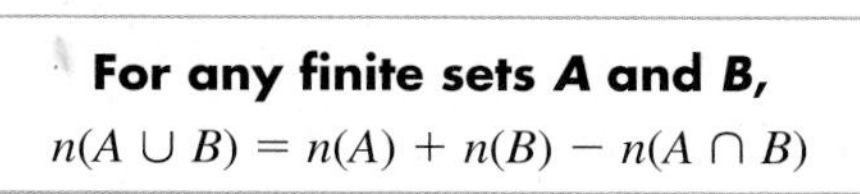

For any finite sets *A* and *B*,
$n(A \cup B) = n(A) + n(B) - n(A \cap B)$

EXAMPLE 7 ***How Many Houses Are White or Have a Chimney?***

On Wilcott Drive, there are 20 white houses, there are 12 houses that have a chimney, and there are 7 white houses that have a chimney. How many houses are either white or have a chimney?

SOLUTION If we let set A be the set of white houses and set B be the set of houses that have a chimney, then we need to determine $n(A \cup B)$. We can use the above formula to find $n(A \cup B)$.

$$n(A \cup B) = n(A) + n(B) - n(A \cap B)$$
$$n(A \cup B) = 20 + 12 - 7$$
$$= 25$$

Thus, there are 25 houses that are either white or have a chimney.

EXAMPLE 8 ***The Number of Elements in Set A or Set B***

Set A contains six letters and five numbers. Set B contains four letters and nine numbers. Two letters and one number are common to both sets A and B. Find the number of elements in set A or set B.

SOLUTION You are asked to find the number of elements in set A or set B, which is $n(A \cup B)$. Because $n(A \cup B) = n(A) + n(B) - n(A \cap B)$, if you can determine $n(A)$, $n(B)$, and $n(A \cap B)$, you can solve the problem. Set A contains 6 letters and 5 numbers, so $n(A) = 11$. Set B contains 4 letters and 9 numbers, so $n(B) = 13$. Because 2 letters and 1 number are common to both sets, $n(A \cap B) = 3$.

$$\begin{aligned} n(A \cup B) &= n(A) + n(B) - n(A \cap B) \\ &= 11 + 13 - 3 = 21 \end{aligned}$$

Thus, the number of elements in set A or set B is 21. ▲

SECTION 2.3 EXERCISES

CONCEPT/WRITING EXERCISES

In Exercises 1–5, use Fig. 2.2 as a guide to draw a Venn diagram that illustrates the situation described.

1. Set A and set B are disjoint sets.

2. $A \subset B$

3. $B \subset A$

4. $A = B$

5. Set A and set B are overlapping sets.

6. If we are given set A, how do we obtain A complement, A'?

7. How do we obtain the union of two sets A and B, $A \cup B$?

8. How do we obtain the intersection of two sets A and B, $A \cap B$?

9. a) Which set operation is the word *or* generally interpreted to mean?

b) Which set operation is the word *and* generally interpreted to mean?

10. Give the relationship between $n(A \cup B)$, $n(A)$, $n(B)$, and $n(A \cap B)$.

11. When constructing a Venn diagram with two sets, which region of the diagram do we generally complete first?

12. When constructing a Venn diagram with two sets, which region of the diagram do we generally complete last?

PRACTICE THE SKILLS

13. For the sets U, A, and B, construct a Venn diagram and place the elements in the proper regions.

$$\begin{aligned} U &= \{1, 2, 3, 4, 5, 6, 7, 8, 9, 10\} \\ A &= \{1, 4, 6, 7, 9, 10\} \\ B &= \{2, 3, 4, 7, 8\} \end{aligned}$$

14. For the sets U, A, and B, construct a Venn diagram and place the elements in the proper regions.

$$\begin{aligned} U &= \{\text{a, b, c, d, e, f, g, h, i, j}\} \\ A &= \{\text{a, b, c, d, f, h}\} \\ B &= \{\text{b, c, e, f, i}\} \end{aligned}$$

15. *Racing Standings* The following table shows the 1997 Winston Cup Final Standings (NASCAR racing). Let the people in the table represent the universal set.

Final standings

Driver	Car	Pts
Jeff Gordon	Chevy	4710
Dale Jarrett	Ford	4696
Mark Martin	Ford	4681
Jeff Burton	Ford	4285
Dale Earnhardt	Chevy	4216
Terry Labonte	Chevy	4177
Bobby Labonte	Pontiac	4101
Bill Elliot	Ford	3836
Rusty Wallace	Ford	3598
Ken Schrader	Chevy	3576

Source: *1999 Sports Illustrated Almanac*

Let A = the set of drivers whose car was a Ford.

Let B = the set of drivers whose points were 4200 or more.

Construct a Venn diagram illustrating the sets.

16. *Championship Golfers* The chart shows the number of times selected professional golfers have won the Masters and U.S. Open golf tournaments. Let these individuals represent the universal set, U.

	Masters	U.S. Open
Jack Nicklaus[a]	6	4
Bobby Jones	0	4
Walter Hagen	0	2
Ben Hogan	2	4
Gary Player[a]	3	1
John Ball	0	0
Arnold Palmer[a]	4	1
Tom Watson[b]	2	1
Gene Sarazen	1	2
Sam Snead	3	0

[a]Active Senior PGA player.
[b]Active PGA player.
Source: *1999 Sports Illustrated Almanac*

Let A = the set of golfers that won the Masters 3 or more times.

Let B = the set of golfers that won the U.S. Open 2 or more times.

Construct a Venn diagram that illustrates this information.

17. Let U represent the set of U.S. senators. Let set A represent the set of senators who voted in favor of the Hartley–Domingo bill. Describe A'.

18. Let U represent the set of marbles in a box. Let set B represent the set of marbles that contain some blue coloring. Describe B'.

In Exercises 19–24,

U is the set of universities in the United States.

A is the set of universities in the United States that have the word State in their name.

B is the set of universities in the United States that have the word South in their name.

Describe the following sets in words.

19. A' **20.** B'

21. $A \cup B$ **22.** $A \cap B$

23. $A \cap B'$ **24.** $A \cup B'$

In Exercises 25–30,

U is the set of U.S. corporations.

A is the set of U.S. corporations whose headquarters are in the state of New York.

B is the set of U.S. corporations whose chief executive officer is a woman.

C is the set of U.S. corporations that employ at least 100 people.

Describe the following sets.

25. $A \cap B$ **26.** $A \cup C$

27. $B \cap C'$ **28.** $A \cup B \cup C$

29. $A \cap B \cap C$ **30.** $A' \cup C'$

In Exercises 31–38, use the Venn diagram in Fig. 2.11 to list the set of elements in roster form.

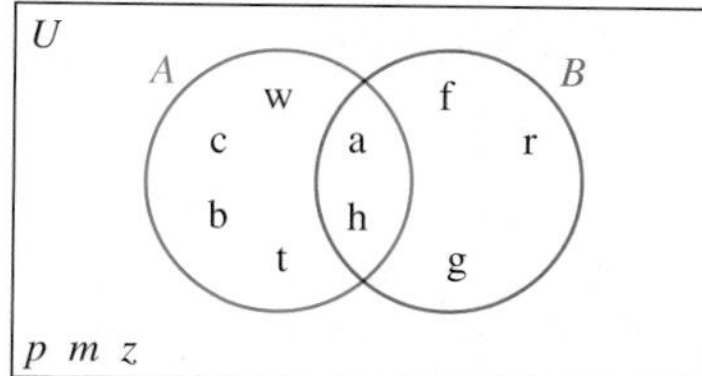

Figure 2.11

31. B **32.** A

33. U **34.** $A \cap B$

35. $A \cup B$ **36.** $(A \cup B)'$

37. $(A \cap B)'$ **38.** $A' \cap B$

In Exercises 39–46, use the Venn diagram in Fig. 2.12 to list the set of elements in roster form.

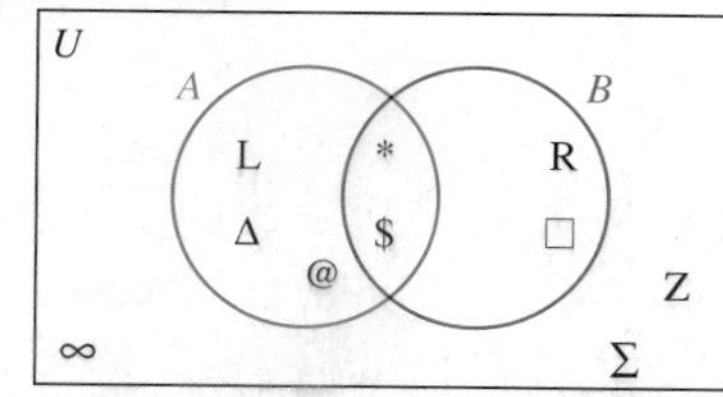

Figure 2.12

39. A **40.** B

41. U **42.** $A \cup B$

43. $A \cap B$ **44.** $A \cup B'$

45. $A' \cap B$ **46.** $(A \cup B)'$

In Exercises 47–56, let

$$U = \{1, 2, 3, 4, 5, 6, 7, 8\}$$
$$A = \{1, 2, 4, 5, 8\}$$
$$B = \{2, 3, 4, 6\}$$

Determine the following.

47. $A \cup B$ **48.** $A \cap B$
49. B' **50.** $A \cup B'$
51. $(A \cup B)'$ **52.** $A' \cap B'$
53. $(A \cup B)' \cap B$ **54.** $(A \cup B) \cap (A \cup B)'$
55. $(B \cup A)' \cap (B' \cup A')$ **56.** $A' \cup (A \cap B)$

In Exercises 57–66, let

$$U = \{\text{a, b, c, d, e, f, g, h, i, j, k}\}$$
$$A = \{\text{a, c, d, f, g, i}\}$$
$$B = \{\text{b, c, d, f, g}\}$$
$$C = \{\text{a, b, f, i, j}\}$$

Determine the following.

57. A' **58.** $B \cup C$
59. $A \cap C$ **60.** $A' \cup B$
61. $(A \cap C)'$ **62.** $(A \cap C) \cup B$
63. $A \cup (C \cap B)'$ **64.** $A \cup (C' \cup B')$
65. $A' \cap (B \cap C)$ **66.** $(C \cap B) \cap (A' \cap B)$

PROBLEM SOLVING

In Exercises 67–80, let

$$U = \{x \mid x \in N \text{ and } x < 10\}$$
$$A = \{x \mid x \in N \text{ and } x \text{ is odd and } x < 10\}$$
$$B = \{x \mid x \in N \text{ and } x \text{ is even and } x < 10\}$$
$$C = \{x \mid x \in N \text{ and } x < 6\}$$

Determine the following.

67. $A \cap B$ **68.** $A \cup B$
69. $A' \cup B$ **70.** $(B \cup C)'$
71. $A \cap C'$ **72.** $A \cap B'$
73. $(B \cap C)'$ **74.** $(A \cup B) \cap C$
75. $(C \cap B) \cup A$ **76.** $(C \cup A) \cap B$
77. $(A' \cup C) \cap B$ **78.** $(A \cap B') \cup C$
79. $(A \cup B)' \cap C$ **80.** $(A \cap C)' \cap B$

81. When will a set and its complement be disjoint? Explain and give an example.

82. When will $n(A \cap B) = 0$? Explain and give an example.

83. *Baseball and Football* At Washington High School, 16 students played on the baseball team, 35 students played on the football team, and 7 students played on both the baseball and football teams. How many students played on either the baseball or football team?

84. *Visiting California* The results of a survey of visitors in Hollywood, California, showed that 27 visited the Hollywood Bowl, 38 visited Disneyland, and 16 visited both the Hollywood Bowl and Disneyland. How many people visited either the Hollywood Bowl or Disneyland?

85. Consider the formula

$$n(A \cup B) = n(A) + n(B) - n(A \cap B)$$

a) Show that this relation holds for $A = \{a, b, c, d\}$ and $B = \{b, d, e, f, g, h\}$.
b) Make up your own sets A and B, each consisting of at least six elements. Using these sets, show that the relation holds.
c) Use a Venn diagram and explain why the relation holds for any two sets A and B.

86. The Venn diagram in Fig. 2.13 shows a technique of labeling the regions to indicate membership of elements in a particular region. Define each of the four regions with a set statement. (*Hint:* $A \cap B'$ defines region I.)

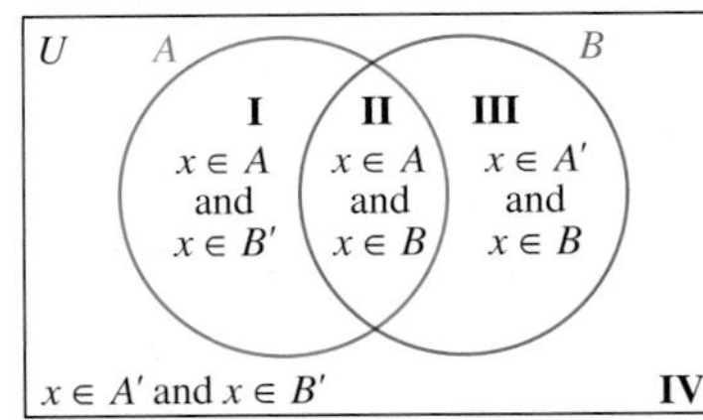

Figure 2.13

In Exercises 87–96, let $U = \{0, 1, 2, 3, 4, 5, \ldots\}$, $A = \{1, 2, 3, 4, \ldots\}$, $B = \{4, 8, 12, 16, \ldots\}$, *and* $C = \{2, 4, 6, 8, \ldots\}$. *Determine the following.*

87. $A \cup B$ **88.** $A \cap B$
89. $B \cap C$ **90.** $B \cup C$
91. $A \cap C$ **92.** $A' \cap C$
93. $B' \cap C$ **94.** $(B \cup C)' \cup C$
95. $(A \cap C) \cap B'$ **96.** $U' \cap (A \cup B)$

CHALLENGE EXERCISES/GROUP ACTIVITIES

In Exercises 97–104, determine whether the answer is $\emptyset$, *A, or U. (Assume* $A \neq \emptyset$, $A \neq U$.)

97. $A \cup A'$ **98.** $A \cap A'$ **99.** $A \cup \emptyset$
100. $A' \cup U$ **101.** $A \cap \emptyset$ **102.** $A \cup U$
103. $A \cap U$ **104.** $A \cup U'$

In Exercises 105–110, determine the relationship between set A and B if

105. $A \cap B = B$. **106.** $A \cup B = B$. **107.** $A \cap B = \emptyset$.

108. $A \cup B = A$. **109.** $A \cap B = A$. **110.** $A \cup B = \emptyset$.

Another set operation is the **difference of two sets.** The difference of two sets A and B, symbolized $A - B$, is defined as

$$A - B = \{x \mid x \in A \text{ and } x \notin B\}$$

Thus, $A - B$ is the set of elements that belong to set A but not to set B. For example, if $U = \{1, 2, 3, 4, 5, 6, 7, 8, 9, 10\}$, $A = \{2, 4, 5, 9, 10\}$, and $B = \{1, 3, 4, 5, 6, 7\}$, then $A - B = \{2, 9, 10\}$ and $B - A = \{1, 3, 6, 7\}$.

In Exercises 111–114, let $U = \{a, b, c, d, e, f, g, h, i, j, k\}$, $A = \{b, c, e, f, g, h\}$, and $B = \{a, b, c, g, i\}$. Determine the following.

111. $A - B$ **112.** $B - A$

113. $A' - B$ **114.** $A - B'$

In Exercises 115–120, let $U = \{1, 2, 3, 4, 5, 6, 7, 8, 9, 10, 11, 12, 13, 14, 15\}$, $A = \{2, 4, 5, 7, 9, 11, 13\}$, and $B = \{1, 2, 4, 5, 6, 7, 8, 9, 11\}$. Determine the following.

115. $A - B$ **116.** $B - A$

117. $(A - B)'$ **118.** $A - B'$

119. $(B - A)'$ **120.** $A \cap (A - B)$

2.4 VENN DIAGRAMS WITH THREE SETS AND VERIFICATION OF EQUALITY OF SETS

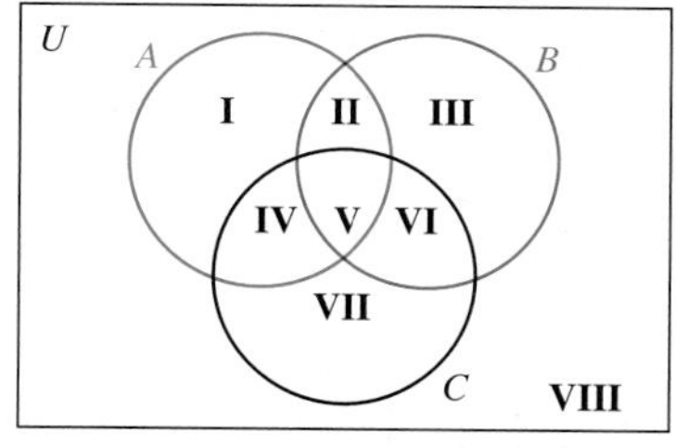

Figure 2.14

Venn diagrams can be used to illustrate three or more sets. For three sets, A, B, and C, the diagram is drawn so the three sets overlap (Fig. 2.14), creating eight regions. The diagrams in Fig. 2.15 emphasize selected regions of three intersecting sets. *When constructing Venn diagrams with three sets, we generally start with region V and work outward,* as explained in the following procedure.

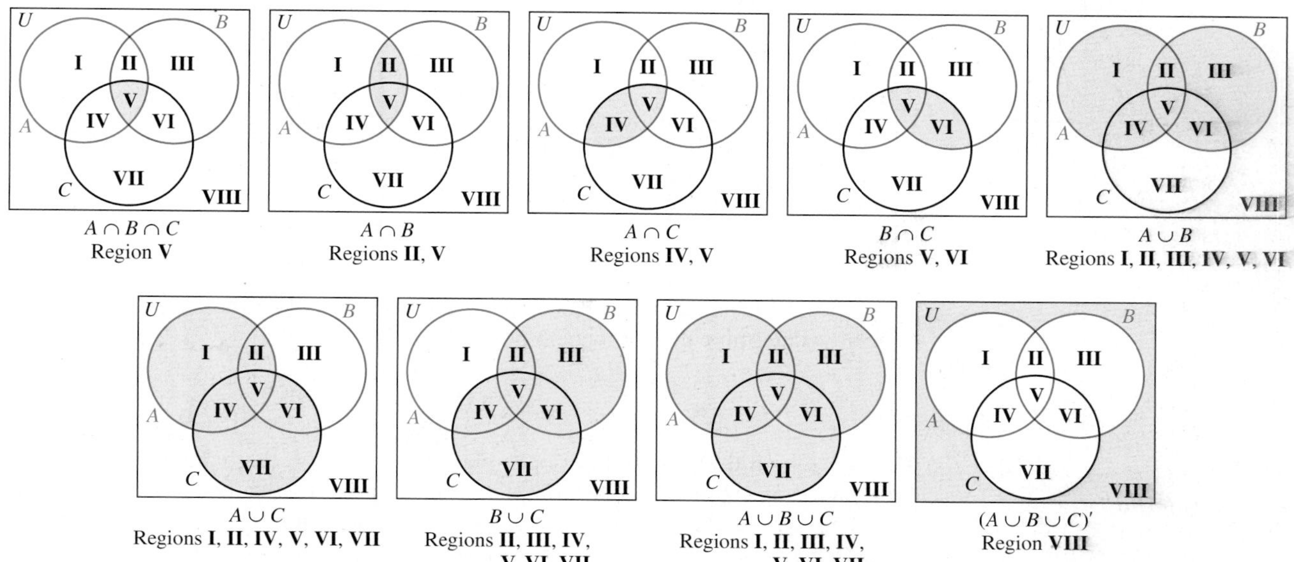

Figure 2.15

General Procedure for Constructing Venn Diagrams with Three Sets, A, B, and C

1. Determine the elements to be placed in region V by finding the elements that are common to all three sets, $A \cap B \cap C$.
2. Determine the elements to be placed in region II. Find the elements in $A \cap B$. The elements in this set belong in regions II and V. Place the elements in the set $A \cap B$ that are not listed in region V in region II. The elements in regions IV and VI are found in a similar manner.
3. Determine the elements to be placed in region I by determining the elements in set A that are not in regions II, IV, and V. The elements in regions III and VII are found in a similar manner.
4. Determine the elements to be placed in region VIII by finding the elements in the universal set that are not in regions I through VII.

Example 1 illustrates the general procedure.

EXAMPLE 1 ***Construct a Venn Diagram for Three Sets***

Construct a Venn diagram illustrating the following sets.

$U = \{1, 2, 3, 4, 5, 6, 7, 8, 9, 10, 11, 12, 13, 14, 15\}$
$A = \{1, 2, 3, 4, 7, 9, 11\}$
$B = \{2, 3, 4, 5, 10, 12, 14\}$
$C = \{1, 2, 4, 8, 9\}$

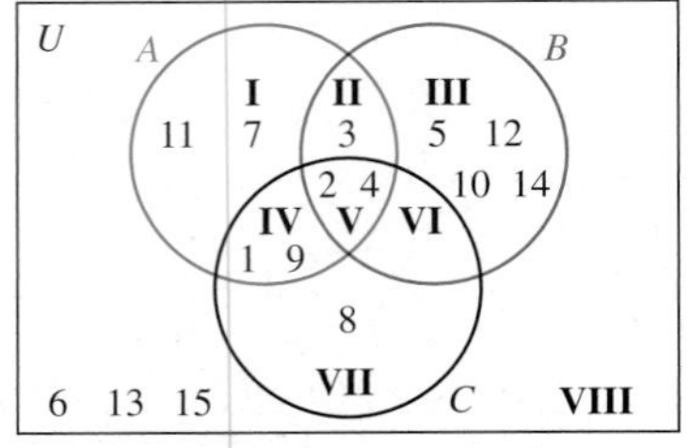

Figure 2.16

SOLUTION First find the intersection of all three sets. Because the elements 2 and 4 are in all three sets, $A \cap B \cap C = \{2, 4\}$. The elements 2 and 4 are placed in region V in Fig. 2.16. Next complete region II by determining the intersection of sets A and B.

$$A \cap B = \{2, 3, 4\}$$

$A \cap B$ consists of regions II and V. The elements 2 and 4 have already been placed in region V, so 3 must be placed in region II.

Now determine what numbers go in region IV.

$$A \cap C = \{1, 2, 4, 9\}$$

Since 2 and 4 have already been placed in region V, place the 1 and 9 in region IV. Now determine the numbers to go in region VI.

$$B \cap C = \{2, 4\}$$

Since both the 2 and 4 have been placed in region V, there are no numbers to be placed in region VI. Now complete set A. The only elements of set A that have not previously been placed in regions II, IV, or V are 7 and 11. Therefore, place the elements 7 and 11 in region I. The elements in region I are only in set A. Using set B, complete region III using the same general procedure used to determine the numbers in region I. Using set C, complete region VII by using the same procedure used to complete regions I and III. To determine the elements in region VIII, find the elements in U that have not been placed in regions I–VII. The elements 6, 13, and 15 have not been placed in regions I–VII, so place them in region VIII. ▲

Venn diagrams can be used to illustrate and analyze many everyday problems. One example follows.

EXAMPLE 2 ***Blood Types***

Human blood is classified (typed) according to the presence or absence of the specific antigens A, B, and Rh in the red blood cells. Antigens are highly specified proteins and carbohydrates that will trigger the production of antibodies in the blood to fight infection. Blood lacking the Rh antigen is labeled negative and blood lacking both A and B antigens is type O. Sketch a Venn diagram with three sets A, B, and Rh and place each type of blood listed in the proper region. A person has only one type of blood.

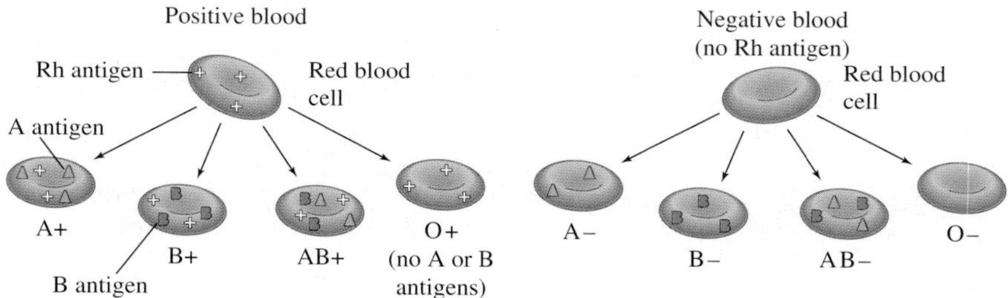

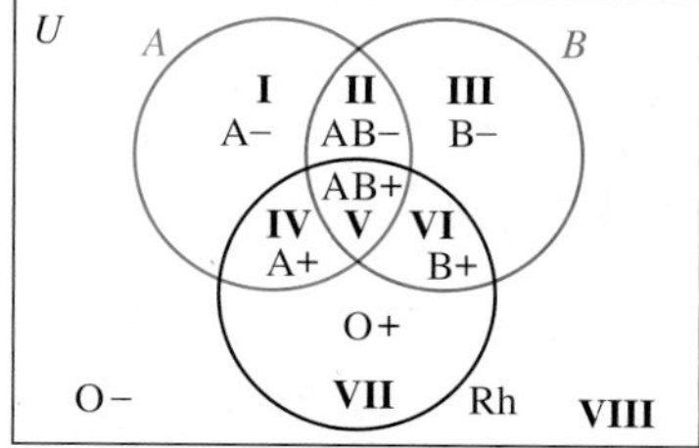

Figure 2.17

SOLUTION As illustrated in Chapter 1, the first thing to do is to read the question carefully and make sure you understand what is given and what you are asked to find. There are three antigens A, B, and Rh. Therefore, begin by naming the three circles in a Venn diagram with the three antigens; see Fig. 2.17.

Any blood containing the Rh antigen is positive, and any blood not containing the Rh antigen is negative. Therefore, all blood in the Rh circle is positive, and all blood outside the Rh circle is negative. The intersection of all three sets, region V, is AB+. Region II contains only antigens A and B and is therefore AB−. Region I is A− because it contains only antigen A. Region III is B−, region IV is A+, and region VI is B+. Region VII is O+, containing only the Rh antigen. Region VIII, which lacks all three antigens, is O−. ▲

VERIFICATION OF EQUALITY OF SETS

In this chapter, for clarity we may refer to operations on sets, such as $A \cup B'$ or $A \cap B \cap C$, as *statements involving sets* or simply as *statements.* Now we discuss how to determine if two statements involving sets are equal.

Consider the question, Is $A' \cup B = A' \cap B$ *for all sets A and B*? For the specific sets $U = \{1, 2, 3, 4, 5\}$, $A = \{1, 3\}$, and $B = \{2, 4, 5\}$, is $A' \cup B = A' \cap B$? To answer the question, we do the following.

Find $A' \cup B$	Find $A' \cap B$
$A' = \{2, 4, 5\}$	$A' = \{2, 4, 5\}$
$A' \cup B = \{2, 4, 5\}$	$A' \cap B = \{2, 4, 5\}$

For these sets, $A' \cup B = A' \cap B$, because both sets are equal to $\{2, 4, 5\}$. At this point you may believe that $A' \cup B = A' \cap B$ for all sets A and B.

If we select the sets $U = \{1, 2, 3, 4, 5\}$, $A = \{1, 3, 5\}$, and $B = \{2, 3\}$, we see that $A' \cup B = \{2, 3, 4\}$ and $A' \cap B = \{2\}$. For this case, $A' \cup B \neq A' \cap B$. Thus, we have proved that $A' \cup B \neq A' \cap B$ for all sets A and B by using a *counterexample.* A counterexample, as explained in Chapter 1, is an example that shows a statement is not true.

In Chapter 1, we explained that proofs involve the use of deductive reasoning. Recall that deductive reasoning begins with a general statement and works to a specific conclusion. To verify, or determine whether set statements are equal for any two sets selected, we use deductive reasoning with Venn diagrams. Venn diagrams are used because they can illustrate general cases. To determine if statements that contain sets, such as $(A \cup B)'$ and $A' \cap B'$, are equal for all sets A and B, we use the regions of Venn diagrams. If both statements represent the same regions of the Venn diagram, then the statements are equal for all sets A and B. See Example 3.

EXAMPLE 3 ***Equality of Sets***

Determine whether $(A \cup B)' = A' \cap B'$ for all sets A and B.

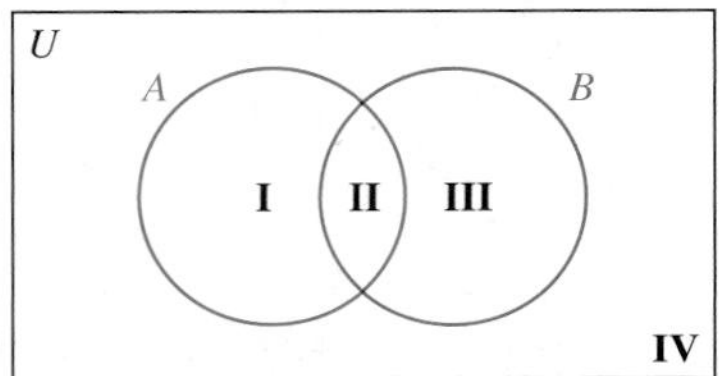

Figure 2.18

SOLUTION Draw a Venn diagram with two sets A and B, as in Fig. 2.18. Label the regions as indicated.

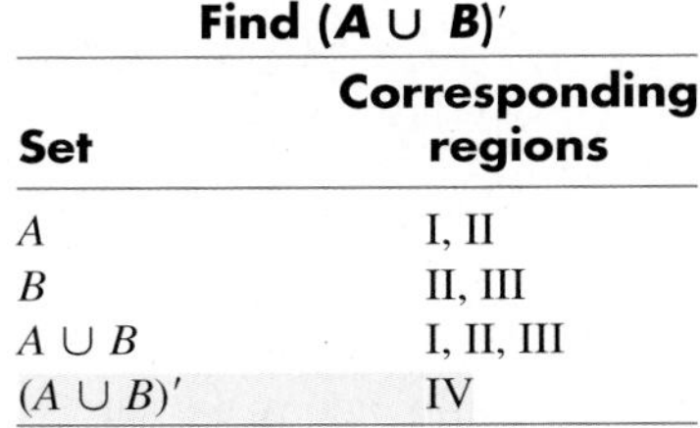

Find $(A \cup B)'$

Set	Corresponding regions
A	I, II
B	II, III
$A \cup B$	I, II, III
$(A \cup B)'$	IV

Find $A' \cap B'$

Set	Corresponding regions
A'	III, IV
B'	I, IV
$A' \cap B'$	IV

Both statements are represented by the same region, IV, of the Venn diagram. Thus, $(A \cup B)' = A' \cap B'$ for all sets A and B. ▲

In Example 3, when we proved that $(A \cup B)' = A' \cap B'$, we started with two general sets and worked to the specific conclusion that both statements represented the same regions of the Venn diagram. We showed that $(A \cup B)' = A' \cap B'$ *for all sets A and B.* No matter what sets we choose for A and B, this statement will be true. For example, let $U = \{1, 2, 3, 4, 5, 6, 7, 8, 9, 10\}$, $A = \{3, 4, 6, 10\}$, and $B = \{1, 2, 4, 5, 6, 8\}$.

$$\begin{aligned} (A \cup B)' &= A' \cap B' \\ \{1, 2, 3, 4, 5, 6, 8, 10\}' &= \{1, 2, 5, 7, 8, 9\} \cap \{3, 7, 9, 10\} \\ \{7, 9\} &= \{7, 9\} \end{aligned}$$

We can also use Venn diagrams to prove statements involving three sets.

EXAMPLE 4 ***Equality of Sets***

Determine whether $A \cap (B \cup C) = (A \cap B) \cup (A \cap C)$ for all sets, A, B, and C.

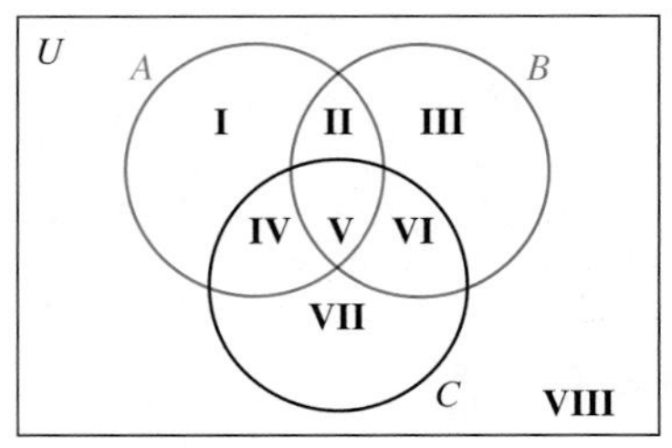

Figure 2.19

SOLUTION Because the statements include three sets, A, B, and C, three circles must be used. The Venn diagram illustrating the eight regions is shown in Fig. 2.19.

First we will find the regions that correspond to $A \cap (B \cup C)$, and then we will find the regions that correspond to $(A \cap B) \cup (A \cap C)$. If both answers are the same, the statements are equal.

Find $A \cap (B \cup C)$

Set	Corresponding regions
A	I, II, IV, V
$B \cup C$	II, III, IV, V, VI, VII
$A \cap (B \cup C)$	II, IV, V

Find $(A \cap B) \cup (A \cap C)$

Set	Corresponding regions
$A \cap B$	II, V
$A \cap C$	IV, V
$(A \cap B) \cup (A \cap C)$	II, IV, V

The regions that correspond to $A \cap (B \cup C)$ are II, IV, and V, and the regions that correspond to $(A \cap B) \cup (A \cap C)$ are also II, IV, and V.

The results show that both statements are represented by the same regions, namely, II, IV, and V, and therefore $A \cap (B \cup C) = (A \cap B) \cup (A \cap C)$ for all sets A, B, and C. ▲

In Example 4, we proved that $A \cap (B \cup C) = (A \cap B) \cup (A \cap C)$ for all sets A, B, and C. Show that this statement is true for the specific sets $U = \{1, 2, 3, 4, 5, 6, 7, 8, 9, 10\}$, $A = \{1, 2, 3, 7\}$, $B = \{2, 3, 4, 5, 7, 9\}$, and $C = \{1, 4, 7, 8, 10\}$.

DE MORGAN'S LAWS

In set theory, logic, and other branches of mathematics, a pair of related theorems known as De Morgan's laws make it possible to transform statements and formulas into alternative and often more convenient forms. In set theory, **De Morgan's laws** are symbolized as follows.

De Morgan's Laws

1. $(A \cup B)' = A' \cap B'$
2. $(A \cap B)' = A' \cup B'$

Law 1 was verified in Example 3. We suggest that you verify law 2 at this time. The laws were expressed verbally by William of Ockham in the fourteenth century. In the nineteenth century, Augustus De Morgan expressed them mathematically. De Morgan's laws will be discussed more thoroughly in Chapter 3, Logic.

SECTION 2.4 EXERCISES

CONCEPT/WRITING EXERCISES

1. When constructing a Venn diagram with three sets, which region do you generally complete first?
2. When constructing a Venn diagram with three sets, after completing region V, which regions do you generally complete next?
3. A Venn diagram contains three sets, A, B, and C, as in Fig. 2.14. If region V contains 6 elements and there are 10 elements in $A \cap B$, how many elements belong in region II? Explain.
4. A Venn diagram contains three sets, A, B, and C, as in Fig. 2.14. If region V contains 4 elements and there are 12 elements in $B \cap C$, how many elements belong in region VI? Explain.
5. Give De Morgan's laws.

6. **a)** For $U = \{1, 2, 3, 4, 5\}$, $A = \{1, 4, 5\}$, and $B = \{1, 4, 5\}$, does $A \cup B = A \cap B$?
   b) By observing the answer to part (a), can we conclude that $A \cup B = A \cap B$ for all sets A and B? Explain.
   c) Determine if $A \cup B = A \cap B$ for all sets A and B.

PRACTICE THE SKILLS

7. Construct a Venn diagram illustrating the following sets.

$U = \{1, 2, 3, 4, 5, 6, 7, 8, 9, 10\}$
$A = \{3, 4, 5, 7, 8, 9\}$
$B = \{1, 3, 4, 7\}$
$C = \{3, 6, 9, 10\}$

8. Construct a Venn diagram illustrating the following sets.

U = {Jan, Feb, March, April, May, June, July, Aug, Sept, Oct, Nov, Dec}
A = {Jan, April, Aug, Sept, Oct, Dec}
B = {Feb, Aug, Oct, Nov, Dec}
C = {Feb, March, June, Dec}

9. Construct a Venn diagram illustrating the following sets. The elements of the sets are the days Amy, Peter, and Carlos work at Al's Sandwich Shop.

U = {Sunday, Monday, Tuesday, Wednesday, Thursday, Friday, Saturday}
Amy = {Monday, Tuesday, Wednesday, Thursday}
Peter = {Sunday, Monday, Tuesday, Wednesday, Thursday, Friday}
Carlos = {Tuesday, Sunday}

10. Construct a Venn diagram illustrating the following sets.

U = {football, basketball, baseball, gymnastics, lacrosse, soccer, tennis, volleyball, swimming, wrestling, cross-country, track, golf, fencing}
A = {football, basketball, soccer, lacrosse, volleyball}
B = {baseball, lacrosse, tennis, golf, volleyball}
C = {swimming, gymnastics, fencing, basketball, volleyball}

11. Construct a Venn diagram illustrating the following sets.

U = {Louis Armstrong, Glenn Miller, Stan Kenton, Charlie Parker, Duke Ellington, Benny Goodman, Count Basie, Jon Coltrane, Dizzy Gillespie, Miles Davis, Thelonius Monk}
A = {Stan Kenton, Count Basie, Dizzy Gillespie, Duke Ellington, Thelonius Monk}
B = {Louis Armstrong, Glenn Miller, Count Basie, Duke Ellington, Miles Davis}
C = {Count Basie, Miles Davis, Stan Kenton, Charlie Parker, Duke Ellington}

12. Construct a Venn diagram illustrating the following sets.

U = {peach, pear, banana, apple, grape, melon, carrot, corn, orange, spinach}
A = {pear, grape, melon, carrot}
B = {peach, pear, banana, spinach, corn}
C = {pear, banana, apple, grape, melon, spinach}

13. *Popular TV Shows* Let U = {*ER, Seinfeld, Suddenly Susan, Home Improvement, 60 Minutes, Friends, Veronica's Closet, NFL Monday Night Football, Touched by an Angel, The Naked Truth, Caroline in the City*}. Let A be the set of the five most popular shows on television in 1997–1998. Let B be the five most popular shows on television in 1996–1997, and let C be the five most popular shows on television in 1995–1996 (according to the 1999 *People Almanac*). Then

A = {*Seinfeld, ER, Veronica's Closet, Friends, NFL Monday Night Football*}
B = {*ER, Seinfeld, Suddenly Susan, Friends, The Naked Truth*}
C = {*ER, Seinfeld, Friends, Caroline in the City, NFL Monday Night Football*}

Construct a Venn diagram illustrating the sets.

14. *Olympic Medals* Consider the chart, which shows teams that won at least 10 medals in the 1998 winter Olympics. Let the teams shown in the chart represent the universal set.

	Gold	Silver	Bronze	Total
Germany	12	9	8	29
Norway	10	10	5	25
Russia	9	6	3	18
Austria	3	5	9	17
Canada	6	5	4	15
United Sates	6	3	4	13
Finland	2	4	6	12
Netherlands	5	4	2	11
Japan	5	1	4	10
Italy	2	6	2	10

Source: *1999 World Almanac*

Let A = set of teams that won at least 15 medals.
Let B = set of teams that won at least 6 gold medals.
Let C = set of teams that won at least 6 bronze medals.

Construct a Venn diagram that illustrates this information.

Men's Brand Loyalty The chart on top left of page 65, taken from the 1999 Wall Street Journal Almanac shows men's loyalty to specific brands. In Fig. 2.20, the set indicated as 1994 represents the set of brands listed in the table under 1994, and so on.

Top men's brands ranked by repurchase intent

1994	1996	1998
1. Levi's	1. Levi's	1. Levi's
2. Starter	2. Land's End	2. Dockers
3. Dockers	3. L.L. Bean	3. Starter
4. Russell Athletic	4. Nike	4. London Fog
5. Hanes	5. Gold Toe	4. Timberland
5. Gold Toe	5. London Fog	4. Tommy Hilfiger
7. Fruit of the Loom	5. Reebok	7. Nike
8. Lee	8. Champion	7. Reebok
8. Reebok	9. Arizona	9. Fruit of the Loom
10. Champion	9. Disney	9. Lee
	9. Hanes	

Source: Kurt Salmon Associates and NPD Group Inc.

Indicate in Fig. 2.20 in which region, I through VIII, each of the following brands belongs.

15. Levi's **16.** Tommy Hilfiger

17. Nike **18.** London Fog

19. Fruit of the Loom **20.** Reebok

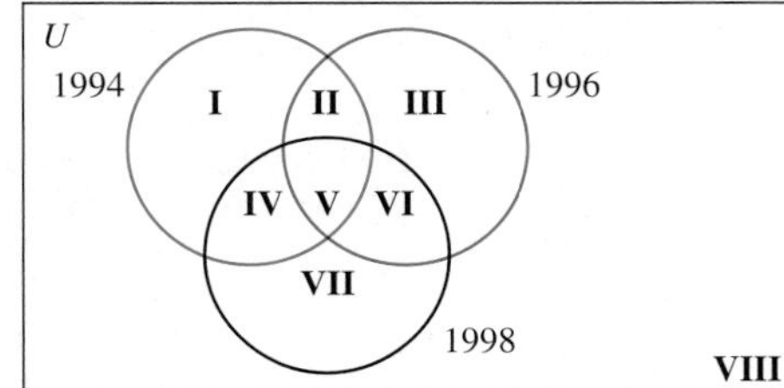

Figure 2.20

Women's Brand Loyalty *The following chart taken from the 1999 Wall Street Journal Almanac shows women's loyalty to specific brands. In Fig. 2.21, the set indicated as 1994 represents the set of brands listed in the table under 1994, and so on.*

Top women's brands ranked by repurchase intent

1994	1996	1998
1. Levi's	1. Levi's	1. Tommy Hilfiger
2. London Fog	2. Hanes	2. Hanes
3. Hanes Her Way	3. Reebok	3. Victoria's Secret
4. Hanes	4. Hanes Her Way	4. L.L. Bean
5. Reebok	5. L.L. Bean	5. Eddie Bauer
6. Dockers	6. Arizona	5. Hanes Her Way
6. Fruit of the Loom	7. Fruit of the Loom	5. Timberland
8. Just My Size	7. Lee	8. Alfred Dunner
9. Nike	7. London Fog	8. Dockers
10. Lee	10. Disney	8. Levi's
		8. Nike

Source: Kurt Salmon Associates and NPD Group Inc.

Indicate in Fig. 2.21 in which region, I through VIII, each of the following brands belongs.

21. Victoria's Secret **22.** Hanes Her Way

23. L.L. Bean **24.** Nike

25. Levi's **26.** Tommy Hilfiger

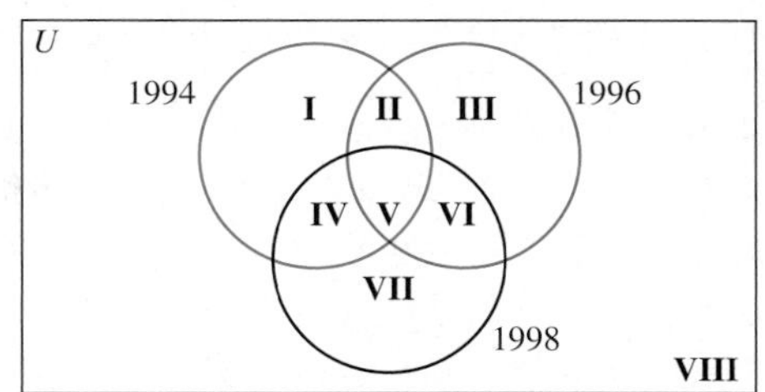

Figure 2.21

In Exercises 27–38, indicate in Fig. 2.22 the region in which each of the figures would be placed.

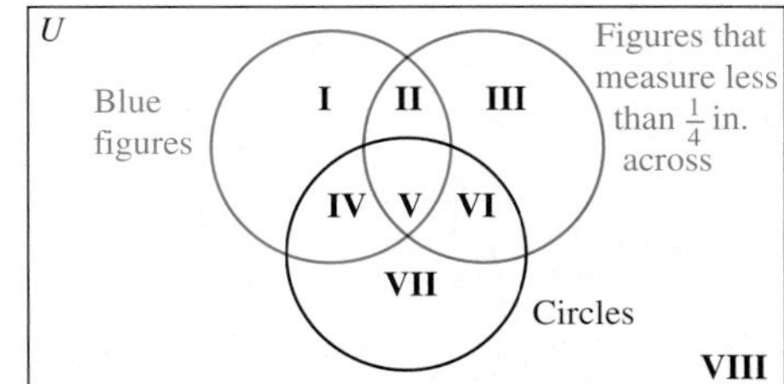

Figure 2.22

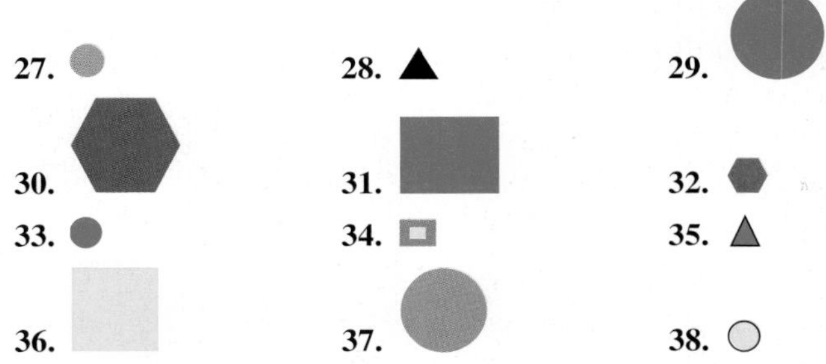

Senate Bills *During a session of the U.S. Senate, three bills were voted on. The votes of six senators are shown in the table. Determine in which region of Fig. 2.23 each senator would be placed. The set labeled bill 1 represents the set of senators who voted yes on bill 1, and so on.*

Senator	Bill 1	Bill 2	Bill 3
39. Grump	yes	no	no
40. Happi	no	no	yes
41. Turwilliger	no	no	no
42. Dillinger	yes	yes	yes
43. Isaitere	no	yes	yes
44. Smith	no	yes	no

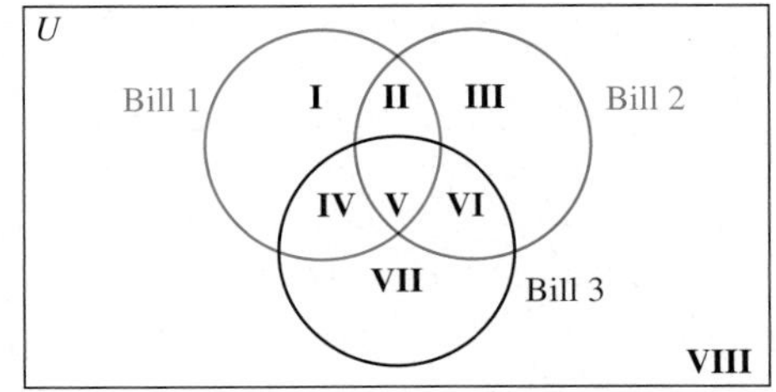

Figure 2.23

In Exercises 45–58, use the Venn diagram in Fig. 2.24 to list the sets in roster form.

45. A	**46.** B	**47.** C
48. U	**49.** $A \cap B$	**50.** $A \cap C$
51. $(B \cap C)'$	**52.** $A \cap B \cap C$	**53.** $A \cup B$
54. $B \cup C$	**55.** $(A \cup C)'$	**56.** $A \cup B \cup C$
57. A'	**58.** $(A \cup B \cup C)'$	

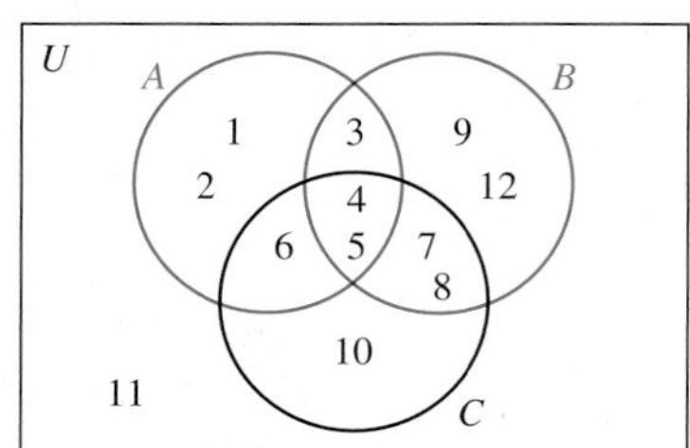

Figure 2.24

In Exercises 59–66, use Venn diagrams to determine whether the following statements are equal for all sets A and B.

59. $(A \cup B)'$, $A' \cap B'$

60. $(A \cup B)'$, $A' \cap B$

61. $A' \cup B'$, $A \cap B$

62. $(A \cup B)'$, $(A \cap B)'$

63. $A' \cup B'$, $(A \cup B)'$

64. $A \cap B'$, $A' \cup B$

65. $(A \cap B')'$, $A' \cup B$

66. $A' \cap B'$, $(A' \cap B')'$

In Exercises 67–76, use Venn diagrams to determine whether the following statements are equal for all sets A, B, and C.

67. $A \cup (B \cap C)$, $(A \cup B) \cap C$

68. $A \cup (B \cap C)$, $(B \cap C) \cup A$

69. $A \cap (B \cup C)$, $(B \cup C) \cap A$

70. $A' \cup (B \cap C)$, $A \cap (B \cup C)'$

71. $A \cap (B \cup C)$, $(A \cap B) \cup (A \cap C)$

72. $A \cup (B \cap C)$, $(A \cup B) \cap (A \cup C)$

73. $A \cap (B \cup C)'$, $A \cap (B' \cap C')$

74. $(A \cup B) \cap (B \cup C)$, $B \cup (A \cap C)$

75. $(A \cup B)' \cap C$, $(A' \cup C) \cap (B' \cup C)$

76. $(C \cap B)' \cup (A \cap B)'$, $A \cap (B \cap C)$

PROBLEM SOLVING

77. Let

$$U = \{1, 2, 3, 4, 5, 6, 7, 8, 9, 10\}$$
$$A = \{1, 2, 3, 4\}$$
$$B = \{3, 6, 7\}$$
$$C = \{6, 7, 9\}$$

a) Show that $(A \cup B) \cap C = (A \cap C) \cup (B \cap C)$ for these sets.

b) Make up your own sets A, B, and C. Verify that $(A \cup B) \cap C = (A \cap C) \cup (B \cap C)$ for your sets A, B, and C.

c) Use Venn diagrams to verify that $(A \cup B) \cap C = (A \cap C) \cup (B \cap C)$ for all sets A, B, and C.

78. Let

$$U = \{a, b, c, d, e, f, g, h, i\}$$
$$A = \{a, c, d, e, f\}$$
$$B = \{c, d\}$$
$$C = \{a, b, c, d, e\}$$

a) Determine whether $(A \cup C)' \cap B = (A \cap C)' \cap B$ for these sets.

b) Make up your own sets, A, B, and C. Determine whether $(A \cup C)' \cap B = (A \cap C)' \cap B$ for your sets.

c) Determine whether $(A \cup C)' \cap B = (A \cap C)' \cap B$ for all sets A, B, and C.

79. *Blood Types* A hematology text gives the following information on percentages of the different types of blood worldwide.

Type	Positive blood, %	Negative blood, %
A	37	6
O	32	6.5
B	11	2
AB	5	0.5

Construct a Venn diagram similar to the one in Example 2 and place the correct percent in each of the eight regions.

80. Define each of the eight regions in Fig. 2.25 using sets A, B, and C and a set operation. (*Hint:* $A \cap B' \cap C'$ defines region I.)

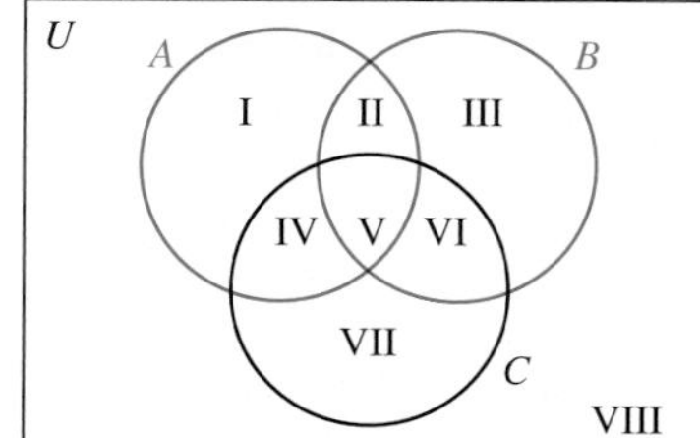

Figure 2.25

81. *Categorizing Personnel* The personnel department at a Gatorade bottling plant wants to classify its employees by their gender, education, and salary. They let set A be the set of males, set B be the set of employees who have a college degree, and set C be the set of employees with a salary greater than \$30,000.

a) Draw a Venn diagram that can be used to categorize the employees of the company according to the listed criteria.
b) Determine the region of the diagram that contains male employees whose salary is greater than \$30,000 and who have a college degree. Describe the region using sets A, B, and C with the operations union, intersection, and complement.
c) Determine the region of the diagram that contains female employees whose salary is greater than \$30,000 and who have a college degree. Describe the region using sets A, B, and C with the operations union, intersection, and complement.
d) Determine the region of the diagram that contains male employees who do not have a college degree and whose salary is less than or equal to \$30,000. Describe the region using sets A, B, and C with the operations union, intersection, and complement.

CHALLENGE EXERCISES/GROUP ACTIVITIES

82. a) Construct a Venn diagram illustrating four sets, A, B, C, and D. (*Hint:* Four circles cannot be used, and you should end up with 16 *distinct* regions.) Have fun!
b) Label each region with a set statement (see Exercise 80). Check all 16 regions to make sure that *each is distinct.*

83. You were able to determine the number of elements in the union of two sets with the formula

$$n(A \cup B) = n(A) + n(B) - n(A \cap B)$$

Can you determine a formula for finding the number of elements in the union of three sets? In other words, write a formula to determine $n(A \cup B \cup C)$. [*Hint:* The formula will contain each of the following: $n(A)$, $n(B)$, $n(C)$, $n(A \cap B \cap C')$, $n(A \cap B' \cap C)$, $n(A' \cap B \cap C)$, and $2n(A \cap B \cap C)$.]

RESEARCH ACTIVITY

84. The two Venn diagrams illustrate what happens when colors are added or subtracted. Do research in an art text, an encyclopedia, or another source and write a report explaining the creation of the colors in the Venn diagrams, using such terms as union of colors and subtraction (or difference) of colors.

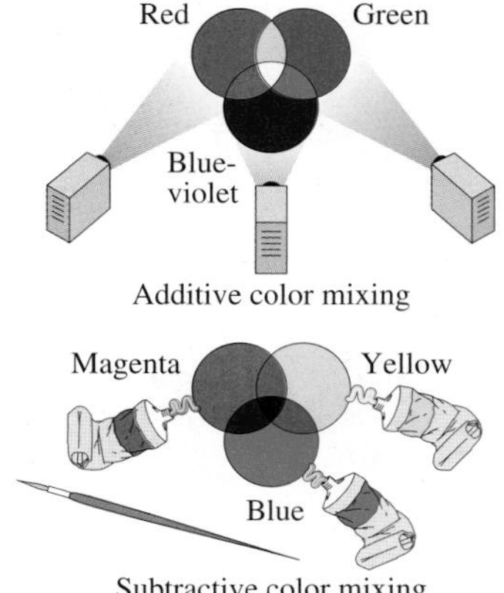

Additive color mixing

Subtractive color mixing

2.5 APPLICATIONS OF SETS

We can solve practical problems involving sets by using the problem-solving process discussed in Chapter 1: Understand the problem, devise a plan, carry out the plan, and then examine and check the results. First determine: What is the problem? or What am I looking for? To devise the plan list all the facts that are given and how they are related. *Look for key words or phrases* like: "only set A," "set A and set B," "set A or set B," "set A and set B and not set C." Remember that *and* means intersection, *or* means union, and *not* means complement. The problems we solve in this section contain two or three sets of elements, which can be represented in a Venn diagram. Our plan will generally include drawing a Venn diagram, labeling the diagram, and filling in the regions of the diagram.

Whenever possible, follow the procedure in Section 2.4 for completing the Venn diagram and then answer the questions. Remember, when drawing Venn diagrams, we generally start with the intersection of the sets and work outward.

EXAMPLE 1 *Voting*

A U.S. senator hired a firm, Rigley and Company, to determine the preference of registered voters in her state on two bills she will soon need to vote on. One is related to education (E) and the other to health care (H). The firm took polls of registered voters in shopping malls and other areas. The following are the results of the polls.

965 registered voters were surveyed.
605 favor the education bill.
427 favor the health care bill.
138 favor both bills.

a) Of those surveyed, how many are not in favor of either the education bill or the health care bill?
b) How many of those surveyed favor the education bill but not the health care bill?
c) How many of those surveyed favor the health care bill but not the education bill?
d) How many of those surveyed favor either the education bill or the health care bill?

SOLUTION The problem provides the following information.

The number of people surveyed is 965: $n(U) = 965$.
The number of people who favor the education bill is 605: $n(E) = 605$.
The number of people who favor the health care bill is 427: $n(H) = 427$.
The number of people who favor both the education and health care bill: $n(E \cap H) = 138$.

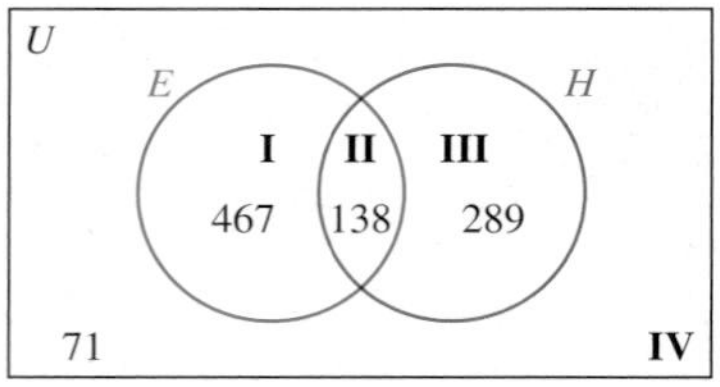

Figure 2.26

We illustrate this information on the Venn diagram shown in Fig. 2.26. We already know that $E \cap H$ corresponds to region II. As $n(E \cap H) = 138$, we write 138 in region II. Set E consists of regions I and II. We know that set E, the people who favor the education bill, contains 605 people. Therefore, region I contains $605 - 138$, or 467 people. We write the number 467 in region I. Set H consists of regions II and III. As $n(H) = 427$, the total in these two regions must be 427. Region II contains 138, leaving 289 for region III.

The total number of people who favor the education bill or the health care bill is found by adding the numbers in regions I, II, and III. Therefore, $n(E \cup H) = 467 + 138 + 289 = 894$. The number of people in region IV is the difference between $n(U)$ and $n(E \cup H)$. There are $965 - 984$, or 71, people in region IV.

a) The people who are not in favor of either the education bill or the health care bill are those members of the universal set who are not contained in set E or set H. The 71 people in region IV are not in favor of either bill.
b) The 467 people in region I are those who favor the education bill but not the health care bill.
c) The 289 people in region III are those who favor the health care bill but not the education bill.
d) The people in regions I, II, or III favor either the education bill or the health care bill. Thus, $467 + 138 + 289$ or 894 people favor either of the bills. Notice that the 138 people in region II who favor both bills are included in those who favor either of the bills. ▲

Similar problems involving three sets can be solved, as illustrated in Example 2.

EXAMPLE 2 *New Cereals*

General Mills is considering producing three new cereals. To determine how the public will respond to the cereals, it sends samples of the new cereals to many households and offers to pay a small amount if the household will complete a small questionnaire. The new cereals are Creepy Crawlers, Mighty Mucks, and Sweet Treats. The following information was obtained from the 970 households that returned the questionnaire.

440 would purchase Creepy Crawlers.
520 would purchase Mighty Mucks.
501 would purchase Sweet Treats.
297 would purchase Creepy Crawlers and Mighty Mucks.
253 would purchase Creepy Crawlers and Sweet Treats.
204 would purchase Mighty Mucks and Sweet Treats.
156 would purchase all three cereals.

Use a Venn diagram to answer the following questions. How many households would purchase:

a) none of these cereals?
b) only Sweet Treats?
c) at least one of the cereals?
d) exactly two of the cereals?

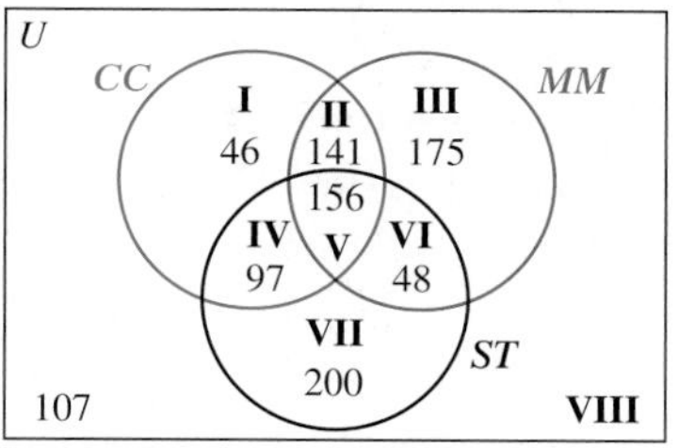

Figure 2.27

SOLUTION Begin by constructing a Venn diagram with three overlapping circles. One circle represents Creepy Crawlers, another Mighty Mucks, and the third Sweet Treats. See Fig. 2.27. Label the eight regions.

Whenever possible, work from the center of the diagram outwards. First fill in region V. Since 156 households would purchase all three cereals, we place 156 in region V. Next determine the number to be placed in region II. Regions II and V together represent the households who would purchase both Creepy Crawlers and Mighty Mucks. Since 297 households would purchase both of these cereals, the sum of the numbers in these regions must be 297. Since 156 have already been placed in region V, $297 - 156 = 141$ must be placed in region II. Now we determine the number to be placed in region IV. Since 253 households would purchase both Creepy Crawlers and Sweet Treats, the sum of the numbers in regions IV and V must be 253. Therefore, $253 - 156 = 97$ must be placed in region IV. Now determine the number to be placed in region VI. A total of 204 households would purchase Mighty Mucks and Sweet Treats. The numbers in regions V and VI must total 204. Since 156 have already been placed in region V, the number to be placed in region VI is $204 - 156 = 48$.

Now that we have determined the numbers for regions V, II, IV, and VI, we can determine the numbers to be placed in regions I, III, and VII. We are given that 440 households would purchase Creepy Crawlers. The sum of the numbers in regions I, II, IV, and V must be 440. To determine the number to be placed in region I, subtract the amounts in regions II, IV, and V from 440. There must be $440 - 141 - 97 - 156 = 46$ in region I. Determine the numbers to be placed in regions III and VII in a similar manner.

$$\text{Region III} = 520 - 141 - 156 - 48 = 175$$
$$\text{Region VII} = 501 - 97 - 156 - 48 = 200$$

Now that we have determined the numbers in regions I through VII, we can determine the number to be placed in region VIII. Adding the numbers in regions I through VII yields a sum of 863. The difference between the total number of responses, 970, and the sum of the numbers in regions I through VII must be placed in region VIII.

$$\text{Region VIII} = 970 - 863 = 107$$

Now that we have completed the Venn diagram, we can answer the questions.

a) One hundred seven households would not purchase any of the cereals. These households are indicated in region VIII.
b) Region VII represents the households who would purchase only the Sweet Treats. Thus 200 households would purchase only Sweet Treats.
c) The words *at least one* mean "one or more." All those in regions I through VII will purchase at least one of the cereals. The sum of the numbers in regions I through VII is 863, so 863 households would purchase at least one of the cereals.
d) The households in regions II, IV, and VI would purchase exactly two of the cereals. Summing the numbers in these regions, 141 + 97 + 48, we find that 286 households would purchase exactly two of the cereals. Notice that we did not include the households in region V. Those in this region would purchase all three cereals.

The procedure to work problems like those given in Example 2 is generally the same. Start by completing region V. Next complete regions II, IV, and VI. Then complete regions I, III, and VII. Finally, complete region VIII. When you are constructing Venn diagrams, be sure to check your work carefully. *The most common mistake made by students is forgetting to subtract the number in region V from the respective values in determining the numbers to be placed in regions II, IV, and VI.*

EXAMPLE 3 *Birds at the Feeders*

In a bird sanctuary, 41 different species of birds are being studied. Three large bird feeders are constructed, each providing a different type of bird feed. One feeder has sunflower seeds. A second feeder has a mixture of seeds and the third feeder has small pieces of fruit. The following information was obtained.

20 species ate sunflower seeds.
22 species ate the mixture.
11 species ate the fruit.
10 species ate the sunflower seeds and the mixture.
4 species ate the sunflower seeds and the fruit.
3 species ate the mixture and the fruit.
1 species ate all three.

Use a Venn diagram to answer the following questions. How many species ate

a) none of the foods?
b) the sunflower seeds, but neither of the other two foods?
c) the mixture *and* the fruit, but not the sunflower seeds?
d) the mixture *or* the fruit, but not the sunflower seeds?
e) exactly one of the foods?

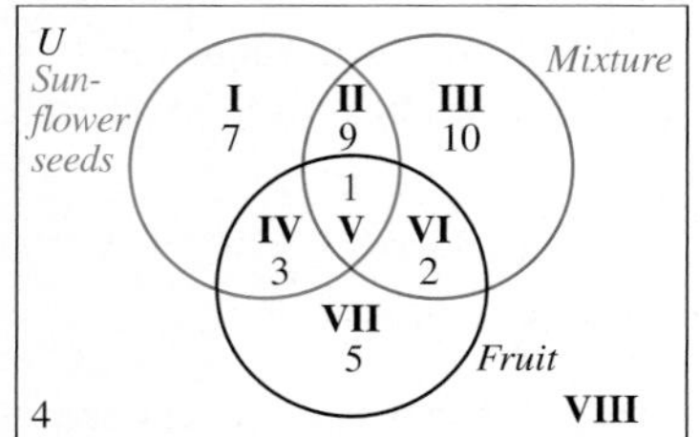

Figure 2.28

SOLUTION The Venn diagram is constructed using the procedure we outlined in Example 2. The diagram is illustrated in Fig. 2.28. We suggest you construct the diagram by yourself now and check your diagram with Fig. 2.28.

a) Four species did not eat any of the food (see region VIII).
b) Seven species (see region I) ate the sunflower seeds but neither of the other two foods.
c) Those in region VI ate both the mixture and fruit but not the sunflower seeds. Therefore, two species satisfy the criteria.
d) The word *or* in this type of problem means one or the other or both. All the species in regions II, III, IV, V, VI, and VII ate the mixture or the fruit or both. Those in regions II, IV, and V also ate the sunflower seeds. The species who ate the mixture or fruit, but not the sunflower seeds, are found by adding the numbers in regions III, VI, and VII. There are $10 + 2 + 5 = 17$ species that satisfy the criteria.
e) Those species indicated in regions I, III, or VII ate exactly one of the foods. Therefore, $7 + 10 + 5 = 22$ species ate exactly one of the three types of bird food. ▲

SECTION 2.5 EXERCISES

PRACTICE THE SKILLS/PROBLEM SOLVING

In Exercises 1–15, draw a Venn diagram to obtain the answers.

1. *Study Locations* At a local college, a survey was taken to determine where students studied on campus. Of 160 students surveyed, it was determined that

79 studied in the library.

65 studied in the student lounge.

43 studied in both the library and the student lounge.

Of those interviewed,

a) how many studied in only the library?
b) how many studied in only the student lounge?
c) how many did not study in either location?

2. *Travel Preferences* A travel agent interviewed 180 people to determine whether they preferred traveling long distances by airplane or by automobile and learned that

120 preferred to travel by airplane.

90 preferred to travel by automobile.

50 preferred to travel by both airplane and automobile.

Of those interviewed,

a) how many preferred to travel only by airplane?
b) how many preferred to travel only by automobile?
c) how many preferred not to travel by airplane or automobile?

3. *Toothpaste Taste Test* A drug company is considering manufacturing a new toothpaste. They are considering two flavors, regular and mint. In a sample of 120 people, it was found that

74 liked the regular.

62 liked the mint.

35 liked both types.

a) How many liked only the regular?
b) How many liked only the mint?
c) How many liked either one or the other or both?

4. *Favoring Legislation* A congressional representative polled her constituents regarding two bills. The results showed that of 200 persons in the sample

55 favored bill I.

100 favored bill II.

40 favored both bills.

a) How many favored only bill I?
b) How many were not in favor of either bill?
c) Did the majority of those surveyed favor bill II?

5. *Overnight Delivery Services* In San Diego, California, a sample of 444 businesses was surveyed to determine which overnight mailing services they used. The following information was determined.

189 used Federal Express.

205 used United Parcel Service.

122 used Airborne.

57 used Federal Express and United Parcel Service.

34 used Federal Express and Airborne.

30 used United Parcel Service and Airborne.

22 used all three.

How many used

a) none of these services?
b) only Airborne?
c) exactly one of the services?
d) exactly two of the services?
e) Federal Express or United Parcel Service?

6. *Raising Grain* The three major grain crops raised in the world are wheat, maize, and rice. A survey of 40 countries that raise grain yielded the following results.

    18 countries raised wheat.
    16 countries raised maize.
    12 countries raised rice.
    9 raised wheat and maize.
    3 raised maize and rice.
    3 raised wheat and rice.
    2 raised all three crops.

    How many countries raised
    a) none of the three crops?
    b) exactly one of the three crops?
    c) exactly two of the three crops?
    d) wheat and maize, but not rice?
    e) maize or rice, but not wheat?

7. *Professional Teams* Thirty-three U.S. cities with large populations were surveyed to determine whether they had a professional baseball team, a professional football team, or a professional basketball team. The following information was determined.

    16 had baseball.
    17 had football.
    15 had basketball.
    11 had baseball and football.
    7 had baseball and basketball.
    9 had football and basketball.
    5 had all three teams.

    How many had
    a) only a football team?
    b) baseball and football, but not basketball?
    c) baseball or football?
    d) baseball or football, but not basketball?
    e) exactly two teams?

8. *Jobs at the Supermarket* A supermarket compiled the following information regarding 30 of its employees.

    8 worked in the produce department.
    9 stacked food on shelves.
    18 operated the cash register.
    4 worked in the produce department and stacked food at different times.
    5 worked in the produce department and operated the cash register at different times.
    3 stacked food and operated the cash register at different times.
    2 did all three jobs at different times.

    How many of the employees
    a) did at least two of these jobs?
    b) worked in produce only?
    c) did only one of the jobs?
    d) stacked shelves or worked the cash register?
    e) stacked shelves and worked the cash register but did not work in produce?

9. *Celebrities* The *People Almanac* performs an annual survey on celebrities. Subscribers are given a list of celebrities and asked to indicate which of the following categories each celebrity belongs to: most powerful, most liked, most trustworthy. From 80 celebrities considered, the following information is determined.

    26 were placed on the most powerful list.
    32 were placed on the most liked list.
    29 were placed on the most trustworthy list.
    7 were placed on the most powerful list and the most liked list.
    8 were placed on the most powerful list and the most trustworthy list.
    12 were placed on the most liked list and the most trustworthy list.
    4 were placed on all three lists.

    How many were on
    a) only the most trustworthy list?
    b) exactly one of the lists?
    c) at least one of the lists?
    d) exactly two of the lists?
    e) none of the lists?

    (Note that in 1999, the two most powerful celebrities were Tom Hanks and Mel Gibson, the two who were most liked were Mel Gibson and Tom Hanks, and the two who were most trustworthy were Andy Griffith and Katharine Hepburn.)

10. *Office Equipment* A survey of 63 business people found that:

30 had desktop computers.
22 had laptop computers.
39 had a fax machine.
15 had a desktop computer and a laptop computer.
18 had a desktop computer and a fax machine.
14 had a laptop computer and a fax machine.
12 had all three.

a) How many had only a laptop computer?
b) How many had only a fax machine?
c) How many had a fax machine and laptop, but not a desktop computer?
d) How many had a fax machine or a laptop, but not a desktop computer?
e) How many had none of the three items?

11. *Blood Types* In a recent blood drive, the following data on donors were recorded (see Example 2 in Section 2.4).

243 had the A antigen.
93 had the B antigen.
325 had the Rh antigen.
28 had both the A and B antigens.
80 had both the B and Rh antigens.
210 had both the A and Rh antigens.
25 had all three antigens.
33 had none of the antigens.

How many donors had
a) A+ blood?
b) B− blood?
c) AB+ blood?
d) How many donors are represented in the blood drive?

12. *Ice Cream Survey* Dreamy Cream Ice Cream hired Molly to find out what kind of ice cream people liked. She surveyed 100 people, with the following results: 78 liked hard ice cream, 61 liked soft ice cream, and 40 liked both hard ice cream and soft ice cream. Every person interviewed liked one or the other or both kinds of ice cream. Does this result seem right? Explain your answer.

13. *Discovering an Error* An immigration agent sampled cars going from the United States into Canada. In his report, he indicated that of the 85 cars sampled,

35 cars were driven by women.
53 cars were driven by U.S. citizens.
43 cars had two or more passengers.
27 cars were driven by women who are U.S. citizens.
25 cars were driven by women and had two or more passengers.
20 cars were driven by U.S. citizens and had two or more passengers.
15 cars were driven by women who are U.S. citizens and had two or more passengers.

After his supervisor reads the report, she explains to the agent that he made a mistake. Explain how his supervisor knew that the agent's report contained an error.

CHALLENGE EXERCISES/GROUP ACTIVITIES

14. *Rodeo Events* At a rodeo, 27 contestants entered at least one of the following events: calf roping, steer wrestling, bronco riding. There were 8 who roped cows but did not wrestle steers. There were 11 who wrestled steers but did not rope cows. Of the 11 who entered the bronco riding event, 3 entered only that event. There were 6 who entered only the calf roping event. One contestant entered all three events.

How many contestants entered
a) only the steer-wrestling event?
b) only one event?
c) the steer-wrestling or the calf-roping or the bronco-riding events?
d) the calf-roping and the bronco-riding events, or the steer-wrestling event?

15. *Surveying Farmers* A survey of 500 farmers in a midwestern state showed the following.

125 grew only wheat.
110 grew only corn.
90 grew only oats.
200 grew wheat.
60 grew wheat and corn.
50 grew wheat and oats.
180 grew corn.

Find the number who
a) grew at least one of the three.
b) grew all three.
c) did not grow any of the three.
d) grew exactly two of the three.

RESEARCH ACTIVITY

16. On page 49, we discussed the ladder of life. Do research and indicate all the different classifications in the Linnaean system, from most general to the most specific, in which a koala belongs.

2.6 INFINITE SETS

On page 41, we state that a finite set is a set in which the number of elements is zero or can be expressed as a natural number. On page 42, we define a one-to-one correspondence. To determine the number of elements in a finite set, we can place it in a one-to-one correspondence with a subset of the set of counting numbers. For example, the set $A = \{\#, ?, \$\}$ can be placed in one-to-one correspondence with set $B = \{1, 2, 3\}$, a subset of the set of counting numbers.

$$\begin{array}{rl} A = & \{\#,\ ?,\ \$\} \\ & \ \ \downarrow\ \ \downarrow\ \ \downarrow \\ B = & \{1,\ 2,\ 3\} \end{array}$$

Because the cardinal number of set B is 3, the cardinal number of set A is also 3. Any two sets such as A and B that can be placed in a one-to-one correspondence must have the same number of elements (therefore the same cardinality) and must be equivalent sets. Note that $n(A)$ and $n(B)$ both equal 3.

The German mathematician Georg Cantor (1845–1918), known as the father of set theory, thought about sets that were not bounded. He called an unbounded set an *infinite set* and provided the following definition.

> An **infinite set** is a set that can be placed in a one-to-one correspondence with a proper subset of itself.

In Example 1, we use Cantor's definition of an infinite set to show that the set of counting numbers is infinite.

EXAMPLE 1 *The Set of Natural Numbers*

Show that $N = \{1, 2, 3, 4, 5, \ldots, n, \ldots\}$ is an infinite set.

SOLUTION To show that the set N is infinite, we establish a one-to-one correspondence between the counting numbers and a proper subset of itself. By removing the first element from the set of counting numbers, we get the set $\{2, 3, 4, 5, \ldots\}$, which is a proper subset of the set of counting numbers. Now we establish the one-to-one correspondence.

$$\begin{array}{ll} \text{Counting numbers} & = \{\,1,\ 2,\ 3,\ 4,\ 5, \ldots,\ \ n\ \ , \ldots\} \\ & \quad\ \ \downarrow\ \ \downarrow\ \ \downarrow\ \ \downarrow\ \ \downarrow \qquad\ \ \downarrow \\ \text{Proper subset} & = \{\,2,\ 3,\ 4,\ 5,\ 6, \ldots, n+1, \ldots\} \end{array}$$

Note that for any number, n, in the set of counting numbers, its corresponding number in the proper subset is one greater, or $n + 1$. We have now shown the desired one-to-one correspondence, and thus the set of counting numbers is infinite. ▲

Note in Example 1 that we showed the pairing of the general terms $n \rightarrow (n + 1)$. Showing a one-to-one correspondence of infinite sets requires showing the pairing of the general terms in the two infinite sets.

In the set of counting numbers, n represents the general term. For any other set of numbers, the general term will be different. A general term should be written in terms of n such that when 1 is substituted for n in the general term, we get the first number in the set; when 2 is substituted for n in the general term, we get the second number in the set; when 6 is substituted for n in the general term, we get the sixth number in the set; and so on.

Consider the set {4, 9, 14, 19, . . .}. Suppose we want to write a general term for this set (or sequence) or numbers. What would the general term be? The numbers differ by 5, so the general term will be of the form $5n$ plus or minus some number. Substituting 1 for n yields 5(1), or 5. Because the first number in the set is 4, we need to subtract 1 from the 5. Thus, the general term is $5n - 1$. Note that when $n = 1$, the value is $5(1) - 1$ or 4; when $n = 2$, the value is $5(2) - 1$ or 9; when $n = 3$, the value is $5(3) - 1$ or 14; and so on. Therefore we write the set of numbers with a general term as

$$\{4, 9, 14, 19, \ldots, 5n - 1, \ldots\}$$

Now that you are aware of how to determine the general term of a set of numbers, we can do some more problems involving sets.

PROFILE IN MATHEMATICS

Leopold Kronecker

Mathematician Leopold Kronecker, Cantor's former mentor, ridiculed Cantor's theories and prevented Cantor from gaining a position at the University of Berlin. Although Cantor's work on infinite sets is now considered a masterpiece, it generated heated controversy when originally published. Cantor's claim that the infinite set was unbounded offended the religious views of the time that God had created a complete universe, which could not be wholly comprehended by humans. Eventually Cantor was given the recognition due to him, but by then the criticism had taken its toll on his health. He had several nervous breakdowns and spent his last days in a mental hospital.

EXAMPLE 2 ***The Set of Even Numbers***

Show that the set of even counting numbers $\{2, 4, 6, \ldots, 2n, \ldots\}$ is an infinite set.

SOLUTION First create a proper subset of the set of even counting numbers by removing the first number from the set. Then establish a one-to-one correspondence.

$$\begin{array}{ll} \text{Even counting numbers:} & \{\,2,\ 4,\ 6,\ 8, \ldots,\ \ 2n\ \ , \ldots\} \\ & \ \ \downarrow\ \ \downarrow\ \ \downarrow\ \ \downarrow \qquad\quad \downarrow \\ \text{Proper subset:} & \{\,4,\ 6,\ 8, 10, \ldots, 2n + 2, \ldots\} \end{array}$$

A one-to-one correspondence exists between the two sets, so the set of even counting numbers is infinite. ▲

EXAMPLE 3 ***The Set of Multiples of Five***

Show that the set $\{5, 10, 15, 20, \ldots, 5n, \ldots\}$ is an infinite set.

SOLUTION

$$\begin{array}{ll} \text{Given set:} & \{\ 5, 10, 15, 20, 25, \ldots,\ \ 5n\ \ , \ldots\} \\ & \ \ \downarrow\ \ \ \downarrow\ \ \ \downarrow\ \ \ \downarrow\ \ \ \downarrow \qquad\quad \downarrow \\ \text{Proper subset:} & \{10, 15, 20, 25, 30, \ldots, 5n + 5, \ldots\} \end{array}$$

Therefore, the given set is an infinite set. ▲

COUNTABLE SETS

In his work with infinite sets, Cantor developed ideas on how to determine the cardinal number of an infinite set. He called the cardinal number of infinite sets "transfinite cardinal numbers" or "transfinite powers." He defined a set as **countable** if it is finite or if it can be placed in a one-to-one correspondence with the set of counting numbers. All infinite sets that can be placed in a one-to-one correspondence with the set of counting numbers have cardinal number, **aleph-null**, symbolized $\aleph_0$ (the first Hebrew letter, aleph, with a zero subscript, read "null").

EXAMPLE 4 ***The Cardinal Number of the Set of Even Numbers***

Show that the set of even counting numbers has cardinal number $\aleph_0$.

SOLUTION In Example 2, we showed that a set of even counting numbers is infinite by setting up a one-to-one correspondence between the set and a proper subset of itself.

Now we will show that it is countable and has cardinality $\aleph_0$ by setting up a one-to-one correspondence between the set of counting numbers and the set of even counting numbers.

$$\begin{array}{ll} \text{Counting numbers:} & N = \{1, 2, 3, 4, \ldots, n, \ldots\} \\ & \quad\;\;\;\downarrow\;\downarrow\;\downarrow\;\downarrow \qquad\;\downarrow \\ \text{Even counting numbers:} & O = \{2, 4, 6, 8, \ldots, 2n, \ldots\} \end{array}$$

For each number n in the set of counting numbers, its corresponding number is $2n$. Since we found a one-to-one correspondence, the set of even counting numbers is not only infinite, it is also countable. Thus, the cardinal number of the set of even counting numbers is $\aleph_0$; that is, $n(E) = \aleph_0$. ▲

> Any set that can be placed in a one-to-one correspondence with the set of counting numbers has cardinality $\aleph_0$ and is countable.

EXAMPLE 5 ***The Cardinal Number of the Set of Odd Numbers***

Show that the set of odd counting numbers has cardinality $\aleph_0$.

SOLUTION To show that the set of odd counting numbers has cardinality $\aleph_0$, we need to show a one-to-one correspondence between the counting numbers and the odd counting numbers.

$$\begin{array}{ll} \text{Counting numbers:} & N = \{1, 2, 3, 4, 5, \ldots, n, \ldots\} \\ & \quad\;\;\;\downarrow\;\downarrow\;\downarrow\;\downarrow\;\downarrow \qquad\;\downarrow \\ \text{Odd counting numbers:} & O = \{1, 3, 5, 7, 9, \ldots, 2n-1, \ldots\} \end{array}$$

Since there is a one-to-one correspondence, the odd counting numbers have cardinality $\aleph_0$; that is, $n(O) = \aleph_0$. ▲

We have shown that both the odd and even counting numbers have cardinality $\aleph_0$. Merging the odd counting numbers with the even counting numbers gives the set of counting numbers, and we may reason that

$$\aleph_0 + \aleph_0 = \aleph_0$$

This result may seem strange, but it is true. What could such a statement mean? Well, consider a hotel with infinitely many rooms. If all the rooms are occupied, then the hotel is, of course, full. If more guests appear wanting accommodations, will they be turned away? The answer is *no,* for if the room clerk were to reassign each guest to a new room with a room number twice that of the present room, then all the odd-numbered rooms would become unoccupied and there would be space for more guests!

. . . where there's always room for one more. . .

In Cantor's work, he showed that there are different orders of infinity. Sets that are countable and have cardinal number $\aleph_0$ are the lowest order of infinity. Cantor showed that the set of integers and the set of rational numbers (fractions of the form p/q, where $q \neq 0$) are infinite sets with cardinality $\aleph_0$. He also showed that the set of real numbers (discussed in Chapter 5) could not be placed in a one-to-one correspondence with the set of counting numbers and that they have a higher order of infinity, aleph-one, $\aleph_1$.

SECTION 2.6 EXERCISES

CONCEPT/WRITING EXERCISES

1. What is an infinite set as defined in this section?
2. How can we determine if a given set has cardinality $\aleph_0$ and is countable?

PRACTICE THE SKILLS

In Exercises 3–12, show that the set is infinite by placing it in a one-to-one correspondence with a proper subset of itself. Be sure to show the pairing of the general terms in the sets.

3. $\{4, 5, 6, 7, 8, \ldots\}$ **4.** $\{3, 4, 5, 6, 7, \ldots\}$

5. $\{6, 8, 10, 12, 14, \ldots\}$ **6.** $\{3, 5, 7, 9, 11, \ldots\}$

7. $\{4, 7, 10, 13, 16, \ldots\}$ **8.** $\{4, 8, 12, 16, 20, \ldots\}$

9. $\{6, 11, 16, 21, 26, \ldots\}$ **10.** $\{1, \frac{1}{2}, \frac{1}{3}, \frac{1}{4}, \frac{1}{5}, \ldots\}$

11. $\{1, \frac{1}{3}, \frac{1}{5}, \frac{1}{7}, \frac{1}{9}, \ldots\}$ **12.** $\{\frac{5}{8}, \frac{6}{8}, \frac{7}{8}, \frac{8}{8}, \frac{9}{8}, \ldots\}$

In Exercises 13–22, show that the set has cardinal number $\aleph_0$ by establishing a one-to-one correspondence between the set of counting numbers and the given set. Be sure to show the pairing of the general terms in the sets.

13. $\{3, 6, 9, 12, 15, \ldots\}$

14. $\{100, 101, 102, 103, 104, \ldots\}$

15. $\{4, 6, 8, 10, 12, \ldots\}$ **16.** $\{0, 2, 4, 6, 8, \ldots\}$

17. $\{2, 5, 8, 11, 14, \ldots\}$ **18.** $\{4, 9, 14, 19, 24, \ldots\}$

19. $\{5, 8, 11, 14, 17, \ldots\}$ **20.** $\{\frac{1}{2}, \frac{1}{4}, \frac{1}{6}, \frac{1}{8}, \ldots\}$

21. $\{\frac{1}{3}, \frac{1}{4}, \frac{1}{5}, \frac{1}{6}, \frac{1}{7}, \ldots\}$ **22.** $\{\frac{1}{2}, \frac{2}{3}, \frac{3}{4}, \frac{4}{5}, \frac{5}{6}, \ldots\}$

CHALLENGE EXERCISES/GROUP ACTIVITIES

In Exercises 23–26, show that the set has cardinality $\aleph_0$ by establishing a one-to-one correspondence between the set of counting numbers and the given set.

23. $\{1, 4, 9, 16, 25, 36, \ldots\}$

24. $\{2, 4, 8, 16, 32, \ldots\}$

25. $\{3, 9, 27, 81, 243, \ldots\}$

26. $\{\frac{1}{3}, \frac{1}{6}, \frac{1}{12}, \frac{1}{24}, \frac{1}{48}, \ldots\}$

RESEARCH ACTIVITIES

27. Do research to explain how Cantor proved that the set of rational numbers has cardinal number $\aleph_0$.
28. Do research to explain how it can be shown that the real numbers do not have cardinal number $\aleph_0$.

CHAPTER 2 SUMMARY

IMPORTANT FACTS

Or is generally interpreted to mean *union.*

And is generally interpreted to mean *intersection.*

DE MORGAN'S LAWS

$$(A \cup B)' = A' \cap B'$$

$$(A \cap B)' = A' \cup B'$$

For any sets A and B,

$$n(A \cup B) = n(A) + n(B) - n(A \cap B).$$

Number of distinct subsets of a finite set with n elements is 2^n.

Symbol	Meaning
$\in$	is an element of
$\notin$	is not an element of
$n(A)$	number of elements in set A
$\emptyset$ or $\{\ \}$	the empty set
U	the universal set
$\subseteq$	is a subset of
$\not\subseteq$	is not a subset of
$\subset$	is a proper subset of
$\not\subset$	is not a proper subset of
$'$	complement
$\cup$	union
$\cap$	intersection
$\aleph_0$	aleph-null

CHAPTER 2 REVIEW EXERCISES

2.1, 2.2, 2.3, 2.6

In Exercises 1–14, state whether each is true or false. If false, give a reason.

1. The set of U.S. presidents who were born in the state of New Jersey is a well-defined set.
2. The set of the three best brands of computers is a well-defined set.
3. $\triangle \in \{\triangle, \square, \bigcirc, ⬠\}$
4. $\{\ \} \subset \varnothing$
5. $\{3, 6, 9, 12, \ldots\}$ and $\{2, 4, 6, 8, \ldots\}$ are disjoint sets.
6. $\{a, b, c, d, e\}$ is an example of a set in roster form.
7. {computer, calculator, pencil} = {calculator, computer, diskette}
8. $\{R, P, L, K\}$ is equivalent to $\{1, 4, 9, 3\}$.
9. If $A = \{\text{a, e, i, o, u}\}$, then $n(A) = 5$.
10. $A = \{1, 4, 9, 16, \ldots\}$ is a countable set.
11. $A = \{1, 4, 7, 10, \ldots, 31\}$ is a finite set.
12. $\{3, 6, 7\} \subseteq \{7, 6, 3, 5\}$.
13. $\{x \mid x \in N \text{ and } 3 < x \leq 5\}$ is a set in set-builder notation.
14. $\{x \mid x \in N \text{ and } 5 < x \leq 15\} \subseteq \{1, 2, 3, 4, 5, \ldots, 20\}$

In Exercises 15–18, express each set in roster form.

15. Set A is the set of odd natural numbers between 5 and 16.
16. Set B is the set of states that border Oklahoma.
17. $C = \{x \mid x \in N \text{ and } x < 297\}$
18. $D = \{x \mid x \in N \text{ and } 8 < x \leq 96\}$

In Exercises 19–22, express each set in set-builder notation.

19. Set A is the set of natural numbers between 72 and 100.
20. Set B is the set of natural numbers greater than 85.
21. Set C is the set of natural numbers less than 3.
22. Set D is the set of natural numbers between 23 and 41, inclusive.

In Exercises 23–26, express each set with a written description.

23. $A = \{x \mid x$ is a letter of the English alphabet from E through M inclusive$\}$
24. $B =$ {penny, nickel, dime, quarter, half-dollar}
25. $C = \{x, y, z\}$
26. $D = \{x \mid 3 \leq x < 9\}$

In Exercises 27–32, let

$$U = \{1, 3, 5, 7, 9, 11, 13, 15\}$$
$$A = \{1, 3, 5, 7\}$$
$$B = \{5, 7, 9, 13\}$$
$$C = \{1, 7, 13\}$$

Determine the following.

27. $A \cap B$
28. $A \cup B'$
29. $A' \cap B$
30. $(A \cup B)' \cup C$
31. The number of subsets of set B
32. The number of proper subsets of set A
33. For the following sets, construct a Venn diagram and place the elements in the proper region.

 $U =$ {Bob, Carol, Diane, Eunice, Frank, George, Heather, Izzy, Janice, Karl}
 $A =$ {Bob, Carol, Eunice, Izzy, Janice, Karl}
 $B =$ {Carol, George, Izzy, Janice, Karl}
 $C =$ {Bob, Carol, Eunice, Karl}

In Exercises 34–39, use Fig. 2.29 to determine the sets.

34. $A \cup B$
35. $A \cap B'$
36. $A \cup B \cup C$
37. $A \cap B \cap C$
38. $(A \cup B) \cap C$
39. $(A \cap B) \cup C$

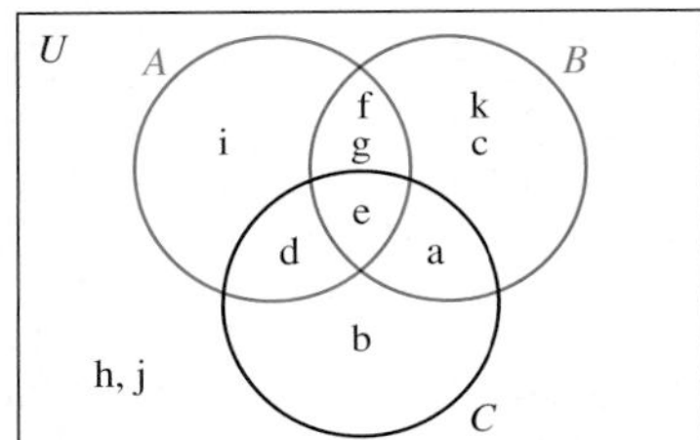

Figure 2.29

2.4

Construct a Venn diagram to determine whether the following statements are true.

40. $(A' \cup B')' = A \cap B$
41. $(A \cup B') \cup (A \cup C') = A \cup (B \cap C)'$

2.5

42. *Pizza Survey* A pizza chain was willing to pay \$1 to each person interviewed about his or her likes and dislikes of types of pizza crust. Of the people interviewed, 200 liked thin crust, 270 liked thick crust, 70 liked both, and 50 did not like pizza at all. What was the total cost of the survey?
43. *Cookie Preferences* The Cookie Shoppe conducted a survey to determine its customers' preferences.

 200 people liked chocolate chip cookies.
 190 people liked peanut butter cookies.
 210 people liked sugar cookies.
 100 people liked chocolate chip cookies and peanut butter cookies.

150 people liked peanut butter cookies and sugar cookies.

110 people liked chocolate chip cookies and sugar cookies.

70 people liked all three.

5 people liked none of these cookies.

Draw a Venn diagram, then determine how many people

a) completed the survey?
b) liked only peanut butter cookies?
c) liked peanut butter cookies and chocolate chip cookies, but not sugar cookies?
d) liked peanut butter cookies or sugar cookies, but not chocolate chip cookies?

44. *Store Survey* In a survey at a shopping mall, 120 people were surveyed to determine their preferences for stores shopped.

62 shopped in Dillards.
71 shopped in J.C. Penney's.
67 shopped in Sears.
37 shopped in Dillards and J.C. Penney's.
32 shopped in Dillards and Sears.
40 shopped in J.C. Penney's and Sears.
12 shopped in all three.

Determine the number of people who shopped in

a) only Dillards.
b) exactly one of the three stores.
c) exactly two of the three stores.
d) Sears and J.C. Penney's, but not Dillards.
e) Sears or J.C. Penney's, but not Dillards.

2.6

In Exercises 45 and 46, show that the sets are infinite by placing each set in a one-to-one correspondence with a proper subset of itself.

45. $\{2, 4, 6, 8, 10, \ldots\}$ **46.** $\{3, 5, 7, 9, 11, \ldots\}$

In Exercises 47 and 48, show that each set has cardinal number $\aleph_0$ by setting up a one-to-one correspondence between the set of counting numbers and the given set.

47. $\{5, 8, 11, 14, 17, \ldots\}$ **48.** $\{4, 9, 14, 19, 24, \ldots\}$

CHAPTER 2 TEST

In Exercises 1–9, state whether each is true or false. If the statement is false, explain why.

1. $\{p, a, r, g\}$ is equivalent to $\{1, a, b, 4\}$.
2. $\{4, 3, 7, p\} = \{p, 3, 7, 2\}$
3. $\{g, h, i\} \subset \{a, r, g, h, i, p\}$
4. $\{7\} \subseteq \{x \mid x \in N \text{ and } x < 7\}$
5. $\{\ \} \not\subseteq \{0\}$
6. $\{p, q, r\}$ has seven subsets.
7. If $A \cap B = \{\ \}$, then A and B are disjoint sets.
8. For any set A, $A \cup A' = \{\ \}$.
9. For any set A, $A \cap U = A$.

In Exercises 10 and 11, use set

$$A = \{x \mid x \in N \text{ and } x < 8\}.$$

10. Write set A in roster form.
11. Write a description of set A.

In Exercises 12–15, use the following information.

$$U = \{3, 5, 7, 9, 11, 13, 15\}$$
$$A = \{3, 5, 7, 9\}$$
$$B = \{7, 9, 11, 13\}$$
$$C = \{3, 11, 15\}$$

Determine the following.

12. $A \cap B$
13. $A \cup C'$
14. $A \cap (B \cap C)'$
15. $n(A \cap B')$
16. Using the sets provided for Exercises 12–15, draw a Venn diagram illustrating the relationship among sets.
17. Use a Venn diagram to determine whether

$$A \cap (B \cup C') = (A \cap B) \cup (A \cap C')$$

for all sets A, B, and C. Show your work.

18. *TV Choices* *Television News* sent a letter to selected subscribers asking which of the following three shows they watched on a regular basis. The three shows they asked about were *ER, The Practice,* and *Ally McBeal.* The results of the 420 questionnaires that were returned showed that

223 selected *ER.*
192 selected *The Practice.*
151 selected *Ally McBeal.*
86 selected *ER* and *The Practice.*
54 selected *ER* and *Ally McBeal.*
68 selected *The Practice* and *Ally McBeal.*
24 selected all three.

a) Construct a Venn diagram and record the appropriate numbers in each region.
b) How many enjoyed watching exactly one of these shows?
c) How many enjoyed watching none of the shows?
d) How many enjoyed watching exactly two of these shows?
e) How many enjoyed watching at least two of these shows?
f) How many watched *ER* or *The Practice,* but not *Ally McBeal?*
g) How many watched *ER* and *The Practice,* but not *Ally McBeal?*

19. Show that the following set is infinite by setting up a one-to-one correspondence between the set and a proper subset of itself.

$$\{7, 8, 9, 10, \ldots\}$$

20. Show that the following set has cardinal number $\aleph_0$ by setting up a one-to-one correspondence between the set of counting numbers and the set.

$$\{1, 3, 5, 7, \ldots\}$$

GROUP PROJECTS

SELECTING A FAMILY PET

1. The Wilcox family is considering buying a dog. They have established several criteria for the family dog: It must be one of the breeds listed in the table, must not shed, must be less than 16 in. tall, and must be good with children.

   a) Using the information in the table*, construct a Venn diagram in which the universal set is the dogs listed. Indicate the set of dogs to be placed in each region of the Venn diagram.
   b) From the Venn diagram constructed in part (a) determine which dogs will meet the criteria set by the Wilcox family. Explain.

Breed	Sheds	Less than 16 in.	Good with children
Airedale	no	no	no
Basset hound	yes	yes	yes
Beagle	yes	yes	yes
Border terrier	no	yes	yes
Cairn terrier	no	yes	no
Cocker spaniel	yes	yes	yes
Collie	yes	no	yes
Dachshund	yes	yes	no
Poodle, miniature	no	yes	no
Schnauzer, miniature	no	yes	no
Scottish terrier	no	yes	no
Wirehaired fox terrier	no	yes	no

*The information is a collection of the opinions of an animal psychologist, Dr. Daniel Tortora, and a group of veterinarians.

CLASSIFICATION OF THE DOMESTIC CAT

2. Read the Did You Know feature on page 49. Do research and indicate the name of the following groupings to which the domestic cat belongs.

a) Kingdom
b) Phylum
c) Class
d) Order
e) Family
f) Genus
g) Species

WHO LIVES WHERE

3. On Diplomat Row, a suburb of Washington, D.C., there are five houses. Each owner is a different nationality, each has a different pet, each has a different favorite food, a different favorite drink, and each house is painted a different color.

The green house is directly to the right of the ivory house.

The Senegalese has the red house.

The dog belongs to the Spaniard.

The Afghanistani drinks tea.

The person who eats cheese lives next door to the fox.

The Japanese eats fish.

Milk is drunk in the middle house.

Apples are eaten in the house next to the horse.

Ale is drunk in the green house.

The Norwegian lives in the first house.

The peach eater drinks whiskey.

Apples are eaten in the yellow house.

The banana eater owns a snail.

The Norwegian lives next door to the blue house.

For each house find

a) the color.
b) the nationality of the occupant.
c) the owner's favorite food.
d) the owner's favorite drink.
e) the owner's pet.
f) Finally, the crucial question is: Does the zebra's owner drink vodka or ale?

CHAPTER 3

Logic

The ancient Greeks were the first people to analyze systematically the way people think and arrive at a conclusion. Aristotle, whose study of logic is presented in a work called *Organon, is* called the father of logic. Since Aristotle's time, the study of logic has been continued by other great mathematicians.

Although most people believe that logic deals with the way people think, it does not. In the study of logic, we use deductive reasoning to analyze complicated situations and come to a reasonable conclusion from a given set of information.

If human thought does not always follow the rules of logic, then why do we study it? Logic enables us to communicate effectively, to make more convincing arguments, and to develop patterns of reasoning for decision making. The study of logic also prepares an individual to better understand other areas of mathematics, computer programming and design, and in general the thought process involved in learning any subject. ■

Logical reasoning can tell us whether a conclusion follows from a set of premises, but not whether those premises are true. For example, Greek astronomers, using the assumption that the planets revolved around Earth, correctly predicted the positions of the planets even though their premise was false.

3.1 STATEMENTS AND LOGICAL CONNECTIVES

HISTORY

The ancient Greeks were the first people to systematically analyze the way humans think and arrive at conclusions. Aristotle (384–322 B.C.) organized the study of logic for the first time in a work called *Organon*. As a result of his work, Aristotle is called the father of logic. The logic from this period, called **Aristotelian logic**, has been taught and studied for more than 2000 years.

Since Aristotle's time, the study of logic has been continued by other great philosophers and mathematicians. Gottfried Wilhelm Leibniz (1646–1716) had a deep conviction that all mathematical and scientific concepts could be derived from logic. As a result, he became the first serious student of **symbolic logic**. One difference between symbolic logic and Aristotelian logic is that in symbolic logic, as its name implies, symbols (usually letters) represent written statements. The forms of the statements in the two types of logic are different. The self-educated English mathematician George Boole (1815–1864) is considered to be the founder of symbolic logic because of his impressive work in this area. Among Boole's publications are *The Mathematical Analysis of Logic* (1847) and *An Investigation of the Law of Thought* (1854). Mathematician Charles Dodgson, better known as Lewis Carroll, incorporated many interesting ideas from logic into his books *Alice's Adventures in Wonderland* and *Through the Looking Glass* and his other children's stories.

Logic has been studied through the ages to exercise the mind's ability to reason. Understanding logic will enable you to think clearly, communicate effectively, make more convincing arguments, and develop patterns of reasoning that will help you in making decisions. It will also help you to detect the fallacies in the reasoning or arguments of others. Studying logic has practical applications as well, such as helping you to understand wills, contracts, and other legal documents.

The study of logic is also good preparation for other areas of mathematics. If you preview Chapter 12, on probability, you will see formulas for the probability of a or b and the probability of a and b, symbolized as $P(A \text{ or } B)$ and $P(A \text{ and } B)$, respectively. Special meanings of common words such as *or* and *and* apply to all areas of mathematics. The meaning of these and other special words are discussed in this chapter.

DID YOU KNOW

Playing on Words

Boole, De Morgan, and other mathematicians of the nineteenth century were anxious to make logic an abstract science that would operate like algebra but be applicable to all fields. One of the problems logicians faced was that verbal language could be ambiguous and could easily lead to confusion and contradiction. Abbott and Costello had fun with the ambiguity of language in their skit about the baseball players: "Who's on first, What's on second, I Don't Know is on third—Yeah, but who's on first?" (in the film *The Naughty Nineties, 1945).*

LOGIC AND THE ENGLISH LANGUAGE

In reading, writing, and speaking, we use many words such as *and, or,* and *if . . . then* . . . to connect thoughts. In logic we call these words **connectives**. How are these words interpreted in daily communication? A judge announces to a convicted offender, "I hereby sentence you to five months of community service *and* a fine of \$100." In this case, we normally interpret the word *and* to indicate that *both* events will take place. That is, the person must do community service and must also pay a fine.

Now suppose a judge states, "I sentence you to six months in prison *or* 10 months of community service." In this case, we interpret the connective *or* as meaning the convicted person must either spend the time in jail or do community service, but not both. The word *or* in this case is the **exclusive or**. When the exclusive or is used, one or the other of the events can take place, but *not both*.

In a restaurant a waiter asks, "May I interest you in a cup of soup or a sandwich?" This question offers three possibilities: You may order soup, you may order a sandwich, or you may order both soup and a sandwich. The *or* in this case is the **inclusive or**. When the inclusive or is used, one or the other, *or both* events can take place. **In this chapter, when we use the word *or* in a logic statement, it will mean the *inclusive or* unless stated otherwise.**

If–then statements are often used to relate two ideas, as in the bank policy statement "If the average daily balance is greater than $500, then there will be no service charge." If–then statements are also used to emphasize a point or add humor, as in the statement "If the Cubs win, then I will be a monkey's uncle."

Now let's look at logic from a mathematical point of view.

STATEMENTS AND LOGICAL CONNECTIVES

A sentence that can be judged either true or false is called a **statement**. Labeling a statement true or false is called *assigning a truth value*. Here are some examples of statements.

1. The Golden Gate Bridge goes over San Francisco Bay.
2. Disney World is in Georgia.
3. The Mississippi River is the longest river in the United States.

In each case, we can say that the sentence is either true or false. Statement 1 is true because the Golden Gate Bridge does go over San Francisco Bay. Statement 2 is false. Disney World is in Florida. By looking at a map or reading an almanac, we can determine that the Mississippi River is the longest in the United States, and therefore, statement 3 is true.

The three sentences are examples of **simple statements** because they convey one idea. Sentences combining two or more ideas that can be assigned a truth value are called **compound statements**. Compound statements are discussed shortly.

QUANTIFIERS

Sometimes it is necessary to change a statement to its opposite meaning. To do so, we use the **negation** of a statement. For example, the negation of the statement "Emily is at home" is "Emily is not at home." The negation of a true statement is always a false statement, and the negation of a false statement is always a true statement. We must use special caution when negating statements containing the words ***all***, ***none*** (or ***no***), and ***some***. These words are referred to as **quantifiers**.

Consider the statement "All lakes contain fresh water." We know this statement is false because the Great Salt Lake in Utah contains salt water. Its negation must therefore be true. We may be tempted to write its negation as "No lake contains fresh water," but this statement is also false because Lake Superior contains fresh water. Therefore, "No lakes contain fresh water" is not the negation of "All lakes contain fresh water." The correct negation of "All lakes contain fresh water" is "Not all lakes contain fresh water" or "At least one lake does not contain fresh water" or "Some lakes do not contain fresh water." These statements all imply that at least one lake does not contain fresh water, which is a true statement.

Now consider the statement "No birds can swim." This statement is false, since at least one bird, the penguin, can swim. Therefore, the negation of this statement must be true. We may be tempted to write the negation as "All birds can swim," but because this statement is also false it cannot be the negation. The correct negation of the statement is "Some birds can swim" or "At least one bird can swim," which are true statements.

Now let's consider statements involving the quantifier *some,* as in "Some students have a driver's license." This is a true statement, meaning that at least one student has a driver's license. The negation of this statement must therefore be false. The negation is "No student has a driver's license," which is a false statement.

Consider the statement "Some students do not ride motorcycles." This statement is true because it means "At least one student does not ride a motorcycle." The negation of this statement must therefore be false. The negation is "All students ride motorcycles," which is a false statement.

The negation of quantified statements is summarized as follows:

Form of statement	**Form of negation**
All are.	Some are not.
None are.	Some are
Some are.	None are.
Some are not.	All are.

The following diagram might help you to remember the statements and their negations:

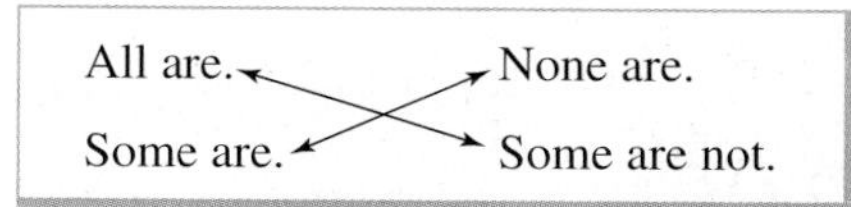

The quantifiers diagonally opposite each other are the negations of each other.

EXAMPLE 1 ***Write Negations***

Write the negation of each statement.

a) Some calculators are solar powered.
b) All judges are elected.

SOLUTION

a) Since *some* means "at least one," the statement "Some calculators are solar powered" is the same as "At least one calculator is solar powered." Because it is a true statement, its negation must be false. The negation is "No calculators are solar powered," which is a false statement.
b) The statement "All judges are elected" is false since some judges are appointed. Its negation must therefore be true. The negation may be written as "Some judges are not elected" or "Not all judges are elected" or "At least one judge is not elected." Each of these statements is true. ▲

COMPOUND STATEMENTS

Statements consisting of two or more simple statements are **compound statements**. The connectives often used to join two simple statements are

and, or, if, . . . then. . . , if and only if

In addition, we consider a simple statement that has been negated to be a compound statement. The word *not* is generally use to negate a statement.

To reduce the amount of writing in logic, it is common to represent each simple statement with a lowercase letter and each connective with a symbol. It is customary to use the letters p, q, r, and s to represent simple statements, but other letters may be used instead. Let's now look at the connectives used to make compound statements.

NOT STATEMENTS

The negation is symbolized by $\sim$ and read "not." For example, the negation of the statement "Steve is a college student" is "Steve is not a college student." If p represents the simple statement "Steve is a college student," then $\sim p$ represents the compound statement "Steve is not a college student." For any statement p, $\sim(\sim p) = p$. For example, the negation of the statement "Steve is not a college student" is "Steve is a college student."

Consider the statement "Maria is not at home." This statement contains the word *not,* which indicates that it is a negation. To write this statement symbolically, we let p represent "Maria *is* at home." Then $\sim p$ would be "Maria is not at home." *We will use this convention of letting letters such as p, q, or r represent statements that are not negated. We will represent negated statements with the negation symbol,* $\sim$.

DID YOU KNOW

Opposing Forces

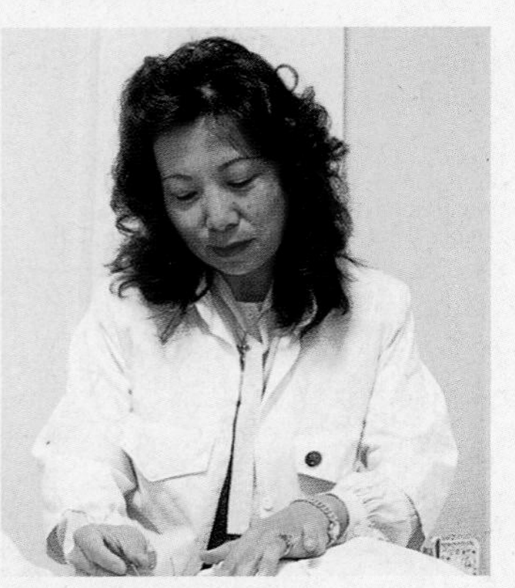

An important symbol in Eastern culture is that of Yin Yang. It suggests a duality of nature—light/dark—and represents the belief that the world arises out of the interplay of opposing forces. Within the human body, traditional Chinese physicians believe, Yin and Yang forces must be balanced. Acupuncture, the insertion of needles into specific spots on the body to relieve pain or disease, a practice developed many centuries ago in China, is believed to "negate," or unblock, the pathways in the body to restore balance and health.

AND STATEMENTS

The **conjunction** is symbolized by $\wedge$ and read "and." The $\wedge$ looks like an A (for And) with the bar missing. Let p and q represent the simple statements.

p: You will perform five months of community service.
q: You will pay a \$100 fine.

Then the following is the conjunction written in symbolic form.

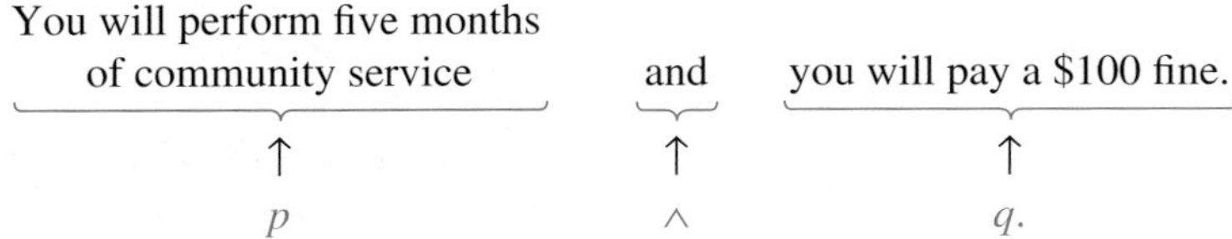
You will perform five months of community service
and
you will pay a $100 fine.
p
∧
q.

The conjunction is generally expressed as *and*. Other words sometimes used to express a conjunction are *but, however,* or *nevertheless*.

EXAMPLE 2 *Write A Conjunction*

Write the following conjunction in symbolic form. The dish is heavy, but the dish in not hot.

SOLUTION Let d and h represent the simple statements.

d: The dish is heavy.
h: The dish is hot.

In symbolic form, the compound statement is $d \wedge \sim h$. ▲

In Example 2, the compound statement is "The dish is heavy, but the dish is not hot." This statement could also be repesented as "The dish is heavy, but *it* is not hot." In this problem, it should be clear that the word *it* means *the dish*. Therefore, the statement, "The dish is heavy, but it is not hot" would also be symbolized as $d \wedge \sim h$.

OR STATEMENTS

The **disjunction** is symbolized by $\vee$ and read "or." The *or* we use in this book (except where indicated in the exercise sets) is the *inclusive or* described on page 84.

EXAMPLE 3 *Write a Disjunction*

Let p: Bill will go to the baseball game.

q: Bill will go to the football game.

Write the following statements in symbolic form.

a) Bill will go to the baseball game or Bill will go to the football game.
b) Bill will go to the football game or Bill will not go to the baseball game.
c) Bill will not go to the baseball game or Bill will not go to the football game.

SOLUTION

a) $p \vee q$ b) $q \vee \sim p$ c) $\sim p \vee \sim q$

Because *or* represents the inclusive or, the statement "Bill will go to the baseball game or Bill will go to the football game" in Example 3(a) may mean that Bill will go to the baseball game, or that Bill will go to the football game, or that Bill will go to both the baseball game and the football game. The statement in Example 3(a) could also be written as "Bill will go to the baseball game or the football game".

When a compound statement contains more that one connective, a comma can be used to indicate which simple statements are to be grouped together. When we write the compound statement symbolically, *the simple statements on the same side of the comma are to be grouped together within parentheses.*

For example, "Phil Collins is a singer (p) or Goldie Hawn is an actress (g), and Dallas is in Texas (d)", is written $(p \vee g) \wedge d$. Note that the p and g are both on the same side of the comma. They are therefore grouped together within parentheses. The statement "Phil Collins is a singer, or Goldie Hawn is an actress and Dallas is in Texas" is written $p \vee (g \wedge d)$. In this case, g and d are on the same side of the comma and are therefore grouped together within parentheses.

EXAMPLE 4 *Understand How Commas Are Used to Group Statements*

Let p: Dinner includes soup.

q: Dinner includes salad.

r: Dinner includes the vegetable of the day.

Write the following statements in symbolic form.

a) Dinner includes soup, and salad or the vegetable of the day.
b) Dinner includes soup and salad, or the vegetable of the day.

◆ PROFILE IN MATHEMATICS

George Boole

The self-taught English mathematician George Boole (1815–1864) took the operations of algebra and used them to extend Aristotelian logic. He used symbols like x and y to represent particular qualities or objects in question. For example, if x represents all butterflies, then $1 - x$ represents everything else except butterflies. If y represents the color yellow, then $(1 - x)(1 - y)$ represents everything except butterflies and things that are yellow, or yellow butterflies. This added a computational dimension to logic that provided a basis for twentieth-century work in the field of computing.

SOLUTION

a) The comma tells us to group the statement "Dinner includes salad" with the statement "Dinner includes the vegetable of the day." Note that both statements are on the same side of the comma. The statement in symbolic form is $p \wedge (q \vee r)$.

In mathematics, we always evaluate the information within the parentheses first. Since the conjunction, $\wedge$, is outside the parentheses and is evaluated *last,* this statement is considered a *conjunction.*

b) The comma tells us to group the statement "Dinner includes soup" with the statement "Dinner includes salad." Note that both statements are on the same side of the comma. The statement in symbolic form is $(p \wedge q) \vee r$. Since the disjunction, $\vee$, is outside the parentheses and is evaluated *last,* this statement is considered a *disjunction.* ▲

The information provided in Example 4 is summarized below.

Statement	Symbolic representation	Type of statement
Dinner includes soup, and salad or the vegetable of the day.	$p \wedge (q \vee r)$	conjunction
Dinner includes soup and salad, or the vegetable of the day.	$(p \wedge q) \vee r$	disjunction

An important point to remember is that a negation has the effect of negating only the statement that directly follows it. To negate a compound statement, we must use parentheses. When a negation symbol is placed in front of a statement in parentheses, it negates the entire statement in parentheses. The negation symbol in this case is read, "It is not true that . . . " or "It is false that . . . "

EXAMPLE 5 ***Change Symbolic Statements to Words***

Let p: Susan is using the camcorder.
q: Walter is taking photos.

Write the following symbolic statements in words.

a) $\sim p \wedge \sim q$ b) $\sim(p \vee q)$

SOLUTION

a) Susan is not using the camcorder and Walter is not taking photos.

b) It is false that Susan is using the camcorder or Walter is taking photos. ▲

Part (a) of Example 5 is a conjunction, since it can be written $(\sim p) \wedge (\sim q)$. Part (b) is a negation, since it negates the entire statement. The similarity of these two statements is discussed in Section 3.4.

Occasionally, we come across a **neither–nor** statement, such as "John is neither handsome nor rich." This statement means that John is not handsome *and* John is not rich. If p represents "John is handsome" and q represents "John is rich," this statement is symbolized by $\sim p \wedge \sim q$.

IF–THEN STATEMENTS

The **conditional** is symbolized by $\rightarrow$ and is read "if–then." The statement $p \rightarrow q$ is read "If p, then q."* The conditional statement consists of two parts: the part that precedes the arrow is the **antecedent**, and the part that follows the arrow is the **consequent**.[†] In the conditional statement $p \rightarrow q$, the p is the antecedent and the q is the consequent.

In the conditional statement $\sim(p \vee q) \rightarrow (p \wedge q)$, the antecedent is $\sim(p \vee q)$ and the consequent is $(p \wedge q)$. An example of a conditional statement is "If you drink your milk, then you will grow up to be healthy." A conditional symbol may be placed between any two statements even if the statements are not related.

Sometimes the word *then* in a conditional statement is not explicitly stated. For example, the statement "If you pass this course, I will buy you a car" is a conditional statement because it actually means "If you pass this course, then I will buy you a car."

EXAMPLE 6 *Write Conditional Statements*

Let

p: I go into the bookstore.
q: I will spend money.

Write the following statements symbolically.

a) If I go into the bookstore, then I will spend money.
b) If I do not go into the bookstore, then I will not spend money.
c) It is false that if I go into the bookstore then I will spend money.

SOLUTION

a) $p \rightarrow q$ b) $\sim p \rightarrow \sim q$ c) $\sim(p \rightarrow q)$

EXAMPLE 7 *Use Commas When Writing a Symbolic Statement in Words*

Let

p: Jorge is enrolled in calculus.
q: Jorge's major is criminal justice.
r: Jorge's major is engineering.

Write the following symbolic statements in words and indicate whether the statement is a negation, conjunction, disjunction, or conditional.

a) $(q \rightarrow \sim p) \vee r$ b) $q \rightarrow (\sim p \vee r)$

SOLUTION The parentheses indicate where to place the commas in the sentences.

a) "If Jorge's major is criminal justice then Jorge is not enrolled in calculus, or Jorge's major is engineering." This statement is a disjunction because $\vee$ is outside the parentheses.

---

*Some books indicate that $p \rightarrow q$ may also be read "p implies q." However, many higher-level mathematics books, indicate that $p \rightarrow q$ may be read "p implies q" only under certain conditions. Implications are discussed in Section 3.3.

[†]Some books refer to the antecedent as the hypothesis or premise and the consequent as the conclusion.

b) "If Jorge's major is criminal justice, then Jorge is not enrolled in calculus or Jorge's major is engineering." This is a conditional statement because $\rightarrow$ is outside the parentheses. ▲

IF AND ONLY IF STATEMENTS

The **biconditional** is symbolized by $\leftrightarrow$ and is read "if and only if." The phrase *if and only if* is sometimes abbreviated as "iff." The statement $p \leftrightarrow q$ is read "p if and only if q."

EXAMPLE 8 *Write Statements Using the Biconditional*

Let p: The cow is brown.

q: The milk is chocolate.

Write the following symbolic statements in words.

a) $q \leftrightarrow p$ b) $\sim(p \leftrightarrow \sim q)$

SOLUTION

a) The milk is chocolate if and only if the cow is brown.

b) It is false that the cow is brown if and only if the milk is not chocolate. ▲

You will learn later that $p \leftrightarrow q$ means the same as $(p \rightarrow q) \wedge (q \rightarrow p)$. Therefore, the statement "I will go to college if and only if I can pay the tuition" has the same logical meaning as "If I go to college then I can pay the tuition and If I can pay the tuition then I will go to college."

The following is a summary of the connectives discussed in this section.

Formal name	Symbol	Read	Symbolic form
Negation	$\sim$	"Not"	$\sim p$
Conjunction	$\wedge$	"And"	$p \wedge q$
Disjunction	$\vee$	"Or"	$p \vee q$
Conditional	$\rightarrow$	"If-then"	$p \rightarrow q$
Biconditional	$\leftrightarrow$	"If and only if"	$p \leftrightarrow q$

DOMINANCE OF CONNECTIVES

What is the answer to the problem $2 + 3 \times 4$? Some of you might say 20, but others might say 14. If you evaluate $2 + 3 \times 4$ on a calculator by pressing

2 [+] 3 [×] 4

some may give you the answer 14, whereas others may give you the answer 20. Which is the correct answer? In mathematics, unless otherwise changed by parentheses or some other grouping symbol, multiplication is *always* performed before addition. Thus,

$$2 + 3 \times 4 = 2 + (3 \times 4) = 14$$

The calculators that gave the incorrect answer of 20 are basic calculators that are not programmed according to the order of operations used in mathematics.

Just as an order of operations exists in the evaluation of arithmetic expressions, a dominance of connectives is used in the evaluation of logic statements. How do we evaluate a symbolic logic statement when no parentheses are used? For example, does $p \vee q \rightarrow r$ mean $(p \vee q) \rightarrow r$, or does it mean $p \vee (q \rightarrow r)$? If we are given a symbolic logic statement for which grouping has not been indicated by parentheses or a written logic statement for which grouping has not been indicated by a comma, then we use the dominance of connectives shown in Table 3.1. Note that *the least dominant connective is the negation and that the most dominant is the biconditional.*

Table 3.1 Dominance of Connectives

Least dominant	1. Negation, $\sim$	Evaluate first
	2. Conjunction, $\wedge$; disjunction, $\vee$	
	3. Conditional, $\rightarrow$	
Most dominant	4. Biconditional, $\leftrightarrow$	Evaluate last

As indicated in Table 3.1, the conjunction and disjunction have the same level of dominance. Thus, to determine whether the symbolic statement $p \wedge q \vee r$ is a conjunction or a disjunction, we have to use grouping symbols (parentheses). When evaluating a symbolic statement that does not contain parentheses, we *evaluate the least dominant connective first and the most dominant connective last.* For example,

Statement	Most dominant connective used	Statement means	Type of statement
$\sim p \vee q$	$\vee$	$(\sim p) \vee q$	Disjunction
$p \rightarrow q \vee r$	$\rightarrow$	$p \rightarrow (q \vee r)$	Conditional
$p \wedge q \rightarrow r$	$\rightarrow$	$(p \wedge q) \rightarrow r$	Conditional
$p \rightarrow q \leftrightarrow r$	$\leftrightarrow$	$(p \rightarrow q) \leftrightarrow r$	Biconditional
$p \vee r \leftrightarrow r \rightarrow \sim p$	$\leftrightarrow$	$(p \vee r) \leftrightarrow (r \rightarrow \sim p)$	Biconditional
$p \rightarrow r \leftrightarrow s \wedge p$	$\leftrightarrow$	$(p \rightarrow r) \leftrightarrow (s \wedge p)$	Biconditional

EXAMPLE 9 *Use the Dominance of Connectives*

Use the dominance of connectives to add parentheses to each statement. Then indicate whether each statement is a negation, conjunction, disjunction, conditional, or biconditional.

a) $p \rightarrow q \vee r$ b) $\sim p \wedge q \leftrightarrow r \vee p$

SOLUTION

a) The conditional has greater dominance than the disjunction, so we place parentheses around $q \vee r$, as follows:

$$p \rightarrow (q \vee r)$$

It is a conditional statement because the conditional symbol is outside the parentheses.

b) The biconditional has the greatest dominance, so we place parentheses as follows:

$$(\sim p \wedge q) \leftrightarrow (r \vee p)$$

It is a biconditional statement because the biconditional symbol is outside the parentheses. ▲

EXAMPLE 10 *Identify the Type of Statement*

Use the dominance of connectives and parentheses to write each statement symbolically. Then indicate whether each statement is a negation, conjunction, disjunction, conditional, or biconditional.

a) If you are late in paying your rent or you have damaged the apartment then you may be evicted.
b) You are late in paying your rent, or if you have damaged the apartment then you may be evicted.

SOLUTION

a) Let p: You are late in paying your rent.
q: You have damaged the apartment.
r: You may be evicted.

No commas appear in the sentence, so we will evaluate it by using the dominance of connectives. Because the conditional has higher dominance than the disjunction, the conditional statement will be evaluated last. Thus, the statements "You are late in paying your rent" and "You have damaged the apartment" are to be grouped together. The statement written symbolically with parentheses is

$$(p \vee q) \rightarrow r$$

This statement is a conditional.

b) A comma is used in this statement to indicate grouping, just as parentheses do in arithmetic. The placement of the comma indicates that the statements "You have damaged the apartment" and "You may be evicted" are to be grouped together. Therefore, this statement written symbolically is

$$p \vee (q \rightarrow r)$$

This statement is a disjunction. Note that the comma overrides the dominance of connectives and tells us to evaluate the conditional statement before the disjunction. ▲

SECTION 3.1 EXERCISES

1. What is a simple statement?
2. List the words identified as quantifiers.
3. Write the general form of the negation for statements of the form
   a) none are.
   b) some are not.
   c) all are.
   d) some are.
4. What are compound statements?
5. Represent the statement "The ink is not purple" symbolically. Explain your answer.
6. Draw the symbol used to represent the
   a) conditional.
   b) disjunction.
   c) conjunction.
   d) negation.
   e) biconditional.
7. Explain how a comma is used to indicate the grouping of simple statements.
8. List the dominance of connectives from the most dominant to the least dominant.

In Exercises 9–22, indicate whether the statement is a simple or a compound statement. If it is a compound statement, indicate whether it is a negation, conjunction, disjunction, conditional, or biconditional by using both the word and its appropriate symbol (for example, "a negation," ~).

9. The cat is in the house and the dog is in the yard.

10. The printer is not out of paper.

11. The figure is a triangle if and only if it has three sides.

12. If the phone rings, then you will answer it

13. Jill Bos is playing tennis or she is shopping.

14. The leather jacket is neither tan nor red.

15. The *World Almanac* has 1000 pages.

16. John Montera will go to college if and only if he passes English.

17. It is false that Emily Falzon has a driver's license and a pilot's license.

18. If the oven warms up, then the dough will rise.

19. Everybody studied in the library and everybody passed the test.

20. We decided not to go to the restaurant, but we called to cancel the reservation.

21. It is false that if there is a loss of power then the security system will not work.

22. If the car is in the garage, then the Johnsons have two cars.

In Exercises 23–34, write the negation of the statement.

23. Some flowers are yellow.

24. No money grows on trees.

25. All fish swim.

26. All bowling balls are round.

27. Some dogs do not have fleas.

28. No doctors make house calls.

29. No books are round.

30. All cows give milk.

31. Some rain forests are not being destroyed.

32. No one likes a bully.

33. Some students maintain an A average.

34. Some people who earn money do not pay taxes.

In Exercise 35–40, write the statement in symbolic form. Let

p: The motor vehicle department office is open.

q: The line of people extends out the door.

35. The motor vehicle department office is not open.

36. The motor vehicle department office is open and the line of people extends out the door.

37. The line of people does not extend out the door or the motor vehicle department office is not open.

38. The line of people does not extend out the door if and only if the motor vehicle department office is not open.

39. If the motor vehicle department office is not open, then the line of people does not extend out the door.

40. The line of people does not extend out the door, however the motor vehicle department office is open.

In Exercises 41–46, write the statement in symbolic form. Let

p: The charcoal is hot.

q: The chicken is on the grill.

41. The chicken is not on the grill if and only if the charcoal is not hot.

42. If the chicken is not on the grill then the charcoal is not hot.

43. Neither is the charcoal hot nor is the chicken on the grill.

44. The charcoal is not hot, but the chicken is on the grill.

45. It is false that if the chicken is on the grill then the charcoal is not hot.

46. It is false that the charcoal is hot and the chicken is on the grill.

In Exercises 47–56, write the compound statement in words. Let

p: Firemen work hard.

q: Firemen wear red suspenders.

47. $\sim q$

48. $\sim p$

49. $q \vee p$

50. $p \wedge q$

51. $\sim p \rightarrow q$

52. $\sim p \leftrightarrow \sim q$

53. $\sim(q \vee p)$

54. $\sim p \vee \sim q$

55. $\sim p \wedge \sim q$

56. $\sim(p \wedge q)$

In Exercises 57–66, write the statements in symbolic form. Let

p: The temperature is 90°.
q: The air conditioner is working.
r: The apartment is hot.

57. The temperature is 90° and the air conditioner is working, or the apartment is hot.

58. If the temperature is 90° or the air conditioner is not working, then the apartment is hot.

59. The apartment is hot if and only if the temperature is not 90°, or the air conditioner is not working.

60. If the temperature is 90°, then the air conditioner is working or the apartment is not hot.

61. If the apartment is hot and the air conditioner is working, then the temperature is 90°.

62. The temperature is not 90° if and only if the air conditioner is not working, or the apartment is not hot.

63. The apartment is hot if and only if the air conditioner is working, and the temperature is 90°.

64. It is false that if the apartment is hot then the air conditioner is not working.

65. If the air conditioner is working, then the temperature is 90° if and only if the apartment is hot.

66. The apartment is hot or the air conditioner is not working, if and only if the temperature is 90°.

In Exercises 67–76, write each symbolic statement in words. Let

p: The water is 70°.
q: The sun is shining.
r: We go swimming.

67. $(p \wedge q) \vee r$
68. $(p \vee q) \wedge \sim r$
69. $(q \rightarrow p) \vee r$
70. $\sim p \wedge (q \vee r)$
71. $\sim r \rightarrow (q \wedge p)$
72. $(q \wedge r) \rightarrow p$
73. $(q \rightarrow r) \wedge p$
74. $\sim p \rightarrow (q \vee r)$
75. $(q \leftrightarrow p) \wedge r$
76. $q \rightarrow (p \leftrightarrow r)$

Dinner Menu *In Exercises 77–80, use the following information to arrive at your answers. Many restaurant dinner menus include statements such as the following. All dinners are served with a choice of: Soup or Salad, and Potatoes or Pasta, and Carrots or Peas. Which of the following selections are permissible? If a selection is not permissible, explain why. See the discussion of the* exclusive or *on page 83.*

77. Soup, salad, and peas
78. Salad, pasta, and carrots
79. Soup, potatoes, pasta, and peas
80. Soup, pasta, and potatoes

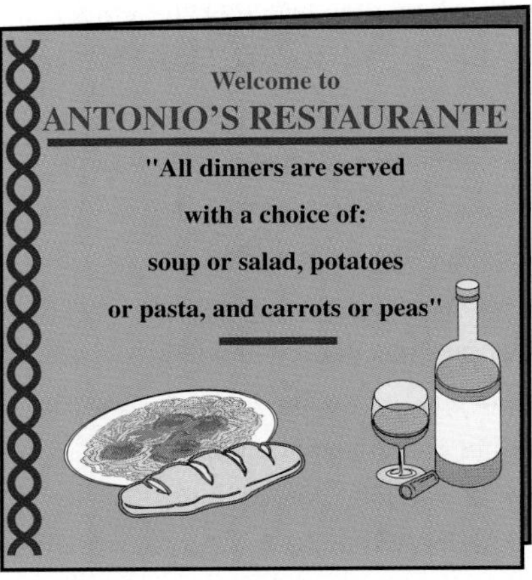

Welcome to
ANTONIO'S RESTAURANTE
"All dinners are served
with a choice of:
soup or salad, potatoes
or pasta, and carrots or peas"

In Exercises 81–94, (a) add parentheses by using the dominance of connectives and (b) indicate whether the statement is a negation, conjunction, disjunction, conditional, or biconditional (see Example 9).

81. $\sim p \rightarrow q$
82. $\sim p \wedge r \leftrightarrow \sim q$
83. $\sim q \wedge \sim r$
84. $\sim p \vee q$
85. $p \vee q \rightarrow r$
86. $q \rightarrow p \wedge \sim r$
87. $r \rightarrow p \vee q$
88. $q \rightarrow p \leftrightarrow p \rightarrow q$
89. $\sim p \leftrightarrow \sim q \rightarrow r$
90. $\sim q \rightarrow r \wedge p$
91. $r \wedge \sim q \rightarrow q \wedge \sim p$
92. $\sim [p \rightarrow q \vee r]$
93. $\sim [p \wedge q \leftrightarrow p \vee r]$
94. $\sim [r \wedge \sim q \rightarrow q \wedge r]$

In Exercises 95–103, (a) select letters to represent the simple statements and write each statement symbolically by using parentheses and (b) indicate whether the statement is a negation, conjunction, disjunction, conditional, or biconditional.

95. Carol Britz visited Nancy Hart or Carol Britz did not drive to Maine.

96. If the moon is out then it is evening or the sun is not shining.

97. It is false that if you pay your taxes then you will not get audited.

98. If the moon is out then it is evening, or the sun is not shining.

99. If the fruit is red or the vegetables are carrots, then you will eat well.

100. If today is Tuesday then tomorrow is Wednesday if and only if today is not Tuesday.

101. The store is open if and only if today is not Sunday or it is before 5 P.M.

102. If a number is divisible by 2 and the number is divisible by 3 then the number is divisible by 6.

103. The store is empty if and only if today is Sunday, or it is after 5 P.M.

PROBLEM SOLVING/GROUP ACTIVITIES

104. *An Ancient Question* If Zeus could do anything, could he build a wall that he could not jump over? Explain your answer.

In Exercises 105 and 106, place parentheses in the statement according to the dominance of connectives. Indicate whether the statement is a negation, conjunction, disjunction, conditional, or biconditional.

105. $\sim q \rightarrow r \vee p \leftrightarrow \sim r \wedge q$

106. $\sim[\sim r \rightarrow p \wedge q \leftrightarrow \sim p \vee r]$

107. **a)** We cannot place parentheses in the statement $p \vee q \wedge r$. Explain why.
b) Make up three simple statements and label them p, q, and r. Then write compound statements to represent $(p \vee q) \wedge r$ and $p \vee (q \wedge r)$.
c) Do you think that the statements for $(p \vee q) \wedge r$ and $p \vee (q \wedge r)$ mean the same thing? Explain.

RESEARCH ACTIVITIES

108. *Legal Documents* Obtain a legal document such as a will or rental agreement and copy one page of the document. Circle every connective used. Then list the number of times each connective appeared. Be sure to include conditional statements from which the word *then* was omitted from the sentence. Give the page and your listing to your instructor.

109. Write a report on the life and accomplishments of George Boole, who was an important contributor to the development of logic. In your report, indicate how his work eventually led to development of the computer. References include encyclopedias, history of mathematics books, and the Internet.

3.2 TRUTH TABLES FOR NEGATION, CONJUNCTION, AND DISJUNCTION

A **truth table** is a device used to determine when a compound statement is true or false. Five basic truth tables are used in constructing other truth tables. Three are discussed in this section (Tables 3.2, 3.4, and 3.7), and two are discussed in the next section. Section 3.5 uses truth tables in determining whether a logical argument is valid or invalid.

NEGATION

Table 3.2 Negation

	p	$\sim p$
Case 1	T	F
Case 2	F	T

The first truth table is for *negation*. If p is a true statement, then the negation of p, "not p," is a false statement. If p is a false statement, then "not p" is a true statement. For example, if the statement "The shirt is blue" is true, then the statement "The shirt is not blue" is false. These relationships are summarized in Table 3.2. For a simple statement, there are exactly two true–false cases, as shown.

Table 3.3

	p	q
Case 1	T	T
Case 2	T	F
Case 3	F	T
Case 4	F	F

If a compound statement consists of two simple statements p and q there are four possible cases, as illustrated in Table 3.3. Consider the statement "The test is today and the test covers Chapter 5." The simple statement "The test is today" has two possible truth values, true or false. The simple statement "The test covers Chapter 5" also has two truth values, true or false. Thus for these two simple statements there are four distinct possible true–false arrangements. Whenever we construct a truth table for a compound statement that consists of two simple statements, we begin by listing the four true–false cases shown in Table 3.3.

CONJUNCTION

To illustrate the conjunction, consider the following situation. You have recently purchased a new house. To decorate it, you ordered a new carpet and new furniture from the same store. You explain to the salesperson that the carpet must be delivered before the furniture. He promises that the carpet will be delivered on Thursday and that the furniture will be delivered on Friday.

Table 3.4 Conjunction

	p	q	$p \wedge q$
Case 1	T	T	T
Case 2	T	F	F
Case 3	F	T	F
Case 4	F	F	F

To help determine whether the salesperson kept his promise, we assign letters to each simple statement. Let p be "The carpet will be delivered on Thursday" and q be "The furniture will be delivered on Friday." The salesperson's statement written in symbolic form is $p \wedge q$. There are four possible true–false situations to be considered. (Table 3.4).

CASE 1: p is true and q is true. The carpet is delivered on Thursday and the furniture is delivered on Friday. The salesperson has kept his promise and the compound statement is true.

CASE 2: p is true and q is false. The carpet is delivered on Thursday but the furniture is not delivered on Friday. Since the furniture was not delivered as promised, the compound statement is false.

CASE 3: p is false and q is true. The carpet is not delivered on Thursday but the furniture is delivered on Friday. Since the carpet was not delivered on Thursday as promised, the compound statement is false.

CASE 4: p is false and q is false. The carpet is not delivered on Thursday and the furniture is not delivered on Friday. Since the carpet and furniture were not delivered as promised, the compound statement is false.

Examining the four cases, we see that in only one case did the salesperson keep his promise: in case 1. Therefore, case 1 (T, T) is true. In cases 2, 3, and 4, the salesperson did not keep his promise and the compound statement is false. The results are summarized in Table 3.4, the truth table for the conjunction.

The **conjunction** $p \wedge q$ is true only when both p and q are true.

EXAMPLE 1 *Construct a Truth Table*

Construct a truth table for $p \wedge \sim q$.

SOLUTION Because there are two statements, p and q, construct a truth table with four cases; see Table 3.5(a). Then write the truth values under the p in the compound statement and label this column 1, as in Table 3.5(b). Copy these

Table 3.5

(a)

	p	q	$p \wedge \sim q$
Case 1	T	T	
Case 2	T	F	
Case 3	F	T	
Case 4	F	F	

(b)

p	q	$p \wedge \sim q$
T	T	T
T	F	T
F	T	F
F	F	F
		1

(c)

p	q	p	$\wedge$	$\sim q$
T	T	T		T
T	F	T		F
F	T	F		T
F	F	F		F
		1		2

(d)

p	q	p	$\wedge$	$\sim$	q
T	T	T		F	T
T	F	T		T	F
F	T	F		F	T
F	F	F		T	F
		1		3	2

(e)

p	q	p	$\wedge$	$\sim$	q
T	T	T	F	F	T
T	F	T	T	T	F
F	T	F	F	F	T
F	F	F	F	T	F
		1	4	3	2

truth values directly from the p column on the left. Write the corresponding truth values under the q in the compound statement and call this column 2, as in Table 3.5(c). Copy the truth values for column 2 directly from the q column on the left. Now find the truth values of $\sim q$ by negating the truth values in column 2 and call this column 3, as in Table 3.5(d). Use the conjunction table, Table 3.4, and the entries in columns 1 and 3 to complete column 4, as in Table 3.5(e). The results in column 4 are obtained as follows:

Row 1: $\text{T} \wedge \text{F}$ is F. Row 2: $\text{T} \wedge \text{T}$ is T.
Row 3: $\text{F} \wedge \text{F}$ is F. Row 4: $\text{F} \wedge \text{T}$ is F.

The answer is always the last column completed. Columns 1, 2, and 3 are only aids in arriving at the answer in column 4. ▲

The statement $p \wedge \sim q$ in Example 1 actually means $p \wedge (\sim q)$. In the future, instead of listing a column for q and a separate column for its negation, we will make one column for $\sim q$, which will have the opposite values of those in the q column on the left. Similarly, when we evaluate $\sim p$, we will use the opposite values of those in the p column on the left. This procedure is illustrated in Example 2.

In Example 1, we spoke about *cases* and also *columns*. Consider Table 3.5(e). This table has four cases indicated by the four different rows of the two left hand (unnumbered) columns. The four *cases* are TT,TF, FT, and FF. In every truth table with two letters, we list the four cases (the first two columns) first. Then we complete the remaining columns in the truth table. In Table 3.5(e), after completing the two left-hand columns, we complete the remaining columns in the order indicated by the numbers below the columns. We will continue to place numbers below the columns to show the order in which the columns are completed.

EXAMPLE 2 *Construct and Interpret a Truth Table*

a) Construct a truth table for the following statement.

The furnace is not on and we are not wasting energy.

b) Under which conditions will the compound statement be true?
c) Suppose "The furnace is on" is a false statement and "We are wasting energy" is a true statement. Is the compound statement given in part (a) true or false?

SOLUTION

a) First write the simple statements in symbolic form by using simple non-negated statements.

Let p: The furnace is on.
q: We are wasting energy.

Table 3.6

p	q	$\sim p$	$\wedge$	$\sim q$
T	T	F	F	F
T	F	F	F	T
F	T	T	(F)	F
F	F	T	(T)	T
		1	3	2

Therefore, the compound statement may be written $\sim p \wedge \sim q$. Now construct a truth table with four cases, as shown in Table 3.6.

Fill in the column labeled 1 by negating the truth values under p on the far left. Fill in the column labeled 2 by negating the values under q in the second column from the left. Fill in the column labeled 3 by using the columns labeled 1 and 2 and the definition of conjunction.

In the first row, to determine the entry for column 3, we use false for $\sim p$ and false for $\sim q$. Since false $\wedge$ false is false (see case 4 of Table 3.4), we place an F in column 3, row 1. In the second row, we use false for $\sim p$ and true for $\sim q$. Since false $\wedge$ true is false (see case 3 of Table 3.4), we place an F in column 3, row 2. In the third row, we use a true for $\sim p$ and a false for $\sim q$. Since true $\wedge$ false is false (see case 2 of Table 3.4), we place an F in column 3, row 3. In the fourth row, we use true for $\sim p$ and true for $\sim q$. Since true $\wedge$ true is true (see case 1 of Table 3.4), we place a T in column 3, row 4.

b) The compound statement in part (a) will be true only in case 4 (circled in blue) when both simple statements, p and q, are false, that is, when the furnace is not on and we are not wasting energy.

c) We are told that "The furnace is on," p, is a false statement and that "We are wasting energy," q, is a true statement. From the truth table (Table 3.6), we can determine that when p is false and q is true, case 3, the compound statement, is false (circled in red).

DISJUNCTION

Consider the job description that contains the following requirements.

Civil Technician

Municipal program for redevelopment seeks on-site technician. **The applicant must have a two-year college degree in civil technology or five years of related experience.** Interested candidates please call 555-1234.

Who qualifies for the job? To help analyze the statement, translate it into symbolic form. Let p be "A requirement for the job is a two-year college degree in civil technology" and q be "A requirement for the job is five years of related experience." The statement in symbolic form is $p \vee q$. For the two simple statements, there are four distinct cases (see Table 3.7).

Table 3.7 Disjunction

p	q	$p \vee q$
T	T	T
T	F	T
F	T	T
F	F	F

CASE 1: p is true and q is true. A candidate has a two-year college degree in civil technology and five years of related experience. The candidate has both requirements and qualifies for the job. Consider qualifying for the job as a true statement and not qualifying as a false statement.

CASE 2: p is true and q is false. A candidate has a two-year college degree in civil technology but does not have five years of related experience. The candidate still qualifies for the job with the two-year college degree.

CASE 3: p is false and q is true. The candidate does not have a two-year college degree in civil technology but does have five years of related experience. The candidate qualifies for the job with the five years of related experience.

CASE 4: p is false and q is false. The candidate does not have a two-year college degree in civil technology and does not have five years of related experience. The candidate does not meet either of the two requirements and therefore does not qualify for the job.

In examining the four cases, we see that there is only one case in which the candidate does not qualify for the job: case 4. As this example indicates, an *or* statement will be true in every case, except when both simple statements are false. The results are summarized in Table 3.7, the truth table for the disjunction.

The **disjunction**, $p \vee q$, is true when either p is true, q is true, or both p and q are true.

The disjunction $p \vee q$ is false only when p and q are both false.

EXAMPLE 3 *Truth Table With a Disjunction*

Construct a truth table for $\sim(q \vee \sim p)$.

Table 3.8

p	q	~	(q	∨	~p)
T	T	F	T	T	F
T	F	T	F	F	F
F	T	F	T	T	T
F	F	F	F	T	T
		4	1	3	2

SOLUTION First construct the standard truth table listing the four cases. Then work within parentheses. The order to be followed is indicated by the numbers below the columns (see Table 3.8). In column 1, copy the values from the q column on the left. Under $\sim p$, column 2, write the negation of the p column on the left. Next complete the *or* column, column 3, using columns 1 and 2 and the truth table for the disjunction. The *or* column is false only when both statements are false, as in case 2. Finally, negate the values in the *or* column and place these negated values in column 4. By examining the truth table you can see that the compound statement $\sim(q \vee \sim p)$ is true only in case 2, that is, when p is true and q is false. ▲

A General Procedure for Constructing Truth Tables

1. Study the compound statement and determine whether it is a negation, conjunction, disjunction, conditional, or biconditional statement, as was done in Section 3.1. The answer to the truth table will appear under $\sim$ if the statement is a negation, under $\wedge$ if the statement is a conjunction, under $\vee$ if the statement is a disjunction, under $\rightarrow$ if the statement is a conditional, and under $\leftrightarrow$ if the statement is a biconditional.
2. Complete the columns under the simple statements, p, q, r, and their negations, $\sim p$, $\sim q$, $\sim r$, within parentheses. If there are nested parentheses (one pair of parentheses within another pair), work with the innermost pair first.
3. Complete the column under the connective within the parentheses. You will use the truth values of the connective in determining the final answer in step 5.
4. Complete the column under any remaining statements and their negations.
5. Complete the column under any remaining connectives. Recall that the answer will appear under the column determined in step 1. If the statement is a conjunction, disjunction, conditional, or biconditional, you will obtain the truth values for the connective by using the last column completed on the left side and on the right side of the connective. If the statement is a negation, you will obtain the truth values by negating the truth values of the last column completed within the grouping symbols on the right side of the negation. Be sure to circle or highlight your answer column or number the columns in the order they were completed.

Table 3.9

p	q	$(\sim p$	$\vee$	$q)$	$\wedge$	$\sim p$
T	T	F	T	T	F	F
T	F	F	F	F	F	F
F	T	T	T	T	T	T
F	F	T	T	F	T	T
		1	3	2	5	4

EXAMPLE 4 *Use the General Procedure to Construct a Truth Table*

Construct a truth table for the statement $(\sim p \vee q) \wedge \sim p$.

SOLUTION We will follow the general procedure outlined in the box. This statement is a conjunction, so the answer will be under the conjunction symbol. Complete columns under $\sim p$ and q within the parentheses and call these columns 1 and 2, respectively (see Table 3.9). Complete the column under the disjunction, $\vee$, using the values in columns 1 and 2, and call this column 3. Next complete the column under $\sim p$, and call this column 4. The answer, column 5, is determined from the definition of the conjunction and the truth values in column 3, the last column completed on the left side of the conjunction, and column 4. ▲

Table 3.10

	p	q	r
Case 1	T	T	T
Case 2	T	T	F
Case 3	T	F	T
Case 4	T	F	F
Case 5	F	T	T
Case 6	F	T	F
Case 7	F	F	T
Case 8	F	F	F

So far, all the truth tables we have constructed have contained at most two simple statements. Now we will explain how to construct a truth table that consists of three simple statements, such as $(p \wedge q) \wedge r$. When a compound statement consists of three simple statements, there are eight different true–false possibilities, as illustrated in Table 3.10. To begin such a truth table, write four Ts and four Fs in the column under p. Under the second statement, q, pairs of Ts alternate with pairs of Fs. Under the third statement, r, T alternates with F. This technique is not the only way of listing the cases, but it ensures that each case is unique and that no cases are omitted.

EXAMPLE 5 *Construct a Truth Table with Eight Cases*

a) Construct a truth table for the statement "Allen is home and he is not at his desk, or he is sleeping".
b) Suppose that "Allen is home" is a false statement, that "Allen is at his desk" is a true statement, and that "Allen is sleeping" is a true statement. Is the compound statement in part (a) true or false?

SOLUTION a) First we will translate the statement into symbolic form.

Let p: Allen is home.
q: Allen is at his desk.
r: Allen is sleeping.

In symbolic form, the statement is $(p \wedge \sim q) \vee r$.

Since the statement is composed of three simple statements, there are eight cases. Begin by listing the eight cases in the three left-hand columns; see Table 3.11. By examining the statement, you can see that it is a disjunction.

Table 3.11

p	q	r	$(p$	$\wedge$	$\sim q)$	$\vee$	r
T	T	T	T	F	F	T	T
T	T	F	T	F	F	F	F
T	F	T	T	T	T	T	T
T	F	F	T	T	T	T	F
F	T	T	F	F	F	(T)	T
F	T	F	F	F	F	F	F
F	F	T	F	F	T	T	T
F	F	F	F	F	T	F	F
			1	3	2	5	4

Therefore, the answer will be under the $\vee$ column. Fill out the truth table by working in parentheses first. Place values under p, column 1, and $\sim q$, column 2. Then find the conjunctions of columns 1 and 2 to obtain column 3. Place the values of r in column 4. To obtain the answer, column 5, use columns 3 and 4 and your knowledge of the disjunction.

b) We are given the following:

p: Allen is home—false.

q: Allen is at his desk—true.

r: Allen is sleeping—true.

Therefore, we need to find the truth value of the following case: false, true, true. In case 5 of the truth table, p, q, and r are F, T, and T, respectively. Therefore, under these conditions, this statement is true (as circled in the table). ▲

We have learned that a truth table with one simple statement has two cases, a truth table with two simple statements has four cases, and a truth table with three simple statements has eight cases. In general, *the number of distinct cases in a truth table with n distinct simple statements is 2^n*. The compound statement $(p \vee q) \vee (r \wedge \sim s)$ has four simple statements, p, q, r, s. Thus, a truth table for this compound statement would have 2^4, or 16, distinct cases.

When we construct a truth table, we determine the truth values of a compound statement for every possible case. If we want to find the truth value of the compound statement for any specific case when we know the truth values of the simple statements, we do not have to develop the entire table. For example, to determine the truth value for the statement

$$2 + 3 = 5 \quad \text{and} \quad 1 + 1 = 3$$

we let p be $2 + 3 = 5$ and q be $1 + 1 = 3$. Now we can write the compound statement as $p \wedge q$. We know that p is a true statement and q is a false statement. Thus, we can substitute T for p and F for q and evaluate the statement:

$$\begin{array}{c} p \wedge q \\ \text{T} \wedge \text{F} \\ \text{F} \end{array}$$

Therefore, this compound statement is false.

EXAMPLE 6 *Determine the Truth Value of a Compound Statement*

Determine the truth value for each simple statement. Then, using these truth values, determine the truth value of the compound statement.

a) 3 is greater than or equal to 2 ($3 \geq 2$).
b) Canada is south of Mexico and New York City is east of the Mississippi River, or Dallas is not a city in Texas.
c) An article in the June 21, 1999, issue of *Newsweek* indicates that the longer a person stays with an Internet provider, the more junk mail (or "spam") a

DID YOU KNOW

Applications of Logic

George Boole provided the key that would unlock the door to modern computing. Not until 1938, however, did Claude Shannon, in his master's thesis at MIT, propose uniting the on–off capability of electrical switches with Boole's two-value system of 0's and 1's. The operations AND, OR, and NOT and the rules of logic laid the foundation for computer gates. Such gates determine whether the current will pass. The closed switch (current flow) is represented as 1, and the open switch (no current flow) is represented as 0. For more information, see the group projects on page 145.

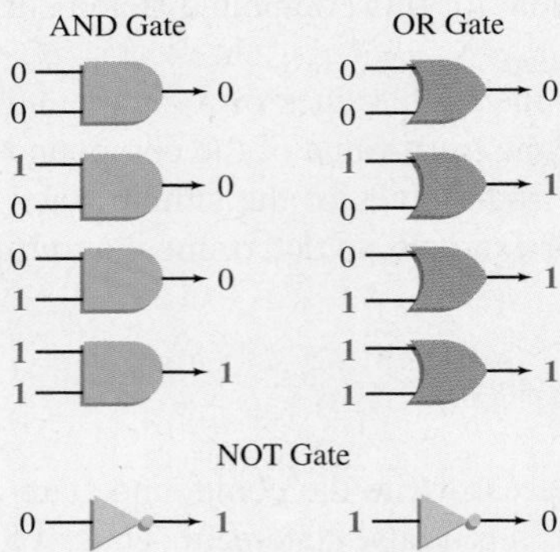

As you can see from the accompanying diagrams, the gates function in essentially the same way as a truth table. The gates shown here are the simplest ones, representing simple statements. There are other gates, such as the NAND and NOR gates, that are combinations of NOT, AND, and OR gates. The microprocessing unit of a computer uses thousands of these switches.

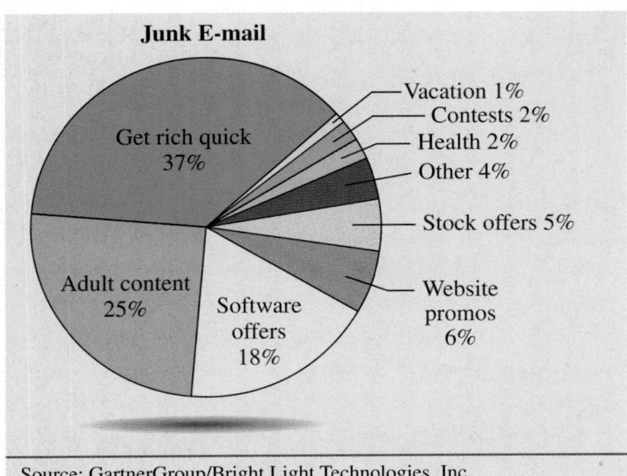

Figure 3.1 Source: GartnerGroup/Bright Light Technologies, Inc.

person receives. The graph in Fig. 3.1 shows the percents of the types of spam the average person gets. Use Fig. 3.1 to determine the truth value of the statement "The average person gets more junk mail dealing with stock offers than software offers or 10% of the junk mail received does not deal with health."

SOLUTION

a) Let p: 3 is greater than 2.
q: 3 is equal to 2.

The statement "3 is greater than or equal to 2" means that 3 is greater than 2 or 3 is equal to 2. The compound statement can be expressed as $p \vee q$. We know that p is a true statement and that q is a false statement. Substitute T for p and F for q and evaluate the statement:

$$\begin{array}{c} p \vee q \\ \mathrm{T} \vee \mathrm{F} \\ \mathrm{T} \end{array}$$

Therefore, the compound statement "3 is greater than or equal to 2" is a true statement.

b) Let p: Canada is south of Mexico.
q: New York City is east of the Mississippi River.
r: Dallas is a city in Texas.

The compound statement can be written in symbolic form as $(p \wedge q) \vee \sim r$. We know that p is a false statement because Canada is north of Mexico and that q is a true statement because New York City is east of the Mississippi River. Dallas is a city in Texas, so r is a true statement. Its negation, $\sim r$, is therefore a false statement. Substitute F for p, T for q, and F for $\sim r$ and then evaluate the statement:

$$\begin{array}{l} (p \wedge q) \vee \sim r \\ (\mathrm{F} \wedge \mathrm{T}) \vee \mathrm{F} \\ \quad \mathrm{F} \quad\;\; \vee \mathrm{F} \\ \qquad\quad \mathrm{F} \end{array}$$

Therefore, the compound statement is a false statement.

c) Let p: The average person gets more junk mail dealing with stock offers than software offers.

q: 10% of the junk mail received deals with health.

The given statement may therefore be represented as $p \vee (\sim q)$. From Fig. 3.1, we see that statement p is false since 18% of junk mail was for software offers and only 5% was for stock offers. Statement q is false since only 2% of the junk mail received on the Internet deals with health. Since q is false, $\sim q$ is true. Therefore,

$$p \vee \sim q$$
$$\text{F} \vee \text{T}$$
$$\text{T}$$

Therefore, the statement given in part (c) is true.

SECTION 3.2 EXERCISES

1. a) How many distinct cases must be listed in a truth table that contains two simple statements?
b) List all the cases.

2. a) How many distinct cases must be listed in a truth table that contains three simple statements?
b) List all the cases.

3. a) Construct the truth table for the conjunction, $p \wedge q$.
b) Under what circumstances is the *and* table true?

4. a) Construct the truth table for the disjunction, $p \vee q$.
b) Under what circumstances is the *or* table false?

In Exercises 5–20, construct a truth table for the statement.

5. $p \wedge \sim p$
6. $p \vee \sim p$
7. $q \vee \sim p$
8. $p \wedge \sim q$
9. $\sim p \vee \sim q$
10. $\sim(p \vee \sim q)$
11. $\sim(p \wedge \sim q)$
12. $\sim(\sim p \wedge \sim q)$
13. $(p \wedge r) \vee \sim q$
14. $(p \vee \sim q) \wedge r$
15. $r \vee (p \wedge \sim q)$
16. $(r \wedge q) \wedge \sim p$
17. $\sim q \wedge (r \vee \sim p)$
18. $\sim p \wedge (q \vee r)$
19. $(\sim q \wedge r) \vee p$
20. $\sim r \vee (\sim p \wedge q)$

In Exercises 21–30, write the statement in symbolic form and construct a truth table.

21. Driving is fun and walking is good exercise.

22. The calculator is solar powered but the sun is not out.

23. The sock fits but the shoe does not fit.

24. It is false that at least 200 tickets must be sold or the concert will be canceled.

25. Ricardo will take mathematics or history, but he will not take psychology.

26. The shirt is green and the suit is blue, however they go well together.

27. Karen uses America Online and Yahoo!, but she does not use Microsoft Explorer.

28. The dog sheds, but I wanted a collie and I wanted a large dog.

29. The password is pistachio, and the gate is open or the gate is not open.

30. The pen is out of ink, or the ink is white or I am sleepy.

In Exercises 31–36, if p is true, q is false, and r is true, determine the truth value of the statement.

31. $\sim p \vee (q \wedge r)$
32. $(\sim p \wedge r) \wedge q$
33. $(\sim q \wedge \sim p) \vee \sim r$
34. $(\sim p \vee \sim q) \vee \sim r$
35. $(p \wedge \sim q) \vee r$
36. $(p \vee \sim q) \wedge \sim(p \wedge \sim r)$

In Exercises 37–42, if p is false, q is true, and r is false, determine the truth value of the statement.

37. $(\sim r \wedge p) \vee q$
38. $\sim q \vee (r \wedge p)$
39. $(\sim q \vee \sim p) \wedge r$
40. $(\sim r \vee \sim p) \vee \sim q$
41. $(\sim p \vee \sim q) \vee (\sim r \vee q)$
42. $(\sim r \wedge \sim q) \wedge (\sim r \vee \sim p)$

In Exercises 43–50, determine the truth value for each simple statement. Then use these truth values to determine the truth value of the compound statement. (You may have to use a reference source such as an almanac.)

43. $4 + 3 = 7$ or $6 + 4 = 12$

44. $9 - 6 = 15$ and $7 - 4 = 3$

45. Elvis Presley was born in Tupelo, Mississippi or the giraffe has only two legs.

Elvis' birthplace

46. The capital of Texas is San Antonio or Kentucky is east of the Mississippi River, and Los Angeles is not the capital of California.

47. Algebra and geometry are mathematics courses, but Shakespearean literature is not a mathematics course.

48. Rome is in France or Paris is not in Germany, and London is in England.

49. Marco Polo played football or John Glenn built houses, and George Washington is on a U.S. one-dollar bill.

50. Mars is a planet and the sun is a star, or the moon is not a star.

Food Consumption *In Exercises 51 and 52, use the chart to determine the truth value of each simple statement. Then determine the truth value of the compound statement.*

Americans are eating more than ever before. Annual per capita consumption in pounds:

	1909	**1997**
Red meat	99	111
Poultry	11	64
Fish	11	15
Cheese	4	28
Fats and oils[a]	38	66
Sweeteners[b]	86	154

[a]Added fats & oils
[b]Caloric sweeteners (sugars, honey, corn syrup).
Source: *Newsweek*

51. Twenty-eight pounds of cheese were consumed by the average American in 1909, and the average American did not consume 154 pounds of sweeteners in 1997.

52. The per capita consumption of red meat was less for the average American in 1997 than it was in 1909 or the per capita consumption of poultry was greater for the average American in 1997 than it was in 1909.

Sleep Time *In Exercises 53 and 54, use the graph, which shows the number of hours Americans sleep, to determine the truth value of each simple statement. Then determine the truth value of the compound statement.*

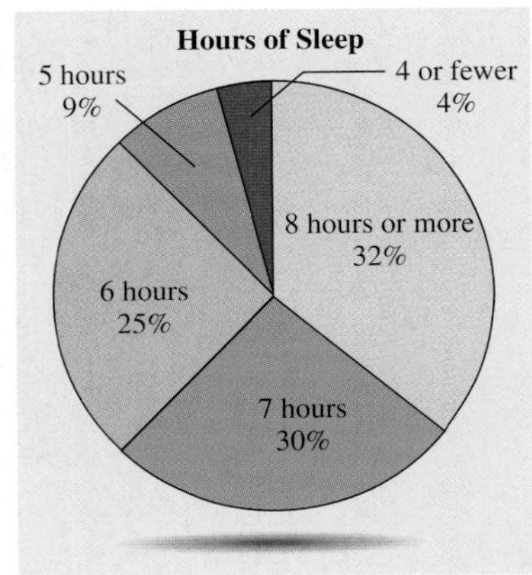

53. It is false that 30% of Americans get 6 hours of sleep each night and 9% get 5 hours of sleep each night.

54. Twenty-five percent of Americans get 6 hours of sleep each night, and 30% get 7 hours of sleep each night or 9% do not get 5 hours of sleep each night.

In Exercises 55–58, let

p: Tanisha owns a convertible.
q: Joan owns a Volvo.

Translate each statement into symbols. Then construct a truth table for each and indicate under what conditions the compound statement is true.

55. Tanisha does not own a convertible, but Joan owns a Volvo.

56. Tanisha owns a convertible and Joan does not own a Volvo.

57. Tanisha owns a convertible or Joan does not own a Volvo.

58. Tanisha does not own a convertible or Joan does not own a Volvo.

In Exercises 59–62, let

p: The house is owned by an engineer.
q: The heat is solar generated.
r: The car is run by electric power.

Translate each statement into symbols. Then construct a truth table for each and indicate under what conditions the compound statement is true.

59. The house is owned by an engineer and heat is solar generated, or the car is run by electric power.

60. The car is run by electric power or the heat is solar generated, but the house is owned by an engineer.

61. The heat is solar generated, or the house is owned by an engineer and the car is not run by electric power.

62. The house is not owned by an engineer, and the car is not run by electric power and the heat is solar generated.

Obtaining a Loan *In Exercises 63 and 64, read the requirements and each applicant's qualifications for obtaining a loan.*

a) Identify which of the applicants would qualify for the loan.

b) For the applicants who do not qualify for the loan, explain why.

63. To qualify for a loan of $40,000, an applicant must have a gross income of $28,000 if single, $46,000 combined income if married, and assets of at least $6,000.

Mr. Furman, married with three children, makes $42,000 on his job. Mrs. Furman does not have an income. The Furmans have assets of $42,000.

Ms. Duncan is not married, works in sales, and earns $31,000. She has assets of $9000.

Mrs. Tuttle and her husband have total assets of $43,000. One earns $35,000, and the other earns $23,500.

64. To qualify for a loan of $45,000, an applicant must have a gross income of $30,000 if single, $50,000 combined income if married, and assets of at least $10,000.

Mr. Argento, married with two children, makes $37,000 on his job. Mrs. Argento earns $15,000 at a part-time job. The Argentos have assets of $25,000.

Ms. McVey, single, has assets of $19,000. She works in a store and earns $25,000.

Mr. Henke earns $24,000 and Mrs. Henke earns $28,000. Their assets total $8,000.

65. *Airline Special Fares* An airline advertisement states, "To get the special fare you must purchase your tickets between January 1 and February 15 and fly round trip between March 1 and April 1. You must depart on a Monday, Tuesday, or Wednesday, and return on a Tuesday, Wednesday, or Thursday, and stay over at least one Saturday."

a) Determine which of the following individuals will qualify for the special fare.

b) If the person does not qualify for the special fare, explain why.

Michael Bolinder plans to purchase his ticket on January 15, depart on Monday, March 3, and return on Tuesday, March 18.

Gina Vela plans to purchase her ticket on February 1, depart on Wednesday, March 10, and return on Thursday, April 2.

Laura Griffin Heller plans to purchase her ticket on February 14, depart on Tuesday, March 5, and return on Monday, March 18.

Christos Giakoumopoulos plans to purchase his ticket on January 4, depart on Monday, March 8, and return on Thursday, March 11.

Alex Chang plans to purchase his ticket on January 1, depart on Monday, March 3, and return on Monday, March 10.

PROBLEM SOLVING/GROUP ACTIVITIES

In Exercises 66–67, construct a truth table for the symbolic statement.

66. $\sim[(\sim(p \vee q)) \vee (q \wedge r)]$

67. $[(q \wedge \sim r) \wedge (\sim p \vee \sim q)] \vee (p \vee \sim r)$

68. On page 101, we indicated that a compound statement consisting of n simple statements had 2^n distinct true–false cases.

a) How many distinct true–false cases does a truth table containing simple statements p, q, r, and s have?

b) List all possible true–false cases for a truth table containing the simple statements p, q, r, and s.

c) Use the list in part (b) to construct a truth table for $(q \wedge p) \vee (\sim r \wedge s)$.

d) Construct a truth table for $(\sim r \wedge \sim s) \wedge (\sim p \vee q)$.

69. Must $(p \wedge \sim q) \vee r$ and $(q \wedge \sim r) \vee p$ have the same number of trues in their answer columns? Explain.

RESEARCH ACTIVITIES

70. Digital computers use gates that work like switches to perform calculations. Information is fed into the gates and information leaves the gates, according to the type of gate. The three basic gates used in computers are the NOT gate, the AND gate, and the OR gate. Do research on the three types of gates.

a) Explain how each gate works.

b) Explain the relationship between each gate and the corresponding logic connectives *not, and,* and *or*.

c) Illustrate how two or more gates can be combined to form a more complex gate.

3.3 TRUTH TABLES FOR THE CONDITIONAL AND BICONDITIONAL

CONDITIONAL

In Section 3.1, we mentioned that the statement preceding the conditional symbol is called the *antecedent* and that the statement following the conditional symbol is called the *consequent*. For example, consider $(p \vee q) \rightarrow [\sim(q \wedge r)]$. In this statement, $(p \vee q)$ is the antecedent and $[\sim(q \wedge r)]$ is the consequent.

Now we will look at the truth table for the conditional. Consider the statement "If you get an A in class, then I will buy you a car." Assume this statement is true except when I have actually broken my promise to you.

Let p: You get an A.
q: I buy you a car.

Translated into symbolic form, the statement becomes $p \rightarrow q$. Let's examine the four cases shown in Table 3.12.

Table 3.12 Conditional

p	q	$p \rightarrow q$
T	T	T
T	F	F
F	T	T
F	F	T

CASE 1: (T, T) You get an A, and I buy a car for you. I have met my commitment, and the statement is true.

CASE 2: (T, F) You get an A, and I do not buy a car for you. I have broken my promise, and the statement is false.

What happens if you don't get and A? If you don't get an A, I no longer have a commitment to you, and therefore I cannot break my promise.

CASE 3: (F, T) You do not get an A, and I buy you a car. I have not broken my promise, and therefore the statement is true.

CASE 4: (F, F) You do not get an A, and I don't buy you a car. I have not broken my promise, and therefore the statement is true.

The conditional statement is false when the antecedent is true and the consequent is false. In every other case the conditional statement is true.

> The **conditional statement** $p \rightarrow q$ is true in every case except when p is a true statement and q is a false statement.

EXAMPLE 1 *A Truth Table with a Conditional*

Construct a truth table for the statement $p \rightarrow \sim q$.

SOLUTION Since this is a conditional statement, the answer will lie under the $\rightarrow$. Fill out the truth table by placing the appropriate values under p, column 1, and under $\sim q$, column 2 (Table 3.13). Then, using the information given in the truth table for the conditional and the truth values in columns 1 and 2, determine the solution, column 3. In row 1, the antecedent, p, is true and the consequent, $\sim q$, is false. Row 1 is T $\rightarrow$ F, which according to row 2 of Table 3.12, is F. Row 2 is T $\rightarrow$ T, which is T. Row 3 is F $\rightarrow$ F, which is T. Row 4 is F $\rightarrow$ T, which is T. ▲

Table 3.13

p	q	p	$\rightarrow$	$\sim q$
T	T	T	F	F
T	F	T	T	T
F	T	F	T	F
F	F	F	T	T
		1	3	2

EXAMPLE 2 *A Conditional Truth Table with Three Simple Statements*

Construct a truth table for the statement $p \rightarrow (q \wedge \sim r)$.

SOLUTION Since this is a conditional statement, the answer will lie under the $\rightarrow$. Work within the parentheses first. Place truth values under q, column 1, and $\sim r$, column 2 (Table 3.14). Then take the conjunction of columns 1 and 2 to obtain column 3. Place the values of p in column 4. To obtain the answer, column 5, use columns 3 and 4 and your knowledge of the conditional statement. Column 4 represents the truth values of the antecedent, and column 3 represents the truth values of the consequent. Remember that the conditional is false only when the antecedent is true and the consequent is false, as in cases 1, 3, and 4 of column 5.

Table 3.14

p	**q**	**r**	**p**	$\rightarrow$	**(q**	$\wedge$	**~r)**
T	T	T	T	F	T	F	F
T	T	F	T	T	T	T	T
T	F	T	T	F	F	F	F
T	F	F	T	F	F	F	T
F	T	T	F	T	T	F	F
F	T	F	F	T	T	T	T
F	F	T	F	T	F	F	F
F	F	F	F	T	F	F	T
			4	5	1	3	2

EXAMPLE 3 *Evaluate a Tire Advertisement*

An advertisement in an automobile magazine makes the following claim: "If you use Goodtread tires, then you will get a smooth ride and you will not spend a lot of money."

Translate the statement into symbolic form and construct a truth table.

SOLUTION Let p: You use Goodtread tires.

q: You will get a smooth ride.

r: You will spend a lot of money.

In symbolic form, the statement is

$$p \rightarrow (q \wedge \sim r)$$

This symbolic statement is identical to the statement in Table 3.14, and the truth tables are the same. Column 3 represents the truth values of $(q \wedge \sim r)$, which corresponds to the statement "You will get a smooth ride and you will not spend a lot of money." Note that column 3 is true in cases (rows) 2 and 6. In case 2, since p is true, Goodtread tires are used. In case 6, however, since p is false, Goodtread tires are not used. From this information, we can conclude that it is possible to get a smooth ride and not spend a lot of money without using Goodtread tires (case 6).

A truth table alone cannot tell us whether a statement is true of false. It can, however, be used to examine the various possibilities.

BICONDITIONAL

The *biconditional statement,* $p \leftrightarrow q$, means that $p \rightarrow q$ and $q \rightarrow p$, or, symbolically, $(p \rightarrow q) \wedge (q \rightarrow p)$. To determine the truth table for $p \leftrightarrow q$, we will construct the truth table for $(p \rightarrow q) \wedge (q \rightarrow p)$ (Table 3.15). Table 3.16 shows the truth values for the biconditional statement.

Table 3.15

p	q	$(p$	$\rightarrow$	$q)$	$\wedge$	$(q$	$\rightarrow$	$p)$
T	T	T	T	T	T	T	T	T
T	F	T	F	F	F	F	T	T
F	T	F	T	T	F	T	F	F
F	F	F	T	F	T	F	T	F
		1	3	2	7	4	6	5

Table 3.16 Biconditional

p	q	$p \leftrightarrow q$
T	T	T
T	F	F
F	T	F
F	F	T

The **biconditional statement**, $p \leftrightarrow q$, is true only when both p and q have the same truth value, that is, when both are true or both are false.

EXAMPLE 4 *A Truth Table Using a Biconditional*

Construct a truth table for the statement $p \leftrightarrow (q \rightarrow \sim r)$.

SOLUTION Since there are three letters, there must be eight cases. The parentheses indicate that the answer must be under the biconditional (Table 3.17). Use columns 3 and 4 to obtain the answer in column 5. When columns 3 and 4 have the same truth values, place a T in column 5. When columns 3 and 4 have different truth values, place an F in column 5.

Table 3.17

p	q	r	p	$\leftrightarrow$	$(q$	$\rightarrow$	$\sim r)$
T	T	T	T	F	T	F	F
T	T	F	T	T	T	T	T
T	F	T	T	T	F	T	F
T	F	F	T	T	F	T	T
F	T	T	F	T	T	F	F
F	T	F	F	F	T	T	T
F	F	T	F	F	F	T	F
F	F	F	F	F	F	T	T
			4	5	1	3	2

In the preceding section, we showed that finding the truth value of a compound statement for a specific case does not require constructing an entire truth table. Example 5 illustrates this technique for the conditional and the biconditional.

EXAMPLE 5 *Determine the Truth Value of a Compound Statement*

Determine the truth value of the statement $(q \leftrightarrow r) \rightarrow (\sim p \wedge r)$ when p is true, q is false, and r is true.

SOLUTION Substitute the truth value for each simple statement:

$$\begin{array}{c} (q \leftrightarrow r) \rightarrow (\sim p \wedge r) \\ (\text{F} \leftrightarrow \text{T}) \rightarrow \quad (\text{F} \wedge \text{T}) \\ \text{F} \quad \rightarrow \quad \text{F} \\ \text{T} \end{array}$$

For this specific case, the statement is true.

EXAMPLE 6 ***Determine the Truth Value of a Compound Statement***

Determine the truth value for each simple statement. Then use the truth values to determine the truth value of the compound statement.

a) If 1 + 1 = 3, then 3 × 5 = 12.
b) The Rock and Roll Hall of Fame is in Cleveland and the Country Music Hall of Fame is in Tulsa, if and only if the Baseball Hall of Fame is in Cooperstown.
c) The graphs in Fig. 3.2(a) show how one-family homes were financed in 1970 and 1997. The graph in Fig. 3.2(b) shows the heating fuel used in 1970 and 1997. Use Fig. 3.2 to determine the truth value of the following statement: "If 56% of one-family homes were purchased with conventional mortgages in 1997 or 63% of one-family homes were heated with gas in 1970, then 26% of one-family homes were not heated with electricity in 1997."

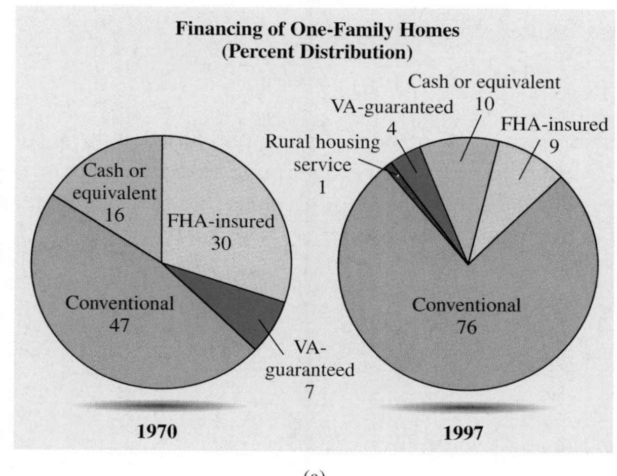

(a)

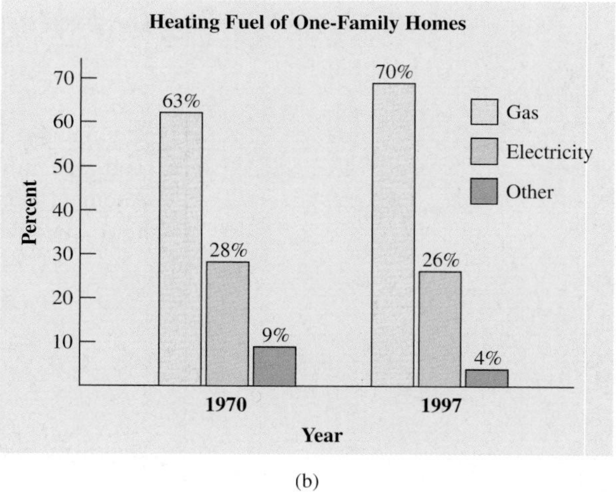

(b)

Figure 3.2

SOLUTION

a) Let p: $1 + 1 = 3$

q: $3 \times 5 = 12$

Then the statement "If 1 + 1 = 3, then 3 × 5 = 12" can be written $p \rightarrow q$. We know that p is a false statement and q is a false statement. Substitute F for p and F for q and evaluate the statement:

$$\begin{array}{c} p \rightarrow q \\ \text{F} \rightarrow \text{F} \\ \text{T} \end{array}$$

Therefore, "If 1 + 1 = 3, then 3 × 5 = 12" is a true statement.

b) Let p: The Rock and Roll Hall of Fame is in Cleveland.

q: The Country Music Hall of Fame is in Tulsa.

r: The Baseball Hall of Fame is in Cooperstown.

The compound statement can be written $(p \wedge q) \leftrightarrow r$. By doing a little research, we can determine that statement p is true since the Rock and Roll Hall of Fame is in Cleveland, Ohio. Statement q is false since the Country Music Hall of Fame is in Nashville, Tennessee. Statement r is true since the Baseball Hall of Fame is in Cooperstown, New York. Substitute T for p, F for q, and T for r and evaluate the statement:

$$\begin{array}{c} (p \wedge q) \leftrightarrow r \\ (\mathrm{T} \wedge \mathrm{F}) \leftrightarrow \mathrm{T} \\ \mathrm{F} \leftrightarrow \mathrm{T} \\ \mathrm{F} \end{array}$$

Therefore, the given compound statement is false.

c) Let p: 56% of one-family homes were purchased with conventional mortgages in 1997.

q: 63% of one-family homes were heated with gas in 1970.

r: 26% of one-family homes were heated with electricity in 1997.

The given statement written symbolically is:

$$(p \vee q) \rightarrow \sim r$$

Statement p is false since 77% of homes were purchased with a conventional mortgage in 1997. From Fig. 3.2(b) we determine that statement q is true and statement r is true. Now substitute these truth values into the statement and evaluate. If statement r is true then $\sim r$ is false.

$$\begin{array}{c} (p \vee q) \rightarrow \sim r \\ (\mathrm{F} \vee \mathrm{T}) \rightarrow \mathrm{F} \\ \mathrm{T} \rightarrow \mathrm{F} \\ \mathrm{F} \end{array}$$

Therefore, the given statement is false. ▲

SELF-CONTRADICTIONS, TAUTOLOGIES, AND IMPLICATIONS

Two special situations can occur in the truth table of the compound statement: The statement may always be true, or the statement may always be false. We give such statements special names.

A **self-contradiction** is a compound statement that is always false.

When every truth value in the answer column of the truth table is false, then the statement is a self-contradiction.

Table 3.18

p	q	$(p \leftrightarrow q)$	$\wedge$	$(p$	$\leftrightarrow$	$\sim q)$
T	T	T	F	T	F	F
T	F	F	F	T	T	T
F	T	F	F	F	T	F
F	F	T	F	F	F	T
		1	5	2	4	3

EXAMPLE 7 *All Falses, a Self-Contradiction*

Construct a truth table for the statement $(p \leftrightarrow q) \wedge (p \leftrightarrow \sim q)$.

SOLUTION See Table 3.18. In this example, the truth values are false in each case of column 5. This is an example of a self-contradiction or a *logically false statement.*

A **tautology** is a compound statement that is always true.

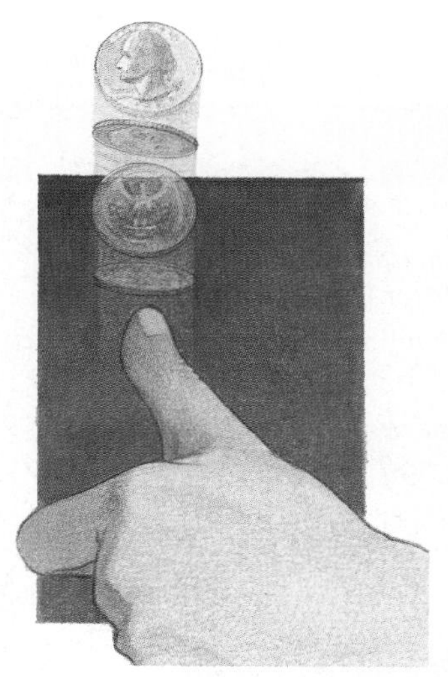

"*Heads I win,* tails you lose." Do you think that this statement is a tautology, self-contradiction, or neither? See Problem-Solving Exercise 81.

When every truth value in the answer column of the truth table is true, the statement is a tautology.

EXAMPLE 8 *All Trues, a Tautology*

Construct a truth table for the statement $(p \wedge q) \rightarrow (p \vee r)$.

SOLUTION The answer is given in column 3 of Table 3.19. The truth values are true in every case. Thus, the statement is an example of a tautology or a *logicallly true statement.*

Table 3.19

p	q	r	$(p \wedge q)$	$\rightarrow$	$(p \vee r)$
T	T	T	T	T	T
T	T	F	T	T	T
T	F	T	F	T	T
T	F	F	F	T	T
F	T	T	F	T	T
F	T	F	F	T	F
F	F	T	F	T	T
F	F	F	F	T	F
			1	3	2

The conditional statement $(p \wedge q) \rightarrow (p \vee q)$ is a tautology. Conditional statements that are tautologies are called *implications.* In Example 8, we can say that $p \wedge q$ implies $p \vee q$.

An **implication** is a conditional statement that is a tautology.

In any implication the antecedent of the conditional statement implies the consequent. In other words, if the antecedent is true, then the consequent must also be true. That is, the consequent will be true whenever the antecedent is true.

EXAMPLE 9 *An Implication?*

Determine whether the conditional statement $[(p \wedge q) \wedge p] \rightarrow q$ is an implication.

SOLUTION If the conditional statement is a tautology, the conditional statement is an implication. Because the conditional statement is a tautology (see Table 3.20), the conditional statement is an implication. The antecedent $[(p \wedge q) \wedge p]$

implies the consequent q. Note that the antecedent is true only in case 1 and that the consequent is also true in case 1.

Table 3.20

p	q	$[(p \wedge q)$	$\wedge$	$p]$	$\rightarrow$	q
T	T	T	T	T	T	T
T	F	F	F	T	T	F
F	T	F	F	F	T	T
F	F	F	F	F	T	F
		1	3	2	5	4

SECTION 3.3 EXERCISES

1. a) Construct the truth table for the conditional statement, $p \rightarrow q$.
   b) Explain when the conditional statement is true and when it is false.
2. a) Construct the truth table for the biconditional statement, $p \leftrightarrow q$.
   b) Explain when the biconditional statement is true and when it is false.
3. a) Explain the procedure to determine the truth value of a compound statement when specific truth values are provided for the simple statements.
   b) Follow the procedure in part (a) and determine the truth value of the symbolic statement

$$[(p \leftrightarrow q) \vee (\sim r \rightarrow q)] \rightarrow \sim r$$

when p = true, q = false, and r = true.

4. What is a self-contradiction?
5. What is a tautology?
6. What is an implication?

In Exercises 7–16, construct a truth table for the statement.

7. $p \rightarrow \sim q$
8. $\sim q \rightarrow \sim p$
9. $\sim(p \leftrightarrow q)$
10. $\sim(q \rightarrow p)$
11. $\sim q \leftrightarrow p$
12. $(p \leftrightarrow q) \rightarrow p$
13. $p \leftrightarrow (q \vee p)$
14. $(\sim q \wedge p) \rightarrow \sim q$
15. $q \rightarrow (p \rightarrow \sim q)$
16. $(p \vee q) \leftrightarrow (p \wedge q)$

In Exercises 17–26, construct a truth table for the statement.

17. $p \rightarrow (q \vee r)$
18. $r \wedge (\sim q \rightarrow p)$
19. $q \leftrightarrow (r \wedge p)$
20. $(q \leftrightarrow p) \wedge \sim r$
21. $(q \vee \sim r) \leftrightarrow \sim p$
22. $(p \wedge r) \rightarrow (q \vee r)$
23. $(\sim r \vee \sim q) \rightarrow p$
24. $[r \wedge (q \vee \sim p)] \leftrightarrow \sim p$
25. $(p \rightarrow q) \leftrightarrow (\sim q \rightarrow \sim r)$
26. $(\sim p \leftrightarrow \sim q) \rightarrow (\sim q \leftrightarrow r)$

In Exercises 27–32, write the statement in symbolic form. Then construct a truth table for it.

27. If I cut the grass, then I will need to rake and I will need to bag the clippings.

28. The pizza will be delivered if and only if the driver finds the house, or the pizza will not be hot.
29. The cable is out if and only if we cannot watch TV, or we will use the antenna.
30. If it rains then the roof will leak, and if the sun shines then the roof will not leak.
31. If the computer is not being used then we can use the telephone, or we can use the fax machine.
32. It is false that if Paige Dunbar went to a movie, then she did not go to the party and she went to school.

In Exercises 33–38, determine whether the statement is a tautology, a self-contradiction, or neither.

33. $\sim q \rightarrow p$
34. $(p \wedge q) \leftrightarrow p$

35. $\sim q \wedge (q \wedge p)$

36. $(\sim p \vee \sim q) \rightarrow p$

37. $(\sim p \rightarrow q) \vee \sim p$

38. $[(p \wedge q) \wedge \sim r] \leftrightarrow [(p \vee q) \wedge r]$

In Exercises 39–44, determine whether the statement is an implication.

39. $(p \vee q) \rightarrow (q \wedge p)$

40. $q \rightarrow (p \vee q)$

41. $(q \wedge p) \rightarrow (p \wedge q)$

42. $(p \vee q) \rightarrow (p \vee \sim r)$

43. $[(p \rightarrow q) \wedge (q \rightarrow p)] \rightarrow (p \leftrightarrow q)$

44. $[(p \vee q) \wedge r] \rightarrow (p \vee q)$

In Exercises 45–56, if p is true, q is false, and r is true, find the truth value of the statement.

45. $\sim p \rightarrow (q \wedge \sim r)$

46. $\sim p \rightarrow (q \vee \sim r)$

47. $(q \wedge \sim p) \leftrightarrow \sim r$

48. $p \leftrightarrow (\sim q \wedge r)$

49. $(\sim p \wedge \sim q) \vee \sim r$

50. $\sim[p \rightarrow (q \wedge r)]$

51. $(p \wedge r) \leftrightarrow (p \vee \sim q)$

52. $(\sim p \vee q) \rightarrow \sim r$

53. $(\sim p \leftrightarrow r) \vee (\sim q \leftrightarrow r)$

54. $(r \rightarrow \sim p) \wedge (q \rightarrow \sim r)$

55. $\sim[(p \vee q) \leftrightarrow (p \rightarrow \sim r)]$

56. $[(\sim r \rightarrow \sim q) \vee (p \wedge \sim r)] \rightarrow q$

In Exercises 57–64, determine the truth value for each simple statement. Then, using the truth values, determine the truth value of the compound statement.

57. If $2 + 3 = 5$, then $4 + 2 = 6$.

58. If $1 + 1 = 2$ and $2 = 2 + 3$, then $5 - 1 = 3$.

59. $2 + 3 = 6$ or $4 + 5 = 8$, and $2 = 1 + 1$.

60. The Verrazano Bridge is in New York and the Golden Gate Bridge is in San Francisco, or the Sears Tower is in Cincinnati.

61. A pound contains 16 ounces, if and only if a foot is equal to 15 inches or a yard is equal to 3 feet.

62. Steven Spielberg is a motion picture director, or if Julia Roberts is not an actress then Tom Cruise is an actor.

63. July 4 is Independence Day or two dimes have the same value as one quarter, and one dollar has the same value as 100 pennies.

64. Wednesday follows Sunday or Wednesday follows Tuesday, if and only if Monday follows Wednesday.

In Exercises 65–68, use the information provided about the moons for the planets Jupiter and Saturn to determine the truth values of the simple statements. Then determine the truth value of the compound statement given.

65. *Jupiter's Moons* Io has a diameter of 1000–3161 miles or Thebe may have water, and Io may have atmosphere.

66. *Moons of Saturn* Titan may have water and Titan may have atmosphere, if and only if Janus may have water.

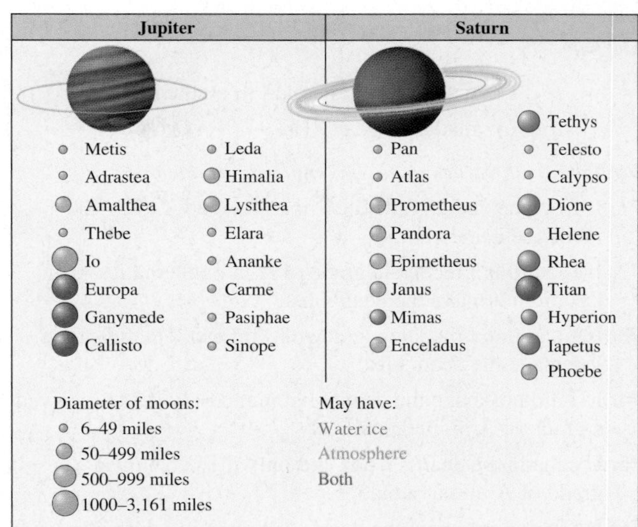

Source: *Time* Magazine

67. *Saturn's Moons* Phoebe has a larger diameter than Rhea if and only if Callisto may have water ice, and Calypso has a diameter of 6–49 miles.

68. *Moon Comparisons* If Jupiter has 16 moons or Saturn does not have 18 moons, then Saturn has 7 moons that may have water ice.

In Exercises 69 and 70, use the graphs to determine the truth values of each simple statement. Then determine the truth value of the compound statement given.

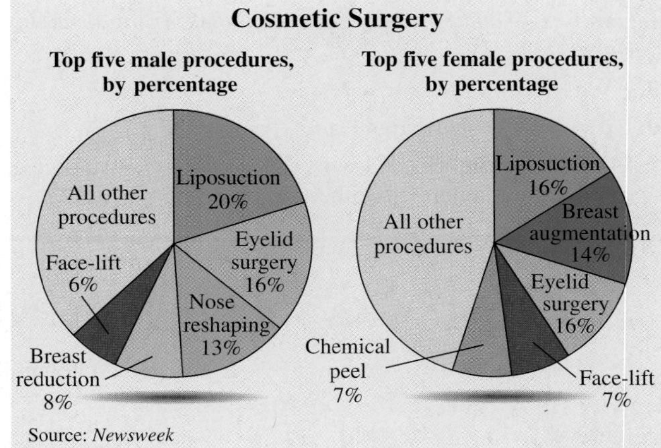

Source: *Newsweek*

69. *Most Common Cosmetic Surgery* The most common cosmetic surgery procedure for females is liposuction or the most common procedure for males is eyelid surgery, and 20% of male cosmetic surgery is for nose reshaping.

70. *Face-lifts and Eyelid Surgeries* 7% of female cosmetic surgeries are for face-lifts and 10% of male cosmetic surgeries are for face-lifts, if and only if males have a higher percent of eyelid surgeries than females.

In Exercises 71–76, suppose both of the following statements are true.

p: Liz received a grade of A in mathematics.

q: Liz made the dean's list.

Find the truth values of each compound statement.

71. If Liz received a grade of A in mathematics, then Liz made the dean's list.

72. If Liz did not receive a grade of A in mathematics, then Liz did not make the dean's list.

73. If Liz did not receive a grade of A in mathematics, then Liz made the dean's list.

74. Liz did not make the dean's list if and only if Liz received a grade of A in mathematics.

75. Liz made the dean's list if and only if Liz received a grade of A in mathematics.

76. If Liz did not make the dean's list, then Liz did not receive a grade of A in mathematics.

77. *A New Car* Your father makes the statement "If I get a raise in salary, then I will buy a new car." At the end of the semester you come home and there is a new car in the driveway. Can you conclude that your father got a raise? Explain.

78. *Job Interview* Consider the statement "If your interview goes well, then you will be offered the job." If you are interviewed and then offered the job, can you conclude that your interview went well? Explain.

PROBLEM SOLVING/GROUP ACTIVITIES

In Exercises 79 and 80, construct truth tables for the symbolic statement.

79. $p \vee q \rightarrow \sim r \leftrightarrow p \wedge \sim q$

80. $[(r \rightarrow \sim q) \rightarrow \sim p] \vee (q \leftrightarrow \sim r)$

81. Is the statement "Heads I win, Tails you lose" a tautology, a self-contradiction, or neither? Explain your answer.

82. *The Youngest Triplet* The Barr triplets have an annoying habit: Whenever a question is asked of the three of them, two tell the truth and the third lies. When I asked them which of them was born last, they replied as follows.

Mary: Katie was born last.

Katie: I am the youngest.

Annie: Mary is the youngest.

Which of the Barr triplets was born last?

83. *Cat Puzzle* Solve the following puzzle. The Joneses have four cats. The parents are Tiger and Boots, and the kittens are Sam and Sue. Each cat insists on eating out of its own bowl. To complicate matters, each cat will eat only its own brand of cat food. The colors of the bowls are red, yellow, green and blue. The different types of cat food are Whiskas, Friskies, Nine Lives, and Meow Mix. Tiger will eat Meow Mix if and only if it is in a yellow bowl. If Boots is to eat her food, then it must be in a yellow bowl. Mrs. Jones knows that the label on the can containing Sam's food is the same color as his bowl. Boots eats Whiskas. Meow Mix and Nine Lives are packaged in a brown paper bag. The color of Sue's bowl is green if and only if she eats Meow Mix. The label on the Friskies can is red. Match each cat with its food and the bowl of the correct color.

84. Construct a truth table for

a) $(p \vee q) \rightarrow (r \wedge s)$.

b) $(q \rightarrow \sim p) \vee (r \leftrightarrow s)$.

RESEARCH ACTIVITY

85. Select an advertisement from a newspaper or magazine that makes or implies a conditional statement. Analyze the advertisement to determine whether the consequent necessarily follows from the antecedent. Explain you answer. (See Example 3.)

3.4 EQUIVALENT STATEMENTS

Equivalent statements are an important concept in the study of logic.

> Two statements are **equivalent**, symbolized $\Leftrightarrow$,* if both statements have exactly the same truth values in the answer columns of the truth tables.

Sometimes the words *logically equivalent* are used in place of the word *equivalent.*

To determine whether two statements are equivalent, construct a truth table for each statement and compare the answer columns of the truth tables. If the answer

*The symbol $\equiv$ is also used to indicate equivalent statements.

columns are identical, the statements are equivalent. If the answer columns are not identical, the statements are not equivalent.

EXAMPLE 1 *Equivalent Statements*

Show that the following two statements are equivalent.

$$[p \vee (q \vee r)] \qquad [(p \vee q) \vee r]$$

SOLUTION Construct a truth table for each statement (see Table 3.21).

Table 3.21

p	q	r	[p	∨	(q ∨ r)]	[(p ∨ q)	∨	r]
T	T	T	T	T	T	T	T	T
T	T	F	T	T	T	T	T	F
T	F	T	T	T	T	T	T	T
T	F	F	T	T	F	T	T	F
F	T	T	F	T	T	T	T	T
F	T	F	F	T	T	T	T	F
F	F	T	F	T	T	F	T	T
F	F	F	F	F	F	F	F	F
			1	3	2	1	3	2

Because the truth tables have the same answer (column 3 for both tables), the statements are equivalent. Thus, we can write

$$[p \vee (q \vee r)] \Leftrightarrow [(p \vee q) \vee r].$$

EXAMPLE 2 *Are the Following Equivalent Statements?*

Determine whether the following statements are equivalent.

a) If you study and go to sleep by 11 P.M., then you will get an A on the test.

b) If you do not study or do not go to sleep by 11 P.M., then you will not get an A on the test.

SOLUTION Write each statement in symbolic form, and then construct a truth table for each statement. If the answer columns of both tables are identical, the statements are equivalent. If the answer columns are not identical, the statements are not equivalent.

Let
p: You study.
q: You go to sleep by 11 P.M.
r: You will get an A on the test.

In symbolic form, the statements are

a) $(p \wedge q) \rightarrow r$. b) $(\sim p \vee \sim q) \rightarrow \sim r$.

The truth tables for these statements are given in Tables 3.22 and 3.23, respectively. The answers in the columns labeled 5 are not identical, so the statements are not equivalent.

Table 3.22

p	q	r	$(p$	$\wedge$	$q)$	$\rightarrow$	r
T	T	T	T	T	T	T	T
T	T	F	T	T	T	F	F
T	F	T	T	F	F	T	T
T	F	F	T	F	F	T	F
F	T	T	F	F	T	T	T
F	T	F	F	F	T	T	F
F	F	T	F	F	F	T	T
F	F	F	F	F	F	T	F
			1	3	2	5	4

Table 3.23

p	q	r	$(\sim p$	$\vee$	$\sim q)$	$\rightarrow$	$\sim r$
T	T	T	F	F	F	T	F
T	T	F	F	F	F	T	T
T	F	T	F	T	T	F	F
T	F	F	F	T	T	T	T
F	T	T	T	T	F	F	F
F	T	F	T	T	F	T	T
F	F	T	T	T	T	F	F
F	F	F	T	T	T	T	T
			1	3	2	5	4

EXAMPLE 3 *Which Statements Are Logically Equivalent?*

Select the statement that is logically equivalent to "It is not true that the plane is both overbooked and departing late."

a) If the plane is not departing late, then the plane is not overbooked.
b) The plane is not overbooked or the plane is not departing late.
c) The plane is not departing late and the plane is not overbooked.
d) If the plane is not overbooked, then the plane is not departing late.

SOLUTION To determine whether any of the choices are equivalent to the given statement, write the given statement and the choices in symbolic form. Then construct and compare their truth tables.

Let p: The plane is overbooked.
q: The plane is departing late.

The given statement may be written "It is not true that the plane is overbooked and the plane is departing late." The statement is expressed in symbolic form as $\sim(p \wedge q)$. Using p and q as indicated, choices (a) through (d) may be expressed symbolically as

a) $\sim q \rightarrow \sim p$. b) $\sim p \vee \sim q$. c) $\sim q \wedge \sim p$. d) $\sim p \rightarrow \sim q$.

Now construct a truth table for the given statement (Table 3.24) and each possible choice, given in Table 3.25(a) through (d). By examining the truth tables, we see that the given statement, $\sim(p \wedge q)$, is logically equivalent to choice (b), $\sim p \vee \sim q$. Therefore, the correct answer is "The plane is not overbooked or the plane is not departing late." This is logically equivalent to "It is not true that the plane is both overbooked and departing late."

Table 3.24

p	q	$\sim$	$(p$	$\wedge$	$q)$
T	T	F	T	T	T
T	F	T	T	F	F
F	T	T	F	F	T
F	F	T	F	F	F
		4	1	3	2

Table 3.25

		(a)			(b)			(c)			(d)		
p	q	$\sim q$	$\rightarrow$	$\sim p$	$\sim p$	$\vee$	$\sim q$	$\sim q$	$\wedge$	$\sim p$	$\sim p$	$\rightarrow$	$\sim q$
T	T	F	T	F	F	F	F	F	F	F	F	T	F
T	F	T	F	F	F	T	T	T	F	F	F	T	T
F	T	F	T	T	T	T	F	F	F	T	T	F	F
F	F	T	T	T	T	T	T	T	T	T	T	T	T

In the preceding section, we showed that $p \leftrightarrow q$ has the same truth table as $(p \rightarrow q) \wedge (q \rightarrow p)$. Therefore, these statements are equivalent, a useful fact for Example 4.

EXAMPLE 4 ***Write an Equivalent Biconditional Statement***

Write the following statement as an equivalent biconditional statement: "If the figure is a triangle then the figure has three angles and if the figure has three angles then the figure is a triangle."

SOLUTION An equivalent statement is "A figure is a triangle if and only if the figure has three angles." ▲

DE MORGAN'S LAWS

Example 3 showed that a statement of the form $\sim(p \wedge q)$ is equivalent to a statement of the form $\sim p \vee \sim q$. Thus, we may write $\sim(p \wedge q) \Leftrightarrow \sim p \vee \sim q$. This equivalent statement is one of two special laws called De Morgan's laws. The laws, named after Augustus De Morgan, an English mathematician, were first introduced in Section 2.4, where they applied to sets.

De Morgan's Laws

1. $\sim(p \wedge q) \Leftrightarrow \sim p \vee \sim q$

2. $\sim(p \vee q) \Leftrightarrow \sim p \wedge \sim q$

You can demonstrate that De Morgan's second law is true by constructing and comparing truth tables for $\sim(p \vee q)$ and $\sim p \wedge \sim q$. Do so now.

When using De Morgan's laws, if it becomes necessary to negate an already negated statement, use of the fact that $\sim(\sim p)$ is equivalent to p. For example, the negation of the statement "Today is not Monday" is "Today is Monday."

EXAMPLE 5 ***Use De Morgan's Laws***

Select the statement that is logically equivalent to "The sun is not shining but it is not raining."

a) It is not true that the sun is shining and it is raining.
b) It is not raining or the sun is not shining.
c) The sun is shining or it is raining.
d) It is not true that the sun is shining or it is raining.

SOLUTION To determine which statement is equivalent, write each statement in symbolic form.

Let p: The sun is shining.
q: It is raining.

The statement "The sun is not shining but it is not raining" written symbolically is $\sim p \wedge \sim q$. Recall that the word *but* means the same as the word *and*. Now, write parts (a) through (d) symbolically.

a) $\sim(p \wedge q)$ b) $\sim q \vee \sim p$ c) $p \vee q$ d) $\sim(p \vee q)$

DID YOU KNOW

Adventures in Logic

One of the more interesting and well-known students of logic was Charles Dodgson (1832–1898), better known to us as Lewis Carroll, the author of *Alice's Adventures in Wonderland* and *Through the Looking-Glass*. Although the books have a child's point of view, many would argue that the audience best equipped to enjoy them is an adult one. Dodgson, a mathematician, logician, and photographer (among other things), uses the naïveté of a 7-year-old girl to show what can happen when the rules of logic are taken to absurd extremes.

"You should say what you mean," the March Hare went on.

"I do," Alice hastily replied; "at least–at least I mean what I say–that's the same thing, you know."

"Not the same thing a bit!" said the Hatter. "You might as well say that 'I see what I eat' is the same thing as 'I eat what I see'!"

De Morgan's law shows that $\sim p \wedge \sim q$ is equivalent to $\sim(p \vee q)$. Therefore, the answer is (d): "It is not true that the sun is shining or it is raining." ▲

PROFILE IN MATHEMATICS

Augustus De Morgan

Augustus De Morgan (1806–1871), the son of a member of the East India Company, was born in India and educated at Trinity College, Cambridge. One of the great reformers of logic in the nineteenth century, De Morgan made his greatest contribution to the subject by realizing that logic as it had come down from Aristotle was narrow in scope and could be applied to a wider range of arguments. His work laid the foundation for modern, symbolic logic.

EXAMPLE 6 *Write a Logically Equivalent Statement*

Write a statement that is logically equivalent to "It is not true that today is Sunday or you have to go to work."

SOLUTION

Let p: Today is Sunday.
q: You have to go to work.

This statement is of the form $\sim(p \vee q)$. An equivalent statement, using De Morgan's laws, is $\sim p \wedge \sim q$. Therefore, an equivalent statement is "Today is not Sunday and you do not have to go to work." ▲

Consider $\sim(p \wedge q) \Leftrightarrow \sim p \vee \sim q$, one of De Morgan's laws. To go from $\sim(p \wedge q)$ to $\sim p \vee \sim q$, we negate both the p and the q within parentheses; change the conjunction, $\wedge$, to a disjunction, $\vee$; and remove the negation symbol preceding the left parentheses and the parentheses themselves. We can use a similar procedure to obtain equivalent statements. For example,

$$\sim(\sim p \wedge q) \Leftrightarrow p \vee \sim q$$
$$\sim(p \wedge \sim q) \Leftrightarrow \sim p \vee q$$

We can use a similar procedure to obtain equivalent statements when a disjunction is within parentheses. Note that

$$\sim(\sim p \vee q) \Leftrightarrow p \wedge \sim q$$
$$\sim(p \vee \sim q) \Leftrightarrow \sim p \wedge q$$

EXAMPLE 7 *Write a Logically Equivalent Statement*

Use De Morgan's laws to write a statement logically equivalent to "Omaha is not the capital of Nebraska but Omaha is a city in Nebraska."

SOLUTION

Let p: Omaha is the capital of Nebraska.
q: Omaha is a city in Nebraska.

The statement written symbolically is $\sim p \wedge q$. Earlier we showed that

$$\sim p \wedge q \Leftrightarrow \sim(p \vee \sim q)$$

Therefore, the statement "It is false that Omaha is the capital of Nebraska or Omaha is not a city in Nebraska" is logically equivalent to the given statement. ▲

There are strong similarities between the topics of sets and logic. We can see them by examining De Morgan's laws for sets and logic.

De Morgan's laws: set theory	**De Morgan's laws: logic**
$(A \cap B)' = A' \cup B'$	$\sim(p \wedge q) \Leftrightarrow \sim p \vee \sim q$
$(A \cup B)' = A' \cap B'$	$\sim(p \vee q) \Leftrightarrow \sim p \wedge \sim q$

The complement in set theory, $'$, is similar to the negation, $\sim$, in logic. The intersection, $\cap$, is similar to the conjunction, $\wedge$; and the union, $\cup$, is similar to the disjunction, $\vee$. If we were to interchange the set symbols with the logic symbols, De Morgan's laws would remain, but in a different form.

Both $'$ and $\sim$ can be interpreted as *not*.

Both $\cap$ and $\wedge$ can be interpreted as *and*.

Both $\cup$ and $\vee$ can be interpreted as *or*.

For example, the set statement $A' \cup B$ can be written as a statement in logic as $\sim a \vee b$.

Statements containing connectives other than *and* and *or* may have equivalent statements. To illustrate this point, construct truth tables for $p \rightarrow q$ and for $\sim p \vee q$. The truth tables will have the same answer columns and therefore the statements are equivalent. That is,

$$p \rightarrow q \Leftrightarrow \sim p \vee q$$

With these equivalent statements, we can write a conditional statement as a disjunction or a disjunction as a conditional statement. For example the statement "If the game is polo, then you ride a horse" can be equivalently stated as "The game is not polo or you ride a horse."

To change a conditional statement to a disjunction, negate the antecedent, change the conditional symbol to a disjunction symbol, and keep the consequent the same. To change a disjunction statement to a conditional statement, negate the first statement, change the disjunction symbol to a conditional symbol, and keep the second statement the same.

EXAMPLE 8 *Write a Logically Equivalent Statement*

Write a conditional statement that is logically equivalent to "The Gators will win or the Flames will lose." Assume the negation of winning is losing.

SOLUTION

Let p: The Gators will win.

q: The Flames will win.

The statement may be written symbolically as $p \vee \sim q$. To write an equivalent statement negate the first statement, p, and change the disjunction symbol to a conditional symbol. Symbolically, the statement is $\sim p \rightarrow \sim q$. The equivalent statement is "If the Gators lose, then the Flames will lose." ▲

NEGATION OF THE CONDITIONAL STATEMENT

Now we will discuss how to negate a conditional statement. To negate a statement we use the fact that $p \rightarrow q \Leftrightarrow \sim p \vee q$ and De Morgan's laws. Examples 9 and 10 show the process.

EXAMPLE 9 *The Negation of a Conditional Statement*

Determine a statement equivalent to $\sim(p \rightarrow q)$.

SOLUTION Begin with $p \rightarrow q \Leftrightarrow \sim p \vee q$, negate both statements, and use De Morgan's laws.

$$p \rightarrow q \Leftrightarrow \sim p \vee q$$

$$\sim(p \rightarrow q) \Leftrightarrow \sim(\sim p \vee q) \quad \text{Negate both statements}$$

$$\Leftrightarrow p \wedge \sim q \quad \text{De Morgan's laws}$$

Therefore, $\sim(p \rightarrow q)$ is equivalent to $p \wedge \sim q$. ▲

EXAMPLE 10 *Write an Equivalent Statement*

Write a statement equivalent to

"It is false that if the dog is snoring then the dog cannot sleep in our bedroom."

SOLUTION

Let p = the dog is snoring

q = the dog can sleep in our room

Then the given statement can be represented symbolically as $\sim(p \rightarrow \sim q)$. Using the procedure illustrated in Example 9 we can determine that $\sim(p \rightarrow \sim q)$ is equivalent to $p \wedge q$. Verify this yourself now. Therefore, an equivalent statement is "The dog is snoring and the dog can sleep in our bedroom." ▲

VARIATIONS OF THE CONDITIONAL STATEMENT

We know that $p \rightarrow q$ is equivalent to $\sim p \vee q$. Are any other statements equivalent to $p \rightarrow q$? Yes, there are many. Now let's look at the variations of the conditional statement to determine whether any are equivalent to the conditional statement. The variations of the conditional statement are made by switching and/or negating the antecedent and the consequent of a conditional statement. The variations of the conditional statement are the **converse** of the conditional, the **inverse** of the conditional, and the **contrapositive** of the conditional.

Listed here are the variations of the conditional with their symbolic form and the words we say to read each one.

Variations of the Conditional Statement

Name	Symbolic form	Read
Conditional	$p \rightarrow q$	"If p, then q"
Converse of the conditional	$q \rightarrow p$	"If q, then p"
Inverse of the conditional	$\sim p \rightarrow \sim q$	"If not p, then not q"
Contrapositive of the conditional	$\sim q \rightarrow \sim p$	"If not q, then not p"

To write the converse of the conditional statement, switch the order of the antecedent and the consequent. To write the inverse, negate both the antecedent and the consequent. To write the contrapositive, switch the order of the antecedent and the consequent and then negate both of them.

Are any of the variations of the conditional statement equivalent? To determine the answer, we can construct a truth table for each variation, as shown in Table 3.26. It reveals that the conditional statement is equivalent to the contrapositive statement and that the converse statement is equivalent to the inverse statement.

Table 3.26

p	q	Conditional $p \to q$	Contrapositive $\sim q \to \sim p$	Converse $q \to p$	Inverse $\sim p \to \sim q$
T	T	T	T	T	T
T	F	F	F	T	T
F	T	T	T	F	F
F	F	T	T	T	T

DID YOU KNOW

In the Service of Science

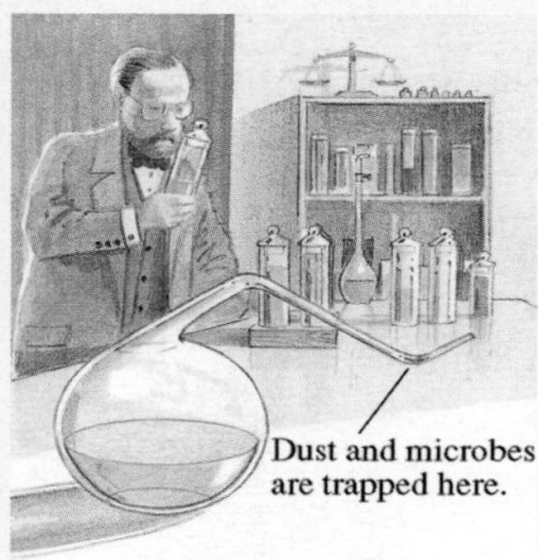
Dust and microbes are trapped here.

Logical reasoning is often used by scientists as they develop theories. Logical reasoning often also plays a part in developing procedures to test theories. Sometimes, their theories are supported by experimental evidence and are accepted. Other times, their theories are contradicted by the evidence and are rejected. For example, Aristotle observed that bugs appeared in spoiling meat and reasoned that life arose spontaneously from nonliving matter. With advances in technology, scientist have had more means at their disposal to test their theories. To refute Aristotle's claim of spontaneous generation of life, Louis Pasteur in 1862 conducted an experiment by isolating some meat broth in a sterile flask to demonstrate that the bugs Aristotle observed grew from microscopic life forms too small to be seen.

EXAMPLE 11 *The Converse, Inverse and Contrapositive*

For the conditional statement "If Paul is in the nursing program, then he will learn the metric system," write the

a) converse. b) inverse. c) contrapositive.

SOLUTION

a) Let

p: Paul is in the nursing program.
q: Paul will learn the metric system.

The conditional statement is of the form $p \to q$, so the converse must be of the form $q \to p$. Therefore, the converse is "If Paul learns the metric system, then he is in the nursing program."

b) The inverse is of the form $\sim p \to \sim q$. Therefore, the inverse is "If Paul is not in the nursing program, then he will not learn the metric system."
c) The contrapositive is of the form $\sim q \to \sim p$. Therefore, the contrapositive is "If Paul does not learn the metric system, then Paul is not in the nursing program."

EXAMPLE 12 *Determine the Truth Values*

Let

p: The number is divisible by 9.
q: The number is divisible by 3.

Write the following statements and determine which are true.

a) The conditional statement, $p \to q$
b) The converse of $p \to q$
c) The inverse of $p \to q$
d) The contrapositive of $p \to q$

SOLUTION

a) *Conditional statement:* $(p \rightarrow q)$
If the number is divisible by 9, then the number is divisible by 3. This statement is true. A number divisible by 9 must also be divisible by 3, since 3 is a divisor of 9.

b) *Converse of the conditional:* $(q \rightarrow p)$
If the number is divisible by 3, then the number is divisible by 9. This statement is false. For instance, 6 is divisible by 3, but 6 is not divisible by 9.

c) *Inverse of the conditional:* $(\sim p \rightarrow \sim q)$
If the number is not divisible by 9, then the number is not divisible by 3. This statement if false. For instance, 6 is not divisible by 9, but 6 is divisible by 3.

d) *Contrapositive of the conditional:* $(\sim q \rightarrow \sim p)$
If the number is not divisible by 3, then the number is not divisible by 9. The statement is true, since any number that is divisible by 9 must be divisible by 3. ▲

EXAMPLE 13 ***Use the Contrapositive***

Use the contrapositive to write a statement logically equivalent to "If the boat is 24 ft long, then it will not fit into the boathouse."

SOLUTION

Let p: The boat is 24 ft long.
q: The boat will fit into the boathouse.

The given statement written symbolically is

$$p \rightarrow \sim q$$

The contrapositive of the statement is

$$q \rightarrow \sim p$$

Therefore, an equivalent statement is "If the boat will fit into the boathouse, then the boat is not 24 ft long. ▲

The contrapositive of the conditional is very important in mathematics. Consider the statement "If a^2 is not a whole number, then a is not a whole number." Is this statement true? You may find this question difficult to answer. Writing the statement's contrapositive may enable you to answer the question. The contrapositive is "If a is a whole number, then a^2 is a whole number." Since the contrapositive is a true statement, the original statement must also be true.

EXAMPLE 14 ***Which Are Equivalent?***

Determine which, if any, of the following statements are equivalent. You may use De Morgan's laws, the fact that $p \rightarrow q \Leftrightarrow \sim p \vee q$, information from the variations of the conditional, or truth tables.

a) If you leave by 9 A.M., then you will get to your destination on time.
b) You do not leave by 9 A.M. or you will get to your destination on time.
c) It is false that you get to your destination on time or you did not leave by 9 A.M.
d) If you do not get to your destination on time, then you did not leave by 9 A.M.

SOLUTION

Let p: You leave by 9 A.M.

q: You will get to your destination on time.

In symbolic form, the four statements are

a) $p \rightarrow q$.
b) $\sim p \vee q$.
c) $\sim(q \vee \sim p)$.
d) $\sim q \rightarrow \sim p$.

Which of these statements are equivalent? Earlier in this section, you learned that $p \rightarrow q$ is equivalent to $\sim p \vee q$. Therefore, statements (a) and (b) are equivalent. Statement (d) is the contrapositive of statement (a). Therefore, statement (d) is also equivalent to statement (a) and statement (b). All these statements have the same truth table (Table 3.27).

Table 3.27

		(a)	(b)	(d)
p	q	$p \rightarrow q$	$\sim p \vee q$	$\sim q \rightarrow \sim p$
T	T	T	T	T
T	F	F	F	F
F	T	T	T	T
F	F	T	T	T

Now let's look at statement (c). If we use De Morgan's laws on statement (c), we get

$$\sim(q \vee \sim p) \Leftrightarrow \sim q \wedge p$$

If $\sim q \wedge p$ was one of the other statements, then $\sim(q \vee \sim p)$ would be equivalent to that statement. Because $\sim q \wedge p$ does not match any of the other choices, it does not necessarily mean that $\sim(q \vee \sim p)$ is not equivalent to the other statements. To determine whether $\sim(q \vee \sim p)$ is equivalent to the other statements, we will construct its truth table (Table 3.28) and compare the answer column with the answer columns in Table 3.27.

Table 3.28

		(c)			
p	q	$\sim$	$(q$	$\vee$	$\sim p)$
T	T	F	T	T	F
T	F	T	F	F	F
F	T	F	T	T	T
F	F	F	F	T	T
		4	1	3	2

The answer columns of the truth tables are not the same, so $\sim(q \vee \sim p)$ is not equivalent to any of the other statements. Therefore, statements (a), (b), and (d) are equivalent to each other. ▲

DID YOU KNOW

Fuzzy Logic

Modern computers, like truth tables, work with only two values, 1 or 0 (equivalent to true or false in truth tables). This constraint prevents a computer from being able to reason as the human brain can and prevents a computer from being able to evaluate items involving vagueness or value judgments that so often occur in real-world situations. For example, a binary computer will have difficulty evaluating the subjective statement "the air is warm."

Fuzzy logic uses the concept: Everything is a matter of degree. Fuzzy logic manipulates vague concepts such as *bright* and *fast* by assigning values between 0 and 1 to each item. For example, suppose *bright* is assigned a value of 0.80; then *not bright* is assigned a value of $1 - 0.8 = 0.20$. As the value assigned to *bright* changes, so does the value assigned to *not bright*. Not p is always $1 - p$, where $0 < p < 1$. Fuzzy logic is used to operate cameras, air conditioners, subways, and many other devices where the change in one condition changes another condition. For example, when it is bright outside, the camera's lens aperture opens less, and when it is overcast, the camera's lens aperture opens more. How many other devices can you name that may use fuzzy logic? See Problem-Solving Exercises 87 and 88.

SECTION 3.4 EXERCISES

CONCEPT/WRITING EXERCISES

1. What symbol is used to represent logically equivalent?
2. What are equivalent statements?
3. Explain how you can determine whether two statements are equivalent.
4. Suppose two statements are connected with the biconditional and the truth table is constructed. If the answer column of the truth table has all trues, what must be true about these two statements? Explain.
5. Write De Morgan's laws for logic.
6. For a statement of the form if p, then q, symbolically indicate the form of the
   a) converse. **b)** inverse. **c)** contrapositive.
7. Which of the following are equivalent statements?
   a) The converse **b)** The contrapositive
   c) The inverse **d)** The conditional
8. Write a disjunctive statement that is logically equivalent to $p \rightarrow q$.

PRACTICE THE SKILLS

In Exercises 9–18, use De Morgan's laws to determine whether the two statements are equivalent.

9. $\sim(p \vee q),\ \sim p \wedge \sim q$
10. $\sim p \vee \sim q,\ \sim(p \wedge q)$
11. $\sim(p \wedge q),\ \sim p \wedge q$
12. $\sim(p \wedge q),\ p \vee \sim q$
13. $p \wedge q,\ \sim(\sim p \vee \sim q)$
14. $\sim(p \wedge q),\ \sim(q \vee \sim p)$
15. $(\sim p \vee \sim q) \rightarrow r,\ \sim(p \wedge q) \rightarrow r$
16. $q \rightarrow \sim(p \wedge \sim r),\ q \rightarrow \sim p \vee r$
17. $\sim(p \rightarrow \sim q),\ p \wedge q$
18. $\sim(\sim p \rightarrow q),\ \sim p \wedge \sim q$

In Exercises 19–28, use a truth table to determine whether the two statements are equivalent.

19. $\sim p \rightarrow q,\ p \wedge q$
20. $p \rightarrow q,\ \sim p \vee q$
21. $p \rightarrow q,\ \sim q \rightarrow \sim p$
22. $(p \wedge q) \wedge r,\ p \wedge (q \wedge r)$
23. $(p \vee q) \vee r,\ p \vee (q \vee r)$
24. $p \vee (q \wedge r),\ \sim p \rightarrow (q \wedge r)$
25. $q \leftrightarrow (p \wedge \sim r),\ q \rightarrow (p \vee r)$
26. $\sim(q \rightarrow p) \vee r,\ (p \vee q) \wedge \sim r$
27. $(p \rightarrow q) \wedge (q \rightarrow r),\ (p \rightarrow q) \rightarrow r$
28. $\sim q \rightarrow (p \wedge r),\ \sim(p \vee r) \rightarrow q$
29. Show that $(p \rightarrow q) \wedge (q \rightarrow p) \Leftrightarrow (p \leftrightarrow q)$.
30. Determine whether $[\sim(p \rightarrow q)] \wedge [\sim(q \rightarrow p)] \Leftrightarrow \sim(p \leftrightarrow q)$.

PROBLEM SOLVING

In Exercise 31–38, use De Morgan's laws to write an equivalent statement for the sentence.

31. It is false that the boat is at the dock or the boat will depart.

32. It is false that the ink is red and the pen has a ball point.

33. The house does not have one phone line or the house does not have two phone lines.

34. David teaches English but he does not teach chemistry.

35. The novel is neither written by Dinya Floyd nor is it well illustrated.

36. Felicia ate the ice cream or Felicia did not eat the yogurt.

37. If we go to Cozumel, then we will go snorkeling or we will not go to Senior Frogs.

38. If we visit Aunt Lilly, then she will not go on vacation but she will make her famous pancakes.

In Exercises 39–44, use the fact that $p \rightarrow q$ is equivalent to $\sim p \vee q$ to write an equivalent form of the statement.

39. If you drink milk, then your bones will be strong.

40. The roller-coaster ride was exciting or the roller-coaster was out of service.

41. John painted the picture or Ada did not purchase the picture.

42. If we do not renew the subscription, then the magazine will stop coming.

43. It is false that if the noise is too loud then the police will come.

44. Stone Mountain is in Georgia, or the Empire State Building is not in Austin, Texas.

In Exercises 45–48, use the fact that $(p \rightarrow q) \wedge (q \rightarrow p)$ is equivalent to $p \leftrightarrow q$ to write the statement in an equivalent form.

45. If you are 18 years old then you are eligible to vote, and if you are eligible to vote then you are 18 years old.

46. If a number is even then it is divisible by 2, and if a number is divisible by 2 then the number is even.

47. An animal is a mammal if and only if it is warm blooded.

48. You need to pay taxes if and only if you receive income.

In Exercises 49–56, write the converse, inverse, and contrapositive of the statement. (For Exercises 55 and 56, use De Morgan's laws.)

49. If the fish are biting, then we will go fishing.

50. If the computer goes on sale, then we will buy the computer.

51. If the phone bill is large, then you will have to give up your phone.

52. If there are clothes on the floor, then you have not cleaned your room.

53. If the dog is not friendly, then I will not get out of the car.

54. If the nut will not turn, then Sean will need to use Liquid Wrench.

55. If the sun is shining, then we will go down to the marina and we will take out the sailboat.

56. If the apple pie is baked, then we will eat a piece of pie and we will save some pie for later.

In Exercises 57–64, write the contrapositive of the statement. Use the contrapositive to determine whether the conditional statement is true or false.

57. If the triangle is not isosceles, then two angles of the triangle are not equal.

58. If the number is not divisible by 3, then the sum of the digits is not divisible by 3.

59. If 2 does not divide the counting number, then 2 does not divide the units digit of the counting number.

60. If $1/n$ is not a natural number, then n is not a natural number.

61. If two lines do not intersect in at least one point, then the two lines are parallel.

62. If $\frac{m \cdot a}{m \cdot b} \neq \frac{a}{b}$, then m is not a counting number.

63. If the sum of the interior angles of a polygon do not measure 360°, then the polygon is not a quadrilateral.

64. If a and b are not both even counting numbers, then the product of a and b is not an even counting number.

In Exercises 65–80, determine which, if any, of the three statements are equivalent (see Example 14).

65. a) Maria has not retired or Maria is still working.
b) If Maria is still working, then Maria has not retired.
c) If Maria has retired, then Maria is not still working.

66. a) If today is Monday, then tomorrow is not Wednesday.
b) It is false that today is Monday and tomorrow is not Wednesday.
c) Today is not Monday or tomorrow is Wednesday.

67. a) The car is not reliable and the car is noisy.
b) If the car is not reliable, then the car is not noisy.
c) It is false that the car is reliable or the car is not noisy.

68. a) The sales tax is 6% if and only if you live in Kings County.
b) If you live in Kings County then the sales tax is 6%, and if the sales tax is 6% then you live in Kings County.
c) You do not live in Kings County and the sales tax is not 6%.

69. a) The house is not made of wood or the shed is not made of wood.
b) If the house is made of wood, then the shed is not made of wood.
c) It is false that the shed is made of wood and the house is not made of wood.

70. a) It is false that if you do not leave your lights on then your electric bill will be higher.
b) Your electric bill will be higher if and only if you leave your lights on.
c) It is false that if you leave your lights on then your electric bill will be higher.

71. a) You will not get a speeding ticket if and only if you do not speed.
b) It is false that you will get a speeding ticket if and only if you do not speed.
c) If you get a speeding ticket, then you were speeding.

72. a) Today is not Sunday or the library is open.
b) If today is Sunday, then the library is not open.
c) If the library is open, then today is not Sunday.

73. a) If you are fishing at 1 P.M., then you are driving a car at 1 P.M.
b) You are not fishing at 1 P.M. or you are driving a car at 1 P.M.
c) It is false that you are fishing at 1 P.M. and you are not driving a car at 1 P.M.

74. a) The grass grows and the trees are blooming.
b) If the trees are blooming, then the grass does not grow.
c) The trees are not blooming or the grass does not grow.

75. a) If the pay is good and today is Monday, then I will take the job.
b) If I do not take the job, then it is false that the pay is good or today is Monday.
c) The pay is good and today is Monday, or I will take the job.

76. a) If you are 18 years old and a citizen of the United States, then you can vote in the presidential election.
b) You can vote in the presidential election, if and only if you are a citizen of the United States and you are 18 years old.
c) You cannot vote in the presidential election, or you are 18 years old and you are not a citizen of the United States.

77. a) The package was sent by Federal Express, or the package was not sent by United Parcel Service but the package arrived on time.
b) The package arrived on time, if and only if it was sent by Federal Express or it was not sent by United Parcel Service.
c) If the package was not sent by Federal Express, then the package was not sent by United Parcel Service but the package arrived on time.

78. a) If we put the dog outside or we feed the dog, then the dog will not bark.
b) If the dog barks, then we did not put the dog outside and we did not feed the dog.
c) If the dog barks, then it is false that we put the dog outside or we feed the dog.

79. a) The car needs oil, and the car needs gas or the car is new.
b) The car needs oil, and it is false that the car does not need gas and the car is not new.
c) If the car needs oil, then the car needs gas or the car is not new.

80. a) The mortgage rate went down, if and only if Tim purchased the house and the down payment was 10%.
b) The down payment was 10%, and if Tim purchased the house then the mortgage rate went down.
c) If Tim purchased the house, then the mortgage rate went down and the down payment was not 10%.

81. Can a conditional statement and its converse both be false statements? Explain your answer with an example.

82. Can a conditional statement be true if its converse is false? Explain your answer with an example.

83. Can a conditional statement and its contrapositive both be false? Explain your answer with an example.

84. Can a conditional statement be true if its contrapositive is false? Explain.

CHALLENGE PROBLEMS/GROUP ACTIVITIES

85. We learned that $p \rightarrow q \Leftrightarrow \sim p \vee q$. Determine a conjunctive statement that is equivalent to $p \rightarrow q$. (*Hint:* There are many answers.)

86. Determine whether $\sim[\sim(p \vee \sim q)] \Leftrightarrow p \vee \sim q$. Explain the method(s) you used to determine your answer.

87. In an appliance or device that uses fuzzy logic, a change in one condition causes a change in a second condition. For example, in a camera, if the brightness increases, the lens aperture automatically decreases to get the proper exposure on the film. Name at least 10 appliances or devices that make use of fuzzy logic and explain how fuzzy logic is used in each appliance or device. See the Did You Know on page 124.

88. In symbolic logic, a statement is either true or false (consider true to have a value of 1 and false a value of 0). In fuzzy logic, nothing is true or false, but everything is a matter of degree. For example, consider the statement "The sun is shining." In fuzzy logic, this statement may have a value between 0 and 1 and may be constantly changing. For example, if the sun is partially blocked by clouds, the value of this statement may be 0.25. In fuzzy logic, the values of connective statements are found as follows for statements p and q.

Not p has a truth value of $1 - p$.

$p \wedge q$ has a truth value equal to the lesser of p and q.

$p \vee q$ has a truth value equal to the greater of p and q.

$p \rightarrow q$ has a truth value equal to the lesser of 1 and $1 - p + q$.

$p \leftrightarrow q$ has a truth value equal to $1 - |p - q|$, that is, 1 minus the absolute value* of p minus q.

Suppose the statement "p: The sun is shining" has a truth value of 0.25 and the statement "q: Mary is getting a tan" has a truth value of 0.20. Find the truth value of

a) $\sim p$. **b)** $\sim q$.
c) $p \wedge q$. **d)** $p \vee q$.
e) $p \rightarrow q$. **f)** $p \leftrightarrow q$.

RESEARCH ACTIVITIES

89. Do research and write a report on fuzzy logic.

90. Read one of Lewis Carroll's books and write a report on how he used logic in the book. Give at least five specific examples.

91. Do research and write a report on the life and achievements of Augustus De Morgan. Indicate in your report his contributions to sets and logic.

3.5 SYMBOLIC ARGUMENTS

In the preceding sections of this chapter, we used symbolic logic to determine the truth value of a compound statement. We now extend those basic ideas to determine whether symbolic arguments are valid or invalid.

Consider the statements:

If Jason is a singer, then he is well known.
Jason is a singer.

If you accept these two statements as true, then a conclusion that necessarily follows is that

Jason is well known.

*Absolute values are discussed in Section 13.8.

These three statements in the following form constitute a symbolic argument.

Premise 1: If Jason is a singer, then he is well known.
Premise 2: Jason is a singer.
Conclusion: Therefore, Jason is well known.

A **symbolic argument** consists of a set of **premises** and a **conclusion**. It is called a symbolic argument because we generally write it in symbolic form to determine its validity.

> An **argument is valid** when its conclusion necessarily follows from a given set of premises.
> An **argument is invalid** or a **fallacy** when the conclusion does not necessarily follow from the given set of premises.

DID YOU KNOW

The Dog Did Nothing

In the case of the disappearance of the racehorse Silver Blaze, Sherlock Holmes demonstrated that sometimes the absence of a clue is itself a clue. The local police inspector asked him, "Is there any point to which you would wish to draw my attention?" Holmes replied, "To the curious incident of the dog in the nighttime." The inspector, confused, asked: "The dog did nothing in the nighttime." "That was the curious incident," remarked Sherlock Holmes. From the lack of the dog's bark, Holmes concluded that the horse had been "stolen" by a stablehand. How did Holmes reach his conclusion?

An argument that is not valid is invalid. The argument just presented is an example of a valid argument, as the conclusion necessarily follows from the premises. Now we will discuss a procedure to determine whether an argument is valid or invalid. We begin by writing the argument in symbolic form. To write the argument in symbolic form, we let p and q be

p: Jason is a singer.
q: Jason is well known.

Symbolically, the argument is written

Premise 1: $p \to q$
Premise 2: p
Conclusion: $\therefore q$ (The three-dot triangle is read "therefore.")

Write the argument in the following form.

If [*premise 1* **and** *premise 2*] **then** *conclusion*
$[\,(p \to q) \quad \wedge \quad p \quad] \to \quad q$

Then construct a truth table for the statement $[(p \to q) \wedge p] \to q$ (Table 3.29). *If the truth table answer column is true in every case, then the statement is a tautology, and the argument is valid. If the truth table is not a tautology, then the argument is invalid.* Since the statement is a tautology, (see column 5), the conclusion necessarily follows from the premises and the argument is valid.

Table 3.29

p	q	$[(p \to q)$	$\wedge$	$p]$	$\to$	q
T	T	T	T	T	T	T
T	F	F	F	T	T	F
F	T	T	F	F	T	T
F	F	T	F	F	T	F
		1	3	2	5	4

Once we have demonstrated that an argument in a particular form is valid, all arguments with exactly the same form will also be valid. In fact, many of these forms have been assigned names. The argument form just discussed,

$$\begin{array}{l} p \rightarrow q \\ \underline{p \quad\quad} \\ \therefore q \end{array}$$

is called the **law of detachment.**

Consider the following argument.

If the water is warm, then the moon is made of cheese.
The water is warm.
∴ The moon is made of cheese.

Now translate the argument into symbolic form.

Let w: The water is warm.
m: The moon is made of cheese.

In symbolic form the argument is

$$\begin{array}{l} w \rightarrow m \\ \underline{w \quad\quad} \\ \therefore m \end{array}$$

This argument is also the law of detachment, and therefore it is a valid argument.

Note that the argument is valid even though the conclusion, "The moon is made of cheese," is a false statement. It is also possible to have an invalid argument in which the conclusion is a true statement. When an argument is valid, the conclusion necessarily follows from the premises. It is not necessary for the premises or the conclusion to be true statements in an argument.

Procedure to Determine Whether an Argument Is Valid

1. Write the argument in symbolic form.
2. Compare the form of the argument with forms that are known to be valid or invalid. If there are no known forms to compare it with, or you do not remember the forms, go to step 3.
3. If the argument contains two premises, write a conditional statement of the form

$$[(\text{premise 1}) \wedge (\text{premise 2})] \rightarrow \text{conclusion}$$

4. Construct a truth table for the statement in step 3.
5. If the answer column of the truth table has all trues, the statement is a tautology, and the argument is valid. If the answer column does not have all trues, the argument is invalid.

Examples 1 through 3 contain two premises. When an argument contains more that two premises, step 3 of the procedure will change slightly, as explained shortly.

EXAMPLE 1 *Valid or Invalid?*

Determine whether the following argument is valid or invalid.

If today is Tuesday, then the furniture will be delivered.
The furniture will not be delivered.
∴ Today is not Tuesday.

SOLUTION We first write the argument in symbolic form.

Let p: Today is Tuesday.
q: The furniture will be delivered.

In symbolic form, the argument is

$$\begin{array}{l} p \rightarrow q \\ \underline{\sim q} \\ \therefore \sim p \end{array}$$

As we have not tested an argument in this form, we will construct a truth table to determine whether it is valid or invalid. We write the argument in the form $[(p \rightarrow q) \wedge \sim q] \rightarrow \sim p$, and construct a truth table (Table 3.30). Since the answer, column 5, has all T's, the argument is valid.

Table 3.30

p	q	$[(p \rightarrow q)$	$\wedge$	$\sim q]$	$\rightarrow$	$\sim p$
T	T	T	F	F	T	F
T	F	F	F	T	T	F
F	T	T	F	F	T	T
F	F	T	T	T	T	T
		1	3	2	5	4

The argument form in Example 1 is an example of the **law of contraposition.**

EXAMPLE 2 *Another Symbolic Argument*

Determine whether the following argument is valid or invalid.

The grass is green or the grass is full of weeds.
The grass is full of weeds.
∴ The grass is green.

SOLUTION

Let p: The grass is green.
q: The grass is full of weeds.

In symbolic form, the argument is

$$\begin{array}{l} p \vee q \\ \underline{q} \\ \therefore p \end{array}$$

As this is not one of the forms we are familiar with, we will construct a truth table. We write the argument in the form $[(p \vee q) \wedge q] \to p$. Next we construct a truth table, as shown in Table 3.31. The answer to the truth table, column 5, is not true in *every case*. Therefore, the statement is not a tautology, and the argument is invalid, or is a fallacy.

Table 3.31

p	q	$[(p \vee q)$	$\wedge$	$q]$	$\to$	p
T	T	T	T	T	T	T
T	F	T	F	F	T	T
F	T	T	T	T	F	F
F	F	F	F	F	T	F
		1	3	2	5	4

DID YOU KNOW

Guilty as Charged?

In America law, a jury must reach unanimous agreement about a verdict. Generally, fewer than 5% of the 120,000 jury trials conducted each year in the United States result in a "hung jury" in which a jury cannot reach a unanimous agreement. Juries use logical reasoning to determine if there are contradictions in the evidence or testimony when deciding their verdict. Most nations of the world rely solely on a judge's opinion to decide guilt or innocence.

Standard forms of commonly used arguments are given in the following chart.

Standard Forms of Arguments

Valid Arguments	*Law of Detachment*	*Law of Contraposition*	*Law of Syllogism*	*Disjunctive Syllogism*
	$p \to q$	$p \to q$	$p \to q$	$p \vee q$
	p	$\sim q$	$q \to r$	$\sim p$
	$\therefore q$	$\therefore \sim p$	$\therefore p \to r$	$\therefore q$
Invalid Arguments	*Fallacy of the Converse*	*Fallacy of the Inverse*		
	$p \to q$	$p \to q$		
	q	$\sim p$		
	$\therefore p$	$\therefore \sim q$		

EXAMPLE 3

Determine whether the following argument is valid or invalid.

If I drink hot milk, then I can sleep.
If I can sleep, then I can dream.
∴ If I drink hot milk, then I can dream.

SOLUTION

Let p: I drink hot milk.
q: I can sleep.
r: I can dream.

In symbolic form, the argument is

$$\begin{array}{l} p \to q \\ q \to r \\ \hline \therefore p \to r \end{array}$$

The argument is in the form of the law of syllogism. Therefore, the argument is valid, and there is no need to construct a truth table.

DID YOU KNOW

Spotty Advertising

Too many artificial colors in your dog's food?

Advertisements often use humorous twists of logic to grab the reader's interest. For example, consider this ad. Can you write a conditional statement that is implied by the ad?

Now we consider an argument that has more than two premises. When an argument contains more than two premises, the statement we test, using a truth table, is formed by taking the conjunction of all the premises as the antecedent and the conclusion as the consequent. For example, if an argument is of the form

$$\begin{array}{l} p_1 \\ p_2 \\ \underline{p_3} \\ \therefore c \end{array}$$

We evaluate the truth table for $[p_1 \wedge p_2 \wedge p_3] \rightarrow c$. When we evaluate $[p_1 \wedge p_2 \wedge p_3]$, it makes no difference whether we evaluate $[(p_1 \wedge p_2) \wedge p_3]$, or $[p_1 \wedge (p_2 \wedge p_3)]$ because both give the same answer. In Example 4, we evaluate $[p_1 \wedge p_2 \wedge p_3]$ from left to right, that is, $[(p_1 \wedge p_2) \wedge p_3]$.

EXAMPLE 4 ***An Argument with Three Premises***

Use a truth table to determine whether the following argument is valid or invalid.

If Donna has a pet, then Donna owns a snail.
Donna owns a snail or Donna drives a truck.
Donna drives a truck or Donna has a pet.
∴ Donna has a pet.

SOLUTION This argument contains three simple statements.

Let
p: Donna has a pet.
q: Donna owns a snail.
r: Donna drives a truck.

In symbolic form, the argument is

$$\begin{array}{l} p \rightarrow q \\ q \vee r \\ \underline{r \vee p} \\ \therefore p \end{array}$$

Write the argument in the form

$$[(p \rightarrow q) \wedge (q \vee r) \wedge (r \vee p)] \rightarrow p.$$

Now construct the truth table (Table 3.32). The answer, column 7, is not true in every case. Thus, the argument is a fallacy.

Table 3.32

p	q	r	[(p → q)	∧	(q ∨ r)	∧	(r ∨ p)]	→	p
T	T	T	T	T	T	T	T	T	T
T	T	F	T	T	T	T	T	T	T
T	F	T	F	F	T	F	T	T	T
T	F	F	F	F	F	F	T	T	T
F	T	T	T	T	T	T	T	F	F
F	T	F	T	T	T	F	F	T	F
F	F	T	T	T	T	T	T	F	F
F	F	F	T	F	F	F	F	T	F
			1	3	2	5	4	7	6

Let's now investigate how we can arrive at a valid conclusion from a given set of premises.

EXAMPLE 5 ***Determine a Logical Conclusion***

Determine a logical conclusion that follows from the given statements. "If you own a house, then you will pay property tax. You own a house. Therefore, . . . "

SOLUTION If you recognize a specific form of an argument, you can use you knowledge of that form to draw a logical conclusion.

Let p: You own a house.

q: You will pay property tax.

$$\begin{array}{l} p \rightarrow q \\ \underline{p \qquad} \\ \therefore ? \end{array}$$

If the question mark is replaced with a q, this argument is of the form of the law of detachment. A logical conclusion is "Therefore, you will pay property tax." ▲

SECTION 3.5 EXERCISES

CONCEPT/WRITING EXERCISES

1. What does it mean when an argument is valid?
2. Is it possible for an argument to be valid if its conclusion is false? Explain your answer.
3. Is it possible for an argument to be invalid if the premises are all true? Explain your answer.
4. Is it possible for an argument to be valid if the premises are all false? Explain your answer.
5. Explain how to determine whether an argument with premises p_1 and p_2 and conclusion c is a valid or invalid argument.
6. What is another name for an invalid argument?

In Exercises 7–10, (a) indicate the form of the valid argument and (b) write an original argument in words for each form.

7. Law of detachment
8. Law of syllogism
9. Law of contraposition
10. Disjunctive syllogism

In Exercises 11 and 12, (a) indicate the form of the fallacy, and (b) write an original argument in words for each form.

11. Fallacy of the converse
12. Fallacy of the inverse

PRACTICE THE SKILLS

In Exercises 13–32, determine whether the argument is valid or invalid. You may compare the argument to a standard form or use a truth table.

13. $\begin{array}{l} p \rightarrow q \\ \underline{\sim p \qquad} \\ \therefore q \end{array}$

14. $\begin{array}{l} p \wedge \sim q \\ \underline{q \qquad} \\ \therefore \sim p \end{array}$

15. $\begin{array}{l} p \rightarrow q \\ \underline{p \qquad} \\ \therefore q \end{array}$

16. $\begin{array}{l} \sim p \vee q \\ \underline{q \qquad} \\ \therefore p \end{array}$

17. $\begin{array}{l} \sim p \\ \underline{p \vee q} \\ \therefore \sim q \end{array}$

18. $\begin{array}{l} p \rightarrow q \\ \underline{\sim q \quad} \\ \therefore \sim p \end{array}$

19. $\begin{array}{l} q \rightarrow p \\ \underline{\sim q \quad} \\ \therefore \sim p \end{array}$

20. $\begin{array}{l} p \vee q \\ \underline{\sim q \quad} \\ \therefore p \end{array}$

21. $\begin{array}{l} \sim p \rightarrow q \\ \underline{\sim q \qquad} \\ \therefore \sim p \end{array}$

22. $\begin{array}{l} q \wedge \sim p \\ \underline{\sim p \qquad} \\ \therefore q \end{array}$

23. $\begin{array}{l} p \rightarrow q \\ \underline{q \rightarrow r} \\ \therefore p \rightarrow r \end{array}$

24. $\begin{array}{l} q \wedge p \\ \underline{q \qquad} \\ \therefore \sim p \end{array}$

25. $\begin{array}{l} p \leftrightarrow q \\ \underline{q \wedge r \quad} \\ \therefore p \vee r \end{array}$

26. $\begin{array}{l} p \leftrightarrow q \\ \underline{q \rightarrow r \qquad} \\ \therefore \sim r \rightarrow \sim p \end{array}$

27. $\begin{array}{l} r \leftrightarrow p \\ \underline{\sim p \wedge q} \\ \therefore p \wedge r \end{array}$

28. $\begin{array}{l} p \vee q \\ \underline{r \wedge p} \\ \therefore q \end{array}$

29. $\begin{array}{l} p \rightarrow q \\ q \vee r \\ \underline{r \vee p \quad} \\ \therefore p \end{array}$

30. $\begin{array}{l} p \rightarrow q \\ q \rightarrow r \\ \underline{r \rightarrow p \quad} \\ \therefore q \rightarrow p \end{array}$

31. $\begin{array}{l} p \rightarrow q \\ r \rightarrow \sim p \\ \underline{p \vee r \qquad} \\ \therefore q \vee \sim p \end{array}$

32. $\begin{array}{l} p \leftrightarrow q \\ p \vee r \\ \underline{q \rightarrow r \quad} \\ \therefore q \vee r \end{array}$

PROBLEM SOLVING

In Exercises 33–44, translate the argument into symbolic form. Determine whether each argument is valid or invalid. You may compare the argument to a standard form or use a truth table.

33. If today is Tuesday, then we will play bingo.
Today is Tuesday.
∴ We will play bingo.

34. If the game is televised, then people will watch the game.
People will watch the game.
∴ The game is televised.

35. The sweater is white or the sweater is red.
The sweater is not red.
∴ The sweater is white.

36. If the canteen is full, then we can go for a walk.
We can go for a walk and we will not get thirsty.
∴ If we go for a walk, then the canteen is not full.

37. Bryce Canyon National Park is in Utah or Bryce Canyon National Park is in Arizona.
If Bryce Canyon National Park is in Arizona, then it is not in Utah.
∴ Bryce Canyon National Park is not in Arizona.

38. If you cook the meal, then I will vacuum the rug.
I will not vacuum the rug.
∴ You will not cook the meal.

39. The package is not more than 2 pounds.
If the package is more than 2 pounds, then we can mail the package.
∴ We can mail the package.

40. It is snowing and I am going skiing.
If I am going skiing, then I will wear a coat.
∴ If it is snowing, then I will wear a coat.

41. The garden has vegetables or the garden has flowers.
If the garden does not have flowers, then the garden has vegetables.
∴ The garden has flowers or the garden has vegetables.

42. If the house has electric heat, then the Flynns will buy the house.
If the price is not less than \$100,000, then the Flynns will not buy the house.
∴ If the house has electric heat, then the price is less than \$100,000.

43. If the children are young, then we will get a dog.
We will get a dog if and only if we will not get a white carpet.
∴ If the children are young, then we will not get a white carpet.

44. If there is an atmosphere, then there is gravity.
If an object has weight, then there is gravity.
∴ If there is an atmosphere, then an object has weight.

In Exercises 45–54, translate the argument into symbolic form. Then determine whether the argument is valid or invalid.

45. If the cat is in the room, then the mice are hiding. The mice are not hiding. Therefore, the cat is not in the room.

46. The television is on or the plug is not plugged in. The plug is plugged in. Therefore, the television is on.

47. If Bonnie passes the bar exam, then she will practice law. Bonnie will not practice law. Therefore, Bonnie did not pass the bar exam.

48. The test was easy and I received a good grade. The test was not easy or I did not receive a good grade. Therefore, the test was not easy.

49. The baby is crying but the baby is not hungry. If the baby is hungry then the baby is crying. Therefore, the baby is hungry.

50. If the car is new, then the car has air conditioning. The car is not new and the car has air conditioning. Therefore, the car is not new.

51. If the football team wins the game, then Dave played quarterback. If Dave played quarterback, then the team is not in second place. Therefore, if the football team wins the game, then the team is in second place.

52. The engineering courses are difficult and the chemistry labs are long. If the chemistry labs are long, then the art tests are easy. Therefore, the engineering courses are difficult and the art tests are not easy.

53. If the lights are on, then we can play ball. If the umpires are present, then we can play ball. Therefore, if the lights are on, then the umpires are present.

54. You fertilize the shrubs or the shrubs will not grow. If the shrubs do not grow, then you will not have privacy. You have privacy. Therefore, you fertilized the shrubs.

In Exercises 55–61, using the standard forms of arguments and other information you have learned, supply what you believe is a logical conclusion to the argument. Verify that the argument is valid for the conclusion you supplied.

55. If you do your homework every day, then you will get an A.
You do your homework every day.
Therefore, . . .

56. If you have lunch, then you must pay the bill.
You did not pay the bill.
Therefore, . . .

57. John reads the history assignment or he fixes the car.
John does not read the history assignment.
Therefore, . . .

58. If the electric bill is too high, then I am not able to pay the bill.
If I am not able to pay the bill, then my electricity will be shut off.
Therefore, . . .

59. If you close the deal, then you will get a commission.
You did not get a commission.
Therefore, . . .

60. If you do not read a lot, then you will not gain knowledge.
You do not read a lot.
Therefore, . . .

61. If you do not pay off your credit card bill, then you will have to pay interest.
If you have to pay interest, then the bank makes money.
Therefore, . . .

CHALLENGE PROBLEMS/GROUP ACTIVITIES

62. Determine whether the argument is valid or invalid.

If Lynn wins the contest or strikes oil, then she will be rich.
If Lynn is rich, then she will stop working.
___
∴ If Lynn does not stop working, she did not win the contest.

63. Is it possible for an argument to be invalid if the conjunction of the premises is false in every case of the truth table? Explain you answer.

RESEARCH ACTIVITIES

64. Show how logic is used in advertising. Discuss several advertisements and show how logic is used to persuade the reader.

65. Find examples of valid (or invalid) arguments in printed matter such as newspaper or magazine articles. Explain why the arguments are valid (or invalid).

3.6 EULER DIAGRAMS AND SYLLOGISTIC ARGUMENTS

In the preceding section, we showed how to determine the validity of *symbolic arguments* using truth tables and comparing the arguments to standard forms. This section presents another form of argument called a **syllogistic argument**, better known by the shorter name **syllogism**. The validity of a syllogistic argument is determined by using Euler diagrams, as is explained shortly.

Syllogistic logic, a deductive process of arriving at a conclusion, was developed by Aristotle in about 350 B.C. Aristotle considered the relationships among the four types of statements that follow.

All ______ are ______.
No ______ are ______.
Some ______ are ______.
Some ______ are not ______.

Examples of these statements are: *All doctors are tall. No doctors are tall. Some doctors are tall. Some doctors are not tall.* Since Aristotle's time, other types of statements have been added to the study of syllogistic logic, two of which are

______ is a ______.
______ is not a ______.

Examples of these statements are: *Maria is a doctor. Maria is not a doctor.*

Leonhard Euler

Swiss mathematician Leonhard Euler (1707–1783) produced such a large volume of work that all of it has not yet been compiled. It has been estimated that a complete edition of his work would make up over 60 volumes. Although his greatest contribution was made in the field of modern analysis, his work also explored physics, astronomy, trigonometry, and calculus. Euler also had a great influence on the notation and symbolism of mathematics, and it was through his work that the symbols e and π came into common use. His facility for mental calculation and his uncommon memory allowed him to continue his prolific career after he lost his vision in both eyes.

The difference between a symbolic argument and a syllogistic argument can be seen in the following chart. Symbolic arguments use the connectives *and*, *or*, *not*, *if–then*, and *if and only if*. Syllogistic arguments use the quantifiers *all*, *some*, and *none*, which were discussed in Section 3.1

Symbolic Arguments Versus Syllogistic Arguments

	Words or phrases used	*Method of determining validity*
Symbolic argument	and, or, not, if–then, if and only if	Truth tables or by comparison with standard forms of arguments
Syllogistic argument	all are, some are, none are, some are not	Euler diagrams.

As with symbolic logic, the premises and the conclusion together form an argument. An example of a syllogistic argument is

All German shepherds are dogs.
All dogs bark.
$\therefore$ All German shepherds bark.

This is an example of a valid argument. Recall from the previous section that an argument is *valid* when its conclusion necessarily follows from a given set of premises. Recall that an argument in which the conclusion does not necessarily follow from the given premises is said to be an *invalid argument* or a *fallacy*.

Before we give another example of a syllogism, let's review the Venn diagrams discussed in Section 2.3 in relationship with Aristotle's four statements.

All *A*s are *B*s	No *A*s are *B*s	Some *A*s are *B*s	Some *A*s are not *B*s
	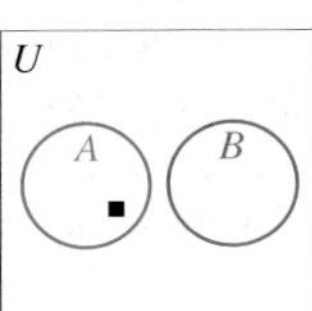	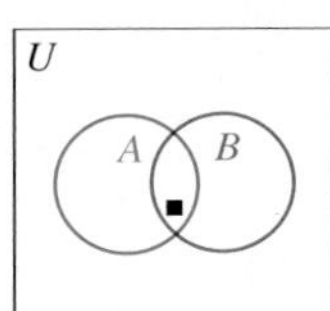	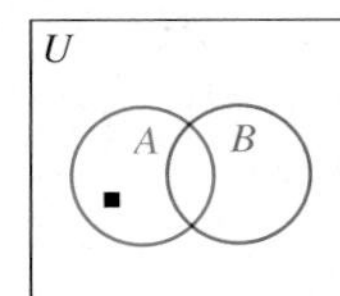
If an element is in set *A*, then it is in set *B*.	If an element is in set *A*, then it is not in set *B*.	There is at least one element that is in both set *A* and set *B*.	There is at least one element that is in set *A* that is not in set *B*

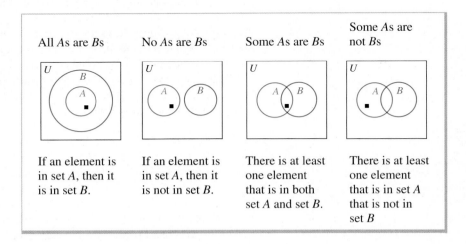

One method used to determine whether an argument is valid or is a fallacy is by means of an **Euler diagram**, named after Leonhard Euler (1707–1783) who used circles to represent sets in syllogistic arguments. The technique of using Euler diagrams is illustrated in Example 1.

EXAMPLE 1 ***Using a Euler Diagram***

Determine whether the following syllogism is valid or is a fallacy.

All horses are brown.
All brown animals have fur.
∴ All horses have fur.

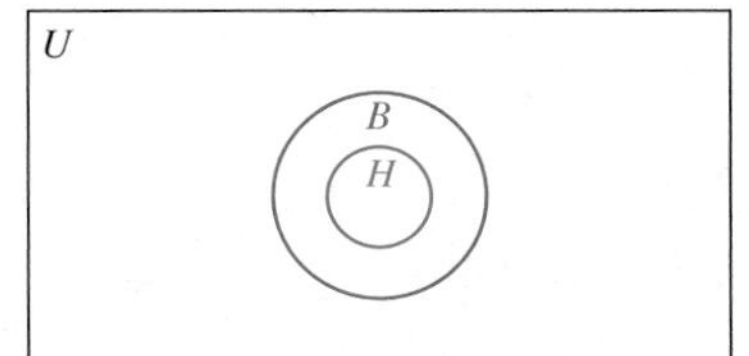

Figure 3.3

SOLUTION The statement "All horses are brown" may be interpreted as "If an animal is a horse, then it is brown." Construct the diagram and represent the first premise "All horses are brown," as shown in Fig. 3.3. The outer circle represents all brown animals, and the inner circle represents all horses. Now illustrate the second premise "All brown animals have fur," as shown in Fig. 3.4. The set containing all animals that have fur must contain all the brown animals, as illustrated in the diagram. Note that the premises force the set of horses to be within the set of animals that have fur. Therefore, the argument is valid, since the conclusion "All horses have fur" necessarily follows from the set of premises. ▲

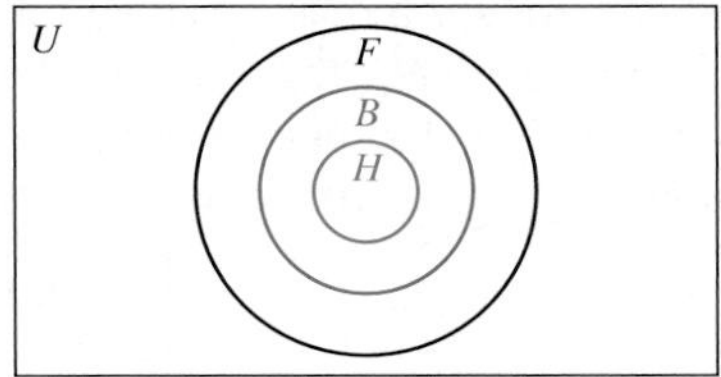

Figure 3.4

The argument in Example 1 is valid even though the conclusion "All horses have fur" is obviously a false statement. Similarly, an argument can be a fallacy even if the conclusion is a true statement.

When we determine the validity of an argument, we are determining whether the conclusion necessarily follows from the premises. When we say that an argument is valid, we are saying that if all the premises are true statements, then the conclusion must also be a true statement.

The form of the argument determines its validity, not the particular statements. For example, consider the syllogism

All earth people have two heads.
All people with two heads can fly.
∴ All earth people can fly.

The form of this argument is the same as that of the previous valid argument. Therefore, this argument is also valid.

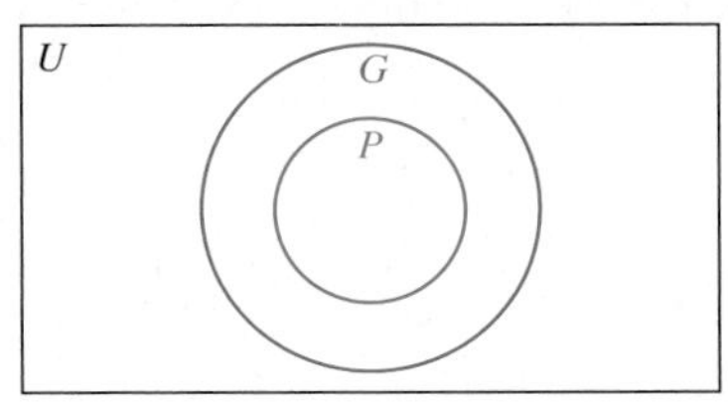

Figure 3.5

EXAMPLE 2 ***Is the Syllogism Valid?***

Determine whether the following syllogism is valid or is a fallacy.

All pilots have good vision.
Kaitlyn is a pilot.
∴ Kaitlyn has good vision.

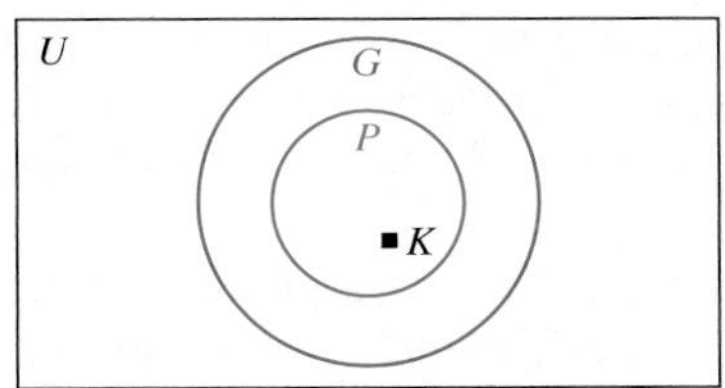

Figure 3.6

SOLUTION The statement "All pilots have good vision" is illustrated in Fig. 3.5. The second premise, "Kaitlyn is a pilot," tells us that Kaitlyn must be placed in the inner circle (see Fig. 3.6). The Euler diagram illustrates that we must accept the conclusion "Kaitlyn has good vision" as true (when we accept the premises as true). Therefore, the argument is valid. ▲

In both Example 1 and Example 2, we had no choice as to where the second premise was to be placed in the Euler diagram. In Example 1, the set of brown animals had to be placed inside the set of animals with fur. In Example 2, Kaitlyn had to be placed inside the set of people with good vision. Often when determining the truth value of a syllogism, a premise can be placed in more than one area in the diagram. *We always try to draw the Euler diagram so that the conclusion does not necessarily follow from the premises. If this can be done, then the conclusion does not necessarily follow from the premises and the argument is invalid.* If we cannot show that the argument is invalid, only then do we accept the argument as valid. We illustrate this process in Example 3.

DID YOU KNOW

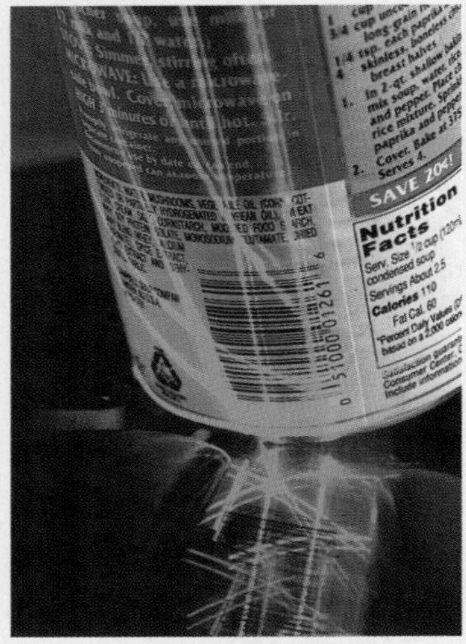

Logic, or Boolean algebra, affects each of us daily. Nowadays, many aspects of our lives are computer related. There are computer chips in our cars, watches, radios, cameras, supermarket checkouts, and most appliances, to name just a few uses. The outcomes of these computer chips are based on the *not*, *and*, and *or* truth tables we have discussed. Sometimes abstract mathematics is not applied until many years after its development, which is exactly what happened with the abstract Boolean algebra concepts considered in the seventeenth and eighteenth centuries.

EXAMPLE 3 *Is Christine a Football Player?*

Determine whether the following syllogism is valid or is a fallacy.

All football players are strong.
Christine is strong.
$\therefore$ Christine is a football player.

SOLUTION The statement "All football players are strong" is illustrated in Fig. 3.7(a). The next premise, "Christine is strong," tells us that Christine must be placed in the set of strong people. Two diagrams in which both premises are satisfied are shown in Figs. 3.7(b) and 3.7(c). By examining Fig. 3.7(b), however, we see that Christine is not a football player. Therefore, the conclusion "Christine is a football player" does not necessarily follow from the set of premises. Thus, the argument is a fallacy.

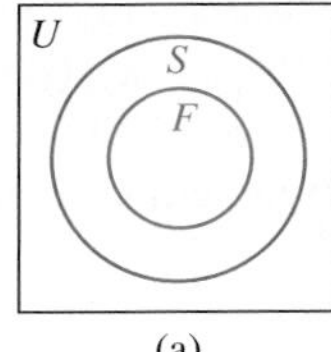

(a)

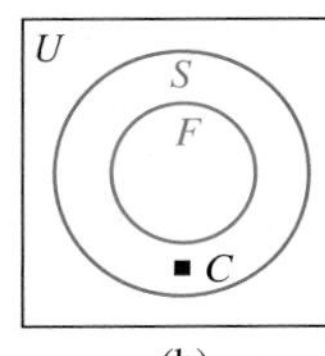

(b)

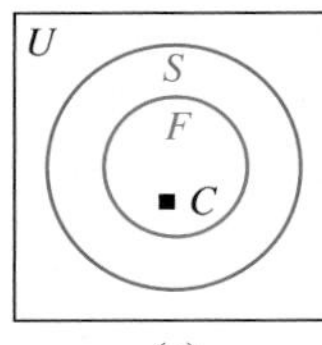

(c)

Figure 3.7

EXAMPLE 4 *Parrots and Chickens*

Determine whether the following syllogism is valid or is a fallacy.

No parrots eat chicken.
Fletch does not eat chicken.
$\therefore$ Fletch is a parrot.

SOLUTION The diagram in Fig. 3.8 satisfies the two given premises and also shows that Fletch is not a parrot. Therefore, the argument is invalid, or is a fallacy.

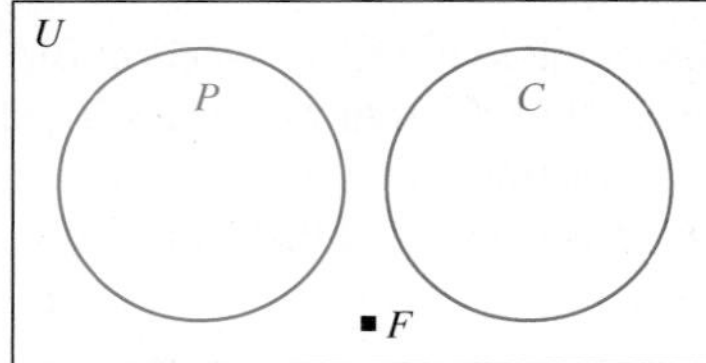

Figure 3.8

Note that in Example 4 if we placed Fletch in circle P, the argument would appear to be valid. Remember, *whenever testing the validity of an argument, always try to show that the argument is invalid.* If there is any way of showing that the conclusion does not necessarily follow from the premises, then the argument is invalid.

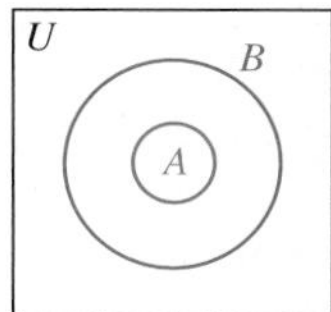

Figure 3.9

EXAMPLE 5 *A Syllogism Involving the Word Some*

Determine whether the following syllogism is valid or invalid.

All *A*s are *B*s.
Some *B*s are *C*s.
∴ Some *A*s are *C*s.

SOLUTION The statement "All *A*s are *B*s" is illustrated in Fig. 3.9. The statement "Some *B*s are *C*s" means that there is at least one *B* that is a *C*. We can illustrate this set of premises in four ways, as illustrated in Fig. 3.10.

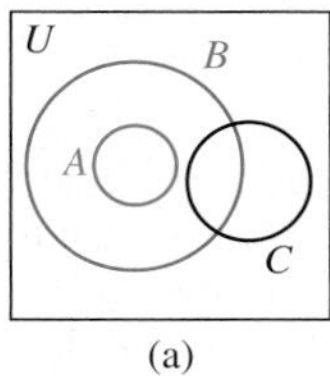

(a)

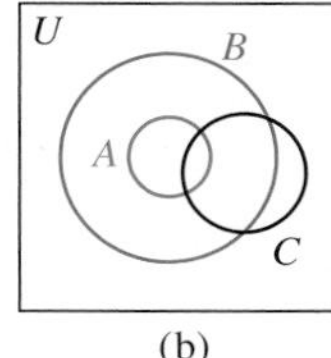

(b)

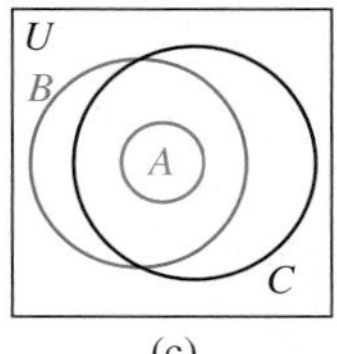

(c)

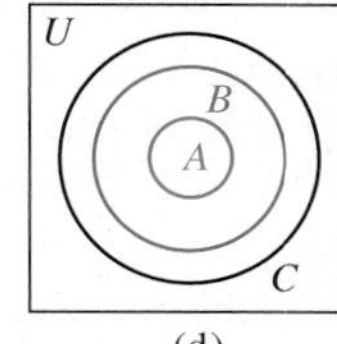

(d)

Figure 3.10

In all four illustrations, we see that (1) all *A*s are *B*s and (2) some *B*s are *C*s. The conclusion is "Some *A*s are *C*s." Since at least one of the illustrations, Fig. 3.10(a), shows that the conclusion does not necessarily follow from the given premises, the argument is invalid. ▲

EXAMPLE 6 *Neurosurgeons and Ice Cream*

Determine whether the following syllogism is valid or invalid.

No neurosurgeons eat ice cream.
All college graduates eat ice cream.
∴ No neurosurgeons are college graduates.

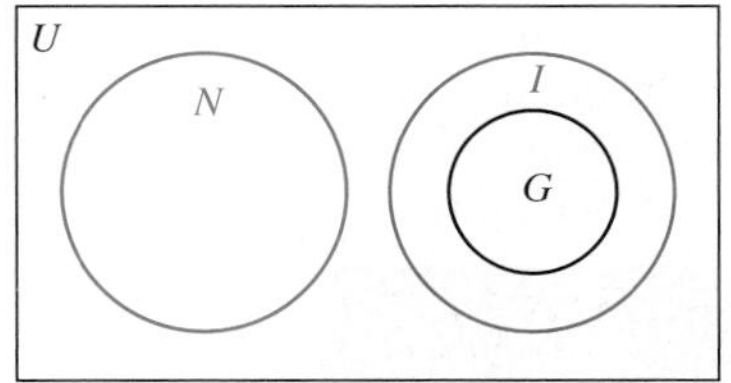

Figure 3.11

SOLUTION The first premise tells us that the neurosurgeons and the people who eat ice cream are disjoint sets, as illustrated in Fig. 3.11. The second premise tells us that the set of college graduates is a subset of the people who eat ice cream.

The set of college graduates and neurosurgeons cannot be made to intersect without violating a premise. Thus, no neurosurgeons can be college graduates, and the syllogism is valid. Note that we did not say that this conclusion was true, only that the argument was valid. ▲

SECTION 3.6 EXERCISES

CONCEPT/WRITING EXERCISES

1. What is another name for a syllogistic argument?
2. What four types of statements did Aristotle consider when he developed syllogistic arguments?
3. What does it mean when we determine that an argument is valid?
4. Explain the differences between a symbolic argument and a syllogistic argument.
5. Can an argument be valid if the conclusion is a false statement? Explain your answer.
6. Can an argument be invalid if the conclusion is a true statement? Explain.

PRACTICE THE SKILLS/PROBLEM SOLVING

In Exercises 7–30, use a Euler diagram to determine whether the syllogism is valid or invalid.

7. All parrots talk.
Chicklet is a parrot.
∴ Chicklet talks.

8. All kangaroos jump.
All things that jump have wings.
∴ All kangaroos have wings.

9. No dogs are cats.
All poodles are dogs.
∴ No poodles are cats.

10. All *A*s are *B*s.
All *B*s are *C*s.
∴ All *A*s are *C*s.

11. All living things breathe.
An oak tree breathes.
∴ An oak tree is a living thing.

12. All doctors have college degrees.
Tong has a college degree.
∴ Tong is a doctor.

13. No sharks bite.
Hammerheads are sharks.
∴ Hammerheads do not bite.

14. No basketball players are greater than 8 feet tall.
Pete is not a basketball player.
∴ Pete is greater than 8 feet tall.

15. Some animals are dangerous.
A lion is an animal.
∴ A lion is dangerous.

16. Some politicians are stuffy.
Todd Hall is a politician.
∴ Todd Hall is not stuffy.

17. Some medical researchers have made important discoveries.
Some people who have made important discoveries are famous.
∴ Some medical researchers are famous.

18. Some professional golfers give golf lessons.
All people who belong to the PGA are professional golfers.
∴ All people who belong to the PGA give golf lessons.

19. No tennis players are wrestlers.
Allison is not a wrestler.
∴ Allison is a tennis player.

20. Some soaps float.
All things that float are lighter than water.
∴ Some soaps are lighter than water.

21. Some people love mathematics.
All people who love mathematics love physics.
∴ Some people love physics.

22. Some desks are made of wood.
All paper is made of wood.
∴ Some desks are made of paper.

23. No *x*s are *y*s.
No *y*s are *z*s.
∴ No *x*s are *z*s.

24. All pilots can fly.
All astronauts can fly.
∴ Some pilots are astronauts.

25. Some dogs wear glasses.
Fido wears glasses.
∴ Fido is a dog.

26. All rainy days are cloudy.
Today it is cloudy.
∴ Today is a rainy day.

27. All sweet things taste good.
All things that taste good are fattening.
All things that are fattening put on pounds.
∴ All sweet things put on pounds.

28. All books have red covers.
All books that have red covers contain 200 pages.
Some books that contain 200 pages are novels.
∴ All books that contain 200 pages are novels.

29. All country singers play the guitar.
All country singers play the drums.
Some people who play the drums are rock singers.
∴ Some country singers are rock singers.

30. Some hot dogs are made of turkey.
All things made of turkey are edible.
Some things that are made of beef are edible.
∴ Some hot dogs are made of beef.

CHALLENGE PROBLEM/GROUP ACTIVITY

31. Statements in logic can be translated into set statements: for example, $p \wedge q$ is similar to $P \cap Q$; $p \vee q$ is similar to $P \cup Q$; and $p \rightarrow q$ is equivalent to $\sim p \vee q$, which is similar to $P' \cup Q$. Euler diagrams can also be used to show that arguments similar to those discussed in Section 3.5 are valid or invalid. Use Euler diagrams to show that the symbolic argument is invalid.

$$\begin{array}{l} p \rightarrow q \\ \underline{p \vee q} \\ \therefore \sim p \end{array}$$

RESEARCH ACTIVITY

32. Leonhard Euler is considered one of the greatest mathematicians of all time. Do research and write a report on Euler's life. Include information on his contributions to sets and to logic. Also indicate other areas of mathematics in which he made important contributions. References include encyclopedias, history of mathematics books, and the Internet.

CHAPTER 3 SUMMARY

IMPORTANT FACTS

Quantifiers

Form of statement	Form of negation
All are.	Some are not.
None are.	Some are.
Some are.	None are.
Some are not.	All are.

Summary of connectives

Formal name	Symbol	Read	Symbolic form
Negation	$\sim$	not	$\sim p$
Conjunction	$\wedge$	and	$p \wedge q$
Disjunction	$\vee$	or	$p \vee q$
Conditional	$\rightarrow$	if–then	$p \rightarrow q$
Biconditional	$\leftrightarrow$	if and only if	$p \leftrightarrow q$

Basic truth tables

Negation

p	$\sim p$
T	F
F	T

p	q	Conjunction $p \wedge q$	Disjunction $p \vee q$	Conditional $p \rightarrow q$	Biconditional $p \leftrightarrow q$
T	T	T	T	T	T
T	F	F	T	F	F
F	T	F	T	T	F
F	F	F	F	T	T

De Morgan's laws

$\sim(p \wedge q) \Leftrightarrow \sim p \vee \sim q$

$\sim(p \vee q) \Leftrightarrow \sim p \wedge \sim q$

Other equivalent forms

$p \rightarrow q \Leftrightarrow \sim p \vee q$

$\sim(p \rightarrow q) \Leftrightarrow p \wedge \sim q$

$p \leftrightarrow q \Leftrightarrow [(p \rightarrow q) \wedge (q \rightarrow p)]$

Variations of the conditional statement

Name	Symbolic form	Read
Conditional	$p \rightarrow q$	If p, then q.
Converse of the conditional	$q \rightarrow p$	If q, then p.
Inverse of the conditional	$\sim p \rightarrow \sim q$	If not p, then not q.
Contrapositive of the conditional	$\sim q \rightarrow \sim p$	If not q, then not p.

Standard forms of arguments
Valid arguments

Law of detachment	Law of contra-position	Law of syllogism	Disjunctive syllogism
$p \rightarrow q$	$p \rightarrow q$	$p \rightarrow q$	$p \vee q$
p	$\sim q$	$q \rightarrow r$	$\sim p$
$\therefore q$	$\therefore \sim p$	$\therefore p \rightarrow r$	$\therefore q$

Invalid arguments

Fallacy of the converse	Fallacy of the inverse
$p \rightarrow q$	$p \rightarrow q$
q	$\sim p$
$\therefore p$	$\therefore \sim q$

Symbolic argument vs. syllogistic argument

	Words or phrases used	Method of determining validity
Symbolic argument	and, or, not, if–then, if and only if	Truth tables or by comparison with standard forms of arguments
Syllogistic argument	all are, some are, none are, some are not	Euler diagrams

CHAPTER 3 REVIEW EXERCISES

3.1

In Exercises 1–6, write the negation of the statement.

1. Some people drink milk.
2. Some dogs do not have fleas.
3. No butterflies bite.
4. Some locks are keyless.
5. All pens use ink.
6. No rabbits wear glasses.

In Exercises 7–12, write each compound statement in words.

p: The coffee is Maxwell House.
q: The coffee is hot.
r: The coffee is strong.

7. $p \vee q$
8. $\sim q \wedge r$
9. $p \leftrightarrow \sim r$
10. $q \rightarrow (r \wedge \sim p)$
11. $(p \vee \sim q) \wedge \sim r$
12. $\sim p \leftrightarrow (r \wedge \sim q)$

3.2

In Exercises 13–18, use the statements for p, q, and r as in Exercises 7–12 to write the statement in symbolic form.

13. If the coffee is Maxwell House, then it is strong.
14. The coffee is strong and the coffee is hot.
15. If the coffee is strong then the coffee is hot, or the coffee is not Maxwell House.
16. The coffee is hot if and only if the coffee is Maxwell House, and the coffee is not strong.
17. The coffee is strong and the coffee is hot, or the coffee is not Maxwell House.
18. It is false that the coffee is strong and the coffee is hot.

In Exercises 19–24, construct a truth table for the statement.

19. $(p \vee q) \wedge \sim p$
20. $q \leftrightarrow (p \vee \sim q)$
21. $p \wedge (\sim q \vee r)$
22. $p \rightarrow (q \wedge \sim r)$
23. $(p \vee q) \leftrightarrow (p \vee r)$
24. $(p \wedge q) \rightarrow \sim r$

3.2, 3.3

In Exercises 25–28, determine the truth value of the statement.

25. If $4 - 1 = 3$, then $2 + 2 = 3$.
26. The St. Louis arch is in St. Louis or Abraham Lincoln is buried in Grant's Tomb.

27. If George Washington was the first president of the United States or all mushrooms are edible, then Florida is south of Central America.
28. $3 + 7 = 11$ or $6 + 5 = 11$, and $7 \cdot 6 = 42$.

In Exercises 29 and 30, use the following chart to determine the truth value of each simple statement. Then determine the truth value of the component statement.

National Basketball Association

Most points, career

	Points	Years	Average
Michael Jordan	5,987	13	33.4
Kareem Abdul-Jabbar	5,762	18	24.3
Jerry West	4,457	13	29.1
Larry Bird	3,897	12	23.8
John Havlicek	3,776	13	22.0
Magic Johnson	3,701	13	19.5
Karl Malone	3,691	13	26.9
Hakeem Olajuwon	3,674	13	27.0
Elgin Baylor	3,623	12	27.0
Wilt Chamberlain	3,607	13	22.5

Source: *1999 Sports Illustrated Almanac*

29. Michael Jordan averaged 33.4 points per game if and only if Jerry West averaged 29.1 points per game, or Wilt Chamberlain played for 10 years.

30. Magic Johnson played for 15 years, or if Kareem Abdul-Jabbar scored 6000 points then Elgin Baylor played for 12 years.

3.3

In Exercises 31–34, determine the truth value of the statement when p is T, q is F, and r is F.

31. $(p \vee q) \leftrightarrow (\sim r \wedge p)$

32. $(p \rightarrow \sim r) \vee (p \wedge q)$

33. $\sim r \leftrightarrow [(p \vee q) \leftrightarrow \sim p]$

34. $\sim[(q \wedge r) \rightarrow (\sim p \vee r)]$

3.4

In Exercises 35–38, determine whether the pairs of statements are equivalent. You may use De Morgan's laws, the fact that $(p \rightarrow q) \Leftrightarrow (\sim p \vee q)$, truth tables, or equivalent forms of the conditional statement.

35. $\sim p \rightarrow \sim q \qquad p \vee \sim q$

36. $\sim p \vee \sim q \qquad \sim p \leftrightarrow q$

37. $\sim p \vee (q \wedge r) \qquad (\sim p \vee q) \wedge (\sim p \vee r)$

38. $(\sim q \rightarrow p) \wedge p \qquad \sim(\sim p \leftrightarrow q) \vee p$

In Exercises 39–43, use De Morgan's laws or the fact that $(p \rightarrow q) \Leftrightarrow (\sim p \vee q)$ to write an equivalent statement for the given statement.

39. The stapler is empty or the stapler is jammed.

40. The boy sang bass and the girl sang alto.

41. It is not true that *Newsweek* is a comic book or *Time* is not an almanac.

42. If there is not water in the vase, then the flowers will wilt.

43. I did not go to the party and I did not finish my special report.

In Exercises 44–47, write the contrapositive for the statement.

44. If the railroad crossing light is flashing red, then you must stop.

45. If John is having difficulty seeing, then John's eyes must be checked.

46. If today is not a holiday, then I will be at work.

47. If the carpet is Scotch-Guarded and properly cared for, then it will not stain.

48. Write the converse, inverse, and contrapositive of the conditional statement "If I study, then I will get a passing grade."

In Exercises 49–52, determine which, if any, of the three statements are equivalent.

49. a) If the temperature is over 80°, then the air conditioner will come on.
b) The temperature is not over 80° or the air conditioner will come on.
c) It is false that the temperature is over 80° and the air conditioner will not come on.

50. a) The screwdriver is on the workbench if and only if the screwdriver is not on the counter.
b) If the screwdriver is not on the counter, then the screwdriver is not on the workbench.
c) It is false that the screwdriver is on the counter and the screwdriver is not on the workbench.

51. a) If $2 + 3 = 6$, then $3 + 1 = 5$.
b) $2 + 3 = 6$ if and only if $3 + 1 \neq 5$.
c) If $3 + 1 \neq 5$, then $2 + 3 \neq 6$.

52. a) If the sale is on Tuesday and I have money, then I will go to the sale.
b) If I go to the sale, then the sale is on Tuesday and I have money.
c) I go to the sale, or the sale is on Tuesday and I have money.

3.5, 3.6

In Exercises 53–58, determine whether the argument is valid.

53.
$$\begin{array}{l} p \rightarrow q \\ \sim p \\ \hline \therefore q \end{array}$$

54.
$$\begin{array}{l} p \wedge q \\ q \rightarrow r \\ \hline \therefore p \rightarrow r \end{array}$$

55. Nicole is in the hot tub or she is in the shower.
Nicole is in the hot tub.
∴ Nicole is not in the shower.

56. If the car has a sound system, then Rick will buy the car. If the price is not less than $18,000, then Rick will not buy the car. Therefore, if the car has a sound system, then the price is less than $18,000.

57. All grasshoppers are green.
Some crickets are green.
∴ Some crickets are grasshoppers.

58. Some driveways are yellow.
All umbrellas are yellow.
∴ Some umbrellas are driveways .

CHAPTER 3 TEST

In Exercises 1–3, write the statement in symbolic form.

p: Celion is the president.

q: Sheldon is the vice president.

r: Ron is the secretary.

1. Celion is the president but Ron is the secretary, or Sheldon is not the vice president.

2. If Ron is the secretary then Sheldon is the vice president, or Celion is not the president.

3. It is false that Ron is the secretary if and only if Sheldon is not the vice president.

In Exercises 4 and 5, use p, q, and r as above to write each symbolic statement in words.

4. $\sim(p \rightarrow \sim r)$ **5.** $p \leftrightarrow (q \wedge r)$

In Exercises 6 and 7, construct a truth table for the given statement.

6. $[\sim(p \rightarrow r)] \wedge q$ **7.** $(q \leftrightarrow \sim r) \vee p$

In Exercises 8 and 9, find the truth value of the statement.

8. $2 + 6 = 8$ or $7 - 12 = 5$.

9. A scissor can cut paper or a dime has the same value as two nickels, if and only if Louisville is a city in Kentucky.

In Exercises 10 and 11, given that p is true, q is false, and r is true, determine the truth value of the statement.

10. $[\sim(r \rightarrow \sim p)] \wedge (q \rightarrow p)$

11. $(r \vee q) \leftrightarrow (p \wedge \sim q)$

12. Determine whether the pair of statements are equivalent.

$$\sim p \vee q \qquad \sim(p \wedge \sim q)$$

In Exercises 13 and 14, determine which, if any, of the three statements are equivalent.

13. a) If the bird is red, then it is a cardinal.
b) The bird is not red or it is a cardinal.
c) If the bird is not red, then it is not a cardinal.

14. a) It is not true that the test is today or the concert is tonight.
b) The test is not today and the concert is not tonight.
c) If the test is not today, then the concert is not tonight.

15. Translate the following argument into symbolic form. Determine whether the argument is valid or invalid by comparing the argument to a recognized form or by using a truth table.

If the soccer team wins the game, then Sue played fullback. If Sue played fullback, then the team is in second place. Therefore, if the soccer team wins the game, then the team is in second place.

16. Use a Euler diagram to determine whether the syllogism is valid or is a fallacy.

All cars have engines.
Some things with engines use gasoline.
∴ Some cars use gasoline.

In Exercises 17 and 18, write the negation of the statement.

17. All leopards are spotted.

18. Some people are funny.

19. The conditional statement "If the apple is red, then it is a delicious apple" is given. Write the inverse, converse, and contrapositive of the statement.

20. Is it possible for an argument to be valid when the conclusion is a false statement? Explain your answer.

GROUP PROJECTS

SWITCHING CIRCUITS

1. An application of logic is *switching circuits.* There are two basic types of electric circuits: *series circuits* and *parallel circuits.* In a series circuit, the current can flow in only one path; see Fig. 3.12. In a parallel circuit the current can flow in more than one path; see Fig. 3.13.

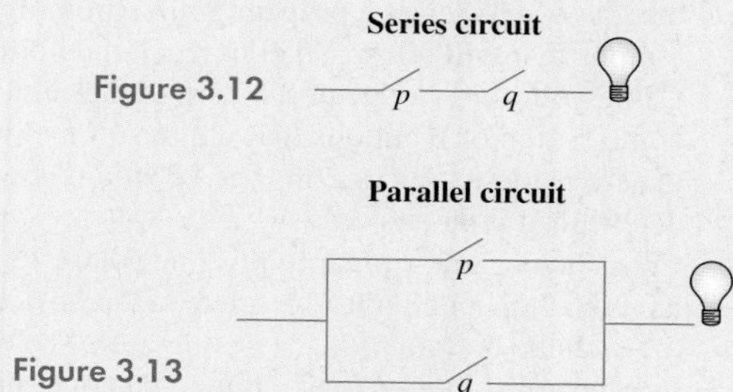

Figure 3.12

Figure 3.13

In Figs. 3.12 and 3.13, the p and q represent switches that may be opened or closed. In the series circuit in Fig. 3.12, if both switches are closed, the current will reach the bulb and the bulb will light. In the parallel circuit in Fig. 3.13, if either switch p or switch q is closed, or if both switches are closed, the current will reach the bulb and the bulb will light.

a) How many different open/closed arrangements of the two switches in Fig. 3.12 are possible? List all the possibilities.

b) Series circuits are represented using conjunctions. The circuit in Fig. 3.12 may be represented as $p \wedge q$. Construct a four-row truth table to represent the series circuit. Construct the table with columns for p, q, and $p \wedge q$. The statement $p \wedge q$ represents the outcome of the circuit (either the bulb lighting or the bulb not lighting). Represent a closed switch with the number 1, an open switch with the number 0, the bulb lighting with the number 1, and the bulb not lighting with the number 0. For example, if both switches are closed, the bulb will light, and so we write the first row of the truth table as

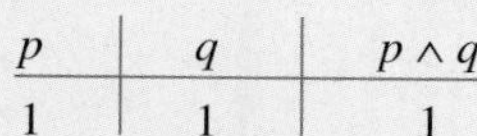

p	q	$p \wedge q$
1	1	1

c) How is the truth table determined in part (b) similar to the truth table for $p \wedge q$ discussed in earlier sections of this chapter?

d) Parallel circuits are represented using disjunctions. The circuit in Fig. 3.13 may be represented as $p \vee q$. Construct a truth table to represent the parallel circuit. Construct the table with columns for p, q, and $p \vee q$. The statement $p \vee q$ represents the outcome of the circuit (either the bulb lighting or the bulb not lighting). Use 1's and 0's as indicated in part (b).

e) How is the truth table determined in part (d) similar to the truth table for $p \vee q$ discussed in earlier sections of this chapter?

f) Represent the following circuit as a symbolic logic statement using parentheses. Explain how you determined your answer.

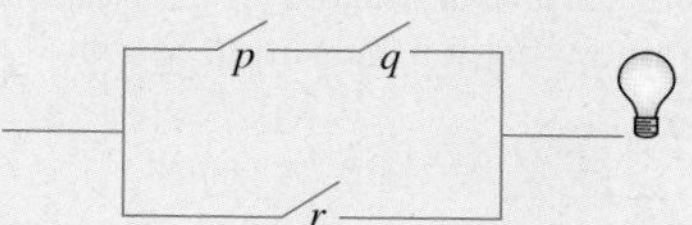

g) Draw a circuit to represent the logic statement $p \wedge (q \vee r)$. Explain how you determined your answer.

COMPUTER GATES

2. Gates in computers work on the same principles as switching circuits. The three basic types of gate are the NOT gate, the AND gate, and the OR gate. Each is illustrated along with a table that indicates current flow entering and exiting the gate. If current flows into a NOT gate, then no current exits, and vice versa. Current exits an AND gate only when both inputs have a current flow. Current exits an OR gate if current flows through either, or both, inputs. In the table, a 1 represents a current flow and a 0 indicates no current flow. For example, in the AND gate, if there is a current flow in input A (I_a has a value of 1) and no current flow in input B (I_b has a value of 0), there is no current flow in the output (O has a value of 0); see row 2 of the AND Gate table.

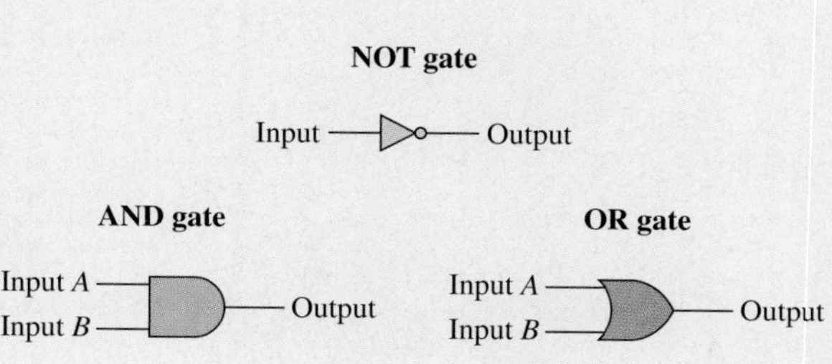

NOT gate

I	O
1	0
0	1

AND gate

I_a	I_b	O
1	1	1
1	0	0
0	1	0
0	0	0

OR gate

I_a	I_b	O
1	1	1
1	0	1
0	1	1
0	0	0

a) If 1 is considered true and 0 is considered false, explain how these tables are similar to the *not, and,* and *or* truth tables.

For the inputs indicated in the following figures determine whether the output is 1 or 0.

b)

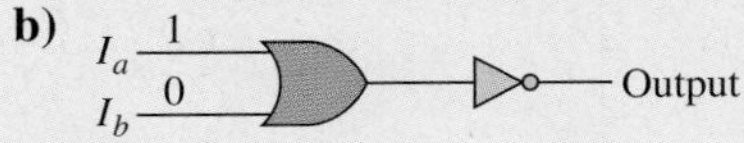

c)

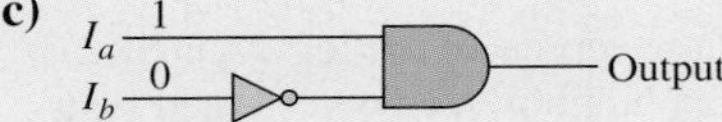

d)

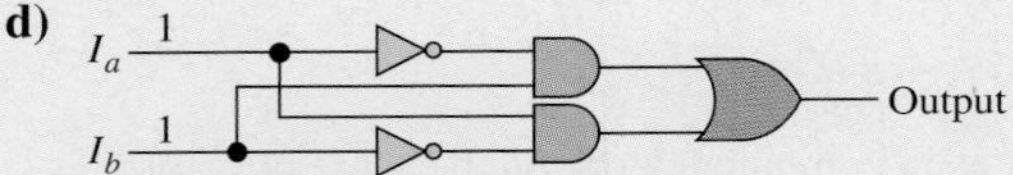

e) What values for I_a and I_b will give an output of 1 in the figure in part (d)? Explain how you determined your answer.

f) Construct a truth table using 1's and 0's for the following gate. Your truth table should have columns I_a, I_b, and O and should indicate the four possible cases for the inputs and each corresponding output.

LOGIC GAME

3. **a)** Shown is a photograph of a logic game at the Ontario Science Centre. There are 12 balls on top of the game board, numbered from left to right, with ball 1 on the extreme left and ball 12 on the extreme right. On the platform in front of the players are 12 buttons, one corresponding to each of the balls. When 6 buttons are pushed, the 6 respective balls are released. When 1 or 2 balls reach an *and* gate or an *or* gate, a single ball may or may not pass through the gate. The object of the game is to select a proper combination of 6 buttons that will allow 1 ball to reach the bottom. Using your knowledge of *and* and *or,* select a combination of 6 buttons that will result in a win. (There is more than one answer.) Explain how you determined your answer.

b) Construct a game similar to this one where 15 balls are at the top and 8 must be selected to allow 1 ball to reach the bottom.

c) Indicate all solutions to the game you constructed in part (c).

CHAPTER 4

Systems of Numeration

The number system we use—called the Hindu–Arabic system—seems to be a permanent, unchanging means of communicating quantities. However, just as languages evolve over time, so do numerical symbols that represent numbers.

Mathematics began with the practical problem of counting and record keeping. People had to count their herds, the passage of days, and objects of barter. They used physical objects—stones, shells, fingers—to represent the objects counted.

As primitive cultures grew from villages to cities, the complexity of human activities increased. Now people needed better ways of recording and communicating. It was a revolutionary step when people started using physical objects to represent not only specific objects like sheep and grain, but also the concept of pure quantity. The later invention of counting boards, the abacus, and wooden or ivory rods made it easier to work with large numbers.

Through the course of human history, the evolution of numeration systems has expanded our knowledge and abilities for record keeping, communication, and computation. As a society's numeration system changes, so do the capabilities of that society. Without an understanding of the binary number system, the computer as we know it today could not exist. Without the computer, our lifestyle would not be as it is today. ■

One of the earliest reasons that human beings needed numbers was for reckoning time, marking off days in the lunar month, so the seasonal changes that dictated human activity could be anticipated. The Mayans, the Egyptians, and the ancient Britons constructed monumental stone observatories that enabled them to mark the passage of the seasons, especially the summer solstice, using the alignment of the sun as a guide.

4.1 ADDITIVE, MULTIPLICATIVE, AND CIPHERED SYSTEMS OF NUMERATION

Just as the first attempts to write were made long after the development of speech, the first representation of numbers by symbols came long after people had learned to count. A tally system using physical objects, such as scratch marks in the soil or on a stone, notches on a stick, pebbles, or knots on a vine, was probably the earliest method of recording numbers.

In primitive societies, such a tally system adequately served the limited need for recording livestock, agriculture, or whatever was counted. As civilization developed, however, more efficient and accurate methods of calculating and keeping records were needed. Because tally systems are impractical and inefficient, societies developed symbols to replace them. For example, the Egyptians used the symbol ∩ and the Babylonians used the symbol ≺ to represent the number we symbolize by 10.

A **number** is a quantity, and it answers the question "How many?" A **numeral** is a symbol such as ∩, ≺, and 10 used to represent the number. We think a number but write a numeral. The distinction between number and numeral will be made here only if it is helpful to the discussion.

In language, relatively few letters of the alphabet are used to construct a large number of words. Similarly, in arithmetic, a small variety of numerals can be used to represent all numbers. In general, when writing a number, we use as few numerals as possible. One of the greatest accomplishments of humankind has been the development of systems of numeration, whereby all numbers are "created" from a few symbols. Without such systems, mathematics would not have developed to its present level.

> A **system of numeration** consists of a set of numerals and a scheme or rule for combining the numerals to represent numbers.

Four types of numeration systems used by different cultures are the topic of this chapter. They are additive (or repetitive), multiplicative, ciphered, and place-value systems. You do not need to memorize all the symbols, but you should understand the principles behind each system. By the end of this chapter, we hope that you better understand the system we use, the **Hindu–Arabic system**, and its relationship to other types of systems.

DID YOU KNOW

An Important Discovery

Archaeologist Denise Schmandt-Besserat made a breakthrough discovery about early systems of numeration. She realized that the little clay geometric objects that had been found in many archaeological sites had actually been used by people to account for their goods. Later in history, these tokens were impressed on a clay tablet to represent quantities, the beginning of writing.

ADDITIVE SYSTEMS

An additive system is one in which the number represented by a particular set of numerals is simply the sum of the values of the numerals. The additive system of numeration is one of the oldest and most primitive types of numeration systems. One of the first additive systems, the Egyptian hieroglyphic system, dates back to about 3000 B.C. The Egyptians used symbols for the powers of 10: 10^0 or 1, 10^1 or 10, 10^2 or $10 \cdot 10$, 10^3 or $10 \cdot 10 \cdot 10$, and so on. Table 4.1 lists the Egyptian hieroglyphic numerals with the equivalent Hindu–Arabic numerals.

To write the number 600 in Egyptian hieroglyphics, we write the numeral for 100 six times: 999999.

DID YOU KNOW

The Rhind Papyrus

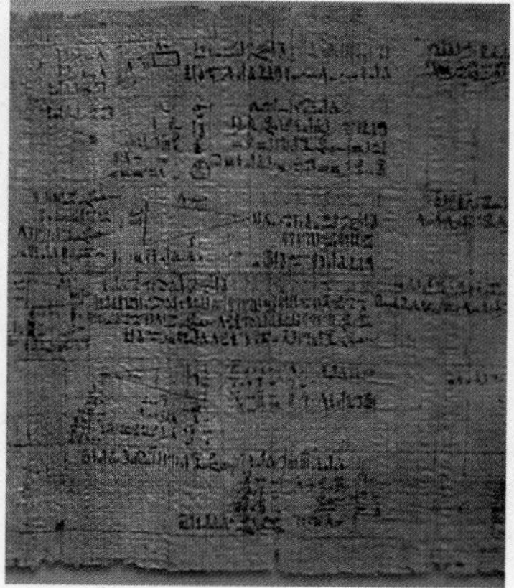

Much of our knowledge of Egyptian mathematics and numeration comes from a roll of papyrus measuring 18 ft by 13 in. It was discovered in a shop in Luxor, Egypt, in 1858 by Henry Rhind, a Scottish lawyer turned archaeologist. Unlike the straightforward accounting of property and events common to Egyptian tombs, the Rhind Papyrus has inscribed on it 85 mathematical problems and solutions involving addition, subtraction, multiplication, division, and geometry.

The key to translation of the papyrus was the Rosetta Stone, which had been discovered some 60 years earlier by one of Napoleon's officers. Both of these treasures now reside at the British Museum.

Table 4.1 Egyptian Hieroglyphics

Hindu–Arabic numerals	Egyptian numerals	Description
1	|	Staff (vertical stroke)
10	∩	Heel bone (arch)
100	𓍢	Scroll (coiled rope)
1,000	𓆼	Lotus flower
10,000	𓂭	Pointing finger
100,000	𓆐	Tadpole (or whale)
1,000,000	𓁨	Astonished person

EXAMPLE 1 ***From Egyptian to Hindu–Arabic Numerals***

Write the following numeral as a Hindu–Arabic numeral.

𓂭𓂭𓍢𓍢𓍢|

SOLUTION 10,000 + 10,000 + 100 + 100 + 100 + 1 = 20,301 ▲

EXAMPLE 2 ***From Hindu–Arabic to Egyptian Numerals***

Write 43,628 as an Egyptian numeral.

SOLUTION

43,628 = 40,000 + 3,000 + 600 + 20 + 8

𓂭𓂭𓂭𓂭𓆼𓆼𓆼𓍢𓍢𓍢𓍢𓍢𓍢∩∩|||||||| ▲

In this system, the order of the symbols is not important. For example, 𓆐𓍢𓍢||∩ and ||𓍢𓍢𓆐∩ both represent 100,212.

Users of additive systems easily accomplished addition and subtraction by combining or removing symbols. Multiplication and division were more difficult; they were performed by a process called *duplation and mediation* (see Section 4.5). The Egyptians had no symbol for zero, but they did have an understanding of fractions. The symbol 𓂋 was used to take the reciprocal of a number; thus, 𓂋||| meant $\frac{1}{3}$ and 𓂋∩| was $\frac{1}{11}$. Writing large numbers in the Egyptian system would have taken longer than in other systems because so many symbols had to be listed. For example, 45 symbols are needed to represent the number 99,999.

The Roman numeration system, a second example of an additive system, was developed later than the Egyptian system. Roman numerals (Table 4.2) were used in most European countries until the eighteenth century. They are still commonly seen on buildings, on clocks, and in books. Roman numerals are selected letters of the Roman alphabet.

Table 4.2 Roman Numerals

Roman numerals	I	V	X	L	C	D	M
Hindu–Arabic numerals	1	5	10	50	100	500	1000

The Roman system has two advantages over the Egyptian system. The first is that it uses the subtraction principle as well as the addition principle. Starting from the left, we add each numeral unless its value is smaller than the value of the numeral to its

DID YOU KNOW

Roman Numerals

Roman numerals remained popular on large clock faces long after their disappearance from daily transactions because they are easier to read from a distance than Hindu–Arabic numerals.

right. In that case, we subtract it from that numeral. Only the numbers 1, 10, 100, 1000, . . . can be subtracted, and they can only be subtracted from the next two higher numbers. For example, C (100) can be subtracted only from D (500) or M (1000). The symbol DC represents 500 + 100, or 600, and CD represents 500 − 100, or 400. Similarly, MC represents 1000 + 100, or 1100, and CM represents 1000 − 100, or 900.

EXAMPLE 3 *A Roman Numeral*

Write MCCLXII as a Hindu–Arabic numeral.

SOLUTION Since each numeral is larger than the one on its right, no subtraction is necessary.

$$\text{MCCLXII} = 1000 + 100 + 100 + 50 + 10 + 1 + 1 = 1262$$

▲

EXAMPLE 4 *A Roman Numeral Involving a Subtraction*

Write DCXLVI as a Hindu–Arabic numeral.

SOLUTION Checking from left to right, we see that X (10) has a smaller value than L (50). Therefore, XL represents 50 − 10, or 40.

$$\text{DCXLVI} = 500 + 100 + (50 - 10) + 5 + 1 = 646$$

▲

EXAMPLE 5 *Writing a Roman Numeral*

Write 289 as a Roman numeral.

SOLUTION

$$289 = 200 + 80 + 9 = 100 + 100 + 50 + 10 + 10 + 10 + 9$$

(Nine is treated as 10 − 1.)

$$289 = \text{CCLXXXIX}$$

▲

In the Roman numeration system, a symbol does not have to be repeated more than three consecutive times. For example, the number 646 would be written DCXLVI instead of DCXXXXVI.

The second advantage of the Roman numeration system over the Egyptian system is that it makes use of the multiplication principle for numbers over 1000. A bar above a symbol or group of symbols indicates that the symbol or symbols are to be multiplied by 1000. Thus, $\overline{\text{V}} = 5 \times 1000 = 5000$, $\overline{\text{X}} = 10 \times 1000 = 10{,}000$, and $\overline{\text{CD}} = 400 \times 1000 = 400{,}000$. This greatly reduces the number of symbols needed to write large numbers. Still, it requires 19 symbols, including the bar, to write the number 33,888.

MULTIPLICATIVE SYSTEMS

Multiplicative numeration systems are more similar than additive systems to our Hindu–Arabic system. The number 642 in a multiplicative system might be written (6) (100) (4) (10) (2) or

6
100
4
10
2

Note that no addition signs are needed to represent the number. From this illustration, try to formulate a rule explaining how multiplicative systems work.

The principle example of a multiplicative system is the traditional Chinese system. The numerals used in this system are given in Table 4.3.

Table 4.3 Traditional Chinese Numerals

Traditional Chinese numerals	一	二	三	四	五	六	七	八	九	十	百	千
Hindu–Arabic numerals	1	2	3	4	5	6	7	8	9	10	100	1000

Chinese numerals are always written vertically. The number on top will be a number from 1 to 9 inclusive. This number is to be multiplied by the power of 10 below it. The number 20 is written

$$\left.\begin{array}{c}\text{二}\\ \text{十}\end{array}\right\} 2 \times 10 = 20$$

The number 400 is written

$$\left.\begin{array}{c}\text{四}\\ \text{百}\end{array}\right\} 4 \times 100 = 400$$

EXAMPLE 6 ***A Traditional Chinese Numeral***

Write 428 as a Chinese numeral.

SOLUTION

$$428 = \left\{\begin{array}{lcl} 400 = & \left\{\begin{array}{r}4\\100\end{array}\right. & \begin{array}{c}\text{四}\\ \text{百}\end{array} \\ 20 = & \left\{\begin{array}{r}2\\10\end{array}\right. & \begin{array}{c}\text{二}\\ \text{十}\end{array} \\ 8 = & \;\;8 & \text{八} \end{array}\right.$$

Note that in Example 6 the units digit, the 8, is not multiplied by a power of the base.

Have you noticed that there is no symbol for zero in the Chinese system? Why is a symbol for zero not needed?

The number system used today in China is different from the traditional system. The present-day system is a positional-value system rather than a multiplicative system and uses the symbol 0 for zero.

DID YOU KNOW

The Hindu–Arabic system of numeration is used throughout the world. In many countries, numbers are displayed in Hindu–Arabic numerals as well as those used in the country itself.

CIPHERED SYSTEMS

A ciphered numeration system is one in which there are numerals for numbers up to and including the base and for multiples of the base. The numbers represented by a particular set of numerals is the sum of the values of the numerals.

Ciphered numeration systems require the memorization of many different symbols but have the advantage that numbers can be written in a compact form. The ciphered numeration system that we discuss is the Ionic Greek system (see Table 4.4).

Table 4.4 Ionic Greek Numerals

1	α	alpha	60	ξ	xi
2	β	beta	70	ο	omicron
3	γ	gamma	80	π	pi
4	δ	delta	90	Q	koph*
5	ϵ	epsilon	100	ρ	rho
6	ϛ	vau*	200	σ	sigma
7	ζ	zeta	300	τ	tau
8	η	eta	400	υ	upsilon
9	θ	theta	500	φ	phi
10	ι	iota	600	χ	chi
20	κ	kappa	700	ψ	psi
30	λ	lambda	800	ω	omega
40	μ	mu	900	ϡ	sampi*
50	ν	nu			

*Taken from the Phoenician alphabet.

The Ionic Greek system was developed in about 3000 B.C., and it used letters of their alphabet for numerals. Other ciphered systems include the Hebrew, Coptic, Hindu, Brahmin, Syrian, Egyptian Hieratic, and early Arabic systems.

Since the Greek alphabet contains 24 letters but 27 symbols were needed, the Greeks borrowed the symbols ϛ, Q, and ϡ from the Phoenician alphabet.

The number 24 = 20 + 4. When 24 is written as a Greek numeral, the plus sign is omitted:

$$24 = \kappa\delta$$

The number 996 written as a Greek numeral is ϡ Q ϛ.

When a prime (′) is placed above a number, it multiplies that number by 1000. For example,

$$\beta' = 2 \times 1000 = 2000$$

$$\sigma' = 200 \times 1000 = 200{,}000$$

EXAMPLE 7 ***The Ionic Greek System: A Ciphered System***

Write $\psi\,\nu\,\gamma$ as a Hindu–Arabic numeral.

SOLUTION $\psi = 700$, $\nu = 50$, and $\gamma = 3$. Adding these numbers gives 753. ▲

EXAMPLE 8 ***Writing an Ionic Greek Numeral***

Write 9432 as an Ionic Greek numeral.

SOLUTION

$$\begin{aligned} 9432 &= 9000 + 400 + 30 + 2 \\ &= (9 \times 1000) + 400 + 30 + 2 \\ &= \theta' \qquad \upsilon \quad \lambda \quad \beta \\ &= \theta'\,\upsilon\lambda\,\beta \end{aligned}$$

▲

SECTION 4.1 EXERCISES

CONCEPT/WRITING EXERCISES

1. What is the difference between a number and a numeral?
2. List four numerals given in this section that may be used to represent the number ten.
3. List four numerals given in this section that may be used to represent the number one hundred.
4. What is a system of numeration?
5. What is the name of the system of numeration that we presently use?
6. Explain how numbers are represented in an additive numeration system.
7. Explain how numbers are represented in a multiplicative numeration system.
8. Explain how numbers are represented in a ciphered numeration system.

PRACTICE THE SKILLS

In Exercises 9–14, write the numeral as a Hindu–Arabic numeral.

9. 𓍢𓍢𓎆𓎆𓎆𓏺𓏺 **10.** 𓎆𓍢𓎆𓏺𓏺

11. 𓆼𓆼𓍢𓍢𓍢𓍢𓎆𓎆𓏺𓏺𓏺 **12.** 𓂭𓂭𓂭𓆼𓍢𓍢𓎆

13. 𓆐𓆐𓆐𓆐𓂭𓂭𓂭𓆼𓆼𓆼𓆼𓍢𓍢𓎆𓏺𓏺𓏺𓏺𓏺 **14.** 𓁨𓁨𓁨𓆐𓆐𓆐𓍢𓍢𓍢𓍢𓎆𓎆𓎆𓏺

In Exercises 15–20, write the numeral as an Egyptian numeral.

15. 437 **16.** 752

17. 2045 **18.** 1812

19. 173,845 **20.** 2,315,132

In Exercises 21–32, write the numeral as a Hindu–Arabic numeral.

21. XIV **22.** XVI

23. DXLVII **24.** CLXII

25. MCDXCII **26.** MCMXVIII

27. MCMXLV **28.** MDCCXLVI

29. $\overline{\text{X}}$MMDCLXVI **30.** $\overline{\text{L}}$MCMXLIV

31. $\overline{\text{IX}}$CDLXIV **32.** $\overline{\text{V}}$MCCCXXXIII

In Exercises 33–44, write the numeral as a Roman numeral.

33. 47 **34.** 94

35. 164 **36.** 269

37. 2000 **38.** 3564

39. 4793 **40.** 6274

41. 9999 **42.** 14,315

43. 20,644 **44.** 99,999

In Exercises 45–50, write the numeral as a Hindu–Arabic numeral.

45. 九十四 **46.** 六十二 **47.** 四千八十一

48. 三千二十九 **49.** 七千六百五十 **50.** 三千四百八十七

In Exercises 51–56, write the numeral as a traditional Chinese numeral.

51. 47 **52.** 178 **53.** 378

54. 2001 **55.** 3570 **56.** 6905

In Exercises 57–62, write the numeral as a Hindu–Arabic numeral.

57. $\sigma\,\xi\,\delta$ **58.** $\chi\,o\,\eta$ **59.** $\kappa'\,\beta'\,\phi\,\epsilon$

60. $\rho'\,\nu'\,\omega\,\iota\,\gamma$ **61.** $\theta'\chi\,\zeta$ **62.** α' Π Q Θ

In Exercises 63–68, write the numeral as an Ionic Greek numeral.

63. 47 **64.** 178 **65.** 726

66. 2001 **67.** 35,704 **68.** 690,540

In Exercises 69–71, compare the advantages and disadvantages of a ciphered system of numeration with those of the named system.

69. An additive system
70. A multiplicative system
71. The Hindu–Arabic system

In Exercises 72–75, write the numeral as numerals in the indicated systems of numeration.

72. 𓆼𓎆𓎆𓏺 in Hindu–Arabic, Roman, Chinese, and Greek

73. MCMXXXVI in Hindu–Arabic, Egyptian, Greek, and Chinese

74. 五百二十七 in Hindu–Arabic, Egyptian, Roman, and Greek

75. $\upsilon\kappa\beta$ in Hindu–Arabic, Egyptian, Roman, and Chinese

CHALLENGE PROBLEMS/GROUP ACTIVITIES

76. Write the Roman numeral for 999,999.

77. Write the Ionic Greek numeral for 999,999.

4.2 PLACE-VALUE OR POSITIONAL-VALUE NUMERATION SYSTEMS

The eighteenth-century mathematician Pierre Simon, Marquis de Laplace, speaking of the positional principle, said: "The idea is so simple that this very simplicity is the reason for our not being sufficiently aware of how much attention it deserves."

Today the most common type of numeration system is the place-value system. The Hindu–Arabic numeration system, used in the United States and many other countries, is an example of a place-value system. In a **place-value system**, which is also called a **positional value system**, the value of the symbol depends on its position in the representation of the number. For example, the 2 in 20 represents 2 tens, and the 2 in 200 represents 2 hundreds. A true positional-value system requires a **base** and a set of symbols, including a symbol for zero and one for each counting number less than the base. Although any number can be written in any base, the most common positional system is the base 10 system (the decimal number system).

The Hindus in India are credited with the invention of zero and the other symbols used in our system. The Arabs, who traded regularly with the Hindus, also adopted the system, thus the name Hindu–Arabic. Not until the middle of the fifteenth century, however, did the Hindu–Arabic numerals take the form we know today.

Babylonian Numerals

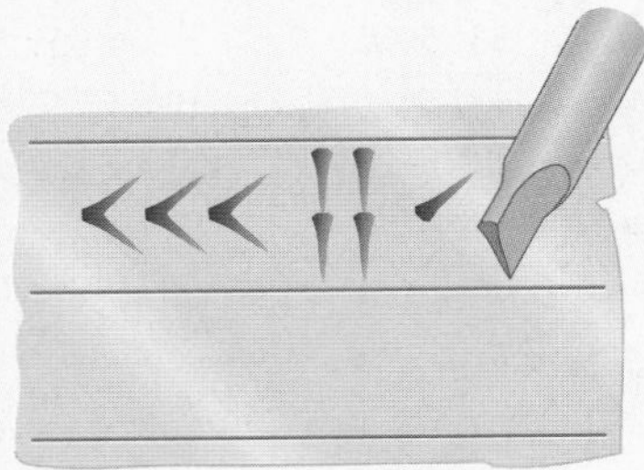

The form Babylonian numerals took is directly related to their writing materials. Babylonians used a reed (later a stylus) to make their marks in wet clay. The end could be used to make a thin wedge, 𒁹, which represents a unit, or a wider wedge, 𒌋, which represents 10 units. The clay dried quickly, so the writings tended to be short but extremely durable.

The Hindu–Arabic numerals and the positional system of numeration revolutionized mathematics by making addition, subtraction, multiplication, and division much easier to learn and very practical to use. Merchants and traders no longer had to depend on the counting board or abacus. The first group of mathematicians, who computed with the Hindu–Arabic system rather than with pebbles or beads on a wire, were known as the "algorists."

In the Hindu–Arabic system, the symbols 0, 1, 2, 3, 4, 5, 6, 7, 8, and 9 are called **digits**. The base 10 system was developed from counting on fingers, and the word *digit* comes from the Latin word for fingers.

The positional values in the Hindu–Arabic system are

$$\ldots, (10)^5, (10)^4, (10)^3, (10)^2, 10, 1$$

To evaluate a number in the Hindu–Arabic system, we multiply the first digit on the right by 1. We multiply the second digit from the right by the base, 10. We multiply the third digit from the right by the base squared, 10^2 or 100. We multiply the fourth digit from the right by the base cubed, 10^3 or 1000, and so on. In general, we multiply the digit n places from the right by 10^{n-1}. Therefore, we multiply the digit eight places from the right by 10^7. Using the place-value rule, we can write a number in **expanded form**. The number 1234 written in expanded form is

$$1234 = (1 \times 10^3) + (2 \times 10^2) + (3 \times 10^1) + (4 \times 1)$$

or

$$(1 \times 1000) + (2 \times 100) + (3 \times 10) + 4$$

The oldest known numeration system that resembled a place-value system was developed by the Babylonians in about 2500 B.C. Their system resembled a place-value system with a base of 60, a sexagesimal system. It was not a true place-value system because it lacked a symbol for zero. The lack of a symbol for zero led to a great deal of ambiguity and confusion. Table 4.5 gives the Babylonian numerals.

Table 4.5 Babylonian Numerals

Babylonian Numerals	𒁹	𒌋
Hindu–Arabic numerals	1	10

The positional values in the Babylonian system are

$$\ldots, (60)^3, (60)^2, 60, 1$$

In a Babylonian numeral, a gap is left between the characters to distinguish between the various place values. From right to left, the sum of the first group of numerals is multiplied by 1. The sum of the second group is multiplied by 60. The sum of the third group is multiplied by $(60)^2$, and so on.

EXAMPLE 1 ***The Babylonian System: A Positional Value System***

Write ◀◀ ◀◀▼▼▼▼ as a Hindu–Arabic numeral.

SOLUTION

$$\underbrace{\text{◀◀}}_{60\text{'s}} \qquad \underbrace{\text{◀◀▼▼▼▼}}_{\text{units}}$$

$$\underbrace{10 + 10}_{60\text{'s}} \qquad \underbrace{10 + 10 + 1 + 1 + 1 + 1}_{\text{units}}$$

$$(20 \times 60) + (24 \times 1)$$
$$1200 + 24 = 1224$$

The Babylonians used the symbol ▼̅ to indicate subtraction. The numeral ◀▼̅▼▼ represents $10 - 2$, or 8. The numeral ◀◀◀▼▼▼▼▼▼̅◀▼▼ represents $35 - 12$, or 23 in decimal notation.

EXAMPLE 2 ***From Babylonian to Hindu–Arabic Numerals***

Write ▼▼ ◀▼ ◀◀▼̅▼▼ as a Hindu–Arabic numeral.

SOLUTION The place value of these three groups of numerals from left to right is

$$(60)^2, \quad 60, \quad 1$$
$$\text{or} \quad 3600, \quad 60, \quad 1$$

The numeral in the group on the right has a value of $20 - 2$, or 18. The numeral in the center has a value of $10 + 1$, or 11. The numeral on the left represents $1 + 1$, or 2. Multiplying each group by its positional value gives

$$\begin{aligned} &(2 \times 60^2) + (11 \times 60) + (18 \times 1) \\ &= (2 \times 3600) + (11 \times 60) + (18 \times 1) \\ &= 7200 + 660 + 18 \\ &= 7878 \end{aligned}$$

To explain the procedure used to convert from a Hindu–Arabic numeral to a Babylonian numeral, we will consider a length of time. How can we change 9820 seconds into hours, minutes, and seconds? Since there are 3600 seconds in an hour

DID YOU KNOW

Counting Boards

One of the earliest counting devices, used in most ancient civilizations, was the counting board. On such a board, each column represents a positional value. The number of times a value occurs is represented by markers (beads, stones, sticks) in the column. An empty column signifies "no value." The widespread use of counting boards meant that Europeans were already long accustomed to working with positional values when they were introduced to Hindu–Arabic numerals in the fifteenth century. Some people in China, Japan, the former Soviet Union, and Eastern Europe still commonly use a type of counting board known as the abacus to perform routine computations.

(60 seconds to a minute and 60 minutes to an hour), we can find the number of hours in 9820 seconds by dividing 9820 by 60^2, or 3600.

$$\begin{array}{r l} 2 & \leftarrow \text{Hours} \\ 3600\overline{)9820} & \\ \underline{7200} & \\ 2620 & \leftarrow \text{Remaining seconds} \end{array}$$

Now we can determine the number of minutes by dividing the remaining seconds by 60, the number of seconds in a minute.

$$\begin{array}{r l} 43 & \leftarrow \text{Minutes} \\ 60\overline{)2620} & \\ \underline{2400} & \\ 220 & \\ \underline{180} & \\ 40 & \leftarrow \text{Remaining seconds} \end{array}$$

Since the remaining number of seconds, 40, is less than the number of seconds in a minute, our task is complete.

$$9820 \text{ sec} = 2 \text{ hr, } 43 \text{ min, and } 40 \text{ sec}$$

The same procedure is used to convert a decimal (base 10) number to a Babylonian number or any number in a different base.

On the High Seas

Vestiges of the Babylonian sexagesimal system are still with us today, especially in navigation. Navigators use degrees, minutes, and seconds of longitude and latitude to chart their course. In the past, merchants and explorers had to have an easy means of computing angles as they guided their ships by the stars. Base 60 easily divides into halves, thirds, fourths, fifths, sixths, tenths, twelfths, fifteenths, twentieths, and thirtieths, making such computations easier. Today's navigators use sophisticated electronic instruments to achieve the same ends.

EXAMPLE 3 ***From Hindu–Arabic to Babylonian Numerals***

Write 1602 as a Babylonian numeral.

SOLUTION The Babylonian numeration system has positional values of

$$\ldots, 60^3, 60^2, 60, 1$$

which can be expressed as

$$\ldots, 216000, 3600, 60, 1$$

The largest positional value less than or equal to 1602 is 60. To determine how many groups of 60 are in 1602, divide 1602 by 60.

$$\begin{array}{r l} 26 & \leftarrow \text{Groups of 60} \\ 60\overline{)1602} & \\ \underline{120} & \\ 402 & \\ \underline{360} & \\ 42 & \leftarrow \text{Units remaining} \end{array}$$

Thus, $602 \div 60 = 26$ with remainder 42. There are 26 groups of 60 and 42 units remaining. Because the remainder, 42, is less than the base, 60, no further division is necessary. The remainder represents the number of units when the number is written in expanded form. Therefore, 1602 = (26 × 60) + (42 × 1). When written as a Babylonian numeral, 1602 is

𒌋𒌋𒁹𒁹𒁹𒁹𒁹𒁹 𒌋𒌋𒌋𒌋𒁹𒁹 ▲

EXAMPLE 4 ***Using Division to Determine a Babylonian Numeral***

Write 6270 as a Babylonian numeral.

SOLUTION Divide 6270 by the largest positional value less than or equal to 6270. That value is 3600.

$$6270 \div 3600 = 1 \text{ with remainder } 2670$$

There is one group of 3600 in 6270. Next divide the remainder 2670 by 60 to determine the number of groups of 60 in 2670.

$$2670 \div 60 = 44 \text{ with remainder } 30$$

There are 44 groups of 60 and 30 units remaining.

$$6270 = (1 \times 60^2) + (44 \times 60) + (30 \times 1)$$

Thus, 6270 written as a Babylonian numeral is

𒁹 𒌋𒌋𒌋𒌋𒁹𒁹𒁹𒁹 𒌋𒌋𒌋 ▲

DID YOU KNOW

Sacred Mayan Glyphs

3

4

In addition to their base 20 numerals the Mayans had a holy numeration system used by priests to create and maintain calendars. They used a special set of hieroglyphs that consisted of pictograms of Mayan gods. For example, the number 3 was represented by the god of wind and rain, the number 4 by the god of sun.

Another place-value system is the Mayan numeration system. The Mayans, who lived on the Yucatan Peninsula, developed a sophisticated numeration system based on their religious and agricultural calendar. The numbers in this system are written vertically rather than horizontally, with the units position on the bottom. In the Mayan system, the number in the bottom row is to be multiplied by 1. The number in the second row from the bottom is to be multiplied by 20. The number in the third row is to be multiplied by 18 × 20, or 360. You probably expected the third row to be multiplied by 20^2 rather than 18 × 20. It is believed that the Mayans used 18 × 20 so that their numeration system would conform to their calendar of 360 days. The positional values above 18 × 20 are 18×20^2, 18×20^3, and so on.

Positional values in the Mayan system

$\ldots 18 \times (20)^3$,	$18 \times (20)^2$,	18×20,	20,	1
or . . . 144,000,	7200,	360,	20,	1

The digits 0, 1, 2, 3, . . . , 19 of the Mayan systems are formed by a simple grouping of dots and lines, as shown in Table 4.6.

Table 4.6 Mayan Numerals

0	1	2	3	4	5	6	7	8	9
⬭	•	••	•••	••••	—	• —	•• —	••• —	•••• —
10	**11**	**12**	**13**	**14**	**15**	**16**	**17**	**18**	**19**
— —	• — —	•• — —	••• — —	•••• — —	— — —	• — — —	•• — — —	••• — — —	•••• — — —

> **DID YOU KNOW**
>
> **Numerals of the World**
>
> 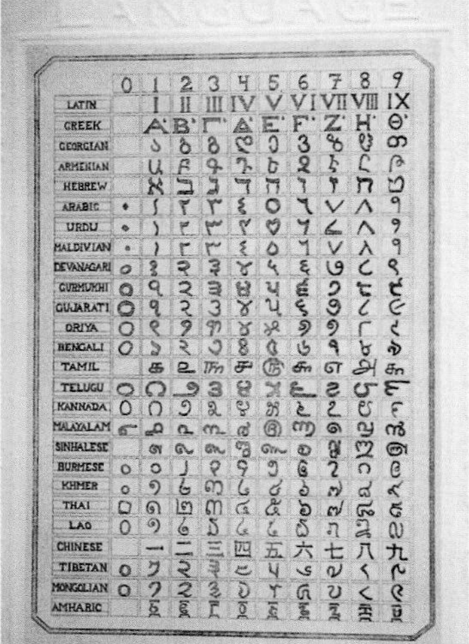
>
>
> Although most countries presently use a place value (or positional value) numeration system with base 10, the numerals used for the digits differ by country. The photo of the artwork entitled *Numbers* by Jan Fleck shows numerals currently used in many countries of the world. For example, in Burmese, the numeral ၃ has a value of 3.

EXAMPLE 5 ***The Mayan System: A Positional Value System***

Write

••
—
••
•••
═

as a Hindu–Arabic numeral.

SOLUTION In the Mayan numeration system, the first three positional values are

$$\begin{array}{c} 18 \times 20 \\ 20 \\ 1 \end{array}$$

••/—	$= 7 \times (18 \times 20)$	$= 2520$
••	$= 2 \times 20$	$= 40$
•••/═	$= 13 \times 1$	$= 13$
		2573

EXAMPLE 6 ***From Mayan to Hindu–Arabic Numerals***

Write

•••
—
•
═
••••

as a Hindu–Arabic numeral.

SOLUTION

•••/—	$= 8 \times (18 \times 20)$	$= 2880$
•/═	$= 11 \times 20$	$= 220$
••••	$= 4 \times 1$	$= 4$
		3104

EXAMPLE 7 ***From Hindu–Arabic to Mayan Numerals***

Write 4025 as a Mayan numeral.

SOLUTION To convert from a Hindu–Arabic to a Mayan numeral, we use a procedure similar to the one used to convert to a Babylonian numeral. The Mayan positional values are . . . , 7200, 360, 20, 1. The greatest positional value less than or equal to 4025 is 360. Divide 4025 by 360.

$$4025 \div 360 = 11 \text{ with remainder } 65$$

There are 11 groups of 360 in 4025. Next divide the remainder, 65, by 20.

$$65 \div 20 = 3 \text{ with remainder } 5$$

There are 3 groups of 20 with five units remaining.

$$4025 = (11 \times 360) + (3 \times 20) + (5 \times 1)$$

4025 written as a Mayan numeral is

$$\left.\begin{array}{c} 11 \times 360 \\ 3 \times 20 \\ 5 \times 1 \end{array}\right\} = \begin{array}{c} \overset{\bullet}{═} \\ \bullet\bullet\bullet \\ — \end{array}$$

SECTION 4.2 EXERCISES

CONCEPT/WRITING EXERCISES

1. What is the most common type of numeration system used in the world today?
2. Consider the numbers 30 and 300 in the Hindu–Arabic numeration system. What does the 3 represent in each number?
3. In a true positional-value system, what symbols are required?
4. a) What is the base in the Hindu–Arabic numeration system?
   b) What are the digits in the Hindu–Arabic numeration system?
5. Explain how to write a number in expanded form in a positional-value numeration system.
6. Why was the Babylonian system not a true place-value system?
7. Consider the Babylonian number represented by <▼. Give two numbers in Hindu–Arabic numerals this number may represent. Explain your answer.
8. a) What problem did not having a symbol for zero cause in the Babylonian system?
   b) Write the numbers 133 and 7980 as Babylonian numerals.
9. List the first five positional values, starting with the units position, for the Mayan numeration system.
10. Describe two ways that the Mayan place-value system differs from the Hindu–Arabic place-value system.

PRACTICE THE SKILLS

In Exercises 11–22, write the Hindu–Arabic numeral in expanded form.

11. 57	**12.** 86	**13.** 359
14. 562	**15.** 897	**16.** 3769
17. 5262	**18.** 23,468	**19.** 10,732
20. 125,678	**21.** 346,861	**22.** 3,765,934

In Exercises 23–28, write the Babylonian numeral as a Hindu–Arabic numeral.

23. <<▼▼▼▼ **24.** <<<▼▼▼▼▼

25. <▼▼▼ ▼▼▼▼ **26.** <▼ <<▼▼▼

27. ▼ <<▼ <▼▼▼ **28.** < <<▼▼▼▼ ▼▼

In Exercises 29–34, write the numeral as a Babylonian numeral.

29. 76	**30.** 86	**31.** 121
32. 512	**33.** 3685	**34.** 3030

In Exercises 35–40, write the Mayan numeral as a Hindu–Arabic numeral.

35. ••
•̲ ═

36. ═
—

37. •• ═
⬭
•

38. •• —
•••• —
•• —

39. • ═
••
⬭

40. ••••
═
—

In Exercises 41–46, write the numeral as a Mayan numeral.

41. 18	**42.** 257	**43.** 300
44. 406	**45.** 3181	**46.** 1978

47. *Comparisons of Systems* Compare the advantages and disadvantages of a place-value system with those of (a) additive numeration systems, (b) multiplicative numeration systems, and (c) ciphered numeration systems.
48. *Your Own System* Create your own place-value system. Write 1996 in your system.

In Exercises 49 and 50, write the numeral in the indicated systems of numeration.

49. <<<▼▼▼ in Hindu–Arabic and Mayan

50. — in Hindu–Arabic and Babylonian
•• —
••••

In Exercises 51 and 52, suppose a place value numeration system has base ⊡, with digits represented by the symbols △, ϕ, □, and ⊠. Write each expression in expanded form.

51. ϕ□△

52. ⊠△ϕ□

CHALLENGE PROBLEMS/GROUP ACTIVITIES

53. a) Is there a largest number in the Babylonian numeration system? Explain.
    b) Write the Babylonian numeral for 999,999.
54. a) Is there a largest number in the Mayan numeration system? Explain.
    b) Write the Mayan numeral for 999,999.

In Exercises 55–58, first convert each numeral to a Hindu–Arabic numeral and then perform the indicated operation. Finally, convert the answer back to a numeral in the original numeration system.

55. ▼▼ <<▼▼▼ + <<▼▼▼ **56.** ▼▼▼ <<<▼▼▼ − <<<▼▼

57. (•• — / • — / ≡) + (• — / •• — / ••• ═)

58. (•• — / • — / ≡) − (• — / •• — / ••• ═)

RESEARCH ACTIVITIES

59. Investigate and write a report on the development of the Hindu–Arabic system of numeration. Start with the earliest records of this system in India.

60. The Arabic numeration system currently in use is a base 10 positional-value system, which uses different symbols than the Hindu–Arabic numeration system. Write the symbols used in the Arabic system of numeration and their equivalent symbols in the Hindu–Arabic numeration system. Write 54, 607, and 2000 in Arabic numerals.

4.3 OTHER BASES

The positional values in the Hindu–Arabic numeration system are

$$\ldots, (10)^4, (10)^3, (10)^2, 10, 1$$

The positional values in the Babylonian numeration system are

$$\ldots, (60)^4, (60)^3, (60)^2, 60, 1$$

The numbers 10 and 60 are called the **bases** of the Hindu–Arabic and Babylonian systems, respectively.

Any counting number greater than 1 may be used as a base for a positional-value numeration system. If a positional-value system has a base b, then its positional values will be

$$\ldots, b^4, b^3, b^2, b, 1$$

The positional values in a base 8 system are

$$\ldots, 8^4, 8^3, 8^2, 8, 1$$

and the positional values in a base 2 system are

$$\ldots, 2^4, 2^3, 2^2, 2, 1$$

As we indicated earlier, the Mayan numeration system is based on the number 20. It is not however, a true base 20 positional-value system. Why?

This is reason for the almost universal acceptance of base 10 numeration systems. Human beings have ten fingers. Even so, there are still some positional-value numeration systems that use bases other than 10. Some societies are still using a base 2 numeration system. They include some groups of people in Australia, New Guinea, Africa, and South America. Bases 3 and 4 are also used in some areas of South America. Base 5 systems were used by some primitive tribes in Bolivia, but the tribes are now extinct. The pure base 6 system occurs only sparsely in Northwest Africa. Base 6 also occurs in other systems in combination with base 12, the *duodecimal system*.

We continue to see remains of other base systems in many countries. For example, there are 12 inches in a foot, 12 months in a year. Base 12 is also evident in the dozen, the 24-hour day, and the gross (12 × 12). English uses the word *score* to mean 20, as in "Four score and seven years ago." Other traces are found in pre-

DID YOU KNOW

Beyond Their Fingers

The Kewa people of Papua, New Guinea, have gone well beyond counting on their fingers–they use the entire upper body. Going from the little finger of one hand, down the elbow, to shoulder to head to shoulder, down the other elbow, to the little finger of the opposite hand provides them with a count of 68.

DID YOU KNOW

The I Ching

One of the most influential books in Chinese history is the *I ching,* or *Book of Changes,* said to have been written by the emperor Fu Hsi in the twenty-ninth century B.C. The Chinese used the 64 hexagrams of the *I ching* to make predictions. These were formed from eight trigrams said to encompass all that happens in heaven and on earth: the creative, the receptive, the arousing, the abysmal, keeping still, the gentle, the clinging, the joyous. German mathematician Gottfriéd Leibniz (1646–1716) recognized in the *I ching's* patterns of broken and unbroken lines a system similar to binary arithmetic. If you imagine the broken line to be a zero and the unbroken line to be a one, each hexagram can be interpreted as a binary number.

English Celtic, Gaelic, Danish, and Welsh. Remains of base 60 are found in measurements of time (60 seconds to a minute, 60 minutes to an hour) and angles (60 seconds to one minute, 60 minutes to one degree).

The base 2, or **binary system**, has become very important because it is the internal language of the computer. For example, when the grocery store's cash register computer records the price of your groceries by using a scanning device, the bar codes it scans on the packages are in binary form. Computers use a two-digit "alphabet" that consists of the numerals 0 and 1. Every character on a standard keyboard can be represented by a combination of those two numerals. A single numeral such as 0 or 1 is called a **bit**. Other bases that computers make use of are base 8 and base 16. A group of eight bits is called a **byte**. In the American Standard Code for Information Interchange (ASCII) code, used in most computers, the byte 01000001 represents the character A, 01100001 represents the character a, 00110000 represents the character 0, and 00110001 represents the character 1.

A place-value system with base b must have b distinct symbols, one for zero and one for each number less than the base. A base 6 system must have symbols for the numbers 0, 1, 2, 3, 4, and 5. All numbers in base 6 are constructed from these 6 symbols. A base 8 system must have symbols for 0, 1, 2, 3, 4, 5, 6, and 7. All numbers in base 8 are constructed from these 8 symbols, and so on.

A number in a base other than base 10 will be indicated by a subscript to the right of the number. Thus, 123_5 represents a number in base 5. The number 123_6 represents a number in base 6. The value of 123_5 is not the same as the value of 123_{10}, and the value of 123_6 is not the same as the value 123_{10}. A base 10 number may be written without a subscript. For example 123 means 123_{10} and 456 means 456_{10}. For clarity in certain problems, we will use the subscript 10 to indicate a number in base 10.

Remember the symbols that represent the base itself, in any base b, are 10_b. For example, in base 5, the symbols 10_5 represent the number 5. The symbols 10_5 mean one group of 5 and no units. In base 6, the symbols 10_6 represent the number 6. The symbols 10_6 represent one group of 6 and no units, and so on.

To change a number in a base other than 10 to a base 10 number, we follow the same procedure we used in Section 4.2 to change the Babylonian and Mayan numbers to base 10 numbers. Multiply each digit in the number by its respective positional value. Then find the sum of the products.

EXAMPLE 1 ***Converting from Base 6 to Base 10***

Convert 524_6 to base 10.

SOLUTION In base 6, the positional values are . . . , $6^3, 6^2, 6, 1$. In expanded form,

$$\begin{aligned} 524_6 &= (5 \times 6^2) + (2 \times 6) + (4 \times 1) \\ &= (5 \times 36) + (2 \times 6) + (4 \times 1) \\ &= \quad 180 \quad + \quad 12 \quad + \quad 4 \\ &= 196 \end{aligned}$$

In Example 1, the units digit in 524_6 is 4. Notice that 4_6 has the same value as 4_{10} since both are equal to 4 units. That is, $4_6 = 4_{10}$. If n is a digit less than the base b, and the base b is less than or equal to 10, then $n_b = n_{10}$.

EXAMPLE 2 ***Converting from Base 8 to Base 10***

Convert 3615_8 to base 10.

SOLUTION

$$\begin{aligned} 3615_8 &= (3 \times 8^3) + (6 \times 8^2) + (1 \times 8) + (5 \times 1) \\ &= (3 \times 512) + (6 \times 64) + (1 \times 8) + (5 \times 1) \\ &= 1536 + 384 + 8 + 5 \\ &= 1933 \end{aligned}$$

A base 12 system must have 12 distinct symbols. In this text, we use the symbols 0, 1, 2, 3, 4, 5, 6, 7, 8, 9, T, and E, where T represents ten and E represents eleven. Why will the numerals 10_{12} and 11_{12} have different meanings than 10 and 11? The number 10_{12} represents 1 group of twelve plus 0 units, or twelve. The number 11_{12} represents 1 group of twelve plus 1 unit, or 13.

EXAMPLE 3 ***Converting from Base 12 to Base 10***

Convert $12\text{T}6_{12}$ to base 10.

SOLUTION

$$\begin{aligned} 12\text{T}6_{12} &= (1 \times 12^3) + (2 \times 12^2) + (\text{T} \times 12) + (6 \times 1) \\ &= (1 \times 1728) + (2 \times 144) + (10 \times 12) + (6 \times 1) \\ &= 1728 + 288 + 120 + 6 \\ &= 2142 \end{aligned}$$

EXAMPLE 4 ***Converting from Base 2***

Convert 101101_2 to base 10.

SOLUTION

$$\begin{aligned} 101101_2 &= (1 \times 2^5) + (0 \times 2^4) + (1 \times 2^3) + (1 \times 2^2) + (0 \times 2) + (1 \times 1) \\ &= 32 + 0 + 8 + 4 + 0 + 1 \\ &= 45 \end{aligned}$$

To change a number from a base 10 system to a different base, we will use the procedure explained in Section 4.2. Divide the base 10 number by the highest power of the new base that is less than or equal to the given number. Record this quotient. Then divide the remainder by the next smaller power of the new base and record this quotient. Repeat this procedure until the remainder is a number less than the new base. The answer is the set of quotients listed from left to right, with the remainder on the far right. This procedure is illustrated in Examples 5 through 7.

EXAMPLE 5 ***Convert to Base 8***

Convert 406 to base 8.

SOLUTION We are converting a number in base 10 to a number in base 8. The positional values in the base 8 system are . . . , 8^3, 8^2, 8, 1, or . . . , 512, 64, 8, 1.

The highest power of 8 that is less than or equal to 406 is 8^2, or 64. Divide 406 by 64.

First digit in answer
↓
$$406 \div 64 = 6 \text{ with remainder } 22$$

Therefore, there are 6 groups of 8^2 in 406. Next divide the remainder, 22, by 8.

Second digit in answer
↓
$$22 \div 8 = 2 \text{ with remainder } 6$$
↑
Third digit in answer

There are two groups of 8 in 22 and 6 units remaining. Since the remainder, 6, is less than the base, 8, no further division is required.

$$\begin{aligned} &= (6 \times 64) + (2 \times 8) + (6 \times 1) \\ &= (6 \times 8^2) + (2 \times 8) + (6 \times 1) \\ &= 626_8 \end{aligned}$$

Notice that we placed the subscript 8 to the right of 626 to show that it is a base 8 number.

EXAMPLE 6 *Convert to Base 3*

Convert 273 to base 3.

SOLUTION The place values in the base 3 system are . . . , $3^6, 3^5, 3^4, 3^3, 3^2$, 3, 1, or . . . , 729, 243, 81, 27, 9, 3, 1. The highest power of the base that is less than 273 is 3^5, or 243. Successive divisions by the powers of the base give the following result.

$$\begin{aligned} 273 \div 243 &= 1 \text{ with remainder of } 30 \\ 30 \div 81 &= 0 \text{ with remainder } 30 \\ 30 \div 27 &= 1 \text{ with remainder } 3 \\ 3 \div 9 &= 0 \text{ with remainder } 3 \\ 3 \div 3 &= 1 \text{ with remainder } 0 \end{aligned}$$

The remainder, 0, is less than the base, 3, so no further division is necessary. To obtain the answer, list the quotients from top to bottom followed by the remainder in the last division.

The number 273 can be represented as one group of 243, no groups of 81, one group of 27, no groups of 9, one group of 3, and no units.

$$\begin{aligned} 273 &= (1 \times 243) + (0 \times 81) + (1 \times 27) + (0 \times 9) + (1 \times 3) + (0 \times 1) \\ &= (1 \times 3^5) + (0 \times 3^4) + (1 \times 3^3) + (0 \times 3^2) + (1 \times 3) + (0 \times 1) \\ &= 101010_3 \end{aligned}$$

EXAMPLE 7 ***Convert to Base 12***

Convert 558 to base 12.

SOLUTION The place values in base 12 are . . . , 12^3, 12^2, 12, 1, or . . . , 1728, 144, 12, 1.

$$558 \div 144 = 3 \text{ with remainder } 126$$
$$126 \div 12 = \text{T with remainder } 6$$

(Remember that T is used to represent ten in base 12.)

$$558 = (3 \times 12^2) + (\text{T} \times 12) + (6 \times 1) = 3\text{T}6_{12}$$

SECTION 4.3 EXERCISES

CONCEPT/WRITING EXERCISES

1. In your own words, explain how to change a number in a base other than base 10 to base 10.
2. In your own words, explain how to change a number in base 10 to a base other than base 10.

PRACTICE THE SKILLS

In Exercises 3–20, convert the numeral to a numeral in base 10.

3. 7_8 **4.** 50_7 **5.** 23_5
6. 101_2 **7.** 1011_2 **8.** 1101_2
9. 84_{12} **10.** 21021_3 **11.** 465_7
12. 654_7 **13.** 20432_5 **14.** 101111_2
15. 4003_6 **16.** 123E_{12} **17.** 123_8
18. 1043_8 **19.** 14705_8 **20.** 67342_9

In Exercises 21–36, convert the base 10 numeral to a numeral in the base indicated.

21. 8 to base 2 **22.** 16 to base 2
23. 22 to base 2 **24.** 243 to base 6
25. 435 to base 7 **26.** 908 to base 4
27. 2061 to base 12 **28.** 100 to base 3
29. 529 to base 8 **30.** 64 to base 2
31. 2867 to base 12 **32.** 4312 to base 6
33. 1011 to base 2 **34.** 1589 to base 7
35. 2307 to base 8 **36.** 13,469 to base 8

In Exercises 37–40, assume that a base 16 positional-value system uses the numerals 0, 1, 2, 3, 4, 5, 6, 7, 8, 9, A, B, C, D, E, and F, where A through F represent 10 through 15, respectively. Convert the numeral to a numeral in base 10.

37. 826_{16} **38.** 581_{16}
39. $6\text{D}3\text{B}7_{16}$ **40.** 24FEA_{16}

In Exercises 41–44, convert the numeral to a numeral in base 16.

41. 412 **42.** 349
43. 5478 **44.** 34,721

In Exercises 45–50, convert 2001 to a numeral in the base indicated.

45. 2 **46.** 3 **47.** 5
48. 7 **49.** 12 **50.** 16

In Exercises 51–54, if any numerals are written incorrectly, explain why.

51. 5013_5 **52.** 1203_3
53. 674_8 **54.** 1206_{12}

PROBLEM SOLVING

In Exercises 55–58, assume the numerals given are in a base 5 numeration system. The numerals in this system and their equivalent Hindu–Arabic numerals are

◯ = 0 ◯ = 1 ◯ = 2 ◯ = 3 ◯ = 4

Write the Hindu–Arabic numerals equivalent to each of the following.

55. ◯◯$_5$ **56.** ◯◯$_5$ **57.** ◯◯◯$_5$ **58.** ◯◯◯$_5$

In Exercises 59–62, write the Hindu–Arabic numerals in the numeration system discussed in Exercises 55–58.

59. 17 **60.** 23 **61.** 74 **62.** 87

In Exercise 63–66, suppose colors as indicated below represent numerals in a base 4 numeration system.

● = 0 ● = 1 ● = 2 ● = 3

Write the Hindu–Arabic numerals equivalent to each of the following.

63. ●●$_4$ **64.** ●●$_4$ **65.** ●●●$_4$ **66.** ●●●$_4$

In Exercises 67–70, write the Hindu–Arabic numerals in the base 4 numeration system discussed in Exercises 63–66. You will need to use a variety of the colors indicated on page 164 to write the answer.

67. 11 **68.** 15 **69.** 60 **70.** 56

71. *Another Conversion Method* There is an alternative method for changing a number in base 10 to a different base. This method will be used to convert 328 to base 5. Dividing 328 by 5 gives a quotient of 65 and a remainder of 3. Write the quotient below the dividend and the remainder on the right, as shown.

	remainder
5⌊328	
65	3

Continue this process of division by 5.

	remainder	
5⌊328		
5⌊ 65	3 ↑	
5⌊ 13	0 Answer	(Since the dividend, 2, is smaller than the divisor, 5, the quotient is 0 and the remainder is 2.)
5⌊ 2	3	
0	2	

Note that the division continues until the quotient is zero. The answer is read from the bottom number to the top number in the remainder column. Thus, $328 = 2303_5$.

a) Explain why this procedure results in the proper answer.

b) Convert 683 to base 5 by this method.

c) Convert 763 to base 8 by this method.

CHALLENGE PROBLEMS/GROUP ACTIVITIES

72. a) Use the numerals 0, 1, and 2 to write the first 20 numbers in the base 3 numeration system.

b) What is the next number after 222_3?

73. a) *Your Own Numeration System* Make up your own base 20 positional-value numeration system. Indicate the 20 numerals you will use to represent the 20 numbers less than the base.

b) Write the numbers 468 and 5293 in your base 20 numeration system.

74. *Computer Code* The ASCII code used by most computers uses the last seven positions of an eight-bit byte to represent all the characters on a standard keyboard. How many different orderings of 0s and 1s (or how many different characters) can be made by using the last seven positions of an eight-bit byte?

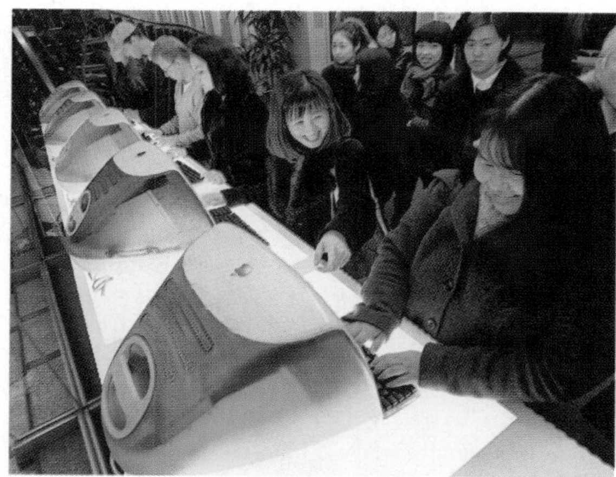

75. Find b if $111_b = 43$.

RESEARCH ACTIVITIES

76. Investigate and write a report on how digital computers use the binary number system.

77. We mention at the beginning of this section that some societies still use a base 2 and base 3 numeration system. These societies are in Australia, New Guinea, Africa, and South America. Write a report on these societies, covering the symbols they use and how they combine these symbols to represent numbers in their numeration system.

4.4 COMPUTATION IN OTHER BASES

ADDITION

When computers perform calculations, they do so in base 2, the binary system. In this section, we explain how to perform calculations in base 2 and other bases.

In a base 2 system, the only digits are 0 and 1, and the place values are

$$\ldots,\ 2^4,\ 2^3,\ 2^2,\ 2,\ 1$$

or

$$\ldots,\ 16,\ 8,\ 4,\ 2,\ 1$$

Suppose we want to add $1_2 + 1_2$. The subscript 2 indicates that we are adding in base 2. Remember the answer to $1_2 + 1_2$ must be written by using only the digits 0 and 1.

The sum of $1_2 + 1_2$ is 10_2, which represents 1 group of two and 0 units in base 2. Recall that 10_2 means $1(2) + 0(1)$.

If we wanted to find the sum of $10_2 + 1_2$, we would add the digits in the right-hand, or units, column. Since $0_2 + 1_2 = 1_2$, the sum of $10_2 + 1_2 = 11_2$.

We are going to work additional examples and exercises in base 2, so rather than performing individual calculations in every problem, we can construct and use an addition table, Table 4.7, for base 2 (just as we used an addition table in base 10 when we first learned to add in base 10).

Table 4.7 Base 2 Addition Table

+	0	1
0	0	1
1	1	10

EXAMPLE 1 *Adding in Base 2*

Add
$$\begin{array}{r} 1101_2 \\ \underline{111_2} \end{array}$$

SOLUTION Begin by adding the numbers in the right-hand, or units, column. From previous discussion, and as can be seen in Table 4.7, $1_2 + 1_2 = 10_2$. Place the 0 under the units column and carry the 1 to the 2's column, the second column from the right.

Place value of columns

$$\begin{array}{cccc} 2^3 & 2^2 & 2 & 1 \\ \downarrow & \downarrow & \downarrow & \downarrow \\ 1 & 1 & {}^{1}0 & 1 \\ & 1 & 1 & 1 \\ \hline & & & 0_2 \end{array}$$

Now add the three digits in the twos column, $1_2 + 0_2 + 1_2$. Treat this as $(1_2 + 0_2) + 1_2$. Therefore, add $1_2 + 0_2$ to get 1_2, then add $1_2 + 1_2$ to get 10_2. Place the 0 under the twos column and carry the 1 to the 2^2 column (the third column from the right).

$$\begin{array}{cccc} 1 & {}^{1}1 & {}^{1}0 & 1 \\ & 1 & 1 & 1 \\ \hline & & 0 & 0_2 \end{array}$$

Now add the three 1s in the 2^2 column to get $(1_2 + 1_2) + 1_2 = 10_2 + 1_2 = 11_2$. Place the 1 under the 2^2 column and carry the 1 to the 2^3 column (the fourth column from the right).

$$\begin{array}{cccc} {}^{1}1 & {}^{1}1 & {}^{1}0 & 1 \\ & 1 & 1 & 1 \\ \hline & 1 & 0 & 0_2 \end{array}$$

Now add the two 1s in the 2^3 column, $1_2 + 1_2 = 10_2$. Place the 10 as follows.

$$\begin{array}{ccccc} & {}^{1}1 & {}^{1}1 & {}^{1}0 & 1 \\ & & 1 & 1 & 1 \\ \hline 1 & 0 & 1 & 0 & 0_2 \end{array}$$

Therefore, the sum is 10100_2.

Let's now look at addition in a base 5 system. In base 5, the only digits are 0, 1, 2, 3, and 4, and the positional values are

$$\ldots,\ 5^4,\ 5^3,\ 5^2,\ 5,\ 1$$
$$\text{or} \quad \ldots,\ 625,\ 125,\ 25,\ 5,\ 1$$

Table 4.8 Base 5 Addition Table

+	0	1	2	3	4
0	0	1	2	3	4
1	1	2	3	4	10
2	2	3	4	10	11
3	3	4	10	11	12
4	4	10	11	(12)	13

What is the sum of $4_5 + 3_5$? We can consider this to mean $(1 + 1 + 1 + 1) + (1 + 1 + 1)$. We can regroup the seven 1s into one group of five and two units as $(1 + 1 + 1 + 1 + 1) + (1 + 1)$. Thus the sum of $4_5 + 3_5 = 12_5$ (circled in Table 4.8). Recall that 12_5 means $1(5) + 2(1)$. We can use this same procedure in obtaining the values in the base 5 addition table.

EXAMPLE 2 ***Use the Base 5 Addition Table***

Add 32_5
$\underline{33_5}$

SOLUTION First determine that $2_5 + 3_5$ is 10_5 from Table 4.8. Record the 0 and carry the 1 to the fives column.

$$\begin{array}{r} {}^{1}3\ 2_5 \\ \underline{3\ 3_5} \\ 0_5 \end{array}$$

Add the numbers in the second column, $(1_5 + 3_5) + 3_5 = 4_5 + 3_5 = 12_5$. Record the 12.

$$\begin{array}{r} {}^{1}3\ 2_5 \\ \underline{3\ 3_5} \\ 1\ 2\ 0_5 \end{array}$$

The sum is 120_5.

DID YOU KNOW

Speaking to Machines

For the past 600 years, we have used the Hindu–Arabic system of numeration without change. Our base 10 numeration system seems so obvious to us, perhaps because of our 10 fingers and 10 toes, but it would be rash to think that numbers in other bases are not useful. In fact, one of the most significant numeration systems is the binary system, or base 2. This system, with its elemental simplicity, is what is used by computers to process information and "talk" to one another. When a computer receives a command or data, every character in the command or data must first be converted into a binary number for the computer to understand and use it. Because of the ever-expanding number of computers in use, the users of the binary number system may soon outnumber the users of base 10!

Almost all packaged goods we buy today are marked with a universal product code (UPC), a black-and-white bar code. An optical scanner "reads" the pattern of black and white, thick and thin, and converts it to a binary code that is sent to the scanner's computer, which then calls up the appropriate price and records the sale for inventory purposes.

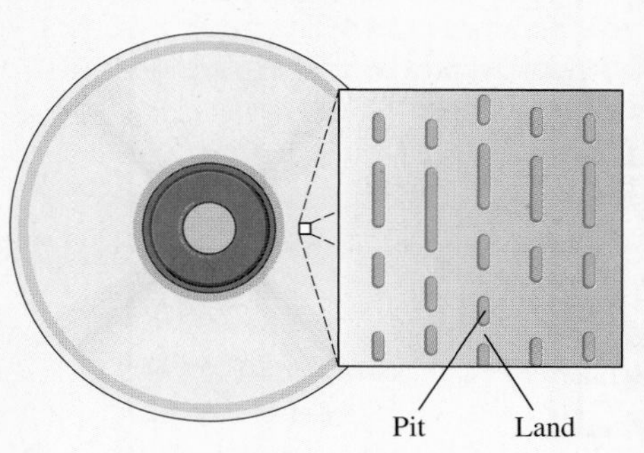

On a compact disc, music is digitally encoded on the underside of the disc in a binary system of pits and "lands" (nonpits). To play the disc, a laser beam tracks along the spiral and is reflected when it hits a land (signal sent = 1), but it is not reflected by the pits (no signal = 0). The binary sequence is then converted into music.

EXAMPLE 3 ***Add in Base 5***

Add 1234_5
$\underline{2042_5}$

SOLUTION

$$\begin{array}{r} 1\ \ {}^{1}2\ \ {}^{1}3\ \ 4_5 \\ 2\ \ 0\ \ 4\ \ 2_5 \\ \hline 3\ \ 3\ \ 3\ \ 1_5 \end{array}$$

▲

You can develop an addition table for any base and use it to add in that base. However, as you get more comfortable with addition in other bases, you may prefer to add numbers in other bases by using mental arithmetic. To do so, convert the sum of the numbers being added from the given base to base 10 and then convert the base 10 number back into the given base. You must clearly understand how to convert from base 10 to the given base, as discussed in Section 4.3. For example, to add $7_9 + 8_9$, add 7 + 8 in base 10 to get 15_{10} and then mentally convert 15_{10} to 16_9 using the procedure given earlier. Remember, 16_9 when converted to base 10 becomes 1(9) + 6(1), or 15. Addition using this procedure is illustrated in Examples 4 and 5.

EXAMPLE 4 ***Adding in Base 10; Converting to Base 3***

Add 1022_3
$\underline{2121_3}$

SOLUTION To solve this problem, make the necessary conversions by using mental arithmetic. $2 + 1 = 3_{10} = 10_3$. Record the 0 and carry the 1.

$$\begin{array}{r} 1\ \ 0\ \ {}^{1}2\ \ 2_3 \\ 2\ \ 1\ \ 2\ \ 1_3 \\ \hline 0_3 \end{array}$$

$1 + 2 + 2 = 5_{10} = 12_3$. Record the 2 and carry the 1.

$$\begin{array}{r} 1\ \ {}^{1}0\ \ {}^{1}2\ \ 2_3 \\ 2\ \ 1\ \ 2\ \ 1_3 \\ \hline 2\ \ 0_3 \end{array}$$

$1 + 0 + 1 = 2_{10} = 2_3$. Record the 2.

$$\begin{array}{r} 1\ \ {}^{1}0\ \ {}^{1}2\ \ 2_3 \\ 2\ \ 1\ \ 2\ \ 1_3 \\ \hline 2\ \ 2\ \ 0_3 \end{array}$$

$1 + 2 = 3_{10} = 10_3$. Record the 10.

$$\begin{array}{r} 1022_3 \\ 2121_3 \\ \hline 10220_3 \end{array}$$

▲

EXAMPLE 5 *Adding in Base 10; Converting to Base 5*

Add 444_5
244_5
143_5
214_5

SOLUTION Adding the digits in the right-hand column gives $4 + 4 + 3 + 4 = 15_{10} = 30_5$. Record the 0 and carry the 3. Adding the 3 with the digits in the next column yields $3 + 4 + 4 + 4 + 1 = 16_{10} = 31_5$. Record the 1 and carry the 3. Adding the 3 with the digits in the left-hand column gives $3 + 4 + 2 + 1 + 2 = 12_{10} = 22_5$. Record both digits. The sum of these four numbers is 2210_5.

$$\begin{array}{r} {}^{3}4\ {}^{3}4\ 4_5 \\ 2\ 4\ 4_5 \\ 1\ 4\ 3_5 \\ 2\ 1\ 4_5 \\ \hline 2\ 2\ 1\ 0_5 \end{array}$$

▲

SUBTRACTION

Subtraction can also be performed in other bases. Always remember that when you "borrow," you borrow the amount of the base given in the subtraction problem. For example, if subtracting in base 5, when you borrow, you borrow 5. If subtracting in base 12, when you borrow, you borrow 12.

EXAMPLE 6 *Subtacting in Base 5*

Subtract 3032_5
$- 1004_5$

SOLUTION We will perform the subtraction in base 10 and convert the results to base 5. Since 4 is greater than 2, we must borrow one group of 5 from the preceding column. This action gives a sum of $5 + 2$, or 7 in base 10. Now we subtract 4 from 7; the difference is 3. We complete the problem in the usual manner. The 3 in the second column becomes a 2, $2 - 0 = 2$, $0 - 0 = 0$, and $3 - 1 = 2$.

$$\begin{array}{r} 3032_5 \\ -\ 1004_5 \\ \hline 2023_5 \end{array}$$

▲

EXAMPLE 7

Subtract 468_{12}
$- 295_{12}$

SOLUTION $8 - 5 = 3$. Next we must subtract 9 from 6. Since 9 is greater than 6, borrowing is necessary. We must borrow one group of 12 from the preceding column. We then have a sum of $12 + 6 = 18$ in base 10. Now we subtract 9 from 18, and the difference is 9. The 4 in the left column becomes 3, and $3 - 2 = 1$.

$$\begin{array}{r} 468_{12} \\ -\ 295_{12} \\ \hline 193_{12} \end{array}$$

▲

MULTIPLICATION

Multiplication can also be performed in other bases. Doing so is helped by forming a multiplication table for the base desired. Suppose we want to determine the product of $4_5 \times 3_5$. In base 10, 4×3 means there are four groups of three units. Similarly, in a base 5 system, $4_5 \times 3_5$ means there are four groups of three units, or

$$(1 + 1 + 1) + (1 + 1 + 1) + (1 + 1 + 1) + (1 + 1 + 1)$$

Regrouping the 12 units above into groups of five gives

$$(1 + 1 + 1 + 1 + 1) + (1 + 1 + 1 + 1 + 1) + (1 + 1)$$

or two groups of five, and two units. Thus, $4_5 \times 3_5 = 22_5$.

We can construct other values in the base 5 multiplication table in the same way. You may, however, find it easier to multiply the values in the base 10 system and then change the product to base 5 by using the procedure discussed in Section 4.3. Multiplying 4×3 in base 10 gives 12, and converting 12 from base 10 to base 5 gives 22_5.

The product of $4_5 \times 3_5$ is circled in Table 4.9, the base 5 multiplication table. The other values in the table may be found by either method discussed.

Table 4.9 Base 5 Multiplication Table

×	0	1	2	3	4
0	0	0	0	0	0
1	0	1	2	3	4
2	0	2	4	11	13
3	0	3	11	14	22
4	0	4	13	(22)	31

EXAMPLE 8 *Using the Base 5 Multiplication Table*

Multiply

$$\begin{array}{r} 13_5 \\ \times\ 3_5 \\ \hline \end{array}$$

SOLUTION Multiply as you would in base 10, but use the base 5 multiplication table to find the products. When the product consists of two digits, record the right digit and carry the left digit. Multiplying gives $3_5 \times 3_5 = 14_5$. Record the 4 and carry the 1.

$$\begin{array}{r} {}^{1}13_5 \\ \times\ 3_5 \\ \hline 4 \end{array}$$

$(3_5 \times 1_5) + 1_5 = 4_5$. Record the 4.

$$\begin{array}{r} {}^{1}13_5 \\ \times\ 3_5 \\ \hline 44_5 \end{array}$$

The product is 44_5.

Constructing a multiplication table is often tedious, especially when the base is large. To multiply in a given base without the use of a table, multiply in base 10 and convert the products to the appropriate base number before recording them. This procedure is illustrated in Example 9.

DID YOU KNOW

Messages to Outer Space

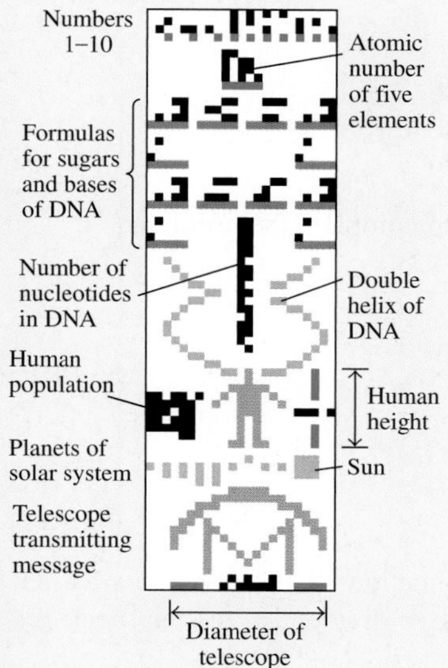

In 1974, radio astronomer Frank Drake beamed a radio message out toward the constellation Hercules. Guessing that alien intelligence might understand binary notation, Drake encoded the numbers 1 through 10 (top), followed by the atomic numbers of the five elements essential to terrestrial life, the chemical formula of the DNA molecule, and the numbers for the average human height and the world's population. If the receivers decode the message pictorially, they will see images of the human form, the solar system, and the transmitting radio dish.

EXAMPLE 9 ***Multiplying in Base 7***

Multiply
$$\begin{array}{r} 43_7 \\ \times 25_7 \\ \hline \end{array}$$

SOLUTION $5 \times 3 = 15_{10} = 2(7) + 1(1) = 21_7$. Record the 1 and carry the 2.

$$\begin{array}{r} {}^{2}43_7 \\ \times\ 25_7 \\ \hline 1 \end{array}$$

$(5 \times 4) + 2 = 20 + 2 = 22_{10} = 3(7) + 1(1) = 31_7$. Record the 31.

$$\begin{array}{r} {}^{2}43_7 \\ \times\ 25_7 \\ \hline 311 \end{array}$$

$2 \times 3 = 6_{10} = 6_7$. Record the 6.

$$\begin{array}{r} {}^{2}43_7 \\ \times\ 25_7 \\ \hline 311 \\ 6 \end{array}$$

$2 \times 4 = 8_{10} = 1(7) + 1(1) = 11_7$. Record the 11. Now add in base 7 to determine the answer. Remember, in base 7, there are no digits greater than 6.

$$\begin{array}{r} {}^{2}43_7 \\ \times\ 25_7 \\ \hline 311 \\ 116 \\ \hline 1501_7 \end{array}$$

DIVISION

Division is performed in much the same manner as long division in base 10. A detailed example of a division in base 5 is illustrated in Example 10. The same procedure is used for division in any other base.

EXAMPLE 10 ***Dividing in Base 5***

Divide $2_5\overline{)143_5}$.

SOLUTION Using Table 4.9, the multiplication table for base 5, we list the multiples of the divisor, 2.

$$\begin{aligned} 2_5 \times 1_5 &= 2_5 \\ 2_5 \times 2_5 &= 4_5 \\ 2_5 \times 3_5 &= 11_5 \\ 2_5 \times 4_5 &= 13_5 \end{aligned}$$

DID YOU KNOW

Evolution of Numbers

12th century	ɩ	ƨ	3	ጸ	y	Ꮆ	Λ	ȣ	ʔ	o
13th century	ȷ	ㄱ	3	ጸ	९	Ꮆ	Λ	8	9	ȣ
About 1300	ı	ㄱ	3	ጸ	կ	Ꮆ	Λ	8	9	ɵ
About 1429	ɩ	2	3	ጸ	Ϥ	Ꮆ	Λ	8	7	ø
15th century	1	2	3	4	5	6	7	8	9	0
20th century	1	2	3	4	5	6	7	8	9	0

During the Middle Ages, Western Europeans were reluctant to give up Roman numerals in favor of Hindu–Arabic numerals. But the rapid expansion of trade and commerce during the fifteenth century caused the need for quicker systems of calculation. The invention of movable type in 1450 also ensured a certain consistency in the way numerals were depicted, yet we still find ways to alter them.

Since $2_5 \times 4_5 = 13_5$, which is less than 14_5, 2_5 goes into 14_5 four times.

$$\begin{array}{r} 4 \\ 2_5\overline{)143_5} \\ \underline{13} \\ 1 \end{array}$$

Subtract 13_5 from 14_5. The difference is 1_5. Record the 1. Now bring down the 3 as when dividing in base 10.

$$\begin{array}{r} 4 \\ 2_5\overline{)143_5} \\ \underline{13} \\ 13 \end{array}$$

We see that $2_5 \times 4_5 = 13_5$. Use this information to complete the problem.

$$\begin{array}{r} 44_5 \\ 2_5\overline{)143} \\ \underline{13} \\ 13 \\ \underline{13} \\ 0 \end{array}$$

Therefore, $143_5 \div 2_5 = 44_5$ with remainder zero. ▲

A division problem can be checked by multiplication. If the division was performed correctly, (quotient × divisor) + remainder = dividend. We can check Example 10 as follows.

$$(44_5 \times 2_5) + 0_5 = 143_5$$

$$\begin{array}{r} 44_5 \\ \times\ 2_5 \\ \hline 143_5 \end{array} \quad \text{Check}$$

EXAMPLE 11 ***Dividing in Base 6***

Divide $4_6\overline{)2430_6}$.

SOLUTION The multiples of 4 in base 6 are

$$\begin{aligned} 4_6 \times 1_6 &= 4_6 \\ 4_6 \times 2_6 &= 12_6 \\ 4_6 \times 3_6 &= 20_6 \\ 4_6 \times 4_6 &= 24_6 \\ 4_6 \times 5_6 &= 32_6 \end{aligned}$$

$$\begin{array}{r} 404_6 \\ 4_6\overline{)2430_6} \\ \underline{24} \\ 03 \\ \underline{00} \\ 30 \\ \underline{24} \\ 2 \end{array}$$

Thus, the quotient is 404_6, with remainder 2_6.

Be careful when subtracting! When subtracting 4 from 0, you will need to borrow. Remember that you borow 10_6, which is the same as 6 in base 10.

Check: Does $(404_6 \times 4_6) + 2_6 = 2430_6$?

$$\begin{array}{r} 404_6 \\ \times \quad 4_6 \\ \hline 2424_6 \end{array} + 2_6 = 2430_6 \quad \text{True}$$

SECTION 4.4 EXERCISES

CONCEPT/WRITING EXERCISES

1. a) What are the first five positional values, from right to left, in base b?

b) What are the first five positional values, from right to left, in base 6?

2. In the addition

$$\begin{array}{r} 463_7 \\ +\ 24_7 \\ \hline \end{array}$$

What are the positional values of the first column on the right, the second column from the right, and the third column from the right? Explain how you determined your answer.

3. In your own words, explain how to add two numbers in a given base. In your explanation, answer the question, "What happens when the sum of the numbers in a column is greater than the base?"

4. In your own words, explain how to subtract two numbers in a given base. Include in your explanation what you do when, in one column, you must subtract a larger number from a smaller number.

PRACTICE THE SKILLS

In Exercises 5–16, add in the indicated base.

5. $\begin{array}{r} 32_5 \\ 41_5 \\ \hline \end{array}$ **6.** $\begin{array}{r} 33_7 \\ 65_7 \\ \hline \end{array}$ **7.** $\begin{array}{r} 3031_4 \\ 232_4 \\ \hline \end{array}$

8. $\begin{array}{r} 101_2 \\ 11_2 \\ \hline \end{array}$ **9.** $\begin{array}{r} 799_{12} \\ 218_{12} \\ \hline \end{array}$ **10.** $\begin{array}{r} 222_3 \\ 22_3 \\ \hline \end{array}$

11. $\begin{array}{r} 1112_3 \\ 1011_3 \\ \hline \end{array}$ **12.** $\begin{array}{r} 470_{12} \\ 347_{12} \\ \hline \end{array}$ **13.** $\begin{array}{r} 14631_7 \\ 6040_7 \\ \hline \end{array}$

14. $\begin{array}{r} 1341_8 \\ 341_8 \\ \hline \end{array}$ **15.** $\begin{array}{r} 1011_2 \\ 110_2 \\ \hline \end{array}$ **16.*** $\begin{array}{r} 43A_{16} \\ 496_{16} \\ \hline \end{array}$

In Exercises 17–28, subtract in the indicated base.

17. $\begin{array}{r} 312_4 \\ -\ 103_4 \\ \hline \end{array}$ **18.** $\begin{array}{r} 426_7 \\ -\ 134_7 \\ \hline \end{array}$ **19.** $\begin{array}{r} 2432_5 \\ -\ 1243_5 \\ \hline \end{array}$

20. $\begin{array}{r} 1011_2 \\ -\ 101_2 \\ \hline \end{array}$ **21.** $\begin{array}{r} 782_{12} \\ -\ 13T_{12} \\ \hline \end{array}$ **22.** $\begin{array}{r} 1221_3 \\ -\ 202_3 \\ \hline \end{array}$

23. $\begin{array}{r} 1001_2 \\ -\ 110_2 \\ \hline \end{array}$ **24.** $\begin{array}{r} 1E41_{12} \\ -\ 345_{12} \\ \hline \end{array}$ **25.** $\begin{array}{r} 4223_7 \\ -\ 304_7 \\ \hline \end{array}$

26. $\begin{array}{r} 4232_5 \\ -\ 2341_5 \\ \hline \end{array}$ **27.** $\begin{array}{r} 2100_3 \\ -\ 1012_3 \\ \hline \end{array}$ **28.*** $\begin{array}{r} 4E7_{16} \\ -\ 189_{16} \\ \hline \end{array}$

In Exercises 29–40, multiply in the indicated base.

29. $\begin{array}{r} 34_5 \\ \times\ 2_5 \\ \hline \end{array}$ **30.** $\begin{array}{r} 123_5 \\ \times\ 4_5 \\ \hline \end{array}$ **31.** $\begin{array}{r} 342_7 \\ \times\ 5_7 \\ \hline \end{array}$

32. $\begin{array}{r} 101_2 \\ \times\ 11_2 \\ \hline \end{array}$ **33.** $\begin{array}{r} 512_6 \\ \times\ 23_6 \\ \hline \end{array}$ **34.** $\begin{array}{r} 124_{12} \\ \times\ 6_{12} \\ \hline \end{array}$

35. $\begin{array}{r} 234_9 \\ \times\ 25_9 \\ \hline \end{array}$ **36.** $\begin{array}{r} 6T3_{12} \\ \times\ 24_{12} \\ \hline \end{array}$ **37.** $\begin{array}{r} 111_2 \\ \times\ 101_2 \\ \hline \end{array}$

38. $\begin{array}{r} 584_9 \\ \times\ 24_9 \\ \hline \end{array}$ **39.** $\begin{array}{r} 316_7 \\ \times\ 16_7 \\ \hline \end{array}$ **40.** $\begin{array}{r} 8T_{12} \\ \times\ 2T_{12} \\ \hline \end{array}$

In Exercises 41–52, divide in the indicated base.

41. $1_2 \overline{)110_2}$ **42.** $4_6 \overline{)231_6}$

43. $4_5 \overline{)143_5}$ **44.** $7_8 \overline{)466_8}$

45. $2_4 \overline{)312_4}$ **46.** $6_{12} \overline{)431_{12}}$

47. $3_4 \overline{)232_4}$ **48.** $5_6 \overline{)214_6}$

49. $3_5 \overline{)224_5}$ **50.** $4_6 \overline{)210_6}$

51. $6_7 \overline{)404_7}$ **52.** $3_7 \overline{)2101_7}$

*For Exercises 16, 21, and 28, see Exercises 37–40 in Section 4.3.

PROBLEM SOLVING

In Exercises 53–56, the numerals in base 5 numeration system are as illustrated with their equivalent Hindu–Arabic numerals.

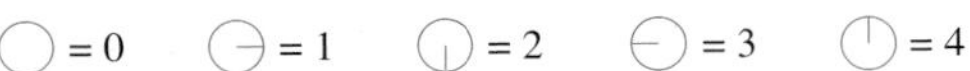

Add the following base 5 numbers.

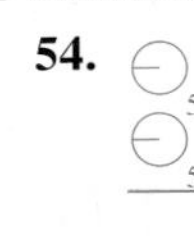

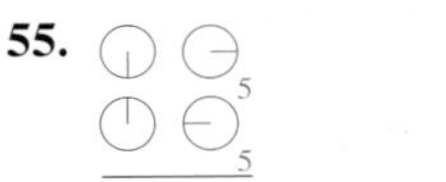

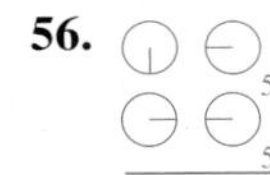

In Exercises 57–64, assume the numerals given are in a base 4 numeration system. In this system, suppose colors are used as numerals, as indicated below.

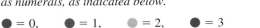

Add the following base 4 numbers. Your answers will contain a variety of the colors indicated.

Subtract the following in base 4. Your answer will contain a variety of the colors indicated.

61. ●●$_4$ − ●●$_4$
62. ●●$_4$ − ●●$_4$
63. ●●●$_4$ − ●●●$_4$
64. ●●●$_4$ − ●●●$_4$

For Exercises 65 and 66, study the pattern in the boxes. The number in the bottom row of each box represents the value of each dot in the box directly above it. For example, the following figure represents $(3 \times 7^2) + (2 \times 7) + (4 \times 7^0)$, or the number 324_7. This number in base 10 is 165.

• • •	• •	• • • •
7^2	7^1	7^0

65. Determine the base 5 number represented by the dots in the top row of the boxes. Then convert the base 5 number to a number in base 10.

• •	• • •		• •
5^3	5^2	5^1	5

66. Fill in the correct amount of dots in the columns above the base values if the number represented by the dots is to equal 327 in base 10.

9^2	9^1	9^0

CHALLENGE PROBLEMS/GROUP ACTIVITIES

Divide in the indicated base.

67. $14_5\overline{)242_5}$ **68.** $20_4\overline{)223_4}$

69. Consider the multiplication

$$\begin{array}{r} 462_8 \\ \times\ 35_8 \\ \hline \end{array}$$

a) Multiply the numbers in base 8.
b) Convert 462_8 and 35_8 to base 10.
c) Multiply the base 10 numbers determined in part (b).
d) Convert the answer obtained in base 8 in part (a) to base 10.
e) Are the answers obtained in parts (c) and (d) the same? Why or why not?

70. If $1304_b = 204$, determine b.

RESEARCH ACTIVITIES

71. Investigate and write a report on the use of the duodecimal (base 12) system as a system of numeration. You might contact the Duodecimal Society, Nassau Community College, Garden City, NY 11530 for information.

72. One method used by computers to perform subtraction is the "end around carry method." Do research and write a report explaining, with specific examples, how a computer performs subtraction by using the end around carry method.

4.5 EARLY COMPUTATIONAL METHODS

Our present procedures for multiplying and dividing numbers are the most recent to be developed. Early civilizations used various methods for multiplying and dividing. Multiplication was performed by *duplation and mediation,* by the *galley method,* and by *Napier rods.* Following is an explanation of each method.

DUPLATION AND MEDIATION

EXAMPLE 1 *A Pairing Technique for Multiplying*

Multiply 17 × 30 using duplation and mediation.

SOLUTION Write 17 and 30 with a dash between to separate them. Divide the number on the left, 17, in half, drop the remainder, and place the quotient, 8, under the 17. Double the number on the right, 30, obtaining 60, and place it under the 30. You will then have the folowing paired lines.

17—30
8—60

Continue this process, taking one-half the number in the left-hand column, disregarding the remainder, and doubling the number in the right-hand column, as shown below. When a 1 appears in the left-hand column, stop.

17—30
8—60
4—120
2—240
1—480

Cross out all the even numbers in the left-hand column and the corresponding numbers in the right-hand column.

17–30
~~8–60~~
~~4–120~~
~~2–240~~
1–480

Now add the remaining numbers in the right-hand column, obtaining 30 + 480 = 510, which is the product you want. If you check, you will find that 17 × 30 = 510.

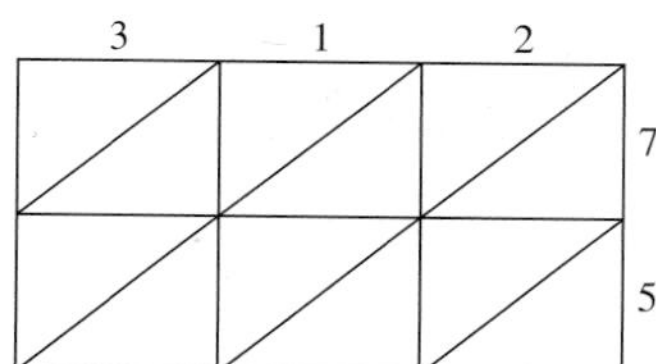

Figure 4.1

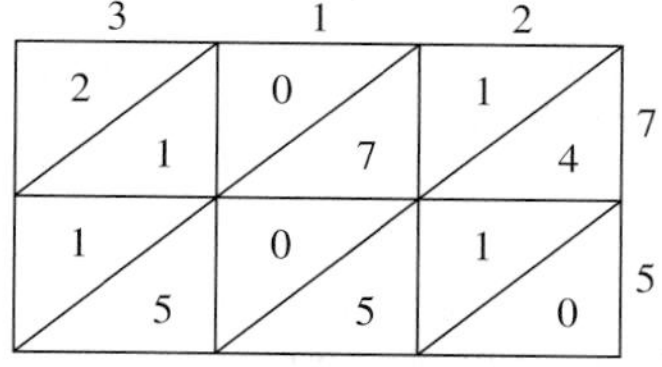

Figure 4.2

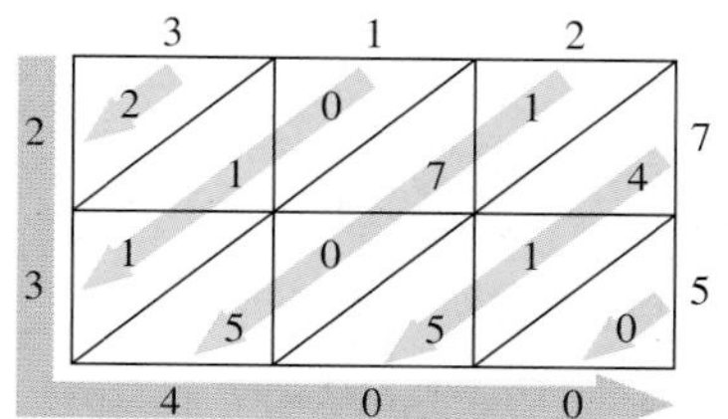

Figure 4.3

THE GALLEY METHOD

The galley method (sometimes referred to as the Gelosia method) was developed after duplation and mediation. To multiply 312 × 75 using the galley method, you construct a rectangle consisting of three columns (one for each digit of 312) and two rows (one for each digit of 75).

Place the digits 3, 1, 2 above the boxes and the digits 7, 5 on the right of the boxes, as shown in Fig. 4.1. Then place a diagonal in each box.

Complete each box by multiplying the number on top of the box by the number to the right of the box (Fig. 4.2). Place the units below the diagonal and the tens above.

Add the numbers along the diagonals, as shown in Fig. 4.3, starting with the bottom right diagonal. If the sum in a diagonal is 10 or greater, record the units digit below the rectangle and carry the tens digit to the next diagonal to the left.

For example, when adding 4, 1, and 5 (along the second blue diagonal from the right), the sum is 10. Record the 0 below the rectangle and carry the 1 to the next blue diagonal. The sum of 1 + 1 + 7 + 0 + 5 is 14. Record the 4 and carry the 1. The sum of the numbers in the next blue diagonal is 1 + 0 + 1 + 1 or 3.

The answer is read down the left-hand column and along the bottom, as shown by the purple arrow in Fig. 4.3. The answer is 23,400.

PROFILE IN MATHEMATICS

John Napier

During the seventeenth century, the growth of scientific fields such as astronomy required the ability to perform often unwieldy calculations. The English mathematician John Napier (1550–1617) made great contributions toward solving the problem of computing these numbers. His inventions include simple calculating machines and a device for performing multiplication and division known as Napier bones. Napier also developed the theory of logarithms.

NAPIER RODS

The third method was developed from the galley method by John Napier in the seventeenth century. His method of multiplication, known as Napier rods, proved to be one of the forerunners of the modern-day computer. Napier developed a system of separate rods numbered from 0 through 9 and an additional strip for an index, numbered vertically 1 through 9 (Fig. 4.4). Each rod is divided into 10 blocks, and each block contains a multiple of the top number. Units are placed to the right and tens to the left. Example 2 explains how Napier rods are used to multiply numbers.

INDEX	0	1	2	3	4	5	6	7	8	9
1	0/0	0/1	0/2	0/3	0/4	0/5	0/6	0/7	0/8	0/9
2	0/0	0/2	0/4	0/6	0/8	1/0	1/2	1/4	1/6	1/8
3	0/0	0/3	0/6	0/9	1/2	1/5	1/8	2/1	2/4	2/7
4	0/0	0/4	0/8	1/2	1/6	2/0	2/4	2/8	3/2	3/6
5	0/0	0/5	1/0	1/5	2/0	2/5	3/0	3/5	4/0	4/5
6	0/0	0/6	1/2	1/8	2/4	3/0	3/6	4/2	4/8	5/4
7	0/0	0/7	1/4	2/1	2/8	3/5	4/2	4/9	5/6	6/3
8	0/0	0/8	1/6	2/4	3/2	4/0	4/8	5/6	6/4	7/2
9	0/0	0/9	1/8	2/7	3/6	4/5	5/4	6/3	7/2	8/1

Figure 4.4

EXAMPLE 2 *Using Napier Rods*

Multiply 8 × 365, using Napier rods.

SOLUTION To multiply 8 × 365, line up the rods 3, 6, and 5 opposite the index, as shown in Fig. 4.5. Use the blocks that are shaded in blue to obtain the answer. Add along the diagonals as in the galley method.

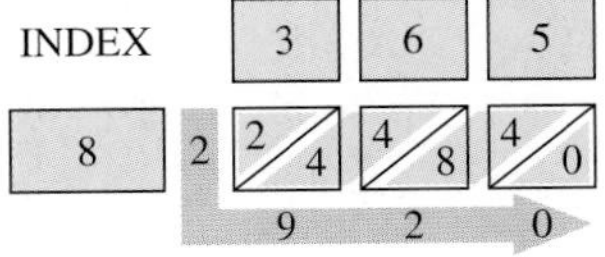

Thus, 8 × 365 = 2920.

INDEX	3	6	5
1	0/3	0/6	0/5
2	0/6	1/2	1/0
3	0/9	1/8	1/5
4	1/2	2/4	2/0
5	1/5	3/0	2/5
6	1/8	3/6	3/0
7	2/1	4/2	3/5
8	2/4	4/8	4/0
9	2/7	5/4	4/5

Figure 4.5

Example 3 illustrates the procedure to follow to multiply numbers containing more than one digit, using Napier rods.

EXAMPLE 3 ***Using Napier Rods to Multiply Two- and Three-Digit Numbers***

Multiply 48 × 365, using Napier rods.

SOLUTION $48 \times 365 = (40 + 8) \times 365$

Write $(40 + 8) \times 365 = (40 \times 365) + (8 \times 365)$. To find 40 × 365, determine 4 × 365 and multiply the product by 10. To evaluate 4 × 365, set up Napier rods for 3, 6, and 5 with index 4, and then evaluate along the diagonals, as indicated.

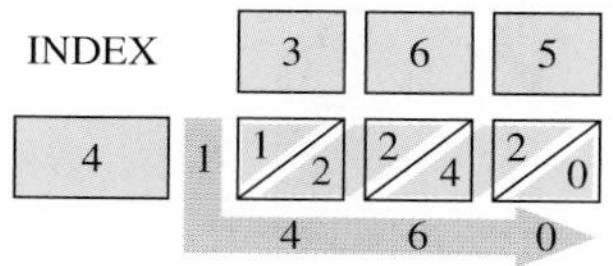

Therefore, $4 \times 365 = 1460$. Then $40 \times 365 = 10 \times 1460 = 14{,}600$.

$$\begin{aligned} 48 \times 365 &= (40 \times 365) + (8 \times 365) \\ &= 14{,}600 + 2920 \\ &= 17{,}520 \end{aligned}$$

8 × 365 = 2920 from Example 2

SECTION 4.5 EXERCISES

CONCEPT/WRITING EXERCISES

1. What are the three early computational methods discussed in this section?
2. **a)** Explain in your own words how multiplication by duplation and mediation is performed.
   b) Using the procedure given in part (a), multiply 267 × 193.
3. **a)** Explain in your own words how multiplication by the galley method is performed.
   b) Using the procedure given in part (a), multiply 423 × 27.
4. **a)** Explain in your own words how multiplication using Napier rods is performed.
   b) Multiply 25 × 6 using Napier rods.

PRACTICE THE SKILLS

In Exercises 5–12, multiply using duplation and mediation.

5. 17 × 29 **6.** 35 × 23
7. 9 × 162 **8.** 182 × 93
9. 35 × 236 **10.** 96 × 53
11. 85 × 85 **12.** 49 × 124

In Exercises 13–20, multiply using the galley method.

13. 7 × 365 **14.** 8 × 365
15. 4 × 583 **16.** 7 × 125
17. 75 × 12 **18.** 39 × 296
19. 314 × 652 **20.** 634 × 832

In Exercises 21–28, multiply using Napier rods.

21. 4 × 52 **22.** 7 × 63
23. 6 × 73 **24.** 7 × 125
25. 5 × 125 **26.** 75 × 125
27. 9 × 6742 **28.** 7 × 3456

PROBLEM SOLVING

In Exercises 29 and 30, we show multiplications using the galley method. (a) Determine the numbers being multiplied. Explain how you determine your answer. (b) Find the product.

29.

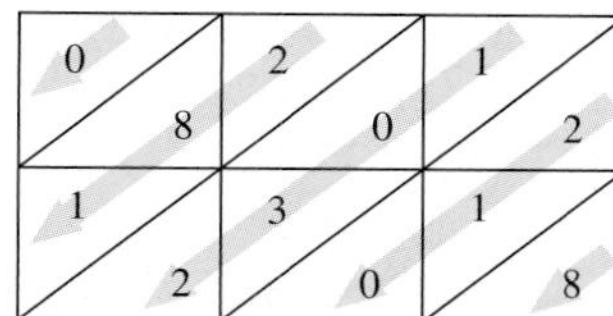

30. 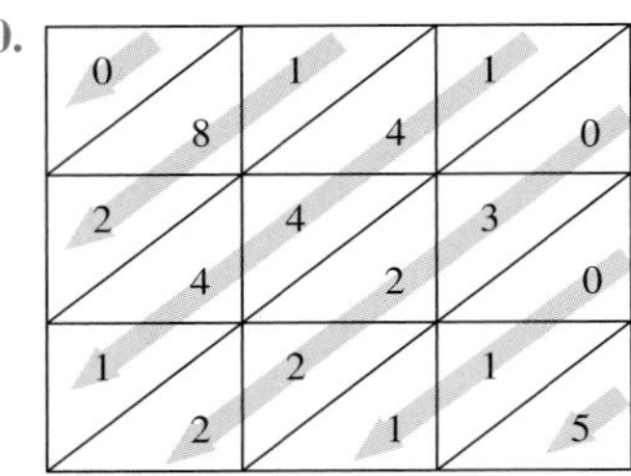

In Exercises 31 and 32, we solve a multiplication problem using Napier rods. (a) Determine the numbers being multiplied. Each empty box contains a single digit. Explain how you determined your answer. (b) Find the product.

31.

32.

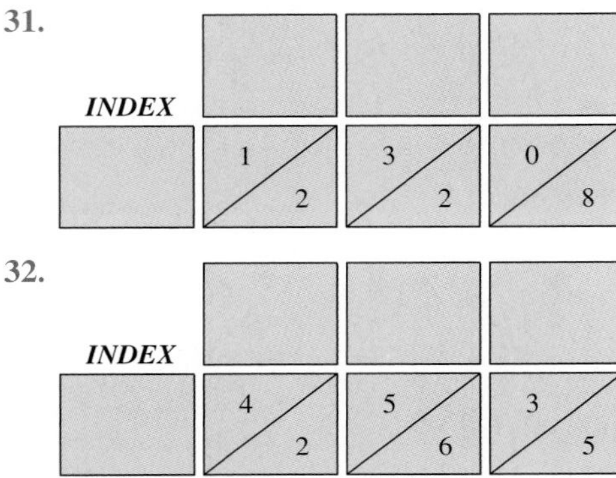

CHALLENGE PROBLEMS/GROUP ACTIVITIES

In Exercises 33 and 34, use the method of duplation and mediation to perform the multiplication. Write the answer in the numeration system in which the exercise is given.

33. (∩|||) • (∩∩||)

34. (XXVI) • (LXVII)

In Exercises 35 and 36, use the galley method to perform the multiplication. Write the answer in the base in which the exercise is given.

35. $12_3 \times 121_3$

36. $24_5 \times 234_5$

37. Develop a set of Napier rods that can be used to multiply numbers in base 5. Illustrate how your rods can be used to multiply $3_5 \times 21_5$.

RESEARCH ACTIVITIES

38. In addition to Napier rods, John Napier is credited with making other important contributions to mathematics. Write a report on John Napier and his contributions to mathematics.

39. Do research on the multiplication method of duplation and mediation and write a paper explaining why the technique works.

CHAPTER 4 SUMMARY

IMPORTANT FACTS

TYPES OF NUMERATION SYSTEMS

Additive (Egyptian hieroglyphics, Roman)
Multiplicative (traditional Chinese)
Ciphered (Ionic Greek)
Place-value (Babylonian, Mayan, Hindu–Arabic)

EARLY COMPUTATIONAL METHODS

Duplation and mediation
The galley method
Napier rods

CHAPTER 4 REVIEW EXERCISES

4.1, 4.2

In Exercises 1–6, assume an additive numeration system in which a = 1, b = 10, c = 100, and d = 1000. Find the value of the numeral.

1. *ddcaa* **2.** *abcda* **3.** *bcccad*
4. *cbdadaaa* **5.** *dddccbaaaa* **6.** *ccbaddac*

In Exercises 7–12, assume the same additive numeration system as in Exercises 1–6. Write the numeral in terms of a, b, c, and d.

7. 67 **8.** 125
9. 293 **10.** 2001
11. 6851 **12.** 2314

In Exercises 13–18, assume a multiplicative numeration system in which a = 1, b = 2, c = 3, d = 4, e = 5, f = 6, g = 7, h = 8, i = 9, x = 10, y = 100, and z = 1000. Find the value of the numeral.

13. *cxe* **14.** *bxg* **15.** *gydxi*

16. *dzfxh* **17.** *ezfydxh* **18.** *dzhyi*

In Exercises 19–24, assume the same multiplicative numeration system as in Exercises 13–18. Write the Hindu–Arabic numeral in that system.

19. 74 **20.** 295 **21.** 862

22. 2020 **23.** 6004 **24.** 2001

In Exercises 25–36, use the following ciphered numeration system.

Decimal	1	2	3	4	5	6	7	8	9
Units	a	b	c	d	e	f	g	h	i
Tens	j	k	l	m	n	o	p	q	r
Hundreds	s	t	u	v	w	x	y	z	A
Thousands	B	C	D	E	F	G	H	I	J
Ten thousands	K	L	M	N	O	P	Q	R	S

Convert the numeral to a Hindu–Arabic numeral.

25. pf **26.** uh **27.** woh

28. NGzqc **29.** Nqb **30.** Pwki

Write the numeral in the ciphered numeration system.

31. 42 **32.** 675 **33.** 493

34. 1997 **35.** 53,467 **36.** 75,496

In Exercises 37–42, convert 1462 to a numeral in the indicated numeration system.

37. Egyptian **38.** Roman **39.** Chinese

40. Ionic Greek **41.** Babylonian **42.** Mayan

In Exercises 43–48, convert the numeral to a Hindu–Arabic numeral.

43. 𓆐𓍢𓆼𓆼𓎆𓎆|||||

44. 八
千
二
百
五
十
四

45. $\phi\ \pi\ \epsilon$

46. MCMXCI

47. 𒌋𒌋𒁹 𒌋𒌋𒁹𒁹𒁹𒁹

48. —
•••
—
•
═

4.3

In Exercises 49–54, convert the numeral to a Hindu–Arabic numeral.

49. 54_9 **50.** 101_2 **51.** 130_4

52. 2746_8 **53.** T0E_{12} **54.** 20220_3

In Exercises 55–60, convert 463 to a numeral in the base indicated.

55. base 4 **56.** base 3 **57.** base 2

58. base 5 **59.** base 12 **60.** base 8

4.4

In Exercises 61–66, add in the base indicated.

61. $42_6 + 55_6$ **62.** $10110_2 + 11001_2$ **63.** $\text{TE}_{12} + 87_{12}$

64. $234_7 + 456_7$ **65.** $3024_5 + 4023_5$ **66.** $1407_8 + 7014_8$

In Exercises 67–72, subtract in the base indicated.

67. $4032_7 - 321_7$ **68.** $1001_2 - 101_2$ **69.** $\text{4TE}_{12} - \text{E7}_{12}$

70. $4321_5 - 442_5$ **71.** $1713_8 - 1243_8$ **72.** $2021_3 - 212_3$

In Exercises 73–78, multiply in the base indicated.

73. $22_5 \times 4_5$ **74.** $23_4 \times 21_4$ **75.** $126_{12} \times 47_{12}$

76. $221_3 \times 22_3$ **77.** $1011_2 \times 101_2$ **78.** $476_8 \times 23_8$

In Exercises 79–84, divide in the base indicated.

79. $1_2\overline{)1011_2}$ **80.** $2_4\overline{)320_4}$ **81.** $3_5\overline{)140_5}$

82. $4_6\overline{)3020_6}$ **83.** $3_6\overline{)2034_6}$ **84.** $6_8\overline{)5072_8}$

4.5

85. Multiply 142 × 24, using the duplation and mediation method.

86. Multiply 142 × 24, using the galley method.

87. Multiply 142 × 24, using Napier rods.

CHAPTER 4 TEST

1. Explain the difference between a numeral and a number.

In Exercises 2–7, convert the numeral to a Hindu–Arabic numeral.

2. MMDCXLVII
3. 𒌋𒌋𒁹 𒌋𒁹𒁹𒁹𒁹𒁹
4. 八
千
九
十
5. ••
•
═
••••
6. 𓂭𓂭𓆼𓆼𓍢𓎆𓎆𓎆𓎆𓏺𓏺
7. θ′ π Q θ

In Exercises 8–12, convert the number written in base 10 to a numeral in the numeration system indicated.

8. 352 to Egyptian
9. 2476 to Ionic Greek
10. 1512 to Mayan
11. 1596 to Babylonian
12. 2378 to Roman

In Exercises 13–16, describe briefly each of the systems of numeration. Explain how each type of numeration system is used to represent numbers.

13. Additive system
14. Multiplicative system
15. Ciphered system
16. Place-value system

In Exercises 17–20, convert the numeral to a numeral in base 10.

17. 37_8
18. 403_5
19. 101101_2
20. 368_9

In Exercises 21–24, convert the base 10 numeral to a numeral in the base indicated.

21. 36 to base 2
22. 84 to base 5
23. 2356 to base 12
24. 2938 to base 7

In Exercises 25–28, perform the indicated operations.

25. $\begin{array}{r} 133_5 \\ +\,434_5 \\ \hline \end{array}$

26. $\begin{array}{r} 425_7 \\ -\,154_7 \\ \hline \end{array}$

27. $\begin{array}{r} 45_6 \\ \times\,23_6 \\ \hline \end{array}$

28. $3_5\overline{)1210_5}$

29. Multiply 14 × 28, using duplation and mediation.
30. Multiply 43 × 196, using the galley method.

GROUP PROJECTS

U.S. POSTAL SERVICE BAR CODES

Wherever we look nowadays, we see bar codes. We find them on items we buy at grocery stores and department stores and on many pieces of mail we receive. There are various types of bar codes, but each can be considered a type of numeration system. Although bar codes may vary in design, most are made up of a series of long and short bars. (New bar codes now being developed use a variety of shapes). In this group project, we explain how postal codes are used.

The U.S. Postal Service introduced a bar coding system for zip codes in 1976. The system became known as Postnet (*post*al *n*umeric *e*ncoding *t*echniqe), and it has been refined over the years. Our basic zip code consists of five digits. The post office would like us to use the basic zip code followed by a dash and four additional digits. The post office refers to this nine-digit zip code as "zip + 4."

The Postnet bar code uses a series of long and short bars. A bar code may contain either 52 or 62 bars. The code designates the location to which the letter is being sent. The following bar code, with 52 bars, is for an address in Pittsburgh, Pennsylvania.

15250-7406 (Pittsburgh, PA)

In bar codes, each short bar represents 0 and each long bar represents 1. Each code starts and ends with a long bar that is *not* used in determining the zip + 4. If the code contains 52 bars, the code represents the zip + 4 and an extra digit referred to as a check digit. If the code contains 62 bars, it contains the zip + 4, the last two digits of the address number, and a check digit. If the code contains 52 bars, the sum of the zip + 4 and the check digit must equal a number that is divisible by 10. If the code contains 62 bars, the sum of the zip + 4, the last two digits of the address number, and the check digit must equal a number that is divisible by

10. The check digit is added to make each sum divisible by 10.

In a postal bar code, each of the digits 0 through 9 is represented by a series of five digits containing zeros and ones:

11000 (0)	00011 (1)	00101 (2)	00110 (3)	01001 (4)
01010 (5)	01100 (6)	10001 (7)	10010 (8)	10100 (9)

Consider the postal code from Pittsburgh given above. If you disregard the bar on the left, the next five bars are ıııII. Since each small bar represents a 0 and each large bar represents a 1, these five bars can be represented as 00011. From the chart, we see that this represents the number 1. The first five bars (after the bar on the far left has been excluded) tells the region of the country in which the address is located on the map shown below. Notice that Pennsylvania is located, along with New York, in the region marked 1 on the map.

National zip code areas

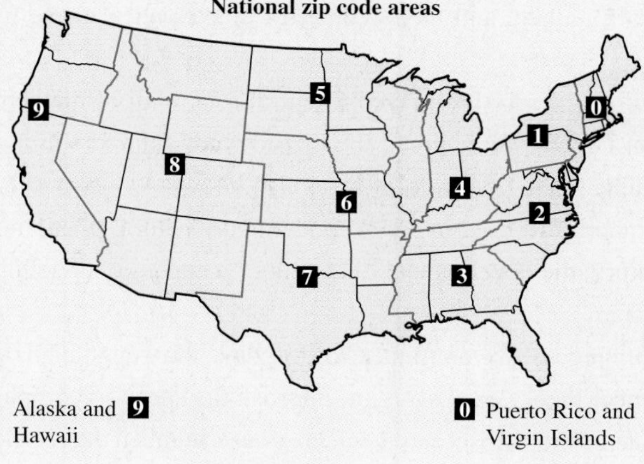

The second five lines in the bar code, ıIıIı represents 01010 and has a value of 5. The other digits in the zip + 4 are determined in a similar manner. This code has 52 bars. The 45 bars, after the first bar, give the zip + 4. If you add the digits in the zip + 4, you get 1 + 5 + 2 + 5 + 0 + 7 + 4 + 0 + 6 = 30. Since 30 is divisible by 10, the five bars to the left of the bar on the very right should be 0. Note that IIııı is represented as 11000 and has a value of 0. If, for example, the sum of the nine digits in the zip + 4 was 36, then the last five digits would need to represent the number 4, to make the sum of the digits divisible by 10. The five bars to the left of the last bar on the right are always used as a check.

Now lets work some problems.

a) For the Postnet code

IIIııııIIıııIııIIıııIIIıııIıIııIııIıııIIıIıIıııııIII

determine the zip + 4 and the check digit. Then check by adding the zip + 4 and the check digit. Is the sum divisible by 10?

b) For each of the following Postnet codes, determine the zip + 4, the last two numbers of the address number (if applicable), and the check digit.

i) IıIıIııııIIıIıIıIıIııııIIıIııIIıIııIIıIııIIıııII

ii) IıııIIıIıIıIıIıııIıIIıııIIıIıIıIıııIıIıIıIIıııIıIıIıIIıI

c) Construct the Postnet code of long and short bars for each of the numbers. The numbers represent the zip + 4 and the last two digits of the address number. Do not forget the check digit.

i) 32226-8600-34 **ii)** 20794-1063-50

d) Construct the 52-bar Postnet code for your college's zip + 4. Don't forget to include the check digit.

CHAPTER 5

Number Theory and the Real Number System

It is impossible to live in our modern world without encountering numbers on a regular basis. In addition to playing a role in our everyday lives, numbers are used to describe the natural world, to communicate vast quantities of information, and to model problems facing scientists and researchers. *Number theory,* the study of numbers and their properties, makes all of these roles possible.

Throughout history, number theory has played a central role to all of mathematics and science. The ancient Greeks were among the first to study numbers in a systematic fashion. In the Middle Ages, Leonardo of Pisa, also known as Fibonacci, introduced a sequence of numbers still used today to model many natural phenomena. During the seventeenth century, the development of calculus enabled the modeling of even more natural phenomena.

This central role of number theory continues to this day. Mathematicians and computer scientists use number theory extensively to improve the speed of computers in our homes, businesses, and schools. Research scientists use number theory along with another branch of mathematics known as **knot theory** to conduct DNA research, new drug research, and to study how infectious diseases spread. Engineers use number theory in satellite technology used in virtually every modern means of communication.

Perhaps more relevant to our everyday lives is the recent focus of the media on *innumeracy,* or the inability to use and understand mathematical information. Literacy of numbers, or *numeracy,* will be important to your career in the twenty-first century. ■

Number Theory Plays an important role in computer science, medical research and satellite technology.

5.1 NUMBER THEORY

This chapter introduces **number theory**, the study of numbers and their properties. The numbers we use to count are called the **counting numbers** or **natural numbers**. Since we begin counting with the number 1, the set of natural numbers begins with 1. The set of natural numbers is frequently denoted by N:

$$N = \{1, 2, 3, 4, 5, \ldots\}$$

Any natural number can be expressed as a product of two or more natural numbers. For example, $8 = 2 \times 4$, $16 = 4 \times 4$, and $19 = 1 \times 19$. The natural numbers that are multiplied together are called factors of the product. For example,

$$\underset{\uparrow}{2} \times \underset{\uparrow}{4} = 8$$

Factors

A natural number may have many factors. For example, what pairs of numbers have a product of 18?

$$1 \cdot 18 = 18$$
$$2 \cdot 9 = 18$$
$$3 \cdot 6 = 18$$

The numbers 1, 2, 3, 6, 9, and 18 are all factors of 18. Each of these numbers divides 18 without a remainder.

If a and b are natural numbers, we say that a is a **divisor** of b or a **divides** b, symbolized $a \mid b$, if the quotient of b divided by a has a remainder of 0. If a divides b, then b is **divisible** by a. For example, 4 divides 12, symbolized $4 \mid 12$, since the quotient of 12 divided by 4 has a remainder of 0. Note that 12 is divisible by 4. The notation $7 \nmid 12$ means that 7 does not divide 12. Note that every factor of a natural number is also a divisor of the natural number. *Caution:* Do not confuse the symbols, $a \mid b$ and a / b; $a \mid b$ means "a divides b" and a / b means "a divided by b" ($a \div b$). The symbols a / b and $a \div b$ indicate that the operation of division is to be performed, and b may or may not be a divisor of a.

> **DID YOU KNOW**
>
> **Lucky Numbers**
>
> 
>
> Triskaidekaphobia is the technical name for the fear of the number 13. Certain numbers are associated with certain beliefs. For example, 13 is considered by some to be unlucky, whereas 7 is considered lucky. The Greek mathematician Pythagoras, in the fifth century B.C., believed the number 1 was the originator of all numbers and signified reason. Two, the first female number, signified diversity of opinion. Three, the first male number, signified harmony (being composed of unity and diversity). Four stood for justice and retribution (the product of equals). Five, being the union of 2 and 3, stood for marriage.

PRIME AND COMPOSITE NUMBERS

Every natural number greater than 1 can be classified as either a prime number or a composite number.

> A **prime number** is a natural number greater than 1 that has exactly two factors (or divisors), itself and 1.

The number 5 is a prime number because it is divisible only by the factors 1 and 5. The first eight prime numbers are 2, 3, 5, 7, 11, 13, 17, and 19. The number 2 is the only even prime number. All other even numbers have at least three divisors: 1, 2, and the number itself.

A **composite number** is a natural number that is divisible by a number other than itself and 1.

Any natural number greater than 1 that is not prime is composite. The first eight composite numbers are 4, 6, 8, 9, 10, 12, 14, and 15.

The number 1 is neither prime nor composite; it is called a **unit**. The number 38 has at least three divisors, 1, 2, and 38, and hence is a composite number. In contrast, the number 23 is a prime number since its only divisors are 1 and 23.

More than 2000 years ago, the ancient Greeks developed a technique for determining which numbers are prime numbers and which are not. This technique is named the **sieve of Eratosthenes**, for the Greek mathematician Eratosthenes of Cyrene who first used it.

~~1~~	(2)	(3)	~~4~~	(5)	~~6~~	(7)	~~8~~	~~9~~	~~10~~
(11)	~~12~~	(13)	~~14~~	~~15~~	~~16~~	(17)	~~18~~	(19)	~~20~~
~~21~~	~~22~~	(23)	~~24~~	~~25~~	~~26~~	~~27~~	~~28~~	(29)	~~30~~
(31)	~~32~~	~~33~~	~~34~~	~~35~~	~~36~~	(37)	~~38~~	~~39~~	~~40~~
(41)	~~42~~	(43)	~~44~~	~~45~~	~~46~~	(47)	~~48~~	~~49~~	~~50~~

Figure 5.1

To find the prime numbers less than or equal to any natural number, say, 50, using this method, list the first 50 counting numbers (Fig. 5.1). Cross out 1 since it is not a prime number. Circle 2, the first prime number. Then cross out all the multiples of 2: 2, 4, 6, 8, . . . , 50. Circle the next prime number, 3. Cross out all multiples of 3 that are not already crossed out. Continue this process until you reach the prime number p, such that $p \cdot p$, or p^2, is greater than the last number listed, in this case 50. Next circle 5 and cross out its multiples. Then circle 7 and cross out its multiples. The next prime number is 11, and $11 \cdot 11$, or 121, is greater than 50, so you are done. At this point, circle all the remaining numbers to obtain the prime numbers less than or equal to 50. The prime numbers less than or equal to 50 are 2, 3, 5, 7, 11, 13, 17, 19, 23, 29, 31, 37, 41, 43, and 47.

Now we turn our attention to composite numbers and their factors. The rules of divisibility given in the chart on page 185 are helpful in finding divisors (or factors) of composite numbers.

The test for divisibility by 6 is a particular case of the general statement that the product of two prime divisors of a number is a divisor of the number. Thus, for example, if both 3 and 7 divide a number, then 21 will also divide the number.

Note that the chart does not list rules of divisibility for the number 7. There is a rule for 7, but it is difficult to remember. The easiest way to check divisibility by 7 is just to perform the division.

DID YOU KNOW

Cole's Factorization

In 1903, a mathematician named F. N. Cole received a standing ovation at a meeting of the American Mathematical Society (AMS). That was the first and only time that anyone has ever received a standing ovation at a meeting of the AMS. And he did so without saying a single word.

Cole walked to the chalkboard, raised 2 to the sixty-seventh power, and subtracted 1: $2^{67} - 1$. He then moved to a second chalkboard and wrote

$$\begin{array}{r} 761{,}838{,}257{,}287 \\ \times \quad 193{,}707{,}721 \\ \hline \end{array}$$

He then multiplied it out and came up with the same result as on the first chalkboard. Cole had factored a number long thought to be prime! And he found this factorization of the number $2^{67} - 1$ long before the age of the computer.

DID YOU KNOW

No More Odd Days!

The date November 19th, 1999 was a very special day. It was a rare "odd day". The numerical format is 11-19-1999 (or some may write it as 19-11-1999), which contains only odd digits. The next odd day will be January 1, 3111, which would be written 1-1-3111, and is over a thousand years away—a date we certainly will never see.

Days such as 4-13-1998 have both odd and even digits. Thus, it is neither an odd day nor an even day. The first of many even days in the year 2000 was 2-2-2000, the first one since 8-28-888.

So now you have a reason to celebrate, since you have seen your last odd day on earth!

Rules of Divisibility

Divisible by	Test	Example
2	The number is even.	924 is divisible by 2 since 924 is even.
3	The sum of the digits of the number is divisible by 3.	924 is divisible by 3 since the sum of the digits, 9 + 2 + 4 = 15, and 15 is divisible by 3.
4	The number formed by the last two digits of the number is divisible by 4.	924 is divisible by 4 since the number formed by the last two digits, 24, is divisible by 4.
5	The number ends in 0 or 5.	265 is divisible by 5 since the number ends in 5.
6	The number is divisible by both 2 and 3.	924 is divisible by 6 since it is divisible by both 2 and 3.
8	The number formed by the last three digits of the number is divisible by 8.	5824 is divisible by 8 since the number formed by the last three digits, 824, is divisible by 8.
9	The sum of the digits of the number is divisible by 9.	837 is divisible by 9 since the sum of the digits, 18, is divisible by 9.
10	The number ends in 0.	290 is divisible by 10 since the number ends in 0.

EXAMPLE 1 *Using the Divisibility Rules*

Determine whether 511,848 is divisible by

a) 2 b) 3 c) 4 d) 5 e) 6 f) 8 g) 9 h) 10

SOLUTION

a) Since 511,848 is even, it is divisible by 2.
b) The sum of the digits of 511,848 is 5 + 1 + 1 + 8 + 4 + 8 = 27. Since 27 is divisible by 3, the number 511,848 is divisible by 3.
c) The number formed by the last two digits is 48. Since 48 is divisible by 4, the number 511,848 is divisible by 4.
d) Since 511,848 does not have 0 or 5 as the last digit, it is not divisible by 5.
e) Since 511,848 is divisible by both 2 and 3, it is divisible by 6.
f) The number formed by the last three digits is 848. Since 848 is divisible by 8, the number 511,848 is divisible by 8.
g) The sum of the digits of 511,848 is 27. Since 27 is divisible by 9, the number 511,848 is divisible by 9.
h) Since 511,848 does not have zero as its last digit, the number is not divisible by 10.

Every composite number can be expressed as a product of prime numbers. The process of breaking a given number down into a product of prime numbers is called **prime factorization**. The prime factorization of 18 is $3 \times 3 \times 2$. No other natural number listed as a product of primes will have the same prime factorization as 18. The *fundamental theorem of arithmetic* states this concept formally. (A **theorem** is a statement or proposition that can be proven true.)

The Fundamental Theorem of Arithmetic
Every composite number can be expressed as a *unique* product of prime numbers.

In writing the prime factorization of a number, the order of the factors is immaterial. For example, we may write the prime factors of 18 as $3 \times 3 \times 2$ or $2 \times 3 \times 3$ or $3 \times 2 \times 3$.

A number of techniques can be used to find the prime factorization of a number. Two methods are illustrated.

METHOD 1: BRANCHING

To find the prime factorization of a number, select any two numbers whose product is the number to be factored. If the factors are not prime numbers, then continue factoring each composite number until all numbers are prime.

EXAMPLE 2 *Prime Factorization by Branching*

Write 450 as a product of primes.

SOLUTION Select any two numbers whose product is 450. Among the many choices, two possibilities are $45 \cdot 10$ and $9 \cdot 50$. Let us consider $45 \cdot 10$. Now find any two numbers whose product is 45 and any two numbers whose product is 10. Continue branching as shown in Fig. 5.2 until the numbers in the last row are all prime numbers. To determine the answer, write the product of all the prime factors. The branching diagram is sometimes called a *factor tree.*

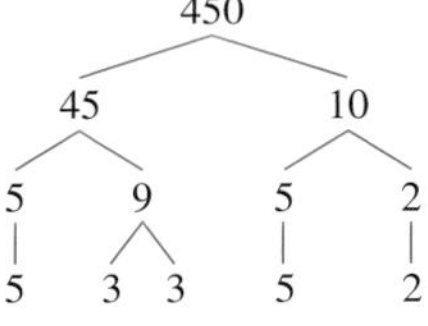

Figure 5.2

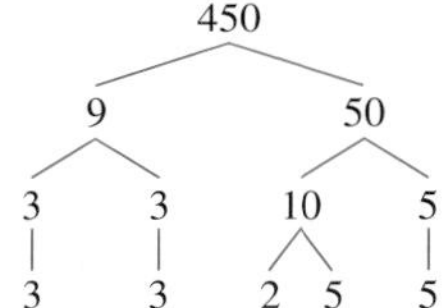

Figure 5.3

We see that the numbers in the last row of factors in Fig. 5.2 are all prime numbers. Thus, the prime factorization of 450 is $5 \cdot 3 \cdot 3 \cdot 5 \cdot 2 = 2 \cdot 3 \cdot 3 \cdot 5 \cdot 5 = 2 \cdot 3^2 \cdot 5^2$. Note from Fig. 5.3 that had we chosen 9 and 50 as the first pair of factors, the prime factorization would still be $2 \cdot 3^2 \cdot 5^2$. ▲

METHOD 2: DIVISION

To obtain the prime factorization of a number by this method, divide the given number by the smallest prime number by which it is divisible. Place the quotient under the given number. Then divide the quotient by the smallest prime number by which it is divisible and again record the quotient. Repeat this process until the quotient is a prime number. The prime factorization is the product of all the prime divisors and the prime (or last) quotient. This procedure is illustrated in Example 3.

EXAMPLE 3 *Prime Factorization by Division*

Write 450 as a product of prime numbers.

SOLUTION Because 450 is an even number, the smallest prime number that divides it is 2. Divide 450 by 2. Place the quotient, 225, below the 450. Repeat this process of dividing each quotient by the smallest prime number that divides it.

DID YOU KNOW

The Long-Lost Factoring Machine

In 1989, Jeffrey Shallit of the University of Waterloo came across an article in a 1920 French journal regarding a machine that was built in 1914 for factoring numbers. Eugene Oliver Carissan, an amateur mathematician, who had invented the machine wrote the article. After reading the article, Shallit wondered whatever became of the factoring machine.

After considerable searching, Shallit found the machine in a drawer of an astronomical observatory in Floirac, France. The machine was in good condition, and it still worked. By rotating the machine by a hand crank at two revolutions per minute, an operator could process 35 to 40 numbers per second. Carissan needed just 10 minutes to prove that 708,158,977 is a prime number, an amazing feat in precomputer times. The machine is now housed at the Conservatoire National des Arts et Métiers in Paris.

PROFILE IN MATHEMATICS

Srinivasa Ramanujan

One of the most interesting mathematicians of modern times is Srinivasa Ramanujan (1887–1920). Born to an impoverished middle-class family in India, he virtually taught himself higher mathematics. He went to England to study with the number theorist G. H. Hardy. Hardy tells the story of a taxicab ride he took to visit Ramanujan. The cab had the license plate number 1729, and he challenged the young Indian to find anything interesting in that. Without hesitating, Ramanujan pointed out that it was the smallest positive integer that could be represented in two different ways as the sum of two cubes: $1^3 + 12^3$ and $9^3 + 10^3$.

2	450
3	225
3	75
5	25
	5

The final quotient, 5, is a prime number, so we stop. The prime factorization is

$$450 = 2 \cdot 3 \cdot 3 \cdot 5 \cdot 5 = 2 \cdot 3^2 \cdot 5^2.$$

Note that, despite the different methods used in Examples 2 and 3, the answer is the same.

GREATEST COMMON DIVISOR

The discussion in Section 5.3 of how to reduce fractions makes use of the greatest common divisor (GCD). One technique of finding the GCD is to use prime factorization.

The **greatest common divisor (GCD)** of a set of natural numbers is the largest natural number that divides (without remainder) every number in that set.

What is the GCD of 12 and 18? One way to determine it is to make a list of the divisors (or factors) of 12 and 18:

Divisors of 12	{**1**, **2**, **3**, 4, **6**, 12}
Divisors of 18	{**1**, **2**, **3**, **6**, 9, 18}

The common divisors are 1, 2, 3, and 6, and therefore the greatest common divisor is 6.

If the numbers are large, this method of finding the GCD is not practical. The GCD can be found more efficiently by using prime factorization.

To Find the Greatest Common Divisor of Two or More Numbers

1. Determine the prime factorization of each number.
2. Find each prime factor with the smallest exponent that appears in each of the prime factorizations.
3. Determine the product of the factors found in step 2.

Example 4 illustrates this procedure.

EXAMPLE 4 ***Using Prime Factorization to Find the GCD***

Find the GCD of 54 and 90.

SOLUTION The branching method of finding the prime factors of 54 and 90 is illustrated in Fig. 5.4.

DID YOU KNOW

Friendly Numbers

The ancient Greeks often thought of numbers as having human qualities. For example, the numbers 220 and 284 were considered "friendly" or "amicable" numbers because each number was the sum of the other number's proper factors. (A proper factor is any factor of a number other than the number itself.) If you sum all the proper factors of 284 (1 + 2 + 4 + 71 + 142), you get the number 220, and if you sum all the proper factors of 220 (1 + 2 + 4 + 5 + 10 + 11 + 20 + 22 + 44 + 55 + 110), you get 284.

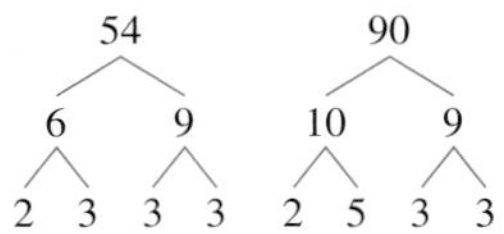

Figure 5.4

a) The prime factorization of 54 is $2 \cdot 3^3$, and the prime factorization of 90 is $2 \cdot 3^2 \cdot 5$.
b) The prime factors with the smallest exponents that appear in each of the factorizations of 54 and 90 are 2 and 3^2.
c) The product of the factors found in step 2 is $2 \cdot 3^2 = 2 \cdot 9 = 18$. The GCD of 54 and 90 is 18. Eighteen is the largest number that divides both 54 and 90.

EXAMPLE 5 *Finding the GCD*

Find the GCD of 225 and 525.

SOLUTION

a) The prime factorization of 225 is $3^2 \cdot 5^2$, and the prime factorization of 525 is $3 \cdot 5^2 \cdot 7$ (you should verify these using the branching method or the division method).
b) The prime factors with the smallest exponents that appear in each of the factorizations of 225 and 525 are 3 and 5^2.
c) The product of the factors found in step 2 is $3 \cdot 5^2 = 3 \cdot 25 = 75$. The GCD of 225 and 525 is 75.

Two numbers with a GCD of 1 are said to be **relatively prime**. The numbers 9 and 14 are relatively prime, since the GCD is 1.

LEAST COMMON MULTIPLE

To perform addition and subtraction of fractions (Section 5.3), we use the least common multiple (LCM). One technique of finding the LCM is to use prime factorization.

The **least common multiple (LCM)** of a set of natural numbers is the smallest natural number that is divisible (without remainder) by each element of the set.

What is the least common multiple of 12 and 18? One way to determine the LCM is to make a list of the multiples of each number:

Multiples of 12 {12, 24, **36**, 48, 60, **72**, 84, 96, **108**, 120, 132, **144**, . . .}
Multiples of 18 {18, **36**, 54, **72**, 90, **108**, 126, **144**, 162, . . .}

Some common multiples of 12 and 18 are 36, 72, 108, and 144. The least common multiple, 36, is the smallest number that is divisible by both 12 and 18. Usually, the most efficient method of finding the LCM is to use prime factorization.

DID YOU KNOW

Keeping Secrets

Prime numbers are essential in the protection of databases used by government and business. Data have to be *encrypted* so that lawful users have access but unlawful users do not. One means of doing this is the Rivest–Shamir–Adleman (RSA) *cryptoalgorithm.* The key to entering the database is a large composite number believed to be over 300 digits long. This number, n, is the product of two very large prime numbers, $p \cdot q$. The number n is publicly available, so information can be entered into the system. The numbers p and q, however, are given only to those who have the right to access the data. The database is considered secure because currently the largest composite number ever factored with a computer is 167 digits long. This 167-digit number was found to be the product of an 80-digit number and an 87-digit number by Samuel Wagstaff of Purdue University. To accomplish this feat, he used a computer and a factoring method known as the number-field sieve. The process took 100,000 hours of computing time.

To Find the Least Common Multiple of Two or More Numbers

1. Determine the prime factorization of each number.
2. List each prime factor with the greatest exponent that appears in any of the prime factorizations.
3. Determine the product of the factors found in step 2.

Example 6 illustrates this procedure.

EXAMPLE 6 ***Using Prime Factorization to Find the LCM***

Find the LCM of 54 and 90.

SOLUTION

a) Find the prime factors of each number. In Example 4, we determined that

$$54 = 2 \cdot 3^3 \qquad \text{and} \qquad 90 = 2 \cdot 3^2 \cdot 5$$

b) List each prime factor with the greatest exponent that appears in either of the prime factorizations: $2, 3^3, 5$.
c) Determine the product of the factors found in step 2:

$$2 \cdot 3^3 \cdot 5 = 2 \cdot 27 \cdot 5 = 270$$

Thus, 270 is the LCM of 54 and 90. It is the smallest number that is divisible by both 54 and 90.

EXAMPLE 7 ***Finding the LCM***

Find the LCM of 225 and 525.

SOLUTION

a) Find the prime factors of each number. In Example 5, we determined that

$$225 = 3^2 \cdot 5^2 \qquad \text{and} \qquad 525 = 3 \cdot 5^2 \cdot 7$$

b) List each prime factor with the greatest exponent that appears in either of the prime factorizations: $3^2, 5^2, 7$.
c) Determine the product of the factors found in step 2:

$$3^2 \cdot 5^2 \cdot 7 = 9 \cdot 25 \cdot 7 = 1575$$

Thus, 1575 is the least common multiple of 225 and 525. It is the smallest number divisible by both 225 and 525.

THE SEARCH FOR LARGER PRIME NUMBERS

More than 2000 years ago, the Greek mathematician Euclid proved that there is no largest prime number. Mathematicians, however, continue to strive to find larger and larger prime numbers.

DID YOU KNOW

GIMPS

Hajratwala

The Great Internet Mersenne Prime Search (GIMPS) was started by George Woltman to provide free software and access to a large database to coordinate the efforts of thousands of people interested in seeking new Mersenne prime numbers. Scott Kurowski provided the technology and services needed to make it simple for anyone to join the project via the Internet. GIMPS allows participants to use their computers to search for prime numbers while their computers are idle. A computer communicates, via the specialized software, with the database. The largest Mersenne prime to date was found by Nayan Hajratwala in conjunction with GIMPS. Hajratwala's home computer used 111 days of idle time to find this number.

Marin Mersenne (1588–1648), a seventeenth-century monk, found that numbers of the form $2^n - 1$ are often prime numbers when n is a prime number. For example,

$$2^2 - 1 = 4 - 1 = 3 \qquad 2^3 - 1 = 8 - 1 = 7$$
$$2^5 - 1 = 32 - 1 = 31 \qquad 2^7 - 1 = 128 - 1 = 127$$

Numbers of the form $2^n - 1$ that are prime are referred to as **Mersenne primes**. The first 10 Mersenne primes occur when $n = 2, 3, 5, 7, 13, 17, 19, 31, 61, 89$. The first time the expression $2^n - 1$ does not generate a prime number, for prime number n, is when n is 11. The number $2^{11} - 1$ is a composite number (see Exercise 86).

Scientists frequently use Mersenne primes in their search for larger and larger primes. The largest prime found to date was discovered in June 1999 by Nayan Hajratwala of Plymouth, Michigan, in conjunction with the Great Internet Mersenne Prime Search (GIMPS; see the Did You Know). The number is the Mersenne prime $2^{6,972,593} - 1$. This record prime is the 38th known Mersenne prime, and when written out, it is 2,098,960 digits long, enough to fill almost 80 newspaper pages!

MORE ABOUT PRIME NUMBERS

Another mathematician who studied prime numbers was Pierre de Fermat (1601–1665). A lawyer by profession, Fermat became interested in mathematics as a hobby. He became one of the finest mathematicians of the seventeenth century. Fermat conjectured that each number of the form $2^{2^n} + 1$, now referred to as the **Fermat number**, was prime for each natural number n. Recall that a *conjecture* is a supposition that has not been proved nor disproved. In 1732, Leonhard Euler proved that for $n = 5$, $2^{32} + 1$ was a composite number, thus disproving Fermat's conjecture.

Since Euler's time, mathematicians have only been able to evaluate the sixth, seventh, eighth, ninth, tenth, and eleventh Fermat numbers to determine whether they are prime or composite. Each of these numbers has been shown to be composite. The eleventh Fermat number was factored by Richard Brent and François Morain in 1988. The sheer magnitude of the numbers involved makes it difficult to test these numbers, even with supercomputers.

In 1742, Christian Goldbach conjectured in a letter to Euler that every even number greater than or equal to 4 can be represented as the sum of two (not necessarily distinct) prime numbers (for example, $4 = 2 + 2$, $6 = 3 + 3$, $8 = 3 + 5$, $10 = 5 + 5$, $12 = 5 + 7$). This conjecture became known as **Goldbach's conjecture**, and it remains unproven to this day. The **twin prime conjecture** is another famous long-standing conjecture. **Twin primes** are primes of the form p and $p + 2$ (for example, 3 and 5, 5 and 7, 11 and 13). This conjecture states that there are an infinite number of pairs of twin primes. Currently, the largest twin primes are of the form $361{,}700{,}055 \cdot 2^{39{,}020}$ plus or minus 1, which were discovered by Henri Lifchitz in 1999.

SECTION 5.1 EXERCISES

CONCEPT/WRITING EXERCISES

1. What is number theory?
2. What does "a and b are factors of c" mean?
3. a) What does "a divides b" mean?
   b) What does "a is divisible by b" mean?
4. What is a prime number?

5. What is a composite number?

6. What does the fundamental theorem of arithmetic state?

7. a) What is the greatest common divisor of a set of natural numbers?
   b) In your own words, explain how to find the GCD of a set of natural numbers by using prime factorization.
   c) Find the GCD of 16 and 40 by using the procedure given in part (b).

8. a) What is the least common multiple of a set of natural numbers?
   b) In your own words, explain how to find the LCM of a set of natural numbers by using prime factorization.
   c) Find the LCM of 16 and 40 by using the procedure given in part (b).

9. What are Mersenne primes?

10. What is a conjecture?

11. What is Goldbach's conjecture?

12. What are twin primes?

PRACTICE THE SKILLS

13. Use the sieve of Eratosthenes to find the prime numbers up to 75.

14. Use the sieve of Eratosthenes to find the prime numbers up to 100.

In Exercises 15–26, determine whether the statement is true or false. Modify each false statement to make it a true statement.

15. 7 is a factor of 42.

16. $7 \mid 56$

17. Forty-two is a multiple of 7.

18. Seven is a divisor of 42.

19. Seven is divisible by 42.

20. Seven is a multiple of 42.

21. If a number is not divisible by 5, then it is not divisible by 10.

22. If a number is not divisible by 10, then it is not divisible by 5.

23. If a number is divisible by 3, then every digit of the number is divisible by 3.

24. If every digit of a number is divisible by 3, then the number itself is divisible by 3.

25. If a number is divisible by 2 and 3, then the number is divisible by 6.

26. If a number is divisible by 3 and 4, then the number is divisible by 12.

In Exercises 27–32, determine whether the number is divisible by each of the following numbers: 2, 3, 4, 5, 6, 8, 9, and 10.

27. 48,324 28. 529,200

29. 2,763,105 30. 3,126,120

31. 1,882,320 32. 3,941,221

33. Determine a number that is divisible by 2, 3, 4, 5, and 6.

34. Determine a number that is divisible by 3, 4, 5, 9, and 10.

In Exercises 35–46, find the prime factorization of the number.

35. 44 36. 51 37. 72

38. 150 39. 303 40. 400

41. 513 42. 663 43. 1336

44. 1313 45. 2001 46. 3190

In Exercises 47–56, find a) the greatest common divisor (GCD) and b) the least common multiple (LCM).

47. 15 and 18 48. 15 and 44

49. 42 and 56 50. 52 and 65

51. 40 and 900 52. 120 and 240

53. 96 and 212 54. 240 and 285

55. 24, 48, and 128 56. 18, 78, and 198

PROBLEM SOLVING

57. Find the next two sets of twin primes that follow the set 11, 13.

58. The primes 2 and 3 are consecutive natural numbers. Is there another pair of consecutive natural numbers both of which are prime? Explain.

59. Show that Goldbach's conjecture is true for the even numbers 4 through 20.

60. Find the first five Mersenne prime numbers.

61. Find the first three Fermat numbers and determine whether they are prime or composite.

62. *Airport Activity* O'Hare International Airport in Chicago has a flight leaving for New York City every 45 minutes and a flight leaving for Atlanta every 60 minutes. If a flight to New York City and a flight to Atlanta leave at the same time, how many minutes will it be before a flight to New York City and a flight to Atlanta again leave at the same time?

63. *Tool Sale Frequency* Sears has a sale on mechanics' tool chests every 40 days and a sale on mechanics' tool sets every 60 days. If both are on sale today, how long will it be before they are on sale together again?

64. *Work Schedules* Sara and Harry both work the 3:00 P.M. to 11:00 P.M. shift. Sara has every fifth night off and Harry has every sixth night off. If they both have tonight off, how many days will pass before they have the same night off again?

65. *Restaurant Service* Bill Leonard runs a professional accounting service for restaurants. Bill goes to Arturo's Family Restaurant every 15 days, and he goes to Xang's Great Wall Restaurant every 18 days. If on October 1 Bill visits both restaurants, how many days would it be before he visited both restaurants on the same day again?

66. *Stacking Trading Cards* John collects trading cards. He has 432 baseball cards and 360 football cards. He wants to make stacks of cards on a table so that each stack contains the same number of cards and each card belongs to one stack. If the baseball and football cards must not be mixed in the stacks, what is the largest number of cards that he can have in a stack?

67. *Toy Car Collection* Karin DeJamaer collects Matchbox® and HotWheels® toy cars. She has 288 red cars and 192 blue cars. She wants to line up her cars in groups so that each group has the same number of cars and each group contains only red cars or only blue cars. What is the largest number of cars she can have in a group?

68. *U.S. Senate Committees* The U.S. Senate consists of 100 members. Senate committees are to be formed so that each of the committees contains the same number of senators and each senator is a member of exactly one committee. The committees are to have more than 2 members but fewer than 50 members. There are various ways that these committees can be formed.

a) What size committees are possible?

b) How many committees are there for each size?

69. Consider the first eight prime numbers greater than 3. The numbers are 5, 7, 11, 13, 17, 19, 23, and 29.

a) Determine which of these prime numbers differs by 1 from a multiple of the number 6.

b) Use inductive reasoning and the results obtained in part (a) to make a conjecture regarding prime numbers.

c) Select a few more prime numbers and determine whether your conjecture appears to be correct.

70. State a procedure that defines a divisibility test for 15.

Another method that can be used to find the greatest common divisor is known as the ***Euclidean algorithm.*** *We illustrate this procedure by finding the GCD of 60 and 220.*

First divide 220 by 60. Disregard the quotient 3 and then divide 60 by the remainder 40. Continue this process of dividing the divisors by the remainders until you obtain a remainder of 0. The divisor in the last division, in which the remainder is 0, is the GCD.

$$\begin{array}{r} 3 \\ 60\overline{)220} \\ \underline{180} \\ 40 \end{array} \qquad \begin{array}{r} 1 \\ 40\overline{)60} \\ \underline{40} \\ 20 \end{array} \qquad \begin{array}{r} 2 \\ 20\overline{)40} \\ \underline{40} \\ 0 \end{array}$$

Since 40/20 had a remainder of 0, the GCD is 20.

In Exercises 71–76, use the Euclidean algorithm to find the GCD.

71. 35, 75 **72.** 20, 160

73. 18, 112 **74.** 96, 115

75. 150, 180 **76.** 210, 560

A number whose proper factors *(factors other than the number itself) add up to the number is called a* perfect number. *For example, 6 is a perfect number because its proper factors are 1, 2, and 3, and $1 + 2 + 3 = 6$. Determine which, if any, of the following numbers are perfect.*

77. 12 **78.** 28

79. 496 **80.** 48

CHALLENGE PROBLEMS/GROUP ACTIVITIES

81. The following procedure can be used to determine the *number of factors (or divisors)* of a composite number. Write the number in prime factorization form. Examine the exponents on the prime numbers in the prime factorization. Add 1 to each exponent and then find the product of these numbers. This product gives the number of positive divisors of the composite number.

a) Use this procedure to determine the number of divisors of 60.

b) To check your answer, list all the divisors of 60. You should obtain the same number of divisors found in part (a).

82. Recall that if a number is divisible by both 2 and 3, then the number is divisible by 6. If a number is divisible by both 2 and 4, is the number necessarily divisible by 8? Explain your answer.

83. The product of any three consecutive natural numbers is divisible by 6. Explain why.

84. A number in which each digit except 0 appears exactly three times is divisible by 3. For example, 888,444,555 and 714,714,714 are both divisible by 3. Explain why this outcome must be true.

85. Use the fact that if $a \mid b$ and $a \mid c$, then $a \mid (b + c)$ to determine whether 36,018 is divisible by 18. (*Hint:* Write 36,018 as 36,000 + 18.)

86. Show that the $2^n - 1$ is a (Mersenne) prime for $n = 2, 3, 5$, and 7 but composite for $p = 11$.

87. Goldbach also conjectured in his letter to Euler that *every* integer greater than 5 is the sum of three prime numbers. For example, $6 = 2 + 2 + 2$ and $7 = 2 + 2 + 3$. Show that this conjecture is true for integers 8 through 20.

RESEARCH ACTIVITIES

88. Do research and explain what *deficient numbers* and *abundant numbers* are. Give an example of each type of number. References include history of mathematics books, encyclopedias, and the Internet.

89. Conduct an Internet search on the GIMPS project. Write a report describing the history and development of the project. Include a current update of the project's findings.

5.2 THE INTEGERS

In Section 5.1, we introduced the natural or counting numbers:

$$N = \{1, 2, 3, 4, \dots\}$$

Another important set of numbers, the **whole numbers**, help to answer the question "How many?"

$$\text{Whole numbers} = \{0, 1, 2, 3, 4, \dots\}$$

Note that the set of whole numbers contains the number 0 but that the set of counting numbers does not. If a farmer were asked how many chickens were in a coop, the answer would be a whole number. If the farmer had no chickens, he or she would answer zero. Although we use the number 0 daily and take it for granted, the number 0 as we know it was not used and accepted until the sixteenth century.

If the temperature is 12°F and drops 20°, the resulting temperature is −8°F. This type of problem shows the need for negative numbers. The set of **integers** consists of the negative integers, 0, and the positive integers.

$$\text{Integers} = \{\underbrace{\dots, -4, -3, -2, -1}_{\text{Negative integers}}, 0, \underbrace{1, 2, 3, \dots}_{\text{Positive integers}}\}$$

The term *positive integers* is yet another name for the natural numbers or counting numbers.

An understanding of addition, subtraction, multiplication, and division of the integers is essential in understanding algebra (Chapter 6). To aid in our explanation of addition and subtraction of integers, we introduce the real number line (Fig. 5.5). To construct the real number line, arbitrarily select a point for zero to serve as the starting point. Place the positive integers to the right of 0, equally spaced from one another. Place the negative integers to the left of 0, using the same spacing. The real number line contains the integers and all of the other real numbers that are not integers. Some examples of real numbers that are not integers are indicated in Fig. 5.5. We discuss real numbers that are not integers in the next two sections.

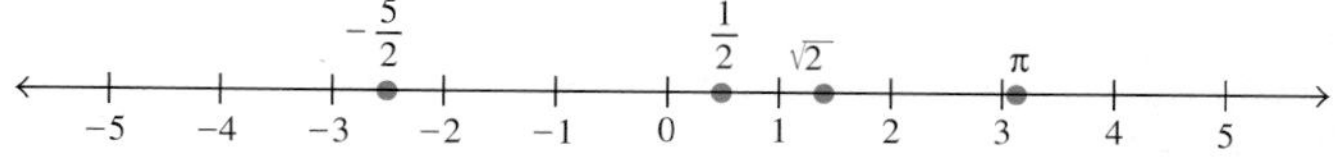

Figure 5.5

Virgil's Party

The poet Virgil was born in 70 B.C. In 1930, some highly respected scholars decided to celebrate his 2000th birthday. The only problem was that the accounting of time does not include a year 0 between B.C. and A.D. So Virgil would have turned 70 in our year 1 A.D. and, had he lived so long, 2000 in 1931. This fact was pointed out to the celebrants, but only after the party was well under way.

The arrows at the ends of the real number line indicate that the line continues indefinitely in both directions. Note that for any natural number, n, on the number line, the *opposite of* that number, $-n$, is also on the number line. This real number line was drawn horizontally, but it could just as well have been drawn vertically. In fact, in the next chapter, we show that the axes of a graph are the union of two number lines, one horizontal and the other vertical.

The number line can be used to determine the order of two integers. On the number line, the numbers increase from left to right. The number 3 is greater than 2, written $3 > 2$. Observe that 3 is to the right of 2. Similarly, we can see that $0 > -1$ by observing that 0 is to the right of -1 on the number line.

Instead of stating that 3 is greater than 2, we could state that 2 is less than 3, written $2 < 3$. Note that 2 is to the left of 3 on the number line. We can also see that $-1 < 0$ by observing that -1 is to the left of 0. The inequality symbol always points to the smaller of the two numbers when the inequality is true.

EXAMPLE 1 *Writing an Inequality*

Insert either $>$ or $<$ in the shaded area between the paired numbers to make the statement correct.

a) -3 ■ 1 b) -3 ■ -5 c) -6 ■ -4 d) 0 ■ -7

SOLUTION

a) $-3 < 1$ since -3 is to the left of 1 on the number line.
b) $-3 > -5$ since -3 is to the right of -5 on the number line.
c) $-6 < -4$ since -6 is to the left of -4 on the number line.
d) $0 > -7$ since 0 is to the right of -7 on the number line.

ADDITION OF INTEGERS

Addition of integers can be represented geometrically with a number line. To do so, begin at 0 on the number line. Represent the first addend by an arrow starting at 0. Draw the arrow to the right if the addend is positive. If the addend is negative, draw the arrow to the left. From the tip of the first arrow, draw a second arrow to represent the second addend. Draw the second arrow to the right or left, as just explained. The sum of the two integers is found at the tip of the second arrow.

EXAMPLE 2 *Adding Integers*

Evaluate the following using the number line.

a) $3 + (-5)$ b) $-1 + (-4)$ c) $-6 + 4$ d) $3 + (-3)$

SOLUTION

a)

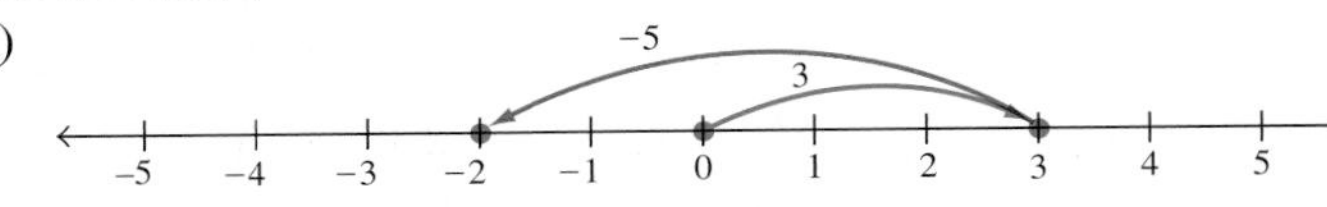

Thus, $3 + (-5) = -2$

b)

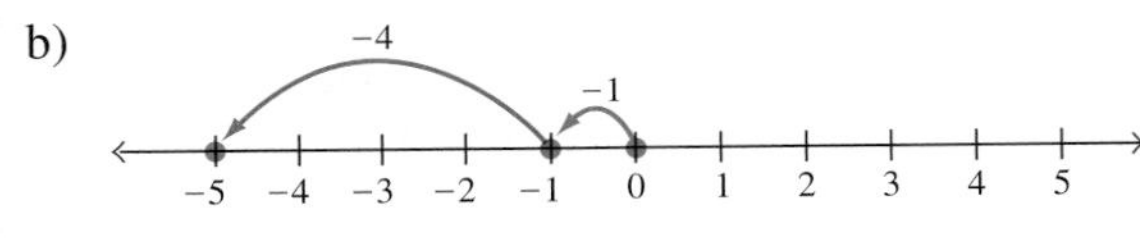

Thus, $-1 + (-4) = -5$

c)

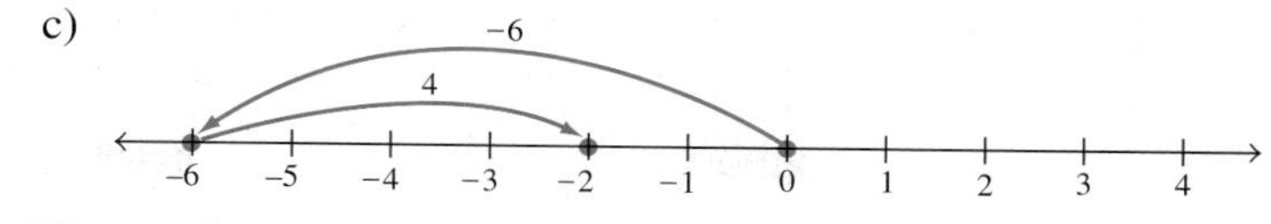

Thus, $-6 + 4 = -2$

d)

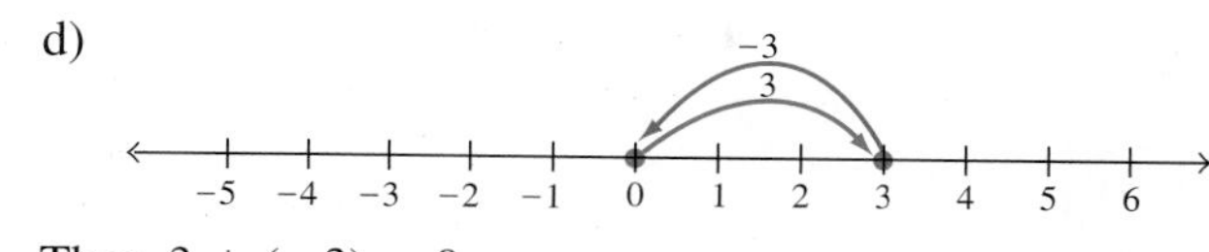

Thus, $3 + (-3) = 0$.

In Example 2(d), the number -3 is said to be the additive inverse of 3 and 3 is the additive inverse of -3, because their sum is 0. In general, the **additive inverse** of the number n is $-n$, since $n + (-n) = 0$. Inverses are discussed more formally in Chapter 10.

SUBTRACTION OF INTEGERS

Any subtraction problem can be rewritten as an addition problem. To do so, we use the following definition of subtraction.

Subtraction

$$a - b = a + (-b)$$

The rule for subtraction indicates that to subtract b from a, *add* the additive inverse of b to a. For example,

$$3 - 5 = 3 + (-5)$$

↑ Subtraction ↑ Addition ↑ Additive inverse of 5

Now we can determine the value of $3 + (-5)$.

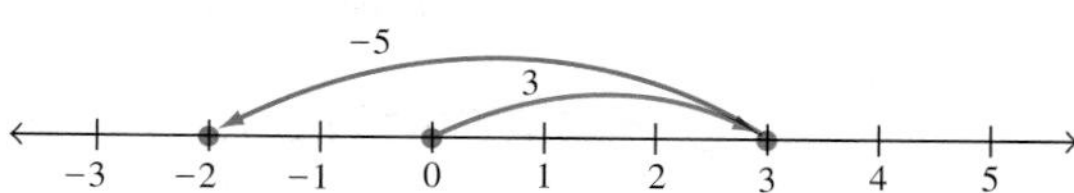

Thus, $3 - 5 = 3 + (-5) = -2$.

EXAMPLE 3 *Subtracting Integers*

Evaluate $-4 - (-3)$ using the number line.

SOLUTION We are subtracting -3 from -4. The additive inverse of -3 is 3; therefore, we add 3 to -4. We now add $-4 + 3$ on the number line to obtain the answer -1.

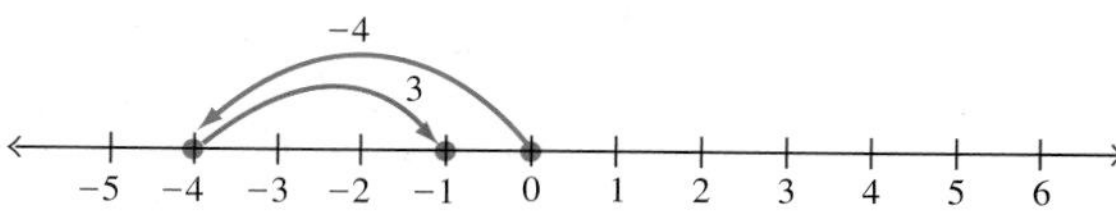

Thus, $-4 - (-3) = -4 + 3 = -1$.

Sparsely Populated

More people have made it into outer space than have stood on the top of Nepal's Mount Everest. Fewer people still have visited the Pacific Ocean's Mariana Trench. In 1960, Jacques Piccard and Don Walsh were the first to make the journey in a vessel designed to withstand the immense pressure: 7 tons per square inch. The diving time was $8\frac{1}{2}$ hours, of which only 20 minutes were spent on the ocean floor.

In Example 3, we found that $-4 - (-3) = -4 + 3$. In general, $a - (-b) = a + b$. As you get more proficient in working with integers, you should be able to answer questions involving them without drawing a number line.

EXAMPLE 4 ***Subtracting: Adding the Inverse***

Evaluate a) $-3 - 5$ b) $4 - (-3)$

SOLUTION

a) $-3 - 5 = -3 + (-5) = -8$ b) $4 - (-3) = 4 + 3 = 7$ ▲

EXAMPLE 5 ***Elevation Difference***

The highest point on Earth is Mount Everest, in the Himalayas, at a height of 29,035 ft above sea level. The lowest point on Earth is the Mariana Trench, in the Pacific Ocean, at a depth of 36,198 ft below sea level (−36,198 ft). Find the vertical height difference between Mount Everest and the Mariana Trench.

SOLUTION We obtain the vertical difference by subtracting the lower elevation from the higher elevation.

$$29{,}035 - (-36{,}198) = 29{,}035 + 36{,}198 = 65{,}233$$

The vertical difference is 65,233 ft. ▲

MULTIPLICATION OF INTEGERS

The multiplication property of zero is important in our discussion of multiplication of integers. It indicates that the product of 0 and any number is 0.

Multiplication Property of Zero

$$a \cdot 0 = 0 \cdot a = 0$$

We will develop the rules for multiplication of integers using number patterns. The four possible cases are

1. positive integer × positive integer,
2. positive integer × negative integer,
3. negative integer × positive integer, and
4. negative integer × negative integer.

CASE 1: POSITIVE INTEGER × POSITIVE INTEGER The product of two positive integers can be defined as repeated addition of a positive integer. Thus, $3 \cdot 2$ means $2 + 2 + 2$. This sum will always be positive. Thus, a positive integer times a positive integer is a positive integer.

CASE 2: POSITIVE INTEGER × NEGATIVE INTEGER Consider the following patterns

$$3(3) = 9$$
$$3(2) = 6$$
$$3(1) = 3$$

Note that each time the second factor is reduced by 1, the product is reduced by 3. Continuing the process gives

$$3(0) = 0$$

What comes next?

$$3(-1) = -3$$
$$3(-2) = -6$$

The pattern indicates that a positive integer times a negative integer is a negative integer.

We can confirm this result by using the number line. The expression $3(-2)$ means $(-2) + (-2) + (-2)$. Adding $(-2) + (-2) + (-2)$ on the number line, we obtain a sum of -6.

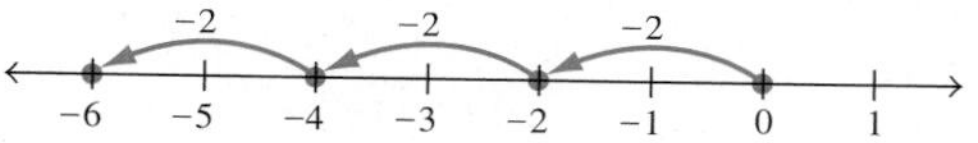

CASE 3: NEGATIVE INTEGER × POSITIVE INTEGER A procedure similar to that used in case 2 will indicate that a negative integer times a positive integer is a negative integer.

CASE 4: NEGATIVE INTEGER × NEGATIVE INTEGER We have illustrated that a positive integer times a negative integer is a negative integer. We make use of this fact in the following pattern.

$$4(-4) = -16$$
$$3(-4) = -12$$
$$2(-4) = -8$$
$$1(-4) = -4$$

In this pattern, each time the first term is decreased by 1, the product is increased by 4. Continuing this process gives

$$0(-4) = 0$$
$$(-1)(-4) = 4$$
$$(-2)(-4) = 8$$

This pattern illustrates that a negative integer times a negative integer is a positive integer.

The examples were restricted to integers. However, the rules for multiplication can be used for any numbers. We summarize them as follows.

Rules for Multiplication

1. The product of two numbers with *like signs* (positive × positive or negative × negative) is a *positive number.*
2. The product of two numbers with *unlike signs* (positive × negative or negative × positive) is a *negative number.*

EXAMPLE 6 ***Multiplying Integers***

Evaluate a) $5(-6)$ b) $-7(3)$ c) $(-7)(-9)$

SOLUTION

a) $5(-6) = -30$ b) $-7(3) = -21$ c) $(-7)(-9) = 63$ ▲

DIVISION OF INTEGERS

You may already realize that a relationship exists between multiplication and division.

$$6 \div 2 = 3 \quad \text{means that} \quad 3 \cdot 2 = 6$$

$$\frac{20}{10} = 2 \quad \text{means that} \quad 2 \cdot 10 = 20$$

These examples demonstrate that division is the reverse process of multiplication.

Division

For any a, b, and c where $b \neq 0$, $\frac{a}{b} = c$ means that $c \cdot b = a$.

We discuss the four possible cases for division, which are similar to those for multiplication.

CASE 1: POSITIVE INTEGER ÷ POSITIVE INTEGER A positive integer divided by a positive integer is positive.

$$\frac{6}{2} = 3 \quad \text{since} \quad 3(2) = 6$$

CASE 2: POSITIVE INTEGER ÷ NEGATIVE INTEGER A positive integer divided by a negative integer is negative.

$$\frac{6}{-2} = -3 \quad \text{since} \quad (-3)(-2) = 6$$

CASE 3: NEGATIVE INTEGER ÷ POSITIVE INTEGER A negative integer divided by a positive integer is negative.

$$\frac{-6}{2} = -3 \quad \text{since} \quad (-3)(2) = -6$$

CASE 4: NEGATIVE INTEGER ÷ NEGATIVE INTEGER A negative integer divided by a negative integer is positive.

$$\frac{-6}{-2} = 3 \quad \text{since} \quad 3(-2) = -6$$

The examples were restricted to integers. However, the rules for division can be used for any numbers. You should realize that division of integers does not always result in an integer. The rules for division are summarized as follows.

Rules for Division

1. The quotient of two numbers with *like signs* (positive ÷ positive or negative ÷ negative) is a *positive number.*
2. The quotient of two numbers with *unlike signs* (positive ÷ negative or negative ÷ positive) is a *negative number.*

EXAMPLE 7 *Dividing Integers*

Evaluate a) $\frac{21}{-3}$ b) $\frac{-100}{25}$ c) $\frac{-60}{-12}$

SOLUTION

(a) $\frac{21}{-3} = -7$ b) $\frac{-100}{25} = -4$ c) $\frac{-60}{-12} = 5$

In the definition of division, we stated that the denominator could not be 0. Division by 0 is not allowed. The quotient of a number divided by zero is said to be **undefined**. For example, 7/0 is undefined.

SECTION 5.2 EXERCISES

CONCEPT/WRITING EXERCISES

1. Explain how to add numbers using a number line.
2. Explain how to rewrite a subtraction problem as an addition problem.
3. Explain the rule for multiplication of real numbers.
4. Explain the rule for division of real numbers.
5. What is the product of 0 times any real number?
6. What do we call the quotient of 3 ÷ 0? Explain your answer.

PRACTICE THE SKILLS

In Exercises 7–16, evaluate the expression.

7. $-4 + 7$ **8.** $6 + (-11)$

9. $(-3) + 9$ **10.** $(-7) + (-7)$

11. $[6 + (-11)] + 0$ **12.** $(2 + 5) + (-4)$

13. $[(-3) + (-4)] + 9$ **14.** $[8 + (-3)] + (-2)$

15. $[(-23) + (-9)] + 11$ **16.** $[5 + (-13)] + 18$

In Exercises 17–26, evaluate the expression.

17. $5 - 8$ **18.** $-6 - 3$

19. $-7 - 6$ **20.** $4 - (-6)$

21. $-5 - (-3)$ **22.** $-4 - 4$

23. $14 - 20$ **24.** $8 - (-3)$

25. $[5 + (-3)] - 4$ **26.** $6 - (8 + 6)$

In Exercises 27–36, evaluate the expression.

27. $-3 \cdot 6$ **28.** $6(-8)$

29. $(-7)(-7)$ **30.** $8(-8)$

31. $[(-8)(-2)] \cdot 6$ **32.** $4(-5)(-6)$

33. $(5 \cdot 6)(-2)$ **34.** $(-9)(-1)(-2)$

35. $[(-3)(-6)] \cdot [(-5)(8)]$ **36.** $[(-8 \cdot 4) \cdot 5](-2)$

In Exercises 37–46, evaluate the expression.

37. $-27 \div (-9)$ **38.** $-24 \div 6$

39. $13 \div (-13)$ **40.** $-48 \div 3$

41. $56/-8$ **42.** $-75/15$

43. $-210/14$ **44.** $186/-6$

45. $144 \div (-3)$ **46.** $(-900) \div (-4)$

In Exercises 47–56, determine whether the statement is true or false. Modify each false statement to make it a true statement.

47. Every integer is a natural number.

48. Every natural number is an integer.

49. The difference of any two negative integers is a negative integer.

50. The sum of any two negative integers is a negative integer.

51. The product of any two positive integers is a positive integer.

52. The difference of a positive integer and a negative integer is always a negative integer.

53. The quotient of a negative integer and a positive integer is always a negative number.

54. The quotient of any two negative integers is a negative number.

55. The sum of a positive integer and a negative integer is always a positive integer.

56. The product of a positive integer and a negative integer is always a positive integer.

In Exercises 57–66, evaluate the expression.

57. $(6 + 8) \div 2$

58. $[-6(5)] + 7$

59. $[6(-2)] - 5$

60. $[(-5)(-6)] - 3$

61. $(4 - 8)(3)$

62. $[18 \div (-2)](-3)$

63. $[2 + (-17)] \div 3$

64. $(5 - 9) \div (-4)$

65. $[(-22)(-3)] \div (2 - 13)$

66. $[15(-4)] \div (-6)$

In Exercises 67–70, write the numbers in increasing order from left to right.

67. $-9, 7, -5, -3, -1, 0$

68. $-10, 8, -6, 4, -2, 0$

69. $-5, -2, -3, -1, -4, -6$

70. $106, 33, -47, -108, 72, -76$

PROBLEM SOLVING

71. *Temperature Difference* On January 5, 1999, Dubuque, Iowa, had an overnight low temperature of −11°F. One week later the overnight low temperature was 15°F. How much warmer was the second temperature than the first?

72. *Elevation Difference* Mount Whitney, in the Sierra Nevada mountains of California, is the highest point in the contiguous United States. It is 14,495 ft above sea level. Death Valley, in California and Nevada, is the lowest point in the United States, 282 ft below sea level. Find the vertical height difference between Mount Whitney and Death Valley.

73. *Vertical Distance Traveled* A helicopter drops a package from a height of 842 ft above sea level. The package lands in the ocean and settles at a point 927 ft below sea level. What was the vertical distance the package traveled?

74. *Football Yardage* In the first four plays of the game, the Eagles gained 8 yd, lost 5 yd, gained 3 yd, and gained 4 yd. What is the total number of yards gained in the first four plays? Did the Eagles make a first down? (Ten yards are needed for a first down.)

75. *Tracking a Stock* Yvette records the closing price of her favorite stock every Friday and the change in price from the previous Friday. The changes in price for the last five Fridays were: gained 7 points, lost 2 points, lost 3 points, lost 2 points, and gained 1 point. What was the net change in the stock's value for the five Fridays?

76. *Stock Market Average Changes* On August 4, 1999, the NASDAQ® average opened at 2540. During that day, it gained 26 points, and on the next day, it lost 91 points. What was the closing NASDAQ average on August 5, 1999?

77. *Time Zone Calculations* Part of a World Standard Time Zones chart used by airlines and the United States Navy is shown. The scale along the bottom is just like a number line with the integers −12, −11, . . . , 11, 12 on it.

a) Find the difference in time between Amsterdam (zone +1) and Los Angeles (zone −8).

b) Find the difference in time between Boston (zone −5) and Puerto Vallarta (zone −7).

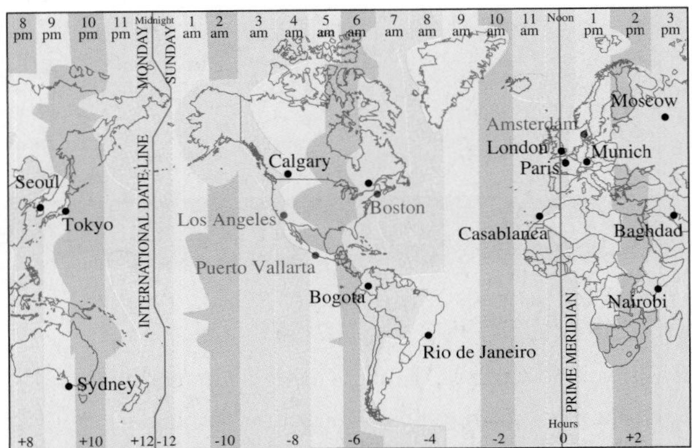

78. Explain why $\dfrac{a}{b} = \dfrac{-a}{-b}$.

CHALLENGE PROBLEMS/GROUP ACTIVITIES

79. Find the quotient:

$$\frac{-1 + 2 - 3 + 4 - 5 + \cdots - 99 + 100}{1 - 2 + 3 - 4 + 5 - \cdots + 99 - 100}$$

80. Triangular numbers and square numbers were introduced in the Section 1.1 Exercises. There are also **pentagonal numbers**, which were also studied by the Greeks. Four pentagonal numbers are 1, 5, 12, and 22.

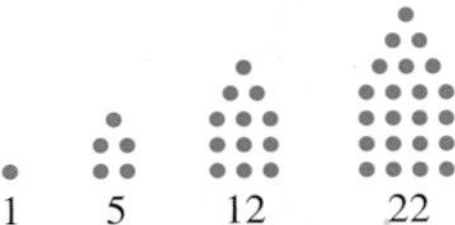

a) Determine the next three pentagonal numbers.

b) Describe a procedure to determine the next five pentagonal numbers without drawing the figures.

c) Is 72 a pentagonal number? Explain how you determined your answer.

81. Place the appropriate plus or minus signs between each digit so that the total will equal 1.

$$0 \quad 1 \quad 2 \quad 3 \quad 4 \quad 5 \quad 6 \quad 7 \quad 8 \quad 9 = 1$$

RESEARCH ACTIVITY

82. Do research and write a report on the history of the number 0 in the Hindu–Arabic numeration system.

5.3 THE RATIONAL NUMBERS

"When you can measure what you are talking about and express it in numbers, you know something about it."
Lord Kelvin

We introduced the number line in Section 5.1 and discussed the integers in Section 5.2. The numbers that fall between the integers on the number line are either rational or irrational numbers. In this section, we discuss the rational numbers, and in Section 5.4, we discuss the irrational numbers.

Any number that can be expressed as a quotient of two integers (denominator not 0) is a rational number.

The set of **rational numbers**, denoted by Q, is the set of all numbers of the form p/q, where p and q are integers and $q \neq 0$.

The following numbers are examples of rational numbers:

$$\frac{1}{3}, \quad \frac{3}{4}, \quad -\frac{7}{8}, \quad 1\frac{2}{3}, \quad 2, \quad 0, \quad \frac{15}{7}$$

The integers 2 and 0 are rational numbers because each can be expressed as the quotient of two integers: $2 = \frac{2}{1}$ and $0 = \frac{0}{1}$. In fact, every integer n is a rational number, since it can be written in the form of $\frac{n}{1}$.

Numbers such as $\frac{1}{3}$ and $-\frac{7}{8}$ are also called **fractions**. The number above the fraction line is called the **numerator**, and the number below the fraction line is called the **denominator**.

DID YOU KNOW

Triangular Numbers

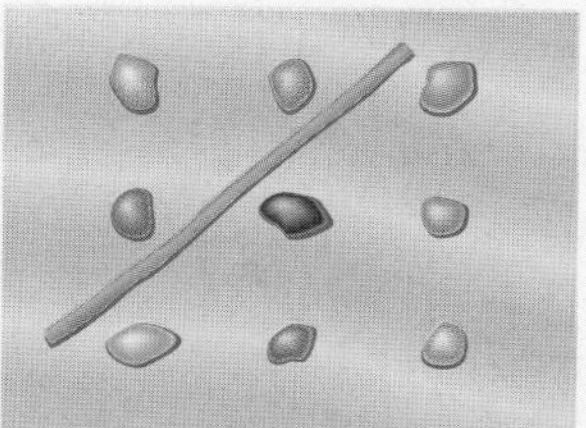

One very simple reason that the ancient Greek mathematicians thought of mathematics in terms of whole numbers and their ratios (the rational numbers) was that they were still working with numbers that were represented by objects, such as squares and triangles, not number symbols. Still, this did not prevent them from drawing some conclusions about number theory. Consider two consecutive triangular numbers, for instance 3 and 6. You can see from the diagram that the sum of the two triangular numbers is equal to the square number 9, the square of a side of the larger triangle.

REDUCING FRACTIONS

Sometimes the numerator and denominator in a fraction have a common divisor (or common factor). For example, both the numerator and denominator of the fraction $\frac{6}{10}$ have the common divisor 2. When a numerator and denominator have a common divisor, we can *reduce the fraction to its lowest terms.*

A fraction is said to be in its lowest terms (or reduced) when the numerator and denominator are relatively prime (that is, have no common divisors other than 1). To reduce a fraction to its lowest terms, divide both the numerator and the denominator by the greatest common divisor. Recall that a procedure for finding the greatest common divisor was discussed in Section 5.1.

The fraction $\frac{6}{10}$ is reduced to its lowest terms as follows.

$$\frac{6}{10} = \frac{6 \div 2}{10 \div 2} = \frac{3}{5}$$

EXAMPLE 1 ***Reducing a Fraction to Lowest Terms***

Reduce $\frac{54}{90}$ to its lowest terms.

SOLUTION In Example 4 of Section 5.1, we determined that the GCD of 54 and 90 is 18. Divide the numerator and the denominator by GCD, 18.

$$\frac{54}{90} = \frac{54 \div 18}{90 \div 18} = \frac{3}{5}$$

Since there are no common divisors of 3 and 5 other than 1, this fraction is in its lowest terms.

MIXED NUMBERS AND IMPROPER FRACTIONS

Consider the number $2\frac{3}{4}$. This is an example of a **mixed number**. It is called a mixed number because it consists of an integer, 2, and a fraction, $\frac{3}{4}$. The mixed number $2\frac{3}{4}$ means $2 + \frac{3}{4}$. The mixed number $-4\frac{1}{4}$ means $-(4 + \frac{1}{4})$. Rational numbers greater than 1 or less than -1 that are not integers may be represented as mixed numbers, or as **improper fractions**. An improper fraction is a fraction whose numerator is greater than its denominator. An example of an improper fraction is $\frac{8}{5}$. Figure 5.6 shows both mixed numbers and improper fractions indicated on a number line. In this section, we show how to convert mixed numbers to improper fractions and vice versa.

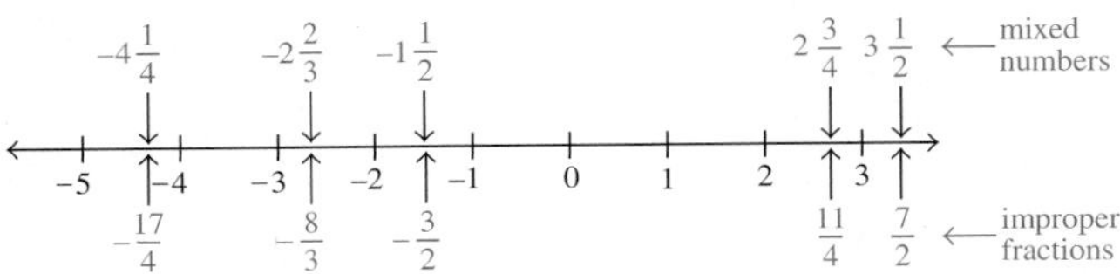

Figure 5.6

We begin by limiting our discussion to positive mixed numbers and positive improper fractions.

Converting a Positive Mixed Number to an Improper Fraction

1. Multiply the denominator of the fraction in the mixed number by the integer preceding it.
2. Add the product obtained in step 1 to the numerator of the fraction in the mixed number. This sum is the numerator of the improper fraction we are seeking. The denominator of the improper fraction we are seeking is the same as the denominator of the fraction in the mixed number.

EXAMPLE 2 *From Mixed Number to Improper Fraction*

Convert the following mixed numbers to improper fractions.

a) $3\frac{1}{2}$ b) $2\frac{3}{4}$

SOLUTION

a) $3\frac{1}{2} = \frac{2 \cdot 3 + 1}{2} = \frac{6 + 1}{2} = \frac{7}{2}$

b) $2\frac{3}{4} = \frac{4 \cdot 2 + 3}{4} = \frac{8 + 3}{4} = \frac{11}{4}$

Note that both $\frac{7}{2}$ and $\frac{11}{4}$ have numerators larger than their denominators and that both are improper fractions.

Converting a Positive Improper Fraction to a Mixed Number

1. Divide the numerator by the denominator. Identify the quotient and the remainder.
2. The quotient obtained in step 1 is the integer part of the mixed number. The remainder is the numerator of the fraction in the mixed number. The denominator in the fraction of the mixed number will be the same as the denominator in the original fraction.

EXAMPLE 3 ***From Improper Fraction to Mixed Number***

Convert the following improper fractions to mixed numbers.

a) $\frac{8}{5}$ b) $\frac{225}{8}$

SOLUTION

a) Divide the numerator, 8, by the denominator, 5.

$$\begin{array}{rl} 1 & \leftarrow \text{Quotient} \\ \text{Divisor} \rightarrow 5\overline{)8} & \leftarrow \text{Dividend} \\ \underline{5} & \\ 3 & \leftarrow \text{Remainder} \end{array}$$

Therefore,

$$\frac{8}{5} = \underset{\text{Quotient}}{1}\frac{3}{5} \begin{array}{l} \leftarrow \text{Remainder} \\ \leftarrow \text{Divisor} \end{array}$$

The mixed number is $1\frac{3}{5}$.

b) Divide the numerator, 225, by the denominator, 8.

$$\begin{array}{rl} 28 & \leftarrow \text{Quotient} \\ \text{Divisor} \rightarrow 8\overline{)225} & \leftarrow \text{Dividend} \\ \underline{16} & \\ 65 & \\ \underline{64} & \\ 1 & \leftarrow \text{Remainder} \end{array}$$

Therefore,

$$\frac{225}{8} = \underset{\text{Quotient}}{28}\frac{1}{8} \begin{array}{l} \leftarrow \text{Remainder} \\ \leftarrow \text{Divisor} \end{array}$$

The mixed number is $28\frac{1}{8}$.

Up to this point, we have only worked with positive mixed numbers and positive improper fractions. When converting a negative mixed number to an improper fraction, or a negative improper fraction to a mixed number, it is best to ignore the negative sign temporarily. Perform the calculation as described earlier and then reattach the negative sign.

EXAMPLE 4 *Negative Mixed Numbers and Improper Fractions*

a) Convert $-2\frac{3}{4}$ to an improper fraction.

b) Convert $-\frac{225}{8}$ to a mixed number.

SOLUTION

a) First, ignore the negative sign and examine $2\frac{3}{4}$. We learned in Example 2(b) that $2\frac{3}{4} = \frac{11}{4}$. Now to convert $-2\frac{3}{4}$ to an improper fraction, we reattach the negative sign.

Thus, $-2\frac{3}{4} = -\frac{11}{4}$.

b) We learned in Example 3(b) that $\frac{225}{8} = 28\frac{1}{8}$. Therefore, $-\frac{225}{8} = -28\frac{1}{8}$.

TERMINATING OR REPEATING DECIMAL NUMBERS

Note the following important property of the rational numbers.

> Every *rational number* when expressed as a decimal number will be either a terminating or a repeating decimal number.

Examples of terminating decimal numbers are 0.5, 0.75, and 4.65. Examples of repeating decimal numbers are 0.333 . . . , 0.2323 . . . , and 8.13456456. . . . One way to indicate that a number or group of numbers repeat is to place a bar above the number or group of numbers that repeat. Thus, 0.333 . . . may be written $0.\overline{3}$, 0.2323 . . . may be written $0.\overline{23}$, and 8.13456456 . . . may be written $8.13\overline{456}$.

EXAMPLE 5 *Terminating Decimal Numbers*

Show that the following rational numbers are terminating decimal numbers.

a) $\frac{3}{4}$ b) $\frac{6}{5}$ c) $\frac{5}{8}$

SOLUTION To express the rational number in decimal form, divide the numerator by the denominator. If you use a calculator, or use long division, you will obtain the following results.

a) $3 \div 4 = 0.75$ b) $6 \div 5 = 1.2$ c) $5 \div 8 = 0.625$

Note that in each part of Example 5, each quotient has a final nonzero digit. Thus, each number is a terminating decimal number.

EXAMPLE 6 *Repeating Decimal Numbers*

Show that the following rational numbers are repeating decimal numbers.

a) $\frac{2}{3}$ b) $\frac{14}{99}$ c) $1\frac{4}{33}$

SOLUTION If you use a calculator, or use long division, you will see that each fraction results in a repeating decimal number.

a) $2 \div 3 = 0.6666\ldots$ or $0.\overline{6}$

b) $14 \div 99 = 0.141414\ldots$ or $0.\overline{14}$

c) $1\frac{4}{33} = \frac{37}{33} = 1.121212\ldots$ or $1.\overline{12}$

Note that in each part of Example 6, the quotient has no final digit and continues indefinitely. Each number is a repeating decimal number.

When a fraction is converted to a decimal number, the maximum number of digits that can repeat is $n - 1$, where n is the denominator of the fraction. For example, when $\frac{2}{7}$ is converted to a decimal number, the maximum number of digits that can repeat is $7 - 1$, or 6.

CONVERTING DECIMAL NUMBERS TO FRACTIONS

We can convert a terminating or repeating decimal number into a quotient of integers. The explanation of the procedure will refer to the positional values to the right of the decimal point, as illustrated here:

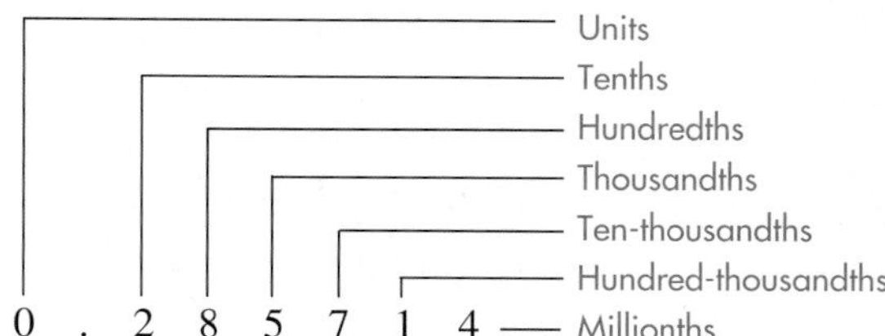

Example 7 demonstrates how to convert from a decimal number to a fraction.

EXAMPLE 7 *Converting a Decimal Number into a Fraction*

Convert the following terminating decimal numbers to a quotient of integers.

a) 0.4 b) 0.62 c) 0.062 d) 1.37

SOLUTION When converting a terminating decimal number to a quotient of integers, we observe the last digit to the right of the decimal point. The position of this digit will indicate the denominator of the quotient of integers.

a) $0.4 = \frac{4}{10}$ because the 4 is in the tenths position.

b) $0.62 = \frac{62}{100}$ because the digit on the right, 2, is in the hundredths position.

c) $0.062 = \frac{62}{1000}$ because the digit on the right, 2, is in the thousandths position.

d) $1.37 = \frac{137}{100}$ because the digit on the right, 7, is in the hundredths position.

Converting a repeating decimal number to a quotient of integers is more difficult. To do so, we must "create" another repeating decimal number with the same repeating digits so that when one repeating decimal number is subtracted from the other repeating decimal number, the difference will be a whole number. To create a number with the same repeating digits, multiply the original repeating decimal number by 10 if one digit repeats, by 100 if two digits repeat, by 1000 if three digits repeat, and so on. Examples 8 through 10 demonstrate this procedure.

DID YOU KNOW

The Latin Fraction

The words *numerator* and *denominator* are derived from the Latin for "the numberer" and "the namer," respectively (*nomin* is Latin for "name"). The ancient Chinese looked at the relationship in a more fanciful way: They called the numerator "the son" and the denominator "the mother."

EXAMPLE 8 *Converting a Repeating Decimal Number into a Fraction*

Convert $0.\overline{3}$ to a quotient of integers.

SOLUTION $0.\overline{3} = 0.3\overline{3} = 0.33\overline{3}$, and so on.

Let the original repeating decimal number be n; thus, $n = 0.\overline{3}$. Because one digit repeats, we multiply both sides of the equation by 10, which gives $10n = 3.\overline{3}$. Then we subtract.

$$\begin{array}{rl} 10n &= 3.\overline{3} \\ -\quad n &= 0.\overline{3} \\ \hline 9n &= 3.0 \end{array}$$

Note that $10n - n = 9n$ and $3.\overline{3} - 0.\overline{3} = 3.0$.

Next, we solve for n by dividing both sides of the equation by 9.

$$\frac{9n}{9} = \frac{3.0}{9}$$

$$n = \frac{3}{9} = \frac{1}{3}$$

Therefore, $0.\overline{3} = \frac{1}{3}$.

EXAMPLE 9 *Converting a Repeating Decimal Number into a Fraction*

Convert $0.\overline{35}$ to a quotient of integers.

SOLUTION Let $n = 0.\overline{35}$. Since two digits repeat, multiply both sides of the equation by 100. Thus, $100n = 35.\overline{35}$. Now we subtract n from $100n$.

$$\begin{array}{rl} 100n &= 35.\overline{35} \\ -\quad n &= 0.\overline{35} \\ \hline 99n &= 35 \end{array}$$

Finally, we divide both sides of the equation by 99.

$$\frac{99n}{99} = \frac{35}{99}$$

$$n = \frac{35}{99}$$

Therefore, $0.\overline{35} = \frac{35}{99}$.

EXAMPLE 10 *Converting a Repeating Decimal Number into a Fraction*

Convert $12.14\overline{2}$ to a quotient of integers.

SOLUTION This problem is different from the two preceding examples in that the repeating digit, 2, is not directly to the right of the decimal point. When this situation arises, move the decimal point to the right until the repeating terms are directly to its right. For each place the decimal point is moved, the number is multiplied by 10. In this example, the decimal point must be moved two places to the right. Thus, the number must be multiplied by 100.

DID YOU KNOW

Where Do These Fractions Come From?

Until recently, new investors in U.S. stock markets were often puzzled by the practice of quoting share prices in fractions of a dollar instead of in dollars and cents. For example, a stock quote of $20\frac{1}{8}$ meant \$20.125.

In the late 1700s, most stock traders were merchants who were involved in foreign trade and currency exchange. The Spanish dollar coin, or piece of eight, was a widely held and stable currency used for trading both stocks and goods. With a hammer and chisel, these dollars could be divided into halves, quarters, and eighths. Eighths were also known as bits. It is from this practice that the synonym of "two bits" for 25 cents comes.

Up until recently, trading in eighths remained part of the securities industry culture. The New York Stock Exchange recently converted to a decimal pricing system. Other stock markets in the United States are expected to follow suit.

$$n = 12.14\overline{2}$$
$$100n = 100 \times 12.14\overline{2} = 1214.\overline{2}$$

Now proceed as in the previous two examples. Since one digit repeats, multiply both sides by 10.

$$100n = 1214.\overline{2}$$
$$10 \times 100n = 10 \times 1214.\overline{2}$$
$$1000n = 12142.\overline{2}$$

Now subtract $100n$ from $1000n$ so that the repeating part will drop out.

$$\begin{array}{rcl} 1000n &=& 12142.\overline{2} \\ -\ 100n &=& 1214.\overline{2} \\ \hline 900n &=& 10928 \end{array}$$
$$n = \frac{10{,}928}{900} = \frac{2732}{225}$$

Therefore, $12.14\overline{2} = \frac{2732}{225}$.

MULTIPLICATION AND DIVISION OF FRACTIONS

The product of two fractions is found by multiplying the numerators together and multiplying the denominators together.

Multiplication of Fractions

$$\frac{a}{b} \cdot \frac{c}{d} = \frac{a \cdot c}{b \cdot d} = \frac{ac}{bd}, \qquad b \neq 0, \quad d \neq 0$$

EXAMPLE 11 *Multiplying Fractions*

Evaluate the following.

a) $\frac{3}{5} \cdot \frac{7}{8}$ b) $\left(\frac{-2}{3}\right)\left(\frac{-4}{9}\right)$ c) $\left(1\frac{7}{8}\right)\left(2\frac{1}{4}\right)$

SOLUTION

a) $\frac{3}{5} \cdot \frac{7}{8} = \frac{3 \cdot 7}{5 \cdot 8} = \frac{21}{40}$

b) $\left(\frac{-2}{3}\right)\left(\frac{-4}{9}\right) = \frac{(-2)(-4)}{(3)(9)} = \frac{8}{27}$

c) $\left(1\frac{7}{8}\right)\left(2\frac{1}{4}\right) = \frac{15}{8} \cdot \frac{9}{4} = \frac{135}{32} = 4\frac{7}{32}$

The **reciprocal** of any number is 1 divided by that number. The product of a number and its reciprocal must equal 1. Examples of some numbers and their reciprocals follow.

Number		Reciprocal		Product
3	$\cdot$	$\frac{1}{3}$	$=$	1
$\frac{3}{5}$	$\cdot$	$\frac{5}{3}$	$=$	1
-6	$\cdot$	$-\frac{1}{6}$	$=$	1

To find the quotient of two fractions, multiply the first fraction by the reciprocal of the second fraction.

Division of Fractions

$$\frac{a}{b} \div \frac{c}{d} = \frac{a}{b} \cdot \frac{d}{c} = \frac{ad}{bc}, \qquad b \neq 0, \quad d \neq 0, \quad c \neq 0$$

EXAMPLE 12 *Dividing Fractions*

Evaluate the following

a) $\frac{2}{3} \div \frac{5}{7}$ b) $\frac{-3}{5} \div \frac{5}{7}$

SOLUTION

a) $\frac{2}{3} \div \frac{5}{7} = \frac{2}{3} \cdot \frac{7}{5} = \frac{2 \cdot 7}{3 \cdot 5} = \frac{14}{15}$

b) $\frac{-3}{5} \div \frac{5}{7} = \frac{-3}{5} \cdot \frac{7}{5} = \frac{-3 \cdot 7}{5 \cdot 5} = \frac{-21}{25}$ or $-\frac{21}{25}$

ADDITION AND SUBTRACTION OF FRACTIONS

Before we can add or subtract fractions, the fractions must have a common denominator. A common denominator is another name for a common multiple of the denominators. The **lowest common denominator (LCD)** is the least common multiple of the denominators.

To add or subtract two fractions with a common denominator, we add or subtract their numerators and retain the common denominator.

Addition and Subtraction of Fractions

$$\frac{a}{c} + \frac{b}{c} = \frac{a+b}{c}, \quad c \neq 0; \qquad \frac{a}{c} - \frac{b}{c} = \frac{a-b}{c}, \quad c \neq 0$$

EXAMPLE 13 *Adding and Subtracting Fractions with a Common Denominator*

Evaluate the following.

a) $\frac{2}{9} + \frac{5}{9}$ b) $\frac{6}{7} - \frac{2}{7}$

SOLUTION

a) $\frac{2}{9} + \frac{5}{9} = \frac{2+5}{9} = \frac{7}{9}$ b) $\frac{6}{7} - \frac{2}{7} = \frac{6-2}{7} = \frac{4}{7}$

▶ DID YOU KNOW

Mathematical Music

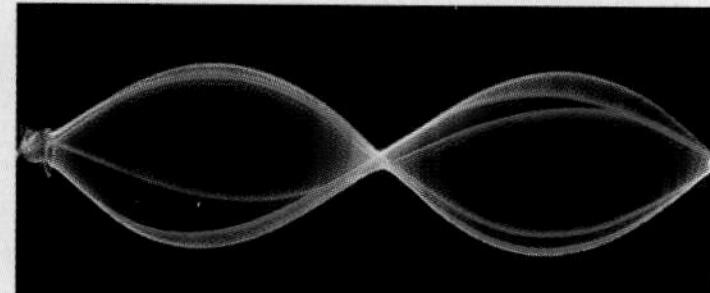

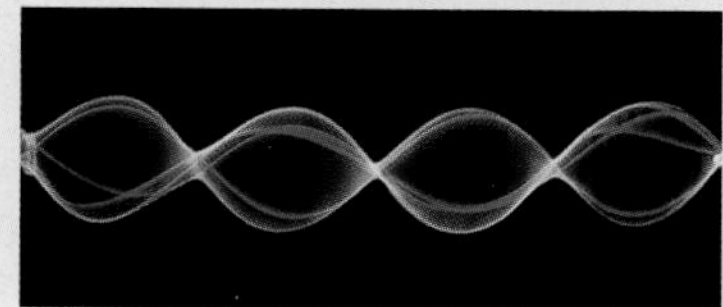

When the string of a musical instrument is plucked or bowed, it moves in a wavelike pattern like the strings shown here. The vibration this creates in the surrounding air is what your eardrums detect as sound.

The ancient Greeks believed that in nature, all harmony and everything of beauty could be explained with rational numbers. This belief was reinforced by the discovery that the sound of plucked strings could be quite pleasing if the strings plucked were in the ratio of 1 to 2 (an octave), 2 to 3 (a fifth), 3 to 4 (a fourth), and so on. Thus, the secret of harmony lies in the rational numbers such as $\frac{1}{2}$, $\frac{2}{3}$, and $\frac{3}{4}$.

The theory of vibrating strings has applications today that go well beyond music. How materials vibrate, and hence the stress they can absorb, is a vital matter in the construction of rockets, buildings, and bridges.

Note that in Example 13, the denominators of the fractions being added or subtracted were the same; that is, they have a common denominator. *When adding or subtracting two fractions with unlike denominators, first rewrite each fraction with a common denominator. Then add or subtract the fractions.*

Writing fractions with a common denominator is accomplished with the *fundamental law of rational numbers.*

Fundamental Law of Rational Numbers

If a, b, and c are integers, with $b \neq 0$ and $c \neq 0$, then

$$\frac{a}{b} = \frac{a}{b} \cdot \frac{c}{c} = \frac{a \cdot c}{b \cdot c}$$

The terms $\frac{a}{b}$ and $\frac{a \cdot c}{b \cdot c}$ are called **equivalent fractions**. For example, since $\frac{5}{12} = \frac{5 \cdot 5}{12 \cdot 5} = \frac{25}{60}$, the fractions $\frac{5}{12}$ and $\frac{25}{60}$ are equivalent fractions. We will see the importance of equivalent fractions in the next two examples.

EXAMPLE 14 ***Subtracting Fractions with Unlike Denominators***

Evaluate $\frac{5}{12} - \frac{3}{10}$.

SOLUTION Using prime factorization (Section 5.1), we find that the LCM of 12 and 10 is 60. We will therefore express each fraction with a denominator of 60. Sixty divided by 12 is 5. Therefore, the denominator, 12, must be multiplied by 5 to get 60. If the denominator is multiplied by 5, the numerator must also be multiplied by 5 so that the value of the fraction remains unchanged. Multiplying both numerator and denominator by 5 is the same as multiplying by 1.

We follow the same procedure for the other fraction, $\frac{3}{10}$. Sixty divided by 10 is 6. Therefore, we multiply both the denominator, 10, and the numerator, 3, by 6 to obtain an equivalent fraction with a denominator of 60.

$$\begin{aligned}\frac{5}{12} - \frac{3}{10} &= \left(\frac{5}{12} \cdot \frac{5}{5}\right) - \left(\frac{3}{10} \cdot \frac{6}{6}\right)\\ &= \frac{25}{60} - \frac{18}{60}\\ &= \frac{7}{60}\end{aligned}$$

▲

EXAMPLE 15 ***Adding Fractions with Unlike Denominators***

Evaluate $\frac{1}{54} + \frac{1}{90}$.

SOLUTION In Example 6 of Section 5.1, we determined that the LCM of 54 and 90 is 270. Rewrite each fraction as an equivalent fraction using the LCM as the common denominator.

$$\begin{aligned}\frac{1}{54} + \frac{1}{90} &= \left(\frac{1}{54} \cdot \frac{5}{5}\right) + \left(\frac{1}{90} \cdot \frac{3}{3}\right)\\ &= \frac{5}{270} + \frac{3}{270}\\ &= \frac{8}{270}\end{aligned}$$

Now we reduce $\frac{8}{270}$ by dividing both 8 and 270 by 2, their greatest common factor.

$$\frac{8}{270} = \frac{8 \div 2}{270 \div 2} = \frac{4}{135}$$

EXAMPLE 16 *Rice Preparation*

Following are the instructions given on a box of Minute Rice. Determine the amount of (a) rice and water, (b) salt, and (c) butter or margarine needed to make 3 servings of rice.

Directions

1. Bring water, salt, and butter (or margarine) to a boil.
2. Stir in rice. Cover; remove from heat. Let stand 5 minutes. Fluff with fork.

To make	Rice & water (equal measures)	Salt	Butter or margarine (if desired)
2 servings	$\frac{2}{3}$ cup	$\frac{1}{4}$ tsp	1 tsp
4 servings	$1\frac{1}{3}$ cups	$\frac{1}{2}$ tsp	2 tsp

SOLUTION Since 3 is halfway between 2 and 4, we can find the amount of each ingredient by finding the average of the amount for 2 and 4 servings. To do so, we add the amounts for 2 servings and 4 servings and divide the sum by 2.

a) Rice and water: $\dfrac{\frac{2}{3} + 1\frac{1}{3}}{2} = \dfrac{\frac{2}{3} + \frac{4}{3}}{2} = \dfrac{\frac{6}{3}}{2} = \dfrac{2}{2} = 1$ cup

b) Salt: $\dfrac{\frac{1}{4} + \frac{1}{2}}{2} = \dfrac{\frac{1}{4} + \frac{2}{4}}{2} = \dfrac{\frac{3}{4}}{2} = \dfrac{3}{4} \cdot \dfrac{1}{2} = \dfrac{3}{8}$ tsp

c) Butter or margarine: $\dfrac{1 + 2}{2} = \dfrac{3}{2}$, or $1\frac{1}{2}$ tsp

The solution to Example 16 can be found in other ways. Suggest two other procedures for solving the same problem.

SECTION 5.3 EXERCISES

CONCEPT/WRITING EXERCISES

1. Describe the set of rational numbers.
2. a) Explain how to write a terminating decimal number as a fraction.
   b) Write 0.013 as a fraction.
3. a) Explain how to reduce a fraction to lowest terms.
   b) Reduce $\frac{12}{45}$ to lowest terms by using the procedure in part (a).
4. Explain how to convert a mixed number into an improper fraction.
5. Explain how to convert an improper fraction into a mixed number.
6. a) Explain how to multiply two fractions.
   b) Multiply $\frac{24}{15} \cdot \frac{7}{36}$ by using the procedure in part (a).
7. a) Explain how to determine the reciprocal of a number.
   b) Using the procedure in part (a), determine the reciprocal of -5.
8. a) Explain how to divide two fractions.
   b) Divide $\frac{12}{19} \div \frac{3}{8}$ by using the procedure in part (a).
9. a) Explain how to add or subtract two fractions having a common denominator.
   b) Add $\frac{13}{27} + \frac{8}{27}$ by using the procedure in part (a).

10. a) Explain how to add or subtract two fractions having unlike denominators.
    b) Using the procedure in part (a), add $\frac{5}{12} + \frac{4}{9}$.
11. In your own words, state the fundamental law of rational numbers.
12. Are $\frac{4}{7}$ and $\frac{20}{35}$ equivalent fractions? Explain your answer.

PRACTICE THE SKILLS

In Exercises 13–22, reduce each fraction to lowest terms.

13. $\frac{14}{21}$ **14.** $\frac{22}{55}$ **15.** $\frac{63}{98}$

16. $\frac{36}{56}$ **17.** $\frac{525}{800}$ **18.** $\frac{13}{221}$

19. $\frac{112}{176}$ **20.** $\frac{120}{135}$ **21.** $\frac{45}{495}$

22. $\frac{124}{148}$

In Exercises 23–28, convert each mixed number into an improper fraction.

23. $2\frac{5}{8}$ **24.** $3\frac{7}{9}$ **25.** $-2\frac{3}{4}$

26. $-7\frac{1}{5}$ **27.** $-4\frac{15}{16}$ **28.** $11\frac{9}{16}$

In Exercises 29–32, write the number of inches indicated by the arrows as an improper fraction.

29.

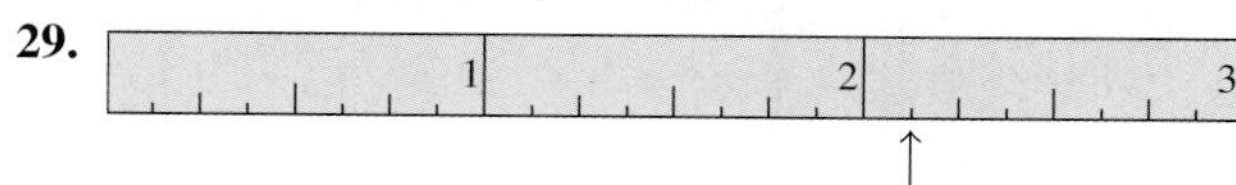

30.

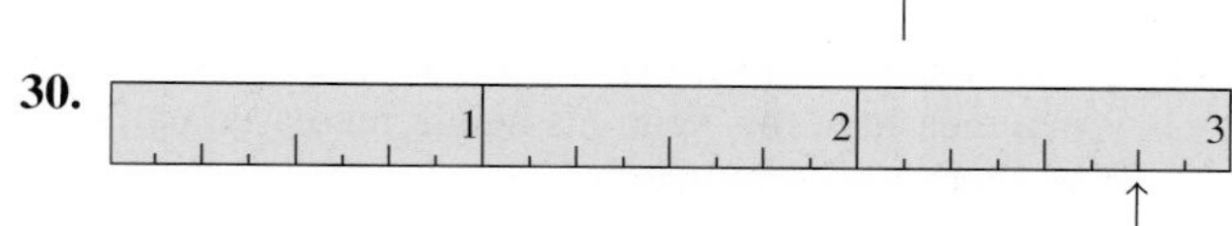

31.

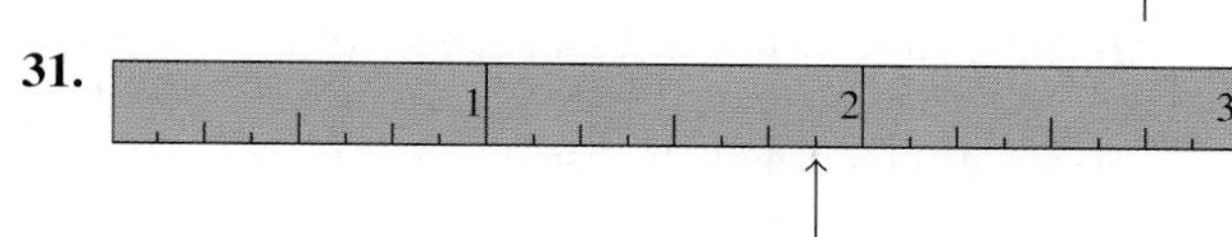

32.

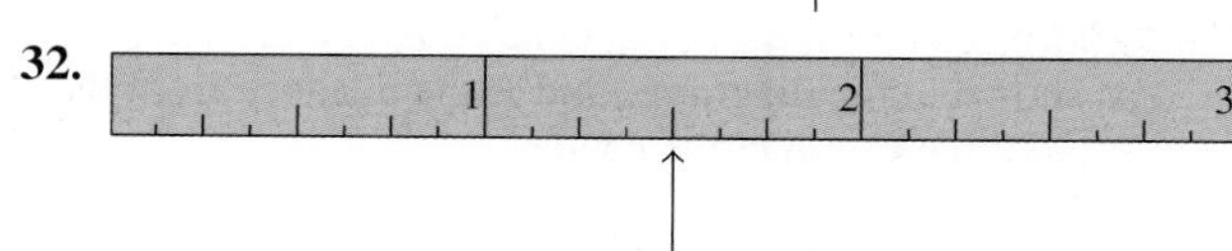

In Exercises 33–38, convert each improper fraction into a mixed number.

33. $\frac{31}{16}$ **34.** $\frac{45}{14}$ **35.** $-\frac{213}{5}$

36. $-\frac{457}{11}$ **37.** $-\frac{878}{15}$ **38.** $\frac{1028}{21}$

In Exercises 39–48, express each rational number as terminating or repeating decimal number.

39. $\frac{1}{4}$ **40.** $\frac{3}{7}$ **41.** $\frac{4}{7}$

42. $\frac{5}{6}$ **43.** $\frac{3}{8}$ **44.** $\frac{23}{7}$

45. $\frac{13}{3}$ **46.** $\frac{115}{15}$ **47.** $\frac{85}{15}$

48. $\frac{1002}{11}$

In Exercises 49–58, express each terminating decimal number as a quotient of two integers.

49. 0.6 **50.** 0.88 **51.** 0.052

52. 0.0125 **53.** 6.2 **54.** 7.25

55. 1.452 **56.** 1.2345 **57.** 3.0001

58. 4.2535

In Exercises 59–68, express each repeating decimal number as a quotient of two integers.

59. $0.\overline{3}$ **60.** $0.\overline{4}$ **61.** $2.\overline{9}$

62. $0.\overline{51}$ **63.** $1.\overline{36}$ **64.** $0.\overline{135}$

65. $1.0\overline{2}$ **66.** $2.\overline{49}$ **67.** $3.4\overline{78}$

68. $5.2\overline{39}$

In Exercises 69–78, evaluate each expression.

69. $\frac{2}{7} \div \frac{5}{3}$ **70.** $\frac{5}{7} \cdot \frac{6}{11}$

71. $\left(\frac{-3}{8}\right)\left(\frac{-16}{15}\right)$ **72.** $\left(-\frac{3}{5}\right) \div \frac{10}{21}$

73. $\frac{7}{8} \div \frac{8}{7}$ **74.** $\frac{3}{7} \div \frac{3}{7}$

75. $\left(\frac{3}{5} \cdot \frac{4}{7}\right) \div \frac{1}{3}$ **76.** $\left(\frac{4}{7} \div \frac{4}{5}\right) \cdot \frac{1}{7}$

77. $\left[\left(\frac{-3}{4}\right)\left(\frac{-2}{7}\right)\right] \div \frac{3}{5}$ **78.** $\left(\frac{3}{8} \cdot \frac{5}{9}\right) \cdot \left(\frac{4}{7} \div \frac{5}{8}\right)$

In Exercises 79–88, perform the indicated operation and reduce your answer to lowest terms.

79. $\frac{1}{5} + \frac{1}{6}$ **80.** $\frac{1}{3} - \frac{2}{15}$ **81.** $\frac{2}{11} + \frac{3}{110}$

82. $\frac{5}{12} + \frac{7}{36}$ **83.** $\frac{5}{9} - \frac{7}{54}$ **84.** $\frac{13}{30} - \frac{17}{120}$

85. $\frac{1}{12} + \frac{1}{48} + \frac{1}{72}$ **86.** $\frac{3}{5} + \frac{7}{15} + \frac{9}{75}$

87. $\frac{1}{30} - \frac{3}{40} - \frac{7}{50}$ **88.** $\frac{4}{25} - \frac{9}{100} - \frac{7}{40}$

PROBLEM SOLVING

Alternative methods for adding and subtracting two fractions are shown. These methods may not result in a solution in its lowest terms.

$$\frac{a}{b} + \frac{c}{d} = \frac{ad + bc}{bd} \quad \text{and} \quad \frac{a}{b} - \frac{c}{d} = \frac{ad - bc}{bd}$$

In Exercises 89–94, use one of the two formulas to evaluate the expression.

89. $\frac{2}{3} + \frac{3}{4}$ **90.** $\frac{5}{7} - \frac{1}{12}$ **91.** $\frac{5}{7} + \frac{3}{4}$

92. $\frac{7}{3} - \frac{5}{12}$ **93.** $\frac{3}{8} + \frac{5}{12}$ **94.** $\left(\frac{2}{3} + \frac{1}{4}\right) - \frac{3}{5}$

In Exercises 95–100, evaluate each expression.

95. $\left(\frac{1}{5} \cdot \frac{1}{4}\right) + \frac{1}{3}$ **96.** $\left(\frac{3}{5} \div \frac{2}{10}\right) - \frac{1}{3}$

97. $\left(\frac{1}{2} + \frac{3}{10}\right) \div \left(\frac{1}{5} + 2\right)$ **98.** $\left(\frac{1}{9} \cdot \frac{3}{5}\right) + \left(\frac{2}{3} \cdot \frac{1}{5}\right)$

99. $\left(3 - \frac{4}{9}\right) \div \left(4 + \frac{2}{3}\right)$ **100.** $\left(\frac{2}{5} \div \frac{4}{9}\right)\left(\frac{3}{5} \cdot 6\right)$

101. *Crop Storage* Todd Hall has a silo on his farm in which he can store silage made from his various crops. He currently has a silo that is $\frac{1}{4}$ full of corn silage, $\frac{2}{5}$ full of hay silage, and $\frac{1}{3}$ full of oats silage. What fraction of Todd's silo is currently in use?

102. *College Majors* A sociology class is $\frac{2}{5}$ math majors, $\frac{1}{4}$ English majors, and $\frac{1}{10}$ chemistry majors; the remaining members of the class are art majors. What fraction of the class is art majors?

103. *Department Budget* Mary Deininger is chair of the humanities department at Santa Fe Community College. Mary has a budget in which $\frac{1}{2}$ of the money is for photocopying, $\frac{2}{5}$ of the money is for computer-related expenses, and the rest of the money is for student tutors in the foreign languages lab. What fraction of Mary's budget is for student tutors?

104. *Stairway Height* A stairway consists of 12 stairs, each $10\frac{3}{4}$ in. high. What is the vertical height of the stairway?

105. *Proofreading a Textbook* To help proofread her new textbook, Chris assigns three students to proofread $\frac{1}{4}$, $\frac{1}{5}$, and $\frac{1}{2}$ of the book, respectively. She decides to proofread the rest of the book herself. If the book has 540 pages, how many pages must Chris proofread herself?

106. *Cutting a Pipe* If a plumber cuts three $2\frac{5}{16}$ ft lengths of pipe from an 8 ft section, how much remains?

107. *Weekly Stock Decrease* According to the *Wall Street Journal,* during the week ending August 7, 1999, Disney stock opened at $27\frac{7}{8}$ and closed at $25\frac{1}{2}$. How many points did the stock decrease during this week?

108. *Modifying a Recipe* A recipe for six servings of stew calls for $1\frac{1}{4}$ tsp of worcestershire sauce. How much worcestershire sauce should be used for eight servings?

109. *Art Supplies* Denise Viale teaches kindergarten and is buying supplies for her class to make papier-mâché piggy banks. Each piggy bank to be made requires $1\frac{1}{4}$ cups of flour. If Denise has 15 students who are going to make piggy banks, how much flour does Denise need to purchase?

110. *Height of a Computer Stand* The instructions for assembling a computer stand include a diagram illustrating its dimensions. Find the total height of the stand.

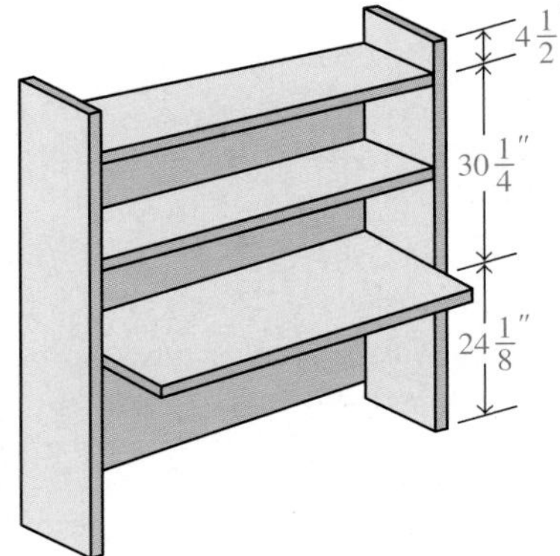

111. *Height Increase* Last year, Mark's height was $46\frac{3}{4}$ in. When measured this year, his height increased by $3\frac{5}{16}$ in. Find Mark's present height.

112. *Cutting Lumber* A piece of wood measures $15\frac{3}{8}$ in.

a) How far from one end should you cut the wood if you want to cut the length in half?

b) What is the length of each piece after the cut? You must allow $\frac{1}{8}$ in. for the saw cut.

113. *Width of a Picture* The width of a picture is $24\frac{7}{8}$ in., as shown in the diagram. Find x, the distance from the edge of the frame to the center.

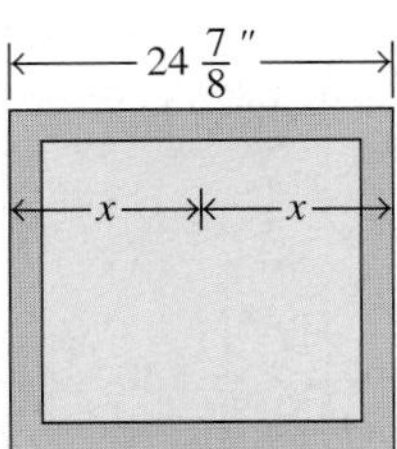

114. *Floor Molding* Rafela wants to place $\frac{1}{2}$ in. molding along the floor around the perimeter of her room (excluding door openings). She finds that she needs lengths of $26\frac{1}{2}$ in., $105\frac{1}{4}$ in., $53\frac{1}{4}$ in., and $106\frac{5}{16}$ in. How much molding will she need?

CHALLENGE PROBLEMS/GROUP ACTIVITIES

115. *Cutting Lumber* If a piece of wood $8\frac{3}{4}$ ft long is to be cut into four equal pieces, find the length of each piece. (Allow $\frac{1}{8}$ in. for each saw cut.)

116. *Increasing a Book Size* The dimensions of the cover of a book have been increased from $8\frac{1}{2}$ in. by $9\frac{1}{4}$ in. to $8\frac{1}{2}$ in. by $10\frac{1}{4}$ in. By how many square inches has the surface area increased? Use area = length × width.

117. *Dimensions of a Room* A rectangular room measures 8 ft 3 in. by 10 ft 8 in. by 9 ft 2 in. high.

a) Determine the perimeter of the room in feet.

b) Calculate the area of the floor of the room in square feet.

c) Calculate the volume of the room in cubic feet.

118. *Hanging a Picture* The back of a framed picture that is to be hung is shown. A nail is to be hammered into the wall, and the picture will be hung by the wire on the nail.

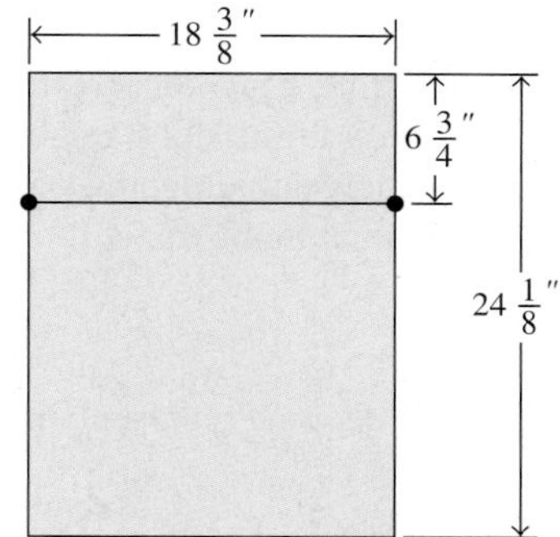

a) If the center of the wire is to rest on the nail and a side of the picture is to be 20 in. from the window, how far from the window should the nail be placed?

b) If the top of the frame is to be $26\frac{1}{4}$ in. from the ceiling, how far from the ceiling should the nail be placed? (Assume the wire will not stretch.)

c) Repeat part (b) if the wire will stretch $\frac{1}{4}$ in. when the picture is hung.

A set of numbers is said to be a dense set *if between any two distinct members of the set there exists a third distinct member of the set. The set of integers is not dense, since between any two consecutive integers, there is not another integer. For example, between 1 and 2 there are no other integers. The set of rational numbers is dense because between any two distinct rational numbers there exists a third distinct rational number. For example, we can find a rational number between 0.243 and 0.244. The number 0.243 can be written as 0.2430, and 0.244 can be written as 0.2440. There are many numbers between these two. Some of them are 0.2431, 0.2435, and 0.243912. In Exercises 119–126, find a rational number between the two numbers in each pair.*

119. 0.25 and 0.26 **120.** 4.003 and 4.3

121. −2.176 and −2.175 **122.** 1.3457 and 1.34571

123. 3.12345 and 3.123451 **124.** 0.4105 and 0.4106

125. 4.872 and 4.873 **126.** −3.7896 and −3.7895

To find a rational number halfway between any two rational numbers given in fraction form, add the two numbers together and divide their sum by 2. In Exercises 127–134, find a rational number halfway between the two fractions in each pair.

127. $\frac{3}{5}$ and $\frac{4}{5}$ **128.** $\frac{1}{9}$ and $\frac{2}{9}$ **129.** $\frac{1}{20}$ and $\frac{1}{10}$

130. $\frac{7}{13}$ and $\frac{8}{13}$ **131.** $\frac{1}{4}$ and $\frac{1}{5}$ **132.** $\frac{1}{3}$ and $\frac{2}{3}$

133. $\frac{1}{10}$ and $\frac{1}{100}$ **134.** $\frac{1}{2}$ and $\frac{2}{3}$

135. *Cooking Oatmeal* Following are the instructions given on a box of oatmeal. Determine the amount of water (or milk) and oats needed to make $1\frac{1}{2}$ servings by:

a) Adding the amount of each ingredient needed for 1 serving to the amount needed for 2 servings and dividing by 2.

b) Adding the amount of each ingredient needed for 1 serving to half the amount needed for 1 serving.

Directions

1. Boil water or milk and salt (if desired).
2. Stir in oats.
3. Stirring occasionally, cook over medium heat for 5 minutes.

Servings	1	2
Water (or milk)	1 cup	$1\frac{3}{4}$ cup
Oats	$\frac{1}{2}$ cup	1 cup
Salt (optional)	dash	$\frac{1}{8}$ tsp

136. Consider the rational number $0.\overline{9}$.

a) Use the method from Example 8 to convert $0.\overline{9}$ to a quotient of integers.

b) Find a number halfway between $0.\overline{9}$ and 1 by adding the two numbers and dividing by 2.

c) Find $\frac{1}{3} + \frac{2}{3}$. Express $\frac{1}{3}$ and $\frac{2}{3}$ as repeating decimals. Now find the same sum using the repeating decimal representation of $\frac{1}{3}$ and $\frac{2}{3}$.

d) What conclusion can you draw from parts (a), (b), and (c)?

RESEARCH ACTIVITY

137. The ancient Greeks are often considered the first true mathematicians. Write a report summarizing the ancient Greeks' contributions to rational numbers. Include in your report what they learned and believed about the rational numbers. References include encyclopedias, history of mathematics books, and Internet web sites.

5.4 THE IRRATIONAL NUMBERS AND THE REAL NUMBER SYSTEM

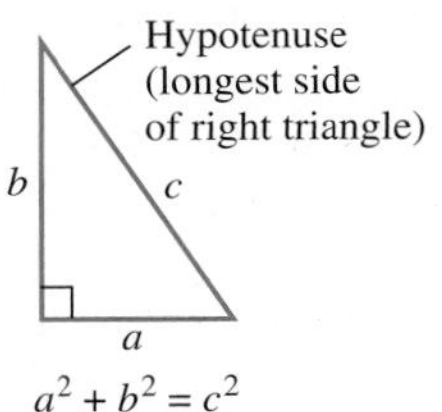

Figure 5.7

Pythagoras (ca. 585–500 B.C.), a Greek mathematician, is credited with providing a written proof that in any *right triangle* (a triangle with a 90° angle; see Fig. 5.7), the square of the length of one side (a^2) added to the square of the length of the other side (b^2) equals the square of the length of the hypotenuse (c^2). The formula $a^2 + b^2 = c^2$ is now known as the **Pythagorean theorem**.* Pythagoras found that the solution of the formula, where $a = 1$ and $b = 1$, is not a rational number.

$$\begin{aligned} a^2 + b^2 &= c^2 \\ 1^2 + 1^2 &= c^2 \\ 1 + 1 &= c^2 \\ 2 &= c^2 \end{aligned}$$

There is no rational number that when squared will equal 2. This prompted a need for a new set of numbers, the irrational numbers.

In Section 5.2, we introduced the real number line. The points on the real number line that are not rational numbers are referred to as irrational numbers. Recall that every rational number is either a terminating or a repeating decimal number. Therefore, irrational numbers, when represented as decimal numbers, will be nonterminating, nonrepeating decimal numbers.

> An **irrational number** is a real number whose decimal representation is a nonterminating, nonrepeating decimal number.

Nonrepeating number patterns can be used to indicate irrational numbers. For example, 6.1011011101111 . . . and 0.525225222 . . . are both irrational numbers.

The expression $\sqrt{2}$ is read "the square root of 2" or "radical 2." The symbol $\sqrt{\ }$ is called the **radical sign**, and the number or expression inside the radical sign is called the **radicand**. In $\sqrt{2}$, 2 is the radicand.

The square roots of some numbers are rational, whereas the square roots of other numbers are irrational. The **principal** (or **positive**) **square root** of a number n, written $\sqrt{n}$, is the positive number that when multiplied by itself, gives n. Whenever we mention the term "square root" in this text, we mean the principal square root. For example,

$$\sqrt{9} = 3 \quad \text{since} \quad 3 \cdot 3 = 9$$
$$\sqrt{36} = 6 \quad \text{since} \quad 6 \cdot 6 = 36$$

Both $\sqrt{9}$ and $\sqrt{36}$ are examples of numbers that are rational numbers because their square roots, 3 and 6, are terminating decimal numbers.

Returning to the problem faced by Pythagoras: If $c^2 = 2$, then c has a value of $\sqrt{2}$, but what is $\sqrt{2}$ equal to? The $\sqrt{2}$ is an irrational number, and it cannot be expressed as a terminating or repeating decimal number. It can only be approximated

---

*The Pythagorean theorem is discussed in more detail in Section 9.3.

PROFILE IN MATHEMATICS

Pythagoras of Samos

Pythagoras of Samos founded a philosophical and religious school in southern Italy in the sixth century B.C. The scholars at the school, known as Pythagoreans, produced important works of mathematics, astronomy, and theory of music. Although the Pythagoreans are credited with proving the Pythagorean theorem, it was known to the ancient Babylonians 1000 years earlier. The Pythagoreans were a secret society that formed a model for many secret societies in existence today. One practice was that students were to spend their first three years of study in silence, while their master, Pythagoras, spoke to them from behind a curtain. Among other philosophical beliefs was "that at its deepest level, reality is mathematical in nature."

by a decimal number: $\sqrt{2}$ is approximately 1.4142135 (to seven decimal places). Later in this section, we will discuss using a calculator to approximate irrational numbers.

Other irrational numbers include $\sqrt{3}$, $\sqrt{5}$, and $\sqrt{37}$. Another important irrational number used to represent the ratio of a circle's circumference to its diameter is pi, symbolized π. Pi is approximately 3.1415926.

We have discussed procedures for performing the arithmetic operations of addition, subtraction, multiplication, and division with rational numbers. We can perform the same operations with the irrational numbers. Before we can proceed, however, we must understand the numbers called perfect squares. Any number that is the square of a natural number is said to be a **perfect square**.

Natural numbers	1,	2,	3,	4,	5,	6, . . .
Squares of the natural numbers	1^2,	2^2,	3^2,	4^2,	5^2,	6^2, . . .
or perfect squares	1,	4,	9,	16,	25,	36, . . .

The numbers 1, 4, 9, 16, 25, and 36 are some of the perfect square numbers. Can you determine the next two perfect square numbers? How many perfect square numbers are there? The square root of a perfect square number will be a natural number. For example, $\sqrt{1} = 1$, $\sqrt{4} = 2$, $\sqrt{9} = 3$, $\sqrt{16} = 4$, $\sqrt{25} = 5$, and so on.

The number that multiplies a radical is called the radical's **coefficient**. For example, in $3\sqrt{5}$, the 3 is the coefficient of the radical.

Some irrational numbers can be simplified by determining whether there are any perfect square factors in the radicand. If there are, the following rule can be used to simplify the radical.

Product Rule for Radicals

$$\sqrt{a \cdot b} = \sqrt{a} \cdot \sqrt{b}, \qquad a \geq 0, \quad b \geq 0$$

To simplify a radical, write the radical as a product of two radicals. One of the radicals should contain the greatest perfect square that is a factor of the radicand in the original expression. Then simplify the radical containing the perfect square factor.

For example,

$$\sqrt{18} = \sqrt{9 \cdot 2} = \sqrt{9} \cdot \sqrt{2} = 3 \cdot \sqrt{2} = 3\sqrt{2}$$

and

$$\sqrt{75} = \sqrt{25 \cdot 3} = \sqrt{25} \cdot \sqrt{3} = 5 \cdot \sqrt{3} = 5\sqrt{3}$$

EXAMPLE 1 ***Simplifying Radicals***

Simplify a) $\sqrt{20}$ b) $\sqrt{32}$

SOLUTION

a) Since 4 is a perfect square factor of 20, we write

$$\sqrt{20} = \sqrt{4 \cdot 5} = \sqrt{4} \cdot \sqrt{5} = 2 \cdot \sqrt{5} = 2\sqrt{5}$$

Since 5 has no perfect square factors, $\sqrt{5}$ cannot be simplified.

A Piece of the Pi

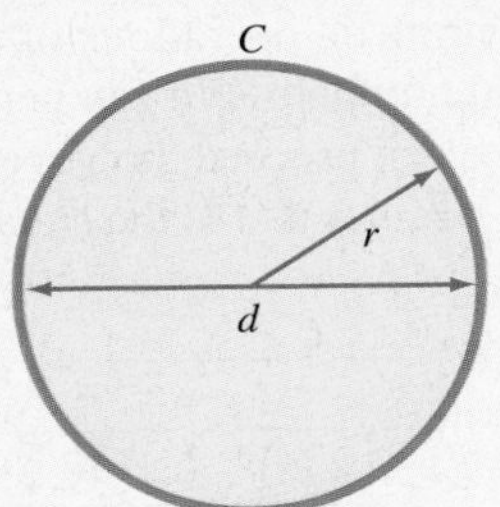

$C = \pi d$
or $\pi = \frac{C}{d}$

The American philosopher and psychologist William James wrote in 1909: "The thousandth decimal of π sleeps there though no one may every try to compute it." It sleeps no more. We now know π, the ratio of a circle's circumference to its diameter, to over 206 billion digits.

In September 1999, Yasumasa Kanada and Daisuke Takahashi of the University of Tokyo announced that they had calculated π to 206,158,430,000 (about $3 \cdot 2^{36}$) decimal places. To check their results, the researchers used two distinct algorithms. These two algorithms required 37 hours and 46 hours, respectively, of computation time on a Hitachi SR8000 computer.

This record, like the record for the largest prime number, will most likely be broken in the near future (it might already be broken as you read this). Mathematicians and computer scientists continue to improve both their computers and their methods used to find numbers like the most accurate approximation for π or the largest prime number.

b) Since 16 is a perfect square factor of 32, we write

$$\sqrt{32} = \sqrt{16 \cdot 2} = \sqrt{16} \cdot \sqrt{2} = 4\sqrt{2}$$

In Example 1(b), you can obtain the correct answer if you start out factoring differently:

$$\sqrt{32} = \sqrt{4 \cdot 8} = \sqrt{4} \cdot \sqrt{8} = 2\sqrt{8}$$

Note that 8 has 4 as a perfect square factor.

$$2\sqrt{8} = 2\sqrt{4 \cdot 2} = 2\sqrt{4} \cdot \sqrt{2} = 2 \cdot 2 \cdot \sqrt{2} = 4\sqrt{2}$$

The second method will eventually give the same answer, but it requires more work. It is best to try to factor out the *largest* perfect square factor from the radicand.

ADDITION AND SUBTRACTION OF IRRATIONAL NUMBERS

To add or subtract two or more square roots with the same radicand, add or subtract their coefficients. The answer is the sum or difference of the coefficients multiplied by the common radical.

EXAMPLE 2 ***Adding and Subtracting Radicals with the Same Radicand***

Simplify a) $3\sqrt{5} + 4\sqrt{5}$ b) $5\sqrt{3} - 2\sqrt{3} + \sqrt{3}$

SOLUTION

a) $3\sqrt{5} + 4\sqrt{5} = (3 + 4)\sqrt{5} = 7\sqrt{5}$

b) $5\sqrt{3} - 2\sqrt{3} + \sqrt{3} = (5 - 2 + 1)\sqrt{3} = 4\sqrt{3}$

Note that $\sqrt{3} = 1\sqrt{3}$.

EXAMPLE 3 ***Adding and Subtracting Radicals with Different Radicands***

Simplify $5\sqrt{3} - \sqrt{12}$.

SOLUTION These radicals cannot be added in their present form because they contain different radicands. When this occurs, determine whether one or more of the radicals can be simplified so that they have the same radicand.

$$\begin{aligned} 5\sqrt{3} - \sqrt{12} &= 5\sqrt{3} - \sqrt{4 \cdot 3} \\ &= 5\sqrt{3} - \sqrt{4} \cdot \sqrt{3} \\ &= 5\sqrt{3} - 2\sqrt{3} \\ &= (5 - 2)\sqrt{3} = 3\sqrt{3} \end{aligned}$$

MULTIPLICATION OF IRRATIONAL NUMBERS

When multiplying irrational numbers, we again make use of the product rule for radicals. After the radicands are multiplied, simplify the remaining radical when possible.

DID YOU KNOW

Is the Computer Always Right?

We often assume so, on the basis of the history and performance of the computer language and programs we use. Obviously, we cannot check calculations that are being made in billionths of a second. We could run a computation twice or on two different systems to be sure we get the same result to a specified number of decimal places. That's what is done with the big supercomputers. Their computation of the value of π often serves as a test of the program's ability to compute accurately.

EXAMPLE 4 ***Multiplying Radicals***

Simplify a) $\sqrt{3} \cdot \sqrt{27}$ b) $\sqrt{3} \cdot \sqrt{7}$ c) $\sqrt{6} \cdot \sqrt{10}$

SOLUTION

a) $\sqrt{3} \cdot \sqrt{27} = \sqrt{3 \cdot 27} = \sqrt{81} = 9$

b) $\sqrt{3} \cdot \sqrt{7} = \sqrt{3 \cdot 7} = \sqrt{21}$

c) $\sqrt{6} \cdot \sqrt{10} = \sqrt{6 \cdot 10} = \sqrt{60} = \sqrt{4 \cdot 15} = \sqrt{4} \cdot \sqrt{15} = 2\sqrt{15}$ ▲

DIVISION OF IRRATIONAL NUMBERS

To divide irrational numbers, use the following rule. After performing the division, simplify when possible.

Quotient Rule for Radicals

$$\frac{\sqrt{a}}{\sqrt{b}} = \sqrt{\frac{a}{b}}, \quad a \geq 0, \quad b > 0$$

EXAMPLE 5 ***Dividing Radicals***

Divide a) $\dfrac{\sqrt{8}}{\sqrt{2}}$ b) $\dfrac{\sqrt{96}}{\sqrt{2}}$

SOLUTION

a) $\dfrac{\sqrt{8}}{\sqrt{2}} = \sqrt{\dfrac{8}{2}} = \sqrt{4} = 2$

b) $\dfrac{\sqrt{96}}{\sqrt{2}} = \sqrt{\dfrac{96}{2}} = \sqrt{48} = \sqrt{16 \cdot 3} = \sqrt{16} \cdot \sqrt{3} = 4\sqrt{3}$ ▲

RATIONALIZING THE DENOMINATOR

A denominator is **rationalized** when it contains no radical expressions. To rationalize a denominator that contains only a square root, multiply both the numerator and denominator of the fraction by a number that will result in the radicand in the denominator becoming a perfect square. (This action is the equivalent of multiplying the fraction by 1 because the value of the fraction does not change.) Then simplify the fractions when possible.

EXAMPLE 6 ***Rationalizing the Denominator***

Rationalize the denominator of

a) $\dfrac{5}{\sqrt{2}}$ b) $\dfrac{5}{\sqrt{12}}$ c) $\dfrac{\sqrt{5}}{\sqrt{10}}$

SOLUTION

a) Multiply the numerator and denominator by a number that will make the radicand a perfect square.

$$\frac{5}{\sqrt{2}} = \frac{5}{\sqrt{2}} \cdot \frac{\sqrt{2}}{\sqrt{2}} = \frac{5\sqrt{2}}{\sqrt{4}} = \frac{5\sqrt{2}}{2}$$

Note that the 2's in the answer cannot be divided out because one 2 is a radicand and the other is not.

b) $\dfrac{5}{\sqrt{12}} = \dfrac{5}{\sqrt{12}} \cdot \dfrac{\sqrt{3}}{\sqrt{3}} = \dfrac{5\sqrt{3}}{\sqrt{36}} = \dfrac{5\sqrt{3}}{6}$

You could have obtained the same answer to this problem by multiplying both the numerator and denominator by $\sqrt{12}$ and then simplifying. Try to do so now.

c) Write $\dfrac{\sqrt{5}}{\sqrt{10}}$ as $\sqrt{\dfrac{5}{10}}$ and reduce the fraction to obtain $\sqrt{\dfrac{1}{2}}$. By the quotient rule for radicals, $\sqrt{\dfrac{1}{2}} = \dfrac{\sqrt{1}}{\sqrt{2}}$ or $\dfrac{1}{\sqrt{2}}$. Now rationalize $\dfrac{1}{\sqrt{2}}$.

$$\frac{1}{\sqrt{2}} = \frac{1}{\sqrt{2}} \cdot \frac{\sqrt{2}}{\sqrt{2}} = \frac{\sqrt{2}}{2}$$

APPROXIMATING SQUARE ROOTS ON A SCIENTIFIC CALCULATOR

Consider the irrational number the square root of two. We use the symbol $\sqrt{2}$ to represent the *exact value* of this number. Although exact values are important, approximations are also important, especially when working with application problems. We can use a scientific calculator to obtain approximations for square roots. Scientific calculators generally have one of the following square root keys:*

[$\sqrt{\ }$] or [$\sqrt{x}$]

For simplicity, we will refer to the square root key with the [$\sqrt{\ }$] symbol.

To approximate $\sqrt{2}$, perform the following keystrokes:

2 [$\sqrt{\ }$]

or, depending on your model of calculator, you may have to do the following:

[$\sqrt{\ }$] [2] [ENTER]

The display on your calculator may read 1.414213562. Your calculator may display more or fewer digits. It is important to realize that 1.414213562 is a rational number *approximation* for the irrational number $\sqrt{2}$. The symbol $\approx$ means *is approximately equal to,* and we write

$$\sqrt{2} \approx 1.414213562$$

Exact value (irrational number) ↗ ↖ Approximation (rational number)

EXAMPLE 7 *Approximating Square Roots*

Use a scientific calculator to approximate the following square roots. Round your answers to two decimal places.

a) $\sqrt{17}$ b) $\sqrt{31}$ c) $\sqrt{73}$ d) $\sqrt{198}$

SOLUTION

a) $\sqrt{17} \approx 4.12$ b) $\sqrt{31} \approx 5.57$ c) $\sqrt{73} \approx 8.54$ d) $\sqrt{198} \approx 14.07$

*If your calculator has the $\sqrt{\ }$ symbol printed *above* the key instead of on the face of the key, you can access the square root function by first pressing the "2nd" or the "inverse" key.

SECTION 5.4 EXERCISES

CONCEPT/WRITING EXERCISES

1. Explain the difference between a rational number and an irrational number.
2. What is the principal square root of a number?
3. What is a perfect square?
4. In your own words state the product rule for radicals.
5. **a)** Explain how to add square roots that have the same radicand.
   b) Using the procedure in part (a), add $7\sqrt{3} - 2\sqrt{3} + 3\sqrt{3}$.
6. In your own words, state the quotient rule for radicals.
7. What does it mean to rationalize a denominator?
8. **a)** Explain how to rationalize a denominator that contains a square root.
   b) Using the procedure in part (a), rationalize $\dfrac{3}{\sqrt{5}}$.

PRACTICE THE SKILLS

In Exercises 9–18, determine whether the number is rational or irrational.

9. $\sqrt{49}$ **10.** $\sqrt{10}$ **11.** $\dfrac{4}{7}$

12. 0.212112111 . . . **13.** 3.575775777 . . .

14. π **15.** $\dfrac{22}{7}$ **16.** 3.14159

17. 3.14159 . . . **18.** $\dfrac{\sqrt{5}}{\sqrt{5}}$

In Exercises 19–28, evaluate the expression.

19. $\sqrt{81}$ **20.** $\sqrt{121}$ **21.** $\sqrt{49}$ **22.** $-\sqrt{144}$

23. $-\sqrt{169}$ **24.** $\sqrt{25}$ **25.** $-\sqrt{225}$ **26.** $-\sqrt{36}$

27. $-\sqrt{100}$ **28.** $\sqrt{256}$

In Exercises 29–38, classify the number as a member of one or more of the following sets: the rational numbers, the integers, the natural numbers, the irrational numbers.

29. 10 **30.** -5

31. $\sqrt{25}$ **32.** $\dfrac{4}{5}$

33. 0.040040004 **34.** 2.718

35. $-\dfrac{7}{8}$ **36.** 0.123123123

37. $0.\overline{123}$ **38.** 0.123112311123 . . .

In Exercises 39–48, simplify the radical.

39. $\sqrt{12}$ **40.** $\sqrt{27}$ **41.** $\sqrt{52}$

42. $\sqrt{60}$ **43.** $\sqrt{63}$ **44.** $\sqrt{75}$

45. $\sqrt{80}$ **46.** $\sqrt{90}$ **47.** $\sqrt{162}$

48. $\sqrt{300}$

In Exercises 49–58, perform the indicated operation.

49. $3\sqrt{5} + 4\sqrt{5}$

50. $\sqrt{11} + 5\sqrt{11}$

51. $3\sqrt{7} - 5\sqrt{7}$

52. $2\sqrt{5} + 3\sqrt{20}$

53. $4\sqrt{12} - 7\sqrt{27}$

54. $2\sqrt{7} + 5\sqrt{28}$

55. $5\sqrt{3} + 7\sqrt{12} - 3\sqrt{75}$

56. $13\sqrt{2} + 2\sqrt{18} - 5\sqrt{32}$

57. $\sqrt{8} - 3\sqrt{50} + 9\sqrt{32}$

58. $\sqrt{63} + 13\sqrt{98} - 5\sqrt{112}$

In Exercises 59–68, perform the indicated operation. Simplify the answer when possible.

59. $\sqrt{3}\sqrt{2}$ **60.** $\sqrt{8}\sqrt{10}$ **61.** $\sqrt{9}\sqrt{15}$

62. $\sqrt{3}\sqrt{6}$ **63.** $\sqrt{10}\sqrt{20}$ **64.** $\sqrt{11}\sqrt{33}$

65. $\dfrac{\sqrt{8}}{\sqrt{4}}$ **66.** $\dfrac{\sqrt{125}}{\sqrt{5}}$ **67.** $\dfrac{\sqrt{72}}{\sqrt{8}}$

68. $\dfrac{\sqrt{136}}{\sqrt{8}}$

In Exercises 69–78, rationalize the denominator.

69. $\dfrac{7}{\sqrt{2}}$ **70.** $\dfrac{5}{\sqrt{11}}$ **71.** $\dfrac{\sqrt{5}}{\sqrt{13}}$

72. $\dfrac{\sqrt{3}}{\sqrt{10}}$ **73.** $\dfrac{\sqrt{20}}{\sqrt{3}}$ **74.** $\dfrac{\sqrt{50}}{\sqrt{14}}$

75. $\dfrac{\sqrt{9}}{\sqrt{2}}$ **76.** $\dfrac{\sqrt{15}}{\sqrt{3}}$ **77.** $\dfrac{\sqrt{10}}{\sqrt{6}}$

78. $\dfrac{8}{\sqrt{8}}$

PROBLEM SOLVING

The following diagram shows a 16 in. ruler marked using $\frac{1}{2}$ inches.

1 2 3 4 5 6 7 8 9 10 11 12 13 14 15

In Exercises 79–84, without using a calculator, indicate between which two markers each of the following irrational numbers will fall. Explain how you obtained your answer. Support your answer by obtaining an approximation with a calculator.

79. $\sqrt{15}$ in. **80.** $\sqrt{43}$ in.

81. $\sqrt{107}$ in. **82.** $\sqrt{135}$ in.

83. $\sqrt{170}$ in. **84.** $\sqrt{200}$ in.

In Exercises 85–90, determine whether the statement is true or false. Rewrite each false statement to make it a true statement.

85. $\sqrt{p}$ is irrational for any prime number p.

86. The sum of any two irrational numbers is a rational number.

87. The sum of any two irrational numbers is an irrational number.

88. The product of any two rational numbers is always a rational number.

89. The product of an irrational and a rational number is always an irrational number.

90. The product of any two irrational numbers is always an irrational number.

In Exercises 91–94, give an example to show that the stated case can occur.

91. The sum of two irrational numbers may be an irrational number.

92. The sum of two irrational numbers may be a rational number.

93. The product of two irrational numbers may be an irrational number.

94. The product of two irrational numbers may be a rational number.

95. Without doing any calculations, determine whether $\sqrt{3} = 1.732$. Explain your answer.

96. Without doing any calculations, determine whether $\sqrt{11} = 3.31\overline{6}$. Explain.

97. Give an example to show that $\sqrt{a+b} \neq \sqrt{a} + \sqrt{b}$.

98. The number π is an irrational number. Often the values 3.14 or $\frac{22}{7}$ are used for π. Does π equal either 3.14 or $\frac{22}{7}$? Explain your answer.

99. *A Swinging Pendulum* The time T required for a pendulum to swing back and forth may be found by the formula

$$T = 2\pi\sqrt{\frac{l}{g}}$$

where l is the length of the pendulum and g is the acceleration of gravity. Find the time in seconds if $l = 35$ cm and $g = 980$ cm/sec^2.

100. *Estimating Speed of a Vehicle* The speed a vehicle was traveling, s, in miles per hour, when the brakes were first applied, can be estimated using the formula $s = \sqrt{\frac{d}{0.04}}$ where d is the length of the vehicle's skid marks, in feet.

a) Determine the speed of a car that made skid marks 4 ft long.

b) Determine the speed of a car that made skid marks 16 ft long.

c) Determine the speed of a car that made skid marks 64 ft long.

d) Determine the speed of a car that made skid marks 256 ft long.

101. *Dropping an Object* The formula $t = \frac{\sqrt{d}}{4}$ can be used to estimate the time, t, in seconds it takes for an object dropped to travel d feet.

a) Determine the time it takes for an object to drop 100 ft.

b) Determine the time it takes for an object to drop 400 ft.

c) Determine the time it takes for an object to drop 900 ft.

d) Determine the time it takes for an object to drop 1600 ft.

CHALLENGE PROBLEMS/GROUP ACTIVITIES

102. a) If a radical expression is evaluated on a calculator, explain how you can determine whether the expression is a rational or irrational number.

b) Is $\sqrt{0.04}$ rational or irrational? Explain.

c) Is $\sqrt{0.7}$ rational or irrational? Explain.

103. One way to find a rational number between two distinct rational numbers is to add the two distinct rational numbers and divide by 2. Do you think that this method will work for finding an irrational number between two distinct irrational numbers? Explain.

RESEARCH ACTIVITIES

In Exercises 104 and 105, references include history of mathematics books, encyclopedias, and Internet web sites.

104. Do the necessary research and write a two-page report on the history of the development of the irrational numbers.

105. Do the necessary research and write a two-page report on the history of pi. In your report, indicate when the symbol π was first used and list the first 10 digits of π.

5.5 REAL NUMBERS AND THEIR PROPERTIES

Now that we have discussed both the rational and irrational numbers, we can discuss the real numbers and the properties of the real number system. The union of the rational numbers and the irrational numbers is the **set of real numbers**, symbolized by $\mathbb{R}$.

Figure 5.8 illustrates the relationship among various sets of numbers. It shows that the natural numbers are a subset of the whole numbers, the integers, the rational numbers, and the real numbers. For example, since the number 3 is a natural or counting number, it is also a whole number, an integer, a rational number, and a real number. Since the rational number $\frac{1}{4}$ is outside the set of integers, it is not an integer, a whole number, or a natural number. The number $\frac{1}{4}$ is a real number, however, as is the irrational number $\sqrt{2}$. Note that the real numbers are the union of the rational numbers and the irrational numbers.

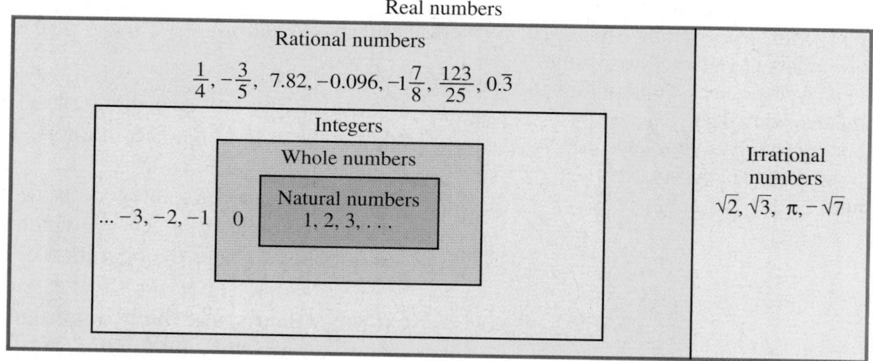

Figure 5.8

The relationship between the various sets of numbers in the real number system can also be illustrated with a tree diagram, as in Fig. 5.9.

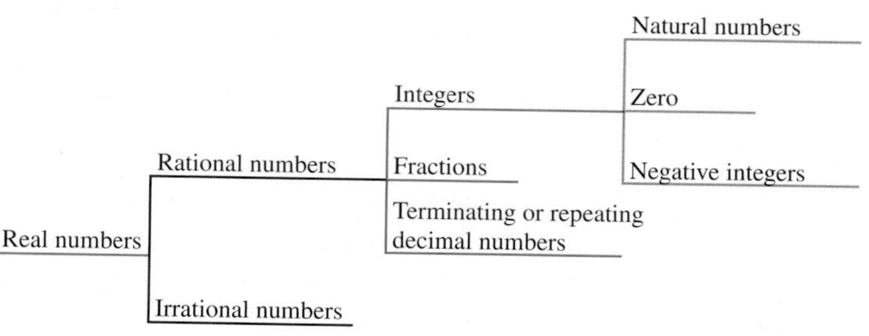

Figure 5.9

Figure 5.9 shows that, for example, the natural numbers are a subset of the integers, the rational numbers, and the real numbers. We can also see, for example, the natural numbers, zero, and the negative integers together form the integers.

PROPERTIES OF THE REAL NUMBER SYSTEM

We are now prepared to consider the properties of the real number system. The first property that we will discuss is *closure*.

DID YOU KNOW

An Important Number

Some years ago, a group of French mathematicians who worked under the collective pseudonym of "Monsieur Nicholas Bourbaki" embarked on the development of an encyclopedic description of all mathematics. They devoted 200 pages simply to introduce the innocent-looking concept, the number 1.

If an operation is performed on any two elements of a set and the result is an element of the set, we say that the set is **closed** under that given operation.

Is the sum of any two natural numbers a natural number? The answer is yes. Thus, we say that the natural numbers are closed under the operation of addition.

Are the natural numbers closed under the operation of subtraction? If we subtract one natural number from another natural number, must the difference always be a natural number? The answer is no. For example, $3 - 5 = -2$, which is not a natural number. Therefore, the natural numbers are not closed under the operation of subtraction.

EXAMPLE 1 ***Closure of Sets***

Determine whether the integers are closed under the operations of (a) multiplication and (b) division.

SOLUTION

a) If we multiply any two integers, will the product always be an integer? The answer is yes. Thus, the integers are closed under the operation of multiplication.

b) If we divide any two integers, will the quotient be an integer? The answer is no. For example, $6 \div 5 = \frac{6}{5}$, which is not an integer. Therefore, the integers are not closed under the operation of division. ▲

Next we will discuss three important properties: the commutative property, the associative property, and the distributive property. A knowledge of these properties is essential for the understanding of algebra. We begin with the commutative property.

Commutative Property

	Addition	***Multiplication***
	$a + b = b + a$	$a \cdot b = b \cdot a$

for any real numbers a and b.

The commutative property states that the *order* in which two numbers are added or multiplied is not important. For example, $4 + 5 = 5 + 4 = 9$ and $3 \cdot 6 = 6 \cdot 3 = 18$. Note that the commutative property does not hold for the operations of subtraction or division. For example,

$$4 - 7 \neq 7 - 4 \quad \text{and} \quad 9 \div 3 \neq 3 \div 9$$

Associative Property

	Addition	***Multiplication***
	$(a + b) + c = a + (b + c)$	$(a \cdot b) \cdot c = a \cdot (b \cdot c)$

for any real numbers a, b, and c.

The associative property states that when adding or multiplying three real numbers, we may place parentheses around any two adjacent numbers. For example,

$$\begin{aligned}(3 + 4) + 5 &= 3 + (4 + 5) \\ 7 + 5 &= 3 + 9 \\ 12 &= 12\end{aligned} \qquad \begin{aligned}(3 \cdot 4) \cdot 5 &= 3 \cdot (4 \cdot 5) \\ 12 \cdot 5 &= 3 \cdot 20 \\ 60 &= 60\end{aligned}$$

The associative property does not hold for the operations of subtraction and division. For example,

$$(10 - 6) - 2 \neq 10 - (6 - 2) \quad \text{and} \quad (27 \div 9) \div 3 \neq 27 \div (9 \div 3)$$

Note the difference between the commutative property and the associative property. The commutative property involves a change in *order,* whereas the associative property involves a change in *grouping* (or the *association* of numbers that are grouped together).

Another property of the real numbers is the distributive property of multiplication over addition.

Distributive Property of Multiplication over Addition

$$a \cdot (b + c) = a \cdot b + a \cdot c$$

for any real numbers a, b, and c.

For example, if $a = 3$, $b = 4$, and $c = 5$, then

$$\begin{aligned}3 \cdot (4 + 5) &= (3 \cdot 4) + (3 \cdot 5) \\ 3 \cdot 9 &= 12 + 15 \\ 27 &= 27\end{aligned}$$

This result indicates that, when using the distributive property, you may either add first and then multiply or multiply first and then add. Note that the distributive property involves two operations, addition and multiplication. Although positive integers were used in the example, any real numbers could have been used.

We frequently use the commutative, associative, and distributive properties without realizing that we are doing so. To add $13 + 4 + 6$, we may add the $4 + 6$ first to get 10. To this sum we then add 13 to get 23. Here we have done the equivalent of placing parentheses around the $4 + 6$. We can do so because of the associative property of addition.

To multiply 102×11 in our heads, we might multiply $100 \times 11 = 1100$ and $2 \times 11 = 22$ and add these two products to get 1122. We are permitted to do so because of the distributive property.

$$\begin{aligned}102 \times 11 = (100 + 2) \times 11 &= (100 \times 11) + (2 \times 11) \\ &= 1100 + 22 = 1122\end{aligned}$$

EXAMPLE 2 ***Identifying properties of Real Numbers***

Name the property illustrated.

a) $7 + 8 = 8 + 7$
b) $(7 + 8) + 9 = 7 + (8 + 9)$
c) $(x \cdot y) \cdot z = x \cdot (y \cdot z)$
d) $7(z + 9) = 7 \cdot z + 7 \cdot 9$
e) $a + (b + c) = a + (c + b)$
f) $7(8 \cdot x) = (8 \cdot x)7$

SOLUTION

a) Commutative property of addition
b) Associative property of addition
c) Associative property of multiplication
d) Distributive property of multiplication over addition
e) The only change between the left and right sides of the equal sign is the order of the b and c within the parentheses. The order is changed from $b + c$ to $c + b$ using the commutative property of addition.
f) The order of 7 and $(8 \cdot x)$ is changed by using the commutative property of multiplication.

EXAMPLE 3 ***Simplifying by Using the Distributive Property***

Use the distributive property to simplify

a) $2(3 + \sqrt{5})$ b) $\sqrt{2}(7 + \sqrt{3})$

SOLUTION

a) $2(3 + \sqrt{5}) = (2 \cdot 3) + (2 \cdot \sqrt{5})$
$= 6 + 2\sqrt{5}$

b) $\sqrt{2}(7 + \sqrt{3}) = (\sqrt{2} \cdot 7) + (\sqrt{2} \cdot \sqrt{3})$
$= 7\sqrt{2} + \sqrt{6}$

Note that $\sqrt{2} \cdot 7$ is written $7\sqrt{2}$.

EXAMPLE 4 ***Multiplying by Using the Distributive Property***

Use the distributive property to multiply $4(x + 5)$. Then simplify the result.

SOLUTION $4(x + 5) = 4 \cdot x + 4 \cdot 5$
$= 4x + 20$

We summarize the properties mentioned in this section as follows, where a, b, and c are any real numbers.

Commutative property of addition	$a + b = b + a$
Commutative property of multiplication	$a \cdot b = b \cdot a$
Associative property of addition	$(a + b) + c = a + (b + c)$
Associative property of multiplication	$(a \cdot b) \cdot c = a \cdot (b \cdot c)$
Distributive property of multiplication over addition	$a \cdot (b + c) = a \cdot b + a \cdot c$

SECTION 5.5 EXERCISES

CONCEPT/WRITING EXERCISES

1. What are the real numbers?
2. What symbol is used to represent the set of real numbers?
3. What does it mean if a set is closed under a given operation?
4. Give the commutative property of addition, explain what it means, and give an example illustrating it.
5. Give the commutative property of multiplication, explain what it means, and give an example illustrating it.
6. Give the associative property of addition, explain what it means, and give an example illustrating it.

7. Give the associative property of multiplication, explain what it means, and give an example illustrating it.
8. Give the distributive property of multiplication over addition, explain what it means, and give an example illustrating it.

PRACTICE THE SKILLS

In Exercises 9–12, determine whether the natural numbers are closed under the given operation.

9. Subtraction **10.** Addition
11. Multiplication **12.** Division

In Exercises 13–16, determine whether the integers are closed under the given operation.

13. Addition **14.** Subtraction
15. Multiplication **16.** Division

In Exercises 17–20, determine whether the rational numbers are closed under the given operation.

17. Subtraction **18.** Addition
19. Division **20.** Multiplication

In Exercises 21–24, determine whether the irrational numbers are closed under the given operation.

21. Subtraction **22.** Addition
23. Division **24.** Multiplication

In Exercises 25–28, determine whether the real numbers are closed under the given operation.

25. Subtraction **26.** Addition
27. Multiplication **28.** Division

29. Does $6 + (7 + 8) = 6 + (8 + 7)$ illustrate the commutative property or the associative property? Explain your answer.
30. Does $(1 + 2) + 3 = 3 + (1 + 2)$ illustrate the commutative property or the associative property? Explain your answer.
31. Give an example to show that the commutative property of multiplication may be true for the negative integers.
32. Give an example to show that the commutative property of addition may be true for the negative integers.
33. Does the commutative property hold for the rational numbers under the operation of division? Give an example to support your answer.
34. Does the commutative property hold for the integers under the operation of subtraction? Give an example to support your answer.
35. Give an example to show that the associative property of multiplication may be true for the negative integers.
36. Give an example to show that the associative property of addition may be true for the negative integers.
37. Does the associative property hold for the integers under the operation of division? Give an example to support your answer.
38. Does the associative property hold for the integers under the operation of subtraction? Give an example to support your answer.
39. Does the associative property hold for the real numbers under the operation of division? Give an example to support your answer.
40. Does $a + (b \cdot c) = (a + b) \cdot (a + c)$? Give an example to support your answer.

In Exercises 41–56, state the name of the property illustrated.

41. $5 + [7 + (-3)] = (5 + 7) + (-3)$
42. $s + t = t + s$
43. $s \cdot t = t \cdot s$
44. $5 \cdot z = z \cdot 5$
45. $(x + 2) + 8 = x + (2 + 8)$
46. $3(5 + 7) = 3 \cdot 5 + 3 \cdot 7$
47. $(\frac{2}{3} + \frac{4}{3}) + \frac{1}{3} = \frac{2}{3} + (\frac{4}{3} + \frac{1}{3})$
48. $(2 + 3) + (5 + 7) = (5 + 7) + (2 + 3)$
49. $5 + (10 + 15) = (10 + 15) + 5$
50. $3(a + b) = 3a + 3b$
51. $(1 + 10) + 100 = (10 + 1) + 100$
52. $(r + s) + t = t + (r + s)$
53. $(r + s) \cdot t = (r \cdot t) + (s \cdot t)$
54. $g \cdot (h + i) = (h + i) \cdot g$
55. $(p + q) + (r + s) = (r + s) + (p + q)$
56. $(a \cdot b) + (c \cdot d) = (b \cdot a) + (c \cdot d)$

In Exercises 57–64, use the distributive property to multiply. Then, if possible, simplify the resulting expression.

57. $3(y + 4)$ **58.** $7(x + 4)$
59. $\sqrt{2}(3 + \sqrt{6})$ **60.** $\sqrt{3}(\sqrt{8} + 4)$
61. $\sqrt{5}(x + \sqrt{5})$ **62.** $x(y + z)$
63. $\sqrt{3}(\sqrt{3} - \sqrt{6})$ **64.** $x(\sqrt{3} - \sqrt{5})$

PROBLEM SOLVING

In Exercises 65 and 66, name the property used to go from step to step.

65. $5(x + 4) + 6 = (5 \cdot x + 5 \cdot 4) + 6$
$= (5x + 20) + 6$
66. $= 5x + (20 + 6)$
$= 5x + 26$

In Exercises 67–70, name the property used to go from step to step.

67. $2(x + 4) + 3x = (2 \cdot x + 2 \cdot 4) + 3x$
$= (2x + 8) + 3x$
68. $= 2x + (8 + 3x)$
69. $= 2x + (3x + 8)$

70. $= (2x + 3x) + 8$

$= 5x + 8$

In Exercises 71–74, name the property used to go from step to step.

71. $2 + 3(x + 4) + 5x = 2 + (3 \cdot x + 3 \cdot 4) + 5x$

$= 2 + (3x + 12) + 5x$

72. $= 2 + (12 + 3x) + 5x$

73. $= (2 + 12) + 3x + 5x$

$= 14 + 8x$

74. $= 8x + 14$

In Exercises 75–80, determine whether the activity can be used to illustrate the commutative property. For the property to hold, the end result must be identical, regardless of the order in which the actions are performed.

75. Turning on a computer and typing a term paper on the computer

76. Putting on shoes and socks

77. Putting sugar and cream in coffee

78. Writing on the blackboard and erasing the blackboard

79. Turning on the lamp and reading a book

80. Washing clothes and drying clothes

In Exercises 81–88, determine whether the activity can be used to illustrate the associative property. For the property to hold, doing the first two actions followed by the third would produce the same end result as doing the second and third actions followed by the first.

81. Mowing the lawn, trimming the bushes, and removing dead limbs from trees

82. Brushing your teeth, washing your face, and combing your hair

83. Cracking an egg, pouring out the egg, and cooking the egg

84. Removing the gas cap, putting the nozzle in the tank, and turning on the gas pump

85. A coffee machine dropping the cup, dispensing the coffee, and then adding the sugar

86. Taking a bath, brushing your teeth, and taking your vitamins

87. Starting a car, moving the stick lever to drive, and then stepping on the gas

88. While making meatloaf, mixing in the milk, mixing in the spices, and mixing in the bread crumbs

89. The man in the photographs below is demonstrating that he can remove his sweater without removing his jacket. Can removal of the sweater and jacket be used to illustrate the commutative property? Explain your answer.

CHALLENGE PROBLEMS/GROUP ACTIVITIES

90. **a)** Consider the three words *man eating tiger*. Does (*man eating*) *tiger* mean the same as *man* (*eating tiger*)?

b) Does (*horse riding*) *monkey* mean the same as *horse* (*riding monkey*)?

c) Can you find three other nonassociative word triples?

91. Does $0 \div a = a \div 0$ (assume $a \neq 0$)? Explain.

RESEARCH ACTIVITY

92. Write a report on *complex numbers*. Include their relationship to real numbers.

5.6 RULES OF EXPONENTS AND SCIENTIFIC NOTATION

An understanding of exponents is important in solving problems in algebra. In the expression 5^2, the 2 is referred to as the **exponent** and the 5 is referred to as the **base**. We read 5^2 as 5 to the second power, or 5 squared, which means

$$5^2 = \underbrace{5 \cdot 5}_{\text{2 factors of 5}}$$

The number 5 to the third power, or 5 cubed, written 5^3, means

$$5^3 = \underbrace{5 \cdot 5 \cdot 5}_{\text{3 factors of 5}}$$

In general, the number b to the nth power, written b^n, means

$$b^n = \underbrace{b \cdot b \cdot b \cdot \cdots \cdot b}_{n \text{ factors of } b}$$

EXAMPLE 1 ***Evaluating the Power of a Number***

Evaluate the following.

a) 4^2 b) $(-5)^2$ c) 5^3 d) 1^{1000} e) 7^1

SOLUTION

a) $4^2 = 4 \cdot 4 = 16$
b) $(-5)^2 = (-5)(-5) = 25$
c) $5^3 = 5 \cdot 5 \cdot 5 = 125$
d) $1^{1000} = 1$. (The number 1 times itself any number of times equals 1.)
e) $7^1 = 7$. (Any number with an exponent of 1 equals the number itself.)

EXAMPLE 2 ***The Importance of Parentheses***

Evaluate the following.

a) -3^4 b) $(-3)^4$

SOLUTION

a) -3^4 means $-(3)^4$, which can also be written $-1(3)^4$.

$$-3^4 = -1(3)^4 = -1(3)(3)(3)(3) = -1(81) = -81$$

b)
$$\begin{aligned}(-3)^4 &= (-3)(-3)(-3)(-3)\\ &= (9)(-3)(-3)\\ &= (-27)(-3)\\ &= 81\end{aligned}$$

From Example 2, we can see that $-x^n \neq (-x)^n$, where n is an even natural number.

RULES OF EXPONENTS

Now that we know how to evaluate powers of numbers we can discuss the rules of exponents. Consider

$$2^2 \cdot 2^3 = \underbrace{2 \cdot 2}_{\text{2 factors}} \cdot \underbrace{2 \cdot 2 \cdot 2}_{\text{3 factors}} = 2^5$$

This example illustrates the product rule for exponents.

Product Rule for Exponents

$$a^m \cdot a^n = a^{m+n}$$

Therefore, by using the product rule, $2^2 \cdot 2^3 = 2^{2+3} = 2^5$.

EXAMPLE 3 ***Using the Product Rule for Exponents***

Use the product rule to simplify.

a) $3^4 \cdot 3^5$ b) $7^2 \cdot 7^6$

SOLUTION

a) $3^4 \cdot 3^5 = 3^{4+5} = 3^9$

b) $7^2 \cdot 7^6 = 7^{2+6} = 7^8$

Consider

$$\frac{2^5}{2^2} = \frac{2 \cdot 2 \cdot 2 \cdot 2 \cdot 2}{2 \cdot 2} = 2 \cdot 2 \cdot 2 = 2^3$$

This example illustrates the quotient rule for exponents.

Quotient Rule for Exponents

$$\frac{a^m}{a^n} = a^{m-n}, \qquad a \neq 0$$

Therefore, $\frac{2^5}{2^2} = 2^{5-2} = 2^3$.

EXAMPLE 4 ***Using the Quotient Rule for Exponents***

Use the quotient rule to simplify.

a) $\frac{5^8}{5^5}$ b) $\frac{8^{12}}{8^5}$

SOLUTION

a) $\frac{5^8}{5^5} = 5^{8-5} = 5^3$ b) $\frac{8^{12}}{8^5} = 8^{12-5} = 8^7$

DID YOU KNOW

A Very Large Number

170,141,183,460,469,231,731,687, 303,715,884,105,727 is a very large number. How would you read this number? Take a breath.

170 undecillion, 141 decillion, 183 nonillion, 460 octillion, 469 septillion, 231 sextillion, 731 quintillion, 687 quadrillion, 303 trillion, 715 billion, 884 million, 105 thousand, 727.

Often numbers this large can be represented with an approximation involving scientific notation. This number, however, is a prime number, and its *exact* representation is very important. Mathematicians frequently represent such large numbers by using exponents. Here, another exact, and more efficient, representation of this number is $2^{127} - 1$.

Consider $2^3 \div 2^3$. The quotient rule gives

$$\frac{2^3}{2^3} = 2^{3-3} = 2^0$$

But $\frac{2^3}{2^3} = \frac{8}{8} = 1$. Therefore, 2^0 must equal 1. This example illustrates the zero exponent rule.

Zero Exponent Rule

$$a^0 = 1, \qquad a \neq 0$$

Note that 0^0 is not defined by the zero exponent rule.

EXAMPLE 5 ***The Zero Power***

Use the zero exponent rule to simplify.

a) 3^0 b) $(-7)^0$

SOLUTION

a) $3^0 = 1$ b) $(-7)^0 = 1$

Consider $2^3 \div 2^5$. The quotient rule yields

$$\frac{2^3}{2^5} = 2^{3-5} = 2^{-2}$$

But $\frac{2^3}{2^5} = \frac{\cancel{2} \cdot \cancel{2} \cdot \cancel{2}}{\cancel{2} \cdot \cancel{2} \cdot \cancel{2} \cdot 2 \cdot 2} = \frac{1}{2^2}$. Since $\frac{2^3}{2^5}$ equals both 2^{-2} and $\frac{1}{2^2}$, then 2^{-2} must equal $\frac{1}{2^2}$. This example illustrates the negative exponent rule.

Negative Exponent Rule

$$a^{-m} = \frac{1}{a^m}, \qquad a \neq 0$$

EXAMPLE 6 ***Using the Negative Exponent Rule***

Use the negative exponent rule to simplify.

a) 5^{-2} b) 8^{-1}

SOLUTION

a) $5^{-2} = \frac{1}{5^2} = \frac{1}{25}$ b) $8^{-1} = \frac{1}{8^1} = \frac{1}{8}$

Consider $(2^3)^2$:

$$(2^3)^2 = (2^3)(2^3) = 2^{3+3} = 2^6$$

DID YOU KNOW

Large and Small Numbers

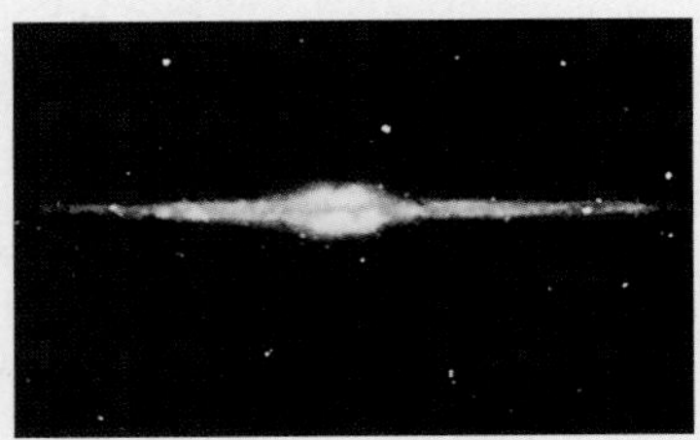

1×10^5 light-years

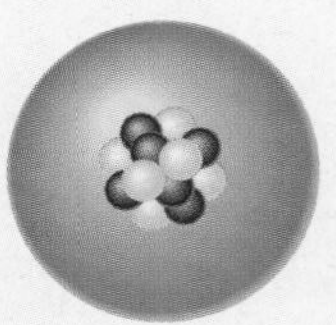

Diameter
1×10^{-10} meters

Our everyday activities don't require us to deal with quantities much above those in the thousands: \$5.95 for lunch, 100 meters to a lap, a \$12,000 car loan, and so on. Yet as modern technology has developed, so has our ability to study all aspects of the universe we live in, from the very large to the very small. Computing values when you have to deal with 10 or 20 digits at a time isn't easy. The rules of exponents and scientific notation, however, allow us to compute with numbers that are out of this world in a very down-to-earth way.

This example illustrates the power rule for exponents.

Power Rule for Exponents

$$(a^m)^n = a^{m \cdot n}$$

Thus, $(2^3)^2 = 2^{3 \cdot 2} = 2^6$.

EXAMPLE 7 ***Evaluating a Power Raised to Another Power***

Use the power rule to simplify.

a) $(5^4)^3$ b) $(7^2)^5$

SOLUTION

a) $(5^4)^3 = 5^{4 \cdot 3} = 5^{12}$ b) $(7^2)^5 = 7^{2 \cdot 5} = 7^{10}$

Summary of the Rules of Exponents

$a^m \cdot a^n = a^{m+n}$		Product rule for exponents
$\dfrac{a^m}{a^n} = a^{m-n}$,	$a \neq 0$	Quotient rule for exponents
$a^0 = 1$,	$a \neq 0$	Zero exponent rule
$a^{-m} = \dfrac{1}{a^m}$,	$a \neq 0$	Negative exponent rule
$(a^m)^n = a^{m \cdot n}$		Power rule for exponents

SCIENTIFIC NOTATION

Often scientific problems deal with very large and very small numbers. For example, the distance from the Earth to the sun is about 93,000,000 miles. The wavelength of a yellow color of light is about 0.0000006 meter. Because working with many zeros is difficult, scientists developed a notation that expresses such numbers with exponents. For example, consider the distance from the Earth to the sun, 93,000,000 miles.

$$\begin{aligned} 93{,}000{,}000 &= 9.3 \times 10{,}000{,}000 \\ &= 9.3 \times 10^7 \end{aligned}$$

The wavelength of a yellow color of light is about 0.0000006 meter.

$$\begin{aligned} 0.0000006 &= 6.0 \times 0.0000001 \\ &= 6.0 \times 10^{-7} \end{aligned}$$

The numbers 9.3×10^7 and 6.0×10^{-7} are written in a form called **scientific notation**. Each number written in scientific notation is written as a number greater than or equal to 1 and less than 10 ($1 \leq a < 10$) multiplied by some power of 10.

Some examples of numbers in scientific notation are

$$3.7 \times 10^3, \quad 2.05 \times 10^{-3}, \quad 5.6 \times 10^8, \quad \text{and} \quad 1.00 \times 10^{-5}$$

The following is a procedure for writing a number in scientific notation.

To Write a Number in Scientific Notation

1. Move the decimal point in the original number to the right or left until you obtain a number greater than or equal to 1 and less than 10.
2. Count the number of places you have moved the decimal point to obtain the number in step 1. If the decimal point was moved to the left, the count is to be considered positive. If the decimal point was moved to the right, the count is to be considered negative.
3. Multiply the number obtained in step 1 by 10 raised to the count found in step 2. (Note that the count determined in step 2 is the exponent on the base 10.)

EXAMPLE 8 *Converting from Decimal Notation to Scientific Notation*

Write each number in scientific notation.

a) In 1999, the world population was about 6,008,000,000 people.
b) In 1999, the U.S. federal government debt was about $5,650,000,000,000.
c) The diameter of a hydrogen atom nucleus is about 0.0000000000011 millimeter.
d) The wavelength of an X ray is about 0.000000000492 meters.

SOLUTION

a) $6{,}008{,}000{,}000 = 6.008 \times 10^9$
b) $5{,}650{,}000{,}000{,}000 = 5.65 \times 10^{12}$
c) $0.0000000000011 = 1.1 \times 10^{-12}$
d) $0.000000000492 = 4.92 \times 10^{-10}$

To convert from a number given in scientific notation to decimal notation we reverse the procedure.

To Change a Number in Scientific Notation to Decimal Notation

1. Observe the exponent on the 10.
2. **a)** If the exponent is positive, move the decimal point in the number to the right the same number of places as the exponent. Adding zeros to the number might be necessary.
   b) If the exponent is negative, move the decimal point in the number to the left the same number of places as the exponent. Adding zeros might be necessary.

EXAMPLE 9 *Converting from Scientific Notation to Decimal Notation*

Write each number in decimal notation.

a) The average distance from the Earth to the sun is about 9.3×10^7 miles.
b) The half-life of uranium 235 is about 4.5×10^9 years.
c) The average grain size in siltstone is 1.35×10^{-3} inch.
d) A *millimicron* is a unit of measure used for very small distances. One millimicron is about 3.94×10^{-8} inch.

SOLUTION

a) $9.3 \times 10^7 = 93{,}000{,}000$
b) $4.5 \times 10^9 = 4{,}500{,}000{,}000$
c) $1.35 \times 10^{-3} = 0.00135$
d) $3.94 \times 10^{-8} = 0.0000000394$

In scientific journals and books, we occasionally see numbers like 10^{15} and 10^{-6}. We interpret these numbers as 1×10^{15} and 1×10^{-6}, respectively, when converting the numbers to decimal form.

EXAMPLE 10 ***Multiplying Numbers in Scientific Notation***

Multiply $(4.3 \times 10^6)(2 \times 10^{-4})$. Write the answer in decimal notation.

SOLUTION

$$\begin{aligned}(4.3 \times 10^6)(2 \times 10^{-4}) &= (4.3 \times 2)(10^6 \times 10^{-4})\\ &= 8.6 \times 10^2\\ &= 860\end{aligned}$$

▲

EXAMPLE 11 ***Dividing Numbers Using Scientific Notation***

Divide $\frac{0.000000000048}{24{,}000{,}000{,}000}$. Write the answer in scientific notation.

SOLUTION First write each number in scientific notation.

$$\begin{aligned}\frac{0.000000000048}{24{,}000{,}000{,}000} = \frac{4.8 \times 10^{-11}}{2.4 \times 10^{10}} &= \left(\frac{4.8}{2.4}\right)\left(\frac{10^{-11}}{10^{10}}\right)\\ &= 2 \times 10^{-11-10}\\ &= 2 \times 10^{-21}\end{aligned}$$

▲

SCIENTIFIC NOTATION ON THE SCIENTIFIC CALCULATOR

One of the advantages of using scientific notation when working with very large and very small numbers is the ease with which you can perform operations. Performing these operations is even easier with the use of a scientific calculator. Most scientific calculators have a scientific notation key labeled "Exp," "EXP," or "EE." We will refer to the scientific notation key as [EXP]. The following keystrokes can be used to enter the number 4.3×10^6

Keystroke(s)	Calculator Display
4.3	4.3
[EXP]	4.3 ^{00}
6	4.3 ^{06}

Your calculator may have some slight variations to the display shown here. The display 4.3 ^{06} means 4.3×10^6. We now will use our calculators to perform some computations using scientific notation.

EXAMPLE 12 ***Use Scientific Notation on a Calculator to Find a Product***

Multiply $(4.3 \times 10^6)(2 \times 10^{-4})$ using a scientific calculator. Write the answer in decimal notation.

SOLUTION Our sequence of keystrokes is as follows

Keystroke(s)	Display	
4.3	4.3	
EXP	4.3 ^{00}	
6	4.3 ^{06}	
×	4300000	4.3×10^6 is now entered in the calculator. Most calculators will convert to decimal notation (if the display can show the number)
2	2.	
EXP	2. ^{00}	
4	2. ^{04}	Enter the positive form of the exponent
+/−	2. $^{-04}$	Make the exponent negative*
=	860.	Press = to obtain the answer of 860†

EXAMPLE 13 ***Scientific Notation on a Calculator to Find a Quotient***

Divide $\dfrac{0.000000000048}{24{,}000{,}000{,}000}$ using a scientific calculator. Write the answer in scientific notation.

SOLUTION We first rewrite the numerator and denominator using scientific notation. See Example 11.

$$\frac{0.000000000048}{24{,}000{,}000{,}000} = \frac{4.8 \times 10^{-11}}{2.4 \times 10^{10}}$$

Next we use a scientific calculator to perform the computation. The keystrokes are as follows

4.8 EXP 11 +/− ÷ 2.4 EXP 10 =

The display on the calculator is 2. $^{-21}$, which means 2.0×10^{-21}.

EXAMPLE 14 ***U.S. Debt per Person***

On August 28, 1999, the U.S. Department of the Treasury estimated the U.S. federal debt to be about \$5.65 trillion. On this same day, the U.S. Bureau of the Census estimated the U.S. population to be about 273 million people. Determine the average debt, per person, by dividing the U.S. federal debt by the U.S. population.

SOLUTION First we will write the numbers involved using decimal notation and then convert them to scientific notation.

$$5.65 \text{ trillion} = 5{,}650{,}000{,}000{,}000 = 5.65 \times 10^{12}$$
$$273 \text{ million} = 273{,}000{,}000 = 2.73 \times 10^{8}$$

*Some calculators will require you to enter the negative sign before entering the exponent.

†Some calculators will display the answer in scientific notation. In this case, the display will show 8.6 ^{02}, which means 8.6×10^2 and equals 860.

DID YOU KNOW

What's the Difference Between Debt and Deficit?

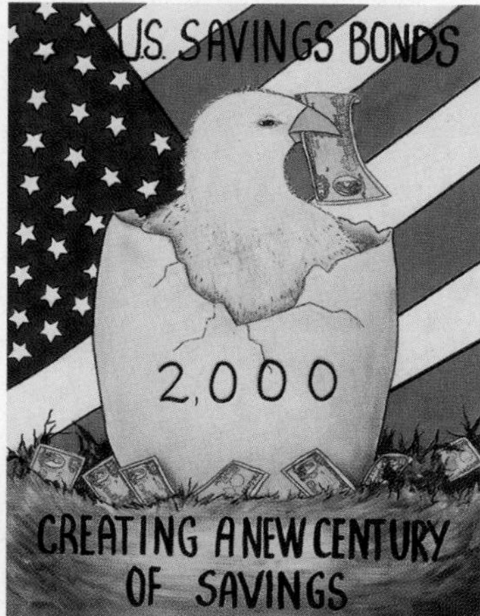

We often hear economists and politicians talk about such things as revenue, expenditures, deficit, surplus, and national debt. Revenue is the money the government collects annually, mostly through taxes. Expenditures are the money the government spends annually. If revenue exceeds expenditures, a surplus occurs; if expenditures exceed revenue, a deficit occurs. The national debt is the total of all the budget deficits and (the few) surpluses encountered by the federal government for over 200 years. How big is the national debt? Search the Internet for the latest figures, but as of August 3, 1999, the national debt was \$5,642,129,525,649.20. This amount is owed to investors worldwide—perhaps including you—who own government bonds.

Now we divide 5.65×10^{12} by 2.73×10^{8} using a scientific calculator. The keystrokes are

5.65 [EXP] 12 [÷] 2.73 [EXP] 8 [=]

The display shows 20,695.9707. This indicates that on August 28, 1999 the U.S. government owed about $20,695.97 per man, woman, and child living in the United States.

SECTION 5.6 EXERCISES

CONCEPT/WRITING EXERCISES

1. In the expression 4^6, what is the name given to the 4, and what is the name given to the 6?
2. Explain the meaning of b^n.
3. a) Explain the product rule for exponents.
   b) Use the product rule to simplify $3^3 \cdot 3^5$.
4. a) Explain the quotient rule for exponents.
   b) Use the quotient rule to simplify $\frac{4^5}{4^3}$.
5. a) Explain the negative exponent rule.
   b) Use the negative exponent rule to simplify 3^{-5}.
6. a) Explain the zero exponent rule.
   b) Use the zero exponent rule to simplify 5^0.
7. a) Explain the power rule for exponents
   b) Use the power rule to simplify $(4^4)^3$.
8. Explain how you can simplify the expression 1^{500}.
9. a) In your own words, explain how to change a number in decimal notation to scientific notation.
   b) Using the procedure in part (a) change 0.000426 to scientific notation.
10. a) In your own words, explain how to change a number in scientific notation to decimal notation.
    b) Change 5.76×10^{-4} to decimal notation.
11. A number is given in scientific notation. What does it indicate about the number when the exponent on the ten is (a) positive, (b) zero, and (c) negative?
12. a) How is the number 10^5 interpreted in scientific notation?
    b) Write 10^5 as a number in decimal notation.

PRACTICE THE SKILLS

In Exercises 13–44, evaluate the expression.

13. 3^2 **14.** 4^3 **15.** $(-5)^2$

16. $(-2)^5$ **17.** -2^5 **18.** $\left(\frac{1}{3}\right)^2$

19. $\left(\frac{4}{5}\right)^2$ **20.** 2^4 **21.** $(-2)^4$

22. -2^4 **23.** $2^3 \cdot 3^2$ **24.** $\frac{15^2}{3^2}$

25. $\frac{5^7}{5^5}$ **26.** $3^3 \cdot 3^4$ **27.** $\frac{7}{7^3}$

28. $3^4 \cdot 7^0$ **29.** $(-13)^0$ **30.** $(-3)^4$

31. 3^4 **32.** -3^4 **33.** 3^{-2}

34. 3^{-3} **35.** $(2^3)^4$ **36.** $(1^{12})^{13}$

37. $\frac{11^{25}}{11^{23}}$ **38.** $5^2 \cdot 5$ **39.** $(-4)^2$

40. 4^{-2} **41.** -4^2 **42.** $(4^3)^2$

43. $(2^2)^{-3}$ **44.** $3^{-3} \cdot 3$

In Exercises 45–60, express the number in scientific notation.

45. 120,000 **46.** 9,751,000 **47.** 45

48. 0.000421 **49.** 0.053 **50.** 0.0000561

51. 19,000 **52.** 1,260,000,000 **53.** 0.000186

54. 0.0003 **55.** 0.00000423 **56.** 54,000

57. 711 **58.** 0.02 **59.** 0.153

60. 416,000

In Exercises 61–76, express the number in decimal notation.

61. 8.4×10^4 **62.** 2.71×10^{-3} **63.** 1.2×10^{-2}

64. 5.19×10^5 **65.** 2.13×10^{-5} **66.** 2.74×10^{-7}

67. 3.12×10^{-1} **68.** 4.6×10^1 **69.** 9×10^6

70. 7.3×10^4 **71.** 2.31×10^2 **72.** 1.04×10^{-2}

73. 3.5×10^4 **74.** 2.17×10^{-6} **75.** 1×10^4

76. 1×10^{-3}

In Exercises 77–86, (a) perform the indicated operation without the use of a calculator and express each answer in decimal notation. (b) Confirm your answer from part (a) by using a scientific calculator to perform the operations. If the calculator displays the answer in scientific notation, convert the answer to decimal notation.

77. $(4 \times 10^2)(3 \times 10^5)$ **78.** $(2 \times 10^{-3})(3 \times 10^2)$

79. $(5.1 \times 10^1)(3 \times 10^{-4})$ **80.** $(1.6 \times 10^{-2})(4 \times 10^{-3})$

81. $\dfrac{6.4 \times 10^5}{2 \times 10^3}$ **82.** $\dfrac{8 \times 10^{-3}}{2 \times 10^1}$ **83.** $\dfrac{8.4 \times 10^{-6}}{4 \times 10^{-3}}$

84. $\dfrac{25 \times 10^3}{5 \times 10^{-2}}$ **85.** $\dfrac{4 \times 10^5}{2 \times 10^4}$ **86.** $\dfrac{16 \times 10^3}{8 \times 10^{-3}}$

In Exercises 87–96, (a) perform the indicated operation without the use of a calculator and express each answer in scientific notation. (b) Confirm your answer from part (a) by using a scientific calculator to perform the operations. If the calculator displays the answer in decimal notation, convert the answer to scientific notation.

87. (700,000)(6,000,000) **88.** (0.0006)(5,000,000)

89. (0.003)(0.00015) **90.** (230,000)(3000)

91. $\dfrac{1,400,000}{700}$ **92.** $\dfrac{20,000}{0.0005}$ **93.** $\dfrac{0.00004}{200}$

94. $\dfrac{0.0012}{0.000006}$ **95.** $\dfrac{150,000}{0.0005}$ **96.** $\dfrac{24,000}{8,000,000}$

PROBLEM SOLVING

In Exercises 97–100, list the numbers from smallest to largest.

97. 5.8×10^5; 3.2×10^{-1}; 4.6; 8.3×10^{-4}

98. 8.5×10^{-5}; 8.2×10^3; 1.3×10^{-1}; 6.2×10^4

99. 40,000; 4.1×10^3; 0.00079; 8.3×10^{-5}

100. 267,000,000; 3.14×10^7; 1,962,000; 4.79×10^6

In Exercises 101–107, express your answer (a) using decimal notation and (b) using scientific notation. You may use a scientific calculator to perform the necessary operations.

101. *World Population* According to the U.S. Bureau of the Census, the population of the world in 1999 was approximately 6.008×10^9 and the population of China was about 1.256×10^9. How many people lived outside China?

102. *Traveling to Jupiter* The distance from Earth to the planet Jupiter is approximately 4.5×10^8 mi. If a spacecraft traveled at a speed of 25,000 mph, how many hours would the spacecraft need to travel from Earth to Jupiter? Use distance = rate × time.

103. *Traveling to the Moon* The distance from Earth to the moon is approximately 239,000 mi. If a spacecraft travels at a speed of 20,000 mph, how many hours would the spacecraft need to travel from Earth to the moon? Use distance = rate × time.

104. *Bucket Full of Molecules* A drop of water contains about 40 billion molecules. If a bucket has half a million drops of water in it, how many molecules of water are in the bucket?

105. *Blood Cells in a Cubic Millimeter* If a cubic millimeter of blood contains 5,800,000 red blood cells, how many red blood cells are contained in 50 cubic millimeters of blood?

106. *Radioactive Isotopes* The half-life of a radioactive isotope is the time required for half the quantity of the isotope to decompose. The half-life of uranium 238 is 4.5×10^9 years, and the half-life of uranium 234 is 2.5×10^5 years. How many times greater is the half-life of uranium 238 than uranium 234?

107. *1950 Niagara Treaty* The 1950 Niagara Treaty between the United States and Canada requires that during the tourist season a minimum of 100,000 cubic feet of water per second (ft^3/sec) flow over Niagara Falls (another 130,000–160,000 ft^3/sec are diverted for power generation). Find the minimum amount of water that will flow over the falls in a 24 hour period during the tourist season.

108. *U.S. Debt per Person: 1992 versus 1999* In Example 14, the U.S. government debt was discussed. It was found that the debt in 1999 was $20,695.97 per person. In 1992, the U.S. government debt was about $4.65 trillion, and the population of the United States was about 257 million people.

a) Determine the amount of debt per person in the United States in 1992.

b) How much more per person did the U.S. government owe in 1999 than in 1992?

109. *Disposable Diaper Quantity* Laid end to end, the 18 billion disposable diapers thrown away in the United States each year would reach the moon and back seven times.

a) Write 18 billion in scientific notation.

b) If the distance from the Earth to the moon is 2.38×10^5 miles, what is the length of all these diapers placed end to end? Write the answer in decimal notation.

110. *Mutual Fund Manager* Lauri Mackey is the fund manager for the Mackey Mutual Fund. This mutual fund has total assets of \$1.2 billion. Lauri wants to maintain the investments in this fund according to the following pie chart.

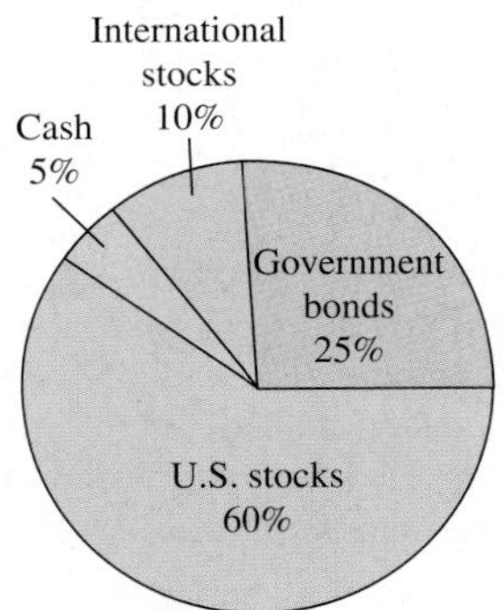

Source: Ibbotson Associates

a) How much of the \$1.2 billion should be invested in U.S. stocks?

b) How much should be invested in government bonds?

c) How much should be invested in international stocks?

d) How much should remain in cash?

111. *Another Mutual Fund Manager* Susan Dratch is the fund manager for the Dratch Mutual Fund. This mutual fund has total assets of \$3.4 billion. Susan wants to maintain the investments in this fund according to the following pie chart.

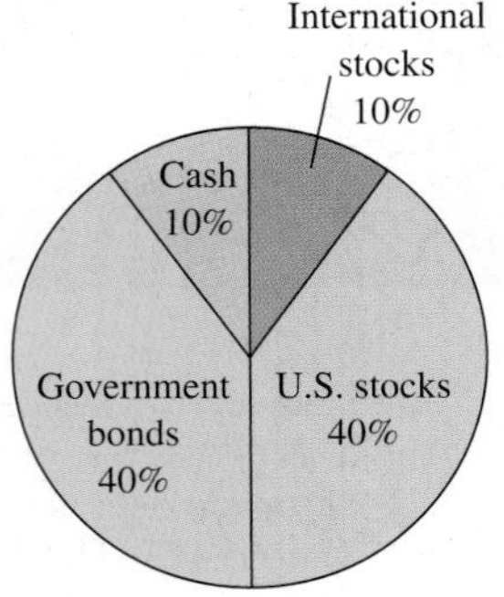

Source: Ibbotson Associates

a) How much of the \$3.4 billion should be invested in U.S. stocks?

b) How much should be invested in government bonds?

c) How much should be invested in international stocks?

d) How much should remain in cash?

112. *Metric System Comparison* In the metric system, 1 meter = 10^3 millimeters. How many times greater is a meter than a millimeter? Explain how you determined your answer.

113. In the metric system, 1 gram = 10^3 milligrams and 1 gram = 10^{-3} kilograms. What is the relationship between milligrams and kilograms? Explain how you determined your answer.

114. *Earth to Sun Comparison* The mass of the sun is approximately 2×10^{30} kilograms, and the mass of the Earth is approximately 6×10^{24} kilograms. How many times greater is the mass of the sun than the mass of the Earth? Write your answer in decimal notation.

115. *The Day of Six Billion* The United Nations declared October 12, 1999, the *Day of Six Billion.* On this day, the Earth's population was estimated to reach 6 billion. Currently, the Earth's population is doubling about every 35 years.

a) Using this figure, estimate the world's population in the year 2034.

b) Assuming 365 days in a year, estimate the average number of additional people added to the Earth's population each day between 1999 and 2034.

116. *Computer Calculation Speed* The IBM Blue Pacific computer is capable of operating at a peak speed of about 3.9 trillion (3,900,000,000,000) calculations per second. At this rate, how long would it take to perform a task requiring 897 quadrillion (897,000,000,000,000,000) calculations?

CHALLENGE PROBLEMS/GROUP ACTIVITIES

117. *Comparing a Million to a Billion* Many people have no idea of the difference in size between a million (1,000,000), a billion (1,000,000,000), and a trillion (1,000,000,000,000).

a) Write a million, a billion, and a trillion in scientific notation.

b) Determine how long it would take to spend a million dollars if you spent \$1000 a day.

c) Repeat part (b) for a billion dollars.

d) Repeat part (b) for a trillion dollars.

e) How many times greater is a billion dollars than a million dollars?

118. *Speed of Light*

a) Light travels at a speed of 1.86×10^5 mi/sec. A *light-year* is the distance that light travels in 1 year. Determine the number of miles in a light year.

b) The Earth is approximately 93,000,000 mi from the sun. How long does it take light from the sun to reach the Earth?

119. *Bacteria in a Culture* The exponential function $E(t) = 2^{10} \cdot 2^t$ approximates the number of bacteria in a certain culture after t hours.

a) The initial number of bacteria is determined when $t = 0$. What is the initial number of bacteria?

b) How many bacteria are there after $\frac{1}{2}$ hour?

RESEARCH ACTIVITIES

120. Obtain data from the U.S. Department of the Treasury and from the U.S. Bureau of the Census to calculate the current U.S. government debt per person. Write a report in which you compare your figure with those obtained in Exercise 108 and Example 14. Include in your report definitions of the following terms: revenues, expenditures, deficit, and surplus.

121. John Allen Paulos of Temple University has written many entertaining books about mathematics for nonmathematicians. Included among these are *Mathematics and Humor* (1980); *I Think, Therefore I Laugh* (1985); *Innumeracy—Mathematical Illiteracy and Its Consequences* (1989); *Beyond Numeracy—Ruminations of a Numbers Man* (1991); *A Mathematician Reads the Newspaper* (1995); and *Once Upon a Number* (1998). Read one of Paulos's books and write a 500-word report on it.

Dr. John Allen Paulos

122. Find an article in a newspaper or magazine that contains scientific notation. Write a paragraph explaining how scientific notation was used. Attach a copy of the article to your report.

5.7 ARITHMETIC AND GEOMETRIC SEQUENCES

Now that you can recognize the various sets of real numbers and know how to add, subtract, multiply, and divide real numbers, we can discuss sequences. A **sequence** is a list of numbers that are related to each other by a rule. The numbers that form the sequence are called its **terms**. If your salary increases or decreases by a fixed amount over a period of time, the listing of the amounts, over time, would form an arithmetic sequence. When interest in a savings account is compounded at regular intervals, the listing of the amounts in the account over time will be a geometric sequence.

ARITHMETIC SEQUENCES

A sequence in which each term after the first term differs from the preceding term by a constant amount is called an **arithmetic sequence**. The amount by which each pair of successive terms differs is called the **common difference**, d. The common difference can be found by subtracting any term from the term that directly follows it.

Examples of arithmetic sequences	**Common differences**
$1, 5, 9, 13, 17, \ldots$	$d = 5 - 1 = 4$
$-7, -5, -3, -1, 1, \ldots$	$d = -5 - (-7) = -5 + 7 = 2$
$\frac{5}{2}, \frac{3}{2}, \frac{1}{2}, -\frac{1}{2}, \ldots$	$d = \frac{3}{2} - \frac{5}{2} = -\frac{2}{2} = -1$

PROFILE IN MATHEMATICS

Carl Friedrich Gauss

Carl Friedrich Gauss (1777–1855), often called the "Prince of Mathematicians," made significant contributions to the fields of algebra, geometry, and number theory. Gauss was only 22 years old when he proved the fundamental theorem of algebra for his doctoral dissertation.

When Gauss was only 10, his mathematics teacher gave him the problem of finding the sum of the first 100 natural numbers, thinking that this would keep him busy for a while. Gauss recognized a pattern in the sequence of numbers when he considered the sum of the following numbers.

$$\begin{array}{r} 1 + \ \ 2 + \ \ 3 + \cdots + \ 99 + 100 \\ 100 + \ 99 + \ 98 + \cdots + \ \ 2 + \ \ \ 1 \\ \hline 101 + 101 + 101 + \cdots + 101 + 101 \end{array}$$

He had the required answer in no time at all. When he added, he had one hundred 101's. Therefore, the sum is $\frac{1}{2}(100)(101) = 5050$.

EXAMPLE 1 ***Writing the First Five Terms of an Arithmetic Sequence***

Write the first five terms of the arithmetic sequence with first term 3 and a common difference of 5.

SOLUTION The first term is 3. The second term is $3 + 5$ or 8. The third term is $8 + 5$ or 13. The fourth term is $13 + 5$ or 18. The fifth term is $18 + 5$ or 23. Thus, the first five terms of the sequence are 3, 8, 13, 18, 23.

EXAMPLE 2 ***An Arithmetic Sequence with a Negative Difference***

Write the first four terms of the arithmetic sequence with first term 3 and a common difference of -2.

SOLUTION $3, 1, -1, -3$

When discussing a sequence, we often represent the first term as a_1 (read "a sub 1"), the second term as a_2, the fifteenth term as a_{15}, and so on. We use the notation a_n to represent the general or nth term of a sequence. Thus a sequence may be symbolized as

$$a_1, a_2, a_3, a_4, \ldots, a_n, \ldots$$

For example, in the sequence 2, 5, 8, 11, 14, . . . , we have

$$a_1 = 2, a_2 = 5, a_3 = 8, a_4 = 11, a_5 = 14, \ldots.$$

When we know the first term of an arithmetic sequence and the common difference, we can use the following formula to find the value of any specific term.

General or *n*th Term of an Arithmetic Sequence

$$a_n = a_1 + (n - 1)d$$

EXAMPLE 3 ***Finding the Seventh Term of an Arithmetic Sequence***

Find the seventh term of the arithmetic sequence whose first term is 3 and whose common difference is -6.

SOLUTION To find the seventh term, or a_7, replace n in the formula with 7, a_1 with 3, and d with -6.

$$\begin{aligned} a_n &= a_1 + (n - 1)d \\ a_7 &= 3 + (7 - 1)(-6) \\ &= 3 + (6)(-6) \\ &= 3 - 36 \\ &= -33 \end{aligned}$$

The seventh term is -33. As a check, we have listed the first seven terms of the sequence: $3, -3, -9, -15, -21, -27, -33$.

EXAMPLE 4 ***Finding the nth Term of an Arithmetic Sequence***

Write an expression for the general or nth term, a_n, for the sequence 1, 6, 11, 16, ...

SOLUTION In this sequence, the first term a_1, is 1, and the common difference, d, is 5. We substitute these values into $a_n = a_1 + (n - 1)d$ to obtain an expression for the nth term, a_n.

$$\begin{aligned} a_n &= a_1 + (n-1)d \\ &= 1 + (n-1)5 \\ &= 1 + 5n - 5 \\ &= 5n - 4 \end{aligned}$$

Note that when $n = 1$, the first term is $5(1) - 4 = 1$. When $n = 2$, the second term is $5(2) - 4 = 6$, and so on. ▲

We can use the following formula to find the sum of the first n terms in an arithmetic sequence.

Sum of the First *n* Terms in an Arithmetic Sequence

$$s_n = \frac{n(a_1 + a_n)}{2}$$

In this formula, s_n represents the sum of the first n terms, a_1 is the first term, a_n is the nth term, and n is the number of terms in the sequence from a_1 to a_n.

EXAMPLE 5 ***Finding the Sum of a Sequence***

Find the sum of the first 25 natural numbers.

SOLUTION The sequence we are discussing is

$$1, 2, 3, 4, 5, \ldots, 25$$

In this sequence, $a_1 = 1$, $a_{25} = 25$, and $n = 25$. Thus, the sum of the first 25 terms is

$$\begin{aligned} s_n &= \frac{n(a_1 + a_n)}{2} \\ s_{25} &= \frac{25(1 + 25)}{2} \\ &= \frac{25(26)}{2} = 325 \end{aligned}$$

Thus, the sum of the terms $1 + 2 + 3 + 4 + \cdots + 25$ is 325. ▲

GEOMETRIC SEQUENCES

The next type of sequence we will discuss is the geometric sequence. A **geometric sequence** is one in which the ratio of any term to the term that directly precedes it is a constant. This constant is called the **common ratio**. The common ratio, r, can be found by taking any term except the first and dividing that term by the preceding term.

Examples of geometric sequences	Common ratios
$2, 4, 8, 16, 32, \ldots$	$r = 4 \div 2 = 2$
$-3, 6, -12, 24, -48, \ldots$	$r = 6 \div (-3) = -2$
$\frac{2}{3}, \frac{2}{9}, \frac{2}{27}, \frac{2}{81}, \ldots$	$r = \frac{2}{9} \div \frac{2}{3} = \left(\frac{2}{9}\right)\left(\frac{3}{2}\right) = \frac{1}{3}$

To construct a geometric sequence when the first term, a_1, and common ratio are known, multiply the first term by the common ratio to get the second term. Then multiply the second term by the common ratio to get the third term, and so on.

EXAMPLE 6 ***Writing the First Five Terms of a Geometric Sequence***

Write the first five terms of the geometric sequence whose first term, a_1, is 5 and whose common ratio, r, is 2.

SOLUTION The first term is 5. The second term, found by multiplying the preceding term by 2 is $5 \cdot 2$ or 10. The third term is $10 \cdot 2$ or 20. The fourth term is $20 \cdot 2$ or 40. The fifth term is $40 \cdot 2$ or 80. Therefore, the first five terms of the sequence are 5, 10, 20, 40, 80. ▲

When we know the first term of a geometric sequence and the common ratio, we can use the following formula to find the value of the general or nth term, a_n.

General or *n*th Term of a Geometric Sequence

$$a_n = a_1 r^{n-1}$$

EXAMPLE 7 ***Finding the Seventh Term of a Geometric Sequence***

Find the seventh term of the geometric sequence whose first term is -3 and whose common ratio is -2.

SOLUTION In this sequence, $a_1 = -3$, $r = -2$, and $n = 7$. Substituting the values, we obtain

$$\begin{aligned} a_n &= a_1 r^{n-1} \\ a_7 &= -3(-2)^{7-1} \\ &= -3(-2)^6 \\ &= -3(64) \\ &= -192 \end{aligned}$$

As a check, we have listed the first seven terms of the sequence: $-3, 6, -12, 24, -48, 96, -192$. ▲

EXAMPLE 8 ***Finding the nth Term of a Geometric Sequence***

Write an expression for the general or nth term, a_n, of the sequence 2, 6, 18, 54, ...

DID YOU KNOW

A Heavenly Sequence

The sequence 0, 3, 6, 12, 24, 48, 96, 192, . . . , which resembles a geometric sequence, is known collectively as the **Titius–Bode law**. The sequence, discovered in 1766 by the two German astronomers, was of great importance to astronomy in the eighteenth and nineteenth centuries. When 4 was added to each term and the result was divided by 10, the sequence closely corresponded with the observed mean distance from the sun to the known principal planets of the solar system (in astronomical units, or au): Mercury (0.4), Venus (0.7), Earth (1.0), Mars (1.6), missing planet (2.8), Jupiter (5.2), Saturn (10.0), missing planet (19.6). The exciting discovery of Uranus in 1781 with a mean distance of 19.2 astronomical units was in near agreement with the Titius–Bode law and stimulated the search for an undiscovered planet at a predicted 2.8 astronomical units. This effort led to the discovery of Ceres and other members of the asteroid belt. Although the Titius–Bode law broke down after discoveries of Neptune (30.1 au) and Pluto (39.5 au), many scientists still believe that other applications of the Titius–Bode law will emerge in the future.

SOLUTION In this sequence, $a_1 = 2$ and $r = 3$. We substitute these values into $a_n = a_1 r^{n-1}$ to obtain an expression for the nth term, a_n.

$$\begin{aligned} a_n &= a_1 r^{n-1} \\ &= 2(3)^{n-1} \end{aligned}$$

Note than when $n = 1$, $a_1 = 2(3)^0 = 2(1) = 2$. When $n = 2$, $a_2 = 2(3)^1 = 6$, and so on.

We can use the following formula to find the sum of the first n terms of a geometric sequence.

Sum of the First *n* Terms of a Geometric Sequence

$$s_n = \frac{a_1(1 - r^n)}{1 - r}, \qquad r \neq 1$$

EXAMPLE 9 ***Adding the First n Terms of a Geometric Sequence***

Find the sum of the first five terms in the geometric sequence whose first term is 4 and whose common ratio is 2.

SOLUTION In this sequence, $a_1 = 4$, $r = 2$, and $n = 5$. Substituting these values into the formula, we get

$$\begin{aligned} s_n &= \frac{a_1(1 - r^n)}{1 - r} \\ s_5 &= \frac{4[1 - (2)^5]}{1 - 2} \\ &= \frac{4(1 - 32)}{-1} \\ &= \frac{4(-31)}{-1} = \frac{-124}{-1} = 124 \end{aligned}$$

The sum of the first five terms of the sequence is 124. The first five terms of the sequence are 4, 8, 16, 32, 64. If you add these five numbers, you will obtain the sum 124.

EXAMPLE 10 ***Pounds and Pounds of Silver***

As a reward for saving his kingdom from a band of thieves, a king offered a knight one of two options. The knight's first option was to be paid 100,000 pounds of silver all at once. The second option was to be paid over the course of a month. On the first day, he would receive one pound of silver. On the second day, he would receive two pounds of silver. On the third day, he would receive four pounds of silver, and so on, each day receiving double the amount given on the previous day. Assuming the month is 30 days, which option would pay the knight more silver?

SOLUTION The first option pays the knight 100,000 pounds of silver. The second option pays according to the geometric sequence 1, 2, 4, 8, 16, ... In this sequence, $a_1 = 1$, $r = 2$, and $n = 30$. The sum of this sequence can be found by substituting these values into the formula to obtain

$$s_n = \frac{a_1(1 - r^n)}{1 - r}$$

$$s_{30} = \frac{1(1 - 2^{30})}{1 - 2}$$

$$= \frac{1 - 1{,}073{,}741{,}824}{-1}$$

$$= \frac{-1{,}073{,}741{,}823}{-1}$$

$$= 1{,}073{,}741{,}823$$

Thus, the knight would get paid 1,073,741,823 pounds of silver with the second option. The second option pays 1,073,641,823 more pounds of silver than the first option.

SECTION 5.7 EXERCISES

CONCEPT/WRITING EXERCISES

1. State the definition of *sequence* and give an example.
2. What are the numbers that make up a sequence called?
3. State the definition of *arithmetic sequence* and give an example.
4. Explain what the common difference in an arithmetic sequence is.
5. State the definition of *geometric sequence* and give an example.
6. Explain what the common ratio in a geometric sequence is.

PRACTICE THE SKILLS

In Exercises 7–14, write the first five terms of the arithmetic sequence with the first term, a_1, and common difference, d.

7. $a_1 = 2, d = 4$
8. $a_1 = 3, d = 5$
9. $a_1 = -3, d = 3$
10. $a_1 = -4, d = 2$
11. $a_1 = 5, d = -2$
12. $a_1 = -3, d = -4$
13. $a_1 = \frac{1}{2}, d = \frac{1}{2}$
14. $a_1 = \frac{5}{2}, d = -\frac{3}{2}$

In Exercises 15–22, find the indicated term for the arithmetic sequence with the first term, a_1, and common difference, d.

15. Find a_5 when $a_1 = 4, d = 3$
16. Find a_8 when $a_1 = -4, d = -5$
17. Find a_9 when $a_1 = -5, d = 2$
18. Find a_{12} when $a_1 = 7, d = -3$
19. Find a_{20} when $a_1 = \frac{4}{5}, d = -1$
20. Find a_{15} when $a_1 = -\frac{1}{2}, d = -2$
21. Find a_{11} when $a_1 = 4, d = \frac{1}{2}$
22. Find a_{15} when $a_1 = \frac{4}{3}, d = \frac{1}{3}$

In Exercises 23–30, write an expression for the general or nth term, a_n, for the arithmetic sequence.

23. 2, 4, 6, 8, . . .
24. 7, 3, −1, − 5, . . .
25. 6, 16, 26, 36, . . .
26. −2, −5, −8, −11, . . .
27. $-\frac{5}{3}, -\frac{4}{3}, -1, -\frac{2}{3}, \ldots$
28. −15, −10, −5, 0, . . .
29. $-3, -\frac{3}{2}, 0, \frac{3}{2}, \ldots$
30. −5, −2, 1, 4, . . .

In Exercises 31–38, find the sum of the terms of the arithmetic sequence. The number of terms, n, is given.

31. 1, 2, 3, 4, . . . , 14; $n = 14$
32. 3, 6, 9, 12, . . . , 30; $n = 10$
33. 45, 40, 35, 30, . . . , 5; $n = 9$
34. −4, −7, −10, −13, . . . , −28; $n = 9$
35. 11, 6, 1, −4, . . . , −24; $n = 8$
36. $-9, -\frac{17}{2}, -8, -\frac{15}{2}, \ldots, -\frac{1}{2}$; $n = 18$
37. $\frac{1}{2}, \frac{5}{2}, \frac{9}{2}, \frac{13}{2}, \ldots, \frac{29}{2}$; $n = 8$
38. $\frac{3}{5}, \frac{4}{5}, 1, \frac{6}{5}, \ldots, 4$; $n = 18$

In Exercises 39–46, write the first five terms of the geometric sequence with the first term, a_1, and common ratio, r.

39. $a_1 = 2, r = 4$

40. $a_1 = 4, r = 3$

41. $a_1 = 4, r = -3$

42. $a_1 = 8, r = \frac{1}{2}$

43. $a_1 = -3, r = -1$

44. $a_1 = -6, r = -2$

45. $a_1 = -16, r = -\frac{1}{2}$

46. $a_1 = 5, r = \frac{3}{5}$

In Exercises 47–54, find the indicated term for the geometric sequence with the first term, a_1, and common ratio, r.

47. Find a_6 when $a_1 = 3, r = 4$.

48. Find a_5 when $a_1 = 2, r = 2$.

49. Find a_8 when $a_1 = 5, r = 3$.

50. Find a_9 when $a_1 = -3, r = -2$.

51. Find a_7 when $a_1 = 10, r = -3$.

52. Find a_3 when $a_1 = 3, r = \frac{1}{2}$.

53. Find a_7 when $a_1 = -3, r = -3$.

54. Find a_5 when $a_1 = \frac{1}{2}, r = 2$.

In Exercises 55–62, write an expression for the general or nth term, a_n, for the geometric sequence.

55. 3, 9, 27, 81, . . .

56. 2, 6, 18, 54, . . .

57. $-5, 5, -5, 5, \ldots$

58. $-16, -8, -4, -2, \ldots$

59. $\frac{1}{4}, \frac{1}{2}, 1, 2, \ldots$

60. $-3, 6, -12, 24, \ldots$

61. $9, 3, 1, \frac{1}{3}, \frac{1}{9}, \ldots$

62. $-4, -\frac{8}{3}, -\frac{16}{9}, -\frac{32}{27}, \ldots$

In Exercises 63–70, find the sum of the first n terms of the geometric sequence for the values of a_1 and r.

63. $n = 4, a_1 = 3, r = 2$

64. $n = 5, a_1 = 2, r = 3$

65. $n = 7, a_1 = 5, r = 4$

66. $n = 9, a_1 = -3, r = 5$

67. $n = 11, a_1 = -7, r = 3$

68. $n = 11, a_1 = -5, r = -2$

69. $n = 13, a_1 = -8, r = -3$

70. $n = 14, a_1 = -1, r = 2$

PROBLEM SOLVING

71. Find the sum of the first 50 natural numbers.

72. Find the sum of the first 50 even natural numbers.

73. Find the sum of the first 50 odd natural numbers.

74. Find the sum of the first 20 multiples of 3.

75. *Annual Pay Raises* Donna is given a starting salary of \$20,200 and promised a \$1200 raise per year after each of the next eight years.

a) Determine her salary during her eighth year of work.

b) Determine the total salary she received over the 8 years.

76. *Pendulum Movement* Each swing of a pendulum (from far left to far right) is 3 in. shorter than the preceding swing. The first swing is 8 ft.

a) Find the length of the twelfth swing.

b) Determine the total distance traveled by the pendulum during the first 12 swings.

77. *Decomposing Substance* A certain substance decomposes and loses 20% of its weight each hour. If there are originally 200 g of the substance, how much remains after 6 hr?

78. *A Bouncing Ball* Each time a ball bounces, the height attained by the ball is 6 in. less than the previous height attained. If on the first bounce the ball reaches a height of 6 ft, find the height attained on the eleventh bounce.

79. *Clock Strikes* A clock strikes once at 1 o'clock, twice at 2 o'clock, and so on. How many times does it strike over a 12 hr period?

80. *Salary Increase* If your salary were to increase at a rate of 6% per year, find your salary during your fifteenth year if your original salary is \$20,000.

81. *Samurai Sword Construction* While making a traditional Japanese samurai sword, the master sword maker prepares the blade by heating a bar of iron until it is white hot. He then folds it over and pounds it smooth. Therefore, after each folding, the number of layers of steel is doubled. Assuming the sword maker starts with a bar of one layer and folds it 15 times, how many layers of steel will the finished sword contain?

82. *Another Bouncing Ball* When dropped, a ball rebounds to four-fifths of its original height. How high will the ball rebound after the fourth bounce if it is dropped from a height of 30 ft?

83. *A Baseball Game* During a baseball game, the visiting team scored 1 run in the first inning, 2 runs in the second inning, 3 runs in the third inning, 4 runs in the fourth inning, and so on. The home team scored 1 run in the first inning, 2 runs in the second inning, 4 runs in the third inning, 8 runs in the fourth inning, and so on. What is the score of the game after eight innings?

Inning	1	2	3	4	5	6	7	8	9
Visitors	1	2	3	4					
Home	1	2	4	8					

84. *Value of a Stock* Ten years ago, Nancy Hart purchased \$2,000 worth of shares in RCF, Inc. Since then, the price of the stock has roughly tripled every two years. Approximately how much are Nancy's shares worth today?

CHALLENGE PROBLEMS/GROUP ACTIVITIES

85. *A Wagering Strategy* The following is a strategy used by some people involved in games of chance. A player begins by betting a standard bet, say \$1. If the player wins, the player again bets \$1 in the next round. If the player loses, the player bets \$2 in the next round. Next, if the player wins, the player again bets \$1; if the player loses, the player now bets \$4 in the next round. The process continues as long as the player keeps playing, betting \$1 after a win or doubling the previous bet after a loss.

a) Assume a player is using a \$1 standard bet and loses five times in a row. How much money should the player bet in the sixth round? How much money has the player lost at the end of the fifth round?

b) Assume a player is using a \$10 standard bet and loses five times in a row. How much money should the player bet in the sixth round? How much money has the player lost at the end of the fifth round?

c) Assume a player is using a \$1 standard bet and loses 10 times in a row. How much money should the player bet in the 11th round? How much money has the player lost at the end of the 10th round?

d) Assume a player is using a \$10 standard bet and loses 10 times in a row. How much money should the player bet in the 11th round? How much money has the player lost at the end of the 10th round?

e) Why is this a dangerous strategy?

86. A geometric sequence has $a_1 = 82$ and $r = \frac{1}{2}$; find s_6.

87. *Sums of Internal Angles* The sums of the interior angles of a triangle, a quadrilateral, a pentagon, and a sextagon are 180°, 360°, 540°, and 720°, respectively. Use this pattern to find a formula for the general term, a_n, where a_n represents the sum of the interior angles of an n-sided quadrilateral.

88. *Divisibility by 6* Determine how many numbers between 7 and 1610 are divisible by 6.

89. Find r and a_1 for the geometric sequence with $a_2 = 24$ and $a_5 = 648$.

90. *Total Distance Traveled by a Bouncing Ball* A ball is dropped from a height of 30 ft. On each bounce it attains a height four-fifths of its original height (or of the previous bounce). Find the total vertical distance traveled by the ball after it has completed its fifth bounce (therefore has hit the ground six times).

RESEARCH ACTIVITY

91. A topic generally associated with sequences is *series.*

a) Research *series* in several mathematics texts (intermediate algebra or college algebra texts). Explain what a series is and how it differs from a sequence. Write a formal definition. Give examples of different kinds of series.

b) Write the arithmetic series associated with the arithmetic sequence 1, 4, 7, 10, 13, ...

c) Write the geometric series associated with the geometric sequence 3, 6, 12, 24, 48, ...

d) What is an infinite geometric series?

e) Find the sum of the terms of the infinite geometric series $1 + \frac{1}{2} + \frac{1}{4} + \frac{1}{8} + \frac{1}{16} + \cdots$.

5.8 FIBONACCI SEQUENCE

Our discussion of sequences would not be complete without mentioning a sequence known as the **Fibonacci sequence**. The sequence is named after Leonardo of Pisa, also known as Fibonacci. He was one of the most distinguished mathematicians of the Middle Ages. This sequence is first mentioned in his book *Liber Abacci* (Book of the Abacus), which contained many interesting problems, such as: "A certain man put a pair of rabbits in a place surrounded on all sides by a wall. How many pairs of rabbits can be produced from that pair in a year if it is assumed that every month each pair begets a new pair which from the second month becomes productive?"

The solution to this problem (Fig. 5.10) led to the development of the sequence that bears its author's name: the Fibonacci sequence. The sequence is shown in Table 5.1. The numbers in the columns titled *Pairs of Adults* form the Fibonacci sequence.

Fibonacci

Leonardo of Pisa (1170–1250) is considered one of the most distinguished mathematicians of the Middle Ages. He was born in Italy, but was sent by his father to study mathematics with an Arab master. When he began writing, he referred to himself as Fibonacci, or "son of Bonacci," the name by which he is known today. In addition to the famous sequence bearing his name, Fibonacci is also credited with introducing the Hindu–Arabic number system into Europe. His 1202 book, *Liber Abacci* (Book of the Abacus), explained the use of this number system and emphasized the importance of the number zero.

Table 5.1

Month	Pairs of adults	Pairs of babies	Total pairs
1	1	0	1
2	1	1	2
3	2	1	3
4	3	2	5
5	5	3	8
6	8	5	13
7	13	8	21
8	21	13	34
9	34	21	55
10	55	34	89
11	89	55	144
12	144	89	233

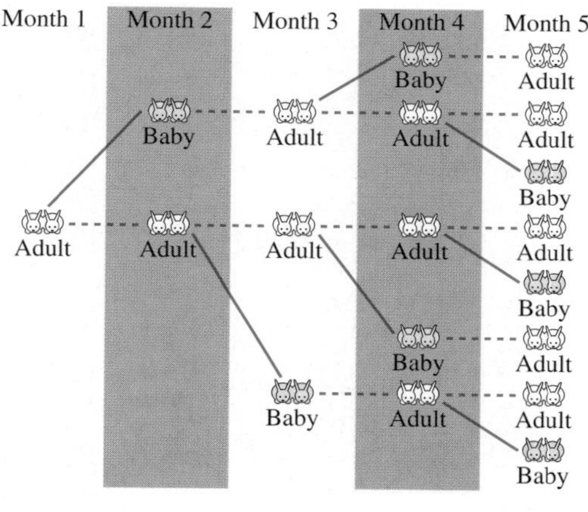

Figure 5.10

Fibonacci Sequence

1, 1, 2, 3, 5, 8, 13, 21, . . .

In the Fibonacci sequence, the first and second terms are 1. The sum of these two terms is the third term. The sum of the second and third terms is the fourth term, and so on.

In the middle of the nineteenth century, mathematicians made a serious study of this sequence and found strong similarities between it and many natural phenomena. Fibonacci numbers appear in the seed arrangement of many species of plants and in the petal counts of various flowers. For example, when the flowering head of the sunflower matures to seed, the seeds' spiral arrangement becomes clearly visible. A typical count of these spirals may give 89 steeply curving to the right, 55 curving more shallowly to the left, and 34 again shallowly to the right. The largest known specimen to be examined had spiral counts of 144 right, 89 left, and 55 right. These numbers, like the other three mentioned, are consecutive terms of the Fibonacci sequence.

On the heads of many flowers, petals (or florets in composite plants) surrounding the central disk generally yield a Fibonacci number. For example, some daisies contain 21 petals, and others contain 34, 55, or 89 petals. (People who use a daisy to play the "love me, love me not" game will likely pluck 21, 34, 55, or 89 petals before arriving at an answer.)

Fibonacci numbers are also observed in the structure of pinecones and pineapples. The tablike or scalelike structures called bracts that make up the main body of the pinecone form a set of spirals that start from the cone's attachment to the branch. Two sets of oppositely directed spirals can be observed, one steep and the other more gradual. A count on the steep spiral will reveal a Fibonacci number, and a count on the gradual one will be the adjacent smaller Fibonacci number, or if not, the next smaller Fibonacci number. One investigation of 4290 pinecones from 10 species of pine trees found in California revealed that only 74 cones, or 1.7%, deviated from this Fibonacci pattern.

Like pinecone bracts, pineapple scales are patterned into spirals, and because they are roughly hexagonal in shape, three distinct sets of spirals can be counted.

Table 5.2

Numbers	Ratio
1, 1	$\frac{1}{1} = 1$
1, 2	$\frac{2}{1} = 2$
2, 3	$\frac{3}{2} = 1.5$
3, 5	$\frac{5}{3} = 1.666\ldots$
5, 8	$\frac{8}{5} = 1.6$
8, 13	$\frac{13}{8} = 1.625$
13, 21	$\frac{21}{13} \approx 1.615$
21, 34	$\frac{34}{21} \approx 1.619$
34, 55	$\frac{55}{34} \approx 1.618$
55, 89	$\frac{89}{55} \approx 1.618$

FIBONACCI NUMBERS AND DIVINE PROPORTIONS

In 1753, while studying the Fibonacci sequence, Robert Simson, a mathematician at the University of Glasgow, noticed that when he took the ratio of any term to the term that immediately preceded it, the value he obtained remained in the vicinity of one specific number. To illustrate this, we indicate in Table 5.2 the ratio of various pairs of sequential Fibonacci numbers.

The ratio of the 50th term to the 49th term is 1.6180. Simson proved that the ratio of the $(n + 1)$ term to the nth term as n gets larger and larger is the irrational number $(\sqrt{5} + 1)/2$, which begins 1.61803. . . . This number was already well known to mathematicians at that time as the **golden number**.

Many years earlier, the Bavarian astronomer and mathematician Johannes Kepler wrote that for him the golden number symbolized the Creator's intention "to create like from like." The golden number $(\sqrt{5} + 1)/2$ is frequently referred to as "phi," symbolized by the Greek letter Φ.

The ancient Greeks, in about the sixth century B.C., sought unifying principles of beauty and perfection, which they believed could be described by using mathematics. In their study of beauty, the Greeks used the term *golden ratio.* To understand the golden ratio, let's consider the line segment AB in Fig. 5.11. When this line segment is divided at a point C, such that the ratio of the whole, AB, to the larger part, AC, is equal to the ratio of the larger part, AC, to the smaller part, CB, then each ratio AB/AC and AC/CB is referred to as a **golden ratio**. The proportion they form, $AB/AC = AC/CB$, is called the **golden proportion**. Furthermore, each ratio in the proportion will have a value equal to the golden number, $(\sqrt{5} + 1)/2$.

Figure 5.11

$$\frac{AB}{AC} = \frac{AC}{CB} = \frac{\sqrt{5} + 1}{2} \approx 1.618$$

The Great Pyramid of Gizeh in Egypt, built about 2600 B.C., is the earliest known example of use of the golden ratio in architecture. The ratio of any of its sides of the square base (775.75 ft) to its altitude (481.4 ft) is about 1.611. Other evidence of the use of the golden ratio appears in other Egyptian buildings and tombs.

The Great Pyramid of Gizeh

In medieval times, people referred to the golden proportion as the **divine proportion**, reflecting their belief in its relationship to the will of God.

The twentieth-century architect Le Corbusier developed a scale of proportions for the human body that he called the Modulor (Fig. 5.12). Note that the navel separates the entire body into golden proportions, as does the neck and knee.

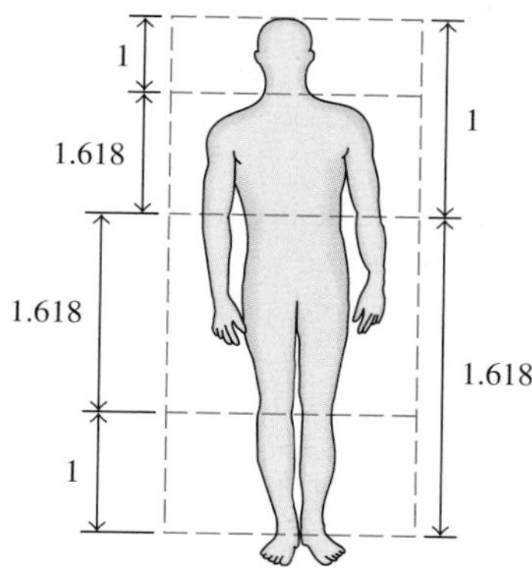

Figure 5.12

From the golden proportion, the **golden rectangle** can be formed, as shown in Fig. 5.13.

$$\frac{\text{Length}}{\text{Width}} = \frac{a+b}{a} = \frac{a}{b} = \frac{\sqrt{5}+1}{2}$$

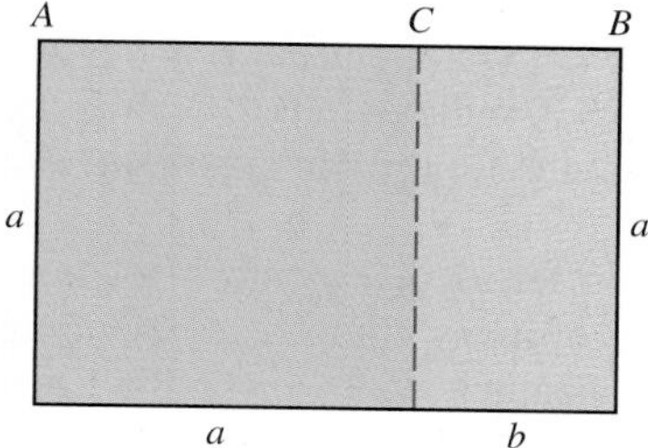

Figure 5.13

Note that when a square is cut off one end of a golden rectangle, as in Fig. 5.13, the remaining rectangle has the same properties as the original golden rectangle (creating "like from like" as Johannes Kepler had written) and is therefore itself a golden rectangle. Interestingly, the curve derived from a succession of diminishing golden rectangles, as shown in Fig. 5.14, is the same as the spiral curve of the chambered nautilus. The same curve appears on the horns or rams and some other animals. It is the same curve observed in the plant structures arranged in Fibonacci sequence mentioned earlier—in sunflowers, other flower heads, pinecones, and pineapples. The curve closely approximates what mathematicians call a *logarithmic spiral.*

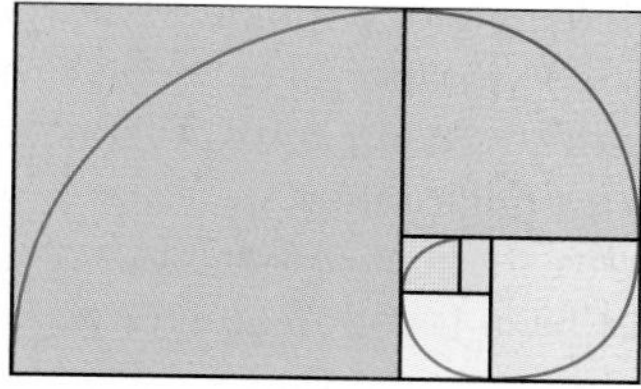

Figure 5.14

The Parthenon

Ancient Greek civilization used the golden rectangle in art and architecture. The main measurements of many buildings of antiquity, including the Parthenon in Athens, are governed by golden ratios and rectangles. Greek statues, vases, urns, and so on also exhibit characteristics of the golden ratio. It is for Phidas, considered the greatest of Greek sculptors, that the golden ratio was named "phi." The proportions can be found abundantly in his work.

The proportions of the golden rectangle can be found in the works of many artists, from the old masters to the moderns. For example, the golden rectangle can be seen in the painting *La Parade* by George Seurat, a French neoimpressionist artist.

La Parade
by George Seurat

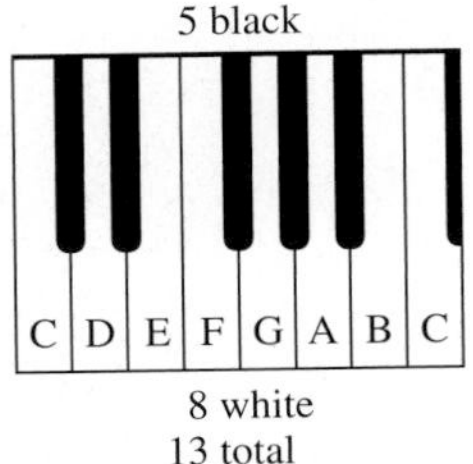

Figure 5.15

Fibonacci numbers are also found in another form of art, namely music. Perhaps the most obvious link between Fibonacci numbers and music can be found on the piano keyboard. An octave (Fig. 5.15) on a keyboard has 13 keys: 8 white, and 5 black (the 5 are in one group of 2 and a group of 3).

In Western music, the most complete scale, the chromatic scale, consists of 13 notes (from C to the next higher C). Its predecessor, the diatonic scale, contains 8 notes (an octave). The diatonic scale was preceded by a 5-note pentatonic scale (*penta* is Greek for "five"). Each number is a Fibonacci number.

The visual arts deal with what is pleasing to the eye, whereas musical composition deals with what is pleasing to the ear. While art achieves some of its goals by using division of planes and area, music achieves some of its goals by a similar division of time, using notes of various duration and spacing. The musical intervals considered by many to be the most pleasing to the ear are the major sixth and minor sixth. A major sixth, for example, consists of the note C, vibrating at about 264* vibrations per second, and note A, vibrating at about 440 vibrations per second. The ratio of 440 to 264 reduces to 5 to 3, or $\frac{5}{3}$, a ratio of two consecutive Fibonacci numbers. An example of a minor sixth would be E (about 330 vibrations per second) and C (about 528 vibrations per second). The ratio 528 to 330 reduces to 8 to 5, or $\frac{8}{5}$, the next ratio of two consecutive Fibonacci numbers. The vibrations of any sixth interval reduce to a similar ratio.

Patterns that can be expressed mathematically in terms of Fibonacci relationships have been found in Gregorian chants and works of many composers, including Bach, Beethoven, and Bartók. A number of twentieth-century works, including Ernst Krened's *Fibonacci Mobile,* have been deliberately structured by using Fibonacci proportions.

A number of studies have tried to explain why the Fibonacci series and related items are linked to so many real-life situations. It appears that the Fibonacci numbers are a part of natural harmony that is pleasing to both the eye and the ear. In the nineteenth century, German physicist and psychologist Gustav Fechner tried to determine which dimensions were most pleasing to the eye. Fechner, along with psychologist Wilhelm Wundt, found that most people do unconsciously favor golden dimensions when purchasing greeting cards, mirrors, and other rectangular objects. This discovery has been widely used by commercial manufacturers in their packaging and labeling designs, by retailers in their store displays, and in other areas of business and advertising.

*Frequencies of notes vary in different parts of the world and change over time.

DID YOU KNOW

Fibonacci and the Male Bee's Ancestors

The most frequent example given to introduce the Fibonacci sequence involves rabbits producing offspring, two at a time. Although this makes for a nice introduction to Fibonacci sequences, it is not at all realistic. A much better example comes from the breeding practices of bees. Female or worker bees are produced when the queen bee mates with a male bee. Male bees are produced from the queen's unfertilized eggs. In essence, then, female bees have two parents, whereas male bees only have one parent. The family tree of a male bee would look like this:

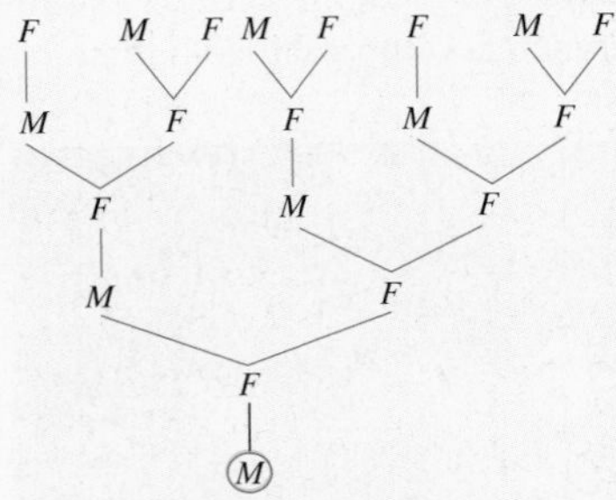

From this tree, we can see that the **1** male bee (circled) has **1** parent, **2** grandparents, **3** great-grandparents, **5** great-great-grandparents, **8** great-great-great-grandparents, and so on. We see the Fibonacci sequence as we move back through the male bees' generations.

SECTION 5.8 EXERCISES

CONCEPT/WRITING EXERCISES

1. Explain how to construct the Fibonacci sequence.
2. a) Find the eighth, ninth, and tenth terms of the Fibonacci sequence.
   b) Divide the ninth term by the eighth term, rounding to the nearest thousandth.
   c) Divide the tenth term by the ninth term, rounding to the nearest thousandth.
   d) Try a few more divisions and then make a conjecture about the result.
3. The ratio of the a_{n+1}/a_n terms of the Fibonacci sequence approaches 1.6180 (rounded to four decimal places) as n increases. Find the ratio of a_n/a_{n+1} rounded to four decimal places as n increases.
4. What is the golden number? Explain the relationship between the golden number and the golden proportion.
5. Select a piece of art and see if you can determine whether the artist used the golden rectangle. Write a brief description of your findings.
6. Explain what is meant by the divine proportion.

PRACTICE THE SKILLS/PROBLEM SOLVING

7. a) To what decimal value is $(\sqrt{5} + 1)/2$ aproximately equal?
   b) To what decimal value is $(\sqrt{5} - 1)/2$ approximately equal?
   c) By how much do the results in parts (a) and (b) differ?
8. The eleventh Fibonacci number is 89. Examine the first six terms in the decimal expression of its reciprocal, $\frac{1}{89}$. What do you find?
9. Find the ratio of the second to the first term of the Fibonacci sequence. Then find the ratio of the third to the second term of the sequence and determine whether this ratio was an increase or decrease from the first ratio. Continue this process for 10 ratios and then make a conjecture regarding the increasing or decreasing values in consecutive ratios.
10. A musical composition is described as follows. Explain why this piece is based on the golden ratio.

Entire Composition

34 measures	55 measures	21 measures	34 measures
Theme	Fast, Loud	Slow	Repeat of theme

11. The greatest common factor of any two consecutive Fibonacci numbers is 1. Show this is true for the first 15 Fibonacci numbers.
12. The sum of any 10 consecutive Fibonacci numbers is always divisible by 11. Select any 10 consecutive Fibonacci numbers and show that for your selection this is true.
13. Twice any Fibonacci number minus the next Fibonacci number equals the second number preceding the original number. Select a number in the Fibonacci sequence and show that this pattern holds for the number selected.
14. For any four consecutive Fibonacci numbers, the difference of the squares of the middle two numbers equals the product of the smallest and largest numbers. Select four consecutive Fibonacci numbers and show that this pattern holds for the numbers you selected.
15. Determine the ratio of the length to width of various photographs and compare these ratios to Φ.
16. Determine the ratio of the length to the width of a 6 inch by 4 inch standard index card, and compare the ratio to Φ.
17. Determine the ratio of the length to width of several picture frames and compare these ratios to Φ.
18. Determine the ratio of the length to the width of your television screen and compare this ratio to Φ.
19. Determine the ratio of the length to width of a desktop in your classroom and compare this ratio to Φ.
20. Determine the ratio of the length to the width of this textbook and compare this ratio to Φ.
21. Determine the ratio of the length to the width of a computer screen and compare this ratio to Φ.
22. Find three physical objects whose dimensions are very close to a golden rectangle.
    a) List the articles and record the dimensions.
    b) Compute the ratios of their lengths to their widths.
    c) Find the difference between the golden ratio and the ratio you obtain in part (b)—to the nearest tenth—for each object.

In Exercises 23–30, determine whether the sequence is a Fibonacci-type sequence (each term is the sum of the two preceding terms). If it is, determine the next two terms of the sequence.

23. 2, 5, 7, 12, 19, 31, . . .
24. 1, 2, 4, 8, 16, 32, . . .
25. 4, 5, 9, 18, 34, 59, . . .
26. 5, 10, 15, 20, 25, . . .
27. 5, 10, 15, 25, 40, 65, . . .
28. $\frac{1}{4}, \frac{1}{4}, \frac{1}{2}, \frac{3}{4}, 1\frac{1}{4}, 2, \ldots$
29. $-5, 3, -2, 1, -1, 0, \ldots$
30. $-4, 5, 1, 6, 7, 13, \ldots$
31. a) Select any two nonzero digits and add them to obtain a third digit. Continue adding the two previous terms to get a Fibonacci-type sequence.
    b) Form ratios of successive terms to show how they will eventually approach the golden number.

32. Repeat Exercise 31 for two different nonzero numbers.

33. a) Select any three consecutive terms of a Fibonacci sequence. Subtract the product of the terms on each side of the middle term from the square of the middle term. What is the difference?

b) Repeat part (a) with three different consecutive terms of the sequence.

c) Make a conjecture about what will happen when you repeat this process for any three consecutive terms of a Fibonacci sequence.

34. One of the most famous number patterns involves **Pascal's triangle**. The Fibonacci sequence can be found by using Pascal's triangle. Can you explain how that can be done? A hint is shown.

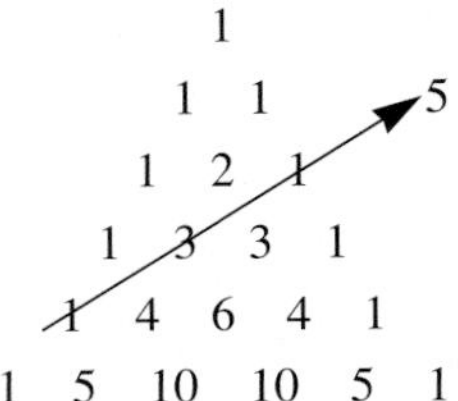

35. a) A sequence related to the Fibonacci sequence is the **Lucas sequence**. The Lucas sequence is formed in a manner similar to the Fibonacci sequence. The first two numbers of the Lucas sequence are 1 and 3. Write the first eight terms of the Lucas sequence.

b) Complete the next two lines of the following chart.

$$1 + 2 = 3$$
$$1 + 3 = 4$$
$$2 + 5 = 7$$
$$3 + 8 = 11$$
$$5 + 13 = 18$$

c) What do you observe about the first column in the chart in part (b)?

CHALLENGE PROBLEMS/GROUP ACTIVITIES

36. The following sequence represents a Fibonacci-type sequence (each term is the sum of the two preceding terms). Here x represents any natural number from 1 to 10:

$$-10, x, -10 + x, -10 + 2x, -20 + 3x, -30 + 5x, \ldots$$

For example, if $x = 2$, the first 10 terms of the sequence would be $-10, 2, -8, -6, -14, -20, -34, -54, -88, -142$.

Write out the first 10 terms of this Fibonacci-type sequence for x equal to:

a) 4 **b)** 5 **c)** 6 **d)** 7 **e)** 8

f) For values of x of 4, 5, and 6, you should have found that each term after the seventh term in the sequence was a negative number. For values of x of 7 and 8, you should have found that each term after the seventh term in the sequence was a positive number. Do you believe that for any value of x greater than or equal to 7, each term after the seventh term of the sequence will always be a positive number? Explain.

37. The divine proportion is $(a + b)/a = a/b$ (see Fig. 5.13). This can be written $1 + (b/a) = a/b$. Now let $x = a/b$, which gives $1 + (1/x) = x$. Multiply both sides of this equation by x to get a quadratic equation and then use the quadratic formula (Section 6.8) to show that one answer is $x = (1 + \sqrt{5})/2$ (the golden ratio).

38. Draw a line of length 5 in. Determine and mark the point on the line that will create the golden ratio. Explain how you determined your answer.

39. A **Pythagorean triple** is a set of three whole numbers, $\{a, b, c\}$, such that $a^2 + b^2 = c^2$. For example, since $6^2 + 8^2 = (10)^2$, $\{6, 8, 10\}$ is a Pythagorean triple. The following steps show how to find Pythagorean triples using any four consecutive Fibonacci numbers. Here we will demonstrate the process with the Fibonacci numbers 3, 5, 8, and 13.

1. Determine the product of 2 and the two inner Fibonacci numbers. We have $2(5)(8) = 80$. This is the first number in the Pythagorean triple. So $a = 80$.
2. Determine the product of the two outer numbers. We have $3(13) = 39$. This is the second number in the Pythagorean triple. So $b = 39$.
3. Determine the sum of the squares of the inner two numbers. We have $5^2 + 8^2 = 25 + 64 = 89$. This is the third number in the Pythagorean triple. So $c = 89$.

This process has produced the Pythagorean triple, $\{80, 39, 89\}$. To verify,

$$(80)^2 + (39)^2 = (89)^2$$
$$6400 + 1521 = 7921$$
$$7921 = 7921$$

Use this process to produce four other Pythagorean triples

40. When two panes of glass are placed face to face, four interior reflective surfaces exist labeled 1, 2, 3, and 4. If light is not reflected, it has just one path through the glass. If it has one reflection, it can be reflected in two ways. If it has two reflections, it can be reflected in three ways. Use this information to answer parts (a) through (c).

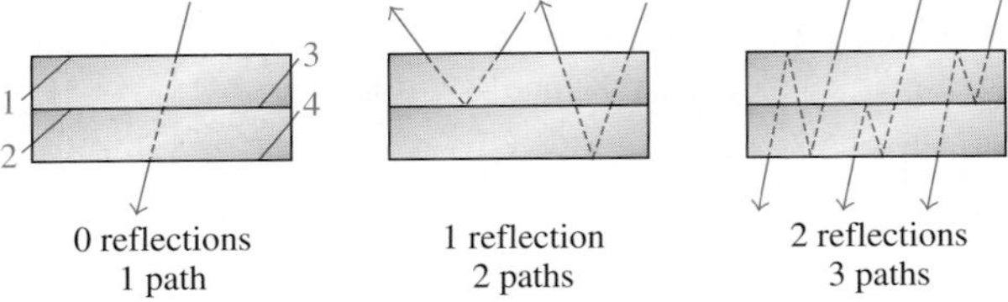

a) If a ray is reflected three times, there are five paths it can follow. Show the paths.

b) If a ray is reflected four times, there are eight paths it can follow. Show the paths.

c) How many paths can a ray follow if it is reflected five times? Explain how you determined your answer.

RESEARCH ACTIVITIES

41. The digits 1 through 9 have evolved considerably since they appeared in Fibonacci's book *Liber Abacci.* Write a report tracing the history of the evolution of the digits 1 through 9 since Fibonacci's time.

42. Write a report on the history and mathematical contributions of Fibonacci.

43. Write a report indicating where the golden ratio and golden rectangle have been used in art and architecture. An art teacher or a staff member of an art museum might be able to give you some information and a list of resources. You may also wish to search the Internet for information on art and architecture related to the golden ratio and Fibonacci sequences.

CHAPTER 5 SUMMARY

IMPORTANT FACTS

Fundamental theorem of arithmetic

Every composite number can be expressed as a unique product of prime numbers.

Sets of numbers

Natural or counting numbers: $\{1, 2, 3, 4, \ldots\}$

Whole numbers: $\{0, 1, 2, 3, 4, \ldots\}$

Integers: $\{\ldots, -3, -2, -1, 0, 1, 2, 3, \ldots\}$

Rational numbers: Numbers of the form p/q, where p and q are integers, $q \neq 0$. Every rational number when expressed as a decimal number will be either a terminating or repeating decimal number.

Irrational number: A real number whose representation is a nonterminating, nonrepeating decimal number (not a rational number).

Definition of subtraction

$$a - b = a + (-b)$$

Fundamental law of rational numbers

$$\frac{a}{b} = \frac{a}{b} \cdot \frac{c}{c} = \frac{ac}{bc}, \qquad b \neq 0, \quad c \neq 0$$

Rules of radicals

Product rule for radicals:

$$\sqrt{a \cdot b} = \sqrt{a} \cdot \sqrt{b}, \qquad a \geq 0, \quad b \geq 0$$

Quotient rule for radicals:

$$\frac{\sqrt{a}}{\sqrt{b}} = \sqrt{\frac{a}{b}}, \qquad a \geq 0, \quad b > 0$$

Properties of real numbers

Commutative property of addition: $a + b = b + a$

Commutative property of multiplication: $a \cdot b = b \cdot a$

Associative property of addition:

$$(a + b) + c = a + (b + c)$$

Associative property of multiplication:

$$(a \cdot b) \cdot c = a \cdot (b \cdot c)$$

Distributive property: $a \cdot (b + c) = ab + ac$

Rules of exponents

Product rule for exponents: $a^m \cdot a^n = a^{m+n}$

Quotient rule for exponents: $\dfrac{a^m}{a^n} = a^{m-n}, \qquad a \neq 0$

Zero exponent rule: $a^0 = 1, \qquad a \neq 0$

Negative exponent rule: $a^{-m} = \dfrac{1}{a^m}, \qquad a \neq 0$

Power rule: $(a^m)^n = a^{m \cdot n}$

Arithmetic sequence

$$a_n = a_1 + (n - 1)d$$

$$s_n = \frac{n(a_1 + a_n)}{2}$$

Geometric sequence

$$a_n = a_1 r^{n-1}$$

$$s_n = \frac{a_1(1 - r^n)}{1 - r}, \qquad r \neq 1$$

Fibonacci sequence

1, 1, 2, 3, 5, 8, 13, 21, . . .

Golden number

$$\frac{\sqrt{5} + 1}{2} \approx 1.618$$

Golden proportion

$$\frac{a + b}{a} = \frac{a}{b}$$

CHAPTER 5 REVIEW EXERCISES

5.1

In Exercises 1 and 2, determine whether the number is divisible by each of the following numbers: 2, 3, 4, 5, 6, 8, 9, and 10.

1. 670,920 **2.** 400,644

In Exercises 3–7, find the prime factorization of the number.

3. 328 **4.** 350 **5.** 840

6. 882 **7.** 1452

In Exercises 8–13, find the GCD and LCM of the numbers.

8. 12, 36 **9.** 72, 52 **10.** 45, 250

11. 840, 320 **12.** 60, 40, 96 **13.** 36, 108, 144

14. *Train Stops* From 1912 to 1971, the Milwaukee Road Railroad Company had a train stop every 15 days in Dubuque, Iowa. During this same period, the same train also stopped in Des Moines, Iowa, every 9 days. If on April 18, 1964, the train made a stop in Dubuque and a stop in Des Moines, how many days was it until the train again stopped in both cities on the same day?

5.2

In Exercises 15–22, use a number line to evaluate the expression.

15. $-5 + 3$ **16.** $7 + (-5)$

17. $4 - 8$ **18.** $-2 + (-4)$

19. $-5 - 4$ **20.** $-3 - (-6)$

21. $(-3 + 7) - 4$ **22.** $-1 + (9 - 4)$

In Exercises 23–30, evaluate the expression.

23. $(-4)(-6)$ **24.** $-3(7)$ **25.** $5(-3)$

26. $\dfrac{-35}{-7}$ **27.** $\dfrac{12}{-6}$

28. $[8 \div (-4)](-3)$

29. $[(-4)(-3)] \div 2$

30. $[(-30) \div (10)] \div (-1)$

5.3

In Exercises 31–39, express the fraction as a terminating or repeating decimal.

31. $\frac{4}{5}$ **32.** $\frac{7}{10}$ **33.** $\frac{12}{16}$

34. $\frac{13}{4}$ **35.** $\frac{3}{7}$ **36.** $\frac{7}{12}$

37. $\frac{3}{8}$ **38.** $\frac{7}{8}$ **39.** $\frac{5}{7}$

In Exercises 40–46, express the decimal number as a quotient of two integers.

40. 0.175 **41.** $0.\overline{3}$ **42.** 5.31

43. $2.\overline{37}$ **44.** 12.083 **45.** 0.0042

46. $2.3\overline{4}$

In Exercises 47–50, express each mixed number as an improper fraction.

47. $5\frac{1}{2}$ **48.** $12\frac{3}{4}$ **49.** $-3\frac{1}{4}$ **50.** $-35\frac{3}{8}$

In Exercises 51–54, express each improper fraction as a mixed number.

51. $\frac{27}{4}$ **52.** $\frac{39}{12}$ **53.** $-\frac{12}{7}$ **54.** $-\frac{136}{5}$

In Exercises 55–63, perform the indicated operation and reduce your answer to lowest terms.

55. $\frac{1}{3} + \frac{1}{7}$ **56.** $\frac{3}{4} - \frac{1}{3}$

57. $\frac{7}{12} + \frac{5}{14}$ **58.** $\frac{2}{3} \cdot \frac{3}{11}$

59. $\frac{5}{9} \div \frac{6}{7}$ **60.** $\left(\frac{4}{5} + \frac{5}{7}\right) \div \frac{4}{5}$

61. $\left(\frac{2}{3} \cdot \frac{1}{7}\right) \div \frac{4}{7}$ **62.** $\left(\frac{1}{5} + \frac{2}{3}\right)\left(\frac{3}{8}\right)$

63. $\left(\frac{1}{5} \cdot \frac{2}{3}\right) + \left(\frac{1}{5} \div \frac{1}{2}\right)$

64. *Cajun Turkey* A recipe for Roasted Cajun Turkey calls for $\frac{1}{8}$ teaspoon of cayenne pepper per pound of turkey. If Jennifer Thornton is preparing a turkey that weighs $17\frac{3}{4}$ pounds, how much cayenne pepper does she need?

5.4

In Exercises 65–80, simplify the expression. Rationalize the denominator when necessary.

65. $\sqrt{20}$ **66.** $\sqrt{32}$

67. $\sqrt{5} + 7\sqrt{5}$ **68.** $\sqrt{3} - 4\sqrt{3}$

69. $\sqrt{8} + 6\sqrt{2}$ **70.** $\sqrt{3} - 7\sqrt{27}$

71. $\sqrt{75} + \sqrt{27}$ **72.** $\sqrt{3} \cdot \sqrt{6}$

73. $\sqrt{8} \cdot \sqrt{6}$ **74.** $\dfrac{\sqrt{18}}{\sqrt{2}}$

75. $\dfrac{\sqrt{56}}{\sqrt{2}}$ **76.** $\dfrac{3}{\sqrt{2}}$

77. $\dfrac{\sqrt{3}}{\sqrt{5}}$ **78.** $5(3 + \sqrt{5})$

79. $\sqrt{3}(4 + \sqrt{6})$ **80.** $\sqrt{3}(\sqrt{6} + \sqrt{15})$

5.5

In Exercises 81–90, state the name of the property illustrated.

81. $y + 5 = 5 + y$

82. $4 \cdot y = y \cdot 4$

83. $(1 + 2) + 3 = 1 + (2 + 3)$

84. $4(y + 3) = 4 \cdot y + 4 \cdot 3$

85. $(1 + 2) + 3 = 3 + (1 + 2)$

86. $(3 + 5) + (4 + 3) = (4 + 3) + (3 + 5)$

87. $(3 \cdot a) \cdot b = 3 \cdot (a \cdot b)$

88. $a \cdot (2 + 3) = (2 + 3) \cdot a$

89. $2(x + 3) = (2 \cdot x) + (2 \cdot 3)$

90. $x \cdot 2 + 6 = 2 \cdot x + 6$

In Exercises 91–96, determine whether the set of numbers is closed under the given operation.

91. Natural numbers, multiplication

92. Integers, subtraction

93. Integers, division

94. Real numbers, subtraction

95. Irrational numbers, multiplication

96. Rational numbers, division

5.6

In Exercises 97–104, evaluate each expression.

97. 2^4 **98.** 2^{-3} **99.** $\dfrac{7^5}{7^4}$

100. $5^2 \cdot 5$ **101.** 7^0 **102.** 4^{-3}

103. $(2^3)^2$ **104.** $(3^2)^2$

In Exercises 105–108, write each number in scientific notation.

105. 230,000 **106.** 0.0000158

107. 0.00275 **108.** 4,950,000

In Exercises 109–112, express each number in decimal notation.

109. 2.5×10^4 **110.** 1.39×10^{-4}

111. 1.75×10^{-4} **112.** 1×10^5

In Exercises 113–116, (a) perform the indicated operation and write your answer in scientific notation. (b) Confirm the result found in part (a) by performing the calculation on a scientific calculator.

113. $(5 \times 10^6)(1.7 \times 10^{-4})$

114. $(4 \times 10^2)(2.5 \times 10^2)$

115. $\dfrac{8.4 \times 10^3}{4 \times 10^2}$

116. $\dfrac{1.5 \times 10^{-3}}{5 \times 10^{-4}}$

In Exercises 117–121, (a) perform the indicated calculation by first converting each number to scientific notation. Write your answer in decimal notation. (b) Confirm the result found in part (a) by performing the calculation on a scientific calculator.

117. (25,000)(600,000)

118. (35,000)(0.00002)

119. $\dfrac{9{,}600{,}000}{3000}$

120. $\dfrac{0.000002}{0.0000004}$

121. At noon, there were 12,000 bacteria in the culture. At 6:00 P.M., there were 300,000 bacteria in the culture. How many times greater is the number of bacteria at 6:00 P.M. than at noon?

122. As a result of a recent water and sewer system improvement, the city of Galena, Illinois, has an outstanding debt of \$20,000,000. If the population of Galena is 3,600 people, how much would each person have to contribute to pay off the outstanding debt?

5.7

In Exercises 123–128, determine whether the sequence is arithmetic or geometric. Then determine the next two terms of the sequence.

123. 1, 6, 11, 16, . . .

124. −3, 9, −27, 81, . . .

125. −3, −6, −9, −12, . . .

126. $\frac{1}{2}, \frac{1}{4}, \frac{1}{8}, \frac{1}{16}, \ldots$

127. 1, 4, 7, 10, 13, . . .

128. 2, −2, 2, −2, 2, . . .

In Exercises 129–134, find the indicated term of the sequence with the given first term, a_1, and common difference, d, or common ratio, r.

129. Find a_6 when $a_1 = -6, d = 3$.

130. Find a_8 when $a_1 = -6, d = -4$.

131. Find a_{10} when $a_1 = -20, d = 5$.

132. Find a_4 when $a_1 = 8, r = 3$.

133. Find a_5 when $a_1 = 4, r = \frac{1}{2}$.

134. Find a_4 when $a_1 = -6, r = 2$.

In Exercises 135–138, find the sum of the arithmetic sequence.

135. 2, 6, 10, 14, . . . , 38

136. $-4, -3\frac{3}{4}, -3\frac{1}{2}, -3\frac{1}{4}, \ldots, -2\frac{1}{4}$

137. 100, 94, 88, 82, . . . , 58

138. 0.5, 0.75, 1.00, 1.25, . . . , 5.25

In Exercises 139–142, find the sum of the first n terms of the geometric sequence for the values of a_1 and r.

139. $n = 3, a_1 = 4, r = 2$

140. $n = 4, a_1 = 2, r = 3$

141. $n = 5, a_1 = 3, r = -2$

142. $n = 6, a_1 = 1, r = -2$

In Exercises 143–148, first determine whether the sequence is arithmetic or geometric; then write an expression for the general or nth term, a_n.

143. 7, 4, 1, −2, . . .

144. 0, 5, 10, 15, . . .

145. $4, \frac{5}{2}, 1, -\frac{1}{2}, \ldots$

146. 3, 6, 12, 24, . . .

147. 4, −4, 4, −4, . . .

148. $5, \frac{5}{3}, \frac{5}{9}, \frac{5}{27}, \ldots$

5.8

In Exercises 149–152, determine whether the sequence is a Fibonacci-type sequence. If so, determine the next two terms.

149. 0, 1, 1, 2, 3, 5, 8, . . .

150. −3, 4, 1, 5, 6, 11, . . .

151. 3, 8, 13, 18, 23, . . .

152. −10, 10, 0, 10, 20, . . .

CHAPTER 5 TEST

1. Which of the numbers 2, 3, 4, 5, 6, 8, 9, and 10 divide 481,248?

2. Find the prime factorization of 420.

3. Evaluate $[(-6) + (-9)] + 8$.

4. Evaluate $-5 - 15$.

5. Evaluate $[(-70)(-5)] \div (8 - 10)$.

6. Convert $4\frac{5}{8}$ to an improper fraction.

7. Convert $\frac{176}{9}$ to a mixed number.

8. Write $\frac{5}{8}$ as a terminating or repeating decimal.

9. Express 6.45 as a quotient of two integers.

10. Evalute $(\frac{5}{16} \div 3) + (\frac{4}{5} \cdot \frac{1}{2})$.

11. Perform the operation and reduce the answer to lowest terms: $\frac{15}{24} - \frac{3}{20}$.

12. Simplify $\sqrt{75} + \sqrt{48}$.

13. Rationalize $\dfrac{\sqrt{5}}{\sqrt{6}}$.

14. Determine whether the integers are closed under the operation of multiplication. Explain your answer.

Name the property illustrated.

15. $(4 + y) + 5 = 4 + (y + 5)$

16. $3(x + y) = 3x + 3y$

Evaluate.

17. $\dfrac{6^7}{6^5}$

18. $4^3 \cdot 4^2$

19. 9^{-2}

20. Perform the operation by first converting the numerator and denominator to scientific notation. Write the answer in scientific notation.

$$\frac{64{,}000}{0.008}$$

21. Write an expression for the general or nth term, a_n, of the sequence −2, −6, −10, −14, ...

22. Find the sum of the terms of the arithmetic sequence −2, −5, −8, −11, . . . , −32.

23. Find a_5 when $a_1 = 3$ and $r = 3$.

24. Find the sum of the first five terms of the sequence when $a_1 = 3$ and $r = 4$.

25. Write an expression for the general or nth term, a_n, of the sequence 3, 6, 12, 24, ...

26. Write the first 10 terms of the Fibonacci sequence.

GROUP PROJECTS

1. *Making Rice* The amount of ingredients needed to make 3 and 5 servings of rice are:

To make	Rice and water	Salt	Butter
3 servings	1 cup	$\frac{3}{8}$ tsp	$1\frac{1}{2}$ tsp
5 servings	$1\frac{2}{3}$ cup	$\frac{5}{8}$ tsp	$2\frac{1}{2}$ tsp

Find the amount of each ingredient needed to make (a) 2 servings, (b) 1 serving, and (c) 29 servings. Explain how you determined your answers.

2. *Finding Areas* **a)** Determine the area of the trapezoid shown by finding the area of the three parts indicated and finding the sum of the three areas. The necessary geometric formulas are given in Chapter 9.

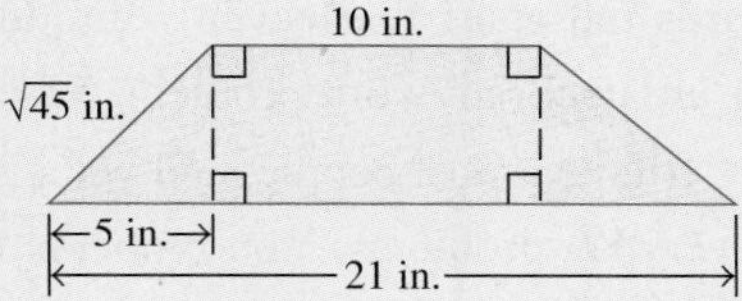

b) Determine the area of the trapezoid by using the formula for the area of a trapezoid given in Chapter 9.
c) Compare your answers from parts (a) and (b). Are they the same? If not, explain why they are different.

3. *Medical Insurance* On a medical insurance policy (such as Blue Cross/Blue Shield), the policyholder may need to make copayments for prescription drugs, office visits, and procedures until the total of all copayments reaches a specified amount. Suppose on the Gattelaro's medical policy that the copayment for prescription drugs is 50% of the cost; the copayment for office visits is $10; and the copayment for all medical tests, X-rays, and other procedures is 20% of the cost. After the family's copayment totals $500 in a calendar year, all medical and prescription bills are paid in full by the insurance company. The Gattelaros had the following medical expenses from January 1 through April 30.

Date	Reason	Cost before copayment
January 10	Office visit	$40
	Prescription	$44
February 27	Office visit	$40
	Medical tests	$188
April 19	Office visit	$40
	X rays	$348
	Prescription	$76

a) How much had the Gattelaros paid in copayments from January 1 through April 30?
b) How much had the medical insurance company paid?
c) What is the remaining copayment that must be paid by the Gattelaros before the $500 copayment limit is reached?

4. *A Branching Plant* A plant grows for two months and then adds a new branch. Each new branch grows for two months and then adds another branch. After the second month, each branch adds a new branch every month. Assume the growth begins in January.
a) How many branches will there be in February?
b) How many branches will there be in May?
c) How many branches will there be after 12 months?
d) How is this problem similar to the problem involving rabbits that appeared in Fibonacci's book *Liber Abacci* (see page 245)?

Algebra, Graphs, and Functions

Gary Larson, creator of *The Far Side* comic strip, touched a common fear in a cartoon titled "Math Phobic's Nightmare." In the cartoon, St. Peter guards the gates to Heaven, admitting only those who can answer an algebraic word problem. Word problems—the very mention of them is enough to frighten most people, and yet algebra is one of the most practical tools for solving everyday problems. You probably use algebra in your daily life without realizing it.

For example, you use a coordinate system when you consult your car map to find directions to a new destination. You solve simple equations when you change a recipe to increase or decrease the number of servings. To evaluate how much interest you will earn on a savings account or to figure out how long it will take you to travel a given distance, you use common formulas that are algebraic equations.

THE FAR SIDE By GARY LARSON

Okay, now listen up. Nobody gets in here without answering the following question: A train leaves Philadelphia at 1:00 p.m. It's traveling at 65 miles per hour. Another train leaves Denver at 4:00... Say, you need some paper?
7-9
Larson

Math phobic's nightmare

The symbolic language of algebra makes it an excellent tool for solving problems. Symbolism has three advantages. First, it allows us to write lengthy expressions in compact form. Second, symbolic language is clear—each symbol has a precise meaning. Finally, symbolism allows us to consider a large or infinite number of separate cases with a common property.

The English philosopher Alfred North Whitehead explained the power of algebra when he stated, "By relieving the brain of all unnecessary work, a good notation sets the mind free to concentrate on more advanced problems and in effect increases the mental power of the race." ■

6.1 ORDER OF OPERATIONS

Algebra is a generalized form of arithmetic. The word *algebra* is derived from the Arabic word *al-jabr* (meaning "reunion of broken parts"), which was the title of a book written by the mathematician Muhammed ibn-Musa al Khwarizmi in about A.D. 825.

Why study algebra? You can solve many problems in everyday life by using arithmetic or by trial and error, but with a knowledge of algebra you can find the solutions with less effort. You can solve other problems, like some we will present in this chapter, only by using algebra.

Algebra uses letters of the alphabet called **variables** to represent numbers. Often the letters x and y are used to represent variables. However, any letter may be used as a variable. A symbol that represents a specific quantity is called a **constant**.

Multiplication of numbers and variables may be represented in several different ways in algebra. Since the "times" sign might be confused with the variable x, a dot between two numbers or variables indicates multiplication. Thus, $3 \cdot 4$ means 3 times 4, and $x \cdot y$ means x times y. Placing two letters or a number and a letter next to one another, with or without parentheses, also indicates multiplication. Thus, $3x$ means 3 times x, xy means x times y, and $(x)(y)$ means x times y.

An **algebraic expression** (or simply an **expression**) is a collection of variables, numbers, parentheses, and operation symbols. Some examples of algebraic expressions are

$$x, \quad x + 2, \quad 3(2x + 3), \quad \frac{3x + 1}{2x - 3}, \quad \text{and} \quad x^2 + 7x + 3$$

Two algebraic expressions joined by an equal sign form an **equation**. Some examples of equations are

$$x + 2 = 4, \quad 3x + 4 = 1, \quad \text{and} \quad x + 3 = 2x$$

The **solution to an equation** is the number or numbers that replace the variable to make the equation a true statement. For example, the solution to the equation $x + 3 = 4$ is $x = 1$. When we find the solution to an equation, we **solve the equation**.

We can determine if any number is a solution to an equation by **checking the solution**. To check the solution, we substitute the number for the variable in the equation. If the resulting statement is a true statement, that number is a solution to the equation. If the resulting statement is a false statement, the number is not a solution to the equation. To check the number $x = 1$ in the equation $x + 3 = 4$, we do the following.

$$x + 3 = 4$$
$$1 + 3 = 4 \quad \text{Substitute 1 for } x.$$
$$4 = 4 \quad \text{True}$$

The same number is obtained on both sides of the equal sign, so the solution is correct. For the equation $x + 3 = 4$, the only solution is $x = 1$. Any other value of x would result in the check being a false statement.

To **evaluate an expression** means to find the value of the expression for a given value of the variable. To evaluate expressions and solve equations, you must have an

DID YOU KNOW

Broken Bones

Imagine yourself walking down a street in Spain during the Middle Ages. You see a sign over a door: "*Algebrista y Sangrador.*" Inside, you would find a person more ready to give you a haircut than help you with your algebra. The sign translates into "Bonesetter and Bloodletter," relatively simple medical treatments administered in barbershops of the day.

The root word *al-jabr,* which the Muslims (Moors) brought to Spain along with some concepts of algebra, suggests the restoring of broken parts. The parts might be bones, or they might be mathematical expressions that are broken into separate parts and the parts moved from one side of an equation to the other and reunited in such a way as to make a solution more obvious.

understanding of exponents. Exponents (Section 5.6) are used to abbreviate repeated multiplication. For example, the expressions 5^2 means $5 \cdot 5$. The 2 in the expression 5^2 is the **exponent**, and the 5 is the **base**. We read 5^2 as "5 to the second power" or "5 squared," and 5^2 means $5 \cdot 5$ or 25.

In general, the number b to the nth power, written b^n, means

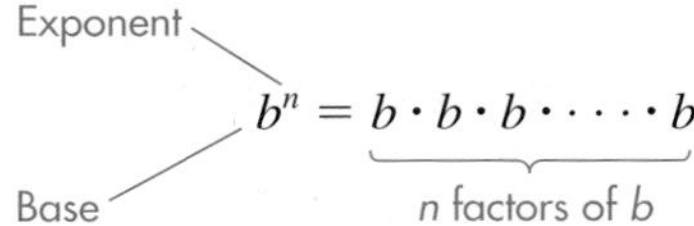

An exponent refers only to its base. In the expression -5^2, the base is 5. In the expression $(-5)^2$, the base is -5.

$$-5^2 = -(5)^2 = -1(5)^2 = -1(5)(5) = -25$$
$$(-5)^2 = (-5)(-5) = 25$$

Note that $-5^2 \neq (-5)^2$ since $-25 \neq 25$.

ORDER OF OPERATIONS

To evaluate an expression or to check the solution to an equation, we need to know the **order of operations** to follow. For example, suppose we want to evaluate the expression $2 + 3x$ when $x = 4$. Substituting 4 for x, we obtain $2 + 3 \cdot 4$. What is the value of $2 + 3 \cdot 4$? Does it equal 20, or does it equal 14? Some standard rules called the order of operations have been developed to ensure that there is only one correct answer. In mathematics, unless parentheses indicate otherwise, always perform multiplication before addition. Thus, the correct answer is 14.

$$2 + 3 \cdot 4 = 2 + (3 \cdot 4) = 2 + 12 = 14$$

The order of operations for evaluating an expression is as follows.

Order of Operations

1. First, perform all operations within parentheses or other grouping symbols (according to the following order).
2. Next, perform all exponential operations (that is, raising to powers or finding roots).
3. Next, perform all multiplications and divisions from left to right.
4. Finally, perform all additions and subtractions from left to right.

Some students use the phrase, "**P**lease **E**xcuse **M**y **D**ear **A**unt **S**ally," or the word "PEMDAS" (**P**arentheses, **E**xponents, **M**ultiplication, **D**ivision, **A**ddition, **S**ubtraction) to remind them of the order of operations. Remember: Multiplication and division are of the same order, and addition and subtraction are of the same order.

EXAMPLE 1 *Evaluating an Expression*

Evaluate the expression $-x^2 + 5x + 29$ for $x = 2$.

SOLUTION Substitute 2 for each x and use the order of operations to evaluate the expression.

$$\begin{aligned} -x^2 + 5x + 29 &= -(2)^2 + 5(2) + 29 \\ &= -4 + 10 + 29 \\ &= 35 \end{aligned}$$

EXAMPLE 2 ***Finding a Temperature***

The temperature T, in degrees Celsius, in a sauna n minutes after the sauna is turned on can be approximated by the expression $-0.04n^2 + 1.6n + 20$ (assuming n is between 0 and 20). Find the temperature after the sauna has been on for 12 minutes.

SOLUTION Substitute 12 for each n.

$$\begin{aligned} &-0.04n^2 + 1.6n + 20 \\ &= -0.04(12)^2 + 1.6(12) + 20 \\ &= -0.04(144) + 19.2 + 20 \\ &= -5.76 + 19.2 + 20 \\ &= 33.44 \end{aligned}$$

The temperature in the sauna is approximately 33°C (or 92°F) after 12 minutes.

EXAMPLE 3 ***Substituting for Two Variables***

Evaluate $-3x^2 + 2xy - 2y^2$ when $x = 2$ and $y = 3$.

SOLUTION Substitute 2 for each x and 3 for each y; then evaluate using the order of operations.

$$\begin{aligned} &-3x^2 + 2xy - 2y^2 \\ &= -3(2)^2 + 2(2)(3) - 2(3)^2 \\ &= -12 + 12 - 18 \\ &= -18 \end{aligned}$$

EXAMPLE 4 ***Is 3 a Solution?***

Determine whether 3 is a solution to the equation $2x^2 + 4x - 9 = 21$.

SOLUTION To determine whether 3 is a solution to the equation, substitute 3 for each x in the equation. Then evaluate the left side of the equation by using the order of operations. If this leads to a 21 on the left side of the equal sign, then both sides of the equation have the same value, and 3 is a solution.

$$\begin{aligned} 2x^2 + 4x - 9 &= 21 \\ 2(3)^2 + 4(3) - 9 &= 21 \\ 2(9) + 12 - 9 &= 21 \\ 18 + 12 - 9 &= 21 \\ 30 - 9 &= 21 \\ 21 &= 21 \quad \text{True} \end{aligned}$$

Because 3 makes the equation a true statement, 3 is a solution to the equation.

DID YOU KNOW

An Important Breakthrough

This painting by Charles Demuth, called *I Saw the Figure 5 in Gold,* depicts the abstract nature of numbers. Mathematician and philosopher Bertrand Russell observed in 1919 that it must have required "many ages to discover that a brace of pheasants and a couple of days were both instances of the number 2." The discovery that numbers could be used not only to count objects such as the number of birds but also to represent abstract quantities represented a breakthrough in the development of algebra.

SECTION 6.1 EXERCISES

CONCEPT/WRITING EXERCISES

1. What is a *variable*?
2. What is a *constant*?
3. What is an algebraic expression? Illustrate with an example.
4. What does *a number is a solution to an equation* mean?
5. **a)** For the term 4^5, identify the base and the exponent.
   b) In your own words, explain how to evaluate 4^5.
6. In your own words, explain the order of operations.
7. Evaluate $6 + 12 \div 2$ using the order of operations.
8. Evaluate $12 + 8 \cdot 3$ using the order of operations.

PRACTICE THE SKILLS

In Exercises 9–28, evaluate the expression for the given value(s) of the variable(s).

9. $x^2, \quad x = -4$

10. $x^2, \quad x = 9$

11. $-x^2, \quad x = -7$

12. $-x^2, \quad x = -5$

13. $-2x^3, \quad x = -7$

14. $-x^3, \quad x = -4$

15. $x - 7, \quad x = 4$

16. $8x - 3, \quad x = \frac{5}{2}$

17. $-5x + 8, \quad x = -4$

18. $x^2 + 6x - 4, \quad x = 5$

19. $-x^2 - 7x + 5, \quad x = -2$

20. $5x^2 + 7x - 11, \quad x = -1$

21. $\frac{1}{2}x^2 - 5x + 2, \quad x = \frac{2}{3}$

22. $x^3 + 4x^2 - 3x + 15, \quad x = 3$

23. $8x^3 - 4x^2 + 7, \quad x = \frac{1}{2}$

24. $-x^2 + 4xy, \quad x = 2, \quad y = 3$

25. $3x^2 - xy + 2y^2, \quad x = 1, \quad y = -2$

26. $3x^2 + \frac{3}{7}xy - \frac{1}{3}y^2, \quad x = 7, \quad y = -3$

27. $4x^2 - 12xy + 9y^2, \quad x = 3, \quad y = 2$

28. $(x + 3y)^2, \quad x = 4, \quad y = -3$

In Exercises 29–38, determine whether the value(s) is (are) a solution to the equation.

29. $6x - 9 = 12, \quad x = 3$

30. $4x - 7 = 15, \quad x = -2$

31. $x + 2y = 0, \quad x = -6, \quad y = 3$

32. $3x + 2y = -2, \quad x = -2, \quad y = 2$

33. $x^2 + 3x - 4 = 5, \quad x = 2$

34. $2x^2 - x - 5 = 0, \quad x = 3$

35. $2x^2 + x = 28, \quad x = -4$

36. $y = x^2 + 3x - 5, \quad x = 1, \quad y = -1$

37. $y = -x^2 + 4x - 1, \quad x = 3, \quad y = 2$

38. $y = x^3 - 3x^2 + 1, \quad x = 2, \quad y = -3$

PROBLEM SOLVING

39. *Pay at Walmart* If Becky Carr earns \$6.55 per hour as a cashier at Walmart, the amount she earns in t hours can be determined by the expression $\$6.55t$. Determine Becky's pay if she works (a) 8 hr and (b) 20 hr.

40. *Sales Tax* If the sales tax on an item is 6%, the sales tax in dollars on an item costing d dollars can be determined by the expression $0.06d$. Determine the sales tax on a lawnmower costing \$325.

41. *Cost of a Car* The total cost of an item in dollars, including a 5% sales tax, can be determined by the expression $x + 0.05x$, where x is the cost of the item before the tax. Determine the total cost of a car whose cost before the tax is \$12,500.

42. *8 Trillion Calculations* If a computer can do a calculation in 0.000002 sec, the time required to do n calculations can be determined by the expression $0.000002n$. Determine the number of seconds needed for the computer to do 8 trillion (8,000,000,000,000) calculations.

43. *Orange Orchard* The number of baskets of oranges that are produced by x trees in a small orchard can be approximated by the expression $25x - 0.2x^2$ (assuming x is no more than 100). Find the number of baskets of oranges produced by 60 trees.

44. *Drying Time* The time, in minutes, needed for clothes hanging on a line outdoors to dry, at a specific temperature and wind speed, depends on the humidity, h.

The time can be approximated by the expression $2h^2 + 80h + 40$, where h is the percent humidity expressed as a decimal number. Find the length of time required for clothing to dry if there is 60% humidity.

45. *Grass Growth* The rate of growth of grass in inches per week depends on a number of factors, including rainfall and temperature. For a certain area, this can be approximated by the expression $0.2R^2 + 0.003RT + 0.0001T^2$, where R is the weekly rainfall, in inches, and T is the average weekly temperature, in degrees Fahrenheit. Find the amount of growth of grass for a week in which the rainfall is 2 in. and the average temperature is 70°F.

CHALLENGE PROBLEMS/GROUP ACTIVITIES

46. Explain why $(-1)^n = 1$ for any even number n.

47. Does $(x + y)^2 = x^2 + y^2$? Complete the table and state your conclusion.

x	y	$(x + y)^2$	$x^2 + y^2$
2	3		
−2	−3		
−2	3		
2	−3		

48. Suppose n represents any natural number. Explain why 1^n equals 1?

RESEARCH ACTIVITY

49. When were exponents first used? Write a paper explaining how exponents were first used and when mathematicians began writing them in the present form. (References include history of mathematics books, encyclopedias, and the Internet.)

6.2 LINEAR EQUATIONS IN ONE VARIABLE

In Section 6.1, we stated that two algebraic expressions joined by an equal sign form an equation. The solution to some equations, such as $x + 3 = 4$, can be found easily by trial and error. However, solving more complex equations, such as $2x - 3 = 4(x + 3)$, requires understanding the meaning of like terms and learning four basic properties.

The parts that are added or subtracted in an algebraic expression are called **terms**. The expression $4x - 3y - 5$ contains three terms, namely $4x$, $-3y$, and -5. The + and − signs that break the expression into terms are a part of the terms. When listing the terms of an expression, however, it is not necessary to include the + sign at the beginning of the term.

The numerical part of a term is called its **numerical coefficient** or, simply, its **coefficient**. In the term $4x$, the 4 is the numerical coefficient. In the term $-4y$, the -4 is the numerical coefficient.

Like terms are terms that have the same variables with the same exponents on the variables. **Unlike terms** have different variables or different exponents on the variables.

Like terms	Unlike terms
$2x$, $7x$ (same variable, x)	$2x$, 9 (only first term has a variable)
$-8y$, $3y$ (same variable, y)	$5x$, $6y$ (different variables)
-4, 10 (both constants)	x, 8 (only first term has a variable)
$-5x^2$, $6x^2$ (same variable with same exponent)	$2x^3$, $3x^2$ (different exponents)

To **simplify an expression** means to combine like terms by using the commutative, associative, and distributive properties discussed in Chapter 5. For convenience, we list these properties on page 262.

Properties of the Real Numbers

$a(b + c) = ab + ac$	Distributive property
$a + b = b + a$	Commutative property of addition
$ab = ba$	Commutative property of multiplication
$(a + b) + c = a + (b + c)$	Associative property of addition
$(ab)c = a(bc)$	Associative property of multiplication

EXAMPLE 1 ***Combining Like Terms***

Combine like terms in each expression.

a) $5x + 3x$ b) $10y - 3y$ c) $15 + x + 9 - 2x$
d) $-x + 6y - 11 + 3y - 4x + 2$

SOLUTION

a) We use the distributive property (in reverse) to combine like terms.

$$\begin{aligned} 5x + 3x &= (5 + 3)x && \text{Distributive property} \\ &= 8x \end{aligned}$$

b) $10y - 3y = (10 - 3)y = 7y$

c) $$\begin{aligned} 15 + x + 9 - 2x &= x - 2x + 15 + 9 && \text{Rearrange terms; place like terms together.} \\ &= -x + 24 && \text{Combine like terms.} \end{aligned}$$

d) $$\begin{aligned} &-x + 6y - 11 + 3y - 4x + 2 \\ &= -x - 4x + 6y + 3y - 11 + 2 && \text{Rearrange terms; place like items together.} \\ &= -5x + 9y - 9 && \text{Combine like terms.} \end{aligned}$$

We are able to rearrange the terms of an expression, as was done in Example 1(c) and (d) by the commutative and associative properties that were discussed in Section 5.5.

The order of the terms in an expression is not crucial. However, when listing the terms of an expression we generally list the terms in alphabetical order with the constant, the term without a variable, last.

SOLVING EQUATIONS

Recall that to solve an equation means to find the value or values for the variable that make(s) the equation true. In this section, we discuss solving **linear (or first degree) equations**. A linear equation in one variable is one in which the exponent on the variable is 1. Examples of linear equations are $5x - 1 = 3$ and $2x + 4 = 6x - 5$.

Equivalent equations are equations that have the same solution. The equations $2x - 5 = 1$, $2x = 6$, and $x = 3$ are all equivalent equations since they all have the same solution, 3. When we solve an equation, we write the given equation as a series of simpler equivalent equations until we obtain an equation of the form $x = c$, where c is some real number.

To solve any equation, we have to **isolate the variable**. That means getting the variable by itself on one side of the equal sign. The four properties of equality that

DID YOU KNOW

In the Beginning

Greek philosopher Diophantus of Alexandria (A.D. 250), who invented notations for powers of a number and for multiplication and division of simple quantities, is thought to have made the first attempts at algebra. But not until the sixteenth century did French mathematician François Viète (1540–1603) use symbols to represent numbers, the foundation of symbolic algebra. However, the work of René Descartes (1596–1660) is considered to be the starting point of modern-day algebra. In 1707, Sir Isaac Newton (1643–1727) gave symbolic mathematics the name *universal arithmetic.*

we are about to discuss are used to isolate the variable. The first is the addition property.

Addition Property of Equality
If $a = b$, then $a + c = b + c$ for all real numbers a, b, and c.

The addition property of equality indicates that the same number can be added to both sides of an equation without changing the solution.

EXAMPLE 2 ***Using the Addition Property of Equality***

Find the solution to the equation $x - 8 = 12$.

SOLUTION To isolate the variable, add 8 to both sides of the equation.

$$\begin{aligned} x - 8 &= 12 \\ x - 8 + 8 &= 12 + 8 \\ x + 0 &= 20 \\ x &= 20 \end{aligned}$$

Check:

$$\begin{aligned} x - 8 &= 12 \\ 20 - 8 &= 12 \quad \text{Substitute 20 for } x. \\ 12 &= 12 \quad \text{True} \end{aligned}$$

In Example 2, we showed the step $x + 0 = 20$. Generally this step is done mentally, and the step is not listed.

Subtraction Property of Equality
If $a = b$, then $a - c = b - c$ for all real numbers a, b, and c.

The subtraction property of equality indicates that the same number can be subtracted from both sides of an equation without changing the solution.

EXAMPLE 3 ***Using the Subtraction Property of Equality***

Find the solution to the equation $x + 7 = 15$.

SOLUTION To isolate the variable, subtract 7 from both sides of the equation.

$$\begin{aligned} x + 7 &= 15 \\ x + 7 - 7 &= 15 - 7 \\ x &= 8 \end{aligned}$$

Note that we did not subtract 15 from both sides of the equation, since this would not result in getting x on one side of the equal sign by itself.

Multiplication Property of Equality
If $a = b$, then $a \cdot c = b \cdot c$ for all real numbers a, b, and c, where $c \neq 0$.

The multiplication property of equality indicates that both sides of the equation can be multiplied by the same nonzero number without changing the solution.

EXAMPLE 4 *Using the Multiplication Property of Equality*

Find the solution to the equation $\frac{x}{5} = 8$.

SOLUTION To solve this equation, multiply both sides of the equation by 5.

$$\frac{x}{5} = 8$$
$$5\left(\frac{x}{5}\right) = 5(8)$$
$$\frac{\overset{1}{\cancel{5}}x}{\underset{1}{\cancel{5}}} = 40$$
$$1x = 40$$
$$x = 40$$

In Example 4, we showed the steps $\frac{5x}{5} = 40$ and $1x = 40$. Generally we will not illustrate these steps.

Division Property of Equality
If $a = b$, then $a/c = b/c$ for all real numbers a, b, and c, $c \neq 0$.

The division property of equality indicates that both sides of an equation can be divided by the same nonzero number without changing the solution. Note that the divisor, c, cannot be 0 because division by 0 is not permitted.

EXAMPLE 5 *Using the Division Property of Equality*

Find the solution to the equation $3x = 15$.

SOLUTION To solve this equation, divide both sides of the equation by 3.

$$3x = 15$$
$$\frac{3x}{3} = \frac{15}{3}$$
$$x = 5$$

An **algorithm** is a general procedure for accomplishing a task. The following general procedure is an algorithm for solving linear (or first-degree) equations. Sometimes the solution to an equation may be found more easily by using a variation of the general procedure. Remember that the primary objective in solving any equation is to isolate the variable.

A General Procedure for Solving Linear Equations

1. If the equation contains fractions, multiply both sides of the equation by the lowest common denominator (or least common multiple). This step will eliminate all fractions from the equation.
2. Use the distributive property to remove parentheses when necessary.
3. Combine like terms on the same side of the equal sign when possible.
4. Use the addition or subtraction property to collect all terms with a variable on one side of the equal sign and all constants on the other side of the equal sign. It may be necessary to use the addition or subtraction property more than once. This process will eventually result in an equation of the form $ax = b$, where a and b are real numbers.
5. Solve for the variable using the division or multiplication property. This will result in an answer in the form $x = c$, where c is a real number.

EXAMPLE 6 *Using the General Procedure*

Solve the equation $2x - 9 = 19$ and check your solution.

SOLUTION Our goal is to isolate the variable; therefore, we start by getting the term $2x$ by itself on one side of the equation.

$$2x - 9 = 19$$

$$2x - 9 + 9 = 19 + 9 \quad \text{Add 9 to both sides of the equation (addition property) (step 4).}$$

$$2x = 28$$

$$\frac{2x}{2} = \frac{28}{2} \quad \text{Divide both sides of the equation by 2 (division property) (step 5).}$$

$$x = 14$$

A check will show that 14 is the solution to $2x - 9 = 19$. ▲

EXAMPLE 7 *Solving a Linear Equation*

Solve the equation $2 = 3 + 5(p + 1)$ for p.

SOLUTION Our goal is to isolate the variable p. To do so, follow the general procedure for solving equations.

$$2 = 3 + 5(p + 1)$$

$$2 = 3 + 5p + 5 \quad \text{Distributive property (step 2)}$$

$$2 = 5p + 8 \quad \text{Combine like terms (step 3)}$$

$$2 - 8 = 5p + 8 - 8 \quad \text{Subtraction property (step 4)}$$

$$-6 = 5p$$

$$-\frac{6}{5} = \frac{5p}{5} \quad \text{Division property (step 5)}$$

$$-\frac{6}{5} = p$$

▲

EXAMPLE 8 ***Solving an Equation Containing Fractions***

Solve the equation $\frac{2x}{3} + \frac{1}{3} = \frac{3}{4}$.

SOLUTION When an equation contains fractions, we generally begin by multiplying each term of the equation by the lowest common denominator, LCD (see Chapter 5). In this example, the LCD is 12, since 12 is the smallest number that is divisible by both 3 and 4.

$$12\left(\frac{2x}{3} + \frac{1}{3}\right) = 12\left(\frac{3}{4}\right) \quad \text{Multiply both sides of the equation by the LCD (step 1).}$$

$$12\left(\frac{2x}{3}\right) + 12\left(\frac{1}{3}\right) = 12\left(\frac{3}{4}\right) \quad \text{Distributive property (step 2)}$$

$$\overset{4}{\cancel{12}}\left(\frac{2x}{\underset{1}{\cancel{3}}}\right) + \overset{4}{\cancel{12}}\left(\frac{1}{\underset{1}{\cancel{3}}}\right) = \overset{3}{\cancel{12}}\left(\frac{3}{\underset{1}{\cancel{4}}}\right) \quad \text{Divide out common factors.}$$

$$8x + 4 = 9$$

$$8x + 4 - 4 = 9 - 4 \quad \text{Subtraction property (step 4)}$$

$$8x = 5$$

$$\frac{8x}{8} = \frac{5}{8} \quad \text{Division property (step 5)}$$

$$x = \frac{5}{8}$$

A check will show that $\frac{5}{8}$ is the solution to the equation. You could have worked the problem without first multiplying both sides of the equation by the LCD. Try it!

EXAMPLE 9 ***Variables on Both Sides of the Equation***

Solve the equation $3x + 4 = 5x + 6$.

SOLUTION Note that the equation has an x on both sides of the equal sign. In equations of this type, you might wonder what to do first. It really does not matter as long as you don't forget the goal of isolating the variable x. Let's collect the terms containing a variable on the left-hand side of the equation.

$$3x + 4 = 5x + 6$$

$$3x + 4 - 4 = 5x + 6 - 4 \quad \text{Subtraction property (step 4)}$$

$$3x = 5x + 2$$

$$3x - 5x = 5x - 5x + 2 \quad \text{Subtraction property (step 4)}$$

$$-2x = 2$$

$$\frac{-2x}{-2} = \frac{2}{-2} \quad \text{Division property (step 5)}$$

$$x = -1$$

DID YOU KNOW

A New Concept

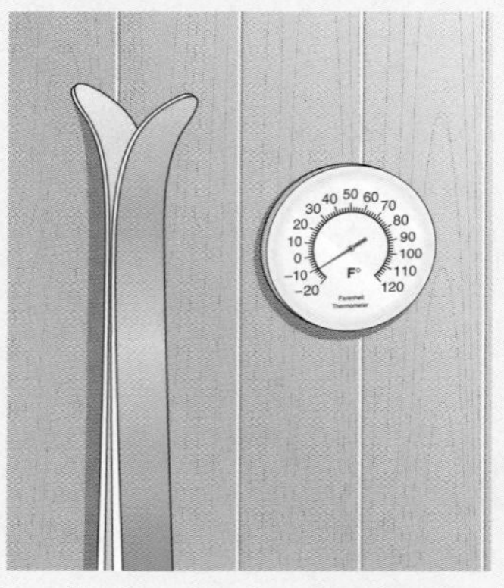

Well into the sixteenth century, mathematicians found it difficult to accept the idea that the solution to a problem (such as Example 9) could be a negative number because negative numbers could not be accepted as physically real. In the early days of algebra, someone working a problem did not isolate a variable by subtracting like terms. Instead, a problem would be put into a form that allowed only positive coefficients and answers. Albert Girard (1595–1637) contributed to the evolution of a correct understanding of negative quantities.

In the solution to Example 9, the terms containing the variable were collected on the left side of the equal sign. Now work Example 9, collecting the terms with the variable on the right side of the equation. If you do so correctly, you will get the same result.

EXAMPLE 10 ***Solving an Equation Containing Decimals***

Solve the equation $4x - 0.48 = 0.8x + 4$ and check your solution.

SOLUTION This problem may be solved with the decimals, or you may multiply each term by 100 and eliminate the decimals. We will solve the problem with the decimals.

$$4x - 0.48 = 0.8x + 4$$

$$4x - 0.48 + 0.48 = 0.8x + 4 + 0.48 \quad \text{Addition property}$$

$$4x = 0.8x + 4.48$$

$$4x - 0.8x = 0.8x - 0.8x + 4.48 \quad \text{Subtraction property}$$

$$3.2x = 4.48$$

$$\frac{3.2x}{3.2} = \frac{4.48}{3.2} \quad \text{Division property}$$

$$x = 1.4$$

Check:

$$4x - 0.48 = 0.8x + 4$$

$$4(1.4) - 0.48 = 0.8(1.4) + 4 \quad \text{Substitute 1.4 for each } x \text{ in the equation.}$$

$$5.6 - 0.48 = 1.12 + 4$$

$$5.12 = 5.12 \quad \text{True}$$

In Chapter 5, we explained that $a - b$ can be expressed as $a + (-b)$. We use this principle in Example 11.

EXAMPLE 11 ***Using the Definition of Subtraction***

Solve $10 = -5 + 3(p - 4)$ for p.

$$10 = -5 + 3(p - 4)$$

$$10 = -5 + 3[p + (-4)] \quad \text{Definition of subtraction}$$

$$10 = -5 + 3(p) + 3(-4) \quad \text{Distributive property}$$

$$10 = -5 + 3p - 12$$

$$10 = 3p - 17 \quad \text{Combine like terms.}$$

$$10 + 17 = 3p - 17 + 17 \quad \text{Addition property}$$

$$27 = 3p \quad \text{Combine like terms.}$$

$$\frac{27}{3} = \frac{3p}{3} \quad \text{Division property}$$

$$9 = p$$

So far, every equation has had exactly one solution. Some equations, however, have no solution and others have more than one solution. Example 12 illustrates an equation that has no solution, and Example 13 illustrates an equation that has an infinite number of solutions.

EXAMPLE 12 ***An Equation with No Solution***

Solve $2(x - 3) + x = 5x - 2(x + 5)$.

SOLUTION

$$
\begin{aligned}
2(x - 3) + x &= 5x - 2(x + 5) & \\
2x - 6 + x &= 5x - 2x - 10 & \text{Distributive property} \\
3x - 6 &= 3x - 10 & \text{Combine like terms.} \\
3x - 3x - 6 &= 3x - 3x - 10 & \text{Subtraction property} \\
-6 &= -10 & \text{False}
\end{aligned}
$$

During the process of solving an equation, if you obtain a false statement like $-6 = -10$, or $4 = 0$, the equation has **no solution**. An equation that has no solution is called an **inconsistent equation**. The equation $2(x - 3) + x = 5x - 2(x + 5)$ is inconsistent and thus has no solution. ▲

EXAMPLE 13 ***An Equation with Infinitely Many Solutions***

Solve $3(x + 2) - 5(x - 3) = -2x + 21$.

SOLUTION

$$
\begin{aligned}
3(x + 2) - 5(x - 3) &= -2x + 21 & \\
3x + 6 - 5x + 15 &= -2x + 21 & \text{Distributive property} \\
-2x + 21 &= -2x + 21 & \text{Combine like terms.}
\end{aligned}
$$

Note that at this point both sides of the equation are the same. Every real number will satisfy this equation. This equation has an infinite number of solutions. An equation of this type is called an **identity**. When solving an equation, if you notice that the same expression appears on both sides of the equal sign, the equation is an identity. The solution to any identity is **all real numbers**. If you continue to solve an equation that is an identity, you will end up with $0 = 0$, as follows.

$$
\begin{aligned}
-2x + 21 &= -2x + 21 & \\
-2x + 2x + 21 &= -2x + 2x + 21 & \text{Addition property} \\
21 &= 21 & \text{Combine like terms.} \\
21 - 21 &= 21 - 21 & \text{Subtraction property} \\
0 &= 0 & \text{True for any value of } x
\end{aligned}
$$

▲

PROPORTIONS

A **ratio** is a quotient of two quantities. An example is the ratio of 2 to 5, which can be written 2:5 or $\frac{2}{5}$ or 2/5.

> A **proportion** is a statement of equality between two ratios.

An example of a proportion is $\frac{a}{b} = \frac{c}{d}$. Consider the proportion

$$\frac{x + 2}{5} = \frac{x + 5}{8}$$

We can solve this proportion by first multiplying both sides of the equation by the least common denominator, 40.

$$\frac{x+2}{5} = \frac{x+5}{8}$$
$$\overset{8}{\cancel{40}}\left(\frac{x+2}{\cancel{5}}\right) = \overset{5}{\cancel{40}}\left(\frac{x+5}{\cancel{8}}\right) \quad \text{Multiplication property}$$
$$8(x+2) = 5(x+5)$$
$$8x + 16 = 5x + 25$$
$$3x + 16 = 25$$
$$3x = 9$$
$$x = 3$$

A check will show that 3 is the solution.

Proportions can often be solved more easily by using cross multiplication.

Cross Multiplication

$$\text{If } \frac{a}{b} = \frac{c}{d}, \text{ then } ad = bc.$$

Let's use cross multiplication to solve the proportion $(x + 2)/5 = (x + 5)/8$.

$$\frac{x+2}{5} = \frac{x+5}{8}$$
$$8(x+2) = 5(x+5) \quad \text{Cross multiplication}$$
$$8x + 16 = 5x + 25$$
$$3x + 16 = 25$$
$$3x = 9$$
$$x = 3$$

Many practical application problems can be solved using proportions.

To Solve Application Problems Using Proportions

1. Represent the unknown quantity by a variable.
2. Set up the proportion by listing the given ratio on the left side of the equal sign and the unknown and other given quantity on the right side of the equal sign. When setting up the right side of the proportion, the same respective quantities should occupy the same respective positions on the left and right. For example, an acceptable proportion might be

$$\frac{\text{miles}}{\text{hour}} = \frac{\text{miles}}{\text{hour}}$$

3. Once the proportion is properly written, drop the units and use cross multiplication to solve the equation.
4. Answer the question or questions asked.

EXAMPLE 14 ***Water Usage***

The cost for water in Orange County is \$1.42 per 750 gallons (gal) of water used. What is the water bill if 30,000 gallons are used?

SOLUTION This problem may be solved by setting up a proportion. One proportion that can be used is

$$\frac{\text{cost of 750 gal}}{\text{750 gal}} = \frac{\text{cost of 30,000 gal}}{\text{30,000 gal}}$$

We want to find the cost for 30,000 gallons of water, so we will call this quantity x. The proportion then becomes

$$\text{Given ratio} \left\{ \frac{1.42}{750} \right. = \frac{x}{30{,}000}$$

Now we solve for x.

$$\begin{aligned} (1.42)(30{,}000) &= 750x \\ 42{,}600 &= 750x \\ \frac{42{,}600}{750} &= \frac{750x}{750} \\ \$56.80 &= x \end{aligned}$$

The cost of 30,000 gallons of water is \$56.80.

EXAMPLE 15 ***Determining the Amount of Insulin***

Insulin comes in 10 cubic centimeter (cc) vials labeled in the number of units of insulin per cubic centimeter of fluid. A vial of insulin marked U40 has 40 units of insulin per cubic centimeter of fluid. If a patient needs 30 units of insulin, how much fluid should be drawn into the syringe from the U40 vial?

SOLUTION The unknown quantity, x, is the number of cubic centimeters of fluid to be drawn into the syringe. Following is one proportion that can be used to find that quantity.

$$\begin{aligned} \text{Given ratio} \left\{ \frac{\text{40 units}}{\text{1 cc}} \right. &= \frac{\text{30 units}}{x\text{ cc}} \\ 40x &= 30(1) \\ 40x &= 30 \\ x &= \frac{30}{40} = 0.75 \end{aligned}$$

The nurse or doctor putting the insulin in the syringe should draw 0.75 cc of the fluid.

SECTION 6.2 EXERCISES

CONCEPT/WRITING EXERCISES

1. Define and give an example of *term.*
2. Define and give an example of *like terms.*
3. Define and give an example of a *numerical coefficient.*
4. Explain how to simplify an expression. Give an example.
5. Define and give an example of a *linear equation.*
6. State the addition property of equality. Give an example.
7. State the subtraction property of equality. Give an example.
8. State the multiplication property of equality. Give an example.
9. State the division property of equality. Give an example.
10. Define *algorithm.*
11. Define and give an example of a *ratio.*
12. Define and give an example of a *proportion.*
13. Are $3x$ and $\frac{1}{2}x$ like terms? Explain.
14. Are $4x$ and $4y$ like terms? Explain.

PRACTICE THE SKILLS

In Exercises 15–38, combine like terms.

15. $3x + 8x$ **16.** $-9x - 6x$

17. $8x + 2x - 11$ **18.** $-8x + 2x + 15$

19. $7x + 3y - 4x + 8y$ **20.** $x - 4x + 3$

21. $-3x + 2 - 5x$ **22.** $-3x + 4x - 2 + 5$

23. $2 - 3x - 2x + 1$ **24.** $-0.2x + 1.7x - 4$

25. $3.7x - 5.8 + 2.6x$ **26.** $\frac{1}{2}x + \frac{3}{4}x + 5$

27. $\frac{1}{3}x - \frac{1}{4}x - 2$

28. $6x + 2y + 8 - 4x - 9y$

29. $5x - 4y - 3y + 8x + 3$

30. $3(p + 2) - 4(p + 3)$

31. $2(s + 3) + 6(s - 4) + 1$

32. $6(r - 3) - 2(r + 5) + 10$

33. $0.2(x - 3) - 1.6(x + 2)$

34. $\frac{1}{4}(x + 3) + \frac{1}{2}x$

35. $\frac{3}{4}x + \frac{3}{5} - \frac{2}{5}x + \frac{1}{2}$

36. $n - \frac{3}{4} + \frac{5}{9}n - \frac{1}{6}$

37. $0.5(2.6x - 4) + 2.3(1.4x - 5)$

38. $\frac{2}{3}(3x + 9) - \frac{1}{4}(2x + 5)$

In Exercises 39–64, solve the equation.

39. $y + 3 = 8$ **40.** $3y - 4 = 11$

41. $16 = 2x + 6$ **42.** $4 = 7 - 3y$

43. $\frac{3}{x} = \frac{7}{8}$ **44.** $\frac{x - 1}{5} = \frac{x + 5}{15}$

45. $\frac{1}{2}x + \frac{1}{3} = \frac{2}{3}$ **46.** $\frac{1}{2}y + \frac{1}{3} = \frac{1}{4}$

47. $0.7x - 0.3 = 1.8$ **48.** $5x + 0.050 = -0.732$

49. $3t - 4 = 2t - 1$ **50.** $\frac{x}{3} + 2x = -\frac{2}{5}$

51. $\frac{x - 5}{4} = \frac{x - 9}{3}$ **52.** $2r + 8 = 5 + 3r$

53. $\frac{x}{15} = 2 + \frac{x}{5}$ **54.** $12x - 1.2 = 3x + 1.5$

55. $2(x + 3) - 4 = 2(x - 4)$

56. $6y + 3(4 + y) = 8$

57. $6(x + 1) = 4x + 2(x + 3)$

58. $\frac{x}{3} + 4 = \frac{2x}{5} - 6$

59. $\frac{1}{4}(x + 4) = \frac{2}{5}(x + 2)$ **60.** $\frac{2}{3}(x + 5) = \frac{1}{4}(x + 2)$

61. $3x + 2 - 6x = -x - 15 + 8 - 5x$

62. $6x + 8 - 22x = 28 + 14x - 10 + 12x$

63. $2(x - 3) + 2 = 2(2x - 6)$

64. $5.7x - 3.1(x + 5) = 7.3$

PROBLEM SOLVING

In Exercises 65 and 66, use the DeKalb County water rate of \$1.95 per 1000 gallons of water used.

65. *Water Bill* What is the water bill if a resident uses 35,300 gal?

66. *Limiting the Cost* How many gallons of water can the customer use if the water bill is not to exceed \$40.68?

67. *Property Tax* In Northhampton County, the property tax rate is \$8.025 per \$1000 of assessed value. If a house and lot have been assessed at \$132,600, determine the amount of tax the owner will have to pay.

68. *Fertilizer Coverage* A 30 lb bag of fertilizer will cover an area of 2500 ft^2.
   a) How many pounds are needed to cover an area of 28,000 ft^2?
   b) How many bags of fertilizer must be purchased to cover an area of 28,000 ft^2?

69. *Toy Making* A machine manufactures 700 toys every 3 hours.
   a) How many toys does it manufacture in 60 hours?
   b) How many hours does it take to manufacture 2800 toys?

70. *Painting a House* A gallon of paint covers 825 ft^2. How much paint is needed to cover a house with a surface area of 5775 ft^2?

71. *Speed Limit* When Jacob Abbott crossed over from Niagara Falls, New York, to Niagara Falls, Canada, he saw a sign that said 50 miles per hour (mph) is equal to 80 kilometers per hour (kph).

   a) How many kilometers per hour are equal to 1 mph?
   b) On a stretch of the Queen Elizabeth Way, the speed limit is 90 kph. What is the speed limit in miles per hour?

72. *The Proper Dosage* A doctor asks a nurse to give a patient 250 milligrams (mg) of the drug Simethicone. The drug is available only in a solution whose concentration is 40 mg Simethicone per 0.6 millimeter (mm) of solution. How many millimeters of solution should the nurse give the patient?

In Exercises 73 and 74, how much insulin (in cc) would be given for the following doses? (Refer to Example 15.)

73. 12 units of insulin from a vial marked U40

74. 35 units of insulin from a vial marked U40

75. a) In your own words, summarize the procedure to use to solve an equation.
   b) Solve the equation $2(x + 3) = 4x + 3 - 5x$ with the procedure you outlined in part (a).

76. a) What is an identity?
   b) When solving an equation, how will you know if the equation is an identity?

77. a) What is an inconsistent equation?
   b) When solving an equation, how will you know if the equation is inconsistent?

CHALLENGE PROBLEMS/GROUP ACTIVITIES

78. The pressure, P, in pounds per square inch (psi), exerted on an object x ft below the sea is given by the formula $P = 14.70 + 0.43x$. The 14.70 represents the weight in pounds of the column of air (from sea level to the top of the atmosphere) standing over a 1 in. by 1 in. square of seawater. The $0.43x$ represents the weight in pounds of a column of water 1 in. by 1 in. by x ft (see Fig. 6.1).

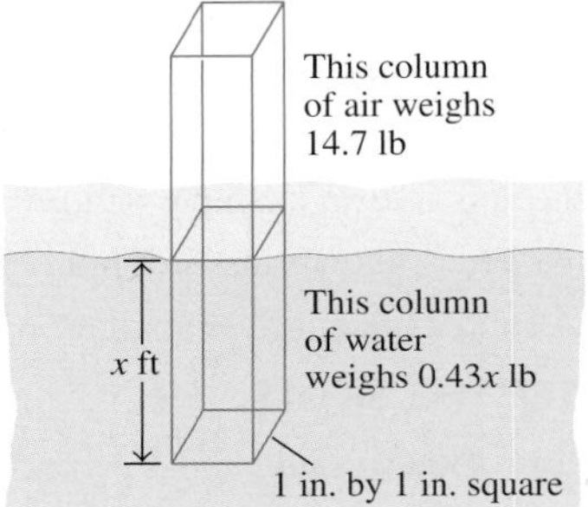

Figure 6.1

   a) A submarine is built to withstand a pressure of 148 psi. How deep can that submarine go?
   b) If the pressure gauge in the submarine registers a pressure of 128.65 psi, how deep is the submarine?

79. a) *Gender Ratios* If the ratio of males to females in a class is 2:3, what is the ratio of males to all the students in the class? Explain your answer?
   b) If the ratio of males to females in a class is $m:n$, what is the ratio of males to all the students in the class?

RESEARCH ACTIVITIES

80. Ratio and proportion are used in many different ways in everyday life. Submit two articles from newspapers or magazines in which ratio and/or proportion are used. Write a brief summary of each article explaining how ratio and/or proportion were used.

81. Write a report explaining how the Egyptians used equations. Include in your discussion the forms of the equations used.

6.3 FORMULAS

A **formula** is an equation that typically has a real-life application. To **evaluate a formula**, substitute the given values for their respective variables and then evaluate using the order of operations given in Section 6.1. Many of the formulas given in this section are discussed in greater detail in other parts of the book.

EXAMPLE 1 ***Simple Interest***

The simple interest formula,* interest = principal × rate × time, or $i = prt$, is used to find the interest you must pay on a simple interest loan when you borrow principal, p, at simple interest rate, r, in decimal form, for time, t. Chris Campbell borrows \$3000 at a simple interest rate of 9% for 3 years.

a) How much will Chris Campbell pay in interest at the end of 3 years?
b) What is the total amount he will repay the bank at the end of 3 years?

SOLUTION

a) Substitute the values of p, r, and t into the formula; then evaluate.

$$\begin{aligned} i &= prt \\ &= 3000(0.09)(3) \\ &= 810 \end{aligned}$$

Thus Chris must pay \$810 interest.

b) The total he must pay at the end of 3 years is the principal, \$3000, plus the \$810 interest, for a total of \$3810. ▲

EXAMPLE 2 ***Finding a Loan Rate***

Use the simple interest formula to find the rate on Josephine's \$8000 loan if the interest is \$1680 after 3 years.

SOLUTION We substitute the appropriate values into the simple interest formula and solve for the desired quantity, r.

$$\begin{aligned} i &= prt \\ 1680 &= (8000)r(3) \\ 1680 &= 24{,}000r \\ \frac{1680}{24{,}000} &= \frac{24{,}000r}{24{,}000} \\ 0.07 &= r \end{aligned}$$

Therefore, the simple interest rate is 7% per year. ▲

In Examples 1 and 2 we used the formula $i = prt$. In these examples we used a mathematical equation to represent real phenomena. When we represent real phenomena, such as finding simple interest, mathematically we say we have created a **mathematical model** or simply a **model** to represent the situation. A model may be a single formula, or equation, or a system of many equations. By using models we gain

---

*The simple interest formula is discussed in Section 10.2.

insight into real life situations, such as how much interest you will accumulate in your savings account. We will use mathematical models throughout this chapter and in Chapter 7. In some exercises in this and the next chapter, when you are asked to determine an equation to represent a real life situation, we will sometimes write the word "model" in the instructions.

Many formulas contain Greek letters, such as μ (mu), σ (sigma), Σ (capital sigma), δ (delta), ϵ (epsilon), π (pi), θ (theta), and λ (lambda). Example 3 makes use of Greek letters.

EXAMPLE 3 ***A Statistics Formula***

A formula used in the study of statistics to find the standard score (or z-score) is

$$z = \frac{\bar{x} - \mu}{\frac{\sigma}{\sqrt{n}}}$$

Find the value of z when $\bar{x}$ (read "x bar") $= 120$, $\mu = 100$, $\sigma = 16$, and $n = 4$.

SOLUTION

$$z = \frac{\bar{x} - \mu}{\frac{\sigma}{\sqrt{n}}} = \frac{120 - 100}{\frac{16}{\sqrt{4}}} = \frac{20}{\frac{16}{2}} = \frac{20}{8} = 2.5$$

Some formulas contain **subscripts**. Subscripts are numbers (or letters) placed below and to the right of variables. They are used to help clarify a formula. For example, if two different amounts are used in a problem, they may be symbolized as A and A_0, or A_1 and A_2. Subscripts are read using the word "sub"; for example, A_0 is read "A sub zero" and A_1 is read "A sub one."

EXPONENTIAL EQUATIONS

Many real-life problems, including population growth, growth of bacteria, and decay of radioactive substances, increase or decrease at a very rapid rate. For example, in Fig. 6.2, which shows global electronic business revenue, in billions of dollars, from 1996 through 1998 and projected to 2003, the graph is increasing rapidly. This is an example where the graph is increasing **exponentially**. The equation of a graph that increases or decreases exponentially is called an **exponential equation** (or **exponential formula**). An exponential equation is of the form $y = a^x$, $a > 0$, $a \neq 1$. We often use exponential equations to model real-life problems. In Section 6.10, we will discuss exponential equations (and exponential formulas) in more detail.

In an exponential formula, letters other than x and y may be used to represent the variables. The following equations are examples of exponential formulas: $y = 2^x$, $A = (\frac{1}{2})^x$, and $A = 2.3^t$. Note in the exponential formula the variable is the exponent of some positive constant that is not equal to 1. In many real-life applications, the variable t will be used to represent time. Problems involving exponential formulas can be evaluated much more easily if you use a calculator containing a $\boxed{y^x}$ or $\boxed{x^y}$ key.

The following formula, referred to as the **exponential growth** or **decay formula**, is used to solve many real-life problems.

$$P = P_0 a^{kt}, \qquad a > 0, \quad a \neq 1$$

DID YOU KNOW

Black Monday

When the world was a simpler place, commercial transactions could be easily computed. In recent decades, however, the capabilities of the computer have speeded up the rate of international commerce to a pace never before imagined. Computers are used to monitor the stock markets electronically and can be programmed to determine algebraically whether the dealer should buy or sell stocks. The stock market crash of October 19, 1987, in which the New York Stock Exchange plunged 508 points, has been partially blamed on the large volume of computer-triggered selling.

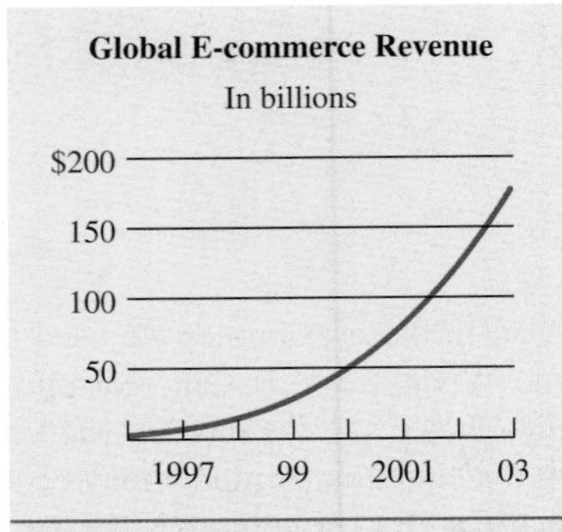

Figure 6.2

In the formula, P_0 represents the original amount present, P represents the amount present after t years, and a and k are constants.

When $k > 0$, P increases as t increases and we have exponential growth. When $k < 0$, P decreases as t increases and we have exponential decay.

EXAMPLE 4 ***Using an Exponential Decay Formula***

Carbon dating is used by scientists to find the age of fossils, bones, and other items. The formula used in carbon dating is

$$P = P_0 2^{-t/5600}$$

where P_0 represents the original amount of carbon 14 (C_{14}) present and P represents the amount of C_{14} present after t years. If 10 mg of C_{14} is present in an animal bone recently excavated, how many milligrams will be present in 3000 years?

SOLUTION Substituting the values in the formula gives

$$P = P_0 2^{-t/5600}$$
$$P = 10(2)^{-3000/5600}$$
$$P \approx 10(2)^{-0.54} \qquad \text{(Recall that } \approx \text{ means "is approximately equal to")}$$
$$P \approx 10(0.69)$$
$$P \approx 6.9 \ mg$$

Thus, in 3000 years, approximately 6.9 mg of the original 10 mg of C_{14} will remain. ▲

In Example 4, we used a calculator to evaluate $(2)^{-3000/5600}$. The steps used to find this quantity on a calculator with a $\boxed{y^x}$ key are

$$2 \ \boxed{y^x} \ \boxed{(} \ 3000 \ \boxed{+/-} \ \boxed{\div} \ 5600 \ \boxed{)} \ \boxed{=} \ 0.689817$$

To evaluate $10(2)^{-3000/5600}$ on a scientific calculator, we can press the following keys.

$$10 \ \boxed{\times} \ 2 \ \boxed{y^x} \ \boxed{(} \ 3000 \ \boxed{+/-} \ \boxed{\div} \ 5600 \ \boxed{)} \ \boxed{=} \ 6.89817$$

When the a in the formula $P = P_0 a^{kt}$ is replaced with the very special letter e, we get the **natural exponential formula**

$$P = P_0 e^{kt}$$

The letter e represents an irrational number whose value is approximately 2.7183. The number e plays an important role in mathematics and is used in finding the solution to many application problems.

EXAMPLE 5 ***Using an Exponential Growth Formula***

Banks often credit compound interest continuously. When that is done, the principal amount in the account, P, at any time t can be calculated by the natural exponential formula $P = P_0 e^{kt}$, where P_0 is the initial principal invested, k is the interest rate in decimal form, and t is the time.

Suppose \$10,000 is invested in a savings account at a 4% interest rate compounded continuously. What will be the balance (or principal) in the account in 5 years?

SOLUTION

$$\begin{aligned} P &= P_0 e^{kt} \\ &= 10{,}000e^{0.04(5)} \\ &= 10{,}000e^{(0.20)} \\ &\approx 10{,}000(1.2214) \\ &\approx 12{,}214 \end{aligned}$$

Thus, after 5 years, the account's value will have grown from \$10,000 to \$12,214, an increase of \$2214. ▲

To evaluate $e^{(0.04)5}$ on a calculator (in Example 5), press*

(.04 × 5) inv ln 1.221402758.

Inverse key — Natural logarithm key

To evaluate $10{,}000e^{(0.04)5}$ on a calculator, press

10,000 × (0.04 × 5) inv ln = 12214.02758.

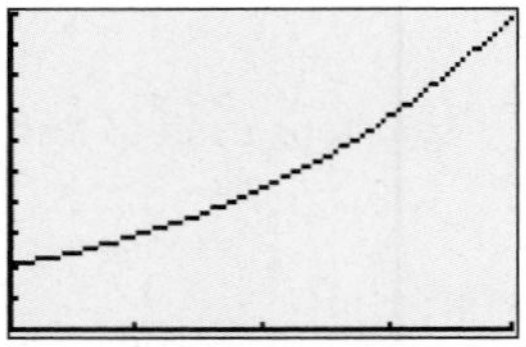

$P = 10{,}000e^{0.04t}$

Figure 6.3

Graphing calculators are a tool that can be used to graph equations. Figure 6.3 shows the graph of $P = 10{,}000e^{0.04t}$ as it appears on the screen (or window) of a Texas Instrument TI-83 graphing calculator. To obtain this screen, the domain (or the x-values) and range (or the y-values) of the window need to be set to selected values. We will speak a little more about graphing calculators shortly. In Section 6.10, we will explain how to graph exponential equations by plotting points.

EXAMPLE 6 *World Population Growth*

The world's population in 1999 was estimated to be 6 billion people. The world's population is continuing to grow exponentially at the rate of about 1.33% per year. The world's expected population, in billions, in t years is given by the formula $P = 6e^{0.0133t}$. Find the expected world population in the year 2099, 100 years after 1999.

SOLUTION

$$\begin{aligned} P &= 6e^{0.0133t} \\ &= 6e^{0.0133(100)} && \text{Substitute 100 for } t. \\ &= 6e^{1.33} \\ &\approx 6(3.78) \\ &\approx 22.68 \end{aligned}$$

Thus, in the year 2099, the world's population is expected to be about 22.68 billion people, or more than three times as great as it was in 1999. ▲

*Keys to press may vary on some calculators.

SOLVING FOR A VARIABLE IN A FORMULA OR EQUATION

Often in mathematics and science courses, you are given a formula or an equation expressed in terms of one variable and asked to express it in terms of a different variable. For example, you may be given the formula $P = i^2r$ and asked to solve the formula for r. To do so, treat each of the variables, except the one you are solving for, as if it were a constant. Then solve for the variable desired, using the properties previously discussed. Examples 7 through 9 show how to do this task.

When graphing equations in Section 6.7, you will sometimes have to solve the equation for the variable y as is done in Example 7.

DID YOU KNOW

~~L~~ = mc²
E = mc²

Many of you recognize the formula $E = mc^2$ used by Albert Einstein in 1912 to describe his groundbreaking theory of relativity. In the formula E is energy, m is mass, and c is the speed of light. In his theory Einstein hypothesized that time was not absolute and that mass and energy were related.

Einstein's 72 page handwritten manuscript, that went on display in 1999 at Jerusalem's Israel Museum, shows that Einstein toyed with using an L rather than an E to represent energy. He wrote $L = mc^2$, then scratched out the L and replaced it with an E, to produce the equation $E = mc^2$. Yisrail Galili at Jerusalem's Hebrew University said the L might have been a Lambda, a Greek letter sometimes used in functions related to energy. He said it might also be related to Joseph Lagrange, an 18th century Italian mathematician.

EXAMPLE 7 ***Solving an Equation Containing More Than One Variable***

Solve the equation $3x + 4y - 8 = 0$ for y.

SOLUTION We need to isolate the term containing the variable y. Begin by moving the constant, -8, and the term $3x$ to the right side of the equation.

$$\begin{aligned}
3x + 4y - 8 &= 0 \\
3x + 4y - 8 + 8 &= 0 + 8 && \text{Addition property} \\
3x + 4y &= 8 \\
-3x + 3x + 4y &= -3x + 8 && \text{Subtraction property} \\
4y &= -3x + 8 \\
\frac{4y}{4} &= \frac{-3x + 8}{4} && \text{Division property} \\
y &= \frac{-3x + 8}{4} \\
y &= -\frac{3x}{4} + \frac{8}{4} \\
y &= -\frac{3}{4}x + 2
\end{aligned}$$

▲

Note that once you have found $y = (-3x + 8)/4$, you have solved the equation for y. The solution can also be expressed in the form $y = -\frac{3}{4}x + 2$. This form of the equation is convenient for graphing equations, as will be explained in Section 6.7. Example 7 can also be solved by moving the y term to the right side of the equal sign. Do so now and note that you obtain the same answer.

EXAMPLE 8 ***Solving for a Variable in a Formula***

An important formula used in statistics is

$$z = \frac{x - \mu}{\sigma}.$$

Solve this formula for x.

SOLUTION To isolate the term x, use the general procedure for solving linear equations given in Section 6.2. Treat each letter, except x, as if it were a constant.

$$z = \frac{x - \mu}{\sigma}$$

$$z \cdot \sigma = \frac{x - \mu}{\cancel{\sigma}} \cdot \cancel{\sigma} \quad \text{Multiplication property}$$

$$z\sigma = x - \mu$$

$$z\sigma + \mu = x - \mu + \mu \quad \text{Add } \mu \text{ to both sides of the equation.}$$

$$z\sigma + \mu = x$$

$$\text{or} \quad x = z\sigma + \mu$$

▲

EXAMPLE 9 ***The Tax Free Yield Formula***

A formula that may be important to you now or sometime in the future is the tax-free yield formula, $T_f = T_a(1 - F)$. This formula can be used to convert a taxable yield, T_a, into its equivalent tax-free yield, T_f, where F is the federal income tax bracket of the individual. A taxable yield is an interest rate for which income tax is paid on the interest made. A tax-free yield is an interest rate for which income tax does not have to be paid on the interest made.

a) For someone in a 28% tax bracket, find the equivalent tax-free yield of an 8% taxable investment.
b) Solve this formula for T_a. That is, write a formula for taxable yield in terms of tax-free yield.

SOLUTION

a) $T_f = T_a(1 - F)$
$= 0.08(1 - 0.28) = 0.08(0.72) = 0.0576, \quad \text{or} \quad 5.76\%$

PROFILE IN MATHEMATICS

Sophie Germain

Because she was a woman, Sophie Germain (1776–1831) was denied admission to the École Polytechnic, the French academy of mathematics and science. Not to be stopped, she obtained lecture notes from courses in which she had an interest, including one taught by Joseph-Louis Lagrange. Under the pen name M. LeBlanc, she submitted a paper on analysis to Lagrange, who was so impressed with the report that he wanted to meet the author and personally congratulate "him." When he found out that the author was a woman, he became a great help and encouragement to her. Lagrange introduced Germain to many of the French scientists of the time.

In 1801, Germain wrote the great German mathematician Carl Friedrich Gauss to discuss Fermat's equation, $x^n + y^n = z^n$. He commended her for showing "the noblest courage, quite extraordinary talents and a superior genius." Germain's interests included work in number theory and mathematical physics. She would have received an honorary doctorate from the University of Göttingen, based on Gauss's recommendation, but died before the honorary doctorate could be awarded.

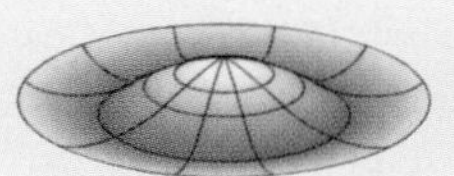
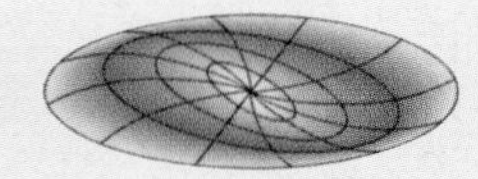

Germain was the first person to devise a formula describing elastic motion. The study of the equations for the elasticity of different materials aided the development of acoustical diaphragms in loudspeakers and telephones.

Thus, a taxable investment of 8% is equivalent to a tax-free investment of 5.76% for a person in a 28% income tax bracket.

b) $T_f = T_a(1 - F)$

$$\frac{T_f}{1 - F} = \frac{T_a\cancel{(1 - F)}}{\cancel{1 - F}}$$ Divide both sides of the equation by $1 - F$.

$$\frac{T_f}{1 - F} = T_a, \quad \text{or} \quad T_a = \frac{T_f}{1 - F}$$

SECTION 6.3 EXERCISES

CONCEPT/WRITING EXERCISES

1. What is a formula?
2. Explain how to evaluate a formula.
3. What are subscripts?
4. What is the simple interest formula?
5. What is an exponential equation?
6. In an exponential equation of the form $y = a^x$, what are the restrictions on a?

PRACTICE THE SKILLS

In Exercises 7–38, use the formula to find the value of the indicated variable for the values given. Use a calculator when one is needed.

7. $A = bh$; find A when $b = 15$ and $h = 4$ (geometry).
8. $P = a + b + c$; find P when $a = 25$, $b = 53$ and $c = 32$ (geometry).
9. $P = 2l + 2w$; find P when $l = 12$ and $w = 16$ (geometry).
10. $F = ma$; find m when $F = 40$ and $a = 5$ (physics).
11. $E = mc^2$; find m when $E = 400$ and $c = 4$ (physics).
12. $p = i^2r$; find r when $p = 62{,}500$ and $i = 5$ (electronics).
13. $m = \dfrac{a + b}{2}$; find b when $m = 55$ and $a = 27$ (statistics).
14. $z = \dfrac{x - \mu}{\sigma}$; find z when $x = 100$, $\mu = 110$, and $\sigma = 5$ (statistics).
15. $z = \dfrac{x - \mu}{\sigma}$; find μ when $z = 2.5$, $x = 42.1$, and $\sigma = 2$ (statistics).
16. $S = B + \frac{1}{2}Ps$; find P when $s = 10$, $S = 300$, and $B = 100$ (geometry).
17. $T = \dfrac{PV}{k}$; find P when $T = 80$, $V = 20$, and $k = 0.5$ (physics).
18. $m = \dfrac{a + b + c}{3}$; find a when $m = 70$, $b = 60$, and $c = 90$ (statistics).
19. $A = P(1 + rt)$; find P when $A = 3600$, $r = 0.04$ and $t = 5$ (economics).
20. $V = \frac{4}{3}\pi r^2 h$; find h when $r = 7$, $\pi = 3.14$, and $V = 1846.32$ (geometry).
21. $V = \frac{1}{2}at^2$; find a when $V = 576$ and $t = 12$ (physics).
22. $\text{F} = \frac{9}{5}\text{C} + 32$; find F when $\text{C} = 7$ (temperature conversion).
23. $\text{C} = \frac{5}{9}(\text{F} - 32)$; find C when $\text{F} = 77$ (temperature conversion).
24. $\text{K} = \dfrac{\text{F} - 32}{1.8} + 273.1$; find K when $\text{F} = 100$ (chemistry).
25. $z = \dfrac{\bar{x} - \mu}{\dfrac{\sigma}{\sqrt{n}}}$; find z when $\bar{x} = 66$, $\mu = 60$, $\sigma = 15$, and $n = 25$ (statistics).
26. $m = \dfrac{y_2 - y_1}{x_2 - x_1}$; find m when $y_2 = 8$, $y_1 = -4$, $x_2 = -3$, and $x_1 = -5$ (mathematics).
27. $S = R - rR$; find R when $S = 186$ and $r = 0.07$ (for determining sale price when an item is discounted).
28. $S = C + rC$; find C when $S = 115$ and $r = 0.15$ (for determining selling price when an item is marked up).
29. $E = a_1p_1 + a_2p_2 + a_3p_3$; find E when $a_1 = 5$, $p_1 = 0.2$, $a_2 = 7$, $p_2 = 0.6$, $a_3 = 10$, and $p_3 = 0.2$ (probability).
30. $x = \dfrac{-b + \sqrt{b^2 - 4ac}}{2a}$; find x when $a = 2$, $b = -5$, and $c = -12$ (mathematics).

31. $x = \dfrac{-b - \sqrt{b^2 - 4ac}}{2a}$; find x when $a = 2$, $b = -5$, and $c = -12$ (mathematics).

32. $R = O + (V - D)r$; find O when $R = 670$, $V = 100$, $D = 10$, and $r = 4$ (economics).

33. $P = \dfrac{f}{1 + i}$; find f when $i = 0.08$ and $P = 3000$ (investment banking).

34. $V = \sqrt{V_x^2 + V_y^2}$; find V when $V_x = 3$ and $V_y = 4$ (physics).

35. $F = \dfrac{Gm_1m_2}{r^2}$; find G when $F = 625$, $m_1 = 100$, $m_2 = 200$, and $r = 4$ (physics).

36. $P = \dfrac{nRT}{V}$; find V if $P = 12$, $n = 10$, $R = 60$, and $T = 8$ (chemistry).

37. $S_n = \dfrac{a_1(1 - r^n)}{1 - r}$; find S_n when $a_1 = 8$, $r = \frac{1}{2}$, and $n = 3$ (mathematics).

38. $A = P\left(1 + \dfrac{r}{n}\right)^{nt}$; find A when $P = 100$, $r = 6\%$, $n = 1$, and $t = 3$ (banking).

In Exercises 39–48, solve the equation for y.

39. $7x - 6y = 15$
40. $4x - 3y = 21$
41. $4x + 7y = 14$
42. $-2x + 4y = 9$
43. $2x - 3y + 6 = 0$
44. $3x + 4y = 0$
45. $-3x - 2y = 20$
46. $3x + 5y - z = 11$
47. $9x + 4z = 7 + 8y$
48. $2x - 3y + 5z = 0$

In Exercises 49–68, solve for the variable indicated.

49. $A = bh$ for b
50. $E = IR$ for R
51. $p = a + b + c$ for a
52. $p = a + b + s_1 + s_2$ for s_1
53. $V = lwh$ for w
54. $p = irt$ for t
55. $C = 2\pi r$ for r
56. $PV = KT$ for T
57. $y = mx + b$ for b
58. $y = mx + b$ for m
59. $P = 2l + 2w$ for w
60. $A = \dfrac{d_1d_2}{2}$ for d_2
61. $A = \dfrac{a + b + c}{3}$ for c
62. $A = \frac{1}{2}bh$ for b
63. $P = \dfrac{KT}{V}$ for T
64. $V = \frac{1}{3}lwh$ for w
65. $F = \frac{9}{5}C + 32$ for C
66. $C = \frac{5}{9}(F - 32)$ for F
67. $S = \pi r^2 + \pi rs$ for s
68. $a_n = a_1 + (n - 1)d$ for n

PROBLEM SOLVING

69. *Savings Account* Janet Jackson deposited \$3500 in a savings account that paid 3% simple interest per year. Determine
- **a)** how much interest was added to her account at the end of one year.
- **b)** the balance in her account at the end of one year.

70. *Interest on a Loan* Jeff Hubbard borrowed \$800 from his brother for 2 years. At the end of 2 years, he repaid the \$800 plus \$128 in interest. What simple interest rate did he pay?

71. *Volume in a Soup Can* Determine the volume of a cylindrical soup can if its diameter is 2.5 in. and its height is 3.75 in. (The formula for the volume of a cylinder is $V = \pi r^2h$. Use your pi key, $\boxed{\pi}$, on your calculator, or 3.14 for π if your calculator does not have a $\boxed{\pi}$ key.)

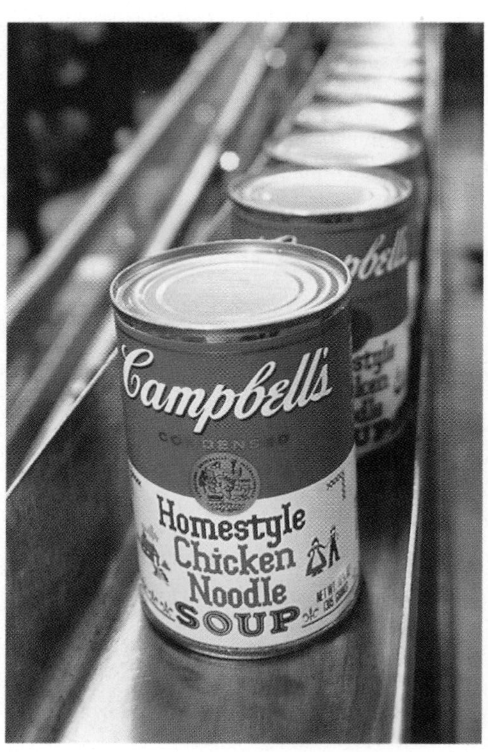

72. *Compound Interest* The formula $A = P(1 + r)^n$ is the **compound interest formula**. When interest is compounded periodically (yearly, monthly, daily), the formula can be used to find the amount, A, in the account after n periods. In the formula, P is the principal, r is the interest rate per compounding period, and n is the number of compounding periods. Suppose Natasha Neuko invests \$10,000 at 8% annual interest compounded quarterly for 6 years. Then $P = 10{,}000$, $r = 8\%/4 = 2\%$, and $n = 6 \cdot 4 = 24$. Find the amount in the account after 6 years.

73. *Carbon Dating* If 20 mg of carbon 14 are originally present in an animal bone, how much will remain at the end of 500 years? (See Example 4.)

74. *Adjusting for Inflation* If P is the price of an item today, the price of the same item n years from today, P_n, is $P_n = P(1 + r)^n$, where r is the constant rate of inflation. Determine the price of a movie ticket 10 years from today if the price today is \$8.00 and the annual rate of inflation is constant at 3%.

75. *Value of New York City* Assume the value of the island of Manhattan has grown at an exponential rate of 8% per year since 1626 when Peter Minuit of the Dutch West India Company purchased the island for \$24. The value of the island, V, at any time, t, in years, can be found by the formula $V = 24e^{0.08t}$. What is the value of the island in 2000, 374 years after Minuit purchased it?

76. *Radioactive Decay* Strontium 90 is a radioactive isotope that decays exponentially at a rate of 2.8% per year. The amount of strontium 90, S, remaining after t years can be found by the formula $S = S_0e^{-0.028t}$, where S_0 is the original amount present. If there are originally 1000 g of strontium 90, find the amount of strontium 90 remaining after 30 years.

CHALLENGE PROBLEMS/GROUP ACTIVITIES

77. Determine the volume of the block shown in Fig. 6.4, excluding the hole.

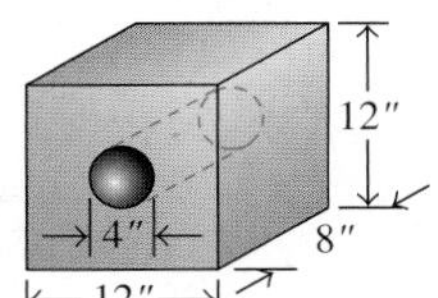

Figure 6.4

78. Joseph Chebet, from Kenya, won the 1999 Boston Marathon, a 26.2 mi course, in a time of 2 hr 9 min and 52 sec. Use the formula, distance = rate · time, to determine Chebet's average speed (or rate), in miles per hour, for the course.

6.4 APPLICATIONS OF LINEAR EQUATIONS IN ONE VARIABLE

> **DID YOU KNOW**
>
> **Survival of the Fittest**
>
> Symbols come and symbols go; the ones that find the greatest acceptance are the ones that survive. The Egyptians used pictorial symbols: a pair of legs walking forward for addition or backward for subtraction. Robert Recorde (1510–1558) used two parallel lines to represent "equals" because "no 2 thynges can be moore equalle." Some symbols evolved from abbreviations, such as the "+" sign, which comes from the Latin *et* meaning "and." The evolution of others is less clear. The invention of the printing press in the fifteenth century led to a greater standardization of symbols already in use.

One of the main reasons for studying algebra is that it can be used to solve everyday problems. In this section, we will do two things: (1) show how to translate a written problem into a mathematical equation and (2) show how linear equations can be used in solving everyday problems. We begin by illustrating how English phrases can be written as mathematical expressions. When writing a mathematical expression, we may use any letter to represent the variable. In the following illustrations, we use the letter x.

Phrase	Mathematical expression
Six more than a number	$x + 6$
A number increased by 3	$x + 3$
Four less than a number	$x - 4$
A number decreased by 9	$x - 9$
Twice a number	$2x$
Four times a number	$4x$

Sometimes the phrase that must be converted to a mathematical expression involves more than one operation.

Phrase	Mathematical expression
Four less than 3 times a number	$3x - 4$
Ten more than twice a number	$2x + 10$
The sum of 5 times a number and 3	$5x + 3$
Eight times a number decreased by 7	$8x - 7$

The word *is* often represents the equal sign.

Phrase	**Mathematical equation**
Six more than a number is 10.	$x + 6 = 10$
Five less than a number is 20.	$x - 5 = 20$
Twice a number decreased by 6 is 12.	$2x - 6 = 12$
A number decreased by 13 is 6 times the number.	$x - 13 = 6x$

The following is a general procedure for solving word problems.

To Solve a Word Problem

1. Read the problem carefully at least twice to be sure that you understand it.
2. If possible, draw a sketch to help visualize the problem.
3. Determine which quantity you are being asked to find. Choose a letter to represent this unknown quantity. Write down exactly what this letter represents.
4. Write the word problem as an equation.
5. Solve the equation for the unknown quantity.
6. Answer the question or questions asked.
7. Check the solution.

This general procedure for solving word problems is illustrated in Examples 1 through 4. In these examples, the equations we obtain are mathematical models of the given situations.

EXAMPLE 1 *How Far Can You Travel?*

Happy Henry's Truck Rental Agency charges \$250 per week plus \$0.21 per mile for a small truck. How far can you travel, to the nearest mile, on a maximum budget of \$500 (disregard the cost of gasoline)?

SOLUTION In this problem, the unknown quantity is the number of miles you can travel. Let's select m to represent the number of miles you can travel. Then we construct an equation using the given information that will allow us to solve for m.

Let m = number of miles you can travel

Then $\$0.21m$ = charge for traveling m miles at 21 cents per mile

$$\begin{array}{ccccc} \text{Rental fee} & + & \text{mileage charge} & = & \text{total amount spent} \\ \$250 & + & \$0.21m & = & \$500 \end{array}$$

Now solve the equation.

$$\begin{aligned} 250 + 0.21m &= 500 \\ 250 - 250 + 0.21m &= 500 - 250 \\ 0.21m &= 250 \\ \frac{0.21m}{0.21} &= \frac{250}{0.21} \\ m &= 1190.48 \end{aligned}$$

Therefore, you can travel 1190 miles. (If you go 1191 miles, you will have to spend more than \$500.)

Check: The check is made with the information given in the original problem.

$$\begin{aligned}\text{Total amount spent} &= \text{rental fee} + \text{mileage charge}\\ &= 250 + 0.21(1190)\\ &= 250 + 249.90\\ &= 499.90\end{aligned}$$

The check shows that the solution is correct. ▲

DID YOU KNOW

Aha!

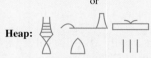

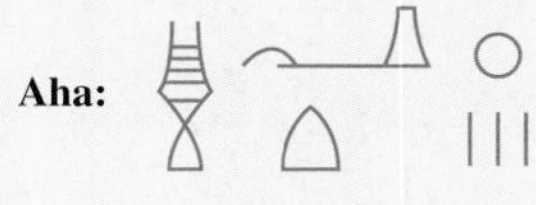

The Egyptians as far back as 1650 B.C. had a knowledge of linear equations. They used the words *aha* or *heap* in place of the variable. Problems involving linear equations can be found in the Rhind Papyrus (Chapter 4).

EXAMPLE 2

Forty hours of overtime must be split among three workers. One worker will be assigned twice the number of hours as each of the other two. How many hours of overtime will be assigned to each worker?

SOLUTION Two workers receive the same amount of overtime, and the third worker receives twice that amount.

Let x = number of hours of overtime for the first worker
x = number of hours of overtime for the second worker
$2x$ = number of hours of overtime for the third worker

Then

$$\begin{aligned}x + x + 2x &= \text{total amount of overtime}\\ x + x + 2x &= 40\\ 4x &= 40\\ x &= 10\end{aligned}$$

Thus, two workers are assigned 10 hours, and the third worker is assigned 2(10), or 20, hours of overtime. A check in the original problem will verify that this answer is correct. ▲

EXAMPLE 3 *Dimensions of a Fenced in Area*

Robert Koch wants to fence in a rectangular region in his backyard for his poodle. He only has 56 ft of fencing to use for the perimeter of the region. What should the dimensions of the region be if he wants the length to be 4 ft greater than the width?

SOLUTION The formula for finding the perimeter of a rectangle is $P = 2l + 2w$, where P is the perimeter, l is the length, and w is the width. A diagram, such as the one shown below, is often helpful in solving problems of this type.

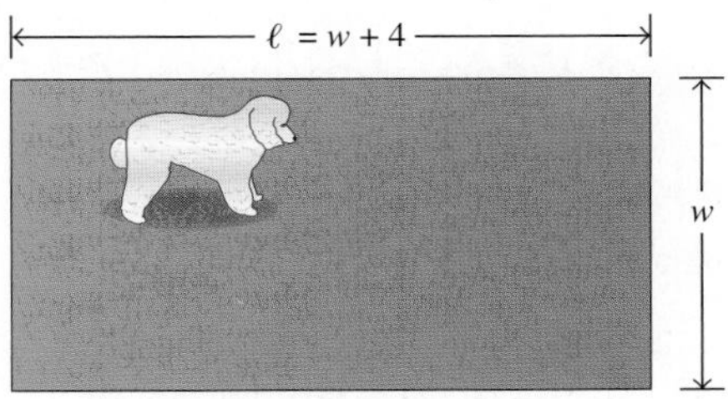

Let w equal the width of the region. The length is 4 ft more than the width, so $l = w + 4$. The total distance around the region P, is 56 ft.

Substitute the known quantities in the formula.

$$\begin{aligned} P &= 2w + 2l \\ 56 &= 2w + 2(w + 4) \\ 56 &= 2w + 2w + 8 \\ 56 &= 4w + 8 \\ 48 &= 4w \\ 12 &= w \end{aligned}$$

The width of the region is 12 ft, and the length of the region is 12 + 4 or 16 ft. ▲

In shopping and other daily activities, we are occasionally asked to solve problems using percents. The word *percent* means "per hundred." Thus, for example, 7% means 7 per hundred, or $\frac{7}{100}$. When $\frac{7}{100}$ is converted to a decimal number, we obtain 0.07. Thus, 7% = 0.07.

Let's look at one example involving percent. (See Section 11.1 for a more detailed discussion of percent.)

EXAMPLE 4 *Price, Including Sales Tax*

Mr. Ramesh is planning to open a hot dog stand in New York City. What should be the price of the hot dog before tax if the total cost of the hot dog, including a 7% sales tax, is to be $2.50?

SOLUTION We are asked to find the cost of the hot dog before sales tax.

Let x = the cost of the hot dog before sales tax.

Then $0.07x$ = 7% of the cost of the hot dog (the sales tax).

$$\begin{aligned} \text{Cost of hot dog before tax} + \text{tax on hot dog} &= 2.50 \\ x + 0.07x &= 2.50 \\ 1.07x &= 2.50 \\ \frac{1.07x}{1.07} &= \frac{2.50}{1.07} \\ x &= \frac{2.50}{1.07} \\ &= 2.336, \qquad \text{or } 2.34 \end{aligned}$$

Thus, the cost of the hot dog before tax is $2.34 to the nearest cent. ▲

SECTION 6.4 EXERCISES

CONCEPT/WRITING EXERCISES

1. What is the difference between a mathematical expression and an equation?

2. Give an example of a mathematical expression and an example of a mathematical equation.

PRACTICE THE SKILLS

In Exercises 3–14, write the phrase as a mathematical expression.

3. 9 decreased by 6 times x
4. 3 times x increased by 7
5. 5 more than 6 times r

6. 10 times s decreased by 13
7. 15 decreased by twice r
8. 2 times m increased by 9
9. 6 more than x
10. 8 increased by 5 times x
11. The sum of 3 and n, divided by 8
12. 15 decreased by t, divided by 4
13. 6 less than the product of 5 times y, increased by 3
14. The quotient of 8 and y, decreased by 3 times x

In Exercises 15–26, write an equation and solve.

15. The sum of a number and 7 is 19.
16. A number decreased by 9 is 5.
17. A number decreased by 10 is 25.
18. A number multiplied by 7 is 42.
19. Twelve increased by 5 times a number is 47.
20. Four times a number decreased by 10 is 42.
21. Sixteen more than 8 times a number is 88.
22. Six more than five times a number is 7 times the number decreased by 18.
23. A number increased by 11 is 1 more than 3 times the number.
24. A number divided by 3 is 4 less than the number.
25. Three more than a number is 5 times the sum of the number and 7.
26. The product of 3 and a number, decreased by 4, is 4 times the number.

PROBLEM SOLVING

In Exercises 27–46, set up an equation that can be used to solve the problem. Solve the equation and find the desired value(s).

27. MODELING - *Investing* Samson receives an inheritance of \$9000. If he wants to invest twice as much in bonds as in mutual funds, how much should he invest in bonds?
28. MODELING - *New Clothing* Miguel purchases two new pairs of pants at The Gap for \$60. If one pair was \$10 more than the other, how much was the more expensive pair?
29. MODELING - *Population Growth* The village of Lexington, which has a population of 6200, is growing by 500 each year. In how many years will the population reach 12,200?
30. MODELING - *Shopping* Budget Warehouse has a plan whereby for a yearly fee of \$80 you save 8% of the price of all items purchased in the store. What is the total Vito will need to spend during the year for his savings to equal the yearly fee?
31. MODELING - *Pet Supplies* PetSmart has a sale offering 10% off of all pet supplies. If Mathew Miller spent \$15.72 on pet supplies before tax, what was the price of the pet supplies he purchased before the discount?

32. MODELING - *Copying* Ronnie pays 8¢ to make a copy of a page at a copy shop. She is considering purchasing a photocopy machine that is on sale for \$250, including tax. How many copies would Ronnie have to make in the copy shop for her cost to equal the purchase price of the photocopy machine she is considering buying?
33. MODELING - *Number of CD's* Samantha and Josie receive 12 free compact discs by joining a compact disc club. How many will each receive if Josie is to have three times as many as Samantha?
34. MODELING - *Recycling* A town recycles 28 tons of newspaper and cardboard each week. The amount of newspaper is three times the amount of cardboard. Determine the number of tons of newspaper and the number of tons of cardboard recycled each week.
35. MODELING - *Truck Rental* Yekcim Esuom budgeted \$100 for renting a truck. How far can he travel in one day if the charges are \$60 a day plus 20¢ per mile?
36. MODELING - *Dimensions of a Deck* Jim is building a rectangular deck and wants the length to be 3 ft greater than the width. What will be the dimensions of the deck if the perimeter is to be 54 ft?
37. MODELING - *Floor Area* The total floor space in three barns is 45,000 ft^2. The two smaller ones have the same area, and the largest one is three times the area of the smaller ones.
    a) Determine the floor space for each barn.
    b) Can merchandise that takes up 8500 ft^2 of floor space fit into either of the smaller barns?
38. MODELING - *Hourly Rate* Mr. Thoms worked a 55 hr week last week. He is paid $1\frac{1}{2}$ times his regular hourly rate for all hours over a 40 hr week. His pay last week was \$500. What is his hourly rate?
39. MODELING - *Moon Weight* A person's weight on the moon is $\frac{1}{6}$ of that on Earth. Assume that the sum of a person's weight on Earth and the moon is 203 lb. What does the person weigh on Earth?

40. MODELING - *Car Purchase* The Gilberts purchased a car. If the total cost, including a 5% sales tax, was $14,512, find the cost of the car before tax.

41. MODELING - *Enclosing Two Pens* Chuck Salvador has 140 ft of fencing in which he wants to build two connecting, adjacent square pens (see the figure). What will be the dimensions if the length of the entire enclosed region is to be twice the width?

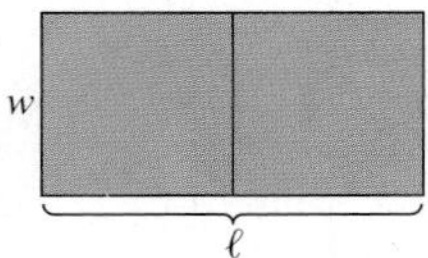

42. MODELING - *Dimensions of a Bookcase* A bookcase with three shelves is to be built by a woodworking student. If the height of the bookcase is to be 2 ft longer than the length of a shelf, and the total amount of wood to be used is 32 ft, find the dimensions of the bookcase.

43. MODELING - *Laundry Cost* The cost of doing the family laundry for a month at a local laundromat is $70. A new washer and dryer cost a total of $760. How many months would it take for the cost of doing the laundry at the laundromat to equal the cost of a new washer and dryer?

44. MODELING - *Health Club Cost* A health club is offering two new membership plans. Plan A costs $56 per month for unlimited use. Plan B costs $20 per month plus $3 for every visit. How many visits to the health club must Doug make for Plan A to result in the same cost as Plan B?

45. MODELING - *Airfare* Rachel has been told that with her half-off airfare coupon, her airfare from New York to San Diego will be $227.00. The $227.00 includes a 7% tax *on the regular fare.* On the way to the airport, Rachel realizes that she has lost her coupon. What will her regular fare be, including tax?

46. MODELING - *Truck Rentals* The cost of renting a small truck at the U-Haul rental agency is $35 per day plus 20¢ a mile. The cost of renting the same truck at the Ryder rental agency is $25 per day plus 32¢ a mile. How far would you have to drive in one day for the cost of renting from U-Haul to equal the cost of renting from Ryder?

CHALLENGE PROBLEMS/GROUP ACTIVITIES

47. *Income Tax* Some states allow a husband and wife to file individual tax returns (on a single form) even though they have filed a joint federal tax return. It is usually to the taxpayers' advantage to do so when both husband and wife work. The smallest amount of tax owed (or the largest refund) will occur when the husband's and wife's taxable incomes are the same.

Mr. McAdams's 2000 taxable income was $24,200, and Mrs. McAdams's taxable income for that year was $26,400. The McAdams's total tax deduction for the year was $3640. This deduction can be divided between Mr. and Mrs. McAdams any way they wish. How should the $3640 be divided between them to result in each individual's having the same taxable income and therefore the greatest tax refund?

48. Write each equation as a sentence. There are many correct answers.

a) $x + 3 = 13$ **b)** $3x + 5 = 8$
c) $3x - 8 = 7$

49. Show that the sum of any three consecutive integers is 3 less than 3 times the largest.

50. *Auto Insurance* A driver education course at the East Lake School of Driving costs $45 but saves those under 25 years of age 10% of their annual insurance premiums until they are 25. Dan has just turned 18, and his insurance costs $600.00 per year.

a) When will the amount saved from insurance equal the price of the course?
b) When Dan turns 25, how much will he have saved?

6.5 VARIATION

In Sections 6.3 and 6.4, we presented many applications of algebra. In this section, we introduce variation, which is an important tool in solving applied problems.

DIRECT VARIATION

Many scientific formulas are expressed as variations. A **variation** is an equation that relates one variable to one or more other variables through the operations of multiplication or division (or both operations). Essentially there are four types of variation problems: direct, inverse, joint, and combined variation.

In **direct variation** the two related variables increase together or decrease together; that is, as one increases so does the other, and as one decreases so does the other.

Consider a car traveling at 40 miles an hour. The car travels 40 miles in 1 hour, 80 miles in 2 hours, and 120 miles in 3 hours. Note that, as the time increases, the distance traveled increases, and, as the time decreases, the distance traveled decreases.

The formula used to calculate distance traveled is

$$\text{Distance} = \text{rate} \cdot \text{time}$$

Since the rate is a constant, 40 miles per hour, the formula can be written

$$d = 40t$$

We say that distance varies directly as time or that distance is directly proportional to time.

The preceding equation is an example of direct variation.

Direct Variation

If a variable y varies directly with a variable x, then

$$y = kx$$

where k is the **constant of proportionality** (or the variation constant).

Examples 1 through 4 illustrate direct variation.

EXAMPLE 1 ***Direct Variation in Geometry***

The circumference of a circle, C, is directly proportional to (or varies directly as) its radius, r. Write the equation for the circumference of a circle if the constant of proportionality, k, is 2π.

SOLUTION

$C = kr$ C varies directly as r.

$C = 2\pi r$ Constant of proportionality is 2π. ▲

EXAMPLE 2 ***Direct Variation in Electricity***

The resistance, R, of a wire varies directly as its length, L.

a) Write this variation as an equation.
b) Find the resistance (measured in ohms) of a 30 ft length of wire, assuming the constant of proportionality for the wire is 0.008.

SOLUTION

a) $R = kL$
b) $R = 0.008(30) = 0.24$

The resistance of the wire is 0.24 ohm. ▲

In certain variation problems, the constant of proportionality, k, may not be known. In such cases, we can often find it by substituting the given values in the variation formula and solving for k.

EXAMPLE 3 ***Finding the Constant of Proportionality***

Suppose w varies directly as the square of y. If w is 60 when y is 20, find the constant of proportionality.

SOLUTION Since w varies directly as the *square of* y, we begin with the formula $w = ky^2$. Since the constant of proportionality is not given, we must first find k using the given information. Substitute 60 for w and 20 for y.

$$\begin{aligned} w &= ky^2 \\ 60 &= k(20)^2 \\ 60 &= 400k \\ \frac{60}{400} &= \frac{400k}{400} \\ 0.15 &= k \end{aligned}$$

Thus, the constant of proportionality is 0.15. ▲

EXAMPLE 4 ***Using the Constant of Proportionality***

The length that a spring will stretch, S, varies directly with the force (or weight), F, attached to the spring. If a spring stretches 4.2 in. when 60 lb is attached, how far will it stretch when 30 lb is attached?

SOLUTION We begin with the formula $S = kF$. Since the constant of proportionality is not given, we must find k using the given information.

$$\begin{aligned} S &= kF \\ 4.2 &= k(60) \\ \frac{4.2}{60} &= k \\ 0.07 &= k \end{aligned}$$

We now use $k = 0.07$ to find S when $F = 30$.

$$\begin{aligned} S &= kF \\ S &= 0.07F \\ S &= 0.07(30) \\ S &= 2.1 \text{ in.} \end{aligned}$$

Thus, a spring will stretch 2.1 in. when a force of 30 lb is attached. ▲

INVERSE VARIATION

A second type of variation is **inverse variation**. When two quantities vary inversely, as one quantity increases, the other quantity decreases, and vice versa.

To explain inverse variation, we use the formula, distance = rate · time. If we solve for time, we get time = distance/rate. Assume the distance is fixed at 100 miles; then

$$\text{Time} = \frac{100}{\text{rate}}$$

At 100 miles per hour it would take 1 hour to cover this distance. At 50 miles an hour, it would take 2 hours. At 25 miles an hour, it would take 4 hours. Note that as the rate (or speed) decreases, the time increases and vice versa.

The preceding equation can be written

$$t = \frac{100}{r}$$

This equation is an example of an inverse variation. The time and rate are inversely proportional. The constant of proportionality is 100.

Inverse Variation

If a variable y varies inversely with a variable x, then

$$y = \frac{k}{x}$$

where k is the constant of proportionality.

Two quantities vary inversely, or are inversely proportional, when as one quantity increases the other quantity decreases and vice versa. Examples 5 and 6 illustrate inverse variation.

EXAMPLE 5 ***Inverse Variation in Illuminance***

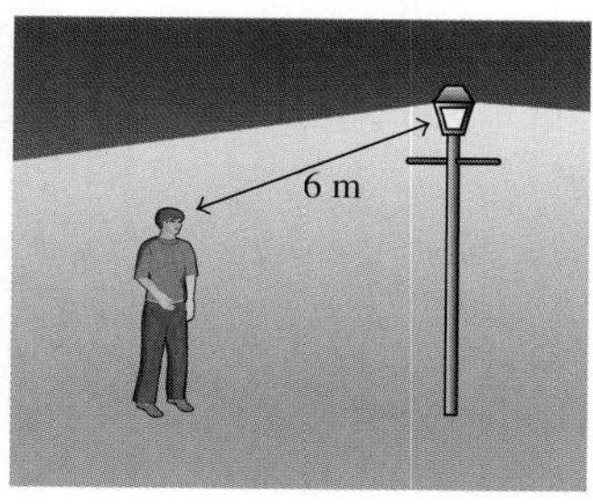

The illuminance, I, of a light source varies inversely as the square of the distance, d, from the source. Assuming that the illuminance is 75 units at a distance of 6 meters, find the equation that expresses the relationship between the illuminance and the distance.

SOLUTION Since the illuminance varies inversely as the *square* of the distance the general form of the equation is

$$I = \frac{k}{d^2}$$

To find k, we substitute the given values for I and d.

$$75 = \frac{k}{6^2}$$

$$75 = \frac{k}{36}$$

$$(75)(36) = k$$

$$2700 = k$$

Thus, the formula is $I = \dfrac{2700}{d^2}$.

EXAMPLE 6 ***Using the Constant of Proportionality***

Suppose y varies inversely as x. If $y = 8$ when $x = 15$, find y when $x = 18$.

SOLUTION First write the inverse variation, then solve for k.

$$y = \frac{k}{x}$$

$$8 = \frac{k}{15}$$

$$120 = k$$

Now substitute 120 for k in $y = k/x$ and find y when $x = 18$.

$$y = \frac{120}{x} = \frac{120}{18} = 6.7 \qquad \text{(to the nearest tenth)}$$

JOINT VARIATION

One quantity may vary directly as a product of two or more other quantities. This type of variation is called **joint variation**.

Joint Variation

The general form of a joint variation, where y varies directly as x and z, is

$$y = kxz$$

where k is the constant of proportionality.

EXAMPLE 7 ***Joint Variation in Geometry***

The area, A, of a triangle varies jointly as its base, b, and height, h. If the area of a triangle is 48 in.^{2} when its base is 12 in. and its height is 8 in., find the area of a triangle whose base is 15 in. and whose height is 20 in.

SOLUTION First write the joint variation, then substitute the known values and solve for k.

$$A = kbh$$

$$48 = k(12)(8)$$

$$48 = k(96)$$

$$\frac{48}{96} = k$$

$$k = \tfrac{1}{2}$$

Now solve for the area of the given triangle.

$$A = kbh$$

$$= \tfrac{1}{2}(15)(20)$$

$$= 150 \text{ in.}^2$$

Summary of Variations

Direct	Inverse	Joint
$y = kx$	$y = \frac{k}{x}$	$y = kxz$

COMBINED VARIATION

Often in real-life situations, one variable varies as a combination of variables. The following examples illustrate the use of **combined variations**.

EXAMPLE 8 *Pretzel Price, Combined Variation*

The owners of the Colonel Mustard Pretzel Shop find that their weekly sales of pretzels, S, varies directly with their advertising budget, A, and inversely with their pretzel price, P. When their advertising budget is \$600 and the price is \$1.20, they sell 6500 pretzels.

a) Write an equation of variation expressing S in terms of A and P. Include the value of the proportionality constant.

b) Find the expected sales if the advertising budget is \$900 and the price is \$1.50.

SOLUTION

a) Since S varies directly as A and inversely as P, we begin with the equation

$$S = \frac{kA}{P}$$

We now find k using the known values.

$$\begin{aligned} 6500 &= \frac{k(600)}{1.20} \\ 6500 &= 500k \\ 13 &= k \end{aligned}$$

Therefore, the equation for the sales of pretzels is $S = \frac{13A}{P}$.

b) $$\begin{aligned} S &= \frac{13A}{P} \\ &= \frac{13(900)}{1.50} = 7800 \end{aligned}$$

They can expect to sell 7800 pretzels.

EXAMPLE 9 *Combined Variation in Electrostatic Force*

The electrostatic force, F, of repulsion between two positive electrical charges is jointly proportional to the two charges, q_1 and q_2, and inversely proportional to the square of the distance, d, between the two charges. Express F in terms of q_1, q_2, and d.

SOLUTION $F = \frac{kq_1q_2}{d^2}$

EXAMPLE 10 ***Combined Variation***

A varies jointly as B and C and inversely as the square of D. If $A = 1$ when $B = 9$, $C = 4$, and $D = 6$, find A when $B = 8$, $C = 12$, and $D = 5$.

SOLUTION We begin with the equation

$$A = \frac{kBC}{D^2}$$

We must first find the constant of proportionality, k, by substituting the known values for A, B, C, and D and solving for k.

$$1 = \frac{k(9)(4)}{6^2}$$

$$1 = \frac{36k}{36}$$

$$1 = k$$

Thus, the constant of proportionality equals 1. Now we find A for the corresponding values of B, C, and D.

$$A = \frac{kBC}{D^2}$$

$$A = \frac{(1)(8)(12)}{5^2} = \frac{96}{25} = 3.84$$

SECTION 6.5 EXERCISES

CONCEPT/WRITING EXERCISES

In Exercises 1–4, use complete sentences to answer the question.

1. Describe direct variation.
2. Describe inverse variation.
3. Describe joint variation.
4. Describe combined variation.

In Exercises 5–20, use your intuition to determine whether the variation between the indicated quantities is direct or inverse.

5. The time required to fill a pool with a hose and the volume of water coming from the hose
6. The distance between two cities on a map and the actual distance between the two cities
7. The speed and the distance traveled by a car in a specified time period
8. A weight and the force needed to lift that weight
9. The interest earned on an investment and the interest rate
10. The volume of a balloon and its radius

See Exercise 5. Photo of Allen R. Angel, author.

11. A person's speed and the time needed for the person to complete the race

12. The time required to cool a room and the temperature of the room

13. The number of painters hired to paint a house and the time required to paint the house

14. The number of pages a person can read in a fixed period of time and his or her reading speed

15. The time required for an ice cube to melt in water and the temperature of the water

16. A person's weight (due to Earth's gravity) and his or her distance from Earth

17. The number of people in the cashier line at the bookstore and the time required to stand in line

18. The light illuminating an object and the distance the light is from the object

19. The cubic-inch displacement, in liters, and the horsepower of the engine

20. The speed of a rider lawn mower and the time it takes to cut a lawn

In Exercises 21 and 22, use Exercises 5–20 as a guide.

21. Name two items that have not been mentioned in this section that have a direct variation.

22. Name two items that have not been mentioned in this section that have an inverse variation.

PRACTICE THE SKILLS

*In Exercises 23–40, **(a)** write the variation and **(b)** find the quantity indicated.*

23. r varies directly as s. Find r when $s = 11$ and $k = 3$.

24. C varies directly as the square of Z. Find C when $Z = 5$ and $k = 2$.

25. y varies inversely as the square of x. Find y when $x = 8$ and $k = 320$.

26. x varies inversely as y. Find x when $y = 25$ and $k = 5$.

27. R varies inversely as W. Find R when $W = 160$ and $k = 8$.

28. D varies directly as J and inversely as C. Find D when $J = 10$, $C = 25$, and $k = 5$.

29. F varies jointly as D and E. Find F when $D = 3$, $E = 10$, and $k = 7$.

30. A varies jointly as R_1 and R_2 and inversely as the square of L. Find A when $R_1 = 120$, $R_2 = 8$, $L = 5$, and $k = \frac{3}{2}$.

31. T varies directly as the square of D and inversely as F. Find T when $D = 8$, $F = 15$, and $k = 12$.

32. x varies jointly as y and z. If x is 72 when $y = 18$ and $z = 2$, find x when $y = 36$ and $z = 1$.

33. Z varies jointly as W and Y. If $Z = 12$ when $W = 9$ and $Y = 4$, find Z when $W = 50$ and $Y = 6$.

34. y varies directly as the square of R. If $y = 4$ when $R = 4$, find y when $R = 8$.

35. H varies directly as L. If $H = 15$ when $L = 50$, find H when $L = 10$.

36. C varies inversely as J. If $C = 7$ when $J = 0.7$, find C when $J = 12$.

37. A varies directly as the square of B. If $A = 245$ when $B = 7$, find A when $B = 9$.

38. F varies jointly as M_1 and M_2 and inversely as the square of d. If $F = 20$ when $M_1 = 5$, $M_2 = 10$, and $d = 0.2$, find F when $M_1 = 10$, $M_2 = 20$, and $d = 0.4$.

39. F varies jointly as q_1 and q_2 and inversely as the square of d. If $F = 8$ when $q_1 = 2$, $q_2 = 8$, and $d = 4$, find F when $q_1 = 28$, $q_2 = 12$, and $d = 2$.

40. S varies jointly as I and the square of T. If $S = 8$ when $I = 20$ and $T = 4$, find S when $I = 2$ and $T = 2$.

PROBLEM SOLVING

41. *Gravity* The gravitational force of attraction, F, in newtons, between an object and the Earth is directly proportional to the mass, m, of the object in kilograms. If the force of attraction is 256 newtons when the object's mass is 8 kg, find the force of attraction when the object's mass is 50 kg.

42. *Finding Interest* The amount of interest earned on an investment, I, varies directly as the interest rate, r. If the interest earned is $40 when the interest rate is 4%, find the amount of interest earned when the interest rate is 6%.

43. *Speaker Loudness* The loudness of a stereo speaker, l, measured in decibels (dB), is inversely proportional to the square of the distance, d, of the listener from the speaker. If the loudness is 20 dB when the listener is 6 ft from the speaker, find the loudness when the listener is 3 ft from the speaker.

44. *Mass and Weight* On Earth, the weight of an object varies directly with its mass. If an object with a weight of 256 lb has a mass of 8 slugs, find the mass of an object weighing 120 lb.

45. *Video Rentals* The weekly videotape rentals, R, at Busterblock Video vary directly with their advertising budget, A, and inversely with the daily rental price, P.

When their advertising budget is \$600 and the rental price is \$3 per day, they rent 4800 tapes per week. How many tapes would they rent per week if they increased their advertising budget to \$700 and raised their rental price to \$3.50?

46. *Weight in Space* The weight, W, of an object in the Earth's atmosphere varies inversely with the square of the distance, d, between the object and the center of the Earth. A 140 lb person standing on Earth is approximately 4000 mi from the Earth's center. Find the weight (or gravitational force of attraction) of this person at a distance 100 mi from the Earth's surface.

47. *Wattage Rating* The wattage rating of an appliance, W, varies jointly as the square of the current, I, and the resistance, R. If the wattage is 1 watt when the current is 0.1 ampere and the resistance is 100 ohms, find the wattage when the current is 0.4 ampere and the resistance is 250 ohms.

48. *Electric Resistance* The electrical resistance of a wire, R, varies directly as its length, L, and inversely as its cross-sectional area, A. If the resistance of a wire is 0.2 ohm when the length is 200 ft and its cross-sectional area is 0.05 in.^{2}, find the resistance of a wire whose length is 5000 ft with a cross-sectional area of 0.01 in^2.

49. *Phone Calls* The number of phone calls between two cities during a given time period, N, varies directly as the populations p_1 and p_2 of the two cities and inversely to the distance, d, between them. If 100,000 calls are made between two cities 300 mi apart and the populations of the cities are 60,000 and 200,000, how many calls are made between two cities with populations of 125,000 and 175,000 that are 450 mi apart?

50. a) If y varies directly as x and the constant of proportionality is 2, does x vary directly or inversely as y? Explain.

b) Give the new constant of proportionality for x as a variation of y.

51. a) If y varies inversely as x and the constant of proportionality is 0.3, does x vary directly or inversely as y? Explain.

b) Give the new constant of proportionality for x as a variation of y.

CHALLENGE PROBLEMS/GROUP ACTIVITIES

52. *Photography* An article in the magazine *Outdoor and Travel Photography* states, "If a surface is illuminated by a point-source of light, the intensity of illumination produced is inversely proportional to the square of the distance separating them. In practical terms, this means that foreground objects will be grossly overexposed if your background subject is properly exposed with a flash. Thus direct flash will not offer pleasing results if there are any intervening objects between the foreground and the subject."

If the subject you are photographing is 4 ft from the flash and the illumination on this subject is $\frac{1}{16}$ of the light of the flash, what is the intensity of illumination on an intervening object that is 3 ft from the flash?

53. *Water Cost* In a specific region of the country, the amount of a customer's water bill, W, is directly proportional to the average daily temperature for the month, T, the lawn area, A, and the square root of F, where F is the family size, and inversely proportional to the number of inches of rain, R.

In one month, the average daily temperature is 78°F and the number of inches of rain is 5.6. If the average family of four who has a thousand square feet of lawn pays \$68.00 for water for that month, estimate the water bill in the same month for the average family of six who has 1500 ft^2 of lawn.

6.6 LINEAR INEQUALITIES

The first four sections of this chapter have dealt with equations. However, we often encounter statements of inequality. The symbols of inequality are as follows.

Symbols of Inequality

$a < b$ means that a is less than b.
$a \leq b$ means that a is less than or equal to b.
$a > b$ means that a is greater than b.
$a \geq b$ means that a is greater than or equal to b.

An **inequality** consists of two (or more) expressions joined by an inequality sign.

Examples of inequalities

$3 < 5, \qquad x < 2, \qquad 3x - 2 \geq 5$

A statement of inequality can be used to indicate a set of real numbers. For example, $x < 2$ represents the set of all real numbers less than 2. Listing all these numbers is impossible, but some are $-\frac{1}{2}$, -1, 0, -2, $\frac{97}{163}$, and -1.234.

A method of picturing all real numbers less than 2 is to graph the solution on the number line. The number line was discussed in Chapter 5.

To indicate the solution set of $x < 2$ on the number line, we draw an open circle at 2 and a line to the left of 2 with an arrow at its end. This technique indicates that all points to the left of 2 are part of the solution set. The open circle indicates that the solution set does not include the number 2.

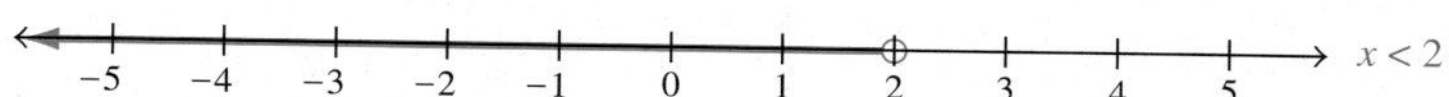

To indicate the solution set of $x \leq 2$ on the number line, we draw a closed (or darkened) circle at 2 and a line to the left of 2 with an arrow at its end. The closed circle indicates that the 2 is part of the solution.

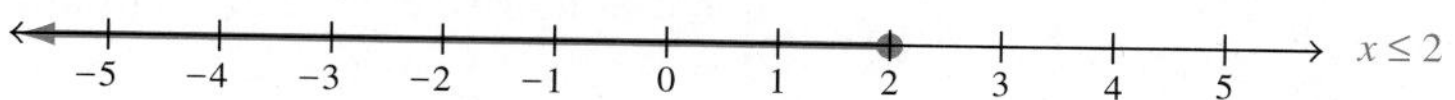

EXAMPLE 1 *Graphing a Less Than or Equal to Inequality*

Graph the solution set of $x \leq -2$, where x is a real number, on the number line.

SOLUTION The numbers less than or equal to -2 are all the points on the number line to the left of -2 and -2 itself. The closed circle at -2 shows that -2 is included in the solution set.

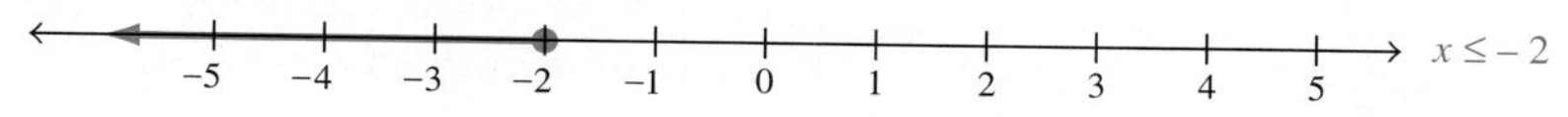

The inequality statements $x < 2$ and $2 > x$ have the same meaning. Note that the inequality symbol points to the x in both cases. Thus, one may be written in place of the other. Likewise, $x > 2$ and $2 < x$ have the same meaning. Note that the inequality symbol points to the 2 in both cases. We make use of this fact in Example 2.

EXAMPLE 2 *Graphing a Less Than Inequality*

Graph the solution set of $3 < x$, where x is a real number, on the number line.

SOLUTION We can restate $3 < x$ as $x > 3$. Both statements have identical solutions. Any number that is greater than 3 satisfies the inequality $x > 3$. The graph includes all the points to the right of 3 on the number line. To indicate that 3 is not part of the solution set, we place an open circle at 3.

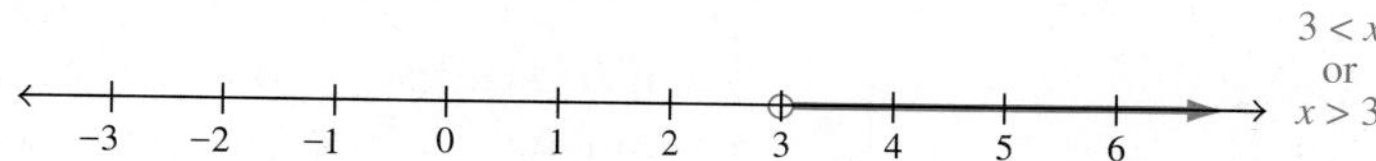

We can find the solution to an inequality by adding, subtracting, multiplying, or dividing both sides of the inequality by the same number or expression. We use the procedure discussed in Section 6.2 to isolate the variable, with one important exception: *When both sides of an inequality are multiplied or divided by a negative number, the direction of the inequality symbol is reversed.* When we change the direction of the inequality symbol, we say we change the **order** (or sense) of the inequality.

EXAMPLE 3 *Multiplying by a Negative Number*

Solve the inequality $-x > 5$ and graph the solution set on the number line.

SOLUTION To solve this inequality, we must eliminate the negative sign in front of the x. To do so, we multiply both sides of the inequality by -1 and change the order of inequality.

$$-x > 5$$

$$-1(-x) < -1(5) \quad \text{Multiply both sides of the inequality by } -1 \text{ and change the order of the inequality.}$$

$$x < -5$$

The solution set is graphed on the number line as follows.

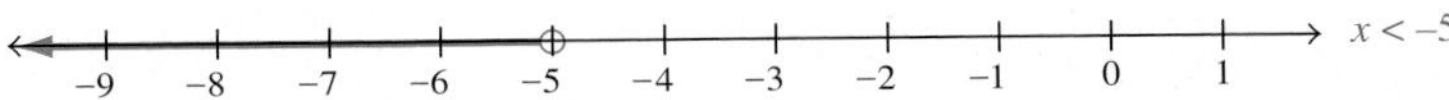

EXAMPLE 4 *Dividing by a Negative Number*

Solve the inequality $-4x < 16$ and graph the solution set on the number line.

SOLUTION Solving the inequality requires making the coefficient of the x term 1. To do so, divide both sides of the inequality by -4 and change the direction of the inequality symbol.

$$-4x < 16$$

$$\frac{-4x}{-4} > \frac{16}{-4} \quad \text{Divide both sides of the inequality by } -4 \text{ and change the direction of the inequality symbol.}$$

$$x > -4$$

The solution set is graphed on the number line as follows.

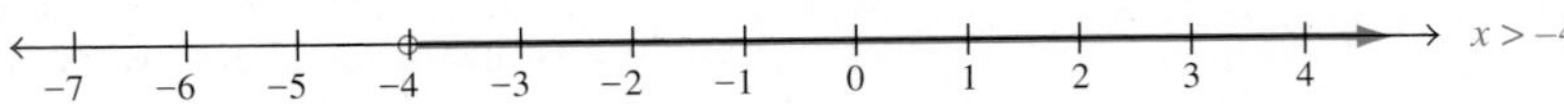

EXAMPLE 5 *Solving an Inequality*

Solve the inequality $2x - 4 < 6$ and graph the solution set on the number line.

SOLUTION To find the solution set, isolate x on one side of the inequality symbol.

$$2x - 4 < 6$$
$$2x - 4 + 4 < 6 + 4 \quad \text{Add 4 to both sides of the inequality.}$$
$$2x < 10$$
$$\frac{2x}{2} < \frac{10}{2} \quad \text{Divide both sides of the inequality by 2.}$$
$$x < 5$$

Thus, the solution set to $2x - 4 < 6$ is all real numbers less than 5.

Note that in Example 5, the *order of the inequality* did not change when both sides of the inequality were divided by the positive number 2.

DID YOU KNOW

What a Bore

In the production of machine parts, engineers must allow a certain tolerance in the way parts fit together. For example, the boring machines that grind cylindrical openings in an automobile's engine block must create a cylinder that allows the piston to move freely up and down, but still fit tightly enough to ensure that compression and combustion are complete. The allowable tolerance between parts can be expressed as an inequality. For example, the diameter of a cylinder may need to be no less than 3.383 in. and no greater than 3.387 in. We can represent the allowable tolerance as $3.383 \le t \le 3.387$.

EXAMPLE 6 *A Solution of Only Integers*

Solve the inequality $x + 4 < 7$, where x is an integer, and graph the solution set on the number line.

SOLUTION

$$x + 4 < 7$$
$$x + 4 - 4 < 7 - 4$$
$$x < 3$$

Since x is an integer and is less than 3, the solution set is the set of integers less than 3, or $\{\ldots -3, -2, -1, 0, 1, 2\}$. To graph the solution set, we make solid dots at the corresponding points on the number line. The three smaller dots to the left of -3 indicate that all the integers to the left of -3 are included.

An inequality of the form $a < x < b$ is called a **compound inequality**. Consider the compound inequality $-3 < x \le 2$, which means that $-3 < x$ *and* $x \le 2$.

EXAMPLE 7 *A Compound Inequality*

Graph the solution set of the inequality $-3 < x \le 2$

a) where x is an integer.
b) where x is a real number.

SOLUTION

a) The solution set is all the integers between -3 and 2, including the 2 but not including the -3, or $\{-2, -1, 0, 1, 2\}$.

b) The solution set consists of all the real numbers between -3 and 2, including the 2 but not including the -3.

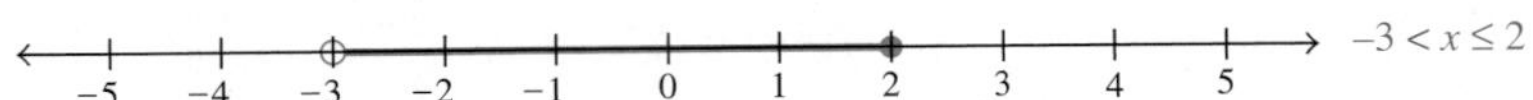

EXAMPLE 8 *Solving a Compound Inequality*

Solve the compound inequality for x and graph the solution set.

$$-4 < \frac{x + 3}{2} \leq 5$$

SOLUTION To solve a compound inequality, we must isolate the x as the middle term. To do so, we use the same principles used to solve inequalities.

$$-4 < \frac{x + 3}{2} \leq 5$$

$$2(-4) < 2\left(\frac{x + 3}{2}\right) \leq 2(5) \qquad \text{Multiply the three terms by 2.}$$

$$-8 < x + 3 \leq 10$$

$$-8 - 3 < x + 3 - 3 \leq 10 - 3 \qquad \text{Subtract 3 from all three terms.}$$

$$-11 < x \leq 7$$

The solution set is graphed on the number line as follows.

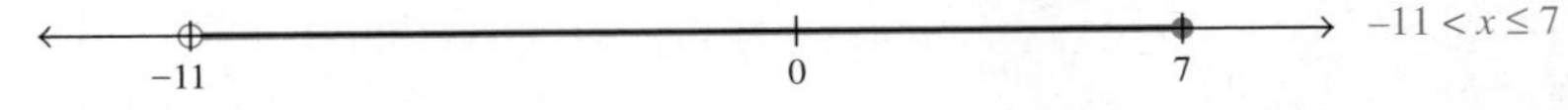

EXAMPLE 9

A student must have an average (the mean) on five tests that is greater than or equal to 80% but less than 90% to receive a final grade of B. Devon's grades on the first four tests were 98%, 76%, 86%, and 92%. What range of grades on the fifth test would give him a B in the course?

SOLUTION The unknown quantity is the range of grades on the fifth test. First construct an inequality that can be used to find the range of grades on the fifth exam. The average (mean) is found by adding the grades and dividing the sum by the number of exams.

Let x = the fifth grade. Then

$$\text{Average} = \frac{98 + 76 + 86 + 92 + x}{5}$$

For Devon to obtain a B, his average must be greater than or equal to 80 but less than 90.

$$80 \le \frac{98 + 76 + 86 + 92 + x}{5} < 90$$

$$80 \le \frac{352 + x}{5} < 90$$

$$5(80) \le 5\left(\frac{352 + x}{5}\right) < 5(90)$$ Multiply the three terms of the inequality by 5.

$$400 \le 352 + x < 450$$

$$400 - 352 \le 352 - 352 + x < 450 - 352$$ Subtract 352 from all three terms

$$48 \le x < 98$$

"In mathematics the art of posing problems is easier than that of solving them."
Georg Cantor

Thus, a grade of 48% up to but not including a grade of 98% on the fifth test will result in a grade of B.

SECTION 6.6 EXERCISES

CONCEPT/WRITING EXERCISES

1. Give the four inequality symbols we use in this section and indicate how each is read.
2. **a)** What is an inequality?
   b) Give an example of three inequalities.
3. When solving an inequality, under what conditions do you need to change the order of the inequality symbol?
4. Does $x > -3$ have the same meaning as $-3 < x$? Explain.
5. Does $x < 2$ have the same meaning as $2 > x$? Explain.
6. When graphing the solution set to an inequality on the number line, when should you use an open circle and when should you use a closed circle?

PRACTICE THE SKILLS

In Exercises 7–24, graph the solution set of the inequality, where x is a real number, on the number line.

7. $x \le 10$
8. $x > 5$
9. $x + 5 < 11$
10. $5x \ge 15$
11. $-3x \le 18$
12. $-4x < 12$
13. $\frac{x}{6} < -2$
14. $\frac{x}{2} > 4$
15. $\frac{-x}{3} \ge 3$
16. $\frac{x}{2} \ge -4$
17. $3x - 8 \le 7$
18. $2x + 9 < 4x + 11$
19. $2(x + 6) \le 15$
20. $-4(x + 2) - 2x > -6x + 2$
21. $3(x + 4) - 2 < 3x + 10$
22. $-2 \le x \le 1$
23. $3 < x - 7 \le 6$
24. $\frac{1}{2} < \frac{x + 4}{2} \le 4$

In Exercises 25–44, graph the solution set of the inequality, where x is an integer, on the number line.

25. $x \ge 3$
26. $-4 < x$
27. $-3x \le 27$
28. $3x \ge 27$
29. $x + 3 < 6$
30. $-7x \le 21$
31. $\frac{x}{6} < -2$
32. $\frac{x}{3} > -3$
33. $-\frac{x}{6} \ge 3$
34. $\frac{2x}{3} \le 4$
35. $-11 < -5x + 4$
36. $2x + 5 < -3 + 6x$
37. $3(x + 4) \ge 4x + 13$
38. $-2(x - 1) < 3(x - 4) + 5$
39. $5(x + 4) - 6 \le 2x + 8$
40. $-3 \le x < 5$
41. $1 > -x > -5$
42. $-2 < 2x + 3 < 6$
43. $0.2 < \frac{x - 3}{10} \le 0.4$
44. $-\frac{1}{3} \le \frac{x - 3}{6} < \frac{1}{2}$

PROBLEM SOLVING

45. *Van Rental* The Berrys need to rent a van for their family vacation. They can rent a van from Jason's Auto Rentals for \$200 per week with no charge for mileage or from Fred's Fine Auto's for \$110 per week plus \$0.25 per mile. Determine the distances the Berrys can drive in the van if the cost of renting from Fred's is to be less than the cost of renting from Jason's.

46. *Salary Options* Greg Brueck has two options for positions at a large manufacturing company. Option A pays \$9.90 per hour plus \$1.10 per unit produced. Option B pays \$12.70 per hour plus \$0.75 per unit produced. How many units must Greg produce in 1 hr to have a higher salary with option A?

47. *Moving Boxes* The janitor must move a large shipment of books from the first floor to the fifth floor. Each box of books weighs 60 lb, and the janitor weighs 180 lb. The sign on the elevator reads, "Maximum weight 1200 lb."
    a) Write a statement of inequality to determine the maximum number of boxes of books the janitor can place on the elevator at one time. (The janitor must ride in the elevator with the books.)
    b) Determine the maximum number of boxes that can be moved in one trip.

48. *Price of a Meal* After Mrs. Franklin is seated in a restaurant, she realizes that she has only \$19.00. If she must pay 7% tax and wants to leave a 15% tip, what is the price range of meals that she can order?

49. *Making a Profit* For a business to realize a profit, its revenue, R, must be greater than its costs, C; that is, a profit will only result if $R > C$ (the company breaks even when $R = C$). A book publishing company has a weekly cost equation of $C = 2x + 200$ and a weekly revenue equation of $R = 12x$, where x is the number of books produced and sold in a week. How many books must be sold weekly for the company to make a profit?

50. *Finding Velocity* The velocity, v, in feet per second, t sec after a tennis ball is projected directly upward is given by the formula $v = 84 - 32t$. How many seconds after being projected upward will the velocity be between 36 ft/sec and 68 ft/sec?

51. *A Grade of B* In Example 9, what range of grades on the fifth test would result in Devon receiving a grade of B if his grades on the first four tests were 78%, 64%, 88%, and 76%?

52. *Speed Limit* The minimum speed for vehicles on a highway is 40 mph, and the maximum speed is 55 mph. If Philip has been driving nonstop along the highway for 4 hr, what range in miles could he have legally traveled?

53. *Day Camp* The cost of running a day camp for one week is \$8000, plus \$175 per person enrolled. If \$18,000 is the minimum amount and \$24,000 is the maximum amount that will be spent running the camp for 1 week, what is the minimum number and the maximum number of people that can be enrolled in day camp?

CHALLENGE PROBLEMS/GROUP ACTIVITIES

54. *Painting a House* J. B. Davis is painting the exterior of his house. The instructions on the paint can indicate that 1 gal covers from 250 to 400 ft^2. The total surface of the house to be painted is 2750 ft^2. Determine the number of gallons of paint he could use and express the answer as an inequality.

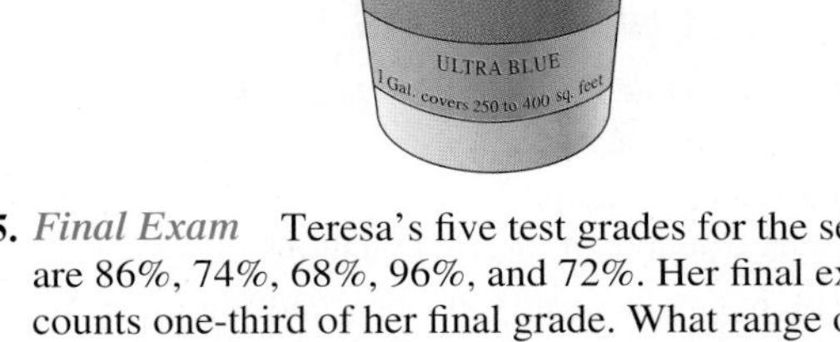

55. *Final Exam* Teresa's five test grades for the semester are 86%, 74%, 68%, 96%, and 72%. Her final exam counts one-third of her final grade. What range of grades on her final exam would result in Teresa receiving a final grade of B in the course? (See Example 9.)

56. A student multiplied both sides of the inequality $-\frac{1}{3}x \leq 4$ by -3 and forgot to reverse the inequality symbol. What is the relation between the student's incorrect solution set and the correct solution set? Is any number in both the correct solution set and the student's incorrect solution set? If so, what is it?

RESEARCH ACTIVITY

57. Find a newspaper or a magazine article that contains the mathematical concept of inequality.
    a) From the information in the article write a statement of inequality.
    b) Summarize the article and explain how you arrived at the inequality statement in part (a).

6.7 GRAPHING LINEAR EQUATIONS

In Section 6.2, we solved equations with a single variable. Real-world problems, however, often involve two or more unknowns. For example, the profit, p, of a company may depend on the sales, s; or the cost, c, of mailing a package may depend on the weight, w, of the package. Thus, it is helpful to be able to work with equations with two variables (for example, $x + 2y = 6$). Doing so requires understanding the

Cartesian (or **rectangular**) **coordinate system**, named after the French mathematician René Descartes (1596–1650).

The rectangular coordinate system consists of two perpendicular number lines (Fig. 6.5). The horizontal line is the **x-axis**, and the vertical line is the **y-axis**. The point of intersection of the x-axis and y-axis is called the **origin**. The numbers on the axes to the right and above the origin are positive. The numbers on the axes on the left and below the origin are negative. The axes divide the plane into four parts: the first, second, third, and fourth **quadrants**.

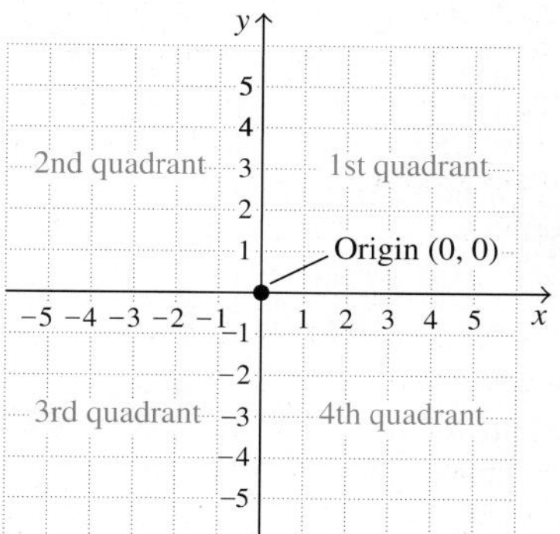

Figure 6.5

We indicate the location of a point in the rectangular coordinate system by means of an **ordered pair** of the form (x, y). The x-coordinate is always placed first and the y-coordinate is always placed second in the ordered pair. Consider the point illustrated in Fig. 6.6). Since the x-coordinate of the point is 5 and the y-coordinate is 3, the ordered pair that represents this point is (5, 3).

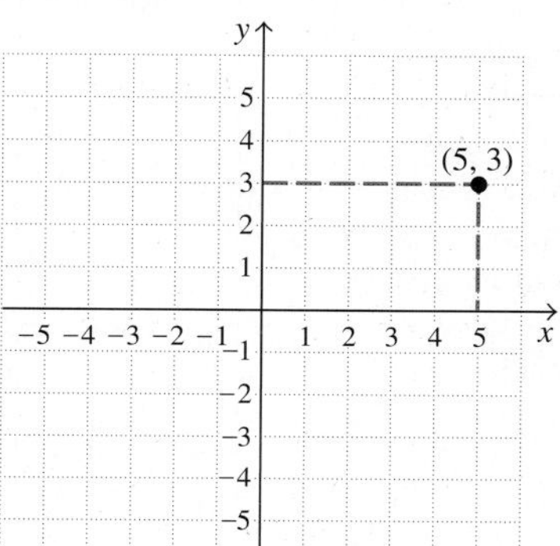

Figure 6.6

The origin is represented by the ordered pair (0, 0). Every point on the plane can be represented by one and only one ordered pair (x, y), and every ordered pair (x, y) represents one and only one point on the plane.

EXAMPLE 1 *Plotting Points*

Plot the points $A(-2, 4)$, $B(3, -4)$, $C(6, 0)$, $D(4, 1)$, and $E(0, 3)$.

SOLUTION Point A has an x-coordinate of -2 and a y-coordinate of 4. Project a vertical line up from -2 on the x-axis and a horizontal line to the left from

4 on the y-axis. The two lines intersect at the point denoted A (Fig. 6.7). The other points are plotted in a similar manner.

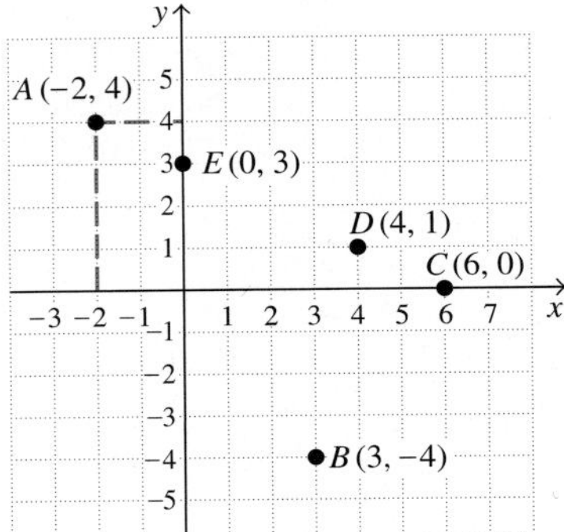

Figure 6.7

PROFILE IN MATHEMATICS

René Descartes

The Mathematician and the Fly

According to legend, the French mathematician and philosopher René Descartes (1596–1650) did some of his best thinking in bed. He was a sickly child, and so the Jesuits who undertook his education allowed him to stay in bed each morning as long as he liked. This practice he carried into adulthood, seldom getting up before noon. One morning as he watched a fly crawl about the ceiling, near the corner of his room, he was struck with the idea that the fly's position could best be described by the connecting distances from it to the two adjacent walls. These became the coordinates of his rectangular coordinate system and were appropriately named after him (Cartesian coordinates) and not the fly.

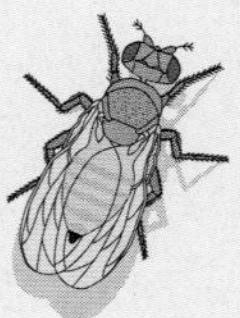

EXAMPLE 2 *A Parallelogram*

The points, A, B, and C are three vertices of a parallelogram with two sides parallel to the x-axis. Plot the three points and find the coordinates of the fourth vertex, D.

$$A(1, 2) \qquad B(2, 4) \qquad C(7, 4)$$

SOLUTION A parallelogram is a figure that has opposite sides that are of equal length and are parallel. (Parallel lines are two lines in the same plane that do not intersect.) The horizontal distance between points B and C is 5 units (see Fig. 6.8). Therefore, the horizontal distance between points A and D must also be 5 units. This problem has two possible solutions, as illustrated in Fig. 6.8.

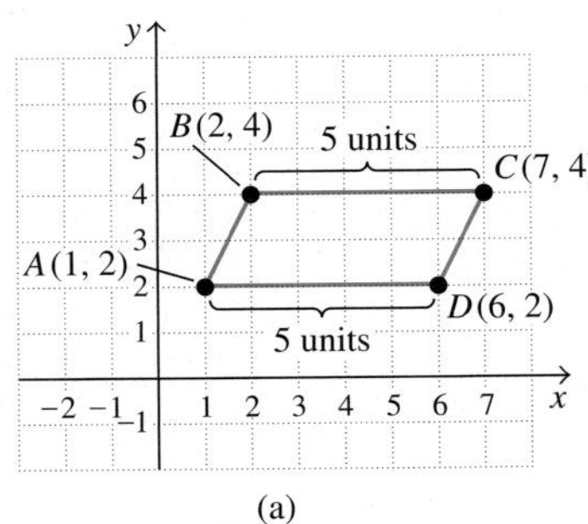

(a)

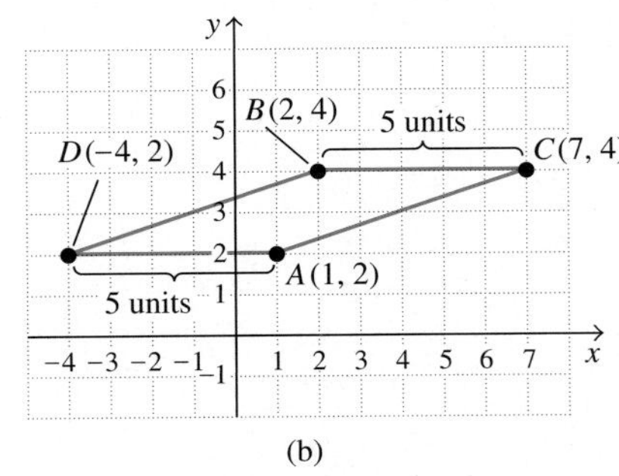

(b)

Figure 6.8

The solutions are the points (6, 2) and (−4, 2).

GRAPHING LINEAR EQUATIONS BY PLOTTING POINTS

Consider the following equation in two variables: $y = x + 1$. Every ordered pair that makes the equation a true statement is a solution to, or satisfies, the equation. We can mentally find some ordered pairs that satisfy the equation $y = x + 1$ by picking some values of x and solving the equation for y. For example, suppose we let $x = 1$; then

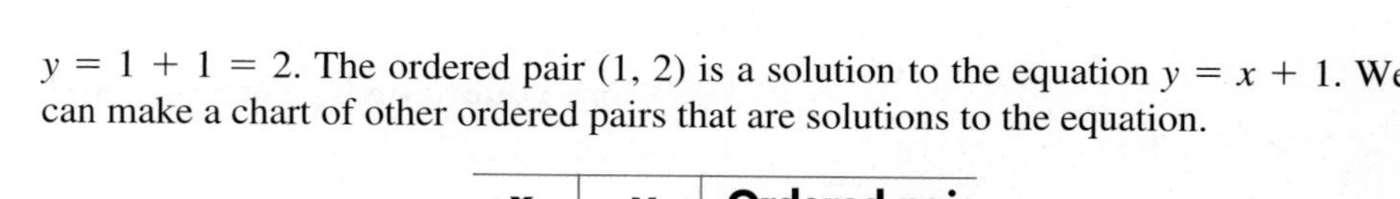

Figure 6.9

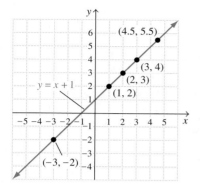

$y = 1 + 1 = 2$. The ordered pair (1, 2) is a solution to the equation $y = x + 1$. We can make a chart of other ordered pairs that are solutions to the equation.

x	y	Ordered pair
1	2	(1, 2)
2	3	(2, 3)
3	4	(3, 4)
4.5	5.5	(4.5, 5.5)
−3	−2	(−3, −2)

How many other ordered pairs satisfy the equation? Infinitely many ordered pairs satisfy the equation. Since we cannot list all the solutions, we show them by means of a graph. A **graph** is an illustration of all the points whose coordinates satisfy an equation.

The points (1, 2), (2, 3), (3, 4), (4.5, 5.5), and (−3, −2) are plotted in Fig. 6.9. With a straightedge we can draw one line that contains all these points. This line, when extended indefinitely in either direction, passes through all the points in the plane that satisfy the equation $y = x + 1$. The arrows on the ends of the line indicate that the line extends indefinitely.

All equations of the form $ax + by = c$, $a \neq 0$, $b \neq 0$, will be straight lines when graphed. Thus, such equations are called **linear equations in two variables**. The exponents on the variables x and y must be 1 for the equation to be linear. Since only two points are needed to draw a line, only two points are needed to graph a linear equation. It is always a good idea to graph a third point as a checkpoint. If no error has been made, all three points will be in a line, or **collinear**. One method that can be used to obtain points is to solve the equation for y, substitute values for x, and find the corresponding values of y.

DID YOU KNOW

Graphing Calculators

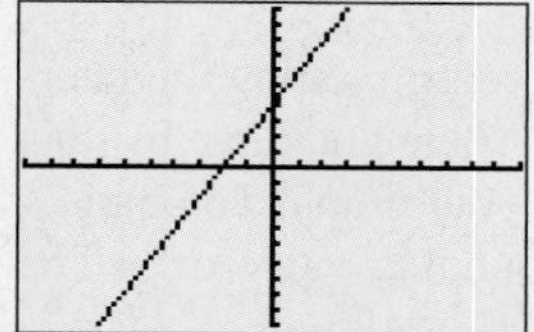

Figure 6.11

On page 276, we illustrated the graph of an exponential equation we obtained from the screen (or window) of a TI-83 graphing calculator. Figure 6.11 shows the graph of the equation $y = 2x + 4$ as illustrated on the screen of that calculator.

Graphing calculators are being used more and more in many mathematics courses. Other makers of graphing calculators include Casio, Sharp, and Hewlett-Packard. Graphing calculators can do a great many more things than a scientific calculator can. If you plan on taking additional mathematics courses, you may find that a graphing calculator is required for those courses.

EXAMPLE 3 *Graphing an Equation by Plotting Points*

Graph $y = 2x + 4$.

SOLUTION Since the equation is already solved for y, select values for x and find the corresponding values for y. The table indicates values arbitrarily selected for x and the corresponding values for y. The ordered pairs are (0, 4), (1, 6), and (−2, 0). The graph is shown in Fig. 6.10.

x	y
0	4
1	6
−2	0

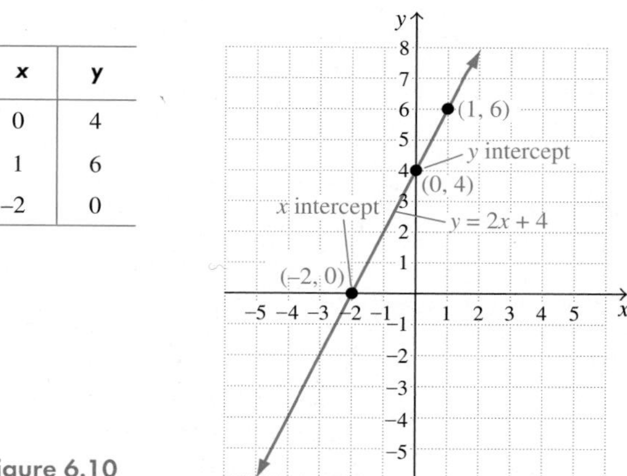

Figure 6.10

DID YOU KNOW

Use of Grids

Grids have long been used in mapping. In archaeological digs, a rectangular coordinate system is used to chart the location of each find.

To Graph Equations by Plotting Points

1. Solve the equation for y.
2. Select at least three values for x and find their corresponding values of y.
3. Plot the points.
4. If the points are in a straight line, draw a line through the set of points and place arrow tips at both ends of the line.

In step 4 of the procedure, if the points are not in a straight line, recheck your calculations and find your error.

GRAPHING BY USING INTERCEPTS

Example 3 contained two special points on the graph, $(-2, 0)$ and $(0, 4)$. At these points, the line crosses the x-axis and the y-axis, respectively. The ordered pairs $(-2, 0)$ and $(0, 4)$ represent the **x-intercept** and the **y-intercept**, respectively. Another method that can be used to graph linear equations is to find the x- and y-intercepts of the graph.

Finding the x- and y-Intercepts

To find the y-intercept, set $x = 0$ and solve the equation for y.

To find the x-intercept, set $y = 0$ and solve the equation for x.

An equation may be graphed by finding the x- and y-intercepts, plotting the intercepts, and drawing a straight line through the intercepts. When graphing by this method, you should always plot a checkpoint before drawing your graph. To obtain a checkpoint, select a nonzero value for x and find the corresponding value of y. The checkpoint should be collinear with the x- and y-intercepts.

EXAMPLE 4 *Graphing Using Intercepts*

Graph $2x + 3y = 6$ by using the x- and y-intercepts.

SOLUTION To find the x-intercept, set $y = 0$ and solve for x.

$$\begin{aligned} 2x + 3y &= 6 \\ 2x + 3(0) &= 6 \\ 2x &= 6 \\ x &= 3 \end{aligned}$$

The x-intercept is $(3, 0)$. To find the y-intercept, set $x = 0$ and solve for y.

$$\begin{aligned} 2x + 3y &= 6 \\ 2(0) + 3y &= 6 \\ 3y &= 6 \\ y &= 2 \end{aligned}$$

The y-intercept is $(0, 2)$. As a checkpoint, try $x = 1$ and find the corresponding value for y.

$$\begin{aligned} 2x + 3y &= 6 \\ 2(1) + 3y &= 6 \\ 2 + 3y &= 6 \\ 3y &= 4 \\ y &= \tfrac{4}{3} \end{aligned}$$

The checkpoint is the ordered pair $(1, \frac{4}{3})$ or $(1, 1\frac{1}{3})$.

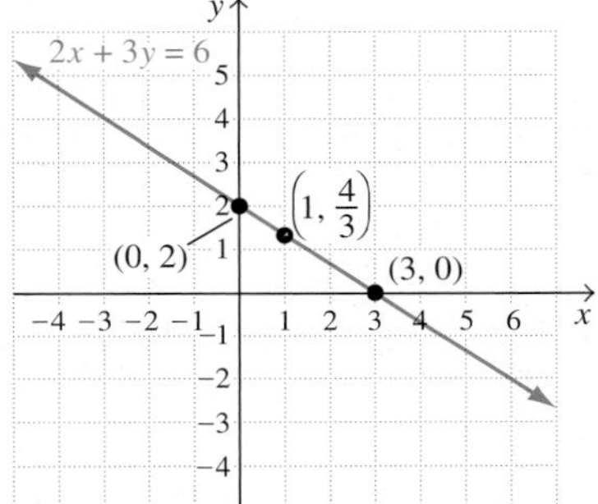

Figure 6.12

Since all three points are collinear in Fig. 6.12, draw a line through the three points to obtain the graph.

DID YOU KNOW

Up Up and Away

Jani Soíninen

Although we may not think much about it, the slope of a line is something we are altogether familiar with. You confront it every time you run up the stairs, late for class, moving 8 inches horizontally for every 6 inches up. The 1998 Olympic gold medalist Jani Soininen of Finland is familiar with the concept of slope. He speeds down 120 meters of a ski ramp at speeds of over 60 mph before he takes flight.

SLOPE

Another useful concept when you are working with straight lines is slope, which is a measure of the "steepness" of a line. The **slope of a line** is a ratio of the vertical change to the horizontal change for any two points on the line. Consider Fig. 6.13. Point A has coordinates (x_1, y_1), and point B has coordinates (x_2, y_2). The vertical change between points A and B is $y_2 - y_1$, and the horizontal change between points A and B is $x_2 - x_1$. Thus, the slope, which is often symbolized with the letter m, can be found as follows.

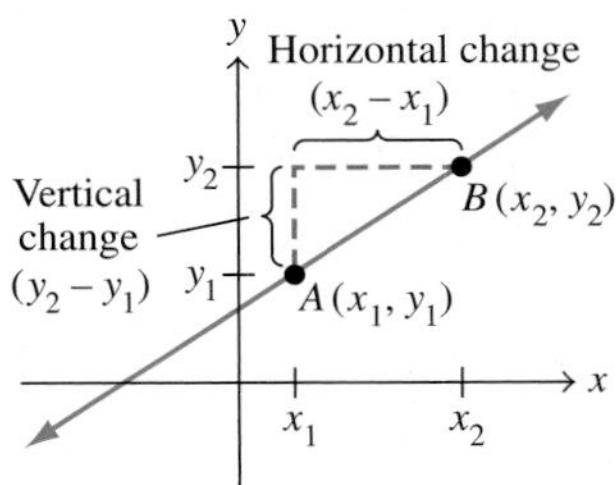

Figure 6.13

$$\textbf{Slope} = \frac{\text{vertical change}}{\text{horizontal change}}$$

$$m = \frac{y_2 - y_1}{x_2 - x_1}$$

The Greek capital letter delta, Δ, is used to represent the words "the change in." Therefore, slope may be defined as

$$m = \frac{\Delta y}{\Delta x}$$

A line may have a positive slope, or a negative slope, or zero slope, as indicated in Fig. 6.14. A line with a positive slope rises from left to right, as shown in Fig. 6.14(a). A line with a negative slope falls from left to right, as shown in Fig. 6.14(b). A horizontal line, which neither rises nor falls, has a slope of zero, as shown in Fig. 6.14(c). Since a vertical line does not have any horizontal change (the x value remains constant) and since we cannot divide by 0, the slope of a vertical line is undefined, as shown in Fig. 6.14(d).

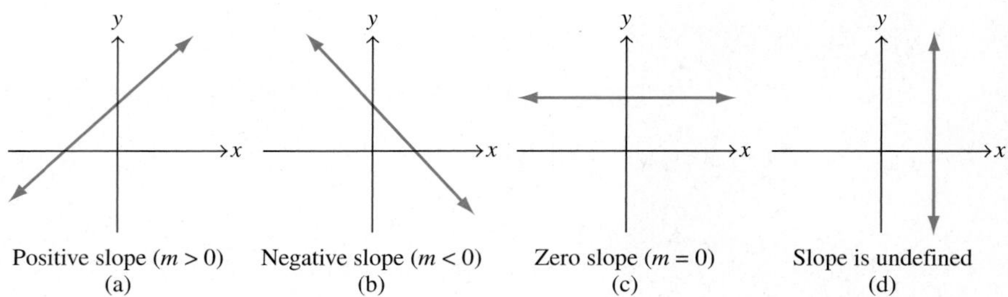

Figure 6.14

EXAMPLE 5 *Finding the Slope of a Line*

Determine the slope of the line that passes through the points $(-1, -3)$ and $(1, 5)$.

SOLUTION Let's begin by drawing a sketch, illustrating the points and the line. See Fig. 6.15(a).

We will let (x_1, y_1) be $(-1, -3)$ and (x_2, y_2) be $(1, 5)$. Then

$$\text{Slope} = \frac{y_2 - y_1}{x_2 - x_1} = \frac{5 - (-3)}{1 - (-1)} = \frac{5 + 3}{1 + 1} = \frac{8}{2} = \frac{4}{1} \quad \text{or} \quad 4$$

The slope of 4 means that there is a vertical change of 4 units for each horizontal change of 1 unit; see Fig. 6.15(b). The slope is positive, and the line rises from left to right. Note that we would have obtained the same results if we let (x_1, y_1) be $(1, 5)$ and (x_2, y_2) be $(-1, -3)$. Try this now and see.

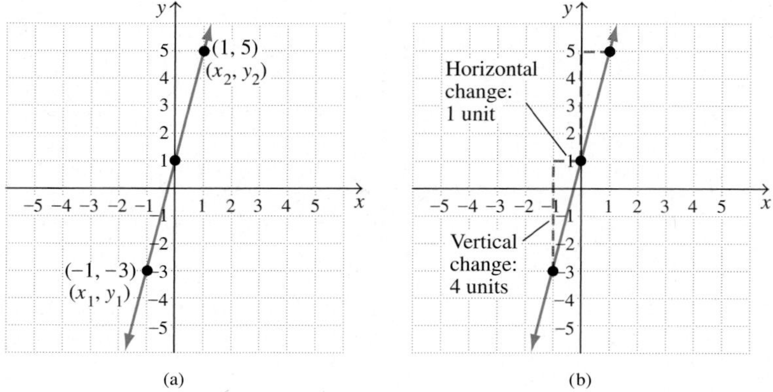

Figure 6.15

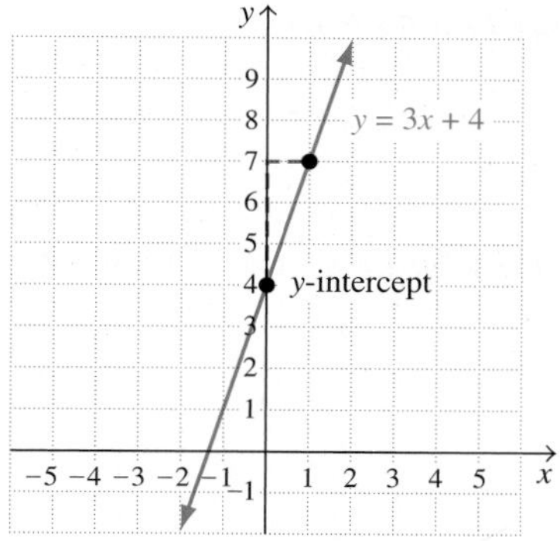

Figure 6.16

GRAPHING EQUATIONS BY USING THE SLOPE AND y-INTERCEPT

A linear equation given in the form $y = mx + b$ is said to be in slope–intercept form.

Slope–Intercept Form of the Equation of a Line

$$y = mx + b$$

where m is the slope of the line and b is the y-intercept of the line.

Consider the graph of the equation $y = 3x + 4$, which appears in Fig. 6.16. By examining the graph, we can see that the y-intercept is 4. We can also see that the graph has a positive slope, since it rises from left to right. Since the vertical change is 3 units for every 1 unit of horizontal change, the slope must be $\frac{3}{1}$ or 3.

We could graph this equation by marking the y-intercept at 4 and then moving *up* 3 units and to the *right* 1 unit to get another point. If the slope were -3, which means $\frac{-3}{1}$, we could start at the y-intercept and move *down* 3 units and to the *right* 1 unit. Thus, if we know the slope and y-intercept of a line, we can graph the line.

To Graph Equations by Using the Slope and y-Intercept

1. Solve the equation for y to place the equation in slope–intercept form.
2. Determine the slope and y-intercept from the equation.
3. Plot the y-intercept.
4. Obtain a second point using the slope.
5. Draw a straight line through the points.

EXAMPLE 6 ***Graphing an Equation Using the Slope and y-Intercept***

Graph $y = -3x + 1$ using the slope–intercept form.

SOLUTION The slope is -3 or $\frac{-3}{1}$ and the y-intercept is 1. Plot (0, 1) on the y-axis. Then plot the next point by moving *down* 3 units and to the *right* 1 unit (see Fig. 6.17). A third point has been plotted in the same way. The graph of $y = -3x + 1$ is the line drawn through these three points.

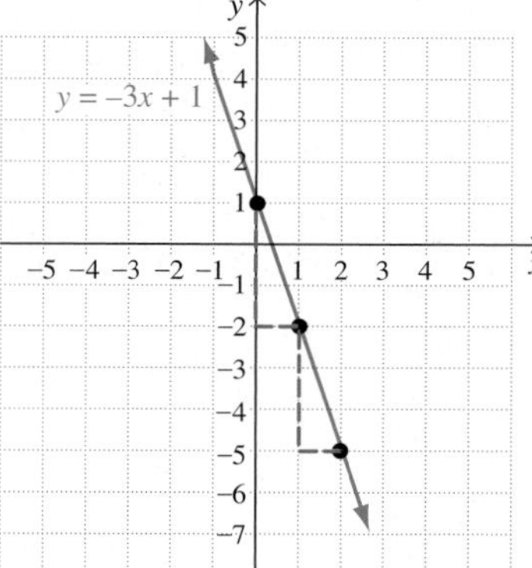

Figure 6.17

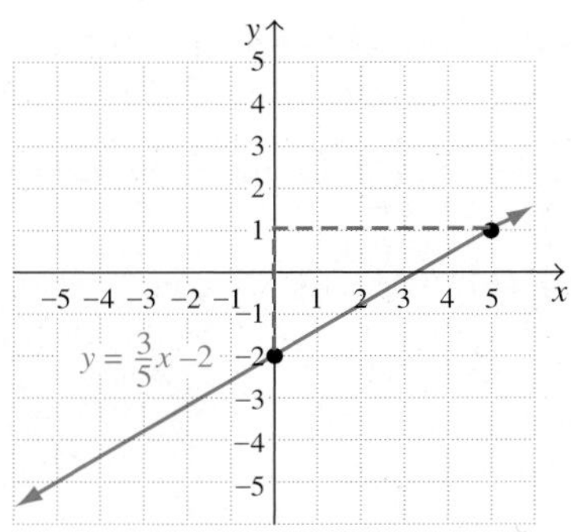

Figure 6.18

EXAMPLE 7 *Write an Equation in Slope–Intercept Form*

a) Write $3x - 5y = 10$ in slope–intercept form.
b) Graph the equation.

SOLUTION

a) To write $3x - 5y = 10$ in slope–intercept form, we solve the given equation for y.

$$3x - 5y = 10$$
$$3x - 3x - 5y = -3x + 10$$
$$-5y = -3x + 10$$
$$\frac{-5y}{-5} = \frac{-3x + 10}{-5}$$
$$y = \frac{-3x}{-5} + \frac{10}{-5} \quad \text{or} \quad y = \frac{3}{5}x - 2$$

Thus, in slope–intercept form, the equation is $y = \frac{3}{5}x - 2$.

b) The y-intercept is -2 and the slope is $\frac{3}{5}$. Place a dot at $(0, -2)$ on the y-axis, then move *up* 3 units and to the *right* 5 units to obtain the second point (see Fig. 6.18). Draw a line through the two points. ▲

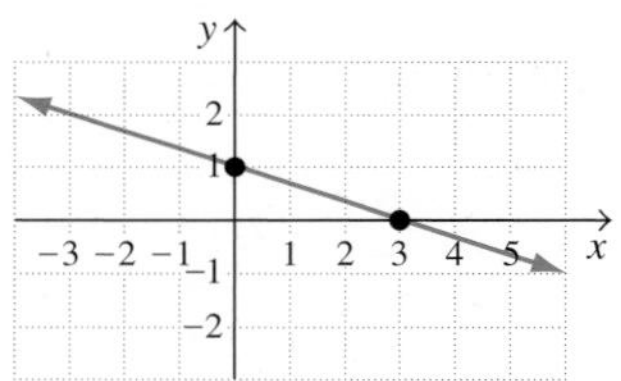

Figure 6.19

EXAMPLE 8 *Determine the Equation of a Line from Its Graph*

Determine the equation of the line in Fig. 6.19.

SOLUTION If we determine the slope and the y-intercept of the line, then we can write the equation using slope–intercept form, $y = mx + b$. We see from the graph that the y-intercept is 1; thus $b = 1$. The slope of the line is negative because the graph falls from left to right. The change in y is one unit for each three-unit change in x. Thus, m, the slope of the line is $-\frac{1}{3}$.

$$y = mx + b$$
$$y = -\frac{1}{3}x + 1$$

The equation of the line is $y = -\frac{1}{3}x + 1$. ▲

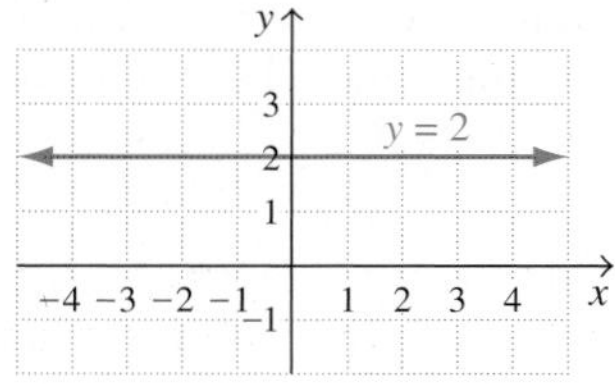

Figure 6.20

EXAMPLE 9 *Horizontal and Vertical Lines*

In the Cartesian coordinate system, graph (a) $y = 2$ and (b) $x = -3$.

SOLUTION

a) For any value of x, the value of y is 2. Therefore, the graph will be a horizontal line through $y = 2$ (Fig. 6.20).
b) For any value of y, the value of x is -3. Therefore, the graph will be a vertical line through $x = -3$ (Fig. 6.21).

Note that the graph of $y = 2$ has a slope of 0. The slope of $x = -3$ is undefined. ▲

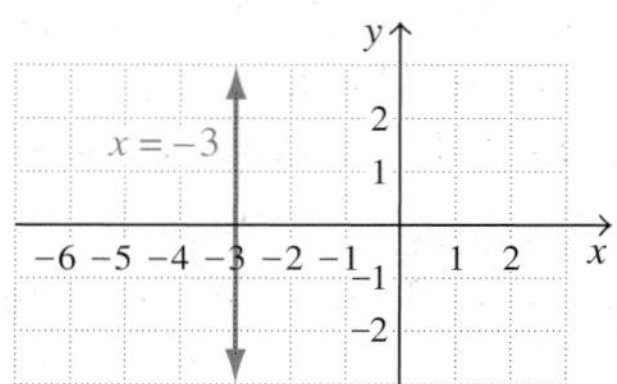

Figure 6.21

We will discuss labeling of the axes of a graph before we look at two applications. In graphing the equations in this section, we labeled the horizontal axis the x-axis and the vertical axis the y-axis. For each equation, we can determine values for y by substituting values for x. Since the value of y depends on the value of x, we refer to y as the **dependent variable** and x as the **independent variable**. We label the *vertical axis* with the *dependent variable* and the *horizontal axis* with the *independent variable*. For the equation $C = 3n + 5$, the C is the dependent variable and n is the independent variable. Thus, to graph this equation, we label the vertical axis C and the horizontal axis n.

In many graphs, the values to be plotted on one axis are much greater than the values to be plotted on the other axis. When that occurs, we can use different scales on the horizontal and the vertical axes, as illustrated in Examples 10 and 11.

EXAMPLE 10 *Using a Graph to Determine Distance*

The Paul Presley Peterson Paving Company has a contract from the state to pave 20 miles of highway. The distance, d, in miles they can pave in t hours can be approximated by the formula $d = 0.2t$.

a) Graph $d = 0.2t$, for $t \leq 300$ hours.
b) Estimate the distance paved in 150 hours.

SOLUTION

a) Since $d = 0.2t$ is a linear equation, its graph will be a straight line. Select three values for t, find the corresponding values for d, and then draw the graph (Fig. 6.22).

$$d = 0.2t$$

Let $t = 0$, $\quad d = 0.2(0) = 0$

Let $t = 100$, $\quad d = 0.2(100) = 20$

Let $t = 300$, $\quad d = 0.2(300) = 60$

t	d
0	0
100	20
300	60

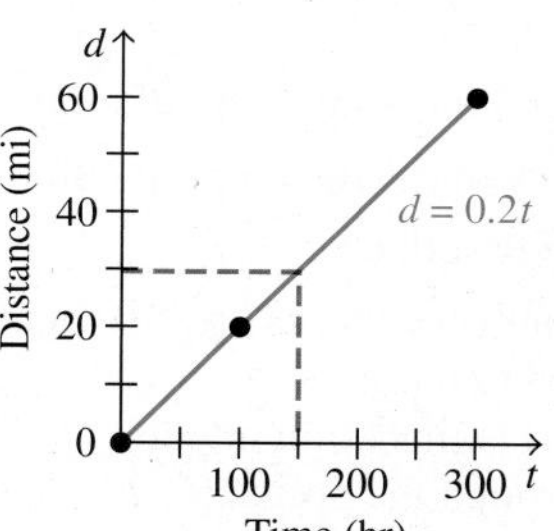

Figure 6.22

b) By drawing a vertical line from $t = 150$ on the time axis up to the graph, and then drawing a horizontal line across to the distance axis, we can determine that the distance covered is about 30 miles. ▲

EXAMPLE 11 *Using a Graph to Determine Profits*

Malik Campbell owns a small business that manufactures compact discs. He believes that the profit (or loss) from each compact disc produced can be estimated by the formula $P = 3.5S - 200{,}000$, where S is the number of compact discs sold.

a) Graph $P = 3.5S - 200{,}000$, for $S \leq 500{,}000$ compact discs.
b) From the graph, estimate the number of compact discs that must be sold for the company to break even.
c) If the profit from selling compact discs is \$1 million, estimate the number of compact discs sold.

SOLUTION

a) Select values for S and find the corresponding values of P.

S	P
0	−200,000
100,000	150,000
500,000	1,550,000

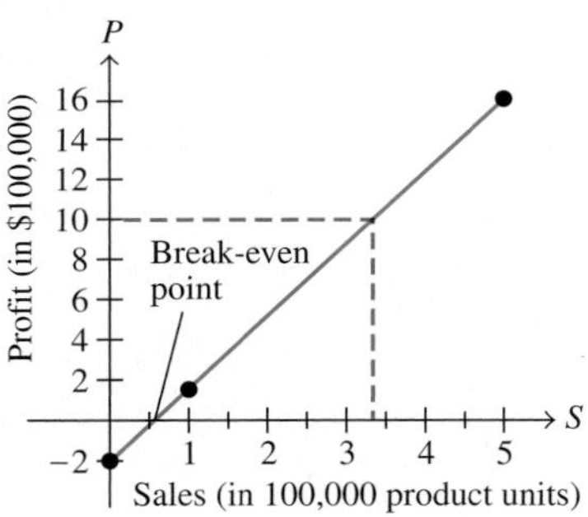

Figure 6.23

b) On the graph (Fig. 6.23), note that the break-even point is about 0.6, or 60,000 compact discs.

c) We can obtain the answer by drawing a horizontal line from 10 on the profit axis. Since the horizontal line cuts the graph at about 3.4 on the S axis, approximately 340,000 compact discs were sold.

SECTION 6.7 EXERCISES

CONCEPT/WRITING EXERCISES

1. What is a graph?
2. Explain how to find the y-intercept of a linear equation.
3. Explain how to find the x-intercept of a linear equation.
4. What is the slope of a line?
5. **a)** Explain in your own words how to find the slope of a line between two points.
   b) Based on your explanation in part (a), find the slope of the line through the points (6, 2) and (−3, 5).
6. Describe the three methods used to graph a linear equation in this section.

PRACTICE THE SKILLS

In Exercises 7–14, plot all the points on the same axes.

7. (−4, 1) **8.** (2, 3)
9. (3, −4) **10.** (−4, 0)
11. (0, 0) **12.** (0, 3)
13. (0, −5) **14.** $(3\frac{1}{2}, 4\frac{1}{2})$

In Exercises 15–22, plot all the points on the same axes.

15. (1, 5) **16.** (0, 2)
17. (−2, −3) **18.** (0, −4)
19. (−3, 0) **20.** (−3, 1)
21. (4, −1) **22.** (4.5, 3.5)

In Exercises 23–32 (indicated on Fig. 6.24), write the coordinates of the corresponding point.

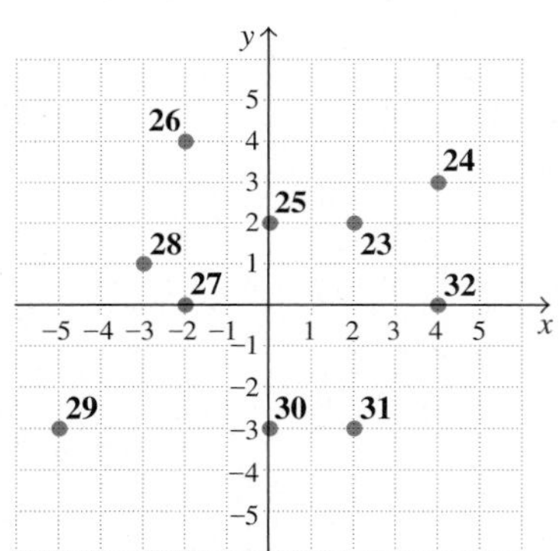

Figure 6.24

In Exercises 33–40, determine which ordered pairs satisfy the given equation.

33. $x - 3y = 8$ (2, −1), (2, −2), (2, −3)
34. $2x + 2y = 10$ (0, 5), (1, 3), (−1, 5)
35. $3x + 2y = 8$ (0, 4), $(1, \frac{5}{2})$, (−1, 3)
36. $2x = 7y + 1$ $(1, \frac{1}{7})$, (3, −1), (4, 1)
37. $7y = 3x - 5$ (1, −1), (−3, −2), (2, 5)
38. $\frac{x}{2} + 3y = 4$ $(0, \frac{4}{3})$, (8, 0), (10, −2)

39. $\frac{x}{2} + \frac{3y}{4} = 2$ $(0, \frac{8}{3})$, $(1, \frac{11}{4})$, $(4, 0)$

40. $2x - 5y = -7$ $(2, 1)$, $(-1, 1)$, $(4, 3)$

In Exercises 41–44, graph the equation and state the slope if it exists (see Example 9).

41. $x = 4$ **42.** $x = -2$

43. $y = 3$ **44.** $y = -5$

In Exercises 45–54, graph the equation by plotting points as in Example 3.

45. $y = x - 2$ **46.** $y = x + 4$

47. $y = -x + 3$ **48.** $y = 2x - 2$

49. $y + 3x = 6$ **50.** $y - 4x = 8$

51. $y = \frac{1}{2}x + 4$ **52.** $3y = 2x - 3$

53. $2y = -x + 6$ **54.** $y = -\frac{3}{4}x$

In Exercises 55–64, graph the equation, using the x- and y-intercepts as in Example 4.

55. $x - y = 5$ **56.** $x + y = 3$

57. $3x + y = 6$ **58.** $4x - 2y = 12$

59. $2x = -4y - 8$ **60.** $y = 4x + 4$

61. $y = -2x - 5$ **62.** $2x + 4y = 6$

63. $-3x + y = 4$ **64.** $5y = 3x + 10$

In Exercises 65–74, find the slope of the line through the given points. If the slope is undefined, so state.

65. (2, 8) and (4, 16)

66. (3, 7) and (7, 3)

67. (−1, −6) and (3, 5)

68. (3, −3) and (−4, 9)

69. (5, 2) and (−3, 2)

70. (−3, −5) and (−1, −2)

71. (8, −3) and (8, 3)

72. (2, 6) and (2, −3)

73. (−2, 3) and (1, −1)

74. (−7, −5) and (5, −6)

In Exercises 75–84, graph the equation using the slope and y-intercept as in Examples 6 and 7.

75. $y = x - 2$

76. $y = 3x + 2$

77. $y = -2x + 1$

78. $y = -x - 4$

79. $y = -\frac{3}{5}x + 3$

80. $y = -x - 2$

81. $7y = 4x - 7$

82. $3x + 2y = 6$

83. $3x - 2y + 6 = 0$

84. $3x + 4y - 8 = 0$

In Exercises 85–88, determine the equation of the graph.

85.

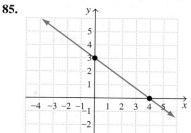

86.

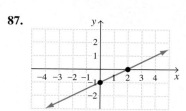

87.

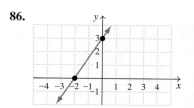

88.

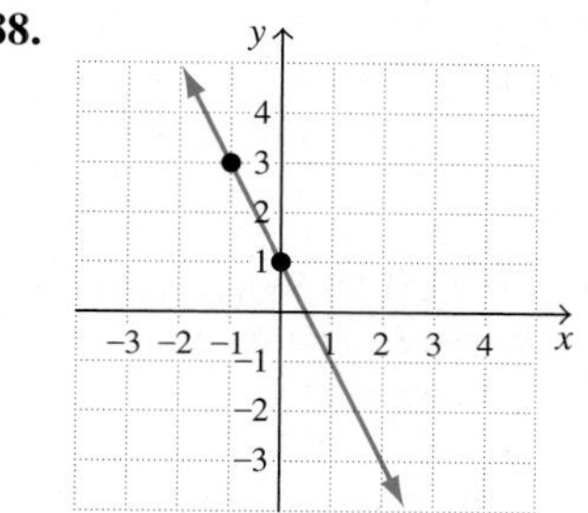

PROBLEM SOLVING

In Exercises 89 and 90, the points A, B, and C are three vertices of a rectangle (the points where two sides meet). Plot the three points. (a) Find the coordinates of the fourth point, D, to complete the rectangle. (b) Find the area of the rectangle; use A = lw.

89. $A(-2, 3)$, $B(4, 3)$, $C(4, -1)$

90. $A(-4, 2)$, $B(7, 2)$, $C(7, 8)$

In Exercises 91 and 92, the points A, B, and C are three vertices of a parallelogram with sides parallel to the x-axis. Plot the three points. Find the coordinates of the fourth point, D, to complete the parallelogram. Note: There are two possible answers for point D.

91. $A(3, 2)$, $B(5, 5)$, $C(9, 5)$

92. $A(-2, 2)$, $B(3, 2)$, $C(6, -1)$

In Exercises 93–96, for what value of b will the line joining the points P and Q be parallel to the indicated axis?

93. $P(-2, 8)$, $Q(6, b)$; x-axis

94. $P(5, 6)$, $Q(b, -2)$; y-axis

95. $P(-6, 2b + 1)$, $Q(2, 7)$; x-axis

96. $P(3b - 1, 5)$, $Q(8, 4)$; y-axis

97. *Photo Processing* The charge, C, for processing a roll of 35-millimeter (mm) film onto a picture compact disc at Scott's Moto Photo is \$8.95 plus \$0.33 per picture, or $C = 8.95 + 0.33n$, where n is the number of pictures printed.

a) Draw a graph of the cost of processing film for up to and including 36 pictures.
b) From the graph estimate the cost of processing a role of 35 mm film containing 20 pictures.
c) If the total cost of processing a roll of 35 mm film is \$20.83, estimate the number of pictures.

98. *VCR Repair* The monthly profit, p, in dollars at Vincent's VCR Repair Shop can be estimated by $p = 30n - 400$, where n is the number of VCRs repaired in a week.
a) Graph $p = 30n - 400$, for $n \leq 100$.
b) Estimate the profit for a week if 50 VCRs are repaired.
c) How many VCRs would need to be repaired in a week for the shop to break even?

99. *A Taxi Business* The weekly cost of operating a taxi, C in dollars, can be approximated by the equation $C = 80 + 0.25n$, where n is the number of miles driven.
a) Graph $C = 80 + 0.25n$, for $n \leq 1000$.
b) Estimate the cost if 600 mi are driven in a week.
c) If the weekly cost is \$180, how many miles were driven in a week?

100. *Earning Simple Interest* When \$1000 is invested in a savings account paying simple interest for a year, the interest, i in dollars, earned can be found by the formula $i = 1000r$, where r is the rate in decimal form.
a) Graph $i = 1000r$, for r up to and including a rate of 15%.
b) If the rate is 6%, what is the simple interest?
c) If the rate is 12%, what is the simple interest?

In Exercises 101 and 102, a set of points is plotted. Also shown is a straight line through the set of points that is called the line of best fit *(or a regression line, as will be discussed in Chapter 13, Statistics.*

101. *Determining a Test Grade* The graph shows the hours studied and the test grades on a biology test for six students. (The two points indicated on the line do not represent any of the six students.) The line of best fit, the red line on the graph, can be used to approximate the test grade the average student receives for the number of hours they study.

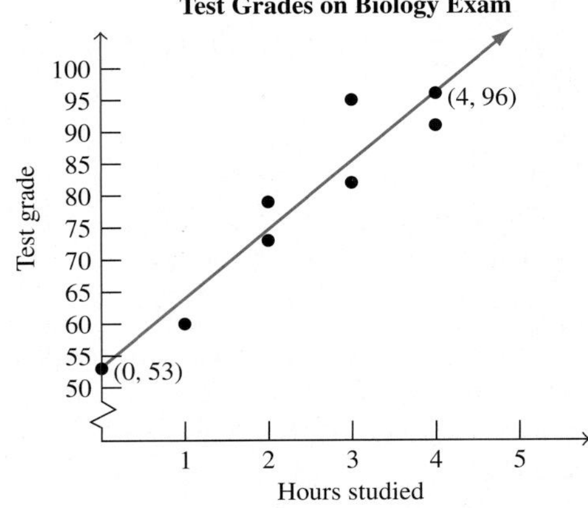

a) Determine the slope of the line of best fit using the two points indicated.
b) Using the slope determined in part (a) and the y-intercept, (0, 53), determine the equation of the line of best fit.
c) Using the equation you determined in part (b), determine the approximate test grade for a student who studied for 3 hours.
d) Using the equation you determined in part (b), determine the amount of time a student would need to study to receive a grade of 80 on the biology test.

102. *Annual Salary* The graph shows the annual salary and the number of years employed for seven workers at Data, Inc. (The two points indicated on the line do not represent any of the seven workers.) The line of best fit, the red line on the graph, can be used to approximate the annual salary of a person employed at the company for a given number of years.

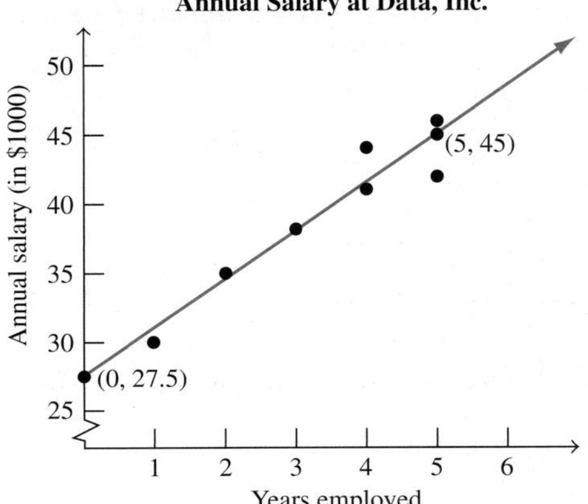

a) Determine the slope of the line of best fit using the two points indicated.

b) Using the slope determined in part (a) and the y-intercept, (0, 27.5), determine the equation of the line of best fit.

c) Using the equation you determined in part (b), approximate the annual salary for an employee that has worked for 4 years at Data, Inc.

d) Using the equation you determined in part (b), approximate the number of years an employee at Data, Inc., needs to work to have a salary of $45,000.

103. *Median Earnings* The graph shows the median weekly earnings of full-time wage and salary workers 25 years old and over, according to the education attained. The black curve represents College, 4 years or more. The straight red dashed line can be used to approximate the median income of people in this category. If we let 0 represent 1980, 1 represent 1981, 2 represent 1982, and so on, then 1997 would be represented by 17. Using the ordered pairs (0, 370) and (17, 790),

a) Determine the slope of the dashed red line.

b) Determine the equation of the dashed red line. Let (0, 370) represent the y-intercept of the graph.

c) Using the equation you determined in part (b), determine the weekly median income in 1986, which would be represented by year 6.

d) Using the equation you determined in part (b), determine the year the median weekly earnings were $600.

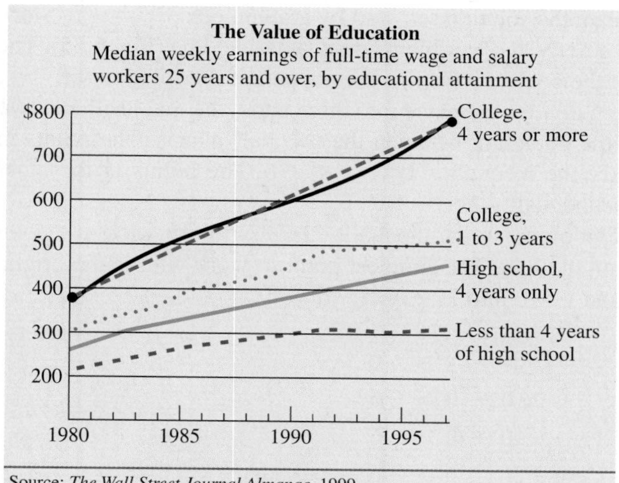

104. *Book Sales* The green graph shows book publishers' net sales, in millions of dollars. The red dashed straight line can be used to approximate the book publishers' net dollar sales. If we let 0 represent 1994, 1 represent 1995, 2 represent 1996, and so on, then 2003 would be represented by 9. Using the ordered pairs (0, 17,000) and (9, 25,000),

a) Determine the slope of the dashed line.

b) Determine the equation of the dashed line using (0, 17,000) as the y-intercept of the graph.

c) Using the equation you determined in part (b), determine the net dollar sales in 1998, which would be represented with year 4.

d) Using the equation you determined in part (b), determine the year that net sales were $20,000 (in millions).

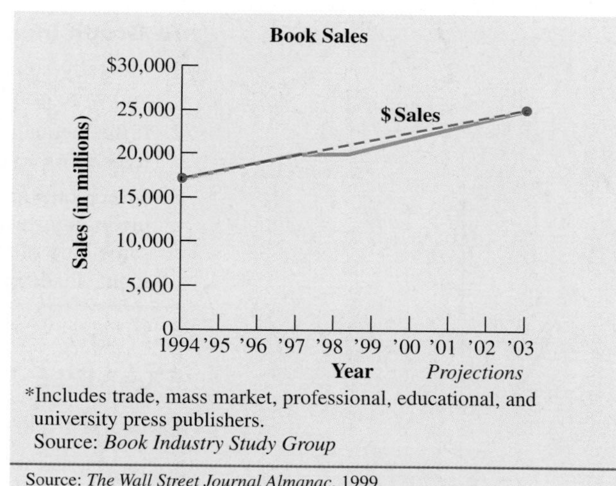

CHALLENGE PROBLEMS/GROUP ACTIVITIES

105. a) Two lines are parallel when they do not intersect no matter how far they are extended. Explain how you can determine, without graphing the equations, whether two equations will be parallel lines when graphed.

b) Determine whether the graphs of the equations $2x - 3y = 6$ and $4x = 6y + 6$ are parallel lines.

106. In which quadrants will the set of points that satisfy the equation $x + y = 1$ lie? Explain.

RESEARCH ACTIVITY

107. René Descartes is known for his contributions to algebra. Write a paper on his life and his contributions to algebra.

6.8 LINEAR INEQUALITIES IN TWO VARIABLES

In Section 6.6, we introduced linear inequalities in one variable. Now we will introduce linear inequalities in two variables. Some examples of linear inequalities in two variables are $2x + 3y \leq 7$, $x + 7y \geq 5$, and $x - 3y < 6$.

The solution set of a linear inequality in one variable may be indicated on a number line. The solution set of a linear inequality in two variables is indicated on a coordinate plane.

An inequality that is strictly less than ($<$) or greater than ($>$) will have as its solution set a **half-plane**. A half-plane is the set of all the points on one side of a line. An inequality that is less than or equal to ($\leq$) or greater than or equal to ($\geq$) will have as its solution set the set of points that consists of a half-plane and a line. To indicate that the line is part of the solution set, we draw a solid line. To indicate that the line is not part of the solution set, we draw a dashed line.

To Graph Inequalities in Two Variables

1. Mentally substitute the equal sign for the inequality sign and plot points as if you were graphing the equation.
2. If the inequality is $<$ or $>$, draw a dashed line through the points. If the inequality is $\leq$ or $\geq$, draw a solid line through the points.
3. Select a test point not on the line and substitute the x- and y-coordinates into the inequality. If the substitution results in a true statement, shade in the area on the same side of the line as the test point. If the test point results in a false statement, shade in the area on the opposite side of the line as the test point.

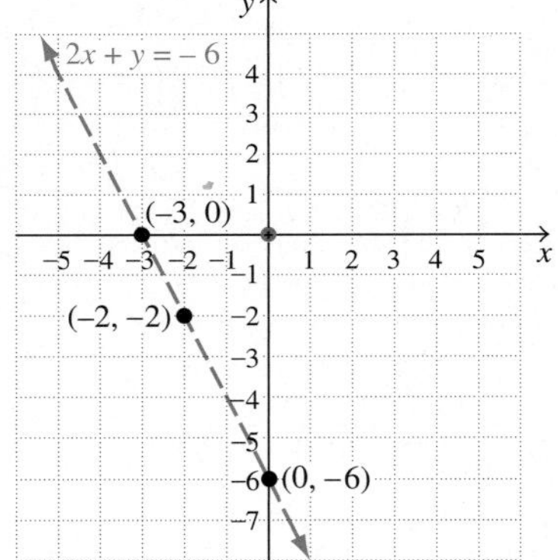

Figure 6.25

EXAMPLE 1 *Graphing an Inequality*

Draw the graph of $2x + y > -6$.

SOLUTION To obtain the solution set, start by graphing $2x + y = -6$. Since the original inequality is strictly "greater than," draw a dashed line (Fig. 6.25). The dashed line indicates that the points on the line are not part of the solution set.

The line $2x + y = -6$ divides the plane into three parts, the line itself and two *half-planes*. The line is the boundary between the two half-planes. The points in one half-plane will satisfy the inequality $2x + y < -6$. The points in the other half-plane will satisfy the inequality $2x + y > -6$.

To determine the solution set to the inequality $2x + y > -6$, pick any point on the plane that is not on the line. The simplest point to work with is the origin, (0, 0). Substitute $x = 0$ and $y = 0$ into $2x + y > -6$.

$$2x + y > -6$$
$$\text{Is } 2(0) + 0 > -6?$$
$$0 + 0 > -6$$
$$0 > -6 \quad \text{True}$$

Since 0 is greater than -6, the point (0, 0) is part of the solution set. All the points on the same side of the graph of $2x + y = -6$ as the point (0, 0) are members of the solution set. We indicate this by shading the half-plane that contains (0, 0). The graph is shown in Fig. 6.26.

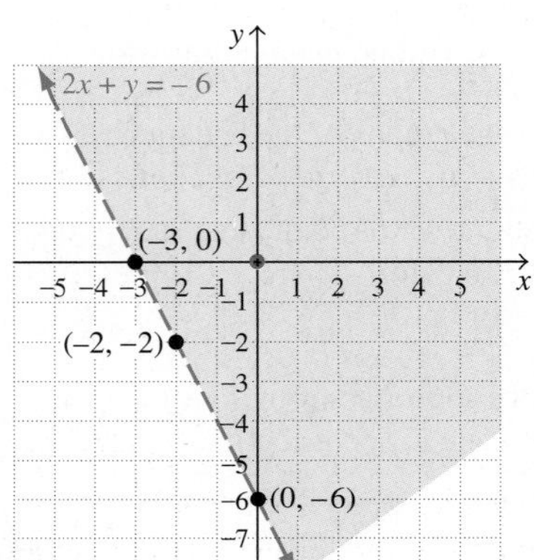

Figure 6.26

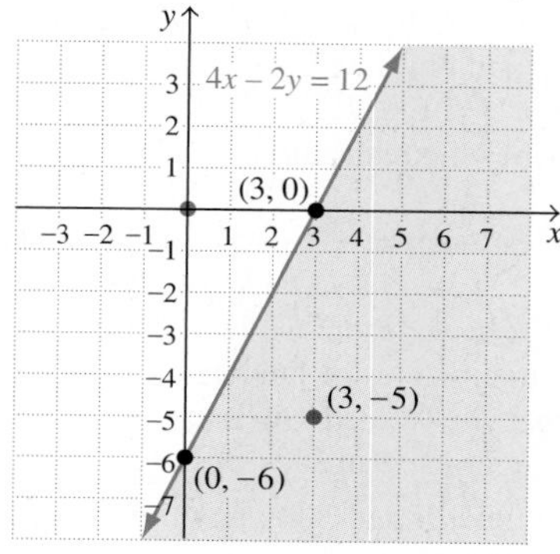

Figure 6.27

EXAMPLE 2 *Graphing an Inequality*

Draw the graph of $4x - 2y \geq 12$.

SOLUTION First draw the graph of the equation $4x - 2y = 12$. Use a solid line because the points on the boundary are included in the solution set. Now pick a point that is not on the line. Take (0, 0) as the test point.

$$4x - 2y \geq 12$$
$$\text{Is } 4(0) - 2(0) \geq 12?$$
$$0 \geq 12 \quad \text{False}$$

Since 0 is not greater than or equal to 12 ($0 \ngeq 12$), the solution set is the line and the half-plane that does not contain the point (0, 0). The graph is shown in Fig. 6.27.

If you had arbitrarily selected the test point (3, −5), you would have found that the inequality would be true: $4(3) - 2(-5) \geq 12$, or $22 \geq 12$. Thus, the point (3, −5) would be in the half-plane containing the solution set.

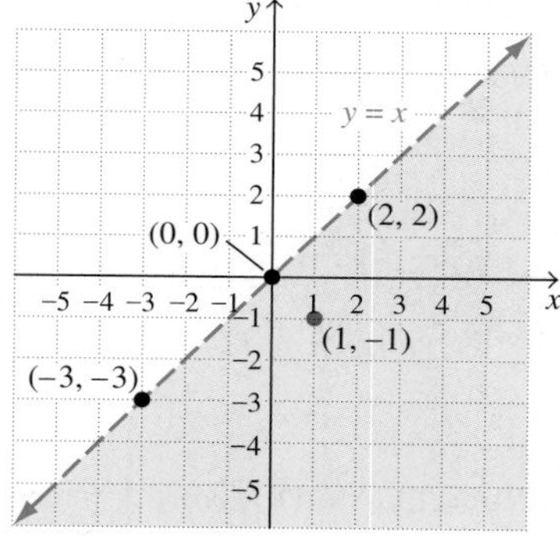

Figure 6.28

EXAMPLE 3 *Graphing an Inequality*

Draw the graph of $y < x$.

SOLUTION The inequality is strictly "less than," so the boundary is not part of the solution set. In graphing the equation $y = x$, draw a dashed line (Fig. 6.28). Since (0, 0) is *on* the line, it cannot serve as a test point. Let's pick the point (1, −1).

$$y < x$$
$$-1 < 1 \quad \text{True}$$

Since $-1 < 1$ is true, the solution set is the half-plane containing the point (1, −1).

SECTION 6.8 EXERCISES

CONCEPT/WRITING EXERCISES

1. Outline the procedure used to graph inequalities in two variables.
2. Explain why we use a solid line when graphing an inequality containing ≤ or ≥ and we use a dashed line when graphing an inequality containing < or >.

PRACTICE THE SKILLS

In Exercises 3–23, draw the graph of the inequality.

3. $x \leq 2$
4. $y \geq 1$
5. $y < x - 2$
6. $y > x + 4$
7. $y \geq 2x - 6$
8. $y < -2x + 2$
9. $3x - 4y > 12$
10. $x + 2y > 4$
11. $3x - 4y \leq 9$
12. $4y - 3x \geq 9$
13. $3x + 2y < 6$
14. $-x + 2y < 2$
15. $x + y > 0$
16. $x + 2y \leq 0$
17. $3y - 2x \leq 0$
18. $y \geq 2x + 7$
19. $3x + 2y > 12$
20. $y \leq 3x - 4$
21. $\frac{2}{5}x - \frac{1}{2}y \leq 1$
22. $0.1x + 0.3y \leq 0.4$
23. $0.2x + 0.5y \leq 0.3$

PROBLEM SOLVING

24. *Telephone Call* Eileen Burke is allowed to talk a maximum of 15 min on the telephone. She has two friends (x and y) with whom she wants to share the time.
    a) State this problem as an inequality in two variables.
    b) Graph the inequality.

25. *Flower Garden* Jim Lawler has 40 ft of landscape edging to place around a new rectangular flower garden.

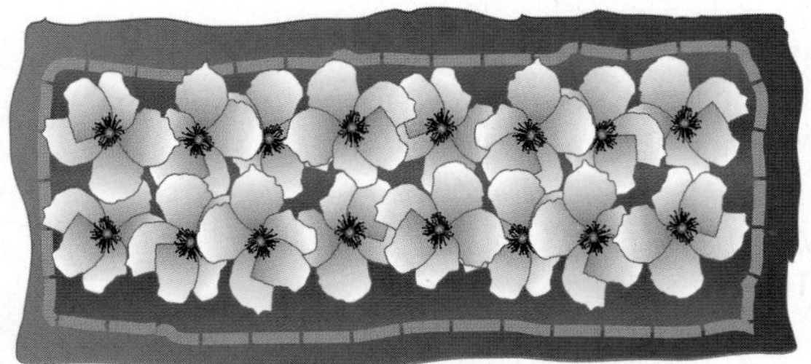

a) Write an inequality illustrating all possible dimensions of the rectangular garden. $P = 2l + 2w$ is the formula for the perimeter of a rectangle.
b) Graph the inequality.

CHALLENGE PROBLEMS

26. *Building a House* Yolanda has \$150,000 to spend on purchasing land and building a new house in the country. She wants at least 1 acre of land but less than 10 acres. If land costs \$1500 per acre and building costs are \$75 per square foot, the inequality $1500x + 75y \leq 150{,}000$, where $1 \leq x < 10$, describes the restriction on her purchase.
a) What quantities do x and y represent in the inequality?
b) Graph the inequality.
c) If Yolanda decides that her house must be at least 1950 ft^2 in size, how many acres of land can she buy?
d) If Yolanda decides that she wants to own at least 5 acres of land, what size house can she afford?

27. *Men's Shirts* The Tommy Hilfiger Company must ship x men's shirts to one outlet and y men's shirts to a second outlet. The maximum number of shirts the manufacturer can produce and ship is 250. We can represent this situation with the inequality $x + y \leq 250$.
a) Can x or y be negative? Explain.
b) Graph the inequality.
c) Write one or two paragraphs interpreting the information that the graph provides.

6.9 SOLVING QUADRATIC EQUATIONS BY USING FACTORING AND BY USING THE QUADRATIC FORMULA

We begin this section by discussing multiplication of binomials and factoring trinomials. After we discuss factoring trinomials, we will explain how to solve quadratic equations using factoring.

We will now look at the **FOIL method** of multiplying two binomials. A **binomial** is an expression that contains two terms, in which each exponent that appears on a variable is a whole number.

Examples of binomials

$x + 3$	$x - 5$
$3x + 5$	$4x - 2$

To multiply two binomials, we can use the FOIL method. The name of the method, FOIL, is an acronym to help its users remember it as a method that obtains the products of the First, Outer, Inner, and Last terms of the binomials.

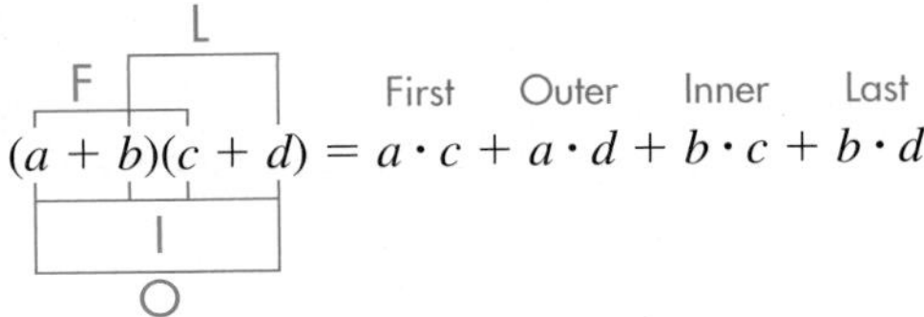

After multiplying the first, outer, inner, and last terms, combine all like terms.

> **DID YOU KNOW**
>
> **Algorithm**
>
> Recall that the series of steps taken to solve a certain type of equation is called an *algorithm*. Because mathematical procedures to solve equations can be generalized, it is possible to program computers to solve problems. For example, the equation $4x^2 + 7x + 3 = 0$ is an equation of the form $ax^2 + bx + c = 0$. If you can find a procedure to solve an equation of the general form, it can be used to solve all equations of that form. One of the first computer languages, FORTRAN (for FORmula TRANslation), was developed specifically to handle scientific and mathematical applications.

EXAMPLE 1 ***Multiplying Binomials***

Multiply $(x + 2)(x + 4)$.

SOLUTION The FOIL method of multiplication yields

$$\begin{aligned}(x + 2)(x + 4) &= \overset{F}{x \cdot x} + \overset{O}{x \cdot 4} + \overset{I}{2 \cdot x} + \overset{L}{2 \cdot 4}\\ &= x^2 + 4x + 2x + 8\\ &= x^2 + 6x + 8\end{aligned}$$

Note that $4x$ and $2x$ were combined to get $6x$.

EXAMPLE 2 ***Multiplying Binomials***

Multiply $(2x - 1)(x + 4)$.

SOLUTION

$$\begin{aligned}(2x - 1)(x + 4) &= \overset{F}{2x \cdot x} + \overset{O}{2x \cdot 4} + \overset{I}{(-1) \cdot x} + \overset{L}{(-1) \cdot 4}\\ &= 2x^2 + 8x - x - 4\\ &= 2x^2 + 7x - 4\end{aligned}$$

FACTORING TRINOMIALS OF THE FORM $x^2 + bx + c$

The expression $x^2 + 6x + 8$ is an example of a trinomial. A **trinomial** is an expression containing three terms in which each exponent that appears on a variable is a whole number.

Example 1 showed that

$$(x + 2)(x + 4) = x^2 + 6x + 8$$

Since the product of $x + 2$ and $x + 4$ is $x^2 + 6x + 8$, we say that $x + 2$ and $x + 4$ are **factors** of $x^2 + 6x + 8$. **To factor an expression** means to write the expression as a product of its factors. For example, to factor $x^2 + 6x + 8$, we write

$$x^2 + 6x + 8 = (x + 2)(x + 4)$$

Let's look at the factors more closely.

$$2 + 4 = 6$$

$$2 \cdot 4 = 8$$

$$x^2 + 6x + 8 = (x + 2)(x + 4)$$

Note that the sum of the two numbers in the factors is $2 + 4$ or 6. The 6 is the coefficient of the x-term. Also note that the product of the numbers in the two factors is $2 \cdot 4$, or 8. The 8 is the constant. In general, when factoring an expression of the form

$x^2 + bx + c$, we need to find two numbers whose product is c and whose sum is b. When we determine the two numbers, the factors will be of the form

$$(x + \square)(x + \square)$$

↑ One number ↑ Other number

EXAMPLE 3 *Factoring a Trinomial, a* = 1

Factor $x^2 + 5x + 6$.

SOLUTION We need to find two numbers whose product is 6 and whose sum is 5. Since the product is +6, the two numbers must both be positive or both be negative. Because the coefficient of the x-term is positive, only the positive factors of 6 need to be considered. Can you explain why? We begin by listing the positive numbers whose product is 6.

Factors of 6	Sum of factors
1(6)	$1 + 6 = 7$
2(3)	$2 + 3 = 5$

Since $2 \cdot 3 = 6$ and $2 + 3 = 5$, 2 and 3 are the numbers we are seeking. Thus, we write

$$x^2 + 5x + 6 = (x + 2)(x + 3)$$

Note that $(x + 3)(x + 2)$ is also an acceptable answer.

To Factor Trinomial Expressions of the Form $x^2 + bx + c$

1. Find two numbers whose product is c and whose sum is b.
2. Write factors in the form

$$(x + \square)(x + \square)$$

↑ One number from step 1 ↑ Other number from step 1

3. Check your answer by multiplying the factors using FOIL.

If, for example, the numbers found in step 1 were 6 and −4, the factors would be written $(x + 6)(x - 4)$.

EXAMPLE 4 *Factoring a Trinomial, a* = 1

Factor $x^2 - 6x - 16$.

SOLUTION We must find two numbers whose product is −16 and whose sum is −6. Begin by listing the factors of −16.

Factors of −16	Sum of factors
−16(1)	$-16 + 1 = -15$
−8(2)	$-8 + 2 = -6$
−4(4)	$-4 + 4 = 0$
−2(8)	$-2 + 8 = 6$
−1(16)	$-1 + 16 = 15$

The table lists all the factors of -16. The only factors listed whose product is -16 and whose sum is -6 are -8 and 2. We listed all factors in this example so that you could see, for example, that $-8(2)$ is a different set of factors than $-2(8)$. Once you find the factors you are looking for, there is no need to go any further. The trinomial can be written in factored form as

$$x^2 - 6x - 16 = (x - 8)(x + 2)$$

▲

FACTORING TRINOMIALS OF THE FORM $ax^2 + bx + c, a \neq 1$

Now we discuss how to factor an expression of the form $ax^2 + bx + c$, where a, the coefficient of the squared term, is not equal to 1.

Consider the multiplication problem $(2x + 1)(x + 3)$.

$$\begin{aligned}(2x + 1)(x + 3) &= 2x \cdot x + 2x \cdot 3 + 1 \cdot x + 1 \cdot 3\\ &= 2x^2 + 6x + x + 3\\ &= 2x^2 + 7x + 3\end{aligned}$$

Since $(2x + 1)(x + 3) = 2x^2 + 7x + 3$, the factors of $2x^2 + 7x + 3$ are $2x + 1$ and $x + 3$.

Let's study the coefficients more closely.

F O + I L

$$2x^2 + 7x + 3 \qquad (2x + 1)(1x + 3)$$

$$\mathbf{F} = 2 \cdot 1 = 2 \qquad \mathbf{O + I} = (2 \cdot 3) + (1 \cdot 1) = 7 \qquad \mathbf{L} = 1 \cdot 3 = 3$$

Note that the product of the coefficient of the first terms in the multiplication of the binomials equals 2, the coefficient of the squared term. The sum of the products of the coefficients of the outer and inner terms equals 7, the coefficient of the x-term. The product of the last terms equals 3, the constant.

A procedure to factor expressions of the form $ax^2 + bx + c$, $a \neq 1$, follows.

To Factor Trinomial Expressions of the Form $ax^2 + bx + c, a \neq 1$

1. Write all pairs of factors of the coefficient of the squared term, a.
2. Write all pairs of factors of the constant, c.
3. Try various combinations of these factors until the sum of the products of the outer and inner terms is bx.
4. Check your answer by multiplying the factors using FOIL.

EXAMPLE 5 *Factoring a Trinomial, a ≠ 1*

Factor $3x^2 + 17x + 10$.

SOLUTION The only factors of 3 are 1 and 3. Therefore, we write

$$3x^2 + 17x + 10 = (3x \quad)(x \quad)$$

The number 10 has both positive and negative factors. However, since both the constant, 10, and the sum of the products of the outer and inner terms, 17, are positive, the two factors must be positive. Why? The positive factors of 10 are 1(10) and 2(5). The following is a list of the possible factors.

Possible factors	Sum of products of outer and inner terms	
$(3x + 1)(x + 10)$	$31x$	
$(3x + 10)(x + 1)$	$13x$	
$(3x + 2)(x + 5)$	$17x$	← Correct middle term
$(3x + 5)(x + 2)$	$11x$	

Thus, $3x^2 + 17x + 10 = (3x + 2)(x + 5)$.

Note that factoring problems of this type may be checked by using the FOIL method of multiplication. We will check the results to Example 5:

$$\begin{aligned}(3x + 2)(x + 5) &= 3x \cdot x + 3x \cdot 5 + 2 \cdot x + 2 \cdot 5\\ &= 3x^2 + 15x + 2x + 10\\ &= 3x^2 + 17x + 10\end{aligned}$$

Since we obtained the expression we started with, our factoring is correct.

EXAMPLE 6 *Factoring a Trinomial, $a \neq 1$*

Factor $6x^2 - 11x - 10$.

SOLUTION The factors of 6 will be either $6 \cdot 1$ or $2 \cdot 3$. Therefore, the factors may be of the form $(6x \quad)(x \quad)$ or $(2x \quad)(3x \quad)$. When there is more than one set of factors for the first term, we generally try the medium-sized factors first. If this does not work, we try the other factors. Thus, we write

$$6x^2 - 11x - 10 = (2x \quad)(3x \quad)$$

The factors of -10 are $(-1)(10)$, $(1)(-10)$, $(-2)(5)$, and $(2)(-5)$. There will be eight different pairs of possible factors of the trinomial $6x^2 - 11x - 10$. Can you list them?

The correct factoring is $6x^2 - 11x - 10 = (2x - 5)(3x + 2)$.

Note that in Example 6 we first tried factors of the form $(2x \quad)(3x \quad)$. If we had not found the correct factors using them, we would have tried $(6x \quad)(x \quad)$.

SOLVING QUADRATIC EQUATIONS BY FACTORING

In Section 6.2, we solved linear, or first-degree, equations. In those equations, the exponent on all variables was 1. Now we deal with the **quadratic equation**. The standard form of a quadratic equation in one variable is shown in the box.

Standard Form of a Quadratic Equation

$$ax^2 + bx + c = 0, \qquad a \neq 0$$

Note that in the standard form of a quadratic equation, the greatest exponent on x is 2 and the right side of the equation is equal to zero. *To solve a quadratic equation* means to find the value or values that make the equation true. In this section, we will solve quadratic equations by factoring and by the quadratic formula.

To solve a quadratic equation by factoring, set one side of the equation equal to 0 and then use the *zero-factor* property.

Zero-Factor Property

If $a \cdot b = 0$, then $a = 0$ or $b = 0$.

The zero-factor property indicates that, if the product of two factors is 0, then one (or both) of the factors must have a value of 0.

EXAMPLE 7 ***Using the Zero-Factor Property***

Solve the equation $(x + 3)(x - 6) = 0$.

SOLUTION When we use the zero-factor property, either $(x + 3)$ or $(x - 6)$ must equal 0 for the product to equal 0. Thus, we set each individual factor equal to 0 and solve each resulting equation for x.

$$(x + 3)(x - 6) = 0$$

$$x + 3 = 0 \quad \text{or} \quad x - 6 = 0$$
$$x = -3 \qquad\qquad x = 6$$

Thus, the solutions are -3 and 6.

Check: $x = -3$

$$(x + 3)(x - 6) = 0$$
$$(-3 + 3)(-3 - 6) = 0$$
$$0(-9) = 0$$
$$0 = 0 \quad \text{True}$$

$x = 6$

$$(x + 3)(x - 6) = 0$$
$$(6 + 3)(6 - 6) = 0$$
$$9(0) = 0$$
$$0 = 0 \quad \text{True}$$

To Solve a Quadratic Equation by Factoring

1. Use the addition or subtraction property to make one side of the equation equal to 0.
2. Factor the side of the equation not equal to 0.
3. Use the zero-factor property to solve the equation.

Examples 8 and 9 illustrate this procedure.

EXAMPLE 8 ***Solving a Quadratic Equation by Factoring***

Solve the equation $x^2 - 8x = -15$.

SOLUTION First add 15 to both sides of the equation to make the right side of the equation equal to 0.

$$x^2 - 8x = -15$$
$$x^2 - 8x + 15 = -15 + 15$$
$$x^2 - 8x + 15 = 0$$

Factor the left side of the equation. The object is to find two numbers whose product is 15 and whose sum is -8. Since the product of the numbers is positive and the sum of the numbers is negative, the two numbers must both be negative. The numbers are -3 and -5. Note that $(-3)(-5) = 15$ and $-3 + (-5) = -8$.

$$x^2 - 8x + 15 = 0$$
$$(x - 3)(x - 5) = 0$$

Now use the zero-factor property to find the solution.

$$x - 3 = 0 \quad \text{or} \quad x - 5 = 0$$
$$x = 3 \qquad\qquad x = 5$$

The solutions are 3 and 5.

EXAMPLE 9 *Solve a Quadratic Equation by Factoring*

Solve the equation $2x^2 - 11x + 12 = 0$.

SOLUTION $2x^2 - 11x + 12$ factors into $(2x - 3)(x - 4)$. Thus, we write

$$2x^2 - 11x + 12 = 0$$
$$(2x - 3)(x - 4) = 0$$
$$2x - 3 = 0 \quad \text{or} \quad x - 4 = 0$$
$$2x = 3 \qquad\qquad x = 4$$
$$x = \frac{3}{2}$$

The solutions are $\frac{3}{2}$ and 4.

SOLVING QUADRATIC EQUATIONS BY USING THE QUADRATIC FORMULA

Not all quadratic equations can be solved by factoring. When a quadratic equation cannot be easily solved by factoring, we can solve the equation with the **quadratic formula**. The quadratic formula can be used to solve any quadratic equation.

Quadratic Formula

For a quadratic equation in standard form, $ax^2 + bx + c = 0$, $a \neq 0$, the quadratic formula is

$$x = \frac{-b \pm \sqrt{b^2 - 4ac}}{2a}$$

In the quadratic formula, the plus or minus symbol, $\pm$, is used. If, for example, $x = 2 \pm 3$, then $x = 2 + 3$ or $x = 2 - 3$.

To use the quadratic formula, first write the quadratic equation in standard form. Then determine the values for a (the coefficient of the squared term), b (the coefficient of the x term), and c (the constant). Finally, substitute the values of a, b, and c into the quadratic formula and evaluate the expression.

DID YOU KNOW

The Mathematics of Motion

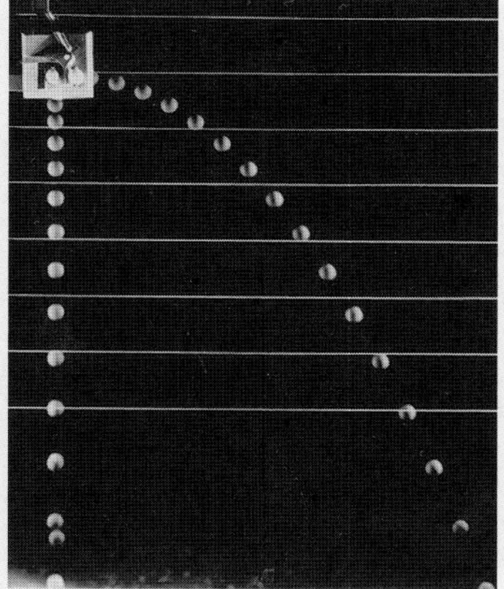

The free fall of an object is something that has interested scientists and mathematicians for centuries. It is described by a quadratic equation. Shown here is a time-lapse photo that shows the free fall of a ball in equal-time intervals. What you see can be described verbally this way: The rate of change in velocity in each interval is the same; therefore, velocity is continuously increasing and acceleration is constant.

EXAMPLE 10 ***Solve a Quadratic Equation Using the Quadratic Formula***

Solve the equation $x^2 + 2x - 15 = 0$ using the quadratic formula.

SOLUTION In this equation, $a = 1$, $b = 2$, and $c = -15$.

$$x = \frac{-b \pm \sqrt{b^2 - 4ac}}{2a} = \frac{-2 \pm \sqrt{2^2 - 4(1)(-15)}}{2(1)}$$

$$= \frac{-2 \pm \sqrt{4 + 60}}{2}$$

$$= \frac{-2 \pm \sqrt{64}}{2}$$

$$= \frac{-2 \pm 8}{2}$$

$$\frac{-2 + 8}{2} = \frac{6}{2} = 3 \quad \text{or} \quad \frac{-2 - 8}{2} = \frac{-10}{2} = -5$$

The solutions are 3 and -5.

Note that Example 10 can also be solved by factoring. We suggest that you do so now.

EXAMPLE 11 ***Irrational Solutions to a Quadratic Equation***

Solve $3x^2 - 6x = 5$ using the quadratic formula.

SOLUTION Begin by writing the equation in standard form by subtracting 5 from both sides of the equation.

$$3x^2 - 6x - 5 = 0$$

$$a = 3, \quad b = -6, \quad c = -5$$

$$x = \frac{-b \pm \sqrt{b^2 - 4ac}}{2a} = \frac{-(-6) \pm \sqrt{(-6)^2 - 4(3)(-5)}}{2(3)}$$

$$= \frac{6 \pm \sqrt{36 + 60}}{6}$$

$$= \frac{6 \pm \sqrt{96}}{6}$$

Since $\sqrt{96} = \sqrt{16}\sqrt{6} = 4\sqrt{6}$ (see Section 5.4), we write

$$\frac{6 \pm \sqrt{96}}{6} = \frac{6 \pm 4\sqrt{6}}{6} = \frac{\overset{1}{\cancel{2}}(3 \pm 2\sqrt{6})}{\underset{3}{\cancel{6}}} = \frac{3 \pm 2\sqrt{6}}{3}$$

The solutions are $\dfrac{3 + 2\sqrt{6}}{3}$ and $\dfrac{3 - 2\sqrt{6}}{3}$.

Note that the solutions to Example 11 are irrational numbers. It is also possible for a quadratic equation to have no real solution. In solving an equation, if the radicand (the expression inside the square root) is a negative number, then the quadratic equation has **no real solution**.

SECTION 6.9 EXERCISES

CONCEPT/WRITING EXERCISES

1. What is a *binomial?* Give two examples of binomials.
2. What is a *trinomial*. Give three examples of trinomials.
3. In your own words, explain the FOIL method used to multiply two binomials.
4. Give the standard form of a quadratic equation.
5. In your own words, state the zero-factor property.
6. Have you memorized the quadratic formula? If not, you need to do so. Without looking at the book, write the quadratic formula.

PRACTICE THE SKILLS

In Exercises 7–22, factor the trinomial. If the trinomial cannot be factored, so state.

7. $x^2 + 10x + 21$ **8.** $x^2 + 4x + 3$
9. $x^2 - 4x - 5$ **10.** $x^2 + 4x - 5$
11. $x^2 + 2x - 24$ **12.** $x^2 - 6x + 8$
13. $x^2 - 2x - 3$ **14.** $x^2 - 5x - 6$
15. $x^2 - 10x + 21$ **16.** $x^2 - 25$
17. $x^2 - 16$ **18.** $x^2 - x - 56$
19. $x^2 + 3x - 28$ **20.** $x^2 + 4x - 32$
21. $x^2 + 2x - 63$ **22.** $x^2 - 2x - 48$

In Exercises 23–34, factor the trinomial. If the trinomial cannot be factored, so state.

23. $2x^2 + 5x + 3$ **24.** $2x^2 + 13x - 7$
25. $3x^2 - 14x - 5$ **26.** $2x^2 - 13x - 7$
27. $5x^2 + 12x + 4$ **28.** $2x^2 - 9x + 10$
29. $5x^2 - 13x - 6$ **30.** $4x^2 + 20x + 21$
31. $5x^2 - 13x + 6$ **32.** $6x^2 - 11x + 4$
33. $3x^2 - 14x - 24$ **34.** $6x^2 + 5x + 1$

In Exercises 35–38, solve each equation, using the zero-factor property.

35. $(x + 2)(x - 5) = 0$ **36.** $(2x - 1)(3x + 2) = 0$
37. $(2x - 3)(3x + 7) = 0$ **38.** $(x - 6)(5x - 4) = 0$

In Exercises 39–58, solve each equation by factoring.

39. $x^2 + 5x + 6 = 0$ **40.** $x^2 - 6x - 7 = 0$
41. $x^2 - 6x + 8 = 0$ **42.** $x^2 + 2x - 15 = 0$
43. $x^2 - 15 = 2x$ **44.** $x^2 - 7x = -6$
45. $x^2 = 4x - 3$ **46.** $x^2 - 13x + 40 = 0$
47. $x^2 - 81 = 0$ **48.** $x^2 - 64 = 0$
49. $x^2 + 5x - 36 = 0$ **50.** $x^2 + 12x + 20 = 0$
51. $3x^2 + 7x = -2$ **52.** $4x^2 + x = 3$
53. $5x^2 + 11x = -2$ **54.** $2x^2 = -5x + 3$
55. $3x^2 - 4x = -1$ **56.** $5x^2 + 16x + 12 = 0$
57. $4x^2 - 9x + 2 = 0$ **58.** $6x^2 + x - 2 = 0$

In Exercises 59–78, solve the equation, using the quadratic formula. If the equation has no real solution, so state.

59. $x^2 - x - 30 = 0$ **60.** $x^2 + 9x + 14 = 0$
61. $x^2 - 3x - 10 = 0$ **62.** $x^2 - 5x - 14 = 0$
63. $x^2 - 8x = 9$ **64.** $x^2 = -8x + 15$
65. $x^2 - 2x + 3 = 0$ **66.** $2x^2 - x - 3 = 0$
67. $x^2 - 4x + 2 = 0$ **68.** $2x^2 - 5x - 2 = 0$
69. $2x^2 - x = 4$ **70.** $5x^2 + 3x - 3 = 0$
71. $4x^2 - x - 1 = 0$ **72.** $4x^2 - 5x - 3 = 0$
73. $2x^2 + 7x + 5 = 0$ **74.** $3x^2 = 9x - 5$
75. $3x^2 - 10x + 7 = 0$ **76.** $4x^2 + 7x - 1 = 0$
77. $4x^2 - 11x + 13 = 0$ **78.** $5x^2 + 9x - 2 = 0$

CHALLENGE PROBLEMS/GROUP ACTIVITIES

79. *Air Conditioning* The yearly profit p of Arnold's Air Conditioning is given by $p = x^2 + 15x - 100$, where x is the number of air conditioners produced and sold. How many air conditioners must be produced and sold to have a yearly profit of \$45,000?

80. **a)** Explain why solving $(x - 4)(x - 7) = 6$ by setting each factor equal to 6 is not correct.
b) Determine the correct solution to $(x - 4)(x - 7) = 6$.

81. The radicand in the quadratic formula, $b^2 - 4ac$, is called the **discriminant**. How many real number solutions will the quadratic equation have if the discriminant is (a) greater than 0, (b) equal to 0, or (c) less than zero? Explain your answer.

RESEARCH ACTIVITY

82. Italian mathematician Girolamo Cardano (1501–1576) is recognized for his skill in solving equations. Write a paper about his life and his contributions to mathematics, in particular his contribution to solving equations.

6.10 FUNCTIONS AND THEIR GRAPHS

The concepts of relations and functions are extremely important in mathematics. A **relation** is any set of ordered pairs. Therefore, every graph will be a relation. A function is a special type of relation. Suppose you are purchasing oranges at a supermarket where each orange costs \$0.20. Then one orange would cost \$0.20, two oranges $2 \times \$0.20 = \$0.40$, three oranges \$0.60, and so on. We can indicate this relation in a table of values.

Number of oranges	Cost
0	0.00
1	0.20
2	0.40
3	0.60
⋮	⋮
10	2.00
⋮	⋮

In general, the cost for purchasing n oranges will be 20 cents times the number of oranges, or $0.20n$. We can represent the cost, c, of n items by the equation $c = 0.20n$. Since the value of c depends on the value of n, we refer to n as the *independent variable* and c as the *dependent variable. Note for each value of the independent variable, n, there is one and only one value of the dependent variable, c.* Such an equation is called a **function**. In the equation $c = 0.20n$, the value of c depends on the value of n, so we say that "c is a function of n."

> A **function** is a special type of relation where each value of the independent variable corresponds to a unique value of the dependent variable.

The set of values that can be used for the independent variable is called the **domain** of the function, and the resulting set of values obtained for the dependent variable is called the **range**. The domain and range for the function $c = 0.20n$ are illustrated in Fig. 6.29.

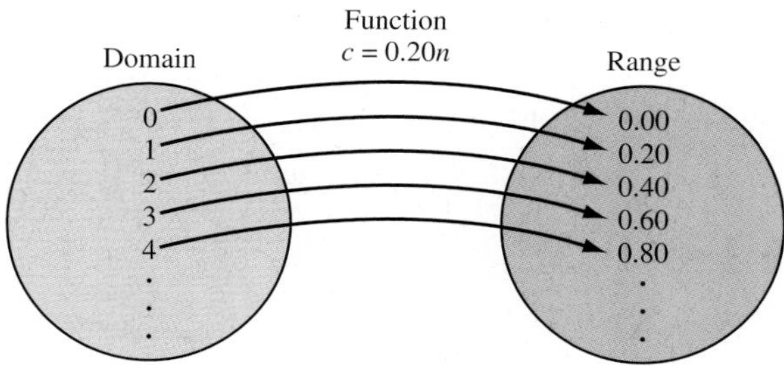

Figure 6.29

When we graphed equations of the form $ax + by = c$ in Section 6.7, we found that they were straight lines. For example, the graph of $y = 2x - 1$ is illustrated in Fig. 6.30.

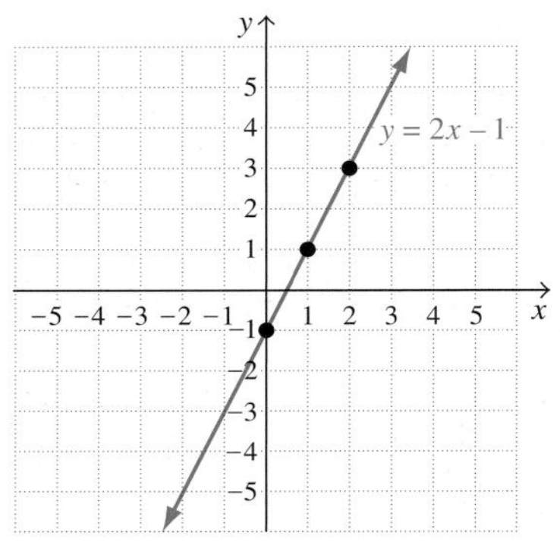

Figure 6.30

Is the equation $y = 2x - 1$ a function? To answer this question, we must ask, "Does each value of x correspond to a unique value of y?" The answer is yes; therefore, this equation is a function.

For the equation $y = 2x - 1$, we say that "y is a function of x" and write $y = f(x)$. The notation $f(x)$ is read "f of x." When we are given an equation that is a function, we may replace the y in the equation with $f(x)$, since $f(x)$ represents y. Thus, $y = 2x - 1$ may be written $f(x) = 2x - 1$.

To evaluate a function for a specific value of x, replace each x in the function with the given value, then evaluate. For example, to evaluate $f(x) = 2x - 1$ when $x = 8$, we do the following.

$$f(x) = 2x - 1$$
$$f(8) = 2(8) - 1 = 16 - 1 = 15$$

Thus, $f(8) = 15$. Since $f(x) = y$, when $x = 8$, $y = 15$. What is the domain and range of $f(x) = 2x - 1$? Because x can be any real number, the domain is the set of real numbers, symbolized $\mathbb{R}$. The range is also $\mathbb{R}$.

We can determine whether a graph represents a function by using the **vertical line test**: If a vertical line can be drawn so that it intersects the graph at more than one point, then each x does not have a unique y and the graph does not represent a function. If a vertical line cannot be made to intersect the graph in at least two different places, then the graph represents a function.

EXAMPLE 1 *Using the Vertical Line Test*

Use the vertical line test to determine which of the graphs in Figure 6.31 represent functions.

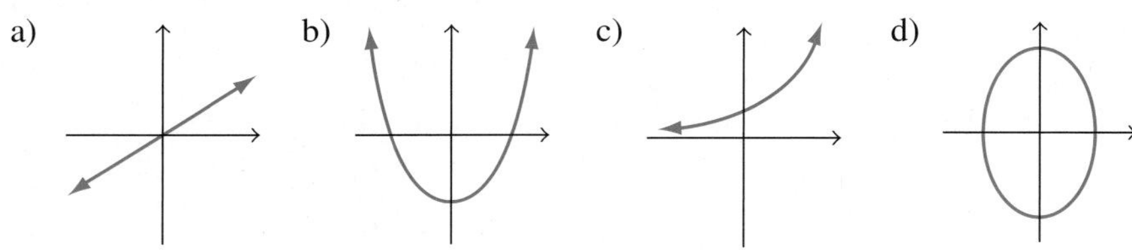

Figure 6.31

SOLUTION (a), (b), and (c) represent functions, but (d) does not.

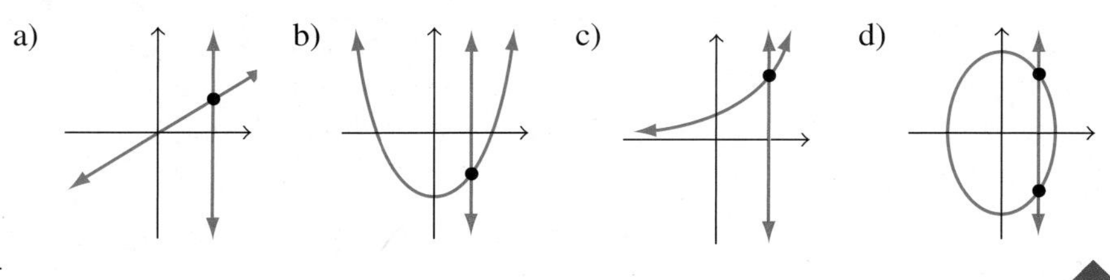

There are many real-life applications of functions. In fact, all the applications illustrated in Sections 6.2 through 6.4 were functions.

In this section, we will discuss three types of functions: linear functions, quadratic functions, and exponential functions.

DID YOU KNOW

Sea of Tranquillity

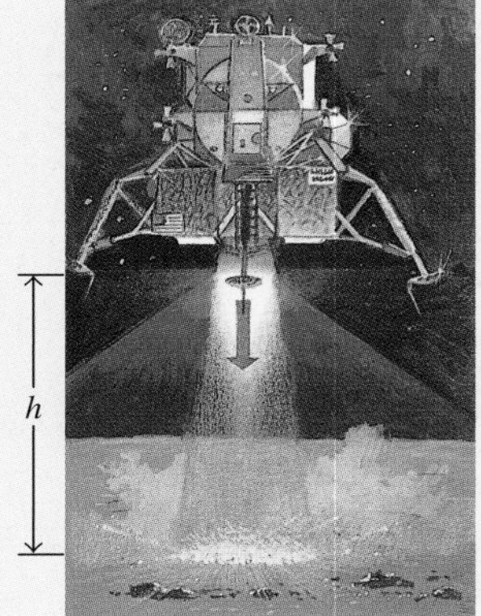

Apollo II touched down at Mare Tranqillitatis, the Sea of Tranquillity. The rock samples taken there placed the age of the rock at 3.5 billion years old, as old as the oldest known Earth rocks.

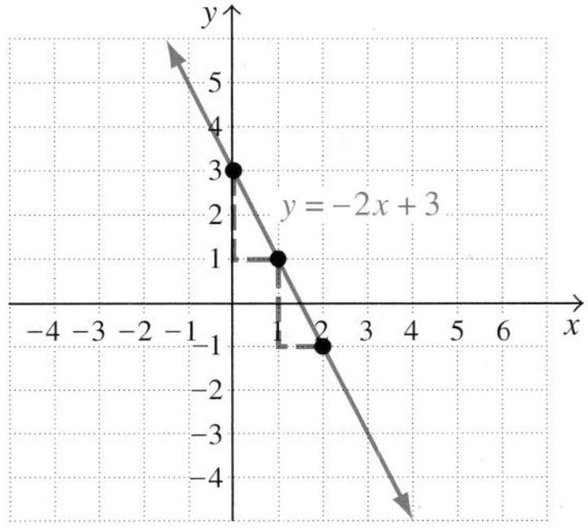

Figure 6.32

LINEAR FUNCTIONS

In Section 6.7, we graphed linear equations. The graph of any linear equation of the form $y = ax + b$ will pass the vertical line test, and so equations of the form $y = ax + b$ are **linear functions**. If we wished, we could write the linear function as $f(x) = ax + b$ since $f(x)$ means the same as y.

EXAMPLE 2 *Salary as a Linear Function*

Donald Karecki is part owner of a newly formed computer software company. His monthly salary, m, is given by the function $m(s) = 500 + 0.10s$, where s is his monthly sales. If Donald's sales for the month of July are \$20,000 what will his salary be for July?

SOLUTION Substitute 20,000 for s in the function.

$$\begin{aligned} m(s) &= 500 + 0.10s \\ m(20{,}000) &= 500 + 0.10(20{,}000) \\ m(20{,}000) &= 500 + 2000 = \$2500 \end{aligned}$$

Thus, if Donald has \$20,000 in sales for the month of July, his monthly salary for July is \$2500. ▲

GRAPHS OF LINEAR FUNCTIONS

The graphs of linear functions are straight lines that will pass the vertical line test. In Section 6.7, we discussed how to graph linear equations. Linear functions can be graphed by plotting points, by using intercepts, or by using the slope and y-intercept.

EXAMPLE 3 *Graphing a Linear Function*

Graph $f(x) = -2x + 3$ by using the slope and y-intercept.

SOLUTION Since $f(x)$ means the same as y, we can rewrite this function as $y = -2x + 3$. From Section 6.7, we know that the slope is -2 and the y-intercept is 3. Plot (0, 3) on the y-axis. Then plot the next point by moving *down* two units and to the *right* 1 unit (see Fig. 6.32). A third point has been plotted in the same way. The graph of $f(x) = -2x + 3$ is the line drawn through these three points. ▲

QUADRATIC FUNCTIONS

The standard form of a quadratic equation is $y = ax^2 + bx + c$, $a \neq 0$. We will learn shortly that graphs of equations of this form always pass the vertical line test and are functions. Therefore, equations of the form $y = ax^2 + bx + c$, $a \neq 0$, may be referred to as **quadratic functions**. We may express quadratic functions using function notation as $f(x) = ax^2 + bx + c$. Two examples of quadratic functions are $y = 2x^2 + 5x - 7$ and $y = -\frac{1}{2}x^2 + 4$.

EXAMPLE 4 *Landing on the Moon*

On July 20, 1969, Neil Armstrong became the first person to walk on the moon. The velocity, v, of his spacecraft, the *Eagle,* in meters per second, was a function of time before touchdown, t, given by

$$v = f(t) = 3.2t + 0.45$$

DID YOU KNOW

Computing Reality with Mathematical Models

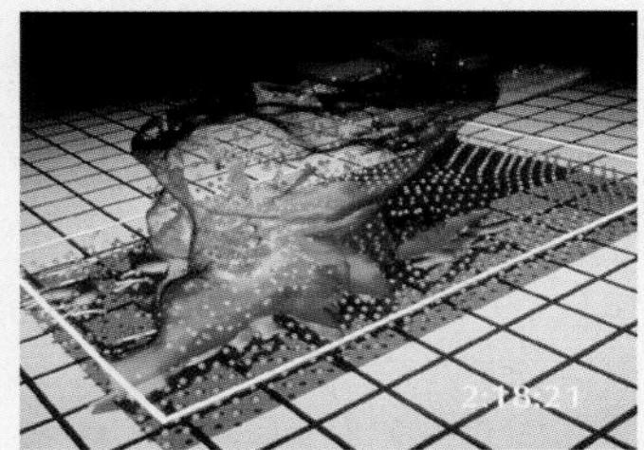

It is a relatively easy matter for scientists and mathematicians to describe and predict a simple motion like that of a falling object. When the phenomenon is complicated, such as making an accurate prediction of the weather, the mathematics becomes much more difficult. The National Weather Service has devised an algorithm that takes the temperature, pressure, moisture content, and wind velocity of more than 250,000 points in the Earth's atmosphere and applies a set of equations that, they believe, will reasonably predict what will happen at each point over time. Researchers at the National Center for Supercomputing Applications are working on a computer model that simulates thunderstorms. Researchers want to know why some thunderstorms can turn severe, even deadly.

The height of the spacecraft, h, above the moon's surface, in meters, was also a function of time before touchdown, given by

$$h = g(t) = 1.6t^2 + 0.45t$$

What was the velocity of the spacecraft and its distance from the surface of the moon

a) at 3 seconds before touchdown? b) at touchdown (0 seconds)?

SOLUTION

a) $v = f(t) = 3.2t + 0.45$, $\quad h = g(t) = 1.6t^2 + 0.45t$

$$f(3) = 3.2(3) + 0.45 = 9.6 + 0.45 = 10.05 \text{ m/sec}$$

$$g(3) = 1.6(3)^2 + 0.45(3) = 1.6(9) + 1.35 = 14.4 + 1.35 = 15.75 \text{ m}$$

The velocity 3 seconds before touchdown was 10.05 meters per second and the height 3 seconds before touchdown was 15.75 meters.

b) $v = f(t) = 3.2t + 0.45$, $\quad h = g(t) = 1.6t^2 + 0.45t$

$$f(0) = 3.2(0) + 0.45 = 0 + 0.45 = 0.45 \text{ m/sec}$$

$$g(0) = 1.6(0)^2 + 0.45(0) = 0 + 0 = 0 \text{ m}$$

The touchdown velocity was 0.45 meters per second. At touchdown, the *Eagle* is on the moon, and therefore the distance from the moon is 0 meters. ▲

GRAPHS OF QUADRATIC FUNCTIONS

The graph of every quadratic function is a parabola. Two parabolas are illustrated in Fig. 6.33. Note that both graphs represent functions since they pass the vertical line test. A parabola opens upward when the coefficient of the squared term, a, is greater than 0, as shown in Figure 6.33(a). A parabola opens downward when the coefficient, a, of the squared term is less than 0, as shown in Fig. 6.33(b).

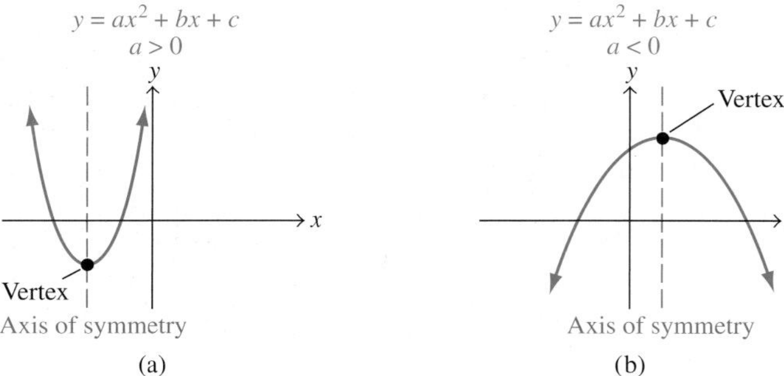

Figure 6.33

The *vertex* of a parabola is the lowest point on a parabola that opens upward and the highest point on a parabola that opens downward. Every parabola is *symmetric* with respect to a vertical line through its vertex. This line is called the *axis of symmetry* of the parabola. The x-coordinate of the vertex and the equation of the axis of symmetry can be found by using the following equation.

DID YOU KNOW

Gravity and the Parabola

Any golf player knows that a golf ball will arch in the same path going down as it did going up. Early gunners knew it too. To hit a distant target, the cannon barrel was pointed skyward, not directly at the target. A cannon fired at a 45° angle will travel the greatest horizontal distance. What the golfer and gunner alike were allowing for is the effect of gravity on a projectile. Projectile motion follows a parabolic path. Galileo was neither a gunner nor a golfer, but he gave us the formula that effectively describes that motion and the distance traveled by an object if it is projected at a specific angle with a specific initial velocity.

Axis of Symmetry of a Parabola

$$x = \frac{-b}{2a}$$

Once the x-coordinate of the vertex has been determined, the y-coordinate can be found by substituting the value found for the x-coordinate into the quadratic equation and evaluating the equation. This procedure is illustrated in Example 5.

EXAMPLE 5 ***Describing the Graph of a Quadratic Equation***

Consider the equation $y = 2x^2 - 4x - 6$.

a) Determine whether the graph will be a parabola that opens upward or downward.
b) Find the equation of the axis of symmetry of the parabola.
c) Find the vertex of the parabola.

SOLUTION

a) Since $a = 2$, which is greater than 0, the parabola opens upward.
b) To find the axis of symmetry, we use the equation $x = -b/2a$. In the equation $y = 2x^2 - 4x - 6$, $a = 2$, $b = -4$, and $c = -6$, so

$$x = \frac{-b}{2a} = \frac{-(-4)}{2(2)} = \frac{4}{4} = 1$$

The equation of the axis of symmetry is $x = 1$.

c) The x-coordinate of the vertex is 1, from part (b). To find the y-coordinate, we substitute 1 for x in the equation and then evaluate.

$$\begin{aligned} y &= 2x^2 - 4x - 6 \\ &= 2(1)^2 - 4(1) - 6 \\ &= 2(1) - 4 - 6 \\ &= 2 - 4 - 6 \\ &= -8 \end{aligned}$$

Therefore, the vertex of the parabola is located at the point $(1, -8)$ on the graph. ▲

General Procedure to Sketch the Graph of a Quadratic Equation

1. Determine whether the parabola opens upward or downward.
2. Determine the equation of the axis of symmetry.
3. Determine the vertex of the parabola.
4. Determine the y-intercept by substituting $x = 0$ into the equation.
5. Determine the x-intercepts (if they exist) by substituting $y = 0$ into the equation and solving for x.
6. Draw the graph, making use of the information gained in steps 1 through 5. Remember the parabola will be symmetric with respect to the axis of symmetry.

In step 5, to determine the x-intercepts, you may use either factoring or the quadratic formula.

EXAMPLE 6 ***Graphing a Quadratic Equation***

Sketch the graph of the equation $y = x^2 - 6x + 8$.

SOLUTION We follow the steps outlined in the general procedure.

1. Because $a = 1$, which is greater than 0, the parabola opens upward.
2. $x = \dfrac{-b}{2a} = \dfrac{-(-6)}{2(1)} = \dfrac{6}{2} = 3$

   Thus, the axis of symmetry is $x = 3$.
3. $y = x^2 - 6x + 8$

   $y = (3)^2 - 6(3) + 8 = 9 - 18 + 8 = -1$

   Thus, the vertex is at $(3, -1)$.
4. $y = x^2 - 6x + 8$

   $y = 0^2 - 6(0) + 8 = 8$

   Thus, the y-intercept is at $(0, 8)$.
5. $0 = x^2 - 6x + 8$, or $x^2 - 6x + 8 = 0$

   We can solve this equation by factoring.

$$(x - 4)(x - 2) = 0$$

$$x - 4 = 0 \qquad \text{or} \qquad x - 2 = 0$$

$$x = 4 \qquad\qquad\qquad x = 2$$

   Thus, the x-intercepts are at $(4, 0)$ and $(2, 0)$.
6. The graph is shown in Fig. 6.34.

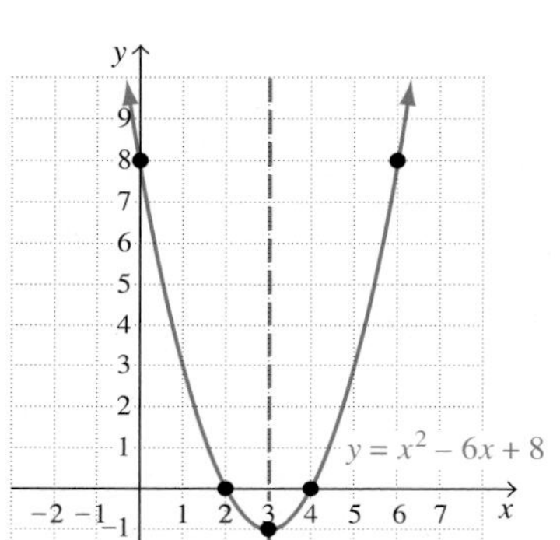

Figure 6.34

Note that the domain of the graph in Example 6, the possible x-values, is the set of all real numbers, $\mathbb{R}$. The range, the possible y-values, is the set of all real numbers greater than or equal to -1. When graphing parabolas, if you feel that you need additional points to graph, you can always substitute values for x and find the corresponding values of y and plot those points. For example, if you substituted 1 for x, the corresponding value of y is 3. Thus, you could plot the point $(1, 3)$.

EXAMPLE 7

a) Sketch the graph of the function $f(x) = -2x^2 + 3x + 4$.

b) Determine the domain and range of the function.

SOLUTION

a) Since $f(x)$ means y, we can replace $f(x)$ with y to obtain $y = -2x^2 + 3x + 4$. Now graph $y = -2x^2 + 3x + 4$ using the steps outlined in the general procedure.

   1. Since $a = -2$, which is less than 0, the parabola opens downward.
   2. Axis of symmetry: $x = \dfrac{-b}{2a} = \dfrac{-(3)}{2(-2)} = \dfrac{-3}{-4} = \dfrac{3}{4}$

3. y-coordinate of vertex:
$$\begin{aligned} y &= -2x^2 + 3x + 4 \\ &= -2\left(\frac{3}{4}\right)^2 + 3\left(\frac{3}{4}\right) + 4 \\ &= -2\left(\frac{9}{16}\right) + \frac{9}{4} + 4 \\ &= -\frac{9}{8} + \frac{9}{4} + 4 \\ &= -\frac{9}{8} + \frac{18}{8} + \frac{32}{8} = \frac{41}{8} \quad \text{or} \quad 5\frac{1}{8} \end{aligned}$$

Thus, the vertex is at $(\frac{3}{4}, 5\frac{1}{8})$.

4. y-intercept:
$$\begin{aligned} y &= -2x^2 + 3x + 4 \\ &= -2(0)^2 + 3(0) + 4 = 4 \end{aligned}$$

The y-intercept is $(0, 4)$.

5. x-intercepts:
$$\begin{aligned} y &= -2x^2 + 3x + 4 \\ 0 &= -2x^2 + 3x + 4 \quad \text{or} \quad -2x^2 + 3x + 4 = 0 \end{aligned}$$

This equation cannot be factored, so we will use the quadratic formula to solve it.

$$a = -2, \qquad b = 3, \qquad c = 4$$

$$\begin{aligned} x &= \frac{-b \pm \sqrt{b^2 - 4ac}}{2a} \\ &= \frac{-3 \pm \sqrt{3^2 - 4(-2)(4)}}{2(-2)} \\ &= \frac{-3 \pm \sqrt{9 + 32}}{-4} \\ &= \frac{-3 \pm \sqrt{41}}{-4} \end{aligned}$$

Since $\sqrt{41} \approx 6.4$,

$$x \approx \frac{-3 + 6.4}{-4} \approx \frac{3.4}{-4} \approx -0.85 \quad \text{or} \quad x \approx \frac{-3 - 6.4}{-4} \approx \frac{-9.4}{-4} \approx 2.35$$

6. Plot the vertex $(\frac{3}{4}, 5\frac{1}{8})$, the y-intercept $(0, 4)$, and the x-intercepts $(-0.85, 0)$ and $(2.35, 0)$. Then sketch the graph (Fig. 6.35).

b) The domain, the values that can be used for x, is the set of all real numbers, $\mathbb{R}$. The range, the values of y, is $y \leq 5\frac{1}{8}$.

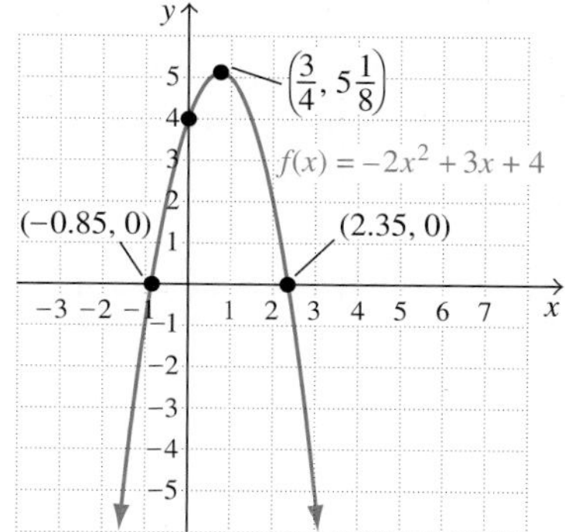

Figure 6.35

When we use the quadratic formula to find the x-intercepts of a graph, if the radicand, $b^2 - 4ac$, is a negative number, the graph has no x-intercepts. The graph will lie totally above or below the x-axis.

EXPONENTIAL FUNCTIONS

In Section 6.3, we discussed exponential equations. Recall that exponential equations are of the form $y = a^x$, $a > 0$, $a \neq 1$. The graph of every exponential equation will pass the vertical line test, and so every exponential equation is also an exponential function. Exponential functions may be written as $f(x) = a^x$, $a > 0$, $a \neq 1$.

In Section 6.3, we also introduced the natural exponential formula $P = P_0e^{kt}$. We can write this formula in function notation as $P(t) = P_0e^{kt}$. This expression is referred to as the **natural exponential function**. In Example 8, we use the natural exponential function.

EXAMPLE 8 *Evaluating an Exponential Decay Function*

The power supply of a satellite is a radioisotope. The power output, p, in watts remaining in the power supply is a function of the time the satellite is in space. If there are originally 100 grams of the isotope, the power remaining after t days is $p(t) = 100e^{-0.001t}$. What will be the remaining power after 1 year (or 365 days) in space?

SOLUTION Substitute 365 days for t in the function, and then evaluate using a calculator as described in Section 6.3.

$$\begin{aligned} p(t) &= 100e^{-0.001t} \\ p(365) &= 100e^{-0.001(365)} \\ &= 100e^{-0.365} \\ &\approx 100(0.694) \\ &\approx 69.4 \text{ watts} \end{aligned}$$

Thus, after 365 days, the power remaining will be about 69.4 watts.

EXAMPLE 9 *Evaluating an Exponential Growth Function*

The number of a certain type of bacteria present in a culture is determined by the function $f(x) = 5000(2^x)$, where x is the number of days the culture has been growing. Find the number of bacteria in a culture that has been growing for 4 days.

SOLUTION Substitute 4 days for x in the function, and then evaluate using a calculator as described in Section 6.3.

$$\begin{aligned} f(x) &= 5000(2^x) \\ f(4) &= 5000(2^4) \\ &= 5000(16) = 80{,}000 \text{ bacteria} \end{aligned}$$

After 4 days, 80,000 bacteria will be present.

GRAPHS OF EXPONENTIAL FUNCTIONS

What does the graph of an **exponential function** of the form $y = a^x$, $a > 0$, $a \neq 1$, look like? Examples 10 and 11 illustrate graphs of exponential functions.

EXAMPLE 10 ***Graphing an Exponential Function, $a > 1$***

a) Graph $y = 2^x$.
b) Determine the domain and range of the function.

SOLUTION

a) Substitute values for x and find the corresponding values of y. The graph is shown in Fig. 6.36.

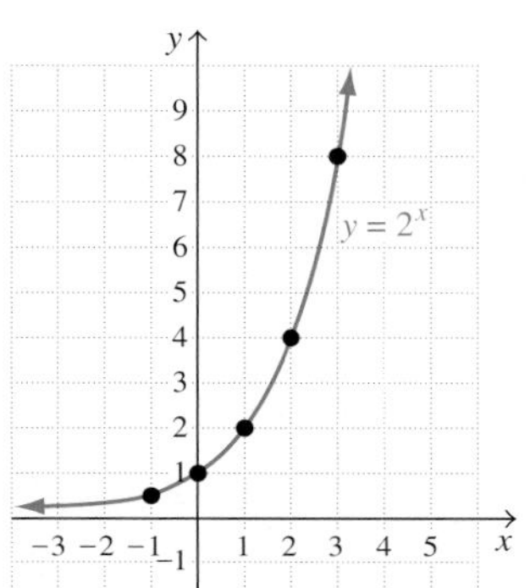

Figure 6.36

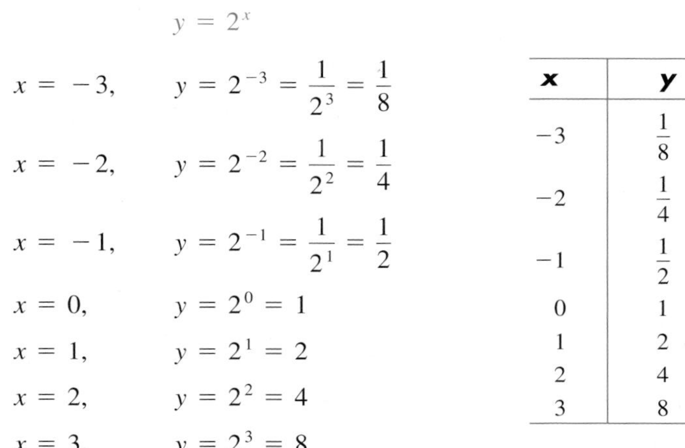

$$y = 2^x$$

$$x = -3, \qquad y = 2^{-3} = \frac{1}{2^3} = \frac{1}{8}$$

$$x = -2, \qquad y = 2^{-2} = \frac{1}{2^2} = \frac{1}{4}$$

$$x = -1, \qquad y = 2^{-1} = \frac{1}{2^1} = \frac{1}{2}$$

$$x = 0, \qquad y = 2^0 = 1$$

$$x = 1, \qquad y = 2^1 = 2$$

$$x = 2, \qquad y = 2^2 = 4$$

$$x = 3, \qquad y = 2^3 = 8$$

x	y
−3	$\frac{1}{8}$
−2	$\frac{1}{4}$
−1	$\frac{1}{2}$
0	1
1	2
2	4
3	8

b) The domain is all real numbers, $\mathbb{R}$. The range is $y > 0$. Note that y can never have a value of 0.

All exponential functions of the form $y = a^x$, $a > 1$, will have the general shape of the graph illustrated in Example 10 (Fig. 6.36). Since $f(x)$ is the same as y, the graphs of functions of the form $f(x) = a^x$, $a > 1$, will also have the general shape of the graph illustrated in Fig. 6.36. Can you now predict the shape of the graph of $y = e^x$? Remember: e has a value of about 2.7183.

EXAMPLE 11 ***Graphing an Exponential Function, $0 < a < 1$***

a) Graph $y = (\frac{1}{2})^x$.
b) Determine the domain and range of the function.

SOLUTION

a) We begin by substituting values for x and calculating values for y. We then plot the ordered pairs and use these points to sketch the graph. To evaluate a fraction with a negative exponent, we use the fact that

$$\left(\frac{a}{b}\right)^{-x} = \left(\frac{b}{a}\right)^{x}$$

For example,

$$\left(\frac{1}{2}\right)^{-3} = \left(\frac{2}{1}\right)^{3} = 8$$

DID YOU KNOW

Population Growth

Population growth during certain time periods can be described by an exponential function. Whether it is a population of bacteria, fish, flowers, or people, the same general trend emerges: a period of rapid (exponential) growth, which is then followed by a leveling-off period.

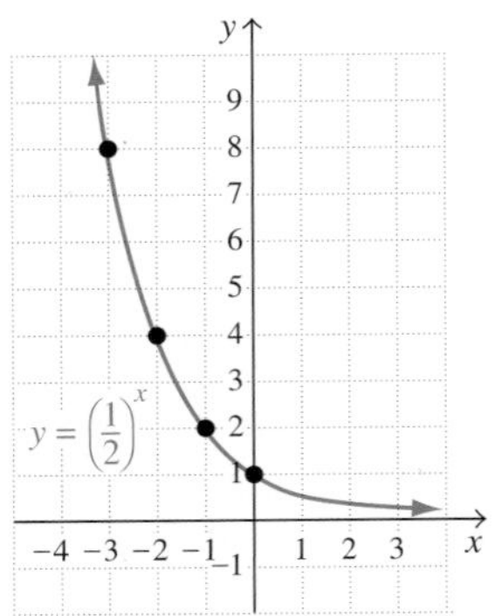

Figure 6.37

Then

$$y = \left(\frac{1}{2}\right)^x$$

$$x = -3, \quad y = \left(\frac{1}{2}\right)^{-3} = 2^3 = 8$$

$$x = -2, \quad y = \left(\frac{1}{2}\right)^{-2} = 2^2 = 4$$

$$x = -1, \quad y = \left(\frac{1}{2}\right)^{-1} = 2^1 = 2$$

$$x = 0, \quad y = \left(\frac{1}{2}\right)^{0} = 1$$

$$x = 1, \quad y = \left(\frac{1}{2}\right)^{1} = \frac{1}{2}$$

$$x = 2, \quad y = \left(\frac{1}{2}\right)^{2} = \frac{1}{4}$$

$$x = 3, \quad y = \left(\frac{1}{2}\right)^{3} = \frac{1}{8}$$

x	y
−3	8
−2	4
−1	2
0	1
1	$\frac{1}{2}$
2	$\frac{1}{4}$
3	$\frac{1}{8}$

The graph is illustrated in Fig. 6.37.

b) The domain is the set of all real numbers, $\mathbb{R}$. The range is $y > 0$.

All exponential functions of the form $y = a^x$ or $f(x) = a^x$, $0 < a < 1$, will have the general shape of the graph illustrated in Example 11 (Fig. 6.37).

DID YOU KNOW

The Electronic Superhighway

Electronic mail, or e-mail, is fast becoming one of the most popular and easiest methods of communicating. The number of e-mails internationally is expected to increase exponentially. According to a survey of office workers* in the United States, Canada, the United Kingdom, and Germany, e-mail ranks second only to the telephone as the most frequent method of communicating. In the United States, an office worker sends and receives, on average, 52 messages daily by telephone compared with 36 by e-mail. Forty-four percent of office workers surveyed prefer e-mail as a method of communication compared with 32% preferring the telephone. The number of e-mails per day in 2002 is expected to be approximately 610.15 million.

*Information from *Pitney Bowes Inc., Gallup Organization,* and the *Institute for the Future.*

EXAMPLE 12 *Is the Growth Exponential?*

The number of U.S. mobile phone users increased tremendously during the 1990s. The graph shows the number of U.S. mobile phone users from 1984 through 1998 projected through 1999.

a) Does the graph approximate the graph of an exponential function?
b) Estimate the number of U.S. mobile phone users in 1999.

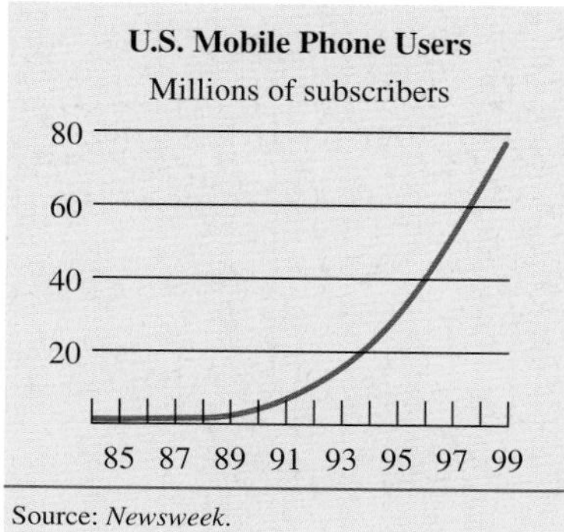

SOLUTION

a) Yes, the graph has the approximate shape of an exponential function. A function that increases rapidly with this general shape, is, or approximates, an exponential function.
b) From the graph, we see that the year marked 1999 corresponds to about 79 million mobile phone users.

SECTION 6.10 EXERCISES

CONCEPT/WRITING EXERCISES

1. What is a function?
2. What is a relation?
3. What is the domain of a function?
4. What is the range of a function?
5. Explain how and why the vertical line test can be used to determine whether a graph is a function.
6. Give three examples of one quantity being a function of another quantity.

PRACTICE THE SKILLS

In Exercises 7–22, determine whether the graph is a function. For each function give the domain and range.

7. 8.

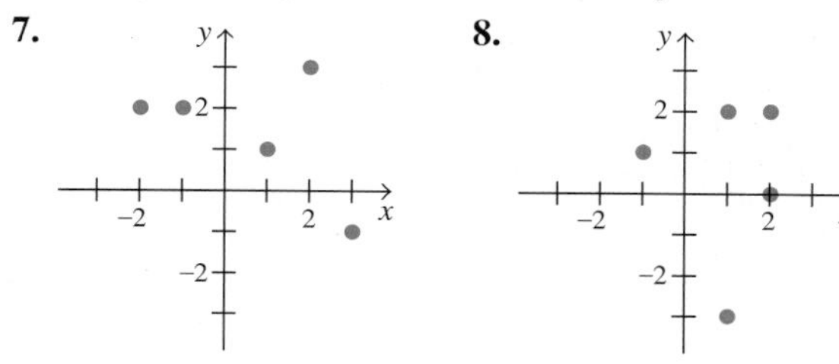

9.

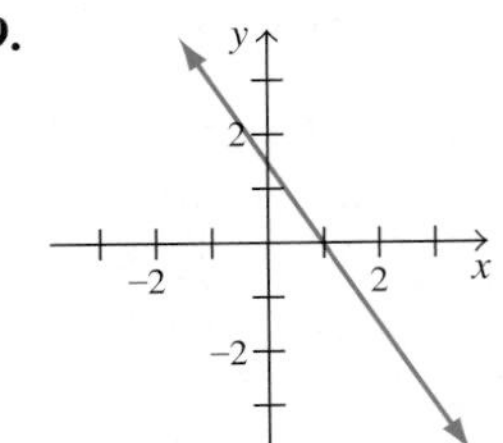

10.

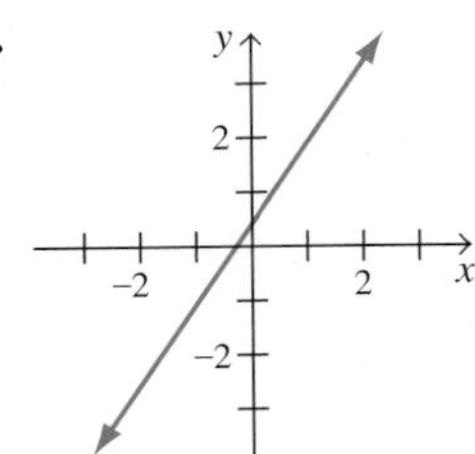

11.

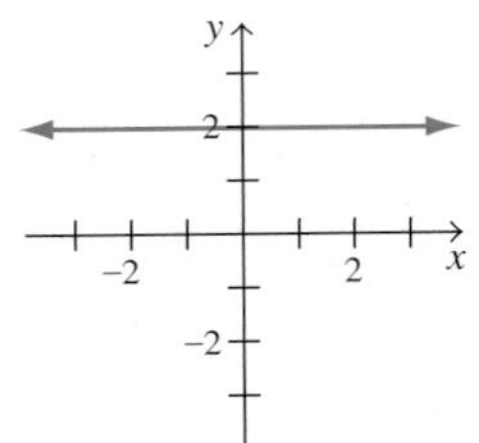

12.

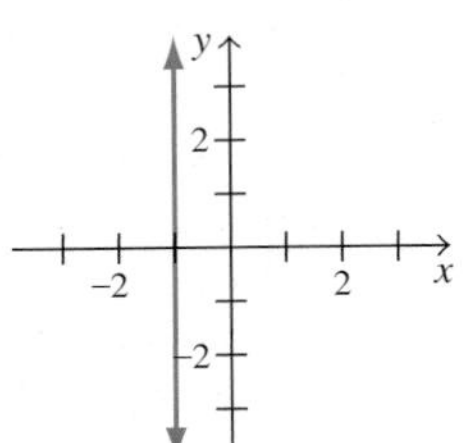

13.

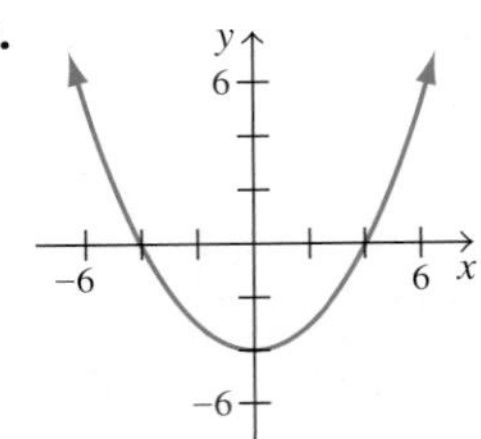

14.

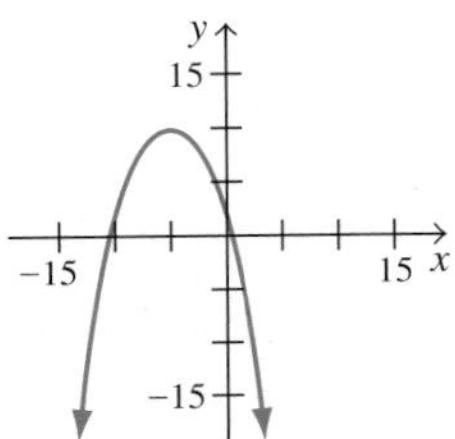

15.

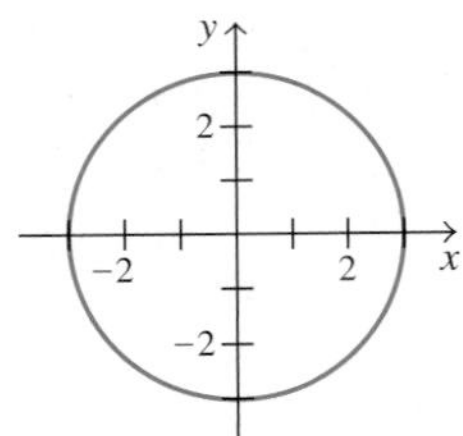

16.

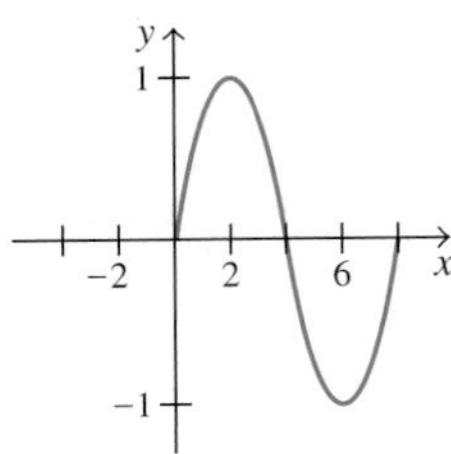

17.

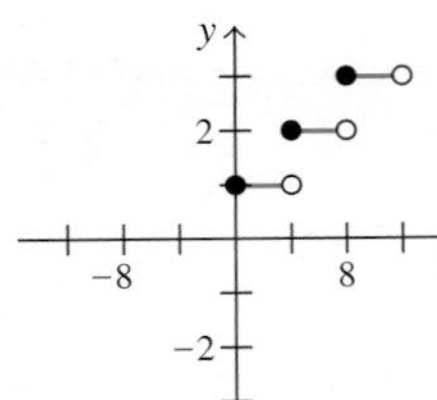

18.

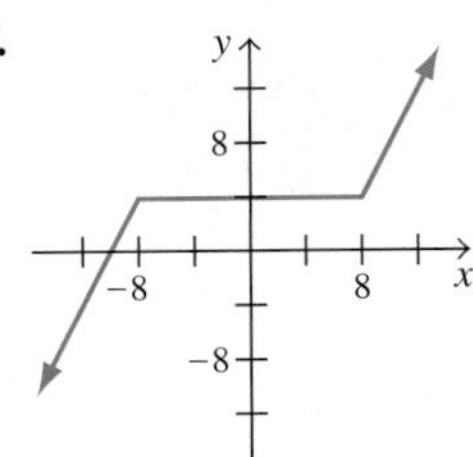

19.

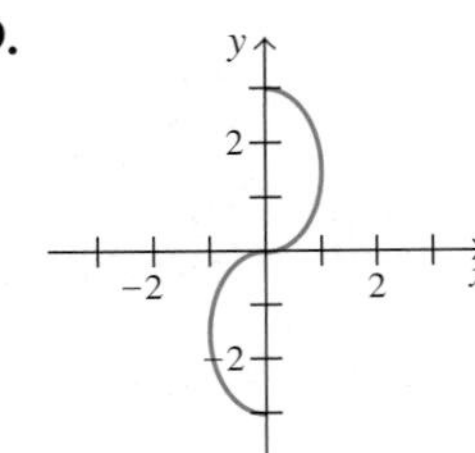

20.

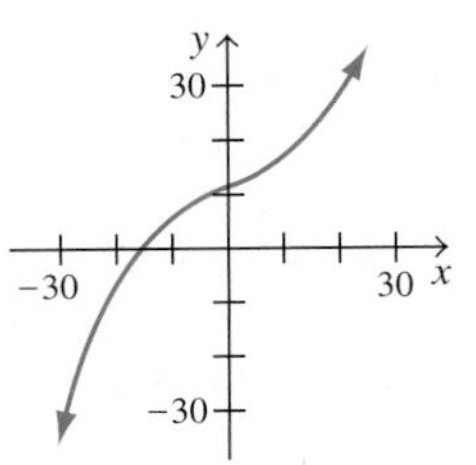

21.

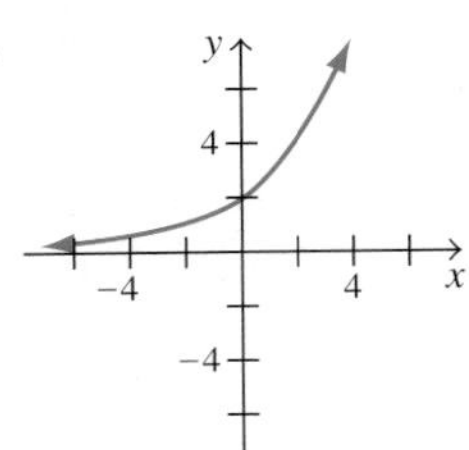

22.

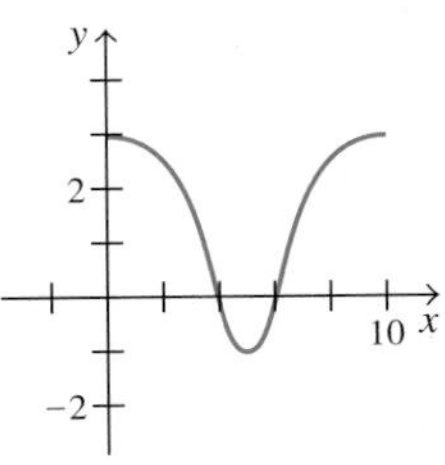

In Exercises 23–28, determine whether the set of ordered pairs is a function.

23. $\{(1, 5), (2, 6), (3, 7)\}$

24. $\{(3, 2), (4, 3), (5, 4)\}$

25. $\{(2, 2), (3, 6), (2, 4)\}$

26. $\{(1, 1), (1, 6), (1, 7)\}$

27. $\{(7, 1), (6, 1), (5, 1)\}$

28. $\{(1, 7), (1, 6), (1, 5)\}$

In Exercises 29–42, evaluate the function for the given value of x.

29. $f(x) = x + 6, \quad x = 9$

30. $f(x) = 3x + 10, \quad x = 2$

31. $f(x) = -2x - 7, \quad x = -4$

32. $f(x) = -5x + 3, \quad x = -1$

33. $f(x) = 10x - 6, \quad x = 0$

34. $f(x) = 7x - 6, \quad x = 4$

35. $f(x) = x^2 + 2x + 4, \quad x = 6$

36. $f(x) = x^2 - 12, \quad x = 6$

37. $f(x) = 2x^2 - 2x - 8, \quad x = -2$

38. $f(x) = -x^2 + 3x + 7, \quad x = 2$

39. $f(x) = -3x^2 + 5x + 4, \quad x = -3$

40. $f(x) = 5x^2 + 2x + 5, \quad x = 4$

41. $f(x) = -6x^2 - 6x - 12, \quad x = -3$

42. $f(x) = -3x^2 + 5x - 9, \quad x = -2$

In Exercises 43–48, graph the function by using the slope and y-intercept.

43. $f(x) = 3x + 1$

44. $f(x) = -x - 6$

45. $f(x) = -4x + 2$

46. $f(x) = 2x + 5$

47. $f(x) = \frac{3}{2}x - 1$

48. $f(x) = -\frac{1}{2}x + 3$

In Exercises 49–64,

a) *determine whether the parabola will open upward or downward.*

b) *find the equation of the axis of symmetry.*

c) *find the vertex.*

d) *find the y-intercept.*
e) *find the x-intercepts if they exist.*
f) *sketch the graph.*
g) *find the domain and range of the function.*

49. $y = x^2 - 1$
50. $y = x^2 - 4$
51. $y = -x^2 + 4$
52. $y = -x^2 + 16$
53. $f(x) = -x^2 - 4$
54. $y = -2x^2 - 8$
55. $y = 2x^2 - 3$
56. $f(x) = -3x^2 - 6$
57. $f(x) = x^2 + 4x + 10$
58. $y = x^2 - 2x + 8$
59. $y = x^2 + 5x + 6$
60. $y = x^2 - 7x - 8$
61. $y = -x^2 + 4x - 6$
62. $y = -x^2 + 8x - 8$
63. $y = -3x^2 + 14x - 8$
64. $y = 2x^2 - x - 6$

In Exercises 65–76, draw the graph of the function and state the domain and range.

65. $y = 3^x$
66. $f(x) = 4^x$
67. $y = (\frac{1}{3})^x$
68. $y = (\frac{1}{4})^x$
69. $f(x) = 2^x + 1$
70. $y = 3^x - 1$
71. $y = 4^x + 1$
72. $y = 2^x - 1$
73. $y = 3^{x-1}$
74. $y = 3^{x+1}$
75. $f(x) = 4^{x+1}$
76. $y = 4^{x-1}$

PROBLEM SOLVING

77. *Finding Distances* The distance a car travels, $d(t)$, at a constant 60 mph is given by the function $d(t) = 60t$, where t is the time in hours. Find the distance traveled in

a) 3 hours
b) 7 hours

78. *Real Estate* Dara Lanier's weekly salary as a real estate agent is given by the function $f(x) = 160 + 0.01x$, where x is her dollar sales for the week. Determine her salary for a specific week if during that week she sells one house for \$90,000.

79. *Room Temperature* A bedroom is at 80°F. The temperature, T, in the bedroom A minutes after the air conditioner is turned on can be approximated by the function

$$T(A) = -0.02A^2 - 0.34A + 80, \qquad 0 \le A \le 15$$

a) Estimate the room temperature 5 minutes after the air conditioner is turned on.
b) Estimate the room temperature 9 minutes after the air conditioner is turned on.

80. *Automobile Stopping Distance* The stopping distance, d, in meters, for a car traveling v kilometers per hour is given by the function $d = f(v) = 0.18v + 0.01v^2$. Determine the stopping distance for the following speeds.

a) 88 km/hr
b) 72 km/hr

81. *Expected Growth* The town of Branchport presently has 4000 residents. The expected future population can be approximated by the function $P(x) = 4000(1.3)^{0.1x}$, where x is the number of years in the future. Find the expected population of Branchport in

a) 10 years.
b) 50 years.

82. *Decay of Plutonium* Plutonium, a radioactive material used in most nuclear reactors, decays exponentially at a rate of 0.003% per year. The amount of plutonium, P, left after t years can be found by the formula $P = P_0e^{-0.00003t}$, where P_0 is the original amount of plutonium present. If there are originally 2000 grams of plutonium, find the amount of plutonium left after 50 years.

83. *Player's Salaries* The average salaries of National Football League (NFL) players are on the rise. The graph shows the average salaries, in thousands of dollars, of NFL players from 1933 through 1997 projected through 1998.

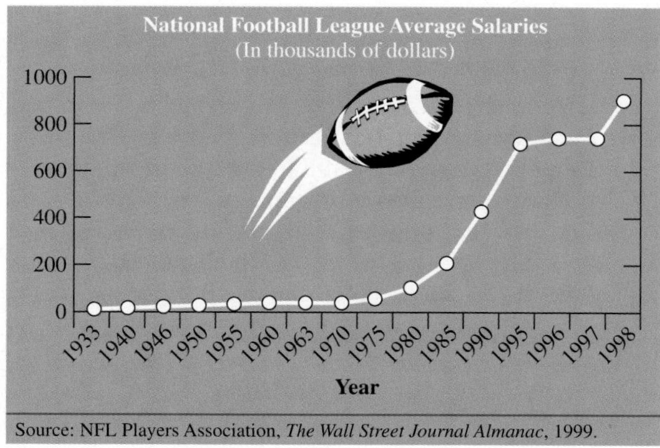

a) Does the graph approximate the graph of an exponential function for the years 1970 through 1995? Explain.
b) Estimate the average player's salary in the NFL in 1995.

84. *Cost of a PC* The graph shows the U.S. average cost of a personal computer (PC), in thousands of dollars, from 1995 through 1998 projected through 2003.

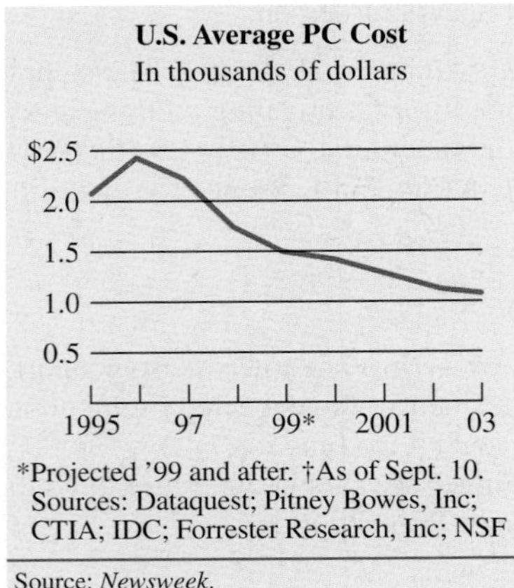

Source: *Newsweek*.

a) From 1995 through 1999 projected through 2003, does the graph approximate the graph of an exponential function? Explain.
b) Estimate the U.S. average cost of a PC in 2002.

85. The spacing of the frets on the neck of a classical guitar is determined from the equation $d = (21.9)(2)^{(20-x)/12}$, where x = the fret number and d = the distance in centimeters of the xth fret from the bridge.

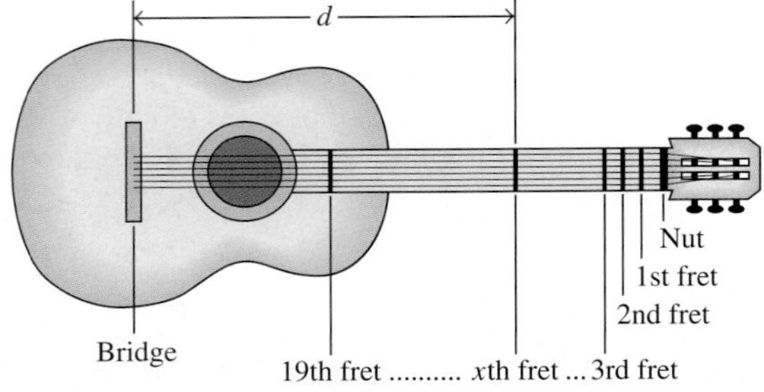

a) Determine how far the 19th fret should be from the bridge (rounded to one decimal place).
b) Determine how far the 4th fret should be from the bridge (rounded to one decimal place).
c) The distance of the nut from the bridge can be found by letting $x = 0$ in the given exponential equation. Find the distance (rounded to one decimal place).

CHALLENGE PROBLEMS/GROUP ACTIVITIES

86. *Appreciation of a House* A house initially cost $85,000. The value, V, of the house after n years if it appreciates at a constant rate of 4% per year can be determined by the function $V = f(n) = \$85{,}000(1.04)^n$.
a) Determine $f(8)$ and explain its meaning.
b) After how many years is the value of the house greater than $153,000? (Find by trial and error.)

87. *Target Heart Rate* While exercising, a person's recommended target heart rate is a function of age. The recommended number of beats per minute, y, is given by the function $y = f(x) = -0.85x + 187$, where x represents a person's age in years. Determine the number of recommended heart beats per minute for the following ages and explain the results.
a) 20
b) 30
c) 50
d) 60
e) How long would a person have to live to have a recommended target rate of 85?

88. *Speed of Light* Light travels at about 186,000 miles per second through space. The distance, d, in miles that light travels in t seconds can be determined by the function $d(t) = 186{,}000t$.
a) Light reaches the moon from the Earth in about 1.3 sec. Determine the approximate distance from the Earth to the moon.
b) Express the distance in miles, d, traveled by light in t minutes as a function of time, t.
c) Light travels from the sun to the Earth in about 8.3 min. Determine the approximate distance from the sun to the Earth.

RESEARCH ACTIVITY

89. The idea of using variables in algebraic equations was introduced by the French mathematician François Viète (1540–1603). Write a paper about his life and his contributions to mathematics. In particular, discuss his work with algebra equations. (References include history of mathematics books, encyclopedias, and the Internet.)

CHAPTER 6 SUMMARY

IMPORTANT FACTS

Properties used to solve equations

If $a = b$, then $a + c = b + c$
Addition property of equality

If $a = b$, then $a - c = b - c$
Subtraction property of equality

If $a = b$, then $ac = bc$
Multiplication property of equality

If $a = b$, then $a/c = b/c$, $c \neq 0$
Division property of equality

Variation

Direct: $y = kx$

Inverse: $y = \dfrac{k}{x}$

Joint: $y = kxz$

Inequality symbols

$a < b$ means that a is less than b.
$a \leq b$ means that a is less than or equal to b.
$a > b$ means that a is greater then b.
$a \geq b$ means that a is greater than or equal to b.

Intercepts

To find the x-intercept, set $y = 0$ and solve the resulting equation for x.

To find the y-intercept, set $x = 0$ and solve the resulting equation for y.

Slope

Slope (m): $m = \dfrac{y_2 - y_1}{x_2 - x_1}$

Equations and formulas

Linear equation in two variables:

$$ax + by = c, \quad a \neq 0, \quad b \neq 0$$

Quadratic equation in one variable:

$$ax^2 + bx + c = 0, \quad a \neq 0$$

Quadratic equation (or function) in two variables:

$$y = ax^2 + bx + c, \quad a \neq 0$$

Exponential equation (or function):

$$y = a^x, \quad a \neq 1, \quad a > 0$$

Exponential growth or decay formula:

$$P = P_0 a^{kt}, \quad a \neq 1, \quad a > 0$$

Quadratic formula:

$$x = \frac{-b \pm \sqrt{b^2 - 4ac}}{2a}$$

Slope–intercept form of a line:

$$y = mx + b$$

Axis of symmetry of a parabola:

$$x = \frac{-b}{2a}$$

Zero-factor property

If $a \cdot b = 0$, then $a = 0$ or $b = 0$.

CHAPTER 6 REVIEW EXERCISES

6.1

In Exercises 1–6, evaluate the expression for the given value(s) of the variable.

1. $x^2 + 7, \quad x = 4$
2. $-x^2 + 8, \quad x = -3$
3. $4x^2 - 2x + 5, \quad x = 2$
4. $-x^2 + 7x - 3, \quad x = \frac{1}{2}$
5. $4x^3 - 7x^2 + 3x + 1, \quad x = -2$
6. $4x^2 - 2xy + 3y^2, \quad x = 2, \quad y = -1$

6.2

In Exercises 7–9, combine like terms.

7. $2x + 5 - 8 - x$
8. $3x + 4(x - 2) + 6x$
9. $2(x - 4) + \frac{1}{2}(2x + 3)$

In Exercises 10–14, solve the equation for the given variable.

10. $-2r - 8 = 20$
11. $3t + 8 = 6t - 13$
12. $\dfrac{x + 5}{6} = \dfrac{x - 3}{3}$
13. $4(x - 2) = 3 + 5(x + 4)$
14. $\dfrac{x}{3} + \dfrac{2}{5} = 4$
15. *Making Oatmeal* A recipe for Hot Oats Cereal calls for 2 cups of water and for $\frac{1}{3}$ cup of dry oats. How many cups of dry oats would be used with 3 cups of water?
16. *Laying Bricks* A mason lays 120 blocks in 1 hr 40 min. How long will it take her to lay 450 blocks?

6.3

In Exercises 17–20, use the formula to find the value of the indicated variable for the values given.

17. $A = lw$
Find A when $l = 13$ and $w = 8$ (geometry).

18. $V = 2\pi R^2 r^2$
Find V when $R = 3$, $r = 1\frac{3}{4}$, and $\pi = 3.14$ (geometry).

19. $z = \dfrac{\bar{x} - \mu}{\dfrac{\sigma}{\sqrt{n}}}$
Find $\bar{x}$ when $z = 2$, $\mu = 100$, $\sigma = 3$, and $n = 16$ (statistics).

20. $K = \frac{1}{2}mv^2$
Find m when $v = 30$ and $k = 4500$ (physics).

In Exercises 21–24, solve for y.

21. $4x - 6y = 12$

22. $5x + 6y = 18$

23. $2x - 3y + 52 = 30$

24. $-3x - 4y + 5z = 4$

In Exercises 25–28, solve for the variable indicated.

25. $A = lw$, for l

26. $P = 2l + 2w$, for w

27. $L = 2(wh + lh)$, for l

28. $a_n = a_1 + (n - 1)d$, for d

6.4

In Exercises 29–32, write the phrase in mathematical terms.

29. 7 decreased by 4 times x

30. 5 times x decreased by 3

31. 10 increased by 3 times r

32. 11 less than 8 divided by q

In Exercises 33–36, write an equation that can be used to solve the problems. Solve the equation and find the desired value(s).

33. Twelve decreased by 3 times a number is 21.

34. The product of 3 and a number, increased by 8, is 6 less than the number.

35. Five times the difference of a number and 4 is 45.

36. Fourteen more than 10 times a number is 8 times the sum of the number and 12.

In Exercises 37–40, write the equation and then find the solution.

37. MODELING - *Income Taxes* On a joint income tax return, Wesley's income was $\frac{1}{3}$ of Marie's income. Their total income was $48,000. Determine Wesley's income.

38. MODELING - *Lawn Chairs* Larry's Lawn Chair Company has fixed costs of $15,000 per month and variable costs of $9.50 per lawn chair manufactured. The company has $95,000 available to meet its total monthly expenditures. What is the maximum number of lawn chairs the company can manufacture in a month? (Fixed costs, like rent and insurance, are those that occur regardless of the level of production. Variable costs, such as those for materials, depend on the level of production.)

39. MODELING - *Restaurant Profit* John Smith owns two restaurants. His profit for a year at restaurant A is $12,000 greater than his profit at restaurant B. The total profit from both restaurants is $68,000. Determine the profit at each restaurant.

40. MODELING - *Renting versus Buying* A saw can be rented for $15 an hour and purchased for $300. For how many hours would the saw have to be rented for the rental cost to equal the cost of purchasing the saw?

6.5

In Exercises 41–44, find the quantity indicated.

41. R is inversely proportional to the square of S. If $R = 8$ when $S = 3$, find R when $S = 6$.

42. m is directly proportional to n. If $m = 80$ when $n = 4$, find m when $n = 12$.

43. W is directly proportional to L and inversely proportional to A. If $W = 80$ when $L = 100$ and $A = 20$, find W when $L = 50$ and $A = 40$.

44. z is jointly proportional to x and y and inversely proportional to the square of r. If $z = 12$ when $x = 20$, $y = 8$, and $r = 8$, find z when $x = 10$, $y = 80$, and $r = 3$.

45. *Map Reading* The scale of a map is 1 in. to 30 mi. What distance on the map represents 120 mi?

46. *Electric Bill* An electric company charges $0.162 per kilowatt-hour (kWh). What is the electric bill if 740 kWh are used in a month?

47. *A Falling Object* The distance, d, an object drops in free fall is directly proportional to the square of the time, t. If an object falls 16 ft in 1 sec, how far will an object fall in 5 sec?

48. *Area of a Circle* The area, A, of a circle varies directly with the square of its radius, r. If the area is 78.5 when the radius is 5, find the area when the radius is 8.

6.6

In Exercises 49–52, graph the solution set for the set of real numbers.

49. $6 + 7x \geq -3x - 4$

50. $3x + 7 \geq 5x + 9$

51. $3(x + 9) \leq 4x + 11$

52. $-3 \leq x + 1 < 7$

In Exercises 53–56, graph the solution set for the set of integers.

53. $2 + 7x > -12$
54. $5x + 13 \geq -22$
55. $-1 < x \leq 7$
56. $-8 \leq x + 2 \leq 7$

6.7

In Exercises 57–60, graph the ordered pair in the Cartesian coordinate system.

57. $(2, 4)$
58. $(-1, 3)$
59. $(-3, -4)$
60. $(6, -7)$

In Exercises 61 and 62, points A, B, and C are vertices of a rectangle. Plot the points. Find the coordinates of the fourth point, D, to complete the rectangle. Find the area of the rectangle.

61. $A(-3, 3)$, $B(2, 3)$, $C(2, -1)$

62. $A(-3, 1)$, $B(-3, -2)$, $C(4, -2)$

In Exercises 63–66, graph the equation by plotting points.

63. $x - y = 4$
64. $2x + 3y = 12$
65. $x = y$
66. $x = 3$

In Exercises 67–70, graph the equation, using the x- and y-intercepts.

67. $x + 4y = 8$
68. $3x - 2y = 6$
69. $4x - 3y = 12$
70. $2x + 3y = 9$

In Exercises 71–74, find the slope of the line through the given points.

71. $(1, 3)$, $(6, 2)$
72. $(3, -1)$, $(5, -4)$
73. $(-1, -4)$, $(5, 3)$
74. $(6, 2)$, $(6, -2)$

In Exercises 75–78, graph the equation by plotting the y-intercept and then plotting a second point by making use of the slope.

75. $y = 2x - 5$
76. $2y - 4 = 3x$
77. $2y + x = 8$
78. $y = -x - 1$

In Exercises 79 and 80, determine the equation of the graph.

79. **80.**

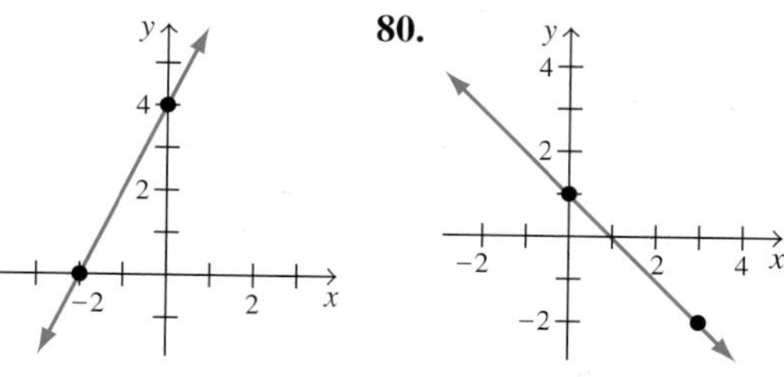

81. *Disability Income* The monthly disability income, I, that Nadja receives is $I = 460 - 0.5m$, where m is her monthly earnings for her part-time job for the previous month.

a) Draw a graph of disability income versus earnings for earnings up to and including \$920.

b) If Nadja earns \$600 in January, how much disability income will she receive in February?

c) If she received \$380 disability income in November, how much did she earn in October?

82. *Business Space Rental* The monthly rental cost, C, in dollars, for space in the Galleria Mall can be approximated by the equation $C = 1.70A + 3000$, where A is the area, in square feet, of space rented.

a) Draw a graph of monthly rental cost versus square feet for up to and including 12,000 ft^2.

b) Determine the monthly rental cost if 2000 ft^2 are rented.

c) If the rental cost is \$10,000 per month, how many square feet are rented?

6.8

In Exercises 83–86, graph the inequality.

83. $6x + 9y \leq 54$
84. $3x + 2y \geq 12$
85. $2x - 3y > 12$
86. $-7x - 2y < 14$

6.9

In Exercises 87–92, factor the trinomial. If the trinomial cannot be factored, so state.

87. $x^2 + 9x + 18$
88. $x^2 + x - 20$
89. $x^2 - 10x + 24$
90. $x^2 - 9x + 20$
91. $2x^2 + x - 21$
92. $3x^2 + 5x - 2$

In Exercises 93–96, solve the equation by factoring.

93. $x^2 + 5x + 6 = 0$
94. $x^2 - 6x = -5$
95. $3x^2 - 17x + 10 = 0$
96. $3x^2 = -7x - 2$

In Exercises 97–100, solve the equation, using the quadratic formula. If the equation has no real solution, so state.

97. $x^2 - 3x - 7 = 0$
98. $x^2 - 3x + 2 = 0$
99. $2x^2 - 3x + 4 = 0$
100. $2x^2 - x - 3 = 0$

6.10

In Exercises 101–104, determine whether the graph is a function. If it is a function give the domain and range.

101. **102.**

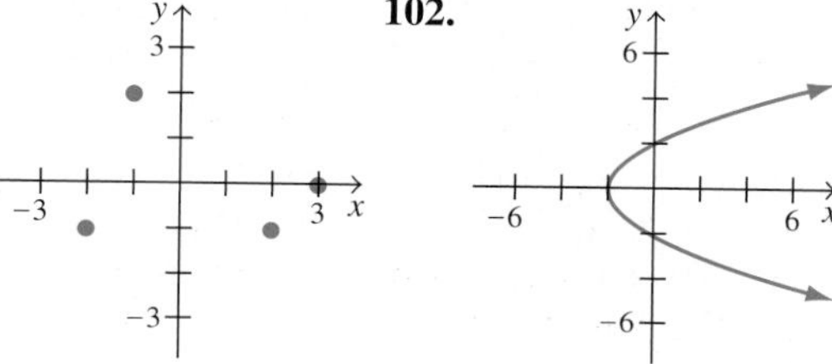

103.

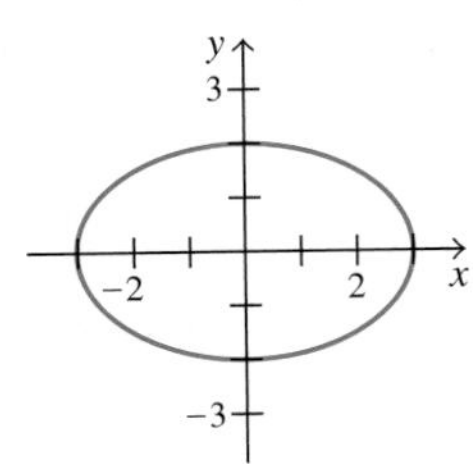

104.

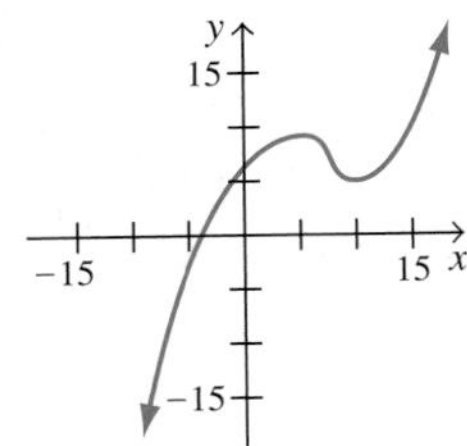

In Exercises 105–108, evaluate f(x) for the given value of x.

105. $f(x) = 2x + 10, \quad x = -3$

106. $f(x) = -3x + 8, \quad x = -2$

107. $f(x) = 2x^2 - 3x + 4, \quad x = 5$

108. $f(x) = -4x^2 + 7x + 9, \quad x = 4$

In Exercises 109 and 110, for each function

a) *determine whether the parabola will open upward or downward.*
b) *find the equation of the axis of symmetry.*
c) *find the vertex.*
d) *find the y-intercept.*
e) *find the x-intercepts if they exist.*
f) *sketch the graph.*
g) *find the domain and range.*

109. $y = -x^2 - 4x + 21$

110. $f(x) = 3x^2 - 24x - 30$

In Exercises 111 and 112, draw the graph of the function and state the domain and range.

111. $y = 2^{2x}$ **112.** $y = (\frac{1}{2})^x$

6.2, 6.3, 6.10

113. *Gas Mileage* The gas mileage, m, of a specific car can be estimated by the equation (or function)

$$m = 30 - 0.002n^2, \qquad 20 \le n \le 80$$

where n is the speed of the car in miles per hour. Estimate the gas mileage when the car travels at 60 mph.

114. *Auto Accidents* The approximate number of accidents in one month, n, involving drivers between 16 and 30 years of age inclusive can be approximated by the equation

$$n = 2a^2 - 80a + 5000, \qquad 16 \le a \le 30$$

where a is the age of the driver. Approximate the number of accidents in one month that involved
a) 18-year-olds **b)** 25-year-olds.

115. *Filtered Light* The percent of light filtering through Swan Lake, p, can be approximated by the function $P(x) = 100(0.92)^x$, where x is the depth in feet. Find the percent of light filtering through at a depth of 4.5 ft.

CHAPTER 6 TEST

1. Evaluate $-3x^2 + 6x + 9$, when $x = 2$.

In Exercises 2 and 3, solve the equation.

2. $3x + 5 = 2(4x - 7)$

3. $-2(x - 3) + 6x = 2x + 3(x - 4)$

In Exercises 4 and 5, write an equation to represent the problem. Then solve the equation.

4. The product of a number and 3, decreased by 10, is 11.

5. *Selling Crafts* Ross can rent a stall in a marketplace for one day for $60. He will sell crafts that cost him $4.35 apiece for $7.75 each. How many items must he sell in 1 day to break even?

6. Evaluate $L = ah + bh + ch$ when $a = 3$, $b = 4$, $c = 5$, and $h = 7$.

7. Solve $5x - 8y = 17$, for y.

8. L varies jointly as M and N and inversely as P. If $L = 12$ when $M = 8$, $N = 3$, and $P = 2$, find L when $M = 10$, $N = 5$, and $P = 15$.

9. For a constant area, the length, l, of a rectangle varies inversely as the width, w. If $l = 15$ ft when $w = 9$ ft, find the length of a rectangle with the same area if the width is 20 ft.

10. Graph the solution set of $-3x + 11 \le 5x + 35$ on the real number line.

11. Determine the slope of the line through the points $(-3, 5)$ and $(10, -26)$.

In Exercises 12 and 13, graph the equation.

12. $y = 3x - 4$ **13.** $2x - 3y = 15$

14. Graph the inequality $3y \ge 5x - 12$.

15. Solve the equation $x^2 - 3x = 28$ by factoring.

16. Solve the equation $3x^2 + 2x = 8$ by using the quadratic formula.

17. Determine whether the graph is a function. Explain your answer.

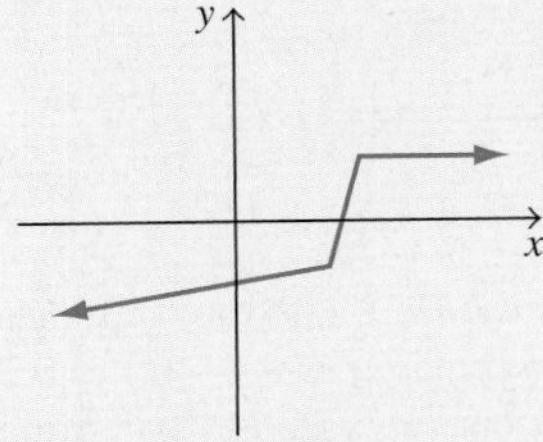

18. Evaluate $f(x) = -4x^2 - 11x + 5$ when $x = -2$.

19. For the equation $y = x^2 - 2x + 4$,
   a) determine whether the parabola will open upward or downward.
   b) find the equation of the axis of symmetry.
   c) find the vertex.
   d) find the y-intercept.
   e) find the x-intercepts if they exist.
   f) sketch the graph.
   g) find the domain and range of the function.

GROUP PROJECTS

ARCHEOLOGY—GAINING INFORMATION FROM BONES

1. Archeologists have developed formulas to predict the height and, in some cases, the age at death of the deceased by knowing the lengths of certain bones in the body. The long bones of the body grow at approximately the same rate. Thus, a linear relationship exists between the length of the bones and the person's height. If the length of one of these major bones—the femur (F), the tibia (T), the humerus (H), and the radius (R)—is known, the height, h, of a person can be calculated with one of the following formulas. The relationship between bone length and height is different for males and females.

Male	**Female**
$h = 2.24\text{F} + 69.09$	$h = 2.23\text{F} + 61.41$
$h = 2.39\text{T} + 81.68$	$h = 2.53\text{T} + 72.57$
$h = 2.97\text{H} + 73.57$	$h = 3.14\text{H} + 64.98$
$h = 3.65\text{R} + 80.41$	$h = 3.88\text{R} + 73.51$

All measurements are in centimeters.

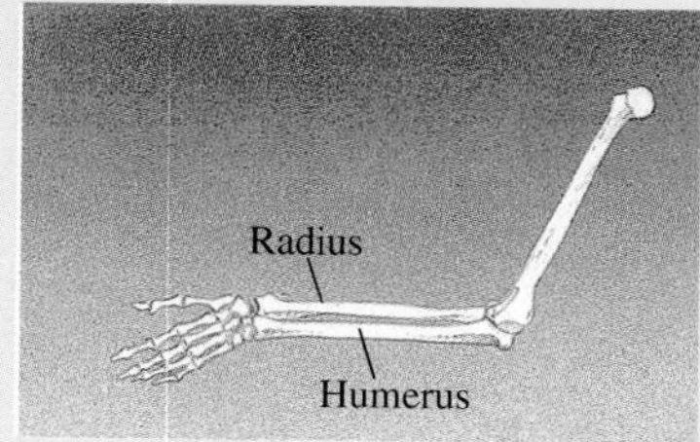

   a) Measure your humerus and use the appropriate formula to predict your height in centimeters. How close is this predicted height to your actual height? (The result is an approximation because measuring a bone covered with flesh and muscle is difficult.)
   b) Determine and describe where the femur and tibia bones are located.
   c) Dr. Juarez, an archeologist, had one female humerus that was 29.42 cm in length. He concluded that the height of the entire skeleton would have been 157.36 cm. Was his conclusion correct?
   d) If a 21-year-old woman is 167.64 cm tall, about how long should her tibia be?
   e) Sometimes the age of a person may be determined by using the fact that the height of a person, and the length of his or her long bones, decreases at the rate of 0.06 cm per year after the age of 30.
      i. At age 30, Jolene is 168 cm tall. Estimate the length of her humerus.
      ii. Estimate the length of Jolene's humerus when she is 60 years old.
   f) Select six people of the same sex and measure their height and one of the bones for which an equation is given (the same bone on each person). Each measurement should be made to the nearest 0.5 cm. For each person, you will have two measurements, which can be considered an ordered pair (bone length, height). Plot the ordered pairs on a piece of graph paper, with the bone length on the horizontal axis and the height on the vertical axis. Start the scale on both axes at zero. Draw a straight line that you feel is the best approximation, or best fit, through these points. Determine where the line crosses the y-axis and the slope of the line. Your y-intercept and slope should be close to the values in the given equation for that bone. (Reference: M. Trotter and G. C. Gleser, "Estimation of Stature from Long Bones of American Whites and Negroes," *American Journal of Physical Anthropology,* 1952, 10:463–514.)

GRAPHING CALCULATOR

2. The functions that we graphed in this chapter can be easily graphed with a graphing calculator (or grapher). If you do not have a graphing calculator, borrow one from your instructor or a friend.
   a) Explain how you would set the domain and range. The calculator key to set the domain and range

may be labeled *range* or *window*. Set the grapher with the following range or window settings: Xmin = −12, Xmax = 12, Xscl = 1, Ymin = −13, Ymax = 6, and Yscl = 1.

b) Explain how to enter a function in the graphing calculator. Enter the function $y = 3x^2 - 7x - 8$ in the calculator.

c) Graph the function you entered in part (b).

d) Learn how to use the *trace feature.* Then use it to estimate the x-intercepts. Record the estimated values for the x-intercepts.

e) Learn how to use the *zoom feature* to obtain a better approximation of the x-intercepts. Use the zoom feature twice and record the x-intercepts each time.

CHAPTER 7

Systems of Linear Equations & Inequalities

Imagine that you and a friend have a great idea for a new T-shirt. First you make a few to give away to friends, using your own money. Then other students see the shirt, and soon everyone on campus wants one. Suddenly, you have entered the T-shirt business. To make a profit in your business, you need to keep track of the cost of your materials, the quantity of shirts sold, and the price at which you sell them, a relatively straightforward calculation.

But now suppose you come up with three other designs, and you want to put them on sweatshirts as well as T-shirts, and you want to offer a variety of colors: black, white, blue, and maroon. The equation for finding the profitability of your venture becomes more complicated because there are more variables. To track your profits, you may need to develop and solve systems of equations.

For most business owners, numerous factors must be considered to determine not only whether the business is profitable, but also how much they should charge their customers, which production method is most efficient, what return they can expect by placing advertisements, and so on. Many small-business owners routinely make these calculations based on their own experience, mathematics, and sometimes computer programs. Larger companies often employ inventory analysts, quality control engineers, and efficiency experts to aid them, along with computers, to keep track of vast quantities of data. ■

Linear programming provides businesses and governments with a mathematical form of decision making that makes the most efficient use of time and resources. Telecommunications companies use it to route calls through satellites, like this one being tested, so that few of their customers will reach a "no circuits available" message.

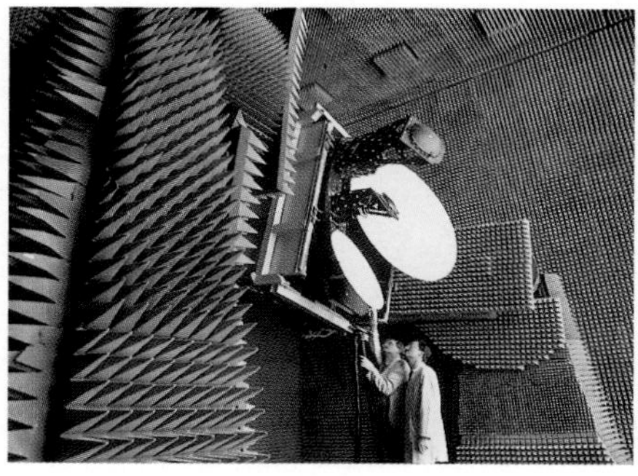

7.1 SYSTEMS OF LINEAR EQUATIONS

In Chapter 6, we discussed linear equations in two variables. In algebra, it is often necessary to find the common solution to two or more such equations. We refer to the equations in this type of problem as a **system of linear equations** or as **simultaneous linear equations**. A **solution to a system of equations** is the ordered pair or pairs that satisfy all equations in the system. A system of linear equations may have exactly one solution, no solution, or infinitely many solutions.

The solution to a system of linear equations may be found by a number of different techniques. In this section, we illustrate how a system of equations may be solved by graphing. In Section 7.2, we illustrate two algebraic methods, the substitution method and the addition method, for solving a system of linear equations.

EXAMPLE 1 *Is the Ordered Pair a Solution?*

Determine which of the ordered pairs is a solution to the following system of equations.

$$\begin{aligned} 2x - y &= 8 \\ 2x + y &= 4 \end{aligned}$$

a) $(0, -8)$ b) $(3, -2)$ c) $(1, 2)$

SOLUTION For the ordered pair to be a solution to the system, it must satisfy each equation in the system.

a)

$$\begin{aligned} 2x - y &= 8 \\ 2(0) - (-8) &= 8 \\ 8 &= 8 \quad \text{True} \end{aligned} \qquad \begin{aligned} 2x + y &= 4 \\ 2(0) + (-8) &= 4 \\ -8 &= 4 \quad \text{False} \end{aligned}$$

Since $(0, -8)$ does not satisfy both equations, it is not a solution to the system.

b)

$$\begin{aligned} 2x - y &= 8 \\ 2(3) - (-2) &= 8 \\ 6 + 2 &= 8 \\ 8 &= 8 \quad \text{True} \end{aligned} \qquad \begin{aligned} 2x + y &= 4 \\ 2(3) + (-2) &= 4 \\ 6 - 2 &= 4 \\ 4 &= 4 \quad \text{True} \end{aligned}$$

Since $(3, -2)$ satisfies both equations, it is a solution to the system.

c)

$$\begin{aligned} 2x - y &= 8 \\ 2(1) - 2 &= 8 \\ 2 - 2 &= 8 \\ 0 &= 8 \quad \text{False} \end{aligned} \qquad \begin{aligned} 2x + y &= 4 \\ 2(1) + 2 &= 4 \\ 2 + 2 &= 4 \\ 4 &= 4 \quad \text{True} \end{aligned}$$

Since $(1, 2)$ does not satisfy both equations in the system, it is not a solution to the system. ▲

To find the solution to a system of linear equations graphically, we graph both of the equations on the same axes. The coordinates of the point or points of intersection of the graphs are the solution or solutions to the system of equations. When two linear equations are graphed, three situations are possible. The two lines may intersect at one point, as in Example 2; or the two lines may be parallel and not intersect, as in Example 3; or the two equations may represent the same line, as in Example 4.

$x + y = 4$	
x	**y**
0	4
1	3
4	0

$2x - y = -1$	
x	**y**
0	1
1	3
−2	−3

Procedure for Solving a System of Equations by Graphing

1. Determine three ordered pairs that satisfy each equation.
2. Plot the ordered pairs and sketch the graphs of both equations on the same axes.
3. The coordinates of the point or points of intersection of the graphs are the solution or solutions to the system of equations.

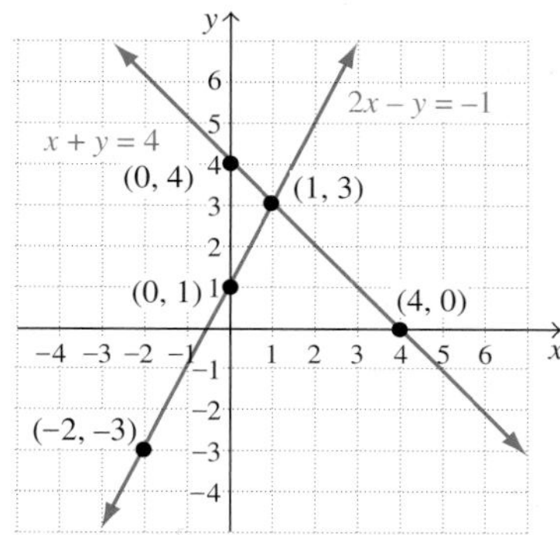

Figure 7.1

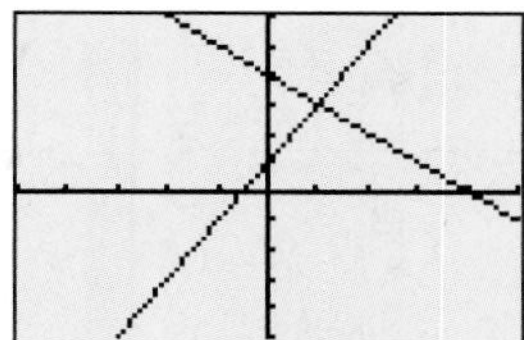
Figure 7.2

Since the solution to a system of equations may not be integer values, you may not be able to obtain the exact solution by graphing.

EXAMPLE 2 ***One Solution***

Find the solution to the following system of equations graphically.

$$x + y = 4$$
$$2x - y = -1$$

SOLUTION To find the solution, graph both $x + y = 4$ and $2x - y = -1$ on the same axes (Fig. 7.1). Three points that satisfy each equation are shown in the tables above Fig. 7.1. Figure 7.2 shows the system $x + y = 4$ and $2x - y = -1$ graphed on a Texas Instrument TI-83 graphing calculator.

The graphs intersect at (1, 3), which is the solution. This is the only point that satisfies *both* equations.

Check:

$x + y = 4$		$2x - y = -1$	
$1 + 3 = 4$		$2(1) - 3 = -1$	
$4 = 4$	True	$2 - 3 = -1$	
		$-1 = -1$	True

The system of equations in Example 2 is an example of a **consistent system of equations**. A consistent system of equations in one that has a solution.

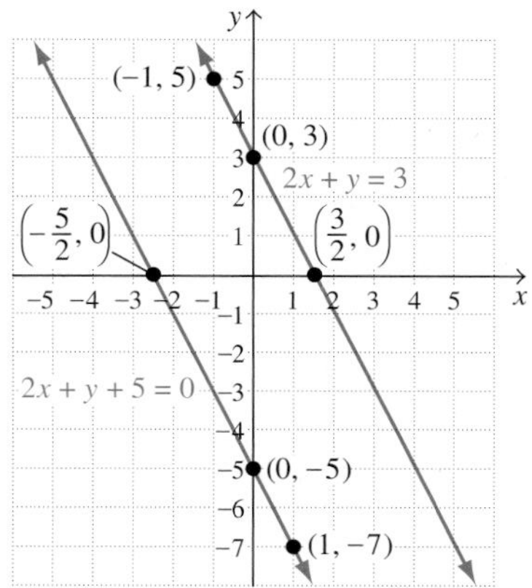

Figure 7.3

EXAMPLE 3 ***No Solution***

Find the solution to the following system of equations graphically.

$$2x + y = 3$$
$$2x + y + 5 = 0$$

SOLUTION Three ordered pairs that satisfy the equation $2x + y = 3$ are $(0, 3)$, $(\frac{3}{2}, 0)$, and $(-1, 5)$. Three ordered pairs that satisfy the equation $2x + y + 5 = 0$ are $(0, -5)$, $(-\frac{5}{2}, 0)$, and $(1, -7)$. The graphs of both equations are given in Fig. 7.3. Since the two lines are parallel, they do not intersect; therefore, the system has *no solution*.

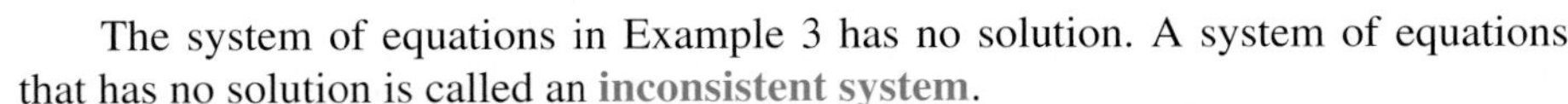

The system of equations in Example 3 has no solution. A system of equations that has no solution is called an **inconsistent system**.

EXAMPLE 4 *An Infinite Number of Solutions*

Find the solution to the following system of equations graphically.

$$x - \frac{1}{2}y = 2$$
$$y = 2x - 4$$

SOLUTION Three ordered pairs that satisfy the equation $x - \frac{1}{2}y = 2$ are $(1, -2)$, $(2, 0)$, and $(-1, -6)$. Three ordered pairs that satisfy the equation $y = 2x - 4$ are $(0, -4)$, $(-2, -8)$, and $(3, 2)$. Graph the equations on the same axes (Fig. 7.4). Because all six points are on the same line, the two equations represent the same line. Therefore, every ordered pair that is a solution to one equation is also a solution to the other equation. Every point on the line satisfies both equations; thus, this system has an *infinite number of solutions*. Solving the first equation for y reveals that the equations are equivalent.

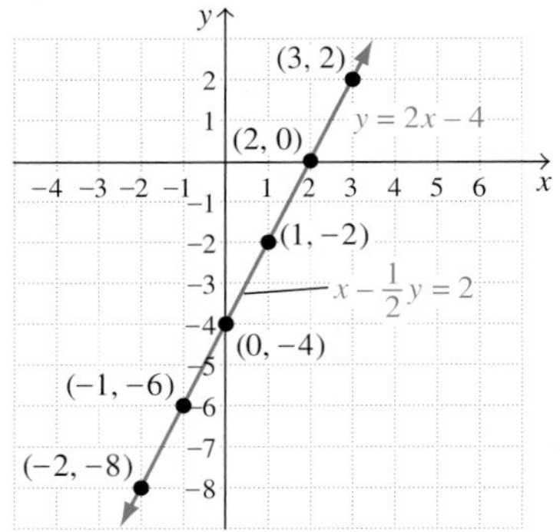

Figure 7.4

When a system of equations has an infinite number of solutions, as in Example 4, it is called a **dependent system**. Note that a dependent system is also a consistent system, since it has a solution.

Figure 7.5 summarizes the three possibilities for a system of linear equations.

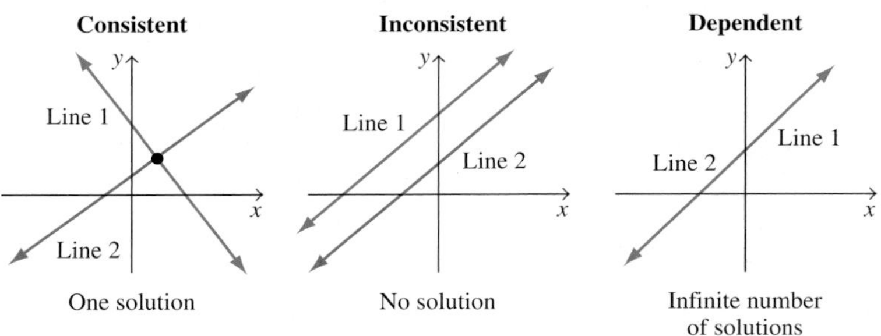

Figure 7.5

In Chapter 6 we introduced *modeling*. Recall that a *mathematical model* is an equation or system of equations that represents a real life situation. In Examples 5 and 6 we develop equations that model the situation.

EXAMPLE 5 MODELING - *A Landscape Service Application*

Tom's Tree and Landscape Service charges a consultation fee of \$200 plus \$50 per hour for labor for landscaping. Lawn Perfect Landscape Service charges a consultation fee of \$300 plus \$25 per hour for labor for landscaping.

a) Write a system of equations to represent the cost, C, of the two landscaping services, each with h hours of labor.
b) Graph both equations on the same axes and determine the number of hours needed for both services to have the same cost.
c) If the Johnsons need 7 hours of landscaping service done at their home, which service is less expensive?

SOLUTION Let h = the number of hours of labor. The total cost of each service is the consultation fee plus the cost of the labor.

a) Tom's Tree and Landscape Service: $C = 200 + 50h$
Lawn Perfect Landscape Service $C = 300 + 25h$

b) We graphed the cost, C, versus the number of hours of labor, h, for 0 to 10 hours (Fig. 7.6). On the graph, the lines intersect at the point (4, 400). Thus, for 4 hours of service, both services would have the same cost, \$400.

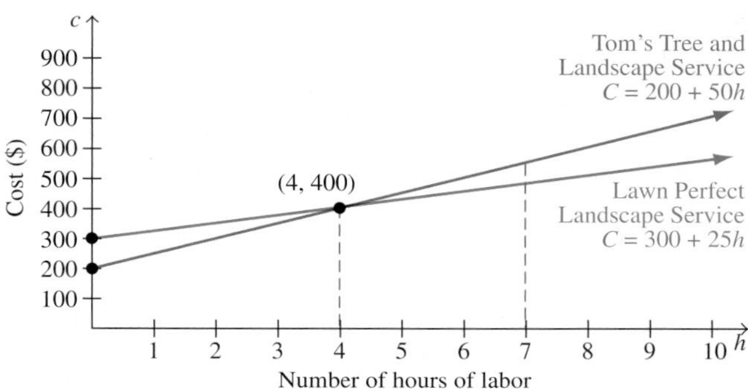

Figure 7.6

c) The graph shows that for more than 4 hours, Lawn Perfect is the least expensive service. Thus, for 7 hours, Lawn Perfect is less expensive than Tom's Tree and Landscape Service.

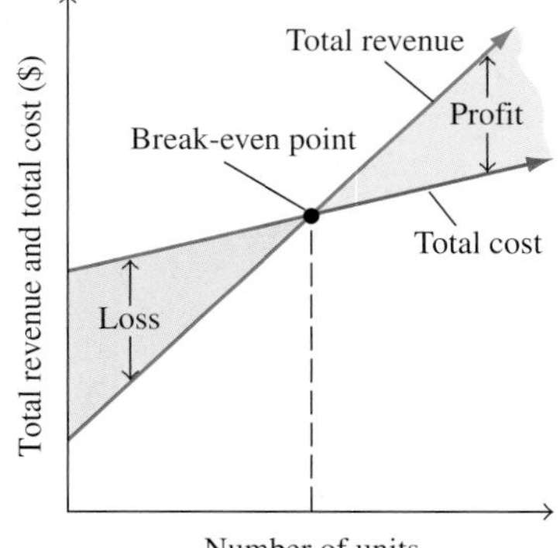

Figure 7.7

Manufacturers use a technique called **break-even analysis** to determine how many units of an item must be sold for the business to "break even," that is, for its total revenue to equal its total cost. Suppose we let the horizontal axis represent the number of units manufactured and sold and the vertical axis represent dollars. Then linear equations for cost, C, and revenue, R, can both be sketched on the same axes (Fig. 7.7). Both C and R are expressed in dollars and both are a function of the number of units.

Initially, the cost graph is higher than the revenue graph because of fixed (overhead) costs. During low levels of production the manufacturer suffers a loss (the cost graph is greater). During higher levels of production the manufacturer realizes a profit (the profit graph is greater). The point at which the two graphs intersect is called the **break-even point**. At that number of units sold revenue equals cost, and the manufacturer breaks even.

EXAMPLE 6 MODELING - *Profit and Loss in Business*

At an art show, Richard can sell personalized caricatures for \$20. The costs for making the caricatures are a fixed cost of \$100 and a production cost of \$10 apiece.

a) How many caricatures must Richard sell to break even?

b) Determine whether Richard makes a profit if he sells 12 caricatures. What is the profit or loss?

c) How many caricatures must Richard sell to make a profit of \$450?

PROFILE IN MATHEMATICS

Hypatia

The Greek mathematician Hypatia (A.D. 370–415) was the first recorded notable woman in mathematics. Daughter of the mathematician and philosopher Theon, she worked in Alexandria (in Egypt) and wrote works on algebra, conic sections, and, it is believed, the construction of scientific instruments. Hypatia actively stood for learning and science at a time in Western history when such learning was associated with paganism.

She paid the ultimate price for her beliefs: she was brutally murdered by religious zealots. Her death led to the departure of many scholars from Alexandria and marked the beginning of the end of the great age of Greek mathematics.

SOLUTION

a) Let x denote the number of caricatures made and sold. The revenue is given by the equation

$$R = 20x \qquad (\$20 \text{ times the number of units})$$

and the cost is given by the equation

$$C = 100 + 10x \qquad (\$100 \text{ plus } \$10 \text{ times the number of units})$$

The break-even point is the point at which the revenue and cost graphs intersect. In Fig. 7.8, the graphs intersect at the point (10, 200), which is the break-even point. Thus, for Richard to break even, he must sell 10 caricatures. When 10 caricatures are made and sold the cost and revenue are both $200.

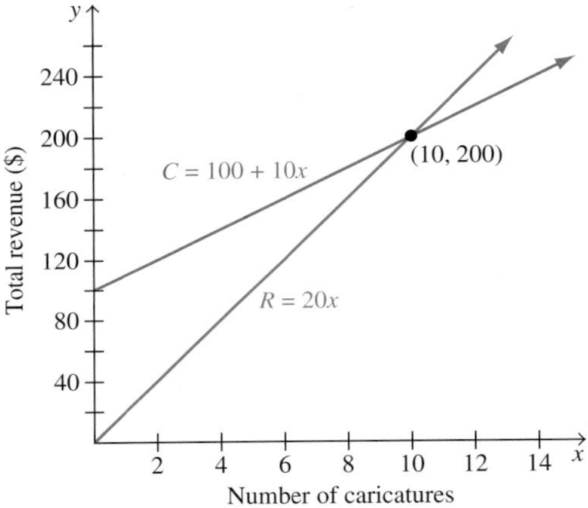

Figure 7.8

b) Examining the graph we can see that if Richard sells 12 caricatures he will have a profit, P, which is the revenue minus the cost. The profit formula is

$$\begin{aligned} P &= R - C \\ &= 20x - (100 + 10x) \\ &= 20x - 100 - 10x \\ &= 10x - 100 \end{aligned}$$

Thus, for 12 caricatures,

$$\begin{aligned} P &= 10x - 100 \\ &= 10(12) - 100 = 20 \end{aligned}$$

Richard has a profit of $20 if he sells 12 caricatures.

c) We can determine the number of caricatures that Richard must sell to have a profit of $450 by using the profit formula. Substituting 450 for P we have

$$P = 10x - 100$$
$$450 = 10x - 100$$
$$550 = 10x$$
$$55 = x$$

Thus, Richard must sell 55 caricatures to make a profit of \$450.

SECTION 7.1 EXERCISES

CONCEPT/WRITING EXERCISES

1. What is a system of linear equations?
2. What is the solution to a system of linear equations?
3. Define a *consistent system of equations*.
4. Define a *dependent system of equations*.
5. Define an *inconsistent system of equations*.
6. Outline the procedure for solving a system of equations by graphing.

PRACTICE THE SKILLS

In Exercises 7 and 8, determine whether the ordered pair is a solution to the given system.

7. $x + 3y = -1$ $\quad(2, -1)$
$2x + y = 2$

8. $x + 2y = 6$ $\quad(-2, 4)$
$x - y = -6$

In Exercises 9–12, solve the system of equations graphically:

9. $x = 2$
$y = 4$

10. $x = -3$
$y = 2$

11. $x = 4$
$y = -3$

12. $x = -5$
$y = -3$

In Exercises 13–28, solve the system of equations graphically. If the system does not have a single ordered pair as a solution, state whether the system is inconsistent or dependent.

13. $y = -x + 3$
$y = 2x - 3$

14. $x + y = 6$
$x - y = 4$

15. $2x - y = 6$
$x + 2y = -2$

16. $3x - y = 3$
$3y - 4x = 6$

17. $x - y = 5$
$-x + y = 2$

18. $x + y = 5$
$x - y = 1$

19. $y = 2x - 4$
$2x + y = 0$

20. $2x + y = 3$
$2y = 6 - 4x$

21. $2x - y = -3$
$2x + y = -9$

22. $y = x + 3$
$y = -1$

23. $x = 1$
$x + y + 3 = 0$

24. $3x + 2y = 6$
$6x + 4y = 12$

25. $2x - 3y = 12$
$3y - 2x = 9$

26. $y = \frac{1}{3}x - 4$
$3y - x = 4$

27. $y = \frac{1}{3}x - 2$
$x - 3y = 6$

28. $2(x - 1) + 2y = 0$
$3x + 2(y + 2) = 0$

29. a) If the two lines in a system of equations have different slopes, how many solutions will the system have? Explain your answer.

b) If the two lines in a system of equations have the same slope but different y-intercepts, how many solutions will the system have? Explain.

c) If the two lines in a system of equations have the same slope and the same y-intercept, how many solutions will the system have? Explain.

30. Indicate whether the graph shown represents a consistent, inconsistent, or dependent system. Explain your answer.

a)

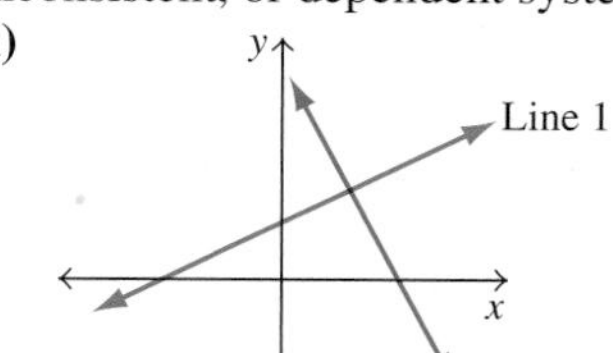

b)

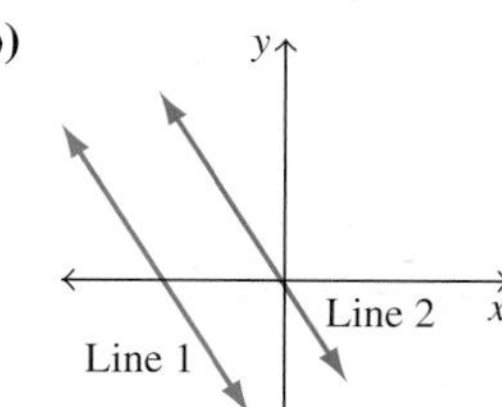

c)

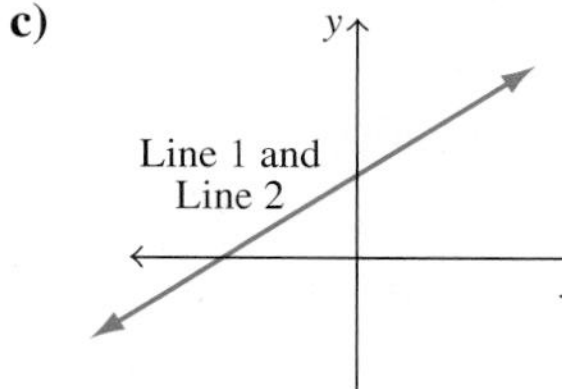

In Exercises 31–42, determine without graphing whether the system of equations has exactly one solution, no solution, or an infinite number of solutions. (Consider your answers to Exercise 29.)

31. $3x + y = 8$
$y = -3x + 8$

32. $2x + 3y = 6$
$6y = -4x + 10$

33. $5x + 10y = 15$
$x - y = 3$

34. $2x - y = 8$
$2y - x = 8$

35. $3x + y = 7$
$y = -3x + 9$

36. $2x - 3y = 6$
$x - \frac{3}{2}y = 3$

37. $x + 4y = 12$
$x = 4y + 3$

38. $3x = 6y + 5$
$y = \frac{1}{2}x - 3$

39. $3y = 6x + 4$
$-2x + y = \frac{4}{3}$

40. $x - 2y = 6$
$x + 2y = 4$

41. $12x - 5y = 4$
$3x + 4y = 6$

42. $4x + 7y = 2$
$4x = 6 + 7y$

PROBLEM SOLVING

Two lines are perpendicular when they meet at a right angle (90° angle). Two lines are perpendicular to each other when their slopes are negative reciprocals of each other. The negative reciprocal of 2 is $-\frac{1}{2}$, the negative reciprocal of $\frac{3}{5}$ is $-1/(\frac{3}{5})$ or $-\frac{5}{3}$, and so on. If a represents any real number, except 0, its negative reciprocal is $-1/a$. Note that the product of a number and its negative reciprocal is -1. In Exercises 43–46, determine, by finding the slope of each line, whether the lines will be perpendicular to each other when graphed.

43. $5y - 2x = 15$
$2y - 5x = 2$

44. $4y - x = 6$
$y = x + 8$

45. $2x + y = 3$
$2y - x = 5$

46. $6x + 5y = 3$
$-10x = 2 + 12y$

In Exercises 47–51 part of the question involves determining a system of equations that models the situation.

47. MODELING - *Landscaping Revisited* In Example 5, assume Tom's Tree and Landscape Service charges $200 for a consultation fee plus $60 per hour for labor and Lawn Perfect Landscape Service charges $305 for a consultation fee plus $25 per hour for labor.

a) Write the system of equations to represent the cost of the two landscaping services.
b) Graph both equations for 0 to 10 hours on the same axes.
c) Determine the number of hours of landscaping that must be used for both services to have the same cost.

48. MODELING - *Security Systems* Tamika Dixon plans to install a security system in her house. She is considering two security companies: ABC Security and SafeHomes Security. ABC's system costs $3380 to install and their monitoring fee is $18 per month. SafeHomes equivalent system costs only $2302 to install but their monitoring fee is $29 per month.
a) Write a system of equations to represent the cost of each system.
b) Graph both equations (for up to and including 180 months) on the same axes.
c) Determine the number of months the service must be used for both companies to have the same cost.
d) If both dealers guarantee not to raise monthly fees for 10 years, and if Tamika plans to use the system for 10 years, which system would be less expensive?

49. MODELING - *Printing Books* The total cost of printing a book consists of a setup charge and an additional fee for material for each book printed. The Sivle Printing Company charges a $1600 setup fee plus $6 per book it prints. The Yelserp Printing Company has a setup fee of $1200 plus $8 per book it prints.
a) Write a system of equations to represent the cost of printing the books with each company.
b) Graph both equations (for up to and including 300 books) on the same axes.
c) Determine the number of books that need to be printed for both companies to have the same cost.
d) If 100 books are to be printed, which is the less expensive printer?

50. MODELING - *Selling Picture Frames* Josh's Picture Framing company can sell decorating picture frames for $25 per frame. The costs for making the frames are a fixed cost of $400 and a production cost of $15 per frame (see Example 6 for an example of cost and revenue equations).
a) Write the cost and revenue equations.
b) Graph both equations, for 0 to 50 frames, on the same axes.
c) How many frames must Josh's Picture Framing company sell to break even?
d) Write the profit formula.

e) Determine whether the company makes a profit or loss if it sells 10 frames. What is the profit or loss?
f) How many frames must the company sell to realize a profit of $1000?

51. MODELING - *Manufacturing Speakers* A manufacturer can sell a certain speaker for $165 per unit. Manufacturing costs consist of a fixed cost of $8400 and a production cost of $95 per unit.
a) Write the cost and revenue equations.
b) Graph both equations, for 0 to 150 units, on the same axes.
c) How many units must the manufacturer sell to break even?
d) Write the profit formula.
e) What is the manufacturer's profit or loss if 100 units are sold?
f) How many units must the manufacturer sell to make a profit of $1250?

52. Explain how you can determine whether a system of two equations will be consistent, dependent, or inconsistent without graphing the equations.

CHALLENGE PROBLEMS/GROUP ACTIVITIES

53. MODELING - *Job Offers* Hubert had two job offers for sales positions. One pays a salary of $300 per week plus a 15% commission on his dollar sales volume. The second position pays a salary of $450 per week with no commission.
a) For each offer, write an equation that expresses the weekly pay.
b) Graph the system of equations and determine the solution.
c) For what dollar sales volume will the two offers result in the same pay?

54. MODELING - *Long-Distance Calling*
a) The Trans America Telephone Company charges 20 cents for the first minute and 10 cents for each additional minute or part thereof for a long-distance call from Happytown to Pleasantville. For the same call, the Pacific Edison Telephone Company charges 26 cents for the first minute and 8 cents for each additional minute or part thereof. Write equations to determine the cost of a long-distance call for the Trans America Telephone Company and for the Pacific Edison Telephone Company. Let x represent the number of additional minutes after the first minute and y the cost of the call in cents.
b) Graph the system of equations and determine the solution.
c) After how many minutes will the cost from the two telephone companies be the same?

55. a) If two lines have different slopes, what is the maximum possible number of points of intersection?
b) If three lines all have different slopes, what is the maximum possible number of points of intersection?
c) If four lines all have different slopes, what is the maximum possible number of points of intersection?
d) If five lines all have different slopes, what is the maximum possible number of points of intersection?
e) Is there a pattern in the number of points of intersection? If so, explain the pattern. Use the pattern to determine the maximum possible number of points of intersection for six lines.

RESEARCH ACTIVITY

56. The Rhind Papyrus indicates that the early Egyptians used linear equations. Do research and write a paper on the use of the symbols used in linear equations and the use of the linear equations by the early Egyptians. (References include history of mathematics books, encyclopedias, and the Internet.)

7.2 SOLVING SYSTEMS OF EQUATIONS BY THE SUBSTITUTION AND ADDITION METHODS

Having solved systems of equations by graphing in Section 7.1, we are ready for two other methods used to solve systems of linear equations: the substitution method and the addition method.

SUBSTITUTION METHOD

Procedure for Solving a System of Equations Using the Substitution Method

1. Solve one of the equations for one of the variables. If possible, solve for a variable with a numerical coefficient of 1. By doing so, you may avoid working with fractions.
2. Substitute the expression found in step 1 into the other equation. This step yields an equation in terms of a single variable.
3. Solve the equation found in step 2 for the variable.
4. Substitute the value found in step 3 into the equation you rewrote in step 1 and solve for the remaining variable.

Examples 1, 2, and 3 illustrate the *substitution method*. These systems of equations are the same as in Examples 2, 3, and 4 in Section 7.1.

EXAMPLE 1 *A Single Solution, by the Substitution Method*

Solve the following system of equations by substitution.

$$\begin{aligned} x + y &= 4 \\ 2x - y &= -1 \end{aligned}$$

SOLUTION The numerical coefficients of the x and y terms in the equation $x + y = 4$ are both 1. Thus, we can solve this equation for either x or y. Let's solve for x in the first equation.

Step 1.

$$\begin{aligned} x + y &= 4 \\ x + y - y &= 4 - y \\ x &= 4 - y \end{aligned}$$

Step 2. Substitute $4 - y$ for x in the other equation.

$$\begin{aligned} 2x - y &= -1 \\ 2(4 - y) - y &= -1 \end{aligned}$$

Step 3. Now solve the equation for y.

$$\begin{aligned} 8 - 2y - y &= -1 \\ 8 - 3y &= -1 \\ 8 - 8 - 3y &= -1 - 8 \\ -3y &= -9 \\ \frac{-3y}{-3} &= \frac{-9}{-3} \\ y &= 3 \end{aligned}$$

Step 4. Substitute $y = 3$ in the equation solved for x and determine the value of x.

$$\begin{aligned} x &= 4 - y \\ x &= 4 - 3 \\ x &= 1 \end{aligned}$$

Thus, the solution is the ordered pair (1, 3). This answer checks with the solution obtained graphically in Section 7.1, Example 2. ▲

EXAMPLE 2 ***No Solution, by the Substitution Method***

Solve the following system of equations by substitution.

$$\begin{aligned} 2x + y &= 3 \\ 2x + y + 5 &= 0 \end{aligned}$$

SOLUTION Solve for y in the first equation.

$$\begin{aligned} 2x + y &= 3 \\ 2x - 2x + y &= 3 - 2x \\ y &= 3 - 2x \end{aligned}$$

Now substitute $3 - 2x$ in place of y in the second equation.

$$\begin{aligned} 2x + y + 5 &= 0 \\ 2x + (3 - 2x) + 5 &= 0 \\ 2x + 3 - 2x + 5 &= 0 \\ 8 &= 0 \quad \text{False} \end{aligned}$$

Since 8 cannot be equal to 0, there is no solution to the system of equations. Thus, the system of equations is inconsistent. This answer checks with the solution obtained graphically in Section 7.1, Example 3. ▲

When solving Example 2, we obtained $8 = 0$ and indicated that the system was inconsistent and that there was no solution. When solving a system of equations, if you obtain a false statement, like $4 = 0$ or $-2 = 0$, the system is *inconsistent* and has *no solution*.

EXAMPLE 3 ***An Infinite Number of Solutions, by the Substitution Method***

Solve the following system of equations by substitution.

$$x - \frac{1}{2}y = 2$$
$$y = 2x - 4$$

SOLUTION The second equation $y = 2x - 4$ is already solved for y, so we will substitute $2x - 4$ for y in the first equation.

$$x - \frac{1}{2}y = 2$$
$$x - \frac{1}{2}(2x - 4) = 2$$
$$x - x + 2 = 2$$
$$2 = 2 \quad \text{True}$$

Since 2 equals 2, the system has an infinite number of solutions. Thus, the system of equations is dependent. This answer checks with the solution obtained in Section 7.1, Example 4. ▲

When solving Example 3, we obtained $2 = 2$ and indicated that the system was dependent and had an infinite number of solutions. When solving a system of equations, if you obtain a true statement, such as $0 = 0$ or $2 = 2$, the system is *dependent* and has an *infinite number of solutions*.

ADDITION METHOD

If neither of the equations in a system of linear equations has a variable with a coefficient of 1, it is generally easier to solve the system by using the **addition** (or **elimination**) **method**.

To solve a system of linear equations by the addition method, it is necessary to obtain two equations whose sum will be a single equation containing only one variable. To achieve this goal, we rewrite the system of equations as two equations where the coefficients of one of the variables are opposites (or additive inverses) of each other. For example, if one equation has a term of $2x$, we might rewrite the other equation so that its x term will be $-2x$. To obtain the desired equations, it might be necessary to multiply one or both equations in the original system by a number. When an equation is to be multiplied by a number, we will place brackets around the equation and place the number that is to multiply the equation before the brackets. For example, $4[2x + 3y = 6]$ means that each term on both sides of the equal sign in the equation $2x + 3y = 6$ is to be multiplied by 4:

$$4[2x + 3y = 6] \qquad \text{gives} \qquad 8x + 12y = 24.$$

This notation will make our explanations much more efficient and easier for you to follow.

DID YOU KNOW

How to Succeed in Business

Economics, a science dependent on mathematics, dates back to just before the Industrial Revolution of the eighteenth century. Technologies were being invented and applied to the manufacture of cloth, iron, transportation, and agriculture. These new technologies led to the development of mathematically based economic models. The French economist Jules Dupuit (1804–1866) suggested a method to calculate the value of railroad bridges; the Irish economist Dionysis Larder (1793–1859) showed railroad companies how to structure their rates so as to increase their profits.

Procedure for Solving a System of Equations by the Addition Method

1. If necessary, rewrite the equations so that the variables appear on one side of the equal sign and the constants on the other side of the equal sign.
2. If necessary, multiply one or both equations by a constant(s) so that when you add the equations, the result will be an equation containing only one variable.
3. Add the equations to obtain a single equation in one variable.
4. Solve the equation in step 3 for the variable.
5. Substitute the value found in step 4 into either of the original equations and solve for the other variable.

EXAMPLE 4 ***Eliminating a Variable by the Addition Method***

Solve the following system of equations by the addition method.

$$\begin{aligned} x + y &= 5 \\ 2x - y &= 7 \end{aligned}$$

SOLUTION Since the coefficients of the y terms, 1 and -1, are additive inverses, the sum of the y terms will be zero when the equations are added. Thus, the sum of the two equations will contain only one variable, x. Add the two equations to obtain one equation in one variable. Then solve for the remaining variable.

$$\begin{aligned} x + y &= 5 \\ \underline{2x - y} &\underline{= 7} \\ 3x \quad\; &= 12 \\ x &= 4 \end{aligned}$$

Now substitute 4 for x in either of the original equations to find the value of y.

$$\begin{aligned} x + y &= 5 \\ 4 + y &= 5 \\ y &= 1 \end{aligned}$$

The solution to the system is (4, 1).

EXAMPLE 5 ***Multiplying by -1 in the Addition Method***

Solve the following system of equations by the addition method.

$$\begin{aligned} x + 3y &= 9 \\ x + 2y &= 5 \end{aligned}$$

SOLUTION We want the sum of the two equations to have only one variable. We can eliminate the variable x by multiplying either equation by -1 and then adding. We will multiply the first equation by -1.

$$\begin{array}{lcl} -1[x + 3y = 9] & \text{gives} & -x - 3y = -9 \\ x + 2y = 5 & & x + 2y = 5 \end{array}$$

We now have a system of equations equivalent to the original system.
Now add the two equations on the right.

$$\begin{aligned} -x - 3y &= -9 \\ x + 2y &= 5 \\ \hline -y &= -4 \\ y &= 4 \end{aligned}$$

Now we solve for x by substituting 4 for y in either of the original equations.

$$\begin{aligned} x + 3y &= 9 \\ x + 3(4) &= 9 \\ x + 12 &= 9 \\ x &= -3 \end{aligned}$$

The solution is $(-3, 4)$. ▲

EXAMPLE 6 ***Multiplying One Equation in the Addition Method***

Solve the following system of equations by the addition method.

$$\begin{aligned} 2x + y &= 6 \\ 3x + 3y &= 9 \end{aligned}$$

SOLUTION We can multiply the top equation by -3 and then add to eliminate the variable y.

$$\begin{array}{lcl} -3[2x + y = 6] & \text{gives} & -6x - 3y = -18 \\ 3x + 3y = 9 & & 3x + 3y = 9 \end{array}$$

$$\begin{aligned} -6x - 3y &= -18 \\ 3x + 3y &= 9 \\ \hline -3x \qquad &= -9 \\ x &= 3 \end{aligned}$$

Now we find y.

$$\begin{aligned} 2x + y &= 6 \\ 2(3) + y &= 6 \\ 6 + y &= 6 \\ y &= 0 \end{aligned}$$

The solution is $(3, 0)$. ▲

Note that in Example 6 we could have eliminated the variable x by multiplying the top equation by 3 and the bottom equation by -2, then adding. Try this method now.

EXAMPLE 7 ***Multiplying Both Equations***

Solve the following system of equations by the addition method.

$$\begin{aligned} 3x - 4y &= 8 \\ 2x + 3y &= 9 \end{aligned}$$

SOLUTION In this system, we cannot eliminate a variable by multiplying only one equation by an integer value and then adding. To eliminate a variable, we can multiply each equation by a different number. To eliminate the variable x, we can multiply the top equation by 2 and the bottom by -3 (or the top by -2 and the bottom by 3) and then add the two equations. If we want, we can instead eliminate the variable y by multiplying the top equation by 3 and the bottom by 4 and then adding the two equations. Let's eliminate the variable x.

$$\begin{aligned} 2[3x - 4y = 8] &\quad \text{gives} \quad 6x - 8y = 16 \\ -3[2x + 3y = 9] &\quad \text{gives} \quad -6x - 9y = -27 \end{aligned}$$

$$\begin{aligned} 6x - 8y &= 16 \\ -6x - 9y &= -27 \\ \hline -17y &= -11 \\ y &= \frac{11}{17} \end{aligned}$$

We could now find x by substituting $\frac{11}{17}$ for y in either of the original equations. Although it can be done, it gets messy. Instead, let's solve for x by eliminating the variable y from the two original equations. To do so, we multiply the first equation by 3 and the second equation by 4.

$$\begin{aligned} 3[3x - 4y = 8] &\quad \text{gives} \quad 9x - 12y = 24 \\ 4[2x + 3y = 9] &\quad \text{gives} \quad 8x + 12y = 36 \end{aligned}$$

$$\begin{aligned} 9x - 12y &= 24 \\ 8x + 12y &= 36 \\ \hline 17x \qquad &= 60 \\ x &= \frac{60}{17} \end{aligned}$$

The solution to the system is $(\frac{60}{17}, \frac{11}{17})$. ▲

When solving a system of linear equations by either the substitution or the addition method, if you obtain the equation $0 = 0$ it indicates that the system is *dependent* (both equations represent the same line; see Fig. 7.4), and there are an infinite number of solutions. When solving, if you obtain an equation such as $0 = 6$, or any other equation that is false, it means that the system is *inconsistent* (the two equations represent parallel lines; see Fig. 7.5) and there is no solution.

EXAMPLE 8 MODELING - *When Are Repair Costs the Same?*

Melinda Melendez needs to purchase a new radiator for her car and have it installed by a mechanic. She is considering two garages: Steve's Repair and Greg's Garage. At Steve's Repair, the parts cost \$200 and the labor cost is \$50 per hour. At Greg's Garage, the parts cost \$375 and the labor cost is \$25 per hour. How many hours would the repair need to take for the total cost at each garage to be the same?

SOLUTION We are asked to find the number of hours the repair would need to take for each garage to have the same total cost, C. First write a system of equations to represent the total cost for each of the garages. The total cost consists of the cost of the parts and the labor cost. The labor cost depends on the number of hours of labor.

Let x = the number of hours of labor.

$$\begin{aligned} \text{Total cost} &= \text{cost of parts} + \text{labor cost} \\ \text{Steve's Repair: } C &= 200 + 50x \\ \text{Greg's Garage: } C &= 375 + 25x \end{aligned}$$

We want to determine when the cost will be the same, so we set the two costs equal to each other (substitution method) and solve the resulting equation.

$$\begin{aligned} 200 + 50x &= 375 + 25x \\ 200 - 200 + 50x &= 375 - 200 + 25x \\ 50x &= 175 + 25x \\ 50x - 25x &= 175 + 25x - 25x \\ 25x &= 175 \\ \frac{25x}{25} &= \frac{175}{25} \\ x &= 7 \end{aligned}$$

Thus, for 7 hours of labor, the cost at both garages would be the same. If we construct a graph (Fig. 7.9) of the two cost equations, the point of intersection is (7, 550). If the repair were to require 7 hours of labor, the total cost at either garage would be \$550.

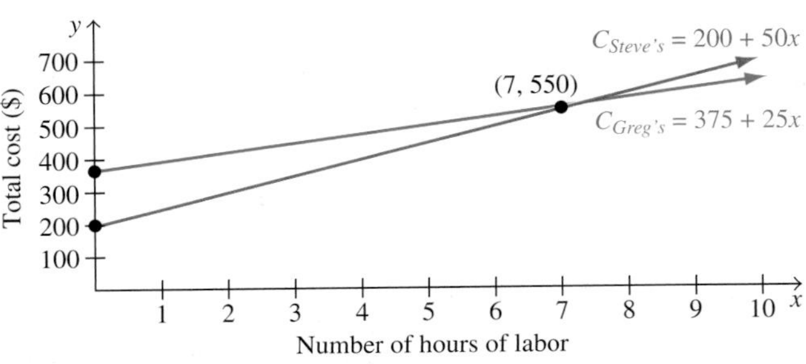

Figure 7.9

EXAMPLE 9 MODELING - *A Mixture Problem*

Florence Nightcap, a pharmacist, needs 100 milliliters (mℓ) of a 10% phenobarbital solution. She has only a 5% solution and a 25% solution available. How many milliliters of each solution should she mix to obtain the desired solution?

SOLUTION First we set up a system of equations. The unknown quantities are the amount of 5% solution and the amount of the 25% solution that must be used. Let

$$x = \text{number of m}\ell \text{ of 5\% solution}$$
$$y = \text{number of m}\ell \text{ of 25\% solution}$$

We know that 100 mℓ of solution are needed. Thus,

$$x + y = 100$$

The total amount of pure phenobarbital in a solution is determined by multiplying the percent of phenobarbital by the number of milliliters of solution. The second equation comes from the fact that

$$\begin{pmatrix}\text{Total amount of}\\ \text{phenobarbital in}\\ \text{5\% solution}\end{pmatrix} + \begin{pmatrix}\text{total amount of}\\ \text{phenobarbital in}\\ \text{25\% solution}\end{pmatrix} = \begin{pmatrix}\text{total amount of}\\ \text{phenobarbital}\\ \text{in 10\% mixture}\end{pmatrix}$$

$$0.05x \quad + \quad 0.25y \quad = \quad 0.10(100)$$

or $\quad 0.05x + 0.25y = 10$

The system of equations is

$$\begin{aligned} x + y &= 100 \\ 0.05x + 0.25y &= 10 \end{aligned}$$

Let's solve this system of equations by using the addition method.

$$\begin{aligned} -5[x + y &= 100] \quad \text{gives} \quad -5x - 5y = -500 \\ 100[0.05x + 0.25y &= 10] \quad \text{gives} \quad 5x + 25y = 1000 \end{aligned}$$

$$\begin{aligned} -5x - 5y &= -500 \\ 5x + 25y &= 1000 \\ \hline 20y &= 500 \\ \frac{20y}{20} &= \frac{500}{20} \\ y &= 25 \end{aligned}$$

Now we find x.

$$\begin{aligned} x + y &= 100 \\ x + 25 &= 100 \\ x &= 75 \end{aligned}$$

Therefore, 75 mℓ of a 5% phenobarbital solution must be mixed with 25mℓ of a 25% phenobarbital solution to obtain 100 mℓ of a 10% phenobarbital solution. ▲

Example 9 can also be solved by using substitution. Try to do so now.

SECTION 7.2 EXERCISES

CONCEPT/WRITING EXERCISES

1. In your own words, explain how to solve a system of equations by using the addition method.
2. In your own words, explain how to solve a system of equations by using the substitution method.
3. How will you know, when solving a system of equations by either the substitution or the addition method, whether the system is inconsistent?
4. How will you know, when solving a system of equations by either the substitution or the addition method, whether the system is dependent?

PRACTICE THE SKILLS

In Exercises 5–22, solve the system of equations by the substitution method. If the system does not have a single ordered pair as a solution, state whether the system is inconsistent or dependent.

5. $y = x - 6$
$y = -x + 4$

6. $y = 5x + 7$
$y = 2x + 1$

7. $2x - 4y = 12$
$2x + y = -3$

8. $x + 3y = 3$
$4y + 3x = -1$

9. $y - x = 4$
$x - y = 3$

10. $x + y = 3$
$y + x = 5$

11. $x = 5y - 12$
$x - y = 0$

12. $3y + 2x = 4$
$3y = 6 - x$

13. $y - 2x = 3$
$2y = 4x + 6$

14. $x = y + 3$
$x = -3$

15. $y = 2$
$y + x + 3 = 0$

16. $x + 2y = 6$
$y = 2x + 3$

17. $y + 3x - 4 = 0$
$2x - y = 7$

18. $x + 4y = 7$
$2x + 3y = 5$

19. $x = 2y + 3$
$y = 3x - 1$

20. $x + 4y = 9$
$2x - y - 6 = 0$

21. $6x - y = 5$
$y = 6x - 3$

22. $x + 3y = 6$
$y = -\frac{1}{3}x + 2$

In Exercises 23–38, solve the system of equations by the addition method. If the system does not have a single ordered pair as a solution, state whether the system is inconsistent or dependent.

23. $4x + y = 9$
$3x - y = 5$

24. $3x + y = 10$
$4x - y = 4$

25. $-x + y = 5$
$x + 3y = 3$

26. $2x - 6y = 8$
$-2x + 4y = -10$

27. $2x - y = -4$
$-3x - y = 6$

28. $x + y = 6$
$-2x + y = -3$

29. $2x + y = 6$
$3x + y = 5$

30. $4x + 3y = -1$
$2x - y = -13$

31. $2x + y = 11$
$x + 3y = 18$

32. $5x - 2y = 11$
$-3x + 2y = 1$

33. $3x - 4y = 11$
$3x + 5y = -7$

34. $4x - 2y = 6$
$4y = 8x - 12$

35. $4x + y = 6$
$-8x - 2y = 13$

36. $2x + 3y = 6$
$5x - 4y = -8$

37. $7x + 8y = 11$
$5x + 6y = 7$

38. $8x + 3y = 7$
$3x + 2y = 9$

PROBLEM SOLVING

In Exercises 39–52, write a system of equations that can be used to solve the problem. Then solve the system and determine the answer.

39. MODELING - *Comparing Salaries* Donald Karecki can make a weekly salary of \$300 plus 4% commission on sales as a sales representative at Appliance City or a weekly salary consisting of a straight 16% commission on sales. Determining the dollar sales necessary for both plans to result in the same salary.

40. MODELING - *Basketball Game* The Tennessee women's basketball team made 45 field goals in a recent game; some were 2-pointers and some were 3-pointers. How many 2-point baskets were made and how many 3-point baskets were made if Tennessee scored 101 points?

41. MODELING - *Hockey Title* Hockey teams receive 2 points when they win a game and 1 point when they tie a game. The Flyers won a division title with 58 points. They won 23 more games than they tied. How many wins and how many ties did the Flyers have?

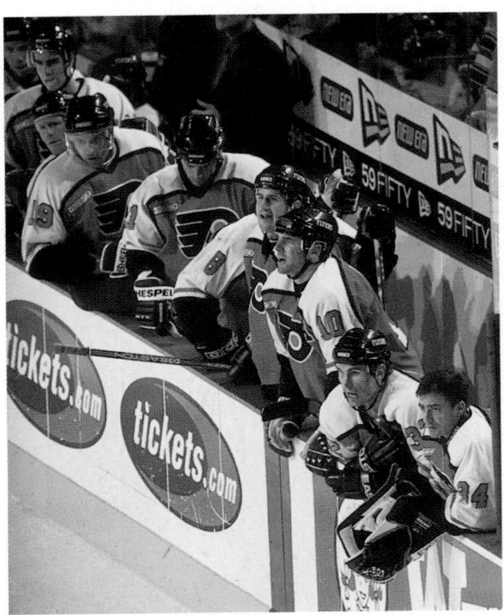

42. MODELING - *Speed Limit* The speed limit is 55 miles per hour in Oxon Hill–Grassmanner, Maryland. The fine for exceeding the speed limit carries a fixed charge plus a charge for each mile per hour over the speed limit. The fine for driving 70 mph is \$175, and the fine for driving 75 mph is \$200. Determine the fixed charge and the charge per mile.

43. MODELING - *A Protein Mix* Soybean meal is 16% protein and corn meal is 7% protein. How many pounds of each should be used to get a 300 lb mixture that is 10% protein?

44. MODELING - *A Milk Mixture* The Guidas own a dairy. They have milk that is 5% butterfat and skim milk without butterfat. How much of the 5% milk and how much of the skim milk should they mix to make 100 gal of milk that is 3.5% butterfat?

45. MODELING - *Choosing a Copy Service* Lori Lanier recently purchased a high-speed copier for her home office and wants to purchase a service contract on the copier. She is considering two sources for the contract. The Economy Sales and Service Company charges \$18 a month plus 2 cents per copy. Office Superstore charges \$24 a month but only 1.5 cents a copy. How many copies would Lori need to make for the monthly costs of both plans to be the same?

46. MODELING - *Cellular Phone Rates* Rich Gratien is considering two cellular phone rate plans. Cellular Two charges \$25 per month plus 5 cents a minute. TGE charges \$15 per month plus 10 cents per minute.

a) How long would Rich have to talk on the phone, in a month, for the two plans to have the same total cost?

b) If Rich talks for 90 minutes a month, which plan would be less expensive for him?

47. MODELING - *Nuts & Pretzels, Mixed* Dave Chwalik wants to purchase 20 pounds of party mix for a total of \$30. To obtain the mixture, he will mix nuts that cost \$3 per pound with pretzels that cost \$1 per pound. How many pounds of each type of mix should he use?

48. MODELING - *School Play Tickets* Jefferson High School sold 250 tickets to their annual school play. Student tickets cost \$2 per ticket and adult tickets cost \$5 per ticket. If \$950 in ticket sales is collected, how many tickets of each type were sold?

49. MODELING - *Laboratory Research* Animals in an experiment are to be kept on a strict diet. Each animal is to receive, among other things, 20 g of protein and 6 g of carbohydrates. The scientist has only two food mixes of the following compositions available.

	Protein (%)	Carbohydrates (%)
Mix *A*	10	6
Mix *B*	20	2

How many grams of each mix should she use to obtain the right diet for a single animal?

50. MODELING - *Two Golf Clubs* Membership in Chippers Country Club costs \$3000 per year and entitles a member to play a round of golf for a greens fee of \$18. At Birdies Country Club, membership costs \$2500 per year and the greens fee is \$20.

a) How many rounds must a golfer play in a year for the costs at the two clubs to be the same?

b) If Sally Sestini planned to play 30 rounds of golf in a year, which club would be the least expensive?

51. MODELING - *Cassettes and CDs* The following graph shows the number of units shipped by record manufacturers from 1988 through 1997. The number of cassettes shipped (the blue curve) can be approximated by the linear equation $y = -29x + 450$, where x is the number of years since 1988. The number of compact discs (CDs) shipped (the dashed red curve) can be approximated by the linear equation $y = 57x + 150$.

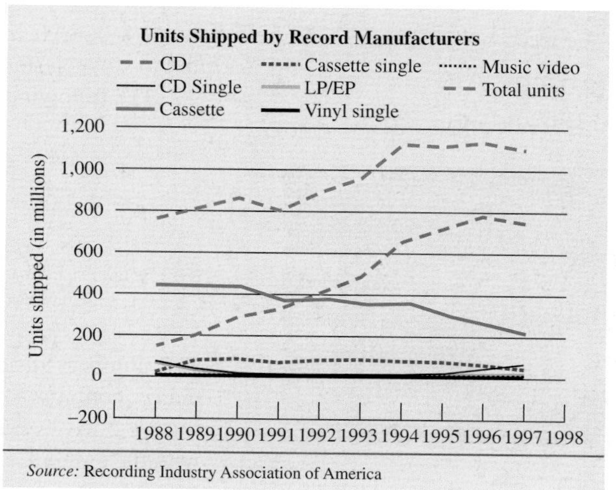

a) Use the substitution method to approximate when the number of cassettes shipped equaled the number of CDs shipped. What was the number of each shipped?

b) Use the graph to verify your answer in part (a).

52. MODELING - *Checking Accounts* The charge for maintaining a checking account at Union Bank is \$6 per month plus 10 cents for each check that is written. The charge at Citrus Bank is \$2 per month and 20 cents per check.

a) How many checks would a customer have to write in a month for the total charges to be the same at both banks?

b) If Brent Pickett planned to write 14 checks per month, which bank would be the least expensive?

CHALLENGE PROBLEMS/GROUP ACTIVITIES

53. Solve the following system of equations for u and v by first substituting x for $1/u$ and y for $1/v$

$$\frac{1}{u} + \frac{2}{v} = 8$$

$$\frac{3}{u} - \frac{1}{v} = 3$$

54. Develop a system of equations that has (6, 5) as its solution. Explain how you developed your system of equations.

55. The substitution or addition methods can also be used to solve a system of three equations in three variables. Consider the following system.

$$\begin{aligned} x + y + z &= 7 \\ x - y + 2z &= 9 \\ -x + 2y + z &= 4 \end{aligned}$$

The ordered triple (x, y, z) is the solution to the system if it satisfies all three equations.

a) Show that the ordered triple (2, 1, 4) is a solution to the system.

b) Use the substitution or addition method to determine the solution to the system. (*Hint:* Eliminate one variable by using two equations. Then eliminate the same variable by using two different equations.)

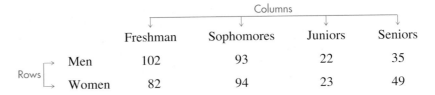

7.3 MATRICES

We have discussed solving systems of equations by graphing, using substitution, and using the addition method. In Section 7.4, we will discuss solving systems of linear equations by using matrices. So that you will become familiar with matrices, in this section, we explain how to add, subtract, and multiply matrices. We also explain how to multiply a matrix by a real number. Matrix techniques are easily adapted to computers.

A **matrix** is a rectangular array of elements. An array is a systematic arrangement of numbers or symbols in rows and columns. In this text, we use brackets to indicate a matrix. Matrices (the plural of matrix) may be used to display information and to solve systems of linear equations.

The following matrix displays information about a survey of 500 college students.

Columns

Rows		Freshman	Sophomores	Juniors	Seniors
→	Men	102	93	22	35
→	Women	82	94	23	49

The numbers inside the brackets are called **elements** of the matrix. The preceding matrix contains 8 elements. Because it has two rows and four columns, it is referred to as a 2 by 4, written 2 × 4, matrix. The **dimensions** of a matrix may be indicated with the symbol $r \times s$, where r is the number of rows and s is the number of columns. A matrix that contains the same number of rows and columns is called a **square matrix**. Following are examples of 2 × 2 and 3 × 3 square matrices.

$$\begin{bmatrix} 2 & 3 \\ 5 & 2 \end{bmatrix} \qquad \begin{bmatrix} 4 & 6 & -1 \\ 2 & 3 & 0 \\ 5 & 2 & 1 \end{bmatrix}$$

Matrices Are Everywhere

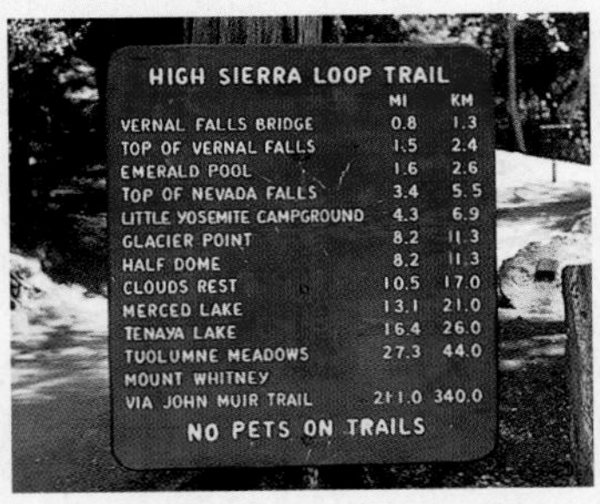

You are already familiar with matrices, although you may not be aware of it. The matrix is a good way to display numerical data, as illustrated on this trail sign in Yosemite National Park.

Two matrices are equal if and only if they have the same elements in the same relative positions.

EXAMPLE 1 ***Equal Matrices***

Given $A = B$, find x and y.

$$A = \begin{bmatrix} 2 & 5 \\ 1 & 8 \end{bmatrix}, \qquad B = \begin{bmatrix} x & 4 \\ 6 & y \end{bmatrix}$$

SOLUTION The corresponding elements must be the same, so $x = 2$ and $y = 8$. ▲

ADDITION OF MATRICES

Two matrices can be added only if they have the same dimensions (same number of rows and same number of columns). To obtain the sum of two matrices with the same dimensions, add the corresponding elements of the two matrices.

EXAMPLE 2

$$A = \begin{bmatrix} 1 & 4 \\ -2 & 6 \end{bmatrix}, \qquad B = \begin{bmatrix} 3 & 8 \\ 6 & 0 \end{bmatrix}. \quad \text{Find } A + B.$$

SOLUTION

$$A + B = \begin{bmatrix} 1 & 4 \\ -2 & 6 \end{bmatrix} + \begin{bmatrix} 3 & 8 \\ 6 & 0 \end{bmatrix}$$

$$= \begin{bmatrix} 1 + 3 & 4 + 8 \\ -2 + 6 & 6 + 0 \end{bmatrix} = \begin{bmatrix} 4 & 12 \\ 4 & 6 \end{bmatrix}$$

▲

EXAMPLE 3 MODELING - *Sales of Bicycles*

Peddler's Bicycle Corporation owns and operates two stores, one in Pennsylvania and one in New Jersey. The number of mountain bicycles, MB, and racing bicycles, RB, sold in each store for the months of January through June and July through December are indicated in the matrices that follow. We will call the matrices A and B.

$$\begin{array}{l} \\ \\ \text{Jan.–June} \\ \text{July–Dec.} \end{array} \begin{array}{c} \textit{Pennsylvania} \\ \begin{array}{cc} MB & RB \end{array} \\ \begin{bmatrix} 515 & 425 \\ 290 & 250 \end{bmatrix} = A \end{array} \qquad \begin{array}{c} \textit{New Jersey} \\ \begin{array}{cc} MB & RB \end{array} \\ \begin{bmatrix} 520 & 350 \\ 180 & 271 \end{bmatrix} = B \end{array}$$

Find the total number of each type of bicycle sold by the corporation during each time period.

SOLUTION To solve the problem, we add matrices A and B.

$$\begin{array}{l} \\ \text{Jan.–June} \\ \text{July–Dec.} \end{array} \begin{array}{c} \begin{array}{cc} MB & RB \end{array} \\ \begin{bmatrix} 515 + 520 & 425 + 350 \\ 290 + 180 & 250 + 271 \end{bmatrix} \end{array} = \begin{array}{c} \begin{array}{cc} MB & RB \end{array} \\ \begin{bmatrix} 1035 & 775 \\ 470 & 521 \end{bmatrix} \end{array}$$

We can see from the sum matrix that during the period from January through June, a total of 1035 mountain bicycles and 775 racing bicycles were sold. During the period from July through December, a total of 470 mountain bicycles and 521 racing bicycles were sold. ▲

The matrix

$$I = \begin{bmatrix} 0 & 0 \\ 0 & 0 \end{bmatrix}$$

is the **additive identity matrix** for 2×2 matrices. We denote this matrix with the letter I. Note that for any 2×2 matrix, A, $A + I = I + A = A$.

SUBTRACTION OF MATRICES

Only matrices with the same dimension may be subtracted. To do so, we subtract each entry in one matrix from the corresponding entry in the other matrix.

EXAMPLE 4 *Subtracting Matrices*

Find $A - B$ if

$$A = \begin{bmatrix} 2 & 6 \\ 3 & -1 \end{bmatrix} \quad \text{and} \quad B = \begin{bmatrix} 3 & -4 \\ 7 & -3 \end{bmatrix}$$

SOLUTION

$$A - B = \begin{bmatrix} 2 & 6 \\ 3 & -1 \end{bmatrix} - \begin{bmatrix} 3 & -4 \\ 7 & -3 \end{bmatrix}$$

$$= \begin{bmatrix} 2-3 & 6-(-4) \\ 3-7 & -1-(-3) \end{bmatrix} = \begin{bmatrix} -1 & 10 \\ -4 & 2 \end{bmatrix}$$ ▲

MULTIPLYING A MATRIX BY A REAL NUMBER

A matrix may be multiplied by a real number by multiplying each entry in the matrix by the real number. Sometimes when we multiply a matrix by a real number, we call that real number a **scalar**.

EXAMPLE 5 *Multiplying a Matrix by a Scalar*

For matrices A and B, find (a) $3A$ and (b) $3A - 2B$.

$$A = \begin{bmatrix} 4 & 6 \\ -3 & 5 \end{bmatrix}, \qquad B = \begin{bmatrix} -1 & 5 \\ 2 & 6 \end{bmatrix}$$

SOLUTION

a) $$3A = 3\begin{bmatrix} 4 & 6 \\ -3 & 5 \end{bmatrix} = \begin{bmatrix} 3(4) & 3(6) \\ 3(-3) & 3(5) \end{bmatrix} = \begin{bmatrix} 12 & 18 \\ -9 & 15 \end{bmatrix}$$

b) We found $3A$ in part (a). Now we find $2B$.

$$2B = 2\begin{bmatrix} -1 & 5 \\ 2 & 6 \end{bmatrix} = \begin{bmatrix} 2(-1) & 2(5) \\ 2(2) & 2(6) \end{bmatrix} = \begin{bmatrix} -2 & 10 \\ 4 & 12 \end{bmatrix}$$

$$3A - 2B = \begin{bmatrix} 12 & 18 \\ -9 & 15 \end{bmatrix} - \begin{bmatrix} -2 & 10 \\ 4 & 12 \end{bmatrix}$$

$$= \begin{bmatrix} 12-(-2) & 18-10 \\ -9-4 & 15-12 \end{bmatrix} = \begin{bmatrix} 14 & 8 \\ -13 & 3 \end{bmatrix}$$

MULTIPLICATION OF MATRICES

Multiplication of matrices is slightly more difficult than addition of matrices. Multiplication of matrices is possible only when the number of *columns* of the first matrix, A, is the same as the number of *rows* of the second matrix, B. We use the notation

$$\underset{3 \times 4}{A}$$

to indicate that matrix A has three rows and four columns. Suppose matrix A is a 3×4 matrix and matrix B is a 4×5 matrix. Then

$$\underset{3 \times 4}{A} \qquad \underset{4 \times 5.}{B}$$

Same

Product matrix 3×5

This indicates that matrix A has four columns and matrix B has four rows. Therefore, we can multiply these two matrices. The product matrix will have the same number of rows as matrix A and the same number of columns as matrix B. Thus, the dimensions of the product matrix are 3×5.

DID YOU KNOW

Spreadsheets

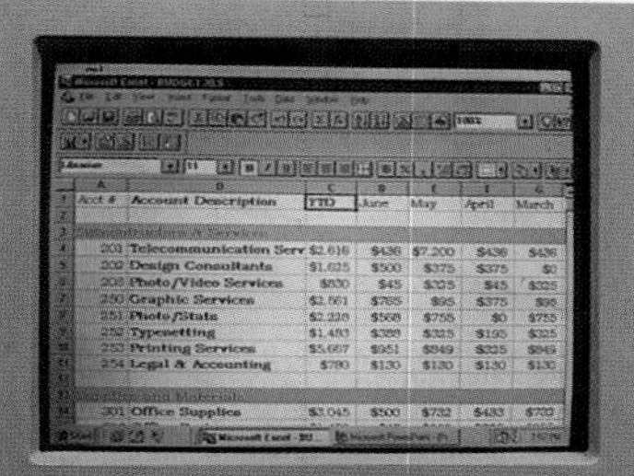

One of the most useful applications of computer software is the spreadsheet. On a computer screen, it looks very much like an accountant's ledger, but there the similarities end. It uses a matrix that assigns an identifying number and letter, in maplike fashion, to each "cell" in the matrix. The data may be dates, numbers, or sums of money. You instruct the computer what operation to perform in each column and row. You can change an entry to see what effect this has on the rest of the spreadsheet. The computer will automatically recompute all affected cells. The computer has the capacity to store information in thousands of cells and can reformat the data in graph form.

EXAMPLE 6 *Can These Matrices Be Multiplied?*

Determine which of the following pairs of matrices can be multiplied.

a) $A = \begin{bmatrix} 3 & 2 \\ 5 & 7 \end{bmatrix}$, $\quad B = \begin{bmatrix} 0 & 6 \\ 4 & 1 \end{bmatrix}$

b) $A = \begin{bmatrix} 2 & 3 \\ 5 & 6 \end{bmatrix}$, $\quad B = \begin{bmatrix} 2 & 4 & -1 \\ 6 & 8 & 0 \end{bmatrix}$

c) $A = \begin{bmatrix} 2 & 1 & 4 \\ 3 & 2 & 8 \end{bmatrix}$, $\quad B = \begin{bmatrix} 2 & 1 & 3 \\ 1 & 0 & -2 \end{bmatrix}$

SOLUTION

a)

$$\underset{2 \times 2}{A} \qquad \underset{2 \times 2}{B}$$

Same

Because matrix A has two columns and matrix B has two rows, the two matrices can be multiplied. The product is a 2×2 matrix.

PROFILE IN MATHEMATICS

James Sylvester, William Rowan Hamilton, and Arthur Cayley

James Sylvester

William Rowan Hamilton

Arthur Cayley

Three mathematicians played important roles in the development of matrix theory: James Sylvester (1814–1897), William Rowan Hamilton (1805–1865), and Arthur Cayley (1821–1895). Sylvester and Cayley cooperated in developing matrix theory. Sylvester was the first to use the term *matrix*. Hamilton, a noted physicist, astronomer, and mathematicians, also used what was essentially the algebra of matrices under the name *linear and vector functions*. The mathematical concept of *vector space* grew out of Hamilton's work on the algebra of vectors.

b)

$$\underset{2\times 2}{A} \qquad \underset{2\times 3}{B}$$

Same

Because matrix A has two columns and matrix B has two rows, the two matrices can be multiplied. The product is a 2×3 matrix.

c)

$$\underset{2\times 3}{A} \qquad \underset{2\times 3}{B}$$

Not same

Because matrix A has three columns and matrix B has two rows, the two matrices cannot be multiplied. ▲

To explain matrix multiplication let's use matrices A and B that follow.

$$A = \begin{bmatrix} 3 & 2 \\ 5 & 7 \end{bmatrix} \quad \text{and} \quad B = \begin{bmatrix} 0 & 6 \\ 4 & 1 \end{bmatrix}$$

Since A contains two rows and B contains two columns, the product matrix will contain two rows and two columns. To multiply two matrices, we use a row–column scheme of multiplying. The numbers in the first row of matrix A are multiplied by the numbers in the first column of matrix B.

$$A \times B = \begin{bmatrix} 3 & 2 \\ 5 & 7 \end{bmatrix}\begin{bmatrix} 0 & 6 \\ 4 & 1 \end{bmatrix}$$

First row First column

$$\begin{bmatrix} 3 & 2 \\ 5 & 7 \end{bmatrix} \qquad \begin{bmatrix} 0 & 6 \\ 4 & 1 \end{bmatrix}$$

$$(3 \times 0) + (2 \times 4) = 0 + 8 = 8$$

The 8 is placed in the first-row, first-column position of the product matrix. The other numbers in the product matrix are obtained similarly, as illustrated in the matrix that follows.

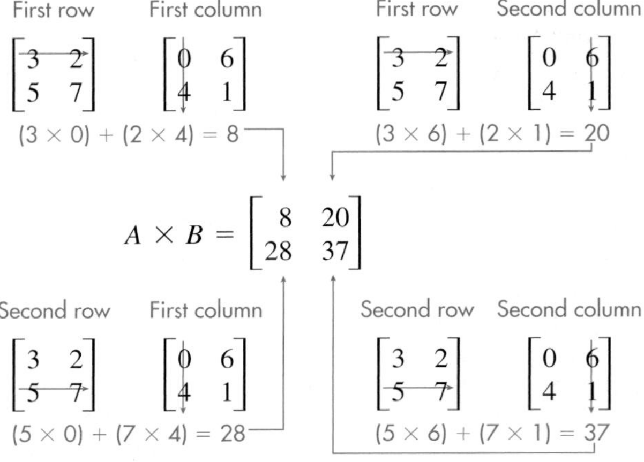

We can shorten the procedure as follows.

$$A \times B = \begin{bmatrix} 3 & 2 \\ 5 & 7 \end{bmatrix} \begin{bmatrix} 0 & 6 \\ 4 & 1 \end{bmatrix}$$

$$= \begin{bmatrix} 3(0) + 2(4) & 3(6) + 2(1) \\ 5(0) + 7(4) & 5(6) + 7(1) \end{bmatrix}$$

$$= \begin{bmatrix} 8 & 20 \\ 28 & 37 \end{bmatrix}$$

In general, if

$$A = \begin{bmatrix} a & b \\ c & d \end{bmatrix} \quad \text{and} \quad B = \begin{bmatrix} e & f \\ g & h \end{bmatrix},$$

then

$$A \times B = \begin{bmatrix} a & b \\ c & d \end{bmatrix} \begin{bmatrix} e & f \\ g & h \end{bmatrix} = \begin{bmatrix} ae + bg & af + bh \\ ce + dg & cf + dh \end{bmatrix}.$$

Let's do one more multiplication.

EXAMPLE 7 ***Multiplying Matrices***

Find $A \times B$, given

$$A = \begin{bmatrix} 2 & 3 \\ 5 & 6 \end{bmatrix} \quad \text{and} \quad B = \begin{bmatrix} 6 & -1 & 3 \\ 2 & 8 & 0 \end{bmatrix}$$

SOLUTION Matrix A contains two columns, and matrix B contains two rows. Thus, the matrices can be multiplied. Since A contains two rows and B contains three columns, the product matrix will contain two rows and three columns.

$$A \times B = \begin{bmatrix} 2 & 3 \\ 5 & 6 \end{bmatrix} \begin{bmatrix} 6 & -1 & 3 \\ 2 & 8 & 0 \end{bmatrix}$$

$$= \begin{bmatrix} 2(6) + 3(2) & 2(-1) + 3(8) & 2(3) + 3(0) \\ 5(6) + 6(2) & 5(-1) + 6(8) & 5(3) + 6(0) \end{bmatrix}$$

$$= \begin{bmatrix} 18 & 22 & 6 \\ 42 & 43 & 15 \end{bmatrix}$$

It should be noted that multiplication of matrices *is not* commutative; that is, $A \times B \neq B \times A$, except in special instances.

We previously discussed the 2×2 additive identity matrix,

$$\begin{bmatrix} 0 & 0 \\ 0 & 0 \end{bmatrix}$$

Square matrices also have a **multiplicative identity matrix**. The multiplicative identity matrices for a 2×2 and a 3×3 matrix, denoted I, follow. Note that in any

DID YOU KNOW

The Prisoner's Dilemma

When two parties pursue conflicting interests, the situation can sometimes be described and modeled in a matrix under a branch of mathematics known as game theory. Consider a famous problem called "the prisoner's dilemma." A pair of criminal suspects, A and B, are being held in separate jail cells and cannot communicate with each other. Each one is told that there are four possible outcomes: If both confess, each receives a 3-year sentence. If A confesses and B does not, A receives a 1-year sentence, whereas B receives a 10-year sentence. If B confesses and A does not, B receives 1 year and A receives 10 years. Finally, if neither confesses, each will be imprisoned for 2 years. (Try arranging this situation in a matrix.)

If neither prisoner knows whether the other will confess, what should each prisoner do? A study of game theory shows that it is in each prisoner's best interest to confess to the crime.

multiplicative identity matrix, 1's go diagonally from top left to bottom right and all other elements in the matrix are 0's.

$$I = \begin{bmatrix} 1 & 0 \\ 0 & 1 \end{bmatrix} \qquad I = \begin{bmatrix} 1 & 0 & 0 \\ 0 & 1 & 0 \\ 0 & 0 & 1 \end{bmatrix}$$

For any square matrix, A, $A \times I = I \times A = A$.

EXAMPLE 8 ***Using the Identity Matrix in Multiplication***

Use the multiplicative identity matrix for a 2×2 matrix and matrix A to show that $A \times I = A$.

$$A = \begin{bmatrix} 4 & 3 \\ 2 & 1 \end{bmatrix}$$

SOLUTION The identity matrix is $I = \begin{bmatrix} 1 & 0 \\ 0 & 1 \end{bmatrix}$.

$$\begin{aligned} A \times I &= \begin{bmatrix} 4 & 3 \\ 2 & 1 \end{bmatrix}\begin{bmatrix} 1 & 0 \\ 0 & 1 \end{bmatrix} \\ &= \begin{bmatrix} 4(1) + 3(0) & 4(0) + 3(1) \\ 2(1) + 1(0) & 2(0) + 1(1) \end{bmatrix} \\ &= \begin{bmatrix} 4 & 3 \\ 2 & 1 \end{bmatrix} = A \end{aligned}$$

▲

Example 9 illustrates an application of multiplication of matrices.

EXAMPLE 9 ***A Manufacturing Application***

The Fancy Frock Company manufactures three types of women's outfits: a dress, a two-piece suit (skirt and jacket), and a three-piece suit (skirt, jacket, and a vest). On a particular day, the firm produces 20 dresses, 30 two-piece suits, and 50 three-piece suits. Each dress requires 4 units of material and 1 hour of work to produce; each two-piece suit requires 3 units of material and 2 hours of work to produce; each three-piece suit requires 5 units of material and 3 hours to produce. Use matrix multiplication to determine the total number of units of material and the total number of hours needed for that day's production.

SOLUTION Let matrix A represent the number of each type of women's outfits produced.

	Dress	Two piece	Three piece
$A =$	[20	30	50]

The units of material and time requirements for each type are indicated in matrix B.

$$B = \begin{array}{c} \begin{array}{cc} \text{Material} & \text{Hours} \end{array} \\ \begin{bmatrix} 4 & 1 \\ 3 & 2 \\ 5 & 3 \end{bmatrix} \end{array} \begin{array}{l} \\ \text{Dress} \\ \text{Two piece} \\ \text{Three piece} \end{array}$$

The product of A and B, or $A \times B$, will give the total number of units of material and the total number of hours of work needed for that day's production.

$$\begin{aligned} A \times B &= [20 \quad 30 \quad 50]\begin{bmatrix} 4 & 1 \\ 3 & 2 \\ 5 & 3 \end{bmatrix} \\ &= [20(4) + 30(3) + 50(5) \quad 20(1) + 30(2) + 50(3)] \\ &= [420 \quad 230] \end{aligned}$$

Thus, a total of 420 units of material and a total of 230 hours of work were required that day.

SECTION 7.3 EXERCISES

CONCEPT/WRITING EXERCISES

1. What is a matrix?
2. Explain how to determine the dimensions of a matrix.
3. What is a square matrix?
4. To add or subtract two matrices, what must be true about the dimensions of those matrices?
5. a) In your own words, explain the procedure used to add matrices.
   b) Use the procedure given in part (a) to add

$$\begin{bmatrix} 5 & 3 & -1 \\ 0 & 2 & 4 \end{bmatrix} \quad \text{and} \quad \begin{bmatrix} 4 & 5 & 6 \\ -1 & 3 & 2 \end{bmatrix}$$

6. a) In your own words, explain the procedure used to subtract matrices.
   b) Use the procedure given in part (a) to subtract

$$\begin{bmatrix} 5 & 3 & -1 \\ 0 & 2 & 4 \end{bmatrix} \quad \text{from} \quad \begin{bmatrix} 4 & 5 & 6 \\ -1 & 3 & 2 \end{bmatrix}$$

7. a) To multiply two matrices, what must be true about the dimensions of those matrices?
   b) What will be the dimensions of the product matrix when multiplying a 2×2 matrix with a 2×3 matrix?
8. a) In your own words, explain the procedure used to multiply matrices.
   b) Use the procedure given in part (a) to multiply

$$\begin{bmatrix} 6 & -1 \\ 5 & 0 \end{bmatrix} \quad \text{by} \quad \begin{bmatrix} 2 & -3 \\ 1 & -4 \end{bmatrix}$$

9. a) What is the multiplicative identity matrix for a 2×2 matrix?
   b) What is the multiplicative identity matrix for a 3×3 matrix?
10. Records are kept each day for a month at a local cinema that houses three movie theaters A, B, and C. The daily average Monday through Sunday receipts for the three theaters are as follows.

    A: \$654, \$785, \$458, \$345, \$1478, \$2109, \$543

    B: \$764, \$778, \$568, \$451, \$1024, \$1689, \$853

    C: \$567, \$764, \$873, \$407, \$2034, \$2432, \$567

    Express this information in the form of a 3×7 matrix.

PRACTICE THE SKILLS

In Exercises 11–14, find A + B.

11. $A = \begin{bmatrix} 2 & 7 \\ 1 & 6 \end{bmatrix}, \quad B = \begin{bmatrix} -3 & -5 \\ 8 & 1 \end{bmatrix}$

12. $A = \begin{bmatrix} 2 & 5 & 1 \\ 6 & 0 & -1 \end{bmatrix}$, $B = \begin{bmatrix} -4 & -3 & 8 \\ 6 & 5 & 0 \end{bmatrix}$

13. $A = \begin{bmatrix} -1 & 0 \\ 0 & 4 \\ 6 & 2 \end{bmatrix}$, $B = \begin{bmatrix} 2 & 3 \\ 5 & 0 \\ 1 & -1 \end{bmatrix}$

14. $A = \begin{bmatrix} 1 & 5 & -3 \\ -1 & -6 & 4 \\ 2 & 0 & 5 \end{bmatrix}$, $B = \begin{bmatrix} -1 & 2 & 1 \\ 8 & -2 & 1 \\ 1 & 3 & 7 \end{bmatrix}$

In Exercises 15–18, find $A - B$.

15. $A = \begin{bmatrix} -2 & 5 \\ 9 & 1 \end{bmatrix}$, $B = \begin{bmatrix} 4 & -2 \\ -3 & 5 \end{bmatrix}$

16. $A = \begin{bmatrix} 1 & 2 \\ 0 & 6 \\ -3 & 9 \end{bmatrix}$, $B = \begin{bmatrix} 1 & 1 \\ 4 & 5 \\ -2 & 8 \end{bmatrix}$

17. $A = \begin{bmatrix} 5 & 3 & -1 \\ 7 & 4 & 2 \\ 6 & -1 & -5 \end{bmatrix}$, $B = \begin{bmatrix} 4 & 3 & 6 \\ -2 & -4 & 9 \\ 0 & -2 & 4 \end{bmatrix}$

18. $A = \begin{bmatrix} -4 & 3 \\ 6 & 2 \\ 1 & -5 \end{bmatrix}$, $B = \begin{bmatrix} -6 & -8 \\ -10 & -11 \\ 3 & -7 \end{bmatrix}$

In Exercises 19–24,

$$A = \begin{bmatrix} 1 & 2 \\ 0 & 5 \end{bmatrix}, \quad B = \begin{bmatrix} 3 & 2 \\ 5 & 0 \end{bmatrix}, \quad \text{and} \quad C = \begin{bmatrix} -2 & 3 \\ 4 & 0 \end{bmatrix}.$$

Find the following.

19. $2B$

20. $-3B$

21. $2B + 3C$

22. $2B + 3A$

23. $3B - 2C$

24. $4C - 2A$

In Exercises 25–30, find $A \times B$.

25. $A = \begin{bmatrix} 2 & 1 \\ 3 & 0 \end{bmatrix}$, $B = \begin{bmatrix} 1 & 4 \\ 2 & 6 \end{bmatrix}$

26. $A = \begin{bmatrix} 1 & -1 \\ 2 & 6 \end{bmatrix}$, $B = \begin{bmatrix} 3 & 2 \\ 5 & -2 \end{bmatrix}$

27. $A = \begin{bmatrix} 2 & 3 & -1 \\ 0 & 4 & 6 \end{bmatrix}$, $B = \begin{bmatrix} 2 \\ 4 \\ 1 \end{bmatrix}$

28. $A = \begin{bmatrix} 1 & 1 \\ 1 & 1 \end{bmatrix}$, $B = \begin{bmatrix} 1 & -1 \\ -1 & 2 \end{bmatrix}$

29. $A = \begin{bmatrix} 5 & 1 & 6 \\ -2 & 3 & 1 \\ 4 & 7 & 2 \end{bmatrix}$, $B = \begin{bmatrix} 1 & 0 & 0 \\ 0 & 1 & 0 \\ 0 & 0 & 1 \end{bmatrix}$

30. $A = \begin{bmatrix} -3 & 1 \\ 2 & 7 \end{bmatrix}$, $B = \begin{bmatrix} 4 & -1 \\ 0 & 6 \end{bmatrix}$

In Exercises 31–36, find $A + B$ and $A \times B$. If an operation cannot be performed, explain why.

31. $A = \begin{bmatrix} 1 & 2 & -2 \\ 3 & 0 & 4 \end{bmatrix}$, $B = \begin{bmatrix} 5 & 1 & 3 \\ 2 & -2 & 1 \end{bmatrix}$

32. $A = \begin{bmatrix} 3 & 5 & -1 \\ 2 & 3 & 4 \end{bmatrix}$, $B = \begin{bmatrix} 0 & 1 \\ 4 & -1 \end{bmatrix}$

33. $A = \begin{bmatrix} 4 & 5 & 3 \\ 6 & 2 & 1 \end{bmatrix}$, $B = \begin{bmatrix} 3 & 2 \\ 4 & 6 \\ -2 & 0 \end{bmatrix}$

34. $A = \begin{bmatrix} 1 & 2 \\ 3 & 4 \\ 5 & 6 \end{bmatrix}$, $B = \begin{bmatrix} 1 & 2 \\ 3 & 4 \\ 5 & 6 \end{bmatrix}$

35. $A = \begin{bmatrix} 1 & 2 \\ 3 & 4 \end{bmatrix}$, $B = \begin{bmatrix} -3 \\ 2 \end{bmatrix}$

36. $A = \begin{bmatrix} 5 & -1 \\ 6 & -2 \end{bmatrix}$, $B = \begin{bmatrix} 1 & 2 \\ 3 & 4 \end{bmatrix}$

In Exercises 37–39, show the commutative property of addition, $A + B = B + A$, holds for matrices A and B.

37. $A = \begin{bmatrix} 1 & 3 \\ 2 & -3 \end{bmatrix}$, $B = \begin{bmatrix} 4 & 5 \\ 6 & 2 \end{bmatrix}$

38. $A = \begin{bmatrix} -3 & 4 \\ 5 & 7 \end{bmatrix}$, $B = \begin{bmatrix} 0 & 6 \\ -1 & 5 \end{bmatrix}$

39. $A = \begin{bmatrix} 0 & -1 \\ 3 & -4 \end{bmatrix}$, $B = \begin{bmatrix} 8 & 1 \\ 3 & -4 \end{bmatrix}$

40. Make up two matrices with the same dimensions, A and B, and show that $A + B = B + A$.

In Exercises 41–43, show that the associative property of addition, $(A + B) + C = A + (B + C)$, holds for the matrices given.

41. $A = \begin{bmatrix} 2 & 3 \\ 1 & 6 \end{bmatrix}$, $B = \begin{bmatrix} -1 & 4 \\ 5 & 0 \end{bmatrix}$, $C = \begin{bmatrix} 3 & 4 \\ -2 & 7 \end{bmatrix}$

42. $A = \begin{bmatrix} -2 & -3 \\ -4 & -5 \end{bmatrix}$, $B = \begin{bmatrix} -9 & 1 \\ -7 & 2 \end{bmatrix}$, $C = \begin{bmatrix} 6 & 3 \\ -3 & 6 \end{bmatrix}$

43. $A = \begin{bmatrix} 7 & 4 \\ 9 & -36 \end{bmatrix}$, $B = \begin{bmatrix} 5 & 6 \\ -1 & -4 \end{bmatrix}$, $C = \begin{bmatrix} -7 & -5 \\ -1 & 3 \end{bmatrix}$

44. Make up three matrices with the same dimensions, A, B, and C, and show that $(A + B) + C = A + (B + C)$.

In Exercises 45–49, determine whether the commutative property of multiplication, $A \times B = B \times A$, holds for the matrices given.

45. $A = \begin{bmatrix} 2 & -1 \\ 4 & -3 \end{bmatrix}$, $B = \begin{bmatrix} 2 & 4 \\ -1 & -3 \end{bmatrix}$

46. $A = \begin{bmatrix} 5 & 1 \\ 4 & 6 \end{bmatrix}$, $B = \begin{bmatrix} 1 & 0 \\ 0 & 1 \end{bmatrix}$

47. $A = \begin{bmatrix} 4 & 2 \\ 1 & -3 \end{bmatrix}$, $B = \begin{bmatrix} 2 & 4 \\ -3 & 1 \end{bmatrix}$

48. $A = \begin{bmatrix} -3 & 2 \\ 6 & -5 \end{bmatrix}$, $B = \begin{bmatrix} -\frac{5}{3} & -\frac{2}{3} \\ -2 & -1 \end{bmatrix}$

49. $A = \begin{bmatrix} 3 & 2 & 1 \\ 4 & 2 & 0 \\ 0 & -2 & 5 \end{bmatrix}$, $B = \begin{bmatrix} 1 & 0 & 0 \\ 0 & 1 & 0 \\ 0 & 0 & 1 \end{bmatrix}$

50. Make up two matrices A and B with the same dimensions, and determine whether $A \times B = B \times A$.

In Exercises 51–55, show that the associative property of multiplication, $(A \times B) \times C = A \times (B \times C)$, holds for the matrices given.

51. $A = \begin{bmatrix} 1 & 2 \\ 4 & 0 \end{bmatrix}$, $B = \begin{bmatrix} 2 & 1 \\ 3 & 0 \end{bmatrix}$, $C = \begin{bmatrix} 4 & 2 \\ 3 & 1 \end{bmatrix}$

52. $A = \begin{bmatrix} -2 & 3 \\ 0 & 4 \end{bmatrix}$, $B = \begin{bmatrix} 4 & 0 \\ 3 & 5 \end{bmatrix}$, $C = \begin{bmatrix} 3 & 4 \\ -2 & 5 \end{bmatrix}$

53. $A = \begin{bmatrix} 4 & 3 \\ -6 & 2 \end{bmatrix}$, $B = \begin{bmatrix} 1 & 2 \\ 0 & 1 \end{bmatrix}$, $C = \begin{bmatrix} 4 & 3 \\ 0 & -2 \end{bmatrix}$

54. $A = \begin{bmatrix} -1 & -2 \\ -3 & -4 \end{bmatrix}$, $B = \begin{bmatrix} 1 & 0 \\ 0 & 1 \end{bmatrix}$, $C = \begin{bmatrix} 0 & 0 \\ 0 & 0 \end{bmatrix}$

55. $A = \begin{bmatrix} 3 & 4 \\ -1 & -2 \end{bmatrix}$, $B = \begin{bmatrix} 0 & 1 \\ 1 & 0 \end{bmatrix}$, $C = \begin{bmatrix} 2 & 0 \\ 3 & 0 \end{bmatrix}$

56. Make up three matrices with the same dimensions, A, B, and C, and show that $(A \times B) \times C = A \times (B \times C)$.

PROBLEM SOLVING

57. MODELING - *Cookie Company Costs* The Original Cookie Factory bakes and sells four types of cookies: chocolate chip, sugar, molasses, and peanut butter. Matrix A shows the number of units of various ingredients used in baking a dozen of each type of cookie.

	Sugar	Flour	Milk	Eggs
Chocolate chip	2	2	$\frac{1}{2}$	1
Sugar	3	2	1	2
Molasses	0	1	0	3
Peanut butter	$\frac{1}{2}$	1	0	0

$$A = \begin{bmatrix} 2 & 2 & \frac{1}{2} & 1 \\ 3 & 2 & 1 & 2 \\ 0 & 1 & 0 & 3 \\ \frac{1}{2} & 1 & 0 & 0 \end{bmatrix}$$

The cost, in cents per cup or per egg, for each ingredient when purchased in small quantities and in large quantities is given in matrix B.

	Small quantities	Large quantities
Sugar	10	12
Flour	5	8
Milk	8	8
Eggs	4	6

$$B = \begin{bmatrix} 10 & 12 \\ 5 & 8 \\ 8 & 8 \\ 4 & 6 \end{bmatrix}$$

Use matrix multiplication to find a matrix representing the comparative cost per item for small and large quantities purchased.

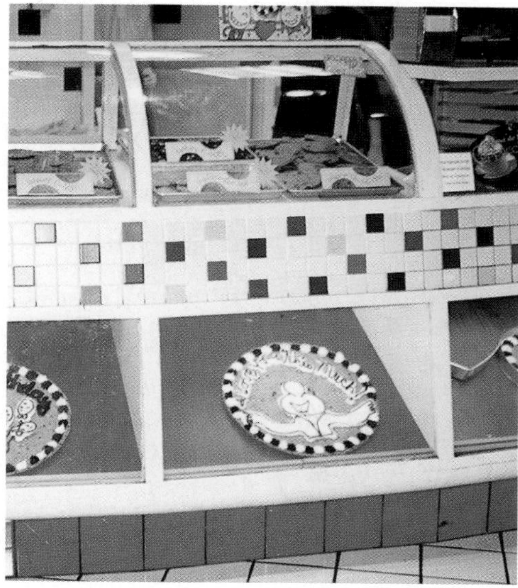

In Exercises 58 and 59, use the information given in Exercise 57. Suppose a typical day's order consists of 40 dozen chocolate chip cookies, 30 dozen sugar cookies, 12 dozen molasses cookies, and 20 dozen peanut butter cookies.

58. a) Express these orders as a 1×4 matrix.
b) Use matrix multiplication to determine the amount of each ingredient needed to fill the day's order.

59. Use matrix multiplication to determine the cost under the two purchase options (small and large quantities) to fill the day's order.

60. MODELING - *Food Prices* Food orders from the Geology Club and the Rugby Club are summarized in matrix A.

	Burger	Fries	Cola
Geology Club	28	32	25
Rugby Club	33	26	31

$$A = \begin{bmatrix} 28 & 32 & 25 \\ 33 & 26 & 31 \end{bmatrix}$$

The prices (in dollars) of a burger, fries, and a cola at three fast-food restaurants are summarized in matrix B.

	McDougal's	Burger Prince	Mendy's
Burger	2.45	2.95	3.15
Fries	1.35	0.99	1.00
Cola	1.40	0.92	1.20

$$B = \begin{bmatrix} 2.45 & 2.95 & 3.15 \\ 1.35 & 0.99 & 1.00 \\ 1.40 & 0.92 & 1.20 \end{bmatrix}$$

a) Multiply the two matrices to form a 2×3 matrix that shows the amount each club would be charged by the fast-food restaurant.
b) Determine which fast-foot restaurant offers each club the best total price.

Two matrices whose sum is the additive identity matrix are said to be **additive inverses**. *That is, if $A + B = B + A = I$, where I is the additive identity matrix, then A and B are additive inverses. In Exercises 61 and 62, determine whether A and B are additive inverses.*

61. $A = \begin{bmatrix} 6 & 3 \\ 4 & -2 \end{bmatrix}, \quad B = \begin{bmatrix} -6 & -3 \\ -2 & 4 \end{bmatrix}$

62. $A = \begin{bmatrix} 4 & 6 & 3 \\ 2 & 3 & -1 \\ -1 & 0 & 6 \end{bmatrix}, \quad B = \begin{bmatrix} -4 & -6 & -3 \\ -2 & -3 & 1 \\ 1 & 0 & -6 \end{bmatrix}$

Two matrices whose product is the multiplicative identity matrix are said to be **multiplicative inverses**. *That is, if $A \times B = B \times A = I$, where I is the multiplicative identity matrix, then A and B are multiplicative inverses. In Exercises 63 and 64, determine whether A and B are multiplicative inverses.*

63. $A = \begin{bmatrix} 5 & -2 \\ -2 & 1 \end{bmatrix}, \quad B = \begin{bmatrix} 1 & 2 \\ 2 & 5 \end{bmatrix}$

64. $A = \begin{bmatrix} 7 & 3 \\ 2 & 1 \end{bmatrix}, \quad B = \begin{bmatrix} 1 & -3 \\ -2 & 7 \end{bmatrix}$

CHALLENGE PROBLEMS/GROUP ACTIVITIES

In Exercises 65 and 66, determine whether the statement is true or false. Give an example to support your answer.

65. $A - B = B - A$, where A and B are any matrices.

66. For scalar a and matrices B and C, $a(B + C) = aB + aC$.

67. MODELING - *Boat-Building Costs* The number of hours of labor required to manufacture one boat of various sizes is summarized in matrix L.

Department: Cutting, Assembly, Packing

$$L = \begin{bmatrix} 1.4 \text{ hr} & 0.7 \text{ hr} & 0.3 \text{ hr} \\ 1.8 \text{ hr} & 1.4 \text{ hr} & 0.3 \text{ hr} \\ 2.7 \text{ hr} & 2.8 \text{ hr} & 0.5 \text{ hr} \end{bmatrix} \begin{matrix} \text{Small} \\ \text{Medium} \\ \text{Large} \end{matrix} \Bigg\} \text{Boat size}$$

The hourly labor rates for cutting, assembly, and packing at the Ames City Plant and at the Bay City Plant are given in matrix C.

Plant: Ames City, Bay City

$$C = \begin{bmatrix} \$14 & \$12 \\ \$10 & \$9 \\ \$7 & \$5 \end{bmatrix} \begin{matrix} \text{Cutting} \\ \text{Assembly} \\ \text{Packaging} \end{matrix} \Bigg\} \text{Department}$$

a) What is the total labor cost for manufacturing a small-sized boat at the Ames City plant?

b) What is the total cost for manufacturing a large-sized boat at the Bay City plant?

c) Determine the product $L \times C$ and explain the meaning of the results.

RESEARCH ACTIVITIES

68. Find an article that shows information illustrated in matrix form. Write a short paper explaining how to interpret the information provided by the matrix. Include the article with your report.

69. Do research and write a paper on the development of matrices. In your paper cover the contributions of James Joseph Sylvester, Arthur Cayley, and William Rowan Hamilton. (References include history of mathematics books, encyclopedias, and the Internet.)

7.4 SOLVING SYSTEMS OF EQUATIONS BY USING MATRICES

In Section 7.3, we introduced matrices. Now we will discuss the procedure to solve a system of linear equations using matrices. We will illustrate how to solve a system of two equations and two unknowns. Systems of equations containing three equations and three unknowns (called third-order systems) and higher-order systems can also be solved by using matrices.

The first step in solving a system of equations using matrices is to represent the system of equations with an **augmented matrix**. An augmented matrix consists of two smaller matrices, one for the coefficients of the variables in the equations and one for the constants in the equations. To determine the augmented matrix, first write each equation in standard form, $ax + by = c$. For the system of equations on the left of the top of the next page, its augmented matrix is shown to its right.

DID YOU KNOW

Early Matrices

The earliest known use of a matrix to solve linear equations appeared in the ancient Chinese mathematical classic *Jiuzhang Suanshu* (*Nine Chapters on the Mathematical Art*) in about 200 B.C. The use of matrices to solve problems did not appear in the West until the nineteenth century. Perhaps the fact that the Chinese used a counting board that took the form of a grid made it easier for them to make the leap to the development and use of matrices. The image shows the same type of counting board as appeared in a book dated 1795.

System of equations	**Augmented matrix**
$a_1x + b_1y = c_1$ $a_2x + b_2y = c_2$	$\left[\begin{array}{cc\|c} a_1 & b_1 & c_1 \\ a_2 & b_2 & c_2 \end{array}\right]$

Following is another example.

System of equations	**Augmented matrix**
$x + 2y = 8$ $3x - y = 7$	$\left[\begin{array}{cc\|c} 1 & 2 & 8 \\ 3 & -1 & 7 \end{array}\right]$

Note that the bar in the augmented matrix separates the numerical coefficients from the constants. The matrix is just a shortened way of writing the system of equations. Thus we can solve a system of equations by using matrices in a manner very similar to solving a system of equations with the addition method.

To solve a system of equations by using matrices, we use *row transformations* to obtain new matrices that have the same solution as the original system. We will discuss three row transformation procedures.

Procedures for Row Transformations

1. Any two rows of a matrix may be interchanged (this is the same as interchanging any two equations in the system of equations).
2. All the numbers in any row may be multiplied by any nonzero real number. (This is the same as multiplying both sides of an equation by any nonzero real number.)
3. All the numbers in any row may be multiplied by any nonzero real number, and these products may be added to the corresponding numbers in any other row of numbers.

We use row transformations to obtain an augmented matrix whose numbers to the left of the vertical bar are the same as in the *multiplicative identity matrix*. From this type of augmented matrix, we can determine the solution to the system of equations. For example, if we get

$$\left[\begin{array}{cc|c} 1 & 0 & 3 \\ 0 & 1 & -2 \end{array}\right],$$

it tells us that $1x + 0y = 3$ or $x = 3$, and $0x + 1y = -2$ or $y = -2$. Thus, the solution to the system of equations that yielded this augmented matrix is $(3, -2)$. Now let's work an example.

EXAMPLE 1 *Using Row Transformations*

Solve the following system of equations by using matrices.

$$\begin{aligned} x + 2y &= 5 \\ 3x - y &= 8 \end{aligned}$$

SOLUTION First we write the augmented matrix.

$$\left[\begin{array}{cc|c} 1 & 2 & 5 \\ 3 & -1 & 8 \end{array}\right]$$

Our goal is to obtain a matrix of the form

$$\left[\begin{array}{cc|c} 1 & 0 & c_1 \\ 0 & 1 & c_2 \end{array}\right]$$

where c_1 and c_2 may represent any real numbers. It is generally easier to work by columns. Therefore, we will try to get the first column of the augmented matrix to be $\substack{1\\0}$ and the second column to be $\substack{0\\1}$. Since the element in the top left position is already a 1, we must work to change the 3 into a 0. We use row transformation procedure 3 to change the 3 into a 0. If we multiply the top row of numbers by -3 and add these products to the second row of numbers, the element in the first column, second row will become a 0:

$$\left[\begin{array}{cc|c} 1 & 2 & 5 \\ 3 & -1 & 8 \end{array}\right]$$

The top row of numbers multiplied by -3 gives

$$1(-3), \qquad 2(-3), \qquad \text{and} \qquad 5(-3)$$

Now add these products to their respective numbers in row 2.

$$\left[\begin{array}{cc|c} 1 & 2 & 5 \\ 3 + 1(-3) & -1 + 2(-3) & 8 + 5(-3) \end{array}\right] = \left[\begin{array}{cc|c} 1 & 2 & 5 \\ 0 & -7 & -7 \end{array}\right]$$

The next step is to obtain a 1 in the second column, second row. At present, -7 is in this position. To change the -7 to a 1, we use row transformation procedure 2. If we multiply -7 by $-\frac{1}{7}$, the product will be 1. Therefore, we multiply all the numbers in the second row by $-\frac{1}{7}$ to get

$$\left[\begin{array}{cc|c} 1 & 2 & 5 \\ 0(-\frac{1}{7}) & -7(-\frac{1}{7}) & -7(-\frac{1}{7}) \end{array}\right] = \left[\begin{array}{cc|c} 1 & 2 & 5 \\ 0 & 1 & 1 \end{array}\right].$$

The next step is to obtain a 0 in the second column, first row. At present, a 2 is in this position. Multiplying the numbers in the second row by -2 and adding the products to the corresponding numbers in the first row gives a 0 in the desired position.

$$\left[\begin{array}{cc|c} 1 + 0(-2) & 2 + 1(-2) & 5 + 1(-2) \\ 0 & 1 & 1 \end{array}\right] = \left[\begin{array}{cc|c} 1 & 0 & 3 \\ 0 & 1 & 1 \end{array}\right]$$

We now have the desired augmented matrix:

$$\left[\begin{array}{cc|c} 1 & 0 & 3 \\ 0 & 1 & 1 \end{array}\right]$$

With this matrix, we see that $1x + 0y = 3$, or $x = 3$, and $0x + 1y = 1$, or $y = 1$. The solution to the system is (3, 1).

Check:

$x + 2y = 5$		$3x - y = 8$	
$3 + 2(1) = 5$		$3(3) - 1 = 8$	
$5 = 5$	True	$8 = 8$	True

To Change an Augmented Matrix to the Form $\left[\begin{array}{cc|c} 1 & 0 & c_1 \\ 0 & 1 & c_2 \end{array}\right]$

1. Change the element in the first column, first row, to a 1.
2. Change the element in the first column, second row, to a 0.
3. Change the element in the second column, second row, to a 1.
4. Change the element in the second column, first row, to a 0.

Generally, when changing an element in the augmented matrix to a 1, we use step 2 in the row transformation box. When changing an element to a 0, we use step 3 in the row transformation box.

EXAMPLE 2 *Using Matrices to Solve a System of Equations*

Solve the following system of equations using matrices.

$$\begin{aligned} 2x + 4y &= 6 \\ 4x - 2y &= -8 \end{aligned}$$

SOLUTION First write the augmented matrix.

$$\left[\begin{array}{cc|c} 2 & 4 & 6 \\ 4 & -2 & -8 \end{array}\right]$$

To obtain a 1 in the first column, first row, multiply the numbers in the first row by $\frac{1}{2}$.

$$\left[\begin{array}{cc|c} 1 & 2 & 3 \\ 4 & -2 & -8 \end{array}\right]$$

To obtain a 0 in the first column, second row, multiply the numbers in the first row by -4 and add the products to the corresponding numbers in the second row.

$$\left[\begin{array}{cc|c} 1 & 2 & 3 \\ 4 + 1(-4) & -2 + 2(-4) & -8 + 3(-4) \end{array}\right] = \left[\begin{array}{cc|c} 1 & 2 & 3 \\ 0 & -10 & -20 \end{array}\right]$$

To obtain a 1 in the second column, second row, multiply the numbers in the second row by $-\frac{1}{10}$.

$$\left[\begin{array}{cc|c} 1 & 2 & 3 \\ 0(-\frac{1}{10}) & -10(-\frac{1}{10}) & -20(-\frac{1}{10}) \end{array}\right] = \left[\begin{array}{cc|c} 1 & 2 & 3 \\ 0 & 1 & 2 \end{array}\right]$$

To obtain a 0 in the second column, first row, multiply the numbers in the second row by -2 and add the products to the corresponding numbers in the first row.

$$\left[\begin{array}{cc|c} 1 + (0)(-2) & 2 + (1)(-2) & 3 + (2)(-2) \\ 0 & 1 & 2 \end{array}\right] = \left[\begin{array}{cc|c} 1 & 0 & -1 \\ 0 & 1 & 2 \end{array}\right]$$

The solution to the system of equations is $(-1, 2)$.

INCONSISTENT AND DEPENDENT SYSTEMS

Assume that you solve a system of two equations and obtain an augmented matrix in which one row of numbers on one side of the vertical line are all zeros but a zero does not appear in the same row on the other side of the vertical line. This indicates that the system is inconsistent and has no solution. For example, a system of equations that yields the following augmented matrix is an inconsistent system.

$$\left[\begin{array}{cc|c} 1 & 2 & 5 \\ 0 & 0 & 4 \end{array}\right] \quad \text{Inconsistent system}$$

The second row of the matrix represents the equation

$$0x + 0y = 4 \qquad \text{or} \qquad 0 = 4$$

which is never true.

If you obtain a matrix in which a 0 appears across an entire row, the system of equations is dependent. For example, a system of equations that yields the following matrix is a dependent system.

$$\left[\begin{array}{cc|c} 1 & 5 & -6 \\ 0 & 0 & 0 \end{array}\right] \quad \text{Dependent system}$$

The second row of the matrix represents the equation

$$0x + 0y = 0 \qquad \text{or} \qquad 0 = 0$$

which is always true.

TRIANGULARIZATION METHOD

Another procedure to solve a system of two equations is to use row transformation procedures to obtain an augmented matrix of the form

$$\left[\begin{array}{cc|c} 1 & a & b \\ 0 & 1 & c \end{array}\right]$$

where a, b, and c represent real numbers. This procedure is called the **triangularization method**, for the ones and zeros form a triangle.

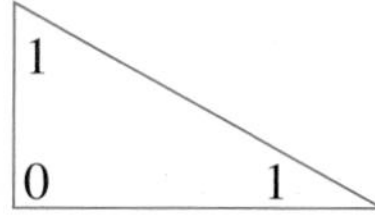

When the matrix is in this form, we can write the following system of equations.

$$\begin{aligned} 1x + ay &= b \\ 0x + 1y &= c \end{aligned} \qquad \text{or} \qquad \begin{aligned} x + ay &= b \\ y &= c \end{aligned}$$

Using substitution, we can easily solve the system.

For example, in Example 2, we obtained the augmented matrix

$$\left[\begin{array}{cc|c} 1 & 2 & 3 \\ 0 & 1 & 2 \end{array}\right]$$

This matrix represents the following system of equations.

$$\begin{aligned} x + 2y &= 3 \\ y &= 2 \end{aligned}$$

To solve for x, we substitute 2 for y in the equation

$$\begin{aligned} x + 2y &= 3 \\ x + 2(2) &= 3 \\ x + 4 &= 3 \\ x &= -1 \end{aligned}$$

Thus, the solution to the system is $(-1, 2)$, as was obtained in Example 2. You may use either method when solving a system of equations with matrices unless your instructor specifies otherwise.

SECTION 7.4 EXERCISES

CONCEPT/WRITING EXERCISES

1. a) What is an augmented matrix?
   b) Determine the augmented matrix for the following system.

$$\begin{aligned} x + 3y &= 7 \\ 2x - y &= 4 \end{aligned}$$

2. In your own words, write the three row transformation procedures.

3. How will you know, when solving a system of equations by using matrices, whether the system is dependent?

4. How will you know, when solving a system of equations by using matrices, whether the system is inconsistent?

PRACTICE THE SKILLS

In Exercises 5–18, use matrices to solve the system of equations.

5. $x + y = 3$
   $2x - y = 9$

6. $x - y = 3$
   $2x - y = 7$

7. $x + 2y = 4$
   $2x - y = 3$

8. $x + y = 1$
   $2x - 5y = 9$

9. $2x - 5y = -6$
   $-4x + 10y = 12$

10. $x + y = 5$
    $3x - y = 3$

11. $x + 3y = 1$
    $-2x + y = 5$

12. $2x - 3y = 10$
    $2x + 2y = 5$

13. $2x - 4y = 0$
    $x - 3y = -1$

14. $4x + 2y = 6$
    $5x + 4y = 9$

15. $-3x + 6y = 5$
    $2x - 4y = 8$

16. $2x - 5y = 10$
    $3x + y = 15$

17. $2x + y = 11$
    $x + 3y = 18$

18. $4x - 3y = 7$
    $-2x + 5y = 14$

PROBLEM SOLVING

In Exercises 19–22, use matrices to solve the problem.

19. MODELING - *Purchasing Supplies* Kym Hampton purchased supplies to decorate her second-grade classroom. She purchased 4 pieces of poster board and 2 marker pens for \$8. Later she purchased 8 pieces of poster board and 5 marker pens for \$18. Find the cost of one piece of poster board and one marker pen.

20. MODELING - *A Geometric Application* The length of a rectangle is 4 ft greater than its width. If its perimeter is 16 ft, find the length and width.

21. MODELING - *Sweets* If Meg buys 2 lb of chocolate-covered cherries and 3 lb of chocolate-covered mints, her total cost is $23. If she buys 1 lb of chocolate-covered cherries and 2 lb of chocolate-covered mints, her total cost is $14. Find the cost of 1 lb of chocolate-covered cherries and 1 lb of chocolate-covered mints.

22. MODELING - *On the Job* Peoplepower, Inc., a daily employment agency, charges $10 per hour for a truck driver and $8 per hour for a laborer. On a certain job, the laborer worked two more hours than the truck driver, and together they cost $144. How many hours did each work?

CHALLENGE PROBLEM/GROUP ACTIVITY

23. MODELING - *Fill in the Missing Information* Pencil World sells two types of mechanical pencils to stationery stores. The nonrefillable pencil that contains 6 pieces of lead sells for $1.50 each, and the refillable pencil sells for $2.00 each. Pencil World received an order for 200 pencils and a check for $337.50 for the pencils. When placing the order, the stationery store clerk failed to specify the number of each type of pencil being ordered. Can Pencil World fill the order with the information given? If so, determine the number of nonrefillable and the number of refillable pencils the clerk ordered.

7.5 SYSTEMS OF LINEAR INEQUALITIES

In earlier sections, we showed how to find the solution to a system of linear equations in two variables. Now we are going to explore the techniques of finding the solution set to a system of linear inequalities in two variables.

The solution set of a system of linear inequalities is the set of points that satisfy all inequalities in the system. The solution set of a system of linear inequalities may consist of infinitely many ordered pairs. To determine the solution set to a system of linear inequalities, graph each inequality on the same axes. The ordered pairs common to all the inequalities are the solution set to the system.

Procedure for Solving a System of Linear Inequalities

1. Select one of the inequalities. Replace the inequality symbol with an equal sign and draw the graph of the equation. Draw the graph with a dashed line if the inequality is $<$ or $>$ and with a solid line if the inequality is $\leq$ or $\geq$.
2. Select a test point on one side of the line and determine whether the point is a solution to the inequality. If so, shade the area on the side of the line containing the point. If the point is not a solution, shade the area on the other side of the line.
3. Repeat steps 1 and 2 for the other inequality.
4. The intersection of the two shaded areas and any solid line common to both inequalities form the solution set to the system of inequalities.

EXAMPLE 1 *Solving a System of Inequalities*

Graph the following system of inequalities and indicate the solution set.

$$x + y < 2$$
$$x - y < 4$$

SOLUTION Graph both inequalities on the same axes. First draw the graph of $x + y < 2$. When drawing the graph, remember to use a dashed line, since the inequality is "less than" (see Fig. 7.10a). If you have forgotten how to graph inequalities, review Section 6.8.

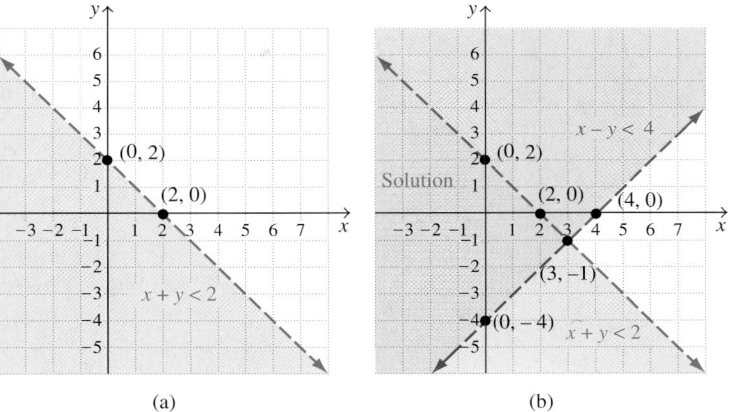

Figure 7.10

Now, on the same axes, shade the half-plane determined by the inequality $x - y < 4$ (see Fig. 7.10b). The solution set consists of all the points common to the two shaded half-planes. These are the points in the region on the graph containing both color shadings. Figure 7.10(b) shows that the two lines intersect at $(3, -1)$. This ordered pair can also be found by any of the algebraic methods discussed in Sections 7.2 and 7.3.

EXAMPLE 2 *Solving a System of Linear Inequalities*

Graph the following system of inequalities and indicate the solution set.

$$2x + 3y \geq 4$$
$$2x - y > -6$$

SOLUTION Graph the inequality $2x + 3y \geq 4$. Remember to use a solid line, because the inequality is "greater than or equal to"; see Fig. 7.11(a).

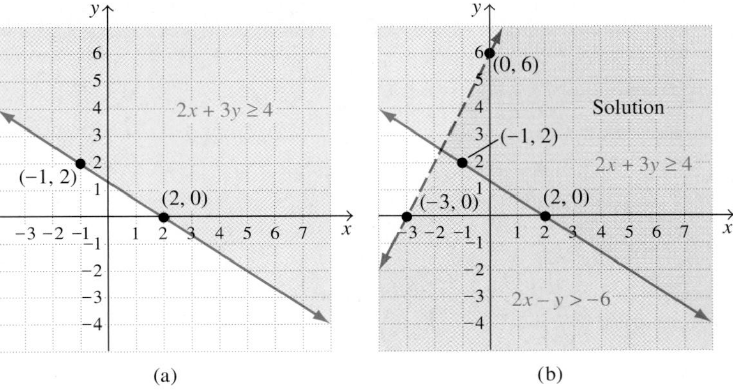

Figure 7.11

On the same set of axes, draw the graph of $2x - y > -6$. Use a dashed line, since the inequality is "greater than"; (see Fig. 7.11(b). The solution is the region of the graph that contains both color shadings and the part of the solid line that satisfies the inequality $2x - y > -6$. Note that the point of intersection of the two lines is not a part of the solution set.

EXAMPLE 3 *Another System of Inequalities*

Graph the following system of inequalities and indicate the solution set.

$$x \leq 5$$
$$y > -1$$

SOLUTION Graph the inequality $x \leq 5$; see Fig. 7.12(a). On the same axes, graph $y > -1$; see Fig. 7.12(b). The solution set is that region of the graph that is

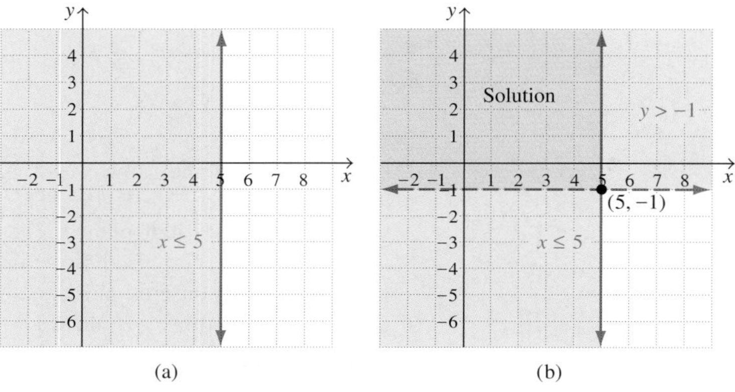

Figure 7.12

shaded in both colors and the part of the solid line that satisfies the inequality $y > -1$. The point of intersection of the two lines, $(5, -1)$, is not part of the solution because it does not satisfy the inequality $y > -1$.

SECTION 7.5 EXERCISES

CONCEPT/WRITING EXERCISES

1. What is the solution set of a system of linear inequalities?
2. In your own words, give the four-step procedure for solving a system of linear inequalities.

PRACTICE THE SKILLS

In Exercises 3–18, graph the system of linear inequalities and indicate the solution set.

3. $y > x + 1$
   $y > 2x$
4. $y \leq 2x - 2$
   $y > -x + 3$
5. $x + y \geq 2$
   $x - y > 5$
6. $2x + 3y \geq 6$
   $x + y < 4$
7. $x - y < 4$
   $x + y < 5$
8. $3x - y \leq 6$
   $x - y > 4$
9. $x - 3y \leq 3$
   $x + 2y \geq 4$
10. $x + 2y \geq 4$
    $3x - y \geq -6$
11. $y \leq 3x$
    $x \geq 3y$
12. $y \leq 4$
    $x - y < 1$
13. $x \geq 1$
    $y \leq 1$
14. $x \leq 0$
    $y \leq 0$
15. $4x + 2y > 8$
    $x \geq y - 1$
16. $5y > 3x + 10$
    $3y < -2x - 3$
17. $3x \geq 2y + 6$
    $x \leq y + 7$
18. $2x - 5y < 7$
    $x > 2y + 1$

CHALLENGE PROBLEMS/GROUP ACTIVITIES

19. MODELING - *Special Diet* Reuben Ricardo's on a diet. He must consume less than 500 calories at a meal that consists of a serving of chicken and a serving of rice. The meal must contain at least 150 calories from each source.

a) Translate the problem into a system of linear inequalities.

b) Solve the system graphically. Graph calories from chicken on the horizontal axis and calories from rice on the vertical axis.

c) There are about 200 calories in 8 oz of rice and about 180 calories in 3 oz of chicken. Select a point in the solution set in part (b). This point represents the number of ounces of chicken and rice to be served at a meal. For the point selected, determine the number of ounces of chicken and rice to be served.

20. Write a system of linear inequalities whose solution is the second quadrant, including the axes.

21. a) Do all systems of linear inequalities have solutions? Explain.

b) Write a system of inequalities that has no solution.

22. Can a system of linear inequalities have a solution set consisting of a single point? Explain.

23. Can a system of linear inequalities have as its solution set all the points on the coordinate plane? Explain your answer, giving an example to support it.

7.6 LINEAR PROGRAMMING

Government, business, and industry often require decision makers to find cost-effective solutions to a variety of problems. Linear programming often serves as a method of expressing the relationships in many of these problems and uses a system of linear inequalities.

The typical linear programming problem has many variables and is generally so lengthy that it is solved on a computer by a technique called the **simplex algorithm**. The simplex method was developed in the 1940s by George B. Dantzig. Linear programming is used to solve problems in the social sciences, health care, land development, nutrition, military, and many other fields.

We will not discuss the simplex method in this textbook. We will merely give a brief introduction to how linear programming works. You can find a detailed explanation in books on finite mathematics.

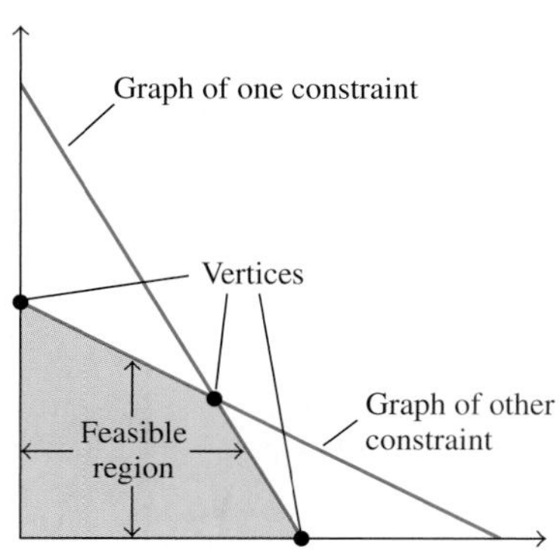

Figure 7.13

In a linear programming problem, there are restrictions called **constraints**. Each constraint is represented as a linear inequality. The list of constraints forms a system of linear inequalities. When the system of inequalities is graphed, we often obtain a region bounded on all sides by line segments (Fig. 7.13). This region is called the **feasible region**. The points where two or more boundaries intersect are called the **vertices** of the feasible region. The points on the boundary of the region and the points inside the feasible region are the solution set for the system of inequalities.

For each linear programming problem, we will obtain a formula of the form $K = Ax + By$, called the **objective function**. The objective function is the formula for the quantity K (or some other variable) that we want to maximize or minimize. The values we substitute for x and y determine the value of K. From the information given in the problem, we determine the real number constants A and B. In a particular linear programming problem, a typical equation that might be used to find the maximum profit, P, is $P = 3x + 7y$. We would find the maximum profit by substituting the ordered pairs (x, y) of the vertices of the feasible region into the formula to see which ordered pair yields the greatest value of P and therefore the maximum profit. The ordered pair that yields the smallest value of P determines the minimum profit.

Linear programming is used to determine which ordered pair will yield the maximum (or minimum) value of the variable that is being maximized (or minimized). The fundamental principle of linear programming provides a rule for finding those maximum and minimum values.

DID YOU KNOW

The Logistics of D-Day

Linear programming was first used to deal with the age-old military problem of logistics: obtaining, maintaining, and transporting military equipment and personnel. George Dantzig developed the simplex method for the Allies of World War II to do just that. Consider the logistics of the Allied invasion of Normandy. Meteorologic experts had settled on three possible dates in June 1944. It had to be a day when low tide and first light would coincide, the winds should not exceed 8 to 13 mph, and visibility had to be not less than 3 miles. A force of 170,000 assault troops was to be assembled and moved to 22 airfields in England where 1200 air transports and 700 gliders would then take them to the coast of France to converge with 5000 ships of the D-Day armada. The code name for the invasion was Operation Overlord, but it is known to most as D-Day.

Winston Churchill called the invasion of Normandy "the most difficult and complicated operation that has ever taken place."

Fundamental Principle of Linear Programming
If the objective function, $K = Ax + By$, is evaluated at each point in a feasible region, the maximum and minimum values of the equation occur at vertices of the region.

Example 1 illustrates how the fundamental principle is used to solve a linear programming problem.

EXAMPLE 1 MODELING - *Using the Fundamental Principle of Linear Programming*

The Ric Shaw Chair company makes two types of rocking chairs: a plain chair and a fancy chair. Each rocking chair must be assembled and then finished. The plain chair takes 4 hours to assemble and 4 hours to finish. The fancy chair takes 8 hours to assemble and 12 hours to finish. The company can provide at most 160 worker-hours of assembling and 180 worker-hours of finishing a day. If the profit on a plain chair is \$10 and the profit on a fancy chair is \$18, how many rocking chairs of each type should the company make per day to maximize profits? What is the maximum profit?

SOLUTION From the information given, we know the following facts.

	Assembly time (hr)	Finishing time (hr)	Profit ($)
Plain chair	4	4	10.00
Fancy chair	8	12	18.00

Let

$$
\begin{aligned}
x &= \text{the number of plain chairs per day} \\
y &= \text{the number of fancy chairs per day} \\
10x &= \text{profit on the plain chairs} \\
18y &= \text{profit on the fancy chairs} \\
P &= \text{the total profit}
\end{aligned}
$$

The total profit is the sum of the profit on the plain chairs and the profit on the fancy chairs. Since $10x$ is the profit on the plain chairs and $18y$ is the profit on the fancy chairs, the profit formula is $P = 10x + 18y$.

The maximum profit, P, is dependent on several conditions, called *constraints*. The number of chairs manufactured each day cannot be a negative amount. This condition gives us the constraints $x \geq 0$ and $y \geq 0$. Another constraint is determined by the total number of hours allocated for assembling. Four hours are needed to assemble the plain chair, so the total number of hours per day to assemble x plain chairs is $4x$. Eight hours are required to assemble a fancy chair, so the total number of hours needed to assemble y fancy chairs is $8y$. The maximum number of hours allocated for assembling is 160 per day. Thus, the third constraint is $4x + 8y \leq 160$. The final constraint is determined by the number of hours allotted for finishing. Finishing a plain chair takes 4 hours, or $4x$ hours to finish x plain chairs. Finishing a fancy chair takes 12 hours, or $12y$ hours to finish y fancy chairs. The total number of hours allotted for finishing is 180 per day. Therefore, the fourth constraint is $4x + 12y \leq 180$. Thus, the four constraints are

$$
\begin{aligned}
x &\geq 0 \\
y &\geq 0 \\
4x + 8y &\leq 160 \\
4x + 12y &\leq 180
\end{aligned}
$$

Figure 7.14

The list of constraints is a system of linear inequalities in two variables. The solution to the system of inequalities is the set of ordered pairs that satisfies all the constraints. These points are plotted in Fig. 7.14. Note that the solution to the system consists of the colored region and the solid boundaries. The points (0, 0), (0, 15), (30, 5), and (40, 0) are the points at which the boundaries intersect. These points can also be found by the addition or substitution method described in Section 7.2.

The goal in this example is to maximize the profit. The objective function is given by the profit formula $P = 10x + 18y$. According to the fundamental principle, the maximum profit will be found at one of the vertices of the feasible region.

Calculate P for each one of the vertices.

$$
\begin{aligned}
& & P &= 10x + 18y \\
&\text{At } (0, 0), & P &= 10(0) + 18(0) = 0 \\
&\text{At } (0, 15), & P &= 10(0) + 18(15) = 270 \\
&\text{At } (30, 5), & P &= 10(30) + 18(5) = 390 \\
&\text{At } (40, 0), & P &= 10(40) + 18(0) = 400
\end{aligned}
$$

The maximum profit is at (40, 0), which means that the company should manufacture 40 plain rocking chairs and no fancy rocking chairs. The maximum profit would be $400. The minimum profit would be at (0, 0), when no rocking chairs of either style were manufactured.

A variation of the problem in Example 1 could be that the company knows that it cannot sell more than 15 plain rocking chairs per day. With this additional constraint, we now have the following set of constraints.

$$\begin{aligned} x &\geq 0 \\ x &\leq 15 \\ y &\geq 0 \\ 4x + 8y &\leq 160 \\ 4x + 12y &\leq 180 \end{aligned}$$

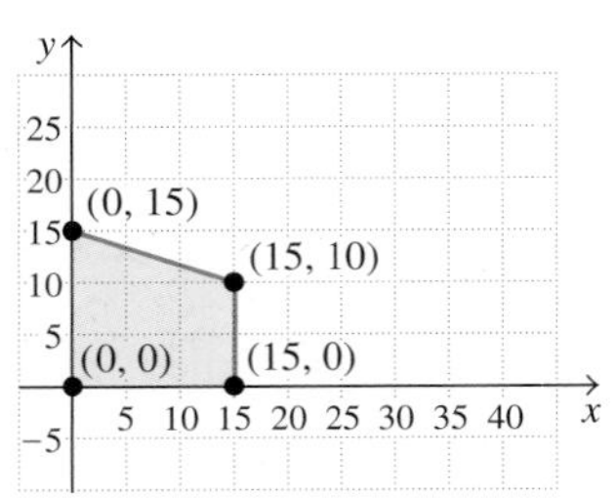

Figure 7.15

The graph of these constraints is shown in Fig. 7.15.

The vertices of the feasible region are (0, 0), (0, 15), (15, 10), and (15, 0). To determine the maximum profit, we calculate P for each of these vertices:

$$P = 10x + 18y$$

At (0, 0),	$P = 10(0) + 18(0) = 0$
At (0, 15),	$P = 10(0) + 18(15) = 270$
At (15, 10),	$P = 10(15) + 18(10) = 330$
At (15, 0),	$P = 10(15) + 18(0) = 150$

This set of constraints gives the maximum profit of $330 when the company manufactures 15 plain rocking chairs and 10 fancy rocking chairs.

EXAMPLE 2 MODELING - *Washers and Dryers, Maximizing Profit*

The Alexander Appliance Company makes washers and dryers. The company must manufacture at least one washer per day to ship to one of its customers. No more than 6 washers can be manufactured due to production restrictions. The number of dryers cannot exceed 7 per day. Also, the number of washers cannot exceed the number of dryers manufactured per day. If the profit on each washer is $20 and the profit on each dryer is $30, how many of each appliance should the company make per day to maximize profits? What is the maximum profit?

SOLUTION

Let

$$\begin{aligned} x &= \text{the number of washers per day} \\ y &= \text{the number of dryers per day} \\ 20x &= \text{the profit on washers} \\ 30x &= \text{the profit on dryers} \\ P &= \text{the total profit} \end{aligned}$$

The maximum profit is dependent on several constraints. The number of appliances manufactured each day cannot be a negative amount. This condition gives us the constraints $x \geq 0$ and $y \geq 0$. The company must produce at least one washer per day; therefore, $x \geq 1$. No more than 6 washers can be manufactured per day; therefore, $x \leq 6$. No more than 7 dryers can be manufactured per day; therefore, $y \leq 7$. The number of washers cannot exceed the number of dryers manufactured per day; therefore, $x \leq y$. Thus, the six constraints are

$$x \geq 0, \quad y \geq 0, \quad x \geq 1, \quad x \leq 6, \quad y \leq 7, \quad x \leq y$$

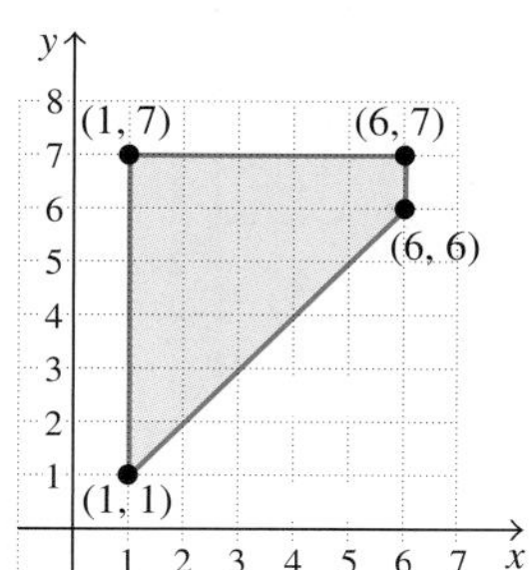

Figure 7.16

Since $20x$ is the profit on x washers and $30y$ is the profit on y dryers, the objective function, the profit formula, is $P = 20x + 30y$. Figure 7.16 shows the feasible region. The feasible region consists of the shaded region and the boundaries. The vertices of the feasible region are the points (1, 1), (1, 7), (6, 7), and (6, 6).

Next we calculate the value of the objective function, P, at each one of the vertices.

$$P = 20x + 30y$$

At (1, 1), $P = 20(1) + 30(1) = 50$

At (1, 7), $P = 20(1) + 30(7) = 230$

At (6, 7), $P = 20(6) + 30(7) = 330$

At (6, 6), $P = 20(6) + 30(6) = 300$

The maximum profit is at (6, 7). This means the company should manufacture 6 washers and 7 dryers to maximize their profit. The maximum profit is $330. ▲

Use the following steps to solve a linear programming problem.

Solving a Linear Programming Problem

1. Determine all necessary constraints.
2. Determine the objective function.
3. Graph the constraints and determine the feasible region.
4. Find the vertices of the feasible region.
5. Find the value of the objective function at each vertex.

The solution is determined by the vertex that yields the maximum or minimum value of the objective function.

SECTION 7.6 EXERCISES

CONCEPT/WRITING EXERCISES

1. What are constraints in a linear programming problem? How are they represented?
2. In a linear programming problem, how is a feasible region formed?
3. What are the points of intersection of the boundaries of the polygonal region called?
4. What is the general form of the objective function?
5. In your own words, state the fundamental principle of linear programming.
6. A profit function is $P = 4x + 6y$ and the vertices of the feasible region are (1, 1), (1, 4), (5, 1), and (7, 1). Determine the maximum profit. Explain how you determined your answer.

PRACTICE THE SKILLS

Exercises 7 and 8 show a feasible region and its vertices. Find the maximum and minimum values of the given objective function.

7. $K = 3x + 4y$

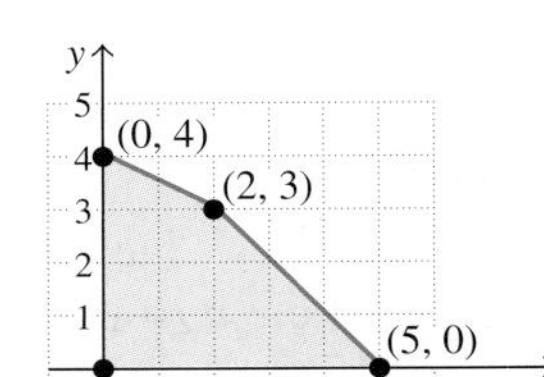

8. $K = 2x + 3y$

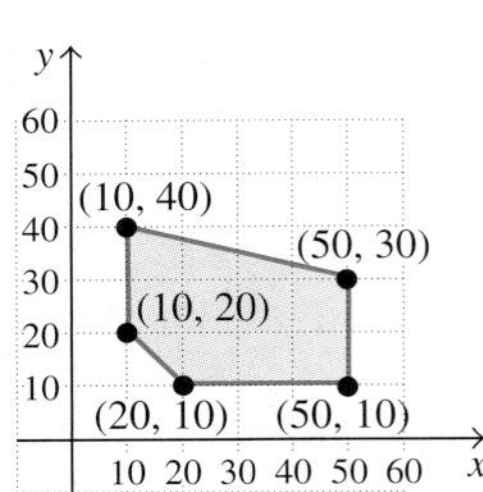

In Exercises 9–14, a set of constraints and a profit formula are given.

a) *Draw the graph of the constraints and find the vertices of the feasible region.*

b) *Use the vertices as obtained in part (a) to determine the maximum and minimum profit.*

9. $x + y \leq 5$
$x + 2y \leq 8$
$x \geq 0$
$y \geq 0$
$P = 3x + 5y$

10. $3x + y \leq 9$
$x + 2y \leq 8$
$x \geq 0$
$y \geq 0$
$P = 5x + 4y$

11. $x + y \leq 4$
$x + 3y \leq 6$
$x \geq 0$
$y \geq 0$
$P = 7x + 6y$

12. $x + y \leq 50$
$x + 3y \leq 90$
$x \geq 0$
$y \geq 0$
$P = 20x + 40y$

13. $4x + 3y \geq 12$
$3x + 4y \leq 36$
$x \geq 2$
$y \leq 5$
$y \geq 1$
$P = 2.20x + 1.65y$

14. $x + 2y \leq 14$
$7x + 4y \geq 28$
$x \geq 2$
$x \leq 10$
$y \geq 1$
$P = 15.13x + 9.35y$

PROBLEM SOLVING

15. MODELING - *On Wheels* The Boards and Blades Company manufactures skateboards and in-line skates. The company can produce a maximum of 20 skateboards and pairs of in-line skates per day. It makes a profit of \$25 on a skateboard and a profit of \$20 on a pair of in-line skates. The company's planners want to make at least 3 skateboards but not more than 6 skateboards per day. To keep customers happy, they must make at least 2 pairs of in-line skates per day.

a) List the constraints.
b) Determine the objective function.
c) Graph the set of constraints.
d) Find the vertices of the feasible region.
e) How many skateboards and pairs of in-line skates should be made to maximize the profit?
f) Find the maximum profit.

16. MODELING - *A Crafts Business* A craftsman makes wooden toy cars and trucks. He wants to make at least 2 cars and 2 trucks but no more than 5 trucks. He can make a total of 15 toys in his spare time in a month. At craft fairs, he makes a profit of \$8 per car and \$10 per truck sold.

a) List the constraints.
b) Determine the objective function.
c) Graph the set of constraints.
d) Find the vertices of the feasible region.
e) How many of each type should he make to maximize his profit?
f) Find the maximum profit.

17. MODELING - *Diet Drinks* A liquid portion of a diet is to provide at least 300 calories, 80 units of vitamin A, and 90 units of vitamin C daily. A cup (8 oz) of dietary drink Trimfit provides 60 calories, 8 units of vitamin A, and 6 units of vitamin C. A cup (8 oz) of dietary drink Usave provides 50 calories, 20 units of vitamin A, and 30 units of vitamin C. Assume Trimfit dietary drink costs \$0.25 per cup and Usave dietary drink costs \$0.32 per cup.

a) List the constraints.
b) Determine the objective function.
c) Graph the set of constraints. Place number of cups of Trimfit on the horizontal axis and number of cups of Usave on the vertical axis.
d) Find the vertices of the feasible region.
e) How many cups of each drink should a person consume each day to minimize the cost and still meet the stated daily requirements?
f) Find the minimum cost.

CHALLENGE PROBLEMS/GROUP ACTIVITIES

18. MODELING - *Hot Dog Profits* To make one package of all-beef hot dogs, a manufacturer uses 1 lb of beef; to make one package of regular hot dogs, the manufacturer uses $\frac{1}{2}$ lb each of beef and pork. The profit on the all-beef hot dogs is 40 cents per pack and the profit on regular hot dogs is 30 cents per pack. If there are 200 lb of beef and 150 lb of pork available, how many packs of all-beef and regular hot dogs should the manufacturer make to maximize the profit? What is the profit?

19. MODELING - *Automobile Engine Profits* Ross's engine reconditioning company works on 4- and 6-cylinder automobile engines. Each 4-cylinder engine requires 1 hr for cleaning, 5 hr for overhauling, and 3 hr for testing. Each 6-cylinder engine requires 1 hr for cleaning, 10 hr for overhauling, and 2 hr for testing. The cleaning station is available for at most 9 hr per week, the overhauling equipment is available for at most 80 hr per week, and the testing equipment is available for at most 24 hr per week. For each reconditioned 4-cylinder engine, the company makes a profit of $150, and for each 6-cylinder engine, it makes a profit of $250. The company can sell all the reconditioned engines it produces. How many of each type should be reconditioned to maximize profit? What is the maximum profit?

RESEARCH ACTIVITY

20. *Operations research* draws on several disciplines, including mathematics, probability theory, statistics, and economics. George Dantzig was one of the key people in developing operations research. Write a paper on Dantzig and his contributions to operations research and linear programming.

CHAPTER 7 SUMMARY

IMPORTANT FACTS

Systems of equations

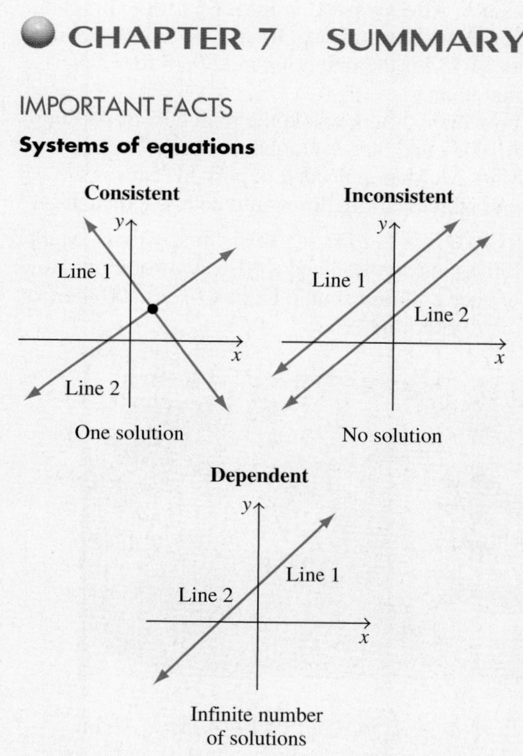

Methods of solving systems of equations

1. Graphing
2. Substitution
3. Addition (or elimination) method

Additive identity matrix

$$\begin{bmatrix} 0 & 0 \\ 0 & 0 \end{bmatrix}$$

Multiplicative identity matrix

$$\begin{bmatrix} 1 & 0 \\ 0 & 1 \end{bmatrix}$$

Fundamental principle of linear programming

If the objective function $K = Ax + By$ is evaluated at each point in a feasible region, the maximum and minimum values of the equation occur at vertices of the region.

CHAPTER 7 REVIEW EXERCISES

7.1

In Exercises 1–4, solve the system of equations graphically. If the system does not have a single ordered pair as a solution, state whether the system is inconsistent or dependent.

1. $x = 2$
$y = 6$

2. $-3x + y = 3$
$3x + y = 3$

3. $x = 3$
$x + y = 5$

4. $x + 2y = 5$
$2x + 4y = 4$

In Exercises 5–8, determine without graphing whether the system of equations has exactly one solution, no solution, or an infinite number of solutions.

5. $y = \frac{2}{3}x + 5$
$3y - 2x = 15$

6. $3y - 2x = 15$
$3y = 2x - 18$

7. $6y - 2x = 20$
$4y + 2x = 10$

8. $2x - 4y = 8$
$-2x + y = 6$

7.2

In Exercises 9–12, solve the system of equations by the substitution method. If the system does not have a single ordered pair as a solution, state whether the system is inconsistent or dependent.

9. $x - 2y = 1$
$2x + y = 7$

10. $x + 2y = -11$
$y = 2x - 3$

11. $2x - y = 4$
$3x - y = 2$

12. $3x + y = 1$
$3y = -9x - 4$

In Exercises 13–18, solve the system of equations by the addition method. If the system does not have a single ordered pair as a solution, state whether the system is inconsistent or dependent.

13. $-x + y = 12$
$x + 2y = -3$

14. $2x + y = 2$
$-3x - y = 5$

15. $x + y = 2$
$x + 3y = -2$

16. $3x + 4y = 6$
$2x - 3y = 4$

17. $3x + 5y = 15$
$2x + 4y = 0$

18. $3x + y = 6$
$-6x - 2y = -12$

7.3

Given $A = \begin{bmatrix} 1 & -3 \\ 2 & 4 \end{bmatrix}$ *and* $B = \begin{bmatrix} -2 & -5 \\ 6 & 3 \end{bmatrix}$, *find the following.*

19. $A + B$

20. $A - B$

21. $2A$

22. $3A - 2B$

23. $A \times B$

24. $B \times A$

7.4

In Exercises 25–30, use matrices to solve the system of equations.

25. $x + 2y = 4$
$x + y = 2$

26. $-x + y = 4$
$x + 2y = 2$

27. $2x + y = 3$
$3x - y = 12$

28. $2x + 3y = 2$
$4x - 9y = 4$

29. $x + 3y = 3$
$3x - 2y = 2$

30. $3x - 6y = 3$
$4x + 5y = 17$

7.1–7.4

31. MODELING - *Minimizing Parking Costs* The cost of parking in All-Day parking lot is $5 for the first hour and $0.50 for each additional hour. Sav-A-Lot parking lot costs $4.25 for the first hour and $0.75 for each additional hour.

a) In how many hours would the total cost of parking at All-Day and Sav-A-Lot be the same?

b) If Mark McMahon needed to park his car for 5 hours, which parking lot would be less expensive?

32. MODELING - *Chemistry* In chemistry class, Mark has an 80% acid solution and a 50% acid solution. How much of each solution should he mix to get 100 liters of a 75% acid solution?

33. MODELING - *A Salesman's Earnings* Abdul Ahmed, an electronics salesman, earns a weekly salary plus a commission on sales. In one week, his salary on sales of $4000 was $660. The next week his salary on sales of $6000 was $740. Find his weekly salary and his commission rate.

34. MODELING - *Quarters and Dimes* Cheryl has a total of 40 coins in her piggy bank. If the coins consist of only quarters and dimes and the total value of the coins is $5.50, what is the number of quarters and the number of dimes?

35. MODELING - *Cool Air* Emily needs to purchase a new air conditioner for the office. Model 1600A costs \$950 to purchase and \$32 per month to operate. Model 6070B, a more efficient unit, costs \$1275 to purchase and \$22 per month to operate.

a) After how many months will the total cost of both units be equal?

b) Which model will be the more cost effective if the life of both units is guaranteed for 10 years?

7.5

In Exercises 36–39, graph the system of linear inequalities and indicate the solution set.

36. $y \le 3x - 1$
$y > -2x + 1$

37. $2x + y < 8$
$y \ge 2x - 1$

38. $x + 3y \le 6$
$2x - 7y \ge 14$

39. $x - y > 5$
$6x + 5y \le 30$

7.6

40. For the following set of constraints and profit formula, graph the constraints and find the vertices. Use the vertices to determine the maximum profit.

$$x + y \le 10$$
$$2x + 1.8y \le 18$$
$$x \ge 0$$
$$y \ge 0$$
$$P = 6x + 5y$$

CHAPTER 7 TEST

1. From a graph, explain how you would identify a consistent system of equations, an inconsistent system of equations, and a dependent system of equations.

2. Solve the system of equations graphically.

$$y = 3x + 4$$
$$-2x + 4y = -4$$

3. Determine without graphing whether the system of equations has exactly one solution, no solution, or an infinite number of solutions.

$$4x + 5y = 6$$
$$-3x + 5y = 13$$

Solve the system of equations by the method indicated.

4. $x - y = 5$
$2x + 3y = -5$
(substitution)

5. $y = 4x + 6$
$y = 2x + 18$
(substitution)

6. $x - y = 4$
$2x + y = 5$
(addition)

7. $4x + 3y = 5$
$2x + 4y = 10$
(addition)

8. $2x + 3y = 4$
$6x + 4y = 7$
(addition)

9. $x + 3y = 4$
$5x + 7y = 4$
(matrices)

For $A = \begin{bmatrix} 2 & -5 \\ 4 & 6 \end{bmatrix}$ *and* $B = \begin{bmatrix} -1 & 3 \\ 2 & 5 \end{bmatrix}$, *find the following.*

10. $A + B$

11. $3A - B$

12. $A \times B$

13. Graph the system of linear inequalities and indicate the solution set.

$$y < -2x + 2$$
$$y > 3x + 2$$

Solve Exercises 14 and 15 by using a system of equations.

14. MODELING - *A Coffee Blend* Louis DiMento plans to mix coffee that sells for \$6 per pound with coffee that sells for \$7.50 per pound to get a 30 lb blend that sells for \$7 per pound. How many pounds of each type should Louis use?

15. MODELING - *Rental Units* The Crossroads apartment building has 20 rental units with one- and two-bedroom units. The rental price for a one-bedroom unit is \$425 per month and for a two-bedroom unit is

$500 per month. If the total rental income from all units is $9100 per month, determine the number of each type of rental unit in the building.

16. The set of constraints and profit formula for a linear programming problem are

$$x + 3y \le 6$$
$$4x + 3y \le 15$$
$$x \ge 0$$
$$y \ge 0$$
$$P = 2x + 3y$$

a) Draw the graph of the constraints and determine the vertices of the feasible region.
b) Use the vertices to determine the maximum and minimum profit.

GROUP PROJECTS

1. Make up three different systems of equations that have (1, 4) as a solution. Explain how you determined your systems.

LINEAR PROGRAMMING

2. MODELING - *Profit from Bookcases* The Bookholder Company manufactures two types of bookcases out of oak and walnut. Model 01 requires 5 board feet of oak and 2 board feet of walnut. Model 02 requires 4 board feet of oak and 3 board feet of walnut. A profit of $75 is made on each Model 01 bookcase, and a profit of $125 is made on each Model 02 bookcase. The company has a supply of 1000 board feet of oak and 600 board feet of walnut. The company has orders for 40 Model 01 bookcases and 50 Model 02 bookcases. These orders indicate the minimum number the company must manufacture of each model.
a) Write the set of constraints.
b) Write the objective function.
c) Graph the set of constraints.
d) Determine the number of bookcases of each type the company should manufacture in order to maximize profits.
e) Determine the maximum profit.

CREATE YOUR OWN WORD PROBLEM

3. a) Write a word problem that can be solved by using a system of two equations with two unknowns.
b) For the problem in part (a), write the system of equations and find the answer.
c) Explain how you developed the problem in part (a).

CHAPTER 8

The Metric System

On a trip to Mexico or France, you might feel uncomfortable asking directions if you do not speak the language of that country. Purchasing clothing or gasoline may also be difficult because their country's measurement system is different than ours. Our neighbors in Mexico and Canada, like almost all countries around the world, use a different measurement system than we use in the United States. Most countries of the world use the *Système international d'unités*, or abbreviated, the *SI system*. In the United States, the SI system is most commonly referred to as the metric system.

The system of measurement most commonly used in the United States is called the U.S. customary system, but metric units are used in many ways in the United States. You can purchase metric tools at your local hardware store. Some clothing is measured in metric units, as are car tires. Drinks are sold in liter bottles, and vitamins and medicines are labeled in milligrams.

As you study this chapter, you will see the many advantages of the metric system. You eventually may even support the movement to change from the U.S. customary system to the metric system. ■

The 2000 Chevrolet Corvette has a standard 5.7-liter engine.

8.1 BASIC TERMS AND CONVERSIONS WITHIN THE METRIC SYSTEM

Most countries of the world use the **Système international d'unités** or **SI system**. The SI system is generally referred to as the **metric system** in the United States. The metric system was named for the Greek word *metron,* meaning "measure." The standard units in the metric system have gone through many changes since the system was first developed in France during the French Revolution. For example, one unit of measure, the meter, was first defined as one ten-millionth of the distance between the North Pole and the equator. Later, the meter was defined as 1,650,763.73 wavelengths of the orange–red line of krypton 86. Since 1893, the meter has been defined as the distance traveled by light in a vacuum in $\frac{1}{299,792,458}$ of a second.

Two systems of weights and measures exist side by side in the United States today, the **U.S. customary system** and the SI system. The SI system is used predominantly in the automotive, construction, farm equipment, computer, and bottling industries and in health-related professions. Many other industries are now in the process of converting to the SI system.

In this chapter, we will discuss metric measurements of length, area, volume, mass, and temperature. Using the metric system has many advantages. Some of them are summarized here.

1. The metric system is the worldwide accepted standard measurement system. All industrial nations that trade internationally, except the United States, use the metric system as the official system of measurement.
2. There is only one basic unit of measurement for each physical quantity. In the customary system, many units are often used to represent the same physical quantity. For example, when discussing length, we use inches, feet, yards, miles, and so on. Converting from one of these units to the other is often a tedious task (consider changing 12 miles to inches). In the metric system, we can make many conversions by simply moving the decimal point.
3. The SI system is based on the number 10, and there is little need for fractions, because most quantities can be expressed as decimals.

DID YOU KNOW

Lost in Space

In September 1999, the United States lost the *Mars Climate Orbiter* as it approached Mars. The loss of the $125 million spacecraft was due to scientists confusing English units and metric units. Two spacecraft teams, one at NASA's Jet Propulsion Laboratory (JPL) in Pasadena, California, and the other at a Lockheed Martin facility in Colorado, where the spacecraft was built, were unknowingly exchanging some vital information in different measurement units.

The spacecraft team in Colorado used English units of *pounds* of force to describe small forces needed to adjust the spacecraft's orbit. The data were shipped via computer, without units, to the JPL, where the navigation team was expecting to receive the information in newtons, a metric measure of force. The mix-up in units led to the JPL scientists giving the spacecraft's computer wrong information, which threw the spacecraft off course. This in turn led to the spacecraft entering the Martian atmosphere, where it burned up. NASA has said it is taking necessary steps to prevent this error from ever happening again.

The missing Mars Climate Orbiter

BASIC TERMS

One- and two-liter bottles.

Because the official definitions of many metric terms are quite technical, we present them informally.

The **meter** (m) is commonly used to measure *length* in the metric system. One meter is a little more than a yard. A door is about 2 meters high.

The **kilogram** (kg) is commonly used to measure *mass*. (The difference between mass and weight is discussed in Section 8.3.) One kilogram is a little more than 2 pounds. A newborn baby may have a mass of about 3 kilograms. The gram (g), a unit of mass derived from the kilogram, is used to measure small amounts. A nickel has a mass of about 5 grams.

The **liter** (ℓ) is commonly used to measure *volume*. One liter is a little more than a quart. The gas tank of a compact car may hold 50 liters of gasoline.

Thus,

$$1 \text{ m} \approx 1 \text{ yd}$$
$$1 \text{ kg} \approx 2.2 \text{ lb}$$
$$1\ \ell \approx 1 \text{ qt}$$

The term **degree Celsius** (°C) is used to measure temperature. The freezing point of water is 0°C, and the boiling point of water is 100°C. The temperature on a warm day may be 30°C.

$$0°\text{C} = 32°\text{F} \quad \text{Water freezes}$$
$$37°\text{C} = 98.6°\text{F} \quad \text{Body temperature}$$
$$100°\text{C} = 212°\text{F} \quad \text{Water boils}$$

PREFIXES

The metric system is based on the number 10 and therefore is a decimal system. Prefixes are used to denote a multiple or part of a base unit. Table 8.1 summarizes the more commonly used prefixes and their meanings. In the table, where we mention "base units" we mean metric units without prefixes, such as meter, gram, or liter. From Table 8.1, we can determine that a *deka*meter represents 10 meters, and a *centi*meter represents $\frac{1}{100}$ of a meter. Also, 1 kiloliter = 1000 liters, 1 kilogram = 1000 grams, 1 centimeter = $\frac{1}{100}$ meter, and 1 milliliter = $\frac{1}{1000}$ liter.

Table 8.1 Metric Prefixes

Prefix	Symbol	Meaning
kilo	k	1000 × base unit
hecto	h	100 × base unit
deka	da	10 × base unit
—	—	base unit
deci	d	$\frac{1}{10}$ of base unit
centi	c	$\frac{1}{100}$ of base unit
milli	m	$\frac{1}{1000}$ of base unit

In the metric system, as used outside the United States, groups of three digits in large numbers are separated by a space, not a comma. For example, the number for

The computer's hard drive can store 30 *giga*bytes (30 GB) of information.

thirty thousand is 30 000, and the number for nine million is 9 000 000. Groups of three digits to the right of the decimal point are also separated by spaces. Commas are not used in the SI system because many countries use the comma as we use the decimal point. For example, 16 millionths is written 0,000 016 in many countries of the world. We will use the decimal point in this book and write 0.000 016. In this section, we will separate groups of three digits using spaces as done outside the United States. Note, however, that the space between groups of three digits is usually omitted if there are only four digits to the left or right of the decimal point. Thus, we will write three thousand as 3000 and five ten-thousandths as 0.0005.

For scientific work, which involves very large and very small quantities, the following prefixes are also used: *mega* (M) is one million times the base unit, *giga* (G) is one billion times the base unit, *tera* (T) is one trillion times the base unit, *micro* (μ, the Greek letter mu) is one millionth of the base unit, *nano* (n) is one billionth of the base unit, and *pico* (p) is one trillionth of the base unit.

In this book, the abbreviations or symbols for units of measure are not pluralized, but full names are. For example, 5 milliliters is symbolized as 5 mℓ, not 5 mℓs. Some countries that use the metric system do not use an "s" in their abbreviations, whereas others do.

CONVERSIONS WITHIN THE METRIC SYSTEM

We will use Table 8.2 to help demonstrate how to change from one metric unit to another metric unit (meters to kilometers and so on).

The meters in Table 8.2 can be replaced by grams, liters, or any other base unit of the metric system. Regardless of which unit we choose, the procedure is the same. For purposes of explanation, we have used the meter.

Table 8.2 Changing Metric Units

Measure of length	kilometer	hectometer	dekameter	meter	decimeter	centimeter	millimeter
Symbol	km	hm	dam	m	dm	cm	mm
Number of meters	1000 m	100 m	10 m	1 m	0.1 m	0.01 m	0.001 m

The table shows that 1 hectometer equals 100 meters and 1 millimeter is 0.001 (or $\frac{1}{1000}$) meter. The millimeter is the smallest unit in the table. A centimeter is 10 times as large as a millimeter, a decimeter is 10 times as large as a centimeter, a meter is 10 times as large as a decimeter, and so on. Because each unit is 10 times as large as the unit on its right, converting from one unit to another is simply a matter of multiplying or dividing by powers of 10.

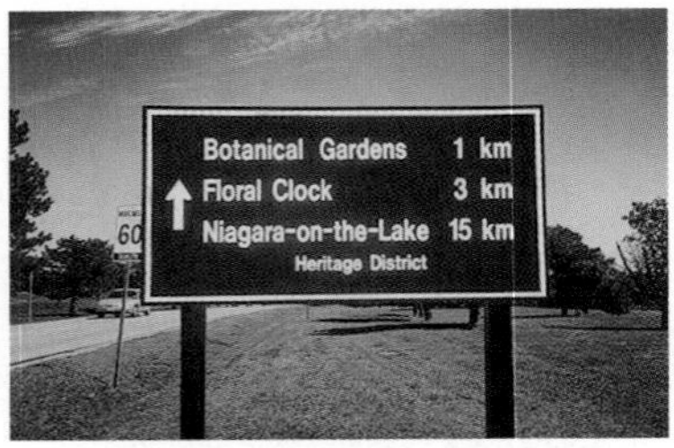

Our neighbors in Canada (and also Mexico) use the metric system. As you will learn shortly, the distance to the Botanical Gardens is about 0.6 mile and the distance to Niagara-on-the-Lake is about 9 miles from the sign.

Changing Units within the Metric System

1. To change from a smaller unit to a larger unit (for example, from meters to kilometers), move the decimal point in the original quantity one place to the left for each larger unit of measurement until you obtain the desired unit of measurement.
2. To change from a larger unit to a smaller unit (for example from kilometers to meters), move the decimal point in the original quantity one place to the right for each smaller unit of measurement until you obtain the desired unit of measurement.

EXAMPLE 1 *Changing Units*

a) Convert 573.2 m to km.
b) Convert 14 g to cg.
c) Convert 0.18 ℓ to mℓ.
d) Convert 240 dℓ to kℓ.

SOLUTION

a) Table 8.2 shows that dekameters, hectometers, and kilometers are all larger units of measurements than meters. Kilometers appear three places to the left of meters in the table. Therefore, to change a measure from meters to kilometers, we must move the decimal point in the given number three places to the left, or

$$573.2 \text{ m} = 0.5732 \text{ km}$$

Note that, since we are changing from a smaller unit of measurement (meter) to a larger unit of measurement (kilometer), the answer will be a smaller number of units.

b) Grams are a larger unit of measurement than centigrams. To convert grams to centigrams, we move the decimal point two places to the right, or

$$14 \text{ g} = 1400 \text{ cg}$$

Note that, since we are changing from a larger unit of measurement (gram) to a smaller unit of measurement (centigram), the answer will be a larger number of units.

c) 0.18 ℓ = 180 mℓ
d) 240 daℓ = 2.40 kℓ

EXAMPLE 2 *Two More Conversions*

a) Convert 305 mm to hectometers.
b) Convert 6.34 dam to decimeters.

SOLUTION

a) Table 8.2 shows that hectometers are five places to the left of millimeters. Therefore, to make the conversion, we must move the decimal point in the given number five places to the left, or

$$305 \text{ mm} = 0.003\,05 \text{ hm}$$

b) Table 8.2 shows that dekameters are two places to the left of decimeters. Therefore, to make the conversion, we must move the decimal point in the given number two places to the right, or

$$6.34 \text{ dam} = 634 \text{ dm}$$

EXAMPLE 3 *A Metric Road Sign*

The sign, from New Zealand, shows that the distance to Queenstown is 48 km and the distance to Invercargill is 141 km.

a) Determine the distance to Queenstown in meters.
b) Determine the distance to Invercargill in decimeters.

SOLUTION

a) We must move the decimal point three places to the right to change from kilometers to meters. Therefore,

$$48 \text{ km} = 48\,000 \text{ m}$$

b) We must move the decimal point four places to the right to change from kilometers to decimeters. Therefore,

$$141 \text{ km} = 1\,410\,000 \text{ dm}$$

▲

EXAMPLE 4 *Comparing Lengths*

Arrange in order from the smallest to largest length: 3.4 m, 3421 mm, and 104 cm.

SOLUTION To be compared, these lengths should all be in the same units of measure. Let's convert all the measures to millimeters, the smallest units of the lengths being compared.

$$3.4 \text{ m} = 3400 \text{ mm} \qquad 3421 \text{ mm} \qquad 104 \text{ cm} = 1040 \text{ mm}$$

The lengths arranged in order from smallest to largest are 104 cm, 3.4 m, and 3421 mm. ▲

SECTION 8.1 EXERCISES

CONCEPT/WRITING EXERCISES

1. What is the name commonly used for the Système international d'unités in the United States?

2. What is the name of the system of measurement primarily used in the United States today?

3. List three advantages of the metric system.

4. What metric unit is commonly used to measure
 a) length?
 b) mass?
 c) volume?
 d) temperature?

5. a) In your own words, explain how to convert from one metric unit of length to a different metric unit of length. Then use this procedure in parts (b) and (c).
 b) Convert 497.2 cm to kilometers.
 c) Convert 30.8 hm to decimeters.

6. What is the name of the prefix that is
 a) a million times the basic unit?
 b) one millionth of the base unit?

7. Without referring to any table, name as many of the metric system prefixes as you can and give their meanings. If you don't already know all the prefixes in Table 8.1, memorize them now.

8. a) How many times greater is 1 hectometer than 1 centimeter?
 b) Convert 1 hm to centimeters.
 c) Convert 1 cm to hectometers.

9. a) How many times greater is 1 dam than 1 dm?
 b) Convert 1 dam to decimeters.
 c) Convert 1 dm to dekameters.

10. a) What is the freezing temperature of water in the metric system?
 b) What is the boiling point of water in the metric system?
 c) What is normal human body temperature in the metric system?

PRACTICE THE SKILLS

In Exercises 11–14, fill in the blank.

11. One kilogram is a little more than ______ pounds.

12. One meter is a little longer than a ______.

13. One nickel has a mass of about ______ grams.

14. The temperature on a warm day may be ______°C.

In Exercises 15–20, match the prefix with the one letter, a–f, that gives the meaning of the prefix.

15. Milli **a)** $\frac{1}{100}$ of base unit

16. Hecto **b)** $\frac{1}{1000}$ of base unit

17. Kilo **c)** 100 times base unit

18. Deci **d)** 1000 times base unit

19. Deka **e)** 10 times base unit

20. Centi **f)** $\frac{1}{10}$ of base unit

In Exercises 21–30, unscramble the word to make a metric unit of measurement.

21. migradec **22.** rteli **23.** magr

24. leritililm **25.** raktileed **26.** terem

27. timenceret **28.** reketolim **29.** greeed sulesic

30. togmeharc

31. Complete the following.
- **a)** 1 dekaliter = ______ liters
- **b)** 1 centiliter = ______ liter
- **c)** 1 milliliter = ______ liter
- **d)** 1 deciliter = ______ liter
- **e)** 1 kiloliter = ______ liters
- **f)** 1 hectoliter = ______ liters

32. Complete the following.
- **a)** 1 hectogram = ______ grams
- **b)** 1 milligram = ______ gram
- **c)** 1 kilogram = ______ grams
- **d)** 1 centigram = ______ gram
- **e)** 1 dekagram = ______ grams
- **f)** 1 decigram = ______ gram

In Exercises 33–38, without referring to any of the tables or your notes, give the symbol and the equivalent in grams for the unit.

33. Milligram **34.** Centigram **35.** Decigram

36. Dekagram **37.** Hectogram **38.** Kilogram

In Exercises 39 and 40, use the photo, which shows the maximum load for an aerial tram in Switzerland. (Notice that in some countries, an "s" is used on the metric abbreviations.)

39. *Aerial Tram Load* What is the maximum load in grams?

40. *Aerial Tram Load* What is the maximum load in milligrams?

In Exercises 41–50, fill in the missing values.

41. 9 m = ______ cm

42. 7 dam = ______ m

43. 35.7 hg = ______ g

44. 0.054 hℓ = ______ ℓ

45. 242.6 cm = ______ hm

46. 1.34 mℓ = ______ ℓ

47. 974 g = ______ hg

48. 14.27 kℓ = ______ ℓ

49. 1.34 hm = ______ cm

50. 0.000 52 kg = ______ cg

In Exercises 51–58, convert the given unit to the unit indicated.

51. 7.3 m to millimeters

52. 102.5 kg to grams

53. 895 ℓ to milliliters

54. 24 dm to kilometers

55. 130 cm to kilometers

56. 6049 mm to meters

57. 8472 mℓ to dekaliters

58. 17 200 mℓ to liters

In Exercises 59–64, arrange the quantities in order from smallest to largest.

59. 514 hm, 62 km, 680 m

60. 5.1 dam, 0.47 km, 590 cm

61. 4.3 ℓ, 420 cℓ, 0.045 kℓ

62. 2.2 kg, 2 400 g, 24 300 dg

63. 0.032 kℓ, 460 dℓ, 48 000 cℓ

64. 2.6 km, 203 000 mm, 52.6 hm

PROBLEM SOLVING

65. If 5 kg are placed on one side of a balance and a 10 lb weight is placed on the other side, which way would the balance tip? Explain.

66. Jim ran 100 m, and Bob ran 100 yd in the same length of time. Who ran faster? Explain.

67. One pump removes 1 daℓ of water in 1 min, and another pump removes 1 dℓ of water in 1 min. Which pump removes water faster? Explain.

68. Would you be walking faster if you walked 1 dam in 10 min or 1 hm in 10 min? Explain.

69. *Framing a Masterpiece* The painting by Picasso, including the frame, measures 74 cm by 99 cm.

a) How many centimeters of framing was needed to frame the painting?
b) How many millimeters of framing was needed to frame the painting?

70. *A Home Run* A baseball diamond is about 27 m along each side.
a) How many meters does a batter run if he hits a home run?
b) How many kilometers?
c) How many millimeters?

71. *Calcium Tablets* Dr. Driscoll recommends that Sean takes two 250 mg chewable calcium tablets each day.
a) How many milligrams of calcium will Sean take in a week?
b) How many grams of calcium will Sean take in a week?

72. *Aerating an Aquarium* The filter pump on an aquarium circulates 360 mℓ of water every minute. If the aquarium holds 30 ℓ of water, how long will it take to circulate all the water?

73. *Gas Consumption* Dale drove 1200 km and used 187 ℓ of gasoline. What was her average rate of gas use for the trip
a) in kilometers per liter?
b) in meters per liter?

74. *Track and Field* The high school track is 400 m long. If Patty runs around the track eight times, how many kilometers has she traveled?

75. *Refreshments* If three 2 ℓ bottles of soda are divided equally among 12 people, how many milliliters will each person receive?

76. *Liters of Soda* A bottle of soda contains 360 mℓ.
a) How many milliliters are contained in a six-bottle carton?
b) How many liters does the amount in part (a) equal?
c) At $2.45 for the carton of soda, what is its cost per liter?

77. *Fill 'er Up* In Europe, gas may cost the equivalent of about 96 cents (American) per liter. What will be the cost of filling the gas tank of a car that has a capacity of 24.3 ℓ?

78. *Roast Beef Dinner* Mr. Gordon ordered a beef roast large enough to serve 14 people. The roast weighs 5.8 kg and costs $6.30/kg.
a) What is the cost of the roast?
b) How many grams of meat are available for each person?

CHALLENGE EXERCISES/GROUP ACTIVITIES

In Exercises 79–82, fill in the blank to make a true statement.

79. 1 gigameter = ______ megameters

80. 1 nanogram = ______ micrograms

81. 1 teraliter = ______ picoliters

82. 1 megagram = ______ nanograms

The recommended daily amount of calcium for an American adult is 0.8 g. In Exercises 83–86, how much of the food indicated must an adult eat to satisfy the entire daily allowance using only that food?

83. Milk: 1 cup contains 288 mg calcium.

84. Eggs: 1 egg contains 27 mg calcium.

85. Raisin bran: 49 g contains 1.6 mg calcium.

86. Broccoli: 1 cup (cooked) contains 195 mg calcium.

One advantage of the metric system is that by using the proper prefix, you can write large and small numbers without large groups of zeros. In Exercises 87–92, write an equivalent measurement with an amount greater than one that does not contain any zeros. For example, you can write 3000 m without zeros as 3 km and 0.0003 hm as 3 cm.

87. 9000 cm

88. 2000 mm

89. 0.0004 km

90. 3000 dm

91. 0.002 km

92. 500 cm

RESEARCH ACTIVITIES

93. Write a report on the development of the metric system in Europe. Indicate who had the most influence in its development.

94. Write a report on why you believe many Americans oppose switching to the metric system. Give your opinion about whether the United States will eventually switch to the metric system, and if so when it might do so.

8.2 LENGTH, AREA, AND VOLUME

This section and the next section are designed to help you *think metric,* that is, to become acquainted with day-to-day usage of metric units. In this section, we consider length, area, and volume.

LENGTH

> **DID YOU KNOW**
>
> **1 Yard = 3 Feet = 36 Inches = . . .**
>
> In the U.S. Customary system, 27 different units of length are used. How many of them can you name? Don't forget rod, mil, paris line, toise, cubit, and light-year. The different units can be found in the *CRC Handbook of Chemistry and Physics.*

The basic unit of length is the meter. In all English-speaking countries except the United States, *meter* is spelled "metre." Until 1960, the meter was officially defined by the length of a platinum bar kept in a vault in France. The modern definition of the meter is based on the speed of light, a constant that has been defined with great precision. Other commonly used units of length are the kilometer, centimeter, and millimeter. The meter, which is a little longer than 1 yard, is used to measure things that we normally measure in yards and feet. A man whose height is about 2 meters is a tall man. A tractor trailer unit (an 18-wheeler) is about 18 meters long.

The kilometer is used to measure what we normally measure in miles. For example, the distance from New York to Seattle is about 5120 kilometers. One kilometer is about 0.6 mile, and 1 mile is about 1.6 kilometers.

Centimeters and millimeters are used to measure what we normally measure in inches. The centimeter is a little less than $\frac{1}{2}$ inch (see Fig. 8.1), and the millimeter is a little less than $\frac{1}{20}$ inch. A millimeter is about the thickness of a dime. A book may measure 20 cm by 25 cm with a thickness of about 3 cm. Millimeters are often used in scientific work and other areas in which small quantities must be measured. The length of a small insect may be measured in millimeters.

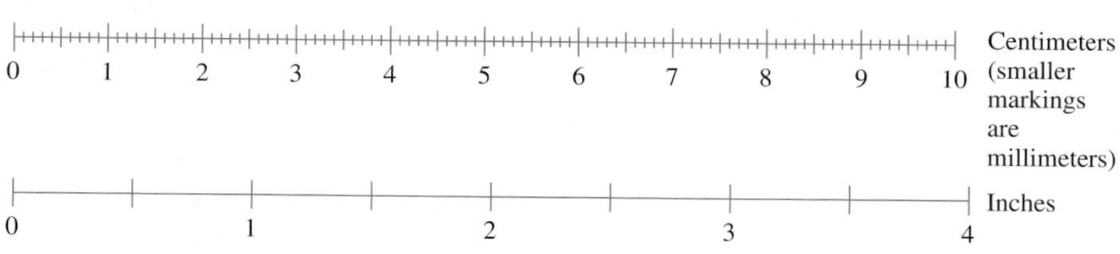

Figure 8.1

EXAMPLE 1 *Choosing an Appropriate Unit of Length*

Determine which unit of length you would use to express the following.

a) The length of a guitar
b) The length of a football field
c) The height of a door
d) The length of an ant
e) The diameter of a half dollar
f) The distance between your house and your school
g) The diameter of the lead in a pencil
h) The diameter of a basketball hoop
i) The thickness of a dime
j) Your height

SOLUTION

a) Meters or centimeters
b) Meters
c) Meters or centimeters
d) Millimeters
e) Centimeters or millimeters
f) Kilometers
g) Millimeters
h) Centimeters
i) Millimeters
j) Meters or centimeters

In some parts of this solution, more than one possible answer is listed. Measurements can often be made by using more than one unit. For example, if someone asks your height, you might answer $5\frac{1}{2}$ feet or 66 inches. Both answers are correct.

AREA

The area enclosed in a square with 1-centimeter sides (Fig. 8.2) is $1 \text{ cm} \times 1 \text{ cm} = 1 \text{ cm}^2$. A square whose sides are 2 cm (Fig. 8.3) has an area of $2 \text{ cm} \times 2 \text{ cm} = 2^2 \text{ cm}^2 = 4 \text{ cm}^2$.

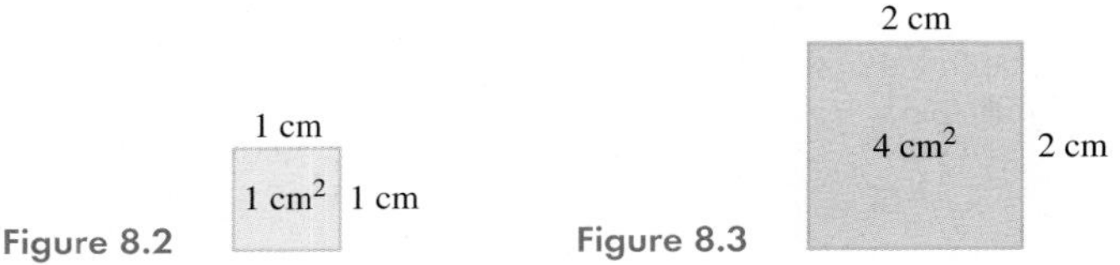

Figure 8.2

Figure 8.3

Areas are always expressed in square units, such as square centimeters, square kilometers, or square meters. When finding areas, be careful that all the numbers being multiplied are expressed in the same units.

In the metric system, the square centimeter replaces the square inch. The square meter replaces the square foot and square yard. In the future, you might purchase carpet or other floor covering by the square meter instead of by the square yard.

For measuring large land areas, the metric system uses a square unit 100 meters on each side (a square hectometer). This unit is called a **hectare** (pronounced "hectair" and symbolized ha). A hectare is about 2.5 acres. One square mile of land contains about 260 hectares. Very large units of area are measured in square kilometers. One square kilometer is about $\frac{4}{10}$ square mile.

EXAMPLE 2 *Choosing an Appropriate Unit of Area*

Determine which unit of area you would use to measure the following.

a) A football field
b) The top of a kitchen table
c) The floor of the classroom
d) A person's property with an average-sized lot
e) The cover of this book
f) A national forest
g) The area of an ice-skating rink
h) A ruler
i) The area of a lens in eyeglasses
j) A dollar bill

SOLUTION

a) Hectares or square meters
b) Square meters
c) Square meters
d) Square meters or hectares
e) Square centimeters
f) Square kilometers or hectares
g) Square meters
h) Square centimeters
i) Square centimeters
j) Square centimeters

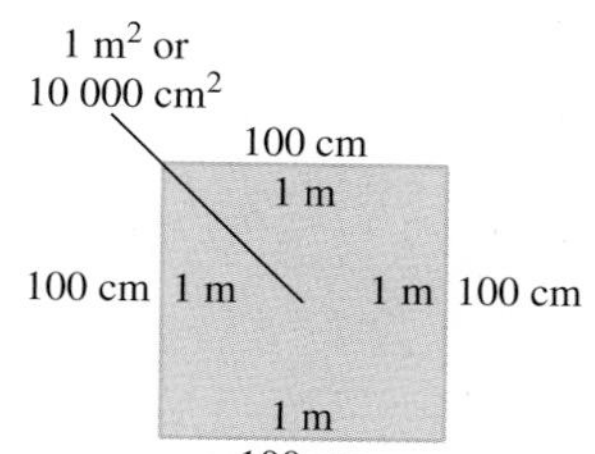

Figure 8.4

EXAMPLE 3 *Same Area, Different Units*

A square meter is how many times as large as a square centimeter?

SOLUTION Since 1 m equals 100 cm, we can replace 1 m with 100 cm (see Fig. 8.4). The area of $1 \text{ m}^2 = 1 \text{ m} \times 1 \text{ m} = 100 \text{ cm} \times 100 \text{ cm} = 10\,000 \text{ cm}^2$. Thus, the area of one square meter is 10 000 times the area of one square centimeter. This technique can be used to convert from any square unit to a different square unit.

VOLUME

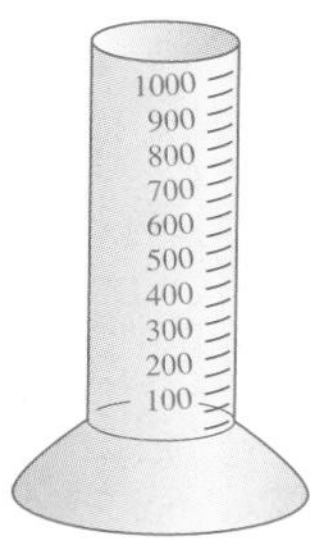

Figure 8.5

When a figure has only two dimensions—length and width—we can find its area. When a figure has three dimensions—length, width, and height—we can find its volume. The volume of an item can be considered the space occupied by the item.

In the metric system, volume may be expressed in terms of liters or cubic meters, depending on what is being measured. In all English-speaking countries except the United States, *liter* is spelled "litre."

The volume of liquids is expressed in liters. A liter is a little larger than a quart. Liters are used in place of pints, quarts, and gallons. A liter can be divided into 1000 equal parts, each of which is called a milliliter. Figure 8.5 illustrates a type of liter container (a 1000 mℓ graduated cylinder) that is often used in chemistry. Milliliters are used to express the volume of very small amounts of liquid. Drug dosages are often expressed in milliliters. An 8 oz cup will hold about 240 mℓ of liquid.

The kiloliter, 1000 liters, is used to represent the volume of large amounts of liquid. Tank trucks carrying gasoline to service stations hold about 10.5 kℓ of gasoline.

Cubic meters are used to express the volume of large amounts of solid material. The volume of a dump truck's load of topsoil is measured in cubic meters. The volume of natural gas used to heat a house may soon be measured in cubic meters instead of cubic feet.

The liquid in a liter container will fit exactly in a cubic decimeter (Fig. 8.6). Note that $1\ \ell = 1000\ \text{m}\ell$ and that $1 \text{ dm}^3 = 1000 \text{ cm}^3$. Because $1\ \ell = 1 \text{ dm}^3$, *1 mℓ*

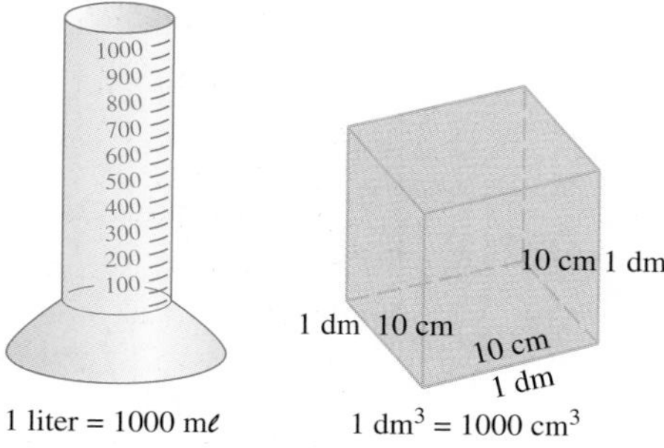

Figure 8.6

must equal 1 cm³. Other useful facts are illustrated in Table 8.3. Thus, within the metric system, conversions are much simpler than in the U.S. customary system. For example, how would you change cubic feet of water into gallons of water?

Table 8.3

Volume in cubic units		**Volume in liters**
1 cm^3	=	1 mℓ
1 dm^3	=	1 ℓ
1 m^3	=	1 kℓ

EXAMPLE 4 *Choosing an Appropriate Unit of Volume*

Determine which unit of volume you would use to measure the following.

a) Air in a large helium balloon (see photo)
b) A carton of milk
c) A truckload of concrete
d) A drug dosage
e) The space available in your refrigerator
f) A handful of sand
g) Water in a drinking glass
h) Water in a water bed
i) Water in an Olympic-sized swimming pool
j) Dirt removed to lay the foundation for a basement

SOLUTION

a) Cubic meters
b) Liters
c) Cubic meters
d) Milliliters
e) Cubic meters
f) Cubic centimeters
g) Milliliters
h) Liters or kiloliters
i) Kiloliters
j) Cubic meters

EXAMPLE 5 *Swimming Pool Volume*

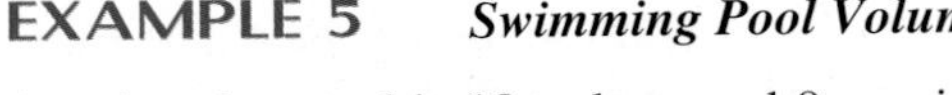

A swimming pool is 18 m long and 9 m wide, and it has a uniform depth of 3 m (Fig. 8.7). Find (a) the volume of the pool in cubic meters and (b) the volume of water in the pool in kiloliters.

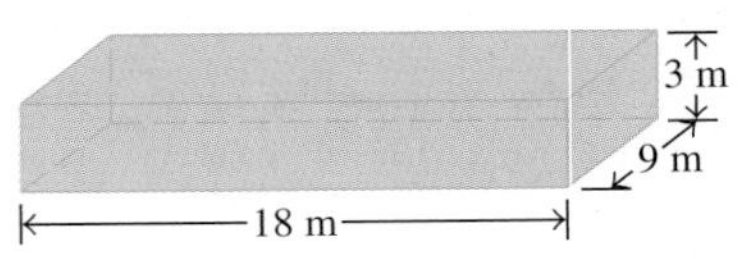

Figure 8.7

SOLUTION

a) To find the volume in cubic meters, we use the formula

$$V = l \times w \times h$$

Substituting, we have

$$V = 18\text{ m} \times 9\text{ m} \times 3\text{ m}$$
$$= 486\text{ m}^3$$

b) Since $1\text{ m}^3 = 1$ kℓ, the pool will hold 486 kℓ of water.

EXAMPLE 6 *Choose an Appropriate Unit*

Select the most appropriate answer. The volume of this textbook is approximately

a) 1500 mm^3 b) 1500 ℓ c) 1500 cm^3.

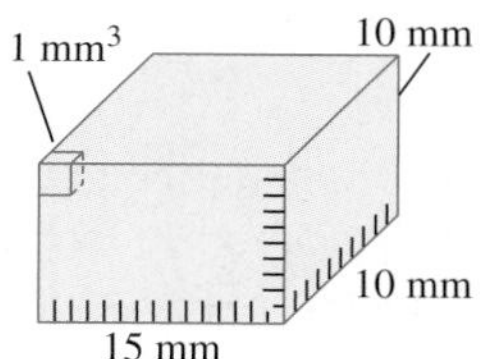

Figure 8.8

SOLUTION This textbook is not a liquid, so its volume is not expressed in liters. Thus, (b) is not the answer. The rectangular solid in Fig. 8.8 is approximately 1500 mm^3, so (a) is not an appropriate answer. This textbook measures about 26 cm × 21 cm × 3 cm, or 1638 cm^3. Therefore, 1500 cm^3 or (c) is the most appropriate answer. ▲

When the volume of a liquid is measured, the abbreviation cc is often used instead of cm^3 to represent cubic centimeters. For example, a nurse may give a patient an injection of 3 cc or 3 mℓ of the drug ampicillin.

EXAMPLE 7 *Measuring Medicine*

A nurse must give a patient 3 cc of the drug Gentamycin mixed in 100 cc of a normal saline solution.

a) How many mℓ of the drug will the nurse administer?
b) What is the total volume of the drug and water in milliliters?

SOLUTION

a) Because 1 cc is equal in volume to 1 mℓ, the nurse will administer 3 mℓ of the drug.
b) The total volume is 3 + 100, or 103, mℓ. ▲

EXAMPLE 8 *A Hot-Water Heater*

A hot-water heater, in the shape of a right circular cylinder, has a radius of 50 cm and a height of 148 cm. What is the capacity, in liters, of the hot water heater?

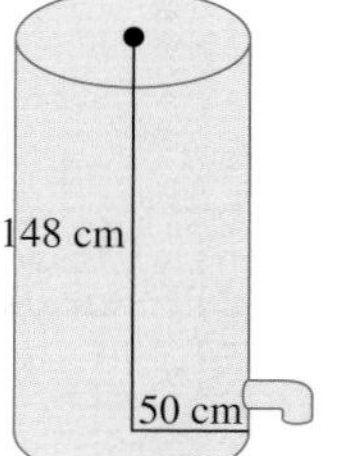

Figure 8.9

SOLUTION The hot-water heater is illustrated in Fig. 8.9. The formula for the volume of a right circular cylinder is $V = \pi r^2 h$, where π is approximately 3.14. If we express all the measurements in meters, the volume will be given in cubic meters. Thus, 50 cm = 0.5 m, and 148 cm = 1.48 m.

$$\begin{aligned} V &= \pi r^2 h \\ &= 3.14(0.5)^2(1.48) \\ &= 3.14(0.25)(1.48) = 1.1618 \text{ m}^3 \end{aligned}$$

We want the volume in liters, so we must change the answer from cubic meters to liters.

$$1 \text{ m}^3 = 1000\ \ell$$

So,

$$1.1618 \text{ m}^3 = 1.1618 \times 1000 = 1161.8\ \ell$$ ▲

DID YOU KNOW

Metrics and Medicine

Both milliliters and cubic centimeters are commonly used in medicine. In the United States, cubic centimeters are commonly denoted cc rather than the cm^3 used in the SI system. A patient's intake and output of fluids and intravenous injections are commonly measured in cubic centimeters. Drug dosage is measured in milliliters.

The following question is from a nursing exam. Can you determine the correct answer?

In caring for a patient after delivery, you are to give 12 units of Pitocin (in 1000 cc of intravenous fluid). The ampule is labeled 10 units per 0.5 mℓ. How much of the solution would you draw and give?

a) 0.6 cc
b) 1.2 cc
c) 6.0 cc
d) 9.6 cc

Answer: (a)

EXAMPLE 9 ***Comparing Volume Units***

a) How many times larger is a cubic meter than a cubic centimeter?
b) How many times larger is a cubic dekameter than a cubic meter?

SOLUTION

a) The procedure used to determine the answer is similar to that used in Example 3 in this section. First we draw a cubic meter, which is a cube 1 m long by 1 m wide by 1 m high. In Fig. 8.10, we represent each meter as 100 centimeters. The volume of the cube is its length times its width times its height, or

$$V = l \times w \times h$$
$$= 100 \text{ cm} \times 100 \text{ cm} \times 100 \text{ cm} = 1\,000\,000 \text{ cm}^3$$

Since $1 \text{ m}^3 = 1\,000\,000 \text{ cm}^3$, a cubic meter is one million times larger than a cubic centimeter.

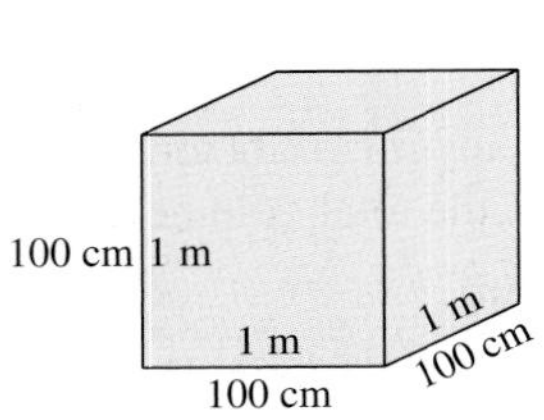

Figure 8.10

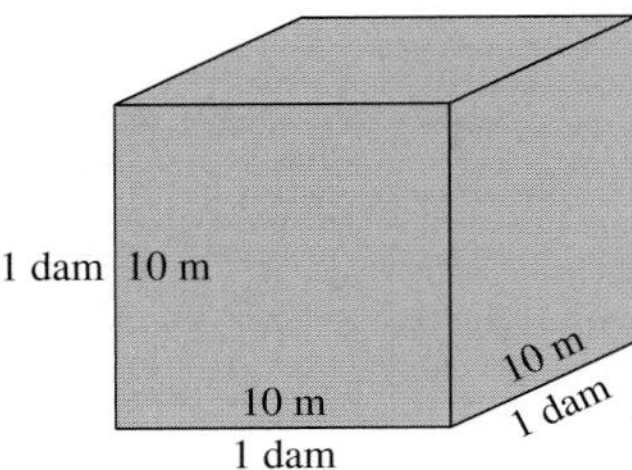

Figure 8.11

b) Work part (b) in a similar manner (Fig. 8.11).

$$V = l \times w \times h$$
$$= 10 \text{ m} \times 10 \text{ m} \times 10 \text{ m} = 1000 \text{ m}^3$$

Since $1 \text{ dam}^3 = 1000 \text{ m}^3$, a cubic dekameter is one thousand times larger than a cubic meter.

SECTION 8.2 EXERCISES

CONCEPT/WRITING EXERCISES

In Exercises 1–12, an object has been measured and the measurement has been written with the unit indicated. Indicate what was measured: length, area, or volume.

1. cm **2.** m^3 **3.** ha **4.** mm
5. cc **6.** ℓ **7.** cm^3 **8.** kℓ
9. m^2 **10.** dℓ **11.** mℓ **12.** cm^2

13. Estimate your height in (a) centimeters and (b) meters.

14. Estimate, in centimeters, the length of this book.

15. Estimate, in square centimeters, the surface area of this book.

16. Estimate, in meters, the length of the classroom in which your mathematics course is held.

17. Estimate, in cubic centimeters, the volume of this book.

18. Estimate, in square centimeters, the surface area of a dollar bill.

19. One liter of liquid has the equivalent volume of which of the following:
a cubic centimeter, a cubic decimeter, or a cubic meter?

20. One cubic meter has the equivalent volume of which of the following liquid measures:
a liter, a milliliter, or a kiloliter?

21. One milliliter of liquid has the equivalent volume of which of the following: a cubic centimeter, a cubic decimeter, or a cubic meter?

22. Is the hectare a measure of length, area, or volume?

23. A hectare has an area of about how many acres: 2.5, 25, or 250?

24. Which metric measurement is used to measure very large areas of land?

PRACTICE THE SKILLS

In Exercises 25–36, indicate the metric unit of measurement that you would use to express the following.

25. The height of a doorway

26. The length of a jump rope

27. The distance between cities

28. The length of a paper clip

29. The width of a Frisbee

30. The length of a newborn infant

31. The diameter of a pencil

32. The diameter of a jump rope

33. The width of a dining room table

34. The length of a photograph

35. The length of a butterfly

36. The distance to the moon

37. Starting with a straight piece of wood of sufficient size, construct a meter stick. Indicate decimeters, centimeters, and millimeters on the meter stick. Use the centimeter measure in Fig. 8.1 as a guide.

38. Construct a metric tape measure from a piece of tape or rope and then determine your waist measurement.

In Exercises 39–46, choose the best answer.

39. A U.S. postage stamp is about how wide and how long?
    a) 2 cm × 3 cm
    b) 2 mm × 3 mm
    c) 2 hm × 3 hm

40. A football field is about how long?
    a) 90 km
    b) 90 m
    c) 90 cm

41. A grown woman is about how tall?
    a) 160 cm
    b) 160 mm
    c) 160 dm

42. The distance between freeway exits could be how long?
    a) 5 mm **b)** 5 m **c)** 5 km

43. The width of a shoelace is about how wide?
    a) 5 cm **b)** 5 mm **c)** 5 dm

44. The diameter of a coffee cup is about which of the following?
    a) 8 mm **b)** 8 cm **c)** 8 dm

45. The Sears Tower is about how tall?
    a) 375 cm **b)** 375 km **c)** 375 m

46. A full-length gown may be how long?
    a) 150 m **b)** 150 mm **c)** 150 cm

In Exercises 47–52, (a) estimate the item in metric units and (b) measure it with a metric ruler. Record your result.

47. The width of a door

48. The width of a card from a deck of cards

49. The length of a car

50. The diameter of a can of soda

51. The height of a milk carton

52. The thickness of a quarter

In Exercises 53–58, replace the customary measure (shown in parentheses) with the appropriate metric measure.

53. Give him a ______ (inch), and he will take a ______ (mile).

54. There was a crooked man and he walked a crooked ______ (mile).

55. One hundred ______ (yard) dash.

56. I wouldn't touch a skunk with a 10-______ (foot) pole.

57. I found ______ (an inch) worm.

58. This is a ______ (mile)stone in my life.

In Exercises 59–68, indicate the metric unit of measurement you would use to express the area of the following.

59. A rug

60. The Great Smoky Mountains National Park

61. The face of a dime
62. The floor of your classroom
63. A baseball field
64. A building lot for a house
65. A postage stamp
66. Washington, D.C.
67. A ceiling tile
68. A professional basketball court

In Exercises 69–76, choose the best answer.

69. A U.S. postage stamp has an area of about
    a) 5 cm^2. **b)** 5 mm^2. **c)** 5 dm^2.
70. The area of a living room floor is about
    a) 20 cm^2. **b)** 20 m^2. **c)** 20 km^2.
71. The area of a city lot is about
    a) 800 m^2. **b)** 800 hm^2. **c)** 800 cm^2.
72. The area of a city lot is about
    a) $\frac{1}{8}$ m^2. **b)** $\frac{1}{8}$ ha. **c)** $\frac{1}{8}$ km^2.
73. The area of a ceiling tile is about
    a) 360 m^2. **b)** 360 km^2. **c)** 360 cm^2.
74. The area of Dinosaur National Park is about
    a) 15 km^2. **b)** 15 ha. **c)** 15 m^2.
75. The area of the face of a dime is about
    a) 2.5 cm^2. **b)** 2.5 m^2. **c)** 2.5 mm^2.
76. The area of the screen of a table top TV is about
    a) 1200 dm^2. **b)** 1200 mm^2. **c)** 1200 cm^2.

In Exercises 77–82, (a) estimate the area of the item in metric units and (b) measure it in metric units and compute its area.

77. A teacher's desk top
78. The cover of this book
79. A five-dollar bill
80. Your kitchen table
81. The end of a 12 oz soda can (Use $A = \pi r^2$.)
82. The face of a penny

In Exercises 83–92, determine the metric unit that would best be used to measure the volume of the following.

83. Water flowing over Niagara Falls per minute

84. Water in a hot-water heater
85. Liquid in an eye dropper
86. Air in a basketball
87. Oil in a can of oil
88. Oil needed to change oil in your car
89. A truckload of ready-mix concrete
90. Asphalt needed to pave a driveway
91. Soda in a soda bottle
92. A truckload of topsoil

In Exercises 93–100, choose the best answer to indicate the volume of the following.

93. A shoebox
    a) 7780 mm^3 **b)** 7780 dm^3 **c)** 7780 cm^3
94. A quarter
    a) 0.5 cm^3 **b)** 0.5 mm^3 **c)** 0.5 dm^3
95. Water in a 24-ft-diameter above-ground circular swimming pool
    a) 55 ℓ **b)** 55 mℓ **c)** 55 kℓ
96. Soda in a can of soda
    a) 355 ℓ **b)** 355 mℓ **c)** 355 m^3
97. Juice that can be squeezed out of an orange
    a) 120 kℓ **b)** 120 mℓ **c)** 120 ℓ
98. A can of vegetables
    a) 550 cm^3 **b)** 550 mm^3 **c)** 550 dm^3
99. Air in a basketball
    a) 14 000 m^3 **b)** 14 000 cm^3 **c)** 14 000 mm^3
100. Air in a balloon with a diameter of 4 meters
    a) 30 m^3 **b)** 30 cm^3 **c)** 30 km^3

In Exercises 101–104, (a) estimate the volume in metric units and (b) compute the actual volume of the item.

101. Water in a water bed that is 2 m long, 1.5 m wide, and 25 cm deep (Use $V = lwh$.)
102. Oil in a barrel that has a height of 1 m and a diameter of 0.5 m (Use $V = \pi r^2 h$.)
103. A new piece of chalk that has a length of 10 cm and a diameter of 1 cm.
104. Water in a cylindrical tank that is 40 cm in diameter and 2 m high

PROBLEM SOLVING

105. Use a metric ruler to measure the length of the sides and the height of the parallelogram. Then compute its perimeter and area. Give your answers in metric units.

106. Use a metric ruler to find the radius of the circle. Then compute the circumference and area of the circle. Give your answers in metric units.

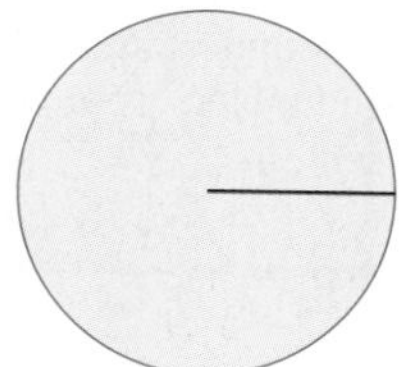

107. *A Mat for a Picture* A framed picture is shown. Find the matted area.

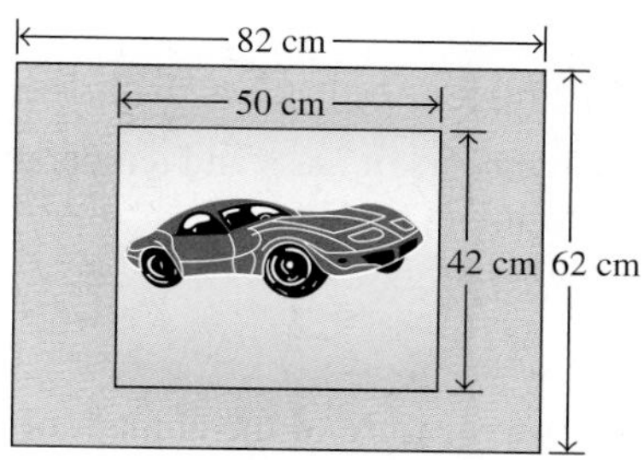

108. *A Walkway* A rectangular building 40 m by 64 m is surrounded by a walk 1.5 m wide.

a) Find the area of the region covered by the building and the walk.

b) Find the area of the walk.

109. *Farmland* Mr. Hershman has purchased a farm that is in the shape of a rectangle. The dimensions of the piece of land are 1.4 km by 3.75 km.

a) How many square kilometers of land did he purchase?

b) If 1 km^2 equals 100 ha, determine the amount of land he purchased in hectares.

110. *Area of a Garden* Guido's garden is 22.5 m by 18.3 m.

a) How large is his garden in square meters?

b) If 1 m^2 equals 0.0001 ha, determine the area of his garden in hectares.

111. a) *Volume of Water* What is the volume of water in a swimming pool that is 18 m long and 10 m wide and has an average depth of 2.5 m? Give your answer in cubic meters.

b) How many kiloliters of water will the pool hold?

112. *Fish Tank Volume* A rectangular fish tank is 70 cm long, 40 cm wide, and 20 cm high.

a) How many cubic centimeters of water will the tank hold?

b) How many milliliters of water will the tank hold?

c) How many liters of water will the tank hold?

113. *Cost of Paint* The first coat of paint for the outside of a building requires 1 ℓ of paint for each 10 m^2. The second coat requires 1 ℓ for every 15 m^2. If the paint costs $4.75 per liter, what will be the cost of two coats of paint for the four outside walls of a building 20 m long, 12 m wide, and 6 m high?

114. *How Much Soup?* A can of Campbell's Home Cookin' chicken vegetable soup has a diameter of 8.0 cm and a height of 12.5 cm. Determine the volume of soup in the can (assume the can is filled with soup).

115. How many times larger is a square kilometer than a square dekameter?

116. How many times larger is a square dekameter than a square meter?

117. How many times larger is a cubic meter than a cubic decimeter?

118. How many times larger is a cubic centimeter than a cubic millimeter?

In Exercises 119–126, replace the question mark with the appropriate value.

119. 1 m^2 = ? mm^2

120. 1 hm^2 = ? cm^2

121. 1 km^2 = ? hm^2

122. 1 cm^2 = ? m^2

123. 1 dm^2 = ? mm^2

124. 1 mm^2 = ? hm^2

125. 1 m^3 = ? cm^3

126. 1 hm^3 = ? km^3

In Exercises 127–130, fill in the blank.

127. 620 cm^3 = ______ mℓ

128. 620 cm^3 = ______ ℓ

129. 76 kℓ = ______ m^3

130. 4.2 ℓ ______ cm^3

CHALLENGE PROBLEMS/GROUP ACTIVITIES

In Exercises 131 and 132, fill in the blank to make a true statement.

131. 6.7 kℓ = ______ dm^3

132. 1.4 ha = ______ cm^2

133. In Example 3, we illustrated how to change an area in a metric unit to an area in a different metric unit

a) Using Example 3 as a guide, change 1 square mile to square inches.

b) Is converting from one unit of area to a different unit of area generally easier in the metric system or the customary system? Explain.

134. In Example 9, we illustrated how to change a volume in one metric unit to a volume in a different metric unit.
   a) Using Example 9 as a guide, change 6 yd^3 (a volume 1 yard by 2 yards by 3 yards) into cubic inches.
   b) Is converting from one unit of volume to a different unit of volume generally easier in the metric system or the customary system? Explain.

RESEARCH ACTIVITIES

135. The definition of the meter has changed several times throughout history. Use the Internet, an encyclopedia, or other reference book and write a one- to two-page report on the history of the meter, from when it was first named to the present.

8.3 MASS AND TEMPERATURE

In this section, we discuss the metric measurements of mass and temperature. As with Section 8.2, the focus of this section is on thinking metric.

MASS

Weight and mass are not the same. **Mass** is a measure of the amount of matter in an object. It is determined by the molecular structure of the object, and it will not change from place to place. Weight is a measure of the gravitational pull on an object. For example, the gravitational pull of the Earth is about six times as great as the gravitational pull of the moon. Thus, a person on the moon weighs about $\frac{1}{6}$ as much as on Earth, even though the person's mass remains the same. In space, where there is no gravity, a person has no weight.

Even on the Earth, the gravitational pull varies from point to point. The closer you are to the center of the Earth, the greater the gravitational pull. Thus, a person weights very slightly less on a mountain than in a nearby valley. Because the mass of an object does not vary with location, scientists generally use mass rather than weight.

Although weight and mass are not the same, on Earth they are proportional to each other (the greater the weight, the greater the mass). Therefore, for our purposes, we can treat weight and mass as the same.

The *kilogram* is the basic unit of mass in the metric system. It is a little more than 2 lb. Recall that the official kilogram is a cylinder of platinum–iridium alloy kept by the International Bureau of Weights and Measures, located in Sèvres, near Paris.

Items that we normally measure in pounds are usually measured in kilograms in other parts of the world. For example, an average-sized man has a mass of about 75 kg.

The *gram* (a unit that is 0.001 kg) is relatively small and is used in place of the ounce. A nickel has a mass of 5 g, a cube of sugar has a mass of about 2 g, and a large paper clip has a mass of about 1 g.

The *milligram* is used extensively in the medical and scientific fields as well as in the pharmaceutical industry. Nearly all bottles of tablets are now labeled in either milligrams or grams.

The *metric tonne* (t) is used to express the mass of heavy items. One metric tonne equals 1000 kg. It is a little larger than our customary ton of 2000 lb. The mass of a large truck may be expressed in metric tonnes.

DID YOU KNOW

The Kilogram

Since 1889, a single platinum–iridium bar has been sealed in an airtight jar in the International Bureau of Weights and Measures in Sèvres, France.

Nicknamed "Le Grand K," this bar constitutes the one and only true kilogram. Of all the standard international units of measure, the kilogram remains the only one whose definition relies on a physical artifact. All the other units have their definitions rooted in constants of nature, such as the speed of light or atomic vibrations.

As part of an international effort, researchers at the U.S. National Institute of Standards and Technology in Washington, D.C., want to redefine the kilogram in a way that will make the standard absolute, unchanging, and accessible to anyone, anywhere.

One problem is that the current standard tends to drift a bit. The kilogram has varied by as much as 0.05 parts per million in the last 100 years. The cause of that variance remains unknown.

EXAMPLE 1 ***Choosing the Appropriate Unit***

Determine the appropriate unit(s) used to express the mass of the following.

a) A teaspoon of sugar
b) A newborn child
c) A car
d) A box of cereal
e) A quarter
f) A fly
g) A steak
h) A refrigerator

SOLUTION

a) Grams
b) Kilograms
c) Kilograms or metric tonnes
d) Kilograms or grams
e) Grams
f) Milligrams
g) Kilograms or grams
h) Kilograms

One kilogram of water has a volume of exactly 1 liter. In fact, a liter is defined to be the volume of 1 kilogram of water at a specified temperature and pressure. Thus, mass and volume are easily interchangeable in the metric system. Converting from weight to volume is not nearly as convenient in our customary system. For example, how would you change pounds of water to cubic feet or gallons of water in our customary system?

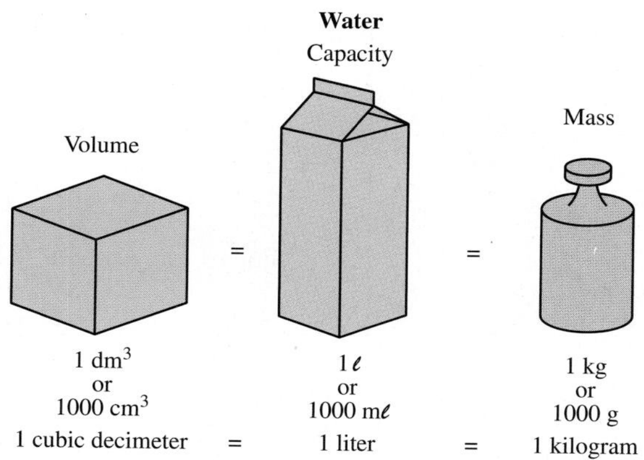

Figure 8.12

Figure 8.12 illustrates the relationship between volume of water in cubic decimeters, capacity in liters, and mass in kilograms. Table 8.4 expands on this relationship between volume and mass of water.

Table 8.4 Volume and Mass of Water

Volume in cubic units		Volume in liters		Mass of water
1 cm^3	=	1 mℓ	=	1 g
1 dm^3	=	1 ℓ	=	1 kg
1 m^3	=	1 kℓ	=	1 t (1000 kg)

EXAMPLE 2 ***A Fish Tank's Capacity***

A fish tank is 1 m long, 50 cm high, and 250 mm wide (Fig. 8.13).

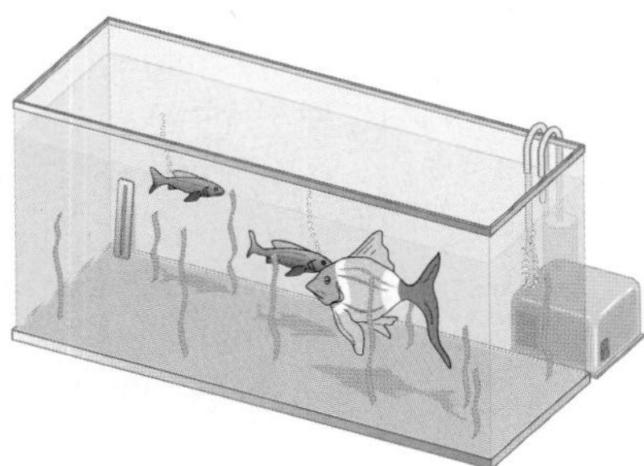

Figure 8.13

a) Determine the number of liters of water the tank holds.
b) What is the mass of the water in kilograms?

SOLUTION

a) We must convert all the measurements to the same units. Let's convert them all to meters: 50 cm is 0.5 m, and 250 mm is 0.25 m.

$$\begin{aligned} V &= l \times w \times h \\ &= 1 \times 0.25 \times 0.5 \\ &= 0.125\ m^3 \end{aligned}$$

Since 1 m^3 of water = 1 kℓ of water,

$$0.125\ \text{m}^3 = 0.125\ \text{k}\ell\text{, or } 125\ \ell \text{ of water}$$

b) Since 1 ℓ = 1 kg, 125 ℓ = 125 kg of water.

To convince yourself of the advantages of the metric system, do a similar problem involving the customary system of measurement, such as Challenge Problems/Group Activities Exercise 69 at the end of this section.

TEMPERATURE

The Celsius scale is used to measure temperatures in the metric system. Figure 8.14 on page 413 shows a thermometer with the Fahrenheit scale on the left and the Celsius scale on the right.

The Celsius scale was named for the Swedish astronomer Anders Celsius (1701–1744), who first devised it in 1742. On the Celsius scale, water freezes at 0°C and boils at 100°C. The Celsius thermometer in the past was called a "centigrade thermometer." Recall that *centi* means $\frac{1}{100}$, and there are 100 degrees between the freezing point of water and the boiling point of water. Thus, 1°C is $\frac{1}{100}$ of this interval. Table 8.5 gives some common temperatures in both Celsius (°C) and Fahrenheit (°F).

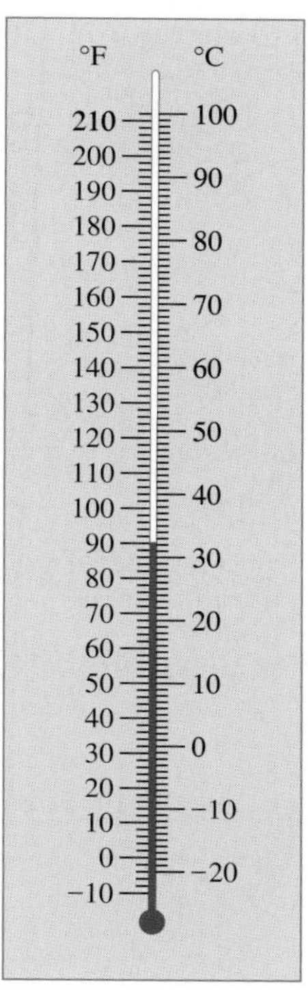

Figure 8.14

Table 8.5

Celsius temperature		Farenheit temperature
−18°C	A very cold day	0°F
0°C	Freezing point of water	32°F
10°C	A warm winter day	50°F
20°C	A mild spring day	68°F
30°C	A warm summer day	86°F
37°C	Body temperature	98.6°F
100°C	Boiling point of water	212°F
177°C	Oven temperature for baking	350°F

At a temperature of −40° the Celsius and Fahrenheit temperatures are the same. That is, −40°C = −40°F. See Exercise 70.

EXAMPLE 3 *Think Metric Temperatures*

Choose the best answer. (Refer to the dual-scale thermometer in Fig. 8.14.)

a) Buffalo, New York, on New Year's Day might have a temperature of

i) −15°C. **ii)** 15°C. **iii)** 40°C.

b) Washington, D.C., on July 4, might have a temperature of

i) 20°C. **ii)** 30°C. **iii)** 40°C.

c) The oven temperature for baking a cake might be

i) 60°C. **ii)** 100°C. **iii)** 175°C.

SOLUTION

a) A temperature of 15°C is possible if it is a very mild winter, but 40°C is much too hot. The best answer for a normal winter is −15°C.

b) The best estimate is 30°C. A temperature of 20°C is too chilly, and 40°C is too hot for July 4.

c) A cake bakes at temperatures well above boiling, so the only reasonable answer is 175°C. ▲

Comparing the temperature scales in Fig. 8.14, we see that the Celsius scale has 100° from boiling to freezing, and the Fahrenheit scale has 180°. Therefore, one Celsius degree represents a greater change in temperature than one Fahrenheit degree does. In fact, one Celsius degree is the same as $\frac{180}{100}$, or $\frac{9}{5}$ Fahrenheit degrees. When converting from one system to the other system, use the following formulas.

From Celsius to Fahrenheit

$$F = \frac{9}{5}C + 32$$

From Fahrenheit to Celsius

$$C = \frac{5}{9}(F - 32)$$

EXAMPLE 4 *Convert to °C*

A typical setting for home thermostats is 68°F. What is the equivalent temperature on the Celsius thermometer?

SOLUTION We use the formula $C = \frac{5}{9}(F - 32)$ to convert from °F to °C. Substituting $F = 68$ gives

$$C = \frac{5}{9}(68 - 32)$$
$$= \frac{5}{9}(36)$$
$$= 20°$$

EXAMPLE 5 ***Convert to °F***

If the temperature outdoors is 22°C, will you need to wear a sweater if going outdoors?

SOLUTION We use the formula $F = \frac{9}{5}C + 32$ to convert from °C to °F. Substituting $C = 22$ yields

$$F = \frac{9}{5}(22) + 32$$
$$= 39.6 + 32$$
$$= 71.6°$$

Since the temperature is about 72°F, you will not need a sweater.

DID YOU KNOW

It's a Metric World

The United States is the only westernized country not presently using the metric system as its primary system of measurement. The only countries in the world besides the United States not presently using or committed to using the metric system are Yemen, Brunei, and a few small islands; see Fig. 8.15.

The European Union (EU) adopted a directive that requires all exporters to EU nations to indicate the dimensions of their products in metric units. Currently, U.S. manufacturers who export goods are doing so. Little by little, the United States is becoming more metric. For example, soft drinks now come in liter bottles and prescription drug dosages are given in metric units. Maybe in the not too distant future gasoline will be measured in liters, not gallons, as it is in Canada and Mexico.

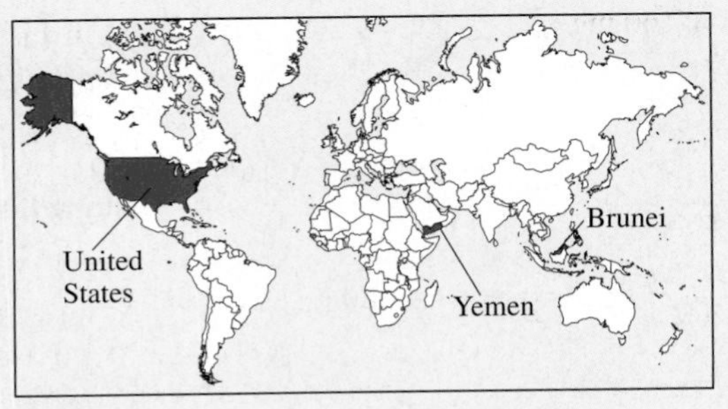

Figure 8.15

SECTION 8.3 EXERCISES

CONCEPT/WRITING EXERCISES

1. What is the basic unit of mass in the metric system?
2. One kilogram is a little more than how many pounds?
3. The mass of a nickel is about how many grams?
4. What unit of mass is used to express the mass of very heavy items?
5. Give an estimate of the temperature, in degrees Celsius, in Florida in August.
6. Give an estimate of the temperature, in degrees Celsius, in North Dakota in February.
7. Give an estimate, in degrees Celsius, of what you would consider an ideal temperature.
8. **a)** Is a person's mass the same in space as on Earth? Explain.
   b) Is a person's weight the same in space as on the Earth? Explain.

PRACTICE THE SKILLS

In Exercises 9–18, what metric unit of measurement would you use to express the mass of the following.

9. A shoe **10.** A man
11. A pair of eyeglasses **12.** A box of cereal
13. A new pencil **14.** A Cadillac
15. A full-grown blue whale **16.** A mosquito
17. A dime **18.** A calculator

In Exercises 19–24, select the best answer.

19. The mass of a 5 lb bag of flour is about how much?
a) 2.26 g **b)** 2.26 kg **c)** 2.26 dag

20. The mass of a newborn baby is about how much?
a) 3.2 g **b)** 3.2 kg **c)** 3.2 dag

21. The mass of a deck of cards is about how much?
a) 60 mg **b)** 60 kg **c)** 60 g

22. The mass of a box of corn flakes is about how much?
a) 0.45 t **b)** 0.45 g **c)** 0.45 kg

23. The mass of a full grown elephant is about how much?
a) 2800 g **b)** 2800 kg **c)** 2800 dag

24. The mass of a full-size car is about how much?
a) 1 962 000 hg **b)** 380 kg **c)** 1.6 t

In Exercises 25–28, estimate the mass of the item. If a scale with metric measure is available, find the mass.

25. Your body **26.** The Yellow Pages
27. A gallon of water **28.** A tomato

In Exercises 29–36, choose the best answer. Use Table 8.5 and Fig. 8.14 to help select your answers.

29. A cup of coffee might have a temperature of
a) 15°C. **b)** 50°C. **c)** 90°C.

30. The thermostat for an air conditioner was set for 80°F. This setting is closest to
a) 2°C. **b)** 27°C. **c)** 57°C.

31. The temperature of the water in a certain lake is 5°C. You could
a) ice fish.
b) dress warmly and walk along the lake.
c) swim in the lake.

32. Freezing rain is most likely to occur at a temperature of
a) −25°C. **b)** 32°C. **c)** 0°C.

33. The weather forecast calls for a high of 15°C. You should plan to wear
a) a down-lined jacket.
b) a sweater.
c) a bathing suit.

34. What might be the temperature of milk from a refrigerator?
a) 30°C **b)** 5°C **c)** 0°C

35. What might be the temperature of an apple pie baking in the oven?
a) 90°C **b)** 100°C **c)** 177°C

36. Upon reentry into the Earth's atmosphere, the space shuttle's tiles protect the shuttle from temperatures as high as 2300°F, or about
a) 195°C. **b)** 620°C. **c)** 1260°C.

In Exercises 37–50, convert each temperature as indicated. Give your answer to the nearest tenth of a degree.

37. 20°C = ______°F **38.** −5°C = ______°F
39. 92°F = ______°C **40.** −25°F = ______°C
41. 350°F = ______°C **42.** 98°F = ______°C
43. 37°C = ______°F **44.** −4°C = ______°F

45. 13°F = ______°C

46. 75°F = ______°C

47. 45°C = ______°F

48. 10°C = ______°F

49. 113°F = ______°C

50. 425°F = ______°C

In Exercises 51–56, use the following graph, which shows the daily low and high temperatures for the week ending January 31, 1998, in Christchurch, New Zealand. The temperatures this week broke many records. Determine the following temperatures in degrees Fahrenheit.

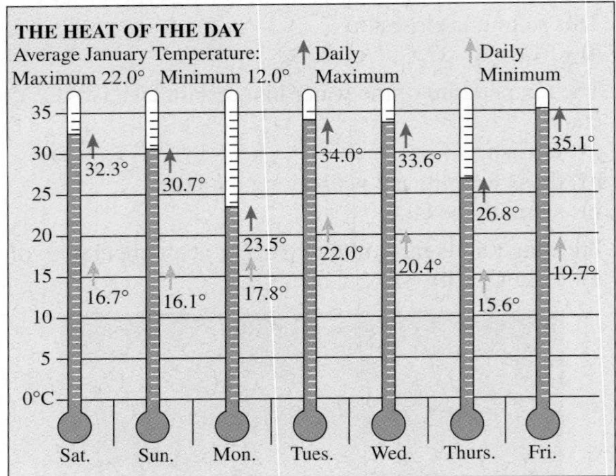

51. The average January maximum temperature

52. The maximum temperature on Friday

53. The maximum temperature for the week

54. The maximum temperature on Saturday

55. The range of temperatures on Monday

56. The range of temperatures on Tuesday

PROBLEM SOLVING

The photo shows the cost of Crest Grower Crumbles and corn at a farm market in Fiji. Use the information provided in the photo to answer Exercises 57 and 58.

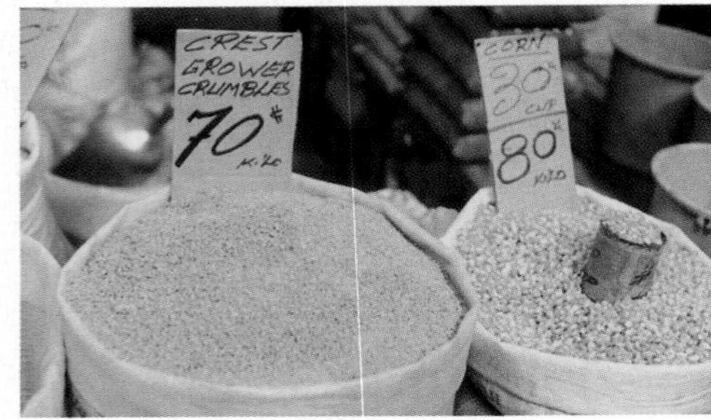

57. Determine the cost of 4.5 kg of Crest Grower Crumbles.

58. Determine the cost of 0.75 kg of corn.

59. *Salt and Soda* A mixture of 45 g of salt and 29 g of baking soda is poured into 370 mℓ of water. What is the total mass of the mixture in grams?

60. *Jet Fuel* A jet can travel about 1 km on 17 kg of fuel. How many metric tonnes of fuel will the jet use flying nonstop between Baltimore and Los Angeles, a distance of about 4320 km?

61. *A Storage Tank* The dimensions of a storage tank are length 16 m, width 12 m, and height 12 m. If the tank is filled with water, determine

a) the volume of water in the tank in cubic meters.

b) the number of kiloliters of water the tank will hold.

c) the mass of the water in metric tonnes.

62. *A Water Heater* A hot-water heater in the shape of a right circular cylinder has a radius of 50 cm and a height of 150 cm. If the tank is filled with water, determine

a) the volume of water in the tank in cubic meters.

b) the number of liters of water the tank will hold.

c) the mass of the water in kilograms.

In Exercises 63–66, convert as indicated

63. 3.6 kg = ______ t

64. 9.52 t = ______ kg

65. 42.6 t = ______ g

66. 1 460 000 mg = ______ t

67. *What's the Problem?* A temperature display at a bank flashes the temperature in degrees Fahrenheit, then flashes the temperature in degrees Celsius. If it flashes 78°F, then 20°C, is there a problem? Explain.

68. *Fever or Chills?* Maria's body temperature is 38.2°C. Should she take an aspirin or put on a sweater? Explain.

CHALLENGE PROBLEMS/GROUP ACTIVITIES

In Example 2, we showed how to find the volume and mass of water in a fish tank. Exercise 69 demonstrates how much more complicated solving a similar problem is in our customary system.

69. A fish tank is 1 yd long by 1.5 ft high by 15 in. wide.

a) Determine the volume of water in the fish tank in cubic feet.

b) Determine the weight of the water in pounds. One cubic foot of water weighs about 62.5 lb.

c) If 1 gal of water weighs about 8.3 lb, how many gallons will the tank hold?

70. Show that $-40°C \approx -40°F$.

RESEARCH ACTIVITIES

71. The SI system is built on a foundation of seven basic units.

a) Do research and list the seven basic units.

b) Which, if any, of these basic units is defined by an actual physical object?

c) For each base unit not defined by a physical object, give its official definition.

72. Do industries in your area export goods? If so, are they training employees to use and understand the metric system? Contact local industries that export goods and write a report on your findings.

8.4 DIMENSIONAL ANALYSIS AND CONVERSIONS TO AND FROM THE METRIC SYSTEM

You may sometimes need to change units of measurement in the metric system to equivalent units in the customary system. To do so, use **dimensional analysis**, which is a procedure used to convert from one unit of measurement to a different unit of measurement. To perform dimensional analysis, you must first understand what is meant by a unit fraction. A **unit fraction** is any fraction in which the numerator and denominator contain different units and the value of the fraction is 1. From Table 8.6, we can obtain many unit fractions involving customary units.

Table 8.6 Customary Units

1 foot = 12 inches
1 yard = 3 feet
1 mile = 5280 feet
1 pound = 16 ounces
1 ton = 2000 pounds
1 cup (liquid) = 8 fluid ounces
1 pint = 2 cups
1 quart = 2 pints
1 gallon = 4 quarts
1 minute = 60 seconds
1 hour = 60 minutes
1 day = 24 hours
1 year = 365 days

Examples of unit fractions

$$\frac{12\text{ in.}}{1\text{ ft}} \qquad \frac{1\text{ ft}}{12\text{ in.}} \qquad \frac{16\text{ oz}}{1\text{ lb}} \qquad \frac{1\text{ lb}}{16\text{ oz}} \qquad \frac{60\text{ min}}{1\text{ hr}} \qquad \frac{1\text{ hr}}{60\text{ min}}$$

In each of these examples, the numerator equals the denominator, so the value of the fraction is 1.

To convert an expression from one unit of measurement to a different unit, multiply the given expression by the unit fraction (or fractions) that will result in the answer having the units you are seeking. When two fractions are being multiplied and the same unit appears in the numerator of one fraction and the denominator of the other fraction, then that common unit may be divided out. For example, suppose we want to convert 30 in. to feet. We consider the following:

$$30\text{ in.} = ?\text{ ft}$$

Since inches are given, we will need to eliminate them. Thus, inches will need to appear in the denominator of the unit fraction. We need to convert to feet, so feet will need to appear in the numerator of the unit fraction. If we multiply a quantity in inches by a unit fraction containing feet/inches, the units will divide out as follows.

$$(\cancel{\text{in.}})\left(\frac{\text{ft}}{\cancel{\text{in.}}}\right) = \text{ft}$$

Thus, the solution to the problem is

$$30 \text{ in.} = (30 \cancel{\text{in.}})\left(\frac{1 \text{ ft}}{12 \cancel{\text{in.}}}\right) = \frac{30}{12} \text{ ft} = 2.5 \text{ ft.}$$

In Examples 1 through 3, we will give examples that do not involve the metric system. After that, we will use dimensional analysis to make conversion to and from the metric system.

EXAMPLE 1 ***Using Dimensional Analysis***

Convert 26 ounces to pounds.

SOLUTION One pound is 16 ounces. Therefore, we write

$$26 \text{ oz} = (26 \cancel{\text{oz}})\left(\frac{1 \text{ lb}}{16 \cancel{\text{oz}}}\right) = \frac{26}{16} \text{ lb} = 1.625 \text{ lb}$$

Thus, 26 oz equals 1.625 lb. ▲

EXAMPLE 2 ***Converting Pesos to Dollars***

On October 1, 1999, \$1 = 9.365 Mexican pesos. What is the amount in dollars of 1000 pesos?

SOLUTION

$$\begin{aligned} 1000 \text{ pesos} &= (1000 \cancel{\text{pesos}})\left(\frac{\$1}{9.365 \cancel{\text{pesos}}}\right) = \frac{\$1000}{9.365} \\ &= \$106.78 \end{aligned}$$

Thus, 1000 pesos have a value of about \$106.78. ▲

If more than one unit needs to be changed, more than one multiplication may be needed, as illustrated in Example 3.

EXAMPLE 3 ***Using Several Unit Fractions***

Convert 60 miles per hour to feet per second.

SOLUTION Let's consider the units given and where we want to end up. We are given $\frac{\text{mi}}{\text{hr}}$ and wish to end with $\frac{\text{ft}}{\text{sec}}$. Thus, we need to change miles into feet and hours into seconds. Because two units need to be changed, we will need to multiply the given quantity by two unit fractions, one for each conversion. First we show how to convert the units of measurement from miles per hour to feet per second:

$$\left(\frac{\cancel{\text{mi}}}{\cancel{\text{hr}}}\right)\left(\frac{\text{ft}}{\cancel{\text{mi}}}\right)\left(\frac{\cancel{\text{hr}}}{\text{sec}}\right) \quad \text{gives an answer in} \quad \frac{\text{ft}}{\text{sec}}$$

DID YOU KNOW

The Tower of Pisa

The Leaning Tower of Pisa, built between 1174 and 1350, actually began leaning before the tower was completed. In 1589 Galileo took advantage of the tower's lean to conduct his experiments on the speed of falling objects.

Throughout history, architects have made many unsuccessful attempts to stop the lean. In 1990, Italian officials recognized that the tower's annual lean increase of 1.5 millimeters at the top had to be stopped to save the tower. The most recent attempt, which began in 1999, involves the slow removal of soil from beneath one side of the base of the tower. As of June 1999, the tower's current lean of about 4.5 meters had been reduced by about 2.5 centimeters. This success has officials optimistic that the lean can be eventually reduced to a total of about 45 centimeters. It is hoped that such a shift would stabilize the marvelous structure. A great variety of mathematics, including problem solving, algebra, geometry, consumer mathematics, and calculus, has gone into the planning and restoration stages of the project.

Now we multiply the given quantity by the appropriate unit fractions to obtain the answer:

$$60\,\frac{\text{mi}}{\text{hr}} = \left(60\,\frac{\cancel{\text{mi}}}{\cancel{\text{hr}}}\right)\left(\frac{5280\text{ ft}}{1\,\cancel{\text{mi}}}\right)\left(\frac{1\,\cancel{\text{hr}}}{3600\text{ sec}}\right) = \frac{(60)(5280)}{(1)(3600)}\,\frac{\text{ft}}{\text{sec}}$$

$$= 88\,\frac{\text{ft}}{\text{sec}}$$

Note that $\left(60\frac{\text{mi}}{\text{hr}}\right)\left(\frac{1\text{ hr}}{3600\text{ sec}}\right)\left(\frac{5280\text{ ft}}{1\text{ mi}}\right)$ will give the same answer. ▲

CONVERSIONS TO AND FROM THE METRIC SYSTEM

Now we will apply dimensional analysis to the metric system.

Table 8.7 is used in making conversions to and from the metric system. A more exact table of conversion factors may be found in many science books at your college's library or on the Internet. However, we can use this table to obtain many unit fractions.

Table 8.7 Conversions Table

Length		
1 inch (in.)	= 2.54	centimeters (cm)
1 foot (ft)	= 30	centimeters (cm)
1 yard (yd)	= 0.9	meter (m)
1 mile (mi)	= 1.6	kilometers (km)
Area		
1 square inch (in^2)	= 6.5	square centimeters (cm^2)
1 square foot (ft^2)	= 0.09	square meter (m^2)
1 square yard (yd^2)	= 0.8	square meter (m^2)
1 square mile (mi^2)	= 2.6	square kilometers (km^2)
1 acre	= 0.4	hectare (ha)
Volume		
1 teaspoon (tsp)	= 5	milliliters (mℓ)
1 tablespoon (tbsp)	= 15	milliliters (mℓ)
1 fluid ounce (fl oz)	= 30	milliliters (mℓ)
1 cup (c)	= 0.24	liter (ℓ)
1 pint (pt)	= 0.47	liter (ℓ)
1 quart (qt)	= 0.95	liter (ℓ)
1 gallon (gal)	= 3.8	liters (ℓ)
1 cubic foot (ft^3)	= 0.03	cubic meter (m^3)
1 cubic yard (yd^3)	= 0.76	cubic meter (m^3)
Weight (Mass)		
1 ounce (oz)	= 28	grams (g)
1 pound (lb)	= 0.45	kilogram (kg)
1 ton (T)	= 0.9	tonne (t)

Table 8.7 shows that 1 in. = 2.54 cm. From this equality, we can write the two unit fractions

$$\frac{1\text{ in.}}{2.54\text{ cm}} \quad \text{or} \quad \frac{2.54\text{ cm}}{1\text{ in.}}$$

Examples of other unit fractions from Table 8.7 are

$$\frac{1\text{ yd}}{0.9\text{ m}},\quad \frac{0.9\text{ m}}{1\text{ yd}},\quad \frac{1\text{ gal}}{3.8\ \ell},\quad \frac{3.8\ \ell}{1\text{ gal}},\quad \frac{1\text{ lb}}{0.45\text{ kg}},\quad \text{and}\quad \frac{0.45\text{ kg}}{1\text{ lb}}$$

To change from a metric unit to a customary unit or vice versa, multiply the given quantity by the unit fraction whose product will result in the units you are seeking. For example, to convert 5 in. to centimeters, multiply 5 in. by a unit fraction with centimeters in the numerator and inches in the denominator.

$$\begin{aligned} 5\text{ in.} &= (5\cancel{\text{in.}})\left(\frac{2.54\text{ cm}}{1\cancel{\text{in.}}}\right) \\ &= 5(2.54)\text{ cm} \\ &= 12.7\text{ cm} \end{aligned}$$

EXAMPLE 4 ***Volume and Length Conversions***

a) A recipe for chicken soup requires $4\frac{1}{2}$ cups of water. How many liters does this amount equal?

b) In New Zealand, a man measures 1.86 m (see photo). What is his height in feet?

SOLUTION

a) In Table 8.7, under the heading of volume, we see that 1 cup = 0.24 ℓ. Thus the unit fractions involving cups and liters are

$$\frac{1\text{ cup}}{0.24\ \ell} \quad \text{or} \quad \frac{0.24\ \ell}{1\text{ cup}}$$

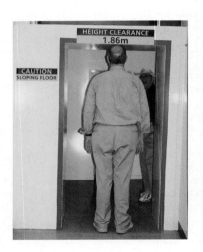

We need to convert from cups to liters. Since $4\frac{1}{2}$ is 4.5 in decimal form, we write

$$4.5\text{ cups} = 4.5\cancel{\text{cups}}\left(\frac{0.24\ \ell}{1\cancel{\text{cup}}}\right) = 1.08\ \ell$$

b) In Table 8.7, under the heading of length, we see that there is no conversion given from meters to feet. However, since 1 inch = 2.54 centimeters, 12 inches or 1 foot = 12 × 2.54 = 30.48 cm or 0.3048 meters. Therefore, we can use the unit fractions.

$$\frac{1\text{ ft}}{0.3048\text{ m}} \quad \text{or} \quad \frac{0.3048\text{ m}}{1\text{ ft}}$$

We want to convert 1.86 m to feet; therefore, we write

$$1.86\text{ m} = (1.86\cancel{\text{m}})\left(\frac{1\text{ ft}}{0.3048\cancel{\text{m}}}\right) = \frac{1.86}{0.3048}\text{ ft} \approx 6.1\text{ ft}$$

Thus, the man is a little taller than 6 ft. ▲

In Example 4(b), a unit fraction involving meters and feet could also be obtained using 1 yard = 0.9 meters. Determine a unit fraction from this conversion factor now.

Land for sale in Fiji, measured in hectares.

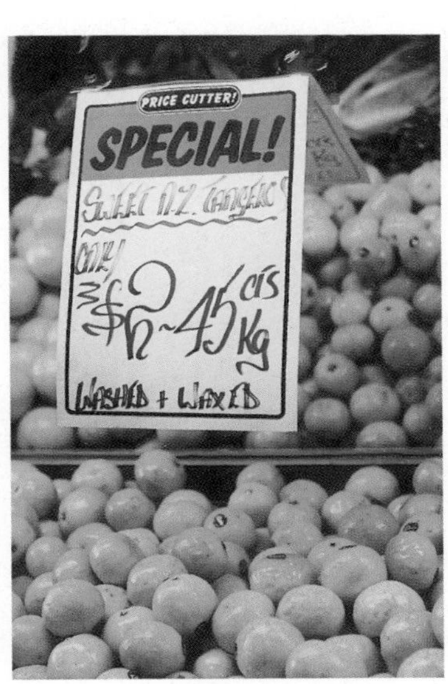

EXAMPLE 5 *Area Conversion*

The photo shows an area of 31.46 hectares for sale. Find the area in acres.

SOLUTION From Table 8.7, we determine that 0.4 ha = 1 acre. Thus,

$$31.46 \text{ ha} = (31.46 \cancel{\text{ha}})\left(\frac{1 \text{ acre}}{0.4 \cancel{\text{ha}}}\right) = \frac{31.46}{0.4} \text{ acres} = 78.65 \text{ acres}$$

▲

EXAMPLE 6 *Weight (Mass) Conversion*

The photo shows that tangelos cost \$2.45 per kilogram. Determine the cost per pound for the tangelos.

SOLUTION From Table 8.7, we see that

$$\frac{1 \text{ lb}}{0.45 \text{ kg}}$$

We use this unit fraction to convert 2.45 kg into pounds.

$$2.45 \text{ kg} = (2.45 \cancel{\text{kg}})\left(\frac{1 \text{ lb}}{0.45 \cancel{\text{kg}}}\right) = \frac{2.45}{0.45} \text{ lb} \approx 5.44 \text{ lb}$$

Therefore, about 5.44 lb of tangelos cost \$2.45. To determine the cost per pound, we divide the cost by the number of pounds.

$$\text{Cost per pound} = \frac{\$2.45}{5.44 \text{ lb}} = \$0.45$$

Therefore, the tangelos cost \$0.45 per pound. ▲

EXAMPLE 7 *Administering a Medicine*

José, a nurse, must administer 4 cc of codeine elixir to a patient.

a) How many milliliters of the drug will he administer?
b) How many ounces is this dosage equivalent to?

SOLUTION

a) Since 1 cc = 1 mℓ, he will administer 4 mℓ of the drug.
b) Since 1 fl oz = 30 mℓ,

$$4 \text{ m}\ell = (4 \cancel{\text{m}\ell})\left(\frac{1 \text{ fl oz}}{30 \cancel{\text{m}\ell}}\right) = \frac{4}{30} \text{ fl oz} \approx 0.13 \text{ fl oz}$$

▲

Suppose we want to convert 150 millimeters to inches. Table 8.7 does not have a conversion factor from millimeters to inches, but it does have one for inches to centimeters. Because 1 inch = 2.54 centimeters and 1 centimeter = 10 millimeters, we can reason that 1 inch = 25.4 millimeters.

$$1 \text{ cm} = 10 \text{ mm}$$

Therefore,

$$2.54 \text{ cm} = 10(2.54) = 25.4 \text{ mm}$$

We can solve the problem as follows.

$$150 \text{ mm} = (150 \cancel{\text{mm}})\left(\frac{1 \text{ in.}}{25.4 \cancel{\text{mm}}}\right) = \frac{150}{25.4} \text{ in.}$$
$$\approx 5.91 \text{ in.}$$

If we wish, we can use dimensional analysis using two unit fractions to make the conversion. The procedure follows:

$$150 \text{ mm} = (150 \cancel{\text{mm}})\left(\frac{1 \cancel{\text{cm}}}{10 \cancel{\text{mm}}}\right)\left(\frac{1 \text{ in.}}{2.54 \cancel{\text{cm}}}\right) = \frac{150}{(10)(2.54)} \text{ in.}$$
$$\approx 5.91 \text{ in.}$$

EXAMPLE 8 ***Converting a Speed***

A road in Cancun, Mexico, has a speed limit of 60 kilometers per hour (kph), see the photo. Determine the speed limit in miles per hour.

SOLUTION In kilometers per hour and miles per hour, the time unit, hour, is the same. Therefore, we just need to convert 60 kilometers to miles. From Table 8.7, we find

$$\frac{1 \text{ mi}}{1.6 \text{ km}} \quad \text{or} \quad \frac{1 \text{ km}}{1.6 \text{ mi}}$$

$$60 \text{ km} = (60 \cancel{\text{km}})\left(\frac{1 \text{ mi}}{1.6 \cancel{\text{km}}}\right) = \frac{60}{1.6} \text{ mi} = 37.5 \text{ mi}$$

Since 60 km equals 37.5 mi, 60 kph is equal to 37.5 mph. ▲

EXAMPLE 9 ***Understanding the Label***

The label on a bottle of Vicks Formula 44D Cough Syrup indicates that the active ingredient is dextromethorphan hydrobromide and that 5 mℓ (1 teaspoon) contains 10 mg of this ingredient. If the recommended dosage for adults is 3 teaspoons, determine the following.

a) How many milliliters of cough medicine will be taken?
b) How many milligrams of the active ingredient will be taken?
c) If the bottle contains 8 fluid ounces of medicine, how many milligrams of the active ingredient are in the bottle?

SOLUTION

a) Since each teaspoon contains 5 mℓ and 3 teaspoons will be taken, 15 mℓ of the cough medicine will be taken.

$$3 \text{ tsp} = (3 \cancel{\text{tsp}})\left(\frac{5 \text{ m}\ell}{1 \cancel{\text{tsp}}}\right) = 15 \text{ m}\ell$$

b) Since each teaspoon contains 10 mg of the active ingredient, 30 mg of the active ingredient will be taken.

$$3 \text{ tsp} = (3 \cancel{\text{tsp}})\left(\frac{10 \text{ mg}}{1 \cancel{\text{tsp}}}\right) = 30 \text{ mg}$$

c) Table 8.7 shows that each fluid ounce contains 30 mℓ. Since each 5 mℓ contains 10 mg of the active ingredient, we can work the problem as follows.

$$8 \text{ fl oz} = (8 \cancel{\text{fl oz}})\left(\frac{30 \cancel{\text{m}\ell}}{1 \cancel{\text{fl oz}}}\right)\left(\frac{10 \text{ mg}}{5 \cancel{\text{m}\ell}}\right) = \frac{8(30)(10)}{5} \text{ mg} = 480 \text{ mg}$$

Therefore, there are 480 mg (or 0.48 g) of the active ingredient in the bottle of cough syrup.

EXAMPLE 10 ***Determining Dosage by Weight***

Drug dosage is often administered according to a patient's weight. For example, 30 mg of the drug Vancomicin is to be given for each kilogram of a person's weight. If Martha, who weighs 136 lb, is to be given the drug, what dosage should she be given?

SOLUTION First we need to convert Martha's weight into kilograms. From Table 8.7, we see that 1 lb = 0.45 kg. We need to determine the number of milligrams of the drug for Martha's weight in kilograms. The answer may be found as follows.

$$136 \text{ lb} = (136 \cancel{\text{lb}})\left(\frac{0.45 \cancel{\text{kg}}}{1 \cancel{\text{lb}}}\right)\left(\frac{30 \text{ mg}}{1 \cancel{\text{kg}}}\right) = (136)(0.45)(30) \text{ mg} = 1836 \text{ mg}$$

Thus, 1836 mg, or 1.836 g, of the drug should be given.

SECTION 8.4 EXERCISES

CONCEPT/WRITING EXERCISES

1. What is dimensional analysis?
2. What is a unit fraction?
3. Give a unit fraction that relates seconds and minutes. Explain how you determined the unit fraction.
4. Give a unit fraction that relates feet and yards. Explain how you determined the unit fraction.
5. When converting from kilograms to pounds, which unit fraction would you use? Explain.

$$\frac{1 \text{ lb}}{0.45 \text{ kg}} \quad \text{or} \quad \frac{0.45 \text{ kg}}{1 \text{ lb}}$$

6. When converting from centimeters to feet, which unit fraction would you use? Explain.

$$\frac{1 \text{ ft}}{30 \text{ cm}} \quad \text{or} \quad \frac{30 \text{ cm}}{1 \text{ ft}}$$

7. When converting from square yards to square meters, which unit fraction would you use? Explain.

$$\frac{1 \text{ yd}^2}{0.8 \text{ m}^2} \quad \text{or} \quad \frac{0.8 \text{ m}^2}{1 \text{ yd}^2}$$

8. When converting from gallons to liters, which unit fraction would you use? Explain.

$$\frac{1 \text{ gal}}{3.8 \ \ell} \quad \text{or} \quad \frac{3.8 \ \ell}{1 \text{ gal}}$$

PRACTICE THE SKILLS

In Exercises 9–24, convert the quantity to the indicated units.

9. 147 km to miles
10. 9 lb to kilograms
11. 4.2 ft to meters
12. 11 in. to centimeters
13. 15 yd^2 to square meters
14. 160 kg to pounds

15. 39 mi to kilometers
16. 765 mm to inches
17. 675 ha to acres
18. 346 g to ounces
19. 10.4 cups to liters
20. 4 T to tonnes
21. 45.6 mℓ to fluid ounces
22. 1.6 km^2 to square miles
23. 120 lb to kilograms
24. 6.2 acres to hectares

In Exercises 25–32, replace the measurement(s) indicated in blue with an equivalent metric measure(s). For example, a foot could be replaced with 30 cm.

25. An ounce of prevention is worth a pound of cure.
26. More bounce to the ounce.
27. Give him an inch and he'll take a mile.
28. He demanded his pound of flesh.
29. Five foot two and eyes of blue.
30. A miss is as good as a mile.
31. First down and 10 yards to go.
32. The longest yard.

In Exercises 33–36, use the part of the score card, which shows the distance in meters for the first four holes of the Millbrook Resort Golf Course in Queenstown, New Zealand. Determine the distances indicated.

HOLE	BLACK Tees	BLUE Tees	HANDICAP	PAR					WHITE Tees	RED Tees
1	505	505	3	5					466	414
2	185	175	15	3					137	91
3	366	357	11	4					344	287
4	396	376	7	4					376	303

33. Hole 1, black tees, in yards
34. Hole 2, blue tees, in yards
35. Hole 3, white tees, in feet
36. Hole 4, red tees, in feet

PROBLEM SOLVING

37. *Distance in Miles* The sign shows an 18 km marker. Determine this distance in miles.

38. *How Far?* Lee Monroe's new car traveled 105 mi on 5 gal of gasoline. How many kilometers can Lee's car travel with the same amount of gasoline?

39. *Buying Carpet* Victoria Montoya is buying outdoor carpet for her lanai, which is 6 yd by 9 yd. The carpeting is sold in square meters. How many square meters of carpeting will she need?

40. *Springfield to St. Louis* The distance from Springfield, Missouri, to St. Louis, Missouri, is about 210 mi. What is the distance in kilometers?

41. *Speed Limit* The speed limit on parts of Route 80 is 70 mph. What is the speed limit in kilometers per hour?

42. *The QEW* Part of the Queen Elizabeth Way (QEW) in Canada has a speed limit of 80 kph. What is the speed in miles per hour?

43. *Milliliters in a Glass* A glass holds 8 fl oz. How many milliliters will it hold?

44. *Louisville to San Antonio* The distance between Louisville, KY and San Antonio, TX is 1776 km. What is the distance in miles between the two cities?

45. *Building a Basement* A basement is to be 50 ft long, 30 ft wide, and 8 ft high. How much dirt will have to be removed when this basement is built? Answer in cubic meters.

46. *Area of Yosemite National Park* Yosemite National Park has an area of 1189 mi^2. What is its area in square kilometers?

47. *Cost of Rice* If rice costs $1.10 per kilogram, determine the cost of a pound of rice.

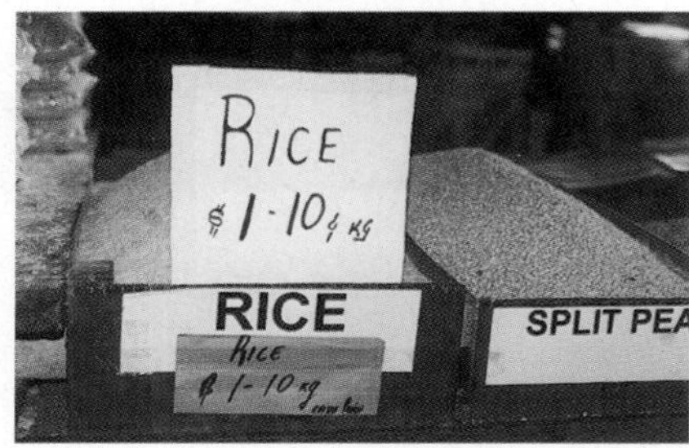

48. *Weight of a Car* A German-made car has a weight of 1.3 t.
    a) How many tons does this weight equal?
    b) How many pounds?

49. *Capacity of a Tank Truck* A tank truck holds 34.5 kℓ of gasoline. How many gallons does it hold?

50. *Cost per Gram* A 0.25 oz bottle of Chanel perfume costs $80. What is the cost per gram?

51. *A Weight in Stones* Some scales in Europe measure a person's weight both in kilograms and in stones. From the photo on top of page 425, we see that a weight of 70 kg is equal to about 11 stones.

a) Using a unit fraction, determine the weight, in kilograms, of a person who weighs 8 stones.
b) Determine the person's weight in pounds.

52. *A Precious Stone* One gram is the same as five carats. Joe's new ring contains a precious stone that is $\frac{1}{8}$ carat. Find the weight of the stone in grams.

53. *Death Valley Elevation* The lowest elevation in the United States is -282 ft at Badwater in Death Valley, California. Determine this elevation in a) centimeters, and b) meters.

54. *Leaning Tower of Pisa* In the Did You Know on page 419 we give four metric measurements related to the restoration of the Leaning Tower of Pisa. Convert each of the given metric measurements to inches. Round answers to the nearest thousandth.

55. *Square Meters to Square Feet* One meter is about 3.3 ft. Use this information to determine
a) the equivalent of one square meter in square feet.
b) the equivalent of one cubic meter in cubic feet.

56. *Square Feet to Square Centimeters* One foot is about 30 cm. Use this information to determine
a) the equivalent of one square foot in square centimeters.
b) the equivalent of one cubic foot in cubic centimeters.

57. *Dosage for a Child* The recommended dosage of the drug codeine for pediatric patients is 1 mg per kilogram of a child's weight. What dosage of codeine should be given to April, who weighs 56 lb?

58. *Dosage for a Man* For each kilogram of a person's weight, 1.5 mg of the antibiotic drug gentamycin is to be administered. If Ron Gigliotti weighs 170 lb, how much of the drug should he receive?

59. *Ampicillin* The recommended dosage of the drug ampicillin for pediatric patients is 200 mg per kilogram of a patient's weight. If Janine weighs 76 lb, how much ampicillin should she receive?

60. *Medicine for a Dog* For each kilogram of weight of a dog, 5 mg of the drug bretylium is to be given. If Blaster, an Irish setter, weighs 82 lb, how much of the drug should be given?

61. *Active Ingredients* The label on the bottle of Triaminic Expectorant indicates that each teaspoon (5 mℓ) contains 12.5 mg of the active ingredient phenylpropanolamine hydrochloride.
a) Determine the amount of the active ingredient in the recommended adult dosage of 2 teaspoons.
b) Determine the quantity of the active ingredient in a 12 oz bottle.

62. *Stomach Ache Remedy* The label on the bottle of Maximum Strength Pepto-Bismol indicates that each tablespoon contains 236 mg of the active ingredient bismuth subsalicyate.
a) Determine the amount of the active ingredient in the recommended dosage of 2 tablespoons.
b) If the bottle contains 8 fl oz, determine the quantity of the active ingredient in the bottle.

63. *Making Cookies* Change all the measurements in the cookie recipe to metric units. Do not forget pan size, temperature, and size of cookies.

Magic Cookie Bar

$\frac{1}{2}$ c graham cracker crumbs

12 oz nuts

8 oz chocolate pieces

$1\frac{1}{3}$ c flaked coconut

$1\frac{1}{3}$ c condensed milk

Coat the bottom of a 9 in. × 13 in. pan with melted margarine. Add rest of ingredients one by one: crumbs, nuts, chocolate, and coconut. Pour condensed milk over all. Bake at 350°F for 25 minutes. Allow to cool 15 minutes before cutting. Makes about two dozen $1\frac{1}{2}$ in. by 3 in. bars.

64. *The Space Shuttle* Write each of the metric units, labeled (*a*) through (*n*), in customary units.

The first human flight, December 17, 1903, was (*a*) *37 m*. Just 66 years later, Neil Armstrong stepped on the moon after journeying (*b*) *370 140 km*. On April 12, 1981, a new era in space flight began when the space shuttle embarked on its maiden voyage.

Here are some characteristics of and facts about the space shuttle. The two solid rocket boosters are jettisoned at (*c*) *44 km*. During reentry, portions of the orbiter's exterior reach temperatures up to (*d*) *1260°C*. The orbiter lands at a speed of (*e*) *335 kph*. It can deliver to orbit up to (*f*) *29 484 kg* of payload in its huge (*g*) *4.5 × 18 m* cargo bay. Propellants can be supplied to the engines at a rate of about (*h*) *171 396 ℓ*/min of hydrogen and (*i*) *63 588 ℓ*/min of oxygen. The external tank is (*j*) *46.89 m* long and (*k*) *8.4 m* in diameter. When fully loaded, the tank contains (*l*) *632 772 kg* of liquid oxygen and (*m*) *106 142 kg* of cold liquid hydrogen at about (*n*) *−251°C*.

CHALLENGE PROBLEMS/GROUP ACTIVITIES

65. *Nursing Question* The following question was selected from a nursing exam. Can you answer it?

In caring for a patient after delivery, you are to give 0.2 mg Ergotrate Maleate. The ampule is labeled $\frac{1}{300}$ grain/mℓ. How much would you draw and give? (60 mg = 1 grain)

a) 15 cc **b)** 1.0 cc **c)** 0.5 cc **d)** 0.01 cc

66. *How Much Beef* Phil is planning a picnic and plans on purchasing 0.18 kg of ground beef for each 100 lb of weight of guests who will be in attendance. If he expects 15 people whose average weight is 130 lb, how many pounds of beef should he purchase?

67. *An Auto Engine* The displacement of automobile engines is measured in liters. A 2000 Chrysler Town and Country minivan has a 3.6 ℓ engine.

a) Determine the displacement of the engine in cubic centimeters.

b) Determine the displacement of the engine in cubic inches.

CHAPTER 8 SUMMARY

IMPORTANT FACTS

Metric units

Prefix	Symbol	Meaning
kilo	k	1000 × base unit
hecto	h	100 × base unit
deka	da	10 × base unit
		base unit
deci	d	$\frac{1}{10}$ of base unit
centi	c	$\frac{1}{100}$ of base unit
milli	m	$\frac{1}{1000}$ of base unit

Water

Volume in cubic units		Volume in liters		Mass of water
1 cm^3	=	1 mℓ	=	1g
1 dm^3	=	1 ℓ	=	1 kg
1 m^3	=	1 kℓ	=	1 t (1000 kg)

Temperature

$$°C = \frac{5}{9}(°F - 32)$$

$$°F = \frac{9}{5}°C + 32$$

CHAPTER 8 REVIEW EXERCISES

8.1

In Exercises 1–6, indicate the meaning of the prefix.

1. Centi **2.** Kilo **3.** Milli

4. Hecto **5.** Deka **6.** Deci

In Exercises 7–12, change the given quantity to that indicated.

7. 80 mg to grams

8. 3.2 ℓ to centiliters

9. 0.197 cm to millimeters

10. 1 000 000 mg to kilograms

11. 4.62 kℓ to liters

12. 192.6 dag to decigrams

In Exercises 13 and 14, arrange the quantities from smallest to largest.

13. 2.67 kℓ, 3000 mℓ, 14 630 cℓ

14. 0.047 km, 4700 m, 47 000 cm

8.2, 8.3

In Exercises 15–24, indicate the metric unit of measurement that would best express the following.

15. The length of a telephone

16. The mass of a telephone

17. The temperature of the water in the ocean

18. The diameter of a quarter

19. The area of a room of a house

20. The volume of a glass of milk

21. The length of an ant

22. The mass of a car

23. The distance from Las Vegas, Nevada, to Jacksonville, Florida

24. The volume of water in a small fish tank

In Exercises 25 and 26, (a) first estimate the following in metric units and then (b) measure with a metric ruler. Record your results.

25. Your height

26. The length of a new pencil

In Exercises 27–32, select the best answer.

27. The length of the distance between Los Angeles and San Francisco is
a) 8000 m **b)** 2000 km **c)** 650 km

28. The mass of a full-grown border collie is about
a) 600 g **b)** 20 kg **c)** 100 kg

29. The volume of a quart of grape juice is about
a) 0.1 kℓ **b)** 0.5 ℓ **c)** 1 ℓ

30. The area of a vegetable garden in a person's yard may be
a) 200 m^2 **b)** 0.5 ha **c)** 0.02 km^2

31. The temperature on a hot summer day in Georgia may be about
a) 34°C **b)** 45°C **c)** 25°C

32. The height of a giant sequoia tree is about
a) 300 m **b)** 3000 cm **c)** 0.3 m

33. Convert 1640 kg to tonnes.

34. Convert 6.3 t to grams.

35. If the temperature outside is 28°C, what is the Fahrenheit temperature?

36. If the room temperature is 68°F, what is the Celsius temperature?

37. If your outdoor thermometer shows a temperature of −6°F, what is the Celsius temperature?

38. If Lynn Colgin's body temperature is 39°C, what is her Fahrenheit temperature?

In Exercises 39 and 40 measure, in centimeters, each of the line segments, then compute the perimeter and the area of the figure.

39.

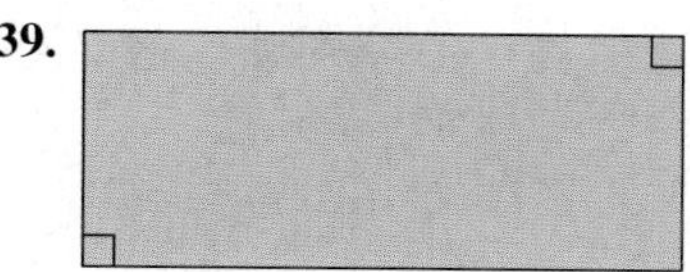

40.

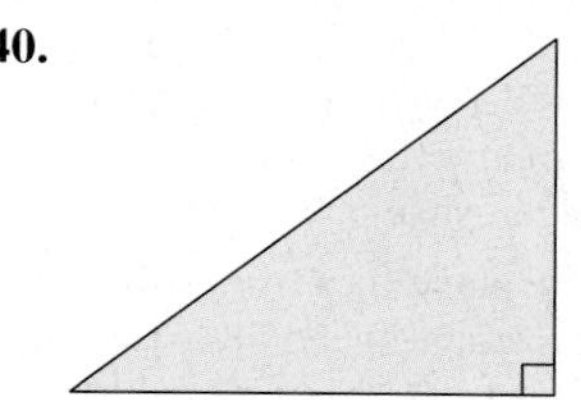

41. a) *A Swimming Pool's Volume* What is the volume of water in a full swimming pool that is 10 m long and 4 m wide and has an average depth of 2 m? Answer in cubic meters.
b) What is the mass of the water in kilograms?

42. *Find the Area* A rectangular lot measures 22 m by 30 m. Find
a) the area in square meters.
b) the area in square kilometers.

43. *Volume of a Fish Tank* A small fish tank measures 80 cm long, 40 cm wide, and 30 cm high.
a) What is its volume in cubic centimeters?
b) What is its volume in cubic meters?
c) How many milliliters of water will the tank hold?
d) How many kiloliters of water will the tank hold?

44. A square kilometer is a square with length and width both 1 km. How many times larger is a square kilometer than a square dekameter?

8.4

In Exercises 45–58, change the given quantity to the indicated quantity.

45. 27 in. = ______ cm

46. 105 kg = ______ lb

47. 83 yd = ______ m

48. 100 m = ______ yd

49. 45 mph = ______ kph

50. 200 lb = ______ kg

51. 15 gal = ______ ℓ

52. 40 m^3 = ______ yd^3

53. 72 lb = ______ kg

54. 4 qt = ______ ℓ

55. 15 yd^3 = ______ m^3

56. 62 mi = ______ km

57. 27 cm = ______ ft

58. $3\frac{1}{4}$ in. = ______ mm

59. *Building a Chimney* John bought 700 bricks to build a chimney. Each brick has a mass of 1.5 kg.
  a) What is the total mass of the bricks in kilograms?
  b) What is the total weight of the bricks in pounds?

60. *Carpeting a Room* Patricia Burgess is buying new carpet for her family room. The room is 15 ft wide and 24 ft long. The carpeting is sold only in square meters. How many square meters of carpeting will she need? Round your answer to the nearest square meter.

61. *Distance Between Cities* A plane flies from Chicago to New York in 2.5 hr at an average speed of 461.6 kph. What is the distance from Chicago to New York in (a) kilometers and (b) miles? (Use distance = rate · time)

62. *The Speed Limit* The speed limit on a certain road is 35 mph. What is the speed limit in
  a) kilometers per hour?
  b) meters per hour?

63. *A Water Tank* A rectangular tank used to test leaks in tires is 90 cm by 70 cm by 40 cm deep.
  a) Determine the number of liters of water the tank holds.
  b) What is the mass of the water in kilograms?

64. If the cost of kiwi fruit is $2.75 per kilogram, determine the cost of 1 lb of kiwi fruit.

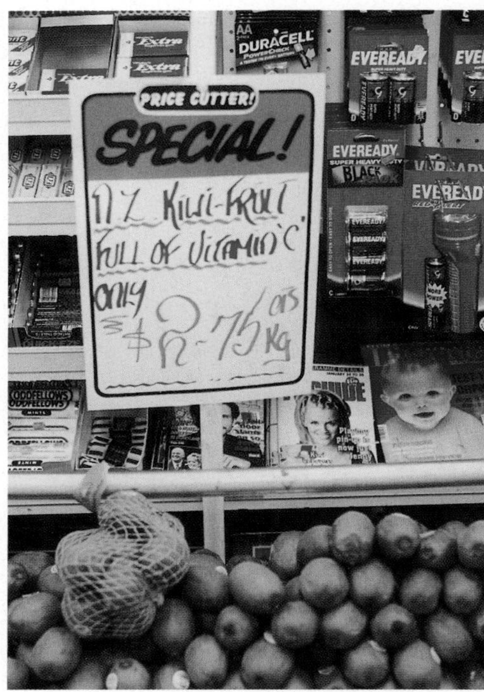

CHAPTER 8 TEST

1. Change 67 km to millimeters.

2. Change 96 cg to hectograms.

3. How many times greater is a kilometer than a dekameter?

4. John's high school track is an oval that measures 300 m around. If John jogs around the track six times, how many kilometers has he gone?

In Exercises 5–9, choose the best answer.

5. The length of this page is about
  a) 10 cm.
  b) 25 cm.
  c) 60 cm.

6. The surface area of the top of a kitchen table is about
  a) 2 m^2.
  b) 200 cm^2.
  c) 2000 cm^2.

7. The volume of a car's gasoline tank is about
  a) 200 ℓ.
  b) 20 ℓ.
  c) 75 ℓ.

8. The mass of a bathroom scale is about
  a) 0.2 t.
  b) 2 kg.
  c) 100 g.

9. The outside temperature on a snowy day is about
  a) 18°C.
  b) −2°C.
  c) −40°C.

10. How many times greater is a square meter than a square centimeter?

11. How many times greater is a cubic meter than a cubic millimeter?

12. Convert 452 in. to centimeters.

13. Change 8 km per hour to miles per hour.

14. Change 50°F to degrees Celsius.

15. Change 50°C to degrees Fahrenheit.

16. A giraffe may be 12 ft tall. How many centimeters is this?

17. *At the Aquarium* A fish tank at an aquarium is 20 m long by 20 m wide by 8 m deep.

a) Determine the volume of the tank in cubic meters.

b) Determine the number of liters of water the tank holds.

c) Determine the weight of the water in kilograms.

18. *Cost of Paint* The first coat of paint for the outside walls of a building requires 1 ℓ of paint for each 10 m^2 of wall surface. The second coat requires 1 ℓ for every 15 m^2. If the paint costs $3.50 per liter, what will be the cost of two coats of paint for the four outside walls of a building 20 m long, 15 m wide, and 6 m high?

GROUP PROJECTS

HEALTH AND MEDICINE

Throughout this chapter, we have shown the importance of the metric system in the medical professions. The following two questions involve applications of metrics to medicine.

1. a) Twenty milligrams of the drug Lincomycin is to be given for each kilogram of a person's weight. The drug is to be mixed with 250 cc of a normal saline solution and the mixture is to be administered intravenously over a 1 hr period. Clyde Dexter, who weighs 196 lb, is to be given the drug. Determine the dosage of the drug he will be given.

b) At what rate per minute should the 250 cc solution be administered?

2. a) At a pharmacy, a parent asks a pharmacist why her child needs such a small dosage of a certain medicine. The pharmacist explains that a general formula may be used to estimate a child's dosage of certain medicines. The formula is

$$\text{Child's dose} = \frac{\left(\begin{array}{c}\text{child's weight}\\ \text{in kilograms}\end{array}\right)}{67.5\text{ kg}} \times \text{adult dose.}$$

What is the amount of medicine you would give a 60 lb child if the adult dosage of the medicine is 70 mg?

b) At what weight, in pounds, would the child receive an adult dose?

TRAVELING TO OTHER COUNTRIES

3. Dale Pollinger is a buyer at Xerox Corporation and travels frequently on business to foreign countries. He always plans ahead and does his holiday shopping overseas where he can purchase items not easily found in the United States.

a) On a trip to Tokyo, he decides to buy a kimono for his sister Kathy. To determine the length of a kimono, one measures, in centimeters, the distance from the bottom of a person's neck to 5 cm above the floor. If the distance from the bottom of Dale's sister's neck to the floor is 5 ft 2 in., calculate the length of the kimono that Dale should purchase.

b) If the conversion rate at the time is 1 U.S. dollar = 102.30 yen, and the kimono cost 8695.5 yen, determine the cost of the kimono in U.S. dollars.

c) On a trip to Mexico City, Mexico, Dale finds a small replica of a Mayan castle that he wants to purchase for his wife Sue. He is going directly from Mexico to Rome, so he wants to mail the castle back to the United States. The mailing rate from Mexico to the United States is 6 pesos per hundred grams. Determine the mailing cost, in U.S. dollars, if the castle weighs 6 lb and the exchange rate is 1 peso = 0.1081 U.S. dollars.

d) Dale takes a personal trip from Rochester, New York, to the Canadian side of Niagara Falls. He notices that in Niagara Falls, New York, the cost of regular gasoline is $1.69 per gallon in U.S. dollars and that in Niagara Falls, Canada, regular gas costs $0.65 per liter in Canadian dollars. His gas tank has a capacity of 18 gallons. If the exchange rate is 1 Canadian dollar = 0.6891 U.S. dollar, is it more expensive to fill the car's empty gas tank in Niagara Falls, Canada, or Niagara Falls, New York, and by how much in U.S. dollars?

CHAPTER 9

Geometry

Many everyday objects in our lives can be described in terms of geometry. The spherical basketball you dribble down a rectangular court, the cylindrical can of soda you drink, and even the rectangular solid shape of this book are all examples of geometric objects that affect our lives. Throughout human history, geometry has played an important role in education, technology, and commerce. This role continues in modern times.

Albert Einstein's use of non-Euclidean geometry in his theory of relativity has enabled mathematicians and scientists to model the universe more accurately. Benoit Mandelbrot's work in fractal geometry has led scientists to discover new ways to describe such intricate and detailed objects as weather systems, air passages in our lungs, and earthquake frequency patterns. A newly discovered form of pure carbon naturally forms molecules whose structure involves hexagons and pentagons in a pattern similar to that of a soccer ball. These molecules hold promise as a microscopic lubricant for a variety of applications from computer chips to new medicines. As the human mind continues to uncover and interact with nature's secrets, geometry undoubtedly will continue to play a vital role. ■

The geometry of fractals now provides mathematicians with a means for describing objects in nature in which a pattern endlessly repeats itself in smaller and smaller versions. This mountain scene is a computer-generated fractal.

9.1 POINTS, LINES, PLANES, AND ANGLES

PROFILE IN MATHEMATICS

Euclid

Euclid (320–275 B.C.) lived in Alexandria, Egypt, and was a teacher and scholar at Alexandria's school called the *Museum.* It was here that Euclid collected and arranged many of the mathematical results known at the time. This collection of works became his 13-volume masterpiece known as the *Elements.* Beginning with a list of definitions, postulates, and axioms, Euclid proved one theorem after another, using only previously proven results. This method of proof became a model of mathematical and scientific investigation that survives today. Next to the *Bible,* Euclid's *Elements* may be the most translated, published, and studied of all the books produced in the Western world.

"It is the glory of geometry that from so few principles, fetched from without, . . . it is able to accomplish so much."

I. Newton

Human beings recognized shapes, sizes, and physical forms long before geometry was developed. Geometry as a science is said to have begun in the Nile Valley of ancient Egypt. The Egyptians used geometry to measure land and to build pyramids and other structures.

The word *geometry* is derived from two Greek words, *ge,* meaning earth, and *metron,* meaning measure. Thus geometry means "earth measure" or "measurement of the earth."

Unlike the Egyptians, the Greeks were interested in more than just the applied aspects of geometry. The Greeks attempted to apply their knowledge of logic to geometry. Thales of Miletus in about 600 B.C. was the first to be credited with using deductive methods to develop geometric concepts. Another outstanding Greek geometer, Pythagoras, continued the systematic development of geometry that Thales had begun.

In about 300 B.C., Euclid collected and summarized much of the Greek mathematics of his time. In a set of 13 books called *Elements,* Euclid laid the foundation for plane geometry, which is also called **Euclidean geometry**.

Euclid is credited with being the first mathematician to use the **axiomatic method** in developing a branch of mathematics. First, Euclid introduced **undefined terms** such as point, line, plane, and angle. He related these to physical space by such statements as "A line is length without breadth," so that we may intuitively understand them. Because such statements play no further role in his system, they constitute primitive or undefined terms.

Second, Euclid introduced certain **definitions**. The definitions are introduced when needed and are often based on the undefined terms. Some terms that Euclid introduced and defined include triangle, right angle, and hypotenuse.

Third, Euclid stated certain primitive propositions called **postulates** (now called **axioms***) about the undefined terms and definitions. The reader is asked to accept these statements as true on the basis of their "obviousness" and their relationship with the physical world. For example, the Greeks accepted all right angles as being equal, which is Euclid's fourth postulate.

Fourth, Euclid proved, using deductive reasoning (Section 1.1), other propositions called **theorems**. One theorem that Euclid proved is known as the Pythagorean theorem: "The sum of the areas of the squares constructed on the arms of a right triangle is equal to the area of the square constructed on the hypotenuse." He also proved that the sum of the angles of a triangle is 180°.

Using only 10 axioms, Euclid deduced 465 propositions (or theorems) in plane and solid geometry, number theory, and Greek geometrical algebra.

POINT AND LINE

Three basic terms in geometry are **point**, **line**, and **plane**. These three terms are not given a formal definition, but we recognize points, lines, and planes when we see them.

Let's consider some properties of a line. Assume that a line means a straight line unless otherwise stated.

*The concept of the axiom has changed significantly since Euclid's time. Now any statement may be designated as an axiom, whether it is self-evident or not. All axioms are *accepted* as true. A set of axioms forms the foundation for a mathematical system.

DID YOU KNOW

Ninth Grader Writes a Theorem

In 1994, the ninth-grade geometry class at Patapsco High School, Maryland, was given the assignment of proving Marion's theorem. The theorem states that, if the sides of a triangle are divided into thirds and lines are drawn from these division points to the opposite vertices, a hexagon is formed inside the triangle whose area is one-tenth the area of the triangle. Ryan Morgan, a 15-year-old student in the class, after finishing the proof, kept looking for "something neat" in the triangle. After much looking, he discovered that Marion's theorem could be extended in the following way. "When the sides of a triangle are divided by an odd number greater than 1, and lines are drawn from the division points on the sides of the triangle to the opposite vertices, there will always be a hexagon in the interior of the triangle. The area of that hexagon will always be a predictable fraction of the area of the larger triangle." The fraction is determined by a complex formula worked out by Ryan. When Ryan first discovered this, he called it Morgan's conjecture. However, now that many mathematicians, even Ryan himself, have proven his conjecture, it is now referred to as Morgan's theorem.

1. A line is a set of points. Each point is on the line and the line passes through each point.
2. Any two distinct points determine a unique line. Figure 9.1(a) illustrates a line containing three points. The arrows at both ends of the line indicate that the line continues in each direction. The line in Fig. 9.1(a) may be symbolized with any two points on the line—for example, $\overleftrightarrow{AB}$, $\overleftrightarrow{BA}$, $\overleftrightarrow{AC}$, $\overleftrightarrow{CA}$, $\overleftrightarrow{BC}$, or $\overleftrightarrow{CB}$.

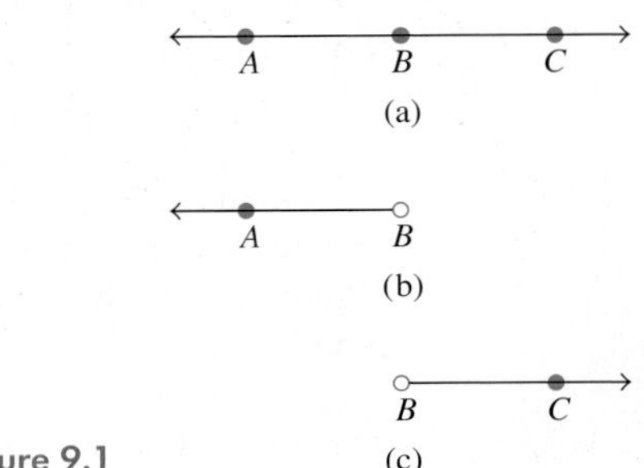

Figure 9.1

3. Any point on a line separates the line into three parts: the point itself and two **half lines** (neither of which includes the point). For example, in Fig. 9.1(a) point B separates the line into the point B and two half lines. Half line BA, symbolized $\overset{\circ\!\!\rightarrow}{BA}$, is illustrated in Fig. 9.1(b). The open circle above the B indicates that point B is not included in the half line. Figure 9.1(c) illustrates half line BC, symbolized $\overset{\circ\!\!\rightarrow}{BC}$.

Look at the half line $\overset{\circ\!\!\rightarrow}{AB}$ in Fig. 9.2(b). If the **end point**, A, is included with the set of points on the half line, the result is called a **ray**. Ray AB, symbolized $\overrightarrow{AB}$, is illustrated in Fig. 9.2(c). Ray BA is illustrated in Fig. 9.2(d).

A **line segment** is that part of a line between two points, including the end points. Line segment AB, symbolized $\overline{AB}$, is illustrated in Fig. 9.2(e).

An open line segment is the set of points on a line between two points, excluding the end points. Open line segment AB, symbolized $\overset{\circ\!-\!\circ}{AB}$, is illustrated in Fig. 9.2(f).

Description	Diagram	Symbol
(a) Line AB	A ●——● B (arrows both ends)	$\overleftrightarrow{AB}$
(b) Half line AB	A ○——● B →	$\overset{\circ\!\!\rightarrow}{AB}$
(c) Ray AB	A ●——● B →	$\overrightarrow{AB}$
(d) Ray BA	← A ●——● B	$\overrightarrow{BA}$
(e) Line segment AB	A ●——● B	$\overline{AB}$
(f) Open line segment AB	A ○——○ B	$\overset{\circ\!-\!\circ}{AB}$
(g) Half open line segments AB	A ●——○ B; A ○——● B	$\overset{-\!\circ}{AB}$; $\overset{\circ\!-}{AB}$

Figure 9.2

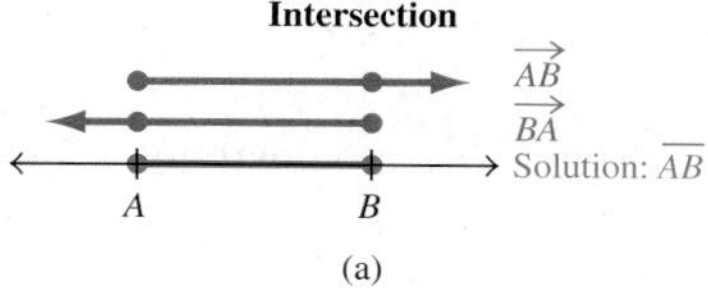

(a)

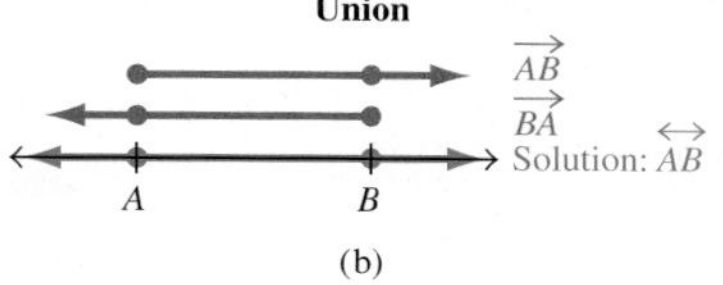

(b)

Figure 9.3

Figure 9.2(g) illustrates two half open line segments, symbolized $\overset{\circ}{\overline{AB}}$ and $\overset{\circ}{\overline{AB}}$.

In Chapter 2 we discussed intersection of sets. Recall that the intersection ($\cap$) of two sets is the set of elements (points in this case) common to both sets.

Consider the lines in Fig. 9.3(a). The intersection of $\overrightarrow{AB}$ and $\overrightarrow{BA}$ is $\overline{AB}$. Thus, $\overrightarrow{AB} \cap \overrightarrow{BA} = \overline{AB}$.

We also discussed the union of two sets in Chapter 2. The union ($\cup$) of two sets is the set of elements (points in this case) that belong to either of the sets or both sets. The union of $\overrightarrow{AB}$ and $\overrightarrow{BA}$ is $\overleftrightarrow{AB}$ (Fig. 9.3b). Thus, $\overrightarrow{AB} \cup \overrightarrow{BA} = \overleftrightarrow{AB}$.

EXAMPLE 1 *Unions and Intersections of Parts of a Line*

Using line *AD*, determine the solution to each part.

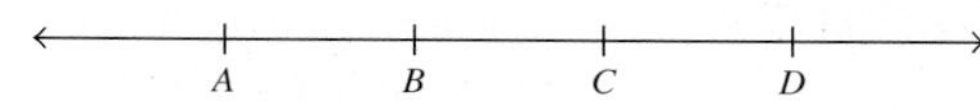

a) $\overrightarrow{AB} \cap \overrightarrow{DC}$ b) $\overrightarrow{AB} \cup \overrightarrow{DC}$ c) $\overline{AB} \cap \overrightarrow{CD}$ d) $\overline{AD} \cup \overset{\circ}{\overrightarrow{CA}}$

SOLUTION

a) $\overrightarrow{AB} \cap \overrightarrow{DC}$

Ray *AB* and ray *DC* are shown below. The intersection of these two rays is that part of line *AD* that is a part of *both* ray *AB and* ray *DC*. The intersection of ray *AB* and ray *DC* is line segment *AD*.

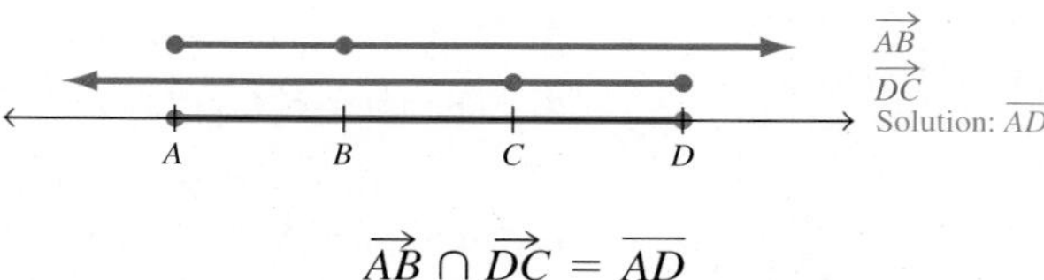

$$\overrightarrow{AB} \cap \overrightarrow{DC} = \overline{AD}$$

b) $\overrightarrow{AB} \cup \overrightarrow{DC}$

Once again ray *AB* and ray *DC* are shown below. The union of these two rays is that part of line *AD* that is part of *either* ray *AB or* ray *DC*. The union of ray *AB* and ray *DC* is the entire line *AD*.

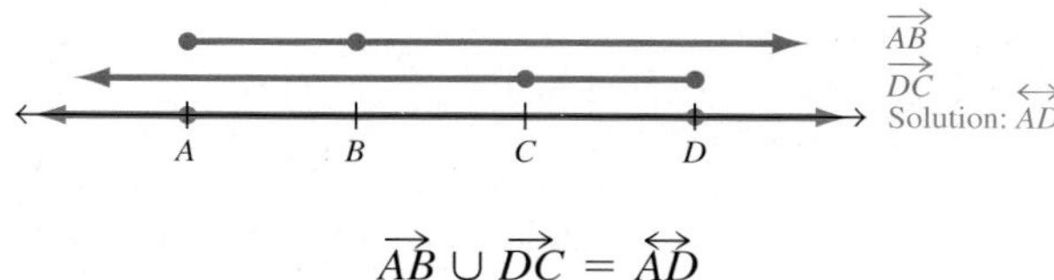

$$\overrightarrow{AB} \cup \overrightarrow{DC} = \overleftrightarrow{AD}$$

c) $\overline{AB} \cap \overrightarrow{CD}$

Line segment *AB* and ray *CD* have no points in common, so their intersection is empty.

$\overline{AB}$
$\overrightarrow{CD}$
Solution: ϕ
A B C D

$$\overline{AB} \cap \overrightarrow{CD} = \varnothing$$

DID YOU KNOW

Let There Be Light

Sunlight travels in straight lines, or rays. When these rays pass through a small hole of a darkened room, they converge and then spread out to form an inverted image of the sun on the opposite wall. In the sixteenth century, artists used lenses to focus images on paper, which could then be traced to create a reproduction. In the eighteenth century, chemists discovered that certain salts of silver would darken when exposed to light, effectively creating a negative image. For the population at large, this all came together in 1888 when 23-year-old George Eastman, using basic geometric concepts, developed the first modern camera intended for the "casual millions"—just $25 apiece.

d) $\overline{AD} \cup \overset{\circ\!\!\rightarrow}{CA}$

The union of line segment AD and half line CA is ray DA (or $\overrightarrow{DB}$ or $\overrightarrow{DC}$).

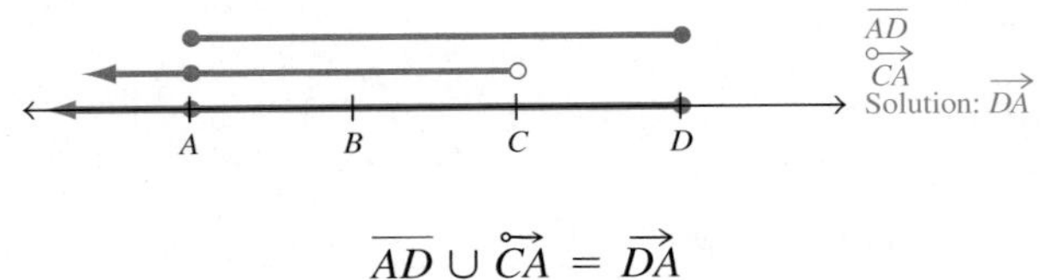

$$\overline{AD} \cup \overset{\circ\!\!\rightarrow}{CA} = \overrightarrow{DA}$$

PLANE

The term *plane* is one of Euclid's undefined terms. For our purposes, we can think of a plane as a two-dimensional surface that extends infinitely in both directions, like an infinitely large blackboard. Euclidean geometry is called **plane geometry** because it is the study of two-dimensional figures in a plane.

Two lines in the same plane that do not intersect are called **parallel lines**. Figure 9.4(a) illustrates two parallel lines in a plane ($\overleftrightarrow{AB}$ is parallel to $\overleftrightarrow{CD}$).

Properties of planes include the following:

1. Any three points that are not on the same line (noncollinear points) determine a unique plane (Fig. 9.4b).
2. A line in a plane divides the plane into three parts—the line and two half planes (Fig. 9.4c).
3. Any line and a point not on the line determine a unique plane.
4. The intersection of two planes is a line (Fig. 9.4d).

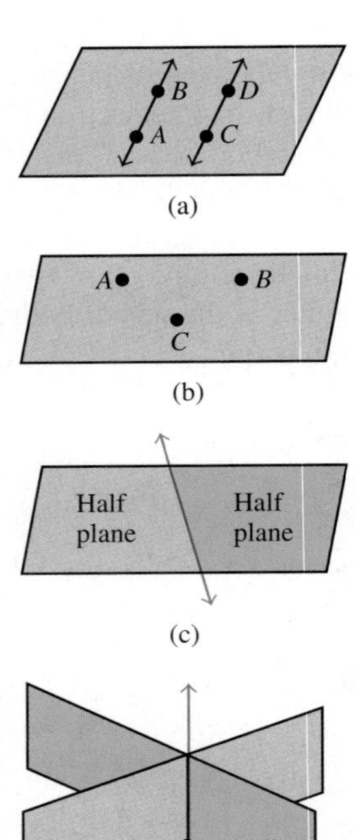

Figure 9.4

Two planes that do not intersect are said to be **parallel planes**. For example, in Fig. 9.5, plane ABE is parallel to plane GHF.

Two lines that do not lie in the same plane and do not intersect are called **skewed lines**. Figure 9.5 illustrates many skewed lines (for example, $\overleftrightarrow{AB}$ and $\overleftrightarrow{CD}$).

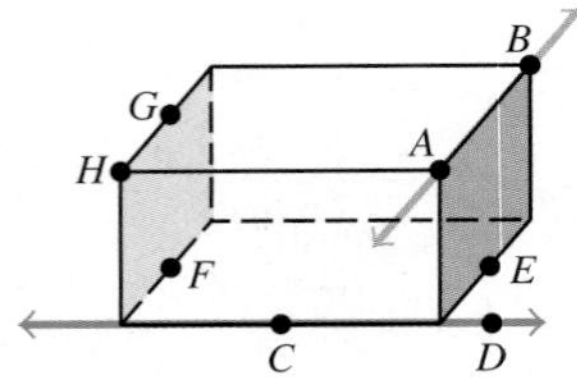

Figure 9.5

ANGLES

An **angle**, denoted $\measuredangle$, is the union of two rays with a common end point (Fig. 9.6),

$$\overrightarrow{BA} \cup \overrightarrow{BC} = \measuredangle ABC \text{ (or } \measuredangle CBA)$$

An angle can be formed by the rotation of a ray about a point. An angle has an initial side and a terminal side: The initial side indicates the position of the ray prior to rotation; the terminal side indicates the position of the ray after rotation. The point common to both rays is called the **vertex** of the angle. The letter designating the vertex is always the middle one of the three letters designating an angle. The rays that make up the angle are called its **sides**.

There are several ways to name an angle. The angle in Fig. 9.6 may be denoted

$$\measuredangle ABC, \quad \measuredangle CBA, \quad \text{or} \quad \measuredangle B$$

An angle divides a plane into three distinct parts: the angle itself, its interior, and its exterior. In Fig. 9.6 the angle is represented by the blue lines, the interior of the angle is shaded pink, and the exterior is shaded green.

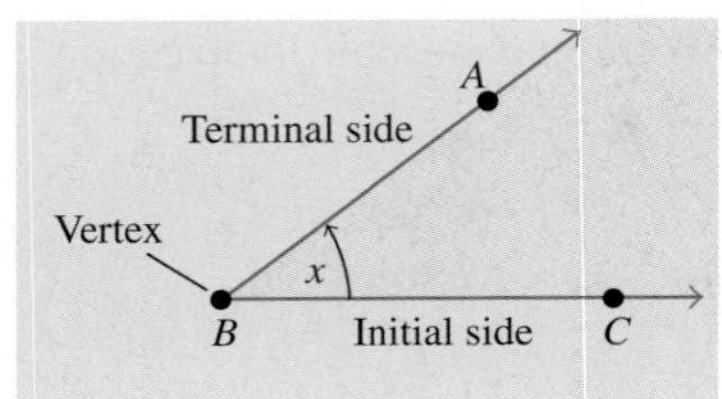

Figure 9.6

DID YOU KNOW

Are We There Yet?

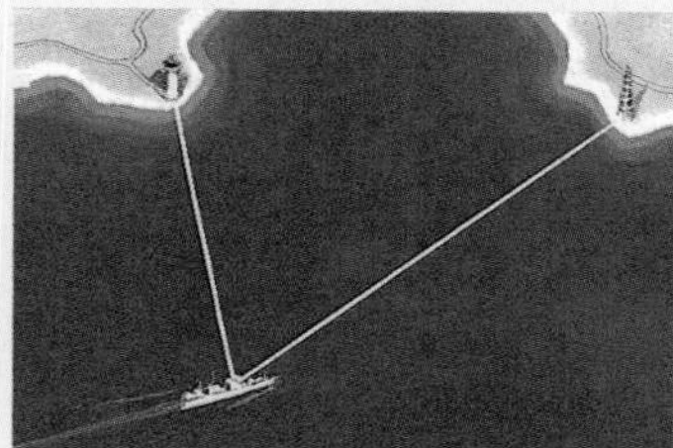

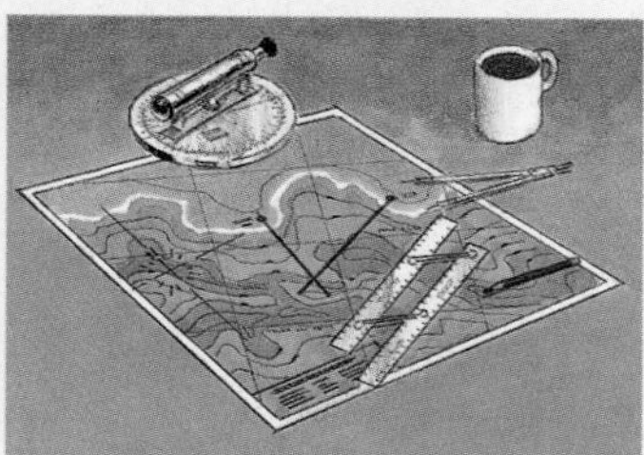

A navigator uses angles to determine a ship's position offshore. A compass fitted with a telescopic site is used to fix the angle, or bearing, of a landmark or buoy relative to a point on the compass, true north for example. Locating the landmark on a map, the navigator draws a line through this point at the given angle. The ship's position is somewhere along that line. A second bearing is taken from a different landmark, and a second line is drawn. The intersection of these two lines shows the ship's position.

EXAMPLE 2 ***Unions and Intersections of Angles***

Refer to Fig. 9.7. Find the following.

a) $\overrightarrow{BF} \cup \overrightarrow{BD}$ b) $\measuredangle ABF \cap \measuredangle FBE$ c) $\overleftrightarrow{AC} \cap \overleftrightarrow{DE}$ d) $\overrightarrow{BA} \cup \overrightarrow{BC}$

SOLUTION

a) $\overrightarrow{BF} \cup \overrightarrow{BD} = \measuredangle FBD$
b) $\measuredangle ABF \cap \measuredangle FBE = \overrightarrow{BF}$
c) $\overleftrightarrow{AC} \cap \overleftrightarrow{DE} = \{B\}$
d) $\overrightarrow{BA} \cup \overrightarrow{BC} = \overleftrightarrow{AC}$

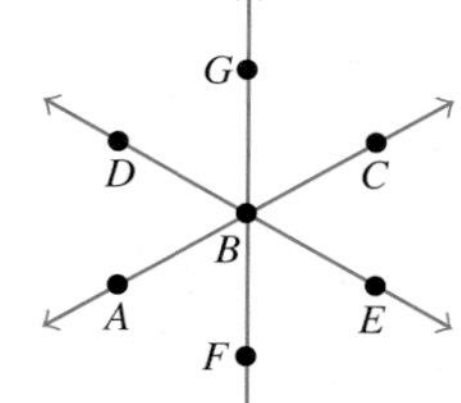

Figure 9.7

The **measure of an angle**, symbolized m, is the amount of rotation from its initial side to its terminal side. In Fig. 9.6, the letter x represents the measure of $\measuredangle ABC$; therefore we may write $m\measuredangle ABC = x$.

Angles can be measured in **degrees**, radians, or gradients. In this text we will discuss only the degree unit of measurement. The symbol for degrees is the same as the symbol for temperature degrees. An angle of 45 degrees is written 45°. A *protractor* is used to measure angles. The angle shown being measured by the protractor in Fig. 9.8 is 50°.

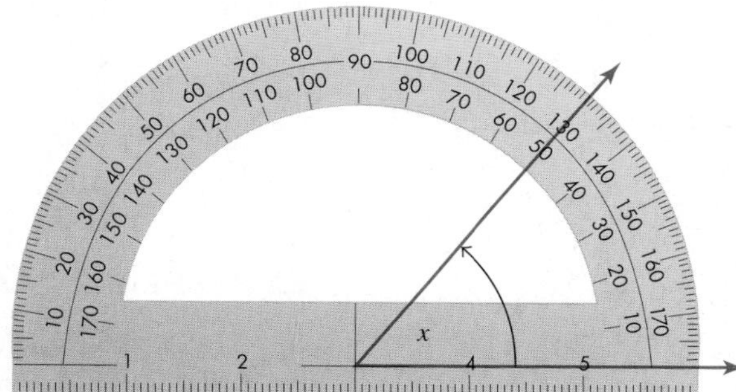

Figure 9.8

Consider a circle whose circumference is divided into 360 equal parts. If we draw a line from each mark on the circumference to the center of the circle, we get 360 wedge-shaped pieces. The measure of an angle formed by the straight sides of each wedge-shaped piece is defined to be 1°.

Angles are classified by their degree measurement, as shown in the following summary. A **right angle** is 90°; an **acute angle** is less than 90°; an **obtuse angle** is greater than 90° but less than 180°; a **straight angle** is 180°.

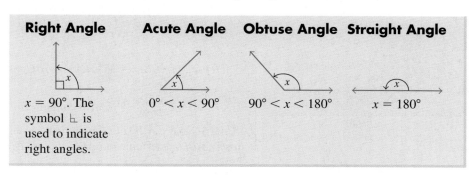

Right Angle	**Acute Angle**	**Obtuse Angle**	**Straight Angle**
$x = 90°$. The symbol ⊾ is used to indicate right angles.	$0° < x < 90°$	$90° < x < 180°$	$x = 180°$

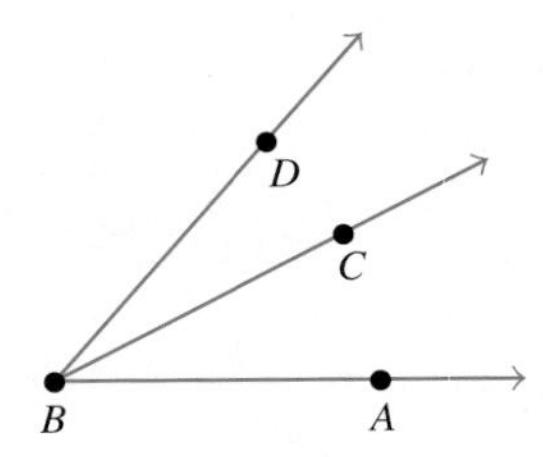

Figure 9.9

Two angles in the same plane are **adjacent angles** when they have a common vertex and a common side but no common interior points. $\measuredangle DBC$ and $\measuredangle CBA$ in Fig. 9.9 are adjacent angles. However, $\measuredangle DBA$ and $\measuredangle CBA$ are not adjacent angles.

Two angles are called **complementary angles** if the sum of their measures is 90°. Two angles are called **supplementary angles** if the sum of their measures is 180°.

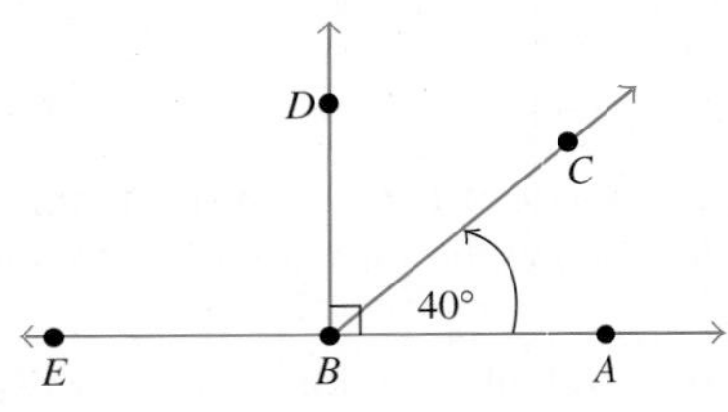

Figure 9.10

EXAMPLE 3 ***Finding Complementary and Supplementary Angles***

In Fig. 9.10 we see $m\measuredangle ABC = 40°$.

a) $\measuredangle ABC$ and $\measuredangle CBD$ are complementary angles, determine $m\measuredangle CBD$.
b) $\measuredangle ABC$ and $\measuredangle CBE$ are supplementary angles, determine $m\measuredangle CBE$.

SOLUTION

a) The sum of two complementary angles must be 90°, so

$$m\measuredangle ABC + m\measuredangle CBD = 90°$$
$$40° + m\measuredangle CBD = 90° \quad \text{Subtract 40° from each side of the equation.}$$
$$m\measuredangle CBD = 90° - 40° = 50°$$

b) The sum of two supplementary angles must be 180°, so

$$m\measuredangle ABC + m\measuredangle CBE = 180°$$
$$40° + m\measuredangle CBE = 180° \quad \text{Subtract 40° from each side of the equation.}$$
$$m\measuredangle CBE = 180° - 40° = 140°$$

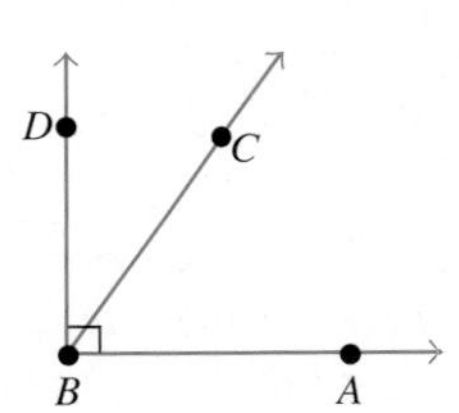

Figure 9.11

EXAMPLE 4 ***Problem-Solving Involving Complementary Angles***

If $\measuredangle ABC$ and $\measuredangle CBD$ are complementary angles and $m\measuredangle ABC$ is 20° greater than $m\measuredangle CBD$, determine the measure of each angle (Fig. 9.11).

SOLUTION Let $m\measuredangle CBD = x$. Then $m\measuredangle ABC = x + 20$ since it is 20° greater than $m\measuredangle CBD$.

Since these angles are complementary, we have

$$m\measuredangle CBD + m\measuredangle ABC = 90°$$
$$x + (x + 20°) = 90°$$
$$2x + 20° = 90°$$
$$2x = 70°$$
$$x = 35°$$

Therefore, $m\measuredangle CBD = 35°$ and $m\measuredangle ABC = 35° + 20°$, or 55°.

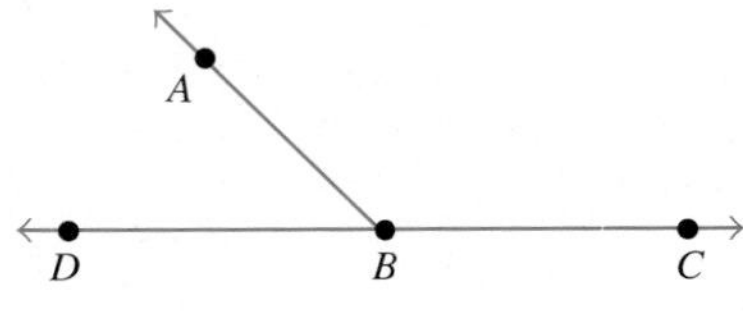

Figure 9.12

EXAMPLE 5 ***Problem-Solving Involving Supplementary Angles***

If $\measuredangle ABC$ and $\measuredangle ABD$ are supplementary angles and $m\measuredangle ABC$ is three times as large as $m\measuredangle ABD$, determine $m\measuredangle ABC$ and $m\measuredangle ABD$ (Fig. 9.12).

SOLUTION Let $m\measuredangle ABD = x$, then $m\measuredangle ABC = 3x$.

$$\begin{aligned} m\measuredangle ABC + m\measuredangle ABD &= 180^\circ \\ 3x + x &= 180^\circ \\ 4x &= 180^\circ \\ x &= 45^\circ \end{aligned}$$

Thus, $m\measuredangle ABD = 45^\circ$, and $m\measuredangle ABC = (3)(45^\circ) = 135^\circ$. ▲

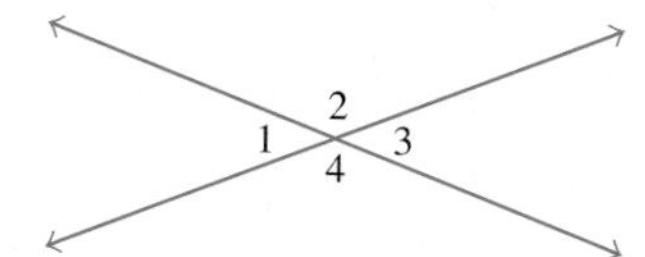

Figure 9.13

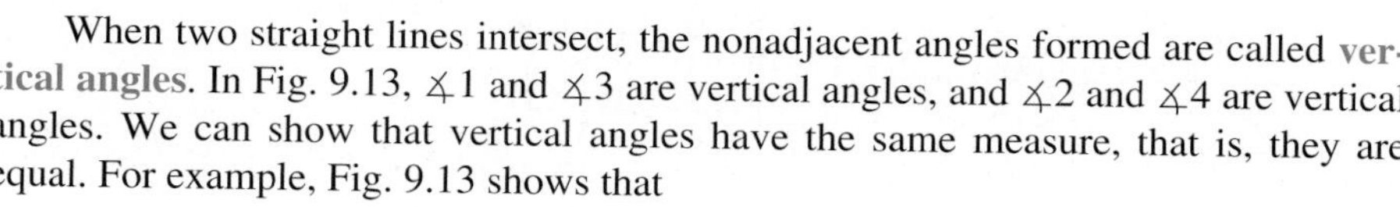
When two straight lines intersect, the nonadjacent angles formed are called **vertical angles**. In Fig. 9.13, $\measuredangle 1$ and $\measuredangle 3$ are vertical angles, and $\measuredangle 2$ and $\measuredangle 4$ are vertical angles. We can show that vertical angles have the same measure, that is, they are equal. For example, Fig. 9.13 shows that

$$m\measuredangle 1 + m\measuredangle 2 = 180^\circ. \qquad \text{Why?}$$
$$m\measuredangle 2 + m\measuredangle 3 = 180^\circ. \qquad \text{Why?}$$

Since $\measuredangle 2$ has the same measure in both cases, $m\measuredangle 1$ must equal $m\measuredangle 3$.

Vertical angles have the same measure.

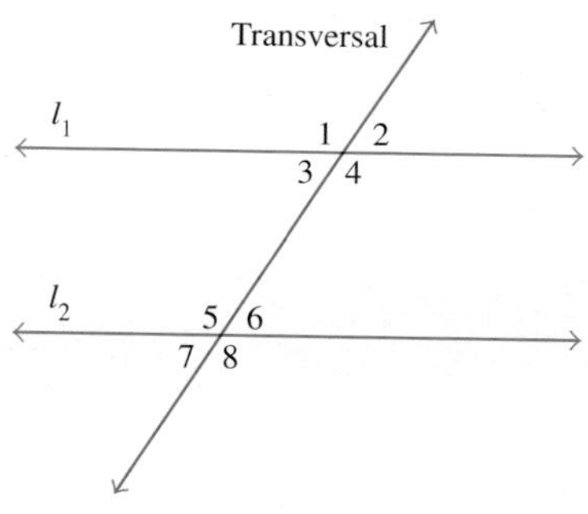

Figure 9.14

A line that intersects two different lines, l_1 and l_2, at two different points is called a **transversal**. Figure 9.14 illustrates that when two parallel lines are cut by a transversal, eight angles are formed. Angles 3, 4, 5, and 6 are called **interior angles**, and angles 1, 2, 7, and 8 are called **exterior angles**. Eight pairs of supplementary angles are formed. Can you list them?

Special names are given to the angles formed by a transversal crossing two parallel lines.

Name	Description	Illustration	Pairs of angles meeting criteria
Alternate interior angles	Interior angles on opposite sides of the transversal	1 2 3 4 5 6 7 8	$\measuredangle 3$ and $\measuredangle 6$ $\measuredangle 4$ and $\measuredangle 5$
Alternate exterior angles	Exterior angles on opposite sides of the transversal	1 2 3 4 5 6 7 8	$\measuredangle 1$ and $\measuredangle 8$ $\measuredangle 2$ and $\measuredangle 7$
Corresponding angles	One interior and one exterior angle on the same side of the transversal	1 2 3 4 5 6 7 8	$\measuredangle 1$ and $\measuredangle 5$ $\measuredangle 2$ and $\measuredangle 6$ $\measuredangle 3$ and $\measuredangle 7$ $\measuredangle 4$ and $\measuredangle 8$

When two parallel lines are cut by a transversal

1. Alternate interior angles have the same measure.
2. Alternate exterior angles have the same measure.
3. Corresponding angles have the same measure.

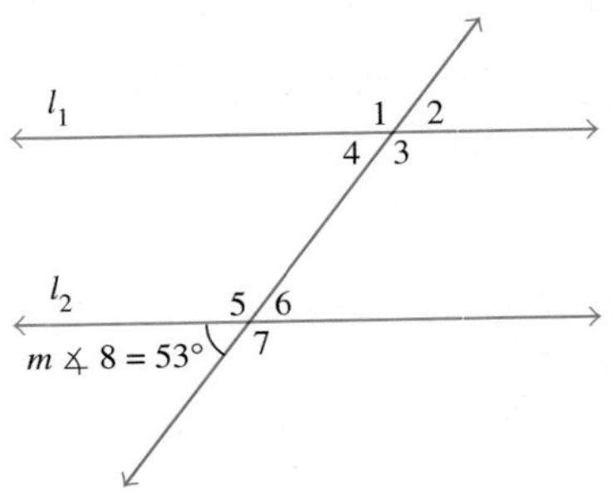

Figure 9.15

EXAMPLE 6 *Finding Angle Measures*

Figure 9.15 shows two parallel lines cut by a transversal. Determine the measure of $\measuredangle 1$ through $\measuredangle 7$.

SOLUTION

$m\measuredangle 6 = 53°$ — $\measuredangle 8$ and $\measuredangle 6$ are vertical angles.

$m\measuredangle 5 = 127°$ — $\measuredangle 8$ and $\measuredangle 5$ are supplementary angles.

$m\measuredangle 7 = 127°$ — $\measuredangle 5$ and $\measuredangle 7$ are vertical angles.

$m\measuredangle 1 = 127°$ — $\measuredangle 1$ and $\measuredangle 7$ are alternate exterior angles.

$m\measuredangle 4 = 53°$ — $\measuredangle 4$ and $\measuredangle 6$ are alternate interior angles.

$m\measuredangle 2 = 53°$ — $\measuredangle 6$ and $\measuredangle 2$ are corresponding angles.

$m\measuredangle 3 = 127°$ — $\measuredangle 3$ and $\measuredangle 1$ are vertical angles.

SECTION 9.1 EXERCISES

CONCEPT/WRITING EXERCISES

1. What is the difference between an axiom and a theorem?
2. a) What are the four key parts in the axiomatic method used by Euclid?
   b) Discuss each of the four parts.
3. What are parallel lines?
4. What are skewed lines?
5. What are adjacent angles?
6. What are complementary angles?
7. What are supplementary angles?
8. What is a right angle?
9. What is an acute angle?
10. What is an obtuse angle?
11. What is a straight angle?
12. Draw two intersecting lines. Identify the two pairs of vertical angles.

PRACTICE THE SKILLS

In Exercises 13–20, identify the figure as a line, half line, ray, line segment, open line segment, or half open line segment. Denote it by its appropriate symbol.

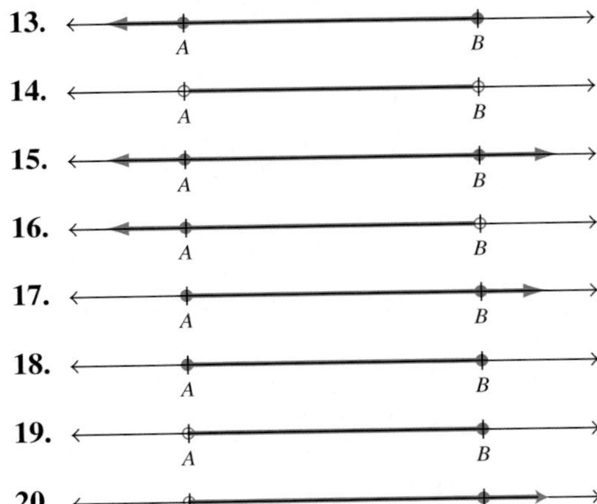

In Exercises 21–32, use the figure to find the following:

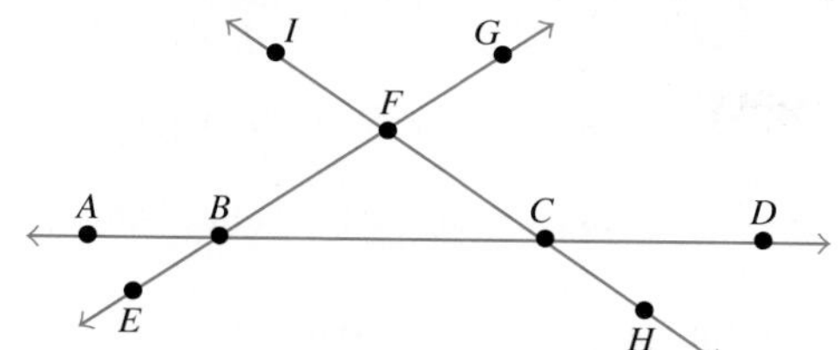

21. $\overset{\circ\!\!\longrightarrow}{FE} \cup \overrightarrow{FG}$
22. $\overrightarrow{FB} \cup \overrightarrow{FC}$
23. $\overleftrightarrow{AD} \cup \overline{BC}$
24. $\overset{\relbar\!\relbar\!\circ}{BC} \cup \overset{\relbar\!\relbar\!\circ}{CD}$
25. $\overleftrightarrow{BA} \cap \overset{\circ\!\!\longrightarrow}{BE}$
26. $\overset{\circ\!\!\longrightarrow}{AB} \cap \overrightarrow{HC}$
27. $\measuredangle HCD \cap \measuredangle ACF$
28. $\overleftrightarrow{AD} \cap \overline{BC}$
29. $\overset{\circ\!\!\longrightarrow}{BD} \cap \overset{\circ\!\!\longrightarrow}{CB}$
30. $\overset{\circ\!\!\longrightarrow}{BD} \cup \overset{\circ\!\!\longrightarrow}{CB}$
31. $\overline{BC} \cup \overline{CF} \cup \overline{FB}$
32. $\{C\} \cap \overset{\circ\!\!\longrightarrow}{CH}$

In Exercises 33–44, use the figure to find each of the following:

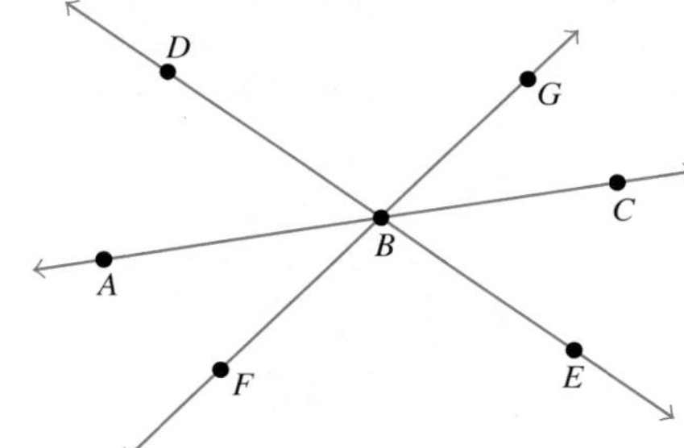

33. $\overrightarrow{BD} \cup \overset{\relbar\!\relbar\!\circ}{BE}$
34. $\overset{\circ\!\!\longrightarrow}{BF} \cap \overrightarrow{BE}$
35. $\overleftrightarrow{DE} \cup \overset{\circ\!\!\longrightarrow}{BE}$
36. $\measuredangle GBC \cap \measuredangle CBE$
37. $\overset{\circ\!\!\longrightarrow}{BF} \cup \overrightarrow{BE}$
38. $\measuredangle ABE \cup \overset{\circ\!\relbar\!\relbar\!\circ}{AB}$
39. $\{B\} \cap \overrightarrow{BA}$
40. $\measuredangle CBE \cap \measuredangle EBC$
41. $\overset{\circ\!\relbar\!\relbar\!\circ}{AC} \cap \overset{\circ\!\relbar\!\relbar\!\circ}{BE}$
42. $\overset{\circ\!\relbar\!\relbar\!\circ}{AC} \cap \overline{AC}$
43. $\overline{GF} \cap \overline{AB}$
44. $\overrightarrow{EB} \cap \overset{\relbar\!\relbar\!\circ}{BE}$

In Exercises 45–52, classify the angle as acute, right, straight, obtuse, or none of these.

45. 

46. 

47. 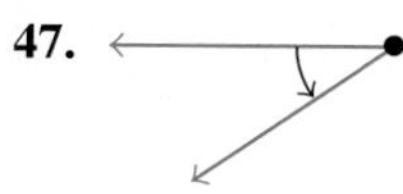

48. 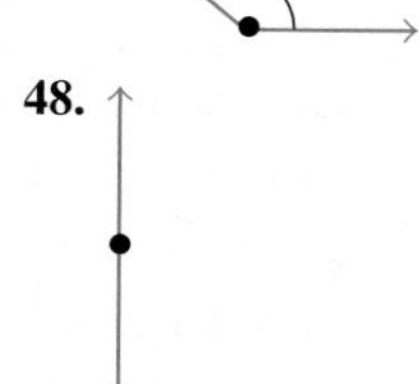

49.  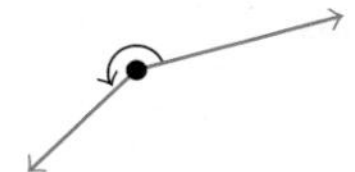

50. 

51. 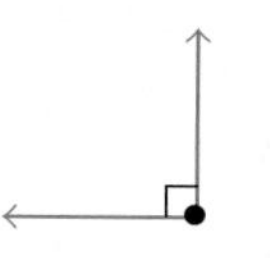

52. 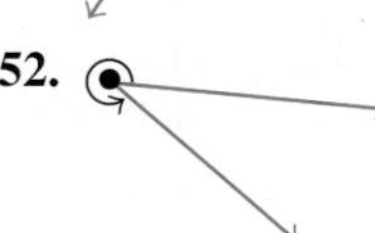

In Exercises 53–58, find the complementary angle of the given angle.

53. 51°
54. 73°
55. $25\frac{1}{2}°$
56. 17.4°
57. 89°
58. $15\frac{1}{8}°$

In Exercises 59–64, find the supplementary angle of the given angle.

59. 76°
60. 19°
61. 135°
62. $2\frac{5}{8}°$
63. $99\frac{1}{5}°$
64. 156.8°

In Exercises 65–70, match the name of the angles with the corresponding figure in parts (a)–(f).

65. Vertical angles
66. Supplementary angles
67. Complementary angles
68. Alternate interior angles
69. Alternate exterior angles
70. Corresponding angles

a)

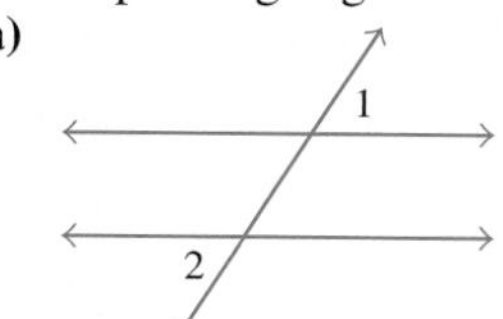

b)

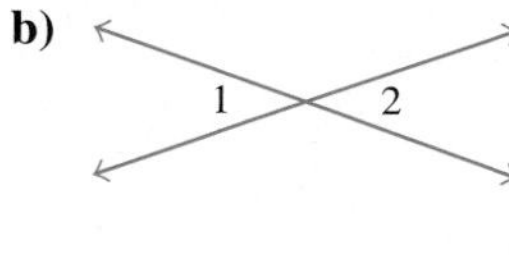

c)

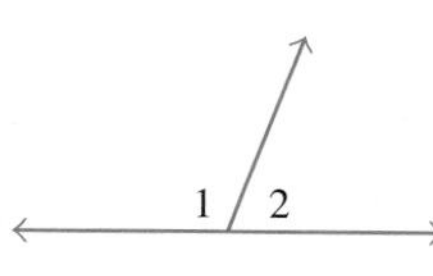

d)

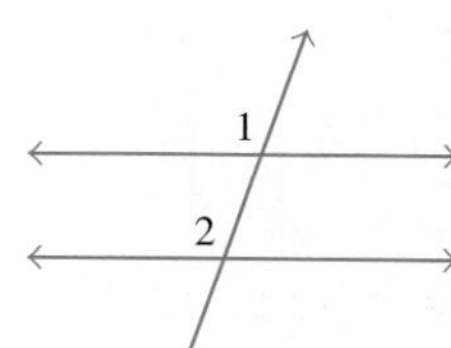

e)

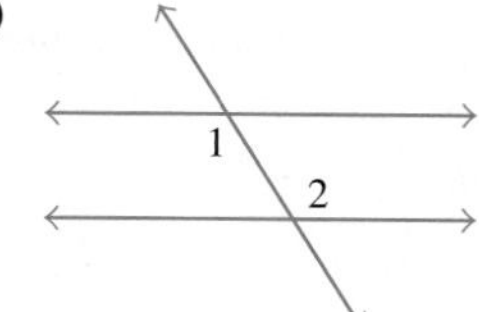

f)

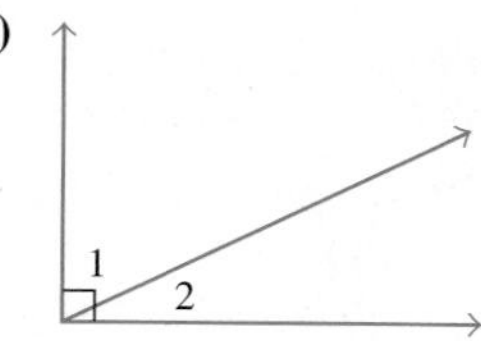

PROBLEM SOLVING

71. MODELING If $\measuredangle 1$ and $\measuredangle 2$ are complementary angles and $\measuredangle 1$ is seven times as large as $\measuredangle 2$, find the measures of $\measuredangle 1$ and $\measuredangle 2$.

72. MODELING The difference between the measures of two complementary angles is 24°. Determine the measures of the two angles.

73. MODELING The difference between the measures of two supplementary angles is 74°. Determine the measures of the two angles.

74. MODELING If $\measuredangle 1$ and $\measuredangle 2$ are supplementary angles and $\measuredangle 1$ is eight times as large as $\measuredangle 2$, find the measures of $\measuredangle 1$ and $\measuredangle 2$.

In Exercises 75–78, parallel lines are cut by the transversal shown. Determine the measures of ∡1 through ∡7.

75.

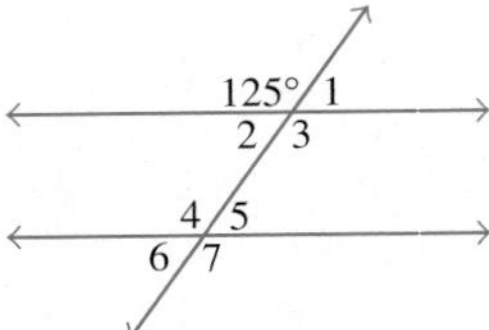

76.

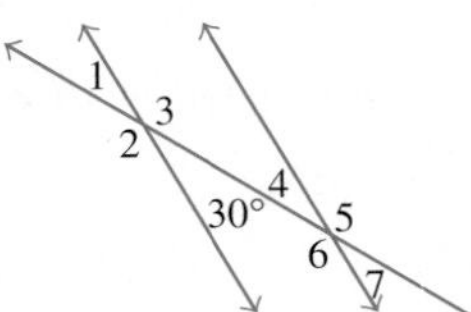

77.

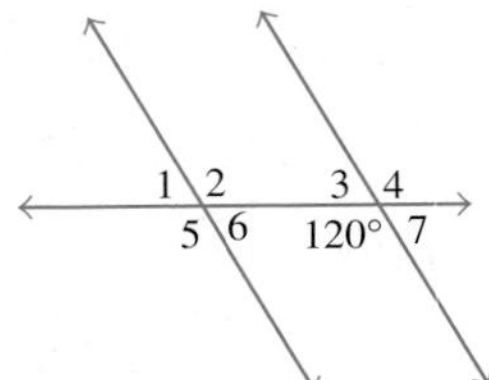

78.

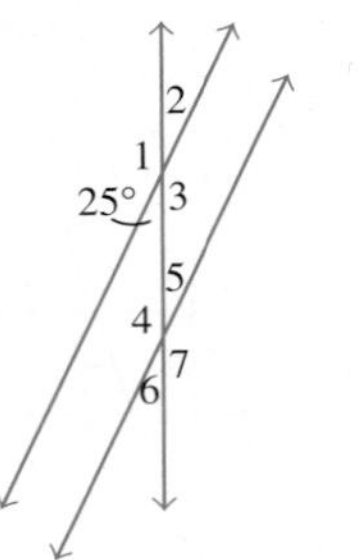

In Exercises 79–82, the angles are complementary angles. Find the measures of ∡1 and ∡2.

79.

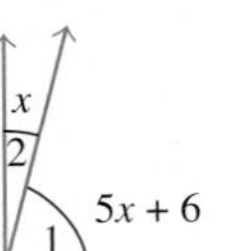

80.

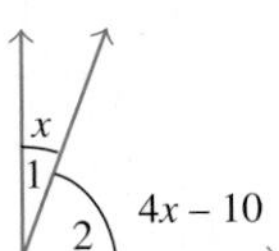

81.

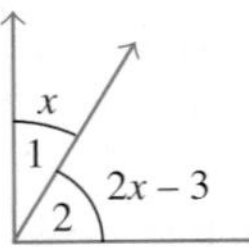

82.

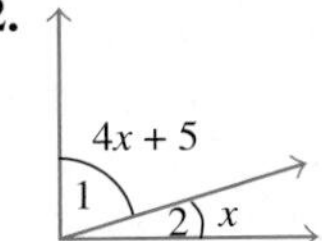

In Exercises 83–86, the angles are supplementary angles. Find the measures of ∡1 and ∡2.

83.

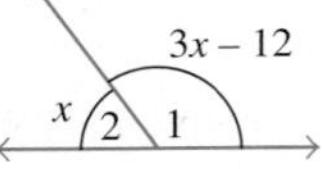

84.

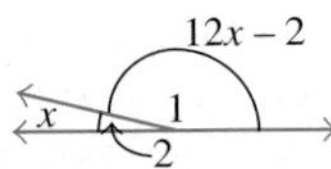

85. 5x – 18 2 1 x

86.

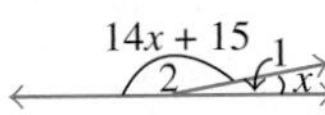

87. **a)** How many lines can be drawn through a given point?
b) How many planes can be drawn through a given point?

88. What is the intersection of two distinct nonparallel planes?

89. How many planes can be drawn through a given line?

90. **a)** Will three noncollinear points A, B, and C always determine a plane? Explain.
b) Is it possible to determine more than one plane with three noncollinear points? Explain.
c) How many planes can be constructed through three collinear points?

The figure suggests a number of lines and planes. The lines may be described by naming two points, and the planes may be described by naming three points. In Exercises 91–96, use the figure to name the following:

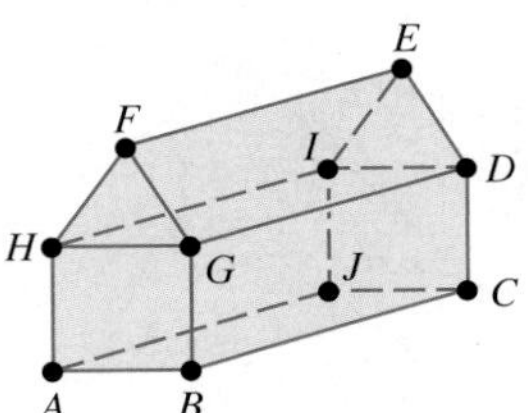

91. A pair of planes whose intersection is a line.

92. Three planes whose intersection is a single point.

93. Three planes whose intersection is a line.

94. A line and a plane whose intersection is a point.

95. A line and a plane whose intersection is a line.

96. Three lines that intersect at a single point.

In Exercises 97–102, determine whether the statement is always true, sometimes true, or never true. Explain your answer.

97. Two lines that are both parallel to a third line must be parallel to each other.

98. A triangle contains two acute angles.

99. Vertical angles are complementary angles.

100. Alternate exterior angles are supplementary angles.

101. Alternate interior angles are complementary angles.

102. A triangle contains two obtuse angles.

CHALLENGE PROBLEMS/GROUP ACTIVITIES

103. Suppose that you have three distinct lines, all lying in the same plane. Find all the possible ways in which the three lines can be related. Sketch each case (four cases).

104. Suppose that you have three distinct planes in space. Find all the possible ways these planes can be related. Sketch each case (five cases).

105. Suppose that you have three distinct lines in space (not necessarily in the same plane). Find all the possible ways in which these three lines can be related. Sketch each case (nine cases).

RESEARCH ACTIVITY

106. Write a paper on Euclid's contributions to geometry.

9.2 POLYGONS

A **polygon** is a closed figure in a plane determined by three or more straight line segments. Examples of polygons are given in Fig. 9.16.

The straight line segments that form the polygon are called its **sides** and a point where two sides meet is called a **vertex** (plural **vertices**). The union of the sides of a polygon and its interior is called a **polygonal region**. A **regular polygon** is one whose sides are all the same length and whose interior angles all have the same measure. Figures 9.16(b) and (d) are regular polygons.

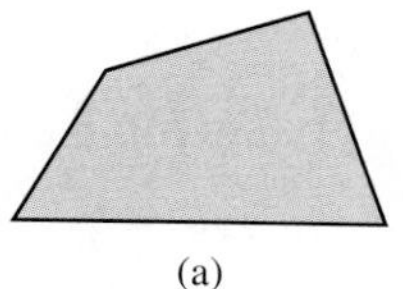
(a)

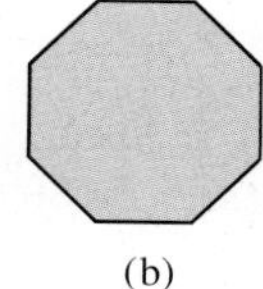
(b)

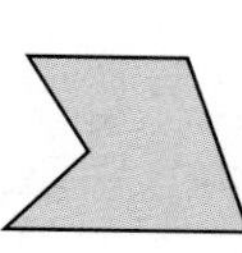
(c)

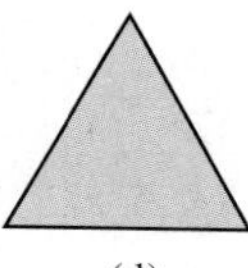
(d)

Figure 9.16

Polygons are named according to their number of sides. The names of some polygons are given in Table 9.1.

Table 9.1

Number of sides	Name	Number of sides	Name
3	Triangle	8	Octagon
4	Quadrilateral	9	Nonagon
5	Pentagon	10	Decagon
6	Hexagon	12	Dodecagon
7	Heptagon	20	Icosagon

One of the most important polygons is the triangle. The sum of the measures of the interior angles of a triangle is 180°. To illustrate, consider triangle *ABC* given in Fig. 9.17. The triangle is formed by drawing two transversals through two parallel lines l_1 and l_2 with the two transversals intersecting at a point on l_1.

In Fig. 9.17, notice that $\measuredangle A$ and $\measuredangle A'$ are corresponding angles. Recall from Section 9.1 that corresponding angles are equal, so $m\measuredangle A = m\measuredangle A'$. Also, $\measuredangle C$ and $\measuredangle C'$ are corresponding angles, therefore, $m\measuredangle C = m\measuredangle C'$. Next, we notice that $\measuredangle B$ and $\measuredangle B'$ are vertical angles. In Section 9.1 we learned that vertical angles are equal, therefore, $m\measuredangle B = m\measuredangle B'$. Figure 9.17 shows $\measuredangle A'$, $\measuredangle B'$, and $\measuredangle C'$ form a straight angle, therefore, $m\measuredangle A' + m\measuredangle B' + m\measuredangle C' = 180°$. Since $m\measuredangle A = m\measuredangle A'$, $m\measuredangle B = m\measuredangle B'$, and $m\measuredangle C = m\measuredangle C'$, we can reason that $m\measuredangle A + m\measuredangle B + m\measuredangle C = 180°$. This illustrates that the sum of the interior angles of a triangle is 180°.

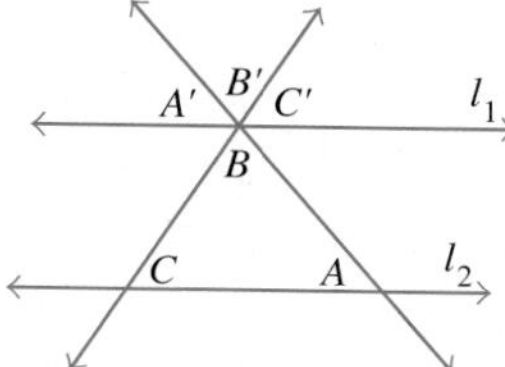

Figure 9.17

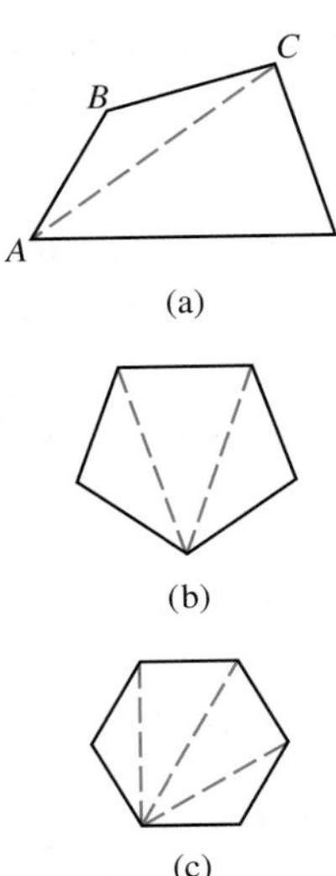

Figure 9.18

Consider the quadrilateral *ABCD* (Fig. 9.18a). Drawing a straight line segment between any two vertices forms two triangles. Since the sum of the measures of the angles of a triangle is 180°, the sum of the measures of the interior angles of a quadrilateral is $2 \cdot 180°$, or 360°.

Now let's examine a pentagon (Fig. 9.18b). We can draw two straight line segments to form three triangles. Thus, the sum of the measures of the interior angles of a five-sided figure is $3 \cdot 180°$, or 540°. Figure 9.18(c) shows that four triangles can be drawn in a six-sided figure. Table 9.2 summarizes this information.

Table 9.2

Sides	Triangles	Sum of the measures of the interior angles
3	1	$1(180°) = 180°$
4	2	$2(180°) = 360°$
5	3	$3(180°) = 540°$
6	4	$4(180°) = 720°$

If we continue this procedure, we can see that for an n-sided polygon the sum of the measures of the interior angles is $(n - 2)180°$.

The **sum** of the measures of the interior angles of an n-sided polygon is $(n - 2)180°$.

EXAMPLE 1 *Angles of a Stop Sign*

A stop sign is in the shape of a regular octagon (Fig. 9.19). Determine (a) the measure of an interior angle and (b) the measure of exterior angle 1.

Figure 9.19

SOLUTION

a) Using the formula $(n - 2)180°$, we can determine the sum of the measures of the interior angles of an octagon, as follows.

$$\begin{aligned} \text{Sum} &= (8 - 2)180° \\ &= 6(180°) \\ &= 1080° \end{aligned}$$

The measure of an interior angle of a regular polygon can be determined by dividing the sum of the interior angles by the number of angles.

The measure of an interior angle of a regular octagon is found as follows:

$$\text{Measure} = \frac{1080°}{8} = 135°$$

b) Since $\measuredangle 1$ is the supplement of an interior angle,

$$m\measuredangle 1 = 180° - 135° = 45°$$

In order to discuss area in the next section we must be able to identify various types of triangles and quadrilaterals. The following is a summary of certain types of triangles and their characteristics.

Acute Triangle	**Obtuse Triangle**	**Right Triangle**
All angles are acute angles.	One angle is obtuse.	One angle is a right angle.
Isosceles Triangle	**Equilateral Triangle**	**Scalene Triangle**
Two equal sides Two equal angles	Three equal sides Three equal angles (60° each)	No two sides are equal in length.

hypotenuse

SIMILAR FIGURES

Figure 9.20

In everyday living we often have to deal with geometric figures that have the "same shape" but are of different sizes. For example, an architect will make a small-scale drawing of a floor plan, or a photographer will make an enlargement of a photograph. Figures that have the same shape but may be of different sizes are called **similar figures**. Two similar figures are illustrated in Fig. 9.20.

Similar figures have *corresponding angles* and *corresponding sides*. In Fig. 9.20 triangle ABC has angles A, B, and C. Their respective corresponding angles in triangle DEF are angles D, E, and F. Sides AB, BC, and AC in triangle ABC have corresponding sides DE, EF, and DF, respectively, in triangle DEF.

> Two polygons are **similar** if their corresponding angles have the same measure and their corresponding sides are in proportion.

In Figure 9.20, $\measuredangle A$ and $\measuredangle D$ have the same measure, $\measuredangle B$ and $\measuredangle E$ have the same measure, and $\measuredangle C$ and $\measuredangle F$ have the same measure. Also, the corresponding sides of similar triangles are in proportion:

$$\frac{\overline{AB}}{\overline{DE}} = \frac{\overline{CB}}{\overline{FE}} = \frac{\overline{AC}}{\overline{DF}}.$$

EXAMPLE 2 *Similar Figures*

Consider the similar figures in Fig. 9.21.

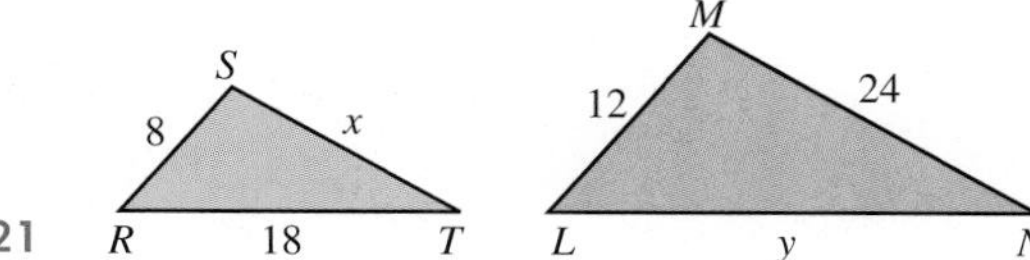

Figure 9.21

Determine the following:

a) The length of side $\overline{ST}$.

b) The length of side $\overline{LN}$.

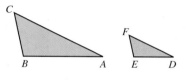

DID YOU KNOW

Polygons in Patterns

An interesting application of polygons is in the design of tiling patterns—patterns that must fit together to cover completely a flat surface. This pattern can be found in the Taj Mahal.

DID YOU KNOW

Islamic Art

Islamic artists perfected the art of ornamentation and mosaic by using geometric forms because Islamic law strictly prohibited the use of the human form in art. Mosaic patterns at the Alhambra, a palace and fortress built in Grenada, Spain, illustrate such mathematical concepts as symmetries, tessellations, reflections, and rotations.

SOLUTION

a) We represent side $\overline{ST}$ with the letter x. Because the corresponding sides must be in proportion, we can write a proportion (as explained in Section 6.2) to find the side $\overline{ST}$. Corresponding sides $\overline{RS}$ and $\overline{LM}$ are known, so we use them as one ratio in the proportion.

$$\frac{\overline{RS}}{\overline{LM}} = \frac{\overline{ST}}{\overline{MN}}$$
$$\frac{8}{12} = \frac{x}{24}$$

Now we solve for x.

$$8 \cdot 24 = 12 \cdot x$$
$$192 = 12x$$
$$16 = x$$

Thus, the length of $\overline{ST} = 16$.

b) We represent $\overline{LN}$ with the letter y.

$$\frac{\overline{RS}}{\overline{LM}} = \frac{\overline{RT}}{\overline{LN}}$$
$$\frac{8}{12} = \frac{18}{y}$$
$$8 \cdot y = 12 \cdot 18$$
$$8y = 216$$
$$y = \frac{216}{8} = 27$$

Thus, the length of $\overline{LN} = 27$.

EXAMPLE 3 *Using Similar Triangles to Find the Height of a Tree*

Saraniti Walker plans to remove a tree from her back yard. She needs to know the height of the tree. Saraniti is 5 ft tall and determines that when her shadow is 8 ft long, the shadow of the tree is 50 ft long (see Fig. 9.22). How tall is the tree?

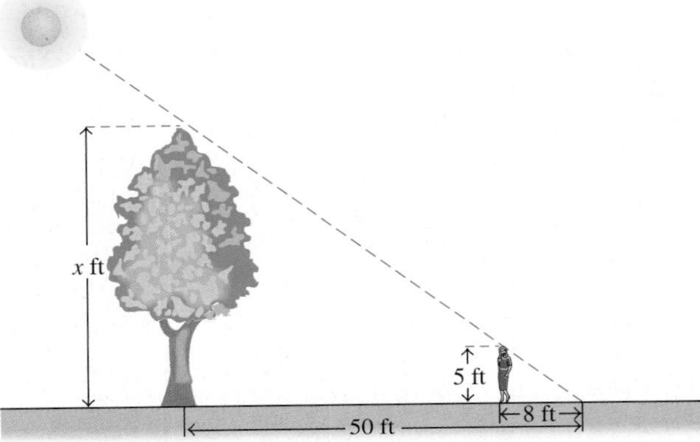

Figure 9.22

SOLUTION We will let x represent the height of the tree. From Fig. 9.22, we can see that the triangle formed by the sun's rays, Saraniti, and her shadow is similar to the triangle formed by the sun's rays, the tree, and its shadow. To find the height of the tree we will set up and solve the following proportion:

$$\frac{\text{Height of the tree}}{\text{Height of Saraniti}} = \frac{\text{length of tree's shadow}}{\text{length of Saraniti's shadow}}$$

$$\frac{x}{5} = \frac{50}{8}$$

$$8x = 250$$

$$x = 31.25$$

Therefore, the tree is 31.25 ft tall.

CONGRUENT FIGURES

If the corresponding sides of two similar figures are the same length, the figures are called **congruent figures**. Corresponding angles of congruent figures have the same measure, and the corresponding sides are equal in length. Two congruent figures coincide when placed one upon the other.

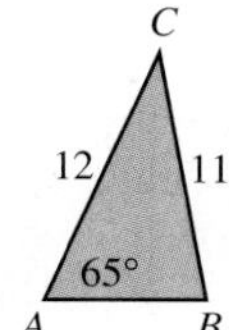

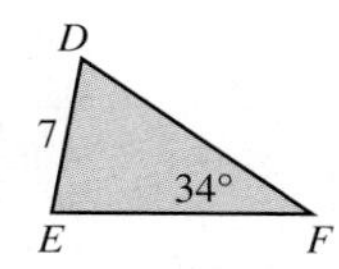

Figure 9.23

EXAMPLE 4 *Congruent Triangles*

Triangles ABC and DEF in Fig. 9.23 are congruent. Find

a) The length of side $\overline{DF}$.
b) The length of side $\overline{AB}$.
c) $m\measuredangle FDE$.
d) $m\measuredangle ACB$.
e) $m\measuredangle ABC$.

DID YOU KNOW

Bucky Balls

The molecular structure of C_{60} resembles the patterns found on a soccer ball.

Buckminsterfullerenes, also known as fullerenes and affectionately known as Bucky Balls, are a class of pure carbon molecules. Along with graphite and diamond, Bucky Balls are the only naturally occurring forms of pure carbon. Named after the American architect-engineer, F. Buckminster Fuller, who designed hemispherical geodesic domes from hexagonal and pentagonal faces, fullerenes are the most spherical molecules known. Discovered in 1985 by Robert Curl, Harold Kroto, and Richard Smalley at Rice University, Bucky Balls are only now beginning to see a wide range of applications. Used primarily as microscopic lubricant, Bucky Balls have potential applications in molecular medical engineering, electrical superconductivity, and in computer chip design. The most common form of buckminsterfullerene contains 60 carbon atoms and has the chemical symbol, C_{60}. The molecular structure of C_{60} contains 12 pentagons and 20 hexagons arranged in a pattern similar to that found on a soccer ball.

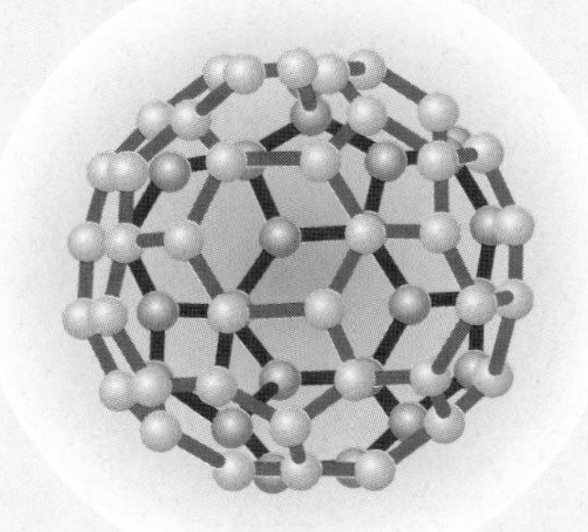

Sketch of a C_{60} molecule (also known as a Bucky Ball)

SOLUTION Because $\triangle ABC$ is congruent to $\triangle DEF$, we know that the corresponding sides and angles are equal.

a) $\overline{DF} = \overline{AC} = 12$
b) $\overline{AB} = \overline{DE} = 7$
c) $m\measuredangle FDE = m\measuredangle CAB = 65°$
d) $m\measuredangle ACB = m\measuredangle DFE = 34°$
e) The sum of the angles of a triangle is 180°. Since $m\measuredangle BAC = 65°$ and $m\measuredangle ACB = 34°$, $m\measuredangle ABC = 180° - 65° - 34° = 81°$. ▲

Earlier we learned that **quadrilaterals** are four-sided polygons, the sum of whose interior angles is 360°. Quadrilaterals may be classified according to their characteristics, as illustrated in the summary box below.

Trapezoid

Two sides are parallel.

Parallelogram

Both pairs of opposite sides are parallel.

Rhombus

Both pairs of opposite sides are parallel. The four sides are equal in length.

Rectangle

Both pairs of opposite sides are parallel. The angles are right angles.

Square

Both pairs of opposite sides are parallel. The four sides are equal in length. The angles are right angles.

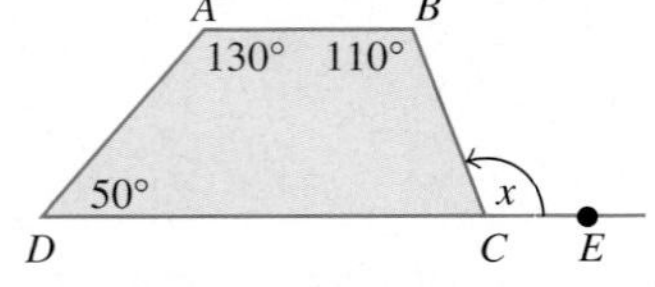

Figure 9.24

EXAMPLE 5 *Angles of a Trapezoid*

Find the measure of the exterior angle, x, of the trapezoid in Fig. 9.24.

SOLUTION Since sides AB and CD are parallel, BC may be considered a transversal. Then by alternate interior angles, $m\measuredangle x = 110°$. ▲

A second method for solving Example 5 is to find the measure of the unknown interior angle of the trapezoid by subtracting the sum of the measures of the three given angles from 360°: $m\measuredangle BCD = 360° - (50° + 130° + 110°) = 360° - 290° = 70°$. Since $\measuredangle DCE$ is a straight angle, $m\measuredangle x$ must be $180° - 70° = 110°$.

SECTION 9.2 EXERCISES

CONCEPT/WRITING EXERCISES

1. What is a polygon?
2. What is a polygonal region?
3. What distinguishes regular polygons from other polygons?
4. In your own words, describe how to find the sum of the measures of the interior angles of a polygon with n sides.
5. What are similar figures?
6. What are congruent figures?

PRACTICE THE SKILLS

In Exercises 7–10, name the polygon.

7. 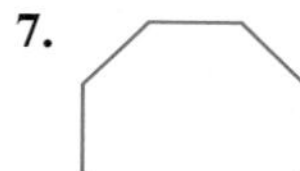

8. 

9. 

10. 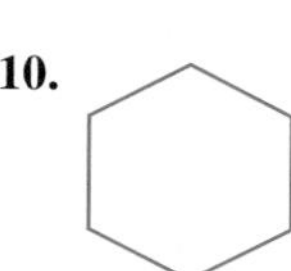

In Exercises 11–16, identify the triangle as scalene, isosceles, or equilateral.

11. 

12. 

13. 

14. 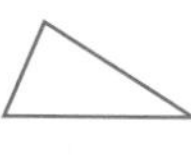

15. 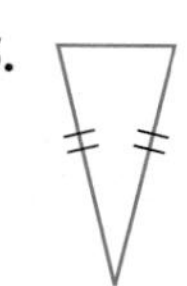

16. 

In Exercises 17–22, identify the triangle as acute, obtuse, or right.

17. 

18. 

19. 

20. 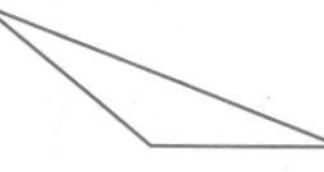

21. 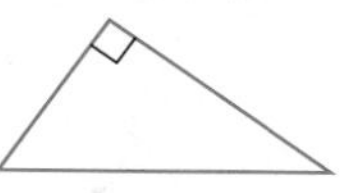

22. 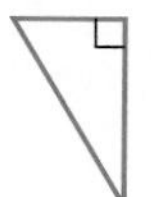

In Exercises 23–28, identify the quadrilateral.

23. 

24. 

25. 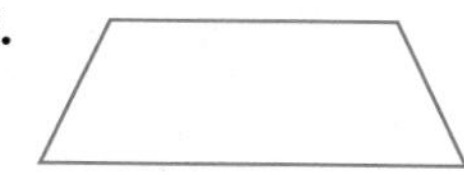

26. 

27. 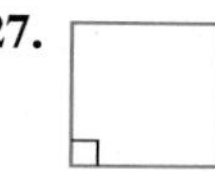

28. 

In Exercises 29–32, find the measure of ∡x.

29. 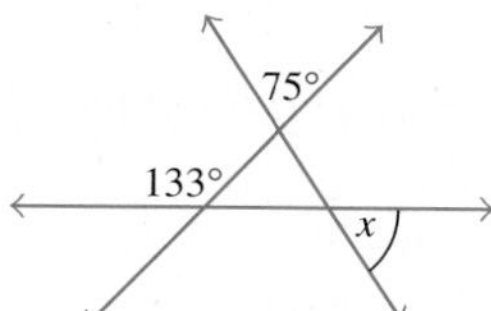

30. 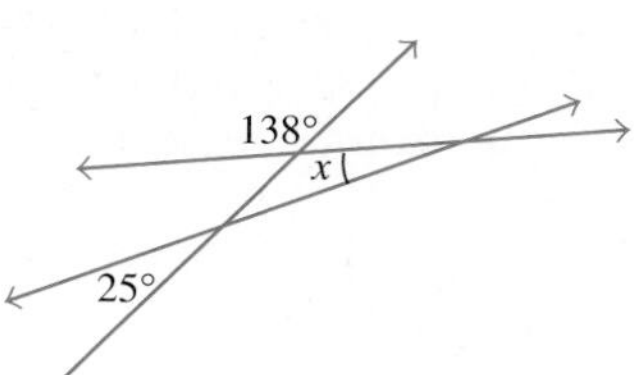

31. 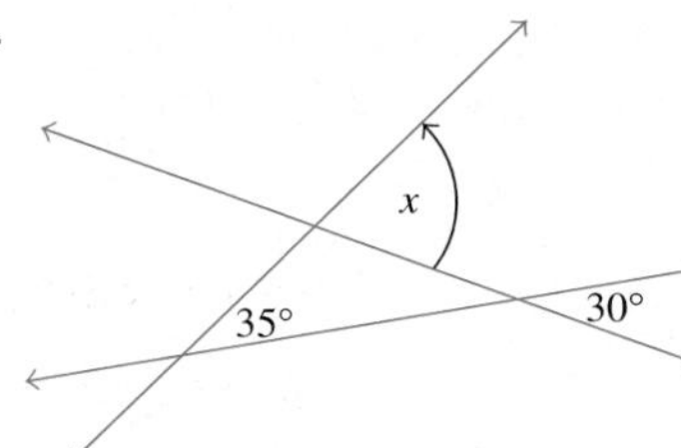

32. 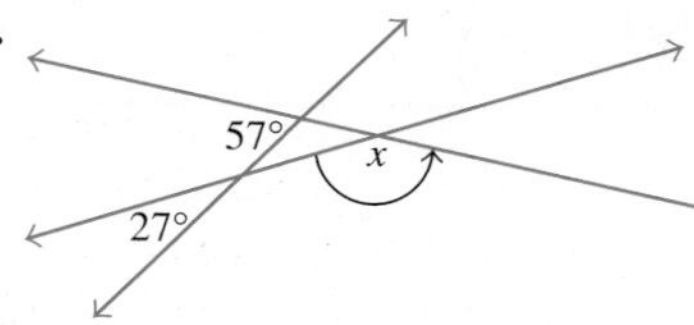

In Exercises 33–34, lines l_1 and l_2 are parallel. Determine the measures of ∡1 through ∡12.

33.

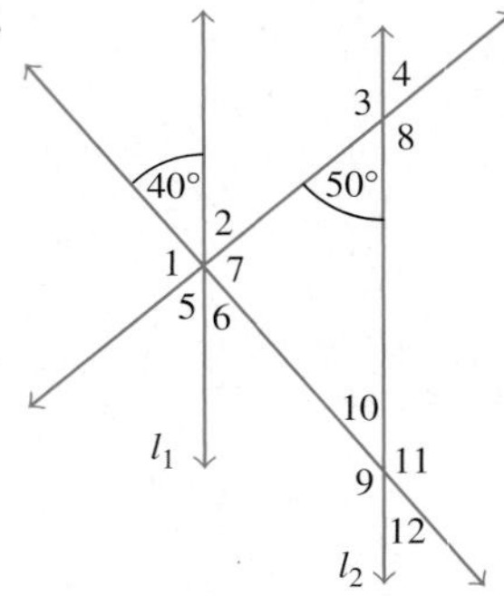

34.

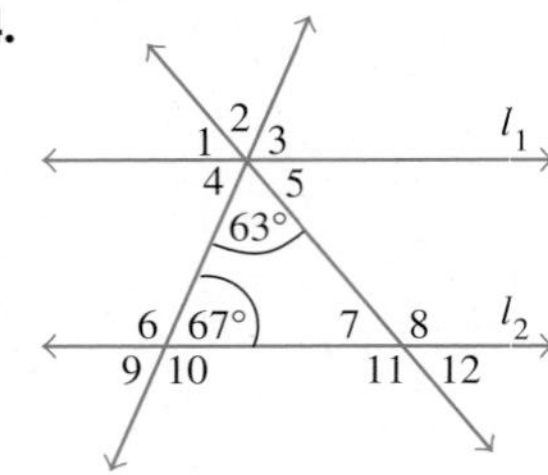

In Exercises 35–38, find the sum of the measures of the interior angles of the named polygon.

35. Octagon

36. Icosagon

37. Heptagon

38. Dodecagon

In Exercises 39–44, find the measure of an interior angle of the regular polygon. If a side of the polygon is extended, find the supplementary angle of an interior angle. See Example 1.

39. Quadrilateral

40. Triangle

41. Pentagon

42. Hexagon

43. Heptagon

44. Octagon

In Exercises 45–48, determine the measure of the angle. In the figure, ∡ABC makes an angle of 125° with the floor and l_1 and l_2 are parallel.

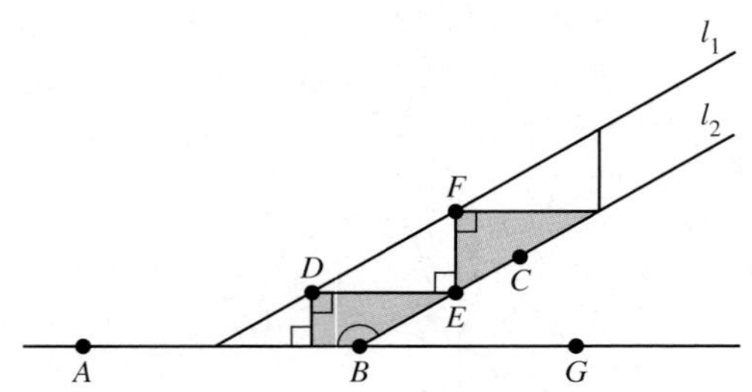

45. ∡*GBC*

46. ∡*EDF*

47. ∡*DFE*

48. ∡*DEC*

In Exercises 49–52, the figures are similar. Find the length of side x and side y.

49.

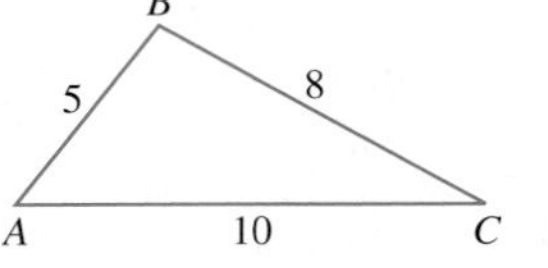

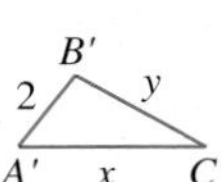

50.

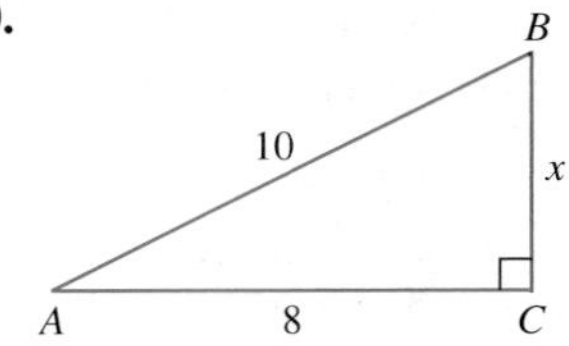

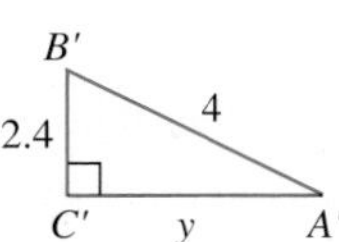

51.

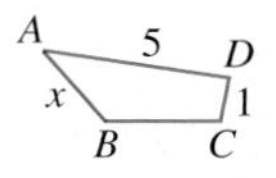

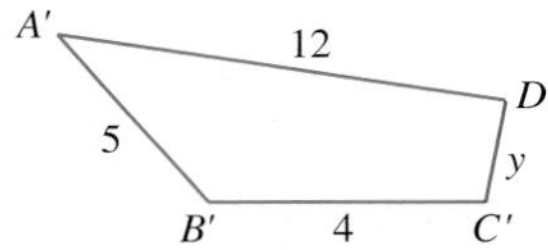

52.

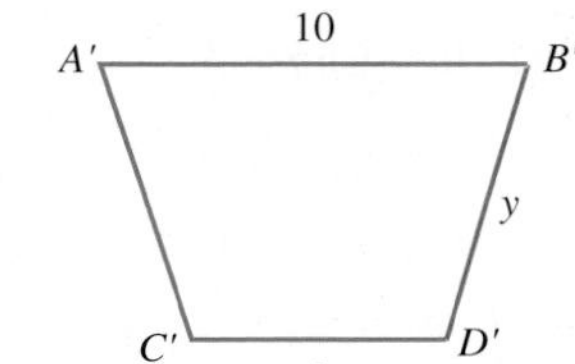

In Exercises 53–56, triangles ABC and DEC are similar figures. Find the length of

53. side $\overline{DC}$.

54. side $\overline{BC}$.

55. side $\overline{BE}$.

56. side $\overline{AD}$

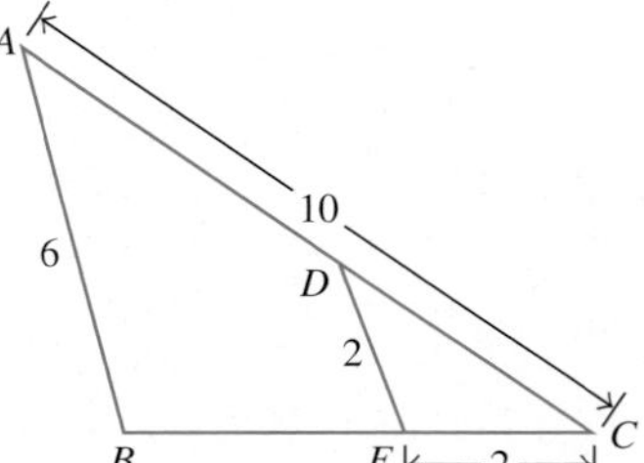

In Exercises 57–62, find the length of the sides and the measures of the angles for the congruent triangles ABC and A′B′C′.

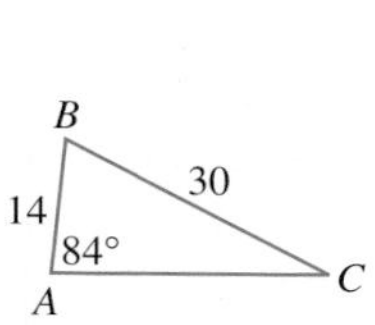

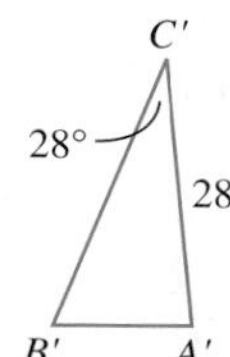

57. The length of side $\overline{A'B'}$

58. The length of side $\overline{B'C'}$

59. The length of side $\overline{AC}$

60. $\measuredangle B'A'C'$

61. $\measuredangle ACB$

62. $\measuredangle ABC$

In Exercises 63–68, find the length of the sides and the measures of the angles for the congruent quadrilaterals ABCD and A′B′C′D′.

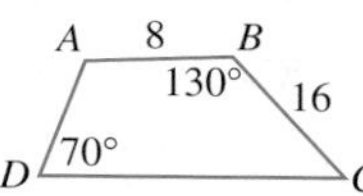

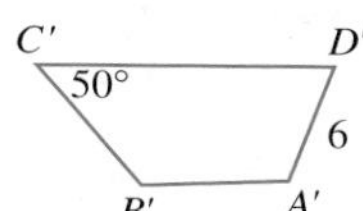

63. The length of side $\overline{A'B'}$

64. The length of side $\overline{AD}$

65. The length of side $\overline{B'C'}$

66. $\measuredangle BCD$

67. $\measuredangle A'D'C'$

68. $\measuredangle DAB$

PROBLEM SOLVING

69. *Angles on a Picnic Table* The legs of a picnic table form an isosceles triangle as indicated in the figure. If $\measuredangle ABC = 80°$, find $\measuredangle x$ and $\measuredangle y$ so that the top of the table will be parallel to the ground.

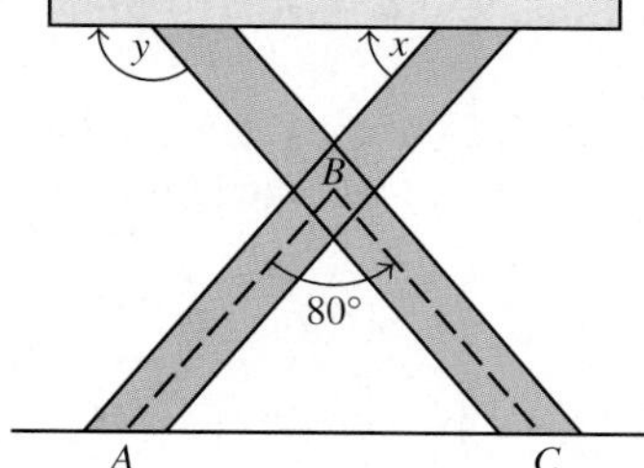

70. *Angles of an Octagon* Sketch an octagon.
 a) By drawing five lines from one vertex, show that the figure can be divided into six triangles.
 b) Use the figure to determine the sum of the interior angles.
 c) Use the formula, sum $= (n - 2)180°$, to verify the results obtained in part (b).

Angles on a Shopping Cart *For the shopping cart illustrated, $\overline{wz}$ is parallel to $\overline{xy}$. Find*

71. $m\measuredangle w$.

72. $m\measuredangle x$.

73. $m\measuredangle y$.

74. $m\measuredangle z$.

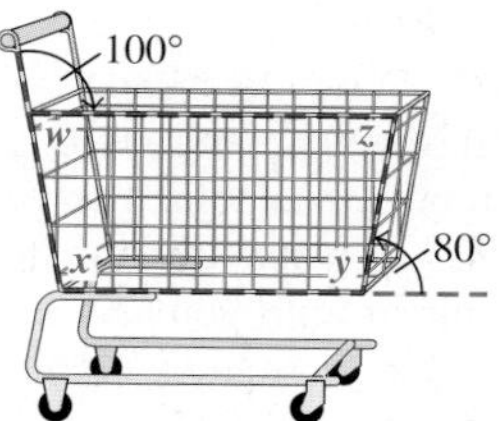

CHALLENGE PROBLEMS/GROUP ACTIVITIES

75. *Distance Across a Lake*
 a) In the figure $m\measuredangle CED = m\measuredangle ABC$. Explain why triangles ABC and DEC must be similar.
 b) Determine the distance across the lake, $\overline{DE}$.

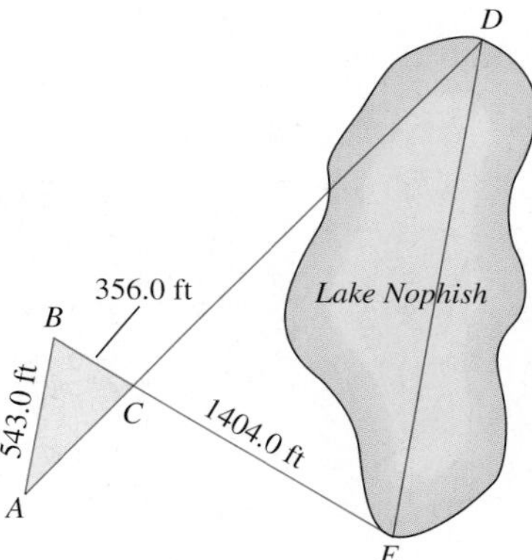

76. *Height of a Tree* As shown, Emily holds a stick 1 ft long at a distance of 2.0 ft from her eye. Her eye is 192 ft from a line that passes through the center of the tree. She holds the stick so that it is parallel to the tree as she sights along the top of the stick to a point A (the top of the tree) and along the bottom of the stick to point B (the base of the tree).
 a) Explain why triangles CDE and CAB are similar figures.
 b) Determine the height of the tree.

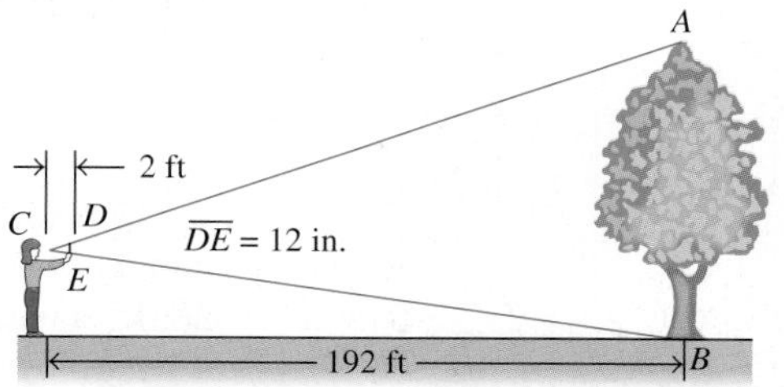

77. In this book the discussion of congruence is limited to polygons. Can three-dimensional objects be congruent? List the criteria that you believe would be necessary for two three-dimensional figures to be congruent. Give several examples of three-dimensional figures that you believe are congruent.

78. *Height of a Wall* You are asked to measure the height of an inside wall of a warehouse. No ladder tall enough to measure the height is available. You borrow a mirror from a salesclerk and place it on the floor. You then move away from the mirror until you can see the reflection of the top of the wall in it, as shown in the figure.
   a) Explain why triangle *HFM* is similar to triangle *TBM*. (*Hint:* In the reflection of light the angle of incidence equals the angle of reflection. Thus, $\measuredangle HMF = \measuredangle TMB$.)
   b) If your eyes are $5\frac{1}{2}$ ft above the floor, you are $2\frac{1}{2}$ ft from the mirror, and the mirror is 20 ft from the wall, how high is the wall?

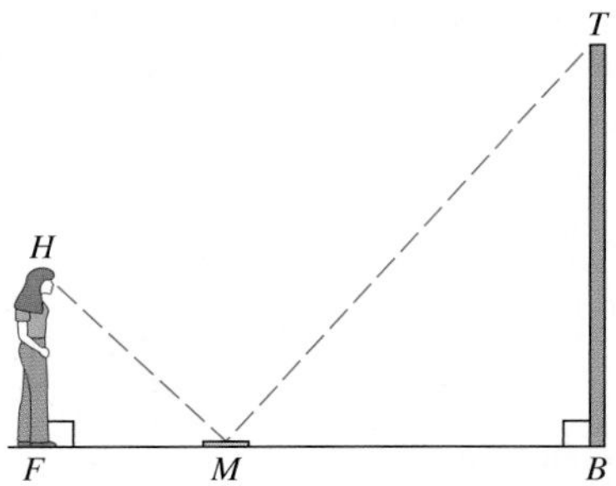

RESEARCH ACTIVITIES

79. Write a paper on transformation geometry. Include a discussion and examples of rotation, reflection, and translation.

80. Write a paper on the history and use of a theodolite, a surveying instrument.

9.3 PERIMETER AND AREA

PERIMETER AND AREA

Figure 9.25

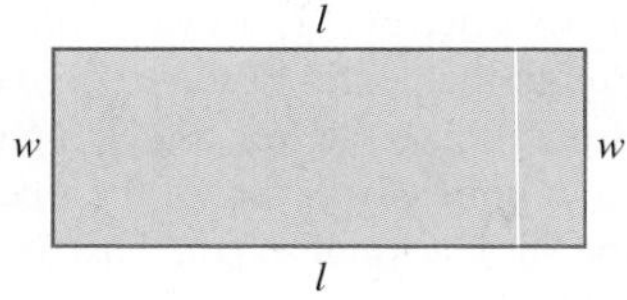

Figure 9.26

Geometric shapes abound in the natural world and the world made by human beings. For example, a basketball court is a rectangle, a basketball is a sphere, a can of food is a cylinder, and a ream of paper is a rectangular solid.

The **perimeter**, P, of a two-dimensional figure, is the sum of the lengths of the sides of the figure. In Figs. 9.25 and 9.26 the sums of the red lines are the perimeters. Perimeters are measured in the same units as the sides. For example, if the sides of a figure are measured in feet, the perimeter will be measured in feet.

The **area**, A, is the region within the boundaries of the figure. The blue color in Figs. 9.25 and 9.26 indicate the areas of the figures. Area is measured in square units. For example, if the sides of a figure are measured in inches, the area of the figure will be measured in square inches (in.^{2}). (See Table 8.7 on page 419 for common units of area in the U.S. customary and metric systems.)

Consider the rectangle in Fig. 9.26. Two sides of the rectangle have length l, and two sides of the rectangle have width w. Thus, if we add the lengths of the four sides to get the perimeter, we find $P = l + w + l + w = 2l + 2w$.

Perimeter of a Rectangle

$$P = 2l + 2w$$

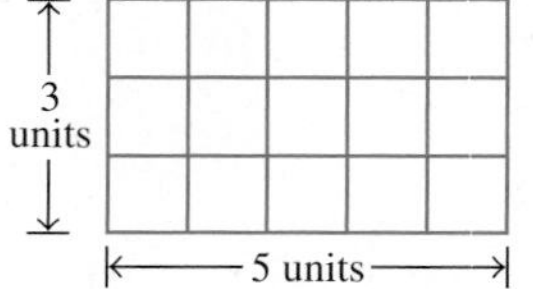

Figure 9.27

Consider a rectangle of length 5 units and width 3 units (Fig. 9.27). Counting the number of 1-unit by 1-unit squares within the figure we obtain the area of the rectangle, 15 square units. The area can also be obtained by multiplying the number of units of length by the number of units of width, or 5 units $\times$ 3 units $=$ 15 square units. We can find the area by the formula Area $=$ length $\times$ width.

Area of a Rectangle

$$A = l \times w$$

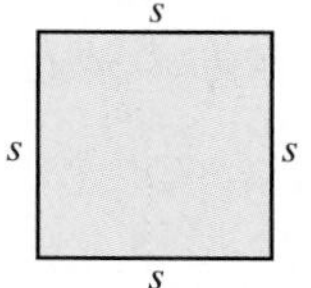

Figure 9.28

Using the formula for the area of a rectangle, we can determine the formulas for the areas of other figures.

A square (Fig. 9.28) is a rectangle that contains four equal sides. Therefore, the length equals the width. If we call both the length and the width of the square s,

$$A = l \times w, \qquad \text{so} \qquad A = s \times s = s^2$$

Area of a Square

$$A = s^2$$

A parallelogram with height h and base b is shown in Fig. 9.29(a).

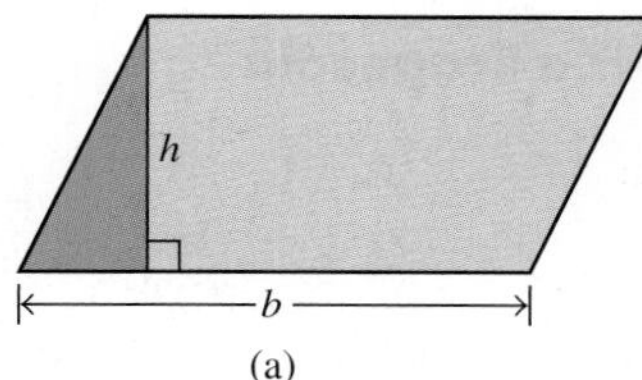

(a)

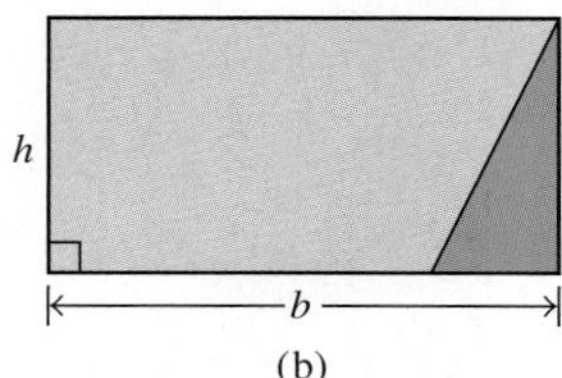

(b)

Figure 9.29

If we were to cut off the red portion of the parallelogram on the left, Fig. 9.29(a), and attach it to the right side of the figure, the resulting figure would be a rectangle, Fig. 9.29(b). Since the area of the rectangle is $b \times h$, the area of the parallelogram is also $b \times h$.

Area of a Parallelogram

$$A = b \times h$$

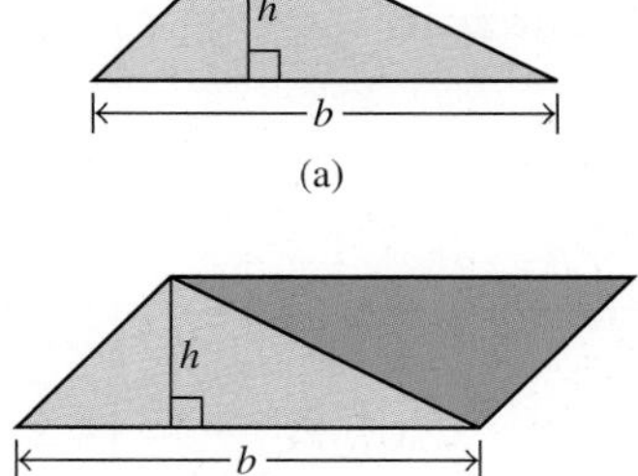

Figure 9.30

Consider the triangle with height, h, and base, b, shown in Fig. 9.30(a). Using this triangle and a second identical triangle, we can construct a parallelogram, Fig. 9.30(b). The area of the parallelogram is bh. The area of the triangle is one-half that of the parallelogram. Therefore, the area of the triangle is $\frac{1}{2}$(base)(height).

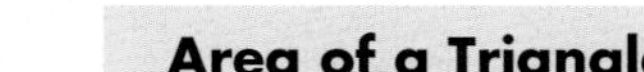

Area of a Triangle

$$A = \tfrac{1}{2}bh$$

Now consider the trapezoid shown in Fig. 9.31(a). We can partition the trapezoid into two triangles by drawing diagonal $\overline{DB}$, as in Fig. 9.31(b). One triangle has base $\overline{AB}$ (called b_2) with height $\overline{DE}$, and the other triangle has base $\overline{DC}$ (called b_1) with height $\overline{FB}$. Note the line used to measure the height of the triangle need not be inside the triangle. Because heights $\overline{DE}$ and $\overline{FB}$ are equal, both triangles have the same height, h. The area of triangles DCB and ADB are $\frac{1}{2}b_1h$ and $\frac{1}{2}b_2h$, respectively. The area of the trapezoid is the sum of the areas of the triangles, $\frac{1}{2}b_1h + \frac{1}{2}b_2h$, which can be written $\frac{1}{2}h(b_1 + b_2)$.

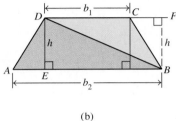

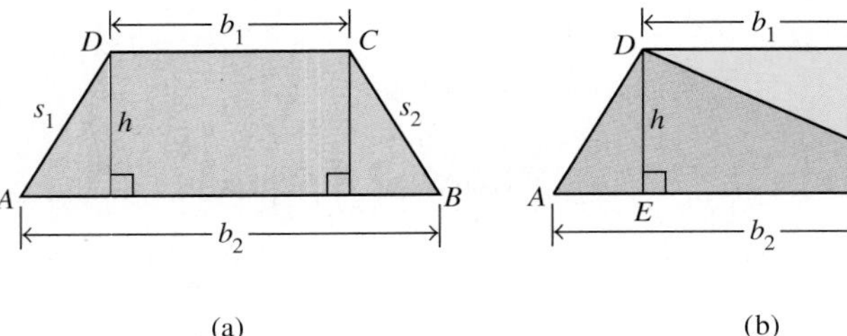

Figure 9.31

DID YOU KNOW

Area Versus Perimeter

Which geographic area, Italy or New Mexico, do you think has the larger land area? As a matter of fact, the land areas of Italy and New Mexico are very similar: Italy's is 116,304 mi^2 and New Mexico's is 121,593 mi^2. However, Italy's perimeter, 5812 mi, is almost five times greater than New Mexico's 1200 mi.

Area of a Trapezoid

$$A = \frac{1}{2}h(b_1 + b_2)$$

Following is a summary of the areas and perimeters of selected figures.

Perimeters and Areas

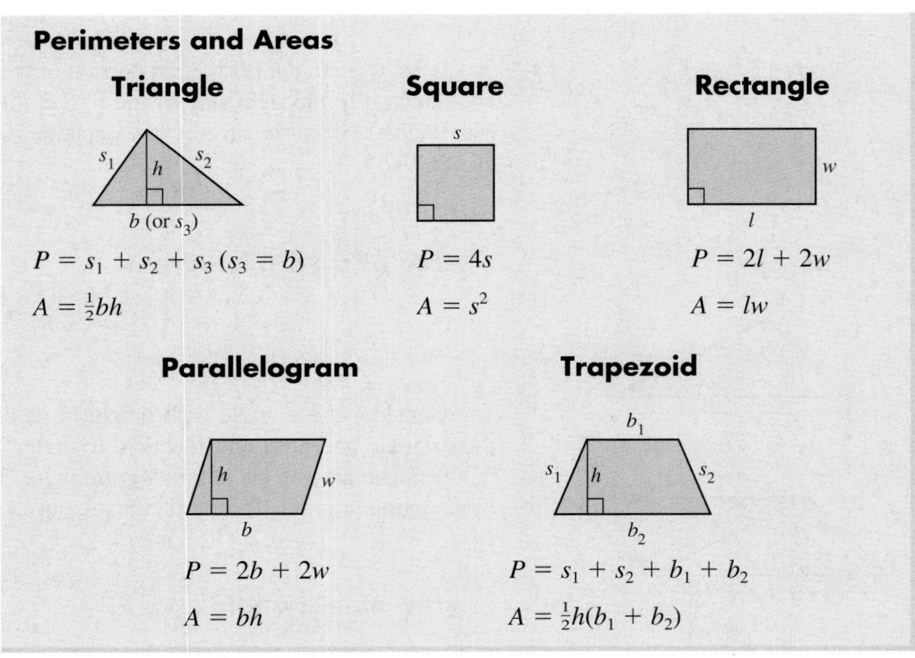

Triangle: $P = s_1 + s_2 + s_3 \ (s_3 = b)$, $A = \frac{1}{2}bh$

Square: $P = 4s$, $A = s^2$

Rectangle: $P = 2l + 2w$, $A = lw$

Parallelogram: $P = 2b + 2w$, $A = bh$

Trapezoid: $P = s_1 + s_2 + b_1 + b_2$, $A = \frac{1}{2}h(b_1 + b_2)$

DID YOU KNOW

Pythagorean Theorem

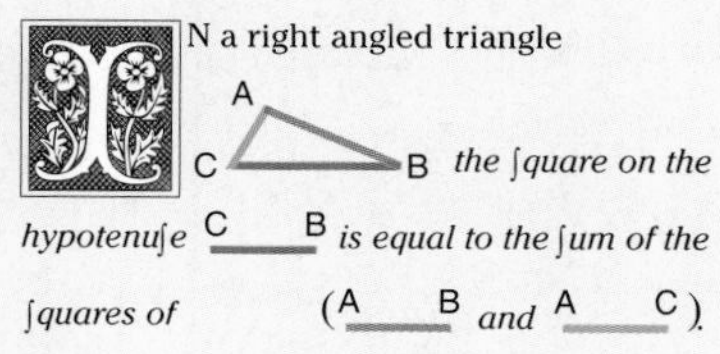

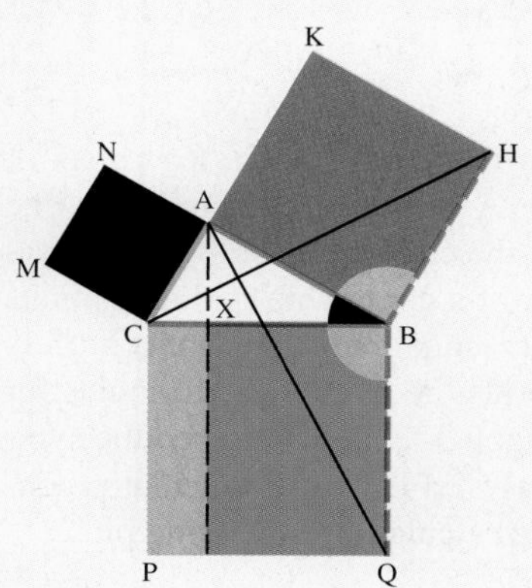

The Pythagorean theorem has probably been proven in more ways than any other theorem in mathematics. The book *Pythagorean Propositions* contains 370 different proofs of the Pythagorean theorem. Another book, the nineteenth-century English-language edition of Euclid's *Elements* by Oliver Byrne described by one critic as "one of the oddest and most beautiful books of the whole [nineteenth] century," uses color and shapes to illustrate a proof of the Pythagorean theorem.

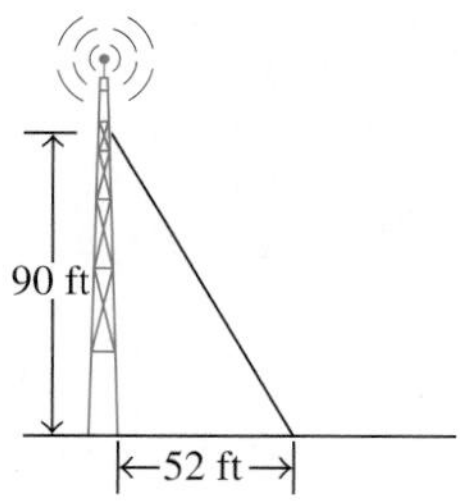

Figure 9.32

EXAMPLE 1 *Waterproofing a Driveway*

Donna Demaree is coating her driveway with Thomson's® Water Seal® waterproofer. One can costs $10.99 and covers 330 ft². If Donna's driveway is 40 ft long and 16 ft wide, determine **(a)** the area of Donna's driveway, **(b)** how many cans of waterproofer she needs, and **(c)** the cost of waterproofing her driveway.

SOLUTION

a) The area of Donna's driveway is

$$A = l \cdot w = 40 \cdot 16 = 640 \text{ ft}^2$$

The area of the driveway is in square feet because both the length and width are measured in feet.

b) To determine the number of cans of waterproofer Donna needs, divide the area of the driveway by the area covered by one can of waterproofer.

$$\frac{\text{Area of driveway}}{\text{Area covered by one can}} = \frac{640}{330} = 1\frac{31}{33} \approx 1.94$$

The waterproofer must be purchased in whole cans, so Donna needs to buy two cans of waterproofer.

c) The cost of the waterproofer is 2 × $10.99, or $21.98. ▲

PYTHAGOREAN THEOREM

We introduced the Pythagorean theorem in Chapter 5. This theorem is an important tool for finding the perimeter and area of triangles. For this reason we restate it here.

Pythagorean Theorem

The sum of the squares of the lengths of the legs of a right triangle equals the square of the length of the hypotenuse.

Symbolically, if a and b represent the lengths of the legs and c represents the length of the hypotenuse (the side opposite the right angle), then $a^2 + b^2 = c^2$.

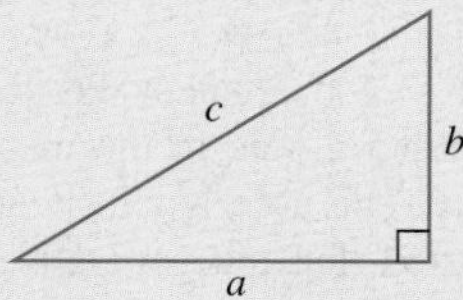

EXAMPLE 2 *Anchoring a Radio Signal Tower*

A 100 ft radio signal tower is being constructed. To steady the tower, guy wires are attached to the tower. One end of the highest guy wire is attached to the tower at a point 90 ft above the ground (see Fig. 9.32). The other end is anchored into the ground at a point 52 ft from the base of the tower. How long is this guy wire?

SOLUTION The tower, the guy wire, and the ground form a right triangle. The tower and the ground form the legs and the guy wire forms the hypotenuse. By the Pythagorean theorem,

$$\begin{aligned} c^2 &= a^2 + b^2 \\ &= (52)^2 + (90)^2 \\ &= 2704 + 8100 \\ &= 10{,}804 \\ c &= \sqrt{10{,}804} \approx 103.9 \text{ ft} \end{aligned}$$

Therefore, the guy wire is about 104 ft long.

CIRCLES

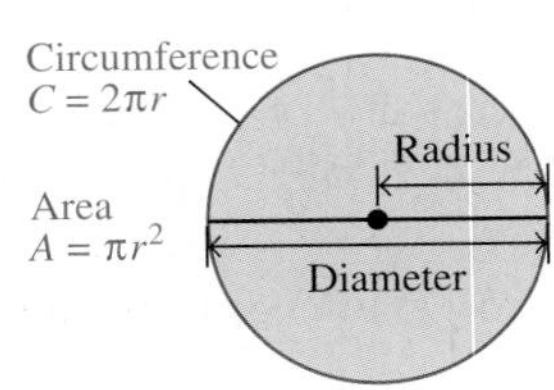

Figure 9.33

A commonly used plane figure that is not a polygon is a *circle*. A **circle** is a set of points equidistant from a fixed point called the center. A **radius**, r, of a circle is a line segment from the center of the circle to any point on the circle (Fig. 9.33). A **diameter**, d, of a circle is a line segment through the center of a circle with both end points on the circle. Note that the diameter of the circle is twice its radius. The **circumference** is the length of the simple closed curve that forms the circle. The formulas for the area and circumference of a circle are given in Fig. 9.33. We introduced the symbol pi, π, in Chapter 5. Recall that π is approximately 3.14. If your calculator contains a $\boxed{\pi}$ key, you should use that key when working calculations involving pi.

EXAMPLE 3 *The Best Pizza Deal*

At Antonio's Pizza, the price for a medium plain pizza is \$7.50, and the price for a large plain pizza is \$14.00. The diameters of the large and medium pizzas are 18 in. and 12 in., respectively, and both pizzas are the same thickness. To get the most for your money, should you buy two medium pizzas or one large pizza?

DID YOU KNOW

Fermat's Last Theorem

Fermat

In 1637 the amateur French mathematician, Pierre de Fermat, scribbled a note in the margin of the book *Arithmetica* by Diophantus. The note would haunt mathematicians for centuries. Fermat stated that the generalized form of the Pythagorean theorem, $a^n + b^n = c^n$, has no positive integer solutions where $n > 2$. Fermat's note concluded with "I have a truly marvelous demonstration of this proposition, which this margin is too narrow to contain." This conjecture became known as Fermat's last theorem. A formal proof of this conjecture escaped mathematicians until on September 19, 1994, Andrew J. Wiles of Princeton University announced he had found a proof. It took Dr. Wiles over 8 years of work—including fixing a flaw in an earlier announced solution—to accomplish the task. Dr. Wiles was awarded the Wolfskehl prize at Göttingen University in Germany in acknowledgement of his achievement.

Wiles

SOLUTION We begin by finding the surface area of the two pizzas. The radius of the larger pizza is 9 in. and the radius of the smaller pizza is 6 in.

Area of a large pizza*

$$A = \pi r^2$$
$$\approx 3.14(9)^2$$
$$\approx 3.14(81) \approx 254.34 \text{ in.}^2$$

Area of a medium pizza

$$A = \pi r^2$$
$$\approx 3.14(6)^2$$
$$\approx 3.14(36) \approx 113.04 \text{ in.}^2$$

The two medium pizzas have an area of about $2(113.04) \approx 226.08$ in.^{2}

	One large pizza	**Two medium pizzas**
Area	254.34 in.^{2}	226.08 in.^{2}
Cost	\$14	\$15

One large pizza has a greater area and a lower cost. Therefore, one large pizza should be purchased.

EXAMPLE 4 *Applying Lawn Fertilizer*

Steve May plans to fertilize his lawn. The shapes and dimensions of his lot, house, driveway, pool, and rose garden are shown in Fig. 9.34. One bag of fertilizer costs \$29.95 and covers 5000 ft^2. Determine how many bags of fertilizer Steve needs and the total cost of the fertilizer.

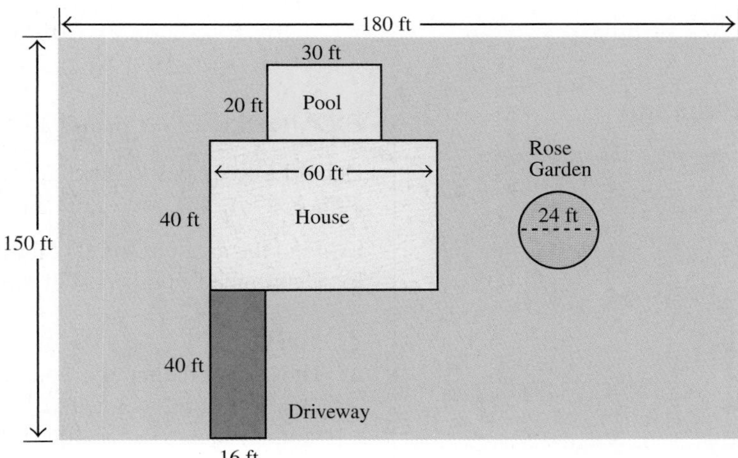

Figure 9.34

SOLUTION The total area of the lot is $150 \cdot 180$, or 27,000 ft^2. To determine the area to be fertilized, subtract the area of house, driveway, pool, and rose garden from the total area.

$$\text{Area of house} = 60 \cdot 40 = 2400 \text{ ft}^2$$
$$\text{Area of driveway} = 40 \cdot 16 = 640 \text{ ft}^2$$
$$\text{Area of pool} = 20 \cdot 30 = 600 \text{ ft}^2$$

*If you use the $\boxed{\pi}$ key on your calculator, your answers will be slightly more accurate.

The diameter of the rose garden is 24 ft, so its radius is 12 ft.

$$\text{Area of rose garden} = \pi r^2 = \pi(12)^2 \approx 3.14(144) \approx 452.16 \text{ ft}^2$$

The total area of the house, driveway, pool, and rose garden is approximately 2400 + 640 + 600 + 452.16, or 4092.16 ft^2. The area to be fertilized is 27,000 − 4092.16 ft^2, or 22,907.84 ft^2. The number of bags of fertilizer is found by dividing the total area to be fertilized by the number of square feet covered per bag.

The number of bags of fertilizer is $\frac{22{,}907.84}{5000}$, or about 4.58 bags. Therefore, Steve needs five bags. At \$29.95 per bag, the total cost is 5 × \$29.95, or \$149.75.

"A man in an ivory tower dreamed of Pi to the 200th power. (This is apt to occur when you mix liquor with clams at a very late hour.)"

John McClellen

EXAMPLE 5 ***Converting Square Feet to Square Inches***

a) Convert 1 ft^2 to square inches.
b) Convert 7.5 ft^2 to square inches.

SOLUTION

a) 1 ft = 12 in.
Therefore, 1 ft^2 = 12 in. × 12 in. = 144 in.^{2}
b) Since 1 ft^2 = 144 in.^{2}, 2 ft^2 = 2(144 in.^{2}) = 288 in.^{2}, then 7.5 ft^2 is 7.5 ft^2 = 7.5 × 144 = 1080 in.^{2}

DID YOU KNOW

City Planning

The Roman poet Virgil tells the story of Queen Dido, who fled to Africa after her brother murdered her husband. There, she begged for some land from King Iarbus, telling him she only needed as much land as the hide of an ox would enclose. Being very clever, she decided that the greatest area would be enclosed if she tore the hide into thin strips and formed them into a circle. On this land she founded the city of Byrsa (the Greek word for "hide"), later known as Carthage in present day Tunisia.

EXAMPLE 6 ***Installing Ceramic Tile***

Debra Levy wishes to purchase ceramic tile for her family room, which measures 30 ft × 27 ft. The cost of the tile, including installation, is \$21 per square yard.

a) Find the area of Debra's family room in square *yards.*
b) Determine Debra's cost of the ceramic tile for her family room.

SOLUTION

a) The area of the family room in square feet is $30 \cdot 27 = 810$ ft^2.
Since 1 yd = 3 ft, 1 yd^2 = 3 ft × 3 ft = 9 ft^2. To find the area of the family room in square yards, divide the area in square feet by 9 ft^2.

$$\text{Area in square yards} = \frac{810}{9} = 90$$

Therefore, the area is 90 yd^2.

b) The cost of 90 yd^2 of ceramic tile, including installation, is $90 \cdot \$21 = \$1890$.

When multiplying units of length, be sure that the units are the same. You can multiply feet by feet to get square feet or yards by yards to get square yards. However, you cannot get a valid answer if you multiply numbers expressed in feet by numbers expressed in yards.

SECTION 9.3 EXERCISES

CONCEPT/WRITING EXERCISES

1. a) Describe in your own words how to determine the *perimeter* of a two-dimensional figure.
   b) Describe in your own words how to determine the *area* of a two-dimensional figure.
   c) Draw a rectangle with a length of 6 units and a width of 2 units. Determine the area and perimeter of this rectangle.
2. What is the relationship between the *radius* and the *diameter* of a circle?

PRACTICE THE SKILLS

In Exercises 3–6, find the area of the triangle.

3.
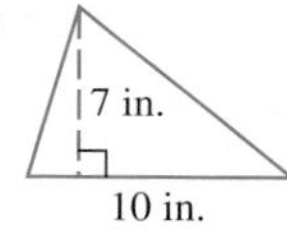

4.
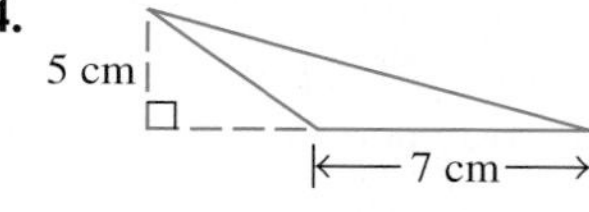

5.
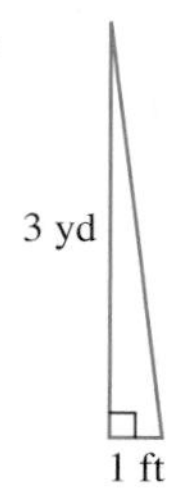

6.
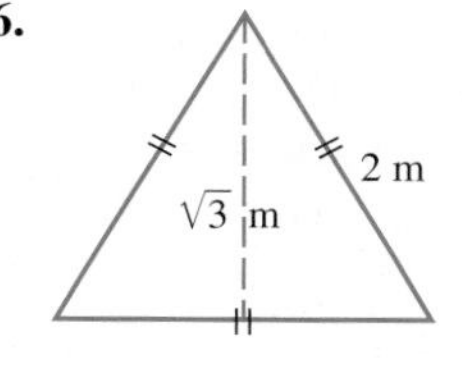

In Exercises 7–12, find the area and perimeter of the quadrilateral.

7.
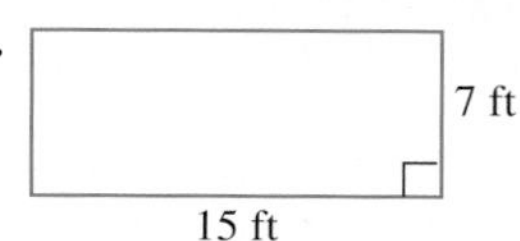

8.
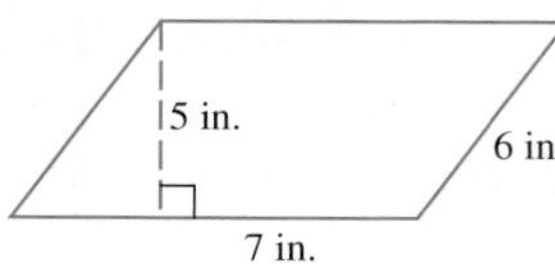

9.
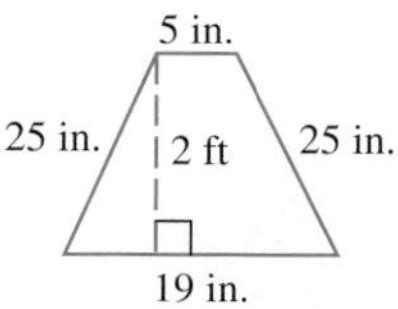

10.
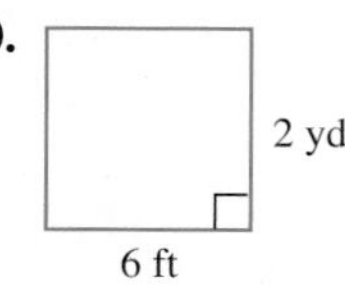

11.
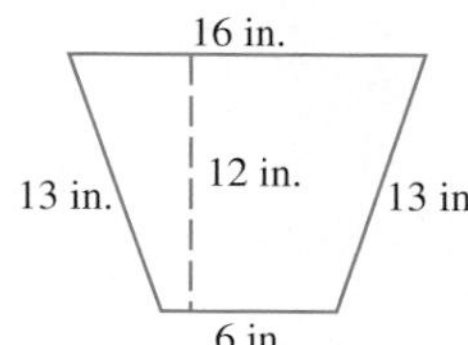

12.
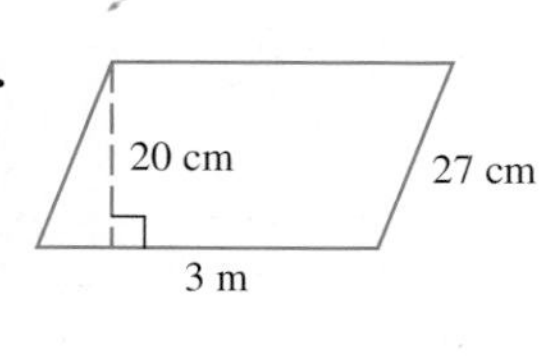

In Exercises 13–16, find the area and circumference of the circle.

13.
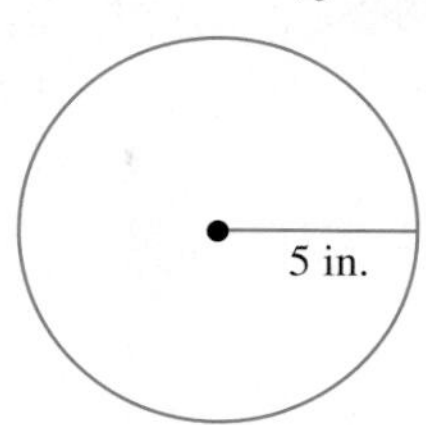

14.
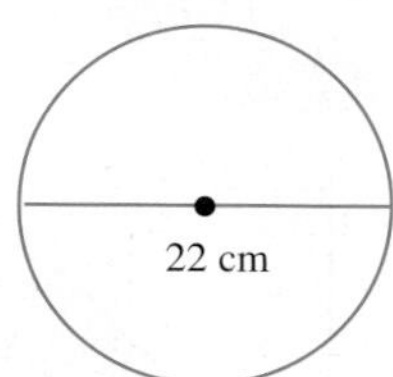

15.
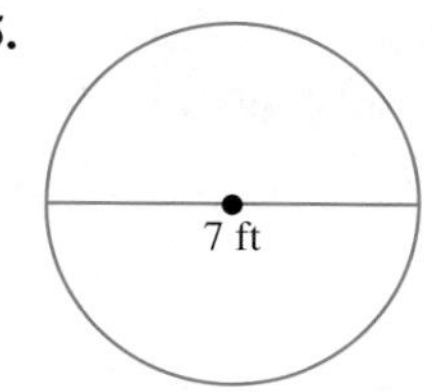

16.
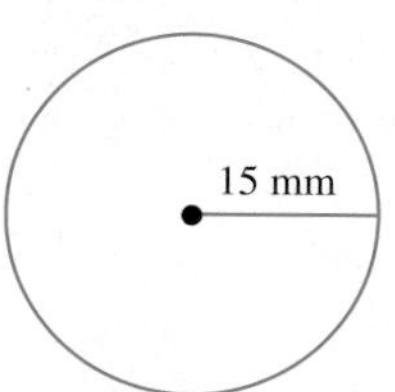

In Exercises 17–20, use the Pythagorean theorem to determine the third side of the triangle.

17.
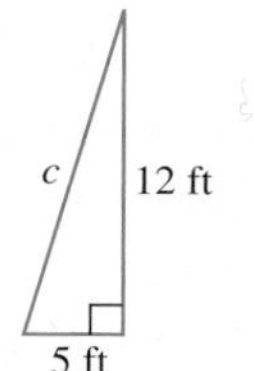

18.
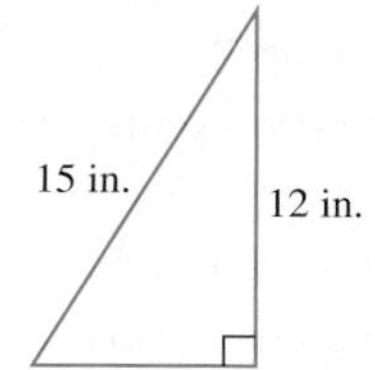

19.
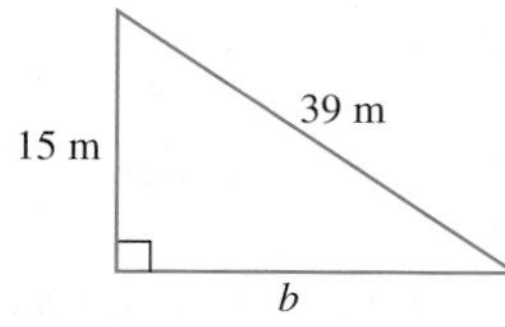

20.
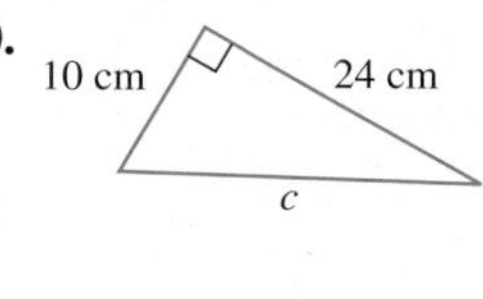

PROBLEM SOLVING

In Exercises 21–28, find the shaded area. Round your answers to hundredths.

21.
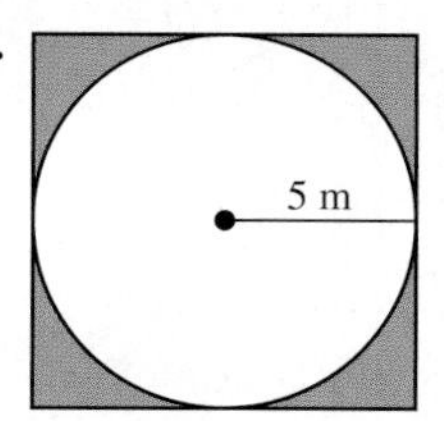

22.
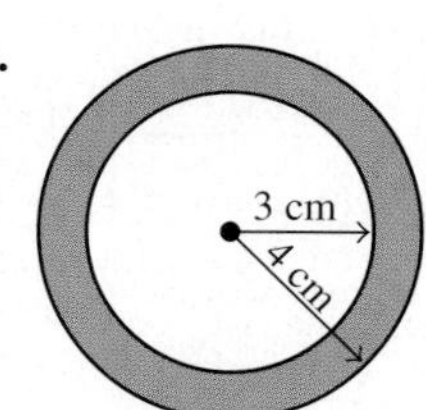

23.
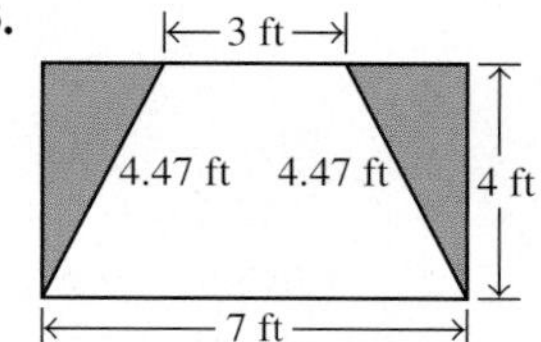

24.
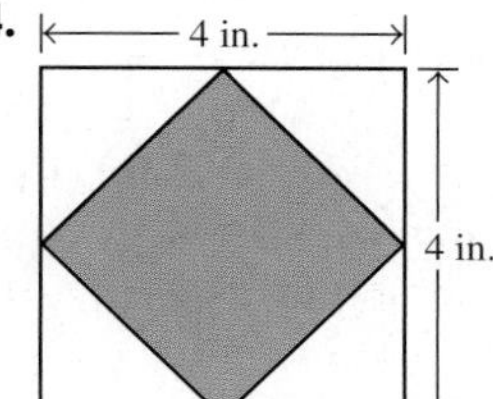

25.

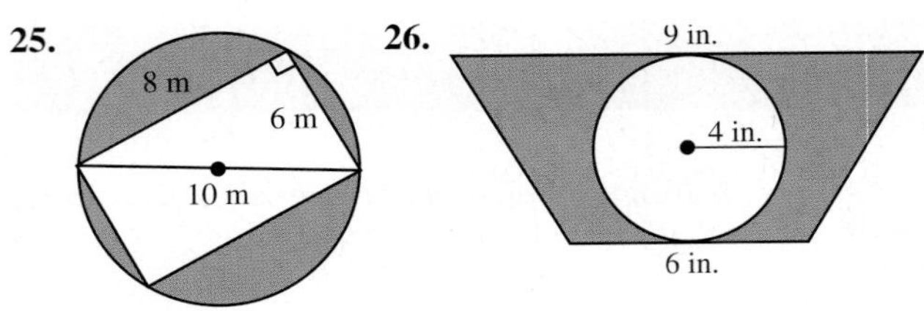

26.

27.

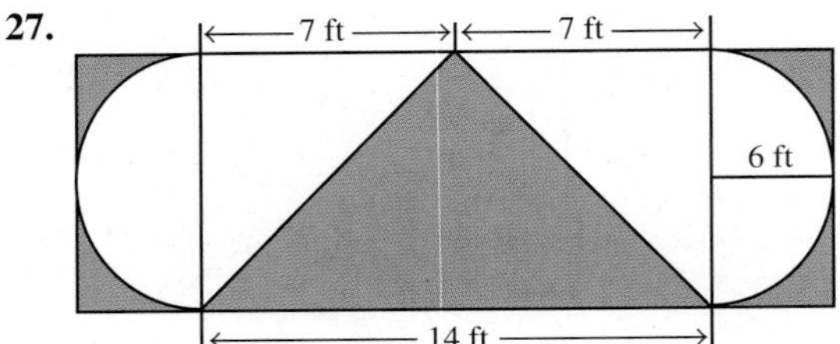

28.

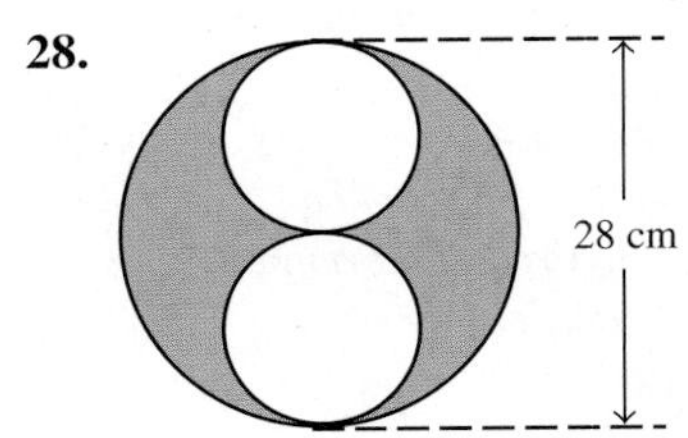

One square yard equals 9 ft². Use this information to convert

29. 15.2 ft^2 to square yards. **30.** 107 ft^2 to square yards.

31. 18.3 yd^2 to square feet. **32.** 14.7 yd^2 to square feet.

One square meter equals 10,000 cm². Use this information to convert

33. 14.7 m^2 to square centimeters

34. 23.4 m^2 to square centimeters

35. 608 cm^2 to square meters

36. 1075 cm^2 to square meters

In Exercises 37–40, use the measurements given on the floor plans to obtain the answer.

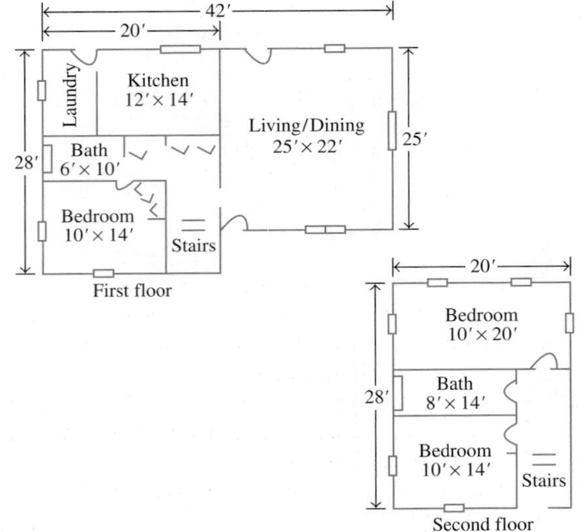

37. *Cost of Building a New House* One method of estimating the cost of building a new house is to use square footage of living space (using exterior dimensions). If the current rate is \$65 per square foot, find the cost of building the house.

38. *Cost of Hardwood Floors* Determine the cost of putting a hardwood floor in the living/dining area of the house. The cost of the hardwood flooring selected is \$4.50 per square foot.

39. *Cost of New Carpeting* Determine the cost of covering the three bedroom floors in the house with carpet. The price of the carpet selected is \$14.95 per square yard. Assume carpeting may be purchased in whole square yards only.

40. *Cost of Ceramic Tile* Determine the total cost of installing ceramic tile in the kitchen and in both bathrooms. The cost of ceramic tile is \$25 per square yard. Assume ceramic tile may be purchased in whole square yards only.

41. *Area of a Swimming Pool Deck* A deck 6 ft wide surrounds a swimming pool 30 ft long by 15 ft wide. Find the area of the deck.

42. *Area of a Picture Frame* A picture frame 3 in. wide surrounds a portrait that is 19 in. wide by 25 in. high. Find the area of the picture frame.

43. *Cost of a Lawn Service* Jim and Wendy Scott's home lot is illustrated here. The Scotts wish to hire a lawn service to cut their lawn. M&M Lawn Service charges \$0.02 per square yard of lawn. How much will it cost the Scotts to have their lawn cut?

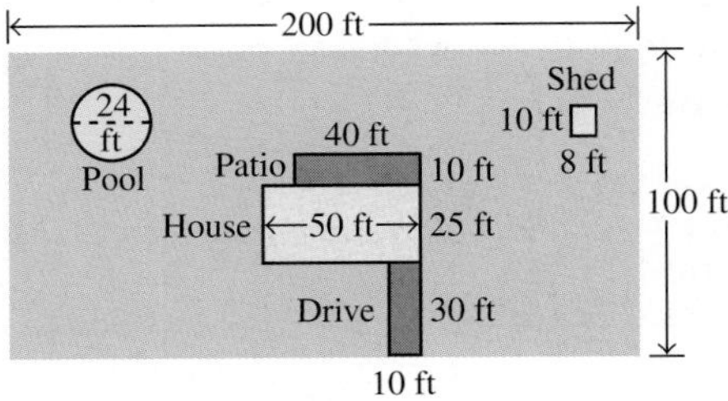

44. *Cost of a Lawn Service* Clarence and Rose Mikelos' home lot is illustrated here. Clarence and Rose wish to hire Picture Perfect Lawn Service to cut their lawn. How much will it cost Clarence and Rose to have their lawn cut if Picture Perfect charges \$0.02 per square yard?

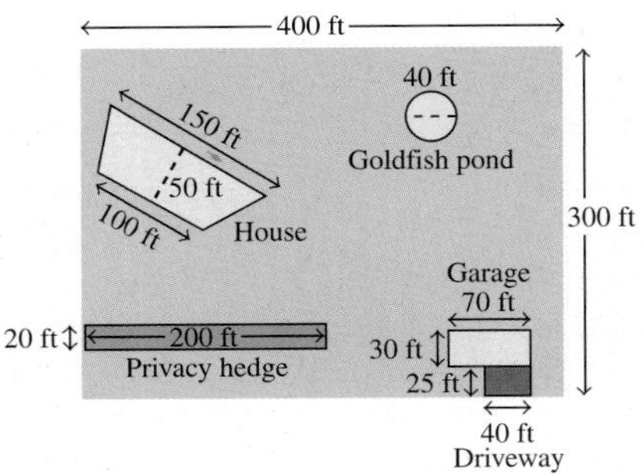

45. *Area of a Fireplace Base* The shape and measurements of the base of a free-standing fireplace are shown. The base will be covered with a special stone. Determine the number of square yards of stone the mason must order.

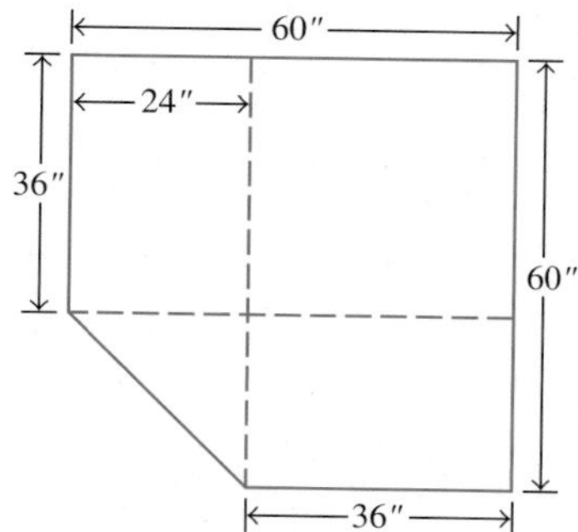

46. *Area of a Garden* Herman's rectangular garden is 11.5 m by 15.4 m.
    a) How large is his garden in square meters?
    b) If 1 hectare (a measurement of area in the metric system) equals 10,000 m^2, how large is his garden in hectares?

47. *Area of an Isosceles Triangle* An isosceles right triangle has an area of 72 cm^2. What are the lengths of the sides?

48. *Hamburger Comparison* Which hamburger has the larger surface area: a square hamburger 3 in. on a side from Wendy's or a $3\frac{1}{2}$-in.-diameter round hamburger from Burger King? Explain your answer and give the difference in their surface areas.

CHALLENGE PROBLEMS/GROUP ACTIVITY

49. *Doubling the Sides of a Square* In the figure below, an original square with sides of length s is shown. Also shown is a larger square with sides double in length, or $2s$.

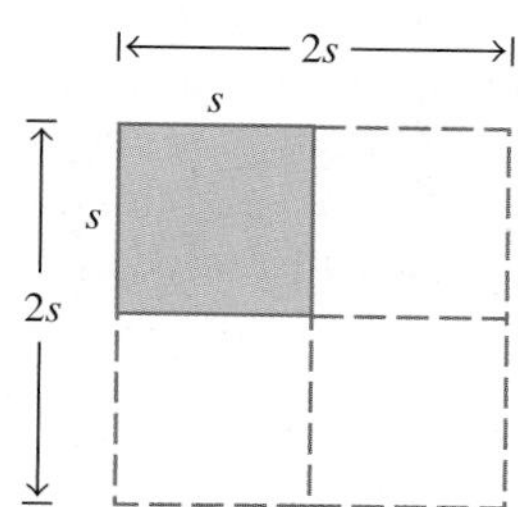

    a) Express the area of the original square in terms of s.
    b) Express the area of the larger square in terms of s.
    c) How many times larger is the area of the square in part (b) than the area of the square in part (a)?

50. *Doubling the Sides of a Triangle* In the figure above and to the right, an original triangle with base b and height h is shown. Also shown is a larger triangle with base and height double in length, or $2b$ and $2h$, respectively.

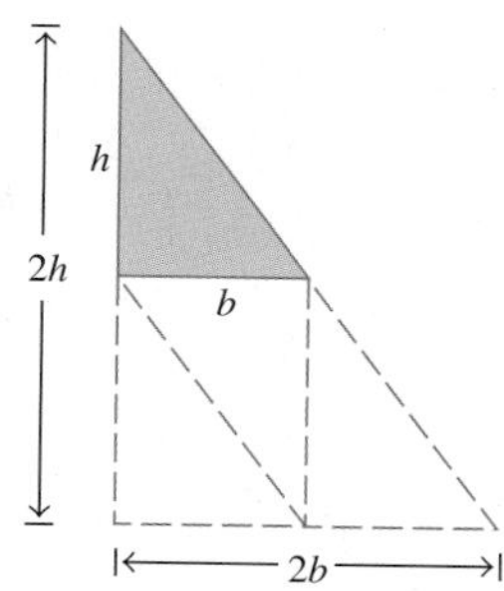

    a) Express the area of the original triangle in terms of b and h.
    b) Express the area of the larger triangle in terms of b and h.
    c) How many times larger is the area of the triangle in part (b) than the area of the triangle in part (a)?

51. *Doubling the Sides of a Parallelogram* In the figure below, an original parallelogram with base b and height h is shown. Also shown is a larger parallelogram with base and height double in length, or $2b$ and $2h$, respectively.

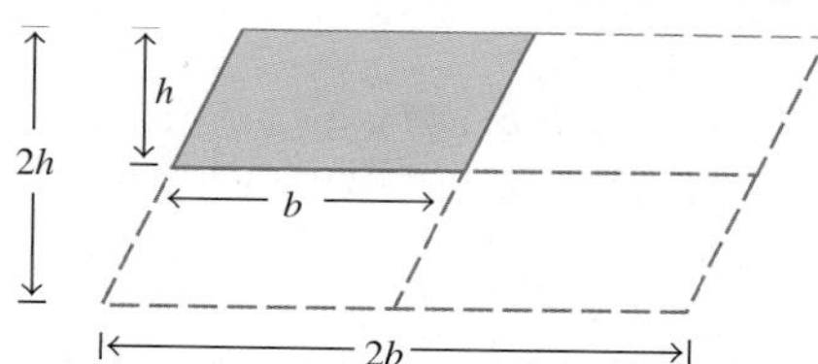

    a) Express the area of the original parallelogram in terms of b and h.
    b) Express the area of the larger parallelogram in terms of b and h.
    c) How many times larger is the area of the parallelogram in part (b) than the area of the parallelogram in part (a)?

52. *Painting a House* The Cunninghams are painting their house this summer. The dimensions of the house are shown.

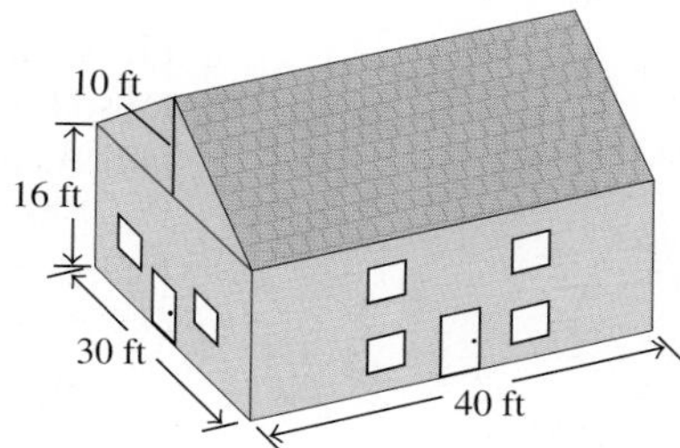

    a) Find the total number of square feet to be painted. (Assume that there are 12 windows, each 2 ft 3 in. × 3 ft 4 in., and two doors, each 80 in. × 36 in. The roof will not be painted.)
    b) Determine the number of gallons needed if two coats of paint are required. For each coat assume that 1 gal of paint covers 500 ft^2.
    c) Determine the cost of the paint if paint sells for \$24.95 per gallon.

53. Find the surface area of the right circular cylinder shown. Use $S = 2\pi rh + 2\pi r^2$.

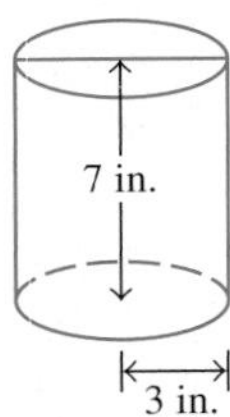

54. Find the surface area of the rectangular solid shown.

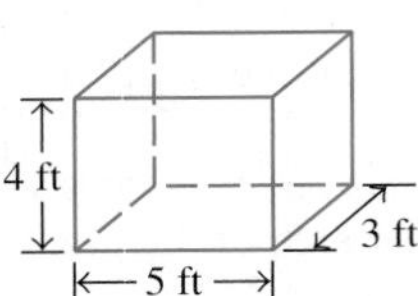

55. A second formula for determining the area of a triangle (called Heron's formula) is

$$A = \sqrt{s(s - a)(s - b)(s - c)}$$

where $s = \frac{1}{2}(a + b + c)$ and a, b, and c are the lengths of the sides of the triangle. Use Heron's formula to determine the area of right triangle ABC (above and to the right) and check your answer using the formula $A = \frac{1}{2}ab$.

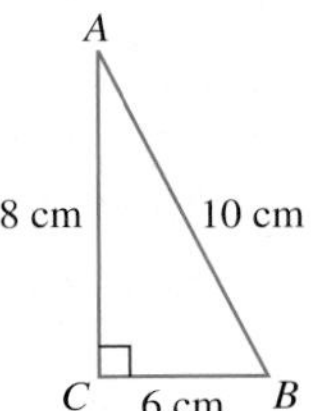

56. *Cost of Land* A developer is offered two triangular pieces of land. The dimensions of the first piece are 15 ft × 25 ft × 35 ft. The dimensions of the second piece are 60 ft × 100 ft × 140 ft. Since the dimensions of the second piece are four times the dimensions of the first, the two triangles are similar. The price of the larger piece of land is eight times the price of the smaller piece of land. Determine which piece of land is the least expensive per square foot.

RESEARCH ACTIVITIES

For Exercises 57 and 58, references include history of mathematics books, encyclopedias, and Internet sites.

57. The early Babylonians and Egyptians did not know about π and had to devise techniques to approximate the area of a circle. Do research and write a paper on the techniques these societies used to determine the area of a circle.

58. Write a paper on the contributions of Heron of Alexandria to geometry.

9.4 VOLUME

When discussing a one-dimensional figure, such as a line, we can find its length. When discussing a two-dimensional figure, such as a rectangle, we can find its area. When discussing a three-dimensional figure, such as a sphere, we can find its volume. **Volume** is a measure of the capacity of a figure. The measure of volume may be confusing because we use different units to measure different types of volumes. For example, water and other liquids may be measured in ounces, quarts, or gallons. A volume of topsoil may be measured in cubic yards. In the metric system, liquid may be measured in liters or milliliters, and topsoil may be measured in cubic meters.

Solid geometry is the study of three-dimensional solid figures (also called space figures). Volumes of three-dimensional figures are measured in cubic units such as cubic feet of cubic meters.

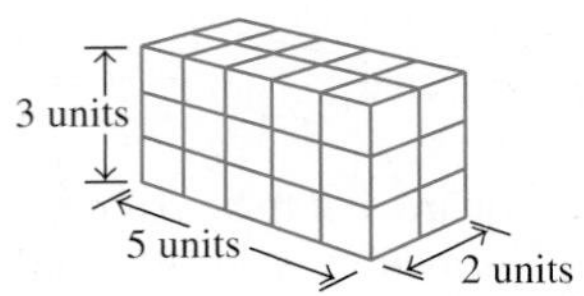

Figure 9.35

We will begin our discussion with the *rectangular solid.* If the length of the solid is 5 units, the width is 2 units, and the height is 3 units, the total number of cubes is 30 (Fig. 9.35). Thus, the volume is 30 cubic units. The volume of a rectangular solid can also be found by multiplying its length times width times height; in this case, 5 units × 2 units × 3 units = 30 cubic units. In general, the volume of any rectangular solid, as shown in part (a) of the summary box on page 461 is $V = l \times w \times h$.

Volume of a Rectangular Solid

$$V = l \times w \times h$$

A *cube* is a rectangular solid with the same length, width, and height (part b of the box below). If we call the side of a cube s and use the formula for a rectangular solid, substituting s for l, w, and h, we obtain $V = s \cdot s \cdot s = s^3$.

Volume of a Cube

$$V = s^3$$

Now consider the right circular cylinder (part c of the box below). The base is a circle with area πr^2. When we add height, h, the figure becomes a cylinder. For the same circular base, the greater the height, the greater is the volume. The volume of the right circular cylinder is found by multiplying the area of the base, πr^2, by the height h.

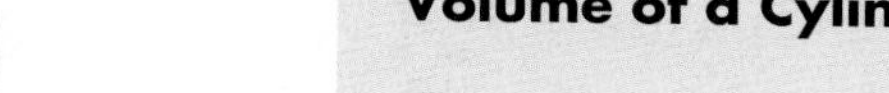

Volume of a Cylinder

$$V = \pi r^2 h$$

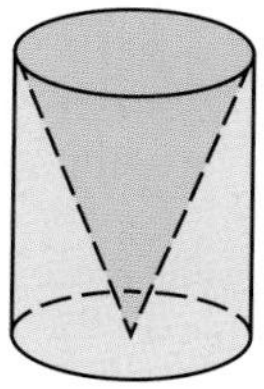

Figure 9.36

A cone is illustrated in part (d) of the box below. Imagine a cone inside a cylinder, sharing the same circle as base. The volume of the cone is less than the volume of the cylinder that has the same base and height (Fig. 9.36). In fact, the volume of the cone is one-third the volume of the cylinder.

Volume of a Cone

$$V = \frac{1}{3}\pi r^2 h$$

The next shape we will discuss in this section is the sphere (part e of the box below). Basketballs, golf balls, and so on have the shape of a sphere. The formula for the volume of a sphere is as follows.

Volume of a Sphere

$$V = \frac{4}{3}\pi r^3$$

The following is a summary of the volumes of selected three-dimensional figures:

Volumes

Rectangular Solid	Cube	Cylinder	Cone	Sphere
$V = l \times w \times h$	$V = s^3$	$V = \pi r^2 h$	$V = \frac{1}{3}\pi r^2 h$	$V = \frac{4}{3}\pi r^3$
(a)	(b)	(c)	(d)	(e)

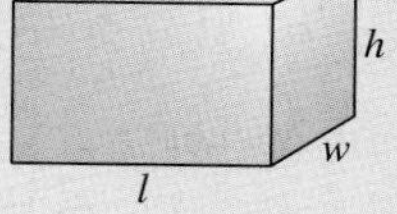

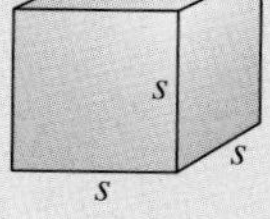

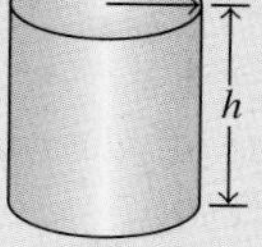

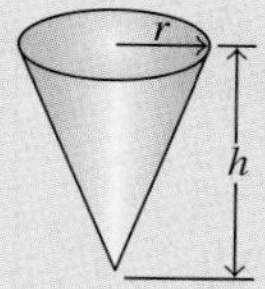

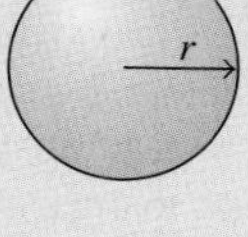

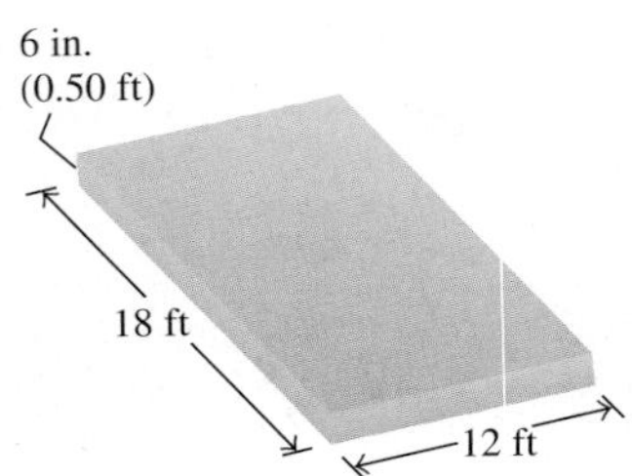

Figure 9.37

DID YOU KNOW

Fractured Volumes

Picasso, Pablo. *Les Demoiselles d'Avignon.* Paris (June–July 1907). Oil on canvas, 8 ft × 7 ft 8 in. (243.9 cm × 233.7 cm). The Museum of Modern Art, New York. Acquired through the Lillie P. Bliss Bequest. Photograph © 1996 The Museum of Modern Art, New York. © Estate of Pablo Picasso.

The cubist painters of the early twentieth century sought to analyze forms as geometric shapes. They were greatly influenced by Cézanne who, in a famous letter, wrote of treating nature in terms of the cylinder, sphere, and cone. In one of the first paintings of the cubist period, *Les Demoiselles d'Avignon,* Picasso dismissed the idea of the human figure as a dynamic unity and instead fractured volumes and planes to present them from multiple angles of vision at one time.

EXAMPLE 1 *Concrete for a Sun Porch*

Johnny Melton is adding a sun porch to his home. The porch will have a concrete floor measuring 18 ft long by 12 ft wide. The concrete will have a uniform thickness of 6 in., see Fig. 9.37. Concrete costs \$61.25 per cubic yard.

a) How many cubic yards of concrete are needed?
b) How much will the concrete cost?

SOLUTION

a) Since we are asked to find the volume in cubic yards, we will convert each measurement to yards. There are 3 ft in a yard. So 18 ft equals $\frac{18}{3}$ or 6 yd and 12 ft equals $\frac{12}{3}$ or 4 yd. There are 36 in. in a yard. So 6 in. equals $\frac{6}{36}$ or $\frac{1}{6}$ yd. The amount of concrete needed is found using the formula for the volume of a rectangular solid:

$$\begin{aligned} V &= l \cdot w \cdot h \\ &= 6 \cdot 4 \cdot \tfrac{1}{6} \\ &= 4 \text{ yd}^3 \end{aligned}$$

Note that since the measurements for length, width, and height are each in terms of yards, the answer is in cubic yards.

b) One cubic yard of concrete costs \$61.25, so 4 yd^3 will cost 4 × \$61.25, or \$245.00.

EXAMPLE 2 *Keeping Soda Cold*

Tobi and Tacinto and their friends are at a picnic at the town park. They have brought a children's wading pool in the shape of a right circular cylinder with a radius of 2 ft and a height of 1 ft into which they will put cold water to keep the soda cold (Fig. 38a). They carry the water from the faucet to the pool in a bucket that is also in the shape of a right circular cylinder, with a diameter of 1 ft and a height of 1 ft (Fig. 9.38b).

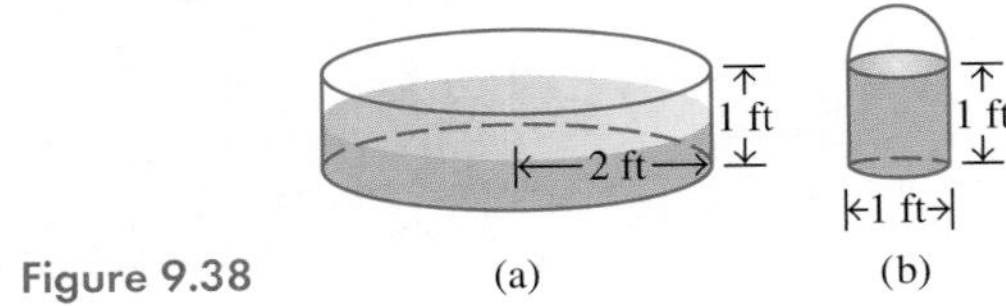

Figure 9.38 (a) (b)

a) How many buckets of water are needed to fill the pool to a height of $\frac{1}{2}$ ft?
b) If 1 ft^3 of water weighs 62.5 lb, what is the weight of the water in the pool?
c) If there are 7.5 gal of water per cubic foot, how many gallons of water are in the pool?

SOLUTION

a) Volume of water in pool

$$\begin{aligned} V &= \pi r^2 h \\ &\approx 3.14(2)^2(0.5) \\ &\approx 3.14(4)(0.5) \approx 6.28 \text{ ft}^3 \end{aligned}$$

Volume of water in bucket

$$\begin{aligned} V &= \pi r^2 h \\ &\approx 3.14(0.5)^2(1) \\ &\approx 3.14(0.25)(1) \approx 0.785 \text{ ft}^3 \end{aligned}$$

To find the number of buckets needed, we divide

$$\frac{6.28}{0.785} = 8.0$$

Thus, eight bucketfuls are needed.

b) Since 1 ft^3 of water weighs 62.5 lb, 6.28 ft^3 of water weighs 6.28×62.5, or 392.5 lb.

c) Since 1 ft^3 of water contains 7.5 gal, 6.28 ft^3 of water contains 6.28×7.5, or 47.1 gal of water. Thus, there are 47.1 gal of water in the pool. ▲

A polyhedron is a closed surface formed by the union of polygonal regions. Figure 9.39 illustrates some polyhedrons.

Polyhedrons

Vertex
Edge
Face

(a) (b) (c) (d)

Figure 9.39

Each polygonal region is called a *face* of the polyhedron. The line segment formed by the intersection of two faces is called an *edge*. The point at which two or more edges intersect is called a *vertex*. In Fig. 9.39(a) there are 6 faces, 12 edges, and 8 vertices. Note that

$$\begin{array}{ccccccc} \text{Number of vertices} & - & \text{number of edges} & + & \text{number of faces} & = & 2 \\ 8 & - & 12 & + & 6 & = & 2 \end{array}$$

This formula, credited to Leonhard Euler, is true for any polyhedron.

Euler's Polyhedron Formula

Number of vertices − number of edges + number of faces = 2

We suggest that you verify this formula holds for Fig. 9.39(b), (c), and (d).

EXAMPLE 3 ***Using Euler's Polyhedron Formula***

A certain polyhedron has 6 vertices and 9 edges. Determine the number of faces on this polyhedron.

SOLUTION Since we are seeking the number of faces, we will let x represent the number of faces on the polyhedron. Next, we will use Euler's polyhedron formula to set up an equation:

DID YOU KNOW

The Shape of Gems?

Gems, like all crystals, are composed of atoms arranged in regular three-dimensional geometric patterns. Many will split naturally into flat, well-defined surfaces, like this rough-cut piece of amethyst. Gems are prized for their beauty, durability, and rarity. They are traditionally cut into certain polyhedral shapes to bring out their beauty.

$$\begin{aligned} \text{Number of vertices} - \text{number of edges} + \text{number of faces} &= 2 \\ 6 - 9 + x &= 2 \\ -3 + x &= 2 \\ x &= 5 \end{aligned}$$

Therefore, the polyhedron has 5 faces.

A **regular polyhedron** is one whose faces are all regular polygons of the same size and shape. Figure 9.39(a) and (b) are regular polyhedrons.

A **prism** is a special type of polyhedron whose bases are congruent polygons and whose sides are parallelograms. These parallelogram regions are called the *lateral faces* of the prism. If all the lateral faces are rectangles, the prism is said to be a **right prism**. Some right prisms are illustrated in Fig. 9.40.

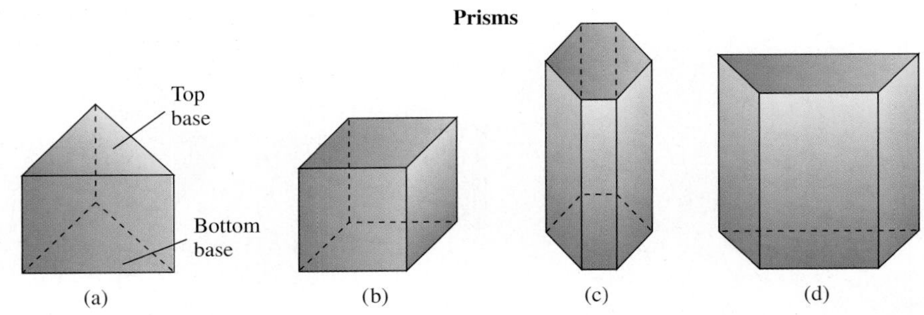

Figure 9.40

The volume of any prism can be found by multiplying the area of the base, B, by the height, h, of the prism.

Volume of a Prism

$$V = Bh$$

where B is the area of a base and h is the height.

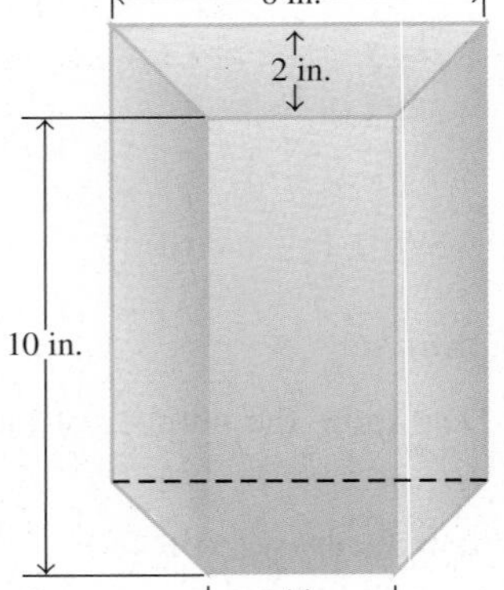

Figure 9.41

EXAMPLE 4 ***Finding the Volume of a Trapezoidal Prism***

Find the volume of the trapezoidal prism shown in Fig. 9.41.

SOLUTION First find the area of the trapezoidal base.

$$\begin{aligned} \text{Area of the base} &= \tfrac{1}{2}h(b_1 + b_2) \\ &= \tfrac{1}{2}(2)(8 + 4) = 12 \text{ in.}^2 \end{aligned}$$

Now find the volume by multiplying the area of the trapezoidal base by the height of the prism.

$$\begin{aligned} V &= B \cdot h \\ &= 12 \cdot 10 = 120 \text{ in.}^3 \end{aligned}$$

EXAMPLE 5 *Volumes Involving Prisms*

Find the volume of the remaining solid after the cylinder, triangular prism, and square prism have been cut from the solid (Fig. 9.42).

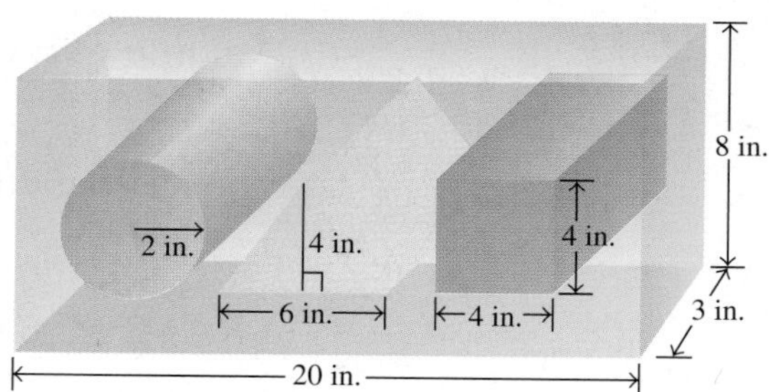

Figure 9.42

SOLUTION To find the volume of the remaining solid, first find the volume of the rectangular solid. Then subtract the volume of the two prisms and the cylinder that were cut out.

$$\text{Volume of rectangular solid} = l \cdot w \cdot h = 20 \cdot 3 \cdot 8 = 480 \text{ in.}^3$$

$$\text{Volume of circular cylinder} = \pi r^2 h \approx (3.14)(2^2)(3) \approx (3.14)(4)(3) \approx 37.68 \text{ in.}^3$$

$$\text{Volume of triangular prism} = \text{area of the base} \cdot \text{height} = \tfrac{1}{2}(6)(4)(3) = 36 \text{ in.}^3$$

$$\text{Volume of square prism} = s^2 \cdot h = 4^2 \cdot 3 = 48 \text{ in.}^3$$

$$\text{Volume of solid} \approx 480 - 37.68 - 36 - 48 \approx 358.32 \text{ in.}^3$$

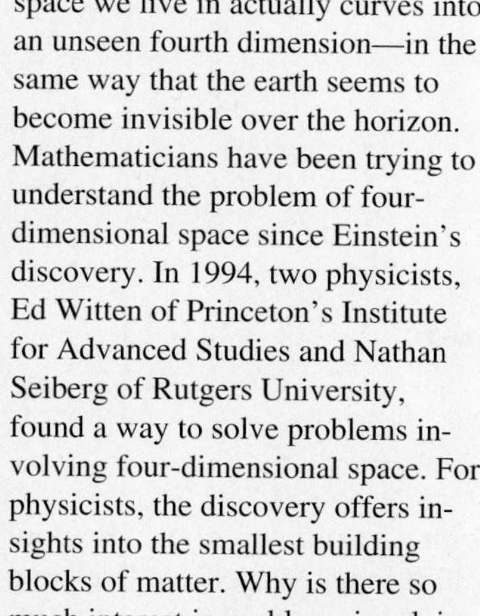

DID YOU KNOW

It's the Space We Live In

Albert Einstein's work showed that the three-dimensional space we live in actually curves into an unseen fourth dimension—in the same way that the earth seems to become invisible over the horizon. Mathematicians have been trying to understand the problem of four-dimensional space since Einstein's discovery. In 1994, two physicists, Ed Witten of Princeton's Institute for Advanced Studies and Nathan Seiberg of Rutgers University, found a way to solve problems involving four-dimensional space. For physicists, the discovery offers insights into the smallest building blocks of matter. Why is there so much interest in problems involving four-dimensional space? One obvious answer, says mathematician Ron Stern of the University of California at Irvine, is that "it's the space we live in."

Another special category of polyhedrons is the **pyramid**. Unlike prisms, pyramids have only one base. Some pyramids are illustrated in Fig. 9.43. Note that all but one face of a pyramid intersect at a common vertex.

Pyramids

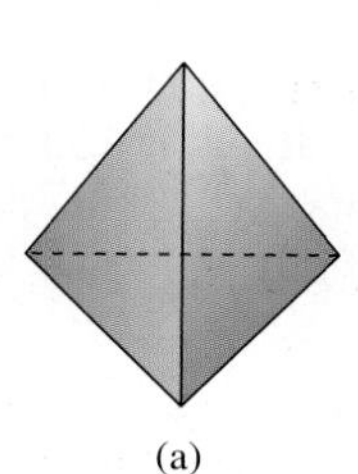
(a)

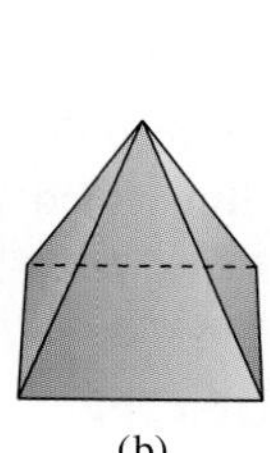
(b)

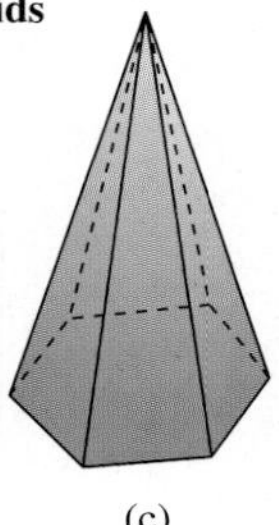
(c)

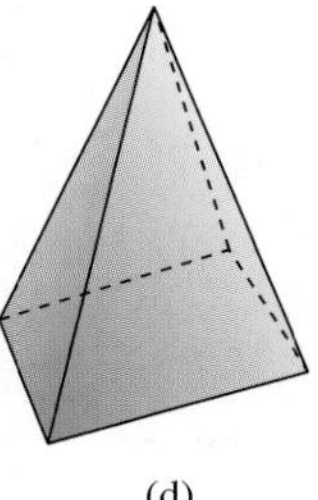
(d)

Figure 9.43

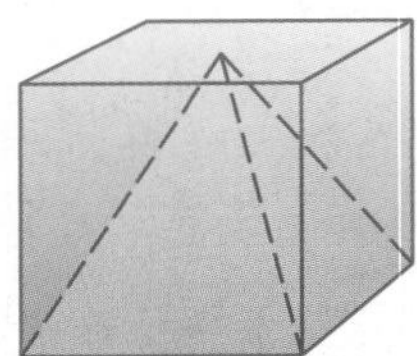
Figure 9.44

If a pyramid is drawn inside a prism, as shown in Fig. 9.44, the volume of the pyramid is less than that of the prism. In fact, the volume of the pyramid is one-third the volume of the prism.

Volume of a Pyramid

$$V = \tfrac{1}{3}Bh$$

where B is the area of the base and h is the height.

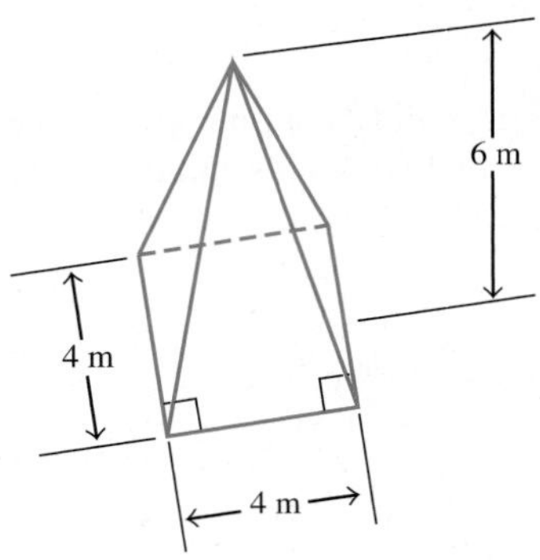

Figure 9.45

EXAMPLE 6 ***Volume of a Pyramid***

Find the volume of the pyramid shown in Fig. 9.45.

SOLUTION First find the area of the base of the pyramid. Since the base of the pyramid is a square

$$\text{Area of base} = s^2 = 4^2 = 16 \text{ m}^2$$

Now use this to find the volume of the pyramid.

$$\begin{aligned} V &= \tfrac{1}{3} \cdot B \cdot h \\ &= \tfrac{1}{3} \cdot 16 \cdot 6 \\ &= 32 \text{ m}^3 \end{aligned}$$

In certain situations converting volume from one cubic unit to a different cubic unit might be necessary. For example, when purchasing topsoil you might have to change the amount of topsoil from cubic feet to cubic yards prior to placing your order.

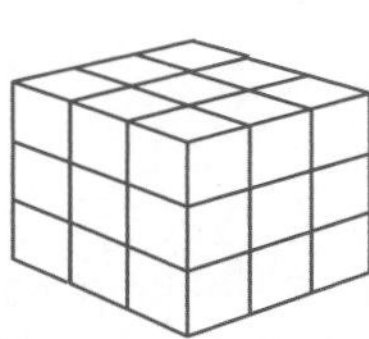
Figure 9.46

EXAMPLE 7 ***Cubic Yards and Cubic Feet***

a) Convert 1 yd^3 to cubic feet. (See Fig. 9.46.)
b) Convert 8.3 yd^3 to cubic feet.

SOLUTION

a) 1 yd = 3 ft
   1 yd^3 = 3 ft · 3 ft · 3 ft = 27 ft^3
b) 1 yd^3 = 27 ft^3
   Thus, 8.3 yd^3 = 8.3 × 27 = 224.1 ft^3.

EXAMPLE 8 ***Filling in a Swimming Pool***

Mary Kaye Leonard recently purchased a home with a swimming pool. The pool has a uniform depth of 4.5 ft and is 15 ft wide by 30 ft long. Mary Kaye lives in a cold climate and so she plans to fill the pool in with dirt to make a flower garden. How many cubic yards of dirt will Mary Kaye have to purchase to fill in the swimming pool?

SOLUTION To find the amount of dirt, we will use the formula for the volume of a rectangular solid:

$$V = lwh$$
$$= (30)(15)(4.5)$$
$$= 2025 \text{ ft}^3$$

Now, we must convert this volume from cubic feet to cubic yards. In Example 7, we learned that 1 $\text{yd}^3 = 27 \text{ ft}^3$. Therefore, $2025 \text{ ft}^3 = \frac{2025}{27} = 75 \text{ yd}^3$. Thus, Mary Kaye needs to purchase 75 yd^3 of dirt to fill in her swimming pool.

SECTION 9.4 EXERCISES

CONCEPT/WRITING EXERCISES

1. In your own words, define *volume.*
2. What is solid geometry?
3. What is the difference between a polyhedron and a regular polyhedron?
4. What is the difference between a prism and a right prism?
5. In your own words, state Euler's polyhedron formula.
6. In your own words, explain the difference between a prism and a pyramid.

PRACTICE THE SKILLS

In Exercises 7–20, find the volume of the solid. Round your answers to hundredths.

7.

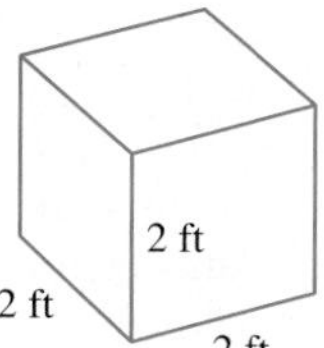

8.

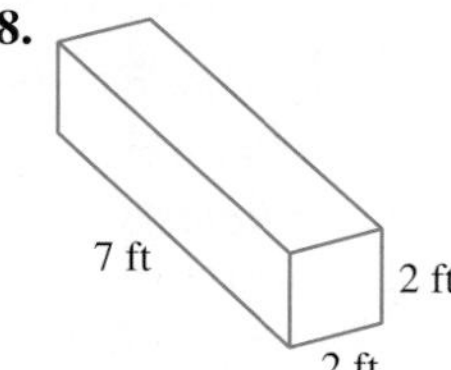

9.

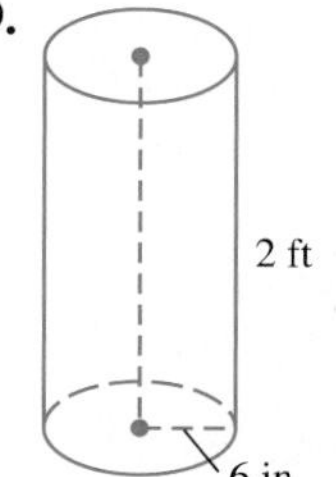

10.

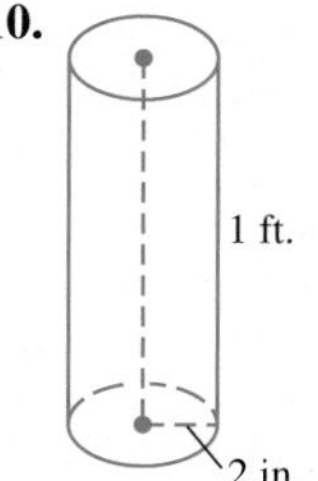

11.

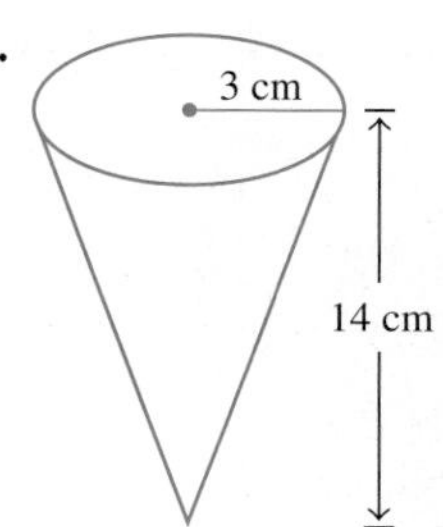

12.

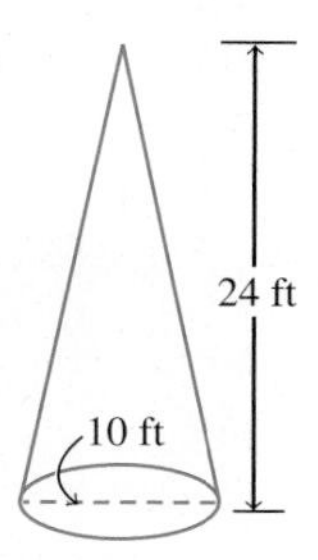

13.

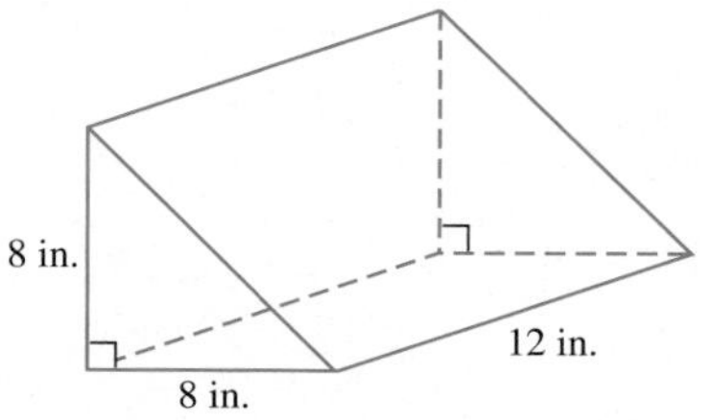

14.

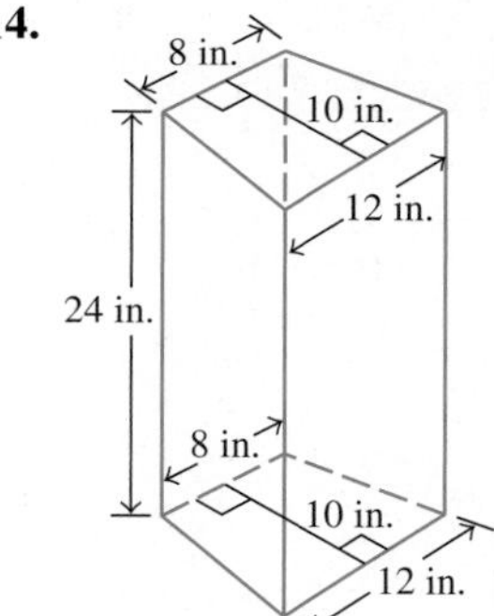

15.

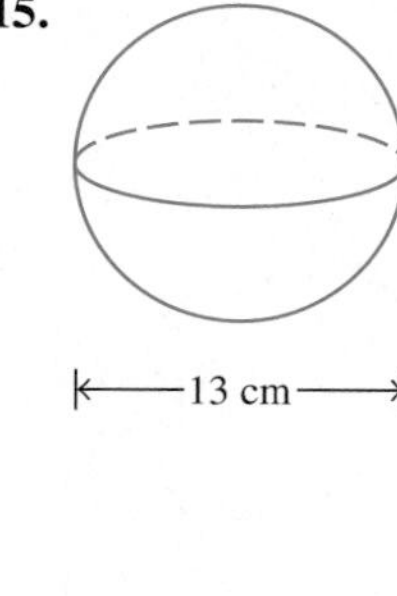

16.

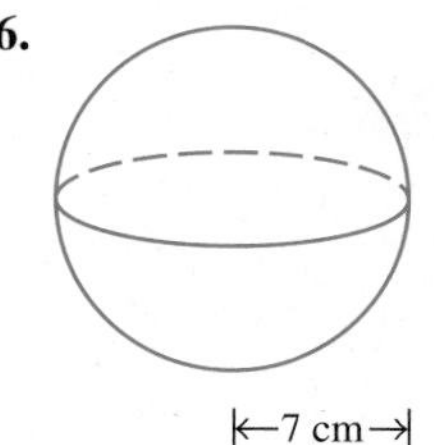

17.

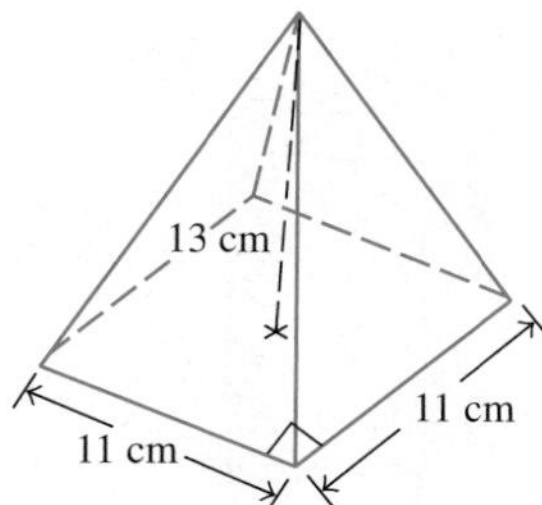

18.

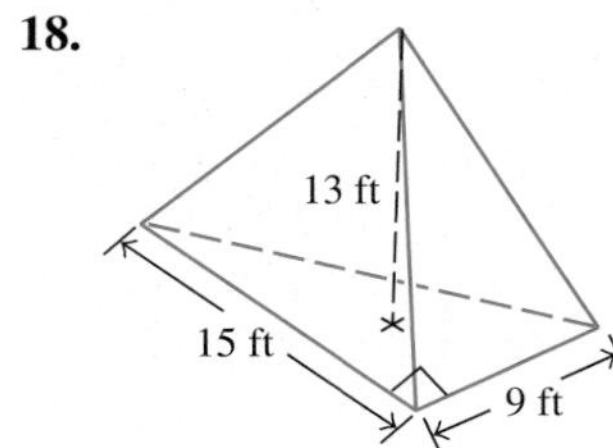

19.

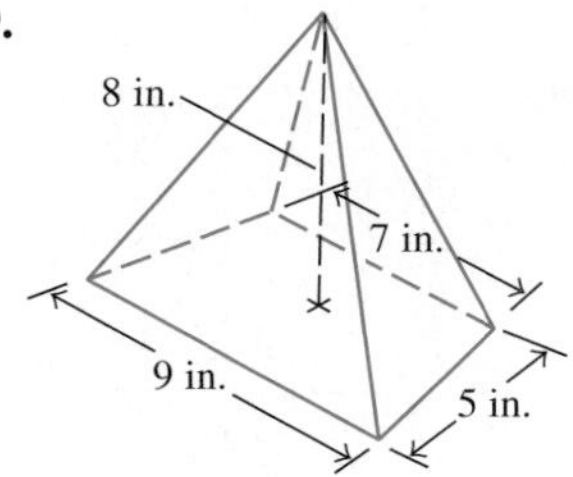

20.

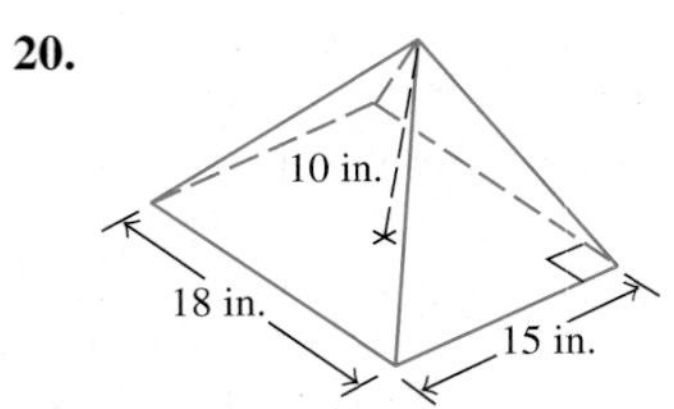

PROBLEM SOLVING

In Exercises 21–28, find the volume of the shaded area. Round your answers to hundredths.

21. 22. 23.

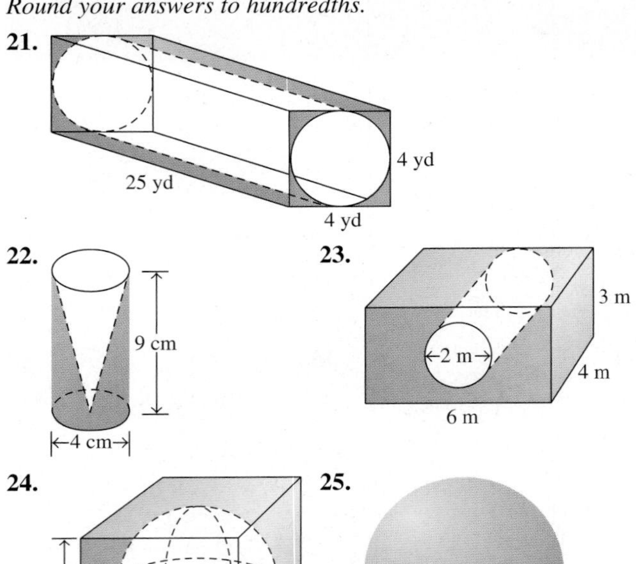

24. 25.

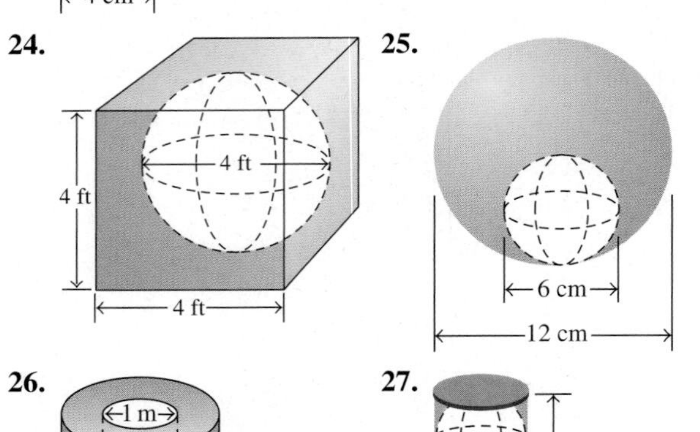

26.

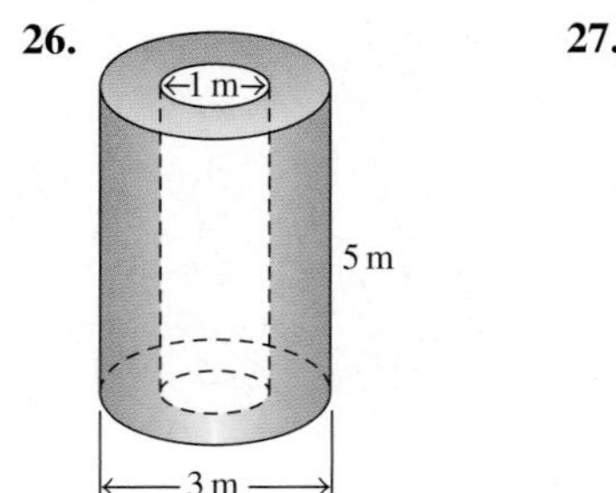

27.

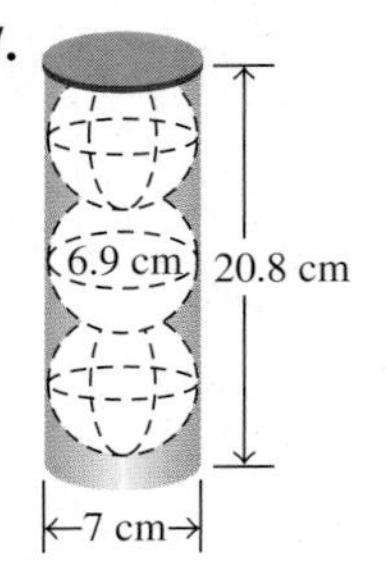

28.

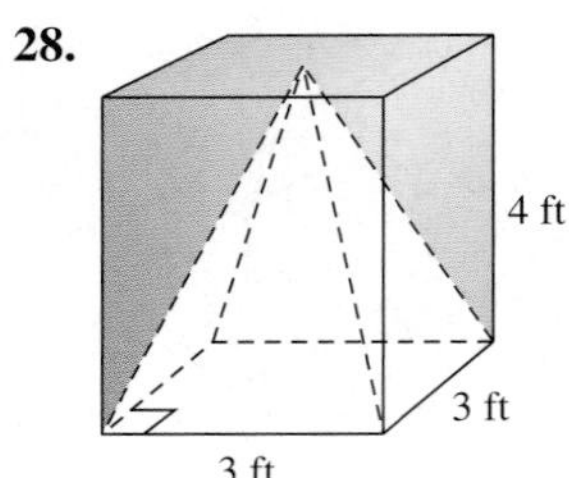

In Exercises 29–32, use the fact that 1 yd³ equals 27 ft³ to make the conversion.

29. 5 yd³ to cubic feet

30. 4.1 yd³ to cubic feet

31. 212 ft³ to cubic yards

32. 84 ft³ to cubic yards

In Exercises 33–36, use the fact that 1 m³ equals 1,000,000 cm³ to make the conversion.

33. 2.7 m³ to cubic centimeters

34. 5.9 m³ to cubic centimeters

35. 4,000,000 cm³ to cubic meters

36. 6,800,000 cm³ to cubic meters

37. *Ice Cream Comparison* The Louisburg Creamery packages its homemade ice cream in tubs and in boxes. The tubs are in the shape of a cylinder with a radius of 3 in. and height of 5 in. The boxes are in the shape of a cube with each side measuring 5 in. Determine the volume of each container.

38. *Volume of a Freezer* The dimensions of the interior of an upright freezer are height 46 in., width 25 in., and depth 25 in. Find its volume

a) In cubic inches. **b)** In cubic feet.

39. *Hamburger Comparison* The dimensions of a square Wendy's hamburger are length and width 4 in. and thickness $\frac{3}{16}$ in. The dimensions of a Magic Burger circular hamburger are diameter $4\frac{1}{2}$ in. and thickness $\frac{1}{4}$ in. Which hamburger has the greater volume? What is the difference in their volumes?

40. *Volume of a Bread Pan* A bread pan is 12 in. × 4 in. × 3 in. How many quarts does it hold, if 1 in.³ ≈ 0.01736 qt?

41. *Gasoline Containers* Bill Savage has two cylindrical containers for storing gasoline. One has a diameter of 10 in. and a height of 12 in. The other has a diameter of 12 in. and a height of 10 in.

a) Which container holds the greater amount of gasoline, the taller one or the one with the greater diameter?

b) What is the difference in volume?

42. *A Fish Tank*

a) How many cubic centimeters of water will a rectangular fish tank hold if the tank is 80 cm long, 50 cm wide, and 30 cm high?

b) If 1 cm³ holds 1 mℓ of liquid, how many milliliters will the tank hold?

c) If 1ℓ = 1000 mℓ, how many liters will the tank hold?

43. *A Swimming Pool*

a) What is the volume of water in a swimming pool that is 15 m long and 9 m wide and has an average depth of 2 m? Give your answer in cubic meters.

b) If 1 $m^3 = 1$ kℓ, how many kiloliters of water will the pool hold?

44. *The Pyramid of Cheops* The Pyramid of Cheops in Egypt has a square base measuring 720 ft on a side. Its height is 480 ft. What is its volume?

45. *Piston Displacement* What is the total piston displacement of a two-cylinder engine if each piston has a diameter of $2\frac{1}{4}$ in. and a $3\frac{1}{2}$ in. stroke?

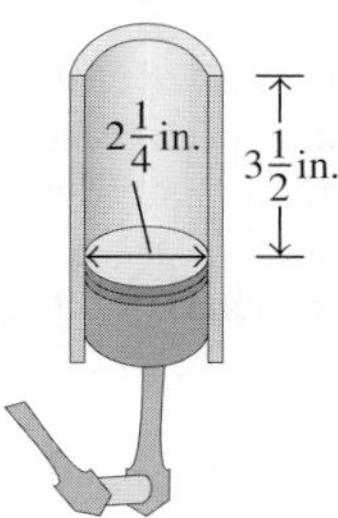

46. *A Ballpoint Pen* The ball at the end of a ballpoint pen has a diameter of 1.2 mm. Find the volume of the ball.

47. *Paving a Driveway* Al Walrich's driveway is 80 ft long by 12.5 ft wide. He plans to lay 4 in. of blacktop on the driveway.

a) How many cubic feet of blacktop does he need?

b) How many cubic yards?

48. *Comparing Cake Pans* In baking a cake you have to choose between a round pan with a 9 in. diameter and a 7 in. × 9 in. rectangular pan.

a) Determine the area of the base of each pan.

b) If both pans are 2 in. deep, determine the volume of each pan.

c) Which pan has the larger volume?

49. *Cake Icing* A bag used to apply icing to a cake is in the shape of a cone with a diameter of 3 in. and a height of 6 in. How much icing will this bag hold when full?

50. *Water Bed* A twin sized waterbed mattress is 6 ft long, 4 ft wide, and 9 in. high.

a) How many cubic feet of water are needed to fill the waterbed mattress?

b) If 1 ft^3 of water weighs 62.5 lb, what is the weight of the water in the mattress?

c) If 1 gal of water weighs 8.3 lb, how many gallons of water does the mattress hold?

51. *The Leaning Tower of Pisa* The Leaning Tower of Pisa was designed to be a vertical bell tower for a cathedral. If the tower were vertical, it would be 60 meters high with a diameter of about 19.6 meters roughly in the shape of a cylinder. Use this information to find the following:

a) the circumference of the tower

b) the volume of tower

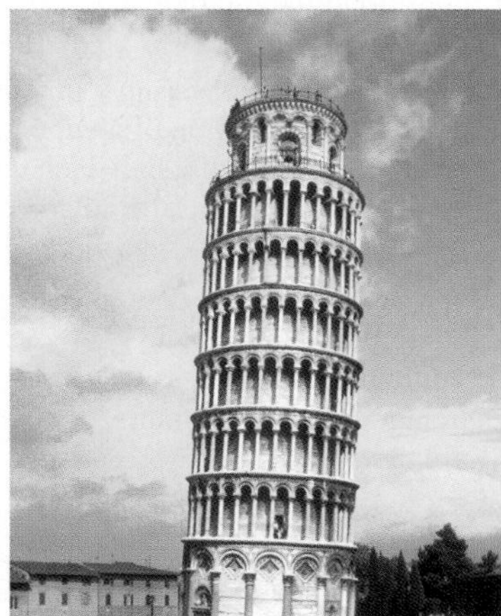

52. *Flower Box* A flower box is 4 ft long, and its ends are in the shape of a trapezoid. The upper and lower bases of the trapezoid measure 12 in. and 8 in., respectively, and the height is 9 in. Find the volume of the flower box

a) in cubic inches.

b) in cubic feet.

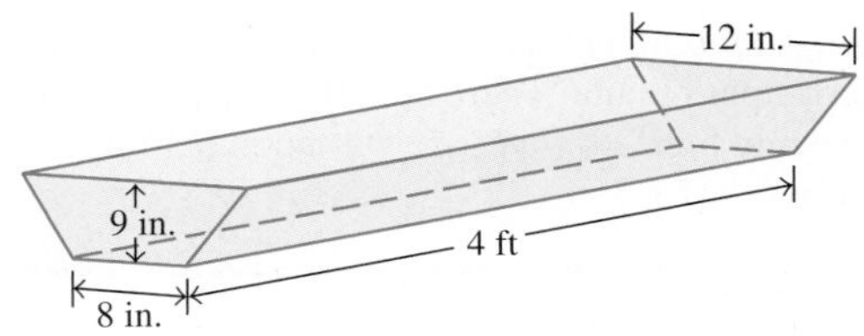

In Exercises 53–58, find the missing value indicated by the question mark. Use the following formula.

$$\begin{pmatrix}\text{Number of}\\ \text{vertices}\end{pmatrix} - \begin{pmatrix}\text{number of}\\ \text{edges}\end{pmatrix} + \begin{pmatrix}\text{number}\\ \text{of faces}\end{pmatrix} = 2$$

	Number of vertices	Number of edges	Number of faces
53.	12	16	?
54.	8	?	3
55.	?	8	4
56.	7	12	?
57.	11	?	5
58.	?	10	4

CHALLENGE PROBLEMS/GROUP ACTIVITIES

59. What effect does doubling the length of each edge of a cube have on its volume? Explain.

60. What effect does doubling the radius of the base of a cylinder have on its volume? Explain.

61. What effect does doubling the radius of a sphere have on its volume? Explain.

62. A half-inch cube is a cube measuring $\frac{1}{2}$ in. on a side. A half cubic inch is half the volume of a cube measuring 1 in. on a side. Is the volume of a half-inch cube the same as the volume of a half cubic inch? Explain.

63. *Comparing Popcorn Containers* A theater decides to change the shape of its popcorn container from a regular cone to a right circular cylinder and charge twice as much. If the containers have the same radius and height, is this new circular cylinder container a better buy for the customer? Explain.

64. *Comparing Grapefruits* Which is a better buy per cubic centimeter, a grapefruit 6 cm in radius that costs \$0.33 or a grapefruit 7 cm in radius that costs \$0.49? Explain.

65. *Packing Orange Juice* A box is packed with six cans of orange juice. The cans are touching each other and the sides of the box, as shown. What percent of the volume of the interior of the box is not occupied by the cans?

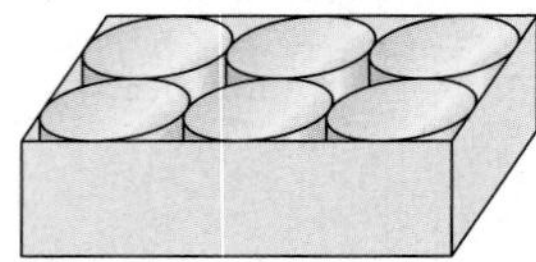

66. *A Dripping Faucet* A faucet is dripping at the rate of 12 drops per minute. There are 20 drops per milliliter. How many liters of water are wasted in a 30-day period? (1000 mℓ = 1 ℓ.)

67. **a)** Explain how to demonstrate, using the cube shown, that

$$(a + b)^3 = a^3 + 3a^2b + 3ab^2 + b^3$$

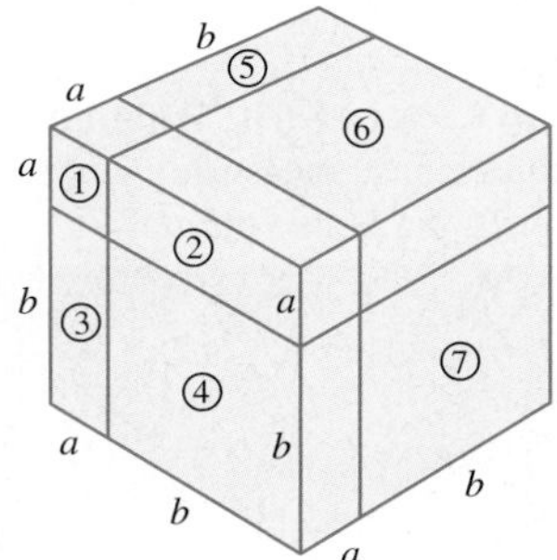

b) What is the volume in terms of a and b of each numbered piece in the figure?

c) An eighth piece is not illustrated. What is its volume?

68. The general procedure for determining the number of rolls of wallpaper needed to wallpaper a given rectangular room is as follows.

i) Determine the room's overall perimeter, in feet, by measuring the distance around the room where the wall meets the floor. Ignore windows and doors at this time.

ii) Multiply the perimeter by the height in feet of the room. This result gives the number of square feet of wall space.

iii) Divide this number by 30, the approximate number of usable square feet in a roll of wallpaper.

iv) From this number, deduct one roll for every two averaged-sized openings (windows, doors, fireplace, and so on).

Helen Latvis plans to wallpaper her kitchen. It is 12 ft long, 10 ft wide, and 8 ft high, and it contains two doorways and two windows. How many rolls of wallpaper will she need for her kitchen?

RESEARCH ACTIVITIES

69. Calculate the volume of the room in which you sleep or study. Go to a store that sells room air conditioners and find out how many cubic feet can be cooled by the different models available. Describe the model that would be the proper size for your room. What is the initial cost? How much does that model cost to operate? If you moved to a room that had twice the amount of floor space and the same height, would the air conditioner you selected still be adequate? Explain.

70. Pappus of Alexandria (ca. A.D. 350) was the last of the well-known ancient Greek mathematicians. Write a paper on his life and his contributions to mathematics.

9.5 THE MÖBIUS STRIP, KLEIN BOTTLE, AND MAPS

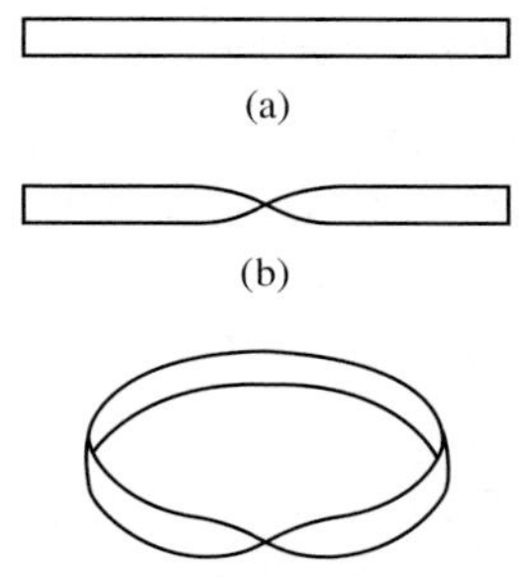

Figure 9.47

A branch of mathematics called **topology** is sometimes referred to as "rubber sheet geometry" because it deals with bending and stretching of geometric figures.

One of the first pioneers of topology was the German astronomer and mathematician August Ferdinand Möbius (1790–1866). A student of Gauss, Möbius was the director of the University of Leipzig's observatory. He spent a great deal of time studying geometry and played an essential part in the systematic development of projective geometry. He is best known for his studies of the properties of one-sided surfaces, including the one called the Möbius strip.

MÖBIUS STRIP

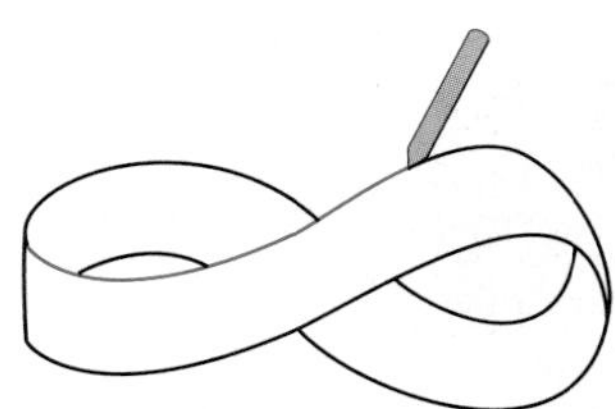
Figure 9.48

If you place a pencil on one surface of a sheet of paper and do not remove it from the sheet, you must cross the edge to get to the other surface. Thus, a sheet of paper has one edge and two surfaces. The sheet retains these properties even when crumpled into a ball. The **Möbius strip**, also called a **Möbius band**, is a one-sided, one-edged surface. You can construct one, as shown in Figure 9.47, by (a) taking a strip of paper, (b) giving one end a half twist, and (c) taping the ends together.

The Möbius strip has some very interesting properties. In order to better understand these properties, perform the following experiments.

Experiment 1 Make a Möbius strip using a strip of paper and tape as illustrated in Fig. 9.47. Place the point of a felt tip pen on the edge of the strip (Fig. 9.48). Pull the strip slowly so that the pen marks the edge—do not remove the pen from the edge. Continue pulling the strip and observe what happens.

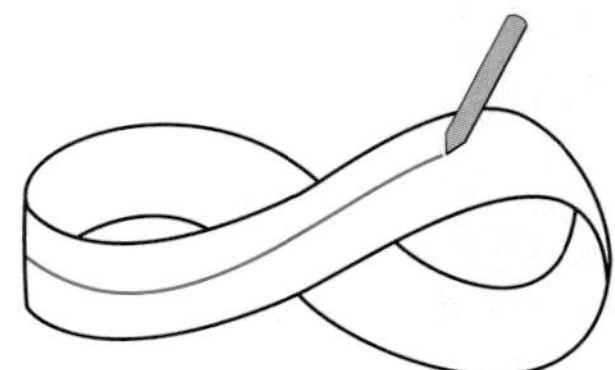
Figure 9.49

Experiment 2 Make a Möbius strip. Place the tip of a felt tip pen on the surface of the strip (Fig. 9.49). Pull the strip slowly so that the pen marks the surface. Continue and observe what happens.

Experiment 3 Make a Möbius strip. With a scissors make a small slit in the middle of the strip. Starting at the slit, cut along the strip, keeping the scissors in the middle of the strip (Fig. 9.50). Continue cutting and observe what happens.

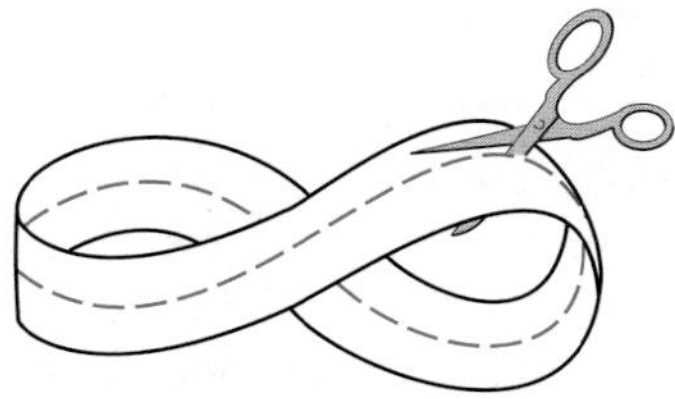
Figure 9.50

Experiment 4 Make a Möbius strip. Make a small slit at a point about one-third of the width of the strip. Cut along the strip, keeping the scissors the same distance from the edge (Fig. 9.51). Continue cutting and observe what happens.

If you give a strip of paper several twists, you get variations on the Möbius strip. To a topologist, the important distinction is between an odd number of twists, which leads to a one-sided surface, and an even number of twists, which leads to a two-sided surface. All strips with an odd number of twists are topologically the same as a Möbius strip and all strips with an even number of twists are topologically the same as an ordinary cylinder, which has no twists.

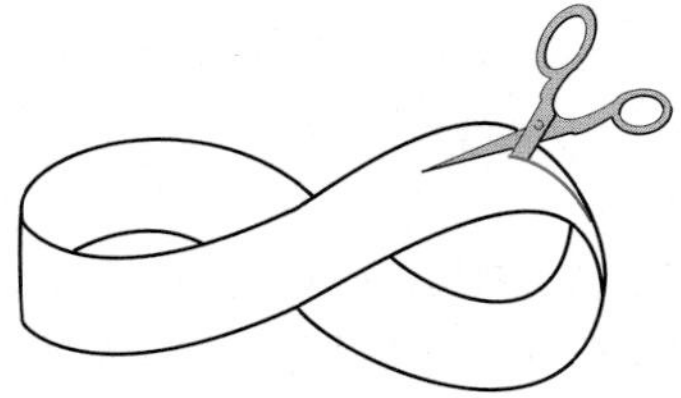
Figure 9.51

KLEIN BOTTLE

Another topological object is the punctured **Klein bottle**, see page 472. This object, named after Felix Klein (1849–1925), resembles a bottle but only has one side.

A punctured Klein bottle can be made by stretching a hollow piece of glass tubing. The neck is then passed through a hole and joined to the base (Fig. 9.52).

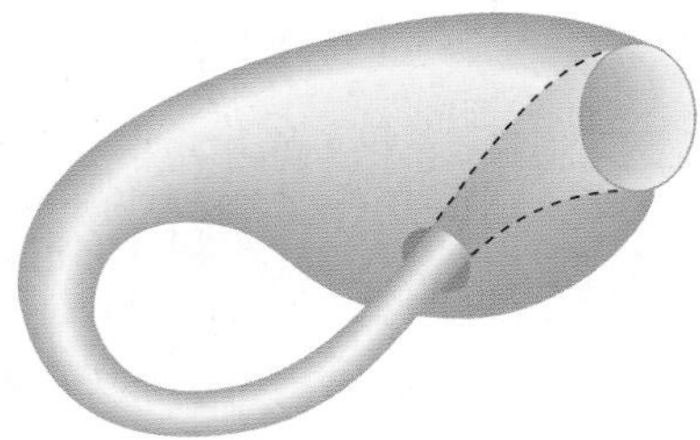

Figure 9.52

Limericks from unknown writers:

"A mathematician confided
That a Möbius band is one-sided,
And you'll get quite a laugh
If you cut one in half
For it stays in one piece when divided."

"A mathematician named Klein
Thought the Möbius band was divine.
He said, 'If you glue
the edges of two
You'll get a weird bottle like mine.' "

Look closely at the model of the Klein bottle shown in Fig. 9.52. The punctured Klein bottle has only one edge and no outside or inside because it has just one side. Figure 9.53 shows a Klein bottle blown in glass by Alan Bennett of Bedford, England.

Klein Bottle, a one-sided surface, blown in glass by Alan Bennett.

Figure 9.53

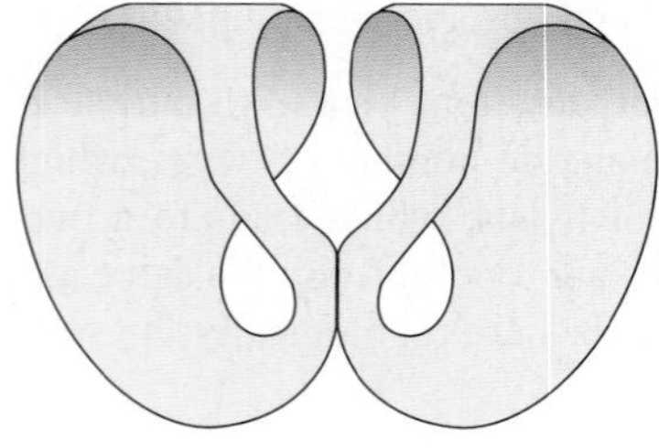

Two Möbius Bands result from cutting a Klein bottle along a curve.

Figure 9.54

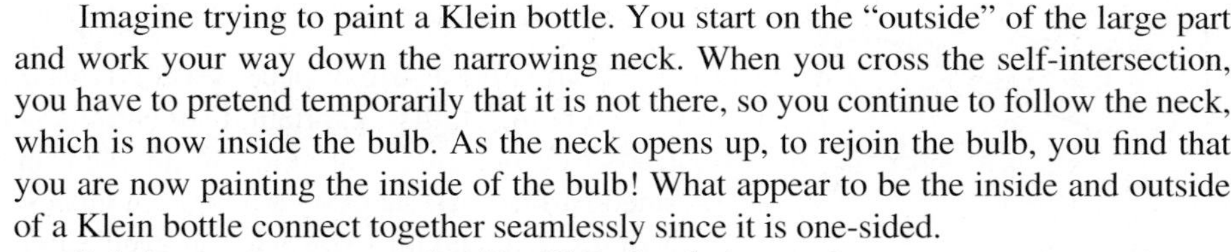

Imagine trying to paint a Klein bottle. You start on the "outside" of the large part and work your way down the narrowing neck. When you cross the self-intersection, you have to pretend temporarily that it is not there, so you continue to follow the neck, which is now inside the bulb. As the neck opens up, to rejoin the bulb, you find that you are now painting the inside of the bulb! What appear to be the inside and outside of a Klein bottle connect together seamlessly since it is one-sided.

It is interesting to note that if a Klein bottle is cut along a curve, the results are two (one-twist) Möbius strips, see Fig. 9.54. Thus, a Klein bottle could also be made by gluing together two Möbius strips along the edges.

MAPS

Maps have fascinated topologists for years because of the many challenging problems they present. Mapmakers have known for a long time that regardless of the complexity of the map and whether it is drawn on a flat surface or a sphere, only four colors are

DID YOU KNOW

His Vision Came in a Dream

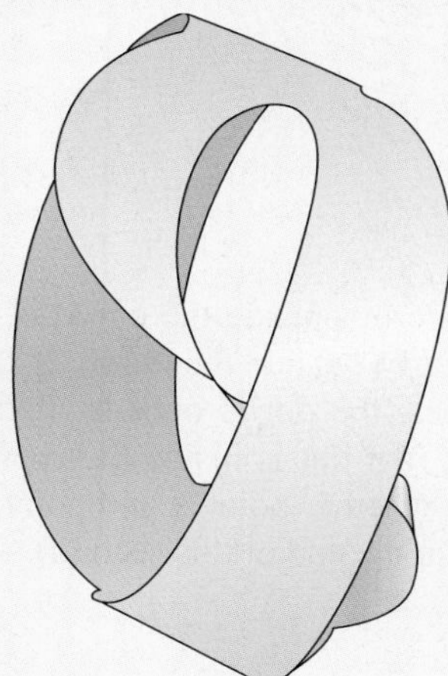

Paper-strip Klein bottle

The Life, the Times, and the Art of Branson Graves Stevenson, by Herbert C. Anderson, Jr. (Janher Publishing, 1979), reports that "in response to a challenge from his son, Branson made his first Klein bottle . . . . He failed in his first try, until the famous English potter, Wedgwood, came to Branson in a dream and showed him how to make the Klein bottle." This was around 50 years ago. Branson's study of claywork and pottery eventually led to the formation of the Archie Bray Foundation in Helena, Montana.

People have made Klein bottles from all kinds of materials. There is a knitting pattern for a woolly Klein bottle and even a paper Klein bottle with a hole.

needed to differentiate each country (or state) from its immediate neighbors. Thus, every map can be drawn by using only four colors, and no two countries with a common border will have the same color. Regions that meet at only one point (such as the states of Arizona, Colorado, Utah, and New Mexico) are not considered to have a common border. In Fig. 9.55(a) no two states with a common border are marked with the same color.

The "four-color" problem was first suggested by a student of Augustus DeMorgan in 1852. In 1976 Kenneth Appel and Wolfgang Haken of the University of Illinois—using their ingenuity, logic, and 1200 hours of computer time—succeeded in proving that only four colors are needed to draw a map. They solved the four-color map problem by reducing any map to a series of points and connecting line segments. They replaced each country with a point. They connected two countries having a common border with a straight line, Fig. 9.55(b). They then showed that the points of any graph in the plane could be colored by using only four colors in such a way that no two points connected by the same line were the same color.

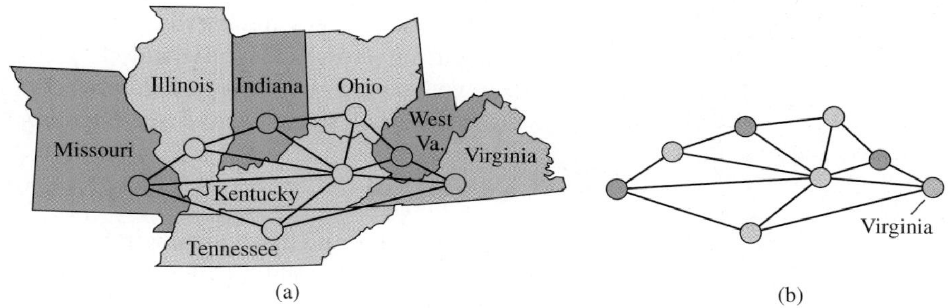

Figure 9.55

Mathematicians have shown that, on different surfaces, more than four colors may be needed to draw a map. For example, a map drawn on a Möbius strip requires a maximum of six colors as in Fig. 9.56(a). A map drawn on a torus (the shape of a doughnut) requires a maximum of seven colors as in Fig. 9.56(b).

Figure 9.56

JORDAN CURVES

A **Jordan curve** is a topological object that can be thought of as a circle twisted out of shape, see Fig. 9.57 (a) to (d). Like a circle, it has an inside and an outside. To get from one side to the other, at least one line must be crossed. Consider the Jordan curve in Fig. 9.57(d). Are points A and B inside or outside the curve?

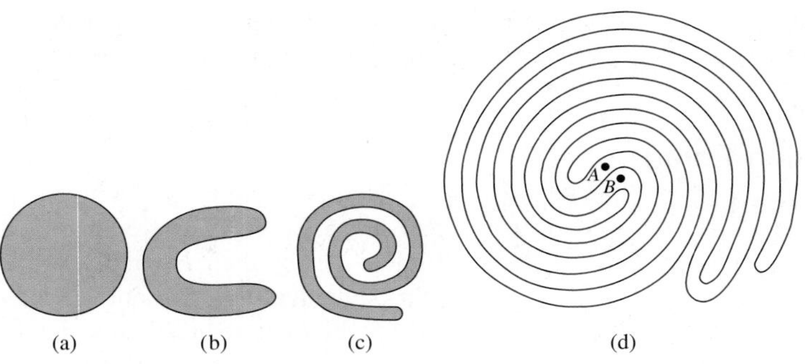

Figure 9.57

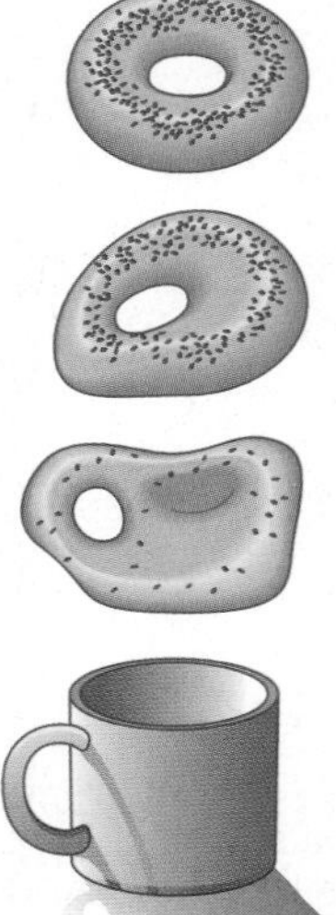

Figure 9.58

A quick way to tell whether the two dots are inside or outside the curve is to draw a straight line from each dot to a point that is clearly outside the curve. If the straight line crosses the curve an even number of times, the dot is outside. If the straight line crosses the curve an odd number of times, the dot is inside the curve. Can you explain why this procedure works? Determine whether point *A* and point *B* are inside or outside the curve (see Exercises 25 and 26 at the end of this section).

TOPOLOGICAL EQUIVALENCE

Someone once said that a topologist is a person who does not know the difference between a doughnut and a coffee cup. Two geometric figures are said to be **topologically equivalent** if one figure can be elastically twisted, stretched, bent, or shrunk into the other figure without puncturing or ripping the original figure. If a doughnut is made of elastic material, it can be stretched, twisted, bent, shrunk, and distorted until it resembles a coffee cup with a handle, as shown in Fig. 9.58. Thus, the doughnut and coffee cup are topologically equivalent.

In topology, figures are classified according to their *genus.* The **genus** of an object is determined by the number of holes in the object. A cup and a doughnut each have one hole and are of genus 1. A kettle and scissors each have two holes and are of genus 2. Figure 9.59 illustrates this type of classification.

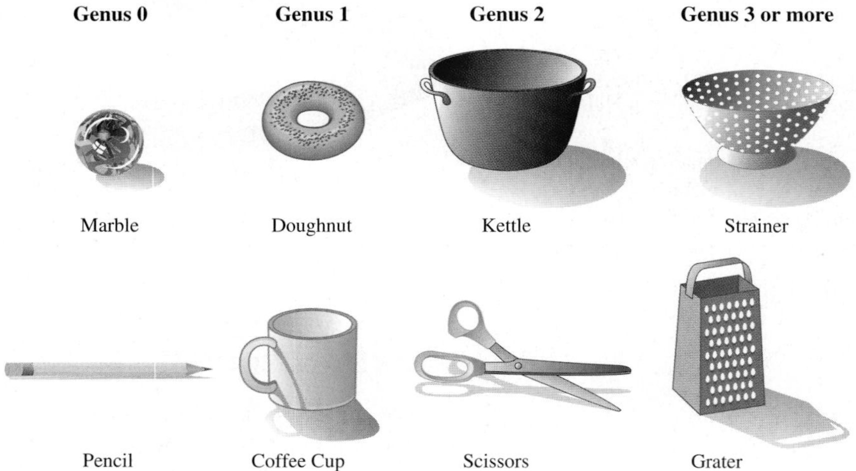

Figure 9.59

SECTION 9.5 EXERCISES

CONCEPT/WRITING EXERCISES

1. What is a Möbius strip?
2. Explain how to make a Möbius strip.
3. What is a Klein bottle?
4. What is the maximum number of colors needed to create a map on a flat surface if no two regions colored the same are to share a common border?
5. What is the maximum number of colors needed to create a map if no two regions colored the same are to share a common border if the surface is a a) Möbius strip b) torus?
6. What is a Jordan curve?
7. When are two figures topologically equivalent?
8. How is the genus of a figure determined?

PRACTICE THE SKILLS

9. Use the result of Experiment 1 to find the number of edges on a Möbius strip.
10. Use the result of Experiment 2 to find the number of surfaces on a Möbius strip.
11. How many separate strips are obtained in Experiment 3?
12. How many separate strips are obtained in Experiment 4?
13. a) Take a strip of paper, give it one full twist, and connect the ends. Is this a Möbius strip with only one side? Explain.
    b) Determine the number of edges, as in Experiment 1.
    c) Determine the number of surfaces, as in Experiment 2.
    d) Cut the strip down the middle. What is the result?
14. Make a Möbius strip. Cut it one-third of the way from the edge, as in Experiment 4. You should get two loops, one going through the other. Determine whether either (or both) of these loops is itself a Möbius strip.
15. Take a strip of paper, make one whole twist and another half twist, and then tape the ends together. Test by a method of your choice to determine whether this has the same properties as a Möbius strip.
16. Can you see any advantage in a Möbius conveyor belt? Explain.

In Exercises 17–22, color the map by using a maximum of four colors so that no two regions with a common border have the same color.

17.

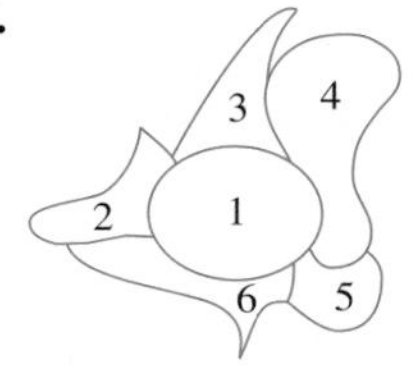

18.

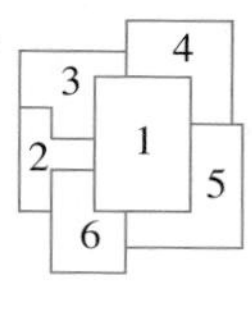

19.

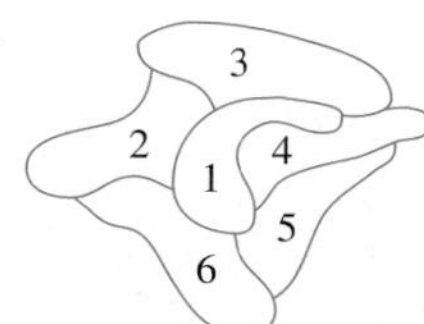

20.

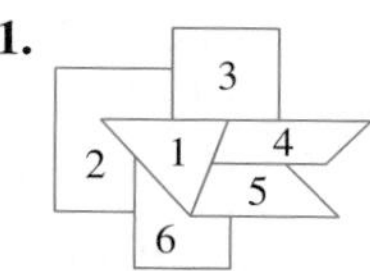

21.

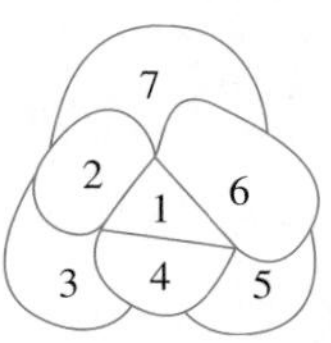

22.

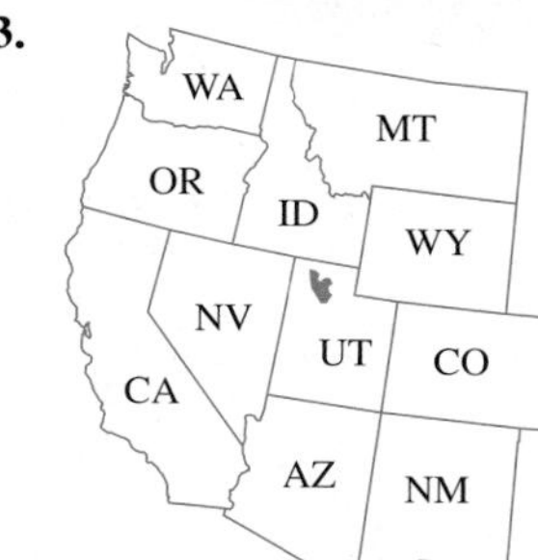

Using the Four-Color Theorem *In Exercises 23 and 24, maps show certain areas of the United States. Shade in the states using a maximum of four colors so that no two states with a common border have the same color.*

23.

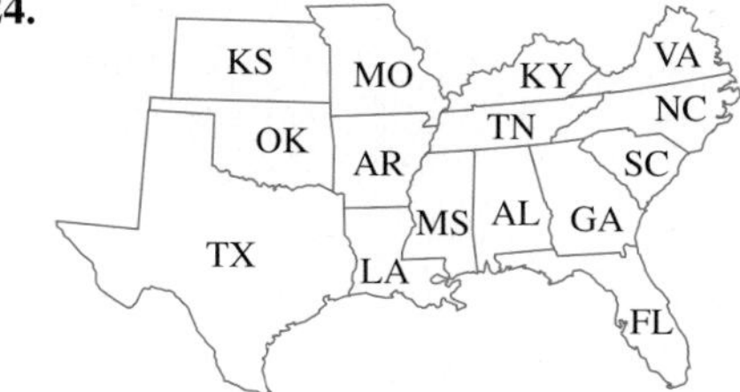

24.

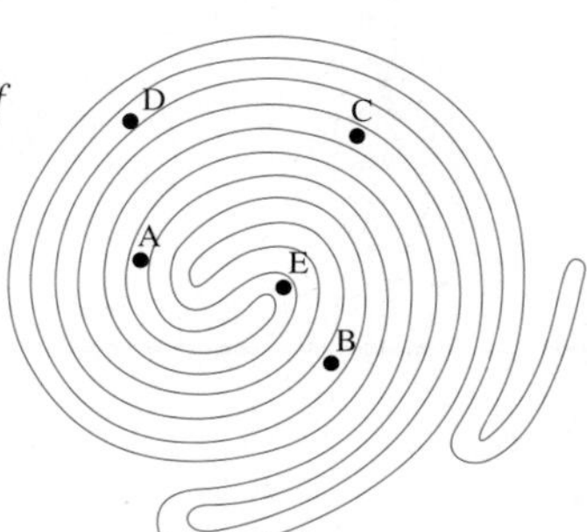

25. Determine whether point A in Fig. 9.57(d) is inside or outside the Jordan curve.
26. Determine whether point B in Fig. 9.57(d) is inside or outside the Jordan curve.

At right is a Jordan curve. In Exercises 27–31, determine if the point is inside or outside of the curve.

27. Point A.
28. Point B.
29. Point C.
30. Point D.
31. Point E.

32. When testing to determine whether a point is inside or outside a Jordan curve, explain why if you count an odd number of lines, the point is inside the curve, and if you count an even number of lines, the point is outside the curve.

In Exercises 33–42, give the genus of the object.

33.

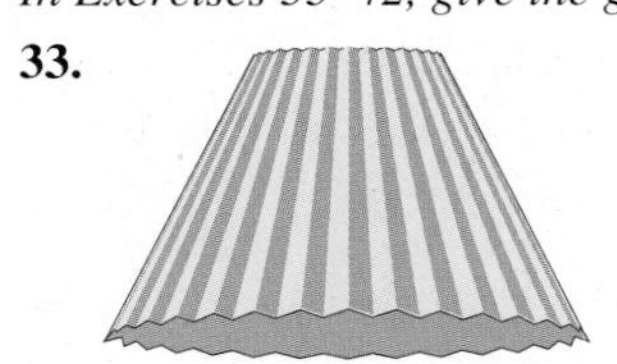

34.

35.

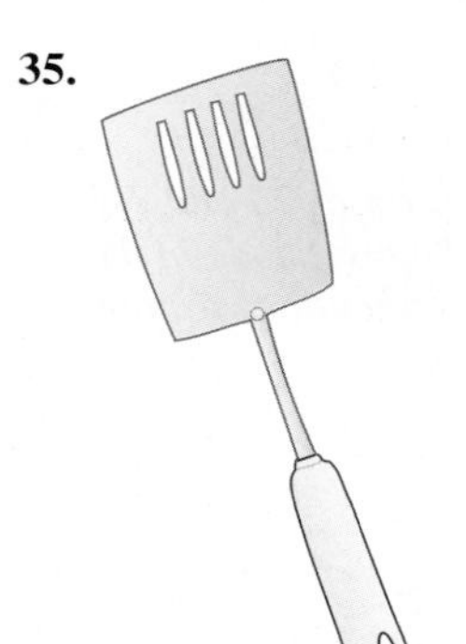

36.

37.

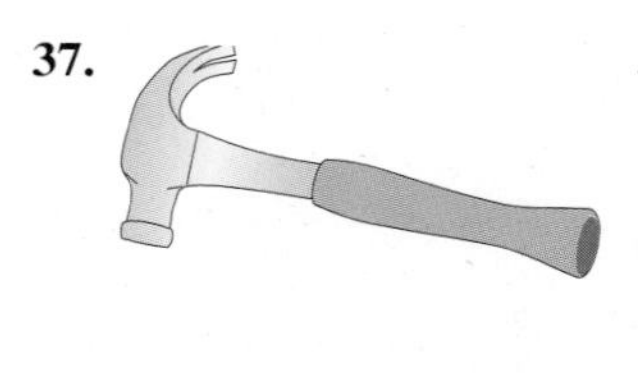

38.

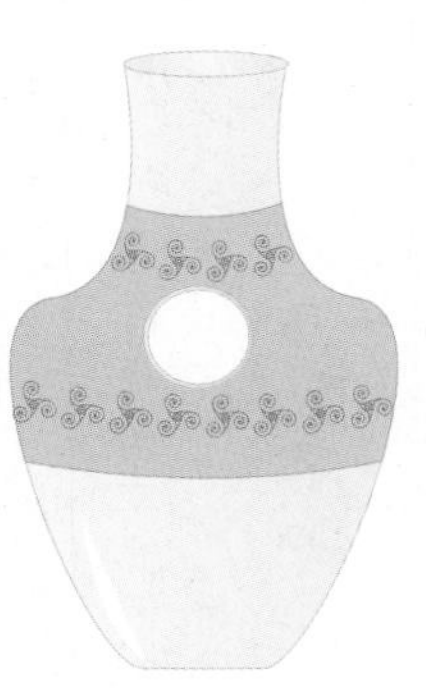

39.

40.

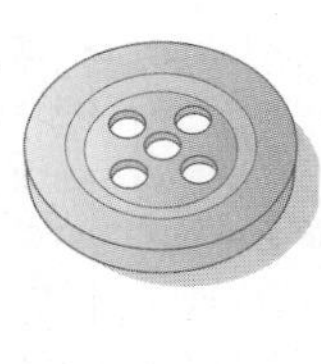

41.

42.

43. Name at least three objects not mentioned in this section that have
   a) Genus 0.
   b) Genus 1.
   c) Genus 2.
   d) Genus 3 or more.

CHALLENGE PROBLEMS/GROUP ACTIVITY

44. Using playdough (or glazing compound) make a doughnut. Without puncturing or tearing the doughnut, reshape it into a topologically equivalent figure, a cup with a handle.

45. Using at most four colors, color the map of South America. Do not use the same color for any two countries that share a common border.

RESEARCH ACTIVITIES

46. Write a paper on the major contributors to the development of topology, starting with Euler.

9.6 NON-EUCLIDEAN GEOMETRY AND FRACTAL GEOMETRY

NON-EUCLIDEAN GEOMETRY

In Section 9.1 we stated postulates or axioms are statements to be accepted as true. In his book *Elements,* Euclid's fifth postulate was "If a straight line falling on two straight lines makes the interior angles on the same side less than two right angles, the two straight lines, if produced indefinitely, meet on that side on which the angles are less than the two right angles."

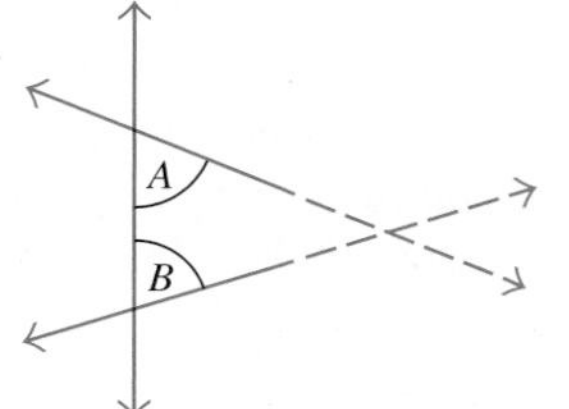

Figure 9.60

Euclid's fifth axiom may be better understood by observing Fig. 9.60. The sum of angles A and B is less than the sum of two right angles (180°). Therefore, the two lines will meet if extended.

John Playfair (1748–1819), a Scottish physicist and mathematician, wrote a geometry book that was published in 1795. In his book, Playfair gave a logically equivalent interpretation of Euclid's fifth postulate. This version is often referred to as Playfair's postulate or the Euclidean parallel postulate.

The Euclidean Parallel Postulate

Given a line and a point not on the line, one and only one line can be drawn through the given point parallel to the given line (Fig. 9.61).

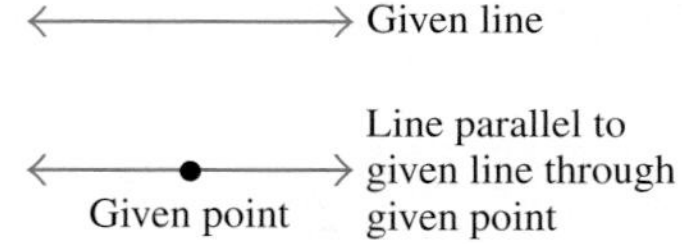

Figure 9.61

The Euclidean parallel postulate may be better understood by looking at Fig. 9.61. Many mathematicians after Euclid believed that this postulate was not as self-evident as the other nine. Others believed this postulate could be proved from the other nine postulates and therefore was not needed at all. Of the many attempts to prove that the fifth postulate was not needed, the most noteworthy one was presented by Girolamo Saccheri (1667–1733), a Jesuit in Italy. In the course of his elaborate chain of deductions, Saccheri proved many of the theorems of what is now called hyperbolic geometry. However, Saccheri did not realize what he had done. He believed that Euclid's geometry was the only "true" geometry and concluded that his own work was in error. Thus, Saccheri narrowly missed receiving credit for a great achievement—the founding of **non-Euclidean geometry**.

Over time, geometers became more and more frustrated at their inability to prove Euclid's fifth postulate. One of them, a Hungarian named Farkos Bolyai, wrote a letter to his son, Janos Bolyai. "I entreat you leave the science of parallels alone. . . . I have traveled past all reefs of this infernal dead sea and have always come back with a broken mast and torn sail." The son, refusing to heed his father's advice, continued to think about parallels until, in 1823, he saw the whole truth and enthusiastically declared, "I have created a new universe from nothing." He recognized that geometry branches in two directions, depending on whether Euclid's fifth postulate is applied. He recognized two different geometries and published his discovery as a 24-page appendix to a textbook written by his father. The famous mathematician George Bruce Halsted called it "the most extraordinary two dozen pages in the whole history of thought." Farkos Bolyai proudly presented a copy of his son's work to his friend Carl Friedrich Gauss, then Germany's greatest mathematician, whose reply to the father had a devastating effect on the son. Gauss wrote, "I am unable to praise this work. . . . To praise it would be to praise myself. Indeed, the whole content of the work, the

Ivanovich Lobachevsky

G. F. Bernhard Riemann

path taken by your son, the results to which he is led, coincides almost entirely with my meditations which occupied my mind partly for the last thirty or thirty-five years." We now know from his earlier correspondence that Gauss had indeed been familiar with **hyperbolic geometry** even before Janos was born. In his letter, Gauss also indicated that it was his intention not to let his theory be published during his lifetime, but to record it so that the theory would not perish with him. It is believed that the reason Gauss did not publish his work was that he feared being ridiculed by other prominent mathematicians of his time.

At about the same time as Bolyai's publication, the Russian Nikolay Ivanovich Lobachevsky published a paper that was remarkably like Bolyai's, although it was quite independent of it. Lobachevsky made a deeper investigation and wrote several books. In marked contrast to Bolyai, who received no recognition during his lifetime, Lobachevsky received great praise and became a professor at the University of Kazan.

After the initial discovery, little attention was paid to the subject until 1854, when G. F. Bernhard Riemann (1826–1866), a student of Gauss, suggested a second type of non-Euclidean geometry, which is now called **spherical**, **elliptical**, or **Riemannian geometry**. The hyperbolic geometry of his predecessors was synthetic; that is, it was not based on or related to any concrete model when it was developed. Riemann's geometry was closely related to the theory of surfaces. A **model** may be considered a physical interpretation of the undefined terms that satisfies the axioms. A model may be a picture or an actual physical object.

The two types of non-Euclidean geometries we have mentioned are elliptical geometry and hyperbolic geometry. The major difference among the three geometries lies in the fifth axiom. The fifth axiom of the three geometries is summarized here.

The Fifth Axiom of Geometry

Euclidean	Elliptical	Hyperbolic
Given a line and a point not on the line, one and only one line can be drawn parallel to the given line through the given point.	Given a line and a point not on the line, no line can be drawn through the given point parallel to the given line.	Given a line and a point not on the line, two or more lines can be drawn through the given point parallel to the given line.

To understand the fifth axiom of the two non-Euclidean geometries, remember the term *line* is undefined. Thus, a line can be interpreted differently in different geometries. A model for Euclidean geometry is a plane, such as a blackboard (Fig. 9.62a). A model for elliptical geometry is a sphere (Fig. 9.62b). A model for hyperbolic geometry is a pseudosphere (Fig. 9.62c). A pseudosphere is similar to two trumpets placed bell to bell. Obviously, a line on a plane cannot be the same as a line

(a) Plane

(b) Sphere

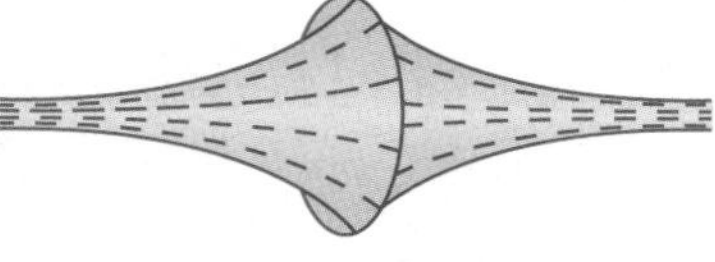

(c) Pseudosphere

Figure 9.62

on either of the other two figures. The curved red lines in Fig. 9.63(a) and both colored lines in Fig. 9.64 are examples of lines in elliptical and hyperbolic geometry, respectively.

ELLIPTICAL GEOMETRY

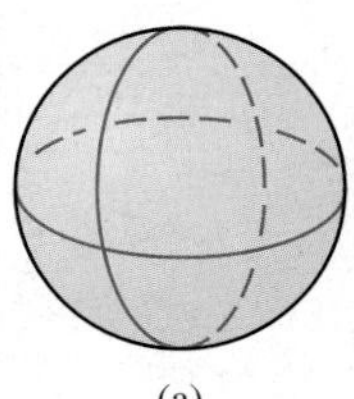

(a)

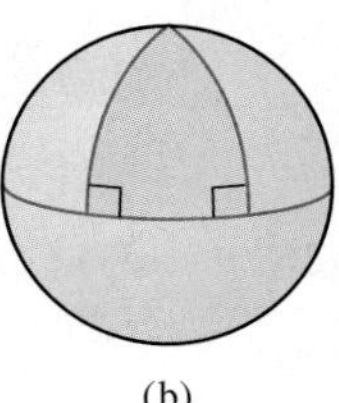

(b)

Figure 9.63

A circle on the surface of a sphere is called a great circle if it divides the sphere into two equal parts. If we were to cut through a sphere along a great circle, we would have two identical pieces. If we interpret a line to be a great circle, we can see the fifth axiom of elliptical geometry is true. Two great circles on a sphere must intersect, hence there can be no parallel lines (Fig. 9.63a).

If we were to construct a triangle on a sphere, the sum of its angles would be greater than 180° (Fig. 9.63b). The theorem, "The sum of the measures of the angles of a triangle is greater than 180°," has been proven by means of the axioms of elliptical geometry. The sum of the measures of the angles varies with the area of the triangle and gets closer to 180° as the area decreases.

HYPERBOLIC GEOMETRY

The lines in hyperbolic geometry are represented by geodesics on the surface of the pseudosphere. A *geodesic* is the shortest and least-curved arc between two points on a surface. Figure 9.64 illustrates two lines on the surface of a pseudosphere.

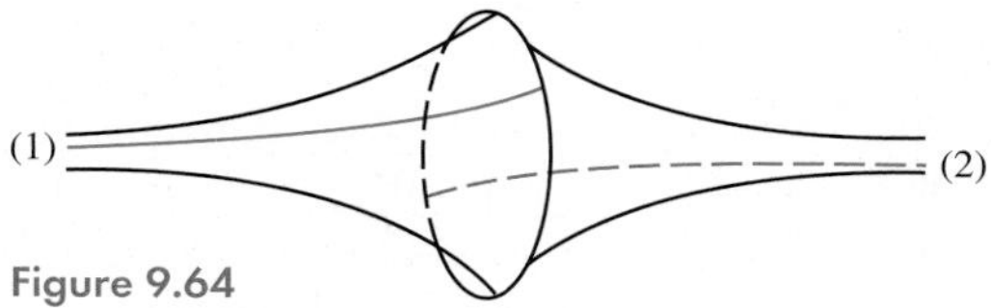

Figure 9.64

Figure 9.65(a) illustrates the fifth axiom of hyperbolic geometry.* Note that, through the given point, two lines are drawn parallel to the given line. If we were to construct a triangle on a pseudosphere, the sum of the measures of the angles would be less than 180° (Fig. 9.65b). The theorem, "The sum of the measures of the angles of a triangle is less than 180°," has been proven by means of the axioms of hyperbolic geometry.

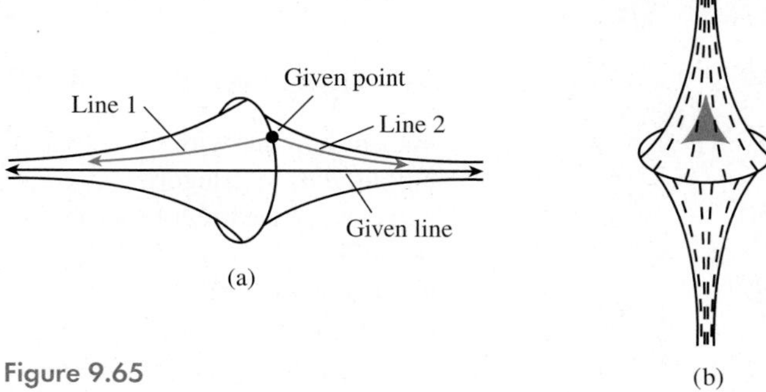

Figure 9.65

*A formal discussion of hyperbolic geometry is beyond the scope of this text.

DID YOU KNOW

It's All Relative

Einstein's general theory of relativity, published in 1916, approached space and time differently from our everyday understanding of them. Einstein's theory unites the three dimensions of space with one of time in a four-dimensional space—time continuum. His theory dealt with the path that light and objects take while moving through space under the force of gravity. Einstein conjectured that mass (such as stars and planets) caused space to be curved. The greater the mass, the greater the curvature. Also, in the region nearer to the mass, the curvature of the space is greater.

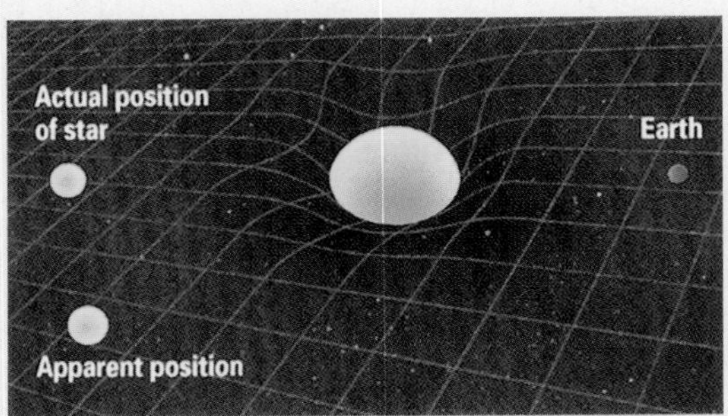

You can visualize Einstein's theory by thinking of space as a rubber sheet pulled taut on which a mass is placed, causing the rubber sheet to bend.

To prove his conjecture, Einstein exposed himself to Riemann's non-Euclidean geometry. Einstein felt that the trajectory of a particle in space represents not a straight line but the straightest curve possible, a geodesic.

Einstein's theory was confirmed by the solar eclipses of 1919 and 1922.

Space–time is now thought to be a combination of three different types of curvature: spherical (described by Riemannian geometry), flat (described by Euclidean geometry), and saddle-shaped (described by hyperbolic geometry).

"The Great Architect of the universe now appears to be a great mathematician."
British physicist Sir James Jeans

We have stated that the sum of the measures of the angles of a triangle is 180°, is greater than 180°, and is less than 180°. Which statement is correct? Each statement is correct *in its own system.* Many theorems hold true for all three geometries—vertical angles still have the same measure, we can uniquely bisect a line segment with a straightedge and compass alone, and so on.

The many theorems based on the fifth postulate may differ in each geometry. It is important for you to realize each theorem proved is true *in its own system* because each is logically deduced from the given set of axioms of the system. No one system is the "best" system. Euclidean geometry may appear to be the one to use in the classroom, where the blackboard is flat. However, in discussions involving the earth as a whole, elliptical geometry may be the most useful, since the earth is a sphere. If the object under consideration has the shape of a saddle or pseudosphere, hyperbolic geometry may be the most useful.

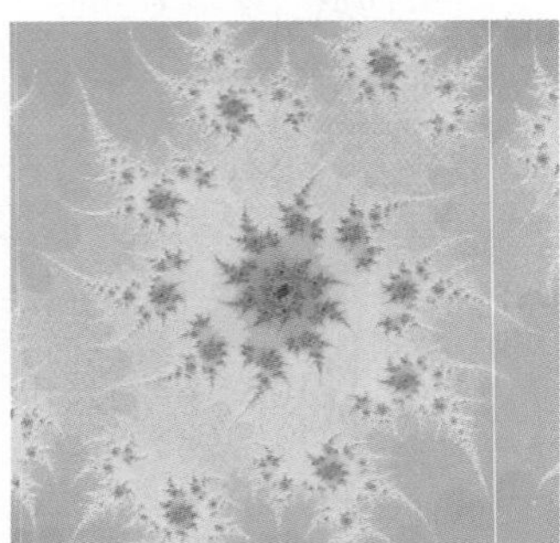

Fractal Images

FRACTAL GEOMETRY

We are familiar with one-, two-, and three-dimensional figures. However, many objects are difficult to categorize as one-, two-, or three-dimensional. For example, how would you classify the irregular shapes we see in nature such as a coast line, or the bark on a tree, or a mountain, or a path followed by lightning? For a long time mathematicians assumed that making realistic geometric models of natural shapes and figures was almost impossible. However, the development of **fractal geometry** now makes it possible. Both color photos on this page were made by using fractal geometry. The discovery and study of fractal geometry has been one of the most popular mathematical topics in recent times.

The word *fractal* (from the Latin word *fractus,* "broken up, fragmented") was first used in the mid-1970s by the mathematician Benoit Mandelbrot to describe

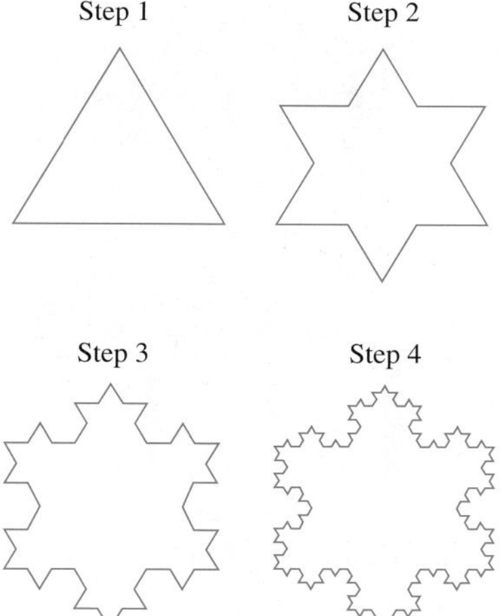

Figure 9.66

shapes that had several common characteristics, including some form of "self-similarity," as will be seen shortly in the Koch snowflake.

Typical fractals are extremely irregular curves or surfaces that "wiggle" enough so that they are not considered one dimensional. Fractals do not have integer dimensions; their dimensions are between 1 and 2. For example, a fractal may have a dimension of 1.26. Fractals are developed by applying the same rule over and over again, with the end point of each simple step becoming the starting point for the next step, a process called **recursion**.

Using the recursive process, we will develop a famous fractal called the **Koch snowflake** after Helga von Koch, the Swedish mathematician who first discovered its remarkable characteristics. The Koch snowflake illustrates a property of all fractals called *self-similarity*—that is, each smaller piece of the curve resembles the whole curve.

To develop the Koch snowflake:

1. start with an equilateral triangle (Step 1, Fig. 9.66), and
2. whenever you see an edge —— replace it with _/_ (Steps 2–4).

What is the perimeter of the snowflake in Fig. 9.66, and what is its area? A portion of the boundary of the Koch snowflake known as the Koch curve or the snowflake curve is represented in Fig. 9.67.

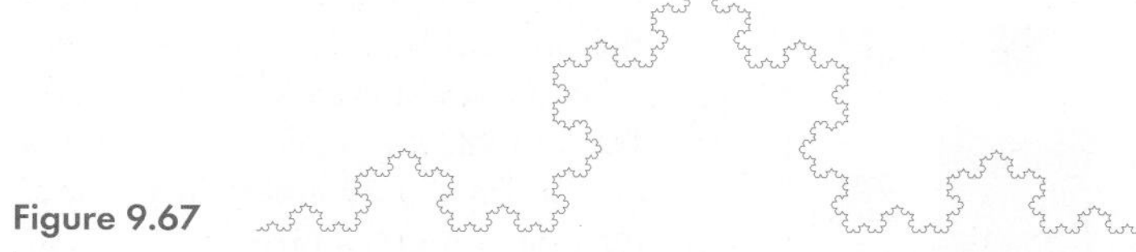

Figure 9.67

Because the Koch curve consists of infinitely many pieces of the form _/_, the perimeter is also infinite. The area of the Koch snowflake is 1.6 times the area of the starting equilateral triangle. Thus the area of the snowflake is finite. The Koch snowflake has a finite area enclosed by an infinite boundary! This fact may seem

DID YOU KNOW

Fractal Antennas

Fractal triangle can act as a miniaturized antenna.

Over the past decade researchers have begun to use antennas made using fractal designs. Many antennas that look like a simple unit, including most radar antennas, are actually arrays of up to thousands of small antennas. Scientists have discovered that a fractal arangement can result in antennas being as powerful as traditional antennas, using only a quarter of the number of elements. Dwight Jasserd of the University of Pennsylvania says "Fractals bridge the gap, they have short-range disorder and long-range order." Fractal antennas are 25% more efficient than the rubbery "stubby" antennas found on most phones. Why do fractal antennas work so well? It has been proven mathematically that for an antenna to work equally well at all frequencies, it must be self-similar, having the same basic appearance at every scale. That is, it has to be fractal!

Hidden inside a cordless phone, a square fractal antenna (center board) replaces the usual rubbery stalk.

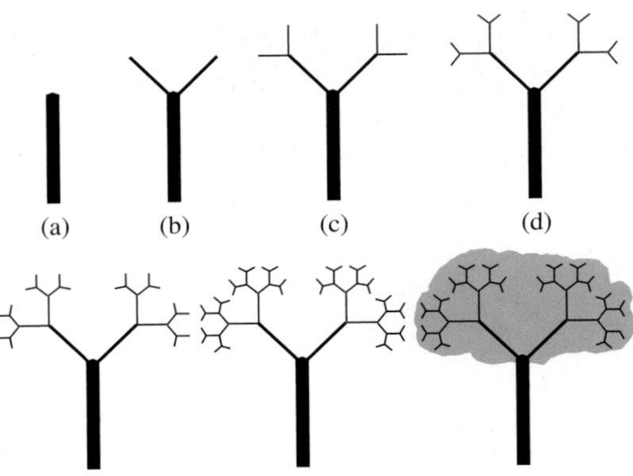

Figure 9.68 The fractal tree

difficult to accept, but it is true. However, the Koch snowflake, like other fractals, is not an everyday run-of-the-mill geometric shape.

Let us look at a few more fractals made using the recursive process. We will now construct what is known as a *fractal tree.* Start with a tree trunk, Fig. 9.68(a). Draw two branches, each one a bit smaller than the trunk, Fig. 68(b). Draw two branches from each of those branches, and continue, see Figs. 9.68(c) and 9.68(d). Ideally we continue the process forever.

If you take a little piece of any branch and zoom in on it, it will look exactly like the original tree. Fractals are *scale independent,* which means you cannot really tell whether you are looking at something very big or something very small because the fractal looks the same whether you are close to it or far from it.

In Figs. 9.69 and 9.70 we develop two other fractals through the process of recursion. Figure 9.69 shows a fractal called the Sierpinski triangle, and Fig. 9.70 shows a fractal called the Sierpinski carpet. Both fractals are named after Waclaw Sierpinski, a Polish mathematician who is best known for his work with fractals and space-filling curves.

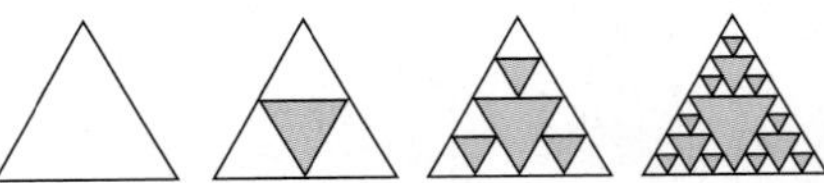
Figure 9.69 Sierpinski triangle

Fractals provide a way to study natural forms such as coastlines, trees, mountains, galaxies, polymers, rivers, weather patterns, brains, lungs, and blood supplies. Fractals also help explain that which appears chaotic. The blood supply in the body is one example. The branching of arteries and veins appear chaotic, but closer inspection reveals the same type of branching occurs for smaller and smaller blood vessels, down to the capillaries. Thus, fractal geometry provides a geometric structure for chaotic processes in nature. The study of chaotic processes is called **chaos theory**.

Fractals nowadays have a potentially important role to play in characterizing weather systems and in providing insight into various physical processes such as the occurrence of earthquakes or the formation of deposits that shorten battery life. Some scientists view fractal statistics as a doorway to unifying theories of medicine, offering a powerful glimpse of what it means to be healthy.

Fractals lie at the heart of current efforts to understand complex natural phenomena. Unraveling their intricacies could reveal the basic design principals at work in our world. Until recently, there was no way to describe fractals. Today, we are beginning to see such features everywhere. Tomorrow, we may look at the entire universe through a fractal lens.

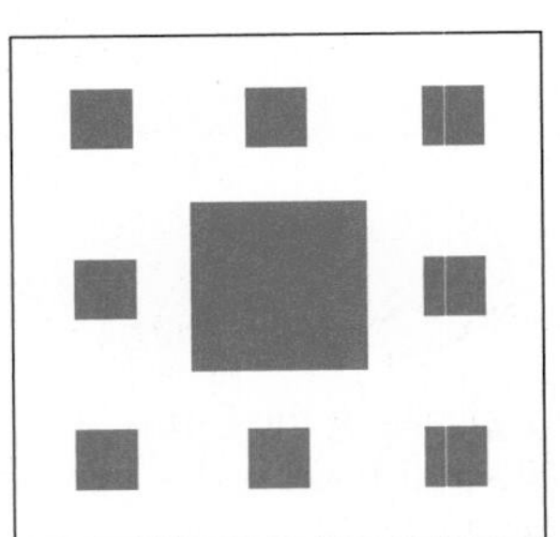
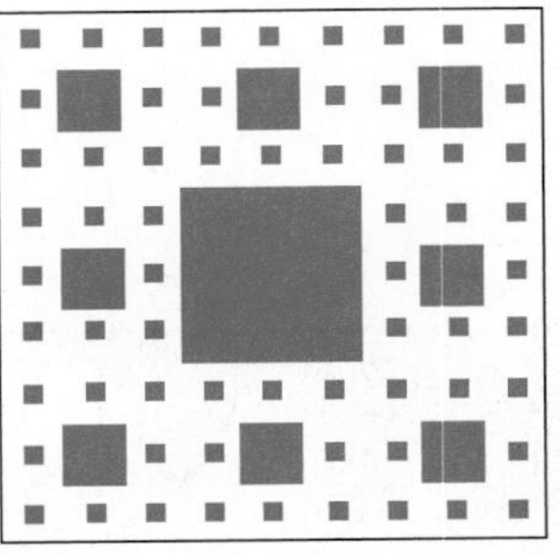
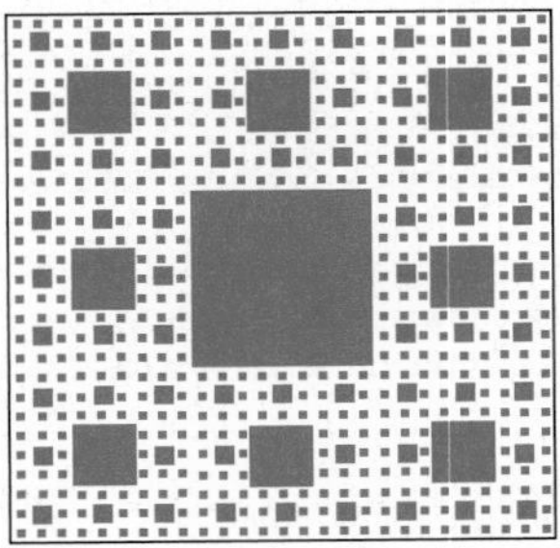
Figure 9.70 Sierpinski carpet

SECTION 9.6 EXERCISES

CONCEPT/WRITING EXERCISES

In Exercises 1–5, list the accomplishments of the mathematician.

1. Janos Bolyai
2. Carl Friedrich Gauss
3. Nikolay Ivanovich Lobachevsky
4. Girolamo Saccheri
5. G. F. Bernhard Riemann
6. State the fifth axiom of
   a) Euclidean geometry. **b)** hyperbolic geometry.
   c) elliptical geometry.
7. State the theorem concerning the sum of the measures of the angles of a triangle in
   a) Euclidean geometry. **b)** hyperbolic geometry.
   c) elliptical geometry.
8. What model is often used in describing and explaining Euclidean geometry?
9. What model is often used in describing and explaining elliptical geometry?
10. What model is often used in describing and explaining hyperbolic geometry?
11. What do we mean when we say that no one axiomatic system of geometry is "best"?
12. What did Einstein conjecture about the relationship between mass and space?
13. List the three types of curvature of space and the types of geometry that correspond to them.
14. In our discussion of the Koch snowflake we indicated that it had a finite area enclosed in an infinite perimeter. Explain how this can be true.

PRACTICE THE SKILLS

In the following we show a fractal-like figure made using a recursive process with the letter "M." Use this fractal-like figure as a guide in constructing fractal-like figures with the letter given in Exercises 15–18. Show three steps, as is done here.

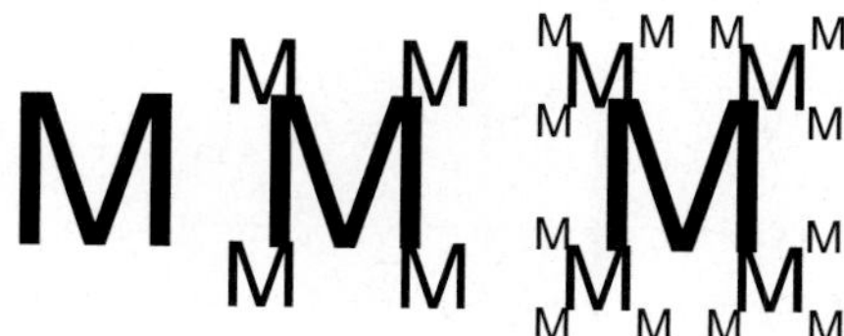

15. V **16.** W **17.** I **18.** H

19. **a)** Develop a fractal by beginning with a square and replacing each side —— with a ⎍. Repeat this process twice.
    b) If you continue this process will the fractal's perimeter be finite or infinite? Explain.
    c) Will the fractal's area be finite or infinite? Explain.

PROBLEM SOLVING/GROUP ACTIVITY

20. In forming the Koch snowflake in Figure 9.66 the perimeter becomes greater at each step in the process. If each side of the original triangle is 1 unit, a general formula for the perimeter, L, of the snowflake at any step, n, may be found by the formula

$$L = 3\left(\frac{4}{3}\right)^{n-1}$$

For example, at the first step when $n = 1$, the perimeter is 3 units, which can be verified by the formula as follows:

$$L = 3\left(\frac{4}{3}\right)^{1-1} = 3\left(\frac{4}{3}\right)^{0} = 3 \cdot 1 = 3.$$

At the second step, when $n = 2$, we find the perimeter as follows:

$$L = 3\left(\frac{4}{3}\right)^{2-1} = 3\left(\frac{4}{3}\right) = 4$$

Thus at the second step the perimeter of the snowflake is 4 units.

a) Use the formula to complete the following table.

Step	Perimeter
1	
2	
3	
4	
5	
6	

b) Use the results of your calculations to explain why the perimeter of the Koch snowflake is infinite.

RESEARCH ACTIVITY

21. The Renaissance painters, for the first time perhaps, worked at giving visual depth to their paintings. To accomplish visual depth the painter had to understand *perspective.* That is, in a painting they had to determine how large a tree or building should be to give the impression that the object was in the background. The branch of geometry that provides this information is called *projective geometry.* Write a paper on the use of projective geometry in art, including specific examples of applications. (References include books on art, encyclopedias, history of mathematics books, and the Internet.)

CHAPTER 9 SUMMARY

IMPORTANT FACTS

The sum of the measures of the angles of a triangle is 180°.

The sum of the measures of the angles of a quadrilateral is 360°.

The sum of the measures of the interior angles of an n-sided polygon is $(n - 2)180°$.

Triangle

$A = \frac{1}{2}bh$

$p = s_1 + s_2 + s_3$

Square

$A = s^2$

$p = 4s$

Rectangle

$A = lw$

$p = 2l + 2w$

Parallelogram

$A = bh$

$p = 2b + 2w$

Trapezoid

$A = \frac{1}{2}h(b_1 + b_2)$

$p = s_1 + s_2 + b_1 + b_2$

Pythagorean theorem

$a^2 + b^2 = c^2$

Circle

$A = \pi r^2$; $C = 2\pi r$ or $C = \pi d$

Cube

$V = s^3$

Rectangular solid

$V = lwh$

Cylinder

$V = \pi r^2 h$

Cone

$V = \frac{1}{3}\pi r^2 h$

Sphere

$v = \frac{4}{3}\pi r^3$

Prism

$V = Bh$, where B is the area of the base

Pyramid

$V = \frac{1}{3}Bh$, where B is the area of the base

Fifth postulate in Euclidean geometry

Given a line and a point not on the line, only one line can be drawn through the given point parallel to the given line.

Fifth postulate in elliptical geometry

Given a line and a point not on the line, no line can be drawn through the given point parallel to the given line.

Fifth postulate in hyperbolic geometry

Given a line and a point not on the line, two or more lines can be drawn through the given point parallel to the given line.

CHAPTER 9 REVIEW EXERCISES

9.1

In Exercises 1–6, use the figure shown to find the following.

1. $\measuredangle EFI \cap \measuredangle BFC$
2. $\overline{BF} \cup \overline{FC} \cup \overline{BC}$
3. $\overleftrightarrow{AB} \cap \overline{BC}$
4. $\overrightarrow{BH} \cup \overrightarrow{HB}$
5. $\overleftrightarrow{HI} \cap \overleftrightarrow{EG}$
6. $\overrightarrow{CF} \cap \overrightarrow{CG}$

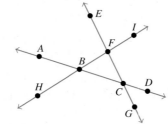

7. $m\measuredangle A = 26.3°$. Find the measure of the complement of $\measuredangle A$.

8. $m\measuredangle B = 105.2°$. Find the measure of the supplement $\measuredangle B$.

9.2

In Exercises 9–12, use the similar triangles ABC and A′B′C shown to find the following.

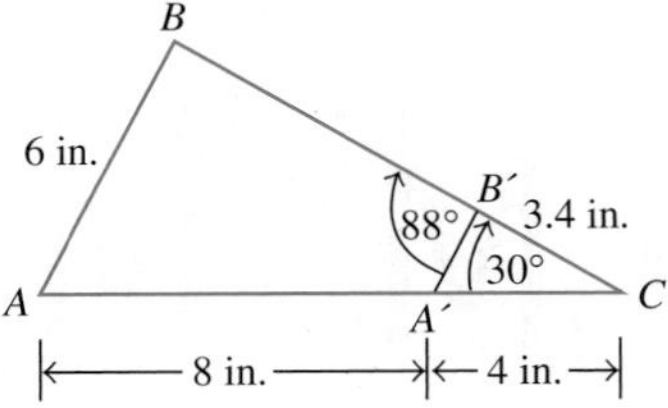

9. The length of $\overline{BC}$
10. The length of $\overline{A'B'}$
11. The measure of $\measuredangle BAC$
12. The measure of $\measuredangle ABC$

13. In the following figure, l_1 and l_2 are parallel lines. Find $m\measuredangle 1$ through $m\measuredangle 6$.

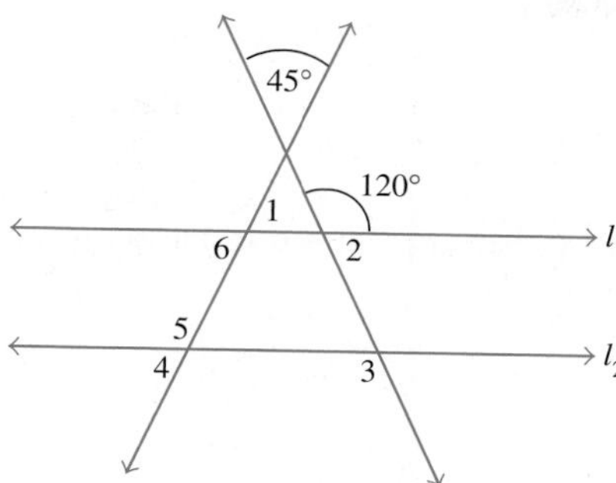

14. Find the sum of the measures of the interior angles of a hexagon.

9.3

In Exercises 15–19, find the area of the figure.

15.
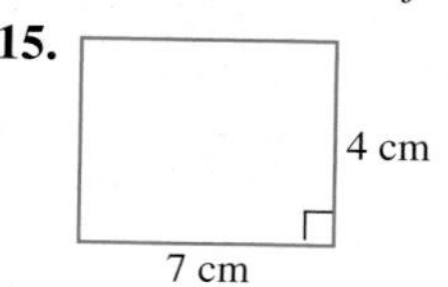

16.
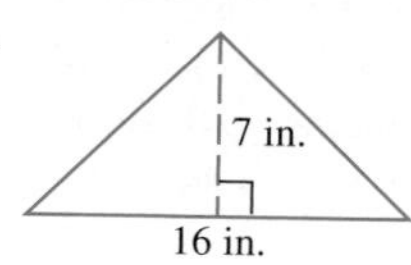

17.
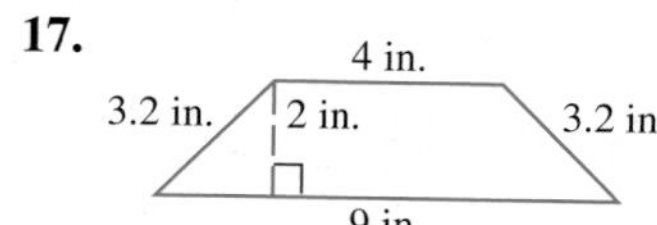

18.
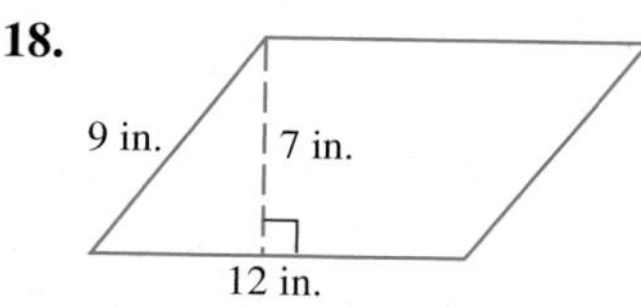

19.
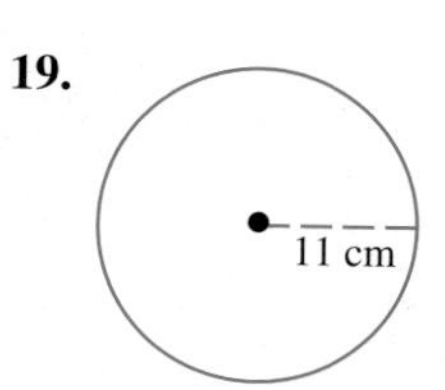

20. *Cost of Kitchen Tile* Determine the total cost of covering a 14 ft × 16 ft kitchen floor with tile. The cost of the tile selected is $18.50 per square yard. Assume tiles may only be purchased in whole square yards.

9.4

In Exercises 21–26, find the volume of the figure. Round your answers to hundredths.

21.
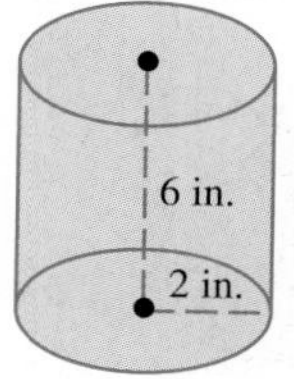

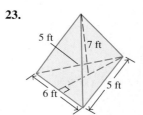

22.
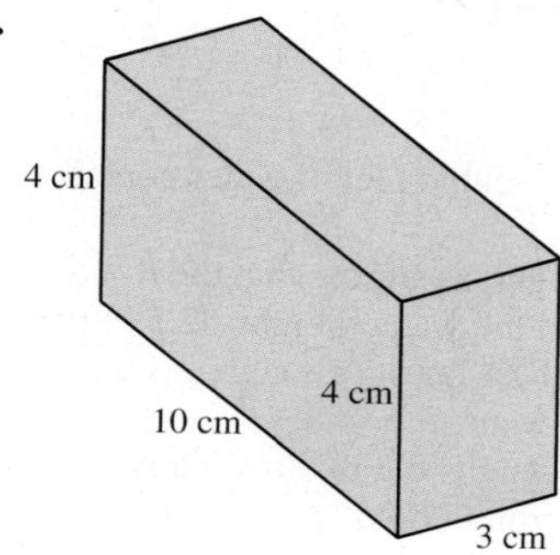

23.

24.
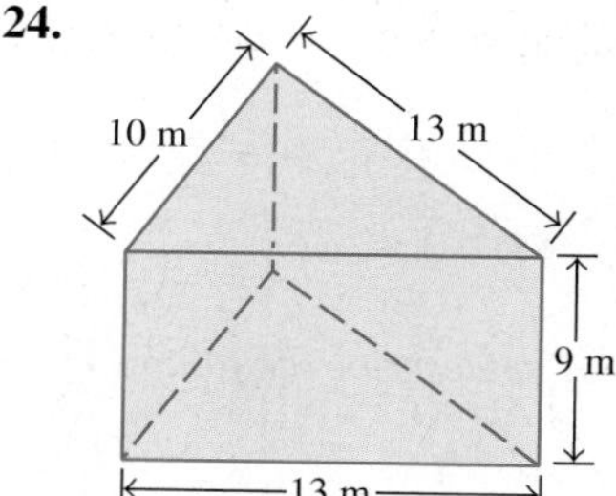

25.
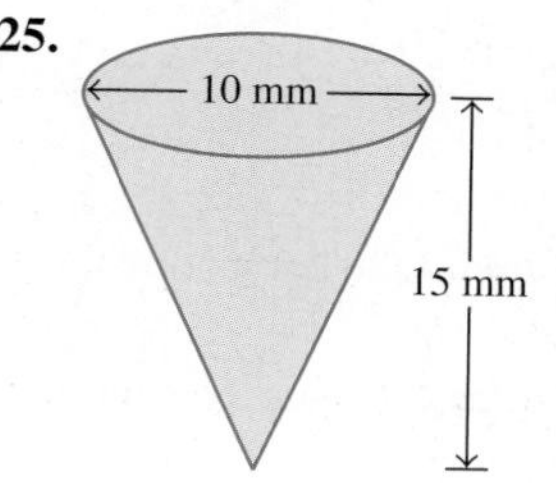

26.
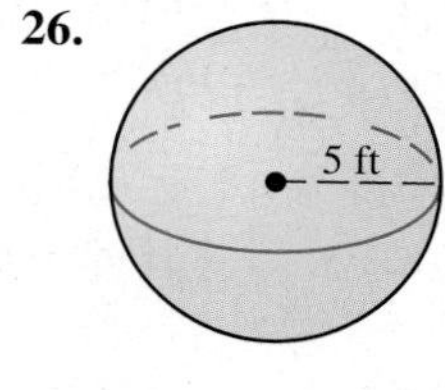

27. *Volume of Water Trough* Steven Dale has a water trough whose ends are trapezoids and whose sides are rectangles, as illustrated. He is afraid that the base it is sitting on will not support the weight of the trough when it is filled with water. He knows that the base will support 4800 lb.

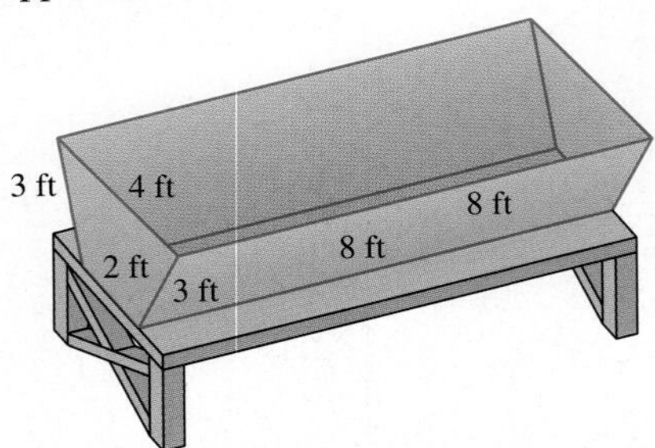

a) Find the number of cubic feet of water contained in the trough.

b) Find the total weight, assuming that the trough weighs 375 lb and the water weighs 62.5 lb per cubic foot. Is the base strong enough to support the trough filled with water?

c) If 1 gal of water weighs 8.3 lb, how many gallons of water will the trough hold?

9.5

28. Give the genus of the object:

29. *Map Coloring* Color the map by using a maximum of four colors so that no two regions with a common border have the same color.

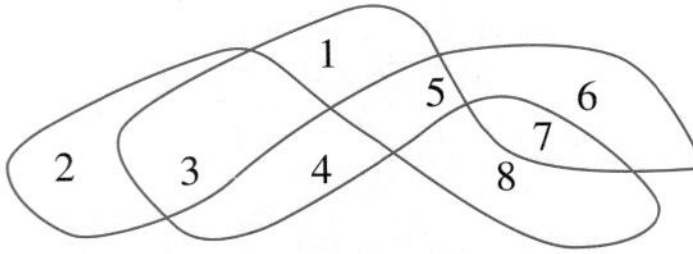

30. Determine whether point A is inside or outside the Jordan curve.

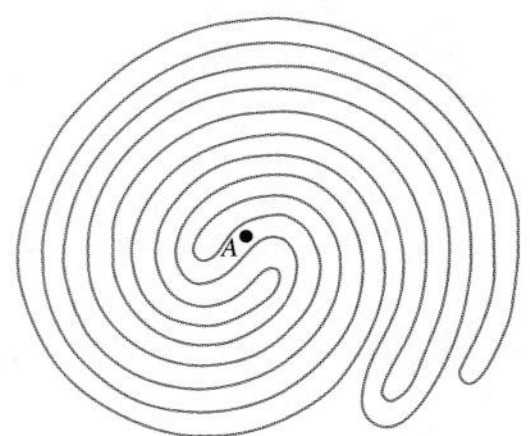

9.6

31. State the fifth axiom of Euclidean, elliptical, and hyperbolic geometry.

32. Develop a fractal by beginning with a square and replacing each side —— with a ⎍. Repeat this process twice.

CHAPTER 9 TEST

In Exercises 1–4, use the figure to describe the following sets of points:

1. $\overset{\circ\!\!\longrightarrow}{AF} \cap \overset{\circ\!\!\longrightarrow}{EF}$

2. $\overline{BC} \cup \overline{CD} \cup \overline{BD}$

3. $\measuredangle EDF \cap \measuredangle BDC$

4. $\overrightarrow{AC} \cup \overrightarrow{BA}$

5. $m\measuredangle A = 17.4°$. Find the measure of the complement of $\measuredangle A$.

6. $m\measuredangle B = 93.6°$. Find the measure of the supplement of $\measuredangle B$.

7. In the figure, find the measure of $\measuredangle x$.

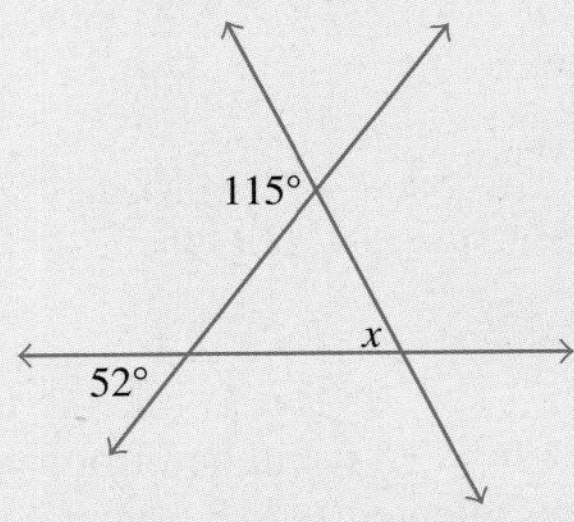

8. Find the sum of the measures of the interior angles of an octagon.

9. Triangles ABC and $A'B'C'$ are similar figures. Find the length of side $\overline{B'C'}$.

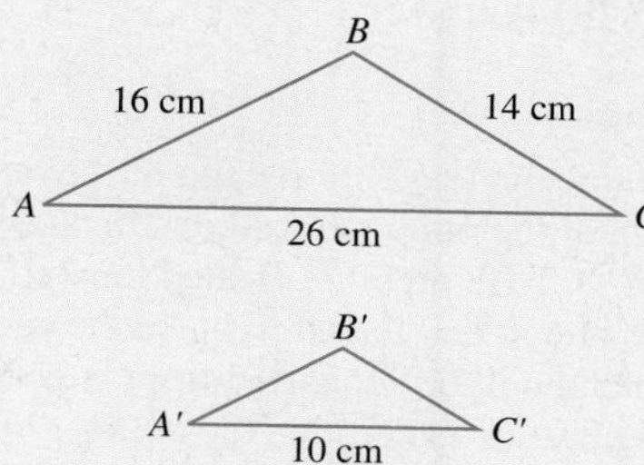

10. Right triangle ABC has one leg of length 5 in. and a hypotenuse of length 13 in.
   a) Find the length of the other leg.
   b) Find the perimeter of the triangle.
   c) Find the area of the triangle.

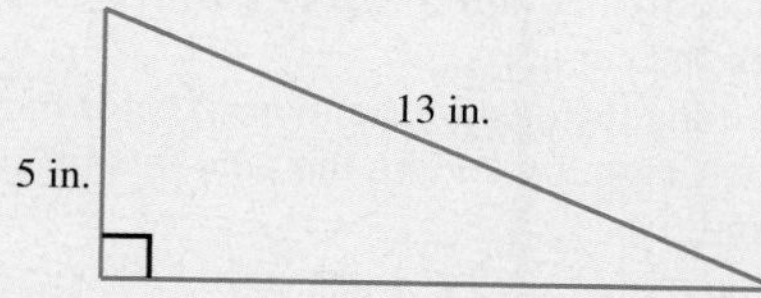

11. Find the volume of a sphere of diameter 16 cm.

12. *Building a Pier* The sketch shows the dimensions of the base of a pier for a bridge with semicircular ends. How many cubic yards of concrete are needed to build a pier 6 ft high?

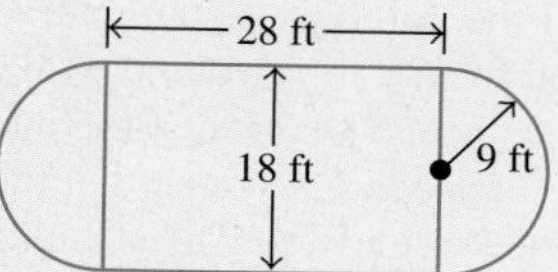

13. Find the volume of the pyramid.

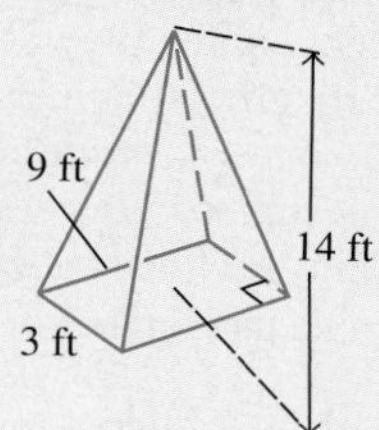

14. Explain how Euclidean geometry, hyperbolic geometry, and elliptical geometry differ.

15. What is a Möbius strip?

16. a) Sketch an object that is of genus 1.
    b) Sketch an object that is of genus 2.

GROUP PROJECTS

SUPPORTING A JACUZZI

1. You are thinking of buying a circular Jacuzzi 12 ft in diameter, 4 ft deep, and weighing 475 lb. You want to place the Jacuzzi on a deck built to support 30,000 lb.
   a) Determine the volume of the water in the Jacuzzi in cubic feet.
   b) Determine the number of gallons of water the Jacuzzi will hold. Note 1 $ft^3 \approx 7.5$ gal.
   c) Determine the weight of the water in the Jacuzzi. (*Hint:* Fresh water weighs about 52.4 lb/ft^3.)
   d) Will the deck support the weight of the Jacuzzi and water?
   e) Will the deck support the weight of the Jacuzzi, water, and four people, whose average weight is 115 lb?

DESIGNING A RAMP

2. The Troutmans are planning to build a ramp so that their front entrance is wheel-chair accessible. The ramp will be 36 in. wide. It will rise 2 in. for each foot of length of horizontal distance. Where the ramp meets the porch, the ramp must be 2 ft high. To provide stability for the ramp, the Troutmans will install a slab of concrete 4 in. thick and 6 in. longer and wider than the ramp (see accompanying figure). The top of the slab will be level with the ground. The ramp may be constructed of concrete or pressure-treated lumber. You are to estimate the cost of materials for constructing the slab, the ramp of concrete, and the ramp of pressure-treated lumber.

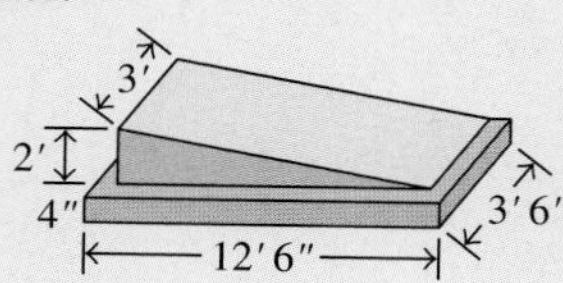

Slab

   a) Determine the length of the base of the ramp.
   b) Determine the dimensions of the concrete slab on which the ramp will set.
   c) Determine the volume of the concrete in cubic yards needed to construct the slab.
   d) If ready-mix concrete costs $45 per cubic yard, determine the cost of the concrete needed to construct the slab.

Concrete Ramp

e) To build the ramp of concrete a form in the shape of the ramp must be framed. The two sides of the form are triangular and the shape of the end, which is against the porch, is rectangular. The form will be framed from $\frac{3}{4}$-in. plywood, which comes in 4 ft × 8 ft sheets. Determine the number of sheets of plywood needed. Assume that the entire sheet(s) will be used to make the sides and the end of the form and that there is no waste.

f) If the plywood costs $18.95 for a 4 ft × 8 ft sheet, determine the cost of the plywood.

g) To brace the form, the Troutmans will need two boards 2 in. × 4 in. × 8 ft (referred to as 8 ft 2 × 4's) and six pieces of lumber 2 in. × 4 in. × 3 ft. These six pieces of lumber will be cut from an 8 ft 2 × 4 boards. Determine the number of 8 ft 2 × 4 boards needed.

h) Determine the cost of the 8 ft 2 × 4 boards needed in part (g) if one board costs $2.14.

i) Determine the volume, in cubic yards, of concrete needed to fill the form.

j) Determine the cost of the concrete needed to fill the form.

k) Determine the total cost of materials for building the ramp of concrete by adding the results in parts (d), (f), (h), and (j).

Wooden Ramp

l) Determine the length of the top of the ramp.

m) The top of the ramp will be constructed of $\frac{5}{4}$-in. × 6-in. × 10-ft pressure-treated lumber. The boards will be butted end to end to make the necessary length and will be supported from underneath by a wooden frame. Determine the number of boards needed to cover the top of the ramp. The boards are laid lengthwise on the ramp.

n) Determine the cost of the boards to cover the top of the ramp if the price of a 10-ft length is $6.47.

o) To support the top of the ramp, the Troutmans will need 10 pieces of 8 ft 2 × 4's. The price of a pressure-treated 8 ft 2 × 4 is $2.44. Determine the cost of the supports.

p) Determine the cost of the materials for building a wooden ramp by adding the amounts from parts (d), (n), and (o).

q) Are the materials for constructing a concrete ramp or a wooden ramp less expensive?

CHAPTER 10

Mathematical Systems

In the last 200 years, much of the focus of scientific study of the fundamental laws of nature has shifted from what things are to how they change in space and time. The mathematics used for this purpose belongs to a branch of mathematics known as group theory. A group is a collection of fundamentally basic elements: It could be numbers, it could be the elementary particles of physics, or it could be a pattern of repeating geometric designs. The way in which the elements change or remain the same when acted upon by some operation or transformation defines membership in the group. A certain underlying pattern and symmetry exists in all this.

To appreciate why group theory is so useful to scientists, consider the physicists who are trying to piece together the history of the universe. They have a clear sense of what the universe is like today, but what about 20 billion years ago when scientists believe the universe exploded into existence in what is called the Big Bang? Group theory may enable them to derive what the initial conditions of the universe may have been from knowledge of what matter is like today. Group theory can also be used to describe number systems as well as the basic building blocks of crystalline solids. It can even be used to define the symmetries that appear in the artifacts of a given culture. No wonder the physicist Sir Arthur Stanley Eddington called group theory the "super-mathematics." ■

Despite the seemingly endless variety of patterns the human imagination can devise, group theory can be used to catalog and define patterns by the way in which the design elements are transformed and positioned.

10.1 GROUPS

We begin our discussion by introducing a mathematical system. As you will learn shortly, you already know and use many mathematical systems.

> A **mathematical system** consists of a set of elements and at least one binary operation.

1 Color + 1 Color = 1 Color

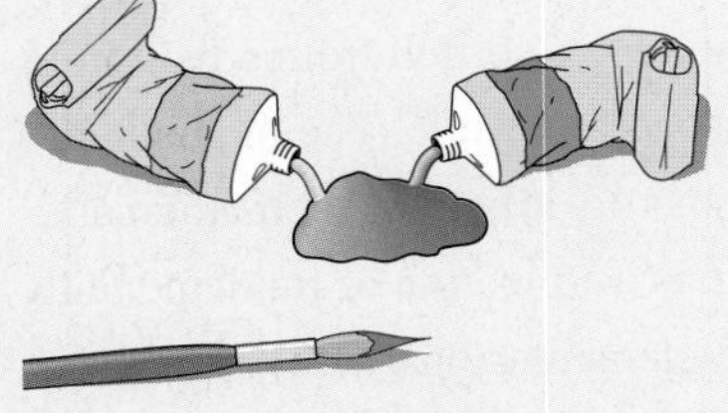

The mixing of two paints is a binary operation. When two colors are mixed—for example, yellow and blue—a single color results, in this case green.

In the above definition we mention binary operation. A **binary operation** is an operation, or rule, that can be performed on two and only two elements of a set. The result is a single element. When we add *two* integers, the sum is *one* integer. When we multiply *two* integers, the product is *one* integer. Thus, addition and multiplication are both binary operations. Is finding the reciprocal of a number a binary operation? No, it is an operation on a single element of a set.

When you learned how to add integers, you were introduced to a mathematical system. The set of elements is the set of integers, and the binary operation is addition. When you learned how to multiply integers, you became familiar with a second mathematical system. The set of integers with the operation of subtraction and the set of integers with the operation of division are two other examples of mathematical systems since subtraction and division are also binary operations.

Some systems are used in solving everyday problems, such as planning work schedules. Others are more abstract and are used primarily in research, chemistry, physical structure, matter, the nature of genes, and other scientific fields.

COMMUTATIVE AND ASSOCIATIVE PROPERTIES

Once a mathematical system is defined, its structure may display certain properties. Consider the set of integers:

$$I = \{\ldots, -3, -2, -1, 0, 1, 2, 3, \ldots\}.$$

Recall that the ellipsis, the three dots, at each end of the set indicates that the set continues in the same manner.

The set of integers can be studied with the operations of addition, subtraction, multiplication, or division as separate mathematical systems. For example, when we study the set of integers under the operations of addition or multiplication, we see the commutative and associative properties hold. The general forms of the properties are shown here.

For Any Elements *a*, *b*, and *c*

	Addition	***Multiplication***
Commutative property	$a + b = b + a$	$a \cdot b = b \cdot a$
Associative property	$(a + b) + c = a + (b + c)$	$(a \cdot b) \cdot c = a \cdot (b \cdot c)$

The integers *are commutative* under the operations of *addition and multiplication.* For example,

$$2 + 4 = 4 + 2 \qquad \text{and} \qquad 2 \cdot 4 = 4 \cdot 2$$
$$6 = 6 \qquad\qquad\qquad 8 = 8$$

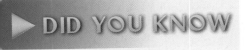

However, the integers *are not commutative* under the operations of *subtraction and division.* For example,

$$4 - 2 \neq 2 - 4 \qquad \text{and} \qquad 4 \div 2 \neq 2 \div 4$$
$$2 \neq -2 \qquad\qquad\qquad 2 \neq \tfrac{1}{2}$$

The integers *are associative* under the operations of *addition and multiplication.* For example,

$$(1 + 2) + 3 = 1 + (2 + 3) \qquad \text{and} \qquad (1 \cdot 2) \cdot 3 = 1 \cdot (2 \cdot 3)$$
$$3 + 3 = 1 + 5 \qquad\qquad\qquad 2 \cdot 3 = 1 \cdot 6$$
$$6 = 6 \qquad\qquad\qquad 6 = 6$$

However, the integers *are not associative* under the operations of *subtraction and division.* See Exercises 17 and 18 at the end of this section.

To say that a set of elements is commutative under a given operation means that the commutative property holds for *any* elements a and b in the set. Similarly, to say that a set of elements is associative under a given operation means that the associative property holds for *any* elements a, b, and c in the set.

Consider the mathematical system consisting of the set of integers under the operation of addition. Because the set of integers is infinite, this mathematical system is an example of an *infinite mathematical system.* We will study certain properties of this mathematical system. The first property that we examine is closure.

CLOSURE

The sum of any two integers is an integer. Therefore, the set of integers is said to be *closed,* or to satisfy the *closure property,* under the operation of addition.

> If a binary operation is performed on any two elements of a set and the result is an element of the set, then that set is **closed** (or has **closure**) under the given binary operation.

Is the set of integers closed under the operation of multiplication? The answer is yes. When any two integers are multiplied, the product will be an integer.

Is the set of integers closed under the operation of subtraction? Again, the answer is yes. The difference of any two integers is an integer.

Is the set of integers closed under the operation of division? The answer is no because two integers may have a quotient that is not an integer. For example, if we select the integers 2 and 3, the quotient of 2 divided by 3 is $\frac{2}{3}$, which is not an integer. Thus, the integers are not closed under the operation of division.

We showed that the set of integers was not closed under the operation of division by finding two integers whose quotient was not an integer. A specific example illustrating that a specific property is not true is called a **counterexample**. Mathematicians and scientists often try to find a counterexample to confirm that a specific property is not always true.

DID YOU KNOW

Secrets of the Universe

A computer simulation of protons and neutrons being smashed together.

Physicists use group theory in their search for the elementary particles that make up all matter. The operations are *transformations*: the coming together or splitting apart of elementary particles as they form larger particles such as protons and neutrons. Some transformations bring two particles together to form a different particle (binary operation), some leave a particle intact (identity), and some follow the rules of associativity and commutativity. In fact, understanding there were distinct "families" of particles related to one another as groups enabled physicists to predict the existence of certain particles before any experimental evidence was available.

IDENTITY ELEMENT

Now we will discuss the identity element for the set of integers under the operation of addition. Is there an element in the set that, when added to any given integer, results in a sum that is the given integer? The answer is yes. The sum of 0 and any integer is the given integer. For example, $1 + 0 = 0 + 1 = 1$, $-4 + 0 = 0 + (-4) = -4$, and so on. For this reason, we call 0 the **additive identity element** for the set of integers. Note that for any integer a, $a + 0 = 0 + a = a$.

> An **identity element** is an element in a set such that when a binary operation is performed on it and any given element in the set, the result is the given element.

Is there an identity element for the set of integers under the operation of multiplication? The answer is yes; it is the number 1. Note that $2 \cdot 1 = 1 \cdot 2 = 2$, $3 \cdot 1 = 1 \cdot 3 = 3$, and so on. For any integer a, $a \cdot 1 = 1 \cdot a = a$. For this reason, 1 is called the **multiplicative identity element** for the set of integers.

INVERSES

What integer, when added to 4, gives a sum of 0; that is, $4 + \square = 0$? The shaded area is to be filled in with the integer -4: $4 + (-4) = 0$. We say that -4 is the additive inverse of 4, and 4 is the additive inverse of -4. Note that the sum of the element and its additive inverse gives the additive identity element 0. What is the additive inverse of 12? Since $12 + (-12) = 0$, -12 is the additive inverse of 12.

Other examples of integers and their additive inverses are

Element	+	Additive Inverse	=	Identity Element
0	+	0	=	0
2	+	(−2)	=	0
−5	+	5	=	0

Note that for the operation of addition, every integer a has a unique inverse, $-a$, such that $a + (-a) = -a + a = 0$.

> When a binary operation is performed on two elements in a set and the result is the identity element for the binary operation, then each element is said to be the **inverse** of the other.

Does every integer have an inverse under the operation of multiplication? For multiplication the product of an integer and its inverse must yield the multiplicative identity element, 1. What is the multiplicative inverse of 2? That is, 2 times what number gives 1?

$$2 \cdot ? = 1 \qquad 2 \cdot \tfrac{1}{2} = 1$$

However, since $\frac{1}{2}$ is not an integer, 2 does not have a multiplicative inverse in the set of integers.

PROFILE IN MATHEMATICS

Niels Abel (1802–1829) Évariste Galois (1811–1832)

Niels Abel

Évariste Galois

Untimely ends: Important contributions to the development of group theory were made by two young men who would not live to see their work gain acceptance: *Niels Abel,* a Norwegian, and *Évariste Galois,* a Frenchman. While their work showed brilliance, neither was noticed in his lifetime by the mathematics community. Abel lived in poverty and died of malnutrition and tuberculosis at the age of 26, just 2 days before a letter arrived with the offer of a teaching position. Galois died of complications from a gunshot wound he received in a duel over a love affair. The 20-year-old man had spent the night before summarizing his theories in the hope that eventually someone would find it "profitable to decipher this mess." After their deaths, both men had important elements of group theory named after them.

GROUP

Let's review what we have learned about the mathematical system consisting of the set of integers under the operation of addition.

1. The set of integers is *closed* under the operation of addition.
2. The set of integers has an *identity element* under the operation of addition.
3. Each element in the set of integers has an *inverse* under the operation of addition.
4. The *associative property* holds for the set of integers under the operation of addition.

The set of integers under the operation of addition is an example of a group. The properties of a group can be summarized as follows.

Properties of a Group

Any mathematical system that meets the following four requirements is called a **group**.

1. The set of elements is *closed* under the given operation.
2. An *identity element* exists for the set.
3. Every element in the set has an *inverse.*
4. The set of elements is *associative* under the given operation.

It is often very time consuming to show that the associative property holds for all cases. In many of the examples that follow we will state that the associative property holds for the given set of elements under the given operation.

COMMUTATIVE GROUP

The commutative property does not need to hold for a mathematical system to be a group. However, if a mathematical system meets the four requirements of a group and is also commutative under the given operation, the mathematical system is a *commutative* (*or abelian*) *group.* The abelian group is named after Niels Abel (see the Profile in Mathematics).

A group that satisfies the commutative property is called a **commutative group** (or **abelian group**).

Because the commutative property holds for the set of integers under the operation of addition, the set of integers under the operation of addition is not only a group, but it is a commutative group.

Properties of a Commutative Group

A mathematical system is a commutative group if all five conditions hold.

1. The set of elements is *closed* under the given operation.
2. An *identity element* exists for the set.
3. Every element in the set has an *inverse.*
4. The set of elements is *associative* under the given operation.
5. The set of elements is *commutative* under the given operation.

DID YOU KNOW

The Enormous Theorem

Mathematicians are attempting to make the world's longest proof shorter. It is called the *enormous theorem.* More than 100 mathematicians toiled for more than 30 years to produce the proof known formally as the classification of finite, simple groups. Completed in the 1980s, it consists of some 500 publications totaling 15,000 pages. The theorem established that all finite, simple groups can fall into 17 infinite families, or 26 so-called sporadic forms. Now an attempt is being made to whittle the theorem down to a mere 3000 pages or so. Those working on it hope that shortening the proof can be completed in a decade or so.

To determine whether a mathematical system is a group under a given operation, check, in the following order, to determine whether (a) the system is closed under the given operation, (b) there is an identity element in the set for the given operation, (c) every element in the set has an inverse under the given operation, and (d) the associative property holds under the given operation. If any of these four requirements is *not* met, stop and state the mathematical system is not a group. If asked to determine whether the mathematical system is a commutative group, then check to determine whether the commutative property holds for the given operation.

EXAMPLE 1 *Is It a Group?*

Determine whether the set of rational numbers under the operation of multiplication forms a group.

SOLUTION Recall from Chapter 5 that the rational numbers are the set of numbers of the form $\frac{p}{q}$ where p and q are integers and $q \neq 0$. All fractions and integers are rational numbers.

1. *Closure:* The product of two rational numbers is a rational number. Therefore, the rational numbers are closed under the operation of multiplication.
2. *Identity Element:* The multiplicative identity element for the set of rational numbers is 1. Note, for example, $3 \cdot 1 = 1 \cdot 3 = 3$, and $\frac{3}{8} \cdot 1 = 1 \cdot \frac{3}{8} = \frac{3}{8}$. For any rational number a, $a \cdot 1 = 1 \cdot a = a$.
3. *Inverse Elements:* For the mathematical system to be a group under the operation of multiplication, *each and every* rational number must have a multiplicative inverse in the set of rational numbers. Remember, for the operation of multiplication, the product of a number and its inverse must give the multiplicative identity element, 1. Let's check a few rational numbers:

Rational number	· inverse	= identity element
3	$\cdot \ \frac{1}{3}$	$= \ 1$
$\frac{2}{3}$	$\cdot \ \frac{3}{2}$	$= \ 1$
$-\frac{1}{5}$	$\cdot \ -5$	$= \ 1$

Looking at these examples you might deduce that each rational number does have an inverse. However, one rational number, 0, does not have an inverse.

$$0 \cdot ? = 1$$

Because there is no rational number that, when multiplied by 0, gives 1, 0 does not have a multiplicative inverse. Since every rational number does not have an inverse, this mathematical system is not a group.

There is no need at this point to check the associative property because we have already shown that the mathematical system of rational numbers under the operation of multiplication is not a group. ▲

SECTION 10.1 EXERCISES

CONCEPT/WRITING EXERCISES

1. What is a binary operation?
2. What are the parts of a mathematical system?
3. Explain why each of the following is a binary operation. Give an example to illustrate each binary operation.
   a) Addition
   b) Subtraction
   c) Multiplication
   d) Division
4. What properties are required for a mathematical system to be a group?
5. What properties are required for a mathematical system to be a commutative group?
6. What is another name for a commutative group?
7. Explain the closure property. Give an example of the property.
8. What is an identity element? Explain. Give the additive and multiplicative identity elements for the set of integers.
9. What is an inverse element? Explain. Give the additive and multiplicative inverse of the number 2 for the set of rational numbers.
10. What is a counterexample?

PRACTICE THE SKILLS

11. Give the associative property of addition and illustrate the property with an example.
12. Give the associative property of multiplication and illustrate the property with an example.
13. Give the commutative property of addition and illustrate the property with an example.
14. Give the commutative property of multiplication and illustrate the property with an example.
15. Give an example to show that the commutative property does not hold for the set of integers under the operation of subtraction.
16. Give an example to show that the commutative property does not hold for the set of integers under the operation of division.
17. Give an example to show that the associative property does not hold for the set of integers under the operation of subtraction.
18. Give an example to show that the associative property does not hold for the set of integers under the operation of division.

PROBLEM SOLVING

In Exercises 19–28, explain your answer.

19. Is the set of positive integers a commutative group under the operation of addition?
20. Is the set of integers a group under the operation of multiplication?
21. Is the set of rational numbers a commutative group under the operation of addition?
22. Is the set of positive integers a group under the operation of subtraction?
23. Is the set of integers a group under the operation of subtraction?
24. For the set of integers list two operations that are not binary.

CHALLENGE PROBLEMS/GROUP ACTIVITIES

25. Is the set of irrational numbers a group under the operation of addition?
26. Is the set of irrational numbers a group under the operation of multiplication?
27. Is the set of real numbers a group under the operation of addition?
28. Is the set of real numbers a group under the operation of multiplication?
29. Create a mathematical system with two binary operations. Select a set of elements and the binary operations so that one binary operation with the set of elements meets the requirements for a group and the other binary operation will not. Explain why the one binary operation with the set of elements is a group. For the other binary operation and the set of elements find counterexamples to show that it is not a group.

RESEARCH ACTIVITY

30. There are other classifications of mathematical systems besides groups. For example, there are *rings* and *fields*. Do research to determine the requirements that must be met for a mathematical system to be (a) a ring and (b) a field. (c) Is the set of real numbers, under the operations of addition and multiplication, a field? Ask your instructor for references to use.

10.2 FINITE MATHEMATICAL SYSTEMS

In the preceding section we presented infinite mathematical systems. In this section we present some finite mathematical systems. A **finite mathematical system** is one whose set contains a finite number of elements.

CLOCK ARITHMETIC

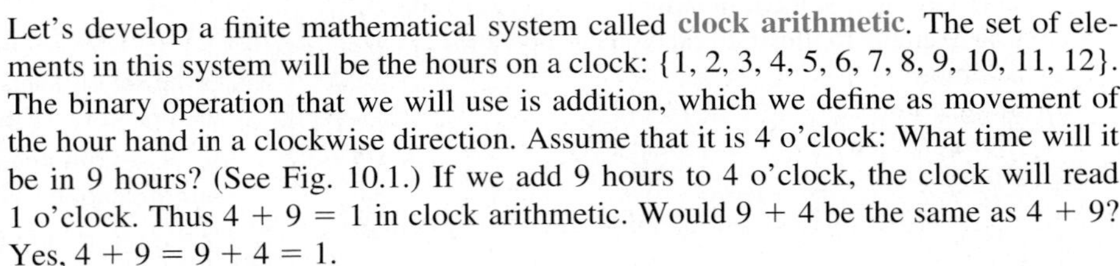

Figure 10.1

Let's develop a finite mathematical system called **clock arithmetic**. The set of elements in this system will be the hours on a clock: {1, 2, 3, 4, 5, 6, 7, 8, 9, 10, 11, 12}. The binary operation that we will use is addition, which we define as movement of the hour hand in a clockwise direction. Assume that it is 4 o'clock: What time will it be in 9 hours? (See Fig. 10.1.) If we add 9 hours to 4 o'clock, the clock will read 1 o'clock. Thus $4 + 9 = 1$ in clock arithmetic. Would $9 + 4$ be the same as $4 + 9$? Yes, $4 + 9 = 9 + 4 = 1$.

Table 10.1 is the addition table for clock arithmetic. Its elements are based on the definition of addition as previously illustrated. For example, the sum of 4 and 9 is 1, so we put a 1 in the table where the row to the right of the 4 intersects the column below the 9. Likewise, the sum of 11 and 10 is 9, so we put a 9 in the table where the row to the right of the 11 intersects the column below the 10.

The binary operation of this system is defined by the table. It is denoted by the symbol +. To determine the value of $a + b$, where a and b are any two numbers in the set, find a in the left-hand column and find b along the top row. Assume there is a horizontal line through a and a vertical line through b; the point of intersection of these two lines is where you find the value of $a + b$. For example $10 + 4 = 2$ has been circled in Table 10.1. Note that $4 + 10$ also equals 2, but this result will not necessarily hold for all examples in this chapter.

Table 10.1 Clock 12 Arithmetic

+	1	2	3	4	5	6	7	8	9	10	11	12
1	2	3	4	5	6	7	8	9	10	11	12	1
2	3	4	5	6	7	8	9	10	11	12	1	2
3	4	5	6	7	8	9	10	11	12	1	2	3
4	5	6	7	8	9	10	11	12	1	2	3	4
5	6	7	8	9	10	11	12	1	2	3	4	5
6	7	8	9	10	11	12	1	2	3	4	5	6
7	8	9	10	11	12	1	2	3	4	5	6	7
8	9	10	11	12	1	2	3	4	5	6	7	8
9	10	11	12	1	2	3	4	5	6	7	8	9
10	11	12	1	(2)	3	4	5	6	7	8	9	10
11	12	1	2	3	4	5	6	7	8	9	10	11
12	1	2	3	4	5	6	7	8	9	10	11	12

EXAMPLE 1 *A Commutative Group?*

Determine whether the clock arithmetic system under the operation of addition is a commutative group.

SOLUTION Check the five requirements that must be satisfied for a commutative group.

1. *Closure:* Is the set of elements in clock arithmetic closed under the operation of addition? Yes, since Table 10.1 contains only the elements in the set

DID YOU KNOW

The Changing Times

Venice clock

The photo shows the famous clock at the Cathedral of St. Marks (called the Basilica) in Piazza San Marco (St. Marks Plaza) in Venice, Italy. Construction of the basilica began in 830, but it has been rebuilt a number of times due to fires. The clock tower was built in the fifteenth century. Note that the clock displays all 24 hours in Roman numerals. Also note the position of the hours. Noon, or XII, is where we expect 9 o'clock to be.

In Florence, Italy, the oldest building is probably the Duomo (The Cathedral House) in the Piazza del Duomo in the city's center. Florentines believe it was constructed in about 1000. The large clock in the Cathedral also displays 24 hours using Roman numerals.

On both clocks 4 o'clock is indicated by IIII, not IV, and 9 o'clock is indicated by VIIII, not IX.

{1, 2, 3, 4, 5, 6, 7, 8, 9, 10, 11, 12}. If Table 10.1 had contained an element other than the numbers 1 through 12, the set would not have been closed under addition.

2. *Identity element:* Is there an identity element for clock arithmetic? If the time is currently 4 o'clock, how many hours have to pass before it is 4 o'clock again? Twelve hours: $4 + 12 = 12 + 4 = 4$. In fact, given any hour, in 12 hours the clock will return to the starting point. Therefore, 12 is the additive identity element in clock arithmetic.

   In examining Table 10.1 we see that the row of numbers next to the 12 in the left-hand column is identical to the row of numbers along the top. We also see that the column of numbers under the 12 in the top row is identical to the column of numbers on the left. The search for such a column and row is one technique for determining whether an identity element exists for a system defined by a table.

3. *Inverse elements:* Is there an inverse for the number 4 in clock arithmetic for the operation of addition? Recall that the identity element in clock arithmetic is 12. What number when added to 4 gives 12; that is, $4 + \square = 12$? Table 10.1 shows that $4 + 8 = 12$ and also that $8 + 4 = 12$. Thus, 8 is the additive inverse of 4, and 4 is the additive inverse of 8.

   To find the additive inverse of 7, find 7 in the left-hand column of Table 10.1. Look to the right of the 7 until you come to the identity element 12. Determine the number at the top of this column. The number is 5. Since $7 + 5 = 5 + 7 = 12$, 5 is the inverse of 7, and 7 is the inverse of 5. The other inverses can be found in the same way, as shown in Table 10.2. Note that each element in the set has an *inverse.*

Table 10.2 Clock 12 Inverses

Element	+	Inverse	=	Identity element
1	+	11	=	12
2	+	10	=	12
3	+	9	=	12
4	+	8	=	12
5	+	7	=	12
6	+	6	=	12
7	+	5	=	12
8	+	4	=	12
9	+	3	=	12
10	+	2	=	12
11	+	1	=	12
12	+	12	=	12

4. *Associative property:* Now consider the associative property. Does $(a + b) + c = a + (b + c)$ for all values a, b, and c of the set? Remember to always evaluate the values within the parentheses first. Let's select some values for a, b, and c. Let $a = 2$, $b = 6$, and $c = 8$. Then

$$(2 + 6) + 8 = 2 + (6 + 8)$$
$$8 + 8 = 2 + 2$$
$$4 = 4 \quad \text{True}$$

Let $a = 5$, $b = 12$, and $c = 9$. Then

$$(5 + 12) + 9 = 5 + (12 + 9)$$
$$5 + 9 = 5 + 9$$
$$2 = 2 \quad \text{True}$$

Randomly selecting *any* elements a, b, and c of the set reveals $(a + b) + c = a + (b + c)$. Thus, the system of clock arithmetic is associative under the operation of addition. Note that if there is just one set of values a, b, and c such that $(a + b) + c \neq a + (b + c)$, the system is not associative. Normally you will not be asked to check every case to determine whether the associative property holds. However, if every element in the set does not appear in every row and column of the table, you need to check the associative property carefully.

5. *Commutative property:* Does the commutative property hold under the given operation? Does $a + b = b + a$ for all elements a and b of the set? Let's randomly select some values for a and b to determine whether the commutative property appears to hold. Let $a = 5$ and $b = 8$; then Table 10.1 shows that

$$5 + 8 = 8 + 5$$
$$1 = 1 \quad \text{True}$$

Let $a = 9$ and $b = 6$; then

$$9 + 6 = 6 + 9$$
$$3 = 3 \quad \text{True}$$

The commutative property holds for these two specific cases. In fact, if we were to select *any* values for a and b, we would find that $a + b = b + a$. Thus, the commutative property of addition is true in clock arithmetic. Note, if there is just one set of values a and b such that $a + b \neq b + a$, the system is not commutative.

This system satisfies the five properties required for a mathematical system to be a commutative group. Thus, clock arithmetic under the operation of addition is a commutative group.

One method that can be used to determine whether a system defined by a table is commutative under the given operation is to determine whether the elements in the table are symmetric about the main diagonal. The main diagonal is the diagonal from the upper left-hand corner to the lower right-hand corner of the table. In Table 10.3 the main diagonal is shaded in color.

If the elements are symmetric about the main diagonal, then the system is commutative. If the elements are not symmetric about the main diagonal, then the system is not commutative. If you examine the system in Table 10.3 you see its elements are symmetric about the main diagonal. Therefore, this mathematical system is commutative.

It is possible to have groups that are not commutative. Such groups are called **noncommutative** or **nonabelian groups**. However, a *noncommutative group defined by a table must be at least a six-element by six-element table.* Nonabelian groups are illustrated in Exercises 69 and 72 at the end of this section.

Table 10.3 Symmetry about the Main Diagonal

+	0	1	2	3	4
0	0	1	2	3	4
1	1	2	3	4	0
2	2	3	4	0	1
3	3	4	0	1	2
4	4	0	1	2	3

Now we will look at another finite mathematical system.

Table 10.4 Four-Element System

$\odot$	1	3	5	7
1	5	7	1	3
3	7	1	3	5
5	1	3	5	7
7	3	5	7	1

EXAMPLE 2 *A Finite System*

Consider the mathematical system defined by Table 10.4. Assume that the associative property holds for the given operation.

a) List the elements in the set of this mathematical system.
b) Identify the binary operation.
c) Determine whether this mathematical system is a commutative group.

SOLUTION

a) The set of elements for this mathematical system consists of the elements found on the top (or left-hand side) of the table: {1, 3, 5, 7}.
b) The binary operation is $\odot$.
c) We must determine whether the five requirements for a commutative group are satisfied.

1. *Closure:* All the elements in the table are in the original set of elements, {1, 3, 5, 7}, so the system is closed.
2. *Identity element:* The identity element is 5. Note that the column of elements under the 5 is identical to the left-hand column *and* the row of elements to the right of the 5 is identical to the top row.
3. *Inverse elements:* When an element operates on its inverse, the result is the identity element. For this example the identity element is 5. To determine the inverse of 1, find the element to replace the question mark:

$$1 \odot ? = 5$$

Since $1 \odot 1 = 5$, 1 is the inverse of 1. Thus, 1 is its own inverse.

To find the inverse of 3, find the element to replace the question mark:

$$3 \odot ? = 5$$

Since $3 \odot 7 = 7 \odot 3 = 5$, 7 is the inverse of 3 (and 3 is the inverse of 7). The elements and their inverses are shown in Table 10.5. Every element has a unique inverse.

Table 10.5 Inverses under $\odot$

Element	$\odot$	Inverse	=	Identity element
1	$\odot$	1	=	5
3	$\odot$	7	=	5
5	$\odot$	5	=	5
7	$\odot$	3	=	5

4. *Associative property:* It is given that the associative property holds for this operation. One example of associative property is

$$(7 \odot 3) \odot 1 = 7 \odot (3 \odot 1)$$
$$5 \odot 1 = 7 \odot 7$$
$$1 = 1 \quad \text{True}$$

5. *Commutative property:* The elements in the table are symmetric about the main diagonal, so the commutative property holds for the operation of $\odot$. One example of the commutative property is

$$3 \odot 5 = 5 \odot 3$$
$$3 = 3 \quad \text{True}$$

The five necessary properties hold. Thus, the mathematical system is a commutative group.

MATHEMATICAL SYSTEMS WITHOUT NUMBERS

Thus far, all the systems we have discussed have been based on sets of numbers. Example 3 illustrates a mathematical system of symbols rather than numbers.

EXAMPLE 3 *Investigating a System of Symbols*

Use the mathematical system defined by Table 10.6 and determine

Table 10.6 A System of Symbols

$\boxdot$	@	$\triangle$	L
@	$\triangle$	L	@
$\triangle$	L	@	$\triangle$
L	@	$\triangle$	L

a) The set of elements.
b) The binary operation.
c) Closure or nonclosure of the system.
d) The identity element.
e) The inverse of @.
f) @ $\boxdot$ $\triangle$ and L $\boxdot$ $\triangle$.
g) (@ $\boxdot$ $\triangle$) $\boxdot$ $\triangle$ and @ $\boxdot$ ($\triangle$ $\boxdot$ $\triangle$).

SOLUTION

a) The set of elements of this mathematical system is {@, $\triangle$, L}.
b) The binary operation is $\boxdot$.
c) Because the table does not contain any symbols other than @, $\triangle$, and L, the system is closed under $\boxdot$.
d) The identity element is L. Note the row next to L in the left-hand column is the same as the top row, and the column under L is identical to the left-hand column. We see that

$$@ \boxdot L = L \boxdot @ = @$$
$$\triangle \boxdot L = L \boxdot \triangle = \triangle$$
$$L \boxdot L = L$$

e) We know that element $\boxdot$ inverse element = identity element, so to find the inverse of @ we write the following.

$$@ \boxdot ? = L$$

To find the inverse of @, we must determine the element to replace the question mark. Since @ $\boxdot$ $\triangle$ = L and $\triangle$ $\boxdot$ @ = L, $\triangle$ is the inverse of @.

f) @ $\boxdot$ $\triangle$ = L and L $\boxdot$ $\triangle$ = $\triangle$

g) $(@ \boxdot \triangle) \boxdot \triangle = L \boxdot \triangle = \triangle$ and $@ \boxdot (\triangle \boxdot \triangle) = @ \boxdot @ = \triangle$

DID YOU KNOW

Group Theory Can Be Fun

Group theory can be fun as well as educational, at least so thought Erno Rubik, a Hungarian teacher of architecture and design. In 1976 he presented the world with a puzzle in the form of a cube with six faces, each of a different color. Each face is divided into nine squares; each row and column of each face can rotate so that in no time at all six faces are transformed into a mix of colors. Your job, should you choose to accept it, is to get the cube back to its original condition.

Table 10.7

$*$	A	B	C	D
A	D	A	B	C
B	A	B	C	D
C	B	C	D	A
D	C	D	A	B

Table 10.8

Element	$*$	Inverse	$=$	Identity element
A	$*$	C	$=$	B
B	$*$	B	$=$	B
C	$*$	A	$=$	B
D	$*$	D	$=$	B

EXAMPLE 4 *Is the System a Commutative Group?*

Determine whether the mathematical system in Table 10.7 in which $*$ is an associative operation is a commutative group.

SOLUTION

1. *Closure:* The system is closed.
2. *Identity element:* The identity element is B.
3. *Inverse elements:* Each element has an inverse as illustrated in Table 10.8.
4. *Associative property:* It is given that the associative property holds. An example illustrating the associative property is

$$\begin{aligned}(D * A) * C &= D * (A * C)\\ C * C &= D * B\\ D &= D \quad \text{True}\end{aligned}$$

5. *Commutative property:* By examining the table we can see that it is symmetric about the main diagonal. Thus, the system is commutative under the given operation. One example of the commutative property is

$$\begin{aligned}D * C &= C * D\\ A &= A \quad \text{True}\end{aligned}$$

All five properties are satisfied. Thus, the system is a commutative group.

Table 10.9

$\ominus$	x	y	z
x	x	z	y
y	z	y	x
z	y	x	z

EXAMPLE 5 *Another System to Study*

Determine whether the mathematical system in Table 10.9 is a commutative group under the operation of $\ominus$.

SOLUTION

1. The system is closed.
2. No row is identical to the top row, so there is no identity element. Therefore, this mathematical system is *not a group.* There is no need to go any further, but for practice, let's look at a few more items.
3. Since there is no identity element, there can be no inverses.
4. The associative property does not hold. The following counterexample illustrates the associative property does not hold for every case.

$$\begin{aligned}(x \ominus y) \ominus z &\neq x \ominus (y \ominus z)\\ z \ominus z &\neq x \ominus x\\ z &\neq x\end{aligned}$$

5. The table is symmetric about the main diagonal. Therefore, the commutative property does hold for the operation of $\ominus$.

Note that the associative property does not hold even though the commutative property does hold. This outcome can occur when there is no identity element and every element does not have an inverse, as in this example.

Table 10.10

*	a	b	c
a	a	b	c
b	b	b	a
c	c	a	c

EXAMPLE 6 *Is the System a Commutative Group?*

Determine whether the mathematical system in Table 10.10 is a commutative group under the operation of $*$.

SOLUTION

1. The system is closed.
2. There is an identity element, a.
3. Each element has an inverse; a is the inverse of a, b is the inverse of c, and c is the inverse of b.
4. Every element in the set does not appear in every row and every column of the table, so we need to check the associative property carefully. There are many specific cases where the associative property does hold. However, the following counterexample illustrates that the associative property does not hold for every case.

$$\begin{aligned} (b * b) * c &\neq b * (b * c) \\ b * c &\neq b * a \\ a &\neq b \end{aligned}$$

5. The commutative property holds because there is symmetry about the main diagonal.

Since we have shown that the associative property does not hold under the operation of $*$, this system is not a group. Therefore, it cannot be a commutative group.

DID YOU KNOW

Creating Patterns by Design

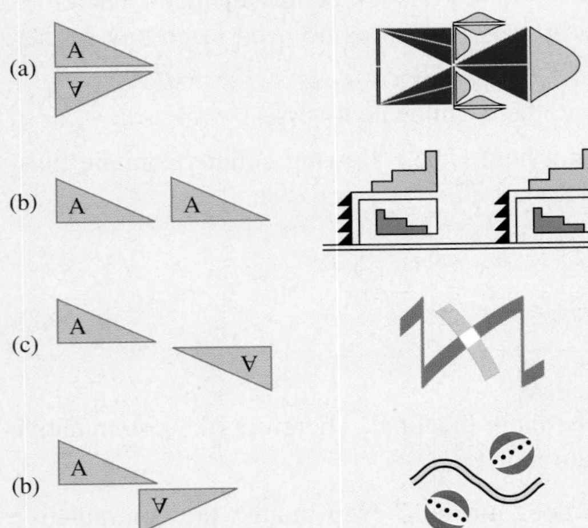

Patterns from *Symmetries of Culture* by Dorothy K. Washburn and Donald W. Crowe (University of Washington Press, 1988)

What makes group theory such a powerful tool is that it can be used to reveal the underlying structure of just about any physical phenomenon that involves symmetry and patterning such as wallpaper or quilt patterns. Interest in the formal study of symmetry in design came out of the Industrial Revolution in the late nineteenth century. The new machines of the Industrial Revolution could vary any given pattern almost indefinitely. Designers needed a way to describe and manipulate patterns systematically. At the same time, explorers were discovering artifacts of other cultures, which stimulated interest in categorizing patterns. Shown here are the four geometric motions that generate all two-dimensional patterns: (a) reflection, (b) translation, (c) rotation, and (d) glide reflection. How these motions are applied, or not applied, is the basis of pattern analysis.

SECTION 10.2 EXERCISES

CONCEPT/WRITING EXERCISES

1. Explain how the clock 12 addition table is formed.
2. What is 12 + 12 in clock 12 arithmetic? Explain how you obtained your answer.
3. a) Explain how to add the numbers (6 + 9) + 5 in clock 12 arithmetic using the addition table.
   b) What is (6 + 9) + 5 in clock 12 arithmetic?
4. a) Explain how to determine a difference of two numbers in clock arithmetic by using the face of a clock.
   b) Determine 5 − 9 in clock 12 arithmetic using the method explained in part (a).
5. a) Explain how to find 3 − 10 in clock 12 arithmetic by adding the number 12 to one of the numbers.
   b) Determine 3 − 10 in clock 12 arithmetic using the method explained in part (a).
   c) Explain why the procedure you give in part (a) works.
6. Explain one method of determining whether a system defined by a table is commutative under the given operation.
7. Is clock 12 arithmetic closed under the operation of addition? Explain.
8. Is there an identity element for addition in clock 12 arithmetic? If so, what is it?
9. Does each element in clock 12 arithmetic have an inverse? If so, give each element and its corresponding inverse.
10. Give an example to illustrate the associative property of addition in clock 12 arithmetic.
11. Is clock 12 arithmetic commutative? Give an example to verify your answer.
12. Is clock 12 arithmetic under the operation of addition a commutative group? Explain.

PRACTICE THE SKILLS

In Exercises 13–24, use Table 10.1 to determine the sum in clock 12 arithmetic.

13. 3 + 9 **14.** 9 + 5
15. 9 + 8 **16.** 10 + 4
17. 5 + 12 **18.** 12 + 12
19. 2 + (9 + 9) **20.** (8 + 7) + 6
21. (6 + 4) + 8 **22.** (10 + 6) + 12
23. (7 + 8) + (9 + 6) **24.** (7 + 11) + (9 + 5)

In Exercises 25–30, determine the difference in clock 12 arithmetic by starting at the first number and counting counterclockwise on the clock the number of units given by the second number.

25. 10 − 4 **26.** 12 − 8 **27.** 4 − 7
28. 3 − 9 **29.** 5 − 10 **30.** 3 − 10

31. Use Fig. 10.2 to develop an addition table for clock 6 arithmetic.

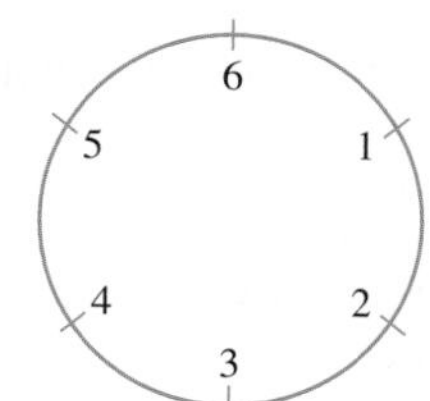

Figure 10.2

In Exercises 32–40, determine the sum or difference in clock 6 arithmetic.

32. 5 + 4 **33.** 3 + 4 **34.** 6 + 4
35. 5 − 2 **36.** 3 − 5 **37.** 2 − 6
38. 3 − 4 **39.** (4 − 5) − 6 **40.** 2 + (1 − 3)

41. Use Fig. 10.3 to develop an addition table for clock 7 arithmetic.

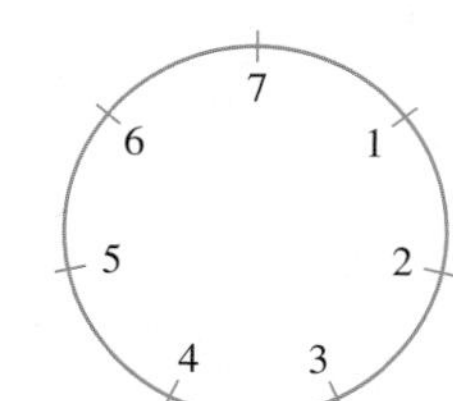

Figure 10.3

In Exercises 42–50, determine the sum or difference in clock 7 arithmetic.

42. 3 + 6 **43.** 4 + 5 **44.** 4 + 4
45. 7 + 6 **46.** 2 − 6 **47.** 3 − 6
48. 2 − 4 **49.** (3 − 5) − 6 **50.** 4 + (2 − 6)

51. Determine whether clock 7 arithmetic under the operation of addition is a commutative group. Explain.
52. A mathematical system is defined by a three-element by three-element table where every element in the set appears in each row and each column. Must the mathematical system be a commutative group? Explain.
53. Consider the mathematical system indicated by the following table in which the operation is associative.

	0	**1**	**2**	**3**
0	0	1	2	3
1	1	2	3	0
2	2	3	0	1
3	3	0	1	2

   a) What are the elements of the set in this mathematical system?

b) What is the binary operation?
c) Is the system closed? Explain.
d) Is there an identity element for the system under the given operation? If so, what is it?
e) Does every element in the system have an inverse? If so, give each element and its corresponding inverse.
f) Give an example to illustrate the associative property.
g) Is the system commutative? Give an example to verify your answer.
h) Is the mathematical system a commutative group? Explain.

In Exercises 54–56, repeat parts (a)–(h) of Exercise 53 for the mathematical system in the table. Assume that the operations are associative.

54.

⅄	15	19	23
15	19	23	15
19	23	15	19
23	15	19	23

55.

⌖	5	8	9	11
5	9	11	5	8
8	11	5	8	9
9	5	8	9	11
11	8	9	11	5

56.

☥	3	5	8	4
3	5	8	4	3
5	8	4	3	5
8	4	3	5	8
4	3	5	8	4

57. a) Is the following mathematical system a group? Explain your answer.
b) Find an example showing that the operation is not associative for the given set of elements.

⌇	1	2	3	4
1	2	3	4	1
2	3	4	1	2
3	4	1	2	3
4	1	4	3	2

58. For the mathematical system

⊛	*f*	*r*	*o*	*m*
f	*f*	*r*	*o*	*m*
r	*r*	*o*	*m*	*f*
o	*o*	*m*	*f*	*r*
m	*m*	*f*	*r*	*o*

determine
a) The elements in the set.
b) The binary operation.
c) Closure or nonclosure of the system.
d) $(r$ ⊛ $o)$ ⊛ f
e) $(f$ ⊛ $r)$ ⊛ m
f) The identity element.
g) The inverse of r.
h) The inverse of m.

In Exercises 59–64, for the mathematical system given, determine which of the five properties of a commutative group do not hold.

59.

◫	○	△	□
○	○	△	□
△	△	○	□
□	□	△	○

60.

∧	*w*	*x*	*y*
w	*w*	*y*	*x*
x	*y*	*x*	*w*
y	*x*	*a*	*y*

61.

∅	~	*	?	L	P
~	L	P	~	*	*
*****	P	~	*	~	L
?	~	*	?	L	P
L	*	~	L	P	?
P	*	L	P	?	~

62.

☺	*a*	*b*	π	*0*	△
a	△	*a*	*b*	π	0
b	*a*	*b*	π	0	△
π	*b*	π	0	△	*a*
0	π	0	△	*a*	*b*
△	0	△	π	*b*	*a*

63.

⇔	*a*	*b*	*c*	*d*	*e*
a	*c*	*d*	*e*	*a*	*b*
b	*d*	*e*	*a*	*b*	*c*
c	*e*	*a*	*b*	*c*	*d*
d	*a*	*b*	*c*	*e*	*d*
e	*b*	*c*	*d*	*e*	*a*

64.

$\triangledown$	0	1	2	3	4	5
0	0	0	0	0	0	0
1	0	1	2	3	4	5
2	0	2	4	0	2	4
3	0	3	0	3	0	3
4	0	4	2	0	4	2
5	0	5	4	3	2	1

PROBLEM SOLVING EXERCISES

65. a) Consider the set consisting of two elements $\{E, O\}$, where E stands for an even number and O stands for an odd number. For the operation of addition, complete the table.

+	*E*	*O*
E		
O		

b) Determine whether this mathematical system forms a commutative group under addition. Explain your answer.

66. a) Let E and O represent even numbers and odd numbers, respectively, as in Exercise 65. Complete the table for the operation of multiplication.

×	*E*	*O*
E		
O		

b) Determine whether this mathematical system forms a commutative group under the operation of multiplication. Explain your answer.

In Exercises 67 and 68, make up your own mathematical system that is a group. List the identity element and the inverses of each element. Do so with sets containing

67. Three elements.

68. Four elements.

69. The following table is an example of a noncommutative or nonabelian group.

∞	1	2	3	4	5	6
1	5	3	4	2	6	1
2	4	6	5	1	3	2
3	2	1	6	5	4	3
4	3	5	1	6	2	4
5	6	4	2	3	1	5
6	1	2	3	4	5	6

a) Show that this system under the operation ∞ is a group. (It would be very time consuming to prove that the associative property holds, but you can give some examples to show that it appears to hold.)

b) Find a counterexample to show that the commutative property does not hold.

CHALLENGE PROBLEMS/GROUP ACTIVITIES

70. If a mathematical system is defined by a four-element by four-element table, how many specific cases must be illustrated to prove the set of elements is associative under the given operation?

71. Repeat Exercise 70 for a five-element by five-element table.

72. *Book Arrangements* Suppose that three books numbered 1, 2, and 3 are placed next to one another on a shelf. If we remove volume 3 and place it before volume 1, the new order of books is 3, 1, 2. Let's call this replacement R. We can write

$$R = \begin{pmatrix} 1 & 2 & 3 \\ 3 & 1 & 2 \end{pmatrix},$$

which indicates the books were switched in order from 1, 2, 3 to 3, 1, 2. Other possible replacements are S, T, U, V, and I, as indicated.

$$S = \begin{pmatrix} 1 & 2 & 3 \\ 2 & 1 & 3 \end{pmatrix} \quad T = \begin{pmatrix} 1 & 2 & 3 \\ 3 & 2 & 1 \end{pmatrix} \quad U = \begin{pmatrix} 1 & 2 & 3 \\ 1 & 3 & 2 \end{pmatrix}$$

$$V = \begin{pmatrix} 1 & 2 & 3 \\ 2 & 3 & 1 \end{pmatrix} \quad I = \begin{pmatrix} 1 & 2 & 3 \\ 1 & 2 & 3 \end{pmatrix}$$

Replacement set I indicates that the books were removed from the shelves and placed back in their original order. Consider the mathematical system with the set of elements, R, S, T, U, V, I, with the operation $*$.

To evaluate $R * S$, write

$$R * S = \overset{R}{\begin{pmatrix} 1 & 2 & 3 \\ 3 & 1 & 2 \end{pmatrix}} * \overset{S}{\begin{pmatrix} 1 & 2 & 3 \\ 2 & 1 & 3 \end{pmatrix}}.$$

As shown in Fig. 10.4, R replaces 1 with 3, and S replaces 3 with 3 (no change), so $R * S$ replaces 1 with 3. R replaces 2 with 1, and S replaces 1 with 2, so $R * S$ replaces 2 with 2 (no change). R replaces 3 with 2, and S replaces 2 with 1, so $R * S$ replaces 3 with 1. $R * S$ replaces 1 with 3, 2 with 2, and 3 with 1.

$$\begin{array}{ccccc} & R & & S & \\ 1 & \longrightarrow & 3 & \longrightarrow & 3 \\ 2 & \longrightarrow & 1 & \longrightarrow & 2 \\ 3 & \longrightarrow & 2 & \longrightarrow & 1 \\ & R & * & S & \end{array}$$

$$R * S = \begin{pmatrix} 1 & 2 & 3 \\ 3 & 2 & 1 \end{pmatrix} = T$$

Figure 10.4

Since this result is the same as replacement set T, we write $R * S = T$.

a) Complete the table for the operation using the procedure outlined.

*	*R*	*S*	*T*	*U*	*V*	*I*
R		*T*				
S						
T						
U						
V						
I						

b) Is this mathematical system a group? Explain.

c) Is this mathematical system a commutative group? Explain.

RESEARCH ACTIVITIES

73. In section 7.3 we introduced matrices. Show that 2 × 2 matrices under the operation of addition form a commutative group.

74. Show that 2 × 2 matrices under the operation of multiplication do not form a commutative group.

10.3 MODULAR ARITHMETIC

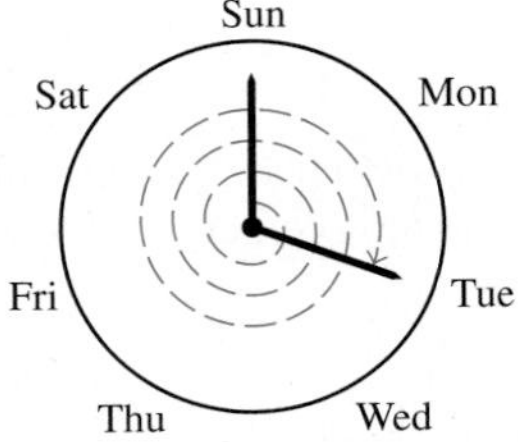

Figure 10.5

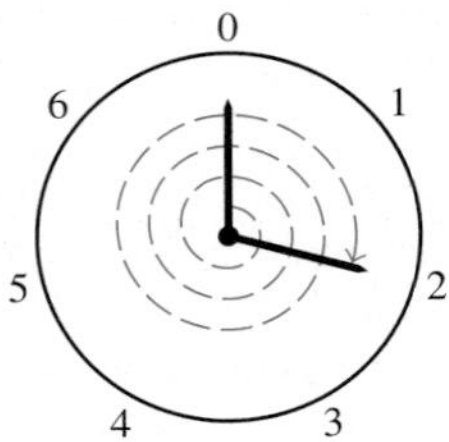

Figure 10.6

The clock arithmetic we discussed in the previous section is similar to modular arithmetic. The set of elements {0, 1, 2, 3, 4, 5, 6, 7, 8, 9, 10, 11} together with the operation of addition is called a modulo 12 or mod 12 system. There is one difference in notation between clock 12 arithmetic and modulo 12 arithmetic. In the modulo 12 system the symbol 12 is replaced with the symbol 0.

A **modulo *m* system** consists of m elements, 0 through $m - 1$, and a binary operation. In this section we will discuss modular arithmetic systems and their properties.

If today is Sunday, what day of the week will it be in 23 days? The answer, Tuesday, is arrived at by dividing 23 by 7 and observing the remainder of 2. Twenty-three days represent 3 weeks plus 2 days. Since we are interested only in the day of the week that the twenty-third day will fall on, the 3-week segment is unimportant to the answer. The remainder of 2 indicates the answer will be 2 days later than Sunday, which is Tuesday.

If we place the days of the week on a clock face as shown in Fig. 10.5, in 23 days the hand would have made three complete revolutions and end on Tuesday. If we replace the days of the week with numbers, a modulo 7 arithmetic system will result. See Fig. 10.6: Sunday = 0, Monday = 1, Tuesday = 2, and so on. If we start at 0 and move the hand 23 places, we will end at 2. Table 10.11 shows a modulo 7 addition table.

Table 10.11 Modular 7 Addition

+	**0**	**1**	**2**	**3**	**4**	**5**	**6**
0	0	1	2	3	4	5	6
1	1	2	3	4	5	6	0
2	2	3	4	5	6	0	1
3	3	4	5	6	0	1	2
4	4	5	6	0	1	2	(3)
5	5	6	0	1	2	3	4
6	6	0	1	2	3	4	5

If we start at 4 and add 6, we end at 3 on the clock in Fig. 10.7. This number is circled in Table 10.11. The other numbers can be obtained in the same way.

A second method of determining the sum of 4 + 6 in modulo 7 arithmetic is to divide the sum, 10, by 7 and observe the remainder.

$$10 \div 7 = 1, \quad \text{remainder } 3$$

The remainder, 3, is the sum of 4 + 6 in a modulo 7 arithmetic system.

The concept of congruence is important in modular arithmetic.

Figure 10.7

> a is **congruent** to b modulo m, written, $a \equiv b \pmod{m}$, if a and b have the same remainder when divided by m.

We can show, for example, that $10 \equiv 3 \pmod 7$ by dividing both 10 and 3 by 7 and observing that we obtain the same remainder in each case.

$$10 \div 7 = 1, \quad \text{remainder } 3 \qquad \text{and} \qquad 3 \div 7 = 0, \quad \text{remainder } 3$$

Since the remainders are the same, 3 in each case, 10 is congruent to 3 modulo 7, and we may write $10 \equiv 3 \pmod{7}$.

Now consider $37 \equiv 5 \pmod{8}$. If we divide both 37 and 5 by 8, each has the same remainder, 5.

In any modulo system we can develop a set of **modulo classes** by placing all numbers with the same remainder in the appropriate modulo class. In a modulo 7 system, every number must have a remainder of either 0, 1, 2, 3, 4, 5, or 6. Thus, a modulo 7 system has seven modulo classes. The seven classes are presented in Table 10.12.

Table 10.12 Modulo 7 Classes

0	**1**	**2**	**3**	**4**	**5**	**6**
0	1	2	3	4	5	6
7	8	9	10	11	12	13
14	15	16	17	18	19	20
21	22	23	24	25	26	27
28	29	30	31	32	33	34
⋮	⋮	⋮	⋮	⋮	⋮	⋮

Every number is congruent to a number from 0 to 6 in mod 7. For example, $24 \equiv 3 \pmod{7}$ because 24 is in the same modulo class as 3.

The solution to a problem in modular arithmetic, if it exists, will always be a number from 0 through $m - 1$, where m is the **modulus** of the system. For example, in a modulo 7 system, because 7 is the modulus, the solution will be a number from 0 through 6.

EXAMPLE 1 *Congruence Modulo 7*

Determine which number, from 0 through 6, the following numbers are congruent to in modulo 7.

a) 62 b) 58 c) 91

SOLUTION We could determine the answer by listing more entries in Table 10.12. Another method of finding the answer is to divide the given number by 7 and observe the remainder. In the solutions we will use a question mark, ?, as a place holder. The ? will represent a number from 0 through 6. When we determine the answer, we will replace the ? with the answer.

a) $62 \equiv ? \pmod{7}$

To determine the value that 62 is congruent to in mod 7, divide 62 by 7 and find the remainder.

$$\begin{array}{r} 8 \\ 7\overline{)62} \\ \underline{56} \\ 6 \end{array} \quad \text{remainder}$$

$$62 \div 7 = 8, \quad \text{remainder } 6$$

Thus, $62 \equiv 6 \pmod{7}$.

b) $58 \equiv ? \pmod{7}$

$$58 \div 7 = 8, \quad \text{remainder } 2$$

Thus, $58 \equiv 2 \pmod{7}$.

c) $91 \equiv ? \pmod{7}$

$$91 \div 7 = 13, \quad \text{remainder } 0$$

Thus, $91 \equiv 0 \pmod{7}$.

DID YOU KNOW

Modulo 100,000

Your car has an odometer that shows your mileage to be a little over 28,000 miles, but this is not the first time you've seen this reading. In fact, this is the third time around. The car's odometer uses a modulo 100,000 system.

EXAMPLE 2 *Congruence Modulo 5*

Evaluate each of the following in mod 5.

a) $4 + 3$ b) $4 - 3$ c) $3 \cdot 4$

SOLUTION In each part the answer will be a number from 0 through 4.

a) $4 + 3 \equiv ? \pmod 5$
$7 \equiv ? \pmod 5$

$$7 \div 5 = 1, \quad \text{remainder } 2$$

Therefore, $4 + 3 \equiv 2 \pmod 5$.

b) $4 - 3 \equiv ? \pmod 5$
$1 \equiv ? \pmod 5$
$1 \equiv 1 \pmod 5$

Remember we want to replace the question mark with a number between 0 and 4, inclusive. Thus, $4 - 3 \equiv 1 \pmod 5$.

c) $3 \cdot 4 \equiv ? \pmod 5$
$12 \equiv ? \pmod 5$

Since $12 \div 5 = 2$, remainder 2, $12 \equiv 2 \pmod 5$. Thus, $3 \cdot 4 \equiv 2 \pmod 5$. ▲

◆ PROFILE IN MATHEMATICS

Amalie Emmy Noether (1882–1935)

Professor Noether was described by Albert Einstein as the "most significant creative mathematical genius thus far produced since the higher education of women began." Because she was a woman, she was never to be granted a regular professorship in her native Germany. In 1933, with the rise of the Nazis to power, she and many other mathematicians and scientists emigrated to the United States. This brought her to Bryn Mawr College, at the age of 51, to her first full-time faculty appointment. She was to die just 2 years later.

Her special interest lay in abstract algebra, which includes such topics as groups, rings, and fields. She developed a number of theories that helped draw together many important mathematical concepts. Her innovations in higher algebra gained her recognition as one of the most creative abstract algebraists of modern times.

Note in Table 10.12 every number in the same modulo class differs by a multiple of the modulo, in this case a multiple of 7. Adding (or subtracting) a multiple of the modulo number to (or from) a given number does not change the modulo class or congruence of the given number. For example, 3, $3 + 1(7)$, $3 + 2(7)$, $3 + 3(7)$, . . . , $3 + n(7)$ are all in the same modulo class, namely, 3. We use this fact in the solution to Example 3.

EXAMPLE 3 *Using Modulo Classes in Subtraction*

Find the replacement for the question mark that makes each of the following true.

a) $3 - 12 \equiv ? \pmod 7$ b) $2 - 4 \equiv ? \pmod 5$ c) $4 - ? \equiv 6 \pmod 8$

SOLUTION

a) In mod 7, adding 7, or a multiple of 7, to a number results in a sum that is in the same modulo class. Thus, if we add 7, 14, 21, . . . to 3, the result will be a number in the same modulo class. We want to replace 3 with an equivalent mod 7 number that is greater than 12. Adding 14 to 3 yields a sum of 17, which is greater than 12.

$$\begin{aligned} 3 - 12 &\equiv ? \pmod 7 \\ (3 + 14) - 12 &\equiv ? \pmod 7 \\ 17 - 12 &\equiv ? \pmod 7 \\ 5 &\equiv 5 \pmod 7 \end{aligned}$$

Therefore, $? = 5$.

b) $2 - 4 \equiv ? \pmod 5$

In mod 5, adding 5, or a multiple of 5, to a number results in a number that is in the same modulo class. Thus, we can add 5 to 2 so that $2 - 4 \equiv ? \pmod 5$ becomes $7 - 4 \equiv ? \pmod 5$.

$$\begin{aligned} 7 - 4 &\equiv ? \pmod 5 \\ 3 &\equiv ? \pmod 5 \\ 3 &\equiv 3 \pmod 5 \end{aligned}$$

Thus, $? = 3$ and we can write $2 - 4 \equiv 3 \pmod 5$. This outcome could be demonstrated with a modulo 5 clock. The minus sign indicates counting counterclockwise.

c) $4 - ? \equiv 6 \pmod 8$
In mod 8, adding 8, or a multiple of 8, to a number results in a sum that is in the same modulo class. Thus, we can add 8 to 4 so that the statement becomes

$$(8 + 4) - ? \equiv 6 \pmod 8$$
$$12 - ? \equiv 6 \pmod 8$$

We can see that $12 - 6 = 6$. Therefore, $? = 6$.

DID YOU KNOW

The Enigma

A cipher, or secret code, actually has group properties: A well-defined operation turns plain text into a cipher and an inverse operation allows it to be deciphered. During World War II, the Germans used an encrypting device based on a modulo 26 system (for the 26 letters of the alphabet) known as the enigma. It has four rotors, each with the 26 letters. The rotors were wired to one another so that if an A were typed into the machine, the first rotor would contact a different letter on the second rotor, which contacted a different letter on the third rotor, and so on. The French secret service obtained the wiring instructions, and Polish and British cryptoanalysts were able to crack the ciphers using group theory. This breakthrough helped to keep the Allies abreast of the deployment of the German navy.

EXAMPLE 4 *Using Modulo Classes in Multiplication*

Find all replacements for the question mark that make the statements true.
a) $4 \cdot ? \equiv 3 \pmod 5$ b) $3 \cdot ? \equiv 0 \pmod 6$ c) $3 \cdot ? \equiv 2 \pmod 6$

SOLUTION

a) One method of determining the solution is to replace the question mark with the numbers 0–4 and then find the equivalent modulo class of the product. We use the numbers 0–4 because we are working in modulo 5.

$$4 \cdot ? \equiv 3 \pmod 5$$
$$4 \cdot 0 \equiv 0 \pmod 5$$
$$4 \cdot 1 \equiv 4 \pmod 5$$
$$4 \cdot 2 \equiv 3 \pmod 5$$
$$4 \cdot 3 \equiv 2 \pmod 5$$
$$4 \cdot 4 \equiv 1 \pmod 5$$

Therefore, $? = 2$ since $4 \cdot 2 \equiv 3 \pmod 5$.

b) Replace the question mark with the numbers 0–5 and follow the procedure used in part (a).

$$3 \cdot ? \equiv 0 \pmod 6$$
$$3 \cdot 0 \equiv 0 \pmod 6$$
$$3 \cdot 1 \equiv 3 \pmod 6$$
$$3 \cdot 2 \equiv 0 \pmod 6$$
$$3 \cdot 3 \equiv 3 \pmod 6$$
$$3 \cdot 4 \equiv 0 \pmod 6$$
$$3 \cdot 5 \equiv 3 \pmod 6$$

Therefore, replacing the question mark with 0, 2, or 4 results in true statements. The answers are 0, 2, and 4.

c) $3 \cdot ? \equiv 2 \pmod 6$
Examining the products in part (b) shows there are no values that satisfy the statement. The answer is "no solution."

Modular arithmetic systems under the operation of addition are commutative groups, as illustrated in Example 5.

EXAMPLE 5 *A Commutative Group*

Construct a mod 5 addition table and show that the mathematical system is a commutative group. Assume that the associative property holds for the given operation.

SOLUTION The set of elements in modulo 5 arithmetic is {0, 1, 2, 3, 4}; the binary operation is +.

+	0	1	2	3	4
0	0	1	2	3	4
1	1	2	3	4	0
2	2	3	4	0	1
3	3	4	0	1	2
4	4	0	1	2	3

For this system to be a commutative group, it must satisfy the five properties of a commutative group.

1. *Closure:* Every entry in the table is a member of the set {0, 1, 2, 3, 4}, so the system is closed under addition.
2. *Identity element:* An easy way to determine whether there is an identity element is to look for a row in the table that is identical to the elements at the top of the table. Note that the row next to 0 is identical to the top of the table. This indicates that 0 *might be* the identity element. Now look at the column under the 0 at the top of the table. If this column is identical to the left-hand column, 0 is the identity element. Since the column under 0 is the same as the left-hand column, 0 is the additive identity element in modulo 5 arithmetic.

Element	+	Identity	=	Element
0	+	0	=	0
1	+	0	=	1
2	+	0	=	2
3	+	0	=	3
4	+	0	=	4

3. *Inverse elements:* Does every element have an inverse? Recall an element plus its inverse must equal the identity element. In this example the identity element is 0. Therefore, for each of the given elements 0, 1, 2, 3, and 4, we must find the element that when added to it results in a sum of zero. These elements will be the inverses.

Element	+	Inverse	=	Identity	
0	+	?	=	0	Since $0 + 0 = 0$, 0 is its own inverse.
1	+	?	=	0	Since $1 + 4 = 0$, 4 is the inverse of 1.
2	+	?	=	0	Since $2 + 3 = 0$, 3 is the inverse of 2.
3	+	?	=	0	Since $3 + 2 = 0$, 2 is the inverse of 3.
4	+	?	=	0	Since $4 + 1 = 0$, 1 is the inverse of 4.

Note that each element has an inverse.

4. *Associative property:* It is given that the associative property holds. One example that illustrates the associative property is

$$(2 + 3) + 4 = 2 + (3 + 4)$$
$$0 + 4 = 2 + 2$$
$$4 = 4 \quad \text{True}$$

5. *Commutative property:* Is $a + b = b + a$ for *all* elements a and b of the given set? The table shows that the system is commutative because the elements are symmetric about the main diagonal. We will give one example to illustrate the commutative property.

$$4 + 2 = 2 + 4$$
$$1 = 1 \quad \text{True}$$

All five properties are satisfied. Thus, modulo 5 arithmetic under the operation of addition forms a commutative group. ▲

SECTION 10.3 EXERCISES

CONCEPT/WRITING EXERCISES

1. What does a *modulo m* system consist of?
2. Explain the meaning of the statement, "*a* is congruent to *b* modulo *m*."
3. In a modulo 5 system, how many modulo classes will there be? Present a table similar to table 10.12 showing elements from each class.
4. In general, for a modulo *m* system, how are modulo classes developed?
5. In a modulo 12 system, how many modulo classes will there be? Explain.
6. In a modulo *n* system, how many modulo classes will there be? Explain.

PRACTICE THE SKILLS

In Exercises 7–12, assume that today is Thursday (day 4). Determine the day of the week it will be at the end of each period. (Assume no leap years.)

7. 20 days **8.** 161 days **9.** 366 days
10. 5 years **11.** 3 years, 34 days **12.** 463 days

In Exercises 13–18, consider 12 months to be a modulo 12 system. Determine the month it will be in the specified number of months. Use the current month as your reference point.

13. 7 months **14.** 36 months
15. 3 years, 5 months **16.** 4 years, 8 months
17. 8 years **18.** 83 months

In Exercises 19–30, determine what number the sum, difference, or product is congruent to in mod 5.

19. $6 + 6$ **20.** $5 + 10$
21. $4 + 7 + 12$ **22.** $9 - 3$
23. $3 - 9$ **24.** $7 \cdot 4$
25. $7 \cdot 9$ **26.** $10 - 15$
27. $4 - 8$ **28.** $3 - 7$
29. $(15 \cdot 4) - 8$ **30.** $(4 - 9) \cdot 7$

In Exercises 31–42, find the modulo class to which each number belongs for the indicated modulo system.

31. 13, mod 4 **32.** 23, mod 7 **33.** 84, mod 12
34. 43, mod 6 **35.** 38, mod 9 **36.** 75, mod 8
37. 34, mod 7 **38.** 53, mod 4 **39.** −6, mod 7
40. −7, mod 4 **41.** −13, mod 11 **42.** −11, mod 13

In Exercises 43–56, find all replacements (less than the modulus) for the question mark that make the statement true.

43. $4 + 4 \equiv ? \pmod 6$ **44.** $? + 9 \equiv 7 \pmod 8$
45. $3 + ? \equiv 4 \pmod 5$ **46.** $4 + ? \equiv 3 \pmod 6$
47. $2 - ? \equiv 5 \pmod 6$ **48.** $4 \cdot 5 \equiv ? \pmod 7$
49. $5 \cdot ? \equiv 7 \pmod 9$ **50.** $3 \cdot ? \equiv 5 \pmod 6$
51. $2 \cdot ? \equiv 3 \pmod 4$ **52.** $3 \cdot ? \equiv 3 \pmod{12}$
53. $3 \cdot ? \equiv 2 \pmod 8$ **54.** $4 - 6 \equiv ? \pmod 8$
55. $5 - 7 \equiv ? \pmod{12}$ **56.** $6 - ? \equiv 8 \pmod 9$

PROBLEM SOLVING

57. *Presidential Elections* The upcoming presidential election years are 2000, 2004, 2008, ...

a) List the next five presidential election years after 2008.
b) What will be the first election year after the year 3000?
c) List the election years between the years 2550 and 2575.

58. *Flight Schedules* A pilot is scheduled to fly for five consecutive days and rest for three consecutive days. If today is the second day of her rest shift, determine whether she will be flying

a) 60 days from today.
b) 90 days from today.
c) 240 days from today
d) Was she flying 6 days ago?
e) Was she flying 20 days ago?

59. *The Weekend Off* A manager of AMC theaters has both Saturday and Sunday off every 7 weeks. This is week 2 of the 7 weeks.

a) Determine the number of weeks before she will have both Saturday and Sunday off.
b) Will she have both Saturday and Sunday off 25 weeks from this week?
c) What is the first week, after 50 weeks from this week, that she will have both Saturday and Sunday off?

60. *Nursing Shifts* A nurse's work pattern at Community Hospital consists of working the 7 A.M.–3 P.M. shift for 3 weeks and then the 3 P.M.–11 P.M. shift for 2 weeks.

a) If this is the third week of the pattern, what shift will the nurse be working 6 weeks from now?
b) If this is the fourth week of the pattern, what shift will the nurse be working 7 weeks from now?
c) If this is the first week of the pattern, what shift will the nurse be working 11 weeks from now?

61. *Restaurant Rotation* A waiter at a restaurant works both day and evening shifts. He works days for 5 consecutive days, then evenings for 3 consecutive days, then days for 4 consecutive days, then evenings for 2 consecutive days. Then the rotation starts again. If this is day 2 of the 5-day consecutive day shift, determine whether he will be working the day or evening shift

a) 20 days from today?
b) 52 days from today?
c) 365 days from today?

62. *A Truck Driver's Schedule* A truck driver's routine is as follows: Drive 3 days from New York to Chicago; rest 1 day in Chicago; drive 3 days from Chicago to Los Angeles; rest 2 days in Los Angeles; drive 5 days to return to New York; rest 3 days in New York. Then the cycle begins again.

If the truck driver is starting his trip to Chicago today, what will he be doing

a) 30 days from today?
b) 70 days from today?
c) 2 years from today?

63. **a)** Construct a modulo 4 addition table.
b) Is the system closed? Explain.
c) Is there an identity element for the system? If so, what is it?
d) Does every element in the system have an inverse? If so, list the elements and their inverses.
e) The associative property holds for the system. Give an example.
f) Does the commutative property hold for the system? Give an example.
g) Is the system a commutative group?
h) Will every modulo system under the operation of addition be a commutative group? Explain.

64. Construct a modulo 8 addition table. Repeat parts (b)–(h) in Exercise 63.

65. **a)** Construct a modulo 4 multiplication table.
b) Is the system closed under the operation of multiplication?
c) Is there an identity element in the system? If so, what is it?
d) Does every element in the system have an inverse? Make a list showing the elements that have a multiplicative inverse and list the inverses.
e) The associative property holds for the system. Give an example.
f) Does the commutative property hold for the system? Give an example.
g) Is this mathematical system a commutative group? Explain.

66. Construct a modulo 7 multiplication table. Repeat parts (b)–(g) in Exercise 65.

CHALLENGE PROBLEMS/GROUP ACTIVITIES

We have not discussed division in modular arithmetic. With what number or numbers, if any, can you replace the question marks to make the statement true?

67. $5 \div 7 \equiv ?\ (\text{mod } 9)$

68. $? \div 5 \equiv 5\ (\text{mod } 9)$

69. $? \div ? \equiv 1\ (\text{mod } 4)$

70. $1 \div 2 \equiv ?\ (\text{mod } 5)$

In Exercises 71–73, solve for x where k is any counting number.

71. $5k \equiv x \pmod{5}$

72. $5k + 4 \equiv x \pmod{5}$

73. $4k - 2 \equiv x \pmod{4}$

74. Find the smallest positive number divisible by 5 to which 2 is congruent in modulo 6.

75. *Deciphering a Code* One important use of modular arithmetic is in coding. One type of coding circle is given in Fig. 10.8. To use it, the person you are sending the message to must know the code key to decipher the code. The code key to this message is *j*. Can you decipher this code? (*Hint:* Subtract the code key from the code numbers.)

23 11 3 18 10 19 2 10 16 4 24

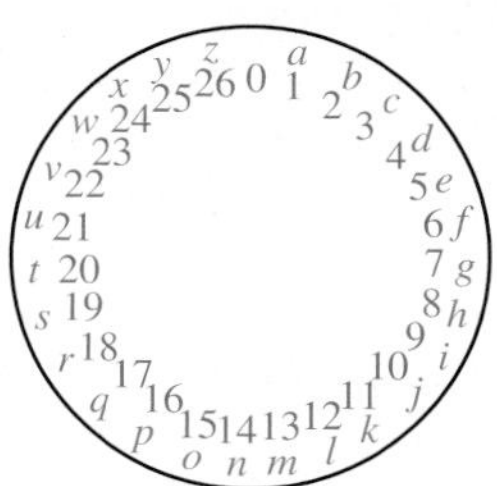

Figure 10.8

RESEARCH ACTIVITY

76. The concept and notation for modular systems were introduced by Carl Friedrich Gauss in 1801. Write a paper on Gauss's contribution to modular systems.

CHAPTER 10 SUMMARY

IMPORTANT FACTS

If the elements in a table are symmetric about the main diagonal, then the system is commutative.

	Addition	**Multiplication**
Commutative property	$a + b = b + a$	$a \cdot b = b \cdot a$
Associative property	$(a + b) + c = a + (b + c)$	$(a \cdot b) \cdot c = a \cdot (b \cdot c)$

a is **congruent** to *b*, written, $a \equiv b \pmod{m}$, if *a* and *b* have the same remainder when divided by *m*.

Properties of a group and a commutative group

A mathematical system is a group if the first four conditions hold and a commutative group if all five conditions hold.

Commutative Group { Group {
1. The set of elements is *closed* under the given operation.
2. An *identity element* exists for the set.
3. Every element in the set has an *inverse*.
4. The set of elements is *associative* under the given operation. }
5. The set of elements is *commutative* under the given operation. }

CHAPTER 10 REVIEW EXERCISES

10.1–10.2

1. List the parts of a mathematical system.

2. What is a binary operation?

3. Are the integers closed under the operation of addition? Explain.

4. Are the natural numbers closed under the operation of subtraction? Explain.

Determine the sum or difference in clock 12 arithmetic.

5. $5 + 10$ **6.** $5 + 12$ **7.** $6 - 10$

8. $6 + 7 + 9$ **9.** $7 - 4 + 6$ **10.** $2 - 8 - 7$

11. List the properties of a group, and explain what each property means.

12. What is an abelian group?

In Exercises 13 and 14, explain your answer.

13. Determine whether the set of integers under the operation of addition forms a group.

14. Determine whether the set of integers under the operation of multiplication forms a group.

15. Determine whether the set of rational numbers under the operation of addition forms a group.

16. Determine whether the set of rational numbers under the operation of multiplication forms a group.

In Exercises 17–19, for the mathematical system, determine which of the five properties of a commutative group do not hold.

17.

⧗	**2**	**c**	**△**
2	2	△	c
c	△	c	2
△	c	2	△

18.

◡	**!**	**?**	**△**	***p***
!	?	△	!	*p*
?	△	*p*	?	!
△	!	?	△	*p*
p	*p*	!	*p*	△

19.

?	**4**	**#**	**L**	**P**
4	P	4	#	L
#	4	#	L	P
L	#	L	P	4
P	L	P	4	L

20. Consider the following mathematical system in which the operation is associative.

⅃	≻	⊙	?	△
≻	≻	⊙	?	△
⊙	⊙	?	△	≻
?	?	△	≻	⊙
△	△	≻	⊙	?

a) What are the elements of the set in this mathematical system?
b) What is the binary operation?
c) Is the system closed? Explain.
d) Is there an identity element for the system under the given operation?
e) Does every element in the system have an inverse? If so, give each element and its corresponding inverse.
f) Give an example to illustrate the associative property.
g) Is the system commutative? Give an example.
h) Is this mathematical system a commutative group? Explain.

10.3

In Exercises 21–30, find the modulo class to which the number belongs for the indicated modulo system.

21. 15, mod 4 **22.** 31, mod 8
23. 27, mod 7 **24.** 59, mod 8
25. 82, mod 13 **26.** 54, mod 4
27. 37, mod 6 **28.** 54, mod 14
29. 97, mod 11 **30.** 42, mod 11

In Exercises 31–40, find all replacements (less than the modulus) for the question mark that make the statement true.

31. $8 + 8 \equiv ? \pmod 9$ **32.** $? - 3 \equiv 0 \pmod 5$
33. $4 \cdot ? \equiv 3 \pmod 6$ **34.** $3 - ? \equiv 5 \pmod 7$
35. $? \cdot 4 \equiv 0 \pmod 8$ **36.** $9 \cdot 7 \equiv ? \pmod{12}$
37. $3 - 5 \equiv ? \pmod 7$ **38.** $? \cdot 7 \equiv 3 \pmod 6$
39. $5 \cdot ? \equiv 3 \pmod 8$ **40.** $7 \cdot ? \equiv 2 \pmod 9$

41. Construct a modulo 6 addition table. Then determine whether the modulo 6 system forms a commutative group under the operation of addition.

42. Construct a modulo 4 multiplication table. Then determine whether the modulo 4 system forms a commutative group under the operation of multiplication.

43. *Work Pattern* Toni's work pattern at the fast food restaurant is as follows: She works 3 evenings, then has 2 evenings off, then she works 2 evenings, and then she has 3 evenings off; then the pattern repeats. If today is the first day of the work pattern,
a) Will Toni be working 18 days from today?
b) Will Toni have the evening off for a party that is being held in 38 days?

CHAPTER 10 TEST

1. What is a mathematical system?
2. List the requirements needed for a mathematical system to be a commutative group.
3. Is the set of the whole numbers a commutative group under the operation of addition? Explain your answer completely.
4. Develop a clock 5 arithmetic addition table.
5. Is clock 5 arithmetic under the operation of addition a commutative group? Assume that the associative property holds. Explain your answer completely.

Determine the following in clock 5 arithmetic.

6. $4 + 3 + 2$

7. $6 - 18$

8. Consider the mathematical system

□	*W*	*S*	*T*	*R*
W	*T*	*R*	*W*	*S*
S	*R*	*W*	*S*	*T*
T	*W*	*S*	*T*	*R*
R	*S*	*T*	*R*	*W*

a) What is the binary operation?
b) Is this system closed? Explain.
c) Is there an identity element for this system under the given operation? Explain.
d) What is the inverse of the element *R*?
e) What is (*T* □ *R*) □ *W*?

In Exercises 9 and 10, determine whether the mathematical system is a commutative group. Explain your answer completely.

9.

*	*a*	*b*	*c*
a	*a*	*b*	*c*
b	*b*	*b*	*a*
c	*c*	*a*	*d*

10.

?	1	2	3
1	3	1	2
2	1	2	3
3	2	3	1

11. Determine whether the mathematical system is a commutative group. Assume that the associative property holds. Explain your answer.

○	@	$	&	%
@	%	@	$	&
$	@	$	&	%
&	$	&	%	@
%	&	%	@	$

In Exercises 12 and 13, determine the modulo class to which the number belongs for the indicated modulo system.

12. 73, mod 9

13. 58, mod 11

In Exercises 14–19, find all replacements for the question mark, less than the modulus, that make the statement true.

14. $6 + 9 \equiv ? \pmod 7$

15. $? - 9 \equiv 4 \pmod 5$

16. $3 - ? \equiv 7 \pmod 9$

17. $4 \cdot 2 \equiv ? \pmod 6$

18. $3 \cdot ? \equiv 2 \pmod 6$

19. $96 \equiv ? \pmod 7$

20. a) Construct a modulo 5 multiplication table.
b) Is this mathematical system a commutative group? Explain your answer completely.

GROUP PROJECTS

1. *Product of Zero* In arithmetic and algebra the statement, "If $a \cdot b = 0$, then $a = 0$ or $b = 0$" is true. That is, for the product of two numbers to be 0, at least one of the factors must be 0. Can the product of two nonzero numbers equal 0 in a specific modulo systems? If so, in what type of modulo systems can this result occur?
a) Construct multiplication tables for modulo systems 3–9.
b) Which, if any, of the multiplication tables in part (a) have products equal to 0 when neither factor is 0?
c) Which, if any, of the multiplication tables in part (a) have products equal to 0 only when at least one factor is 0?
d) Using the results in parts (b) and (c), can you write a conjecture as to which modulo systems have a product of 0 when neither factor is 0.

2. *Conjecture about Multiplication Inverses* Are there certain modulo systems where all numbers have multiplicative inverses?
a) If you have not worked Group Projects Exercise 1, construct multiplication tables for modulo systems 3–9.
b) Which of the multiplication tables in part (a) contain multiplicative inverses for all nonzero numbers?
c) Which, if any, of the multiplication tables in part (a) do not contain multiplicative inverses for all nonzero numbers?
d) Using the results in parts (b) and (c), can you write a conjecture as to which modulo systems contain multiplicative inverses for all nonzero numbers?

3. *Rotating a Square* The square *ABCD* on page 516 has a pin through it at point *P* so that the square can be rotated in a clockwise direction. The mathematical system consists of the set $\{A, B, C, D\}$ and the operation ♣ . The elements of the set represent different rotations of the square clockwise, as follows.

A = rotate about the point P clockwise 90°.

B = rotate about the point P clockwise 180°.

C = rotate about the point P clockwise 270°.

D = rotate about the point P clockwise 360°.

The operation ♣ means *followed by*. For example, A ♣ B means a clockwise rotation of 90° followed by a clockwise rotation of 180°, or a clockwise rotation of 270°. Since C represents a clockwise rotation of 270°, we write A ♣ $B = C$. This and two other results are shown in the following table.

♣	**A**	**B**	**C**	**D**
A		C		
B			A	
C	D			
D				

Complete the table and determine if the mathematical system is a commutative group. Explain.

CHAPTER 11

Consumer Mathematics

Managing your money takes much thought and planning. Your daily needs, such as food and transportation, as well as your monthly and yearly needs, such as car payments, rent or mortgage, and utility bills, must all be paid. At the same time, you should consider your long-term goals such as saving enough for large purchases—a new car or a new house—as well as for unexpected emergencies and for retirement.

The decisions you make on a daily basis affect your ability to reach your long-term goals. To make the best use of your money, you need to be able to make intelligent judgments. This chapter will help provide information on several areas of consumer mathematics, including loans, interest rates, and mortgages. A solid understanding of these and other consumer mathematics topics can help you achieve your financial goals. ■

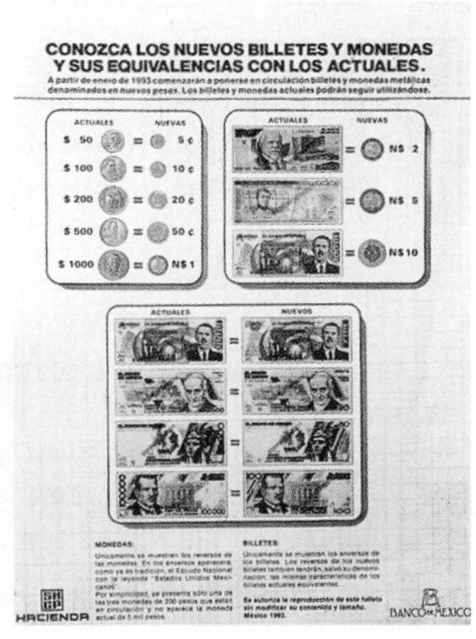

To some extent, money is a central aspect of everyone's life. Being knowledgeable about consumer mathematics can help you reach your financial goals.

11.1 PERCENT

The study of mathematics is crucial to understanding how to make better financial decisions. A basic topic necessary for understanding the material in this chapter is percent. This section will give you a better understanding of the meaning of percent and its use in real-life situations.

The word *percent* comes from the Latin *per centum,* meaning "per hundred." A **percent** is simply a ratio of some number to 100. Thus, $\frac{15}{100} = 15\%$, and $\frac{x}{100} = x\%$.

Percents are useful in making comparisons. Consider Ross, who took two psychology tests. On the first test Ross answered 18 of the 20 questions correctly, and on the second test he answered 23 of 25 questions correctly. On which test did he have the higher score? One way to compare the results is to write a ratio of the correct answers to the number of questions on the test and then convert the ratios to percents. We find the grades in percent for each test by (a) writing a ratio of the number of correct answers to the total number of questions, (b) rewriting these ratios with a denominator of 100, and (c) expressing the ratios as percents.

Test 1 (a) (b) (c)

$$\frac{\text{Number of correct answers}}{\text{Number of questions on the test}} = \frac{18}{20} = \frac{18 \times 5}{20 \times 5} = \frac{90}{100} = 90\%$$

Test 2 (a) (b) (c)

$$\frac{\text{Number of correct answers}}{\text{Number of questions on the test}} = \frac{23}{25} = \frac{23 \times 4}{25 \times 4} = \frac{92}{100} = 92\%$$

By changing the results of both tests to percents, we have a common standard for comparison. The results show that Ross scored 90% on the first test and 92% on the second test. He had a higher score on the second test.

Another procedure to change a fraction to a percent follows.

Procedure to Change a Fraction to a Percent

1. Divide the numerator by the denominator.
2. Multiply the quotient by 100 (which has the effect of moving the decimal point two places to the right).
3. Add a percent sign.

Note that steps 2 and 3 together are the equivalent of multiplying by 100%. Since 100% = 100/100 = 1, we are not changing the *value* of the number, we are simply changing the number to a percent.

EXAMPLE 1 ***Converting a Fraction to a Percent***

Change $\frac{17}{20}$ to a percent.

SOLUTION Follow the steps in the procedure box.

1. $17 \div 20 = 0.85$ **2.** $0.85 \times 100 = 85$ **3.** 85%

Thus, $\frac{17}{20} = 85\%$

DID YOU KNOW

Is It Pure?

99 and 44/100% Pure: It is a slogan that has been around since 1881, ever since Procter & Gamble decided to market its laundry soap, Ivory Soap, as a cosmetic soap. "Pure what?" you may ask. Soap, as opposed to detergent, is a mixture of fatty acids and alkali, so the makeup of Ivory Soap is 99.44% soap and only 0.56% impurities. A semanticist, who studies language and its meaning, might say that's pure baloney. That which is described as pure, by definition, cannot be a mixture at all, and it is either 100% pure or not pure at all.

Procedure to Change a Decimal Number to a Percent

1. Multiply the decimal number by 100.
2. Add a percent sign.

The procedure for changing a decimal number to a percent is equivalent to moving the decimal point two places to the right and adding a percent sign.

EXAMPLE 2 ***Converting a Decimal Number to a Percent***

Change 0.235 to percent.

SOLUTION $0.235 = (0.235 \times 100)\% = 23.5\%$ ▲

Procedure to Change a Percent to a Decimal Number

1. Divide the number by 100.
2. Remove the percent sign.

The procedure for changing a percent to a decimal number is equivalent to moving the decimal point two places to the left and removing the percent sign. Another way to remember this is: percent means per hundred (or to *divide by 100*).

EXAMPLE 3 ***Converting a Percent to a Decimal Number***

a) Change 35% to a decimal number.

b) Change $\frac{1}{2}\%$ to a decimal number.

SOLUTION

a) $35\% = \frac{35}{100} = 0.35$. Thus, $35\% = 0.35$.

b) $\frac{1}{2}\% = 0.5\% = \frac{0.5}{100} = 0.005$. Thus, $\frac{1}{2}\% = 0.005$. ▲

EXAMPLE 4 ***How Old Is Old?***

A Yahoo! Question of the Week asked, "At what age do you consider someone old?" Out of the 3496 people that responded, 1017 said age 80 is old. What percent of respondents feel that age 80 is old?

SOLUTION To find the percent that feel age 80 is old, divide the number of those who responded that age 80 is old by the total number of respondents. Then move the decimal point two places to the right and add a percent sign:

$$\text{Percent who felt age 80 is old} = \frac{1017}{3496} \approx 0.2909 = 29.09\%$$

Thus about 29.1% (to the nearest tenth of a percent) felt that age 80 is old. ▲

All answers in this section will be given to the nearest tenth of a percent. When rounding answers to the nearest tenth of a percent, we carry the division to four places after the decimal point (ten-thousandths) and then round to the nearest thousandths position.

The White House.

EXAMPLE 5 *1996 Presidential Election Results*

In the 1996 U.S. Presidential Election, 96,456,345 people voted. Of these, 47,402,357 people voted for Bill Clinton; 39,198,755 voted for Bob Dole; 8,085,402 voted for Ross Perot, and 1,769,831 voted for other candidates. Find the percent of voters who voted for Clinton, Dole, Perot, and other candidates.

SOLUTION To find each percent, divide the number of votes for each candidate by the total number of votes.

$$\text{Percent voting for Clinton} = \frac{47{,}402{,}357}{96{,}456{,}345} \approx 0.4914 \approx 49.1\%$$

$$\text{Percent voting for Dole} = \frac{39{,}198{,}755}{96{,}456{,}345} \approx 0.4063 \approx 40.6\%$$

$$\text{Percent voting for Perot} = \frac{8{,}085{,}402}{96{,}456{,}345} \approx 0.0838 \approx 8.4\%$$

$$\text{Percent voting for other candidates} = \frac{1{,}769{,}831}{96{,}456{,}345} \approx 0.0183 \approx 1.8\%$$

The sum of the percents, 49.1% + 40.6% + 8.4% + 1.8%, equals 99.9%. Since we rounded each percent, the sum is slightly different from 100%.

The percent increase or decrease, or percent change, over a period of time is found by the following formula:

$$\textbf{Percent change} = \frac{\begin{pmatrix}\text{Amount in}\\ \text{latest period}\end{pmatrix} - \begin{pmatrix}\text{Amount in}\\ \text{previous period}\end{pmatrix}}{\text{Amount in previous period}} \times 100$$

If the latest amount is greater than the previous amount, the answer will be positive and will indicate a percent increase. If the latest amount is smaller than the previous amount, the answer will be negative and will indicate a percent decrease.

Oakland Coliseum.

EXAMPLE 6 *Most Improved Baseball Record*

In 1999, the Major League baseball team with the most improved record was the Oakland Athletics. In 1998, the Athletics won 74 games. In 1999, the Athletics won 87 games. Find the percent increase in number of games won from 1998 to 1999. (*Source: USA Today.*)

SOLUTION The previous period is 1998, and the latest period is 1999.

$$\begin{aligned}\text{Percent change} &= \frac{87 - 74}{74} \times 100\\ &= \frac{13}{74} \times 100\\ &\approx 0.1756 \times 100\\ &\approx 17.6\end{aligned}$$

Therefore, there was about a 17.6% increase.

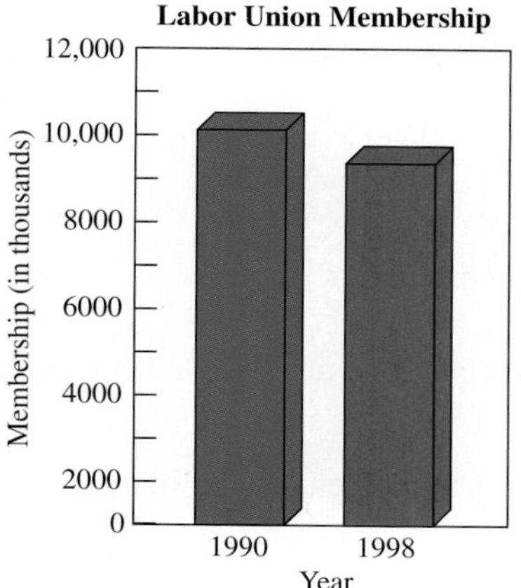

Source: U.S. Bureau of Labor Statistics

EXAMPLE 7 *Labor Union Membership*

In 1990 there were approximately 10,247,000 labor union members in the United States. By 1998 this number had dropped to 9,306,000. Find the percent change in labor union membership from 1990 to 1998.

SOLUTION The previous period is 1990, and the latest period is 1998.

$$\text{Percent change} = \frac{9{,}306{,}000 - 10{,}247{,}000}{10{,}247{,}000} \times 100$$
$$= \frac{-941{,}000}{10{,}247{,}000} \times 100$$
$$\approx -0.0918 \times 100$$
$$\approx -9.2$$

Thus, union membership decreased by about 9.2% over this period.

A similar formula is used to calculate percent markup or markdown on cost. A positive answer indicates a markup and a negative answer indicates a markdown.

$$\textbf{Percent markup on cost} = \frac{\text{Selling price} - \text{Dealer's cost}}{\text{Dealer's cost}} \times 100$$

EXAMPLE 8 *Determining Percent Markup*

Holdren Hardware stores pay \$48.76 for glass fireplace screens. They regularly sell them for \$79.88. At a sale they sell them for \$69.99. Find the following.

a) The percent markup on the regular price
b) The percent markup on the sale price
c) The percent decrease of the sale price from the regular price

SOLUTION

a) We determine the percent markup on the regular price as follows.

$$\text{Percent markup} = \frac{\$79.88 - \$48.76}{\$48.76} \times 100$$
$$\approx 0.6382 \times 100$$
$$\approx 63.8$$

Thus, the percent markup was about 63.8%.

b) We determine the percent markup on the sale price as follows.

$$\text{Percent markup} = \frac{\$69.99 - \$48.76}{\$48.76} \times 100$$
$$\approx 0.4353 \times 100$$
$$\approx 43.5$$

Thus, the percent markup was about 43.5%.

DID YOU KNOW

To Insure Promptness

The practice of tipping began in an English coffeehouse in the eighteenth century. Legend has it that a hurried businessman dropped a few pence on the table as a friendly incentive to the waiter. In fact, the coffeehouses eventually installed small collection boxes marked **T**o **I**nsure **P**romptness. The tip, as it became known, was paid before, not after, the meal. Today, there are many books and Internet web sites devoted to the guidelines of tipping. For example, according to the Original Tipping Page (*www.tipping.com*), it is proper to tip waiters and waitresses 15–20% of the total restaurant bill. Hair stylists should be tipped 15% of the total bill and pizza delivery persons should receive \$1–\$3, depending on the distance driven. Skycaps at the airport should be tipped at least \$1 per bag and newspaper delivery persons should receive \$15–\$25 at holiday time.

c) Based on the regular price, we determine the percent decrease of the sale price.

$$\text{Percent decrease} = \frac{\$69.99 - \$79.88}{\$79.88} \times 100$$
$$\approx -0.1238 \times 100$$
$$\approx -12.4$$

The sale price is about 12.4% lower than the regular price. ▲

In daily life we may need to know how to solve any one of the following three types of problems involving percent:

1. What is a 15% tip on a restaurant bill of \$24.66? The problem can be stated as

   15% of \$24.66 is what number?

2. If Nancy Johnson made a sale of \$500 and received a commission of \$25, what percent of the sale is the commission? The problem can be stated as

   What percent of \$500 is \$25?

3. If the price of a jacket was reduced by 25% or \$12.50, what was the original price of the jacket? The problem can be stated as

   25% of what number is 12.50?

To answer these questions we will write each problem as an equation. The word *is* means "is equal to," or =. In each problem we will represent the unknown quantity with the letter x. Therefore, the preceding problems can be represented as

1. 15% of $\$24.66 = x$ **2.** $x\%$ of $\$500 = \$25$ **3.** 25% of $x = \$12.50$

The word *of* in such problems indicates multiplication. To solve each problem change the percent to a decimal number, and express the problem as an equation; then solve the equation for the variable x. The solutions follow.

1. 15% of $\$24.66 = x$

$$0.15(24.66) = x \quad \text{(15\% is written as 0.15 in decimal form.)}$$
$$3.699 = x$$

Since 15% of \$24.66 is \$3.699, the tip would be \$3.70.

2. $x\%$ of $\$500 = \$25$

$$(0.01x)500 = 25 \quad \text{(}x\text{\% is written as }0.01x\text{ in decimal form.)}$$
$$5x = 25$$
$$\frac{5x}{5} = \frac{25}{5}$$
$$x = 5$$

Since 5% of \$500 is \$25, the commission is 5% of the sale.

3. $25\% \text{ of } x = \$12.50$

$$0.25(x) = 12.50$$

$$\frac{0.25x}{0.25} = \frac{12.50}{0.25}$$

$$x = 50$$

Since 25% of \$50 is \$12.50, the original price of the jacket was \$50.00.

EXAMPLE 9 ***Down Payment on a House***

Melissa Bell wishes to buy a house for \$87,000. To obtain a mortgage, she needs to pay 20% of the selling price as a down payment. Determine the amount of Melissa's down payment.

SOLUTION We want to find the amount of the down payment. Let $x =$ the down payment. Then

$$\begin{aligned} x &= 20\% \text{ of the selling price} \\ &= 20\% \text{ of } \$87{,}000 \\ &= 0.20(87{,}000) \\ &= 17{,}400. \end{aligned}$$

Melissa will have a down payment of \$17,400. ▲

EXAMPLE 10 ***Young Chess Players***

In 1999, about 40,000 out of the 86,000 U.S. Chess Federation (USCF) members were under age 20. What percent of USCF members were under age 20? (*Source:* October/November 1999 issue of *Civilization* magazine.)

SOLUTION We need to determine what percent of 86,000 is 40,000. Let $x =$ the percent of USCF members under age 20. Then

$$\begin{aligned} x\% \text{ of } 86{,}000 &= 40{,}000 \\ 0.01x(86{,}000) &= 40{,}000 \\ 860x &= 40{,}000 \\ x &= \frac{40{,}000}{860} \\ x &\approx 46.5 \end{aligned}$$

Therefore, about 46.5% of USCF members were under age 20. ▲

EXAMPLE 11 ***Population of Iraq***

About 10,340,000, or 47%, of Iraq's population is younger than 15 years old. What is the population of Iraq? (*Source:* 1999 *Wallstreet Journal Almanac.*)

Shrine of Imam El-Hussein at Karbala.

SOLUTION This problem can be stated as, 47% of what number is 10,340,000? Let x = the population of Iraq. Then

$$47\% \text{ of } x = 10{,}340{,}000$$
$$0.47x = 10{,}340{,}000$$
$$x = \frac{10{,}340{,}000}{0.47}$$
$$x = 22{,}000{,}000$$

Therefore, the population of Iraq is about 22,000,000 people.

SECTION 11.1 EXERCISES

CONCEPT/WRITING EXERCISES

1. What is a percent?
2. Explain in your own words how to change a fraction to a percent.
3. Explain in your own words how to change a decimal number to a percent.
4. Explain in your own words how to change a percent to a decimal number.
5. Explain in your own words how to determine percent change.
6. Explain in your own words how to determine percent markup on cost.

PRACTICE THE SKILLS

In Exercises 7–14, change the number to a percent. Express your answer to the nearest tenth of a percent.

7. $\frac{3}{8}$ **8.** $\frac{3}{4}$ **9.** $\frac{5}{8}$

10. $\frac{7}{8}$ **11.** 0.007654 **12.** 0.5688

13. 3.78 **14.** 13.678

In Exercises 15–24, change the percent to a decimal number.

15. 12% **16.** 25.9% **17.** 3.75%

18. 0.0005% **19.** $\frac{1}{4}\%$ **20.** $\frac{3}{8}\%$

21. $\frac{1}{5}\%$ **22.** 135.9% **23.** 1%

24. 0.50%

PROBLEM SOLVING

For Exercises 25–42, round answers to the nearest tenth of a percent.

25. *Cost of Raising a Child* In 1997, the estimated annual expenditures for a child under the age of 2 years old were \$5820. Of this, \$830 was spent on food. What percent of the annual expenditures was for food? (*Source:* U.S. Agriculture Department.)

26. *Toothpaste Sales* In 1997, Proctor & Gamble sold about \$370 million worth of Crest toothpaste. The total sales for all types of toothpaste in 1997 were \$1.5 billion. What percent of all toothpaste sales did Crest account for in 1997? (*Source:* Information Resources, Inc.)

27. *River Pollution* In a study of U.S. rivers, 693,905 miles of river were studied. It was found that 36% of the total miles of rivers studied had impaired water quality. How many miles of rivers had impaired quality? (*Source:* U.S. Environmental Protection Agency.)

28. *Preschool Child Care* In 1999, 18.6% of all preschool age children with working mothers were cared for by their fathers while their mothers worked. If the estimated number of preschool age children with working mothers was 10,398,000, how many children were cared for by their fathers while their mothers worked? (*Source:* U.S. Census Bureau.)

Inventory Shrinkage In Exercises 29 and 30, use the circle graph to answer the questions. In 1998, U.S. companies lost \$43,281.7 million due to "inventory shrinkage." Each sector of the circle graph shows the percent of this total due to each of four sources.

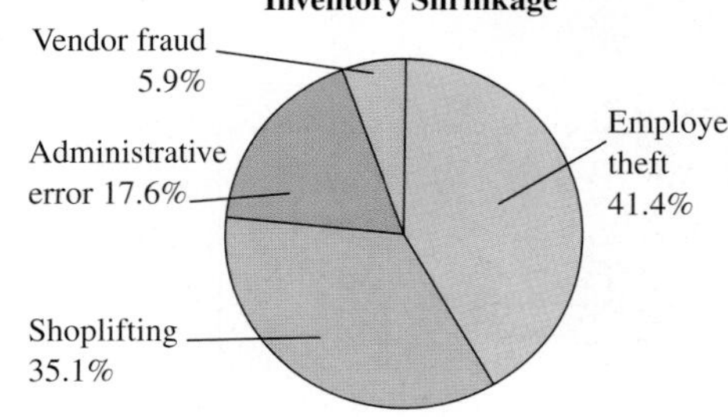

29. Determine the amount lost due to employee theft.

30. Determine the amount lost due to shoplifting.

U.S. Worker Demographics In Exercises 31 and 32, use the circle graph to answer the questions. The projected number of U.S. workers in the year 2006 is 148,847,000. The projected percent of this total by race and ethnicity is shown in the circle graph.

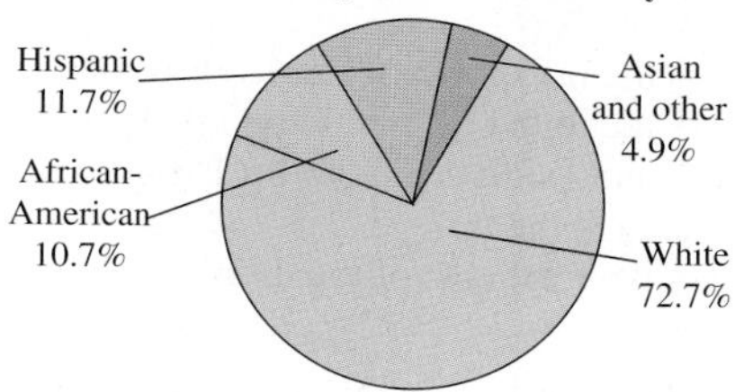

Source: U.S. Bureau of Labor

31. Determine the projected number of African-American workers.

32. Determine the projected number of Hispanic workers.

Top Five Advertisers In Exercises 33 and 34, use the circle graph to answer the questions. The top five advertisers spent $7,534 million in 1997.

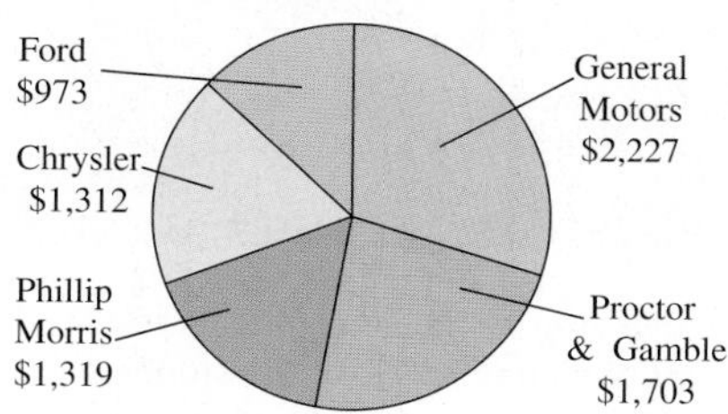

Source: Competitive Media Reporting and Publishers Information Bureau

33. What percent of the $7,534 million did General Motors spend?

34. What percent of the $7,534 million did Proctor & Gamble spend?

Top Five Corn-Producing States In Exercises 35 and 36, use the circle graph to answer the questions. The top five corn-producing states produced $13,815 million worth of corn in 1996.

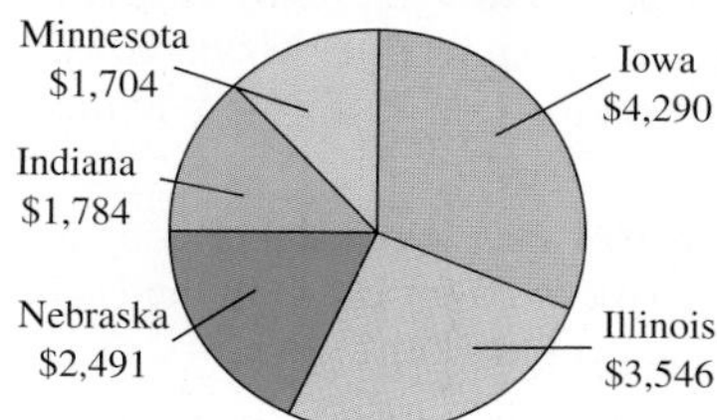

Source: The Wall Street Journal Almanac, 1999

35. What percent of the corn did Nebraska produce?

36. What percent of the corn did Indiana produce?

37. *U.S. Population Increase* The population of the United States rose from approximately 248.7 million in 1990 to approximately 270.3 million in 1998. (*Source:* U.S. Census Bureau.)

a) Find the percent increase in population from 1990 to 1998.

b) What will the population be in the year 2006 if it increases at the same percent as it did from 1990 to 1998?

38. *Decreasing Sales Volume* In 1996 Montgomery Ward had a sales volume of $6.6 billion. In 1997 the sales volume was $5.7 billion. Find the percent decrease from 1996 to 1997. (*Source:* National Retail Federation.)

39. *Median Household Income* The graph shows the median household income (in 1997 dollars) in the United States for the years 1967, 1977, 1987, and 1997.

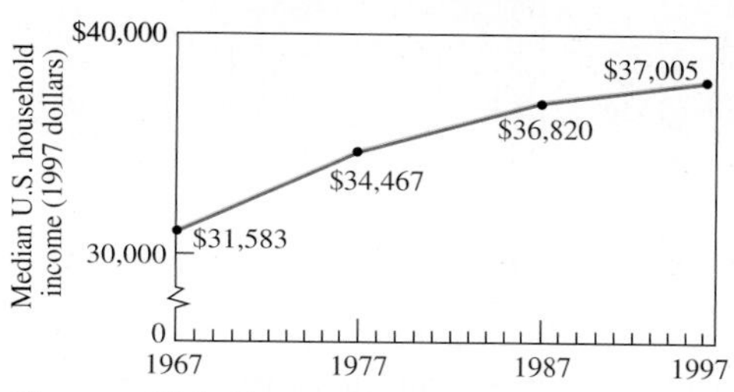

(*Source:* U.S. Census Bureau.)

a) Find the percent increase in median household income from 1967 to 1977.

b) Find the percent increase in median household income from 1977 to 1987.

c) Find the percent increase in median household income from 1987 to 1997.

d) Find the percent increase in median household income from 1967 to 1997.

40. *Hospital Beds* The graph shows the number of hospital beds, in thousands, in the United States in the years 1966, 1976, 1986, and 1996.

a) Find the percent increase in the number of hospital beds from 1966 to 1976.

b) Find the percent decrease in the number of hospital beds from 1976 to 1986.

c) Find the percent decrease in the number of hospital beds from 1986 to 1996.

d) Find the percent decrease in the number of hospital beds from 1966 to 1996.

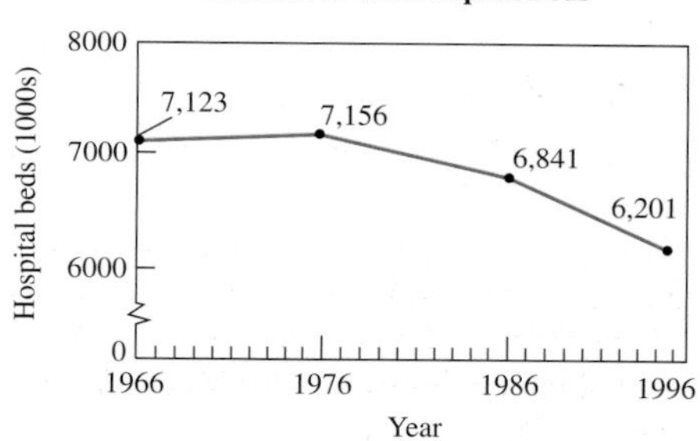

41. *Median Age of First Marriage* The following graph shows the median age of first marriage by gender for the years 1966, 1976, 1986, and 1996.

a) Which 10 year period saw the largest percent increase in median age of first marriage for males? (*Hint:* You will need to calculate the percent increase from 1966 to 1976, from 1976 to 1986, and from 1986 to 1996.)

b) Which 10 year period saw the largest percent increase in median age of first marriage for females?

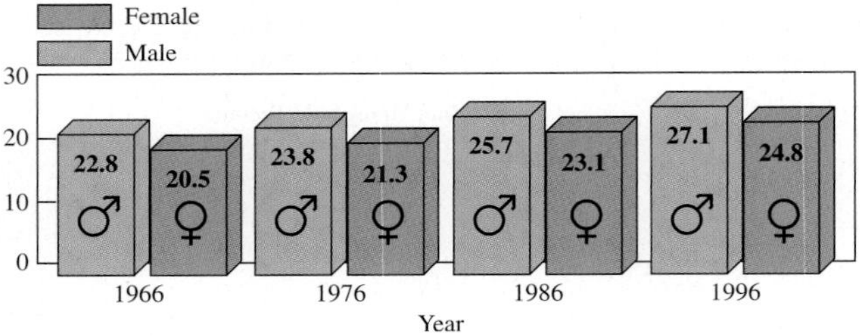

(*Source:* U.S. Census Bureau.)

42. *Mutual Funds* The following graph shows the number of mutual funds available to consumers in the years 1985 through 1997.

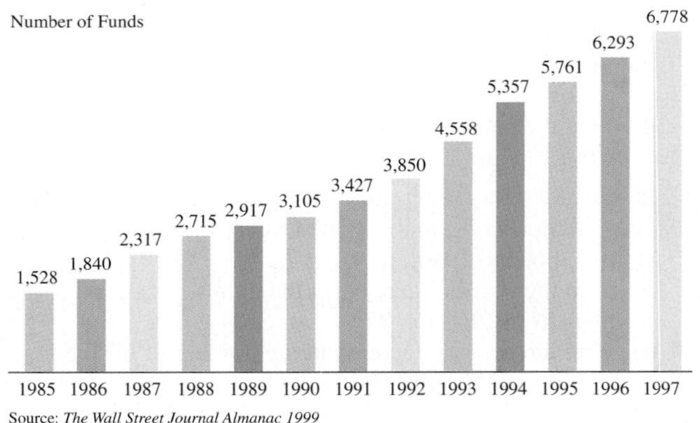

a) Find the percent increase from 1985 to 1990.

b) Find the percent increase from 1990 to 1995.

c) Find the percent increase from 1985 to 1997.

In Exercises 43–48, determine the answer to the question.

43. Fifty-four is 18% of what number?

44. What percent of 75 is 15?

45. What is 7% of 9?

46. What is 134% of 400?

47. What percent of 346 is 86.5?

48. Eleven is 20% of what number?

49. *Tax and Tip* According to the Original Tipping Page (*www.tipping.org*), it is proper to tip waiters and waitresses 15–20% of the total restaurant bill—*including the tax.* Mary and Keith's dinner costs $43.50 before tax, and the tax rate is 6%.

a) What is the tax on Mary and Keith's dinner?

b) What is the total bill, including tax, before the tip?

c) If Mary and Keith decide to tip 15% of the total bill, how much is the tip?

d) What is the total cost of the dinner including tax and tip?

50. *Fishing* The Gordon and Stallard Charter Fishing Boat Company has recently increased the number of crewmembers by 25%, or 10 crewmembers. What was the original number of crewmembers?

51. *Percentage of A's* Eighteen students received an A on the third test, which is 150% of the students who received an A on the second test. How many students received an A on the second test?

52. *Employee Increase* The Fastlock Company hired 57 new employees, which increased its staff by 30%. What was the original number of employees?

53. *Salary Increase* Mr. Brown's present salary is $36,500. He is getting an increase of 7% in his salary next year. What will his new salary be?

54. *Salad Dressing Preference* In a survey of 300 people, 17% prefer Ranch dressing on their salad. How many people in the group prefer Ranch dressing?

55. *Vacuum Cleaner Sales* A Kirby vacuum cleaner dealership sold 430 units in 1998 and 407 units in 1999. Find the percent increase or decrease in the number of units sold.

56. *Vacuum Cleaner Markup* If the Kirby dealership in Exercise 55 pays $320 for each unit and sells it for $699, what is the percent markup?

57. *Poverty in the United States* In the United States in 1993 there were 39.3 million people classified as living in poverty. In 1997 there were 35.6 million people classified as living in poverty. Find the percent decrease in the number of people living in poverty from 1993 to 1997. (*Source: 1999 Wall Street Journal Almanac.*)

58. *More Grandchildren* The Ngs had eight great-grandchildren in 1998. In 1999 they had 12 great-grandchildren. Find the percent increase in the number of great-grandchildren from 1998 to 1999.

59. *Television Sale* The regular price of a Zenith color TV is $539.62. During a sale, Hill TV is selling it for $439. Find the percent decrease in the price of this TV.

60. *Restaurant Markup* The cost of a fish dinner to the owner of the Golden Wharf restaurant is $7.95. The fish dinner is sold for $11.95. Find the percent markup.

61. *Truck Sale Profit* Bonnie sold a truck and made a profit of $675. Her profit was 18% of the sale price. What was the sale price?

62. *Furniture Sale* Kane's Furniture Store advertised a table at a 15% discount. The original price was $115, and the sale price was $100. Was the sale price consistent with the ad? Explain.

63. *Reselling a Car* Quincy purchased a used car for $1000. He decided to sell the car for 10% above his purchase price. Quincy could not sell the car so he reduced his asking price by 10%. If he sells the car at the reduced price, will he have a profit or loss or will he break even? Explain how you arrived at your answer.

CHALLENGE PROBLEMS/GROUP ACTIVITIES

64. *Comparing Markdowns*
    a) A coat is marked down 10%, and the customer is given a second discount of 15%. Is this the same as a single discount of 25%? Explain.
    b) The regular price of a chair is $189.99. Determine the sale price of the chair if the regular price is reduced by 10% and this price is then reduced another 15%.
    c) Determine the sale price of the chair if the regular price of $189.99 is reduced by 25%.
    d) Examine the answers obtained in parts (b) and (c). Does your answer to part (a) appear to be correct? Explain.

65. *Selling Ties* The Tie Shoppe paid $5901.79 for a shipment of 500 ties and wants to make a profit of 40% of the cost on the whole shipment. The store is having two special sales. At the first sale it plans on selling 100 ties for $9.00 each, and at the second special sale it plans on selling 150 ties for $12.50 each. What should be the selling price of the other 250 ties for the Tie Shoppe to make a 40% profit on the whole shipment?

RESEARCH ACTIVITY

66. Find two circle graphs in newspapers or magazines whose data are not given in percents. Redraw the graphs and label them with percents.

11.2 PERSONAL LOANS AND SIMPLE INTEREST

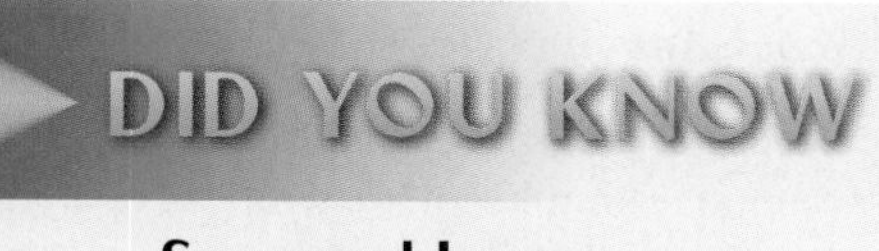

Seasonal Loan

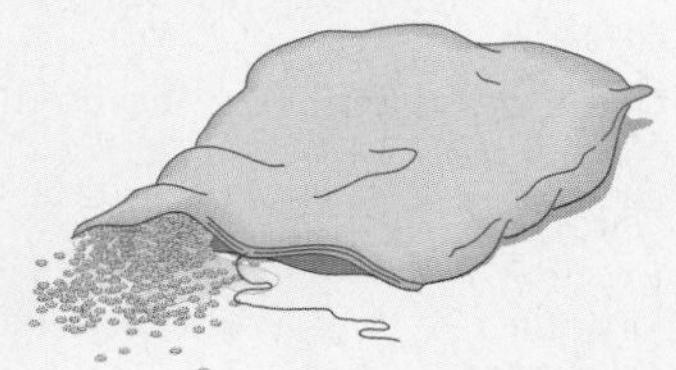

The most common type of loan is the short-term (seasonal) loan. Originally, these were agricultural loans provided to farmers to plant their crops and see them through the period between planting and harvesting. In fact, "seed" loans were first recorded in ancient Babylonia.

Often consumers want to buy clothing, appliances, or furniture but do not have the cash to do so. They then have to determine whether the cost of borrowing the money is worth the convenience or pleasure of having the item today versus waiting several months or years and paying cash. If you are faced with this choice, you may decide that you must have the item today, and may choose to borrow the money from a bank or other lending institution. The money a bank is willing to lend you is called the amount of **credit** extended or the **principal of the loan**.

The amount of credit and the interest rate that you may obtain depend on the assurance that you can give the lender that you will be able to repay the loan. Your credit is determined by your business reputation for honesty, by your earning power, and by what you can pledge as security to cover the loan. **Security** (or *collateral*) is anything of value pledged by the borrower that the lender may sell or keep if the borrower does not repay the loan. Acceptable security may be a business, a mortgage on a property, the title to an automobile, savings accounts, or stocks or bonds. The more marketable the security, the easier it is to obtain the loan, and in some cases marketability may help in getting a lower interest rate.

Bankers sometimes grant loans without security, but they require the signature of one or more other persons, called **cosigners**, who guarantee the loan will be repaid. For either of the two types of loans, the secured loan or the cosigner loan, the borrower (and cosigner, if there is one) must sign an agreement called a **personal note**. This document states the terms and conditions of the loan.

The most common way for individuals to borrow money is through an installment loan or using a credit card. (Installment loans and credit cards are discussed in Section 11.4.)

The concept of simple interest is essential to the understanding of installment buying. **Interest** is the money the borrower pays for the use of the lender's money. One type of interest is called simple interest. **Simple interest** is based on the entire amount of the loan for the total period of the loan. The formula used to find simple interest follows.

Simple Interest Formula

$$\text{Interest} = \text{principal} \times \text{rate} \times \text{time}$$

$$i = prt$$

In the simple interest formula the **principal**, p, is the amount of money lent, the **rate**, r, is the rate of interest expressed as a percent, and the **time**, t, is the number of days, months, or years for which the money will be lent. Time is expressed in the same period as the rate. For example, if the rate is 2% per month, the time must be expressed in months. Typically, rate means the annual rate unless otherwise stated. Principal and interest are expressed in dollars in the United States.

ORDINARY INTEREST

The most common type of simple interest is called *ordinary interest.* For computing ordinary interest, each month has 30 days and a year has 12 months or 360 days. On the due date of a *simple interest note* the borrower must repay the principal plus the interest. (*Note:* Simple interest will mean ordinary interest unless stated otherwise.)

EXAMPLE 1 *Calculating Interest and Payback Amount*

Corliss Bretz needs to borrow \$4200 to have corrective eye surgery. From her credit union, she obtains a 9 month loan with an annual simple interest rate of 8.5%.

a) Calculate the simple interest on the loan.
b) Determine the amount (principal + interest) that Corliss will pay the credit union at the end of the 9 months.

SOLUTION

a) To find the interest on the loan, we use the formula $i = prt$. We know that $p = \$4200$, $r = 8.5\%$ (or converted to a decimal, 0.085), and t in years $= \frac{9}{12} =$ 0.75. We substitute the appropriate values in the formula

$$\begin{aligned} i &= p \times r \times t \\ &= \$4200 \times 0.085 \times 0.75 \\ &= \$267.75 \end{aligned}$$

The simple interest on \$4200 at 8.5% for 9 months is \$267.75.

b) The amount to be repaid is equal to the principal plus interest, or

$$\begin{aligned} A &= p + i \\ &= \$4{,}200 + 267.75 \\ &= \$4{,}467.75 \end{aligned}$$

To pay off her loan, Corliss will pay the credit union \$4,467.75 at the end of 9 months.

EXAMPLE 2 *Determining the Annual Rate of Interest*

Patricia Allaire lent her friend \$300. Six months later her friend repaid the original \$300 plus \$30.00 interest. What annual rate of interest did Patricia receive?

SOLUTION Using the formula $i = prt$, we get

$$\begin{aligned} \$30 &= \$300 \times r \times 0.50 \\ 30 &= 150r \\ \frac{30}{150} &= r \\ 0.2 &= r \end{aligned}$$

The annual rate of interest paid is 20%.

EXAMPLE 3 *A Pawn Loan*

To obtain money for new eyeglasses, Gilbert Kennedy decides to pawn his trumpet. Gilbert borrows \$240 and after 30 days he gets his trumpet back by paying the pawnbroker \$288. What annual rate of interest did Gilbert pay?

SOLUTION Gilbert paid \$288 − \$240 = \$48 in interest, and the length of the loan is for one month or $\frac{1}{12}$ of a year.

Using the formula $i = prt$, we get

$$\begin{aligned} \$48 &= \$240 \times r \times \frac{1}{12} \\ 48 &= 20r \\ \frac{48}{20} &= r \\ r &= 2.4 \end{aligned}$$

The annual rate of interest paid is 240%.

In Examples 1, 2, and 3 we illustrated a simple interest loan, for which the interest and principal are paid on the due date of the note. There is another type of loan, the **discount note**, for which the interest is paid at the time the borrower receives the loan. The interest charged in advance is called the **bank discount**. A Federal Reserve bill is a bank discount note issued by the U.S. government. Example 4 illustrates a discount note.

DID YOU KNOW

Pawn Loans

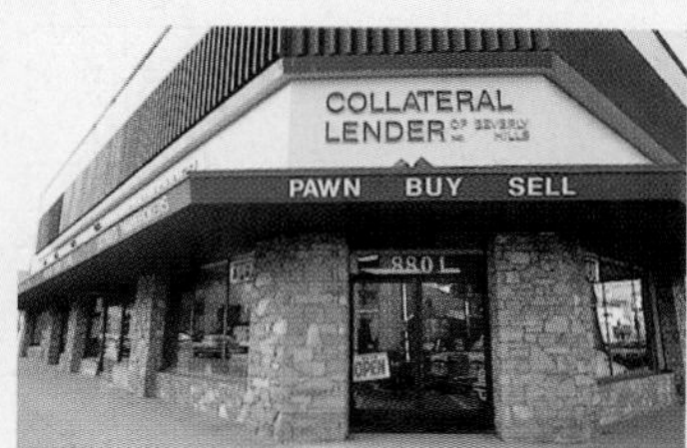

Since ancient times, pawnbroking has been a source of credit for people who cannot obtain loans from other financial institutions. A typical pawn transaction involves the customer presenting an item of value to a pawnbroker. The pawnbroker makes a loan to the customer based on a percentage of the value of the item. The pawnbroker keeps the item as collateral until the customer repays the loan plus the interest. If the customer defaults on the loan, the pawnbroker keeps the item and usually offers it for sale in his or her pawnshop. Due to the 15 to 20% default rate, as well as the need to pay for retail space, pawn loans usually have a significantly higher rate of interest than other financial institutions (see Example 3). (*Source: http://npa.polygon.net/infopawn.html.*)

DID YOU KNOW

Origin of Banks

In ancient Babylonia, as early as 2000 B.C., temples were considered safe depositories for assets. It was believed that these sacred places enjoyed the special protection of the gods and were not likely to be robbed. It is no coincidence that many of the bank buildings built during the early twentieth century resembled ancient temples. In medieval times, monies were kept in vaults protected by armies of the ruling nobility. The word *bank* is derived from the Italian word *banca,* meaning "board." It refers to the counting boards used by merchants. Dishonest moneychangers in the marketplace had their boards smashed to prevent them from continuing in business. The word *bankrupt* is literally "a broken (ruptured) board."

EXAMPLE 4 *True Interest Rate of a Discount Note*

Heather borrowed \$500 on a 10% discount note for a period of 3 months. Find the following:

a) The interest she must pay to the bank on the date she receives the loan.
b) The net amount of money she receives from the bank.
c) The actual rate of interest for the loan.

SOLUTION

a) To find the interest, use the simple interest formula.

$$\begin{aligned} i &= prt \\ &= \$500 \times 0.10 \times \frac{3}{12} \\ &= \$12.50 \end{aligned}$$

b) Since Heather must pay \$12.50 interest when she first receives the loan, the net amount she receives is \$500 − \$12.50 or \$487.50.
c) Calculate the actual rate of interest charged by substituting the net amount she received and the actual interest paid into the simple interest formula.

$$\begin{aligned} i &= prt \\ \$12.50 &= \$487.50 \times r \times \frac{3}{12} \\ 12.50 &= 121.875 \times r \\ \frac{12.50}{121.875} &= r \\ 0.1026 &\approx r \end{aligned}$$

Thus, the actual rate of interest is about 10.3% rather than the quoted 10%. ▲

EXAMPLE 5 *Partial Payments*

Pat wishes to purchase a new racing bicycle but does not have the \$2000 purchase price. Luckily, the bike shop has two payment options. With option 1, Pat can pay \$1000 as a down payment and then pay \$1150 in 6 months. With option 2, Pat can pay \$500 as a down payment and then pay \$1700 in 6 months. Which payment option has a higher annual simple interest rate?

SOLUTION Option 1: To determine the principal, subtract the down payment from the purchase price of the bicycle. Therefore, $p = \$2000 - \$1000 = \$1000$. To determine the interest, subtract the purchase price of the bicycle from the total amount paid. Therefore, $i = (\$1000 + \$1150) - \$2000 = \$150$. Then

$$\begin{aligned} i &= p \times r \times t \\ 150 &= 1000 \times r \times \frac{1}{2} \\ 150 &= 500r \\ \frac{150}{500} &= r \\ 0.3 &= r \end{aligned}$$

Option 1 has a 30% annual simple interest rate.

Option 2: The principal is $p = \$2000 - \$500 = \$1500$, and interest is $i = (\$500 + \$1700) - \$2000 = \$200$. Then

$$\begin{aligned} i &= p \times r \times t \\ 200 &= 1500 \times r \times \frac{1}{2} \\ 200 &= 750r \\ \frac{200}{750} &= r \\ 0.2667 &\approx r \end{aligned}$$

Option 2 has about 26.7% annual simple interest rate. Therefore, option 1 charges a higher annual simple interest rate.

DID YOU KNOW

Tallying Accounts

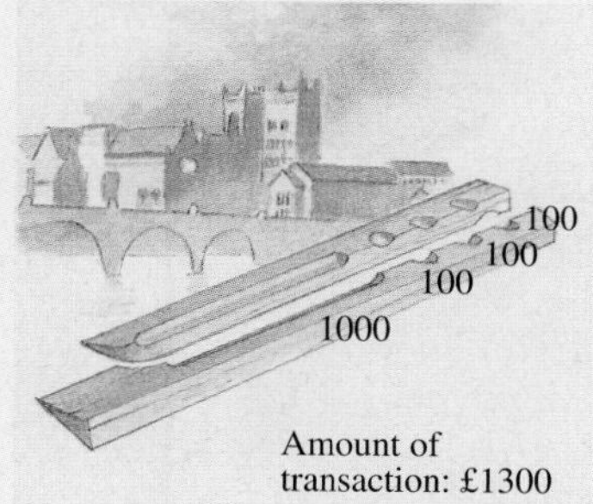

Amount of transaction: £1300

Until 1826, British banks used tally sticks to record the amounts of loans and deposits. The tallies were flat pieces of wood about 8 in. long. The amount was recorded as notches, different values represented by different widths. A hand's width signified £1000, a thumb's width £100. The stick was split in two—the borrower got "the short end of the stick," the money lender held the larger stock ("the stockholder"). Transactions could be verified by putting the sticks together to show the accounts "tallied."

In 1834, as the tallies were being burned in the furnaces of the House of Lords, the fire got out of control and burned the old House of Parliament.

THE UNITED STATES RULE

A loan has a date of maturity, at which time the principal and interest are due. It is possible to make payments on a loan before the date of maturity. A Supreme Court decision specified the method by which these payments are credited. The procedure is called the **United States rule**.

The United States rule states that if a partial payment is made on the loan, interest is computed on the principal from the first day of the loan until the date of the partial payment. The partial payment is used to pay the interest first; then the rest of the payment is used to reduce the principal. The next time a partial payment is made, interest is calculated on the unpaid principal from the date of the previous date of payment. Again, the payment goes first to pay the interest, with the rest of the payment used to reduce the principal. An individual can make as many partial payments as he or she wishes; the procedure is repeated for each payment. The balance due on the date of maturity is found by computing interest due since the last partial payment and adding this interest to the unpaid principal.

The **Banker's rule** is used to calculate simple interest when applying the United States rule. The Banker's rule considers a year to have 360 days, and any fractional part of a year is the exact number of days of the loan.

To determine the exact number of days in a period, we can use Table 11.1 on page 532. Example 6 illustrates how to use the table.

EXAMPLE 6 *Determining the Due Date of a Note*

Use Table 11.1 to find (a) the due date of a loan made on March 15 for 180 days and (b) the number of days from April 18 to July 31.

SOLUTION

a) To determine the due date of the loan, do the following: In Table 11.1 find Day 15 in the left column (with heading Day of Month), then move three columns to the right (heading at the top of the column is March) and find the number 74 (circled in red). Thus, March 15 is the 74th day of the year. Add 180 to 74, since the loan will be due 180 days after March 15:

$$74 + 180 = 254$$

Thus, the due date of the note is the 254th day of the year. Find 254 (circled in blue) in Table 11.1 in the column headed September. The number in the same row as 254 in the left column is Day 11. Thus, the due date of the note is September 11.

b) To determine the number of days from April 18 to July 31, use Table 11.1 to find that April 18 is the 108th day of the year and that July 31 is the 212th day of the year. Then find the difference: 212 − 108 = 104. Thus, the number of days from April 18 to July 31 is 104 days. ▲

Table 11.1

	Days in each month											
	31	28	31	30	31	30	31	31	30	31	30	31
Day of Month	**Jan**	**Feb**	**Mar**	**Apr**	**May**	**Jun**	**Jul**	**Aug**	**Sep**	**Oct**	**Nov**	**Dec**
Day 1	1	32	60	91	121	152	182	213	244	274	305	335
Day 2	2	33	61	92	122	153	183	214	245	275	306	336
Day 3	3	34	62	93	123	154	184	215	246	276	307	337
Day 4	4	35	63	94	124	155	185	216	247	277	308	338
Day 5	5	36	64	95	125	156	186	217	248	278	309	339
Day 6	6	37	65	96	126	157	187	218	249	279	310	340
Day 7	7	38	66	97	127	158	188	219	250	280	311	341
Day 8	8	39	67	98	128	159	189	220	251	281	312	342
Day 9	9	40	68	99	129	160	190	221	252	282	313	343
Day 10	10	41	69	100	130	161	191	222	253	283	314	344
Day 11	11	42	70	101	131	162	192	223	254	284	315	345
Day 12	12	43	71	102	132	163	193	224	255	285	316	346
Day 13	13	44	72	103	133	164	194	225	256	286	317	347
Day 14	14	45	73	104	134	165	195	226	257	287	318	348
Day 15	15	46	74	105	135	166	196	227	258	288	319	349
Day 16	16	47	75	106	136	167	197	228	259	289	320	350
Day 17	17	48	76	107	137	168	198	229	260	290	321	351
Day 18	18	49	77	108	138	169	199	230	261	291	322	352
Day 19	19	50	78	109	139	170	200	231	262	292	323	353
Day 20	20	51	79	110	140	171	201	232	263	293	324	354
Day 21	21	52	80	111	141	172	202	233	264	294	325	355
Day 22	22	53	81	112	142	173	203	234	265	295	326	356
Day 23	23	54	82	113	143	174	204	235	266	296	327	357
Day 24	24	55	83	114	144	175	205	236	267	297	328	358
Day 25	25	56	84	115	145	176	206	237	268	298	329	359
Day 26	26	57	85	116	146	177	207	238	269	299	330	360
Day 27	27	58	86	117	147	178	208	239	270	300	331	361
Day 28	28	59	87	118	148	179	209	240	271	301	332	362
Day 29	29		88	119	149	180	210	241	272	302	333	363
Day 30	30		89	120	150	181	211	242	273	303	334	364
Day 31	31		90		151		212	243		304		365

Add 1 day for leap year if February 29 falls between the two dates under consideration.

DID YOU KNOW

Selling Time

For most of human history the practice of charging interest on money borrowed, or *usury*, was considered not only immoral, but a crime. This attitude arose in ancient times, in part because a peasant or farmer could literally be enslaved if payments could not be made. Charging interest over time was considered "selling time," something not ours to sell. What was especially objectionable was the idea of "pure interest," a fixed payment set at the time of the loan, even when there was no risk to the lender. Much has changed in the past 200 years. Borrowing money has become a fact of life for individuals, businesses, and governments alike. In 1998, the U.S. government paid about $353.5 billion in interest alone on a debt of more than $5.5 trillion.

Today the word usury means lending money at a rate of interest that is excessive or unlawfully high. A person who engages in such loans is often referred to as a "loan shark."

EXAMPLE 7 *Using the Banker's Rule*

Find the simple interest on \$300 at 12% for the period March 3 to May 3 using the Banker's rule.

SOLUTION The exact number of days from March 3 to May 3 is 61. The period of time in years is 61/360. Substituting in the simple interest formula gives

$$\begin{aligned} i &= prt \\ &= \$300 \times 0.12 \times \frac{61}{360} \\ &= \$6.10 \end{aligned}$$

The interest is \$6.10.

The next example illustrates how a partial payment is credited under the United States rule. Making partial payments reduces the amount of interest paid and the cost of the loan.

EXAMPLE 8 *Using the United States Rule*

Cathy Panik is a mathematics teacher and she plans to attend a national conference. To pay for her airfare, on November 1, 2000, Cathy takes out a 120-day loan for \$400 at an interest rate of 12.5%. Cathy uses some birthday gift money to make a partial payment of \$150 on January 5, 2001. She makes a second partial payment of \$100 on February 2, 2001.

a) Determine the due date of the loan.
b) Determine the interest and the amount credited to the principal on January 5.
c) Determine the interest and the amount credited to the principal on February 2.
d) Determine the amount that Cathy must pay on the due date.

SOLUTION

a) Using Table 11.1, we see that November 1 is the 305th day of the year. Next, we note that the sum of 305 and 120 is 425. Since this due date will extend into the next year, we subtract 365 from 425 to get 60. From Table 11.1, we see that the 60th day of the year is March 1. Therefore, the loan due date is March 1, 2001. Had 2001 been a leap year, the due date would have been February 29, 2001.
b) Using Table 11.1, January 5 is the 5th day of the year and November 1 is the 305th day of the year. The number of days from November 1 to January 5 can be computed as follows: $(365 - 305) + 5 = 65$. Then using $i = prt$, and the Banker's rule, we get:

$$\begin{aligned} i &= \$400 \times 0.125 \times \frac{65}{360} \\ &\approx \$9.03 \end{aligned}$$

The interest of \$9.03 that is due January 5, 2001, is deducted from the payment of \$150. The remaining payment of \$150 − \$9.03 or \$140.97 is then credited to the principal. Therefore, the adjusted principal is now \$400 − \$140.97, or \$259.03.

c) We use the Banker's rule to calculate the interest on the unpaid principal for the period from January 5 to February 2. According to Table 11.1, the number of days from January 5 to February 2 is 33 − 5, or 28 days.

$$i = \$259.03 \times 0.125 \times \frac{28}{360}$$
$$\approx \$2.52$$

The interest of \$2.52 that is due February 2, 2001, is deducted from the payment of \$100. The remaining payment of \$100 − \$2.52, or \$97.48, is then credited to the principal. Therefore, the new adjusted principal is now \$259.03 − \$97.48, or \$161.55.

d) The due date of the loan is March 1. Using Table 11.1, we see that there are 60 − 33 or 27 days from February 2 to March 1. The interest is computed on the remaining balance of \$161.51 by using the simple interest formula.

$$i = \$161.55 \times 0.125 \times \frac{27}{360}$$
$$\approx \$1.51$$

Therefore, the balance due on the maturity date of the loan is the sum of the principal and the interest, \$161.55 + \$1.51, or \$163.06. *Note:* The sum of the days in the three calculations, 65 + 28 + 27, equals the total number of days in the loan, 120.

SECTION 11.2 EXERCISES

CONCEPT/WRITING EXERCISES

1. What is interest?
2. What is credit?
3. What is security (or collateral)?
4. What is a cosigner?
5. What is a personal note?
6. Explain what each letter in the simple interest formula, $i = prt$, represents.
7. In your own words, explain the difference between ordinary interest and interest under the Banker's rule.
8. In your own words, explain the United States rule.

In Exercises 9–18, find the simple interest. (The rate is an annual rate unless otherwise noted. Assume 360 days in a year.)

9. $p = \$420$, $r = 9\%$, $t = 3$ years

10. $p = \$520$, $r = 6.5\%$, $t = 4$ years

11. $p = \$875$, $r = 12\%$, $t = 30$ days

12. $p = \$365.45$, $r = 11\frac{1}{2}\%$, $t = 8$ months

13. $p = \$587$, $r = 0.045\%$ per day, $t = 2$ months

14. $p = \$6742.75$, $r = 6.05\%$, $t = 90$ days

15. $p = \$2,756.78$, $r = 10.15\%$, $t = 103$ days

16. $p = \$550.31$, $r = 8.9\%$, $t = 67$ days

17. $p = \$12,752$, $r = 1\frac{1}{2}\%$ per month, $t = 9$ months

18. $p = \$12,752$, $r = 0.055\%$ per day, $t = 120$ days

In Exercises 19–24, use the simple interest formula to find the missing value.

19. $p = \$1300$, $r = ?$, $t = 2$ years, $i = \$104.00$

20. $p = ?$, $r = 6\%$, $t = 60$ days, $i = \$6.00$

21. $p = ?$, $r = 8\%$, $t = 3$ months, $i = \$12.00$

22. $p = \$800.00$, $r = 6\%$, $t = ?$, $i = \$64.00$

23. $p = \$957.62$, $r = 6.5\%$, $t = ?$, $i = \$124.49$

24. $p = \$1650.00$, $r = ?$, $t = 6.5$ years, $i = \$343.20$

PRACTICE THE SKILLS/PROBLEM SOLVING

25. *Personal Loan* As a favor to his friend, Rod lent Sam \$2000 for 9 months at an annual simple interest rate that was 1% lower than the rate Rod could get in his credit union's money market account. If the money market rate is 7.08%, find how much Sam needs to repay Rod at the end of the 9 months.

26. *Bank Loan* Bobby Kunkel borrowed \$1500 from his bank for 60 days at a simple interest rate of $6\frac{1}{2}\%$. He gave the bank stock certificates as security.
    a) How much did he pay for the use of the money?
    b) What is the amount he paid to the bank on the date of maturity?

27. *Bank Personal Note* Kelly Droessler borrowed \$3500 from the bank for 6 months. Her friend Ms. Harris was cosigner of Kelly's personal note. The bank collected $7\frac{1}{2}\%$ simple interest on the date of maturity.
    a) How much did Kelly pay for the use of the money?
    b) Find the amount she repaid to the bank on the due date of the note.

28. *Bank Discount Note* Kwame Adebele borrowed \$2500 for 5 months from his bank, using U.S. government bonds as security. The bank discounted the loan at 8%.
    a) How much interest did Kwame pay the bank for the use of its money?
    b) How much did he receive from the bank?
    c) What was the actual rate of interest he paid?

29. *Bank Discount Note* Julie Jansen borrowed \$3650 from the bank for 8 months. The bank discounted the loan at 7.5%.
    a) How much interest did Julie pay the bank for the use of its money?
    b) How much did she receive from the bank?
    c) What was the actual rate of interest she paid?

30. *Credit Union Loan* Enrico Montoyo wants to borrow \$350 for 6 months from his credit union, using his savings account as security. The credit union's policy is that the maximum amount a person can borrow is 80% of the amount in the person's savings account. The interest rate is 2% higher than the interest rate being paid on the savings account. The current rate on the savings account is $3\frac{1}{4}\%$.
    a) How much money must Enrico have in his account in order to borrow \$350?
    b) What is the rate of interest the credit union will charge for the loan?
    c) Find the amount Enrico must repay in 6 months.

31. *Investing Tuition Payments* Sand Ridge School is requiring parents to pay half of the yearly tuition at the time of registration and half on the date classes begin. Registration is held 5 months prior to the beginning of school, and administrators expect 470 students to register. If annual tuition is \$4500 and if the money paid at the time of registration is placed in an account paying 5.4% simple interest, how much interest will Sand Ridge School earn by the time school begins?

32. *A Pawn Loan* Jeffrey Kowalski wants to take his mother out for dinner on her birthday, but he doesn't get paid until the following week. To borrow money, Jeffrey pawns his watch. Based on the value of the watch, the pawnbroker loans Jeffrey \$75. Fourteen days later Jeffrey gets his watch back by paying the pawnbroker \$80.25. What annual simple interest rate did the pawnbroker charge Jeffrey?

In Exercises 33–38, find the exact time from the first date to the second date. Use Table 11.1.

33. April 4 to October 11
34. May 19 to September 17
35. June 19 to November 5
36. June 14 to January 24
37. August 24 to May 15
38. December 21 to April 28

In Exercises 39–42, determine the due date of the loan, using the exact time, if the loan is made on the given date for the given number of days.

39. April 18 for 90 days
40. June 8 for 120 days
41. November 25 for 120 days (the loan is due in a leap year)
42. July 5 for 210 days

In Exercises 43–50, a partial payment is made on the date indicated. Use the United States rule to determine the balance due on the note at the date of maturity. (The Effective Date is the date the note was written.)

	Prin-cipal	Rate	Effec-tive date	Matu-rity date	Partial payment(s) Amount	Date(s)
43.	$2500	8%	July 1	Oct. 15	$300	Aug. 1
44.	$2400	7%	April 10	July 7	$500	May 5
45.	$8000	9%	May 1	Nov. 1	$2000	June 15
46.	$7500	12%	April 15	Oct. 1	$1000	Aug. 1
47.	$9000	6%	July 15	Feb. 1	$4000	Dec. 27
48.	$1000	12.5%	Jan. 1	Feb. 15	$300	Jan. 15
49.	$1800	15%	Aug. 1	Nov. 1	$500	Sept. 1
					$500	Oct. 1
50.	$5000	14%	Oct. 15	Jan. 1	$800	Nov. 15
					$800	Dec. 15

51. *Company Loan* On March 1 the Zwick Balloon Company signed a $6500 note with simple interest of $10\frac{1}{2}\%$ for 180 days. The company made payments of $1750 on May 1 and $2350 on July 1. How much will the company owe on the date of maturity?

52. *Restaurant Loan* The Sweet Tooth Restaurant borrowed $3000 on a note dated May 15 with simple interest of 11%. The maturity date of the loan is September 1. They made partial payments of $875 on June 15 and $940 on August 1. Find the amount due on the maturity date of the loan.

53. *U.S. Treasury Bills* The U.S. government borrows money by selling Treasury bills. Treasury bills are discounted notes issued by the U.S. government. Previously, Treasury bills were available in minimum purchases of $10,000. But on August 5, 1998, the U.S. Treasury Department announced that all Treasury bills would be available to the public in minimum amounts of $1000, thereby making these investments more accessible to individual investors. On May 5, 1999 Kris purchased a 182-day, $1000 U.S. Treasury bill at a 4.34% discount. On the date of maturity Kris will receive $1000.

a) What is the date of maturity of the Treasury bill?
b) How much did Kris actually pay for the Treasury bill?
c) How much interest did the U.S. goverment pay Kris on the date of maturity?
d) What is the actual rate of interest of the Treasury bill? (Round the answer to the nearest hundredth of a percent.)

54. *U.S. Treasury Bills* On August 31, 1999, Tran purchased a 364-day, $6000 U.S. Treasury bill at a 4.4% discount. (See Exercise 53.)

a) What is the date of maturity of the Treasury bill (the year 2000 is a leap year)?
b) How much did Tran actually pay for the Treasury bill?
c) How much interest did the U.S. government pay Tran on the date of maturity?
d) What is the actual rate of interest of the Treasury bill?

55. *Tax Preparation Loan* Many tax preparation organizations will prepay customers' tax refunds if they pay a one time finance charge. In essence, the customer is borrowing the money (the refund minus the finance charge) from the tax preparer, prepaying the interest (as in a discount note), and then repaying the loan with the tax refund. This allows customers access to their tax refund money without having to wait. Joy Stallard had a tax refund of $743.21 due. She was able to get her tax refund immediately by paying a finance charge of $39.95. What annual simple interest rate is Joy paying for this loan assuming

a) The tax refund check would be available in 5 days?
b) The tax refund check would be available in 10 days?
c) The tax refund check would be available in 20 days?

56. *Prime Interest Rate* Mr. Harrison borrowed $600 for 3 months. The banker said he must repay the loan at the rate of $200 per month plus interest. The bank was charging a rate of 2% above the prime interest rate. The *prime interest rate* is the rate charged to preferred customers of the bank. During the first month the prime rate was $8\frac{1}{2}\%$, during the second month it was 9%, and during the third month it was 10%.

a) Find the amount he paid the bank at the end of the first month; at the end of the second month; at the end of the third month.
b) What was the total amount of interest he paid the bank?

CHALLENGE PROBLEM/GROUP ACTIVITY

57. *U.S. Treasury Bills* Mark purchased a 52-week U.S. Treasury bill (see Exercises 53 and 54) for $93,337. The par value (the value of the bill upon maturity) was $100,000.

a) What was the discount rate?
b) What was the actual interest Mark received?
c) Since U.S. Treasury bills are sold through auctions at Federal Reserve Banks, Mark did not know the purchase price of his Treasury bill until after he was notified by mail. If Mark had sent the Federal Reserve Bank a check for $100,000 to purchase the Treasury bill, how much would the Federal Reserve Bank have to rebate him upon notice of his purchase?
d) On the day he received the rebate that was discussed in part (c) he invested it in a 1-year certificate of deposit yielding 5% interest. What is the total amount of interest he will receive from both investments?

RESEARCH ACTIVITIES

58. *Banking Practices* Three types of simple interest may be calculated with the simple interest formula. They are ordinary, Banker's rule, and exact time.
   a) Visit a local bank to determine how exact time is used in the simple interest formula.
   b) Compare the results for each method on a loan for \$500 at 8% for the period January 2, 2000, to April 2, 2000.

59. Log on to a World Wide Web search engine and search for information on: banks, savings and loans, credit unions, and pawn shops. Then write a report that includes the following information:
   a) Describe the ownership structure of each.
   b) Historically, what need is each fulfilling?
   c) What are the advantages and disadvantages of obtaining a loan from each of those listed?

11.3 COMPOUND INTEREST

Albert Einstein was once asked to name the greatest discovery of man. His reply was, "Compound interest."

In this section we discuss some ways to put your money to work for your benefit.

An **investment** is the use of money or capital for income or profit. We can divide investments into two classes: fixed investments and variable investments. In a **fixed investment** the amount invested as principal is guaranteed, and the interest is computed at a fixed rate. *Guaranteed* means that the exact amount invested will be paid back together with any accumulated interest. Examples of a fixed investment are savings accounts and certificates of deposit. Another fixed investment is a government savings bond. In a **variable investment**, neither the principal nor the interest is guaranteed. Examples of variable investments are stocks, mutual funds, and commercial bonds.

Simple interest, introduced earlier in the chapter, is calculated once for the period of a loan using the formula $i = prt$. The interest paid on savings accounts at most banks is compound interest. A bank computes the interest periodically (for example, daily or quarterly) and adds this interest to the original principal. The interest for the following period is computed by using the new principal (original principal plus interest). In effect, the bank is computing interest on interest, which is called compound interest.

> Interest that is computed on the principal and any accumulated interest is called **compound interest.**

EXAMPLE 1 *Computing Compound Interest*

Marjorie Thrall recently won the \$1000 first prize in a raffle contest. Marjorie deposits the \$1000 in a 1-year certificate of deposit paying 6.0% compounded quarterly. Find the amount, A, to which the \$1000 will grow in 1 year.

SOLUTION Compute the interest for the first quarter using the simple interest formula. Add this interest to the principal to find the amount at the end of the first quarter.

$$i = prt$$
$$= \$1000 \times 0.06 \times 0.25 = \$15.00$$
$$A = \$1000 + \$15 = \$1015$$

DID YOU KNOW

A Penny Saved . . .

"Money makes money and the money that money makes makes more money." So said Benjamin Franklin, who in his will left 1000 pounds sterling to the city of Boston, where he was born. The money was to be lent to young apprentices who would pay $\frac{1}{10}$ of the principal each year plus 5% interest compounded yearly. The money earned was to be used to benefit the inhabitants of Boston. By 1894, the fund's value had increased to 90,000 pounds. Part of this legacy became the Franklin Institute of Boston, a 2-year technical college founded in 1908.

Now repeat the process for the second quarter, this time using a principal of $1015

$$i = \$1015 \times 0.06 \times 0.25 \approx \$15.23$$
$$A = \$1015 + \$15.23 = \$1030.23$$

For the third quarter, use a principal of $1030.23

$$i = \$1030.23 \times 0.06 \times 0.25 \approx \$15.45$$
$$A = \$1030.23 + \$15.45 = \$1045.68$$

For the fourth quarter, use a principal of $1045.68

$$i = \$1045.68 \times 0.06 \times 0.25 \approx \$15.69$$
$$A = \$1045.68 + \$15.69 = \$1061.37$$

Hence, the $1000 grows to a final value of $1061.37. ▲

This example shows the effect of earning interest on interest, or compounding interest. In 1 year, the amount of $1000 has grown to $1061.37, compared to $1060 that would have been obtained with a simple interest rate of 6%. Thus, in 1 year alone the gain was $1.37 more with compound interest than with simple interest.

A simpler and less time-consuming way to calculate compound interest is to use the compound interest formula and a calculator.

Compound Interest Formula

$$A = p\left(1 + \frac{r}{n}\right)^{nt}$$

In this formula, A is the amount, p is the principal, r is the annual rate of interest, t is the time in years, and n is the number of periods per year.

EXAMPLE 2 *Using the Compound Interest Formula*

When Alexander was born, he received several gifts of cash from his relatives and his parents' friends. His father invested this money in a money market account that had a rate of 6.6% compounded monthly. If the amount invested was $410, determine the amount in the account after 5 years.

SOLUTION We will use the formula for compound interest, $A = p\left(1 + \frac{r}{n}\right)^{nt}$.

Since the interest is compounded monthly, there are 12 periods per year. Thus, $n = 12$. Since the money is invested for 5 years, $t = 5$.

$$\begin{aligned} A &= 410\left(1 + \frac{0.066}{12}\right)^{(12)(5)} \\ &= 410\,(1 + 0.0055)^{60} \\ &= 410\,(1.0055)^{60} \\ &= 410\,(1.389711) \\ &= 569.78 \end{aligned}$$

▲

DID YOU KNOW

Cost of Manhattan

In 1626 Peter Du Minuit, representing the Dutch West India Company, traded beads and blankets to the native American inhabitants of Manhattan Island for the island. The trade was valued at 60 Dutch guilders, about $24. If at that time the $24 had been invested at 6% interest compounded annually, the investment would be worth $69,920,552,399.60 in 2000. As a comparison, the simple interest on $24 for 374 years at 6% is only $538.56. If the $24 had been compounded continuously, as is done in many savings accounts today, the investment would be worth approximately $133,591,681,294. These results dramatically illustrate the effect of compounding interest over time.

In Example 2, to find the value of $(1.0055)^{60}$ using a scientific calculator, you can use the $\boxed{y^x}$ key. The key strokes are

$$1.0055 \ \boxed{y^x} \ 60 \ \boxed{=} \ 1.389711$$

Note: Some scientific calculators have an $\boxed{x^y}$ or a $\boxed{\wedge}$ key in place of a $\boxed{y^x}$ key. In either case, the procedure works the same: Enter the base then the key ($\boxed{y^x}$, $\boxed{x^y}$, or $\boxed{\wedge}$), the exponent, and then the $\boxed{=}$ key.

EXAMPLE 3 *Calculating Compound Interest*

Calculate the interest on $650 at 8% compounded semiannually for 3 years, using the compound interest formula.

SOLUTION Since interest is compounded semiannually, there are two periods per year. Thus $n = 2$, $r = 0.08$, and $t = 3$. Substituting into the formula, we find the amount, A.

$$\begin{aligned} A &= p\left(1 + \frac{r}{n}\right)^{nt} \\ &= 650\left(1 + \frac{0.08}{2}\right)^{(2)(3)} \\ &= 650(1.04)^6 \\ &= 650(1.2653190) \\ &= \$822.46 \end{aligned}$$

Since the total amount is $822.46 and the original principal is $650, the interest must be $822.46 − $650, or $172.46.

In Example 3, the interest rate is stated as an annual rate of 8%, but the number of compounding periods per year is two. In applying the compound interest formula, the rate for one period is $r \div n$, which was $8\% \div 2$, or 4% per period.

Now we will calculate the amount, and interest, on $1 invested at 8% compounded semiannually for 1 year. The result is

$$A = 1\left(1 + \frac{0.08}{2}\right)^{(2)(1)} = 1.0816$$

$$\begin{aligned} \text{Interest} &= \text{amount} - \text{principal} \\ i &= 1.0816 - 1 \\ &= 0.0816 \end{aligned}$$

The interest for 1 year is 0.0816. This amount written as a percent, 8.16%, is called the **effective annual yield** (or **effective yield**). If $1 was invested at a simple interest rate of 8.16% and $1 was invested at 8% interest compounded semiannually (equivalent to an effective yield of 8.16%), the interest from both investments would be the same.

Many banks compound interest daily. When computing the effective annual yield, they use 360 for the number of periods in a year. To find the effective annual yield for any interest rate, calculate the amount using the compound interest formula where p is $1. Then subtract $1 from that amount. The difference, written as a percent, is the effective annual yield, as illustrated in Example 4.

On the sign in the margin it shows that the annual rate is 5.25% and the effective annual yield is 5.39%. Show that an interest rate of 5.25% compounded daily has an effective yield of 5.39% now.

DID YOU KNOW

Federal Reserve Banks

Federal Reserve Banks are institutions owned by commecial banks in the district which they serve. The many functions performed by these banks include the following: clearing checks, issuing new currency, retiring damaged currency, making discount loans to banks within their districts, examining member state banks, collecting data on local business conditions, and conducting research to help establish monetary policy, including the setting of the discount interest rate used when banks borrow money from each other. Currently, there are Federal Reserve Banks in Boston, New York, Philadelphia, Richmond, Cleveland, Atlanta, Chicago, St. Louis, Minneapolis, Kansas City, Dallas, and San Francisco. Take a look at any dollar bill and you will see a circular symbol with the location of the Federal Reserve Bank that printed the bill.

EXAMPLE 4 ***Determining Effective Annual Yield***

Determine the effective annual yield for $1 invested for 1 year at

a) 8% compounded daily.
b) 6% compounded quarterly.

SOLUTION

a) With daily compounding $n = 360$.

$$\begin{aligned} A &= p\left(1 + \frac{r}{n}\right)^{nt} \\ &= 1\left(1 + \frac{0.08}{360}\right)^{(1)(360)} \\ &\approx 1.0832774 \approx 1.0833 \\ i &= A - 1 \\ &= 1.0833 - 1 \\ &= 0.0833 \end{aligned}$$

Thus, when the interest is 8% compounded daily, the effective annual yield is 8.33%.

b) With quarterly compounding, $n = 4$.

$$\begin{aligned} A &= p\left(1 + \frac{r}{n}\right)^{nt} \\ &= 1\left(1 + \frac{0.06}{4}\right)^{(1)(4)} \\ &\approx 1.06136355 \approx 1.0614 \\ i &= 1.0614 - 1 \\ &= 0.0614 \end{aligned}$$

Thus, when the interest is 6% compounded quarterly, the effective annual yield is 6.14%. ▲

There are numerous types of savings accounts. Many savings institutions compound interest daily. Some pay interest from the day of deposit to the day of withdrawal, and others pay interest from the first of the month on all deposits made before the tenth of the month. In each of these accounts in which interest is compounded daily, interest is entered into the depositor's account only once each quarter. Some savings banks will not pay any interest on a day-to-day account if the balance falls below a set amount.

PRESENT VALUE

You may wonder about what amount of money you must deposit in an account today to have a certain amount of money in the future. For example, how much must you deposit in an account today at a given rate of interest so it will accumulate to $25,000 to pay your child's college costs in 4 years? The principal, p, which would have to be invested now is called the **present value**. Following is a formula for determining the present value.

Present Value Formula

$$p = \frac{A}{\left(1 + \frac{r}{n}\right)^{nt}}$$

In the formula, A represents the amount to be accumulated in n years. Note that the present value formula is a variation of the compound interest formula.

EXAMPLE 5 ***Savings for College***

Nicholas is currently in the eighth grade and intends to attend college when he finishes high school. Nicholas' parents currently have some money invested in mutual funds, but would like to invest this money in a more secure investment now to pay for college in 4 years. If they will need \$25,000, how much do Nicholas' parents have to invest now in a 48-month Certificate of Deposit that has a rate of 6.5% compounded monthly?

SOLUTION To answer this question, we will use the present value formula.

$$\begin{aligned} p &= \frac{A}{\left(1 + \frac{r}{n}\right)^{nt}} \\ &= \frac{25{,}000}{\left(1 + \frac{0.065}{12}\right)^{(12)(4)}} \\ &= \frac{25{,}000}{(1.005417)^{48}} \\ &= \frac{25{,}000}{1.2960} \\ &= 19{,}290.12 \end{aligned}$$

Nicholas' parents need to invest approximately \$19,290.12 now to have \$25,000 in 4 years.

DID YOU KNOW

The Principal/Principle of the Thing

In 1777, Jacob DeHaven lent George Washington \$450,000 worth of money and supplies. Washington was afraid that without this money and supplies his army would have to either "disband or starve." The army went on to win the Revolutionary War. Supposedly, DeHaven was offered Continental money as repayment but DeHaven demanded gold. No gold was ever paid, and the once-wealthy DeHaven died penniless.

Although historians can find no evidence to support this story, Jacob DeHaven's descendants have throughout history tried to reclaim the original debt—plus interest. The most recent attempt came in 1989 when the DeHaven family decided to sue the U.S. government. Allowing 6% interest compounded daily (the rate offered by the Continental Congress at the time), by 1989 the \$450,000 would have grown to \$150.3 billion. (You can verify this by using the compound interest formula.) After the case was dismissed, the DeHaven's claimed they would be satisfied with a "Thank You" note and a statue at Valley Forge in Jacob DeHaven's honor. There currently are no plans to erect such a statue. (*Source:* Historic Valley Forge Internet Website: www.libertynet.org)

SECTION 11.3 EXERCISES

CONCEPT/WRITING EXERCISES

1. What is an investment?
2. What is a fixed investment?
3. What is a variable investment?
4. What is compound interest?
5. What is effective annual yield?
6. What is meant by present value in the Present Value Formula?

PRACTICE THE SKILLS

In Exercises 7–16, use the compound interest formula

$$A = p\left(1 + \frac{r}{n}\right)^{nt}$$

to compute

a) the total amount.
b) the interest earned on each investment.

7. \$4000 for 2 years at 4.00% compounded annually
8. \$4000 for 2 years at 4.00% compounded semiannually
9. \$3000 for 5 years at 5.00% compounded semiannually
10. \$3000 for 5 years at 5.00% compounded annually
11. \$1500 for 3 years at 4.75% compounded quarterly
12. \$1500 for 4 years at 4.75% compounded quarterly
13. \$2500 for 2 years at 6.25% compounded monthly
14. \$3000 for 2 years at 6.25% compounded monthly
15. \$5000 for 5 years at 6.75% compounded daily (use $n = 360$)
16. \$5000 for 10 years at 6.75% compounded daily (use $n = 360$)

PROBLEM SOLVING

17. *Investing Prize Winnings* Mary Jo and Pat win third prize in the Clearinghouse Sweepstakes and receive a check for \$250,000. After spending \$10,000 on a vacation, Mary Jo and Pat decide to invest the rest in a money market account that pays 5.6% interest compounded monthly. How much money will be in the account after 10 years?

18. *Investing Gifts and Scholarships* Scott just graduated from high school and has received \$800 in gifts of cash from friends and relatives. Additionally, Scott received three scholarships in the amounts of \$150, \$300, and \$1000. If Scott takes all of his gift and scholarship money and invests it in a 24-month CD paying 5.2% interest compounded daily, how much will he have when he cashes in the CD at the end of the 24 months?

19. *Investing a Signing Bonus* Tim just started a new job and has received a \$5000 signing bonus. Tim decides to invest this money now so he can buy a new car in 5 years. If Tim invests in a CD paying 6% interest compounded quarterly, how much money will he receive from his CD in 5 years?

20. *Saving for a Tractor* Buddy wants to invest some money now to buy a new tractor in the future. If he wants to have \$30,000 available in 5 years, how much does he need to invest now in a CD paying 5.15% interest compounded monthly?

21. *A New Water Tower* The village of Kieler completes an exploratory study and finds the current village water tower will need replacement in 10 years at a cost of \$290,000. To finance this amount, the village board will at this time assess its 958 homeowners with a one-time surcharge and then invest this amount in a 10-year CD paying 8.25% interest compounded semiannually.
    a) How much will the village of Kieler need to invest at this time in this CD in order to raise the \$290,000 in 10 years.
    b) What amount should each homeowner pay as a surcharge?

22. *Water Tower Surcharge* (See Exercise 21.) After seeing its neighboring village raise the money to invest in a new water tower, the city board of East Dubuque decides to adopt a similar plan. However, since East Dubuque is a much larger community, they will need to raise $783,000 to build three new water towers in 15 years. At this time the city board plans to assess its 2682 home owners with a one time surcharge, and then invest the money received in a money market account paying 9% interest compounded monthly.
   a) How much money will the city board need to raise at this time to meet the city's water tower needs at the end of 15 years?
   b) Before applying the surcharge, the city board decides to use a $50,000 benefactor gift toward the water tower investment. Taking this gift into account, how much should the surcharge be on each homeowner?

23. *Comparing Loan Sources* Troy needs to borrow $1500 to expand his farm implement maintenance business. He learns that the local bank will lend him the money for 2 years at a rate of 10% compounded quarterly. After hearing this, Troy's grandfather offers to lend him the money for 2 years with a simple interest rate of 7%. How much money will Troy save by borrowing the money from his grandfather?

24. *Savings Account Investment* When Richard Zucker was born, his father deposited $2000 in his name in a savings account. The account was paying 5% interest compounded semiannually.
   a) If the rate did not change, what was the value of the account after 15 years?
   b) If the money had been invested at 5% compounded quarterly, what would the value of the account have been after 15 years?

25. *Savings and Loan Investment* When Lois Martin was born, her father deposited $2000 in her name at a savings and loan association. At the time, that S&L was paying 6% interest compounded semiannually on savings. After 10 years, the S&L changed to an interest rate of 6% compounded quarterly. How much had the $2000 amounted to after 18 years when the money was withdrawn for Lois to use to help pay her college expenses?

26. *Personal Loan* Brent Pickett borrowed $3000 from his brother Dave. He agreed to repay the money at the end of 2 years, giving Dave the same amount of interest that he would have received if the money had been invested at 8% compounded quarterly. How much money did Brent repay his brother?

27. *Forgoing Interest* Rikki Blair borrowed $6000 from her daughter, Lynette. She repaid the $6000 at the end of 2 years. If Lynette had left the money in a bank account that paid $5\frac{1}{4}$% compounded monthly, how much interest would she have accumulated?

28. *Saving for a Down Payment* Bob Dedrick invested $6000 at 8% compounded quarterly. Three years later he withdrew the full amount and used it for the down payment on a house. How much money did he put down on the house?

29. *Investing a Salary* After Karen Estes began her job as a waitress, she invested in a money market account paying 5.6% interest compounded daily. What was the effective annual yield of this account?

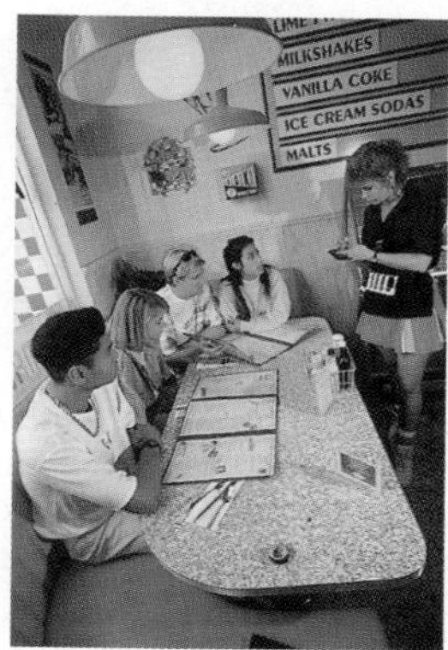

30. Determine the total amount and the interest paid on $1000 with interest compounded semiannually for 2 years at
   a) 2%. **b)** 4%. **c)** 8%.
   d) Is there a predictable outcome in either the amount or the interest when the rate is doubled? Explain.

31. Compute the total amount and the interest paid on the principals (a)–(c) at 12% compounded monthly for 2 years.
   a) $100 **b)** $200 **c)** $400
   d) Is there a predictable outcome in the interest when the principal is doubled? Explain.

32. Compute the total amount and the interest paid on $1000 at 6% compounded semiannually for
   a) 2 years. **b)** 4 years. **c)** 8 years.
   d) Is there a predictable outcome in the amount when the time is doubled? Explain.

33. *Determining Effective Yield* Determine the effective annual yield for $1 invested for 1 year at 7.5% compounded quarterly.

34. *Determining Effective Yield* Determine the effective annual yield for $1 invested for 1 year at 6.5% compounded quarterly.

35. *Comparing Investments* Madeline Bates has a choice of two accounts in which to invest. The first pays 7.00% simple interest rate and the second pays 6.77% interest compounded daily. Calculate the effective annual yield on the second account to determine which account has the better rate.

36. *Comparing Investments* Dave Dudley won a photography contest and received a $1000 cash prize. Will he earn more interest in one year if he invests his

winnings in a simple interest account that pays 5% or in an account that pays 4.75% interest compounded monthly?

37. *Investing for Retirement* The Pearsons are planning on retiring in 20 years and feel that they will need $200,000 in addition to income from their retirement plans. How much must they invest today at 7.5% compounded quarterly to accomplish their goal?

38. *Investment for a Newborn* How much money should parents invest at the birth of their child in order to provide their child with $50,000 at age 18? Assume that the money earns interest at 8% compounded quarterly.

39. *Future Value* How much money must the Nagrockis invest today to have $20,000 in 15 years? Assume that the money earns interest at 7% compounded monthly.

40. *Future Value* How much money must Harry Kim invest today to have $20,000 in 15 years? Assume that the money earns interest at 7% compounded quarterly.

CHALLENGE PROBLEMS/GROUP ACTIVITIES

41. *A Loaf of Bread* If the cost of a loaf of bread was $1.35 in 1999 and the annual average inflation rate is $3\frac{1}{2}$%, what will be the cost of a loaf of bread in 2009?

42. *Finding the Interest Rate* For a total accumulated amount of $3586.58, a principal of $2000, and a time period of 5 years, use the compound interest formula to find r if interest is compounded monthly.

43. *Rule of 72* A simple formula can help you estimate the number of years required to double your money. It's called the *Rule of 72.* You simply divide 72 by the interest rate (without the percent sign). For example, with an interest rate of 4%, your money would double in approximately 72 ÷ 4 or 18 years. In (a)–(d), determine the approximate number of years it will take for $1000 to double at the given interest rate.

a) 3% **b)** 6% **c)** 8% **d)** 12%

e) If $120 doubles in approximately 22 years, estimate the rate of interest.

44. *Finding the Interest Rate* Richard Maruszewski borrowed $2000 from Linda Tonolli. The terms of the loan are as follows: The period of the loan is 3 years, and the rate of interest is 8% compounded semiannually. What rate of simple interest would be equivalent to the rate Linda charged Richard?

45. *Investing with an Annuity* The question may be asked: If I deposit a fixed sum of money monthly, quarterly, semiannually, or annually at a fixed rate of interest, how much money will I have accumulated in x years? This is called an **annuity**. The formula for determining the value of the annuity is

$$S = \frac{R\left[\left(1 + \dfrac{r}{n}\right)^{nt} - 1\right]}{\dfrac{r}{n}}$$

where S is the value of the annuity in t years, R is the amount invested each period, r is the annual rate of interest, n is the number of payments per year, and t is the number of years. After Denisse Brown's birth, her parents invested $500 semiannually (every 6 months) at an annual rate of 5.5% compounded semiannually. What is the value of the annuity after 17 years?

46. *A Retirement Annuity* To supplement her retirement income, Chris Dunn is investing $50 each quarter at 8% compounded quarterly. How many dollars will she accumulate in 30 years? (See Exercise 45.)

47. *Investing in Annuities* Rodney is saving money to send his boys to college by investing in annuities. (See Exercise 45.) For his oldest son, Jacob, Rodney is investing $150 per month in an annuity that pays 5.6% interest compounded monthly. For his second oldest son, Justin, he is investing $900 twice a year in an annuity that pays 5.8% interest compounded semiannually.

a) If Rodney started Jacob's annuity when he was born, how much would the annuity be worth when Jacob turns 18 years old?

b) If Rodney started Justin's annuity when he was born, how much would the annuity be worth when Justin turns 18 years old?

48. *Changing a Payment Plan* Currently, the faculty of Manatee Community College gets paid in 12 equal payments—one payment in each of the months September through April and then four individual payments (all at one time) in May. The Faculty Senate is proposing an alternative plan that would pay the faculty in nine equal payments—one payment in each of the months September through May. Thus, the proposed plan would pay the faculty the same salary, but more would be paid in each of the 8 months from September through April and less would be paid in May (the faculty would receive only one check in May instead of four). Assume Louis Okonkwo's take-home pay for the school year is $26,600.

a) How much does Louis receive monthly from September through April under the current plan?

b) How much would Louis receive monthly from September through May under the proposed plan?

c) Assume Louis can invest the difference in monthly payments between the two plans (each month from

September through April) in a money market account paying an annual interest rate of 5.35% compounded monthly. If Louis left all the money in the account, how much would he earn in interest over the course of the school year? (*Hint:* Perform eight different computations—the computation from September will have $t = \frac{8}{12}$, the computation from October will have $t = \frac{7}{12}$, . . . the computation from April will have $t = \frac{1}{12}$. After each computation, subtract the principal to find the interest.)

RESEARCH ACTIVITIES

49. Imagine you have $4000 to invest and that you need this money to grow to $5000 by investing in a certificate of deposit (CD). Contact a local bank, a savings or loan, and a credit union to obtain CD information—the interest rate, the length of the term, and the number of times per year the CD is compounded. Find out how long it would take for you to reach your goal with the institution selected. Write a report summarizing your findings.

50. Write a paper on the history of simple interest and compound interest. Answer the questions: When was simple interest first charged on loans? When was compound interest first given on investments? (References include the Internet; encyclopedias; and National Council of Teachers of Mathematics, *Historical Topics For the Classroom,* Thirty-First Yearbook.)

11.4 INSTALLMENT BUYING

In Section 11.2, we discussed personal notes and discounted notes. When borrowing money by either of these methods, the borrower normally repays the loan as a single payment at the end of the specified time period. There may be circumstances under which it is more convenient for the borrower to repay the loan on a weekly or monthly basis or to use some other convenient time period. One method of doing so is to borrow money on the *installment plan.*

There are two types of installment loans, open-end and fixed payment. An **open-end installment loan** is a loan on which you can make variable payments each month. Credit cards, such as MasterCard, Visa, and Discover, are actually open-end installment loans, used to purchase items such as clothing, textbooks, and meals. A **fixed installment loan** is one on which you pay a fixed amount of money for a set number of payments. Examples of items purchased with fixed-payment installment loans are college tuition loans, and loans for cars, boats, appliances, and furniture. These loans are generally repaid in 24, 36, 48, or 60 equal monthly payments.

Lenders give any individual wishing to borrow money or purchase goods or services on the installment plan a credit rating to determine if the borrower is likely to repay the loan. The lending institution determines whether the applicant is a good "credit risk" by examining the individual's income, assets, liabilities, and history of repaying debts.

The advantage of installment buying is that the buyer has the use of an article while paying for it. If the article is essential, installment buying may serve a real need. A disadvantage is that some people buy more on the installment plan than they can afford. Another disadvantage is the interest the borrower pays for the loan. The method of determining the interest charged on an installment plan may vary with different lenders.

To provide the borrower with a way to compare interest charged, Congress passed the Truth in Lending Act in 1969. The law requires that the lending institution tell the borrower two things: the annual percentage rate and the finance charge. The **annual percentage rate (APR)** is the true rate of interest charged for the loan. The APR is calculated by using a complex formula, so we use tables to determine the APR. The technique of using the tables to find the APR is illustrated in Example 1. The total **finance charge** is the total amount of money the borrower must pay for its use. The finance charge includes the interest plus any additional fees charged. The additional fees may include service charges, credit investigation fees, mandatory insurance premiums, and so on.

The finance charge a consumer pays when purchasing goods or services on the installment plan is the difference between the total installment price and the cash price. The **total installment price** is the sum of all the monthly payments and the down payment, if any.

> **DID YOU KNOW**
>
> **Give Credit Where Credit Is Due**
>
> 
>
>
> The use of credit cards originated in the United States during the 1920s. The first credit cards were issued by individual firms such as oil companies, department stores, and hotels for use by only the firm that issued them.
>
> The first universal credit card, which could be used at a variety of establishments, was introduced by Diner's Club, Inc., in 1950. The first national plan was Bank-Americard, begun on a statewide basis by the Bank of America in California in 1959 and licensed in other states beginning in 1966. Cooperating merchants throughout the world pay a service charge of 4–7% of the total credit card billings to use a bank credit card system.

FIXED INSTALLMENT LOAN

In Example 1 we learn how to determine the finance charge and the monthly payment on a fixed installment loan.

EXAMPLE 1 *Financing Appliances*

Robyn wishes to buy a new washer and dryer for \$900. The store has an advertised finance option of no down payment and 11% APR for 24 months.

a) Determine the finance charge.
b) Determine the monthly payment.

SOLUTION

a) First look at Table 11.2 to see that the finance charge per \$100 of the amount financed for 24 months at 11% is \$11.86 (circled in red). Since Robyn is financing \$900, we need to multiply \$11.86 by $\frac{900}{100}$ or 9. This gives us

$$\begin{aligned}\text{Total finance charge} &= \$11.86 \times 9\\ &= \$106.74\end{aligned}$$

So Robyn will pay a total finance charge of \$106.74.

b) Next, we calculate the total installment price by adding the finance charge to the purchase price:

$$\begin{aligned}\text{Total installment price} &= \$900 + \$106.74\\ &= \$1006.74\end{aligned}$$

Table 11.2 Annual Percentage Rate Table for Monthly Payment Plans

Number of payments	7.00%	7.50%	8.00%	8.50%	9.00%	9.50%	10.00%	10.50%	11.00%	11.50%	12.00%	12.50%	13.00%
	(Finance charge per \$100 of Amount Financed)												
6	2.05	2.20	2.35	2.49	2.64	2.79	2.93	3.08	3.23	3.38	3.53	3.68	3.83
12	3.83	4.11	4.39	4.66	4.94	5.22	5.50	5.78	6.06	6.34	6.62	6.90	7.18
18	5.63	6.04	6.45	6.86	7.28	7.69	8.10	8.52	8.93	9.35	9.77	10.19	10.61
24	7.45	8.00	8.54	9.09	9.64	10.19	10.75	11.30	11.86	12.42	12.98	13.54	14.10
30	9.30	9.98	10.66	11.35	12.04	12.74	13.43	14.13	14.83	15.54	16.24	16.95	17.66
36	11.16	11.98	12.81	13.64	14.48	15.32	16.16	17.01	17.86	18.71	19.57	20.43	21.30
48	14.94	16.06	17.18	18.31	19.45	20.59	21.74	22.90	24.06	25.23	26.40	27.58	28.77
60	18.81	20.23	21.66	23.10	24.55	26.01	27.48	28.96	30.45	31.96	33.47	34.99	36.52

To determine the monthly payment, we divide the total installment price by the number of payments:

$$\text{Monthly payment} = \frac{\$1006.74}{24}$$
$$= \$41.95$$

Robyn will have 24 monthly payments of $41.95. ▲

Our next example shows how to determine the annual percentage rate on a loan that has a schedule for repaying a fixed amount each period.

EXAMPLE 2 *Determining the APR*

Jim and Bonnie Parker are purchasing a new dining room set for $3600, including taxes. They decide to make a $600 down payment and finance the balance, $3000, through their credit union. The credit union loan officer informs them that their payment will be $101 per month for 36 months.

a) Determine the finance charge. b) Determine the APR.

SOLUTION

a) The total installment price is the down payment plus the total monthly installment payments.

$$\text{Total installment price} = 600 + (36 \times 101)$$
$$= 600 + 3636 = \$4236$$

The finance charge is the total installment price minus the cash price.

$$\text{Finance charge} = 4236 - 3600 = \$636$$

b) To determine the annual percentage rate, use Table 11.2 as follows: First divide the finance charge by the amount financed and multiply the quotient by 100. The result is the finance charge per $100 of the amount financed.

$$\frac{\text{Finance charge}}{\text{Amount financed}} \times 100 = \frac{636}{3000} \times 100 = 0.212 \times 100 = 21.2$$

Thus, the borrower pays $21.20 interest for each $100 being financed. To use Table 11.2, look for 36 in the left column under the heading Number of Payments. Then move across to the right until you find the value closest to $21.20. The value closest to $21.20 is $21.30 (circled in blue). At the top of this column is the value 13.00%. Therefore, the annual percentage rate is approximately 13.00%. ▲

Much more complete APR tables similar to Table 11.2 are available at your local bank.

EXAMPLE 3 *Financing a Restored Car*

Tino Garcia borrowed $9800 in 1999 to purchase a classic 1965 Ford Mustang. He does not recall the APR of the loan but remembers that there are 48 payments of $255.75. If he did not make a down payment on the car, determine the APR.

DID YOU KNOW

Rent-to-Own Transactions

The rent-to-own industry has mushroomed in recent years to a $4.4 billion per year industry. Rent-to-own retailers allow you to rent furniture, appliances, electronics and the like over a period of time and, at a certain point, obtain ownership of the item. Is such a transaction an example of installment buying? It depends. If the consumer returns the item to the retailer prior to obtaining ownership, then the transaction is truly a rental agreement. If, however, the consumer obtains ownership, then the item was actually purchased on an installment plan. Since retailers classify the transaction as a rental agreement, they are not always subject to state mandated limits on the interest they can charge. The result is that often items are purchased at rates higher than even credit cards are allowed to charge. This controversy has come to the attention of lawmakers, and several states have enacted legislation controlling the amount of interest retailers may charge on rent-to-own transactions.

SOLUTION First determine the finance charge by subtracting the cash price from the total amount paid.

$$\begin{aligned}\text{Finance charge} &= (255.75 \times 48) - 9800\\ &= 12{,}276 - 9800\\ &= \$2476\end{aligned}$$

Next divide the finance charge by the amount of the loan and multiply this quotient by 100.

$$\frac{2476}{9800} \times 100 \approx 25.27$$

Next find 48 payments in the left column of Table 11.2. Move to the right until you find the value that is closest to 25.27. The value closest to 25.27 is 25.23 (circled in green). At the top of the column is the APR of 11.50%. ▲

In Example 3 Tino agreed to pay a finance charge of \$2476. If he decides to repay the loan after making 30 payments, must he pay the total finance charge? The answer is no. Two methods are used to determine the finance charge when you repay an installment loan early: the **actuarial method** and the **rule of 78s**. The actuarial method uses the APR tables, whereas the more commonly used method, the rule of 78s, does not.

Actuarial method

$$u = \frac{n \cdot P \cdot V}{100 + V}$$

u = unearned interest

n = number of remaining monthly payments (excluding current payment)

P = monthly payment

V = the value from the APR table that corresponds to the annual percentage rate for the number of remaining payments (excluding current payment)

Rule of 78s

$$u = \frac{f \cdot k(k + 1)}{n(n + 1)}$$

u = unearned interest

f = original finance charge

k = number of remaining monthly payments (excluding current payment)

n = original number of payments

These two formulas determine the unearned interest, u, which is the interest saved by paying off the loan early.

Example 4 illustrates the actuarial method and Example 5 illustrates the rule of 78s.

EXAMPLE 4 *Using the Actuarial Method*

In Example 3, we determined the APR of Tino's loan to be 11.50%. Instead of making his 30th payment of his 48 payment loan, Tino wishes to pay his remaining balance and terminate the loan.

a) Use the actuarial method to determine how much interest Tino will save (the unearned interest, u) by repaying the loan early.
b) What is the total amount due on that day?

SOLUTION

a) After 30 payments have been made, 18 payments remain. Thus, $n = 18$ and $P = \$255.75$. To determine V, use the APR table (Table 11.2). In the Number of Payments column find the number of remaining payments, 18, and then look to the right until you reach the column headed by 11.50%, the APR. This row and column intersect at 9.35. Thus, $V = 9.35$.

$$\begin{aligned} u &= \frac{n \cdot P \cdot V}{100 + V} \\ &= \frac{(18)(255.75)(9.35)}{100 + 9.35} \\ &\approx 393.62 \end{aligned}$$

Tino will save \$393.62 in interest by the actuarial method.

b) Since the remaining payments total 18(\$255.75) = \$4603.50, Tino's remaining balance is

\$4603.50	Total of remaining payments (which includes interest)
− 393.62	Interest saved (unearned interest)
\$4209.88	Balance due

A payment of \$4209.88 plus the 30th monthly payment of \$255.75 will terminate Tino's installment loan. The total amount due is \$4465.63. ▲

EXAMPLE 5 *Using the Rule of 78s*

In Example 3, we determine the APR of Tino's loan to be 11.50%. Instead of making his 30th payment of his 48 payment loan, Tino decides to pay his remaining balance and terminate the loan.

a) Use the rule of 78s to determine how much interest Tino will save by repaying the loan early.
b) What is the total amount due on that day?

SOLUTION

a) After the 30 payments have been made, 18 remain. Thus $k = 18$, $n = 48$. From Example 3 we know the finance charge, $f = \$2476$.

$$\begin{aligned} u &= \frac{f \cdot k(k + 1)}{n(n + 1)} \\ &= \frac{2476 \cdot 18(18 + 1)}{48(48 + 1)} \\ &= \$360.03 \end{aligned}$$

Tino saves \$360.03 in interest by the rule of 78s.

b) Tino's balance is computed in a manner similar to the method in Example 4.

\$4603.50	Total of remaining payments
− 360.03	Interest saved
\$4243.47	Balance due

A payment of \$4243.47 plus a regular monthly payment of \$255.75 will terminate Tino's installment loan. The total amount due is \$4499.22. ▲

Most states use the rule of 78s. However, there is some discussion that the rule of 78s is outdated and is not fair to the borrower.

OPEN-END INSTALLMENT LOAN

A credit card is a popular way of making purchases or borrowing money. Use of a credit card is an example of an open-end installment loan. A typical charge account with a bank may have the terms in Table 11.3.

Typically, credit card monthly statements contain the following information: balance at the beginning of the period, balance at the end of the period (or new balance), the transactions for the period, statement closing date (or billing date), payment due date, and the minimum payment due. For *purchases* there is no finance or interest charge if there is no previous balance due and you pay the entire new balance by the payment due date. However, if you borrow money (*cash advances*) through this account, a finance charge is applied from the date you borrowed the money until the date you repay the money. When you make purchases or borrow money, the minimum monthly payment is sometimes determined by dividing the balance due by 36 and rounding the answer up to the nearest whole dollar, thus ensuring repayment in 36 months. However, if the balance due for any month is less than $360, the minimum monthly payment is typically $10. These general guidelines may vary by bank and store.

Table 11.3 Credit Card Terms

Type of charge	Daily periodic rate*	Annual percentage rate*
Purchases	0.04655%	16.99%
Cash advances	0.05751%	20.99%

**These rates vary with different charge accounts and localities.*

EXAMPLE 6 *Holiday Shopping Charges*

While doing her holiday shopping, Jan Reckard charged the following items to her Discover card: a set of tools for her husband ($250), an original print from a local artist for her daughter ($155), a set of classical music CDs for her best friend ($100), and a gift certificate to a bookstore for her boss ($15). Her balance on December 1 is $250 + $155 + $100+ $15, or $520. The bank requires repayment within 36 months and charges an interest rate of 0.04655% per day.

a) Determine the minimum payment due on December 1.
b) On December 1, Jan makes a payment of $100. Determine the balance due on January 1, assuming that there are no additional charges or cash advances.

SOLUTION

a) To determine the minimum payment, we divide the balance due on December 1 by 36. $520 ÷ 36 = $14.44. Rounding up to the nearest dollar, we determine that the minimum payment due on December 1 is $15.
b) With no additional purchases or cash advances and paying $100 on December 1, the balance on January 1 is $520 − $100 or $420. In addition, she must pay interest on the outstanding balance of $420 for the month of December. December has 31 days. The interest is

$$\begin{aligned} i &= prt \\ &= \$420(0.0004655)(31) \\ &= \$6.06 \end{aligned}$$

Therefore, the balance due on January 1 is $420 + $6.06, or $426.06 ▲

DID YOU KNOW

"Plastic"

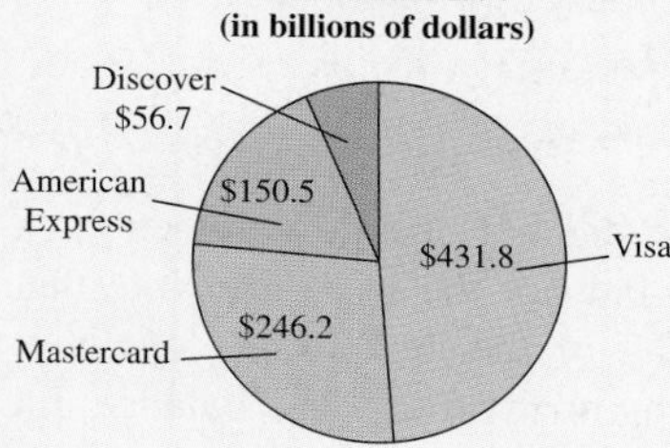

In 1997, Americans used their credit cards or "plastic" to charge $885.2 billion at a rate of about $1.7 million per minute. The four most commonly used cards are Visa, MasterCard, American Express, and Discover. Unlike American Express and Discover, each Visa and MasterCard is associated with one of thousands of different lending institutions such as banks, savings and loans, or credit unions. The institutions that issue the cards are members of the Visa or MasterCard consortium, a system of institutions that use the name and logo and share the electronic database. The individual institutions are responsible for the money charged and reap the benefit of the interest applied. In 1997, the largest Visa and MasterCard issuers in the United States were Citibank, MBNA America, Bank One/First USA, and Chase Manhattan.
(*Source:* CardWeb Inc.'s CardData, web site.)

In Example 6, there were no additional transactions in the account for the period. When additional charges are made during the period, the finance charges on open-end installment loans or credit cards are generally calculated in one of two ways: the *unpaid balance method* or the *average daily balance method.*

With the **unpaid balance method**, the borrower is charged interest or a finance charge on the unpaid balance from the previous charge period. Example 7 illustrates the unpaid balance method.

EXAMPLE 7 *Finance Charges Using the Unpaid Balance Method*

Jim Carlson charged all of the supplies for his Halloween party to his Visa card. On November 5, the billing date, Jim had a balance due of $275. From November 5 through December 4, he did some shopping and charged items totaling $320 and he also made a payment of $145.

a) Find the finance charge due on December 5, using the unpaid balance method. Assume the interest rate is 1.3% per month.
b) Find the new account balance on December 5.

SOLUTION

a) The finance charge is based on the $275 balance due on November 5. To find the finance charge due on December 5, we used the simple interest formula with a time of 1 month.

$$i = \$275 \times 0.013 \times 1 \approx \$3.58$$

The finance charge on December 5 is $3.58.

b) The balance due on December 5 is

$$\$275 + \$320 + \$3.58 - \$145 = \$453.58$$

The balance due on December 5 is $453.58. The finance charge on January 5 is based on $453.58.

Many lending institutions use the **average daily balance method** of calculating the finance charge because they believe it is fairer to the customer.

With the average daily balance method, a balance is determined each day of the billing period for which there is a transaction in the account. The average daily balance method is illustrated in Example 8.

EXAMPLE 8 *Finance Charges Using the Average Daily Balance Method*

The balance on the Margolius's credit card account on July 1, their billing date, was $375.80. They had the following transactions during the month of July.

July 5	Payment	$150.00
July 10	Charge: Toy store	74.35
July 18	Charge: Garage	123.50
July 28	Charge: Restaurant	42.50

a) Find the average daily balance for the billing period.
b) Find the finance charge to be paid on August 1. Assume the interest rate is 1.3% per month.
c) Find the balance due on August 1.

DID YOU KNOW

No More PINs

An emeging technology may make *p*ersonal *i*dentification *n*umbers (PINs) at automatic teller machines (ATMs) a thing of the past. Iris-based identification systems use a series of cameras to identify people by the characteristics of the irises in their eyes. No two irises are identical—not even an individual's left and right irises. A representation of one of your irises would be obtained and stored electronically when you open an account. Each time you used your ATM card, instead of entering your PIN, the system would scan your eye and compare this to the representation on file. Similar systems have been successfully used at highly classified military installations throughout the United States. The system has been successfully tested at ATMs in London and may soon be tested in the United States. The technology also has huge potential to reduce credit card, debit card, and telephone calling card fraud.

(*Source: Technology Review* March/April 1999.)

SOLUTION

a) To determine the average daily balance, we do the following: (i) Find the balance due for each transaction date.

July 1	\$375.80
July 5	\$375.80 − \$150 = \$225.80
July 10	\$225.80 + \$74.35 = \$300.15
July 18	\$300.15 + \$123.50 = \$423.65
July 28	\$423.65 + \$42.50 = \$466.15

(ii) Find the number of days that the balance did not change between each transaction. Count the first day in the period but not the last day. Note that from July 28 through August 1, the beginning of the next billing cycle, is 4 days. (iii) Multiply the balance due by the number of days the balance did not change. (iv) Find the sum of the products.

Date	(i) Balance due	(ii) Number of days balance did not change	(iii) (Balance)(Days)
July 1	\$375.80	4	(\$375.80)(4) = \$ 1503.20
July 5	\$225.80	5	(\$225.80)(5) = \$ 1129.00
July 10	\$300.15	8	(\$300.15)(8) = \$ 2401.20
July 18	\$423.65	10	(\$423.65)(10) = \$ 4236.50
July 28	\$466.25	4	(\$466.15)(4) = \$ 1864.60
		31	(iv) Sum = \$11,134.50

(v) Divide this sum by the number of days in the billing cycle (in the month). The number of days may be found by summing the days in column ii.

$$\frac{\$11{,}134.50}{31} = \$359.18$$

Thus, the average daily balance is \$359.18.

b) The finance charge for the month is found using the simple interest formula with the average daily balance as the principal.

$$i = \$359.18 \times 0.013 \times 1 = \$4.67$$

c) Since the finance charge for the month is \$4.67, the balance owed on August 1 is \$466.15 + \$4.67 or \$470.82.

The calculations in Example 8 are tedious. However, these calculations are made almost instantaneously with computers.

Example 9 illustrates how a credit card is used to borrow money.

EXAMPLE 9 *Using a Credit Card for a Cash Advance*

To obtain money to buy a stereo system at a garage sale, Bobby Bueker obtained a cash advance of \$1500 from his credit card. He borrowed the money on July 10 and repaid it on July 31. If Bobby is charged an interest rate of 0.05751% per day, how much did Bobby pay the credit card company on July 31?

SOLUTION The amount Bobby pays is the original principal plus any accrued interest. Interest on cash advances is generally calculated for the exact number of days of the loan, starting with the day the money is obtained. The time of the loan in this case is 21 days. Using the simple interest formula we get the following:

$$\begin{aligned} i &= prt \\ &= \$1500 \times 0.0005751 \times 21 \\ &= \$18.12 \end{aligned}$$

Therefore, Bobby must repay the credit card company \$1500 + \$18.12 or \$1518.12.

Anyone purchasing a car or other costly items should consider a number of different sources for a loan. Example 10 illustrates one method of making a comparison.

EXAMPLE 10 *Comparing Loan Sources*

Franz Helfenstein purchased carpeting costing \$2400 with his credit card. When the bill comes due on February 1, Franz realizes he can pay \$350 per month until the debt is paid off. His credit card charges 1.5% interest per month.

a) Assuming Franz makes no other purchases with this credit card, how many payments are necessary to retire this debt?
b) What is the total interest Franz will pay?
c) How much money could Franz have saved by obtaining a fixed installment loan with 12% interest and 6 equal monthly payments?

SOLUTION

a) Franz would make his first monthly payment of \$350 on February 1, resulting in a new balance of \$2400 − \$350 = \$2050. His next bill reflects the \$2050 balance plus the monthly interest. He continues to make payments until the debt is retired. For each date indicated, the amount on the far right represents the amount due on that date.

February 1	\$2400 − \$350 = \$2050
March 1	\$2050 + 0.015(\$2050) = \$2080.75; \$2080.75 − \$350 = \$1730.75
April 1	\$1730.75 + 0.015(\$1730.75) = \$1756.71; \$1756.71 − \$350 = \$1406.71
May 1	\$1406.71 + 0.015(\$1406.71) = \$1427.81; \$1427.81 − \$350 = \$1077.81
June 1	\$1077.81 + 0.015(\$1077.81) = \$1093.98; \$1093.98 − \$350 = \$743.98
July 1	\$743.98 + 0.015(\$743.98) = \$755.14; \$755.14 − \$350 = \$405.14
August 1	\$405.14 + 0.015(\$405.14) = \$411.22; \$411.22 − \$350 = \$61.22
September 1	\$61.22 + 0.015(\$61.22) = \$62.14

After eight payments—seven for \$350 and one for \$62.14—Franz has paid off his carpeting.

b) To calculate the total interest paid, we add up all his payments and then subtract the cost of the carpeting:

$$\begin{aligned} \text{Total of all payments} &= 7(\$350) + \$62.14 \\ &= \$2512.14 \\ \text{Total interest} &= \$2512.14 - \$2400 \\ &= \$112.14 \end{aligned}$$

c) To determine the interest that Franz pays, we will use Table 11.2 on page 546. From Table 11.2, we see that a fixed installment loan of 12% for 6 months requires \$3.53 per \$100 of the amount financed. So the interest, or finance charge, is

$$\begin{aligned}\text{Finance charge} &= 3.53\left(\frac{2400}{100}\right)\\ &= 3.53(24)\\ &= \$84.72\end{aligned}$$

Therefore, Franz would save \$112.14 − \$84.72 = \$27.42 in interest by using an installment loan instead of a credit card.

SECTION 11.4 EXERCISES

CONCEPT/WRITING EXERCISES

1. Explain how an installment plan differs from a personal note.
2. Explain in your own words the difference between an open-end installment loan and a fixed installment loan.
3. What is an annual percentage rate (APR)?
4. What is a finance charge?
5. What is a total installment price?
6. Name the two methods used to determine the finance charge when a loan is repaid early.
7. Name the two methods used to determine the finance charge on an open-end installment loan.
8. What is a cash advance?

PRACTICE THE SKILLS/PROBLEM SOLVING

9. *Financing a New Tractor* Stephen Plett purchased a new tractor for \$36,000. He paid 20% as a down payment and financed the balance of the purchase with a 60-month fixed installment loan with an APR of 11%.
   a) Determine Stephen's total finance charge.
   b) Determine Stephen's monthly payment.
10. *Remodeling a Living Room* Oda Lisa Hernandez received a bid of \$4200 to remodel her living room. To finance this amount, her savings and loan requires her to pay 15% down, with the balance being financed with a 24-month installment loan with an APR of 10.5%.
   a) Determine Oda's total finance charge.
   b) Determine Oda's monthly payment.
11. *Financing Student Loans* Joni Gile has a total of \$4000 in student loans that will be paid with a 60-month installment loan with an APR of 7.5%.
   a) Determine Joni's total finance charge.
   b) Determine Joni's monthly payment.
12. *Financing Eye Surgery* Tiger Lucas is planning to have laser eye surgery. The total cost of \$4200 will be paid with a 48-month installment loan with an APR of 9.5%.
   a) Determine Tiger's total finance charge.
   b) Determine Tiger's monthly payment.
13. *Financing a New Business* Cheryl Sisson is a hair designer and wishes to convert her garage into a hair salon to use for her own business. The entire project would have a cash price of \$3200. She decides to finance the project by paying 20% down, with the balance paid in 60 monthly payments of \$53.14.
   a) What finance charge will Cheryl pay?
   b) What is the APR to the nearest half a percent?
14. *Financing a Computer* Gilberto Garza purchased a new laptop computer on a monthly purchase plan. The computer sold for \$2350. He paid \$500 down and \$85.79 a month for 24 months.
   a) What finance charge did Gilberto pay?
   b) What is the APR to the nearest half a percent?
15. *Financing a Canoe* Bob Angel bought a new canoe that had a cash price of \$675. The installment terms include a down payment of \$175 and 12 monthly payments of \$44.66.
   a) What finance charge will Harry pay?
   b) What is the APR to the nearest half a percent?

16. *Financing Furniture* Mr. and Mrs. Chan want to buy furniture that has a cash price of \$3450. On the installment plan they must pay one-third of the cash price as a down payment and make six monthly payments of \$398.
   a) What finance charge will the Chans pay?
   b) What is the APR to the nearest half percent?

17. *Early Repayment Using the Actuarial Method* Ray Flagg took out a 60-month fixed installment loan of $12,000 to open a new pet store. He began making monthly payments of $260.90. Ray's business does better than expected and instead of making his 24th payment, Ray wishes to repay his loan in full.
    a) Determine the APR of the loan.
    b) How much interest will Ray save (use the actuarial method)?
    c) What is the total amount due to pay off the balance?

18. *Early Repayment Using the Actuarial Method* Tina has a 48-month installment loan, with a fixed monthly payment of $193.75. The amount borrowed was $7500. Instead of making her 18th payment, Tina is paying the remaining balance on the loan.
    a) Determine the APR of the installment loan.
    b) How much interest will Tina save (use the actuarial method)?
    c) What is the total amount due to pay off the balance?

19. *Early Repayment Using the Actuarial Method* Catherine Marie buys a new sport utility vehicle for $32,000. She trades in her old truck and receives $10,000, which she uses as a down payment. She finances the balance at 12% APR over 36 months. Before making her 24th payment, she decides to pay off the balance.
    a) Use Table 11.2 to determine the total interest Catherine would pay if all 36 payments were made.
    b) What were Catherine's monthly payments?
    c) How much interest will Catherine save (use the actuarial method)?
    d) What is the total amount due to pay off the balance?

20. *Early Repayment Using the Actuarial Method* The cash price for furniture for Melissa's apartment was $6520. She made a down payment of $3962 and financed the balance on a 24-month fixed payment installment loan. The monthly payments are $119.82. Instead of making her 12th payment, Melissa decides to pay the remaining balance.
    a) Determine the APR on the installment loan.
    b) How much interest will Melissa save (use the actuarial method)?
    c) What is the total amount due to pay off the balance?

21. *Early Repayment Using the Rule of 78s* Rod wishes to purchase new equipment for his landscaping business worth $7345, including taxes. Rod is able to secure a no-money-down, 48-month, 12.5% APR fixed installment loan from his local bank. Before making his 12th payment, Rod decides to pay off the balance with some of his yearly profits.
    a) What was Rod's original finance charge?
    b) What was Rod's original monthly payment?
    c) How much interest will Rod save (use the rule of 78s)?
    d) What is the total amount due to pay off the balance?

22. *Early Repayment Using the Rule of 78s* To pay for remodeling their kitchen, the Leesebergs obtained a no-money down, 36-month, 8.5% fixed installment loan for the amount of $3600. Before making their 12th payment, the Leesebergs decide to pay off the balance.
    a) What was the Leesebergs' original finance charge?
    b) What was the Leesebergs' original monthly payment?
    c) How much interest will the Leesebergs save (use the rule of 78s)?
    d) What is the total amount due to pay off the balance?

23. *Early Repayment Using the Rule of 78s* Tony Gambino is buying a $3000 sound system by making a down payment of $500 and 18 monthly payments of $151.39. Instead of making his 12th payment, Tony decides to pay the remaining balance.
    a) How much interest will Tony save (use the rule of 78s)?
    b) What is the total amount due to pay off the balance?

24. *Early Repayment Using the Rule of 78s* Roger Beaman purchased woodworking tools for $2375. He made a down payment of $850 and financed the balance with a 12-month fixed payment installment loan. Instead of making his sixth monthly payment of 134.71, he decides to pay the remaining balance.
    a) How much interest will Roger save (use the rule of 78s)?
    b) What is the total amount due to pay off the balance?

25. *Travel Expenses* To pay for his trip to Portland, Oregon for a teaching conference in November, David Dean charged the following expenses to his credit card: airfare ($365), hotel ($180), conference fee ($195), and meals ($84). David had a previous balance of zero, he bought no other items with this credit card, and on December 1 made a payment of $200. The bank that issued the card requires repayment within 48 months and charges an interest rate of 1.1% per month.
    a) What is the minimum payment due on December 1?
    b) What is the balance due on January 1?

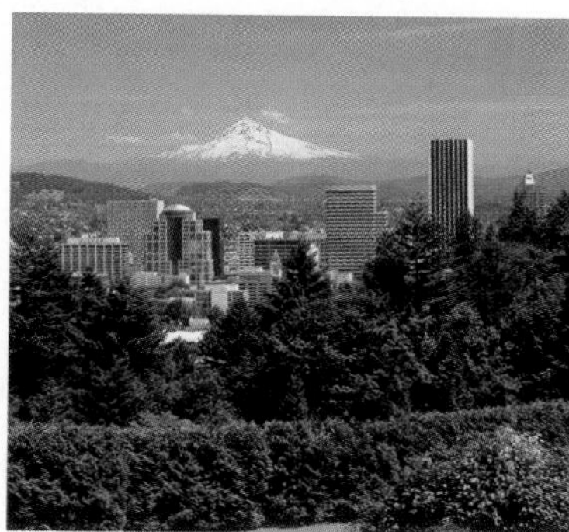

26. *College Expenses* Brian Hickey uses his credit card in August to purchase the following college supplies: books ($425), yearlong bus pass ($175), food service meal ticket ($450), and season tickets to the basketball games ($125). On September 1, he used $650 of his financial aid check to reduce the balance. The issuing bank charges 1.2% interest per month and requires full payment within 36 months. Brian had a previous balance of zero and he makes no other purchases with this card.
   a) What is the minimum payment due September 1?
   b) What is the balance due on October 1?

27. *Business Expenses* In February, Denny Droessler used his credit card to pay for the following business expenses: truck repair ($423), lunch for himself and three of his clients ($36), laundering of business uniforms ($145), and maintenance of equipment ($491). On March 1, he received payment from a client in the amount of $548 which he used as payment on his credit card. The issuing credit union charges 1.1% interest per month and requires full payment within 36 months. Denny had a previous balance of zero and he makes no other purchases with this card.
   a) What is the minimum payment due March 1?
   b) What is the balance due on April 1?

28. *Vacation Expenses* In June, while on vacation, the Greenbergs charged the following expenses to their credit card: airfare ($512), car rental ($172), meals ($190), and hotel ($350). On July 1, the Greenbergs paid $500 to reduce the balance. The issuing savings and loan charges 1.3% interest per month and requires full payment within 48 months. The Greenbergs make no other purchases with this card.
   a) What is the minimum payment due on July 1?
   b) What is the balance due on August 1?

29. *Unpaid Balance Method* On the April 5 billing date, Michaelle Chappell had a balance due of $1097.86 on her credit card. From April 5 through May 4, Michaelle charged an additional $425.79 and makes a payment of $800.
   a) Find the finance charge on May 5, using the unpaid balance method. Assume that the interest rate is 1.8% per month.
   b) Find the new balance on May 5.

30. *Unpaid Balance Method* On September 5, the billing date, Verna had a balance due of $385.75 on her credit card. The transactions during the following month were

September 8	Payment	$275.00
September 21	Charge: Airline ticket	330.00
September 27	Charge: Hotel bill	190.80
October 2	Charge: Clothing	84.75

   a) Find the finance charge on October 5, using the unpaid balance method. Assume that the interest rate is 1.40% per month.
   b) Find the new balance on October 5.

31. *Unpaid Balance Method* On February 3, the billing date, Carol Ann Bluesky had a balance due of $124.78 on her credit card. Her bank charges an interest rate of 1.25% per month. She made the following transactions during the month:

February 8	Charge: Art supplies	$ 25.64
February 12	Payment	100.00
February 14	Charge: Flowers delivered	67.23
February 25	Charge: Music CD	13.90

   a) Find the finance charge on March 3, using the unpaid balance method.
   b) Find the new balance on March 3.

32. *Unpaid Balance Method* On April 15, the billing date, Gabrielle Michaelis had a balance due of $57.88 on her credit card. She is redecorating her apartment and has the following transactions.

April 16	Charge: Paint	$ 64.75
April 20	Payment	45.00
May 3	Charge: Curtains	72.85
May 10	Charge: Chair	135.50

   a) Find the finance charge on May 15, using the unpaid balance method. Assume that the interest rate is 1.35% per month.
   b) Find the new balance on May 15.

33. *Average Daily Balance Method* The balance on the Razazada's credit card on May 12, their billing date, was $378.50. For the period ending June 12, they had the following transactions.

May 13	Charge: Toys	$129.79
May 15	Payment	50.00
June 1	Charge: Clothing	135.85
June 8	Charge: Housewares	37.63

   a) Find the average daily balance for the billing period.
   b) Find the finance charge to be paid on June 12. Assume an interest rate of 1.3% per month.
   c) Find the balance due on June 12.

34. *Average Daily Balance Method* The Levy's credit card statement shows a balance due of $1578.25 on March 23, the billing date. For the period ending April 23, they had the following transactions.

March 26	Charge: Party supplies	$ 79.98
March 30	Charge: Restaurant meal	52.76
April 3	Payment	250.00
April 15	Charge: Clothing	190.52
April 22	Charge: Car repairs	190.85

   a) Find the average daily balance for the billing period.
   b) Find the finance charge to be paid on April 23. Assume an interest rate of 1.3% per month.
   c) Find the balance due on April 23.

35. *Average Daily Balance Method* Refer to Exercise 31. Instead of the unpaid balance method, suppose Carol Ann's bank uses the average daily balance method.
   a) Find Carol Ann's average daily balance for the billing period from February 3 to March 3.
   b) Find the finance charge to be paid on March 3.
   c) Find the balance due on March 3.
   d) Compare these answers to those in Exercise 31.

36. *Average Daily Balance Method* Refer to Exercise 30. Instead of the unpaid balance method, suppose Verna's bank uses the average daily balance method.
   a) Find Verna's average daily balance for the billing period from September 5 to October 5.
   b) Find the finance charge to be paid on October 5.
   c) Find the balance due on October 5.
   d) Compare these answers to those in Exercise 30.

37. *A Cash Advance* Jeff Cole uses his credit card to obtain a cash advance of $1500 for a family emergency. The interest rate charged for the loan is 0.05751% per day and Jeff repays the money plus interest after 24 days.
   a) Determine the interest charged for the cash advance.
   b) What amount did he pay the bank when he repaid the loan?

38. *A Cash Advance* John Richards borrowed $875 against his charge account on September 12 and repaid the loan on October 14 (32 days later). Assume that the interest rate is 0.04273% per day.
   a) How much interest did John pay on the loan?
   b) What amount did he pay the bank when he repaid the loan?

39. *Comparing Loan Options* Penny needs $250. She finds that the ABC Loan Company charges 7.4% simple interest on the amount borrowed for the duration of the loan and requires the loan to be repaid in 6 equal monthly payments. The XYZ Loan Company offers loans of $250 to be repaid in 12 monthly payments of $22.19.
   a) How much interest is charged by the ABC Loan Company?
   b) How much interest is charged by the XYZ Loan Company?
   c) What is the APR, to the nearest half percent, on the ABC Loan Company loan?
   d) What is the APR, to the nearest half percent, on the XYZ Loan Company loan?

40. *Comparing Loan Options* Sara wants to purchase a new television set. The purchase price is $890. If she purchases the set today and pays cash, she must take money out of her savings account. Another option is to charge the TV on her credit card, take the set home today, and pay next month. Next month she will have cash and can pay her credit card balance without paying any interest. The simple interest rate on her savings account is $5\frac{1}{4}$%. How much is she saving by using the credit card instead of taking the money out of her savings account?

CHALLENGE PROBLEMS/GROUP ACTIVITY

41. *Comparing Loans* Suppose the Chans in Exercise 16 use a credit card rather than the installment plan. Assume that they make the same down payment, have no finance charge the first month, make no additional purchases on their credit card, and pay $384 per month plus the finance charge starting with the second month. If the interest rate is 1.3% per month
   a) How many months will it take them to repay the loan?
   b) How much interest will they pay on the loan?
   c) Which method of borrowing will cost the Chans the least amount of interest, the installment loan in Exercise 16 or the credit card?

42. *Determining Purchase Price* Ken bought a new car, but now he cannot remember the original purchase price. His payments are $379.50 per month for 36 months. He remembers that the salesperson said the simple interest rate for the period of the loan was 6%. He also recalls he was allowed $2500 on his old car. Find the original purchase price.

43. *Borrowing Money Interest Free* Martina wants to buy a camera in time for a June 30 family gathering. She knows that she will not have the money to pay for the camera until August 5. The billing date on her credit card is the 25th of the month and she has a 20-day grace period from the billing date to pay the bill with no finance charge. Explain how she can buy the camera before June 30, pay for it after August 5, and pay no interest.

RESEARCH ACTIVITIES

44. Write a brief report giving the advantages and disadvantages of leasing a car. Determine all the individual costs involved with leasing a car. Indicate why you would prefer to lease or purchase a car at the present time.

45. Assume that you are married and have a child. You don't own a washer and dryer and have no money to buy the appliances. Would it be cheaper to borrow money on an installment loan and buy the appliances or to continue

going to the local coin-operated laundry for 5 years until you have saved enough to pay cash for a washer and dryer?

With the aid of parents or friends, establish how many loads of laundry you would be doing each week. Then determine the cost of doing that number of loads at a coin-operated laundry. (Don't forget the cost of transportation to the laundry.) Shop around for a washer and dryer, and determine the total cost on an installment plan. Don't forget to include the cost of gas, electricity, and water. This information can be obtained from a local gas and electric company. With this information, you should be able to make a decision about whether to buy now or wait for 5 years.

46. Do research on the features offered by MasterCard, Visa, Discover and American Express. Include discussions of regular, gold, and platinum cards if they exist. For example, features might include life insurance when traveling by a common carrier, miles toward air travel, discounts on automobiles, cash back at the end of the year, insurance on rental cars, and the like. Determine which card or cards have the most appropriate features for you and explain why you arrived at that conclusion.

47. By reading a financial newspaper or magazine, determine the five specific credit cards that offer the lowest interest rates. List each card and its interest rate and annual fee. Indicate where you obtained this information.

11.5 BUYING A HOUSE WITH A MORTGAGE

Buying a house is the largest purchase of a lifetime for most people. The purchaser will normally be committed to 10, 15, 20, 25, or 30 years of mortgage payments. Before selecting the "dream house," the buyer should consider the following questions: "Can I afford it?" and "Does it suit my needs?"

The question "Can I afford the house?" must be answered carefully and accurately. If a family buys a house beyond its means, it will have a difficult time living within its income. When deciding whether to purchase a particular house, the purchaser must also consider crucial questions such as: "Do I have enough cash for the down payment?" "Can I afford the monthly payments with my current income?" These two items, down payment and mortgage payments over time, constitute the buyer's total cost of buying a house.

Buyers usually seek a *mortgage* from a bank or other lending institution. Before approving a mortgage, which is a long-term loan, the bank will require the buyer to have a specified minimum amount for the down payment. The **down payment** is the amount of cash the buyer must pay to the seller before the lending institution will grant the buyer a mortgage. If the buyer has the down payment and meets the other criteria for the mortgage, the lending institution prepares a written agreement called the mortgage, stating the terms of the loan. The loan specifies the repayment schedule, the duration of the loan, whether the loan can be assumed by another party, and the penalty if payments are late. The party borrowing the money accepts the terms of this agreement and gives the lending institution the title or deed to the property as security.

DID YOU KNOW

Sears Sold Houses

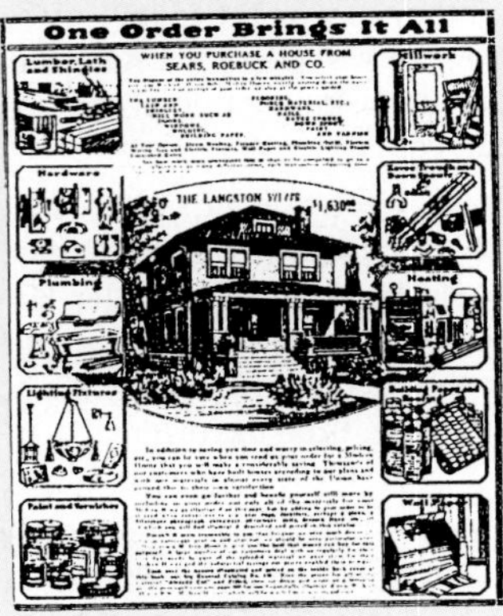

Courtesy of Sears, Roebuck Co.

Sears, Roebuck started its mail-order business in the late 1880s. Its catalog, or "Consumer's Guide," was one of the most eagerly awaited publications of the time. You could order just about anything from the 1908 "Wish Book," including a line of "modern homes." The Langston, a modest two-story home, was offered for \$1630. The houses were shipped in pieces by rail, assembly required.

Homeowner's Mortgage

A long-term loan in which the property is pledged as security for payment of the difference between the down payment and the sale price.

The two most popular types of mortgage loans available today are the **adjustable-rate loan** (or **variable-rate loan**) and the **conventional loan**. The major difference between the two is that the interest rate for a conventional loan is fixed for the duration of the loan, whereas the interest rate for the variable-rate loan may change every period, as specified in the loan. We will first discuss the requirements that are the same for both types of loans.

The size of the down payment required depends on who is lending the money, how old the property is, and whether or not it is easy to borrow money at that particular time. The down payment required by the lending institution can vary from 5% to 50% of the purchase price. A larger down payment is required when money is "tight"—that is, when it is difficult to borrow money. Furthermore, most lending institutions tend to require larger down payments on older homes and smaller down payments on newer homes.

Most lending institutions may require the buyer to pay one or more **points** for their loan at the time of the **closing** (the final step in the sale process). **One point** amounts to 1% of the mortgage money (the amount being borrowed). By charging points, the bank reduces the rate of interest on the mortgage, thus reducing the size of the monthly payments and enabling more people to purchase houses. However, because they charge points, the rate of interest that banks state is not the APR (annual percentage rate) for the loan. The APR would be determined by adding the amount paid for points to the total interest paid and then using an APR table.

CONVENTIONAL LOANS

Example 1 illustrates purchasing a house with a conventional loan.

EXAMPLE 1 *Calculating Down Payment and Points*

Chris and Daryl Cahill want to purchase a house selling for \$85,000. Their bank requires a 15% down payment and a payment of 1 point at the time of closing.

a) Determine the Cahill's down payment.
b) With a 15% down payment, determine the Cahill's mortgage.
c) What is the cost of the point paid by the Cahills on their mortgage.

SOLUTION

a) The down payment is 15% of \$85,000, or

$$0.15 \times \$85{,}000 = \$12{,}750$$

b) The mortgage on the Cahill's new home is the selling price minus the down payment.

$$\$85{,}000 - \$12{,}750 = \$72{,}250$$

c) One point equals 1% of the mortgage amount.

$$0.01 \times \$72{,}250 = \$722.50$$

At the closing, the Cahills will pay the down payment of \$12,750 to the seller and the 1 point, or \$722.50, to their bank. ▲

Banks use a formula to determine the maximum monthly payment that they believe is within the purchaser's ability to pay. A mortgage loan officer first determines the buyer's **adjusted monthly income** by subtracting from the gross monthly income (total income before any deductions) any fixed monthly payments with more than 10 months remaining (such as for a student loan, a car, furniture, or a television). The loan officer then multiplies the adjusted monthly income by 28%. (This percent, and the maximum number of payments remaining on other fixed loans, may vary in different locations.) In general, this product is the maximum monthly payment the lending institution believes the purchaser can afford to pay for a house. This estimated payment must cover principal, interest, property taxes, and insurance. Taxes and insurance are not necessarily paid to the bank; they may be paid directly to the tax collector and the insurance company. Example 2 shows how a bank uses the formula to determine whether a prospective buyer qualifies for a mortgage.

DID YOU KNOW

The Benefits of Home Ownership

Most of us are aware that a big part of the "American dream" is to own your own home. What we may not realize is that in addition to having a place to call your own, several impotant financial benefits occur when you own your own home. First, instead of paying rent to someone else, you make a mortgage payment that builds the equity in your home. *Equity* is the difference between the appraised value of your home and your loan balance, and it increases with each payment you make. As years go by, this equity may also help you qualify for other loans such as college and car loans. Second, the interest and real estate taxes you pay (in most cases) are deductible on your federal income tax. These deductions can add up to significant savings each year, and may result in a larger tax refund. Finally, over time you can typically expect your home to increase in value. Thus, your home not only becomes your place of dwelling; in most cases, it also serves as a wise financial investment.

EXAMPLE 2 *Qualifying for a Mortgage*

Suppose that the Cahill's (see Example 1) gross monthly income is \$3200 and that they have 15 remaining payments of \$185 per month on their car loan and 14 remaining payments of \$35 per month on a loan used to purchase a new washer and dryer. The taxes on the house they want to purchase are \$135 per month and the insurance is \$38 per month.

a) What maximum monthly payment does the bank's loan officer think the Cahills can afford?
b) The Cahills want a 30-year \$72,250 mortgage. If the interest rate is 8%, determine whether the Cahills qualify for the mortgage.

SOLUTION

a) To find the maximum monthly payment the bank's loan officer believes the Cahills can afford, first determine their adjusted monthly income.

\$3200	Gross income
− 220	Monthly payments (car and appliance loans)
\$2980	Adjusted monthly income

Next, find 28% of the adjusted monthly income.

$$0.28 \times \$2980 = \$834.40$$

The loan officer determines that the Cahills can afford a maximum monthly payment—including taxes and insurance—of \$834.40.

b) To determine whether the Cahills qualify for a 30-year conventional mortgage with their current income, calculate the monthly mortgage payment, which includes principal and interest. Lending institutions and lawyers use computer programs or calculators to determine monthly mortgage payments, per thousand dollars, for a specific number of years at a specific interest rate. This information is also available in table form, as shown in Table 11.4 on page 561. With an interest rate of 8%, a 30-year loan would have a monthly payment of \$7.34 per thousand dollars of mortgage (circled in blue in the table).

To determine the Cahill's monthly payment, first divide the mortgage by \$1000.

$$\frac{72{,}250}{1000} = 72.25$$

Then find the monthly mortgage payment by multiplying the number of thousands of dollars of mortgage, 72.25, by the value found in Table 11.4, \$7.34

$$\$7.34 \times 72.25 = \$530.32$$

To the \$530.32, add the monthly cost of real estate taxes, \$135, and insurance of \$38, for a total cost of \$703.32. Since \$703.32 is less than \$834.40, the maximum monthly payment the loan officer determined they can afford, the Cahills will most likely be granted the loan. ▲

Table 11.4 Monthly Payment per \$1000 of Mortgage, Including Principal and Interest

	Number of years				
Rate%	**20**	**25**	**30**	**35**	**40**
6.5	\$ 7.46	\$ 6.75	\$ 6.32	\$ 6.04	\$ 5.85
7	7.75	7.07	6.65	6.39	6.21
7.5	8.06	7.39	6.99	6.74	6.58
8	8.36	7.72	7.34	7.10	6.95
8.5	8.68	8.05	7.69	7.47	7.33
9	9.00	8.40	8.05	7.84	7.72
9.5	9.33	8.74	8.41	8.22	8.11
10	9.66	9.09	8.70	8.60	8.50
10.5	9.98	9.44	9.15	8.98	8.89
11	10.32	9.80	9.52	9.37	9.28
11.5	10.66	10.16	9.90	9.76	9.68
12	11.01	10.53	10.29	10.16	10.08
12.5	11.36	10.90	10.67	10.55	10.49
13	11.72	11.28	11.06	10.95	10.90
13.5	12.07	11.66	11.45	11.35	11.30
14	12.44	12.04	11.85	11.76	11.71
14.5	12.80	12.42	12.25	12.16	12.12
15	13.17	12.81	12.64	12.57	12.53
15.5	13.54	13.20	13.05	12.98	12.94
16	13.91	13.59	13.45	13.38	13.36

What is the effect on the monthly payments when only the period of time has been changed? The total monthly payments for the Cahills in Example 2 would have been \$777.01 for 20 years, \$730.77 for 25 years, \$703.32 for 30 years, \$685.98 for 35 years, and \$675.14 for 40 years. (You should verify these numbers yourself.) Increasing the length of time decreases the monthly payment but increases the total amount of interest paid because the borrower is paying for a longer period of time. The longer the term of a mortgage, the more expensive the total cost of the house.

EXAMPLE 3 *The Total Cost of a House*

The Cahills of Examples 1 and 2 obtained a house selling for \$85,000. They made a 15% down payment and obtained a 30-year \$72,250 conventional mortgage at 8%. They also paid 1 point at closing. Their monthly mortgage payment, including principal and interest, is \$530.32.

a) Determine the total amount the Cahills will pay for their house over 30 years.
b) How much of the cost will be interest?
c) How much of the first payment on the mortgage is applied to the principal?

SOLUTION

a) To find the total amount the Cahills will pay for their house, perform the following computation:

\$530.32	Mortgage payment for 1 month
× 12	Number of months in a year
\$6363.84	Mortgage payments for 1 year
× 30	Number of years
\$190,915.20	Mortgage payments for 30 years
+ 12,750.00	Down payment
+ 722.50	One Point
\$204,387.70	Total cost of the house

Note: The result might not be the exact cost of the house, since the final payment on the mortgage might be slightly more or less than the regular monthly payment.

b) To determine the amount of interest paid over 30 years, subtract the purchase price of the house and the cost of the points from the total cost.

\$204,387.70	Total cost
− 85,000.00	Purchase price
\$119,387.70	Total interest including one point
− 722.50	One Point
\$118,665.20	Total interest on mortgage payments

c) To find the amount of the first payment that is applied to the principal, subtract the amount of interest on the first payment from the monthly mortgage payment. Use the simple interest formula to find the interest on the first payment.

$$\begin{aligned} i &= prt \\ &= \$72{,}250 \times 0.08 \times \frac{1}{12} \\ &= \$481.67 \end{aligned}$$

Now subtract the interest for the first month from the monthly mortgage payment. The difference will be the amount paid on the principal for the first month.

\$530.32	Monthly mortgage payment
− 481.67	Interest paid for the first month
48.65	Principal paid for the first month

Thus, the first payment of \$530.32 consists of \$481.67 in interest and \$48.65 in principal. The \$48.65 is applied to reduce the loan. Thus, the balance due after the first payment is \$72,250 − \$48.65, or \$72,201.35. ▲

By repeatedly using the simple interest formula month-to-month on the unpaid balance, you could calculate the principal and the interest for all the payments—a tedious task. However, a list containing the payment number, payment on the interest, payment on the principal, and balance of the loan can be prepared using a computer. Such a list is called a loan **amortization schedule**. One way to obtain an amortization schedule is by using a spreadsheet program. Another way is to access an amortization "calculator" program on the Internet. A part of the amortization schedule for the Cahill's loan in Example 3 is given in Table 11.5. This schedule was generated from an Internet site called *Mortgage Amortization Calculator* (*www.equitynow.com*). Note that the monthly payment in Table 11.5 ($530.15) is slightly less than the monthly payment we calculated using Table 11.4 ($530.32). This difference is due to round-off error that occurs when estimating the monthly principal and interest payment from Table 11.4.

Table 11.5 Amortization Schedule

Annual % Rate: 8.0 **Monthly Payment: $530.15***
Loan: $72,250 **Term: Years 30, Months 0**
Periods: 360

Payment number	Interest	Principal	Balance of loan
1	$481.67	$48.48	$72,201.52
2	$481.34	$48.81	$72,152.71
3	$481.02	$49.13	$72,103.58
4	$480.69	$49.46	$72,054.12
11	$478.33	$51.81	$71,698.54
12	$477.99	$52.16	$71,646.39
119	$423.95	$106.19	$63,487.07
120	$423.25	$106.90	$63,380.17
239	$294.43	$235.71	$43,929.61
240	$292.86	$237.28	$43,692.33
359	$6.95	$523.20	$519.09
360	$3.46	$519.09	$0.00

*Table obtained from www.equitynow.com

ADJUSTABLE-RATE MORTGAGES

Now let's consider **adjustable-rate mortgages**, ARMs (also called *variable-rate mortgages*). The rules for ARMs vary from state to state and by local bank, so the material presented on ARMs may not apply in your state or at your local lending institution. Generally, with adjustable-rate mortgages, the monthly mortgage payment remains the same for a 1-, 2-, or 5-year period even though the interest rate of the mortgage may change every 3 months, 6 months, or some other predetermined period. The interest rate in an adjustable-rate mortgage may be based on an index that is determined by the Federal Home Loan Bank Association. Or it may be based on the interest rate of a 3-month, 6-month, or 1-year Treasury bill. The interest rate of a 3-month Treasury bill may change every 3 months, a 6-month Treasury bill may change every 6 months, and so on. When the base is a Treasury bill, the actual interest rate charged for the mortgage is often determined by adding 3% to $3\frac{1}{2}$%, called the **add on rate** or **margin** to the rate of the Treasury bill. Thus, if the rate of the Treasury bill is 6% and the add on rate is 3%, the interest rate charged is 9%.

EXAMPLE 4 *An Adjustable Rate Mortgage*

Tony and Keisha Torrence purchased a house for \$115,000 with a down payment of \$23,100. They obtained a 30-year adjustable-rate mortgage with the following terms. The interest rate is based on a 6-month Treasury bill, the effective interest rate is 3% above the rate of the Treasury bill on the date of adjustment (3% is the add on rate), the interest rate is adjusted every 6 months, the interest rate will not change more than 1% (up or down) when the interest rate is adjusted, the maximum interest rate for the duration of the loan is 16%, there is no lower limit on the interest rate, the initial mortgage interest rate is 8%, and the monthly payment (including principal and interest) is adjusted every 5 years.

a) Determine the initial monthly payment.
b) Determine the adjusted interest rate in 6 months if the interest rate on the Treasury bill at that time is 4.5%.

SOLUTION

a) To determine the initial monthly payment of interest and principal, divide the amount of the loan, \$115,000 − \$23,100 = \$91,900, by \$1000. The result is 91.9. Now multiply the number of thousands of dollars of mortgage, 91.9, by the value found in Table 11.4 with $r = 8\%$ for 30 years.

$$\$7.34 \times 91.9 = \$674.55$$

Thus, the initial monthly payment for principal and interest is \$674.55. This amount will not change for the first 5 years of the mortgage.

b) The adjusted interest rate in 6 months will be the Treasury bill rate plus the add on rate.

$$4.5\% + 3\% = 7.5\%$$

In Example 4(b), note that the rate after 6 months is lower than the initial rate of 8%. Since the monthly payment remains the same, the additional money paid the bank is applied to reduce the principal. The monthly interest and principal payment of \$674.55 would pay off the loan in 30 years if the interest remained constant at 8%. What happens if the interest rate drops and stays lower than the initial 8% for the length of the loan? In this case, at the end of each 5-year period the bank reduces the monthly payment so that the loan will be paid off in 30 years. What happens if the interest rates increase above the initial 8% rate? In this case part or, if necessary, all of the principal part of the monthly payment would be used to meet the interest obligation. At the end of the 5-year period, the bank will increase the monthly payment so that the loan can be repaid by the end of the 30-year period. Or the bank may increase the time period of the loan beyond 30 years so that the monthly payment is affordable.

To prevent rapid increases in interest rates, some banks have a rate cap. A **rate cap** limits the maximum amount the interest rate may change. A **periodic rate cap** limits the amount the interest rate may increase in any one period. For example, your mortgage could provide that, even if the index increases by 2% in 1 year, your rate can only go up 1% per year. An **aggregate rate cap** limits the interest rate increase and decrease over the entire life of the loan. If the initial interest rate is 9% and the aggregate rate cap is 5%, the interest rate could go no higher than 14% and no lower than 4% over the life of the mortgage. A **payment cap** limits the amount the monthly payment may change but does not limit changes in interest rates. If interest rates increase rapidly on a loan with a payment cap, the monthly payment may not be large enough to pay the monthly principal and interest on the loan. If that happens, the borrower could end up paying interest on interest.

OTHER TYPES OF MORTGAGES

Conventional mortgages and variable-rate mortgages are not the only methods of financing the purchase of a house. We next briefly describe four other methods, and we briefly discuss home equity loans.

FHA MORTGAGE A house can be purchased with a smaller down payment than with a conventional mortgage if the individual qualifies for a Federal Housing Administration (FHA) loan. The loan application is made through a local bank. The bank's loan officer determines the maximum monthly payment a loan applicant can afford by taking 29% of the adjusted monthly income. The bank provides the money, but the FHA insures the loan. The down payment for an FHA loan is as low as 2.5% of the purchase price, rather than the standard 5 to 50%. Another advantage is FHA loans can be assumed at the original rate of interest on the loan. For example, if you purchase a home today that already has an 8% FHA mortgage, you, the new buyer, can assume that 8% mortgage regardless of the current interest rates. However, to be able to assume a mortgage, the purchaser must be able to make a down payment equal to the difference between the purchase price and the balance due on the original mortgage. The government sets the maximum interest rate the lender may charge.

One drawback to FHA loans is the borrower must pay an FHA insurance premium as part of the monthly mortgage payment. The insurance premium is calculated at a rate of one-half percent (0.5%) of the unpaid balance of the loan on the anniversary date of the loan. Thus, even though the insurance premium decreases each year, it adds to the monthly payments.

VA MORTGAGE A veteran certified by the Department of Veterans Affairs (VA) who wants to purchase a house applies for a mortgage with a bank or lending institution. The individual must meet the requirements set by the bank or lending institution for a mortgage. The VA guarantees a certain percentage of the loan. For example, if the loan is less than $45,000 the VA guarantees 50% of the loan. For loans from $45,000 to $144,000, the VA guarantees the lesser of 40% of the loan or $36,000. This guarantee takes the place of the down payment. A veteran who can make the monthly mortgage payments may therefore obtain a certain mortgage without a down payment.

The government sets the maximum interest rate that the lender may charge. A VA loan is always assumable. With a VA loan, the seller may be asked to pay points, but there is no monthly insurance premium.

GRADUATED PAYMENT MORTGAGE (GPM) A GPM mortgage is designed so that for the first 5 to 10 years the size of the mortgage payment is smaller than the payments for the remaining time of the mortgage. After the 5- to 10-year period, the mortgage payments remain constant for the duration of the mortgage. Depending on the size of the mortgage, the lender might find that the monthly payments made for the first few years may actually be less than the interest owed on the loan for those few years. The interest not paid during the first few years of the mortgage is then added to the original loan. This type of loan is strictly for those who are confident their annual incomes will increase as rapidly as the mortgage payments do.

BALLOON-PAYMENT MORTGAGE (BPM) This type of loan could be for the person who needs time to find a permanent loan. It may work this way: The individual pays the interest for 3 to 8 years, the period of the loan. At the end of the 3- to 8-year period, the buyer must repay the entire principal unless the lender agrees to a loan extension. Balloon-payment loans generally offer lower rates than conventional mortgages. This type of loan may be advantageous if the buyer plans to sell the house before the maturity date of the balloon-payment mortgage.

HOME EQUITY LOANS As you make monthly payments and pay off the principal you owe on your home, you are said to be gaining *equity* in your home. **Equity** is the difference between the appraised value of your home and your loan balance. This equity can be used as collateral in obtaining a loan. Such a loan is referred to as a **home equity loan** or a **second mortgage**. One advantage of home equity loans over other types of loans (such as those discussed earlier in this chapter) is the interest charged on a home equity loan is often tax deductible on federal income taxes. Home equity loans are commonly used for home improvements, bill consolidation, or to pay for children's college education.

For further information about the types of mortgages or loans discussed in this section, consult a loan officer at a local bank, savings and loan, or credit union. Additional information on buying a house may be obtained from the U.S. Government Printing Office in Pueblo, CO 81009. This information is also available on the Internet at *http://www.access.gpo.gov*.

SECTION 11.5 EXERCISES

CONCEPT/WRITING EXERCISES

1. What is a mortgage?
2. What is a down payment?
3. What is the difference between a variable-rate mortgage and a conventional mortgage.
4. **a)** What are points in a mortgage agreement?
   b) Explain how to determine the cost of x points.
5. Explain how to determine a buyer's adjusted monthly income.
6. What is an amortization schedule?
7. What is an add on rate, or margin?
8. Who insures an FHA loan? Who provides the money for an FHA loan?
9. What is equity?
10. What is a home equity loan?

PROBLEM SOLVING

11. *Down Payment and Mortgage Payment* Martha Cuttler wishes to buy a house selling for \$90,000. Her credit union requires her to make a 20% down payment. The current mortgage rate is 7%.
    a) Determine the amount of the required down payment.
    b) Determine the monthly mortgage payment for a 20-year loan with a 20% down payment.
12. *Monthly Payment on a House* Craig Kelling is buying a house selling for \$97,000. The bank is requiring a minimum down payment of 20%. The current mortgage rate is 8.5%.
    a) Determine the amount of the required down payment.
    b) Determine the monthly mortgage payment for a 30-year loan with a minimum down payment.
13. *Monthly Payment on a House* Sergio and Barb Loche are buying a house selling for \$74,000. The bank is requiring a minimum down payment of 20%. The current mortgage rate is 8.5%.
    a) Determine the amount of the required down payment.
    b) Determine the monthly mortgage payment for a 25-year loan with a minimum down payment.

14. *Buying a Condo* Sherrie Nicol is buying a condominium selling for $65,000. The bank is requiring a minimum down payment of 15%. To obtain a 30-year mortgage at 7.5% interest, she must pay 2 points at the time of closing.
    a) What is the required down payment?
    b) With the 15% down payment, what is the amount of the mortgage on the property?
    c) What is the cost of 2 points on the mortgage determined in part (b)?

15. *Buying a House* The Guenthers are buying a house that sells for $93,500. The bank is requiring a minimum down payment of 15%. To obtain a 30-year mortgage at 7.5% interest, they must pay 3 points at the time of closing.
    a) What is the required down payment?
    b) With the 15% down payment, what is the amount of the mortgage on the propety?
    c) What is the cost of 3 points on the mortgage determined in part (b)?

16. *Affordable Payments* Keith Stafford's gross monthly income is $2500. He has 12 remaining payments of $95 on furniture and appliances. The taxes and insurance on the house are $105 per month.
    a) What maximum monthly payment does the bank's loan officer feel that Keith can afford?
    b) Keith would like to get a 30-year, $52,200 mortgage. Does he qualify for this mortgage with an 8% interest rate?

17. *Affordable Payments* The Weigel's gross monthly income is $7200. They have 15 remaining payments of $290 on a new car. The taxes on the house are $350 per month and insurance is $50 per month.
    a) What maximum monthly payment does the bank's loan officer feel that the Weigels can afford?
    b) The Weigels want a 30-year, $140,000 mortgage. Do they qualify for this mortgage with a 6.5% interest rate?

18. *A 30-Year Conventional Mortgage* Ingrid Holzner obtains a 30-year, $63,750 conventional mortgage at 8.5% on a house selling for $75,000. Her monthly payment, including principal and interest, is $490.24.
    a) Determine the total amount Ingrid will pay for her house.
    b) How much of the cost will be interest?
    c) How much of the first payment on the mortgage is applied to the principal?

19. *A 25-Year Conventional Mortgage* Mr. and Mrs. Alan Bell obtain a 25-year, $110,000 conventional mortgage at 10.5% on a house selling for $160,000. Their monthly mortgage payment, including principal and interest, is $1038.40.
    a) Determine the total amount the Bells will pay for their house.
    b) How much of the cost will be interest?
    c) How much of the first payment on the mortgage is applied to the principal?

20. *A 35-Year Conventional Mortgage* The DiMartos purchased a home selling for $58,000 with a 20% down payment and were required to pay 2 points at the time of closing. The period of the conventional mortgage is 35 years, and the rate of interest is 9.5%.
    a) Determine the amount of the down payment.
    b) Determine the cost of the 2 points.
    c) Determine the monthly mortgage payment.
    d) Determine the total cost of the house (including points).
    e) Determine the total interest paid.
    f) Determine how much of the first payment on the loan is applied to the principal.

21. *Evaluating a Loan Request* The Yakomos found a house selling for $113,500. The taxes on the house are $1200 per year, and insurance is $320 per year. They are requesting a conventional loan from the local bank. The bank is currently requiring a 28% down payment and 3 points, and the interest rate is 10%. The Yakomo's monthly income is $4750. They have more than 10 monthly payments remaining on a car, a boat, and furniture. The total monthly payments for these items is $420.
    a) Determine the required down payment.
    b) Determine the cost of the 3 points.
    c) Determine their adjusted monthly income.
    d) Determine the maximum monthly payment the bank's loan officer believes they can afford.
    e) Determine the monthly payments of principal and interest for a 20-year loan.
    f) Determine their total monthly payment, including insurance and taxes.
    g) Determine whether the Yakomos qualify for the 20-year loan.
    h) Determine how much of the first payment on the loan is applied to the principal.

22. *Evaluating a Loan Request* Kathy Morgan wants to buy a condominium selling for $95,000. The taxes on the property are $1500 per year, and insurance is $336 per year. Kathy's gross monthly income is $4000. She has 15 monthly payments of $135 remaining on her van. She has $20,000 in her savings account. The bank is requiring 20% down and is charging 9.5% interest.
    a) Determine the required down payment.
    b) Determine the maximum monthly payment the bank's loan officer believes Kathy can afford.

c) Determine the monthly payment of principal and interest for a 35-year loan.
d) Determine her total monthly payment, including insurance and taxes.
e) Does Kathy qualify for the loan?
f) Determine how much of the first payment on the mortgage is applied to the principal.
g) Determine the total amount she pays for the condominium with a 35-year conventional loan. (Do not include taxes or insurance.)
h) Determine the total interest paid for the 35-year loan.

23. *Comparing Loans* The Riveras are negotiating with two banks for a mortgage to buy a house selling for $105,000. The terms at bank A are a 10% down payment, an interest rate of 10%, a 30-year conventional mortgage, and three points to be paid at the time of closing. The terms at bank B are a 20% down payment, an interest rate of 11.5%, a 25-year conventional mortgage, and no points. Which loan should the Riveras select in order for the total cost of the house to be less?

24. *An Adjustable-Rate Mortgage* The Bretz Family purchased a house for $95,000 with a down payment of $13,000. They obtained a 30-year adjustable-rate mortgage. The terms of the mortgage are as follows: The interest rate is based on a 3-month Treasury bill, the effective interest rate is 3.25% above the rate of the Treasury bill on the date of adjustment, the interest rate is adjusted every 3 months, the interest rate will not change more than 1% (up or down) when the interest rate is adjusted, the maximum interest rate that can be charged for the duration of the loan is 16%, there is no lower limit on the interest rate, the initial mortgage interest rate is 8.5%, and the monthly payment of interest and principal is adjusted annually.
a) Determine the initial monthly payment for interest and principal.
b) Determine the interest rate in 3 months if the interest rate on the Treasury bill at the time is 5.65%.
c) Determine the interest rate in 6 months if the interest rate on the Treasury bill at the time is 4.85%.

CHALLENGE PROBLEMS/GROUP ACTIVITIES

25. *How Much House Can They Afford?* A bank's loan officer determines that the Pappys can afford to make a $950 monthly mortgage payment. If the bank will give them a 25-year conventional mortgage at 9% and requires a 25% down payment, what is
a) the maximum mortgage the bank will grant the Pappys?
b) the highest-priced house they can afford?

26. *Comparing Mortgages* The Hassads are applying for a $90,000 mortgage. They can choose between a conventional mortgage and a variable-rate mortgage. The interest rate on a 30-year conventional mortgage is 9.5%. The terms of the variable-rate mortgage are 6.5% interest rate the first year, an annual cap of 1%, and an aggregate cap of 6%. The interest rates and the mortgage payments are adjusted annually. Assume the interest rates for the variable-rate mortgage increase by the maximum amount each year. Then the monthly mortgage payments for the variable-rate mortgage for years 1–6 are $568.86, $628.05, $688.29, $749.35, $811.02, and $873.11, respectively.
a) Knowing that they will be in the house for only 6 years, which mortgage will be the least expensive for that period?
b) How much will they save by choosing the less expensive mortgage?

27. *An Adjustable-Rate Mortgage* The Simpsons purchased a house for $105,000 with a down payment of $5000. They obtained a 30-year adjustable-rate mortgage. The terms of the mortgage are as follows: The interest rate is based on a 3-month Treasury bill, the effective interest rate is 3.25% above the rate of the Treasury bill on the date of adjustment, the interest rate is adjusted every 3 months, the interest rate will not change more than 1% (up or down) when the interest rate is adjusted, the maximum interest rate that can be charged for the duration of the loan is 16%, there is no lower limit on the interest rate, the initial mortgage interest rate is 9%, and the monthly payment of interest and principal is adjusted semiannually.
a) Determine the initial monthly payment for interest and principal.
b) Determine an amortization schedule for months 1–3.
c) Determine the interest rate for months 4–6 if the interest rate on the Treasury bill at the time is 6.13%.
d) Determine an amortization schedule for months 4–6.
e) Determine the interest rate for months 7–9 if the interest rate on the Treasury bill at the time is 6.21%.

RESEARCH ACTIVITIES

28. *Closing Costs* An important part of buying a house is the closing. The exact procedures for the closing differ with individual cases and in different parts of the country. In any closing, however, both the buyer and the seller have certain expenses. To determine what is involved in the closing of a property in your community, contact a lawyer, a real estate agent, or a banker. Explain that you are a student and your objective is to understand the procedure for closing a real estate purchase and the costs to both buyer and seller. Select a specific piece of property that is for sale. Use the asking price to determine the total closing costs to both buyer and seller. The following is a partial list of the most common costs. Consider them in your research.

a) Fee for title search and title insurance
b) Credit report on buyer
c) Fees to the lender for services in granting the loan
d) Fee for property survey
e) Fee for recording of the deed
f) Appraisal fee
g) Lawyer's fee
h) Escrow accounts (taxes, insurance)
i) Mortgage assumption fee

29. *Finding Your Dream Home* Examine a local newspaper to find your "dream home" and note the asking price. Next, contact a loan officer from your local bank, savings and loan, or credit union. Assuming you can make a 20% down payment, determine the interest rates for a 15-year and a 30-year mortgage. Use an amortization calculator (see page 563) with the data you obtained to print amortizations schedules. Compare the monthly payments with the 15-year mortgage to those of the 30-year mortgage. Compare the total interest costs of the 15-year mortgage to that of the 30-year mortgage. Write a report summarizing your findings.

 CHAPTER 11 SUMMARY

IMPORTANT FACTS

Ordinary interest

When computing ordinary interest, each month is considered to have 30 days and a year is considered to have 360 days.

United States rule

If a partial payment is made on a loan, interest is computed on the principal from the first day of the loan until the date of the partial payment. The partial payment is used to pay the interest first; then the rest of the payment is used to reduce the principal.

Banker's rule

When computing interest with the Banker's rule, a year is considered to have 360 days and any fractional part of a year is the exact number of days.

Simple interest formula

Interest = principal × rate × time or $i = prt$

Compound interest formula $A = p\left(1 + \frac{r}{n}\right)^{nt}$

Present value formula $p = \dfrac{A}{\left(1 + \frac{r}{n}\right)^{nt}}$

Actuarial method $u = \dfrac{u \cdot P \cdot V}{100 + V}$

Rule of 78s $u = \dfrac{f \cdot k(k + 1)}{n(n + 1)}$

Percent change $= \dfrac{\text{amount in latest period} - \text{amount in previous period}}{\text{amount in previous period}} \times 100$

Percent markup on cost $= \dfrac{\text{selling price} - \text{dealer's cost}}{\text{dealer's cost}} \times 100$

CHAPTER 11 REVIEW EXERCISES

11.1

Change the number to a percent. Express your answer to the nearest tenth of a percent.

1. $\frac{1}{4}$ **2.** $\frac{2}{3}$ **3.** $\frac{5}{8}$

4. 0.039 **5.** 0.0098 **6.** 3.141

Change the percent to a decimal number.

7. 26% **8.** 12.1% **9.** 123%

10. $\frac{2}{5}\%$ **11.** $\frac{5}{6}\%$ **12.** 0.00045%

13. *Car Price Increase* The 1999 Chevrolet Cavalier base coupe had a manufacturer's suggested retail price (MSRP) of $11,916. The 2000 Chevrolet Cavalier base coupe had a MSRP of $13,065. Determine the percent increase in the price of the Cavalier from 1999 to 2000. (*Source: Consumer Reports.*)

2000 Chevrolet Cavalier

14. *Salary Increase* Gail Linton had a salary of $46,200 in 1999 and a salary of $51,300 in 2000. Determine the percent increase in Gail's salary from 1999 to 2000.

In Exercises 15–19, solve for the unknown quantity.

15. What percent of 80 is 25?

16. Forty-four is 16% of what number?

17. What is 17% of 540?

18. *Tipping* At Empress Garden Restaurant, Vishnu and Krishna's bill comes to $42.79, including tax. If they wish to leave a 15% tip, how much should they tip the waiter?

19. *Increased Membership* If the number of people in your chess club increased by 20%, or 8, what was the original number of people in the club?

20. *Increased Membership* The Sarasota Wheelers skateboard club had 75 members and increased the number of members to 95. What is the percent increase in the number of members?

11.2

In Exercises 21–24, find the missing quantity by using the simple interest formula.

21. $p = \$2400,\quad r = 7\%,\quad t = 30 \text{ days},\quad i = ?$

22. $p = \$1575,\quad r = ?,\quad t = 100 \text{ days},\quad i = \$41.56$

23. $p = ?,\quad r = 8\frac{1}{2}\%,\quad t = 3 \text{ years},\quad i = \$114.75$

24. $p = \$5500,\quad r = 11\frac{1}{2}\%,\quad t = ?,\quad i = \$316.25$

25. *Home Improvement Loan* George Alexander borrowed $3600 for home improvements. The personal note was for 24 months and it had a simple interest rate of 11.25%. Determine the amount George paid his credit union on the date of maturity.

26. *A Bank Loan* Lori Holdren borrowed $3000 from her bank for 240 days at a simple interest rate of 8.1%.

a) How much interest did she pay for the use of the money?

b) How much did she pay the bank on the date of maturity?

27. *A Bank Loan* Nikos Papadopoulos borrowed $6000 for 24 months from the bank, using stock as security. The bank discounted the loan at $11\frac{1}{2}\%$.

a) How much interest did Nikos pay the bank for the use of the money?

b) How much did he receive from the bank?
c) What was the actual rate of interest?

28. *Savings as Security* Golda Frankl borrowed $800 for 6 months from her bank, using her savings account as security. A bank rule limits the amount that can be borrowed in this manner to 85% of the amount in the borrower's savings account. The rate of interest is 2% higher than the interest rate being paid on the savings account. The current rate on the savings account is $5\frac{1}{2}\%$.
a) What rate of interest will the bank charge for the loan?
b) Find the amount that Golda must repay in 6 months.
c) How much money must she have in her account in order to borrow $800.

11.3

29. *Compound Interest* Determine the amount and the interest when $1500 is invested for 5 years at 6%
a) Compounded annually.
b) Compounded quarterly.
c) Compounded monthly.

30. *Total Amount* Choi deposited $2500 in a savings account that pays 4.75% interest compounded quarterly. What will be the total amount of money in the account 15 years from the day of deposit?

31. *Effective Annual Yield* Determine the effective annual yield of an investment if the interest is compounded daily at an annual rate of 5.6%.

32. *Present Value* How many dollars must you invest today to have $40,000 in 20 years? Assume the money earns 5.5% interest compounded quarterly.

11.4

33. *An Installment Loan* Bill Jordan has a 48-month installment loan, with a fixed monthly payment of $193.75. The amount borrowed was $7500. Instead of making his 24th payment, Bill is paying the remaining balance on the note.
a) Determine the APR of the installment loan.
b) How much interest will Bill save, computed by the actuarial method?
c) What is the total amount due on that day?

34. *Rule of 78s* Carter Fenton is buying a book collection that costs $4000. He is making a down payment of $500 and 24 monthly payments of $163.33. Instead of making his 12th payment, Carter decides to pay the total remaining balance and terminate the loan.
a) How much interest will Carter save, computed by the rule of 78s?
b) What is the total amount due on that day?

35. *Installment Loan* Dara paid $3420 for a new wardrobe. She made a down payment of $860 and financed the balance on a 24-month fixed payment installment loan. The monthly payment are $119.47. Instead of making her 12th payment, Dara decides to pay the total remaining balance and terminate the loan.
a) Determine the APR of the installment loan.
b) How much interest will Dara save, computed by the actuarial method?
c) What is the total amount due on that day?

36. *Unpaid Balance Method* On November 1, the billing date, Arumagam had a balance due of $485.75 on his credit card. The transactions during the month of November are:

November 4	Payment	$375.00
November 8	Charge: Lawn mower	370.00
November 21	Charge: Edger	175.80
November 28	Charge: Trimmer	184.75

a) Find the finance charge on December 1 by using the unpaid balance method. Assume the interest rate is 1.3% per month.
b) Find the new account balance on December 1.

37. *Average Daily Balance Method* On March 5, the billing date, Judith Gersting had a balance due of $185.72 on her credit card. During the month of March, she had the following transactions.

March 8	Charge: Groceries	$ 85.75
March 10	Payment	75.00
March 15	Charge: Dress	72.85
March 21	Charge: Car Repair	275.00

a) Find the finance charge on April 5 by using the average daily balance method. Assume the interest rate is 1.4% per month.
b) Find the new account balance on April 5.

38. *Financing a Car* David Buckley bought a new 2000 Honda Civic for $14,900. He was required to make a 25% down payment. He financed the balance with the dealer on a 48-month payment plan. The salesperson told him the interest rate was 6.5% simple interest on the principal for the duration of the loan.
a) Find the amount of the down payment.
b) Find the amount to be financed.
c) Find the total interest paid.
d) Find the APR.

39. *Financing a Ski Outfit* Lucille Groenke can buy a cross-country skiing outfit for $135. The store is offering the following terms: $35 down and 12 monthly payments of $8.79.
   a) Find the interest paid.
   b) Find the APR.

11.5

40. The Freemans have decided to build a new house. The contractor quoted them a price of $135,700. The taxes on the house will be $3450 per year, and insurance will be $350 per year. They have applied for a conventional loan from a local bank. The bank is requiring a 25% down payment, and the interest rate on the loan is 9.5%. The Freemans' annual income is $64,000. They have more than 10 monthly payments remaining on each of the following: $218 on a car, $120 on new furniture, and $190 on a camper. Determine
   a) The required down payment.
   b) Their adjusted monthly income.
   c) The maximum monthly payment the bank's loan officer believes they can afford.
   d) The monthly payment of principal and interest for a 40-year loan.
   e) Their total monthly payment, including insurance and taxes.
   f) Do the Freemans qualify for the mortgage?

41. The Clars purchased a home selling for $89,900 with a 15% down payment. The period of the mortgage is 30 years and the interest rate is 11.5%. Determine the
   a) Amount of the down payment.
   b) Monthly mortgage payment.
   c) Amount of the first payment applied to the principal.
   d) Total cost of the house.
   e) Total interest paid.

42. The Leglers purchased a house for $105,000 with a down payment of $26,250. They obtained a 30-year adjustable-rate mortgage. The terms of the mortgage are as follows: The interest rate is based on the 6-month Treasury bill, the effective interest rate is 3.00% above the rate of the Treasury bill on the date of adjustment, the interest rate is adjusted every 6 months, the interest rate will not change more than 1% (up or down) when the interest rate is adjusted, the maximum interest rate that can be charged for the duration of the loan is 16%, there is no lower limit on the interest rate, the initial mortgage interest rate is 7.5%, and the monthly payment of interest and principal is adjusted annually. Determine the
   a) Initial monthly payment for interest and principal.
   b) Interest rate in 6 months if the interest rate on the Treasury bill at the time is 5.00%.
   c) Interest rate in 6 months if the interest rate on the Treasury bill at the time is 4.75%

CHAPTER 11 TEST

In Exercises 1 and 2, find the missing quantity by using the simple interest formula.

1. $i = ?$, $p = \$1500$, $r = 8\%$, $t = 8$ months
2. $i = \$288$, $p = \$1200$, $r = 8\%$, $t = ?$

In Exercises 3 and 4, Greg borrowed $5000 from a bank for 18 months. The rate of simple interest charged is 8.5%.

3. How much interest did he pay for the use of the money?
4. What is the amount he repaid to the bank on the due date of the loan?

In Exercises 5–7, a new laser printer sells for $2350. To finance the laser printer through a bank, the bank will require a down payment of 15% and monthly payments of $94.50 for 24 months.

5. How much money will the purchaser borrow from the bank?
6. What finance charge will the individual pay the bank?
7. What is the APR?
8. *Rule of 78s* Sandi Abramowicz purchased a used fishing boat for $6750. She made a down payment of

$1550 and financed the balance with a 12-month fixed-payment installment loan. Instead of making the sixth monthly payment of $465.85, she decides to pay the remaining balance.

a) How much interest will Sandi save (use the rule of 78s)?

b) What is the total amount due on that day?

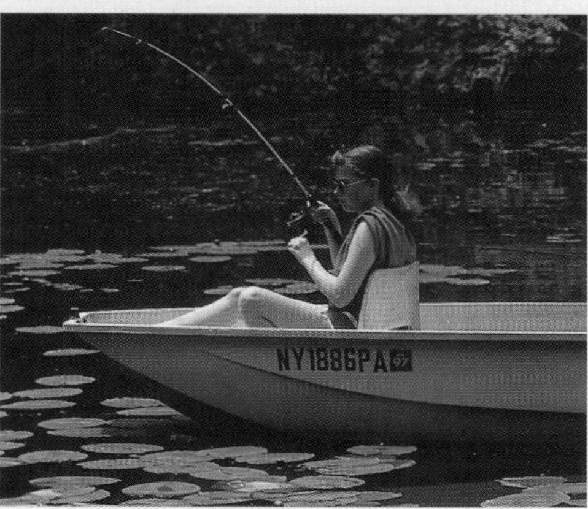

9. *Average Daily Balance Method* The balance on the Garzoni's credit card on May 8, their billing date, was $378.50. For the period ending June 8, they had the following transactions.

May 13	Charge: Car repair	$276.79
May 15	Payment	250.00
May 18	Charge: Compact discs	144.85
May 29	Charge: Restaurant bill	55.63

a) Find the finance charge on June 8 by the average daily balance method. Assume the interest rate is 1.3% per month.

b) Find the new balance on June 8.

10. *Unpaid Balance Method* Michael Murphy's credit card statement shows a balance due of $878.25 on March 23, the billing date. For the period ending on April 23, they had the following transactions.

March 26	Charge: Groceries	$ 95.89
March 30	Charge: Restaurant bill	68.76
April 3	Payment	450.00
April 15	Charge: Clothing	90.52
April 22	Charge: Eye glasses	450.85

a) Find the finance charge due March 23 by using the unpaid balance method. Assume the interest rate is 1.4% per month.

b) Find the new account balance on April 23.

In Exercises 11 and 12, Yolanda Fernandez signed a $5400 note with interest at 12.5% for 90 days on August 1. Yolanda made a payment of $3000 on September 15.

11. How much did she owe the bank on the date of maturity?

12. What total amount of interest did she pay on the loan?

In Exercises 13 and 14, compute the amount and the compound interest.

	Principal	Time	Rate	Compounded
13.	$7500	5 years	8%	Quarterly
14.	$2500	3 years	6.5%	Monthly

Building a House *In Exercises 15–21, the Leungs decided to build a new house. The contractor quoted them a price of $144,500, including the lot. The taxes on the house would be $3200 per year, and insurance would cost $450 per year. They have applied for a conventional loan from a bank. The bank is requiring a 15% down payment, and the interest rate is $10\frac{1}{2}$%. The Leung's annual income is $86,500. They have more than 10 monthly payments remaining on each of the following: $220 for a car, $175 for new furniture, and $210 on a college education loan.*

15. What is the required down payment?

16. Determine their adjusted monthly income.

17. What is the maximum monthly payment the bank's loan officer believes the Leungs can afford?

18. Determine the monthly payments of principal and interest for a 30-year loan.

19. Determine their total monthly payments, including insurance and taxes.

20. Does the bank's loan officer believe that the Leungs meet the requirements for the mortgage?

21. a) Find the total cost of the house (excluding insurance and taxes) after 30 years.

b) How much of the total cost is interest?

GROUP PROJECTS

MORTGAGE LOAN

1. The Young family is purchasing a $130,000 house with a VA mortgage. The bank is offering them a 25-year mortgage with an interest rate of 9.5%. They have $20,000 invested that could be used for a down payment. Since they do not need a down payment, Mr. Young wants to keep the money invested. Mrs. Young feels that they should make a down payment of $20,000.
   a) Determine the total cost of the house with no down payment.
   b) Determine the total cost of the house if they make a down payment of $20,000.
   c) Mr. Young feels that the $20,000 investment will have an annual rate of return of 10% compounded quarterly. Assuming Mr. Young is right, calculate the value of the investment in 25 years. (See Section 11.3.)
   d) If the Youngs use the $20,000 as a down payment, their monthly payments will decrease. Determine the difference of the monthly payments in parts (a) and (b).
   e) Assume that the difference in monthly payments, part (d), is invested each month at a rate of 6% compounded monthly for 25 years. Determine the value of the investment in 25 years. (See Exercise 45 in Section 11.3.)
   f) Use the information from parts (a)–(e) to analyze the problem. Would you recommend that the Youngs make the down payment of $20,000 and invest the difference in their monthly payments as in part (e), or that they do not make the down payment and keep the $20,000 invested as in part (c). Explain.

CREDIT CARD TERMS

2. With each credit card comes a credit agreement (or security agreement) the cardholder must sign. Select two members of your group who have a major credit card (MasterCard, Visa, Discover, or American Express). If possible, one of the credit cards should be a gold card. Obtain a copy of the credit agreement signed by each cardholder and answer questions (a)–(p).
   a) What are the cardholder's responsibilities?
   b) What is the cardholder's maximum line of credit?
   c) What restrictions apply to the use of the credit card?
   d) How many days after the billing date does the cardholder have to make a payment without being charged interest?
   e) What is the minimum monthly payment required, and how is it determined?
   f) What is the interest charged on purchases?
   g) How does the bank determine when to start charging interest on purchases?
   h) Is there an annual fee for the credit card? If so, what is it?
   i) What late charge applies if payments are not made on time?
   j) What information is given on the monthly statement?
   k) How is the finance charge computed?
   l) What other fees, if any, may the bank charge you?
   m) If the card is lost, what is the responsibility of the cardholder?
   n) If the card is lost, what is the liability of the cardholder?
   o) What are the advantages of a gold card over a regular card?
   p) In your opinion which of the two cards is more desirable? Explain.

RETIREMENT PLANS

3. Working men and women often have the benefit of contributing to various retirement plans. Such plans include traditional and Roth IRAs, 403(b) plans, 401(k) plans, IRA–SEP plans, and Keogh plans. Research each of these plans and answer questions (a)–(d).
   a) What type of employees are eligible to use each type of plan?
   b) What is the maximum annual contribution that can be made to each plan?
   c) What type of investments can be made through each of these plans?
   d) What are the tax advantages of each type of plan?

CHAPTER 12

Probability

Millions of Americans each year play state lotteries or play bingo. If you play the lottery, your hope is to beat the odds and be the person with the winning numbers. Mathematicians of the sixteenth, seventeenth, and eighteenth centuries, not satisfied with leaving things to chance, invented the study of probability to use mathematics to determine the likelihood of an event (winning numbers, for example) occurring.

Although the rules of probability were first applied to gaming, they have many other applications. The quality of the food you eat, the pedigree of your cat or dog, the cost of your car insurance—all have to do with probability. In the business of insurance underwriting, the likelihood of an event such as an automobile accident, given the age, sex, and location of the driver, are facts used in determining the cost of the insurance. ■

The laws of probability have applications in the science of genetics. The larger and more diverse the population, the greater is the probability that the species will have the characteristics necessary to adapt to changes in the environment. The cheetah, the world's fastest-running mammal, faces extinction because it lacks the genetic diversity necessary to survive disease. Once found worldwide, the species now lives wild in only a few areas of Africa.

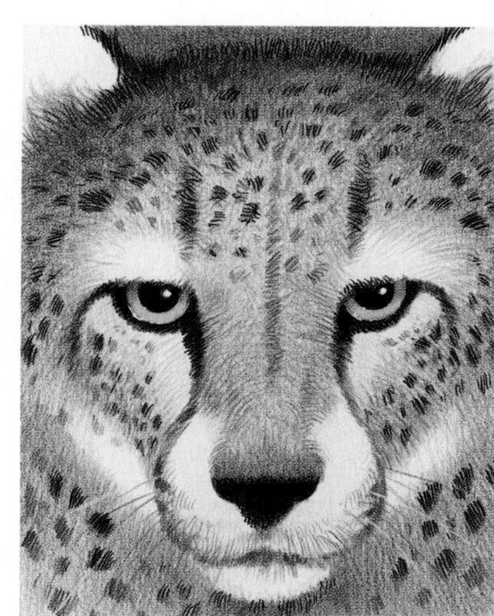

12.1 THE NATURE OF PROBABILITY

PROFILE IN MATHEMATICS

Jacob Bernoulli (1654–1705)

The Swiss mathematician Jacob Bernoulli was considered a pioneer of probability theory. He was part of a famous family of mathematicians and scientists. In *Ars Conjectand,* published posthumously in 1713, he proposed that an increased degree of accuracy can be obtained by increasing the number of trials of an experiment. This theorem, called Bernoulli's theorem (of probability), is also known as the law of large numbers. Bernoulli's theorem of fluid dynamics, used in aircraft wing design, was developed by *Daniel Bernoulli* (1700–1782), Jacob's second son.

HISTORY

Probability is presently used in many areas, including public finance, medicine, insurance, elections, manufacturing, educational tests and measurements, genetics, weather forecasting, investments, opinion polls, the natural sciences, games of chance, and a multitude of others. The study of probability originated from the study of games of chance. Archaeologists have found artifacts used in games of chance in Egypt dating from about 3000 B.C.

Mathematical problems relating to games of chance were studied by a number of mathematicians of the Renaissance. Italy's Girolamo Cardano (1501–1576) in his *Liber de Ludo Aleae* (book on the games of chance) presents one of the first systematic computations of probabilities. Although it is basically a gambler's manual, many consider it the first book ever written on probability. A short time later, two French mathematicians, Blaise Pascal (1623–1662) and Pierre de Fermat (1601–1665), worked together studying "the geometry of the die." In 1657 the Dutch mathematician Christian Huygens (1629–1695) published *De Ratiociniis in Luno Aleae* (on ratiocination in dice games), which contained the first documented reference to the concept of mathematical expectation (see Section 12.4). The Swiss mathematician Jacob Bernoulli (1654–1705), whom many consider the founder of probability theory, is said to have fused pure mathematics with the empirical methods used in statistical experiments. The works of Pierre-Simon de Laplace (1749–1827) dominated probability throughout the nineteenth century.

THE NATURE OF PROBABILITY

Before we discuss the meaning of the word *probability* and learn how to calculate probabilities, we must introduce a few definitions.

> An **experiment** is a controlled operation that yields a set of results.

The process by which medical researchers administer experimental drugs to patients to determine their reaction is one type of experiment.

> The possible results of an experiment are called its **outcomes.**

For example, the possible outcomes from administering an experimental drug may be a favorable reaction, no reaction, or an adverse reaction.

> An **event** is a subcollection of the outcomes of an experiment.

For example, when a die is rolled, the event of rolling a number greater than 2 can be satisfied by any one of four outcomes—3, 4, 5, or 6. The event of rolling a 5 can be satisfied by only one outcome, the 5 itself. The event of rolling an even number can be satisfied by any of three outcomes—2, 4, or 6.

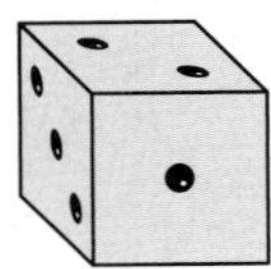

A die (one of a pair of dice) contains six surfaces. Each surface contains a unique number of dots, from 1 to 6. The sum of the dots on opposite surfaces is 7.

Probability is classified as either *empirical* (experimental) or *theoretical* (mathematical). **Empirical probability** is the relative frequency of occurrence of an event and is determined by actual observations of an experiment. **Theoretical probability** is determined through a study of the possible *outcomes* that can occur for the given experiment. We will indicate the probability of an event E by $P(E)$.

In this section we will briefly discuss empirical probability. The emphasis in the remaining sections is on theoretical probability. Following is the formula for computing empirical probability, or relative frequency.

Empirical Probability (Relative Frequency)

$$P(E) = \frac{\text{number of times event } E \text{ has occurred}}{\text{total number of times the experiment has been performed}}$$

The probability of an event, whether empirical or theoretical, is always a number between 0 and 1, inclusive, and may be expressed as a decimal, a fraction, or a percent. An empirical probability of 0 indicates the event has never occurred. An empirical probability of 1 indicates the event has always occurred.

EXAMPLE 1 ***Heads Up!***

In 100 tosses of a fair coin, 53 landed heads up. Find the empirical probability of the coin landing heads up.

SOLUTION Let E be the event that the coin lands heads up. Then

$$P(E) = \frac{53}{100} = 0.53$$

▲

EXAMPLE 2 ***Weight Reduction***

A pharmaceutical company is testing a new drug that is supposed to help with weight reduction. The drug is given to 500 individuals with the following outcomes.

Weight reduced	Weight unchanged	Weight increased
274	93	133

If this drug is given to an individual, find the empirical probability that the person's weight is (a) reduced, (b) unchanged, (c) increased.

SOLUTION

a) Let E be the event that the weight is reduced.

$$P(E) = \frac{274}{500} = 0.548$$

b) Let E be the event that the weight is unchanged.

$$P(E) = \frac{93}{500} = 0.186$$

c) Let E be the event that the weight is increased.

$$P(E) = \frac{133}{500} = 0.266$$

▲

One Blue Eye, One Brown Eye

Seeing a dog or a person, with different colored eyes is unusual, but not that unusual. Our features, such as eye color, are determined genetically, as are certain illnesses. Below is a chart listing some genetic disorders and their rates of occurrence. Genetics, which has its foundation in probability, is used in explaining the cause of these illnesses, and it plays an important role in research to find cures for genetic disorders.

Disorder/Incidence

Cystic fibrosis:
1 out of 2500 white births

Down syndrome:
1 out of 800–1000 births

Duchenne muscular dystrophy:
1 out of 3300 male births

Fragile X syndrome:
1 out of 1500 male births
1 out of 2500 female births

Hemophilia A:
1 out of 8500 male births

Huntington's disease:
4–7 out of 100,000 births

Polycystic kidney disease:
1 out of 3000 births

Prader–Willi syndrome:
1 out of 15,000 births

Sickle-cell anemia:
1 out of 400–600 African-American births

Tay-Sachs disease:
1 out of 3600 Ashkenazi Jewish births

Empirical probability is used when probabilities cannot be theoretically calculated. For example, life insurance companies use empirical probabilities to determine the chance of an individual in a certain profession, with certain risk factors, living to age 65.

EMPIRICAL PROBABILITY IN GENETICS

Using empirical probability, Gregor Mendel (1822–1884) developed the laws of heredity by crossbreeding different types of "pure" pea plants and observing the relative frequencies of the resulting offspring. These laws became the foundation for the study of genetics. For example, when he crossbred a pure yellow pea plant and a pure green pea plant, the resulting offspring (the first generation) were always yellow, see Fig. 12.1(a). When he crossbred a pure round-seeded pea plant and a pure wrinkled-seeded pea plant, the resulting offspring (the first generation) were always round, see Fig. 12.1(b).

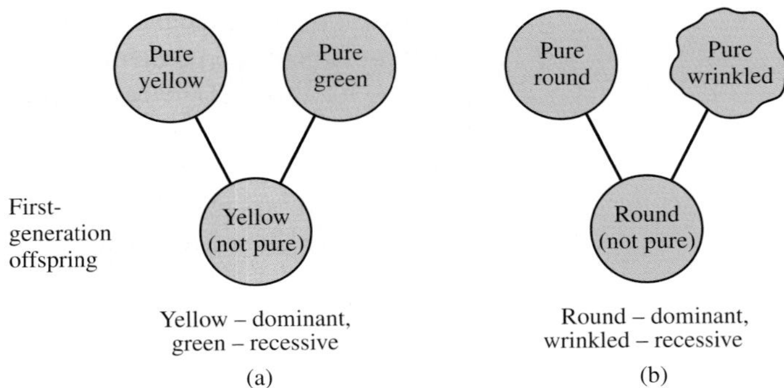

Figure 12.1

Mendel called traits such as yellow color and round seeds **dominant** because they overcame or "dominated" the other trait. The green and wrinkled traits he labeled **recessive**.

Mendel then crossbred the offspring of the first generation. The resulting second generation had both the dominant and the recessive traits of their grandparents, see Fig. 12.2(a) and (b). What's more, these traits always appeared in approximately a 3 to 1 ratio of dominant to recessive.

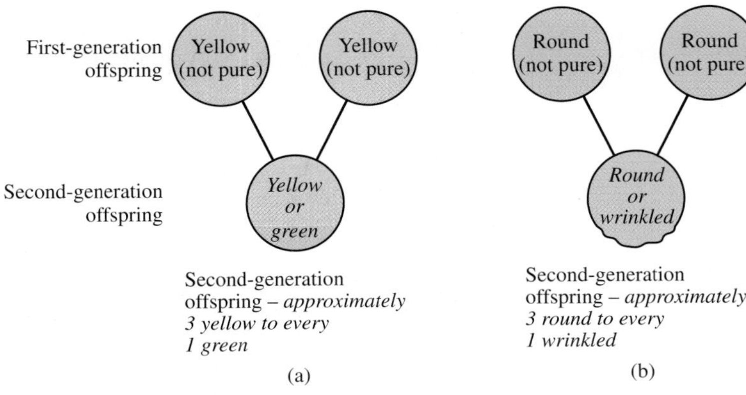

Figure 12.2

Table 12.1 Second-Generation Offspring

Dominant trait	**Number with dominant trait**	**Recessive trait**	**Number with recessive trait**	**Ratio of dominant to recessive**	***P* (dominant trait)**
Yellow seeds	6022	Green seeds	2001	3.01 to 1	$\frac{6022}{8023} = 0.75$
Round seeds	5474	Wrinkled seeds	1850	2.96 to 1	$\frac{5474}{7324} = 0.75$

Table 12.1 lists some of the results of Mendel's experiments with pea plants. Note that the ratio of dominant trait to recessive trait in the second-generation offspring is about 3 to 1 for each of the experiments. The empirical probability of the dominant trait has also been calculated. How would you find the empirical probability of the recessive trait?

From his work Mendel concluded that the sex cells (now called gametes) of the pure yellow (dominant) pea plant carried some factor that caused the offspring to be yellow, and the gametes of the green variety had a variant factor that "induced the development of green plants." In 1909 The Danish geneticist W. Johannsen called these factors "genes." Mendel's work led to the understanding that each pea plant contains

DID YOU KNOW

The Royal Disease

The effect of genetic inheritance is dramatically demonstrated by the occurrence of hemophilia in some of the royal families of Europe. Great Britain's Queen Victoria (1819–1901) was the initial carrier. The disease was subsequently introduced into the royal lines of Prussia, Russia, and Spain through the marriages of her children.

Hemophilia is a disease that keeps blood from clotting. As a result, even a minor bruise or cut can be dangerous. It is also a recessive sex-linked disease. Females have a second gene that enables the blood to clot, which blocks the effects of the recessive carrier gene; males do not. So even though both males and females are carriers, the disease afflicts only males.

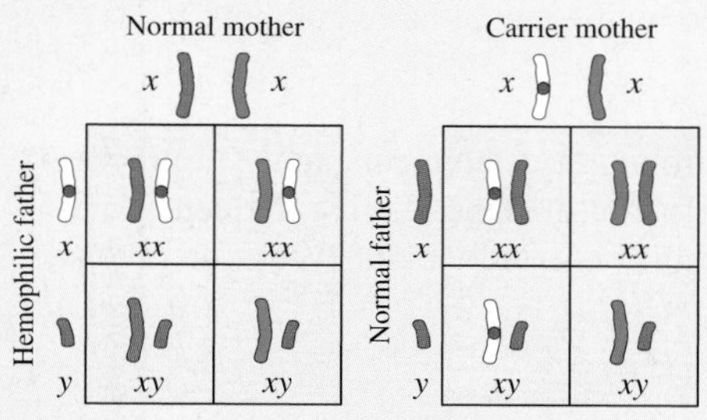

Queen Victoria had 9 children, 26 grandchildren, and 34 greatgrandchildren. Among them 1 son and 9 grandsons were hemophiliacs, and 2 daughters and 4 granddaughters were carriers of the gene for hemophilia. The genetic line of the present-day royal family is free of the disease.

In humans, genes are located on 23 pairs of *chromosomes.* Each parent contributes one member of each pair to a child. The gene that affects blood clotting is carried on the *x* chromosome. Females have two *x* chromosomes; males have one *x* chromosome and one *y* chromosome.

DID YOU KNOW

Batting Averages

If Larry Walker of the Colorado Rockies gets three hits in his first three at bats of the season, he is batting a thousand (1.000). But over the course of the 162 games of the season (with three or four at bats per game), his batting average will fall closer to .379 (his 1999 major league leading batting average). In 1999 out of 438 at bats, Walker had 166 hits, an average above all other players' but much less than 1.000. His batting average is a relative frequency (or empirical probability) of hits to at bats. It is only the long-term average that we take seriously because it is based on the law of large numbers.

"The laws of probability, so true in general, so fallacious in particular."
Edward Gibbon, 1796

two genes for color, one that comes from the mother and the other from the father. If the two genes are alike, for instance both for yellow plants or both for green plants, the plant will be that color. If the genes for color are different, the plant will grow the color of the dominant gene. Thus, if one parent contributes a gene for the plant to be yellow (dominant) and the other parent contributes a gene for the plant to be green (recessive), the plant will be yellow.

THE LAW OF LARGE NUMBERS

Most of us accept the fact that if a "fair coin" is tossed many, many times, it will land heads up approximately half of the time. Intuitively, we can guess that the probability that a fair coin will land heads up is $\frac{1}{2}$. Does this mean that if a coin is tossed twice, it will land heads up exactly once? If a fair coin is tossed 10 times, will there necessarily be five heads? The answer is clearly no. What then does it mean when we state that the probability that a fair coin will land heads up is $\frac{1}{2}$? To answer this question, let's examine Table 12.2, which shows what may occur when a fair coin is tossed a given number of times.

Table 12.2

Number of tosses	Expected number of heads	Actual number of heads observed	Relative frequency of heads
10	5	4	$\frac{4}{10} = 0.4$
100	50	43	$\frac{43}{100} = 0.43$
1,000	500	540	$\frac{540}{1{,}000} = 0.54$
10,000	5,000	4,852	$\frac{4{,}852}{10{,}000} = 0.4852$
100,000	50,000	49,770	$\frac{49{,}770}{100{,}000} = 0.49770$

The last column of Table 12.2, the relative frequency of heads, is a ratio of the number of heads observed to the total number of tosses of the coin. The relative frequency is the empirical probability, as defined earlier. Note as the number of tosses increases, the relative frequency of heads gets closer and closer to $\frac{1}{2}$, or 0.5, which is what we expect.

The nature of probability is summarized by the law of large numbers.

The **law of large numbers** states that probability statements apply in practice to a large number of trials—not to a single trial. It is the relative frequency over the long run that is accurately predictable, not individual events or precise totals.

What does it mean to say that the probability of rolling a 2 on a die is $\frac{1}{6}$? It means that over the long run, on the average, one of every six rolls will result in a 2.

SECTION 12.1 EXERCISES

CONCEPT/WRITING EXERCISES

1. What is an experiment?
2. a) What are outcomes of an experiment?
   b) What is an event?
3. What is empirical probability and how is empirical probability determined?
4. What are theoretical probabilities based on?
5. Explain in your own words the law of large numbers.
6. Explain in your own words why empirical probabilities are used in determining premiums for life insurance policies.
7. *Rain in Denver* A probability statement is often interpreted differently by different people. Consider the statement "The National Weather Service predicts that the probability of rain today in the Denver, Colorado, area is 70%." Below are five possible interpretations of that statement. Which, if any, do you think is its actual meaning?
   a) In 70% of the Denver area there will be rain.
   b) There is a 70% chance that at least any one point in the Denver area will receive rain.
   c) It has rained on 70% of the days with similar weather conditions in the Denver area.
   d) There is a 70% chance that a specific location in the Denver area will receive rain.
   e) There is a 70% chance that the entire Denver area will receive rain.

8. The probability of rolling a 2 on a die is $\frac{1}{6}$. Does this probability mean that, if a die is rolled six times, one 2 will appear? If not, what does it mean?
9. In order to determine premiums, life insurance companies must compute the probable date of death. On the basis of a great deal of research Mr. Reebe, age 36, is expected to live another 42.94 years. Does this determination mean that Mr. Reebe will live until he is 78.94 years old? If not, what does it mean?
10. a) Explain how you would find the empirical probability of rolling a 3 on a die.
    b) What do you believe is the empirical probability of rolling a 3?
    c) Find the empirical probability of rolling a 3 by rolling a die 50 times.

PRACTICE THE SKILLS

11. Flip a coin 40 times and record the results. Find the empirical probability of tossing a
    a) head.
    b) tail.
    c) Does the probability of tossing a head appear to be the same as tossing a tail?
12. Roll a die 50 times and record the results. Find the empirical probability of rolling a
    a) 1.
    b) 6.
    c) Does the probability of rolling a 1 appear to be the same as the probability of rolling a 6? Explain.
13. Roll a pair of dice 60 times and record the sums. Compute the empirical probability of rolling a sum of
    a) 2.
    b) 7.
    c) Does the probability of rolling a sum of 2 appear to be the same as the probability of rolling a sum of 7?
14. Toss two coins 50 times and record the number of times exactly one head was obtained. Compute the empirical probability of tossing exactly one head.

PROBLEM SOLVING

15. *Hair Color* Of the last 60 people who went to the cash register at Kmart 12 had blond hair, 15 had black hair, 28 had brown hair, and 5 had red hair. Determine the empirical probability that the next person to come to the cash register has
    a) red hair.
    b) brown hair.
    c) blond hair.
16. *Birds at a Feeder* The last 20 birds that fed at the Zwicks' bird feeder were 10 finches, 7 cardinals, and 3 blue jays. Use this information to determine the empirical probability that the next bird to feed from the feeder is a
    a) finch.
    b) cardinal.
    c) blue jay.

17. *Fishing* Six friends chartered a deep-sea fishing boat for a day's fishing in the Gulf of Mexico. They caught a total of 62 fish. The following chart provides information about the type and number of fish caught.

Fish	Number caught
Grouper	18
Shark	6
Flounder	30
Kingfish	8

Determine the empirical probability that the next fish caught is a

a) grouper.
b) shark.
c) flounder.

18. *Prader–Willi Syndrome* In a sample of 45,000 births, 3 were found to have Prader–Willi syndrome. Find the empirical probability that a family's first child will be born with this syndrome.

19. *Long Distance Carrier* Five hundred consumers were surveyed to determine the long distance telephone carrier they use. The results of the survey are shown in the figure below.

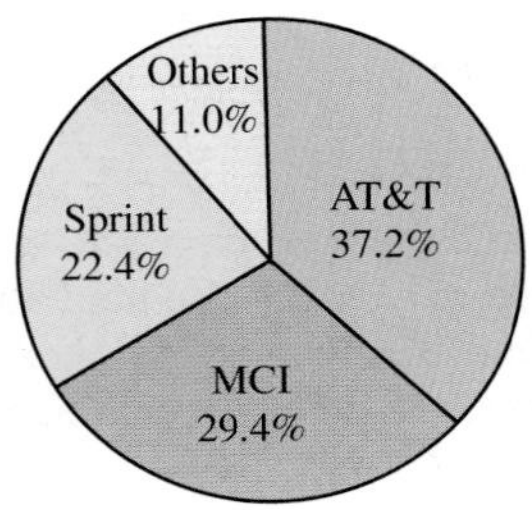

Determine the empirical probability that a person selected at random from the 500 surveyed uses

a) AT&T.
b) MCI.
c) How many of the 500 people surveyed used Sprint?

20. *Air Jordan* The following chart shows Michael Jordan's regular season field goals and field goal attempts from 1984–1998. Also shown are his free throws and free throw attempts during the same period. During this time period, determine the empirical probability of Michael Jordan making a

a) field goal
b) free throw.

Michael Jordan's career stats: regular season

	Games	*Field goals/ Attempts*	*Free throws/ Attempts*
(1984–1998)	930	10,962/21,686	6798/8115

(*Source: Newsweek,* January 25, 1999.)

21. *Dow Jones Gains* The following chart shows the gain made by the Dow–Jones Industrial Average (DJIA) in each year that ends in a 5 since records have been kept.

Years Ending in 5

Year	DJIA return
1885	27.7%
1895	2.3%
1905	38.2%
1915	86.5%
1925	30.0%
1935	38.5%
1945	26.6%
1955	20.8%
1965	10.9%
1975	38.3%
1985	27.7%
1995	36.8%
2005	?

a) What is the empirical probability that the DJIA will increase in a year ending in 5?
b) Is it possible that the DJIA could have a loss in the year 2005? Explain.

22. *Grade Distribution* Mr. Doole's grade distribution over the past 3 years for a course in college algebra is shown in the chart below.

Grade	Number
A	43
B	182
C	260
D	90
F	62
I	8

If Sue Gilligan plans on taking college algebra with Mr. Doole, determine the empirical probability she receives a grade of

a) A
b) C
c) a grade of D or higher

23. *Distribution of Women by Age* The following graph shows the distribution of U.S. women by age in the years 1950 and 1996.

a) If a woman was selected at random in 1950, what is the empirical probability the woman was 15–24 years old?
b) If a woman was selected at random in 1996, what is the empirical probability the woman was 15–24 years old?

c) Repeat parts (a) and (b) for 65+ years of age.

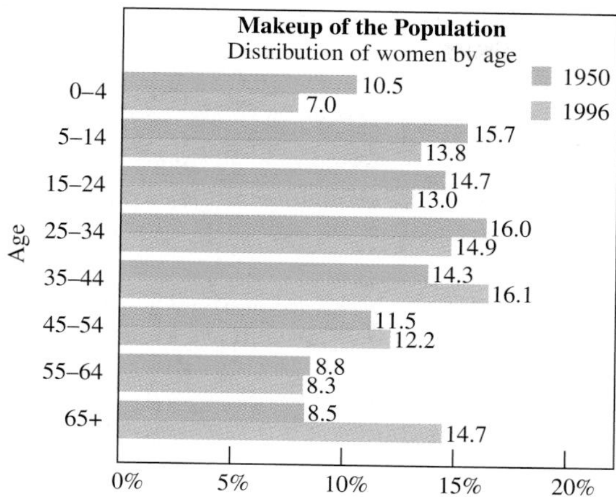

Source: Newsweek Special Edition on Woman's Health, 1999

24. *Repair History of Dishwashers* The following graph from *Consumer Reports* shows the frequency of repair for various brands of dishwashers.

a) Explain why this graph illustrates empirical probabilities.

b) Determine the empirical probability that a person who purchased a KitchenAid dishwasher from 1983 to 1988 had a repair and serious problems with this dishwasher.

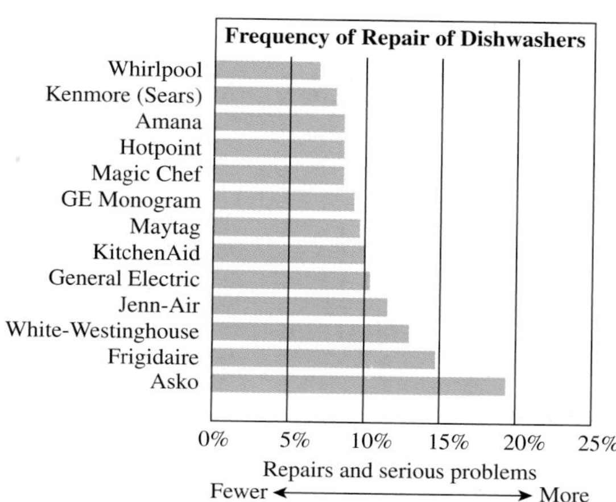

Source: Consumer Reports

25. *Hitting a Bulls-Eye* The pattern of hits shown on the target resulted from a marksman firing 20 rounds. For a single shot

a) find the empirical probability that the marksman hits the 50-point bulls-eye (the center of the target).

b) find the empirical probability that the marksman does not hit the bulls-eye.

c) find the empirical probability that the marksman scores at least 20 points.

d) find the empirical probability that the marksman does not score any points (the area outside the large circle).

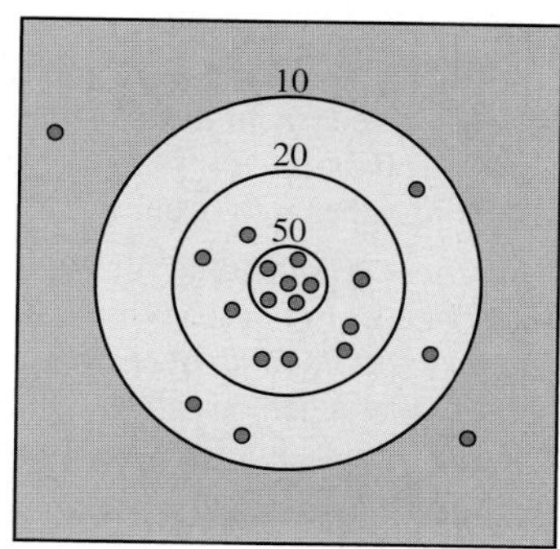

26. *Rock Toss* Jim finds an irregularly shaped five-sided rock. He labels each side and tosses the rock 100 times. The results of his tosses are shown in the table. Find the empirical probability that the rock will land on side 4 if tossed again.

Side	1	2	3	4	5
Frequency	32	18	15	13	22

27. *Cell Biology Experiment* An experimental serum was injected into 500 guinea pigs. Initially, 150 of the guinea pigs had circular cells, 250 had elliptical cells, and 100 had irregularly shaped cells. After the serum was injected, none of the guinea pigs with circular cells were affected, 50 with elliptical cells were affected, and all of those with irregular cells were affected. Find the empirical probability that a guinea pig with (a) circular cells, (b) elliptical cells, and (c) irregular cells will be affected by injection of the serum.

28. *Baby Gender* In the United States more male babies are born than female. In 1998, 2,264,031 males were born and 2,187,419 females were born. Find the empirical probability of an individual being born (a) male and (b) female.

29. *Mendel's Experiment* In one of Mendel's experiments, he crossbred nonpure purple flower pea plants. These purple pea plants had two traits for flowers, purple (dominant) and white (recessive). The result of this crossbreeding was 705 second-generation plants with purple flowers and 224 second-generation plants with white flowers. Find the empirical probability of a second-generation plant having

a) white flowers.

b) purple flowers.

30. *Second-Generation Offspring* In another experiment, Mendel crossbred nonpure tall pea plants. As a result, the second-generation offspring were 787 tall plants and 277 short plants. Find the empirical probability of a second-generation plant being **(a)** tall and **(b)** short.

CHALLENGE PROBLEMS/GROUP ACTIVITIES

31. a) *Design an Experiment* Do you believe that the word *a* or the word *the* is used more frequently?
   b) Design an experiment to determine the empirical probabilities (or relative frequencies) of the words *a* and *the* appearing in a book or newspaper article.
   c) Perform the experiment in part (b) and determine the empirical probabilities.
   d) Which word appears to occur more frequently?
32. *Cola Preference* Can people selected at random distinguish Coke from Pepsi? Which do they prefer?
   a) Design an experiment to determine the empirical probability that a person selected at random can select Coke when given samples of both Coke and Pepsi.
   b) Perform the experiment in part (a) and determine the probability.
   c) Determine the empirical probability that a person selected at random will prefer Coke over Pepsi.

RESEARCH ACTIVITIES

33. Write a paper on how insurance companies use empirical probabilities in determining insurance premiums. An insurance agent may be able to direct you to a source of information.
34. Write a paper on how Gregor Mendel's use of empirical probability led to the development of the science of genetics. You may want to check with a biology professor to determine references to use.

12.2 THEORETICAL PROBABILITY

To be able to do the problems in this section and the remainder of the chapter, you must have a thorough understanding of fractions. If you have forgotten how to use fractions, we strongly suggest that you review Section 5.3 before beginning this section.

Should you spend the 33 cents for a stamp to return the *Reader's Digest* Sweepstakes ticket? What are your chances of winning a lottery? If you go to a carnival, bazaar, or casino which games provide the greatest chance of winning? These and similar questions can be answered once you have an understanding of theoretical probability. *In the remainder of this chapter, the word* probability *will refer to theoretical probability.*

Recall from Section 12.1 that the results of an experiment are called outcomes. When you roll a die and observe the number of points that face up, the possible outcomes are 1, 2, 3, 4, 5, and 6. It is equally likely that you will roll any one of the possible numbers.

If each outcome of an experiment has the same chance of occurring as any other outcome, they are said to be **equally likely outcomes.**

Can you think of a second set of equally likely outcomes when a die is rolled? An odd number is as likely to be rolled as an even number. Therefore, odd and even are another set of equally likely outcomes.

If an event has *equally likely outcomes,* the probability of event *E*, symbolized by $P(E)$, may be calculated with the following formula.

Probability

$$P(E) = \frac{\text{number of outcomes favorable to } E}{\text{total number of possible outcomes}}$$

Example 1 illustrates how to use this formula.

EXAMPLE 1 ***Finding Probabilities***

A die is rolled. Find the probability of rolling

a) a 5. b) an odd number. c) a number greater than 4.
d) a 7. e) a number less than 7.

SOLUTION

a) There are six possible equally likely outcomes: 1, 2, 3, 4, 5, and 6. The event of rolling a 5 can occur in only one way.

$$P(5) = \frac{\text{number of outcomes that will result in a 5}}{\text{total number of possible outcomes}} = \frac{1}{6}$$

b) The event of rolling an odd number can occur in three ways (1, 3, 5).

$$P(\text{odd number}) = \frac{\text{number of outcomes that result in an odd number}}{\text{total number of possible outcomes}} = \frac{3}{6} = \frac{1}{2}$$

A Random Walk Down Wall Street

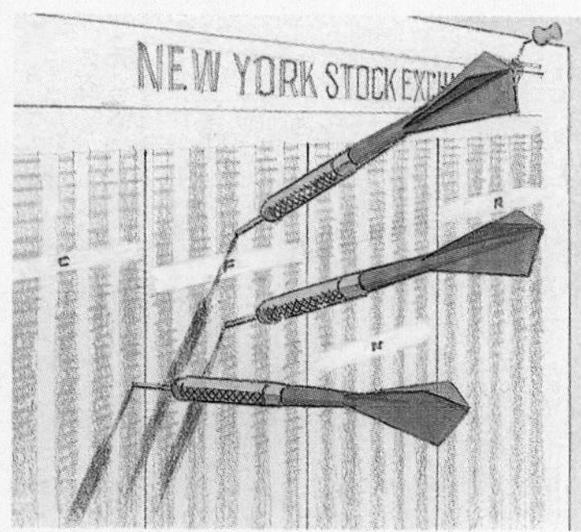

Some theorists have proposed that the best way to make money in the stock market is to randomly select a large, diversified portfolio of stocks. This hypothesis, known as "a random walk down Wall Street," states that the probability of a stock price going up or down is as random as flipping a coin and that commonly available information cannot help you predict the way it will go.

c) Two numbers are greater than 4, namely, 5 and 6.

$$P(\text{number greater than 4}) = \frac{2}{6} = \frac{1}{3}$$

d) No outcomes will result in a 7. Thus, the event cannot occur and the probability is 0.

$$P(7) = \frac{0}{6} = 0$$

e) All the outcomes 1 through 6 are less than 7. Thus, the event must occur and the probability is 1.

$$P(\text{number less than 7}) = \frac{6}{6} = 1$$

Four important facts about probability follow.

Important Facts

1. The probability of an event that cannot occur is 0.
2. The probability of an event that must occur is 1.
3. Every probability is a number between 0 and 1 inclusive; that is, $0 \leq P(E) \leq 1$.
4. The sum of the probabilities of all possible outcomes of an experiment is 1.

EXAMPLE 2 *Choosing One Bird from a List*

The names of 15 birds and their food preferences are listed in Table 12.3. Each of the birds' names is listed on a slip of paper, and the 15 slips are deposited in a bag. One slip is to be selected at random from the bag. Find the probability the slip containing

a) a sparrow is selected.
b) a bird that has a high attractiveness to peanut kernels is selected.
c) a bird that has a low attractiveness to peanut kernels, a low attractiveness to cracked corn, *and* a high attractiveness to black striped sunflower seeds is selected.
d) a bird that has a high attractiveness to either peanut kernels *or* cracked corn (or both) is selected.

Table 12.3 Birds and Their Food Preferences

Bird	**Peanut kernels**	**Cracked corn**	**Black striped sunflower seeds**
American goldfinch	L	L	H
Blue jay	H	M	H
Chickadee	M	L	H
Common grackle	M	H	H
Evening grosbeak	L	L	H
House finch	M	L	H
House sparrow	L	M	M
Mourning dove	L	M	M
Northern cardinal	L	L	H
Purple finch	L	L	H
Scrub jay	H	L	H
Song sparrow	L	L	M
Tufted titmouse	H	L	H
White-crowned sparrow	H	M	H
White-throated sparrow	H	H	H

H = high attractiveness; M = medium attractiveness; L = low attractiveness.

Source: How to Attract Birds (Ortho Books)

SOLUTION

a) Four of the 15 birds listed are sparrows (house sparrow, song sparrow, white-crowned sparrow, and white-throated sparrow).

$$P(\text{sparrow}) = \frac{4}{15}$$

b) Five of the 15 birds listed have a high attractiveness to peanut kernels.

$$P(\text{high attractiveness to peanut kernels}) = \frac{5}{15} = \frac{1}{3}$$

c) Reading across the rows reveals that 4 birds have a low attractiveness to peanut kernels and cracked corn and a high attractiveness to black striped sunflower seeds (American goldfinch, evening grosbeak, northern cardinal, and purple finch).

$$P(\text{low to peanuts, low to corn, and high to black sunflower seeds}) = \frac{4}{15}$$

d) The birds that have a high attractiveness to either peanut kernels or to cracked corn (or both) are the bluejay, common grackle, jay scrub, tufted titmouse, white-crowned sparrow, and white-throated sparrow.

$$P(\text{high to peanut kernels or cracked corn}) = \frac{6}{15} = \frac{2}{5}$$

In any experiment an event must either occur or not occur. *The sum of the probability that an event will occur and the probability that it will not occur is 1.* Thus, for any event A we conclude

$$P(A) + P(\text{not } A) = 1$$
$$\text{or } P(\text{not } A) = 1 - P(A).$$

For example, if the probability that event A will occur is $\frac{5}{12}$, the probability event A will not occur is $1 - \frac{5}{12}$, or $\frac{7}{12}$. Similarly, if the probability that event A will not occur is 0.3, the probability that event A will occur is $1 - 0.3 = 0.7$ or $\frac{7}{10}$. We make use of this concept in Example 3.

EXAMPLE 3 *Selecting One Card from a Deck*

A deck of 52 playing cards is shown. It consists of four suits: hearts, clubs, diamonds, and spades. Each suit has 13 cards, including numbered cards ace (1) through 10 and three picture (or face) cards, the jack, the queen, and the king. Hearts and diamonds are red; clubs and spades are black. There are 12 picture cards, consisting of 4 jacks, 4 queens, and 4 kings. One card is to be selected at random from the deck of cards. Find the probability that the card selected is

a) a 7.
b) not a 7.
c) a diamond.
d) a jack *or* queen *or* king (a picture card).
e) a heart *and* a club.
f) a card greater than 6 *and* less than 9.

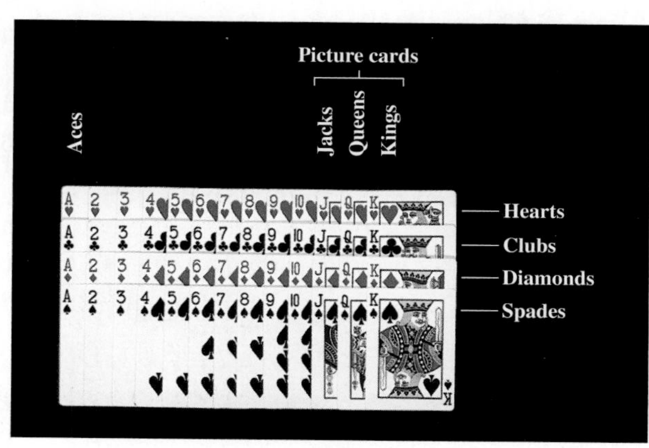

SOLUTION

a) There are four 7s in a deck of 52 cards.

$$P(7) = \frac{4}{52} = \frac{1}{13}$$

b) $P(\text{not a } 7) = 1 - P(7) = 1 - \frac{1}{13} = \frac{12}{13}$

This probability could also have been found by noting that there are 48 cards that are not 7s in a deck of 52 cards.

$$P(\text{not a } 7) = \frac{48}{52} = \frac{12}{13}$$

c) There are 13 diamonds in the deck.

$$P(\text{diamond}) = \frac{13}{52} = \frac{1}{4}$$

d) There are 4 jacks, 4 queens, and 4 kings, or a total of 12 picture cards.

$$P(\text{jack } or \text{ queen } or \text{ king}) = \frac{12}{52} = \frac{3}{13}$$

e) The word *and* means that *both* events must occur. Since it is not possible to select one card that is both a heart and a club, the probability is 0.

$$P(\text{heart and club}) = \frac{0}{52} = 0$$

f) The cards that are both greater than 6 and less than 9 are 7's and 8's. There are four 7's and four 8's, or a total of eight cards.

$$P(\text{greater than 6 } and \text{ less than 9}) = \frac{8}{52} = \frac{2}{13}$$

SECTION 12.2 EXERCISES

CONCEPT/WRITING EXERCISES

1. What are equally likely outcomes?
2. Explain in your own words how to find the theoretical probability of an event.
3. State the relationship that exists for $P(A)$ and $P(\text{not } A)$.
4. How many of each of the following are there in a standard deck of cards?
   a) Total cards
   b) Hearts
   c) Red cards
   d) Fives
   e) Black cards
   f) Picture cards
   g) Aces
   h) Queens
5. Using the definition of probability, explain in your own words why the probability of an event that cannot occur is 0.
6. Using the definition of probability, explain in your own words why the probability of an event that must occur is 1.
7. Between what two numbers (inclusively) will all probabilities lie?

8. What is the sum of all the probabilities of all possible outcomes of an experiment?

PRACTICE THE SKILLS

9. A multiple choice test has five possible answers for each question.
   a) If you guess at an answer, what is the probability that you select the correct answer for one particular question?
   b) If you eliminate one of the five possible answers and guess from the remaining possibilities, what is the probability that you select the correct answer to that question?

10. A TV remote has keys for channels 0 through 9. If you select one key at random,
    a) what is the probability that you press channel 3?
    b) what is the probability that you press a key for an even number?
    c) what is the probability that you press a key for a number less than 7?

11. In a lottery where one number is chosen, determine the probability you would win if you have a choice of 48 numbers to choose from. Explain your answer.

12. In a lottery where one number is chosen, determine the probability you would win if you have a choice of 52 numbers to choose from. Explain your answer.

In Exercises 13–22, one card is selected at random from a deck of cards. Find the probability that the card selected is

13. a 4.
14. a 4 or a 5.
15. not a 4.
16. the queen of hearts.
17. a heart.
18. a red card.
19. a red card or a black card.
20. a red card and a black card.
21. a card greater than 6 and less than 10.
22. a king and a diamond.

In Exercises 23–26, assume that the spinner cannot land on a line. Find the probability that the spinner lands on (a) red, (b) blue, (c) yellow.

23.

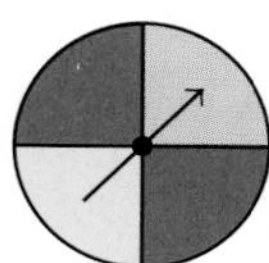

24.

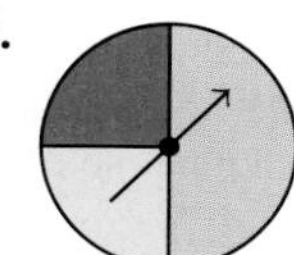

25.

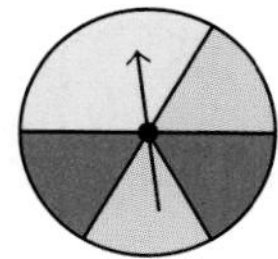

26.

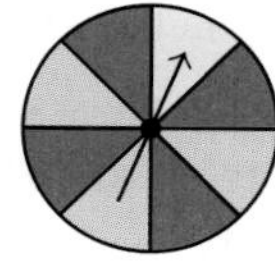

In Exercises 27–30, a bin contains 100 batteries (all size D). Forty are Eveready, 25 are Duracell, 20 are Sony, 10 are Panasonic, and 5 are Rayovac. One battery is selected at random from the bin. Find the probability that the battery selected is a

27. Duracell.
28. Duracell or Eveready.
29. Duracell, Eveready, or Sony.
30. Fuji.

In Exercises 31–34, use the small replica of the Wheel of Fortune.

If the wheel is spun at random, find the probability of the sector indicated stopping under the pointer.

31. \$400
32. Lose a turn or Bankrupt
33. A number greater than \$500
34. \$2500 or Surprise

In Exercises 35–38, 40 tennis balls including 18 Wilson, 12 Penn, and 10 other brand-name balls are on a tennis court. Barry Wood closes his eyes and arbitrarily picks up a ball from the court. Determine the probability the ball selected is

35. a Penn.
36. a Wilson.
37. not a Penn.
38. a Wilson or a Penn.

In Exercises 39–42, a traffic light is red for 30 sec, yellow for 5 sec, and green for 40 sec. What is the probability that when you reach the light,

39. the light is red.
40. the light is yellow.
41. the light is not red.
42. the light is not red or yellow.

In Exercises 43–48, each individual letter of the word "Chautauqua" is placed on a piece of paper, and all 10 pieces of paper are placed in a hat. If one letter is selected at random from the hat, find the probability that

43. the letter a is selected.

44. the letter a is not selected.

45. a vowel is selected.

46. the letter q or u is selected.

47. the letter t is not selected.

48. the letter r is selected.

PROBLEM SOLVING

The Environment *In Exercises 49–52, use the chart below that shows the 8 states in the United States with the greatest number of hazardous waste sites.*

State	Total
New Jersey	110
Pennsylvania	100
California	96
New York	80
Michigan	74
Florida	55
Washington	47
Illinois	41

Source: U.S. Environmental Protection Agency.

If one state from the list is picked at random, determine the probability the state has

49. exactly 80 hazardous waste sites.

50. greater than 80 hazardous waste sites.

If one of the *sites* is selected at random, determine the probability the site is from the state of

51. Florida.

52. New Jersey *or* New York.

In Exercises 53–56, a dart is thrown randomly and sticks on the circular dart board shown.

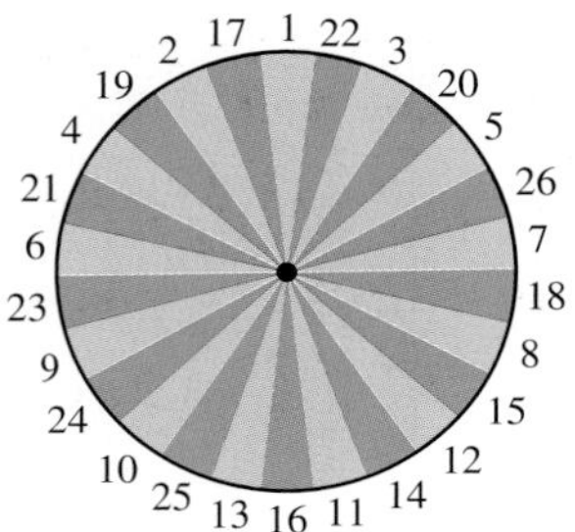

Assuming that the dart cannot land on the black area or on a border between colors, find the probability that the dart lands on

53. the area marked 25.

54. a green area.

55. an area marked with a number greater than or equal to 24.

56. an area marked with a number greater than 6 and less than or equal to 9.

Public or Private College *In Exercises 57–62, refer to the table, which shows preference among a sample of high school students as to the college they plan to attend.*

	Public college	Private college	Total
Males	192	73	265
Females	204	106	310
Total	396	179	575

If one student is selected at random from those surveyed, determine the probability the student is

57. a female.

58. a male.

59. planning on attending a public college.

60. planning on attending a private college.

61. a male who plans on attending a private college.

62. a female who plans on attending a public college.

Peanut Butter Preference *In Exercises 63–68, refer to the following table, which contains information about a sample of shoppers selecting various brands of peanut butter at a grocery store. Assume each shopper purchased exactly one jar of peanut butter.*

Brand	Smooth	Chunky	Total
Peter Pan	30	23	53
Jiff	28	22	50
Skippy	23	16	39
Other	12	5	17
Total	93	66	159

63. Determine the number of shoppers in this sample who purchased peanut butter.

If one shopper is selected at random, determine the probability the shopper selected

64. Jiff peanut butter.

65. Skippy peanut butter.

66. a chunky peanut butter.

67. a smooth peanut butter.

68. a peanut butter other than Peter Pan smooth.

In Exercises 69–73, a bean bag is randomly thrown onto the square table and does not touch a line.

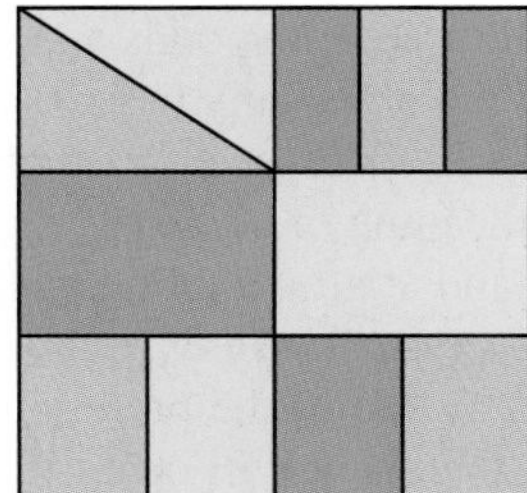

Find the probability that the bean bag lands on

69. a red area.

70. a green area.

71. a yellow area.

72. a red or green area.

73. a yellow or green area.

Before working Exercises 74 and 75, reread the material on genetics in Section 12.1.

74. *Genetics* Cystic fibrosis is an inherited disease that occurs in about 1 in every 2500 Caucasian births in North America and in about 1 in every 250,000 non-Caucasian births in North America. Let's denote the cystic fibrosis gene as c and a disease-free gene as C. Since the disease-free gene is dominant, only a person with cc genes will have the disease. A person who has Cc genes is a carrier of cystic fibrosis but does not actually have the disease. If one parent has CC genes and the other parent has cc genes, find the probability that

a) an offspring will inherit cystic fibrosis, that is, cc genes.

b) an offspring will be a carrier of cystic fibrosis but not contract the disease.

75. *Genetics* Sickle cell anemia is an inherited disease that occurs in about 1 in every 500 African-American births and about 1 in every 160,000 non-African-American births. Unlike cystic fibrosis, in which the cystic fibrosis gene is recessive, sickle cell anemia is *codominant.* In other words, a person inheriting two sickle cell genes will have sickle cell anemia, whereas a person inheriting only one of the sickle cell genes will have a mild version of sickle cell anemia, called *sickle cell trait.* Let's call the disease-free genes s_1 and the sickle cell gene s_2. If both parents have s_1s_2 genes, determine the probability that

a) an offspring will have sickle cell anemia.

b) an offspring will have the sickle cell trait.

c) an offspring will have neither sickle cell anemia nor the sickle cell trait.

CHALLENGE PROBLEMS/GROUP ACTIVITIES

In Exercises 76–78, the solutions involve material that we will discuss in later sections of the chapter. Try to solve them before reading ahead.

76. *Darts* Consider the figure accompanying Exercises 53–56. If one dart is thrown, find the probability that

a) the dart sticks on a red area.

b) the dart sticks on an even number.

c) the dart sticks on a red and even numbered area.

d) the dart sticks on a red area or an even numbered area. (To be discussed in Section 12.6.)

e) Using the probabilities found in parts (a)–(d), can you discover a formula for finding P(dart lands on red *or* an even numbered area)?

77. *Marbles* A bottle contains two red and two green marbles, and a second bottle also contains two red and two green marbles. If you select one marble at random from each bottle, find the probability (to be discussed in Section 12.6) that you obtain

a) two red marbles.

b) two green marbles.

c) a red marble from the first bottle and a green marble from the second bottle.

78. *Birds* Consider Table 12.3. Suppose you are told that one bird's name was selected from the birds listed and the bird selected has a low attractiveness to peanut kernels. Find the probability (to be discussed in Section 12.7) that

a) the bird is a sparrow.

b) the bird has a high attractiveness to cracked corn.

c) the bird has a high attractiveness to black striped sunflower seeds.

RESEARCH ACTIVITY

79. On page 576 we briefly discuss Jacob Bernoulli. The Bernoulli family produced several prominent mathematicians, including Jacob I, Johann I, and Daniel. Write a paper on the Bernoulli family, indicating some of the accomplishments of each of the three Bernoullis named and their relationship. Indicate which Bernoulli the Bernoulli numbers are named after, which Bernoulli the Bernoulli theorem in statistics is named after, and which Bernoulli the Bernoulli theorem of fluid dynamics is named after. References include encyclopedias, history of mathematics books, and the Internet.

12.3 ODDS

The odds against winning a lottery are 7 million to 1; the odds against being audited by the IRS this year are 50 to 1. We see the word *odds* daily in newspapers and magazines and often use it ourselves. Yet there is a widespread misunderstanding of its meaning. In this section we will explain the meaning of odds.

"We figured the odds as best we could, and then we rolled the dice."
Jimmy Carter, on his decision to run for president

The odds given at horse races, at craps, and at all gambling games in Las Vegas and other casinos throughout the world are always *odds against* unless they are otherwise specified. The **odds against** an event is a ratio of the probability that the event will fail to occur (failure) to the probability the event will occur (success). Thus, *in order to find odds you must first know or determine the probability of success and the probability of failure.*

$$\text{Odds against event} = \frac{P(\text{event fails to occur})}{P(\text{event occurs})} = \frac{P(\text{failure})}{P(\text{success})}$$

EXAMPLE 1 *Rolling a 2*

Find the odds against rolling a 2 on one roll of a die.

SOLUTION Before we can determine the odds, we must first determine the probability of rolling a 2 (success) and the probability of not rolling a 2 (failure). When a die is rolled there are six possible outcomes: 1, 2, 3, 4, 5, and 6.

$$P(\text{rolls a 2}) = \frac{1}{6} \qquad P(\text{fails to roll a 2}) = \frac{5}{6}$$

Now that we know the probabilities of success and failure, we can determine the odds against rolling a 2.

$$\text{Odds against rolling a 2} = \frac{P(\text{fails to roll a 2})}{P(\text{rolls a 2})} = \frac{\frac{5}{6}}{\frac{1}{6}} = \frac{5}{\cancel{6}_1}\cdot\frac{\cancel{6}^1}{1} = \frac{5}{1}$$

The ratio $\frac{5}{1}$ is commonly written as 5:1 and is read "5 to 1." Thus the odds against rolling a 2 are 5 to 1.

DID YOU KNOW

Gambling in Ancient Times

Archaeologists have found evidence of gambling in all cultures, from the Stone Age Australian aborigines to the ancient Egyptians, and across cultures touched by the Roman Empire. There is an equally long history of moral and legal opposition to gambling.

Note: The denominators of the probabilities in an odds problem will always divide out.

In Example 1 consider the possible outcomes of the die—1, 2, 3, 4, 5, 6. Over the long run, one of every six rolls will result in a 2, and five of every six rolls will result in a number other than 2. Therefore, for each dollar bet in favor of the rolling of a 2, \$5 should be bet against the rolling of a 2 if it is to be a fair game. The person betting in favor of the rolling of a 2 will either lose \$1 (if a number other than a 2 is rolled) or win \$5 (if a 2 is rolled). The person betting against the rolling of a 2 will either win \$1 (if a number other than a 2 is rolled) or lose \$5 (if a 2 is rolled). If this game is played for a long enough period, each player theoretically will break even.

DID YOU KNOW

Flying Is Safer

Many people fear the risks of air travel, but the odds against dying in an airplane crash are relatively low: approximately 800,000 to 1. The odds against dying in a car accident are approximately 5000 to 1. Thus, the odds in favor of dying in a car accident are much greater than of dying in an airplane accident.

DID YOU KNOW

What Were the Odds on This One?

In the year 1905, only two automobiles were registered in the entire state of Missouri.

In the year 1905, Missouri's only two registered automobiles were involved in a head-on collision!

EXAMPLE 2 *Watch Purchase*

At the Bond Jewelers in the Mall of America in Minneapolis, 4 out of every 13 purchases made are for watches. What are the odds against the next item being purchased at the jewelry store being a watch?

SOLUTION The probability that a watch is the next item purchased is $\frac{4}{13}$. Therefore, the probability that the next item purchased is not a watch is $1 - \frac{4}{13} = \frac{9}{13}$.

$$\text{Odds against next item purchased being a watch} = \frac{P(\text{next item purchased is not a watch})}{P(\text{next item purchased is a watch})}$$

$$= \frac{\frac{9}{13}}{\frac{4}{13}} = \frac{9}{\cancel{13}} \cdot \frac{\cancel{13}}{4} = \frac{9}{4} \quad \text{or } 9:4$$

Thus, the odds against the next item purchased at the jewelers being a watch is 9:4.

Although odds are generally given against an event, at times they may be given in favor of an event. The **odds in favor of** an event are expressed as a ratio of the probability that the event will occur to the probability that the event will fail to occur.

$$\text{Odds in favor of event} = \frac{P(\text{event occurs})}{P(\text{event fails to occur})} = \frac{P(\text{success})}{P(\text{failure})}$$

If the odds *against* an event are $a:b$, the odds *in favor of* the event are $b:a$.

EXAMPLE 3 *Attending the Least Expensive Colleges*

The following circle graph shows that most college students attend the least expensive colleges.

Most Attend Least Expensive Schools

Percent of full-time undergraduates attending schools by cost of tuition and fees, 1999–00:

Tuition and fees plus room and board, books, supplies, transportation

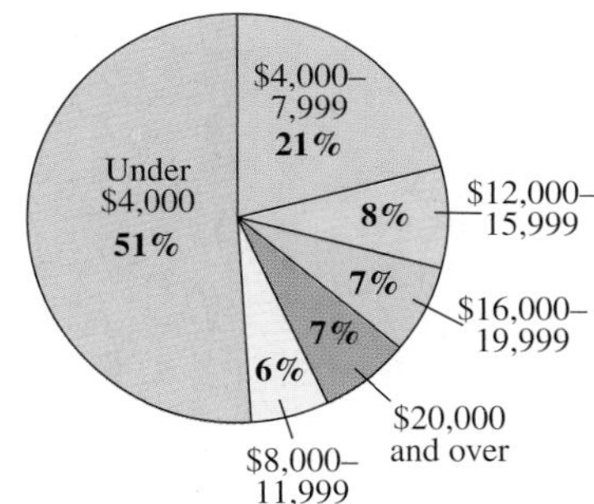

Source: The College Board

Use the graph to determine

a) The odds against a college student selected at random attending a college that costs $4000–$7999.

b) The odds in favor of a college student selected at random attending a college that costs $4000–$7999.

> **DID YOU KNOW**
>
> **One Lucky Man!**
>
> 
>
>
> Before you buy your next lottery ticket, consider the odds. Compare the odds of being struck by lightning (approximately 700,000 to 1) with your chances of winning a lottery. In Colorado, for example, where you must select 6 of 42 numbers, the odds against winning the lottery are about 5.2 million to 1. Thus, you are about 7.5 times more likely to be struck by lightning than to win the lottery. Given those odds, consider the case of a Colorado man, Don Whittman, 29, who won half of a $4 million jackpot on December 23, 1989, and won a $2.2 million jackpot again on October 22, 1991!

SOLUTION

a) The graph shows that 21%, or $\frac{21}{100}$, of all college students attend a college whose costs are \$4000–\$7999. Thus, the probability that a student attends a college whose costs are \$4000–\$7999 is $\frac{21}{100}$. The probability the student's college cost are not \$4000–\$7999 is therefore $1 - \frac{21}{100} = \frac{79}{100}$.

$$\text{Odds against college whose costs are \$4000–\$7999} = \frac{P(\text{college costs not \$4000–\$7999})}{P(\text{college costs \$4000–\$7999})}$$

$$= \frac{\frac{79}{100}}{\frac{21}{100}} = \frac{79}{\cancel{100}} \cdot \frac{\cancel{100}}{21} = \frac{79}{21} \quad \text{or } 79:21$$

Thus, the odds against a college student selected at random attending a college that costs \$4000–\$7999 are 79:21.

b) The odds in favor of a college student selected at random attending a college that costs \$4000–\$7999 is 21:79.

FINDING PROBABILITIES FROM ODDS

When odds are given, either in favor of or against a particular event, it is possible to determine the probabilities of that event. The denominators of the probabilities are found by adding the numbers in the odds statement. The numerators of the probabilities are the numbers given in the odds statements.

EXAMPLE 4 *Determining Probabilities from Odds*

The odds against Robin Murphy being admitted to the college of her choice are 9:2. Find the probability that (a) Robin is admitted and (b) Robin is not admitted.

SOLUTION

a) We have been given odds against and have been asked to find probabilities.

$$\text{Odds against being admitted} = \frac{P(\text{fails to be admitted})}{P(\text{is admitted})}$$

Since the odds statement is 9:2, the denominators of both the probability of success and failure must be 9 + 2 or 11. To get the odds ratio of 9:2, the probabilities must be $\frac{9}{11}$ and $\frac{2}{11}$. Since odds against is a ratio of failure to success, the $\frac{9}{11}$ and $\frac{2}{11}$ represent the probabilities of failure and success, respectively. Thus, the probability that Robin is admitted (success) is $\frac{2}{11}$.

b) The probability that Robin is not admitted (failure) is $\frac{9}{11}$.

Odds and probability statements are sometimes stated incorrectly. For example, consider the statement, "The odds of being selected to represent the district are 1 in 5." Odds are given using the word *to,* not *in.* Thus, there is a mistake in this statement. The correct statement might be, "The odds of being selected to represent the district are 1 to 5" or "The probability of being selected to represent the district is 1 in 5." Without additional information, it is not possible to tell which is the correct interpretation.

SECTION 12.3 EXERCISES

CONCEPT/WRITING EXERCISES

1. Explain in your own words how to determine the odds against an event.
2. Explain in your own words how to determine the odds in favor of an event.
3. Which odds are generally quoted, odds against or odds in favor?
4. Explain how to determine probabilities when you are given an odds statement.
5. The odds in favor of winning the door prize are 2 to 15. Find the odds against winning the door prize.
6. The odds against Speedo winning the horse race are 5:2. Find the odds in favor of Speedo winning.
7. If the odds against an event are 1:1 what is the probability the event will
   a) occur.
   b) fail to occur.
   Explain your answer.
8. If the probability an event will occur is $\frac{1}{2}$, determine
   a) the probability the event will fail to occur.
   b) the odds against the event occurring.
   c) the odds in favor of the event occurring.
   Explain your answer.

PRACTICE THE SKILLS/PROBLEM SOLVING

9. *Dressing Up* Lalo is going to wear a blue sportcoat and is trying to decide what tie he should wear to work. In his closet he has 24 ties, 11 of which he feels go well with the sportcoat. If Lalo selects one tie at random, determine
   a) the probability it goes well with the sportcoat.
   b) the probability it goes not go well with the sportcoat.
   c) the odds against it going well with the sportcoat.
   d) the odds in favor of it going well with the sportcoat.
10. *Making a Donation* In her wallet, Peg Hovde has 12 bills. Six are \$1 bills, two are \$5 bills, three are \$10 bills and one is a \$20 bill. She passes a volunteer seeking donations for the American Red Cross and she decides to select one bill at random from her wallet and give it to the Red Cross. Determine
   a) the probability she selects a \$5 bill.
   b) the probability she does not select a \$5 bill.
   c) the odds in favor of her selecting the \$5 bill.
   d) the odds against her selecting the \$5 bill.

In Exercises 11–14, a die is tossed. Find the odds against rolling

11. a 5.
12. an odd number.
13. a number greater than 4.
14. a number less than 3.

In Exercises 15–18, a card is picked from a deck of cards. Find the odds against and the odds in favor of selecting

15. a 6.
16. a heart.
17. a picture card.
18. a card greater than 5 (ace is low).

In Exercises 19–22, assume that the spinner cannot land on a line. Find the odds against the spinner landing on the color red.

19.
20.
21.
22.

23. One person is selected at random from a class of 15 males and 12 females. Find the odds against selecting
   a) a female.
   b) a male.
24. One million tickets are sold for a lottery.
   a) If you purchase a ticket, find your odds against winning.
   b) If you purchase 10 tickets, find your odds against winning.

In Exercises 25–30, use the rack of 15 billiard balls shown.

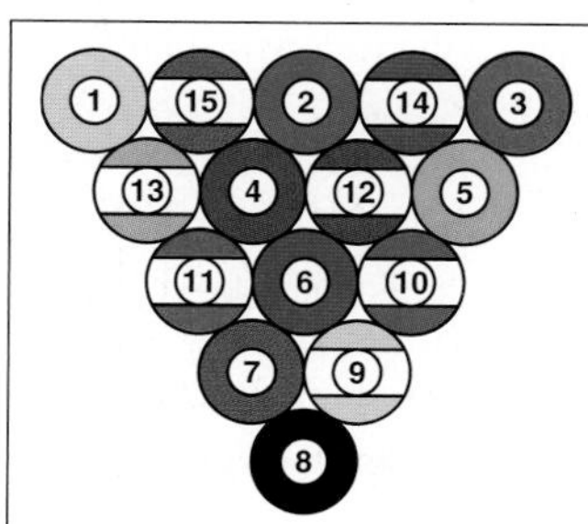

25. If one ball is selected at random, find the odds against it containing a stripe. (Balls numbered 9 through 15 contain stripes.)
26. If one ball is selected at random, find the odds in favor of it being an even-numbered ball.
27. If one ball is selected at random, find the odds in favor of it being a ball other than the 8 ball.
28. If one ball is selected at random, find the odds against it containing any yellow coloring (solid or striped).
29. If one ball is selected at random, find the odds against it containing a number greater than or equal to 9.
30. If one ball is selected at random, find the odds in favor of it containing two digits.

31. *Medical Tests* The results of a medical test show that of 72 people selected at random who were given the test, 70 tested negative and 2 tested positive. Determine the odds against a person selected at random testing negative on the test. Explain how you determined your answer.

32. *A Red Marble* A box contains many marbles. You grab a handful of marbles from the box and get 9 red and 2 blue marbles. Using the handful of marbles selected, determine the odds against selecting a red marble from the box. Explain how you determined your answer.

33. *Spelling Bee* The odds in favor of Carrie Hartmann winning the spelling bee are 7:5. Find the probability that
a) Carrie wins.
b) Carrie does not win.

34. *Tennis Match* The odds in favor of Claire Pearson winning a tennis match are 1:6. Find the probability that Claire will
a) win the match.
b) not win the match.

35. *Getting Promoted* The odds against Jason getting promoted are 5:9. Find the probability that Jason gets promoted.

36. *Winning a Race* The odds against Paul winning the 100 yard dash are 5:2. Find the probability that
a) Paul wins.
b) Paul loses.

Playing Bingo The game of bingo is played in many states. When playing bingo, 75 balls are placed in a bin and balls are selected at random. Each ball is marked with a letter and number as indicated in the following chart.

B	I	N	G	O
1–15	16–30	31–45	46–60	61–75

For example, there are balls marked B1, B2, up to B15; I16, I17, up to I30; and so on. In Exercises 37–42, assume one bingo ball is selected at random, determine

37. the probability it contains the letter *N*.

38. the probability it does not contain the letter *N*.

39. the odds in favor of it containing the letter *N*.

40. the odds against it containing the letter *N*.

41. the odds against it being I27.

42. the odds in favor of it being I27.

Absences from Work *In Exercises 43–48, use the following graphs, which show the reasons for unscheduled absences from work for the years 1995 and 1998.*

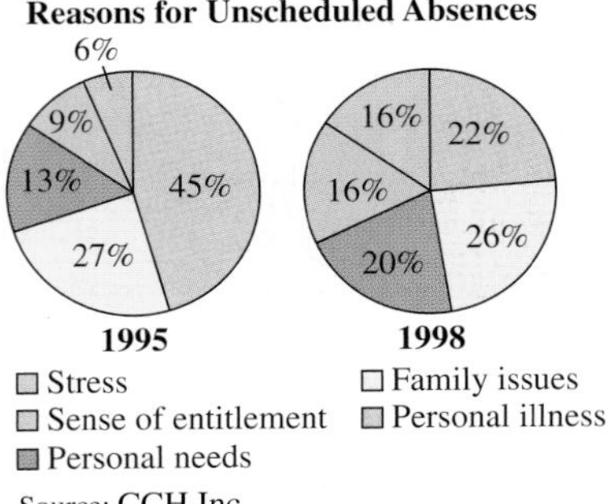

Use the graphs to determine

43. the probability an unscheduled absence was for stress in 1995.

44. the probability an unscheduled absence was not for stress in 1995.

45. the odds against an unscheduled absence being for stress in 1995.

46. the odds in favor of an unscheduled absence being for stress in 1995.

47. the odds in favor of an unscheduled absence being due to a sense of entitlement in 1998.

48. the odds against an unscheduled absence being due to a sense of entitlement in 1998.

49. *Selling a Car* Suppose the probability that you sell your car this week is 0.4. Find the odds against selling your car this week.

50. *Working Overtime* Suppose that the probability that you are asked to work overtime this week is 3/8. Find the odds in favor of your being asked to work overtime.

51. *Car Repair* Suppose that the probability that the mechanic fixes your car right the first time is 0.8. Find the odds against your car being repaired right on the first attempt.

52. *IRS Audit* One in 40 individuals whose salaries range between $10,000 and $40,000 will be randomly selected to have their income tax returns audited for this year. Mr. Frank is in this income tax range. Find
a) the probability that Mr. Frank will be audited.
b) the odds against Mr. Frank being audited.

53. *Arthritis* Gout constitutes about 5% of all systemic arthritis, and it is uncommon in women. The male-to-female ratio of gout is estimated to be 20 to 1.
a) If J. Douglas has gout, what are the odds against J. Douglas's being female?
b) If J. Douglas has gout, what is the probability that J. Douglas is a male?

CHALLENGE PROBLEMS/GROUP ACTIVITIES

54. *Odds Against* Find the odds against an even number or a number greater than 3 being rolled on a die.

55. *Horse Racing* Racetracks quote the approximate odds against each horse winning on a large board called a *tote board.* The odds quoted on a tote board for a race with five horses is as follows.

Horse number	Odds
1	7:2
2	2:1
3	15:1
4	7:5
5	1:1

Find the probability of each horse winning the race. (Do not be concerned that the sum of the probabilities is not 1.)

56. *Roulette* Turn to the roulette wheel illustrated on page 605. If the wheel is spun, find

a) the probability that the ball lands on black.
b) the odds against the ball landing on black.
c) the probability that the ball lands on 0 or 00.
d) the odds in favor of the ball landing on 0 or 00.

RESEARCH ACTIVITIES

57. *State Lottery* Determine whether your state has a lottery. If so, do research and write a paper indicating

a) the probability of winning the grand prize.
b) the odds against winning the grand prize.
c) Explain, using real objects such as pennies or ping pong balls, what these odds really mean.

58. *Casino Advantages* There are many types of games of chance to choose from at casinos. The house has the advantage in each game, but the advantages differ according to the game.

a) List the games available at a typical casino.
b) List those for which the house has the smallest advantage.
c) List those for which the house has the greatest advantage of winning.

Your local library and the Internet have many sources of information available on this topic.

12.4 EXPECTED VALUE (EXPECTATION)

Expected value, also called **expectation,** is often used to determine the expected results of an experiment or business venture *over the long run.* Expectation is used to make important decisions in many different areas. In business, for example, expectation is used to predict future profits of a new product. In the insurance industry, expectation is used to determine how much each insurance policy should cost for the company to make an overall profit. Expectation is also used to predict the expected gain or loss in games of chance such as the lottery, roulette, craps, and slot machines.

Consider the following: Tim tells Barbara that he will give her \$1 if she can roll an even number on a single die. If she fails to roll an even number, she must give Tim \$1. Who would win money in the long run if this game were played many times? We would expect in the long run that half the time Tim would win \$1 and half the time he would lose \$1, therefore, Tim would break even. Mathematically, we could find Tim's expected gain or loss by the following procedure:

$$\text{Tim's expected gain or loss} = P\begin{pmatrix}\text{Tim}\\\text{wins}\end{pmatrix}\begin{pmatrix}\text{amount}\\\text{Tim wins}\end{pmatrix} + P\begin{pmatrix}\text{Tim}\\\text{loses}\end{pmatrix}\begin{pmatrix}\text{amount}\\\text{Tim loses}\end{pmatrix}$$
$$= \frac{1}{2}(\$1) + \frac{1}{2}(-\$1) = \$0$$

DID YOU KNOW

Tough Decision

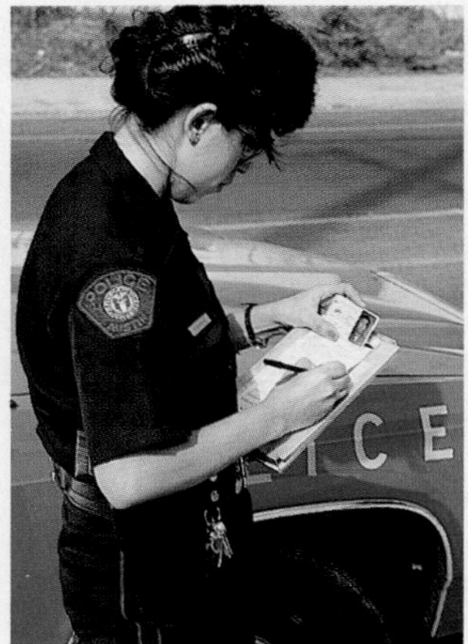

The concept of expected value can be used to help evaluate the consequences of many decisions. You use this concept when you consider whether to double-park your car for a few minutes. The probability of being caught may be low, but the penalty, a parking ticket, may be high. You weigh these factors when you decide whether to spend time looking for a legal parking spot.

Note that the loss is written as a negative number. This procedure indicates that Tim has an expected gain or loss (or expected value) of \$0. The expected value of zero indicates that he would indeed break even, as we had anticipated. Thus, the game is a *fair game*. If his expected value were positive, it would indicate a gain; if negative, a loss.

The expected value, E, is calculated by multiplying the probability of an event occurring by the *net* amount gained or lost if the event occurs. If there are a number of different events and amounts to be considered, we use the following formula.

Expected Value

$$E = P_1A_1 + P_2A_2 + P_3A_3 + \cdots + P_nA_n$$

The symbol P_1 represents the probability that the first event will occur, and A_1 represents the net amount won or lost if the first event occurs. P_2 is the probability of the second event, and A_2 is the net amount won or lost if the second event occurs, and so on. The sum of these products of the probabilities and their respective amounts is the expected value. The expected value is the average (or mean) result that would be obtained if the experiment were performed a great many times.

EXAMPLE 1 *A New Business Venture*

Northeast Airlines is considering adding a route to the city of Austin, Texas. Before the company makes their decision as to whether or not to service Austin, they need to consider many factors, including their potential profits and losses. Factors that may affect their profits and losses include the number of competing airlines, the potential number of customers, the overhead costs, fees they must pay, and so on. After considerable research, the company estimates that if they serve Austin, there is a 60% chance of making an \$800,000 profit, a 10% chance of breaking even, and a 30% chance of losing \$1,200,000. How much can Northeast Airlines "expect" to make on this new route?

SOLUTION The three amounts to be considered are a gain of \$800,000, breaking even at \$0, and a loss of \$1,200,000. The probability of gaining \$800,000 is 0.6, the probability of breaking even is 0.1, and the probability of losing \$1,200,000 is 0.3.

$$\begin{aligned}\text{Northeast's expectation} &= \underbrace{P_1A_1}_{\text{Gain}} + \underbrace{P_2A_2}_{\text{Break even}} + \underbrace{P_3A_3}_{\text{Loss}}\\ &= (0.6)(\$800{,}000) + (0.1)(\$0) + (0.3)(-\$1{,}200{,}000)\\ &= \$480{,}000 + \$0 - \$360{,}000\\ &= \$120{,}000\end{aligned}$$

Northeast Airlines has an expectation, or expected average gain, of \$120,000 for adding this particular service. This means that if they opened routes like this, with these particular probabilities and amounts, in the long run they would have an average gain of \$120,000 per route. However, you must remember that there is a 30% chance they will lose \$1,200,000 on this *particular* route (or any particular route with these probabilities and amounts.)

To obtain the fair price, using gross amounts when more than one amount is awarded, find the sum of the products of the probabilities and their respective amounts. If, for example, two amounts are awarded, we would use the formula

$$\text{Fair price} = P\begin{pmatrix}\text{winning}\\\text{amount 1}\end{pmatrix}\begin{pmatrix}\text{gross}\\\text{amount 1}\end{pmatrix} + P\begin{pmatrix}\text{winning}\\\text{amount 2}\end{pmatrix}\begin{pmatrix}\text{gross}\\\text{amount 2}\end{pmatrix}.$$

EXAMPLE 4 *Lottery Tickets*

One thousand lottery tickets are sold for \$1 each. One grand prize of \$500 and two consolation prizes of \$100 will be awarded. Using *net amounts,* find

a) Fred Zerla's expectation if he purchases one ticket.
b) Fred's expectation if he purchases five tickets.
c) What is the fair price of a ticket?

SOLUTION

a) Three amounts are to be considered: the net gain in winning the grand prize, the net gain in winning the consolation prize, and the loss of the cost of the ticket. If Fred wins the grand prize, his net gain is \$499 (\$500 minus \$1 spent for the ticket). If Fred wins the consolation prize, his net gain is \$99 (\$100 minus \$1). We assume that the winners' names are replaced in the pool after being selected. The probability that Fred wins the grand prize is $\frac{1}{1000}$. Since two consolation prizes will be awarded, the probability that he wins a consolation prize is $\frac{2}{1000}$. The probability that he does not win either prize is $1 - \frac{3}{1000} = \frac{997}{1000}$.

$$\begin{aligned} E &= P_1A_1 + P_2A_2 + P_3A_3 \\ &= \frac{1}{1000}(\$499) + \frac{2}{1000}(\$99) + \frac{997}{1000}(-\$1) \\ &= \frac{499}{1000} + \frac{198}{1000} - \frac{997}{1000} = -\frac{300}{1000} = -\$0.30 \text{ or } -30¢ \end{aligned}$$

b) On average, Fred loses 30¢ on each ticket purchased. On five tickets his expectation is $(-\$0.30)(5)$, or $-\$1.50$.
c) We learned in part (a) that Fred's expectation on one ticket is $-\$0.30$.

$$\begin{aligned} \text{Expectation} &= \text{fair price} - \text{cost} \\ -0.30 &= f - 1.00 \\ 0.70 &= f \end{aligned}$$

The fair price of a ticket is 70¢.

The fair price and expectation of the ticket in Example 4 may also be found using *gross amounts* as follows.

$$\text{Fair price} = \frac{1}{1000}(\$500) + \frac{2}{1000}(\$100) = \frac{500}{1000} + \frac{200}{1000} = \frac{700}{1000} = \$0.70$$

$$\begin{aligned} \text{Expectation} &= \text{fair price} - \text{cost to play} \\ &= \$0.70 - \$1.00 = -\$0.30 \end{aligned}$$

In expectation problems, the amount does not have to be a monetary amount, as is illustrated in Example 5.

EXAMPLE 5 ***Pothole Repairs***

A highway crew repairs 30 potholes a day in dry weather and 12 potholes a day in wet weather. If the weather in this region is wet 40% of the time, find the expected (average) number of potholes that can be repaired per day.

SOLUTION The amounts in this problem are the number of potholes repaired. Since the weather is wet 40% of the time, it will be dry 60% (100% − 40%) of the time:

$$\begin{aligned} E &= P(\text{dry})(\text{amount repaired}) + P(\text{wet})(\text{amount repaired}) \\ &= 0.60(30) + 0.40(12) = 18.0 + 4.8 = 22.8 \end{aligned}$$

Thus, the average, or expected, number of potholes repaired per day is 22.8. ▲

SECTION 12.4 EXERCISES

CONCEPT/WRITING EXERCISES

1. What does the expected value of an experiment or business venture represent?
2. What does an expected value of 0 mean?
3. What is the fair price of a game of chance?
4. Write the formula used to find the expected value of an experiment with
   a) two possible outcomes.
   b) three possible outcomes.
5. Is the fair price to pay for a game of chance the same as the expected value of that game of chance? Explain your answer.
6. If the expected value and cost to play are known for a particular game of chance, explain how you can determine the fair price to pay to play that game of chance.
7. If a particular game costs $1.50 to play and the expectation for the game is −$1.00, what is the fair price to pay to play the game? Explain how you determined your answer.
8. Give the formula for determining the fair price to pay to play a particular game of chance with three possible *gross* amounts that can be won.

PRACTICE THE SKILLS/PROBLEM SOLVING

9. If on a $2 bet Marty's expected value is −$0.40,
   a) what is Marty's expected value on a $10 bet?
   b) how much can Marty expect to win or lose if he places a $10 bet? Explain.
10. If on a $1 bet, Paul's expected value is $0.20,
    a) what is Paul's expected value on a $5 bet?
    b) how much can Paul expect to win or lose if he places a $5 bet? Explain.
11. *A New Business* In a proposed business venture Stephanie Morrison estimates there is a 65% chance she will make $70,000 and a 35% chance she will lose $30,000. Determine Stephanie's expected value.
12. *Seminar Attendance* At an investment tax seminar, Judy Johnson estimates that 20 people will attend if it does not rain and 12 people will attend if it rains. The weather forecast indicates there is a 40% chance it will not rain and a 60% chance it will rain on the day of the seminar. Determine the expected number of people who will attend the seminar.
13. *Basketball* Paige Sauer is a star player for the University of Connecticut women's basketball team. She has injured her ankle and it is doubtful if she will be able to play. If she can play, the coach estimates they will score 78 points. If she is not able to play, the coach estimates they will score 62 points. The team doctor estimates there is a 50–50 chance she will play (the probability of her playing is therefore $\frac{1}{2}$). Determine the number of points the team can expect to score.

EXAMPLE 2 *Test-Taking Strategy*

Maria is taking a multiple choice exam in which there are five possible answers for each question. The instructions indicate that she will be awarded 2 points for each correct response, that she will lose $\frac{1}{2}$ point for each incorrect response, and that no points will be added or subtracted for answers left blank.

a) If Maria does not know the correct answer to a question, is it to her advantage or disadvantage to guess at an answer?
b) If she can eliminate one of the possible choices, is it to her advantage or disadvantage to guess at the answer?

SOLUTION

a) Let's determine the expected value if Maria guesses at an answer. Only one of five possible answers is correct.

$$P(\text{guesses correctly}) = \frac{1}{5} \qquad P(\text{guesses incorrectly}) = \frac{4}{5}$$

$$\begin{aligned}\text{Maria's expectation} &= \underbrace{P_1A_1}_{\text{Guesses correctly}} + \underbrace{P_2A_2}_{\text{Guesses incorrectly}} \\ &= \frac{1}{5}(2) + \frac{4}{5}\left(-\frac{1}{2}\right) \\ &= \frac{2}{5} - \frac{2}{5} = 0\end{aligned}$$

Thus, Maria's expectation is zero when she guesses. Thus, over the long run she will neither gain nor lose points by guessing.

b) If Maria can eliminate one possible choice, one of four answers will be correct.

$$P(\text{guesses correctly}) = \frac{1}{4} \qquad P(\text{guesses incorrectly}) = \frac{3}{4}$$

$$\begin{aligned}\text{Maria's expectation} &= \underbrace{P_1A_1}_{\text{Guesses correctly}} + \underbrace{P_2A_2}_{\text{Guesses incorrectly}} \\ &= \frac{1}{4}(2) + \frac{3}{4}\left(-\frac{1}{2}\right) \\ &= \frac{2}{4} - \frac{3}{8} = \frac{4}{8} - \frac{3}{8} = \frac{1}{8}\end{aligned}$$

Since the expectation is a positive $\frac{1}{8}$, Maria will, on average, gain $\frac{1}{8}$ point each time she guesses when she can eliminate one possible choice. ▲

The amounts in the expectation formula are **net amounts**, or the actual amounts won or lost. Consider Example 3, which illustrates the use of net amounts.

EXAMPLE 3 *Winning a Raffle*

A charity sells 100 tickets for \$2.00 each for a \$50 door prize to be awarded. Susan just purchased one ticket. Find her expectation.

DID YOU KNOW

It Pays to Be Original

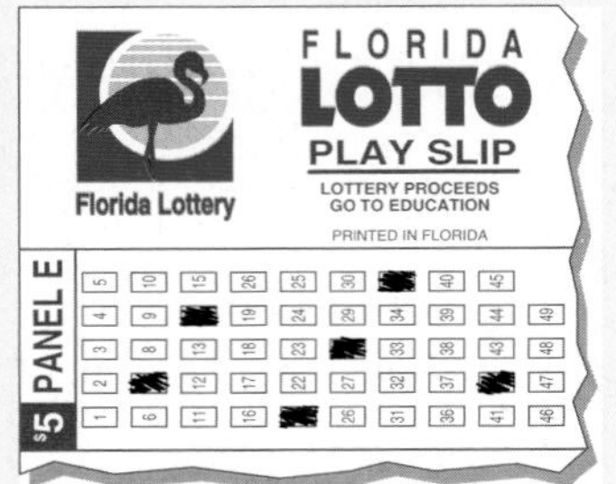

Commonly Selected Numbers

7-14-21-28-35-42
1-2-3-4-5-6
5-10-15-20-25-30
3-8-13-18-23-28

The numbers you select when picking lottery numbers has no effect on your probability of winning. However, your expectation (expected winnings) varies greatly with the numbers you select. Because the jackpot is divided among all the winners, the fewer the winners the more each winner receives. There are some groups of numbers that are commonly selected. Some are illustrated above. In a lottery with a large jackpot, there may be as many as 10,000 people who select the numbers 7 - 14 - 21 - 28 - 35 - 42. If the jackpot was $40 million, and these numbers were selected, each winner would receive about $4000. This is quite a difference from the $40 million that a single winner would receive. Many people use birthdays or other dates when selecting lottery numbers. Therefore, there may be fewer people who select the large numbers.

SOLUTION Two amounts are to be considered: the net amount Susan may win and the cost of the ticket. If Susan wins, her net or actual winnings are $48 (the $50 prize minus her $2 cost of the ticket). If she does not win, she loses the $2 paid for the ticket. Since only one prize will be awarded, Susan's probability of winning is $\frac{1}{100}$. Her probability of losing is, therefore, $\frac{99}{100}$.

$$\begin{aligned} E &= P(\text{Susan wins})(\text{amount won}) + P(\text{Susan loses})(\text{amount lost}) \\ &= \frac{1}{100}(48) + \frac{99}{100}(-2) \\ &= \frac{48}{100} - \frac{198}{100} = -\frac{150}{100} = -\$1.50 \end{aligned}$$

Susan's expectation is $-\$1.50$ per ticket. ▲

The **fair price** of a game of chance is the amount that should be charged for the game to be fair and result in an expectation of 0. If the fair price and cost to play a game are known, the expectation may be found by the formula

Expectation = fair price − cost to play

If any two of the three items in the formula are known, the third may be found. For instance, in Example 3, Susan's expectation is $-\$1.50$ and the cost of a ticket is $2.00; thus, the fair price, f, may be calculated as follows.

$$\begin{aligned} \text{Expectation} &= \text{fair price} - \text{cost to play} \\ -1.50 &= f - 2.00 \\ 2.00 - 1.50 &= f - 2.00 + 2.00 \\ 0.50 &= f \end{aligned}$$

Thus, the fair price of a ticket is $0.50, or 50¢. This price makes sense because, if the price of a ticket were reduced by $1.50 to 50¢, Susan's expectation would be $0.00. Determine Susan's expectation now using a cost to play of $0.50.

Expectation problems requiring money to be paid in advance, as in Example 3, can be solved by an alternative procedure. First, determine the fair price to play by using the formula

$$\textbf{Fair price} = P_1G_1 + P_2G_2 + \cdots + P_nG_n$$

where each P represents the probability of winning and each corresponding G represents the **gross amount** won. The gross amounts do not include the cost to play the game. After determining the fair price, use the formula, Expectation = fair price − cost to play, to determine the expectation.

The expectation of the ticket in Example 3 can be found as follows

$$\text{Fair price} = P_1G_1 = \frac{1}{100}(\$50) = \frac{50}{100} = \$0.50$$

$$\begin{aligned} \text{Expectation} &= \text{fair price} - \text{cost to play} \\ &= \$0.50 - \$2.00 = -\$1.50 \end{aligned}$$

Using either procedure, we see that Susan's expectation is $-\$1.50$.

14. *Tax Law Changes* Alicia Seeway, an investment counselor, is advising her client on a particular investment. She estimates that if the tax law does not change, the client will make \$12,000 but if the tax law changes, the client will lose \$3,000. Find the client's expected value if there is a 70% chance that the tax law will change.

15. *Seattle Greenery* In July in Seattle, the grass grows $\frac{1}{2}$ in. a day on a sunny day and $\frac{1}{4}$ in. a day on a cloudy day. In Seattle in July, 75% of the days are sunny and 25% are cloudy.
 a) Find the expected amount of grass growth on a typical day in July in Seattle.
 b) Find the expected total grass growth in the month of July in Seattle.

16. *Buying Stock* The Palm Coast investment club is considering purchasing a certain stock. After considerable research, the club members determine that there is a 60% chance of making \$8000, a 10% chance of breaking even, and a 30% chance of losing \$6200. Find the expectation of this purchase.

17. *Clothing Sale* At the Crescent Oaks Country Club, they have an annual clothing sale. You select the items you plan to purchase, the cashier rings up your purchase, then you select a slip of paper from a bag which indicates whether you get either 20% or 30% off the original purchase price. The price of your goods is then reduced by the amount on the slip. The probability of selecting a 20% off slip is $\frac{7}{10}$ and the probability of selecting a 30% off slip is $\frac{3}{10}$. If the original price of the goods purchased is \$100, determine
 a) the expected percent to be deducted from your purchase.
 b) the expected dollar amount you will pay for your purchase.

18. *Pick a Card* Mike and Dave play the following game: Mike picks a card from a deck of cards. If he selects a heart, Dave gives him \$5. If not, he gives Dave \$2.
 a) Find Mike's expectation.
 b) Find Dave's expectation.

19. *To Guess or Not to Guess?* A multiple choice exam has four possible answers for each question. For each correct answer, you are awarded 5 points. For each incorrect answer, 2 points are subtracted from your score. For answers left blank, no points are added or subtracted.
 a) If you do not know the correct answer to a particular question, is it to your advantage to guess? Explain.
 b) If you do not know the correct answer but can eliminate one possible choice, is it to your advantage to guess? Explain.

20. *Fortune Cookies* At the Royal Dragon Chinese restaurant, a slip in the fortune cookies indicates a dollar amount that will be subtracted from your total bill. A bag of 10 fortune cookies (each individually wrapped) is given to you from which you will select 1. If 7 fortune cookies contain "\$1 off," 2 contain "\$2 off," and 1 contains "\$5 off," find the expectation of a selection.

21. *Raffle Tickets* One thousand raffle tickets are sold for \$1 each. One prize of \$500 is to be awarded.
 a) Rena purchases one ticket. Find her expected value.
 b) Find the fair price of a ticket.
 c) If the vendor sells all 1000 tickets, how much profit will he or she make?

22. *Raffle Tickets* Ten thousand raffle tickets are sold for \$5 each. Four prizes will be awarded—one for \$10,000, one for \$5000, and two for \$1000. Sidhardt purchases one of these tickets.
 a) Find his expected value.
 b) Find the fair price of a ticket.

In Exercises 23–26, assume that you spin the pointer and are awarded the amount indicated by the pointer.
 a) *Find the fair price to play the game.*
 b) *If it costs \$2 to play the game, find the expectation of a person who plays the game.*

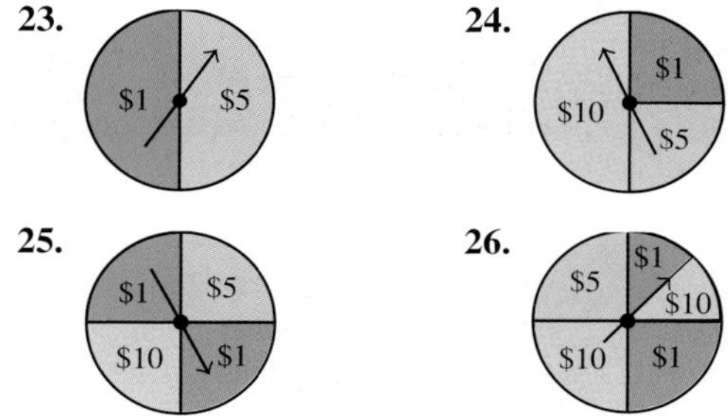

27. *Life Insurance* According to Bristol Mutual Life Insurance's mortality table, the probability that a 20-year-old woman will survive one year is 0.994, and the probability that she will die within one year is 0.006. If she buys a \$10,000 one-year term policy for \$100, what is Bristol Mutual's expected gain or loss?

28. *Reaching Base Safety* Based upon past history, Jim Devias, a minor league professional baseball player, has a 17% chance of reaching first base safely (either a single or a walk), a 10% chance of hitting a double, a 2% chance of hitting a triple, a 8% chance of hitting a home run, and a 63% chance of making an out at his next at bat. Determine Jim's expected number of bases for his next at bat.

29. *Choosing a Colored Chip* In a box there are a total of 10 chips. The chips are orange, green, or yellow, as shown below.

If you select an orange chip you get 4 points, a green chip 3 points, and a yellow chip 1 point.
 a) If you select one chip at random, determine the expected number of points you will get.
 b) If you had to pay, in points, to play this game, what would be the fair number of points to pay?

c) If you select 3 chips with replacement (therefore, each chip selected is returned before the next selection so it is as if you are starting with the original box of chips each time), determine the expected number of points you will get for the 3 selections.

30. Repeat Exercise 29 but assume an orange chip is worth 5 points, a green chip 2 points, and a yellow chip −3 points (3 points are taken away).

31. *Drilling for Oil* It will cost an oil drilling company \$30,000 to sink a test well. If it hits oil, the company will make a net profit of \$500,000. If it hits natural gas, the net profit will be \$100,000. If it hits nothing, it will lose its \$30,000. If the probability of hitting oil is 0.08 and the probability of hitting gas is 0.20, what is the expectation of the oil drilling company? Should it sink the test well? Explain.

32. *Bidding on a Project* An instrumentation contractor is considering making a bid on a water pollution control project. She determines that if her bid is accepted and she completes the project on schedule, her net profit would be \$350,000. If she completes the project between 0 and 3 months after the contract deadline date, her net profit drops to \$140,000. If she completes the project more than 3 months late, her net loss is \$420,000. The probability that she completes the job on schedule is 0.5, the probability that she completes the job between 0 and 3 months late is 0.3, and the probability that she completes the job more than 3 months late is 0.2. Find her expected gain or loss for this bid.

33. *Airline Hiring* American Airlines is planning its staffing needs for next year. On January 1 the Civil Aeronautics Board will inform American Airlines whether it will be granted the new routes it has requested. If the new routes are approved, American will hire 920 new employees. If the new routes are not granted, American will hire only 170 new employees. If the probability that the Board will grant American Airlines' request is 0.36, what is the expected number of new employees to be hired by American Airlines?

34. *Rolling a Die* A die is rolled many times, and the points facing up are recorded. Find the expected (average) number of points facing up over the long run.

35. *Lawsuit* Don Vello is considering bringing a lawsuit against the Dummote Chemical Company. His lawyer estimates that there is a 70% chance he will make \$40,000 (after legal fees), a 10% chance he will break even (the award will equal the legal fees), and a 20% chance they will lose the case and Don will need to pay \$30,000 in legal fees. Estimate Don's expected gain or loss if he proceeds with the lawsuit.

36. *Road Service* On a clear day in Boston, the Automobile Association of American (AAA) makes an average of 110 service calls for motorist assistance, on a rainy day it makes an average of 160 service calls, and on a snowy day it makes an average of 210 service calls. If the weather in Boston is clear 200 days of the year, rainy 100 days of the year, and snowy 65 days of the year, find the expected number of service calls made by the AAA in a given day.

37. *Real Estate* The expenses for Jorge, a realtor, to list, advertise, and attempt to sell a house are \$1000. If Jorge succeeds in selling the house, he will receive 6% of the sales price. If a realtor with a different company sells the house, Jorge still receives 3% of the sales price. If the house is unsold after 3 months, Jorge loses the listing and receives nothing. Suppose the probability that he sells a \$100,000 house is 0.2, the probability another realtor sells the house is 0.5, and the probability that the house is unsold after 3 months is 0.3. Find Jorge's expectation if he accepts this house for listing. Should Jorge list the house? Explain.

38. *Life Insurance* An insurance company has written life insurance policies on people of age 25. There are 200 policies of \$10,000, 400 policies of \$5000, and 1000 policies of \$1000. Experience shows that the probability of an individual dying at age 25 is 0.002. Determine the amount the insurance company can expect to pay out on these policies this year.

In Exercises 39 and 40, assume that you are blindfolded and throw a dart at the dart board shown. Assuming your dart sticks in the dart board, determine

a) *the probabilities that the dart lands on \$1, \$10, \$20, and \$100, respectively.*

b) *If you win the amount of money indicated by the section of the board where the dart lands, find your expectation when you throw the dart.*

c) *If the game is to be fair, how much should you pay to play?*

39.

$1 $20 $1 $100 $10

40.

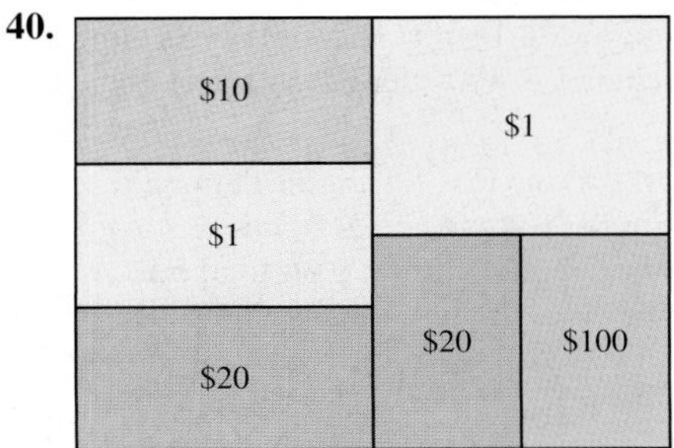

41. *Term Life Insurance* During those years when financial responsibilities are greatest, such as when children are in school or there is a mortgage on the house, a person may choose to buy a term life insurance policy. The insurance company will pay the face value of the policy if the insured person dies during the term of the policy. For how much should an insurance company sell a 10-year term policy with a face value of \$40,000 to a 30-year-old man in order for the company to make a profit? The probability of a 30-year-old man living to age 40 is 0.97. Explain your answer. Remember the customer pays for the insurance before the policy becomes effective.

CHALLENGE PROBLEMS/GROUP ACTIVITIES

Roulette *In Exercises 42 and 43, use the roulette wheel illustrated. A roulette wheel typically contains slots with numbers 1–36 and slots marked 0 and 00. A ball is spun on the wheel and comes to rest in one of the 38 slots. Eighteen numbers are colored red, and 18 numbers are colored black. The 0 and 00 are colored green. If you bet on one particular number and the ball lands on that number the house pays off odds of 35 to 1. If you bet on a red number or black number and win, the house pays 1 to 1 (even money).*

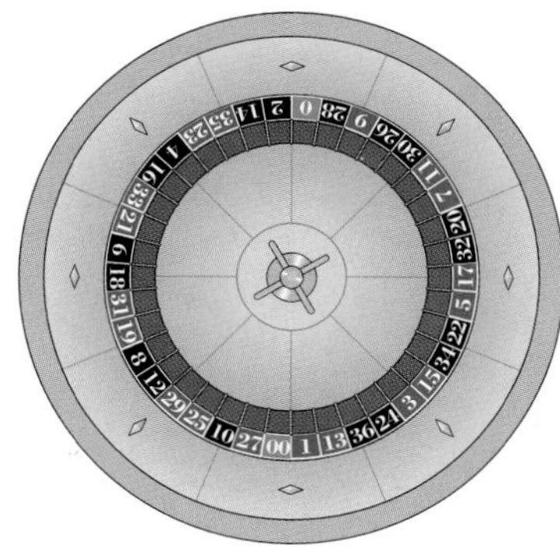

42. Find the expected value of betting \$1 on a particular number.

43. Find the expected value of betting \$1 on red.

44. *A Fair Game?* The dealer shuffles five black cards and five red cards and spreads them out on the table face down. You choose two at random. If both cards are red or both cards are black, you win a dollar. Otherwise, you lose a dollar. Determine whether the game favors you, is fair, or favors the dealer. Explain your answer.

45. *Wheel of Fortune* The following is a miniature version of the Wheel of Fortune. When Dave Salem spins the wheel, he is awarded the amount on the wheel indicated by the pointer. If the wheel points to Bankrupt, he loses the total amount he has accumulated and also loses his turn. Assume that the wheel stops on a position at random and that each position is equally likely to occur.

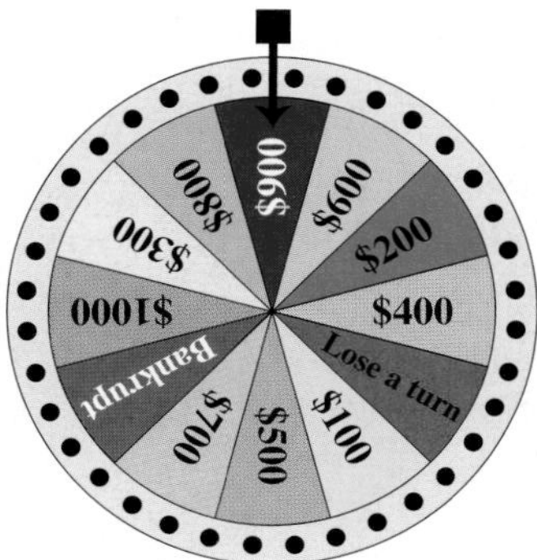

a) Find Mr. Salem's expectation when he spins the wheel at the start of the game (he has no money to lose if he lands on Bankrupt).

b) If Mr. Salem presently has a balance of \$1800, find his expectation when he spins the wheel.

46. *Lottery Ticket* Is it possible to determine your expectation when you purchase a lottery ticket? Explain.

47. *Sweepstakes Ticket* Do not throw away the next sweepstakes ticket you get as part of an advertisement.

a) Use the information provided to determine the expectation of the ticket.

b) If you consider the current cost of a stamp, is it to your benefit to mail in the ticket? Explain.

12.5 TREE DIAGRAMS

We stated earlier that the possible results of an experiment are called its outcomes. In order to solve more difficult probability problems, we must first be able to determine all the possible outcomes of the experiment. The counting principle can be used to determine the number of outcomes of an experiment and is helpful in constructing tree diagrams.

Counting Principle

If a first experiment can be performed in M distinct ways and a second experiment can be performed in N distinct ways, then the two experiments in that specific order can be performed in $M \cdot N$ distinct ways.

DID YOU KNOW

Mix and Match

Fashion experts often suggest you select your wardrobe from pieces of clothing that can be mixed and matched. If, for example, you have 7 shirts, 3 sweaters, and 4 pairs of pants, you actually have 84 possible outfits. If you wear only a sweater or a shirt, but not both, the possibilities go down to 40.

If we wanted to find the number of possible outcomes when a coin is tossed and a die is rolled, we could reason: The coin has two possible outcomes—heads and tails. The die has six possible outcomes—1, 2, 3, 4, 5, and 6. Thus, the two experiments together have $2 \cdot 6$, or 12, possible outcomes.

A list of all the possible outcomes of an experiment is called a **sample space**. Each individual outcome in the sample space is called a **sample point**. **Tree diagrams** are helpful in determining sample spaces.

A tree diagram illustrating all the possible outcomes when a coin is tossed and a die is rolled (see Fig. 12.3) has two initial branches, one for each of the possible outcomes of the coin. Each of these branches will have six branches emerging from them, one for each of the possible outcomes of the die. This will give a total of 12 branches, the same number of possible outcomes found by using the counting principle. We can obtain the sample space by listing all the possible combinations of branches. Note that this sample space consists of 12 sample points.

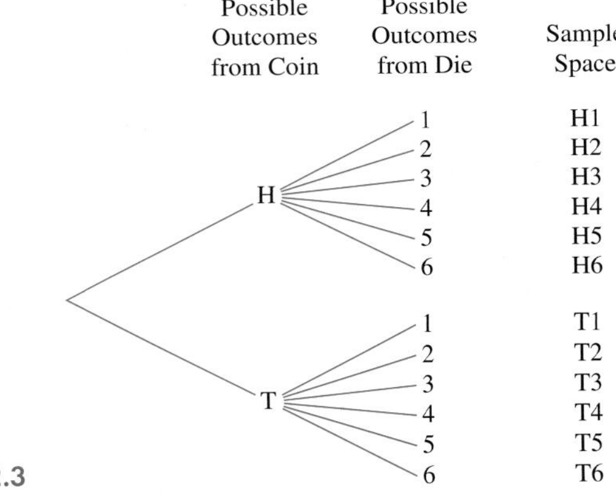

Figure 12.3

Example 1 uses the phrase "without replacement." This phrase tells us that once an item is selected, it cannot be selected again, making it impossible to select the same item twice.

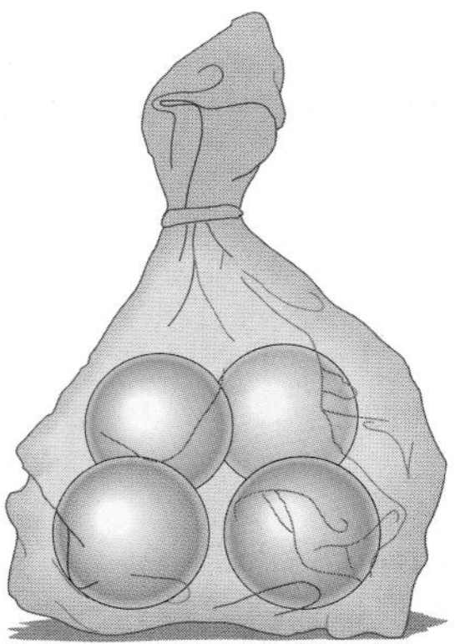

Figure 12.4

EXAMPLE 1 ***Selecting Balls without Replacement***

Two balls are to be selected *without replacement* from a bag that contains one red, one blue, one green, and one orange ball (see Fig. 12.4).

a) Use the counting principle to determine the number of points in the sample space.
b) Construct a tree diagram and list the sample space.
c) Find the probability that one red ball is selected.
d) Find the probability that a green ball followed by a red ball is selected.

SOLUTION

a) The first selection may be any one of the four balls. Once the first ball is selected, only three balls remain for the second selection. Thus, there are $4 \cdot 3$, or 12, sample points in the sample space.

b) The first ball selected can be red, blue, green, or orange. Since this experiment is done without replacement, the same colored ball cannot be selected twice. For example, if the first ball selected is red, the second ball selected must be either blue, green, or orange. The tree diagram and sample space are shown in Fig. 12.5.The sample space contains 12 points. That result checks with the answer obtained with the counting principle.

First Selection	Second Selection	Sample Space
R	B	RB
	G	RG
	O	RO
B	R	BR
	G	BG
	O	BO
G	R	GR
	B	GB
	O	GO
O	R	OR
	B	OB
	G	OG

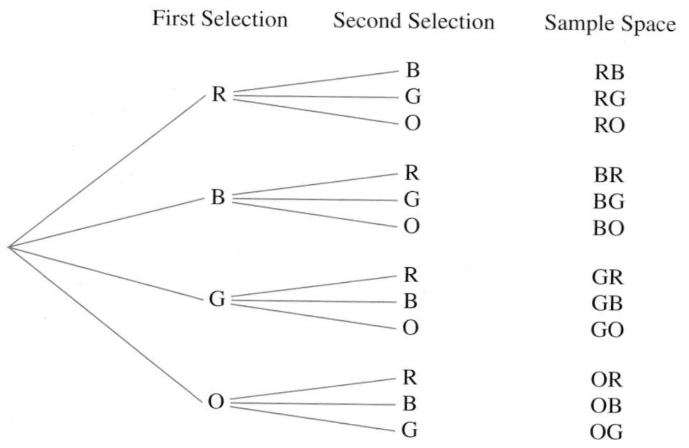

Figure 12.5

c) If we know the sample space, we can compute probabilities using the formula

$$P(E) = \frac{\text{number of outcomes favorable to } E}{\text{total number of outcomes}}$$

The total number of outcomes will be the number of points in the sample space. From Fig. 12.5 we determine that there are 12 possible outcomes. Six outcomes have one red ball: RB, RG, RO, BR, GR, and OR.

$$P(\text{one red ball is selected}) = \frac{6}{12} = \frac{1}{2}$$

d) One possible outcome meets the criteria of a green ball followed by a red ball: GR

$$P(\text{green followed by red}) = \frac{1}{12}$$

The counting principle can be extended to any number of experiments, as illustrated in Example 2.

EXAMPLE 2 *Using the Counting Principle*

The Gilligans are driving from New York to San Francisco and wish to stop in Cleveland and Chicago. They are considering two highways from New York to

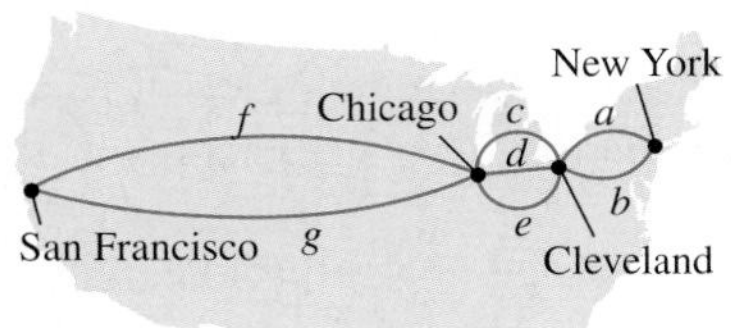

Figure 12.6

Cleveland, three highways from Cleveland to Chicago, and two highways from Chicago to San Francisco, as illustrated in Fig. 12.6.

a) Use the counting principle to determine the number of different routes the Gilligans can take from New York to San Francisco.
b) Use a tree diagram to determine the routes.
c) If a route from New York to San Francisco is selected at random and all routes are considered equally likely, find the probability that both routes a and g are used.
d) Find the probability that neither routes d nor f are used.

SOLUTION

a) Using the counting principle, we can determine that there are $2 \cdot 3 \cdot 2$, or 12, routes from New York to San Francisco.
b) The tree diagram illustrating the 12 possibilities is given in Fig. 12.7.

New York to Cleveland	Cleveland to Chicago	Chicago to San Francisco	Possible Routes
a	c	f	acf
		g	acg
	d	f	adf
		g	adg
	e	f	aef
		g	aeg
b	c	f	bcf
		g	bcg
	d	f	bdf
		g	bdg
	e	f	bef
		g	beg

Figure 12.7

c) Of the 12 possible routes, 3 use both a and g (acg, adg, aeg).

$$P(\text{routes } a \text{ and } g \text{ are both used}) = \frac{3}{12} = \frac{1}{4}$$

d) Of the 12 possible routes, 4 use neither d nor f (acg, aeg, bcg, beg).

$$P(\text{neither } d \text{ nor } f \text{ is used}) = \frac{4}{12} = \frac{1}{3}$$

EXAMPLE 3 *Selecting Employees for Training*

A group of three stock traders at ISI Securities includes the following people: Kim (K), Bob Alger (BA), Bob Parson (BP), and Sarah (S). Two will be selected and sent to Washington for additional training.

a) Use the counting principal to determine the number of points in the sample space.
b) Construct a tree diagram and list the sample space.
c) Determine the probability that neither Bob is selected.
d) Determine the probability that at least one Bob is selected.

SOLUTION

a) The first selection may be any one of the four people. Once the first person is selected, only three people remain for the second selection. Thus, there are $4 \cdot 3$ or 12 sample points in the sample space.

b)

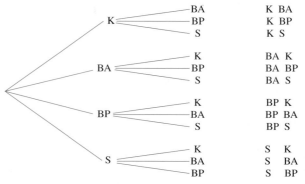

First Selection	Second Selection	Sample Space
K	BA	K BA
	BP	K BP
	S	K S
BA	K	BA K
	BP	BA BP
	S	BA S
BP	K	BP K
	BA	BP BA
	S	BP S
S	K	S K
	BA	S BA
	BP	S BP

c) Of the 12 points in the sample space, two have neither Bob. They are K S (Kim, Sarah) and S K (Sarah Kim).

$$P(\text{neither Bob selected}) = \frac{2}{12} = \frac{1}{6}$$

d) At least one Bob means that one or more Bob is selected. There are 10 points in the sample space with at least one Bob (all those except Kim, Sarah and Sarah, Kim).

$$P(\text{at least one Bob is selected}) = \frac{10}{12} = \frac{5}{6}$$

In Example 3, if you add the probability of no Bob being selected with the probability of at least one Bob being selected, you get $\frac{1}{6} + \frac{5}{6}$, or 1. In any probability problem, the sum of the probabilities of an event not happening or the event happening at least once is 1, since one of those events must always occur. This leads to the following rule.

P(*event happening at least once*) = *1* − *P(event does not happen)*

For example, suppose the probability of not getting any red flowers from 3 seeds that are planted is $\frac{2}{7}$. Then the probability of getting at least one red flower from the 3 seeds that are planted is $1 - \frac{2}{7} = \frac{5}{7}$. We will use this rule in later sections.

SECTION 12.5 EXERCISES

CONCEPT/WRITING EXERCISES

1. Explain the counting principle.
2. **a)** What is a sample space?
   b) What is a sample point?
3. If a first experiment can be performed in 3 distinct ways, and a second experiment can be performed in 5 distinct ways, how many possible ways can the two experiments be performed? Explain your answer.
4. In your own words, describe how to construct a tree diagram.
5. A problem states that two selections are made "without replacement." Explain what this means.

6. If a tree diagram contains a total of 8 branches, how many sample points will be in the corresponding sample space? Explain.

PRACTICE THE SKILLS

7. If two states are selected at random from the 50 states, determine the number of possible outcomes if the states are selected
   a) with replacement.
   b) without replacement.
8. If two dates are selected at random from the 365 days of the year, determine the number of possible outcomes if the dates are selected
   a) with replacement.
   b) without replacement.
9. A bag contains 5 different light bulbs. The bulbs are the same size, but each has a different wattage. If you select 2 bulbs at random, how many sample points will be in the sample space if the bulbs are selected
   a) with replacement.
   b) without replacement.
10. Six hardboiled eggs, labeled 1 through 6, are placed in a bag. If you select 3 eggs at random, how many sample points will be in the sample space if the eggs are selected
   a) with replacement.
   b) without replacement.

In Exercises 11–26 use the counting principal to determine the answer to part (a). Assume that each event is equally likely to occur.

11. *Coin Toss* Two coins are tossed.
   a) Determine the number of points in the sample space.
   b) Construct a tree diagram and list the sample space.
   Find the probability that
   c) no heads are tossed.
   d) exactly one head is tossed.
   e) two heads are tossed.
12. *Face Cards* A bag contains three cards: a jack, a queen, and a king.

   Two cards are to be selected at random with replacement.
   a) Determine the number of points in the sample space.
   b) Construct a tree diagram and determine the sample space.
   Find the probability that
   c) two jacks are selected.
   d) a jack and then a queen are selected.
   e) at least one king is selected.
13. Repeat Exercise 12 without replacement.
14. *Boys and Girls* A couple plans to have two children.
   a) Determine the number of points in the sample space of the possible arrangements of boys and girls.
   b) Construct a tree diagram and list the sample space.
   Find the probability that the family has
   c) two girls.
   d) at least one girl.
   e) a girl and then a boy.
15. *Marble Selection* A hat contains 4 marbles: 1 yellow, 1 red, 1 blue, and 1 green. Two marbles are to be selected at random without replacement from the hat.

   a) Determine the number of points in the sample space.
   b) Construct a tree diagram and list the sample space.
   Find the probability of selecting
   c) exactly 1 red marble.
   d) at least 1 marble that is not red.
   e) no green marbles.
16. *Three Children* A couple plans to have three children.
   a) Determine the number of points in the sample space of possible arrangements of boys and girls.
   b) Construct a tree diagram and list the sample space.
   Find the probability that the family has
   c) no boys.
   d) at least one girl.
   e) either exactly two boys or two girls.
   f) two boys first, then one girl.

PROBLEM SOLVING

17. *Rolling Dice* Two dice are rolled.
   a) Determine the number of points in the sample space.
   b) Construct a tree diagram and list the sample space.
   Find the probability that
   c) a double is rolled.
   d) a sum of 7 is rolled.
   e) a sum of 2 is rolled.
   f) Are you as likely to roll a sum of 2 as you are of rolling a sum of 7? Explain your answer.
18. *Door Prizes* For door prizes, three different CDs will be awarded to three different people. The first person selected gets to choose between Enya, Mariah Carey, and U2. The second person selected chooses between the two remaining CD's. The third person selected is given the left-over CD.
   a) Determine the number of points in the sample space.
   b) Construct a tree diagram and determine the sample space.
   Determine the probability that
   c) the Mariah Carey CD is selected first.
   d) the Enya CD is selected first and Mariah Carey is selected last.
   e) The CDs are selected in this order: Mariah Carey, U2, and Enya.
19. *Voting* At a Homeowners Association, a Board member can vote yes, no, or abstain on a motion. There are three motions the Board is voting on.

a) Determine the number of points in the sample space.
b) Construct a tree diagram and determine the sample space.

Find the probability that a Board member votes

c) no, yes, no in that order.
d) yes on exactly two of the motions.
e) yes on at least one motion.

20. *Vacationing in Florida* Mrs. and Mrs. Peter Collinge are vacationing in the Orlando, Florida, area. They have listed what they consider the major attractions in the area in two groups: Disney attractions and non-Disney attractions. The four Disney attractions include: the Magic Kingdom, Epcot Center, MGM Studios, and Animal Kingdom. The three non-Disney attractions include Sea World, Universal Studios, and Busch Gardens (Tampa). Because of time limitations, they decided to first visit one Disney and then one non-Disney attraction. They will select the name of the one Disney attraction from one hat and the non-Disney attraction from a second hat.

a) Determime the number of points in the sample space.
b) Construct a tree diagram and determine the sample space.

Find the probability that they select

c) the Magic Kingdom or Epcot Center.
d) MGM Studios or Universal Studios.
e) the Magic Kingdom and either Sea World or Busch Gardens.

21. *Planning an Education* Rikki Blair is considering the following colleges for her Bachelors, Masters, and Ph.D. degrees.

Bachelors: University of Texas, State University of New York (SUNY) at Brockport, and Johns Hopkins University

Masters: University of Hawaii, University of Massachusetts, and University of Wisconsin.

Ph.D.: Emory University, University of California at Los Angeles (UCLA).

Assume she selects one of the schools in each category.

a) Determine the number of points in the sample space.
b) Construct a tree diagram and determine the sample space.

Find the probability she selects

c) SUNY at Brockport for her Bachelors degree.
d) the University of Massachusetts or the University of Hawaii for her Masters
e) the University of Texas for her Bachelors and UCLA for her Ph.D.

22. *TV Viewing* Below we list the TV shows shown at 7 A.M., 9 A.M., and 10 A.M. in Tampa, Florida in May 2000.

Network	7 A.M.	9 A.M.	10 A.M.
ABC	"Good Morning America"	"Martha Stewart"	"Martin Short"
CBS	"This Morning"	"Maury"	"Sally Jessy Raphael"
NBC	"Today"	"Later Today"	"Donny & Marie"

Assume one show will be watched during each time slot.

a) Determine the number of points in the sample space.
b) Construct a tree diagram and determine the sample space.

Determine the probability that

c) all three shows on NBC are watched.
d) "Today" and "Martin Short" are watched.
e) "Martha Stewart" is not watched.

23. *A New Computer* You visit Computer City to purchase a new computer system. You are going to purchase a computer, printer, and monitor from among the following brands.

Computer	Printer	Monitor
Compaq	Hewlett-Packard	Omega
IBM	Epson	Toshiba
Apple		
Dell		

a) Determine the number of points in the sample space.
b) Construct a tree diagram and determine the sample space.

Determine the probability of selecting

c) an Apple computer
d) a Hewlett-Packard printer
e) an Apple computer and a Hewlett-Packard printer

24. *Summer School* You decide to take three courses during summer school—an English course, a mathematics course, and a science course. The available courses that you can take are illustrated below.

English	Mathematics	Science
English composition	college algebra	biology
English literature	statistics	geology
	calculus	chemistry
		physics

a) Determine the number of points in the sample space.
b) Construct a tree diagram and determine the sample space.

Determine the probability that

c) geology is elected.
d) either geology or chemistry is selected.
e) calculus is not selected.

25. *Personal Characteristics* An individual can be classified as male or female with red, brown, black, or blonde hair and with brown, blue, or green eyes.
a) How many different classifications are possible?
b) Construct a tree diagram to determine the sample space (for example, male, red-headed, blue-eyed).
c) If each outcome is equally likely, find the probability that the individual will be a male with black hair and blue eyes.
d) Find the probability that the individual will be a female with blonde hair.

26. *Mendel Revisited* A pea plant must have exactly one of each of the following pairs of traits: short (s) or tall (t); round (r) or wrinkled (w) seeds; yellow (y) or green (g) peas; and white (wh) or purple (p) flowers (for example, short, wrinkled, green pea with white flowers).

a) How many different classifications of pea plants are possible?
b) Use a tree diagram to determine all the classifications possible.
c) If each characteristic is equally likely, find the probability that the pea plant will have round peas.
d) Find the probability that the pea plant will be short, have wrinkled seeds, have yellow seeds, and have purple flowers.

CHALLENGE PROBLEMS/GROUP ACTIVITY

27. *Using a Tree Diagram* Use the tree diagram to answer the following questions.

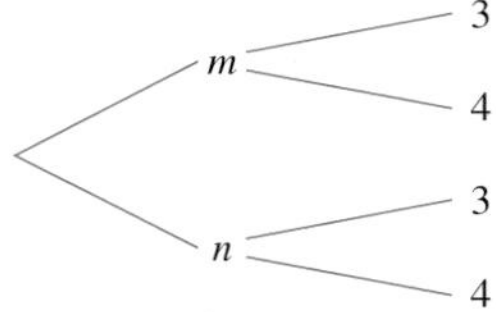

a) What are the possible outcomes of the first experiment?
b) What are the possible outcomes of the second experiment?
c) List the sample space indicated by this tree diagram.
d) From the tree diagram, is it possible to determine the probability that the outcome of the first experiment is m? Explain.
e) From the tree diagram, is it possible to determine whether the probability of getting m followed by 3 is the same as the probability of n followed by 4? Explain.
f) Do your responses to parts (d) and (e) change if you are told that the probabilities of experiment 1 are equally likely and the probabilities of experiment 2 are also equally likely? Explain.

12.6 *OR* and *AND* Problems

In Section 12.5 we showed how to work probability problems by constructing sample spaces. Often it is inconvenient or too time consuming to solve a problem by first constructing a sample space. For example, if an experiment consists of selecting two cards with replacement from a deck of 52 cards, there would be $52 \cdot 52$ or 2704 points in the sample space. Trying to list all these sample points could take hours. In this section we learn how to solve **compound probability** problems that contain the words *and* or *or* without constructing a sample space.

Or Problems

The ***or* probability problem** requires obtaining a "successful" outcome for *at least one* of the given events. For example, suppose that we roll one die and we are interested in finding the probability of rolling an even number *or* a number greater than 4. For this situation rolling either a 2, 4, or 6 (an even number) or a 5 or 6 (a number greater than 4) would be considered successful. Note the number 6 satisfies both criteria. Since 4 out of 6 of the numbers meet the criteria (the 2, 4, 5, and 6), the probability of rolling an even number *or* a number greater than 4 is $\frac{4}{6}$ or $\frac{2}{3}$.

A formula for finding the probability of event A or event B, symbolized $P(A \text{ or } B)$, follows.

$$P(A \text{ or } B) = P(A) + P(B) - P(A \text{ and } B)$$

Since we add (and subtract) probabilities to find $P(A \text{ or } B)$, this formula is sometimes referred to as the *addition formula*. We explain the use of the *or* formula in Example 1.

EXAMPLE 1 *Using the Addition Formula*

Each of the numbers 1, 2, 3, 4, 5, 6, 7, 8, 9, and 10 is written on a separate piece of paper. The 10 pieces of paper are then placed in a hat, and one piece is randomly selected. Find the probability that the piece of paper selected contains an even number or a number greater than 6.

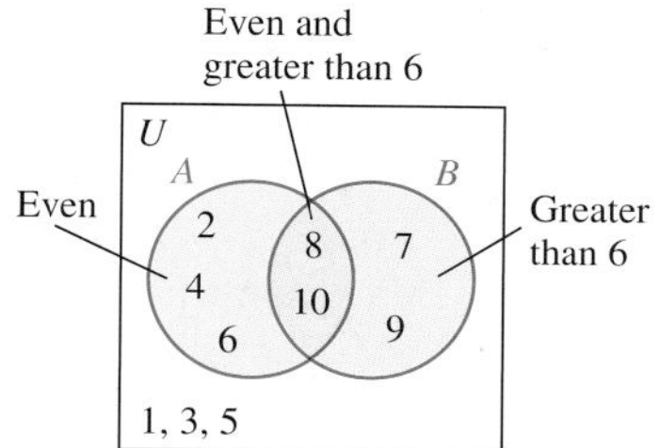

Figure 12.8

SOLUTION We are asked to find the probability the number selected *is even* or *greater than 6*. Let's use set A to represent the statement, "the number is even," and set B to represent the statement, "the number is greater than 6." Figure 12.8 is a Venn diagram, as introduced in Chapter 2, with sets A (even) and B (greater than 6). There are a total of 10 numbers, of which five are even (2, 4, 6, 8, and 10). Thus, the probability of selecting an even number is $\frac{5}{10}$. Four numbers are greater than 6: the 7, 8, 9, and 10. Thus, the probability of selecting a number greater than 6 is $\frac{4}{10}$. Two numbers are both even and greater than 6: the 8 and 10. Thus, the probability of selecting a number that is both even and greater than 6 is $\frac{2}{10}$.

If we substitute the appropriate statements for A and B in the formula, we obtain

$$\begin{aligned} P(A \text{ or } B) &= P(A) + P(B) - P(A \text{ and } B) \\ P\left(\begin{matrix}\text{even or}\\ \text{greater than 6}\end{matrix}\right) &= P(\text{even}) + P\left(\begin{matrix}\text{greater}\\ \text{than 6}\end{matrix}\right) - P\left(\begin{matrix}\text{even and}\\ \text{greater than 6}\end{matrix}\right) \\ &= \frac{5}{10} + \frac{4}{10} - \frac{2}{10} \\ &= \frac{7}{10} \end{aligned}$$

Example 1 illustrates that when finding the probability of A or B, we sum the probabilities of events A and B and then subtract the probability of both events occurring simultaneously.

EXAMPLE 2 *Using the Addition Formula*

Consider the same sample space, the numbers 1 through 10, as in Example 1. If one piece of paper is selected, find the probability that it contains a number less than 4 or a number greater than 6.

SOLUTION Let A represent the statement, "the number is less than 4," and let B represent the statement, "the number is greater than 6." A Venn diagram illustrating these statements is shown in Fig. 12.9.

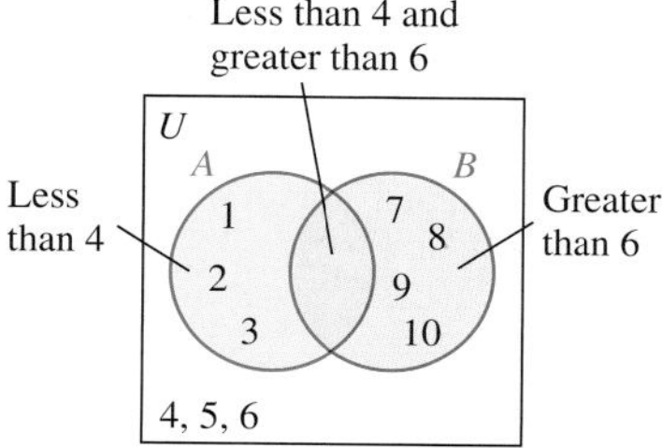

Figure 12.9

$$P(\text{selecting a number less than 4}) = \frac{3}{10}$$

$$P(\text{selecting a number greater than 6}) = \frac{4}{10}$$

Since there are no numbers that are both less than 4 and greater than 6, P(selecting a number less than 4 and greater than 6) $= 0$. Therefore,

$$P\begin{pmatrix}\text{less than 4 or}\\ \text{greater than 6}\end{pmatrix} = P(\text{less than 4}) + P\begin{pmatrix}\text{greater}\\ \text{than 6}\end{pmatrix} - P\begin{pmatrix}\text{less than 4 and}\\ \text{greater than 6}\end{pmatrix}$$

$$= \frac{3}{10} + \frac{4}{10} - 0 = \frac{7}{10}$$

In Example 2, it is impossible to select a number that is both less than 4 *and* greater than 6 when only one number is to be selected. Events such as these are said to be *mutually exclusive.*

Two events A and B are **mutually exclusive** if it is impossible for both events to occur simultaneously.

If events A and B are mutually exclusive, then $P(A \text{ and } B) = 0$, and the addition formula simplifies to $P(A \text{ or } B) = P(A) + P(B)$.

EXAMPLE 3 *Probability of A or B*

One card is selected from a standard deck of playing cards. Determine whether the following pairs of events are mutually exclusive, and find $P(A \text{ or } B)$.

a) $A =$ an ace, $B =$ a king
b) $A =$ an ace, $B =$ a spade
c) $A =$ a red card, $B =$ a black card
d) $A =$ a picture card, $B =$ a red card

SOLUTION

a) It is impossible to select both an ace and a king when only one card is selected. Therefore, these events are mutually exclusive.

$$P(\text{ace or king}) = P(\text{ace}) + P(\text{king}) = \frac{4}{52} + \frac{4}{52} = \frac{8}{52} = \frac{2}{13}$$

b) The ace of spades is both an ace and a spade; therefore, these events are not mutually exclusive.

"The ace of spades is both an ace and a spade."

$$P(\text{ace}) = \frac{4}{52} \qquad P(\text{spade}) = \frac{13}{52} \qquad P(\text{ace and spade}) = \frac{1}{52}$$

$$P(\text{ace or spade}) = P(\text{ace}) + P(\text{spade}) - P(\text{ace and spade})$$

$$= \frac{4}{52} + \frac{13}{52} - \frac{1}{52}$$

$$= \frac{16}{52} = \frac{4}{13}$$

c) It is impossible to select one card that is both a red card and a black card. Therefore, the events are mutually exclusive.

$$P(\text{red or black}) = P(\text{red}) + P(\text{black})$$

$$= \frac{26}{52} + \frac{26}{52} = \frac{52}{52} = 1$$

Therefore, a red card or a black card must be selected.

d) Six picture cards are red: jack, queen, and king of hearts and jack, queen, and king of diamonds. Thus, these events are not mutually exclusive.

$$P\begin{pmatrix}\text{picture card}\\ \text{or red card}\end{pmatrix} = P\begin{pmatrix}\text{picture}\\ \text{card}\end{pmatrix} + P\begin{pmatrix}\text{red}\\ \text{card}\end{pmatrix} - P\begin{pmatrix}\text{picture card}\\ \text{and red card}\end{pmatrix}$$

$$= \frac{12}{52} + \frac{26}{52} - \frac{6}{52}$$

$$= \frac{32}{52} = \frac{8}{13}$$

And Problems

A second type of problem is the ***and* probability problem**, which requires obtaining a favorable outcome in *each* of the given events. For example, suppose that *two* cards are to be selected from a deck of cards and we are interested in the probability of selecting two aces (one ace *and* then a second ace). Only if *both* cards selected are aces would this experiment be considered successful. A formula for finding the probability of events A and B, symbolized $P(A \text{ and } B)$, follows.

$$P(A \text{ and } B) = P(A) \cdot P(B), \text{ assuming that event } A \text{ has occurred*}$$

DID YOU KNOW

They're All Girls

The odds against a family having eight girls in a row is 255 to 1. Thus, for every 256 families with eight children, on average, one family will have all girls and 255 will not. However, should another child be born to that one family, the probability of the ninth child being a girl is still $\frac{1}{2}$!

Since we multiply to find $P(A \text{ and } B)$, this formula is sometimes referred to as the *multiplication formula.* When using the multiplication formula, we will not always write the words "assuming that event A has occurred." However, since this type of problem required obtaining a favorable outcome in *both* of the given events, **we must always assume that event A has occurred before calculating the probability of event B** (even when "assuming that event A has occurred" has not been written).

EXAMPLE 4 *An Experiment with Replacement*

Two cards are to be selected with replacement from a deck of cards. Find the probability that two aces will be selected.

SOLUTION Since the deck of 52 cards contains four aces, the probability of selecting an ace on the first draw is $\frac{4}{52}$. The card selected is then returned to the deck. Therefore, the probability of selecting an ace on the second draw remains $\frac{4}{52}$.

If we let A represent the selection of the first ace and B represent the selection of the second ace, the formula may be written

$$P(A \text{ and } B) = P(A) \cdot P(B)$$

$$P(2 \text{ aces}) = P(\text{ace } \textit{and} \text{ ace}) = P(\text{ace } 1) \cdot P(\text{ace } 2)$$

$$= \frac{4}{52} \cdot \frac{4}{52}$$

$$= \frac{1}{13} \cdot \frac{1}{13} = \frac{1}{169}$$

*$P(B)$, assuming that event A has occurred, may be denoted $P(B \mid A)$, which is read "the probability of B, given A." We will discuss this type of probability (conditional probability) further in Section 12.7.

DID YOU KNOW

The Birthday Problem

Among 24 people chosen at random, what would you guess is the probability that at least 2 of them have the same birthday? It might surprise you to learn it is greater than 0.5. There are 365 days on which the first person selected can have a birthday. That person has a 365/365 chance of having a birthday on one of those days. The probability that the second person's birthday is on any other day is 364/365. The probability that the third person's birthday is on a day different from the first two is 363/365, and so on. The probability that the 24th person has a birthday on any other day than the first 23 people is 342/365. Thus, the probability, P, that of 24 people, no 2 have the same birthday is $(365/365) \times (364/365) \times (363/365) \times \cdots \times (342/365) = 0.462$. Then the probability of at least 2 people of 24 having the same birthday is $1 - P = 1 - 0.462 = 0.538$, or slightly larger than $\frac{1}{2}$.

EXAMPLE 5 ***An Experiment without Replacement***

Repeat Example 4 without replacement.

SOLUTION The probability of selecting an ace on the first draw is $\frac{4}{52}$. When calculating the probability of selecting the second ace, we must assume that the first ace has been selected. Once this first ace has been selected, only 51 cards, including 3 aces, remain in the deck. The probability of selecting an ace on the second draw becomes $\frac{3}{51}$. The probability of selecting two aces without replacement is

$$\begin{aligned} P(\text{2 aces}) &= P(\text{ace 1}) \cdot P(\text{ace 2}) \\ &= \frac{4}{52} \cdot \frac{3}{51} \\ &= \frac{1}{13} \cdot \frac{1}{17} = \frac{1}{221} \end{aligned}$$

Event A and event B are **independent events** if the occurrence of either event in no way affects the probability of occurrence of the other.

Rolling dice and tossing coins are examples of independent events. In Example 4, the events are independent since the first card was returned to the deck. The probability of selecting an ace on the second draw was not affected by the first selection. The events in Example 5 are not independent since the probability of the selection of the second ace was affected by removing the first ace selected from the deck. Such events are called **dependent events**. *Experiments done with replacement will result in independent events, and those done without replacement will result in dependent events.*

EXAMPLE 6 ***Independent or Dependent Events?***

At a dinner to recognize the accomplishments of a local citizen, it is found that of the 100 people in attendance, 45 are Democrats, 35 are Republicans, 12 are Independents, and 8 are Reform Party members. Three of those in attendance will be selected at random and each will be awarded one door prize. Lindsey, a mathematician in attendance, is interested in determining the probability that all three selected will be Democrats. Are the events of picking the Democrats independent or dependent events?

SOLUTION The events are dependent since each time one Democrat is selected, it changes the probability that the next person selected will be a Democrat.

The multiplication formula may be extended to more than two events, as illustrated in Example 7.

EXAMPLE 7 ***Flower Probabilities***

A package of 25 zinnia seeds contain 8 seeds for red flowers, 12 seeds for white flowers, and 5 seeds for yellow flowers. Three seeds are randomly selected and planted. Find the probability of each of the following.

a) All three seeds will produce red flowers.
b) The first seed selected will produce a red flower, the second seed will produce a white flower, and the third seed will produce a red flower.
c) None of the seeds will produce red flowers.
d) At least one will produce red flowers.

SOLUTION Each time a seed is selected and planted, the number of seeds remaining decreases by one.

a) The probability that the first seed selected produces a red flower is $\frac{8}{25}$. If the first seed selected is red, only 7 red seeds in 24 are left. The probability of selecting a second red seed is $\frac{7}{24}$. If the second seed selected is red, only 6 red seeds in 23 are left. The probability of selecting a third red seed is $\frac{6}{23}$.

$$\begin{aligned} P(\text{3 red seeds}) &= P(\text{red seed 1}) \cdot P(\text{red seed 2}) \cdot P(\text{red seed 3}) \\ &= \frac{8}{25} \cdot \frac{7}{24} \cdot \frac{6}{23} = \frac{14}{575} \end{aligned}$$

b) The probability that the first seed selected produces a red flower is $\frac{8}{25}$. Once a seed for a red flower is selected, only 24 seeds are left. Twelve of the remaining 24 will produce white flowers. Thus, the probability that the second seed selected will produce a white flower is $\frac{12}{24}$. After the second seed has been selected, there are 23 seeds left, 7 of which will produce red flowers. The probability that the third seed produces a red flower is therefore $\frac{7}{23}$.

$$\begin{aligned} P\left(\begin{matrix}\text{first red and second} \\ \text{white and third red}\end{matrix}\right) &= P(\text{first red}) \cdot P(\text{second white}) \cdot P(\text{third red}) \\ &= \frac{8}{25} \cdot \frac{12}{24} \cdot \frac{7}{23} = \frac{28}{575} \end{aligned}$$

c) If no flowers are to be red, they must either be white or yellow. Seventeen seeds will not produce red flowers (12 for white and 5 for yellow). The probability that the first seed does not produce a red flower is $\frac{17}{25}$. After the first seed has been selected, 16 of the remaining 24 seeds will not produce red flowers. After the second seed has been selected, 15 of the remaining 23 seeds will not produce red flowers.

$$\begin{aligned} P(\text{none red}) &= P(\text{first not red}) \cdot P(\text{second not red}) \cdot P(\text{third not red}) \\ &= \frac{17}{25} \cdot \frac{16}{24} \cdot \frac{15}{23} = \frac{34}{115} \end{aligned}$$

d) In Section 13.5, we learned that

$$P(\text{event happening at least once}) = 1 - P(\text{event does not happen})$$

In part (c), we found that the probability of selecting no red flowers is $\frac{34}{115}$. Therefore, the probability that at least one of the seeds will produce red flowers can be found as follows:

$$\begin{aligned} P(\text{at least one red flower}) &= 1 - P(\text{no red flowers}) \\ &= 1 - \frac{34}{115} = \frac{115}{115} - \frac{34}{115} = \frac{81}{115} \end{aligned}$$

In newspapers and magazines we often see charts and graphs given in percents. The percents are generally determined from a large sample of people. When using such graphs to find probabilities, since the number of people involved is very large, we assume that each selection is independent of every other selection. Example 8 shows how we can determine probabilities using charts and graphs.

EXAMPLE 8 *Population by Age*

The following graph shows the U.S. population by age in July 1999. (*Source: Newsweek.*)

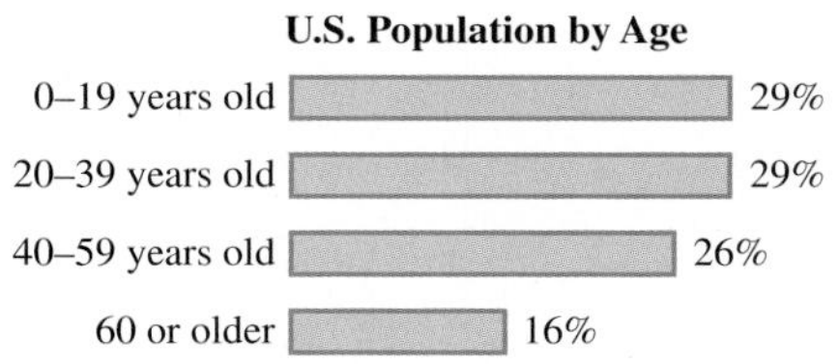

U.S. Population by Age

0–19 years old	29%
20–39 years old	29%
40–59 years old	26%
60 or older	16%

Assuming four Americans are selected at random, find the probability that

a) all four selected are from 0–19 years old.
b) the first selected is 40–59 years old, the second selected is 20–39 years old, the third selected is 60 or older, and the fourth selected is 40–59 years old.

SOLUTION We are given the percents of individuals in each category rather than the actual number in each category. Since the number of Americans is large, we do not need to be concerned about replacement or nonreplacement of people selected. In problems of this type we assume the selections are independent events. The answers we determine will be very close approximations to the actual probabilities.

a) Write the percents as decimal numbers.

$$\begin{aligned} P(\text{all four are from 0–19 years of age}) &= P(\text{1st 0–19}) \cdot P(\text{2nd 0–19}) \cdot \\ &\quad P(\text{3rd 0–19}) \cdot P(\text{4th 0–19}) \\ &= (0.29)(0.29)(0.29)(0.29) \approx 0.0071 \end{aligned}$$

b) $$\begin{aligned} P(\text{40–59, 20–39, 60 or older, 40–59}) &= P(\text{1st 40–59}) \cdot P(\text{2nd 20–39}) \cdot \\ &\quad P(\text{3rd 60 or older}) \cdot P(\text{4th 40–59}) \\ &= (0.26)(0.29)(0.16)(0.26) \approx 0.0031 \end{aligned}$$ ▲

Which formula to use

It is sometimes difficult to determine when to use the *or formula* and when to use the *and formula.* The following information may be helpful in deciding which formula to use.

Or formula

Or problems will almost always contain the word *or* in the statement of the problem. For example, find the probability of selecting a heart *or* a 6. *Or* problems in this book generally involve only *one* selection. For example, "one card is selected" or "one die is rolled."

And formula

And problems often do *not* use the word *and* in the statement of the problem. For example, "find the probability that both cards selected are red," or "find the probability that none of those selected is a banana" are both *and*-type problems. *And* problems in this book will generally involve *more than one* selection. For example, the problem may read "two cards are selected" or "three coins are flipped."

SECTION 12.6 EXERCISES

CONCEPT/WRITING EXERCISES

1. **a)** In $P(A \text{ or } B)$, what does the word *or* indicate?
   b) In $P(A \text{ and } B)$, what does the word *and* indicate?
2. **a)** Give the formula for $P(A \text{ or } B)$.
   b) In your own words, explain how to find $P(A \text{ or } B)$ with the formula.
3. **a)** What are mutually exclusive events? Give an example.
   b) How do you calculate $P(A \text{ or } B)$ when A and B are mutually exclusive?
4. **a)** Give the formula for $P(A \text{ and } B)$.
   b) In your own words, explain how to find $P(A \text{ and } B)$ with the formula.
5. When finding $P(B)$ using the formula $P(A \text{ and } B)$, what do we always assume?
6. What are independent events? Give an example.
7. What are dependent events? Give an example.
8. A family is selected at random. Let event A be the mother likes classical music. Let event B be the daughter likes classical music.
   a) Are events A and B mutually exclusive? Explain.
   b) Are they independent events? Explain.
9. An individual is selected at random. Let event A be the individual is happy. Let event B be the individual is healthy.
   a) Are events A and B mutually exclusive? Explain.
   b) Are they independent events? Explain.
10. A family is selected at random. Let event A be the father is a teacher. Let event B be the mother is a teacher.
    a) Are events A and B mutually exclusive? Explain.
    b) Are they independent events? Explain.
11. **a)** Write a problem that you would use the *or formula* to solve. Solve the problem and give the answer.
    b) Write a problem that you would use the *and formula* to solve. Solve the problem and give the answer.
12. If events A and B are mutually exclusive, explain why the formula $P(A \text{ or } B) = P(A) + P(B) - P(A \text{ and } B)$ can be simplified to $P(A \text{ or } B) = P(A) + P(B)$.

PRACTICE THE SKILLS

In Exercises 13–16, find the indicated probability.

13. If $P(A) = 0.4$, $P(B) = 0.5$, and $P(A \text{ and } B) = 0.2$, find $P(A \text{ or } B)$.
14. If $P(A \text{ or } B) = 0.9$, $P(A) = 0.5$, and $P(B) = 0.6$, find $P(A \text{ and } B)$.
15. If $P(A \text{ or } B) = 0.8$, $P(A) = 0.4$, and $P(A \text{ and } B) = 0.1$, find $P(B)$.
16. If $P(A \text{ or } B) = 0.6$, $P(B) = 0.3$, and $P(A \text{ and } B) = 0.1$, find $P(A)$.

In Exercises 17–20, a single die is rolled one time. Find the probability of rolling

17. a 3 or 5.
18. an odd number or a number greater than 2.
19. a number greater than 5 or less than 3.
20. a number greater than 3 or less than 5.

In Exercises 21–26, one card is selected from a deck of cards. Find the probability of selecting

21. a queen or a king.
22. a jack or a diamond.
23. a picture card or a red card.
24. a heart or a black card.
25. a card less than 9 or a club. (*Note:* The ace is considered a low card.)
26. a card greater than 8 or a black card.

PROBLEM SOLVING

In Exercises 27–34, a board game uses the deck of 20 cards shown.

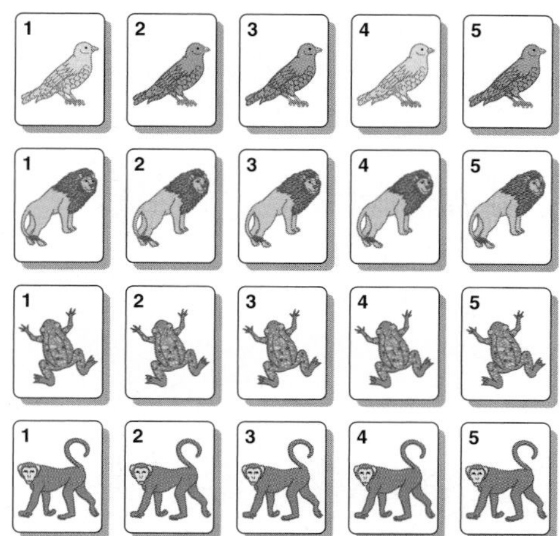

Two cards are selected at random from this deck. Find the probability of the following

a) *with replacement.*
b) *without replacement.*

27. They both show the number 3.
28. They both show frogs.

29. The first shows a monkey, and the second shows a bird.

30. The first shows a 2, and the second shows a 4.

31. The first shows a frog, and the second shows a yellow bird.

32. They both show odd numbers.

33. Neither shows an odd number.

34. The first shows a lion, and the second shows a red bird.

In the deck of cards used in Exercises 27–34, if one card is drawn, find the probability that the card shows

35. a frog or an odd number.

36. a yellow bird or a number greater than 4.

37. a monkey or a 5.

38. a lion or an even number.

In Exercises 39–48, assume that the pointer cannot land on the line and assume that each spin is independent.

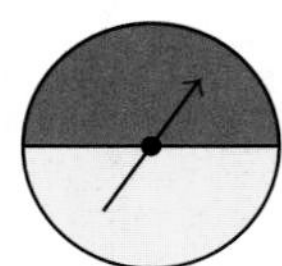

Figure 12.10

If the pointer in Fig. 12.10 is spun twice, find the probability that the pointer lands on

39. yellow on both spins. **40.** red and then yellow.

If the pointer in Fig. 12.11 is spun twice, find the probability that the pointer lands on

41. green and then red **42.** red on both spins.

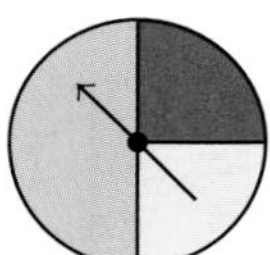

Figure 12.11

If the pointer in Fig. 12.12 is spun twice. Find the probability that the pointer lands on

43. yellow on both spins.

44. a color other than red on both spins.

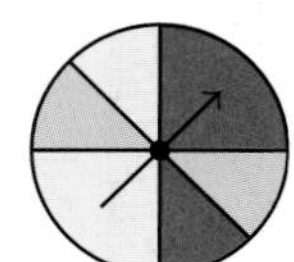

Figure 12.12

In Exercises 45–48, assume that the pointer in Fig. 12.10 is spun and then the pointer in Fig. 12.11 is spun. Find the probability of the pointers landing on

45. yellow on both spins.

46. red on the first spin and yellow on the second spin.

47. a color other than yellow on both spins.

48. yellow on the first spin and a color other than yellow on the second spin.

Having a Family *In Exercises 49–52, a family has three children. Assuming independence and the probability of a boy is $\frac{1}{2}$, find the probability that*

49. all three are girls.

50. all three are boys.

51. the youngest child is a boy and the older children are girls.

52. the youngest child is a girl, the middle child is a boy, and the oldest child is a girl.

53. **a)** *Five Children* The Metzs plan to have five children. Find the probability that all their children will be boys. (Assume that $P(\text{boy}) = \frac{1}{2}$, and assume independence.)

b) If their first five children are boys, and Mrs. Metz is expecting another child, what is the probability it will be a boy?

54. **a)** *The Probability of a Girl* The Bronson's plan to have seven children. Find the probability that all their children will be girls. (Assume that $P(\text{girl}) = \frac{1}{2}$, and assume independence.)

b) If their first seven children are girls, and Mrs. Bronson is expecting another child, what is the probability it will be a girl?

Soda Choices *In Exercises 55–58, a cooler contains 6 cans of soda: 3 colas, 2 orange, and 1 cherry. Two cans will be selected at random. Find the probability of selecting each of the following.*

a) *with replacement*

b) *without replacement*

55. a cola and then an orange.

56. no colas.

57. at least one cola.

58. at least one orange.

Population by Age *In Example 8, we gave a chart for population by age as of July 1999. The same* Newsweek *article provided information on the expected population by age in July 2050. In Exercises 59–62, use the graph to answer the questions.*

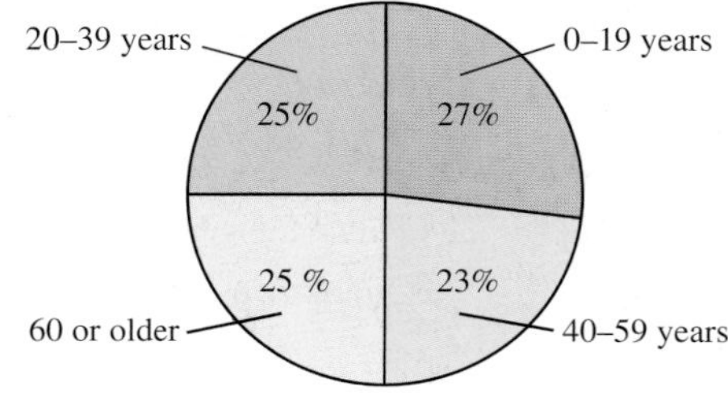

Assuming this chart is correct, if two people were to be selected at random in July 2050, determine the probability that

59. they both are 40–59 years old.

60. the first is 60 or older and the second is 0–19 years old.

61. the first is 0–19 years old and the second is 20–39 years old.

62. neither is 0–19 years old.

Fitness *In Exercises 63–66, use the following chart which shows how often American women exercise.*

How often do you exercise or work at being fit?

20% Daily
31% At least three times a week
15% About once a week
23% Less often than once a week
11% Never

(*Source: Newsweek* (special issue), 1999, Princeton Survey Research Associates.)

If three women are selected at random, determine the probability that

63. they all exercise daily.

64. the first two exercise about once a week, and the third exercises daily.

65. the first exercises at least three times a week, the second exercises daily, and the third never exercises.

66. they all exercise about once a week or more.

An Experimental Drug *In Exercises 67–70, an experimental drug was given to a sample of 100 hospital patients with an unknown sickness. Of the total, 70 patients reacted favorably, 10 reacted unfavorably, and 20 were unaffected by the drug. Assume that this sample is representative of the entire population. If this drug is given to Mr. and Mrs. Rivera and their son Carlos, what is the probability of each of the following? (Assume independence.)*

67. Mrs. Rivera reacts favorably.

68. Mr. and Mrs. Rivera react favorably, and Carlos is unaffected.

69. All three react favorably.

70. None reacts favorably.

Multiple Choice Exam *In Exercises 71–76, each question of a five-question multiple choice exam has four possible answers. Gurshawn picks an answer at random for each question. Find the probability that he selects the correct answer on*

71. any one question.

72. only the first question.

73. only the third and fourth questions.

74. all five questions.

75. none of the questions.

76. at least one of the questions.

A Slot Machine *In Exercises 77–80, consider a slot machine.*

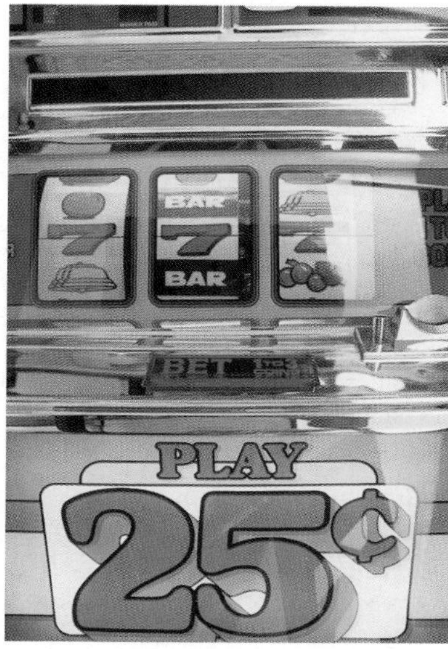

Most people who play slot machines end up losing money because the machines are designed to favor the casino (the house). A slot machine generally has three or four reels that spin independently of each other. When you pull the handle, the reels spin and the pictures on the reels come to rest in a random order. The following is a list of the number of symbols of each type on each of the three reels (for one particular slot machine). Each reel has 22 positions.

Pictures on Reels	Reels 1	2	3
Cherries	2	5	4
Oranges	5	4	5
Plums	6	4	4
Bells	3	4	4
Melons	3	2	3
Bars BAR	2	2	1
7s 7	1	1	1

For this slot machine, find the probability of obtaining

77. an orange on the first reel.

78. oranges on all three reels.

79. no 7s.

80. three 7s.

Two Wheels *In Exercises 81–84, the following double wheel is spun. The outer wheel is spun clockwise and the inner wheel is spun counterclockwise.*

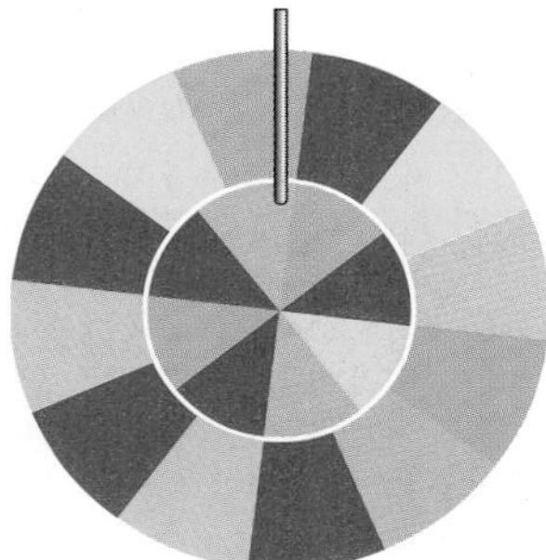

Assuming that the wheels are independent and each outcome is equally likely, find the probability that

81. red on both wheels stop under the pointer.

82. red on the outer wheel and blue on the inner wheel stop under the pointer.

83. red on neither wheel stops under the pointer.

84. red on at least one wheel stops under the pointer.

Hitting a Target *In Exercises 85–88, the probability that a heat-seeking torpedo will hit its target is 0.4. If the first torpedo hits its target, the probability that the second torpedo will hit the target increases to 0.9 because of the extra heat generated by the first explosion. If two heat-seeking torpedoes are fired at a target, find the probability that*

85. neither hits the target.

86. the first hits the target and the second misses the target.

87. both hit the target.

88. the first misses the target and the second hits the target.

89. *Polygenetic Afflictions* Certain birth defects and syndromes are *polygenetic* in nature. Typically, the chance that an offspring will be born with a polygenetic affliction is small. However, once an offspring is born with the affliction, the probability that future offspring of the same parents will be born with the same affliction increases. Let's assume that the probability of a child being born with affliction A is 0.001. If a child is born with this affliction, the probability of a future child being born with the same affliction becomes 0.04.

a) Are the events of the births of two children in the same family with affliction A independent? Explain.

b) A couple plans to have one child. Find the probability that the child will be born with this affliction.

c) A couple plans to have two children. Find the probability that

i) both children will be born with the affliction.

ii) the first will be born with the affliction and the second will not.

iii) the first will not be born with the affliction and the second will.

iv) neither will be born with the affliction.

Chance of an Audit *In Exercises 90–93, use the fact that the Internal Revenue Service claims that 28 in every 1000 people in the $10,000–$50,000 income bracket are audited yearly. Assuming that the returns to be audited are selected at random and that each year's selections are independent of the previous year's selections, find the probability that a person in this income bracket will be audited.*

90. this year.

91. the next two years in succession.

92. this year but not next year.

93. neither this year nor next year.

94. *Lottery Ticket* Ms. Jones has a lottery ticket with a three-digit number in the range 000 to 999. Three balls are to be selected at random with replacement from a bin. An equal number of balls are marked with the digits 0, 1, 2, . . . , 9. Find the probability that Ms. Jones's number is selected.

A Different Die *A die has 1 dot on one side, 2 dots on two sides, and 3 dots on three sides. If the die is rolled once, find the probability of rolling*

95. a 2.

96. a 3.

97. an even number or a number less than 3.

98. an odd number or a number greater than 1.

CHALLENGE PROBLEMS/GROUP ACTIVITIES

99. *Picking Chips* A bag contains five red chips, three blue chips, and two yellow chips. Two chips are selected from the bag without replacement. Find the probability that two chips of the same color are selected.

100. *Ten Yen Coins* Ron has ten coins from Japan: three 1-yen coins, one 10-yen coin, two 20-yen coins, one 50-yen coin, and three 100-yen coins. He selects two coins at random without replacement. Assuming that each coin is equally likely to be selected, find the probability that Ron selects at least one 1-yen coin.

101. *Selecting Stocks* An investment advisor is considering the purchase of stock of 5 drug companies and 10 computer companies. Knowing that each company had the same record of gain over the last year, the investment advisor decides to select 3 stocks at random from the 15 stocks. What is the probability that 2 stocks are for computer companies and 1 is for a drug company?

102. *A Fair Game?* Two playing cards are dealt to you from a well-shuffled deck of 52 cards. If either card is a diamond, or both are diamonds, you win; otherwise, you lose. Determine whether this game favors you, is fair, or favors the dealer. Explain your answer.

103. *Picture Card Probability* You have three cards: an ace, a king, and a queen. A friend shuffles the cards, selects two of them at random, and discards the third. You ask your friend to show you a picture card, and she turns over the king. What is the probability that she also has the queen?

RESEARCH ACTIVITY

104. Girolamo Cardano (1501–1576) wrote *Liber de Ludo Aleae,* which is considered to be the first book on probability. Cardano had a number of different vocations. Do research and write a paper on the life and accomplishments of Girolamo Cardano.

12.7 CONDITIONAL PROBABILITY

In Section 12.6, we indicated that events are independent when the outcome of either event has no effect on the outcome of the other event. For example, selecting two cards from the deck of cards *with replacement* represents independent events. However, not all events are independent events. Consider this problem: Find the probability of selecting two aces from a deck of cards *without replacement.* The probability of selecting the first ace is $\frac{4}{52}$. The probability that the second ace is selected becomes $\frac{3}{51}$, since we assume that an ace was removed from the deck with the first selection. Since the probability of the second ace being selected is affected by the first ace being selected, these two events are *dependent.* Probability problems involving dependent events can be solved by using conditional probability.

DID YOU KNOW

Dice-Y Music

During the eighteenth century, composers were fascinated with the idea of creating compositions by rolling dice. A musical piece attributed by some to Mozart, titled "Musical Dice Game," consists of 176 numbered musical fragments of three-quarter beats each. Two charts show the performer how to order the fragments given a particular roll of two dice. Modern composers can create music by using the same random technique and a computer; however, as in Mozart's time, the music created by random selection of notes or bars is still considered by musicologists to be esthetically inferior to the work of a creative composer.

Conditional Probability

In general, the probability of event E_2 occurring, given that an event E_1 has happened (or will happen—the time relationship does not matter) is called a **conditional probability** and is written $P(E_2|E_1)$.

The symbol $P(E_2|E_1)$, read "the probability of E_2, given E_1," represents the probability of E_2 occurring, assuming that E_1 has already occurred (or will occur).

EXAMPLE 1 *Using Conditional Probability*

A single card is selected from a deck of cards. Find the probability it is a club, given that it is black.

SOLUTION We are told that it is black. Thus, only 26 cards are possible, of which 13 are clubs. Therefore,

$$P(\text{Club}|\text{Black}) \text{ or } P(\text{C}|\text{B}) = \frac{13}{26} = \frac{1}{2}$$

EXAMPLE 2 *Girls in a Family*

Given a family with two children, and assuming that boys and girls are equally likely, find the probability that the family has

a) two girls.
b) two girls if you know at least one of the children is a girl.
c) two girls given that the older child is a girl.

Chance of Showers

At one time or another, you probably have been caught in a downpour on a day the weather forecaster had predicted sunny skies. Short-term (24-hour) weather forecasts are correct nearly 85% of the time, a level of accuracy achieved through the use of conditional probability. Computer models are used to analyze data taken on the ground and in the air and make predictions of atmospheric pressures at some future time, say 10 minutes ahead. Based on these predicted conditions, another forecast is then computed. This process is repeated until a weather map has been generated for the next 12, 24, 36, and 48 hours. Since each new prediction relies on the previous prediction being correct, the margin of error increases as the forecast extends farther into the future.

SOLUTION

a) The sample space of two children, BB, BG, GB, GG, can be determined by a tree diagram (see Fig. 12.13).

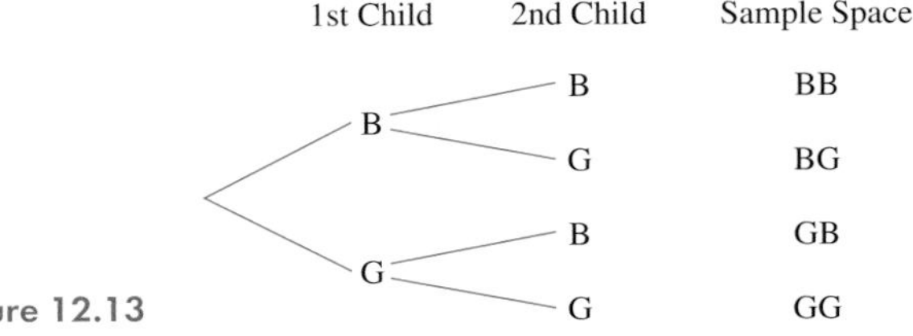

Figure 12.13

Since there are four possibilities, of which only one has two girls,

$$P(2 \text{ girls}) = \frac{1}{4}$$

b) We are told that at least one of the children is a girl. Therefore, for this problem the sample space is BG, GB, GG. Since there are three possibilities, of which only one has two girls,

$$P(\text{both girls} \mid \text{at least one is a girl}) = \frac{1}{3}$$

c) If the older child is a girl, the sample space reduces to GB, GG. Thus,

$$P(\text{both girls} \mid \text{older child is a girl}) = \frac{1}{2}$$

There are a number of formulas that can be used to find conditional probabilities. The one we will use in this book follows.

Conditional Probability

For any two events, E_1 and E_2,

$$P(E_2|E_1) = \frac{n(E_1 \text{ and } E_2)}{n(E_1)}$$

In the formula, $n(E_1$ and $E_2)$ represents the number of sample points common to both event 1 and event 2, and $n(E_1)$ is the number of sample points in event E_1, the given event. Since the intersection of E_1 and E_2, $E_1 \cap E_2$, represents the sample points common to both E_1 and E_2, the formula can also be expressed as

$$P(E_2|E_1) = \frac{n(E_1 \cap E_2)}{n(E_1)}$$

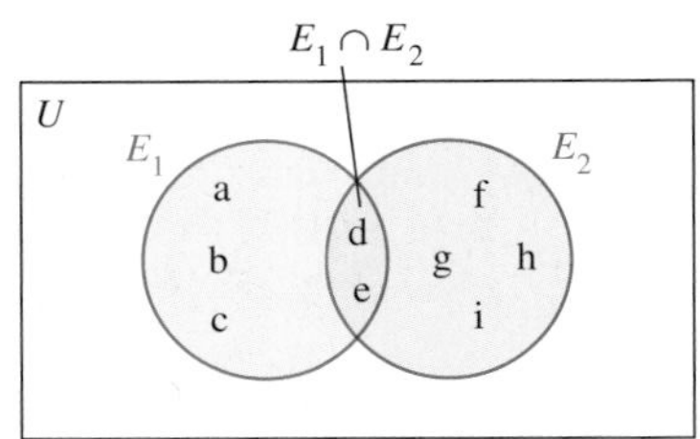

Figure 12.14

Figure 12.14 is helpful in explaining conditional probability.

Here, the number of elements in E_1 is five, the number of elements in E_2 is six, and the number of elements in both E_1 and E_2, or $E_1 \cap E_2$, is two.

$$P(E_2|E_1) = \frac{n(E_1 \text{ and } E_2)}{n(E_1)} = \frac{2}{5}$$

Thus, for this situation, the probability of selecting an element from E_2, given that the element is in E_1, is $\frac{2}{5}$.

EXAMPLE 3 ***Using the Conditional Probability Formula***

Each person in a sample of 200 residents in Lincoln County was asked whether he or she favored having a single countywide police department. The county consists of one large city and a number of small townships. The response of those sampled, with their place of residence specified, is given in the following table.

Residence	**Favor**	**Oppose**	**Total**
Live in city	80	50	130
Live outside city	60	10	70
Total	140	60	200

If one person from the sample is selected at random, find the probability that the person

a) favors a single countywide police force.
b) favors a single countywide police force if the person lives in the city.
c) opposes a single countywide police force if the person lives outside the city.

SOLUTION

a) The total number of respondents is 200, of which 140 favor a single police force.

$$P(\text{favors a single police force}) = \frac{140}{200} = \frac{7}{10}$$

b) Let E_1 be the given information "the person lives in the city." Let E_2 be "the person favors a single countywide police force." We are being asked to find $P(E_2|E_1)$. The number of people who live in the city, $n(E_1)$, is 130. The number of people who live in the city and favor a single countywide police force, $n(E_1 \text{ and } E_2)$, is 80. Thus,

$$P(E_2|E_1) = \frac{n(E_1 \text{ and } E_2)}{n(E_1)} = \frac{80}{130} = \frac{8}{13}$$

c) Let E_1 be the given information "the person lives outside the city." Let E_2 be "the person opposes a single countywide police force." We are asked to find $P(E_2|E_1)$. The number of people who live outside the city, $n(E_1)$, is 70. The number of people who live outside the city and oppose the countywide police force, $n(E_1 \text{ and } E_2)$, is 10. Thus,

$$P(E_2|E_1) = \frac{n(E_1 \text{ and } E_2)}{n(E_1)} = \frac{10}{70} = \frac{1}{7}$$

SECTION 12.7 EXERCISES

CONCEPT/WRITING EXERCISES

1. What does the notation $P(E_2|E_1)$ mean?
2. Give the formula for $P(E_2|E_1)$.
3. If $n(E_1 \cap E_2) = 4$ and $n(E_1) = 12$, find $P(E_2|E_1)$.
4. If $n(E_1 \cap E_2) = 5$ and $n(E_1) = 22$, find $P(E_2|E_1)$.

PRACTICE THE SKILLS

In Exercises 5–10, consider the circles shown.

  3   

Assume that one circle is selected at random and each circle is equally likely to be selected. Find the probability of selecting

5. a 3, given that the circle is orange.
6. a 3, given that the circle is yellow.
7. an even number, given that the number is greater than 2.
8. a number less than 2, given that the number is less than 5.
9. a green number, given that the circle is orange.
10. a number greater than 3, given that the circle is yellow.

In Exercises 11–18, consider the following wheel.

If the wheel is spun and each section is equally likely to stop under the pointer, find the probability that the pointer lands on

11. a four, given that the color is purple.
12. an even number, given that red stops under the pointer.
13. purple, given that the number is odd.
14. a number greater than 4, given that red stops under the pointer.
15. a number greater than 4, given that purple stops under the pointer.
16. an even number, given that red or purple stops under the pointer.
17. purple, given that a number greater than 5 stops under the pointer.
18. yellow, given that a number greater than 10 stops under the pointer.

Money from a Hat *In Exercises 19–22, assume that a hat contains four bills: a \$1, a \$5, a \$10, and a \$20 bill. Two bills are to be selected at random with replacement. Construct a sample space as was done in Example 2 and find the probability that*

19. both bills are \$5 bills.
20. both bills are \$5 if the first selected is a \$5 bill.
21. both bills are \$5 if at least one of the bills is a \$5 bill.
22. both bills are greater than \$5 if the second bill is a \$10 bill.

Two Dice *In Exercises 23–28, two dice are rolled. Construct a sample space and find the probability that the sum of the points on the dice total*

23. 7.
24. 7 if the first die is a 1.
25. 7 if the first die is a 3.
26. an even number if the second die is a 2.
27. a number greater than 7 if the second die is a 5.
28. a 7 or 11 if the first die is a 5.

PROBLEM SOLVING

NY Yankee Payroll *In Exercises 29–34, use the following chart which shows the 1999 annual salaries for the New York Yankees.*

New York Yankee Payroll

Player	Position	Salary
Bernie Williams	of	\$9,857,143
Roger Clemens	p	8,250,000
David Cone	p	8,000,000
Paul O'Neill	of	6,250,000
Chuck Knoblauch	2b	6,000,000
Andy Pettitte	p	5,950,000
Scott Brosius	3b	5,250,000
Derek Jeter	ss	5,000,000

table continued on page 627

Player	Position	Salary
Chili Davis	dh	4,333,333
Tino Martinez	1b	4,300,000
Mariano Rivera	p	4,250,000
Joe Girardi	c	3,400,000
Hideki Irabu	p	3,125,000
Mike Stanton	p	2,017,000
Chad Curtis	of	2,000,000
Jim Leyritz	c	1,900,000
Orlando Hernandez	p	1,850,000
Jeff Nelson	p	1,816,666
Luis Sojo	2b	800,000
Alan Watson	p	750,000
Ramiro Mendoza	p	375,000
Jorge Posada	c	350,000
Ricky Ledee	of	202,850
Clay Bellinger	ss	200,000
Darryl Strawberry	of	200,000
Total: $86,426,992		
Average: $3,457,079		

Source: U.S.A. Today, Oct. 22, 1999.

If one Yankee is selected at random, find the probability that that person

29. earns more than $1 million.

30. earns more than $1 million, given the player is a pitcher, *p*.

31. earns more than $2 million, given the player's last name begins with the letter C.

32. is a pitcher, given the player earns more than $3 million.

33. is a designated hitter, dh, given the player earns less than $1 million.

34. is an outfielder, of, given the player earns less than $300,000.

Restaurant Service *In Exercises 35–38, use the result of a survey of the service at a local restaurant, which is summarized as follows.*

Meals	Service good	Service poor	Total
Lunch	50	15	65
Dinner	45	25	70
Total:	95	40	135

Find the probability that the service was rated

35. good.

36. good, given that the meal was lunch.

37. poor, given that the meal was dinner.

38. poor, given that the meal was lunch.

Psychology Experiment *In Exercises 39–42, use the results of an experiment designed by Allison, a student in a psychology class. She has asked 200 adult men and 200 adult women to place a roll of toilet paper on a toilet paper holder. She is interested in whether they place the toilet paper on the holder in such a manner that the paper hangs down the back of the roll, or comes over the top of the roll and hangs in front of the roll. The results of her findings are indicated in the table.*

	Paper in back	Paper over front	Total
Women	34	166	200
Men	180	20	200
Total	214	186	400

Assume these data are typical of the entire population as a whole. If one adult is selected at random, determine the probability that the person places the paper

39. in the back.

40. over the front.

41. in the back, given the person is a male.

42. over the front, given the person is a female.

Quality Control *In Exercises 43–48, Sally Horsefall, a quality control inspector is checking a sample of light bulbs for defects. The following table summarizes her findings.*

Wattage	Good	Defective	Total
20	80	15	95
50	100	5	105
100	120	10	130
Total	300	30	330

If one of the light bulbs is selected at random, find the probability that the light bulb is

43. good.

44. good, given that it is 50 watts.

45. defective, given that it is 20 watts.

46. good, given that it is 100 watts.

47. good, given that it is 50 or 100 watts.

48. defective, given that it is not 50 watts.

News Survey *In Exercises 49–54, 210 individuals are asked which evening news they watch most often. The results are summarized as follows.*

Viewers	ABC	NBC	CBS	Other	Total
Men	30	20	40	25	115
Women	50	10	20	15	95
Total	80	30	60	40	210

If one of these individuals is selected at random, find the probability that the person watches

49. ABC or NBC.

50. ABC, given that the individual is a woman.

51. ABC or NBC, given that the individual is a man.

52. a station other than CBS, given that the individual is a woman.

53. ABC, NBC, or CBS, given that the individual is a man.

54. NBC or CBS, given that the individual is a woman.

Peter Jennings, ABC News

CHALLENGE PROBLEMS/GROUP ACTIVITIES

In Exercises 55–60, suppose that each circle is equally likely to be selected. One circle is selected at random.

Find the probability indicated.

55. P(green circle $|$+ obtained)

56. P(+$|$ orange circle obtained)

57. P(yellow circle $|$− obtained)

58. P(green +$|$+obtained)

59. P(green or orange circle $|$ green + obtained)

60. P(orange circle with green +$|$+ obtained)

61. Consider the Venn diagram that follows. The numbers in the regions of the circle indicate the number of items that belong to that region. For example, 60 items are in set A but not in set B.

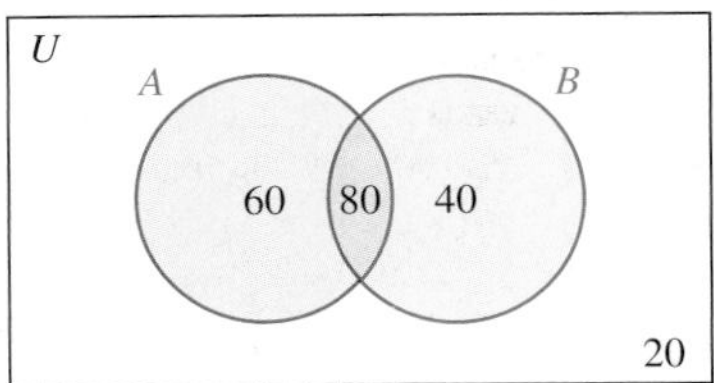

Find **a)** $n(A)$ **b)** $n(B)$ **c)** $p(A)$ **d)** $p(B)$

Use the formula on page 624 to find:

e) $p(A \mid B)$ **f)** $p(B \mid A)$

g) Explain why $p(A \mid B) \neq p(A) \cdot p(B)$.

62. A formula we gave for conditional probability is

$$p(E_2|E_1) = \frac{n(E_1 \text{ and } E_2)}{n(E_1)}$$

This formula may be derived from the formula

$$p(E_2|E_1) = \frac{p(E_1 \text{ and } E_2)}{p(E_1)}$$

Can you explain why? [*Hint:* Consider what happens to the denominators of $p(E_1$ and $E_2)$ and $p(E_1)$ when they are expressed as fractions and the fractions are divided out.]

63. Given that $p(A) = 0.3$, $p(B) = 0.4$, and $p(A \text{ and } B) = 0.12$, use the formula

$$p(E_2|E_1) = \frac{p(E_1 \text{ and } E_2)}{p(E_1)}$$

to find

a) $p(A \mid B)$ **b)** $p(B \mid A)$

c) Are A and B independent? Explain.

12.8 THE COUNTING PRINCIPLE AND PERMUTATIONS

In Section 12.5, we introduced the counting principle, which is repeated here for your convenience.

> **Counting Principle**
> If a first experiment can be performed in M distinct ways and a second experiment can be performed in N distinct ways, then the two experiments in that specific order can be performed in $M \cdot N$ distinct ways.

The counting principle is illustrated in Examples 1 and 2.

DID YOU KNOW

Slot Machines

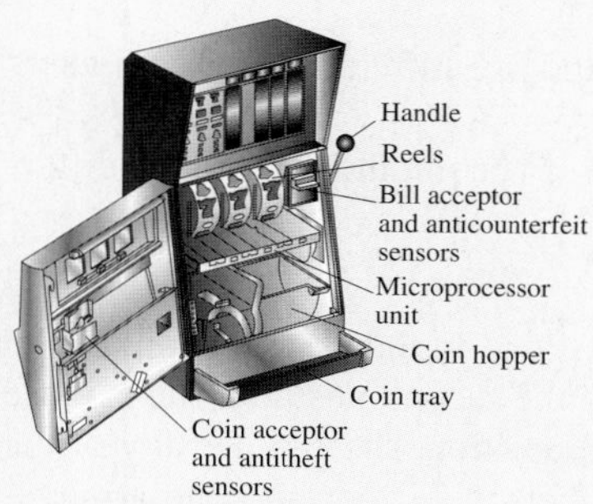

Source: Scientific American

Slot machines may qualify as the world's most user-friendly computer. Nowadays, slot machines are a conglomeration of microprocessors and memory chips. Each of three reels can stop at 22 positions (each one displaying various symbols or a blank space) for a total of 10,648 different arrangements ($22 \times 22 \times 22 = 10{,}648$). The heart of a contemporary slot machine is a microprocessor programmed to generate random numbers that can be assigned to any arrangement of the reels. In effect, the microprocessor generates random numbers that dictates the machine's display and payout. Because many or few random numbers can be assigned to any given arrangement, makers of slot machines can vary the odds as desired. For the top jackpot, the odds of obtaining the single correct arrangement may be set at 10 million to 1.

EXAMPLE 1 ***Counting Principal—License Plates***

A license plate is to consist of two letters followed by three digits. Determine how many different license plates are possible if

a) repetition of letter and digits is permitted.
b) repetition of letters and digits is not permitted.
c) the first letter must be a vowel (a, e, i, o, u) and the first digit cannot be a 0, and repetition of letters and digits is not permitted.

SOLUTION There are 26 letters and 10 digits (0–9). We have five positions to fill, as indicated.

$$\overline{\text{L}}\ \overline{\text{L}}\ \overline{\text{D}}\ \overline{\text{D}}\ \overline{\text{D}}$$

a) Since repetition is permitted, there are 26 possible choices for both the first and second positions. There are 10 possible choices for the third, fourth, and fifth positions.

$$\frac{26}{\text{L}}\ \frac{26}{\text{L}}\ \frac{10}{\text{D}}\ \frac{10}{\text{D}}\ \frac{10}{\text{D}}$$

Since $26 \cdot 26 \cdot 10 \cdot 10 \cdot 10 = 676{,}000$, there are 676,000 different possible arrangements.

b) There are 26 possibilities for the first position. Since repetition of letters is not permitted, there are only 25 possibilities for the second position. The same reasoning is used when determining the number of digits for positions 3 through 5.

$$\frac{26}{\text{L}}\ \frac{25}{\text{L}}\ \frac{10}{\text{D}}\ \frac{9}{\text{D}}\ \frac{8}{\text{D}}$$

Since $26 \cdot 25 \cdot 10 \cdot 9 \cdot 8 = 468{,}000$, there are 468,000 different possible arrangements.

c) Since the first letter must be an a, e, i, o, or u, there are five possible choices for the first position. The second position can be filled by any of the letters except for the vowel selected for the first position. Therefore, there are 25 possibilities for the second position.

Since the first digit cannot be a 0, there are nine possibilities for the third position. The fourth position can be filled by any digit except the one selected for the third position. Thus, there are nine possibilities for the fourth position. Since the last position cannot be filled by any of the two digits previously used, there are eight possibilities for the last position:

$$\frac{5}{\text{L}}\ \frac{25}{\text{L}}\ \frac{9}{\text{D}}\ \frac{9}{\text{D}}\ \frac{8}{\text{D}}$$

Since $5 \cdot 25 \cdot 9 \cdot 9 \cdot 8 = 81{,}000$, there are 81,000 different arrangements that meet the conditions specified. ▲

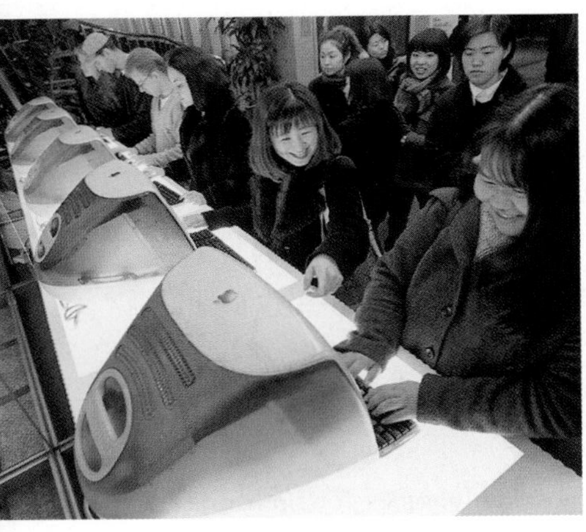

EXAMPLE 2 *Counting Principle—Computer Colors*

At Computer City, they have just received a supply of Apple iMac computers. The computers come in the following colors: tangerine, strawberry, blueberry, grape, and lime. Sadie Bragg, the floor manager, decides to display one of each color computer.

a) In how many different ways can she display the five different color computers on a shelf?
b) If she wants to place the strawberry computer in the middle, how many different ways can she arrange the computers?

SOLUTION

a) There are five positions to fill, using the five colors. In the first position, on the left, she can use any one of the five colors. In the second position, she can use any of the four remaining colors. In the third position, she can use any of the three remaining colors, and so on. The number of distinct possible arrangements is

$$\underline{5} \cdot \underline{4} \cdot \underline{3} \cdot \underline{2} \cdot \underline{1} = 120$$

b) We begin by satisfying the specified requirements stated. In this case, the strawberry computer must be placed in the middle. Therefore, there is only one possibility for the middle position.

$$\underline{\ } \ \underline{\ } \ \underline{1} \ \underline{\ } \ \underline{\ }$$

For the first position, there are now four possibilities. For the second position, there will be three possibilities. For the fourth position, there will be two possibilities. Finally, in the last position, there is only one possibility.

$$\underline{4} \cdot \underline{3} \cdot \underline{1} \cdot \underline{2} \cdot \underline{1} = 24$$

Thus, under the condition stated, there are 24 different possible arrangements.

PERMUTATIONS

Now we introduce the definition of a permutation.

> A **permutation** is any *ordered arrangement* of a given set of objects.

Peter, Paul, Mary and Mary, Paul, Peter represent two different permutations of the same three names. In Example 2(a), there are 120 different ordered arrangements, or permutations, of the five colored computers. In Example 2(b), there are 24 different arrangements, or permutations possible, if the red computer must be displayed in the middle.

When determining the number of permutations possible, we assume that repetition of an item is not permitted. To help you understand and visualize permutations, we illustrate the various permutations possible when a triangle, rectangle, and circle are to be placed in a line.

Six Permutations

Note for this set of three shapes, six different arrangements, or six permutations, are possible. We can obtain the number of permutations by using the counting principle. For the first position, there are three choices. There are then two choices for the second position, and only one choice is left for the third position.

$$\text{Number of permutations} = 3 \cdot 2 \cdot 1 = 6$$

The product $3 \cdot 2 \cdot 1$ is referred to as 3 factorial, and is written 3!. Thus,

$$3! = 3 \cdot 2 \cdot 1 = 6$$

Number of Permutations

The number of permutations of n distinct items is n factorial, symbolized $n!$, where

$$n! = n(n-1)(n-2) \cdots (3)(2)(1)$$

It is important to note that 0! is defined to be 1. Many calculators have the ability to determine factorials. Often to determine factorials you need to press the 2nd or INV key. Read your calculator manual to determine how to find factorials on your calculator.

EXAMPLE 3 *Permutations of Books*

In how many ways can eight different books be arranged on a shelf?

SOLUTION Since there are eight different books, the number of permutations is 8!

$$8! = 8 \cdot 7 \cdot 6 \cdot 5 \cdot 4 \cdot 3 \cdot 2 \cdot 1 = 40{,}320$$

Example 4 illustrates how to use the counting principle to determine the number of permutations possible when only a part of the total number of items is to be selected and arranged.

EXAMPLE 4 *Permutations of Three Out of Five Letters*

Consider the five letters p, q, r, s, t. In how many distinct ways can three letters be selected and arranged if repetition is not allowed?

SOLUTION We are asked to select and arrange only three of the five possible letters. Using the counting principle, we find there are five possible letters for the first choice, four possible letters for the second choice, and three possible letters for the third choice:

$$5 \cdot 4 \cdot 3 = 60.$$

Thus, there are 60 different possible ordered arrangements, or permutations. 

In Example 4, we determined the number of different ways in which we could select and arrange three of the five items. We can indicate this by using the notation ${}_5P_3$. The notation ${}_5P_3$ is read "the number of permutations of five items taken three at a time." The notation ${}_nP_r$ is read "the number of permutations of n items taken r at a time."

We use the counting principle below to evaluate ${}_8P_4$, ${}_9P_3$, and ${}_{10}P_5$. Notice the relationship between the number preceding the P, the number following the P, and the last number in the product.

One more than 8 − 4

$${}_8P_4 = 8 \cdot 7 \cdot 6 \cdot 5$$

One more than 9 − 3

$${}_9P_3 = 9 \cdot 8 \cdot 7$$

One more than 10 − 5

$${}_{10}P_5 = 10 \cdot 9 \cdot 8 \cdot 7 \cdot 6$$

Notice to evaluate ${}_nP_r$ we begin with n and form a product of r consecutive descending factors. For example, to evaluate ${}_{10}P_5$, we start with 10 and form a product of 5 consecutive descending factors (see the preceding illustration).

In general, the number of permutations of n items taken r at a time, ${}_nP_r$, may be found by the formula

One more than $n - r$

$${}_nP_r = n(n-1)(n-2)\cdots(n-r+1)$$

Therefore, when evaluating ${}_{20}P_{15}$, we would find the product of consecutive decreasing integers from 20 to $(20 - 15 + 1)$ or 6, which is written as $20 \cdot 19 \cdot 18 \cdot 17 \cdot \cdots \cdot 6$.

Now let's develop an alternative formula that we can use to find the number of permutations possible when r objects are selected from n objects:

$${}_nP_r = n(n-1)(n-2)\cdots(n-r+1)$$

Now multiply the expression on the right side of the equals sign by $\dfrac{(n-r)!}{(n-r)!}$ which is equivalent to multiplying the expression by 1.

$${}_nP_r = n(n-1)(n-2)\cdots(n-r+1) \times \frac{(n-r)!}{(n-r)!}.$$

For example,

$${}_{10}P_5 = 10 \cdot 9 \cdot \cdots \cdot 6 \times \frac{5!}{5!}$$

or

$${}_{10}P_5 = \frac{10 \cdot 9 \cdot \cdots \cdot 6 \times 5!}{5!}$$

Since $(n - r)!$ means $(n - r)(n - r - 1) \cdots (3)(2)(1)$, the expression for ${}_nP_r$ can be rewritten as

$$ {}_nP_r = \frac{n(n-1)(n-2)\cdots(n-r+1)\overbrace{(n-r)(n-r-1)\cdots(3)(2)(1)}^{(n-r)!}}{(n-r)!} $$

Since the numerator of this expression is $n!$, we can write

$$ {}_nP_r = \frac{n!}{(n-r)!} $$

For example,

$$ {}_{10}P_5 = \frac{10!}{(10-5)!} $$

> The number of permutations possible when r objects are selected from n objects is found by the **permutation formula**
>
> $$ {}_nP_r = \frac{n!}{(n-r)!} $$

In Example 4, we found that when selecting three of five letters, there were 60 permutations. We can obtain the same result using the permutation formula:

$$ {}_5P_3 = \frac{5!}{(5-3)!} = \frac{5!}{2!} = \frac{5 \cdot 4 \cdot 3 \cdot \cancel{2} \cdot \cancel{1}}{\cancel{2} \cdot \cancel{1}} = 60 $$

EXAMPLE 5 *Using the Permutation Formula*

You are among eight people forming a bicycle club. Collectively, you decide to put everyone's name in a hat and to randomly select a president, a vice-president, and a secretary. How many different arrangements or permutations of officers are possible?

SOLUTION Since there are eight people, $n = 8$, of which three are to be selected; thus $r = 3$.

$$ {}_8P_3 = \frac{8!}{(8-3)!} = \frac{8!}{5!} = \frac{8 \cdot 7 \cdot 6 \cdot \cancel{5 \cdot 4 \cdot 3 \cdot 2 \cdot 1}}{\cancel{5 \cdot 4 \cdot 3 \cdot 2 \cdot 1}} = 336 $$

Thus, with eight people there can be 336 different arrangements for president, vice-president, and secretary. ▲

In Example 5, the fraction

$$ \frac{8 \cdot 7 \cdot 6 \cdot \cancel{5 \cdot 4 \cdot 3 \cdot 2 \cdot 1}}{\cancel{5 \cdot 4 \cdot 3 \cdot 2 \cdot 1}} = 336 $$

can be also expressed as

$$\frac{8 \cdot 7 \cdot 6 \cdot \cancel{5!}}{\cancel{5!}} = 336$$

The solution to Example 5, like other permutation problems, can also be obtained using the counting principle.

EXAMPLE 6 *Permutations of Stereo Receivers*

The Stereo Shop's warehouse has 10 different receivers in stock. The owner of the chain calls the warehouse requesting a different receiver be sent to each of its 6 stores. How many ways can the distribution of the receivers to the stores be made?

SOLUTION There are 10 receivers from which 6 are to be selected and distributed. Thus, $n = 10$ and $r = 6$.

$$_{10}P_6 = \frac{10!}{(10 - 6)!} = \frac{10!}{4!} = \frac{10 \cdot 9 \cdot 8 \cdot 7 \cdot 6 \cdot 5 \cdot \cancel{4!}}{\cancel{4!}}$$
$$= 151{,}200$$

There are 151,200 different permutations of receivers possible. ▲

We have worked permutation problems (selecting and arranging, without replacement, r items out of n *distinct* items) by using the counting principle and using the permutation formula. When you are given a permutation problem, unless specified by your instructor, you may use either technique to determine its solution.

PERMUTATIONS OF DUPLICATE ITEMS

So far, all the examples we have discussed in this section have involved arrangements with distinct items. Now we will consider permutation problems in which some of the items to be arranged are duplicates. For example, the name BOB contains three letters, of which the two Bs are duplicates. How many permutations of the letters in the name BOB are possible? If the two Bs were distinguishable (one red and the other blue), there would be six permutations.

BOB	**BBO**	**OBB**
BOB	**BBO**	**OBB**

However, if the Bs are not distinguishable (replacing all colored Bs with black print), we see there are only three permutations

BOB	BBO	OBB

The number of permutations of the letters in BOB can be computed as

$$\frac{3!}{2!} = \frac{3 \cdot \cancel{2 \cdot 1}}{\cancel{2 \cdot 1}} = 3$$

where 3! represents the number of permutations of three letters, assuming that none are duplicates, and 2! represents the number of ways the two items that are duplicates can be arranged (**BB** or **BB**). In general, we have the following rule.

Permutations of Duplicate Objects

The number of distinct permutations of n objects where n_1 of the objects are identical, n_2 of the objects are identical, . . . , n_r of the objects are identical is found by the formula

$$\frac{n!}{n_1!n_2! \cdots n_r!}$$

EXAMPLE 7 *Duplicate Letters*

In how many different ways can the letters of the word TENNESSEE be arranged?

SOLUTION Of the nine letters, four are e's, two are n's, and two are s's. The number of possible arrangements is

$$\frac{9!}{4!2!2!} = \frac{9 \cdot 8 \cdot 7 \cdot 6 \cdot 5 \cdot \cancel{4 \cdot 3 \cdot 2 \cdot 1}}{\cancel{4 \cdot 3 \cdot 2 \cdot 1} \cdot 2 \cdot 1 \cdot 2 \cdot 1} = 3{,}780$$

There are 3,780 different possible arrangements of the letters in the word "Tennessee."

SECTION 12.8 EXERCISES

CONCEPT/WRITING EXERCISES

1. In your own words, state the counting principle.
2. In your own words, describe a permutation.
3. In your own words, explain how to find $n!$ for any whole number n.
4. Give the formula for the number of permutations of n distinct items.
5. How do you read ${}_nP_r$? When you evaluate ${}_nP_r$, what does the outcome represent?
6. Give the formula for the number of permutations when r objects are selected from n objects.
7. Give the formula for the number of permutations of n objects when $n_1, n_2, \ldots, n_r$, of the objects are identical.
8. Does ${}_1P_1 = {}_1P_0$? Explain.

PRACTICE THE SKILLS

In Exercises 9–20, evaluate the expression.

9. $5!$	**10.** $7!$	**11.** $9!$	**12.** ${}_5P_2$
13. $0!$	**14.** ${}_6P_4$	**15.** ${}_8P_0$	**16.** ${}_5P_0$
17. ${}_8P_7$	**18.** ${}_4P_4$	**19.** ${}_8P_8$	**20.** ${}_9P_6$

PROBLEM SOLVING

21. *ATM Codes* To use an automated teller machine, you generally must enter a four-digit code, using the digits 0–9. How many four-digit codes are possible if repetition of digits is permitted?

22. *Daily Double* The daily double at most race tracks consists of selecting the winning horse in both the first and second races. If the first race has seven entries and the second race has eight entries, how many daily double tickets must you purchase to guarantee a win?

23. *Car Door Locks* Some doors on cars can now be opened by pressing the correct sequence of buttons. A display of the five buttons by the door handle of a car follows.*

The correct sequence of five buttons must be pressed to unlock the door. If the same button may be pressed consecutively,

a) how many possible ways can the five buttons be pressed (repetition is permitted)?

b) If five buttons are pressed at random, find the probability that a sequence that unlocks the door will be entered.

*On most cars, although each key lists two numbers, the key acts as a single number. Therefore, if your code is 1, 6, 8, 5, 3, the code 2, 5, 7, 6, 4 will also open the lock.

24. *Social Security Numbers* A social security number consists of nine digits. How many different social security numbers are possible if repetition of digits is permitted?

25. *Selecting a Movie* At many hotels, the guests can pay a fee to watch a current movie. Of the movies available at a specific hotel there are 12 comedies, 7 dramas and 10 action movies. Carlos Sanchez wants to watch the following type of movies, in the specific order given, and he does not want to watch the same movie more than once. Determine the number of different ways he can select the movies.

a) drama, comedy, action
b) comedy, comedy, drama

26. *Purchasing CD's* At a music store Renee is going to purchase one CD by each of the following artists: Britney Spears, Celine Dion, and Ricky Martin. The store has 7 Britney Spears CD's, 5 Celine Dion CD's, and 4 Ricky Martin CD's. In how many ways can Renee select the CD's?

27. *Advertising* The operator of the Sound Great Stereo store is planning a grand opening. He wishes to advertise that the store has many different sound systems available. They stock 8 different CD players, 10 different receivers, and 9 different sets of speakers. Assuming that a sound system will consist of one of each, and all pieces are compatible, how many different sound systems can they advertise?

28. *Winning the Trifecta* The trifecta at most race tracks consists of selecting the first-, second-, and third-place finishers in a particular race in their proper order. If there are seven entries in the trifecta race, how many tickets must you purchase to guarantee a win?

29. *Geometric Shapes* Consider the five figures shown.

In how many different ways can the figures be arranged

a) from left to right?
b) from top to bottom if placed one under the other?
c) from left to right if the triangle is to be placed on the far right?
d) from left to right if the circle is to be placed on the far left and the triangle is to be placed on the far right?

30. *Arranging Pictures* The six pictures shown are to be placed side by side along a wall.

In how many ways can they be arranged from left to right if

a) they can be arranged in any order?
b) the bird must be on the far left?
c) the bird must be on the far left and the giraffe must be next to the bird?
d) a four-legged animal must be on the far right?

31. *Car Rental* At a local car rental agency 8 midsize cars are available and 3 customers want midsize cars. If each customer can choose his or her own car, how many different ways can the cars be selected?

32. *Club Officers* If a club consists of ten members, how many different arrangements of president, vice-president, and secretary are possible?

33. *ISBN Codes* Each book registered in the Library of Congress must have an ISBN code number. For an ISBN number of the form D-DD-DDDDDD-D, where D represents a digit from 0–9, how many different ISBN numbers are possible if repetition of digits is allowed? (See research activity Exercise 66.)

34. *Wedding Reception* At the reception line of a wedding, the bride, the groom, the best man, the maid of honor, the four ushers, and the four bridesmaids must line up to receive the guests.

a) If these individuals can line up in any order, how many arrangements are possible?
b) If the groom must be the last in line and the bride must be next to the groom, and the others can line up in any order, how many arrangements are possible?
c) If the groom is to be last in line, the bride next to the groom, and males and females are to alternate, how many arrangements are possible?

Letter Codes *In Exercises 35–38, an identification code is to consist of two letters followed by four digits. How many different codes are possible if*

35. repetition is permitted?

36. repetition is not permitted?

37. the first letter must be a *A*, *B*, *C*, or *D* and repetition is not permitted?

38. the first two entries must both be the same letter and repetition of the digits is not permitted?

License Plates *In Exercises 39–42, a license plate is to consist of three digits followed by two letters. Determine the number of different license plates possible if*

39. repetition of numbers and letters is permitted.

40. repetition of numbers and letters is not permitted.

41. the first and second digits must be odd, and repetition is not permitted.

42. the first digit cannot be zero, and repetition is not permitted.

43. *Possible Phone Numbers* A telephone number consists of seven digits with the restriction that the first digit cannot be 0 or 1.

a) How many distinct telephone numbers are possible?
b) How many distinct telephone numbers are possible with three-digit area codes preceding the seven-digit number, where the first digit of the area code is not 0 or 1?
c) With the increasing use of cellular phones and paging systems, our society is beginning to run out of usable

phone numbers. Various phone companies are developing phone numbers that use 11 digits instead of 7. How many distinct phone numbers can be made with 11 digits, assuming that the area code remains three digits and the first digit of the area code and the phone number cannot be 0 or 1?

44. *Prizes* A teacher decides to give six different prizes to 6 of the 30 students in her class. In how many ways can she do so?

45. *Selecting Cereal* Mrs. Williams and her 3 children go shopping at a local grocery store. Each of the children will be allowed to select one box of cereal for their own from the 12 different boxes of cereal available (there are no two the same). In how many ways can the selections be made?

46. *Security* A night guard visits 10 different offices every hour. The pattern is varied each night so that the guard will not follow a specific routine. In how many different ways can this pattern be varied?

47. *Color Permutations* Find the number of permutations of the colors in the spectrum that follows.

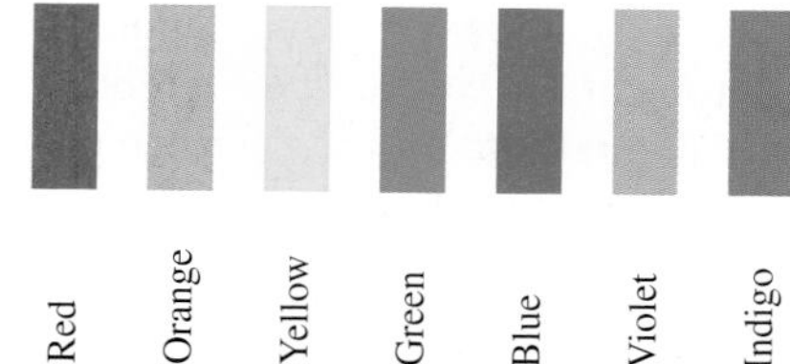

48. *Drive-Through at a Bank* A bank has three drive-through stations. Assuming that each is equally likely to be selected by customers, how many different ways can the next six drivers select a station?

49. *Training Employees* The Tallahassee, Florida, office of ISI Securities employs, among others, 5 sales representatives, 3 product specialists, 12 stock traders, and 4 supervisors. ISI plans to send one person from each of the mentioned categories to Memphis, Tennessee, for additional training. Determine the number of different ways this can be done.

50. *Computer Systems* At a computer store, a customer is considering 5 different computers, 4 different monitors, 7 different printers, and 2 different scanners. Assuming each of the components is compatible with one another, and one of each is to be selected, determine the number of different computer systems possible.

51. Determine the number of permutations of the letters of the word "JACKPOT."

52. Determine the number of permutations of the letters of the word "PUBLISHED."

53. In how many ways can the letters in the word "SASSAFRAS" be arranged?

54. In how many ways can the letters in the word "MISSISSIPPI" be arranged?

55. In how many ways can the digits in the number 4,321,324 be arranged?

56. In how many ways can the digits of the number 1,242,332 be arranged?

57. *Flag Messages* Five different colored flags will be placed on a pole, one beneath another. The arrangement of the colors indicates the message. How many messages are possible if five flags are to be selected from eight different colored flags?

58. *History Test* In one question of a history test the student is asked to match 10 dates with 10 events; each date can only be matched with 1 event. In how many different ways can this question be answered?

59. *Batting Order* In how many ways can the manager of a National League baseball team arrange his batting order of nine players if

a) the pitcher must bat last?

b) there are no restrictions?

60. *Painting Exhibit* Five Monet paintings are to be displayed in a museum.

a) In how many different ways can they be arranged if they must be next to one another?

b) In how many different ways can they be displayed if a specific one is to be in the middle?

CHALLENGE PROBLEMS/GROUP ACTIVITIES

61. *Car Keys* Door keys for a certain automobile are made from a blank key on which five cuts are made. Each cut may be one of five different depths.

a) How many different keys can be made?

b) If 400,000 of these automobiles are made and the same number of each type key is made, how many cars can be opened by a specific key?

c) If one of these cars is selected at random, what is the probability that the key selected at random will unlock the door?

62. *Voting* On a ballot, each committee member is asked to rank three of seven candidates for recommendation for promotion, giving their first, second, and third choices (no ties). What is the minimum number of ballots that must be cast in order to guarantee that at least two ballots are the same?

63. *Scrabble* Nancy Lin, who is playing Scrabble with Dale Grey, has seven different letters. She decides to test each five-letter permutation before her next move. If each permutation takes 5 sec, how long will it take Nancy to check all the permutations?

64. *Scrabble* In Exercise 63, assume, of Nancy's seven letters, three are identical and two are identical. How long will it take Nancy to try all different permutations of her seven letters?

65. Does ${}_nP_r = {}_nP_{(n-r)}$ for all whole numbers, where $n \geq r$? Explain.

RESEARCH ACTIVITY

66. When a book is published it is assigned a 10-digit code number called the International Standard Book Number (ISBN). Do research and write a report on how this coding system works.

12.9 COMBINATIONS

When the order of the selection of the items is important to the final outcome, the problem is a permutation problem. When the order of the selection of the items is unimportant to the final outcome, the problem is a **combination** problem.

Recall from Section 12.8 that permutations are *ordered* arrangements. Thus, for example, *a*, *b*, *c* and *b*, *c*, *a* are two different permutations because the ordering of the three letters is different. The letters *a*, *b*, *c* and *b*, *c*, *a* represent the same combination of letters because the *same letters* are used in each set. However, the letters *a*, *b*, *c* and *a*, *b*, *d* represent two different combinations of letters because the letters contained in each set are different.

Cornering the Market

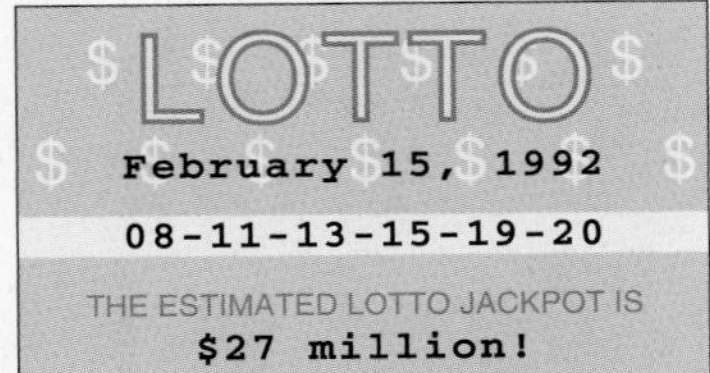

What would it take to guarantee winning the lottery? Buying every possible combination of numbers, of course. In February 1992, an Australian investment group came very close to cornering the market on all possible combinations of six numbers from 1 to 44 in the Virginia State Lottery. The group bought 5 million of a possible 7,059,052 tickets at $1 each, in a lottery with a $27 million jackpot. Only time prevented them from buying the last 2 million combinations. The group was the only winner; their profit: approximately $22,000,000. Since then the Virginia State Lottery Commission (as well as other states) has written regulations to prevent individuals and groups from cornering the market.

A **combination** is a distinct group (or set) of objects without regard to their arrangement.

EXAMPLE 1 ***Permutation or Combination***

Determine whether the situation represents a permutation or combination problem.

a) A group of five friends, Arline, Inez, Judy, Dan, and Eunice, are forming a club. The group will elect a president and a treasurer. In how many different ways can the president and treasurer be selected?
b) Of the five individuals named two will be attending a meeting together. How many different ways can they do so?

SOLUTION

a) Since the president's position is different from the treasurer's position, this problem is a permutation problem. Judy as president with Dan as treasurer is different from Dan as president with Judy as treasurer.
b) Since the order in which the two individuals selected to attend the meeting is not important, this is a combination problem. There is no difference if Judy is selected and then Dan is selected, or if Dan is selected and then Judy is selected.

In Section 12.8, you learned that ${}_nP_r$ represents the number of permutations when r items are selected from n distinct items. *Similarly* ${}_nC_r$ *represents the number of combinations when r items are selected from n distinct items.*

Consider the set of elements $\{a, b, c, d, e\}$. The number of permutations of two letters from the set is represented as ${}_5P_2$ and the number of combinations of two letters from the set is represented as ${}_5C_2$. Twenty permutations of two letters and 10 combinations of two letters are possible from these five letters. Thus, ${}_5P_2 = 20$ and ${}_5C_2 = 10$, as shown.

Permutations	**Combinations**
$\left.\begin{matrix} ab, ba, ac, ca, ad, da, ae, ea, bc, cb, \\ bd, db, be, eb, cd, dc, ce, ec, de, ed \end{matrix}\right\} 20$	$\left.\begin{matrix} ab, ac, ad, ae, bc, \\ bd, be, cd, ce, de \end{matrix}\right\} 10$

When discussing both combination and permutation problems, we always assume that the experiment is performed without replacement. That is why duplicate letters such as *aa* or *bb* are not included in the preceding example.

Note from one combination of two letters two permutations can be formed. For example, the combination *ab* gives the permutations *ab* and *ba*, or twice as many permutations as combinations. Thus, for this example we may write

$${}_5P_2 = 2 \cdot ({}_5C_2)$$

Since $2 = 2!$ we may write

$${}_5P_2 = 2!({}_5C_2)$$

If we repeated this same process for comparing the number of permutations in ${}_nP_r$ with the number of combinations in ${}_nC_r$, we would find that

$${}_nP_r = r!({}_nC_r)$$

Dividing both sides of the equation by $r!$ gives

$${}_nC_r = \frac{{}_nP_r}{r!}$$

Since ${}_nP_r = \dfrac{n!}{(n-r)!}$, the combination formula may be expressed as

$${}_nC_r = \frac{n!/(n-r)!}{r!} = \frac{n!}{(n-r)!r!}$$

> The number of combinations possible when r objects are selected from n objects is found by the **combination formula**
>
> $${}_nC_r = \frac{n!}{(n-r)!r!}$$

EXAMPLE 2 *Exam Question Selection*

An exam consists of six questions. Any four may be selected for answering. In how many ways can this selection be made?

DID YOU KNOW

Poker vs. Bridge

In popular card games there is such a variety of possible combinations of cards that a player rarely gets the same hand twice. The total number of different 5-card poker hands is 2,598,960, and for 13-card bridge hands this number increases to 635,013,559,600.

SOLUTION This is a combination problem because the order in which the four questions are answered is immaterial.

$$_6C_4 = \frac{6!}{(6-4)!4!} = \frac{6!}{2!4!} = \frac{\overset{3}{\cancel{6}} \cdot 5 \cdot \cancel{4 \cdot 3 \cdot 2 \cdot 1}}{\cancel{2} \cdot 1 \cdot \cancel{4 \cdot 3 \cdot 2 \cdot 1}} = 15$$

There are 15 different ways that four of the six questions can be selected. List them now. ▲

EXAMPLE 3 *Airline Seat Availability*

A commercial airline is seating passengers. With 15 minutes to go before take off, there are 9 passengers who are flying standby and wish to get seats, but only 5 seats are available. Determine the number of different ways a group of 5 people can be selected from the 9 to board the plane.

SOLUTION This problem is a combination problem because the order in which the five people are selected is unimportant. There are a total of nine people, so $n = 9$. Five are to be selected, so $r = 5$.

$$_9C_5 = \frac{9!}{(9-5)!5!} = \frac{9!}{4!5!} = \frac{9 \cdot \overset{2}{\cancel{8}} \cdot 7 \cdot \cancel{6} \cdot \cancel{5 \cdot 4 \cdot 3 \cdot 2 \cdot 1}}{\cancel{4} \cdot \cancel{3} \cdot \cancel{2} \cdot 1 \cdot \cancel{5 \cdot 4 \cdot 3 \cdot 2 \cdot 1}} = 126$$

Thus, 126 different combinations are possible when five from a group of nine are to be selected. ▲

EXAMPLE 4 *Dinner Combinations*

At the Royal Dynasty Chinese restaurant, dinner for eight consists of three items from column A, four items from column B, and three items from column C. If columns A, B, and C have five, seven, and six items, respectively, how many different dinner combinations are possible?

SOLUTION For column A, three of five items must be selected. This can be represented as $_5C_3$. For column B, four of seven items must be selected. This can be represented as $_7C_4$. For column C, three of six items must be selected, or $_6C_3$.

$$_5C_3 = 10 \qquad _7C_4 = 35 \quad \text{and} \quad _6C_3 = 20$$

Using the counting principle, we can determine the total number of dinner combinations by multiplying the number of choices from columns A, B, and C:

$$\begin{aligned}\text{Total number of dinner choices} &= {}_5C_3 \cdot {}_7C_4 \cdot {}_6C_3 \\ &= 10 \cdot 35 \cdot 20 = 7000\end{aligned}$$

Therefore, 7000 different combinations are possible under these conditions. ▲

We have presented various counting methods, including the counting principle, permutations, and combinations. You often need to decide which method to use to solve a problem. Table 12.4 may help you in selecting the procedure to use.

Table 12.4 Summary of Counting Methods

Counting Principle	Determining the number of ways of selecting *r* items from *n* items. Repetition not permitted.	
	Permutations	**Combinations**
Counting Principle: If a first experiment can be performed in M distinct ways and a second experiment can be performed in N distinct ways, then the two experiments in that specific order can be performed in $M \cdot N$ distinct ways. May be used with or without repetition. Use when determining the number of different ways that two or more experiments can occur. Also use when there are specific placement requirements, such as the first digit must be a 0 or 1.	Use when order is important. For example, a, b, c and b, c, a are two different permutations of the same three letters. $_nP_r = \frac{n!}{(n-r)!}$ Problems solved with the permutation formula may also be solved by using the counting principle.	Use when order is not important. For example, a, b, c and b, c, a are the same combination of three letters. But a, b, c, and a, b, d are two different combinations of three letters. $_nC_r = \frac{n!}{r!(n-r)!}$

SECTION 12.9 EXERCISES

CONCEPT/WRITING EXERCISES

1. In your own words, explain what is meant by a combination.
2. What does $_nC_r$ mean?
3. Give the formula for finding $_nC_r$.
4. What is the relationship between $_nC_r$ and $_nP_r$?
5. In your own words, explain the difference between a permutation and a combination.
6. Assume that you have five different objects and that you are going to select three without replacement. Will there be more combinations or more permutations of the three items? Explain.

PRACTICE THE SKILLS

In Exercises 7–18, evaluate the expression.

7. $_6C_3$
8. $_6C_2$
9. a) $_7C_3$ b) $_7P_3$
10. a) $_8C_0$ b) $_8P_0$
11. a) $_8C_2$ b) $_8P_2$
12. a) $_{12}C_8$ b) $_{12}P_8$
13. a) $_{10}C_3$ b) $_{10}P_3$
14. a) $_5C_5$ b) $_5P_5$
15. $\frac{_5C_3}{_5P_3}$
16. $\frac{_6C_2}{_6P_2}$
17. $\frac{_8C_5}{_8C_2}$
18. $\frac{_6C_6}{_8C_0}$

PROBLEM SOLVING

19. *Banana Split* An ice cream parlor has 20 different flavors. Cynthia orders a banana split and has to select 3 different flavors. How many different selections are possible?
20. *Car Rental* At a car rental agency, the agent has 9 midsize cars on her lot and 6 people have reserved midsize cars. In how many different ways can the 6 cars to be used be selected?
21. *Test Essays* A student must select and answer four of five essay questions on a test. In how many ways can she do so?
22. *Book Selection* A textbook search committee is considering eight books for possible adoption. The committee has decided to select three of the eight for further consideration. In how many ways can it do so?
23. *Floral Arrangements* Jan Ford makes floral arrangements. She has 12 different cut flowers she can use. She plans to use 8 different flowers. How many different selections of the 8 flowers are possible?
24. *Scholarships* Four of eight finalists will be selected and awarded $10,000 scholarships. In how many ways can this selection be made?
25. *Music CDs* Ruben Scott is shopping for CDs. He decides to purchase 4 Shania Twain CDs. The music store has 8 different Shania CDs in stock. How many different selections of Shania's CDs are possible?
26. *Course Grades* At the beginning of the semester your instructor informs the class that she marks on a curve and exactly 5 of the 26 students in the class will receive a final grade of A. In how many ways can this result occur?

Spin the Wheel *In Exercises 27–30, a double wheel shown is to be spun. The outer wheel is to be spun clockwise and the inner wheel counterclockwise.*

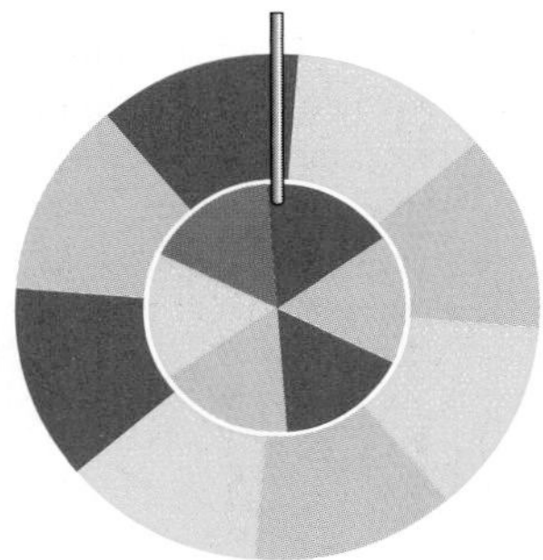

Assuming that each outcome is equally likely, find

27. the number of ways that green on both wheels can stop under the pointer.

28. the number of ways that the outer wheel can stop at red and the inner wheel can stop at green under the pointer.

29. the number of ways that the outer wheel can stop at green and the inner wheel can stop at red under the pointer.

30. the number of different color combinations possible when the two wheels stop under the pointer. Treat, for example, yellow on the outer wheel and red on the inner wheel as the same combination as red on the outer wheel and yellow on the inner wheel. Red on both wheels counts as one combination. List the possible combinations.

31. *Quinella Bet* A quinella bet consists of selecting the first- and second-place winners, in any order, in a particular event. For example, suppose you select a 2–5 quinella. If 2 wins and 5 finishes second, or if 5 wins and 2 finishes second, you win. Mr. Smith goes to a jai alai match. In the match, 8 jai alai teams are entered, and each team plays every other team. How many quinella tickets must Mr. Smith purchase to guarantee a win?

32. *Test Question* On an English test, Tito must write an essay for three of the five questions in Part 1 and four of the six questions in Part 2. How many different combinations of questions can he answer?

33. *Music* The singing group Collective Soul is planning to make a compact disc consisting of 6 fast songs and 4 slow songs. If they have 10 fast songs and 7 slow songs to choose from, how many different possible combinations do they have?

34. *Lottery* In the New York State Lottery you must select the 6 winning numbers (in any order without replacement) out of 51 possible numbers to win the grand prize.

a) How many combinations of the 6 winning numbers can be selected?

b) How many combinations of 6 numbers out of the 51 are possible?

35. *An Editor's Choice* An editor has eight manuscripts for mathematics books and five manuscripts for computer science books. If he is to select five mathematics and three computer science manuscripts for publication, how many different choices does he have?

36. *Selecting Soda* Michael is sent to the store to get 5 different bottles of regular soda and 3 different bottles of diet soda. If there are 10 different types of regular sodas and 7 different types of diet sodas to choose from, how many different choices does Michael have?

37. *Forming a Committee* How many different committees can be formed from 6 teachers and 50 students if the committee is to consist of 2 teachers and 3 students?

38. *Constructing a Test* A teacher is constructing a mathematics test consisting of 10 questions. She has a pool of 28 questions, which are classified by level of difficulty as follows: 6 difficult questions, 10 average questions, and 12 easy questions. How many different 10-question tests can she construct from the pool of 28 questions if her test is to have 3 difficult, 4 average, and 3 easy questions?

39. *Expense Account Dinner* Ashley DuMont, a sales representative for a medical-supply company, wants to take a group of 5 potential customers, out of a possible 12 potential customers, out for dinner on Friday. On Saturday, she wants to take a group of 7 current customers, out of 9 current customers, out for dinner. How many different ways can this be done?

40. *Extra Football Tickets* Mr. Bryson just won 6 tickets for each of two consecutive Dallas Cowboys home football games. For the first game, Mrs. Bryson, his wife, will not be able to attend and so he has 5 extra tickets to give away. He is considering inviting 5 of 9 close friends from where he works. Mr. and Mrs. Bryson will both attend the second game, leaving them with 4 extra tickets. They are considering inviting 4 of 7 different friends from where Mrs. Bryson works. How many different combinations of guests are possible for

a) the first.

b) the second game.

c) both games together.

41. *New Breakfast Cereals* General Mills is testing 6 oat cereals, 5 wheat cereals, and 4 rice cereals. If it plans to market 3 of the oat cereals, 2 of the wheat cereals, and 2 of the rice cereals, how many different combinations are possible?

42. *Catering Service* A catering service is making up trays of hors d'oeuvres. The hors d'oeuvres are categorized as inexpensive, average, and expensive. If the client must select three of the seven inexpensive, five of the eight average, and two of the four expensive hors d'oeuvres, how many different choices are possible?

CHALLENGE PROBLEMS/GROUP ACTIVITIES

43. *Test Answers* Consider a 10-question test in which each question can be answered either correctly or incorrectly.

a) How many different ways are there to answer the questions so that eight are correct and two are incorrect?

b) How many different ways are there to answer the questions so that at least eight are correct?

44. a) *A Dinner Toast* Four people at dinner make a toast. If each person is to tap glasses with each other person one at a time, how many taps will take place?

b) Repeat part (a) with five people.

c) How many taps will there be if there are n people at the dinner table?

45. a) *Combination Lock* To open a combination lock, you must know the lock's three-number sequence in its proper order. Repetition of numbers is permitted. Why is this lock more like a permutation lock than a combination lock? Why is it not a true permutation problem?

b) Assuming that a combination lock has 40 numbers, determine how many different three-number arrangements are possible if repetition of numbers is allowed.

c) Answer the question in part (b) if repetition is not allowed.

46. *Pascal's Triangle* The notation ${}_nC_r$ may be written $\binom{n}{r}$.

a) Use this notation to evaluate each of the combinations in the following array. Form a triangle of the results, similar to the one given, by placing the answer to each combination in the same relative position in the triangle.

$$\binom{0}{0}$$

$$\binom{1}{0} \quad \binom{1}{1}$$

$$\binom{2}{0} \quad \binom{2}{1} \quad \binom{2}{2}$$

$$\binom{3}{0} \quad \binom{3}{1} \quad \binom{3}{2} \quad \binom{3}{3}$$

$$\binom{4}{0} \quad \binom{4}{1} \quad \binom{4}{2} \quad \binom{4}{3} \quad \binom{4}{4}$$

b) Using the number pattern in part (a), find the next row of numbers of the triangle (known as **Pascal's triangle**).

47. *Lottery Combinations* Determine the number of combinations possible in a state lottery where you must select

a) 6 of 46 numbers.

b) 6 of 47 numbers.

c) 6 of 48 numbers.

d) 6 of 49 numbers.

e) Does the number of combinations increase by the same amount going from part (a) to part (b) as from part (b) to part (c)?

48. Prove that ${}_nC_r = {}_nC_{(n-r)}$.

49. a) *Table Seating Arrangements* How many distinct ways can four people be seated in a row?

b) How many distinct ways can four people be seated at a circular table?

12.10 SOLVING PROBABILITY PROBLEMS BY USING COMBINATIONS

In Section 12.9, we discussed combination problems. Now we will use combinations to solve probability problems.

Suppose that we want to find the probability of selecting two picture cards (jacks, queens, or kings) when two cards are selected, without replacement, from a deck of 52 cards. Using the *and* probability formula discussed in Section 12.6, we could reason as follows.

$$P(\text{2 picture cards}) = P(\text{1st picture card}) \cdot P(\text{2nd picture card})$$

$$= \frac{12}{52} \cdot \frac{11}{51} = \frac{132}{2652}, \quad \text{or} \quad \frac{11}{221}$$

Since the order of the two picture cards selected is not important to the final answer, this problem can be considered a combination probability problem.

We can also find the probability of selecting two picture cards, using combinations, by finding the number of possible successful outcomes (selecting two picture cards) and dividing that answer by the total number of possible outcomes (selecting any two cards).

The number of ways in which two picture cards can be selected from the 12 picture cards in a deck is ${}_{12}C_2$, or

$${}_{12}C_2 = \frac{12!}{(12-2)!2!} = \frac{\overset{6}{\cancel{12}} \cdot 11 \cdot \cancel{10!}}{\cancel{10!} \cdot \cancel{2} \cdot 1} = \frac{66}{1} = 66$$

The number of ways in which two cards can be selected from a deck of 52 cards is ${}_{52}C_2$, or

$${}_{52}C_2 = \frac{52!}{(52-2)!2!} = \frac{\overset{26}{\cancel{52}} \cdot 51 \cdot \cancel{50!}}{\cancel{50!} \cdot \cancel{2} \cdot 1} = 1326$$

Thus,

$$P(\text{selecting 2 picture cards}) = \frac{{}_{12}C_2}{{}_{52}C_2} = \frac{66}{1326} = \frac{11}{221}$$

Note that the same answer is obtained with either method. To give you more exposure to counting techniques, we will work the problems in this section using combinations.

The Dead Man's Hand

James Butler "Wild Bill" Hickok (1837–1876), known as "the fastest gun in the west," was a famous scout and federal marshal. Hickok was shot in the back and mortally wounded by Jack McCall while playing poker in a saloon in Deadwood, Dakota Territory (now South Dakota). McCall was hanged for his deed. The cards Hickok was holding when shot are shown here. Since that time, the two pair, aces and eights, have become known as "the dead man's hand" (See Exercise 48). Hickok is buried next to Martha Canary, better known as "Calamity Jane," in a cemetery overlooking the town of Deadwood.

EXAMPLE 1 *Committee of Three Women*

A club consists of four men and five women. Three members are to be selected at random to form a committee. What is the probability that the committee will consist of three women?

SOLUTION

$$P\begin{pmatrix}\text{committee consists} \\ \text{of 3 women}\end{pmatrix} = \frac{\text{number of possible committees with 3 women}}{\text{total number of possible 3-member committees}}$$

The number of possible committees with three women is ${}_5C_3 = 10$.
The total number of possible three-member committees is ${}_9C_3 = 84$.

$$P(\text{committee consists of 3 women}) = \frac{10}{84} = \frac{5}{42}$$

The probability of randomly selecting a committee with three women is $\frac{5}{42}$. ▲

EXAMPLE 2

You are dealt a flush in the game of poker when you are dealt five cards of the same suit. If you are dealt a five-card hand, find the probability you will be dealt a heart flush.

SOLUTION

$$P(\text{heart flush}) = \frac{\text{number of possible 5-card heart flushes}}{\text{total number of possible 5-card hands}}$$

The order in which the five cards are received is immaterial. Thus, the number of possible hands can be found by the combination formula. Since there are 13 hearts in a deck of cards, the number of possible five-card heart flush hands is ${}_{13}C_5 = 1287$. The total number of possible five-card hands in a deck of 52 cards is ${}_{52}C_5 = 2{,}598{,}960$.

$$P(\text{heart flush}) = \frac{{}_{13}C_5}{{}_{52}C_5} = \frac{1287}{2{,}598{,}960} = \frac{33}{66{,}640}$$

The probability of being dealt a heart flush is $\frac{33}{66{,}640}$ or ≈ 0.000495.

EXAMPLE 3 *Employment Assignments*

A temporary employment agency has 6 men and 5 women who wish to be assigned for the day. One employer has requested four employees for security positions and the second employer has requested three employees for moving furniture in an office building. If we assume that each of the potential employees has the same chance of being selected and being assigned at random, and that only seven employees will be assigned, find the probability that

a) three men will be selected for moving furniture.
b) three men will be selected for moving furniture and four women will be selected for security positions.

SOLUTION

a) $$P\begin{pmatrix}\text{3 men selected}\\ \text{for moving furniture}\end{pmatrix} = \frac{\begin{pmatrix}\text{number of possible combinations}\\ \text{of 3 men selected}\end{pmatrix}}{\begin{pmatrix}\text{total number of possible combinations}\\ \text{for selecting 3 people}\end{pmatrix}}$$

The number of possible combinations with 3 men is ${}_6C_3$. The total number of possible selections of three people is ${}_{11}C_3$.

$$P\begin{pmatrix}\text{3 men selected}\\ \text{for moving furniture}\end{pmatrix} = \frac{{}_6C_3}{{}_{11}C_3} = \frac{20}{165} = \frac{4}{33}$$

Thus, the probability that 3 men are selected is $\frac{4}{31}$.

b) The number of ways of selecting 3 men out of 6 is ${}_6C_3$ and the number of ways of selecting 4 women out of 5 is ${}_5C_4$. The total number of possible selections when 7 people are selected from 11 is ${}_{11}C_7$. Since both the 3 men *and* the 4 women must be selected, the probability is calculated as follows:

$$P\begin{pmatrix}\text{3 men and then}\\ \text{4 women selected}\end{pmatrix} = \frac{\begin{pmatrix}\text{number of combinations}\\ \text{of 3 men selected}\end{pmatrix}\cdot\begin{pmatrix}\text{number of combinations}\\ \text{of 4 women selected}\end{pmatrix}}{\text{total number of possible combinations of 7 people}}$$

$$= \frac{{}_6C_3 \cdot {}_5C_4}{{}_{11}C_7} = \frac{20\cdot 5}{330} = \frac{100}{330} = \frac{10}{33}$$

Thus, the probability is $\frac{10}{33}$.

DID YOU KNOW

More Numbers to Choose From

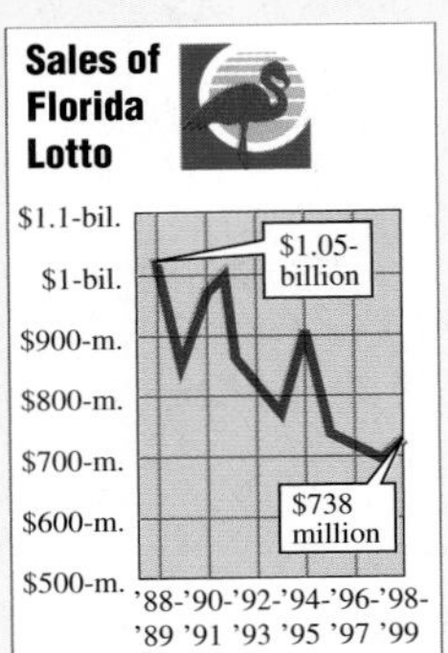

Source: Florida Lottery

Hoping to recoup lost players and increase the jackpot, on October 27, 1999, the Florida Lottery Secretary, David Griffin, increased the numbers that may be selected in the lotto from 49 to 53. Now players choose 6 numbers from 53 (instead of 49). They must select all six numbers to win the grand prize. This change decreased the probability of winning the grand prize from 1 in 13,983,816 or $\left(\frac{{}_6C_6}{{}_{49}C_6}\right)$ to 1 in 22,957,480 or $\left(\frac{{}_6C_6}{{}_{53}C_6}\right)$. He also increased the lottery drawing to twice weekly. "We are looking to increase sales," said Griffin. Time will tell if he is right.

EXAMPLE 4 *New Breakfast Cereals*

The Kellogg Company is testing 12 new cereals for possible production. They are testing 3 oat cereals, 4 wheat cereals, and 5 rice cereals. If we assume that each of the 12 cereals has the same chance of being selected and 4 new cereals will be produced, find the probability that

a) no wheat cereals are selected,
b) at least 1 wheat cereal is selected,
c) 2 wheat cereals and 2 rice cereals are selected.

SOLUTION

a) If no wheat cereals are to be selected, then only oat and rice cereals must be selected. A total of 8 cereals are oat or rice. Thus, the number of ways that 4 oat or rice cereals may be selected from the 8 possible oat or rice cereals is ${}_8C_4$. The total number of possible selections is ${}_{12}C_4$.

$$P(\text{no wheat cereals}) = \frac{{}_8C_4}{{}_{12}C_4} = \frac{70}{495} = \frac{14}{99}$$

b) When 4 cereals are selected, the choice must contain either no wheat cereal or at least 1 wheat cereal. Since one of these outcomes must occur, the sum of the probabilities must be 1, or

$$P(\text{no wheat cereal}) + P(\text{at least 1 wheat cereal}) = 1$$

Therefore,

$$\begin{aligned} P(\text{at least 1 wheat cereal}) &= 1 - P(\text{no wheat cereal}) \\ &= 1 - \frac{14}{99} = \frac{99}{99} - \frac{14}{99} = \frac{85}{99} \end{aligned}$$

Note that the probability of selecting no wheat cereals was found in part (a).

c) The number of ways of selecting 2 wheat cereals out of 4 wheat cereals is ${}_4C_2$, which equals 6. The number of ways of selecting 2 rice cereals out of 5 rice cereals is ${}_5C_2$, which equals 10. The total number of possible selections when 4 cereals are selected from the 12 choices is ${}_{12}C_4$. Since both the 2 wheat *and* the 2 rice cereals must be selected, the probability is calculated as follows.

$$P(\text{2 wheat and 2 rice}) = \frac{{}_4C_2 \cdot {}_5C_2}{{}_{12}C_4} = \frac{6 \cdot 10}{495} = \frac{60}{495} = \frac{4}{33}$$

SECTION 12.10 EXERCISES

CONCEPT/WRITING EXERCISES

In Exercises 1–8, set up the problem as if it were to be solved, but do not solve. Assume each problem is to be done without replacement. Explain why you set up the exercises as you did.

1. Find the probability of selecting 4 red balls at random out of 10 balls of which 6 are red.

2. Find the probability of selecting 3 vowels (*a*, *e*, *i*, *o*, *u*) out of the English alphabet when 3 letters are selected at random.

3. Find the probability of selecting 12 girls at random in a class of 19 girls and 15 boys.

4. A telephone book for a small town has 1206 listed numbers. Two hundred seventy of the names begin with the letter P. If 70 names are selected at random from the phone book, find the probability the names all start with the letter P.

5. A basket of 70 tennis balls contains 22 Wilson balls. If 8 balls are selected at random from the basket, determine the probability they are all Wilson balls.

6. On a horse farm there are 24 horses, of which 18 are palaminos. If 7 horses are selected at random, determine the probability they are all palaminos.

7. In a small area of a forest there are 30 trees, of which 16 are oak tress. Nine trees are to be selected at random. Find the probability that *none* of those selected are oak trees.

8. In a department store 36 different calculators are on display, including 13 Texas Instrument calculators. If 9 are to be selected at random, determine the probability that *none* are Texas Instrument calculators.

PRACTICE THE SKILLS/PROBLEM SOLVING

In Exercises 9–18, the problems are to be done without replacement. Use combinations to determine probabilities.

9. *Blue and Red Balls* A bag contains five red balls and four blue balls. You plan to draw three balls at random. Find the probability of selecting three red balls.

10. *Drawing from a Hat* Each of the numbers 1–6 is written on a piece of paper, and the six pieces of paper are placed in a hat. If two numbers are selected at random, find the probability that both numbers selected are even.

11. *Selecting Batteries* A box contains four good and four defective batteries. If you select three at random, find the probability that you select three good batteries.

12. *Bills of Four Denominations* Bob Dies's wallet contains 8 bills of the following denominations: four $5 bills, two $10 bills, one $20 bill, and one $50 bill. If he selects two bills at random, determine the probability that he selects two $5 bills.

13. *Selecting Digits* Each of the digits 0, 1, 2, 3, 4, 5, 6, 7, 8, and 9 is written on a slip of paper, and the slips are placed in a hat. If three slips of paper are selected at random, find the probability that the three numbers selected are greater than 4.

14. *Forming a Committee* A three-person committee is to be selected at random from four Democrats and three Republicans. Find the probability that all three selected are Democrats.

15. *Bike Riding* A bicycle club has 10 members. Six members ride Huffy bicycles, two members ride Roadmaster bicycles, and two members ride American Flier bicycles. If four of the members are selected at random, determine the probability that they all ride Huffy bicycles.

16. *Faculty-Student Committee* A committee of four is to be randomly selected from a group of seven teachers and eight students. Find the probability that the committee will consist of four students.

17. *Winning the Grand Prize* A lottery consists of 46 numbers. You select 6 numbers and if they match the 6 numbers selected by the lottery commission, you win the grand prize. Find the probability of winning the grand prize.

18. *Red Cards* You are dealt 5 cards from a deck of 52 cards. Find the probability that you are dealt 5 red cards.

TV Game Show In Exercises 19–22, a television game show has five doors, of which the contestant must pick two. Behind two of the doors are expensive cars, and behind the other three doors are consolation prizes. The contestant gets to keep the items behind the two doors she selects. Find the probability that the contestant wins

19. both cars.
20. no cars.
21. at least one car.
22. exactly one car.

Baseball In Exercises 23–26, assume a particular professional baseball team has 10 pitchers, 6 infielders (excluding catchers), and 9 other players. If 3 players names are selected at random, determine the probability that

23. all 3 are pitchers.
24. none of the three is a pitcher.
25. 2 are pitchers and 1 is an infielder.
26. 1 is a pitcher and 2 are players other than pitchers and infielders.

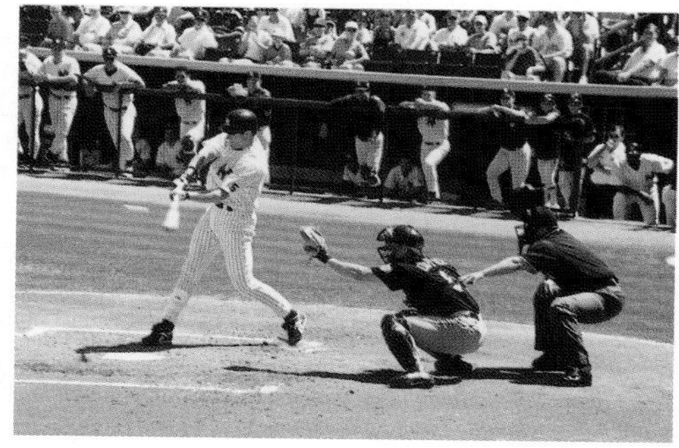

Jury Selection *In Exercises 27–30, a jury pool has 17 men and 22 women, from which 12 will be selected. Assuming that each person is equally likely to be selected and that the jury is selected at random, find the probability the jury consists of*

27. all women.

28. 8 women and 4 men.

29. 6 men and 6 women.

30. at least 1 man.

A Medical Experiment *In Exercises 31–34, drug A is given to 5 patients. Drug B is given to 4 patients, and drug C is given to 6 patients. If 4 of these 15 patients are selected at random, find the probability that*

31. 2 were given drug A and 2 were given drug C.

32. 3 were given drug C and 1 was given drug A.

33. at least 1 patient was given drug C.

34. 1 was given drug A, 2 were given drug B, and 1 was given drug C.

Theater *In Exercises 35–38, five men and six women are going to be assigned to a specific row of seats in the theater. If the 11 tickets for the numbered seats are given out at random, find the probability that*

35. five women are given the first five seats next to the center aisle.

36. at least one woman is in one of the first five seats.

37. exactly one woman is in one of the first five seats.

38. three women are seated in the first three seats and two men are seated in the next two seats.

39. *Work Shift* Among 24 employees who work at Wendy's, three are brothers. If six of the 24 are selected at random to work a late shift, find the probability that the three brothers are selected.

40. *Poker Probability* A full house in poker consists of getting three of one kind and two of another kind in a five-card hand. For example, if a hand contains three kings and two 5s, it is a full house. If 5 cards are dealt at random from a deck of 52 cards, without replacement, find the probability of getting three kings and two 5s.

41. *A Royal Flush* A royal flush consists of the ace, king, queen, jack, and 10 all in the same suit. If 7 cards are dealt at random from a deck of 52 cards, find the probability of getting a

a) royal flush in spades.

b) royal flush in any suit.

CHALLENGE PROBLEMS/GROUP ACTIVITIES

Slot Machine Probabilities *In Exercises 42–45, the first reel of a three-reel slot machine consists of three oranges, four cherries, two lemons, six plums, two bells, and one bar. The second reel consists of two oranges, five cherries, three lemons, five plums, two bells, and one bar. The third reel consists of four oranges, four cherries, three lemons, three plums, two bells, and two bars. When you pull the handle of the slot machine, find the probability of obtaining*

42. cherries on all three reels.

43. cherry, cherry, bell.

44. bars on all three reels.

45. a cherry on at least one reel.

46. *Alternate Seating* If three men and three women are to be assigned at random to six seats in a row at a theater, find the probability they will alternate by sex.

47. *Selecting Officers* A club consists of 15 people including Ali, Kendra, Ted, Alice, Marie, Dan, Linda, and Frank. From the 15 members a president, vice-president, and treasurer will be selected at random, and an advisory committee of 5 other individuals will also be selected at random.

a) Find the probability that Ali is selected president, Kendra is selected vice-president, Ted is selected treasurer, and the other 5 individuals named form the advisory committee.

b) Find the probability that 3 of the 8 individuals named are selected for the three officers' positions and the other 5 are selected for the advisory board.

48. *"Dead Man's Hand"* A pair of aces and a pair of 8s is often referred to as the "dead man's hand." (See the Did You Know on page 644.)

a) Find the probability of being dealt the dead man's hand when dealt 5 cards, without replacement, from a deck of 52 cards.

b) The actual cards Hickok was holding when he was shot were the aces of spades and clubs, the 8s of spades and clubs, and the 9 of diamonds. If you are dealt five cards without replacement, find the probability of being dealt this exact hand.

49. *A Marked Deck* A number is written with a magic marker on each card of a deck of 52 cards. The number 1 is put on the first card, 2 on the second, and so on. The cards are then shuffled and cut. What is the probability that the top 4 cards will be in ascending order? (For example, the top card is 12, the second 22, the third 41, and the fourth 51.)

12.11 BINOMIAL PROBABILITY FORMULA

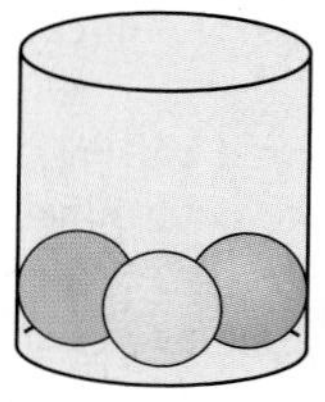

Figure 12.15

Suppose that a basket contains 3 identical balls, except for their color. One is red, one is blue, and one is yellow (Fig. 12.15). Suppose further that we are going to select 3 balls *with replacement* from the basket. We can determine specific probabilities by examining the tree diagram shown in Fig. 12.16. Note that 27 different selections are possible, as indicated in the sample space.

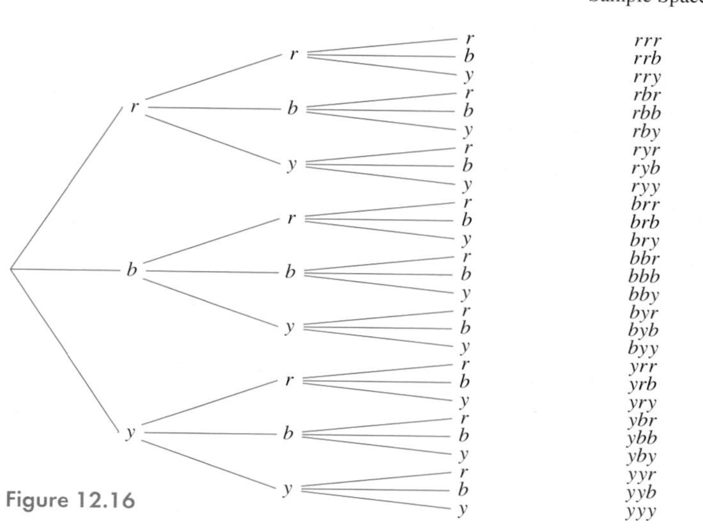

Figure 12.16

Our three selections may yield 0, 1, 2, or 3 red balls. We can determine the probability of selecting exactly 0, 1, 2, or 3 red balls by using the sample space. To determine the probability of selecting 0 red balls, we count those outcomes that do not contain a red ball. There are 8 of them (*bbb*, *bby*, *byb*, *byy*, *ybb*, *yby*, *yyb*, *yyy*). Thus, the probability of obtaining exactly 0 red balls is 8/27. We determine the probability of selecting exactly 1 red ball by counting the sample points that contain exactly 1 red ball. There are 12 of them. Thus, the probability is 12/27, or 4/9.

We can determine the probability of selecting exactly 2 red balls and exactly 3 red balls in a similar manner. The probabilities of selecting exactly 0, 1, 2, and 3 red balls are illustrated in Table 12.5.

Table 12.5 A Probability Distribution

Number of red balls (x)	**Probability of selecting the number of red balls, $P(x)$**
0	$\frac{8}{27}$
1	$\frac{12}{27}$
2	$\frac{6}{27}$
3	$\frac{1}{27}$
	Sum $= \frac{27}{27} = 1$

Note that the sum of the probabilities is 1. This table is an example of a **probability distribution**, which shows the probabilities associated with each specific outcome of an experiment. In a probability distribution every possible outcome must be listed, and the sum of the probabilities must be 1.

Let us specifically consider the probability of selecting 1 red ball in 3 selections. We see from Table 12.5 that this probability is $\frac{12}{27}$ or $\frac{4}{9}$. Can we determine this probability without developing a tree diagram? The answer is yes.

Suppose we consider selecting a red ball success, S, and a non-red ball failure, F. Furthermore, suppose we let p represent the probability of success and q the probability of failure on any trial. Then $p = \frac{1}{3}$ and $q = \frac{2}{3}$. We can obtain 1 success in three selections in the following ways:

SFF FSF FFS

We can compute the probabilities of each of these outcomes using the multiplication formula because each of the selections is independent.

$$P(\text{SFF}) = P(\text{S}) \cdot P(\text{F}) \cdot P(\text{F}) = p \cdot q \cdot q = pq^2 = \frac{1}{3}\left(\frac{2}{3}\right)^2 = \frac{4}{27}$$

$$P(\text{FSF}) = P(\text{F}) \cdot P(\text{S}) \cdot P(\text{F}) = q \cdot p \cdot q = pq^2 = \frac{1}{3}\left(\frac{2}{3}\right)^2 = \frac{4}{27}$$

$$P(\text{FFS}) = P(\text{F}) \cdot P(\text{F}) \cdot P(\text{S}) = q \cdot q \cdot p = pq^2 = \frac{1}{3}\left(\frac{2}{3}\right)^2 = \frac{4}{27}$$

$$\text{Sum} = \frac{12}{27} = \frac{4}{9}$$

We obtained an answer of 4/9, the same answer that was obtained using the tree diagram. Note that each of the 3 sets of outcomes above has one success and two failures. Rather than listing all the possibilities containing 1 success and 2 failures, we can use the combination formula to determine the number of possible combinations of 1 success in 3 trials. To do so, evaluate ${}_3C_1$.

$${}_3C_1 = \frac{3!}{(3-1)!1!} = \frac{3 \cdot 2 \cdot 1}{2 \cdot 1 \cdot 1} = 3$$

Number of trials (the 3 in ${}_3C_1$); Number of sucesses (the 1 in ${}_3C_1$)

Thus, we see that there are 3 ways the 1 success could occur in 3 trials. To compute the probability of 1 success in 3 trials, we can multiply the probability of success in any one trial, $p \cdot q^2$, by the number of ways the 1 success can be arranged among the 3 trials, ${}_3C_1$. Thus, the probability of selecting 1 red ball, $P(1)$, in 3 trials may be found as follows,

$$P(1) = ({}_3C_1)p^1q^2 = 3\left(\frac{1}{3}\right)\left(\frac{2}{3}\right)^2 = \frac{12}{27} = \frac{4}{9}$$

The binomial probability formula, which we introduce shortly, explains how to obtain expressions like $P(1) = ({}_3C_1)p^1q^2$ and is very useful in finding certain types of probabilities.

To use the binomial probability formula, the following two conditions must hold.

To Use the Binomial Probability Formula

1. Each trial has two possible outcomes, *success* and *failure.*
2. There are n repeated independent trials.

Before going further, let's discuss why we can use the binomial probability formula to find the probability of selecting a specific number of red balls when balls are selected with replacement. First, we may consider selecting a red ball as success and selecting any ball of another color as failure. Second, since we are doing each trial *with replacement,* the trials are independent of each other. Now let's discuss the binomial probability formula.

Binomial Probability Formula

The probability of obtaining exactly x successes, $P(x)$, in n independent trials is given by

$$P(x) = ({}_nC_x)p^x q^{n-x}$$

where p is the probability of success on a single trial and q is the probability of failure on a single trial.

In the formula, p will be a number between 0 and 1, inclusive, and $q = 1 - p$. Therefore, if $p = 0.2$, then $q = 1 - 0.2 = 0.8$. If $p = 3/5$, then $q = 1 - 3/5 = 2/5$. Note that $p + q = 1$ and the values of p and q remain the same for each independent trial. The combination ${}_nC_x$ is called the *binomial coefficient.*

In Example 1, we use the binomial probability formula to solve the same problem we recently solved by using a tree diagram.

EXAMPLE 1 *Selecting Colored Balls with Replacement*

A basket contains 3 balls: 1 red, 1 blue, and 1 yellow. Three balls are going to be selected with replacement from the basket. Find the probability that

a) no red balls are selected.
b) exactly 1 red ball is selected.
c) exactly 2 red balls are selected.
d) exactly 3 red balls are selected.

SOLUTION

a) We will consider selecting a red ball a success and selecting a ball of any other color a failure. Since only 1 of the 3 balls is red, the probability of success on any single trial, p, is 1/3. The probability of failure on any single trial, q, is $1 - 1/3 = 2/3$. We do not want any success (red balls), so $x = 0$. There are 3 independent selections (or trials), so $n = 3$. We determine the probability of 0 successes, or $P(0)$, as follows.

$$P(x) = ({}_nC_x)p^x q^{n-x}$$
$$P(0) = ({}_3C_0)\left(\frac{1}{3}\right)^0\left(\frac{2}{3}\right)^{3-0}$$
$$= (1)(1)\left(\frac{2}{3}\right)^3$$
$$= \left(\frac{2}{3}\right)^3 = \frac{8}{27}$$

b) We are finding the probability of obtaining exactly 1 red ball or exactly 1 success in 3 independent selections. Thus, $x = 1$ and $n = 3$. We find the probability of exactly 1 success, or $P(1)$, as follows.

$$P(x) = ({}_nC_x)p^x q^{n-x}$$
$$P(1) = ({}_3C_1)\left(\frac{1}{3}\right)^1\left(\frac{2}{3}\right)^{3-1}$$
$$= 3\left(\frac{1}{3}\right)\left(\frac{2}{3}\right)^2$$
$$= 3\left(\frac{1}{3}\right)\left(\frac{4}{9}\right) = \frac{4}{9}$$

c) We are finding the probability of selecting exactly 2 red balls in 3 independent trials. Thus, $x = 2$ and $n = 3$. We find $P(2)$ as follows.

$$P(x) = ({}_nC_x)p^x q^{n-x}$$
$$P(2) = {}_3C_2\left(\frac{1}{3}\right)^2\left(\frac{2}{3}\right)^{3-2}$$
$$= 3\left(\frac{1}{3}\right)^2\left(\frac{2}{3}\right)^1$$
$$= 3\left(\frac{1}{9}\right)\left(\frac{2}{3}\right) = \frac{2}{9}$$

d) We are finding the probability of selecting exactly 3 red balls in 3 independent trials. Thus, $x = 3$ and $n = 3$. We find $P(3)$ as follows.

$$P(x) = ({}_nC_x)p^x q^{n-x}$$
$$P(3) = ({}_3C_3)\left(\frac{1}{3}\right)^3\left(\frac{2}{3}\right)^{3-3}$$
$$= 1\left(\frac{1}{3}\right)^3\left(\frac{2}{3}\right)^0$$
$$= 1\left(\frac{1}{27}\right)(1) = \frac{1}{27}$$

▲

All the probabilities obtained in Example 1 agree with the answers obtained by using the tree diagram. Whenever you obtain a value for $P(x)$, you should obtain a value between 0 and 1, inclusive. If you obtain a value greater than 1, you have made a mistake.

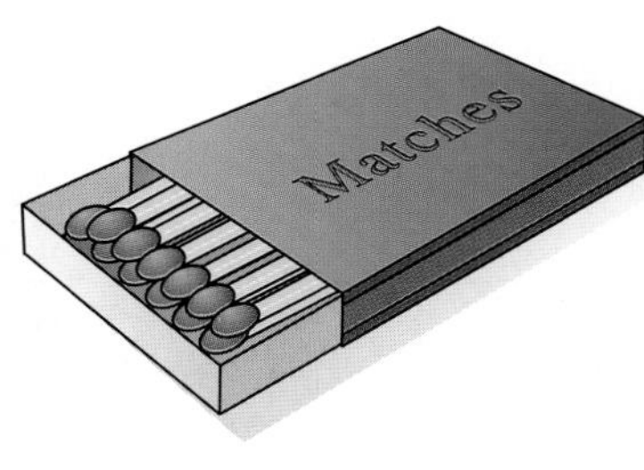

EXAMPLE 2 *Quality Control for Matches*

A manufacturer of matches knows that 0.1% of the matches produced by the company are defective.

a) Write the binomial probability formula that would be used to determine the probability that exactly x matches in a package of n matches are defective.
b) Write the binomial probability formula that would be used to find the probability exactly 3 matches in a 50-match package will be defective.

SOLUTION

a) We want to find the probability that exactly x matches are defective where selecting a defective match is considered success. The probability an individual match is defective is 0.1%, or 0.001 in decimal form. The probability a match is not defective, q, is $1 - 0.001$, or 0.999. The general formula for finding the probability that exactly x matches are defective out of n matches is

$$P(x) = ({}_nC_x)p^xq^{n-x}$$

Substituting 0.001 for p and 0.999 for q, we obtain the formula

$$P(x) = ({}_nC_x)(0.001)^x(0.999)^{n-x}$$

b) We want to determine the probability that exactly 3 matches are defective out of 50. Thus, $x = 3$ and $n = 50$. Substituting these values into the formula in part (a) gives

$$\begin{aligned} P(3) &= ({}_{50}C_3)(0.001)^3(0.999)^{50-3} \\ &= ({}_{50}C_3)(0.001)^3(0.999)^{47} \end{aligned}$$

▲

The answer to part (b) of Example 2 may be obtained using a scientific calculator.

EXAMPLE 3 *Weather Forecast Accuracy*

The local weather person has been accurate in her "within 2 degrees" temperature forecast 80% of the time. Find the probability that she is accurate

a) exactly 3 of the next 5 days.
b) exactly 4 of the next 4 days.

SOLUTION

a) We want to find the probability that the forecaster is successful (within 2°) exactly 3 of the next 5 days. Thus, $x = 3$ and $n = 5$. The probability of success on any one day, p, is 80%, or 0.8. The probability of failure, q, is $1 - 0.8 = 0.2$. Substituting these values into the binomial probability formula yields

$$\begin{aligned} P(x) &= ({}_nC_x)p^xq^{n-x} \\ P(3) &= ({}_5C_3)(0.8)^3(0.2)^{5-3} \\ &= 10(0.512)(0.04) \\ &= 0.2048 \end{aligned}$$

b) We want to find $P(4)$.

$$\begin{aligned} P(x) &= ({}_nC_x)p^x q^{n-x} \\ P(4) &= ({}_4C_4)(0.8)^4(0.2)^{4-4} \\ &= 1(0.8)^4(0.2)^0 \\ &= 1(0.4096)(1) \\ &= 0.4096 \end{aligned}$$

EXAMPLE 4 *Blue Eyes*

The probability that an individual selected at random has blue eyes is 0.4. Find the probability that

a) none of four people selected at random has blue eyes.
b) at least one of four people selected at random has blue eyes.

SOLUTION Success is selecting a person with blue eyes. Thus, $p = 0.4$ and $q = 1 - 0.4 = 0.6$. We want to find the probability of 0 successes in 4 trials. Thus, $x = 0$ and $n = 4$. We find the probability of 0 successes, or $P(0)$, as follows.

$$\begin{aligned} P(x) &= ({}_nC_x)p^x q^{n-x} \\ P(0) &= ({}_4C_0)(0.4)^0(0.6)^{4-0} \\ &= 1(1)(0.6)^4 \\ &= 1(1)(0.1296) \\ &= 0.1296 \end{aligned}$$

b) The probability that at least one person of the four has blue eyes can be found by subtracting from 1 the probability none of the people has blue eyes. We worked problems of this type in earlier sections of the chapter.

In part (a), we determined the probability that none of the people has blue eyes is 0.1296. Thus,

$$\begin{aligned} P(\text{at least 1 has blue eyes}) &= 1 - P(\text{none has blue eyes}) \\ &= 1 - 0.1296 \\ &= 0.8704 \end{aligned}$$

SECTION 12.11 EXERCISES

CONCEPT/WRITING EXERCISES

1. What is a probability distribution?
2. What are the two requirements that must be met to use the binomial probability formula?
3. Write the binomial probability formula.
4. In the binomial probability formula what do p and q represent?

PRACTICE THE SKILLS

In Exercises 5–10, assume that each of the n trials is independent and that p is the probability of success on a given trial. Use the binomial probability formula to find P(x).

5. $n = 4, x = 1, p = 0.1$
6. $n = 3, x = 2, p = 0.6$
7. $n = 5, x = 2, p = 0.4$
8. $n = 3, x = 3, p = 0.9$
9. $n = 4, x = 0, p = 0.5$
10. $n = 5, x = 3, p = 0.4$

PROBLEM SOLVING

11. *A Dozen Eggs* An egg distributor determines that the probability that any individual egg has a crack is 0.15.
    a) Write the binomial probability formula to determine the probability that exactly x of n eggs are cracked.
    b) Write the binomial probability formula to determine the probability that exactly 2 in a one dozen-egg carton are cracked.

12. *Getting Audited* The probability that a family selected at random will be audited by the Internal Revenue Service (IRS) this year is 0.0237.
    a) Write the binomial probability formula to determine the probability that exactly x out of n families selected at random will be audited by the IRS this year.
    b) Write the binomial probability formula to determine the probability that exactly 5 of 20 families selected at random will be audited by the IRS this year.

In Exercises 13–21, use the binomial probability formula to answer the question.

13. *Telephone Booth* Assume that the probability that a telephone in any given telephone booth is defective is 0.2. Determine the probability that if 5 phone booths are examined, exactly 1 has a defective phone.

14. *Rolling a 7* The probability of rolling a sum of 7 when a pair of dice is rolled is 1/6. Find the probability that if a pair of dice is rolled 6 times, exactly one sum of 7 is rolled.

15. *Vision Surgery* Ninety-six percent of Dr. William's LASIK surgery patients end up with 20–30, or better, vision. Find the probability that exactly 2 of her next 3 patients end up with 20–30, or better, vision.

16. *Basketball* Cheryl Miller makes 80% of her free throws in a basketball game. Find the probability that she makes exactly 4 of the next 6 free throws.

17. *Dolphin Drug Care* When treated with the antibiotic resonocyllin, 92% of all dolphins are cured of a particular bacterial infection. If six dolphins with the particular bacterial infection are treated with resonocyllin, find the probability that exactly 4 are cured.

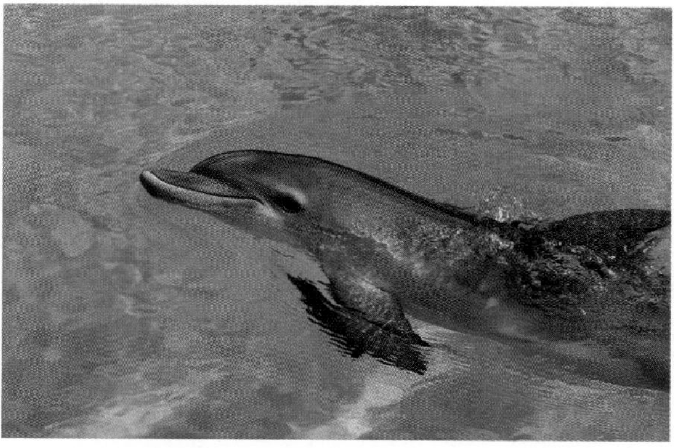

18. *Manufacturing Light Bulbs* A quality control engineer at a GE light bulb plant finds that 1% of its bulbs are defective. Find the probability that exactly 2 of the next 4 bulbs made are defective.

19. *The Roll of a Die* The probability of rolling a number greater than 2 on a die is 2/3. Find the probability that on the next 4 rolls of a die exactly 3 show a number greater than 2.

20. *TV Purchases* At a Best Buy store, 1/4 of those purchasing color televisions purchase a large-screen TV. Find the probability that
    a) none of the next four people who purchase a color television at Best Buy purchases a large-screen TV.
    b) at least one of the next four people who purchase a color television at Best Buy purchases a large-screen TV.

21. *Supporting a Candidate* Sixty percent of the eligible voting residents of a certain community support Ms. Stein, the incumbent candidate. If five of the residents are selected at random, find the probability that
    a) none supports Ms. Stein.
    b) at least one supports Ms. Stein.

CHALLENGE PROBLEMS/GROUP ACTIVITIES

22. *Transportation to Work* In a random sample of working mothers in Duluth, Minnesota, the following data indicating how they get to work were obtained.

Mode of transportation	Number of mothers
Car	40
Bus	20
Bike	16
Other	4

    If this sample is representative of all working mothers in Duluth, find the probability that exactly 3 of 5 working mothers selected at random
    a) take a car to work.
    b) take a bus to work.

23. *Selecting 6 Cards* Six cards are selected from a deck of playing cards with replacement. Find the probability that
    a) exactly 3 picture cards are obtained.
    b) exactly 2 spades are obtained.

24. *Office Visit* The probability that a person visiting Dr. Guillermo Suarez's office is over 60 years old is 0.7. Find the probability that
    a) exactly 3 of the next 5 people visiting the office are over 60 years old.
    b) at least 3 of the next 5 people visiting the office are over 60 years old.

CHAPTER 12 SUMMARY

IMPORTANT FACTS

Empirical probability

$$P(E) = \frac{\text{number of times event } E \text{ has occurred}}{\left(\begin{array}{c}\text{total number of times the}\\ \text{experiment has been performed}\end{array}\right)}$$

The law of large numbers

Probability statements apply in practice to a large number of trials, not to a single trial. It is the relative frequency over the long run that is accurately predictable, not individual events or precise totals.

Theoretical probability

$$P(E) = \frac{\text{number of outcomes favorable to } E}{\text{total number of possible outcomes}}$$

The probability of an event that cannot occur is 0. The probability of an event that must occur is 1. Every probability must be a number between 0 and 1 inclusively; that is

$$0 \leq P(E) \leq 1$$

The sum of the probabilities of all possible outcomes of an event is 1.

$$P(A) + P(\text{not } A) = 1$$

Odds against an event

$$\text{Odds against} = \frac{P(\text{event fails to occur})}{P(\text{event occurs})} = \frac{P(\text{failure})}{P(\text{success})}$$

Odds in favor of an event

$$\text{Odds in favor} = \frac{P(\text{event occurs})}{P(\text{event fails to occur})} = \frac{P(\text{success})}{P(\text{failure})}$$

Expected value

$$E = P_1A_1 + P_2A_2 + P_3A_3 + \cdots + P_nA_n$$

Expected value = fair price − cost to play

Counting principle

If a first experiment can be performed in M distinct ways and a second experiment can be performed in N distinct ways, then the two experiments in that specific order can be performed in $M \cdot N$ distinct ways.

***Or* and *And* problems**

$$P(A \text{ or } B) = P(A) + P(B) - P(A \text{ and } B)$$

$P(A \text{ and } B) = P(A) \cdot P(B)$, assuming that event A has occurred

Conditional probability

$$P(E_2|E_1) = \frac{n(E_1 \text{ and } E_2)}{n(E_1)}$$

The **number of permutations** of n items is $n!$.

$$n! = n(n-1)(n-2)\cdots(3)(2)(1)$$

Permutation formula

$$_nP_r = \frac{n!}{(n-r)!}$$

The number of different permutations of n objects where $n_1, n_2, \ldots, n_r$ of the objects are identical is

$$\frac{n!}{n_1!n_2!\cdots n_r!}$$

Combination formula

$$_nC_r = \frac{n!}{(n-r)!r!}$$

Binomial probability formula

$$P(x) = ({}_nC_x)p^xq^{n-x}$$

CHAPTER 12 REVIEW EXERCISES

12.1–12.11

1. In your own words, explain the law of large numbers.
2. Explain how empirical probability can be used to determine whether a die is "loaded" (not a fair die).
3. Of 60 tosses of a trick coin, 55 resulted in heads. Find the empirical probability of the coin landing heads up.
4. Select a card from a deck of cards 40 times with replacement and compute the empirical probability of selecting a heart.
5. In 100 births at County Hospital, 58 were males. Find the empirical probability that the next baby born at the hospital will be a male.

In Exercises 6–9, each of the digits 0, 1, 2, 3, 4, 5, 6, 7, 8, 9 is written on a piece of paper, and all the pieces of paper are placed in a hat. One number is selected at random. Find the probability that the number selected is

6. odd.
7. even or greater than 4.
8. greater than 2 or less than 5.
9. even and greater than 4.

Soda Preference *In Exercises 10–13, assume that at a supermarket a taste test is given to 50 customers to determine their preference for sodas. The results are summarized in the following chart.*

Brand	Numbers of people
Coke	17
Pepsi	15
Dr. Pepper	10
7-Up	8

If one person who completed the survey is selected at random, find the probability the person selected

10. Coke

11. Pepsi

12. either Dr. Pepper or 7-Up

13. a soda other than Dr. Pepper

14. *Gold Star* Planters Peanuts is having a contest. They indicate that 1 in 10 jars will have a gold star under the cover. Jason purchases one jar. Find the odds
a) against his jar having a star.
b) in favor of his jar having a star.

15. *Vegetable Mix-up* Nicholas, a mischievous little boy, has removed labels on the eight cans of vegetables in the cabinet. Nicholas's father knows that there are three cans of corn, three cans of beans, and two cans of carrots. If the father selects and opens one can at random, find the odds against his selecting a can of corn.

16. *Chess Odds* Assuming that the odds against winning a game of chess are 2:3, find the probability of winning the game.

17. *Restaurant Success* The probability that a new restaurant will succeed in a given location is 0.6. Find the odds in favor of the restaurant succeeding.

18. *Raffle Tickets* A thousand raffle tickets are sold at \$2 each. Three prizes of \$200 and two prizes of \$100 will be awarded.
a) Find the expectation of a person who purchases a ticket.
b) Find the expectation of a person who purchases three tickets.
c) Find the fair price to pay for a ticket.

19. *Expectation of a Card* If Cameron selects a picture card from a deck of cards, Lindsey will give him \$9. If Cameron does not select a picture card, he must give Lindsey \$3.
a) Find Cameron's expectation.
b) Find Lindsey's expectation.
c) If Cameron plays this game 100 times, how much can he expect to lose or gain?

20. *Expected Attendance* If the day is sunny, 1000 people will attend the baseball game. If the day is cloudy, only 500 people will attend. If it rains, only 100 people will attend. The local meteorologist states that the probability of a sunny day is 0.4, of a cloudy day is 0.5, and of a rainy day is 0.1. Find the number of people that are expected to attend.

21. *Club Officers* Tina, Jake, Gina, and Carla form a club. They plan to select a president and a vice-president.
a) Construct a tree diagram showing all the possible outcomes.
b) List the sample space.
c) Find the probability that Gina is selected president and Jake is selected vice-president.

22. *A Coin and a Marble* A coin is flipped and then a marble is to be selected from a bag. The bag contains four marbles: one red, one blue, one green, and one purple.
a) Contruct a tree diagram showing all the possible outcomes.
b) List the sample space.
c) Find the probability that a head is flipped and either a red or purple marble is selected.

Spinning Two Wheels *In Exercises 23–28, the following double wheel is spun. The outer wheel is spun clockwise and the inner wheel is spun counterclockwise.*

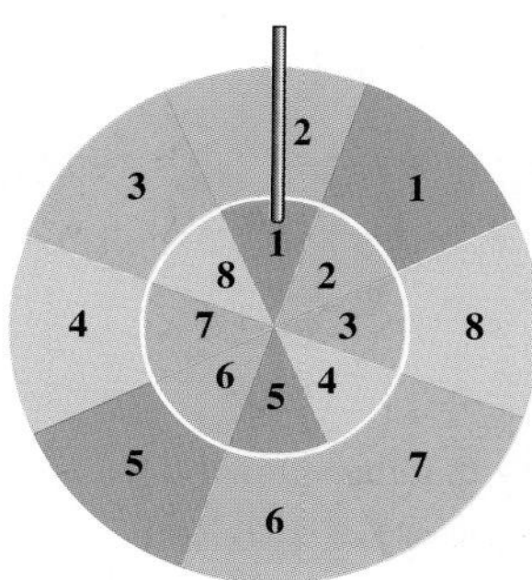

Assuming equally likely outcomes, find the probability that

23. odd numbers on both wheels stop under the pointer.

24. numbers greater than 5 on both wheels stop under the pointer.

25. an odd number on the outer wheel and a number less than 6 on the inner wheel stop under the pointer.

26. an even number or a number less than 6 on the outer wheel stops under the pointer.

27. an even number or a color other than blue on the inner wheel stops under the pointer.

28. blue on the outer wheel and a color other than blue on the inner wheel stop under the pointer.

Candy Selections *In Exercises 29–32, assume that to prepare for Halloween, the Pollingers are going to purchase three large bags of candy. The store has only 12 different bags of candies, including 5 different varieties made by Hershey, 4 different varieties made by Nestle, and 3 different varieties made by H.B. Reese Candy Company. If Mrs. Pollinger selects 3 of these 12 varieties at random, find the probability she selects*

29. 3 varieties of Hershey candies.

30. no varieties of Nestle candies.

31. at least 1 variety of Nestle candy.

32. varieties of Hershey, Hershey, Reese in this order.

Spinner Probabilities *In Exercises 33–36, assume that the spinner cannot land on a line.*

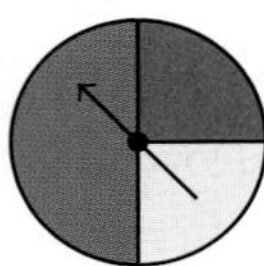

If spun once, find

33. the probability that the spinner lands on red.

34. the odds against and in favor of the spinner landing on red.

35. If you are awarded $5 if the spinner lands on red, find a fair price to play the game.

36. If the spinner is spun twice, find the probability that it lands on red and then green (assume independence).

Spinner Probabilities *In Exercises 37–40, assume that the spinner cannot land on a line.*

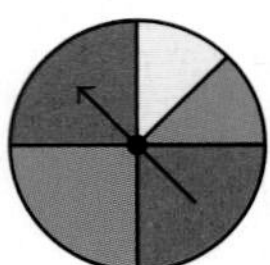

If spun once, find

37. the probability that the spinner does not land on green.

38. the odds in favor of and against the spinner landing on green.

39. A person wins $10 if the spinner lands on green, wins $5 if the spinner lands on red, and loses $20 if the spinner lands on yellow. Find the expectation of a person who plays this game.

40. If the spinner is spun three times, find the probability that at least one spin lands on red.

Automobile Quality Control *In Exercises 41–44, a sample of 180 new cars were checked for defects. The following table shows the results of the suvey.*

Car	Fewer than six defects	Six or more defects	Total
American built	89	17	106
Foreign built	55	19	74
Total	144	36	180

Find the probability that if one car is selected from this sample, the car has

41. fewer than six defects if it is American built.

42. fewer than six defects if it is foreign built.

43. six or more defects if it is foreign built.

44. six or more defects if it is American built.

Neuroscience *In Exercises 45–48, assume that in a neuroscience course the students perform an experiment. Tests are given to determine if people are right brained, left brained, or have no predominance. It is also recorded whether they are right handed or left handed. The following chart shows the results obtained.*

	Right brained	Left brained	No predominance	Total
Right Handed	40	130	60	230
Left Handed	120	30	20	170
Total	160	160	80	400

If one person who completed the survey is selected at random, find the probability the person selected is

45. right handed.

46. left brained, if the person is left handed.

47. right handed, if the person has no predominance.

48. right brained, if the person is left handed.

49. *Finalist* Four finalists remain in a lottery drawing. The prizes to be awarded among the finalists are $100, $500, $1000, and $10,000.

a) In how many different ways can the winners be selected?

b) What is the expectation of a finalist?

50. *Spelling Bee* Five finalists remain in a high school spelling bee. Two will receive $50 each, two will receive $100 each, and one will receive $500. How many different arrangements of prizes are possible?

51. *Selecting a Pet* Each of Mr. Vargas' three children is planning to select his or her own pet rabbit from a litter of eight rabbits. In how many different ways can they do so?

52. *Astronaut Selection* Three of nine astronauts must be selected for a mission. One will be the captain, one will be the navigator, and one will perform scientific experiments. In how many ways can a three-person crew be selected so that each person has a different assignment?

53. *Medicine* Dr. Goldberg has three doses of serum for influenza type A. Six patients in the office require the serum. In how many different ways could Dr. Goldberg dispense the serum?

54. *Dogsled*

a) Ten of 15 huskies are to be selected to pull a dogsled. In how many ways can this selection be made?

b) How many different arrangements of the 10 huskies on a dogsled are possible?

55. *The Big Game* The Big Game is a multistate lottery game offered in Michigan, Georgia, Illinois, Maryland, Massachusetts, New Jersey, and Vermont. To play, you select 5 numbers from 1 through 50 and 1 Big Money Ball number from 1 through 36. If you win the Big Game by matching all 6 numbers, your guaranteed minimum payoff is $5 million. If you match the 5 numbers but do not match the Big Money number, your guaranteed payoff is $150,000.

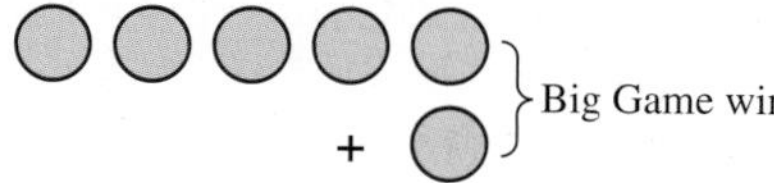

a) What is the probability you match the 5 numbers?

b) What is the probability you have a Big Game win?

56. *Conference Speaker* A newly formed honor society has 4 mathematics majors and 6 members whose major is not mathematics. Two mathematics majors and 3 others from the honor society are to be selected to talk at a conference. How many different combinations are possible?

57. *Selecting Test Subjects* In a psychology research laboratory, one room contains eight men, numbered 1–8, and another room contains five women, numbered 1–5. Three men and two women are to be selected at random to be given a psychological test. How many different combinations of these people are possible?

58. *Choosing Two Aces* Two cards are selected at random, without replacement, from a deck of 52 cards. Find the probability that two aces are selected (use combinations).

Color Chips In Exercises 59–62, a bag contains five red chips, three white chips, and two blue chips. Three chips are to be selected at random, without replacement. Find the probability that

59. all are red.

60. the first two are red and the third is blue.

61. the first is red, the second is white, and the third is blue.

62. at least one is red.

Types of Trees In Exercises 63–66, Agway Lawn and Garden Center carries 5 pine, 6 maple, and 3 birch trees. Gary Egan plans to select 3 trees at random from the 14 mentioned. Assuming that each tree has the same chance of being selected, find the probability that

63. 3 maple trees are selected.

64. 2 pine trees and 1 maple tree are selected.

65. no pine trees are selected.

66. at least one pine tree is selected.

67. *New Homes* In the community of Spring Hill, 60% of the homes purchased cost more than $90,000.

a) Write the binomial probability formula to determine the probability that exactly x of the next n homes purchased in Spring Hill cost more than $90,000.

b) Write the binomial probability formula to determine the probability that exactly 75 of the next 100 home purchases cost more than $90,000.

68. *Long-Stemmed Roses* At the Floyd's Flower Shop, 1/5 of those ordering flowers select long-stemmed roses. Find the probability that exactly 3 of the next 5 customers ordering flowers select long-stemmed roses.

69. *Taking a Math Course* During any semester at City College, 60% of the students are taking a mathematics course. Find the probability that of four students selected at random

a) none is taking a mathematics course this semester.

b) at least one is taking a mathematics course this semester.

CHAPTER 12 TEST

1. *Fishing* Of 20 people who went fishing in Lake Tarpon, 14 were fishing for bass. Find the empirical probability that the next person who goes fishing in Lake Tarpon will be fishing for bass.

In Exercises 2–5, each of the numbers 1–9 is written on a sheet of paper, and the nine sheets of paper are placed in a hat. If one sheet of paper is selected at random from the hat, find the probability that the number selected is

2. greater than 7.
3. odd.
4. even or greater than 4.
5. odd and greater than 4.

In Exercises 6–9, if 2 of the same 9 sheets of paper are selected, without replacement, from the hat, find the probability that

6. both numbers are greater than 5.
7. both numbers are even.
8. the first number is odd and the second number is even.
9. neither of the numbers is greater than 6.
10. One card is selected at random from a deck of cards. Find the probability that the card selected is a red card or a picture card.

In Exercises 11–15, one die is rolled and one letter—a, b, or c—is selected at random.

11. Use the counting principle to determine the number of sample points in the sample space.
12. Construct a tree diagram illustrating all the possible outcomes, and list the sample space.

In Exercises 13–15, by observing the sample space of the same die and three letters, determine the probability of obtaining

13. the number 4 and the letter *a*.
14. the number 4 or the letter *a*.
15. an even number or the letter *b*.
16. *Passwords* A personal password for an internet brokerage account is to consist of a digit, followed by two letters, followed by two digits. Find the number of personal codes possible if the first digit cannot be zero and repetition is permitted.
17. *Selecting a Pet* A litter of dalmation puppies consists of 5 males and 3 females. If one dalmation puppy is selected at random, find the odds against it being a male.

18. *Tennis Odds* The odds against Aimee Calhoun winning the tennis tournament are 5:2. Find the probability that Aimee wins the tournament.
19. *Pick a Card* You get to select one card at random from a deck of cards. If you pick a club, you win \$8. If you pick a heart, you win \$4. If you pick any other suit, you lose \$6. Find your expectation for this game.
20. *Distribution of Squirrels* A biologist is studying the habits of squirrels at two national parks. The following table indicates the number and type of squirrels studied at the two parks.

National park	Gray squirrels	Other types of squirrel	Total
Great Smoky Mountain	82	33	115
Yosemite	60	45	105
Total	142	78	220

If one of the squirrels being studied is spotted at Yosemite, find the probability that it is a gray squirrel.

21. *Awarding Prizes* Three of six people are to be selected and given small prizes. One will be given a book, one will be given a calculator, and one will be given a \$10 bill. In how many different ways can these prizes be awarded?

Quality Control In Exercises 22 and 23, a bin contains a total of 20 batteries, of which 8 are defective. If you select 2 at random, without replacement, find the probability that

22. none is good.
23. at least one is good.

24. *Apples from a Bucket* Five green apples and seven red apples are in a bucket. Five apples are to be selected at random, without replacement. Find the probability that three red apples and two green apple are selected.

25. At the Gross Point Association for the Prevention of Cruelty to Animals (GPAPCA), 3/5 of the people who adopt pets adopt dogs. Find the probability that exactly 3 of the next 4 people who adopt pets from GPAPCA adopt a dog.

GROUP PROJECTS

THE PROBABILITY OF AN EXACT MEASURED VALUE

1. Your car's speedometer indicates that you are traveling at 65 mph. What is the probability that you are traveling at *exactly* 65 mph? Explain your answer.

TAKING A TEST

2. A 10-question multiple choice exam is given, and each question has five possible answers. Pascal takes this exam and guesses at every question. Use the binomial probability formula to find the probability (to 5 decimal places) that
   a) he gets exactly 2 questions correct.
   b) he gets no questions correct.
   c) he gets at least 1 question correct (use the information from part (b) to answer this part).
   d) he gets at least 9 questions correct.
   e) Without using the binomial probability formula, determine the probability that he gets exactly 2 questions correct.
   f) Compare your answers to parts (a) and (e). If they are not the same explain why.

KEYLESS ENTRY

3. Many cars now have keyless entry. To open the lock you may press a 5-digit code on a set of buttons like that illustrated. The code may include repeated digits like 11433 or 55512.

   a) How many different 5-digit codes can be made using the 10 digits if repetition is permitted?
   b) How many different ways are there of pressing 5 buttons if repetition is allowed?
   c) A burglar is going to press 5 buttons at random, with repetition allowed. Find the probability that the burglar hits the sequence to open the door.
   d) Suppose that each button had only one number associated with it as illustrated below. How many different 5-digit codes can be made with the 5 digits if repetition is permitted?

   e) Using the buttons labeled 1–5, how many different ways are there to press 5 buttons if repetition is allowed?
   f) A burglar is going to press 5 buttons of those labeled 1–5 at random with repetition allowed. Find the probability that the burglar hits the sequence to open the door.
   g) Is a burglar more likely, less likely, or does he or she have the same likelihood of pressing 5 buttons and opening the car door if the buttons are labeled as in the first illustration or as in the second illustration? Explain your answer.
   h) Can you see any advantages in labeling the buttons as in the first illustration? Explain.*

*In actuality, in most cars that have key pads like that shown on the bottom left, each key acts as if it contains a single digit. For example, if your code is 7, 9, 5, 1, 3 the code 8, 0, 6, 2, 4 will unlock the door. The extra numbers, in effect, give the owner a false sense of security.

CHAPTER 13

Statistics

Benjamin Disraeli (1804–1881), once Prime Minister of Britain, said that there are three kinds of lies: "Lies, damned lies, and statistics." Do numbers lie? Numbers are, after all, the foundation of all statistical information. The "lie" comes in when, either intentionally or carelessly, a number is used in such a way as to lead us to a conclusion that is unjustified or incorrect.

The first large-sale survey was commissioned in 1086 by William the Conqueror of England to provide a basis for taxation. The first modern census, for use as a basis for government representation, was taken in the United States in August 1790. A census has been taken in the United States every 10 years since then, and is now used for many purposes.

Today there are methods by which statistics obtained from a small sample are used to represent a much larger population. In fact, a sample as small as 1600 may be used to predict the outcome of a national election. Information gathering is conducted by a variety of people, including medical researchers, scientists, advertisers, and political pollsters. You are likely to find examples of statistics every day on the front page of your newspaper.

When evaluating statistical information, remember that you need to judge the sampling methods as well as the numbers given. Ask yourself who conducted the study and whether they have a bias, how large the sample was and whether it was representative, and where the study appeared and whether there was an opposing side. Numbers may not lie, but they can be manipulated and misinterpreted. ■

Now in its 19th edition and weighing over 2 pounds, the *Statistical Abstract of the United States* is a compilation of facts and figures taken from the U.S. Census. In it you'll find many items, including what Americans owe (projected for 2000, 1499 million credit cards were used to charge \$1419 billion worth of goods) and what kinds of pets Americans have (in 1996 31.6% of households had dogs, 27.3% had cats, and 4.6% had birds).

13.1 SAMPLING TECHNIQUES

Statistics is the art and science of gathering, analyzing, and making inferences (predictions) from numerical information obtained in an experiment. This numerical information is referred to as **data**. The use of statistics, originally associated with numbers gathered for governments, has grown significantly and is now applied in all walks of life.

Governments use statistics to estimate the amount of unemployment and the cost of living. Thus, statistics has become an indispensable tool in attempting to regulate the economy. In psychology and education, the statistical theory of tests and measurements has been developed to compare achievements of individuals from diverse places and backgrounds. Another use of statistics with which we are all familiar is the public opinion poll. Newspapers and magazines carry the results of different polls on topics ranging from the president's popularity to the number of cans of soda consumed. In recent years these polls have attained a high degree of accuracy. The A. C. Nielsen rating is a public opinion poll that determines the country's most and least watched TV shows. Statistics is used in scores of other professions; in fact, it is difficult to find one that does not depend on some aspect of statistics.

Tune in Tomorrow

The A. C. Nielsen Company, which measures the viewing population of TV shows, uses a random sample of 5000 households in the United States to draw conclusions about 93.1 million viewers. An electronic measurement system, called the People Meter, is placed on each TV in the sample household. The meter measures what program is being tuned in and who is watching. The People Meter is used to collect audience information for broadcast and cable networks, nationally syndicated programs, and satellite distributors. Each household member is assigned a personal viewing button on the People Meter to keep track of which channels he or she watches and for how long. Nielsen then computes the rating of the show, using the data obtained from their sample. Due in part to the claim by ABC, CBS, and NBC that its statistics are unreliable, Nielsen may soon begin using a new video recorder to monitor viewing habits.
Source: http//www.nielsenmedia.com

Statistics is divided into two main branches: descriptive and inferential. **Descriptive statistics** is concerned with the collection, organization, and analysis of data. **Inferential statistics** is concerned with making generalizations or predictions from the data collected.

Probability and statistics are closely related. Someone in the field of probability is interested in computing the chance of occurrence of a particular event when all the possible outcomes are known. A statistician's interest lies in drawing conclusions about possible outcomes through observations of only a few particular events.

If a probability expert and a statistician find identical boxes, the probability expert might open the box, observe the contents, replace the cover, and proceed to compute the probability of randomly selecting a specific object from the box. The statistician might select a few items from the box without looking at the contents and make a prediction as to the total contents of the box.

The entire contents of the box constitute the **population**. A population consists of all items or people of interest. The statistician often uses a subset of the population, called a **sample**, to make predictions concerning the population. It is important to understand the difference between a population and a sample. A population includes *all* items of interest. A sample includes *some* of the items in the population.

When a statistician draws a conclusion from a sample, there is always the possibility that the conclusion is incorrect. For example, suppose that a jar contains 90 blue marbles and 10 red marbles, as shown in Fig. 13.1. If the statistician selects a random sample of five marbles from the jar and all are blue, he or she may wrongly

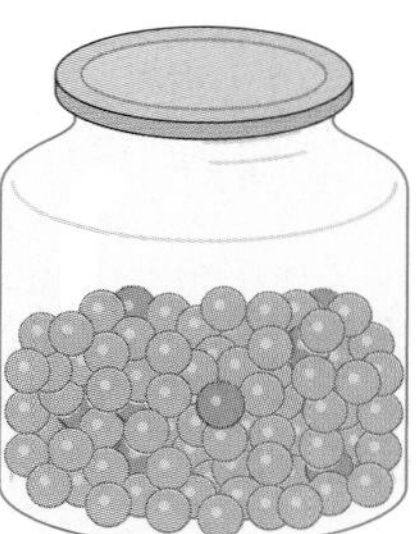

Figure 13.1

conclude that the jar contains all blue marbles. If the statistician takes a larger sample, say, 15 marbles, he or she is likely to select some red marbles. At that point, the statistician may make a prediction about the contents of the jar based on the sample selected. Of course, the most accurate result would occur if every object in the jar, the entire population, were observed. However, in most statistical experiments, observing the entire population is not practical.

Statisticians use samples instead of the entire population for two reasons: (a) it is often impossible to obtain data on an entire population and (b) sampling is less expensive because collecting the data takes less time and effort. For example, suppose that you wanted to determine the number of each species of all the fish in a lake. To do so would be almost impossible without using a sample. If you did try to obtain this information from the entire population, the cost would be astronomical.

Later in this chapter we will discuss statistical measures such as the *mean* and the *standard deviation.* When statisticians calculate the mean and the standard deviation of the entire population they use different symbols and formulas than when they calculate the mean and standard deviation of a sample. The following chart shows the symbols used to represent the mean and standard deviation of a sample and of a population. Note the mean and standard deviation of a population are symbolized by Greek letters.

Measure	Sample	Population
Mean	$\bar{x}$ (read "ex bar")	μ (mu)
Standard deviation	s	σ (sigma)

In this book we will always assume we are working with a sample and so we will use $\bar{x}$ and s. If you take a course in statistics, you will use all four symbols and different formulas for a sample and for a population.

Consider the task of determining the political strength of a certain candidate running in a national election. It is not possible for pollsters to ask each of the approximately 185 million eligible voters his or her preference for a candidate. Thus, pollsters must select and use a sample of the population to obtain their information. How large a sample do you think they use to make predictions about an upcoming national election? You might be surprised to learn that pollsters use only about 1600 registered voters in their national sample. How can a pollster using such a small percentage of the population make an accurate prediction?

The answer lies in the fact that, when pollsters select a sample, they use sophisticated statistical techniques to obtain an unbiased sample. An **unbiased sample** is one that is a small replica of the entire population with regard to income, education, sex, race, religion, political affiliation, age, and so on. The procedures statisticians use to obtain unbiased samples are quite complex. The following sampling techniques will give you a brief idea of how statisticians obtain unbiased samples.

RANDOM SAMPLING

If a sample is drawn in such a way that each time an item is selected, each item in the population has an equal chance of being drawn, the sample is said to be a **random sample**. Under these conditions, one combination of a specified number of items has the same probability of being selected as any other combination. When all the items in the population are similar with regard to the specific characteristic we are interested in, a random sample can be expected to produce satisfactory results. For

The Birth of Inferential Statistics

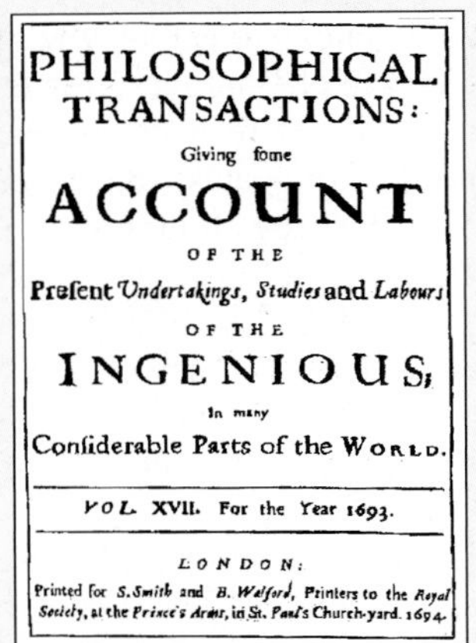

PHILOSOPHICAL TRANSACTIONS: Giving ſome ACCOUNT OF THE Preſent Undertakings, Studies and Labours OF THE INGENIOUS, In many Conſiderable Parts of the WORLD.

VOL. XVII. For the Year 1693.

LONDON: Printed for S. Smith and B. Walford, Printers to the Royal Society, at the Prince's Arms, in St. Paul's Church-yard. 1694.

The London merchant John Gaunt is credited with first making statistical predictions, or inferences, from a set of data rather than basing them simply on the laws of chance. He studied the vital statistics (births, deaths, marriages) contained in the Bills of Mortality published during the years of the Great Plague. He observed that more males were born than females and women lived longer than men. From these observations, he made predictions about life expectancies. The keeping of mortality statistics was stimulated considerably by the growth of the insurance industry.

DID YOU KNOW

Don't Count Your Votes Until They're Cast

A classic instance of faulty sampling occurred in the 1936 presidential election. On the basis of the responses of 2,300,000 voters, selected from automobile owners and telephone subscribers, the *Literary Digest* confidently predicted that the Republican candidate, Alf Landon, would be elected. As it turned out, Franklin D. Roosevelt, the Democratic candidate, won—and by a large margin. The erroneous prediction occurred because the voters used in the sample were not representative of the general voting population. In 1936 telephones and automobiles were unaffordable to the average voter.

example, consider a large container holding 300 tennis balls that are identical except for color. One-third of the balls are yellow, one-third are white, and one-third are green. If the balls can be thoroughly mixed between each draw of a tennis ball so that each ball has an equally likely chance of being selected, randomness is not difficult to achieve. However, if the objects or items are not all the same size, shape, or texture, it might be impossible to obtain a random sample by reaching into a container and selecting an object.

The best procedure for selecting a random sample is to use a random number generator or a table of random numbers. A random number generator is a device, usually a calculator or computer program, that produces a list of random numbers. A random number table is a collection of random digits in which each digit has an equal chance of appearing. To select a random sample, first assign a number to each element in the population. Numbers are usually assigned in order. Then select the number of random numbers needed, which is determined by the sample size. Each numbered element from the population that corresponds to a selected random number becomes part of the sample.

SYSTEMATIC SAMPLING

When a sample is obtained by drawing every nth item on a list or production line, the sample is a **systematic sample**. The first item should be determined by using a random number.

It is important that the list from which a systematic sample is chosen include the entire population being studied. See the Did You Know called "Don't Count Your Votes Until They're Cast." Another problem that must be avoided when this method of sampling is used is the constantly recurring characteristic. For example, on an assembly line, every tenth item could be the work of robot X. If only every tenth item is checked, the work of other robots doing the same job may not be checked and may be defective.

CLUSTER SAMPLING

The **cluster sample** is sometimes referred to as an *area sample* because it is frequently applied on a geographical basis. Essentially, the sampling consists of a random selection of groups of units. For example, geographically we might select blocks of a city or some other geographical subdivision to use as a sample unit. Another example would be to select x boxes of screws from a whole order, count the number of defective screws in the x boxes, and use this number to determine the expected number of defective screws in the whole order.

STRATIFIED SAMPLING

When a population is divided into parts, called strata, for the purpose of drawing a sample, the procedure is known as **stratified sampling**. Stratified sampling involves dividing the population by characteristics called *stratifying factors* such as sex, race, religion, or income. When a population has varied characteristics, it is desirable to separate the population into classes with similar characteristics and then take a random sample from each stratum (or class).

The use of stratified sampling requires some knowledge of the population. For example, to obtain a cross section of voters in a city, we must know where various groups are located and the approximate numbers in each location.

CONVENIENCE SAMPLING

A **convenience sample** uses data that is easily or readily obtained. Occasionally, data that is conveniently obtained may be all that is available. In some cases, some information is better than no information at all. Nevertheless, convenience sampling can be extremely biased. For example, suppose a town wants to raise taxes to build a new elementary school. The local newspaper wants to obtain the opinion of some of the residents and sends a reporter to a senior citizens center. The first 10 people who exit the building are asked if they are in favor of raising taxes to build a new school. This sample could be biased against raising taxes for the new school. Most senior citizens would not have school age children and may not be interested in paying increased taxes to build a new school. Although a convenience sample may be very easy to select, one must be very cautious when using the results obtained from this method.

EXAMPLE 1 *Identifying Sampling Techniques*

Identify the sampling technique used to obtain a sample in the following. Explain your answer.

a) A security agent checks every 20th piece of luggage moving on a conveyor belt at an airport.
b) A $50 gift certificate is given away at the Annual Bankers Convention. Tickets are placed in a bin and the tickets are mixed up. Then the winning ticket is selected by a blindfolded person.
c) A state is divided into regions using zip codes. A random sample of 20 zip code areas is selected.
d) The first 25 people entering a football stadium are surveyed about using taxpayer money to build a new sports stadium.
e) Students at Portland State University are classified according to their major. Then a random sample of 15 students from each major is selected.

SOLUTION

a) Systematic sampling. The sample is obtained by drawing every *n*th item. In this example, every 20th item on a production line is selected.
b) Random sampling. Every ticket has an equal chance of being selected.
c) Cluster sampling. A random sample of geographic areas is selected.
d) Convenience sampling. The sample is selected by picking people that are easily obtained.
e) Stratified sampling. The students are divided into strata based on their majors. Then random samples are selected from each strata. ▲

SECTION 13.1 EXERCISES

CONCEPT/WRITING EXERCISES

1. Explain the difference between descriptive and inferential statistics.
2. Define *statistics* in your own words.
3. When you hear the word *statistics,* what specific words or ideas come to mind?
4. Name five areas other than those mentioned in this section in which statistics is used.
5. Attempt to list at least two professions in which no aspect of statistics is used.
6. Explain the difference between probability and statistics.

7. a) What is a population?
   b) What is a sample?
8. a) What is a random sample?
   b) How might a random sample be selected?
9. a) What is a systematic sample?
   b) How might a systematic sample be selected?
10. a) What is a cluster sample?
    b) How might a cluster sample be selected?
11. a) What is a stratified sample?
    b) How might a stratified sample be selected?
12. a) What is a convenience sample?
    b) How might a convenience sample be selected?
13. What is an unbiased sample?
14. The principal of an elementary school wishes to determine the "average" family size of the children who attend the school. To obtain a sample, the principal visits each room and selects the four students closest to each corner of the room. The principal asks each of these students how many people are in his or her family.
    a) Will this technique result in an unbiased sample? Explain your answer.
    b) If the sample is biased, will the average be greater than or less than the true family size? Explain.

PRACTICE THE SKILLS

In Exercises 15–24, identify the sampling technique used to obtain a sample. Explain your answer.

15. A door prize is given away at a home improvement show. Tickets are placed in a bin and the tickets are mixed up. Then a ticket is selected by a blindfolded person.
16. Every 15th CD player coming off an assembly line is checked for defects.
17. A state is divided into counties. Individuals are randomly selected from five randomly selected counties.
18. A group of people are classified according to age and then random samples of people from each group are taken.
19. Every 25th roll of film coming off the assembly line is selected and checked for defects.
20. At Disney World, everyone whose birthday is October 13 will be used as a sample.
21. The first 40 students leaving the bookstore are asked how much money they spend per day on lunch in the student cafeteria.
22. Bingo balls in a bin are shaken and then balls are selected from the bin.

23. The Food and Drug Administration randomly selects five stores from each of four randomly selected sections of a large city and checks food items for freshness. These stores are used as a representative sample of the entire city.
24. The Student Senate at the University of New Orleans is electing a new president. The first 25 people leaving the library are asked for whom they will vote.

CHALLENGE PROBLEMS/GROUP ACTIVITIES

25. a) *Random Sampling* Select a topic and population of interest to which a random sampling technique can be applied to obtain data.
    b) Explain how you or your group would obtain a random sample for your population of interest.
    c) Actually obtain the sample by the procedure stated in part (b).
26. *Data from Questionnaire* Some subscribers of *Consumers Reports* respond to an annual questionnaire regarding their satisfaction with new appliances, cars, and other items. The information obtained from these questionnaires is then used as a sample from which frequency of repairs and other ratings are made by the magazine. Are the data obtained from these returned questionnaires representative of the entire population or are they biased? Explain your answer.

RESEARCH PROBLEM

27. We have briefly introduced sampling techniques. Using statistics books and Internet web sites as references, select one type of sampling technique (it may be one that we have not discussed in this section) and write a report on how statisticians obtain that type of sample. Also indicate when that type of sampling technique may be preferred. List two examples of when the sampling technique may be used.

13.2 THE MISUSES OF STATISTICS

Statistics, when used properly, is a valuable tool to society. However, many individuals, businesses, and advertising firms misuse statistics to their own advantage. You should examine statistical statements very carefully before accepting them as fact. Two questions you should ask yourself are: Was the sample used to gather the statistical data unbiased and of sufficient size? Is the statistical statement ambiguous; that is, can it be interpreted in more than one way?

Let's examine two advertisements. "Four out of five dentists recommend sugarless gum for their patients who chew gum." In this advertisement, we do not know the sample size and the number of times the experiment was performed to obtain the desired results. The advertisement does not mention that possibly only 1 of 100 dentists recommended gum at all.

In a golf ball commercial, a "type A" ball is hit, and a second ball is hit in the same manner. The type A ball travels farther. We are supposed to conclude that the type A is the better ball. The advertisement does not mention the number of times the experiment was previously performed or the results of the earlier experiments. Possibly sources of bias include (1) wind speed and direction, (2) the fact that no two swings are identical, and (3) the fact that the ball may land on a rough or smooth surface.

Vague or ambiguous words also lead to statistical misuses or misinterpretations. The word *average* is one such culprit. There are at least four different "averages," some of which are discussed in Section 13.5. Each is calculated differently, and each may have a different value for the same sample. During contract negotiations, it is not uncommon for an employer to state publicly that the average salary of its employees is \$35,000, while the employees' union states that the average is \$30,000. Who is lying? Actually, both may be telling the truth. Each will use the average that best suits its needs to arrive at a figure. Advertisers also use the average that most enhances their products. Consumers often misinterpret this average as the one with which they are most familiar.

Another vague word is *largest.* For example, ABC claims that it is the largest department store in the United States. Does that mean largest profit, largest sales, largest building, largest staff, largest acreage, or largest number of outlets?

Still another deceptive technique used in advertising is to state a claim from which the public may draw irrelevant conclusions. For example, a disinfectant manufacturer claims that its product killed 40,760 germs in a laboratory in 5 seconds. "To prevent colds, use disinfectant A." It may well be that the germs killed in the laboratory were not related to any type of cold germ. In another example, company C claims that its paper towels are heavier than its competition's towels. Therefore, they will hold more water. Is weight a measure of absorbency? A rock is heavier than a sponge, yet a sponge is more absorbent.

An insurance advertisement claims that in Duluth, Minnesota, 212 people switched to insurance company Z. One may conclude that this company is offering something special to attract these people. What may have been omitted from the advertisement is that 415 people in Duluth, Minnesota, dropped insurance company Z during the same period.

A foreign car manufacturer claims that 9 of every 10 of a popular model it sold in the United States during the previous 10 years were still on the road. From this statement the public is to conclude that this foreign car is well manufactured and would last for many years. The commercial neglects to state that this model has been selling in the United States for only a few years. The manufacturer could just as well

Creative Displays

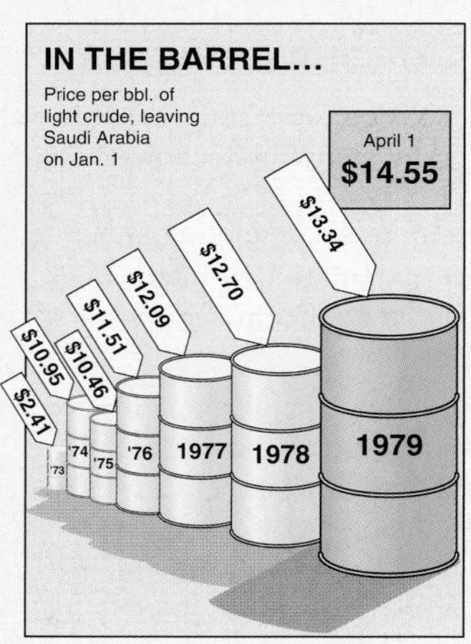

Visual graphics are often used to "dress up" what might otherwise be considered boring statistics. Although visually appealing, such creative displays of numerical data can be misleading. In his book *The Visual Display of Quantitative Information* (Graphics Press, 1983), Edward Tufte cites this misleading graphic taken from *Time,* April 9, 1979. The data shown are the annual increase in oil prices from 1973 to 1979, an overall increase of 454%. Yet what you "see" is a volumetric increase of 27,000% in the relative size of the barrels.

have stated that 9 of every 10 of these cars sold in the United States in the previous 100 years were still on the road.

Charts and graphs can also be misleading or deceptive. In Fig. 13.2 two graphs show performance of two stocks. Which stock would you purchase? Actually, the two graphs present identical information; the only difference is that the vertical scale of the graph for stock B has been exaggerated.

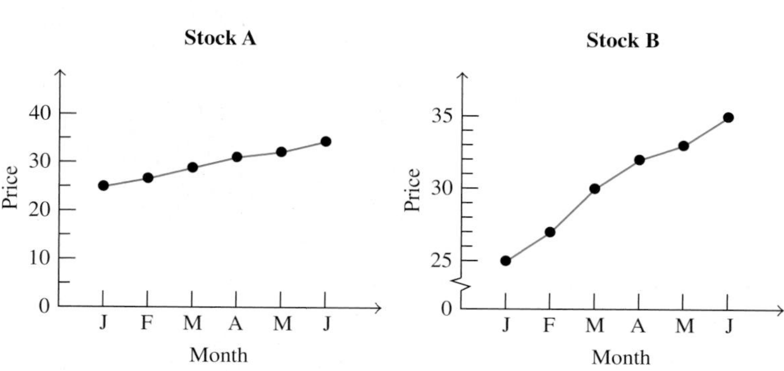

Figure 13.2

The two graphs in Fig. 13.3 show the same change. However, the graph in part (a) appears to show a greater decrease than the graph in part (b), again because of a different scale.

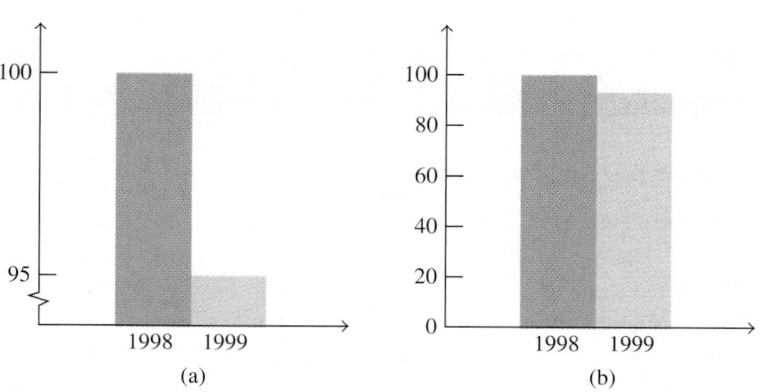

Figure 13.3

Consider a claim that, if you invest $1, by next year you will have $2. This type of claim is sometimes misrepresented, as in Fig. 13.4. Actually, your investment has only doubled, but the area of the square on the right is four times that of the square on the left. By expressing the amounts as cubes (Fig. 13.5 on page 670), you increase the volume eightfold.

The graph in Fig. 13.6 is an example of a circle graph. We will discuss how to construct circle graphs in Section 13.4. In a circle graph, the total circle represents 100%. Therefore, the sum of the parts should add up to 100%. This graph is misleading since the sum of its parts is 183%. A graph other than a circle graph should have been used to display the top six reasons Americans say they use the Internet.

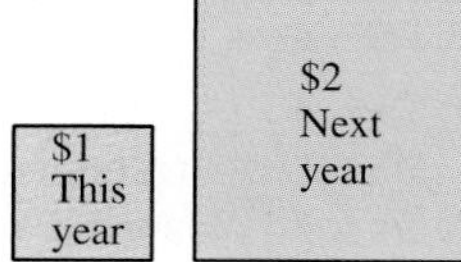

Figure 13.4

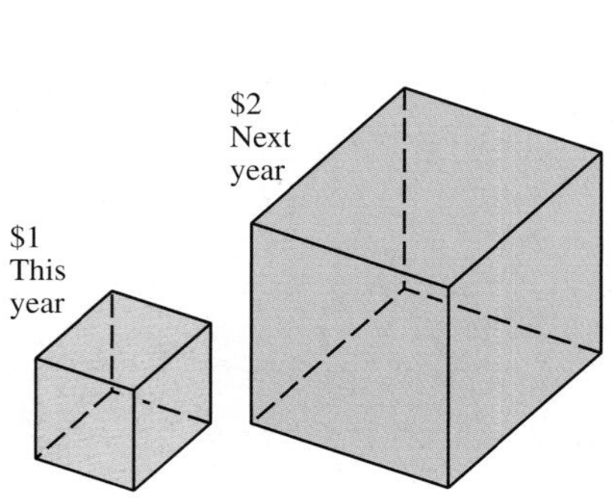

Figure 13.5

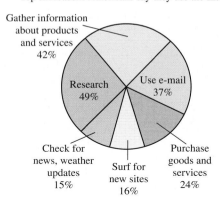

Why People Surf the Web

Top six reasons Americans say they use the Internet

Gather information about products and services 42%

Research 49%

Use e-mail 37%

Check for news, weather updates 15%

Surf for new sites 16%

Purchase goods and services 24%

Figure 13.6

Despite the examples presented in this section, you should not be left with the impression that statistics is used solely for the purpose of misleading or cheating the consumer. As stated earlier, there are many important and necessary uses of statistics. Most statistical reports are accurate and useful. You should realize, however, the importance of being an aware consumer.

SECTION 13.2 EXERCISES

CONCEPT/WRITING EXERCISES

1. Find five advertisements or commercials that may be statistically misleading. Explain why each may be misleading.
2. The following graph appeared in *Newsweek* magazine, July 13, 1998. The graph shows the percentage of adults who are overweight for the period 1962–1994. Is the graph misleading? Explain.

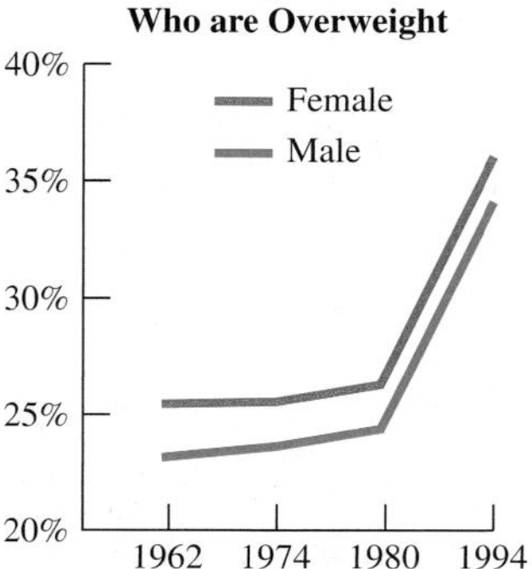

Sources: USDA, Beverage Digest.

PRACTICE THE SKILLS

In Exercises 3–16, discuss the statement and tell what possible misuses or misinterpretations may exist.

3. The average salary for employees of XYZ, Inc., is \$51,000. Therefore, all XYZ, Inc., employees make approximately \$51,000.
4. Healthy Snacks cookies are fat free. So eat as many as you like and you will not gain weight.
5. Morgan's is the largest department store in New York. So shop at Morgan's and save money.
6. Most accidents occur on Saturday night. This means that people do not drive carefully on Saturday night.
7. Eighty percent of all automobile accidents occur within 10 miles of the driver's home. Therefore, it is safer to take long trips.
8. Arizona has the highest death rate for asthma in the country. Therefore, it is unsafe to go to Arizona if you have asthma.
9. Females have a higher average score than males on the English part of the Scholastic Aptitude Test. Therefore,

on this test a particular female selected at random will outperform a particular male selected at random.

10. Four out of five dermatologists recommend Soft and Smooth dry-skin lotion. Therefore, it is the best dry-skin lotion.

11. The average depth of the pond is only 3 ft, so it is safe to go wading.

12. Treadware Tires are the most expensive tires. Therefore, they will last the longest.

13. At West High School, half the students are below average in mathematics. Therefore, the school should receive more federal aid to raise student scores.

14. Florida has the greatest number of orange trees. Therefore, Floridians drink more orange juice than people in any other state.

15. A recent survey showed that 65% of those surveyed preferred Edge shaving gel to Foamy shaving cream. Therefore, more people buy Edge shaving gel than Foamy shaving cream.

16. More men than women are involved in automobile accidents. Therefore, women are better drivers.

17. *Number of Hospitals* The following table shows the number of hospitals in the United States from 1991–1996.

Year	Number of hospitals
1991	6634
1992	6539
1993	6467
1994	6374
1995	6291
1996	6201

Source: American Hospital Association.

Draw a line graph that makes the decrease in the number of hospitals in the United States appear to be
a) small.
b) large.

18. *Health Benefit Costs* The following table shows the total health benefit costs per employee for active and retired workers from 1992–1997.

Year	Health benefit costs (dollars)
1992	3502
1993	3781
1994	3741
1995	3821
1996	3915
1997	3924

(*Source:* Foster Higgins.)

Draw a line graph that makes the increase in health benefits costs per employee appear to be
a) small.
b) large.

In Exercises 19–20, use the following table.

Median Age at First Marriage

Male		Female	
Year	**Age**	**Year**	**Age**
1966	22.8	1966	20.5
1976	23.8	1976	21.3
1986	25.7	1986	23.1
1996	27.1	1996	24.8

Source: U.S. Census Bureau.

19. **a)** Draw a bar graph that appears to show a small increase in the median age at first marriage for males.
b) Draw a bar graph that appears to show a large increase in the median age at first marriage for males.

20. **a)** Draw a bar graph that appears to show a small increase in the median age at first marriage for females.
b) Draw a bar graph that appears to show a large increase in the median age at first marriage for females.

21. *Computer Sales* The following graph shows the annual sales, in billions of dollars, for Gateway Computer in 1997 and 1998.

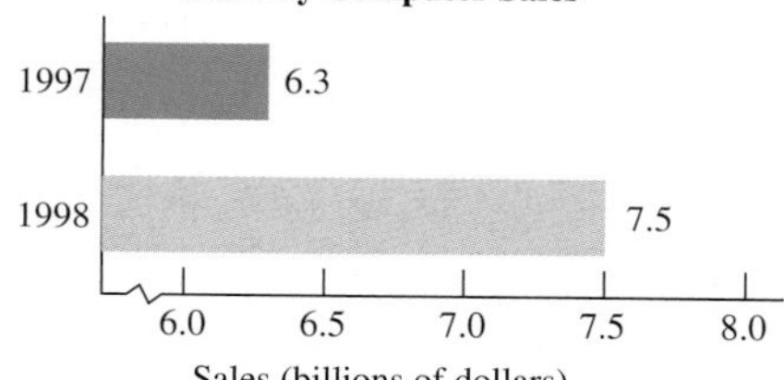

a) Draw a bar graph that shows the entire scale (using intervals of $1 billion) from $0 to $8 billion in sales.
b) Does the new graph give a different impression? Explain.

CHALLENGE PROBLEM/GROUP ACTIVITY

22. Consider the following graph, which shows the U.S. population in 2000 and the projected U.S. population in 2050.

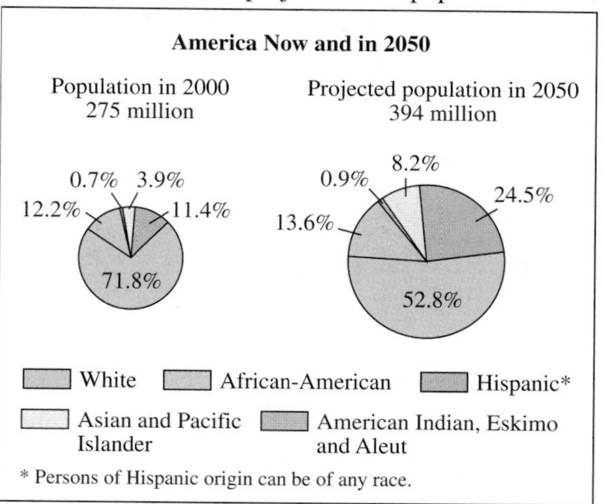

Source: U.S. Census Bureau

a) Compute the projected percent increase in population from 2000 to 2050 by using the formula given on page 520.
b) Measure the radius and then compute the area of the circle representing 2000. Use $A = \pi r^2$.
c) Repeat part (b) for the circle representing 2050.
d) Compute the percent increase in the size of the area of the circle from 2000 to 2050.
e) Are the circle graphs misleading? Explain your answer.

RESEARCH ACTIVITY

21. Read the book *How to Lie with Statistics* by Darrell Huff and write a book report on it. Select three illustrations from the book that show how people manipulate statistics.

13.3 FREQUENCY DISTRIBUTIONS

It is not uncommon for statisticians and others to have to analyze thousands of pieces of data. A **piece of data** is a single response to an experiment. When the amount of data is large, it is usually advantageous to construct a frequency distribution. A **frequency distribution** is a listing of the observed values and the corresponding frequency of occurrence of each value.

"Statistical thinking will one day be as necessary for efficient citizenship as the ability to read and write."

H. G. Wells

EXAMPLE 1 *Frequency Distribution*

The number of children per family is recorded for 64 families surveyed. Construct a frequency distribution of the following data:

0	1	1	2	2	3	4	5
0	1	1	2	2	3	4	5
0	1	1	2	2	3	4	6
0	1	2	2	2	3	4	6
0	1	2	2	2	3	4	7
0	1	2	2	3	3	4	8
0	1	2	2	3	3	5	8
0	1	2	2	3	3	5	9

SOLUTION Listing the number of children (observed values) and the number of families (frequency) gives the following frequency distribution.

Number of children (observed values)	Number of families (frequency)
0	8
1	11
2	18
3	11
4	6
5	4
6	2
7	1
8	2
9	1
	64

Eight families had no children, 11 families had one child, 18 families had two children, and so on. Note that the sum of the frequencies is equal to the original number of pieces of data, 64. ▲

DID YOU KNOW

Can You Count the F's?

Statistical errors often result from careless observations. To see how such errors can occur, consider the statement below. How many F's do you count in the statement? You can find the answer in the answer section.

FINISHED FILES ARE THE RESULT OF YEARS OF SCIENTIFIC STUDY COMBINED WITH THE EXPERIENCE OF YEARS.

Often data are grouped in classes to provide information about the distribution that would be difficult to observe if the data were ungrouped. Graphs called *histograms* and *frequency polygons* can be made of grouped data, as will be explained in Section 13.4. These graphs also provide a great deal of useful information.

When data are grouped in classes, certain rules should be followed.

Rules for Data Grouped by Classes

1. The classes should be of the same "width."
2. The classes should not overlap.
3. Each piece of data should belong to only one class.

In addition, it is often suggested that a frequency distribution should be constructed with 5 to 12 classes. If there are too few or too many classes, the distribution may become difficult to interpret.

To understand these rules, let's consider a set of observed values that go from a low of 0 to a high of 26. Let's assume that the first class is arbitrarily selected to go from 0 through 4. Thus, any of the data with values of 0, 1, 2, 3, 4 would belong in this class. We say that the **class width** is 5, since there are five integral values that belong to the class. This first class ended with 4, so the second class must start with 5. If this class is to have a width of 5, at what value must it end? The answer is 9 (5, 6, 7, 8, 9). The second class is 5–9. Continuing in the same manner, we obtain the following set of classes.

Classes

Lower class limits { 0–4, 5–9, 10–14, 15–19, 20–24, 25–29 } Upper class limits

Classes
0–4
5–9
10–14
15–19
20–24
25–29

We need not go beyond the 25–29 class because the largest value we are considering is 26. The classes meet our three criteria: They have the same width, there is no overlap among the classes, and each of the values from a low of 0 to a high of 26 belongs to one and only one class.

The choice of the first class, 0–4, was arbitrary. If we wanted to have more classes or fewer, we would make the class widths smaller or larger, respectively.

The numbers 0, 5, 10, 15, 20, 25 are called the **lower class limits**, and the numbers 4, 9, 14, 19, 24, 29 are called the **upper class limits**. Each class has a width of 5. Note that the class width, 5, can be obtained by subtracting the first lower class limit from the second lower class limit: $5 - 0 = 5$. The difference between any two consecutive lower class or upper class limits is also 5.

EXAMPLE 2 *A Frequency Distribution of Y2K Expenses*

Table 13.1 on page 674 shows what some of the largest companies in the United States expected to spend to ensure that their computers recognized the year 2000. Numbers are rounded to the nearest million dollars.

Table 13.1

Company	Total costs (millions of dollars)
Citigroup	950
AT&T	756
General Motors	628
General Electric	575
IBM	575
Lucent Technologies	560
MCI WorldCom	552
Philip Morris	550
BankAmerica	550
American Express	530
Merrill Lynch	520
Mobil	465
Bell Atlantic	416
Ford Motor	400
GTE	400
Chase Manhattan	394
DuPont	375
BellSouth	300
J.P. Morgan	300
Sprint	300
Motorola	275
US West	275
SBC Communications	265
Walt Disney	261
Ameritech	244
Exxon	238
Morgan Stanley	223
PG&E	223
AMR	218
Aetna	203
Chevron	200
Johnson & Johnson	200
Xerox	183
American International	175
United Technologies	175
Columbia/HCA	166
Hewlett-Packard	160
American Home Products	155
Cigna	145
Coca-Cola	140
AlliedSignal	135
Allstate	125
Compaq Computer	125
Caterpillar	125
Raytheon	125
Electronic Data Systems	115
PepsiCo	113
Intel	105
United Parcel Service	102
Procter & Gamble	90

Source: Security and Exchange Commission.

Construct a frequency distribution of the data, letting the first class be \$90 million to \$184 million.

SOLUTION Fifty pieces of data are given in *descending order* from highest to lowest. The first class is 90–184. The second class must therefore start at 185. To find the class width, we subtract 90 (the lower class limit of the first class) from 185 (the lower class limit of the second class) to obtain a class width of 95. The upper class limit of the second class is found by adding the class width, 95, to the upper class limit of the first class, 184. The upper class limit of the second class is $184 + 95 = 279$. Thus,

$$\begin{aligned} 90\text{–}184 &= \text{first class} \\ 185\text{–}279 &= \text{second class} \end{aligned}$$

The remaining classes will be 280–374, 375–469, 470–564, 565–659, 660–754, 755–849, 850–944, and 945–1039. Since the highest observed value is 950, there is no need to go any further. Note that each two consecutive lower class limits differ by 95 as do each two consecutive upper class limits. There are 18 pieces of data in the 90–184 class (90, 102, 105, 113, 115, 125, 125, 125, 125, 135, 140, 145, 155, 160, 166, 175, 175, and 183). There are 12 pieces of data in the 185–279 class, 3 in the 280–374 class, 6 in the 375–469 class, 6 in the 470–564 class, 3 in the 565–659 class, 0 in the 660–754 class, 1 in the 755–849 class, 0 in the 850–944 class, and 1 in the 945–1039 class. The complete frequency distribution of nine classes follows. The number of companies totals 50, so we have included each piece of data.

Cost (millions of dollars)	Number of companies
90–184	18
185–279	12
280–374	3
375–469	6
470–564	6
565–659	3
660–754	0
755–849	1
850–944	0
945–1039	1
	50

The **modal class** of a frequency distribution is the class with the greatest frequency. In Example 2, the modal class is 90–184. The **midpoint of a class**, also called the **class mark**, is found by adding the lower and upper class limits and dividing the sum by 2. The midpoint of the first class in Example 2 is

$$\frac{90 + 184}{2} = \frac{274}{2} = 137$$

Note the difference between successive class marks is the class width.

DID YOU KNOW

Cyberspace Is the Place to Be

San Francisco, the "most wired" city.

The Internet, first developed as a communications tool for the government and for research universities, is now being used for everything from research to entertainment to shopping to chatting with friends in chat rooms. In 1998, 47.8% of households in the United States had computers, while 27.3% of households in the United States were online. In 1998, San Francisco was the overall "most wired" city in the United States. Austin, Texas, was the leader in home online usage, with 32.4% of homes online. In San Jose, California, the online office leader, 37.9% of offices were online. (*Source: http://www.zdnet.com.*)

EXAMPLE 3 ***A Frequency Distribution of Family Income***

The following set of data represents the family income (in thousands of dollars, rounded to the nearest hundred) of 15 randomly selected families.

31.5	16.8	30.8	29.7	25.9
50.2	37.4	17.6	38.7	33.8
20.5	25.3	24.8	41.3	35.7

Construct a frequency distribution with a first class of 16.5–22.6.

SOLUTION First rearrange the data from lowest to highest (or highest to lowest) so that the data will be easier to categorize.

16.8	24.8	29.7	33.8	38.7
17.6	25.3	30.8	35.7	41.3
20.5	25.9	31.5	37.4	50.2

The first class goes from 16.5 to 22.6. Since the data are in tenths, the class limits will also be given in tenths. The first class ends with 22.6; therefore, the second class must start with 22.7. The class width of the first class is 22.7 − 16.5, or 6.2. The upper class limit of the second class must therefore be 22.6 + 6.2, or 28.8. The frequency distribution is as follows.

Income ($1000)	Number of families
16.5–22.6	3
22.7–28.8	3
28.9–35.0	4
35.1–41.2	3
41.3–47.4	1
47.5–53.6	1
	15

Note in Example 3 that the class width is 6.2, that the modal class is 28.9–35.0, and that the class mark of the first class is (16.5 + 22.6)/2, or 19.55.

SECTION 13.3 EXERCISES

CONCEPT/WRITING EXERCISES

1. What is a frequency distribution?
2. How can a class width be determined using class limits?
3. Suppose that the first class of a frequency distribution is 9–15.
   a) What is the width of this class?
   b) What is the second class?
   c) What is the lower class limit of the second class?
   d) What is the upper class limit of the second class?
4. Repeat Exercise 3 for a frequency distribution whose first class is 12–20.
5. What is the modal class of a frequency distribution?
6. What is another name for the midpoint of a class? How is the midpoint of a class determined?

PRACTICE THE SKILLS/PROBLEM SOLVING

In Exercises 7 and 8, use the frequency distribution to determine

a) *the total number of observations.*
b) *the width of each class.*
c) *the midpoint of the second class.*
d) *the modal class (or classes).*
e) *the class limits of the next class if an additional class were to be added.*

7.

Class	Frequency
9–13	2
14–18	6
19–23	1
24–28	0
29–33	3
34–38	5

8.

Class	Frequency
40–49	7
50–59	5
60–69	3
70–79	2
80–89	7
90–99	1

9. *Car Stereo Sales* A car stereo supplier is interested in the number of car stereos sold daily. A sample is taken over 40 days to obtain the following data regarding the number of car stereos sold daily. Construct a frequency distribution, letting each class have a width of 1 (as in Example 1).

0	1	1	3	4	5	7	8
0	1	2	3	5	5	7	8
0	1	2	3	5	5	7	9
1	1	2	3	5	6	8	10
1	1	3	4	5	6	8	10

10. *Park Visits by Families* The town of Brighton is planning to improve the local park. The responses of 32 families who were asked how many times per year they visit the park are shown below. Construct a frequency distribution letting each class have a width of 1.

20	21	24	25	26	27	29	32
20	23	24	25	26	27	30	32
20	23	24	26	26	28	31	33
21	23	24	26	26	28	31	34

Note: No one visited the park 22 times per year. However, it is customary to include a missing value as an observed value and assign to it a frequency of 0.

IQ Scores In Exercises 11–14, use the following data, which show the result of 50 sixth-grade I.Q. scores.

80	89	92	95	97	100	102	106	110	120
81	89	93	95	98	100	103	108	113	120
87	90	94	97	99	100	103	108	114	122
88	91	94	97	100	100	103	108	114	128
89	92	94	97	100	101	104	109	119	135

Use this data to construct a frequency distribution with a first class of

11. 78–86. **12.** 80–88.
13. 80–90. **14.** 80–92.

U.S. Presidents In Exercises 15–18, use the following data, which represent the ages of the U.S. presidents at their first inauguration.

57	57	49	52	50	51	51	56	46
61	61	64	56	47	56	60	61	
57	54	50	46	55	55	62	52	
57	68	48	54	54	51	43	69	
58	51	65	49	42	54	55	64	

Use this data to construct a frequency distribution with a first class of

15. 40–45. **16.** 42–47.
17. 42–46. **18.** 40–44.

Company Spending In Exercises 19–22, use the data in Example 2 on pages 673 and 674 to construct a frequency distribution with a first class (in millions of dollars) of

19. 90–199. **20.** 90–195.
21. 50–149. **22.** 50–199.

World Population In Exercises 23–26, use the following data, which represent the 1997 population of the world's 35 largest cities, in millions of people (rounded to the nearest 100,000).

27.5	13.8	11.5	10.3	9.5	7.0	6.4
17.3	12.7	11.5	10.1	9.3	7.0	6.3
17.0	12.3	10.6	9.9	9.1	6.9	6.0
16.5	12.1	10.6	9.7	8.5	6.8	6.0
16.3	11.9	10.5	9.6	7.6	6.5	5.4

Use this data to construct a frequency distribution with a first class of

23. 5.4–7.4 **24.** 5.0–8.0
25. 5.1–7.6 **26.** 5.4–7.9

Bachelor's Degree *In Exercises 27–30, use the data in the following table.*

Percent of Total Population with a Bachelor's Degree

State	%	State	%	State	%
AL	19.3	LA	18.1	OK	20.5
AK	27.5	MA	33.5	OR	24.3
AZ	19.5	MD	32.2	PA	22.9
AR	14.6	ME	20.0	RI	25.7
CA	27.5	MI	21.0	SC	19.2
CO	28.9	MN	28.3	SD	20.1
CT	30.3	MO	22.9	TN	17.1
DE	26.8	MS	20.9	TX	22.4
DC	33.7	MT	25.2	UT	26.7
FL	21.7	NC	22.6	VA	28.0
GA	22.3	ND	20.5	VT	23.7
HI	22.5	NE	21.3	WA	26.1
IA	21.7	NH	27.0	WI	22.4
ID	19.4	NJ	28.5	WV	14.7
IL	25.0	NM	23.6	WY	22.2
IN	16.2	NV	19.9		
KS	27.5	NY	25.8		
KY	17.6	OH	21.5		

Note: District of Columbia (D.C.) is included in the data.
Source: U.S. Census Bureau.

Construct a frequency distribution with a first class (in percents) of

27. 14.6–18.5

28. 14.6–17.5

29. 14.6–18.0

30. 14.6–17.0

13.4 STATISTICAL GRAPHS

Now we will consider four types of graphs: the circle graph, the histogram, the frequency polygon, and the stem-and-leaf display.

Circle graphs (also known as pie charts) are often used to compare parts of one or more components of the whole to the whole. The circle graph in Fig. 13.7 shows how Americans describe their holiday spending. Since the total circle represents 100%, the sum of the percents of the sectors should be 100%, and it is.

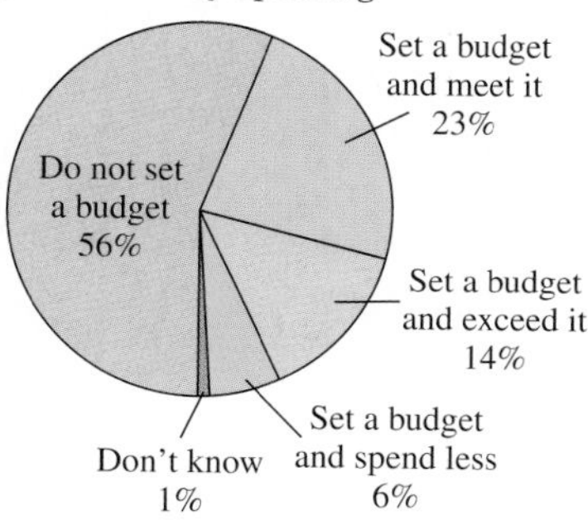

Figure 13.7

EXAMPLE 1 *Software Sales*

The following table indicates the 1997 software sales (in millions of dollars) for software companies.

Company	Sales (millions of dollars)
Microsoft	9.4
Cendant Software	3.8
Intuit	2.6
Other companies	17.1
Total	32.9

Source: PC Data.

Use this information to construct a circle graph illustrating the percent of software sales, in 1997, for Microsoft, Cendant Software, Intuit, and others.

SOLUTION Determine the measure of the corresponding central angle, as illustrated in the following table.

Company	Sales (millions of dollars)	Percent of total (to the nearest tenth of a percent)	Measure of central angle (degrees)
Microsoft	9.4	$\frac{9.4}{32.9} \times 100 = 28.6\%$	$0.286 \times 360 = 103.0°$
Cendant Software	3.8	$\frac{3.8}{32.9} \times 100 = 11.6\%$	$0.116 \times 360 = 41.8°$
Intuit	2.6	$\frac{2.6}{32.9} \times 100 = 7.9\%$	$0.079 \times 360 = 28.4°$
Other	17.1	$\frac{17.1}{32.9} \times 100 = 52.0\%$	$0.52 \times 360 = 187.2°$
Total	32.9	100.1%*	360.4°**

*Due to rounding, we get 100.1%. If the percent were rounded to hundredths, the sum would be exactly 100%.

**Due to rounding, we get 360.4°, not exactly 360°. If the measure of the central angle were rounded to hundredths, the sum would be exactly 360°.

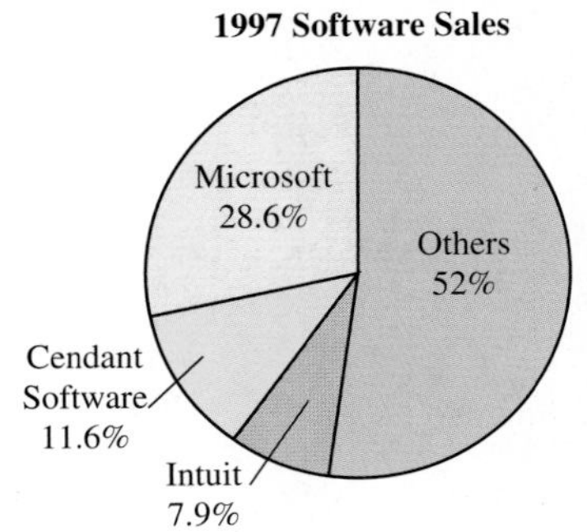

Figure 13.8

Now use a protractor to construct a circle graph and label it properly, as illustrated in Fig. 13.8. The measure of the central angle for Microsoft is about 103°, for Cendant Software it is about 41.8°, for Intuit it is about 28.4°, and for other companies it is about 187.2°

Histograms and frequency polygons are statistical graphs used to illustrate frequency distributions. A **histogram** is a graph with observed values on its horizontal scale and frequencies on its vertical scale. A bar is constructed above each observed value (or class when classes are used), indicating the frequency of that value. The horizontal scale need not start at zero, and the calibrations on the horizontal and vertical scales do not have to be the same. The vertical scale must start at zero. To accommodate large frequencies on the vertical scale, it may be necessary to break the scale. Because histograms and other bar graphs are easy to interpret visually, they are used a great deal in newspapers and magazines.

EXAMPLE 2 *Construct a Histogram*

The frequency distribution developed in Example 1, Section 13.3, is repeated here. Construct a histogram of this frequency distribution.

Number of children (observed values)	Number of families (frequency)
0	8
1	11
2	18
3	11
4	6
5	4
6	2
7	1
8	2
9	1

SOLUTION The vertical scale must extend at least to the number 18, since that is the greatest recorded frequency (see Fig. 13.9). The horizontal scale must include the numbers 0–9, the number of children observed. Eight families have no children. We indicate this by constructing a bar above the number 0 on the horizontal scale extended up to 8 on the vertical scale. Eleven families have one child, so we construct a bar extending to 11 above the number 1 on the horizontal scale. We continue this procedure for each observed value. Both the horizontal and vertical scales should be labeled, the bars should be the same width, and the histogram should have a title. In a histogram the bars should always touch.

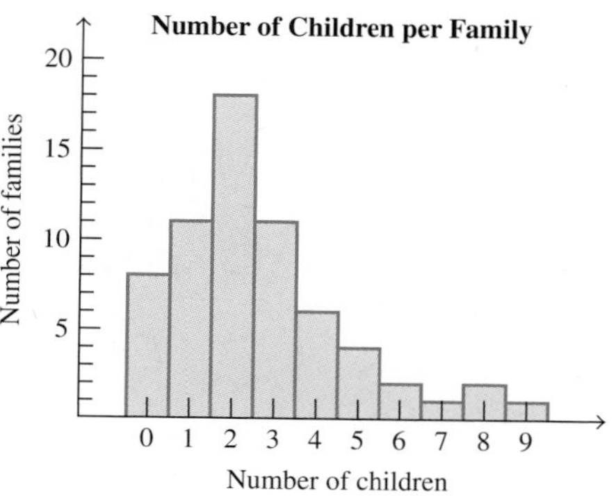

Figure 13.9

Frequency polygons are line graphs with scales the same as those of the histogram; that is, the horizontal scale indicates observed values and the vertical scale indicates frequency. To construct a frequency polygon, place a dot at the corresponding frequency above each of the observed values. Then connect the dots with straight-line segments. When constructing frequency polygons, always put in two additional class marks, one at the lower end and one at the upper end on the horizontal scale (values for these added class marks are not needed on the frequency polygon). Since the frequency at these added class marks is 0, the endpoints of the frequency polygon will always be on the horizontal scale.

EXAMPLE 3 *Construct a Frequency Polygon*

Construct a frequency polygon of the frequency distribution in Example 2.

SOLUTION Since eight families have no children, place a mark above the 0 at 8 on the vertical scale, as shown in Fig. 13.10. Because there are 11 families

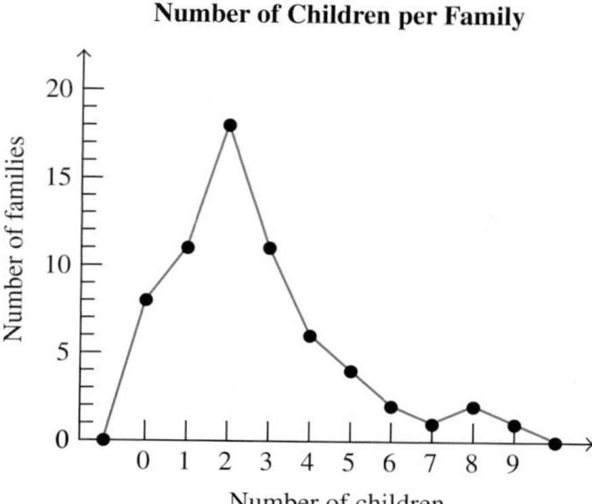

Figure 13.10

with one child, place a mark above the 1 at the 11 on the vertical scale, and so on. Connect the dots by straight-line segments, and bring the endpoints of the graph down to the horizontal scale, as shown.

EXAMPLE 4 *Gas Mileage*

The frequency distribution of average gas mileage for 1999 automobiles is listed in Table 13.2. Construct a histogram and then construct a frequency polygon on the histogram.

Table 13.2

Mileage (mpg)	Number of cars
10–14	5
15–19	31
20–24	36
25–29	8
30–34	5
35–39	4
40–44	1

SOLUTION The histogram can be constructed with either class limits or class marks (class midpoints) on the horizontal scale. Frequency polygons are constructed with class marks on the horizontal scale. Since we will construct a frequency polygon on the histogram, we will use class marks. Recall that class marks are found by adding the lower class limit and upper class limit and dividing the sum by 2. For the first class, the class mark is $(10 + 14)/2$, or 12. Since the class widths are 5 units, the class marks will also differ by 5 units (see Fig. 13.11).

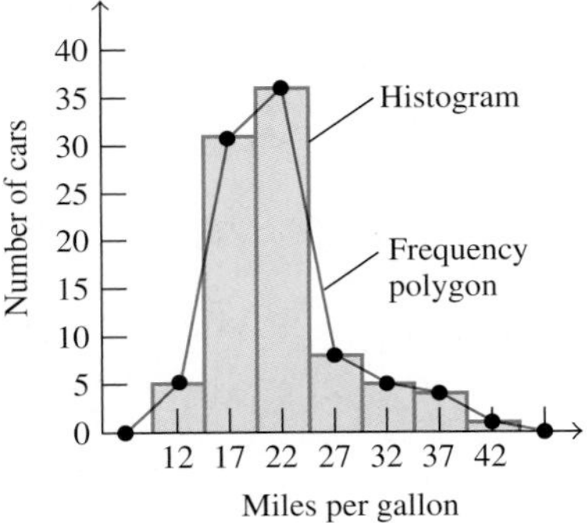

Figure 13.11

EXAMPLE 5 *Carry-on Luggage Weights*

The histogram in Fig. 13.12 shows the weights of selected pieces of carry-on luggage at an airport. Construct the frequency distribution from the histogram in Fig.13.12.

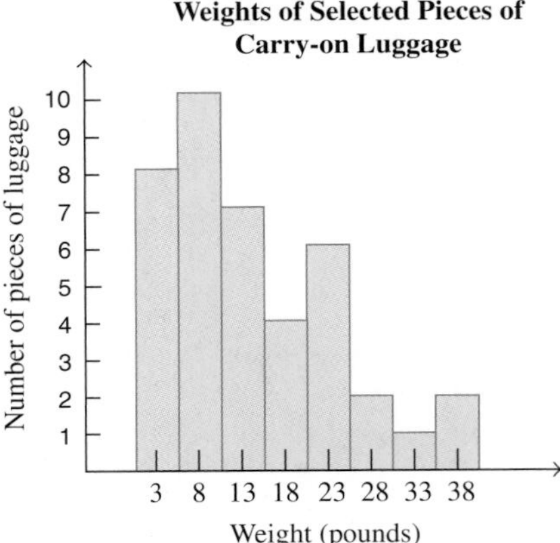

Figure 13.12

Table 13.3

Weight (pounds)	Number of pieces of luggage
1–5	8
6–10	10
11–15	7
16–20	4
21–25	6
26–30	2
31–35	1
36–40	2

SOLUTION There are five units between class midpoints, so each class width must also be five units. Since three is the midpoint of the first class, there must be two units below and two above it. The first class must be 1–5. The second class must therefore be 6–10. The frequency distribution is given in Table 13.3. ▲

Frequency distributions and histograms provide very useful tools to organize and summarize data. However, if the data are grouped, we cannot identify specific data values in a frequency distribution and in a histogram. For example, in Example 5, we know that there are eight pieces of luggage in the class of 1 to 5 pounds. But we don't know the specific weights of those eight pieces of luggage. A **stem-and-leaf display** is a tool that organizes and groups the data while allowing us to see the actual values that make up the data.

To construct a stem-and-leaf display, each value greater than 10 is divided into two groups. The left group of digits is called the *stem.* The remaining group of digits on the right is called the leaf. There is no rule for the number of digits to be included in the stem. Usually the units digit is the leaf and the remaining digits are the stem. For example, the number 53 would be broken up into 5 and 3. The 5 would be the stem and the 3 would be the leaf. The number 417 would be broken up into 41 and 7. The 41 would be the stem and the 7 would be the leaf. With a stem-and-leaf display, the stems are listed, in order, to the left of a vertical line. Then we place each leaf to the right of its corresponding stem, to the right of the vertical line.* Example 6 illustrates this procedure.

EXAMPLE 6 *Stem-and-Leaf Display*

James Fadden tested 20 size-AA batteries for a science project. The table below indicates the lifetime of each battery, to the nearest hour. Construct a stem-and-leaf display using this data.

58	67	62	70	56
45	55	70	68	60
67	72	71	85	91
57	48	63	65	59

SOLUTION By quickly glancing at the data, we can see the battery lifetimes consist of two digit numbers. Let's use the first digit, the tens digit, as our stem and the second digit, the units digit, as the leaf. For example, for a lifetime of 62 hours, the stem is 6, and the leaf is 2. Our values are numbers in the 40's, 50's, 60's, 70's, 80's, and 90's. Therefore, the stems will be 4, 5, 6, 7, 8, 9 as shown below.

4 |
5 |
6 |
7 |
8 |
9 |

Next we place each leaf on its stem. We will do this by placing the second digit of each value next to its stem, to the right of the vertical line. Our first value is 58. The 5 is the stem and the 8 is the leaf. Therefore, we place an 8 next to the stem of 5 and to the right of the vertical line.

5 | 8

*In stem-and-leaf displays the leaves are sometimes listed from lowest digit to greatest digit, but this is not necessary.

The next value is 67. We will place a leaf of 7 next to the stem of 6.

5	8
6	7

The next value is 62. Therefore, we will place a leaf of 2 next to the stem of 6.

5	8
6	7 2

We continue this process until we have listed all of the leaves on the display. The diagram below shows the stem-and-leaf display for the battery lifetimes (in hours). In our display, we will also include a legend to indicate the values represented by the stems and leaves. For example 5 | 6 represents 56.

5 | 6 represents 56

Stem	Leaves
4	5 8
5	8 6 5 7 9
6	7 2 8 0 7 3 5
7	0 0 2 1
8	5
9	1

Every piece of the original data can be seen in a stem-and-leaf display. From the above diagram, we can see that most of the batteries had a lifetime in the range of 60 hours. Only two batteries had a lifetime of more than 72 hours. Two batteries had a lifetime less than 50 hours. Note that the stem-and-leaf display gives the same visual impression as a sideways histogram.

SECTION 13.4 EXERCISES

CONCEPT/WRITING EXERCISES

1. In your own words, explain how to construct a circle graph from a table of values.
2. a) What is listed on the horizontal axis of a histogram and frequency polygon?
   b) What is listed on the vertical axis of a histogram and frequency polygon?
3. In your own words, explain how to construct a histogram from a set of data.
4. In your own words, explain how to construct a frequency polygon from a set of data.
5. a) In your own words, explain how to construct a frequency polygon from a histogram.
   b) Construct a frequency polygon from the histogram below.

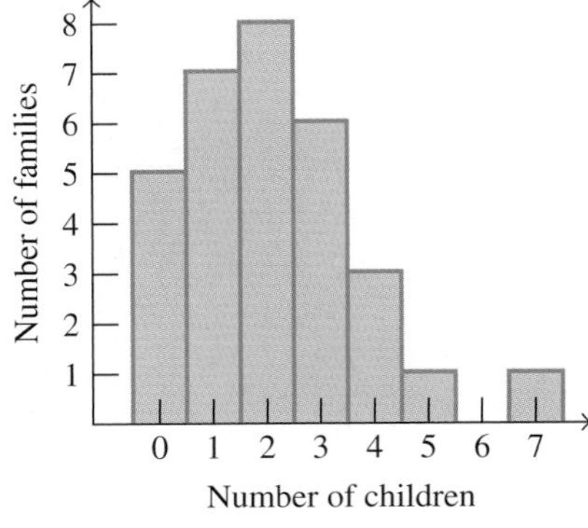

6. **a)** In your own words, explain how to construct a histogram from a frequency polygon.
b) Construct a histogram from the frequency polygon below.

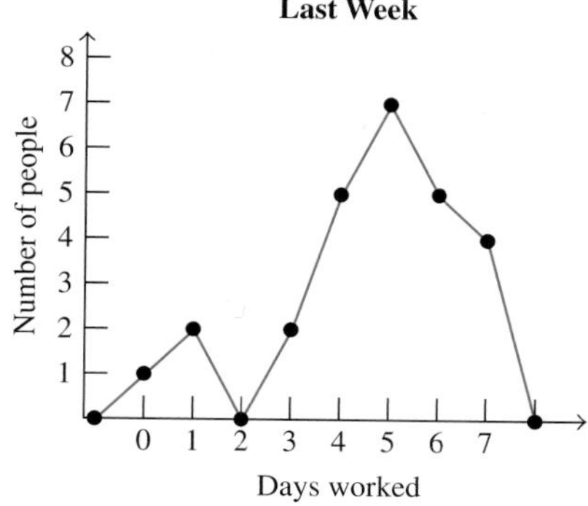

7. **a)** In your own words, explain how to construct a stem-and-leaf display.
b) Construct a frequency distribution, letting each class have a width of 1, from the following stem-and-leaf display.

4 | 5 represents 45

Stem	Leaf
4	5 5 5 7 9
5	0 1 1

8. Construct a frequency distribution, letting each class have a width of 1, from the following stem-and-leaf display.

2 | 3 represents 23

Stem	Leaf
1	7 8 7 9 6
2	3 1 2 2 5 5 4

PRACTICE THE SKILLS

9. *Coffee Drinkers* *Newsweek* magazine surveyed Americans on health issues. One question asked was "How many cups of coffee do you drink on an average day?" The following circle graph shows the percent of the respondents who answered none, one, two, three, or four or more cups of coffee on an average day. If 500 people were surveyed, determine the number of people in each category.

How Many Cups of Coffee Do You Drink on an Average Day?

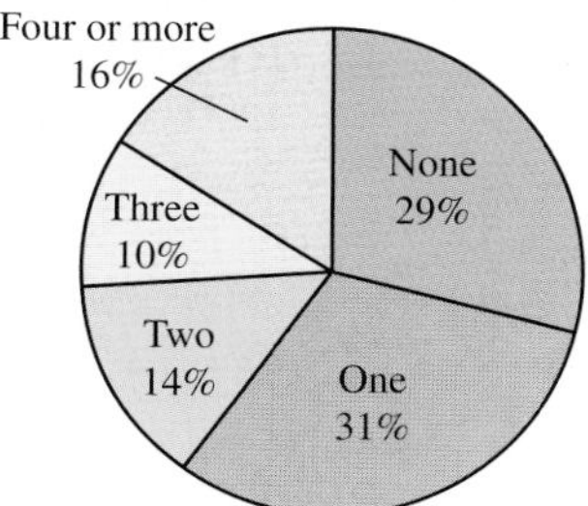

10. *Online Households* In 2000 there are 66.6 million households online worldwide. Of the total, 57% are in North America, as shown in the graph. Estimate the number of households online in each region shown on the graph.

Online Households

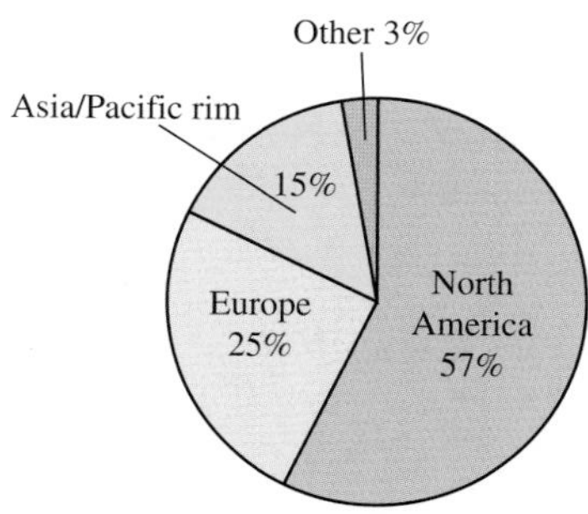

Source: Jupiter Communications

11. A sample of 500 adults at a shopping mall was asked why they came to the mall. Their responses are given in the following table. Construct a circle graph, with sectors given in percent, which illustrates this information. Round percents to tenths.

Reason	Number of responses
Buy clothing	260
Buy items other than clothing	125
To get out of the house	60
Other	55

12. A sample of 600 housing permits for new houses was randomly selected. The number of bedrooms was recorded, as indicated in the following table. Construct a circle graph, with sectors given in percent, which illustrates this information. Round percents to tenths.

Number of bedrooms	Number of permits
2	182
3	230
4	100
5 or more	88

13. *Jazz Concert* The frequency distribution indicates the ages of a group of 45 people attending a jazz concert.

Age	Number of people
17	2
18	5
19	7
20	8
21	0
22	10
23	5
24	8

a) Construct a histogram of the frequency distribution.
b) Construct a frequency polygon of the frequency distribution.

14. *Attendance at a Party* The frequency distribution shown indicates the ages of a group of 40 people attending a party.

Age	Number of people
20	6
21	3
22	0
23	4
24	6
25	3
26	8
27	10

a) Construct a histogram of the frequency distribution.
b) Construct a frequency polygon of the frequency distribution.

15. *Compact Discs* The frequency distribution indicates the number of compact discs owned by a sample of 40 people.

Number of compact discs	Number of people
6–13	4
14–21	5
22–29	10
30–37	11
38–45	6
46–53	3
54–61	1

a) Construct a histogram of the frequency distribution.
b) Construct a frequency polygon of the frequency distribution.

16. *Annual Salaries* The frequency distribution illustrates the annual salaries, in thousands of dollars, of the people in management positions at the Bradley Thomas Corporation.

Salary (in $1000)	Number of people
20–25	4
26–31	6
32–37	8
38–43	9
44–49	8
50–55	5
56–61	3

a) Construct a histogram of the frequency distribution.
b) Construct a frequency polygon of the frequency distribution.

PROBLEM SOLVING

17. *Number of Books Purchased* Use the histogram below to answer the following questions.

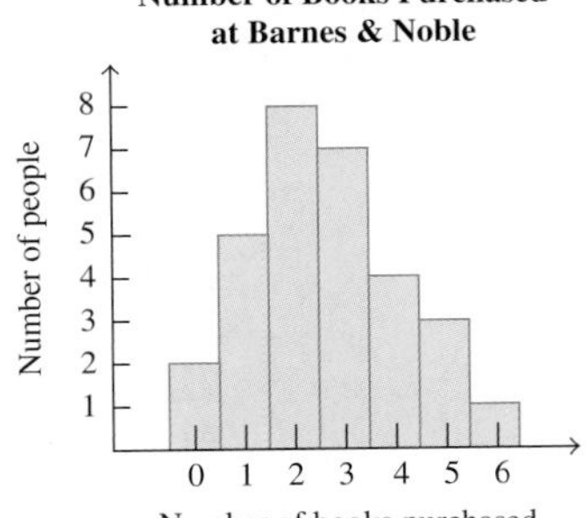

a) How many people were surveyed?
b) How many people purchased four books?
c) What is the modal class?
d) How many books were purchased?
e) Construct a frequency distribution from this histogram.

18. *Student Rent* Use the histogram below to answer the following questions.

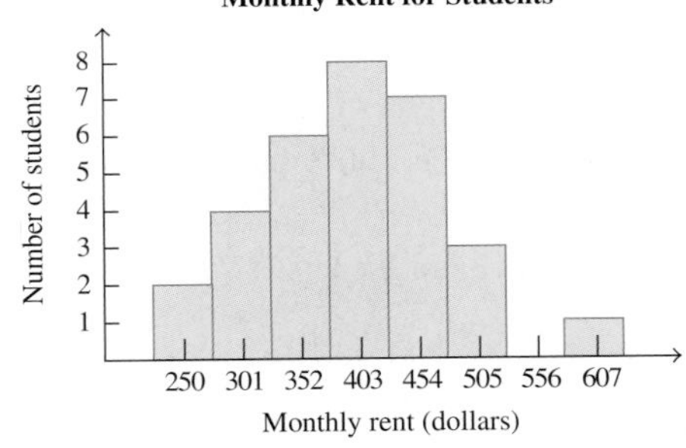

a) How many students were surveyed?
b) What are the lower and upper class limits of the first and second classes?
c) How many students have a monthly rent in the class with a class mark of $352?
d) What is the class mark of the modal class?
e) Construct a frequency distribution from this histogram. Use a first class of 225–275.

19. *Response Time* Use the frequency polygon below to answer the following questions.

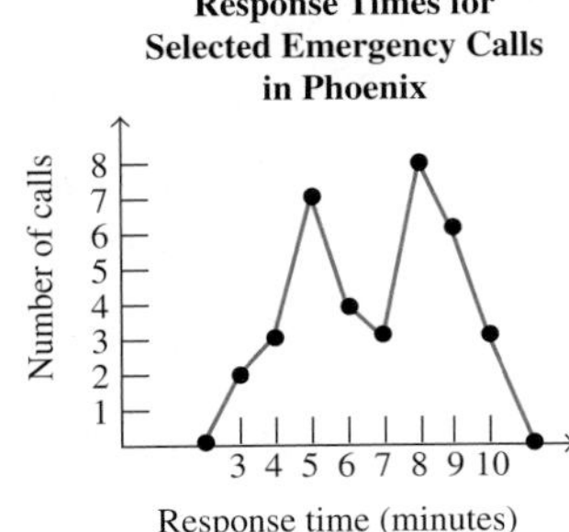

a) How many calls were responded to in 5 minutes?
b) How many calls were responded to in 6 minutes or less?
c) How many calls were included in the survey?
d) Construct a frequency distribution from the frequency polygon.
e) Construct a histogram from the frequency distribution in part (d).

20. *Pacific Science Center* Use the frequency polygon below to answer the following questions.

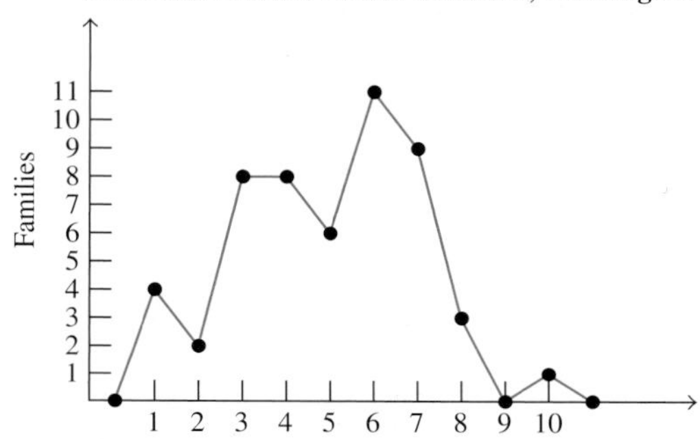

a) How many families visited the Pacific Science Center four times?
b) How many families visited the Pacific Science Center at least six times?
c) How many families were surveyed?
d) Construct a frequency distribution from the frequency polygon.
e) Construct a histogram from the frequency distribution in part (d).

21. Construct a histogram and a frequency polygon from the frequency distribution given in Exercise 7 of Section 13.3. See page 676.

22. Construct a histogram and a frequency polygon from the frequency distribution given in Exercise 8 of Section 13.3. See page 676.

23. *College Credits* Eighteen students in a physical fitness class were asked how many college credits they had earned. The responses are as follows. Construct a stem-and-leaf display.

10	15	24	36	48	45
42	53	60	17	24	30
33	45	48	62	54	60

24. *Distance to Work* Twenty workers at a small company were asked how many miles they drive to work, one way. The responses are as follows. Construct a stem-and-leaf display. For single-digit data, use a stem of 0.

12	18	3	8	12	25	21
33	15	2	5	27	41	22
19	13	23	34	17	16	

25. *Starting Salaries* Starting salaries (rounded to the nearest thousand) for chemical engineers with B.S. degrees and no experience are shown for 25 different companies.

25	26	27	29	31
26	26	27	29	31
26	26	28	30	31
26	27	28	30	32
26	27	28	30	32

a) Construct a frequency distribution.
b) Construct a histogram.
c) Construct a frequency polygon.
d) Construct a stem-and-leaf display.

26. *Broadway Shows* The ages of a random sample of people attending a Broadway show are

20	23	25	30	32	35	39	44
21	23	26	30	33	35	40	45
21	24	27	30	34	35	40	45
22	24	28	31	34	37	40	46
23	25	28	31	34	38	42	47

a) Construct a frequency distribution with a first class of 20–24.
b) Construct a histogram.
c) Construct a frequency polygon.
d) Construct a stem-and-leaf display.

27. *Magazines* The following table shows the 50 leading U.S. magazines in terms of advertising revenue in 1997, rounded to the nearest million dollars.

Magazine	Sales (millions of dollars)
People Weekly	589
Sports Illustrated	549
Time	533
TV Guide	469
Newsweek	408
Better Homes and Gardens	377
PC Magazine	334
Business Week	330
Forbes	244
U.S. News & World Report	239
Woman's Day	228
Fortune	226
Good Housekeeping	219
Reader's Digest	212
Family Circle	206
Cosmopolitan	189
Ladies' Home Journal	184
Vogue	149
Entertainment Weekly	146
Glamour	137

(*continued*)

Magazine	Sales (millions of dollars)
Money	133
Rolling Stone	122
Southern Living	121
Golf Digest	120
McCall's	118
Redbook	113
Martha Stewart Living	111
PC Computing	110
Car and Driver	108
Elle	100
Parents	98
Vanity Fair	96
New Yorker	94
Inc.	90
Country Living	89
Bride's	83
Golf Magazine	79
Windows Magazine	78
W	77
Travel & Leisure	77
Gentleman's Quarterly	76
Seventeen	75
Modern Bride	74
Road & Track	74
Harper's Bazaar	70
Parenting	69
Self	67
National Geographic	63
Mademoiselle	63
House Beautiful	63

(*Source:* Audit Bureau of Circulations and Magazine Publishers of America.)

a) Construct a frequency distribution with the first class \$63 million to \$122 million.
b) Construct a histogram.
c) Construct a frequency polygon.

28. *U.S. Ambassadors* The ages of a random sample of U.S. ambassadors are

40	43	45	50	52	55	59	64
41	43	46	50	53	55	60	65
41	44	47	50	54	55	60	65
42	44	48	51	54	57	60	66
43	45	48	51	54	58	62	67

a) Construct a frequency distribution with the first class 40–44.
b) Construct a histogram.
c) Construct a frequency polygon.

CHALLENGE PROBLEMS/GROUP ACTIVITIES

29. a) *Birthdays* What do you believe a histogram of the months in which the students in your class were born (January is month 1 and December is month 12) would look like? Explain.
b) By asking, determine the month in which the students in your class were born (include yourself).
c) Construct a frequency distribution containing 12 classes.
d) Construct a histogram from the frequency distribution in part (c).
e) Construct a frequency polygon of the frequency distribution in part (c).

30. *Social Security Numbers* Repeat Exercise 29 for the last digit of the students' social security numbers. Include classes for the digits 0–9.

RESEARCH ACTIVITY

31. Over the years many changes have been made in the U.S. Social Security System.
a) Do research and determine the number of people receiving social security benefits for the years 1945, 1950, 1955, 1960, . . . , 2000. Then construct a frequency distribution and histogram of the data.
b) Determine the maximum amount that self-employed individuals had to pay into social security (the FICA tax) for the years 1945, 1950, 1955, 1960, . . . , 2000. Then construct a frequency distribution and a histogram of the data.

13.5 MEASURES OF CENTRAL TENDENCY

Most people have an intuitive idea of what is meant by an "average." The term is used daily in many familiar ways: "This car averages 19 miles per gallon." "The average test grade was 78." "The average height of adult males is 5 feet 8 inches."

An **average** is a number that is representative of a group of data. There are at least four different averages: the mean, the median, the mode, and the midrange. Each is calculated differently and may yield different results for the same set of data. Each will result in a number near the center of the data; for this reason, averages are commonly referred to as **measures of central tendency**.

The **arithmetic mean**, or simply the **mean**, is symbolized either by $\bar{x}$ (read "x bar") or by the Greek letter mu, μ. The symbol $\bar{x}$ is used when the mean of a *sample* of the population is calculated. The symbol μ is used when the mean of the *entire population* is calculated. We will assume that the data featured in this book represent samples, and therefore we will always use $\bar{x}$ for the mean.

The Greek letter sigma, Σ, is used to indicate "summation." The notation Σx, read "the sum of x," is used to indicate the sum of all the data. For example, if there are five pieces of data, 4, 6, 1, 0, 5, then $\Sigma x = 4 + 6 + 1 + 0 + 5 = 16$.

Now we can discuss the procedure for finding the mean of a set of data.

The mean, $\bar{x}$, is the sum of the data divided by the number of pieces of data. The formula for calculating the mean is

$$\bar{x} = \frac{\Sigma x}{n}$$

where Σx represents the sum of all the data and n represents the number of pieces of data.

The most common use of the word *average* is the mean.

EXAMPLE 1 *Find the Mean*

Find the mean age of a group of people in Starbuck's coffee shop on a Friday evening if the ages are 28, 19, 47, 34, and 47.

SOLUTION

$$\bar{x} = \frac{\Sigma x}{n} = \frac{28 + 19 + 47 + 34 + 47}{5} = \frac{175}{5} = 35$$

Therefore, the mean, $\bar{x}$, is 35 years. ▲

The mean represents "the balancing point" of a set of data. For example, if a seesaw were pivoted at the mean and uniform weights were placed at points corresponding to the ages in Example 1, the seesaw would balance. Figure 13.13 shows the five ages given in Example 1 and the calculated mean.

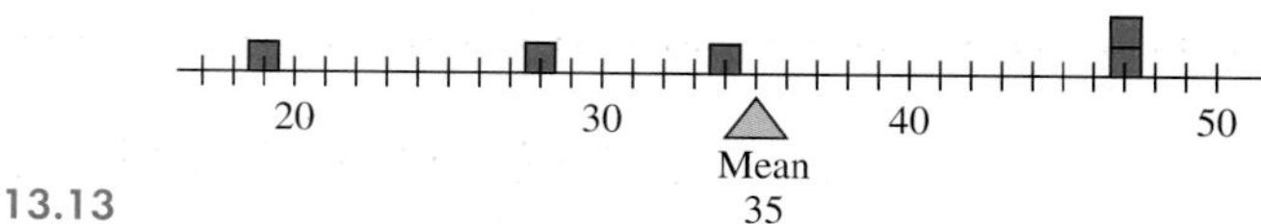

Figure 13.13

A second average is the *median*. To find the median of a set of data, *rank the data* from smallest to largest, or largest to smallest, and determine the value in the middle of the set of *ranked data*. This value will be the median.

The **median** is the value in the middle of a set of *ranked data*.

DID YOU KNOW

Buying the American Dream

San Francisco, CA.

One of the biggest dreams for most people is to own their own home. But depending on where you live, the American dream may be hard to achieve. In 1997, San Francisco had the highest median home price for major metropolitan housing markets, $289,700. Orlando had the lowest median home price for major metropolitan housing markets, $103,300.
Source: Census Bureau, Bureau of Labor Statistics.

Orlando, FL.

EXAMPLE 2 ***Find the Median***

Find the median of the people's ages at Starbuck's in Example 1.

SOLUTION Ranking the data from smallest to largest gives 19, 28, 34, 47, and 47. Since 34 is the value in the middle of this set of ranked data (two pieces of data above it and two pieces below it) 34 years is the median. ▲

If there are an even number of pieces of data, the median will be halfway between the two middle pieces. In this case, to find the median, add the two middle pieces and divide this sum by 2.

EXAMPLE 3 ***Find the Median of an Even Number of Pieces of Data***

Find the median of the following sets of data:

a) 7, 10, 12, 14, 15, 14, 9, 9
b) 7, 8, 8, 8, 9, 10

SOLUTION

a) Ranking the data gives 7, 9, 9, 10, 12, 14, 14, 15. There are eight pieces of data. Therefore, the median will lie halfway between the two middle pieces, the 10 and the 12. The median is $\frac{10 + 12}{2}$ or $\frac{22}{2}$ or 11.
b) There are six pieces of data and they are already ranked. Therefore, the median lies halfway between the two middle pieces. Both middle pieces are 8's. The median is $\frac{8 + 8}{2}$, or $\frac{16}{2}$, or 8. ▲

A third average is the **mode**.

> The **mode** is the piece of data that occurs most frequently.

EXAMPLE 4 ***Find the Mode***

Find the mode of the people's ages at Starbuck's in Example 1.

SOLUTION The ages were 28, 19, 47, 34, and 47. The age 47 is the mode because it occurs twice and the other values occur only once. ▲

If no one piece of data occurs more frequently than every other piece of data, the set of data has no mode. For example, neither of the following two sets of data has a mode.

1, 2, 3, 4, 5 no mode
1, 1, 2, 3, 3, 4, 5, no mode*

The last average that we will discuss is the midrange. The **midrange** is the value halfway between the lowest (L) and highest (H) values in a set of data. It is found by adding the lowest and highest values and dividing the sum by 2. A formula for finding the midrange follows.

*Some textbooks refer to sets of data of this type as bimodal.

DID YOU KNOW

When Babies' Eyes Are Smiling

Experimental psychologists formulate hypotheses about human behavior, design experiments to test them, make observations, and draw conclusions from their data. They use statistical concepts at each stage to help ensure that their conclusions are valid. In one experiment, researchers observed that 2-month-old infants who learned to move their heads in order to make a mobile turn began to smile as soon as the mobile turned. Babies in the control group did not smile as often when the mobile moved independently of their head turning. The researchers concluded that it was not the movement of the mobile that made the infant smile; rather, the infants smiled at their own achievement.

$$\textbf{Midrange} = \frac{\text{lowest value} + \text{highest value}}{2}$$

EXAMPLE 5 ***Find the Midrange***

Find the midrange of the people's ages at Starbuck's given in Example 1.

SOLUTION The lowest age is 19 and the highest age is 47.

$$\text{Midrange} = \frac{\text{lowest} + \text{highest}}{2} = \frac{19 + 47}{2} = \frac{66}{2} = 33 \text{ years}$$

The "average" of the ages 28, 19, 47, 34, 47 can be considered any one of the following values: 35 (mean), 34 (median), 47 (mode), or 33 (midrange). Which average do you feel is most representative of the ages? We will discuss this question later in this section.

EXAMPLE 6 ***Measures of Central Tendency***

The salaries of eight selected chemical engineers rounded to the nearest thousand dollars are 40, 25, 28, 35, 42, 60, 60, and 73. For this set of data, find the (a) mean, (b) median, (c) mode, and (d) midrange, and then (e) rank the measures of central tendency from lowest to highest.

SOLUTION

a) $\bar{x} = \frac{40 + 25 + 28 + 35 + 42 + 60 + 60 + 73}{8} = \frac{363}{8} = 45.375$

b) Listing the data from the smallest to largest gives

25, 28, 35, 40, 42, 60, 60, 73

Since there are an even number of pieces of data, the median is halfway between 40 and 42. The median = (40 + 42)/2 = 82/2 = 41.

c) The mode is the piece of data that occurs most frequently. The mode is 60.

d) The midrange = (L + H)/2 = (25 + 73)/2 = 98/2 = 49.

e) The averages from lowest to highest are the median, mean, midrange, and mode. Their values are 41, 45.375, 49, and 60, respectively.

At this point you should be able to calculate the four measures of central tendency: mean, median, mode, and midrange. Now let's examine the circumstances in which each is used.

The mean is used when each piece of data is to be considered and "weighed" equally. It is the most commonly used average. It is the only average that can be affected by *any* change in the set of data; for this reason, it is the most sensitive of all the measures of central tendency (see Exercise 23).

Occasionally, one or more pieces of data may be much greater or much smaller than the rest of the data. When this occurs, these "extreme" values have the effect of increasing or decreasing the mean significantly so that the mean will not be representative of the set of data. Under these circumstances, the median should be used instead of the mean. The median is often used in describing average family incomes because a relatively small number of families have extremely large incomes. These few families would inflate the mean income, making it nonrepresentative of the millions of families in the population.

Consider a set of exam scores from a mathematics class: 0, 16, 19, 65, 65, 65, 68, 69, 70, 72, 73, 73, 75, 78, 80, 85, 88, 92. Which average would best represent these grades? The mean is 64.06. The median is 71. Since only 3 of the 18 scores fall below the mean, the mean would not be considered a good representative score. The median of 71 probably would be the better average to use.

The mode is the piece of data, if any, that occurs most frequently. Builders planning houses are interested in the most common family size. Retailers ordering shirts are interested in the most common shirt size. An individual purchasing a thermometer might select one with the most common reading. These examples illustrate how the mode may be used.

The midrange is sometimes used as the average when the item being studied is constantly fluctuating. Average daily temperature, used to compare temperatures in different areas, is calculated by adding the lowest and highest temperatures for the day and dividing the sum by 2. The midrange is actually the mean of the high value and the low value of a set of data. Occasionally, the midrange is used to estimate the mean, since it is much easier to calculate.

Sometimes an average itself is of little value, and care must be taken in interpreting its meaning. For example, Jim is told the average depth of Willow Pond is only 3 feet. He is not a good swimmer but decides it is safe to go out a short distance in this shallow pond. After he is rescued, he exclaims, "I thought this pond was only 3 feet deep." Jim didn't realize an average does not indicate extreme values or the spread of the values. The spread of data is discussed in Section 13.6.

DID YOU KNOW

Measures of Location

If you took the Scholastic Aptitude Test (SAT) before applying to college, your score was described as a measure of location rather than a measure of central tendency. **Measures of location** are often used to make comparisons, such as comparing the scores of individuals from different populations, and are generally used when the amount of data is large.

Two measures of location are **percentiles** and **quartiles**. Percentiles divide the set of data into 100 equal parts (see Fig. a). For example, suppose you scored 490 on the math half of the SAT, and the score of 490 was reported to be in the 78th percentile of high school students. This *does not* mean that 78% of your answers were correct; it *does* mean that you outperformed about 78% of all those taking the exam. In general, a score in the *n*th percentile means you outperformed about $n\%$ of the population who took the test, and that $(100 - n)\%$ of the people taking the test performed better than you did.

Quartiles divide data into four equal parts: The first quartile is the value that is higher than about 1/4 or 25% of the population. It is the same as the 25th percentile. The second quartile is the value that is higher than about 1/2 the population and is the same as the 50th percentile, or the median. The third quartile is the value that is higher than about 3/4 of the population and is the same as the 75th percentile (see Fig. b).

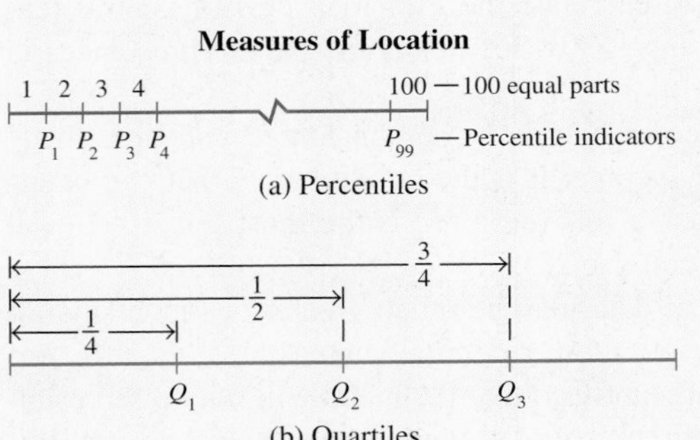

(a) Percentiles

(b) Quartiles

SECTION 13.5 EXERCISES

CONCEPT/WRITING EXERCISES

1. Describe the mean of a set of data and explain how to find it.
2. What is *ranked data*?
3. Describe the median of a set of data and explain how to find it.
4. Describe the midrange of a set of data and explain how to find it.
5. Describe the mode of a set of data and explain how to find it.
6. When might the mode be the preferred average to use? Give an example.
7. When might the median be the preferred average to use? Give an example.
8. When might the midrange be the preferred average to use? Give an example.
9. When might the mean be the preferred average to use? Give an example.
10. **a)** What symbol is used for the sample mean?
    b) What symbol is used for the population mean?

PRACTICE THE SKILLS

In Exercises 11–20, find the mean, median, mode, and midrange of the set of data. Round your answer to the nearest tenth.

11. 2, 2, 3, 6, 8, 11, 11, 11, 36
12. 7, 8, 13, 15, 13, 12, 11, 375, 43, 40
13. 60, 72, 80, 84, 86, 45, 96
14. 5, 3, 6, 6, 6, 9, 11
15. 1, 3, 5, 7, 9, 11, 13, 15
16. 1, 7, 11, 27, 36, 14, 12, 9, 1
17. 40, 50, 30, 60, 90, 100, 140
18. 1, 1, 1, 1, 4, 4, 4, 4, 6, 8, 10, 12, 15, 21
19. 6, 8, 12, 13, 11, 13, 15, 17
20. 5, 15, 5, 15, 5, 15

PROBLEM SOLVING

21. *Selling Hot Dogs* The number of hot dogs sold daily for 1 week at Harry's hot dogs cart is 25, 43, 17, 36, 51, 29, and 31. Find the mean, median, mode, and midrange.
22. *Cholesterol Level* The cholesterol level of 10 patients of Dr. Jackson are 200, 153, 186, 210, 207, 286, 172, 246, 214, and 132. Find the mean, median, mode, and midrange.
23. *Change in the Data* The mean is the "most sensitive" average because it is affected by any change in the data.
    a) Determine the mean, median, mode, and midrange for 1, 2, 3, 5, 5, 7, 11.
    b) Change the 7 to a 10 in part (a). Find the mean, median, mode, and midrange.
    c) Which averages were affected by changing the 7 to a 10?
    d) Which averages will be affected by changing the 11 to a 10 in part (a)?
24. *Life Expectancy* In 1997 the National Center for Health Statistics indicated a new record "average life expectancy" of 76.5 years for the total U.S. population. The average life expectancy for men was 73.6 years, and for women it was 79.4 years. Which "average" do you think the National Center for Health is using? Explain your answer.
25. *A Grade of B* To get a grade of B, a student must have a mean average of 80. Jim has a mean average of 79 for 10 quizzes. He approaches his teachers and asks for a B, reasoning that he missed a B by only one point. What is wrong with Jim's reasoning?
26. *Employee Salaries* The salaries of 10 employees of a small company follow.

$26,000	$62,000
23,000	22,000
29,000	25,000
24,000	79,000
24,000	27,000

    Calculate the
    a) mean.
    b) median.
    c) mode.
    d) midrange.
    e) If the employees wanted to demonstrate the need for a raise, which average would they use to show they are being underpaid: the mean or the median? Explain.
    f) If the management did not want to give the employees a raise, which average would they use: the mean or the median? Explain.
27. *Tossing a Frisbee* Ted tosses a frisbee as far as he can 15 times. The following distances, in feet, were recorded.

190	102	188	270	105
201	95	196	284	200
164	191	110	169	212

Calculate the
a) mean.
b) median.
c) mode.
d) midrange.
e) Which average is the best measure of central tendency for this set of data? Explain.

28. *Visa Charges* Natasha's monthly Visa charges for one year are as follows:

$123.72	$396.81	$127.96
101.95	157.48	161.59
96.27	131.62	145.30
115.82	119.25	106.82

Find
a) mean.
b) median.
c) mode
d) midrange.

29. *Theme Park Attendance* The theme parks with the most attendance in 1998 are listed below.

Park	Attendance (millions of people)
Disney World Magic Kingdom	15.6
Disneyland	13.7
Disney World Epcot Center	10.6
Disney World MGM Studios	9.5
Universal Studios Florida	8.9
Disney World Animal Kingdom	6.0
Universal Studios Hollywood	5.1
Sea World Florida	4.9
Busch Gardens Tampa Bay	4.2
Sea World California	3.7

(*Source:* Amusement Business.)

Determine the
a) mean.
b) median.
c) mode.
d) midrange.

Busch Gardens Tampa Bay

30. *Exam Average* Malcolm's mean average on five exams is 83. Find the sum of his scores.

31. *Exam Average* Jeremy's mean average on six exams is 78. Find the sum of his scores.

32. *Creating a Data Set* Construct a set of five pieces of data in which the mode has a lower value than the median and the median has a lower value than the mean.

33. *Creating a Data Set* Construct a set of six pieces of data with a mean, median, and midrange of 75 and where no two pieces of data are the same.

34. *Creating a Data Set* Construct a set of six pieces of data with a mean of 84 and where no two pieces of data are the same.

35. *Ski Resort* For the 1998–1999 season, 24,000 people skied at Big Snow Ski Resort. The resort was open 120 days for skiing. The highest number of skiers on a single day was 500. The lowest number of skiers on a single day was 20. Determine whether it is possible to find the following with the given information:
a) the mean number of skiers per day.
b) the median number of skiers per day.
c) the mode number of skiers per day.
d) the midrange number of skiers per day.
e) Find all the measures of central tendency that can be found with the information and explain why the others cannot be found.

36. *Determine a Necessary Grade* A mean average of 80 for five exams is needed for a final grade of B. Jorge's first four exam grades are 68, 78, 83, and 80. What grade does Jorge need on the fifth exam to get a B in the course?

37. *Grading Methods* A mean average of 60 on seven exams is needed to pass a course. On her first six exams, Sheryl received grades of 49, 72, 80, 60, 57, and 69.
a) What grade must she receive on her last exam to pass the course?
b) An average of 70 is needed to get a C in the course. Is it possible for Sheryl to get a C? If so, what grade must she receive on the seventh exam?
c) If her lowest grade of the exams already taken is to be dropped, what grade must she receive on her last exam to pass the course?
d) If her lowest grade of the exams already taken is to be dropped, what grade must she receive on her last exam to get a C in the course?

38. Which of the measures of central tendency *must* be an actual piece of data in the distribution? Explain.

39. *Creating a Data Set* Construct a set of six pieces of data such that if only one piece of data is changed, the mean, median, and mode will all change.

40. *Changing One Piece of Data* Consider the set of data 1, 1, 1, 2, 2, 2. If one 2 is changed to a 3, which of the following will change: mean, median, mode, midrange? Explain.

41. *Changing One Piece of Data* Is it possible to construct a set of six different pieces of data such that by changing only one piece of data you cause the mean, median, mode, and midrange to change? Explain.

42. *Grocery Expenses* The Taylor's have recorded their weekly grocery expenses for the past 12 weeks and determined the mean weekly expense was $85.20. Later Mrs. Taylor discovered 1 week's expense of $74 was incorrectly recorded as $47. What is the correct mean?

In Exercises 43–46, refer to the Did You Know on Measures of Location on page 690, and then answer the following questions.

43. *Percentiles* For any set of data, what must be done to the data before percentiles can be determined?

44. *Percentiles* Josie scored in the 73rd percentile on the verbal part of her College Board test. What does that mean?

45. *Percentiles* When a national sample of heights of kindergarten children was taken, Kevin was told that he was in the 35th percentile. Explain what that means.

46. *Percentiles* A union leader is told that, when all workers' salaries are considered, the first quartile is $20,750. Explain what that means.

47. *The 50th Percentile* Give the names of two other statistics that have the same value as the 50th percentile.

48. *College Admissions* Jonathan took an admission test for the University of California and scored in the 85th percentile. The following year Jonathan's sister Kendra took a similar admission test for the University of California and scored in the 90th percentile.
   a) Is it possible to determine which of the two answered the higher percent of questions correctly on their respective exams? Explain your answer.
   b) Is it possible to determine which of the two was in a better relative position with regard to their respective populations? Explain.

49. *Employee Salaries* The following statistics represent weekly salaries at the Midtown Construction Company:

Mean	$510	First quartile	$470
Median	$500	Third quartile	$535
Mode	$490	83rd percentile	$575

   a) What is the most common salary?
   b) What salary did half the employees surpass?
   c) About what percent of employees surpassed $535?
   d) About what percent of employees' salaries were below $470?
   e) About what percent of employees surpassed $575?
   f) If the company has 100 employees, what is the total weekly salary of all employees?

CHALLENGE PROBLEMS/GROUP ACTIVITIES

50. *The Mean of the Means* Consider the following five sets of values.

i)	5	6	7	7	8	9	14
ii)	3	6	8	9			
iii)	1	1	1	2	5		
iv)	6	8	9	12	15		
v)	50	51	55	60	80	100	

   a) Compute the mean of each of the five sets of data.
   b) Compute the mean of the five means in part (a).
   c) Find the mean of the 27 pieces of data.
   d) Compare your answer in part (b) to your answer in part (c). Are the values the same? Does your answer make sense? Explain.

51. *Ruth versus Mantle* The following tables compare the batting performances for selected years for two well-known former baseball players, Babe Ruth and Mickey Mantle.

Babe Ruth
Boston Red Sox 1914–1919
New York Yankees 1920–1934

Year	At bats	Hits	Pct.
1925	359	104	
1930	518	186	
1933	459	138	
1916	136	37	
1922	406	128	
Total	1878	593	

Mickey Mantle
New York Yankees 1951–1968

Year	At bats	Hits	Pct.
1954	543	163	
1957	474	173	
1958	519	158	
1960	527	145	
1962	377	121	
Total	2440	760	

a) For each player, compute the batting average percent (pct.) for each year by dividing the hits by the at bats. Round off to thousandths. Place the answers in the pct. column.
b) Going across each of the five horizontal lines (for example Ruth, 1925, vs. Mantle, 1954), compare the percents (pct.) and determine which is greater in each case.
c) For each player, compute the mean batting average percent for the 5 given years by determining the total at bats and hits over the 5 years and dividing the total hits by the total at bats. Which is greater, Ruth's or Mantle's?
d) Based on your answer in part (b), does your answer in part (c) make sense? Explain.
e) Find the mean percent for each player by adding the five pcts. and dividing by 5.Which is greater, Ruth's or Mantle'?
f) Why do the answers obtained in parts (c) and (e) differ? Explain.
g) Who would you say has the better batting average percent for the 5 years selected? Explain.

52. *Employee Salaries* The following table gives the annual salary distribution for employees at Richardson's Home Improvement. Using the information provided in the table, find the:
a) mean annual salary.
b) median annual salary.
c) mode annual salary.
d) midrange annual salary.
e) Which is the best measure of central tendency for this set of data? Explain your answer.

Annual salary	Number receiving salary
$100,000	1
85,000	2
24,000	6
21,000	4
18,000	5
17,000	7

RESEARCH ACTIVITY

53. Two other measures of location that we did not mention in the Did You Know on page 690 are *stanines* and *deciles*. Use statistics books, books on educational testing and measurements, and Internet web sites to determine what stanines and deciles are and when percentiles, quartiles, stanines, and deciles are used.

13.6 MEASURES OF DISPERSION

The measures of central tendency by themselves do not always give sufficient information to analyze a situation and make decisions. As an example, two manufacturers of airplane engines are being considered for a contract. Manufacturer A's engines have an average (mean) life of 1000 hours of flying time before they must be rebuilt. Manufacturer B's engines have an average life of 950 hours of flying time before they must be rebuilt. If you assume that both cost the same, which engines should be purchased? The average engine life may not be the most important factor. The fact that manufacturer A's engines have an average life of 1000 hours could mean that half will last about 500 hours and the other half will last about 1500 hours. If in fact all of manufacturer B's engines have a life span of between 900 and 1000 hours, then B's engines are more consistent and reliable. If A's engines were purchased, they would all have to be rebuilt every 300 hours or so because it would be impossible to determine which ones would fail first. If B's engines were purchased, they could go much longer before having to be rebuilt. This example is of course an exaggeration used to illustrate the importance of knowing something about the *spread,* or *variability,* of the data.

Measures of dispersion are used to indicate the spread of the data. The range and standard deviation* are the measures of dispersion that will be discussed in this book.

The **range** is the difference between the highest and lowest values; it indicates the total spread of the data.

Range = highest value − lowest value

EXAMPLE 1 *Find the Range*

Find the range of the following mathematics SAT scores:

480, 510, 530, 570, 450, 525, 490

SOLUTION Range = highest value − lowest value = 570 − 450 = 120. The range of this set of mathematics SAT scores is 120.

The second measure of dispersion, the **standard deviation**, measures how much the data *differs from the mean.* It is symbolized either by the letter s or by the Greek letter sigma, σ.[†] The s is used when the standard deviation of a *sample* is calculated. The σ is used when the standard deviation of the entire *population* is calculated. Since we are assuming all data presented here are for samples, we use s to represent the standard deviation in this book (note, however, on the doctors' charts on page 700, σ is used). The larger the spread of the data about the mean, the larger is the standard deviation. Consider the following two sets of data.

5, 8, 9, 10, 12, 13   8, 9, 9, 10, 10, 11

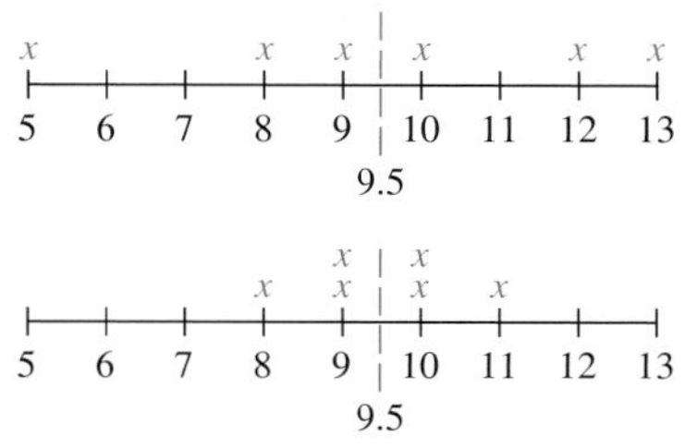

Both have a mean of 9.5. Which set of values on the whole do you feel differs less from the mean of 9.5? Figure 13.14 may make the answer more apparent. The scores in the second set of data are closer to the mean and therefore have a smaller standard deviation. You will soon be able to verify such relationships yourself.

Figure 13.14

Sometimes only a very small standard deviation is desirable or acceptable. Consider a cereal box that is to contain 8 oz of cereal. If the amount of cereal put into the boxes varies too much—sometimes underfilling, sometimes overfilling—the manufacturer will soon be in trouble with consumer groups and government agencies.

At other times, a larger spread of data is desirable or expected. For example, intelligence quotients (IQs) are expected to exhibit a considerable spread about the mean, since everyone is different. The following procedure explains how we determine the standard deviation of a set of data.

**Variance,* another measure of dispersion, is the square of the standard deviation.

[†]Our alphabet uses both uppercase and lowercase letters—for example, *A* and *a*. The Greek alphabet also uses both uppercase and lowercase letters. The symbol Σ is the capital Greek letter sigma, and σ is the lowercase Greek letter sigma.

To Find the Standard Deviation of a Set of Data:

1. Find the mean of the set of data.
2. Make a chart having three columns:

   Data   Data − Mean   (Data − Mean)^2

3. List the data vertically under the column marked Data.
4. Complete the Data − Mean column for each piece of data.
5. Square the values obtained in the Data − Mean column and record these values in the (Data − Mean)^2 column.
6. Find the sum of the values in the (Data − Mean)^2 column.
7. Divide the sum obtained in step 6 by $n - 1$, where n is the number of pieces of data.*
8. Find the square root of the number obtained in step 7. This number is the standard deviation of the set of data.

Example 2 illustrates the procedure to follow to find the standard deviation of a set of data.

EXAMPLE 2 *Find the Standard Deviation*

Find the standard deviation of

$$7, 9, 11, 15, 18$$

SOLUTION First, determine the mean:

$$\bar{x} = \frac{\Sigma x}{n} = \frac{7 + 9 + 11 + 15 + 18}{5} = \frac{60}{5} = 12$$

Next, construct a table with three columns, as illustrated in Table 13.4, and list the data in the first column (it is often helpful to list the data in ascending or descending order). Complete the second column by subtracting the mean, 12 in this case, from each piece of data in the first column.

Table 13.4

Data	Data − mean	(Data − mean)^2
7	$7 - 12 = -5$	
9	$9 - 12 = -3$	
11	$11 - 12 = -1$	
15	$15 - 12 = 3$	
18	$18 - 12 = 6$	
	0	

*To find the standard deviation of a sample, divide the sum of (Data − Mean)^2 column by $n - 1$. To find the standard deviation of a population, divide the sum by n. In this book we assume the set of data represents a sample and divide by $n - 1$. The quotient obtained in step 7 represents a measure of dispersion called the *variance*.

The sum of the values in the Data − Mean column should always be zero; if not, you have made an error. (If the mean is a decimal number, there may be a slight round-off error.)

Next square the values in the second column and place the squares in the third column (Table 13.5).

Table 13.5

Data	Data − mean	(Data − mean)^2
7	−5	$(-5)^2 = (-5)(-5) = 25$
9	−3	$(-3)^2 = (-3)(-3) = 9$
11	−1	$(-1)^2 = (-1)(-1) = 1$
15	3	$(3)^2 = (3)(3) = 9$
18	6	$(6)^2 = (6)(6) = 36$
	0	80

Add the squares in the third column. In this case the sum is 80. Divide this sum by one less than the number of pieces of data ($n - 1$). In this case the number of pieces of data is 5. Therefore, we divide by 4 and get

$$\frac{80}{4} = 20^*$$

Finally, take the square root of this number. The square root of 20 can be found by using a calculator. Since $\sqrt{20} \approx 4.5$, the standard deviation, symbolized s, is 4.5 (to the nearest tenth). ▲

Now we will develop a formula for finding the standard deviation of a set of data. If we call the individual data x and the mean $\bar{x}$, we could write the three column heads Data, Data − Mean, and (Data − Mean)^2 in Table 13.4 as

$$x \qquad x - \bar{x} \qquad (x - \bar{x})^2$$

Let's follow the procedure we used to obtain the standard deviation in Example 2. We found the sum of the (Data − Mean)^2 column, which is the same as the sum of the $(x - \bar{x})^2$ column. We can represent the sum of the $(x - \bar{x})^2$ column by using the summation notation, $\Sigma(x - \bar{x})^2$. Thus, in Table 13.4, $\Sigma(x - \bar{x})^2 = 80$. We then divided this number by 1 less than the number of pieces of data, $n - 1$. Thus, we have

$$\frac{\Sigma(x - \bar{x})^2}{n - 1}$$

Finally, we took the square root of this value to obtain the standard deviation.

Standard Deviation

$$s = \sqrt{\frac{\Sigma(x - \bar{x})^2}{n - 1}}$$

*20 is the variance, symbolized s^2, of this set of data.

EXAMPLE 3 ***Find the Standard Deviation of Stock Prices***

The following are the prices of nine stocks on the New York Stock Exchange. Find the standard deviation.

$15, $28, $32, $36, $50, $52, $68, $74, $104

SOLUTION The mean, $\bar{x}$, is

$$\bar{x} = \frac{\Sigma x}{n} = \frac{15 + 28 + 32 + 36 + 50 + 52 + 68 + 74 + 104}{9} = \frac{459}{9} = 51$$

The mean is $51.

Table 13.6

x	$x - \bar{x}$	$(x - \bar{x})^2$
15	−36	1296
28	−23	529
32	−19	361
36	−15	225
50	−1	1
52	1	1
68	17	289
74	23	529
104	53	2809
	0	6040

Table 13.6 shows us that $\Sigma(x - \bar{x})^2 = 6040$. Since there are nine pieces of data, $n - 1 = 9 - 1$, or 8.

$$s = \sqrt{\frac{\Sigma(x - \bar{x})^2}{n - 1}} = \sqrt{\frac{6040}{8}} = \sqrt{755} \approx 27.5$$

The standard deviation, to the nearest tenth, is 27.5. ▲

Standard deviation will be used in Section 13.7 to find the percent of data between any two values in a normal curve. Standard deviations are also often used in determining norms for a population (see Exercise 28).

SECTION 13.6 EXERCISES

CONCEPT/WRITING EXERCISES

1. Explain how to find the range of a set of data.
2. What does the standard deviation of a set of data measure?
3. Explain how to find the standard deviation of a set of data.
4. What is the standard deviation of a set of data in which all the data is the same? Explain.
5. Why is measuring variation (dispersion) in observed data important?
6. **a)** What symbol is used to represent the sample standard deviation?
   b) What symbol is used to represent the population standard deviation?
7. Can you think of any situations in which a large standard deviation may be desirable? Explain.
8. Can you think of any situations in which a small standard deviation may be desirable? Explain.
9. Without actually doing the calculations, decide which of the following two sets of data will have the greater

standard deviation: 13, 16, 17, 18, 20, 24, or 16, 17, 17, 18, 18, 19. Explain why.

10. Of the following two sets of data, which would you expect to have the larger standard deviation? Explain.

2, 4, 6, 8, 10 or 102, 104, 106, 108, 110

11. By studying the standard deviation formula, explain why the standard deviation of a set of data will always be greater than or equal to 0.

12. Patricia Burgess has two statistics classes, one in the morning and the other in the evening. On the midterm exam, the morning class had a mean of 75.2 and a standard deviation of 5.7. The evening class had a mean of 75.2 and a standard deviation of 12.5.
   a) How do the means compare?
   b) If we compare the set of scores from the first class with those in the second class, how will the distributions of the two sets of scores compare? Explain.

PRACTICE THE SKILLS

In Exercises 13–22, find the range and standard deviation of the set of data.

13. 5, 3, 0, 6, 11

14. 8, 8, 12, 14, 6, 6

15. 150, 151, 152, 153, 154, 155, 156

16. 4, 0, 3, 6, 9, 12, 2, 3, 4, 7

17. 4, 8, 9

18. 9, 9, 9, 9, 9, 9

19. 7, 9, 7, 9, 9, 10, 12

20. 52, 50, 54, 59, 40, 43, 64, 62

21. 3, 4, 5, 9, 3, 7, 4, 4, 9, 2

22. 103, 106, 109, 112, 115, 118, 121

23. *Computer Games* Find the range and standard deviation of the following prices of selected computer games: $28, $28, $50, $45, $30, $45, $48, $18, $45, $23.

24. *Textbook Prices* Nine students in Ms. Elko's chemistry class reported they spent the following amounts on a calculator: $44, $77, $80, $88, $55, $44, $43, $77, and $77. Find the range and standard deviation of the prices.

PROBLEM SOLVING

25. *Count Your Money* Six people were asked to determine the amount of money they were carrying, to the nearest dollar. The results were

$32, $60, $14, $25, $5, $68

   a) Determine the range and standard deviation of the amounts.
   b) Add $10 to each of the six amounts. How do you expect the range and standard deviation of the new set of data to change? Explain your answer.
   c) Determine the range and standard deviation of the new set of data. Do the results agree with your answer to part b)? If not, explain why.

26. a) *Adding to or Subtracting from Each Number* Pick any five numbers. Compute the mean and the standard deviation of this distribution.
   b) Add 20 to each of the numbers in your original distribution and compute the mean and the standard deviation of this new distribution.
   c) Subtract 5 from each number in your original distribution and compute the mean and standard deviation of this new distribution.
   d) What conclusions can you draw about changes in the mean and the standard deviation when the same number is added to or subtracted from each piece of data in a distribution?
   e) How will the mean and standard deviation of the numbers 6, 7, 8, 9, 10, 11, 12 differ from the mean and standard deviation of the numbers 596, 597, 598, 599, 600, 601, 602? Find the mean and standard deviation of both sets of numbers.

27. a) *Multiplying Each Number* Pick any five numbers. Compute the mean and standard deviation of this distribution.
   b) Multiply each number in your distribution by 4 and compute the mean and the standard deviation of this new distribution.
   c) Multiply each number in your original distribution by 9 and compute the mean and the standard deviation of this new distribution.
   d) What conclusions can you draw about changes in the mean and the standard deviation when each value in a distribution is multiplied by the same number?
   e) The mean and standard deviation of the distribution 1, 3, 4, 4, 5, 7 are 4 and 2, respectively. Use the conclusion drawn in part (d) to determine the mean and standard deviation of the distribution

5, 15, 20, 20, 25, 35

28. *Waiting in Line* Consider the following illustrations of two bank-customer waiting systems.

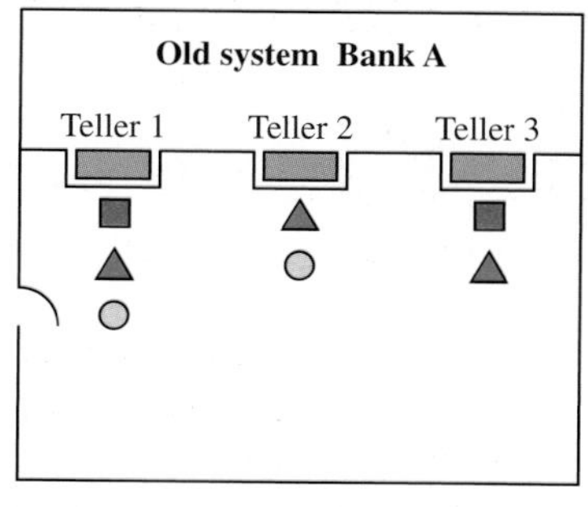

 Customers  Tellers

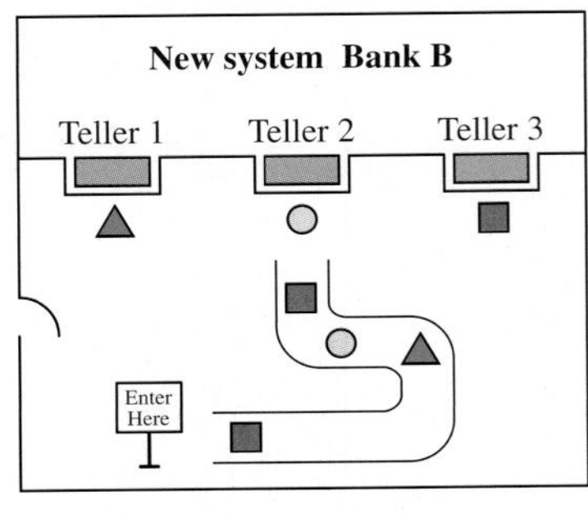

a) How would you expect the mean waiting time in Bank A to compare to the mean waiting time in Bank B? Explain your answer.

b) How would you expect the standard deviation of waiting times in Bank A to compare to the standard deviation of waiting times in Bank B? Explain your answer.

29. *Height and Weight Distribution* The chart shown below uses the symbol σ to represent the standard deviation. Note that 2σ represents the value that is two standard deviations above the mean; -2σ represents the value that is two standard deviations below the mean. The unshaded areas, from two standard deviations below the mean to two standard deviations above the mean, are considered the normal range. For example, the average (mean) 8-year-old boy has a height of about 50 inches. But any heights between approximately 45 inches and 55 inches are considered normal for 8-year-old boys. Refer to the chart below to answer the following questions.

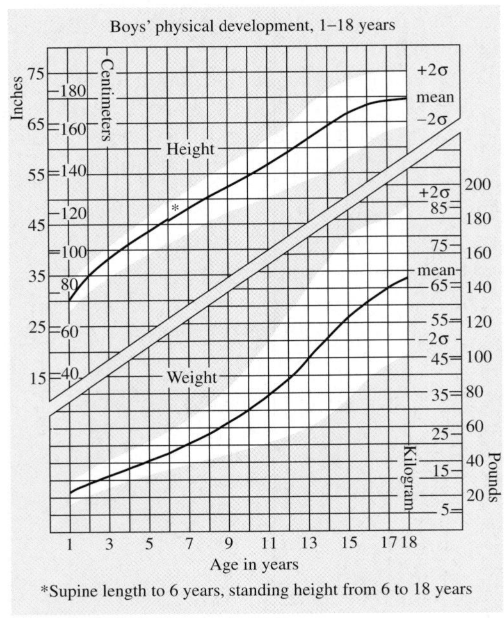

a) What happens to the standard deviation for weights of boys as the age of boys increases? What is the significance of this fact?

b) At age 16, what is the mean weight, in pounds, of boys?

c) What is the approximate standard deviation of boys' weights at age 16?

d) Find the mean weight and normal range for boys at age 13.

e) Find the mean height and normal range for boys at age 13.

f) Assuming this chart was constructed so that approximately 95% of all boys are always in the normal range, determine what percentage of boys are not in the normal range.

CHALLENGE PROBLEMS/GROUP ACTIVITIES

30. *Athletes' Salaries* The following table lists the 10 highest-paid athletes in Major League Baseball and the National Basketball Association.

Major League Baseball (1998 Season)

Player	Salary (millions of dollars)
1. Albert Belle	\$10
2. Gary Sheffield	\$10
3. Greg Maddux	\$9.6
4. Barry Bonds	\$8.9
5. Mark McGwire	\$8.3
6. Roger Clemens	\$8.25
Bernie Williams	\$8.25
8. Andres Galarraga	\$8
Mike Piazza	\$8
Sammy Sosa	\$8

Source: Major League Baseball Players Association.

National Basketball Association (1997–1998 Season)

Player	Salary (millions of dollars)
1. Michael Jordan	\$33.1
2. Patrick Ewing	\$20.5
3. Horace Grant	\$14.3
4. Shaquille O'Neal	\$12.9
5. David Robinson	\$12.4
6. Alonzo Mourning	\$11.26
7. Juwan Howard	\$11.25
8. Hakeem Olajuwon	\$11.16
9. Gary Payton	\$10.5
10. Dikembe Mutumbo	\$9.6

Source: National Basketball Association.

a) Without doing any calculations, do you believe the mean salary of the 10 baseball players or the mean salary of the 10 basketball players is greater? Explain.

b) Without doing any calculations, do you believe the standard deviation of the salary of the 10 baseball players or the standard deviation of the salary of the 10 basketball players is greater? Explain.

c) Compute the mean salary of the 10 baseball players and the mean salary of the 10 basketball players and determine whether your answer in part (a) is correct.

d) Compute the standard deviation of the salary of the 10 baseball players and the standard deviation of the salary of the 10 basketball players and determine whether your answer in part (b) is correct. Round each mean to the nearest tenth to determine the standard deviation.

31. *Selling Shoes* Fred's Footware has franchises in two different parts of the city. The number of footware sold weekly, for 25 weeks, is given below.

East store					West store				
33	59	27	30	42	38	46	38	38	30
19	42	25	22	32	38	38	37	39	31
43	27	57	37	52	39	36	40	37	47
40	67	38	44	43	30	34	42	45	29
15	31	49	41	35	31	46	28	45	48

a) Construct a frequency distribution for each store with a first class of 15–20.

b) Draw a histogram for each store.

c) Using the histogram, determine which store appears to have a greater mean, or do the means appear about the same? Explain.

d) Using the histogram, determine which store appears to have the greater standard deviation? Explain.

e) Calculate the mean for each store and determine whether your answer in part (c) was correct.

f) Calculate the standard deviation for each store and determine whether your answer in part (d) was correct.

RESEARCH PROBLEM

32. Use a calculator with statistical function keys to find the mean and standard deviation of the salaries of the 10 Major League Baseball players and the 10 National Basketball Association players in Exercise 30.

13.7 THE NORMAL CURVE

When examining data using a histogram, we can refer to the overall appearance of the histogram as the *shape* of the distribution of the data. Certain shapes of distributions of data are more common than others. In this section, we will illustrate and discuss a few of the more common ones. In each case the vertical scale is the frequency and the horizontal scale is the observed values.

In a **rectangular distribution** (Fig. 13.15), all the observed values occur with the same frequency. If a die is rolled many times, we would expect the numbers 1–6 to occur with about the same frequency. The distribution representing the outcomes of the die is rectangular.

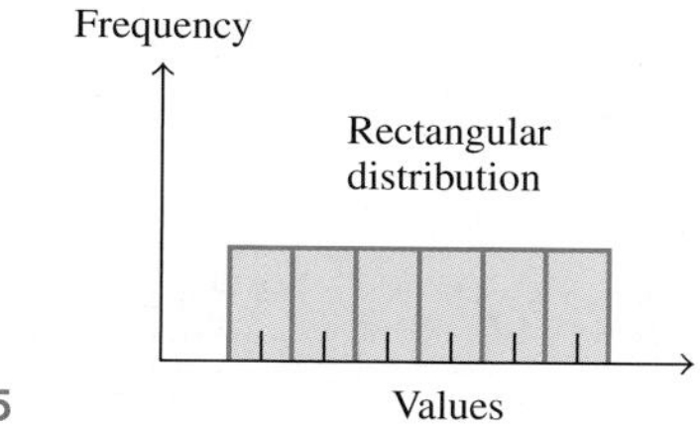

Figure 13.15

In **J-shaped distributions**, the frequency is either constantly increasing (Fig. 13.16a) or constantly decreasing (Fig. 13.16b). The number of hours studied per week by students may have a distribution like that in Fig. 13.16(b). The bars might represent (from left to right) 0–5, 6–10, 11–15 hours, and so on.

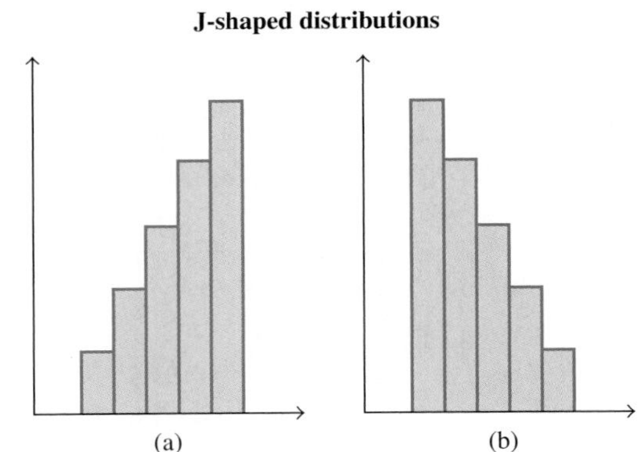

Figure 13.16

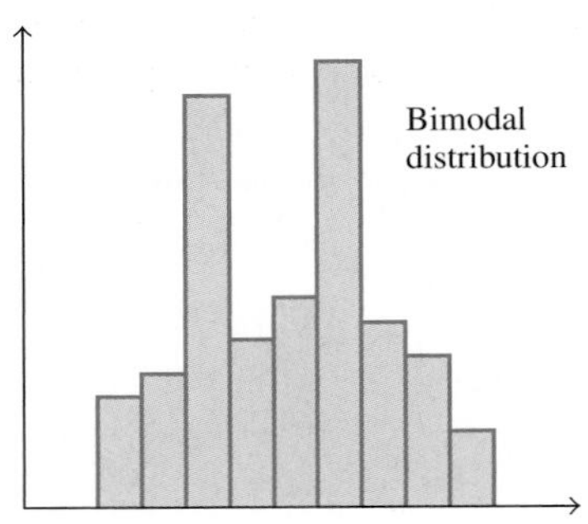

Figure 13.17

A **bimodal distribution** (Fig. 13.17) is one in which two nonadjacent values occur more frequently than any other values in a set of data. For example, if an equal number of men and women were weighed, the distribution of their weights would probably be bimodal, with one mode for the women's weights and the second for the men's weights.

The life expectancy of light bulbs has a bimodal distribution: a small peak very near 0 hours of life, resulting from the bulbs that burned out very quickly because of a manufacturing defect, and a much broader peak representing the nondefective bulbs. A bimodal frequency distribution generally means that you are dealing with two distinct populations, in this case, defective and nondefective bulbs.

Another distribution, called a **skewed distribution**, has more of a "tail" on one side than the other. A skewed distribution with a tail on the right (Fig. 13.18a) is said to be skewed to the right. If the tail is on the left (Fig. 13.18b), the distribution is referred to as skewed to the left.

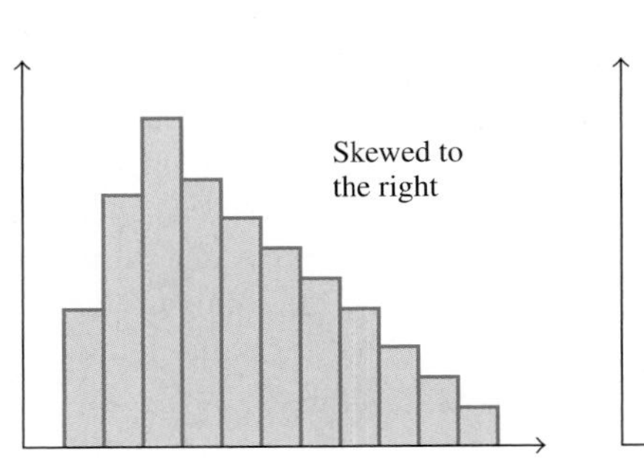

Skewed to
the left

(b)

Figure 13.18

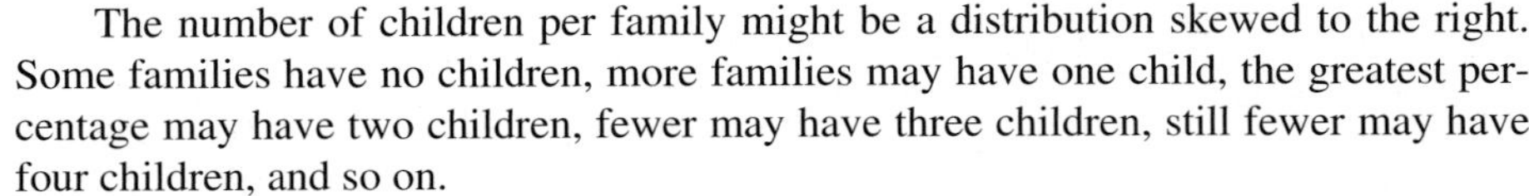

The number of children per family might be a distribution skewed to the right. Some families have no children, more families may have one child, the greatest percentage may have two children, fewer may have three children, still fewer may have four children, and so on.

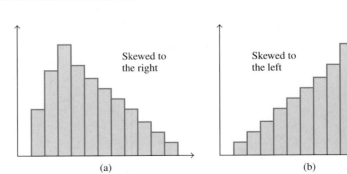

Since few families have high incomes, distributions of family incomes might be skewed to the right.

Smoothing the histograms of the skewed distributions shown in Fig. 13.18 to form curves gives the curves illustrated in Fig. 13.19.

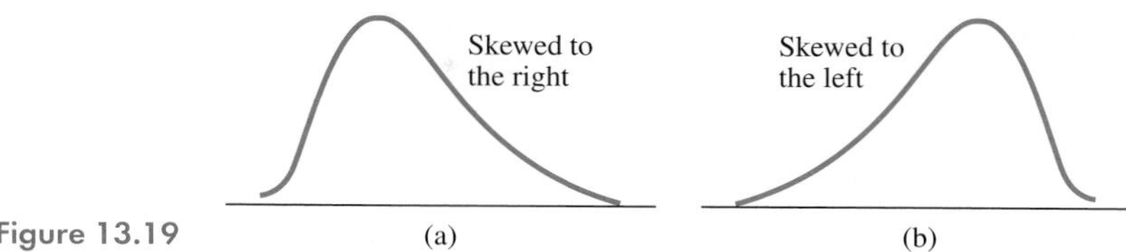

Figure 13.19

In Fig. 13.19(a), the greatest frequency appears on the left side of the curve, and the frequency decreases from left to right. Since the mode is the value with the greatest frequency, the mode would appear on the left side of the curve.

Every value in the set of data is considered in determining the mean. The values on the far right side of the curve in Fig. 13.19(a) would tend to increase the value of the mean. Thus, the value of the mean would be farther to the right than the mode. The median would be between the mode and the mean. The relationship between the mean, median, and mode for curves that are skewed to the right and left is given in Fig. 13.20.

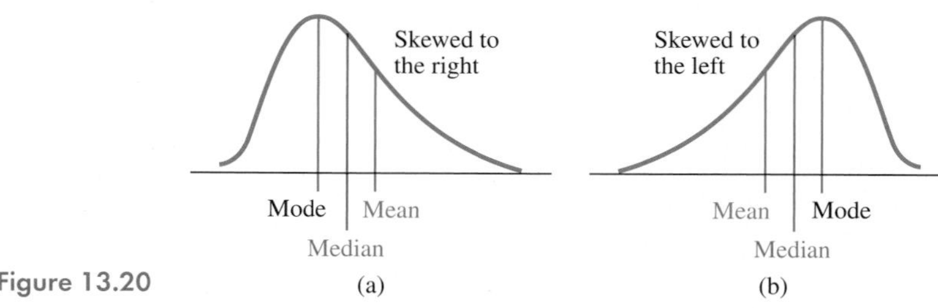

Figure 13.20

Each of these distributions is useful in describing sets of data. However, the most important distribution is the **normal** or **Gaussian distribution**, named for the German mathematician Carl Friedrich Gauss. The histogram of a normal distribution is illustrated in Fig. 13.21.

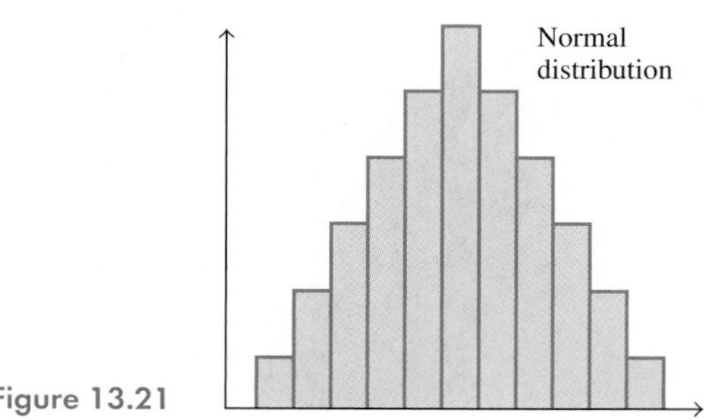

Figure 13.21

The normal distribution is important because many sets of data are normally distributed, or they closely resemble a normal distribution. Such distributions include intelligence quotients, heights and weights of males, heights and weights of females,

DID YOU KNOW

What Conclusions Can You Draw?

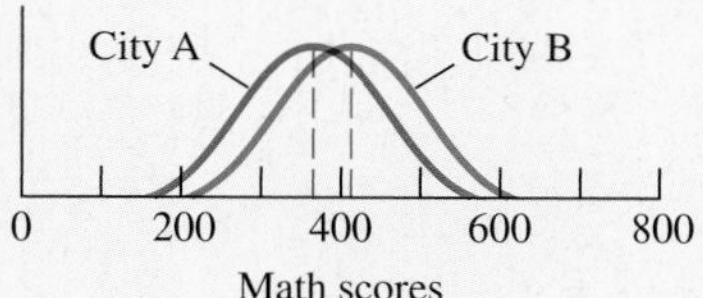

Based on the figure, which shows the distribution of scores on the mathematics part of the SAT test for two different cities, can we say that any given person selected at random from city B has outperformed any given person selected at random from city A? Both distributions appear normal, and the mean of city A is slightly smaller than the mean of city B. But consider two randomly selected students who took this test: Sally from city A, and Kendra, from city B. The graph shows that many students in city A outperformed students from city B, so we cannot conclude that Sally scored higher than Kendra or that Kendra scored higher than Sally.

◆ PROFILE IN MATHEMATICS

David Blackwell (1919–)

David H. Blackwell, (1919–), Professor of Statistics, is the author of more than 90 publications on statistics, probability, game theory, set theory, dynamic programming, and information theory. Dr. Blackwell, past president of the American Statistical Society, is the first African-American elected to the National Academy of Sciences.

His first interests in mathematics were in Geometry as a schoolboy growing up in southern Illinois. When he enrolled at the University of Illinois at the age of 16, he planned to be an elementary school teacher. When he received his Ph.D. in mathematics in 1941, he was just the sixth African-American to receive a doctorate in mathematics in the country.

After spending 2 years as a postdoctoral fellow at the Institute for Advanced Study at Princeton, he taught mathematics for 10 years at Howard University. While at Howard, Blackwell built a strong national reputation as a gifted teacher and creative researcher.

In 1954, he joined the Department of Statistics at the University of California, Berkeley. Dr. Blackwell, who has taught a wide variety of mathematics courses, says "Basically, I'm not interested in doing research and I never have been. I'm interested in understanding, which is quite a different thing."

lengths of full-grown boa constrictors, weights of watermelons, wearout mileage of automobile brakes, and life spans of refrigerators—to name just a few.

The normal distribution is symmetric about the mean. If you were to fold the histogram down the middle, the left side would fit the right side exactly. **In a normal distribution, the mean, median, and mode all have the same value.**

When the histogram of a normal distribution is smoothed to form a curve, the curve is bell-shaped. The bell may be high and narrow or short and wide. Each of the three curves in Fig. 13.22 represents a normal curve. Curve 13.22(a) has the smallest standard deviation (spread from the mean); curve 13.22(c) has the largest.

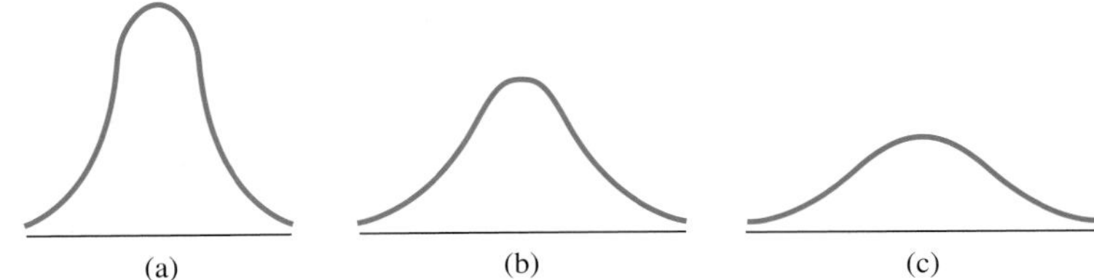

Figure 13.22

Since the curve is symmetric, 50% of the data always falls above (to the right of) the mean and 50% of the data falls below (to the left of) the mean. In addition, every normal distribution has approximately 68% of the data between the value that is one standard deviation below the mean, $\bar{x} - 1s$, and the value that is one standard deviation above the mean, $\bar{x} + 1s$, see Fig. 13.23. Approximately 95% of the data falls between the value that is two standard deviations below the mean, $\bar{x} - 2s$, and the value that is two standard deviations above the mean, $\bar{x} + 2s$.

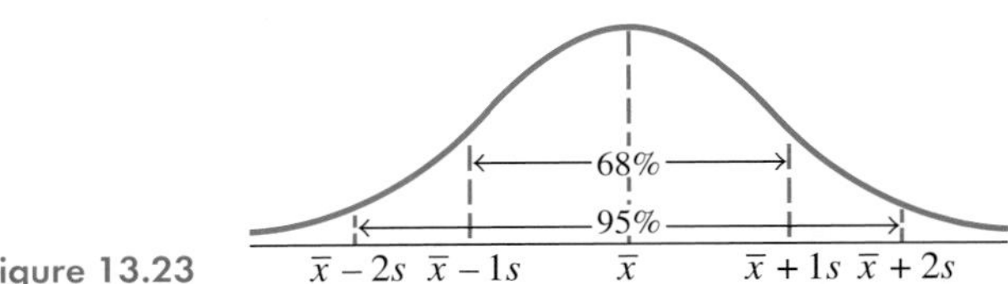

Figure 13.23

Thus, if a normal distribution has a mean of 100 and a standard deviation of 10, then approximately 68% of all the data falls between 100 − 10 and 100 + 10, or between 90 and 110. Approximately 95% of the data falls between 100 − 20 and 100 + 20, or between 80 and 120. In fact, given any normal distribution with a known standard deviation and mean, it is possible through the use of Table 13.6 on page 706 (the z-table) to determine the percent of data between any two given values.

We use **z-scores** (or **standard scores**) to determine how far, in terms of standard deviations, a given score is from the mean of the distribution. For example, a score that has a z-value of 1.5 means the score is 1.5 standard deviations above the mean. The standard or z-score is calculated as follows.

$$z = \frac{\text{value of the piece of data} - \text{mean}}{\text{standard deviation}}$$

Letting x represent the value of the given piece of data, $\bar{x}$ the mean, and s the standard deviation, we can symbolize the formula as follows.

$$z = \frac{x - \bar{x}}{s}$$

In this book the notation z_x represents the z-score, or standard score, of the value x. For example, if a normal distribution has a mean of 86 with a standard deviation of 12, a score of 110 has a standard or z-score of

$$z_{110} = \frac{110 - 86}{12} = \frac{24}{12} = 2$$

Therefore, a value of 110 in this distribution has a z-score of 2. The score of 110 is two standard deviations above the mean.

Data below the mean will always have negative z-scores; data above the mean will always have positive z-scores. The mean will always have a z-score of 0.

EXAMPLE 1 *Finding z-Scores*

A normal distribution has a mean of 100 and a standard deviation of 10. Find z-scores for the following values.

a) 110 b) 115 c) 100 d) 82

SOLUTION

a)

$$z = \frac{\text{value} - \text{mean}}{\text{standard deviation}}$$

$$z_{110} = \frac{110 - 100}{10} = \frac{10}{10} = 1$$

A score of 110 is one standard deviation above the mean.

b)

$$z_{115} = \frac{115 - 100}{10} = \frac{15}{10} = 1.5$$

A score of 115 is 1.5 standard deviations above the mean.

c)

$$z_{100} = \frac{100 - 100}{10} = \frac{0}{10} = 0$$

The mean always has a z-score of 0.

d)

$$z_{82} = \frac{82 - 100}{10} = \frac{-18}{10} = -1.8$$

A score of 82 is 1.8 standard deviations below the mean. ▲

Let's now consider finding areas under the normal curve. The total area under the normal curve is 1.00. Table 13.7 on page 706 will be used to determine the area under the normal curve between any two given points (the values in the table have been rounded). **Table 13.7 gives the area under the normal curve from the mean (a z-value of 0) to a z-value to the right of the mean.**

For example, between the mean and $z = 2.00$, the table shows a value of .477. Thus, there is 0.477 of the total area under the curve between the mean and $z = 2.00$, see Fig. 13.24 on page 706. To change this area of 0.477 to a percent, simply multiply by 100%: $0.477 \times 100\%$ is 47.7%. Thus, 47.7% of all scores will be between the mean and the score that is two standard deviations above the mean.

Table 13.7 Areas under the Standard Normal Curve (the z-table)

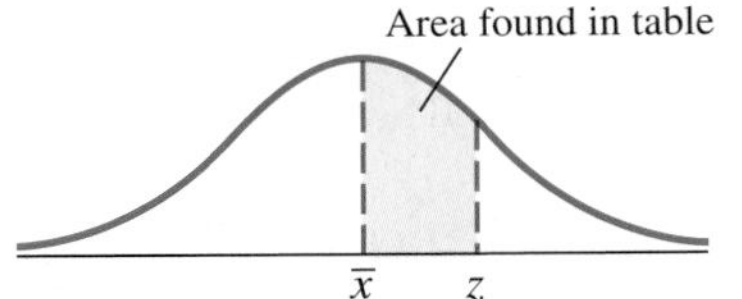

The column under A gives the area under the entire curve that is between $z = 0$ (or the mean) and a positive value of z.

z	A	z	A	z	A	z	A	z	A	z	A	z	A	z	A	z	A
.00	.000	.37	.144	.74	.270	1.11	.367	1.48	.431	1.85	.468	2.22	.487	2.59	.495	2.96	.499
.01	.004	.38	.148	.75	.273	1.12	.369	1.49	.432	1.86	.469	2.23	.487	2.60	.495	2.97	.499
.02	.008	.39	.152	.76	.276	1.13	.371	1.50	.433	1.87	.469	2.24	.488	2.61	.496	2.98	.499
.03	.012	.40	.155	.77	.279	1.14	.373	1.51	.435	1.88	.470	2.25	.488	2.62	.496	2.99	.499
.04	.016	.41	.159	.78	.282	1.15	.375	1.52	.436	1.89	.471	2.26	.488	2.63	.496	3.00	.499
.05	.020	.42	.163	.79	.285	1.16	.377	1.53	.437	1.90	.471	2.27	.488	2.64	.496	3.01	.499
.06	.024	.43	.166	.80	.288	1.17	.379	1.54	.438	1.91	.472	2.28	.489	2.65	.496	3.02	.499
.07	.028	.44	.170	.81	.291	1.18	.381	1.55	.439	1.92	.473	2.29	.489	2.66	.496	3.03	.499
.08	.032	.45	.174	.82	.294	1.19	.383	1.56	.441	1.93	.473	2.30	.489	2.67	.496	3.04	.499
.09	.036	.46	.177	.83	.297	1.20	.385	1.57	.442	1.94	.474	2.31	.490	2.68	.496	3.05	.499
.10	.040	.47	.181	.84	.300	1.21	.387	1.58	.443	1.95	.474	2.32	.490	2.69	.496	3.06	.499
.11	.044	.48	.184	.85	.302	1.22	.389	1.59	.444	1.96	.475	2.33	.490	2.70	.497	3.07	.499
.12	.048	.49	.188	.86	.305	1.23	.391	1.60	.445	1.97	.476	2.34	.490	2.71	.497	3.08	.499
.13	.052	.50	.192	.87	.308	1.24	.393	1.61	.446	1.98	.476	2.35	.491	2.72	.497	3.09	.499
.14	.056	.51	.195	.88	.311	1.25	.394	1.62	.447	1.99	.477	2.36	.491	2.73	.497	3.10	.499
.15	.060	.52	.199	.89	.313	1.26	.396	1.63	.449	2.00	.477	2.37	.491	2.74	.497	3.11	.499
.16	.064	.53	.202	.90	.316	1.27	.398	1.64	.450	2.01	.478	2.38	.491	2.75	.497	3.12	.499
.17	.068	.54	.205	.91	.319	1.28	.400	1.65	.451	2.02	.478	2.39	.492	2.76	.497	3.13	.499
.18	.071	.55	.209	.92	.321	1.29	.402	1.66	.452	2.03	.479	2.40	.492	2.77	.497	3.14	.499
.19	.075	.56	.212	.93	.324	1.30	.403	1.67	.453	2.04	.479	2.41	.492	2.78	.497	3.15	.499
.20	.079	.57	.216	.94	.326	1.31	.405	1.68	.454	2.05	.480	2.42	.492	2.79	.497	3.16	.499
.21	.083	.58	.219	.95	.329	1.32	.407	1.69	.455	2.06	.480	2.43	.493	2.80	.497	3.17	.499
.22	.087	.59	.222	.96	.332	1.33	.408	1.70	.455	2.07	.481	2.44	.493	2.81	.498	3.18	.499
.23	.091	.60	.226	.97	.334	1.34	.410	1.71	.456	2.08	.481	2.45	.493	2.82	.498	3.19	.499
.24	.095	.61	.229	.98	.337	1.35	.412	1.72	.457	2.09	.482	2.46	.493	2.83	.498	3.20	.499
.25	.099	.62	.232	.99	.339	1.36	.413	1.73	.458	2.10	.482	2.47	.493	2.84	.498	3.21	.499
.26	.103	.63	.236	1.00	.341	1.37	.415	1.74	.459	2.11	.483	2.48	.493	2.85	.498	3.22	.499
.27	.106	.64	.239	1.01	.344	1.38	.416	1.75	.460	2.12	.483	2.49	.494	2.86	.498	3.23	.499
.28	.110	.65	.242	1.02	.346	1.39	.418	1.76	.461	2.13	.483	2.50	.494	2.87	.498	3.24	.499
.29	.114	.66	.245	1.03	.349	1.40	.419	1.77	.462	2.14	.484	2.51	.494	2.88	.498	3.25	.499
.30	.118	.67	.249	1.04	.351	1.41	.421	1.78	.463	2.15	.484	2.52	.494	2.89	.498	3.26	.499
.31	.122	.68	.252	1.05	.353	1.42	.422	1.79	.463	2.16	.485	2.53	.494	2.90	.498	3.27	.500
.32	.126	.69	.255	1.06	.355	1.43	.424	1.80	.464	2.17	.485	2.54	.495	2.91	.498	3.28	.500
.33	.129	.70	.258	1.07	.358	1.44	.425	1.81	.465	2.18	.485	2.55	.495	2.92	.498	3.29	.500
.34	.133	.71	.261	1.08	.360	1.45	.427	1.82	.466	2.19	.486	2.56	.495	2.93	.498	3.30	.500
.35	.137	.72	.264	1.09	.362	1.46	.428	1.83	.466	2.20	.486	2.57	.495	2.94	.498	3.31	.500
.36	.141	.73	.267	1.10	.364	1.47	.429	1.84	.467	2.21	.487	2.58	.495	2.95	.498	3.32	.500

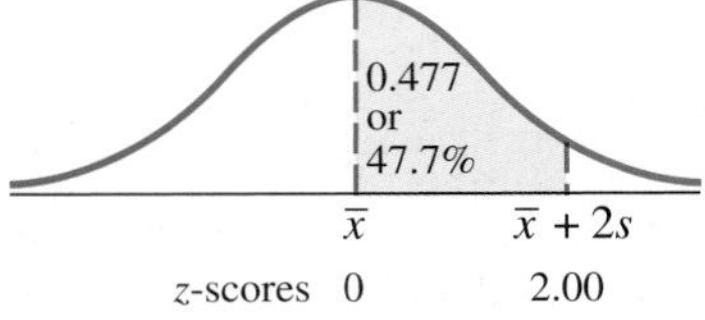

Figure 13.24

When you are finding the area under the normal curve, it is often helpful to draw a picture such as the one in Fig. 13.24, indicating the area or percent to be found.

The normal curve is symmetric about the mean. Thus, the same percent of data is between the mean and a positive z-score as between the mean and the corresponding negative z-score. For example, there is the same area under the normal curve between

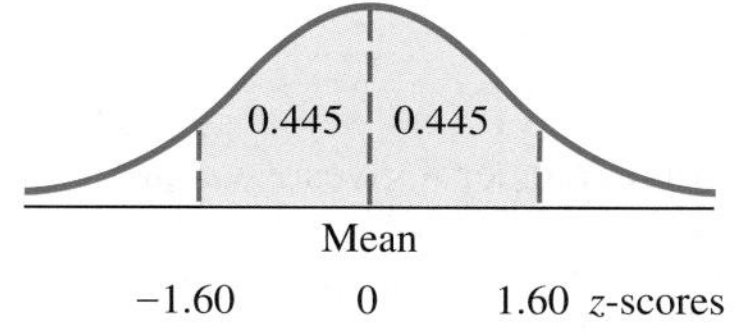

Figure 13.25

a z of 1.60 and the mean as between a z of -1.60 and the mean. Both have an area of 0.445 (Fig. 13.25).

You now have the necessary knowledge to find the percent of data between any two values in a normal distribution.

To Find the Percent of Data Between any Two Values:

1. Draw a diagram of the normal curve, indicating the area or percent to be determined.
2. Use the formula $z = (x - \bar{x})/s$ to convert the given values to z-scores. Indicate these z-scores on the diagram.
3. Look up the percent that corresponds to each z-score in Table 13.7.
4. **a)** When finding the percent of data between two z-scores on the opposite side of the mean (when one z-score is positive and the other is negative), you find the sum of the individual percents.
   b) When finding the percent of data between two z-scores on the same side of the mean (when both z-scores are positive or both are negative), subtract the smaller percent from the larger percent.

EXAMPLE 2 *IQ Scores*

Intelligence quotients are normally distributed with a mean of 100 and a standard deviation of 15. Find the percent of individuals with IQs in the following ranges.

a) Between 100 and 115
b) Between 70 and 100
c) Between 70 and 115
d) Between 115 and 130
e) Below 130
f) Above 122.5

SOLUTION

a) We want to find the area under the normal curve between the values of 100 and 115, as illustrated in Fig. 13.26(a). Converting 100 to a z-score yields a z-score of 0.

$$z_{100} = \frac{100 - 100}{15} = \frac{0}{15} = 0$$

Converting 115 to a z-score yields a z-score of 1.00.

$$z_{115} = \frac{115 - 100}{15} = \frac{15}{15} = 1.00$$

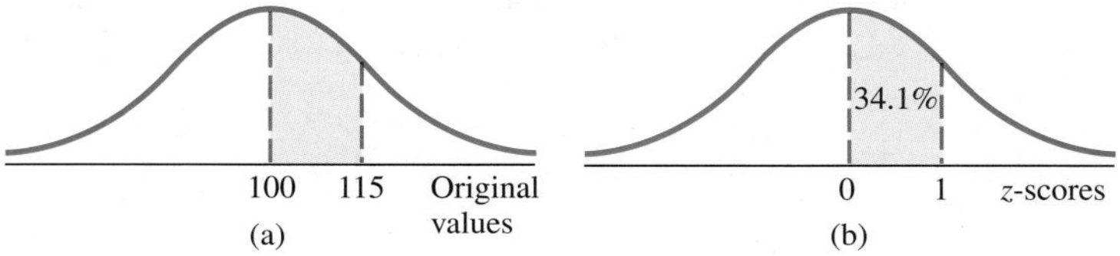

100 115 Original values
(a)

34.1%
0 1 z-scores
(b)

Figure 13.26

The percent of individuals with IQs between 100 and 115 is the same as the percent of data between z-scores of 0 and 1 (Fig. 13.26b).

From Table 13.7, we determine 0.341 of the area, or 34.1% of all the data, is between z-scores of 0 and 1.00. Therefore, 34.1% of individuals have IQs between 100 and 115.

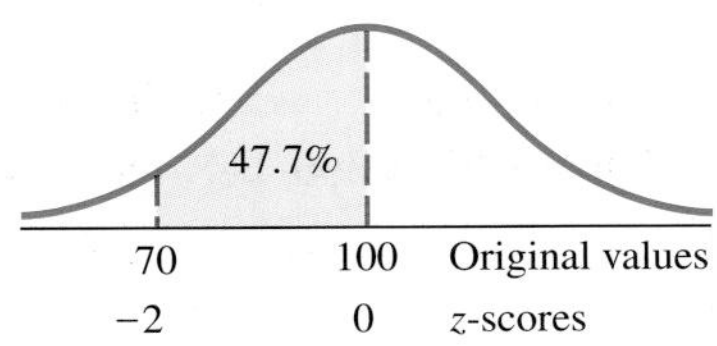

Figure 13.27

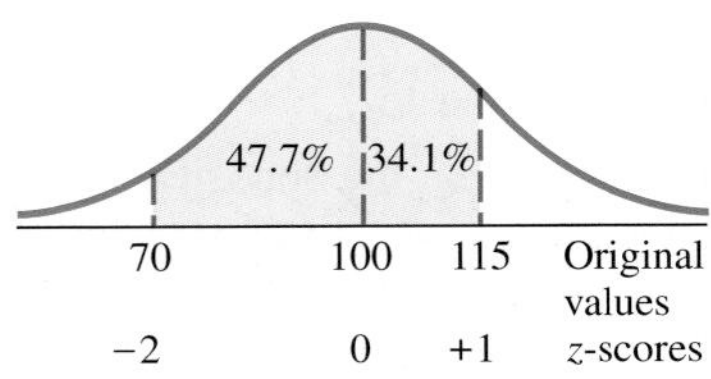

Figure 13.28

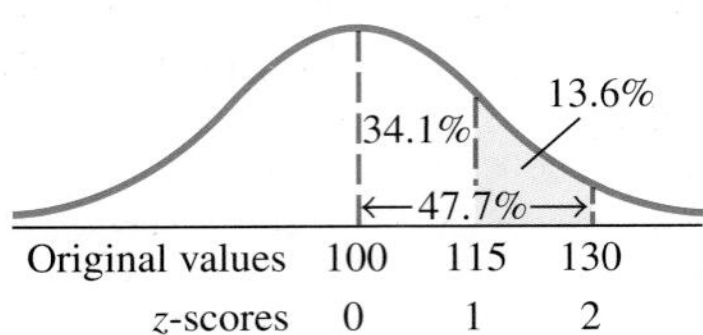

Figure 13.29

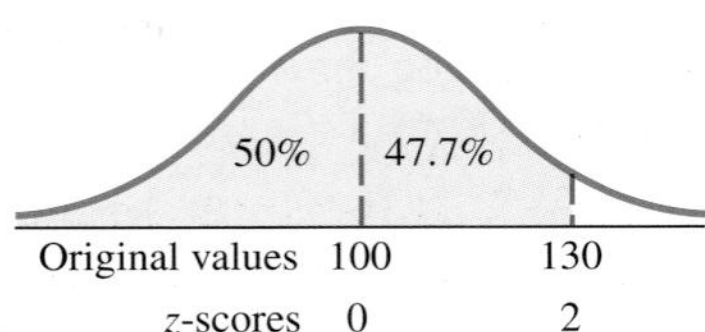

Figure 13.30

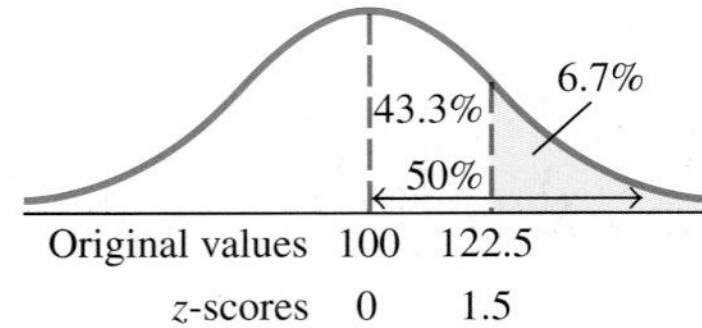

Figure 13.31

b)

$$z_{70} = \frac{70 - 100}{15} = \frac{-30}{15} = -2.00$$

$$z_{100} = 0 \text{ (from part a)}$$

The percent of data between scores of 70 and 100 is the same as the percent between $z = -2$ and $z = 0$ (Fig. 13.27). The percent of data between the mean and two standard deviations below the mean is the same as between the mean and two standard deviations above the mean. From Table 13.7, we determine that 47.7% of the data is between $z = 0$ and $z = 2$. Thus, 47.7% of the data is also between $z = -2$ and $z = 0$. Therefore, 47.7% of all individuals have IQs between 70 and 100.

c) In parts (a) and (b) we determined that $z_{115} = 1.00$ and $z_{70} = -2.00$. Since the values are on opposite sides of the mean, the percent of data between the two values is found by adding the individual percents: 34.1% + 47.7% = 81.8% (Fig. 13.28). Thus, 81.8% of the IQs are between 70 and 115.

d)

$$z_{130} = \frac{130 - 100}{15} = \frac{30}{15} = 2.00$$

$$z_{115} = 1.00 \text{ (from part a)}$$

Since both values are on the same side of the mean (Fig. 13.29), the smaller percent must be subtracted from the larger percent to obtain the percent of data in the shaded area: 47.7% − 34.1% is 13.6%. Thus, 13.6% of all the individuals have IQs between 115 and 130.

e) The percent of IQs below 130 is the same as the percent of data below a z-score of 2. Between $z = 2$ and the mean is 47.7% of the data (Fig. 13.30). To this 47.7%, we add the 50% of the data below the mean to give 97.7%. Thus, 50% +47.7%, or 97.7%, of all IQs are below 130.

f)

$$z_{122.5} = \frac{122.5 - 100}{15} = \frac{22.5}{15} = 1.50$$

The percent of IQs above 122.5 is the same as the percent of data above $z = 1.5$ (Fig. 13.31). Fifty percent of the data is to the right of the mean. Since 43.3% of the data is between the mean and $z = 1.5$, 50% − 43.3%, or 6.7%, of the data is greater than $z = 1.5$. Thus, 6.7% of all IQs are greater than 122.5. ▲

EXAMPLE 3 *Hours Worked by College Students*

Assume the number of hours college students spend working per week is normally distributed with a mean of 18 hours and standard deviation of 4 hours.

a) Find the percent of college students that work at least 18 hours per week.
b) Find the percent of college students that work between 14 and 26 hours per week.
c) Find the percent of college students that work at least 23 hours per week.
d) Find the percent of college students that work less than 11 hours per week.
e) In a random sample of 500 college students, how many work at least 23 hours per week?

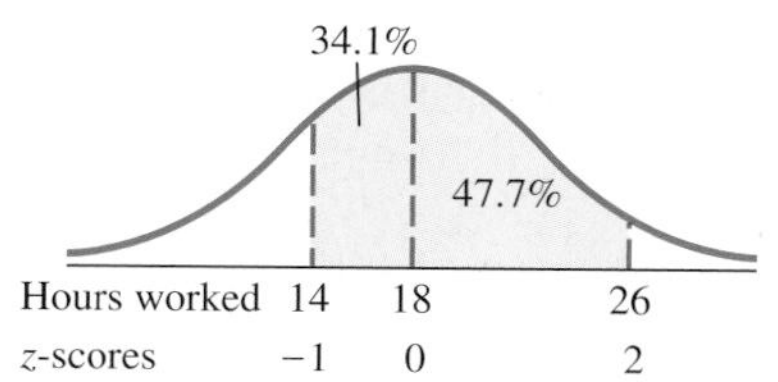

Figure 13.32

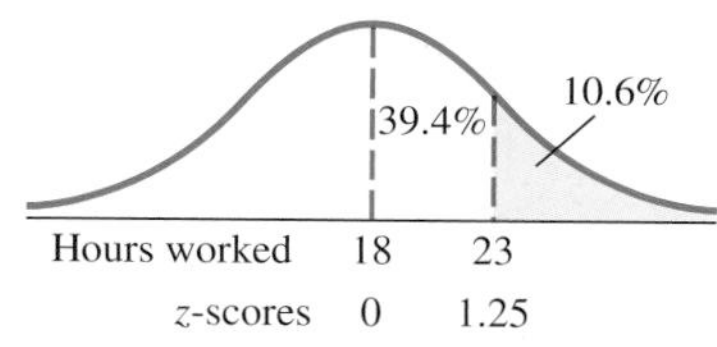

Figure 13.33

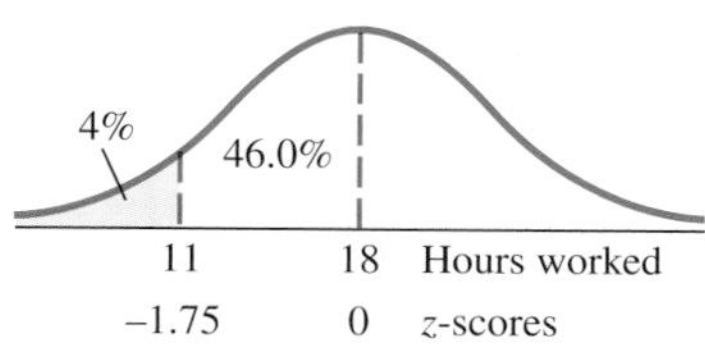

Figure 13.34

SOLUTION

a) In a normal distribution, half the data are always above the mean. Since 18 hours is the mean, half, or 50%, of college students work at least 18 hours per week.

b) Convert 14 hours and 26 hours to z-scores.

$$z_{14} = \frac{14 - 18}{4} = -1.00$$

$$z_{26} = \frac{26 - 18}{4} = 2.00$$

Now look up the areas in Table 13.7. The percent of college students that work between 14 and 26 hours per week is 34.1% + 47.7% or 81.8% (Fig. 13.32).

c)
$$z_{23} = \frac{23 - 18}{4} = 1.25$$

Figure 13.33 shows 39.4% of the data is between the mean and $z = 1.25$. Therefore, the percent of data above $z = 1.25$ is 50% − 39.4% = 10.6%. Thus, 10.6% of college students work at least 23 hours per week.

d)
$$z_{11} = \frac{11 - 18}{4} = -1.75$$

Figure 13.34 shows that 46.0% of the data is between the mean and $z = -1.75$. The percent of data to the left of $z = -1.75$ is found by subtracting 46% from 50% to obtain 4%. Thus, 4% of college students work less than 11 hours per week.

e) In part (c), we determined that 10.6% of all college students work at least 23 hours per week. We now multiply 0.106 times 500 to determine the number of students who work at least 23 hours per week. There are $0.106 \times 500 = 53$ students who work at least 23 hours per week.

SECTION 13.7 EXERCISES

CONCEPT/WRITING EXERCISES

In Exercises 1–6, describe

1. a rectangular distribution.
2. a J-shaped distribution.
3. a bimodal distribution.
4. a distribution that is skewed to the right.
5. a distribution that is skewed to the left.
6. a normal distribution.

In Exercises 7–10, give an example of the type of distribution.

7. rectangular
8. J-shaped
9. skewed
10. bimodal

For the distributions in Exercises 11–14, state whether you think the distribution would be normal, J-shaped, bimodal, rectangular, skewed left, or skewed right. Explain your answers.

11. the numbers resulting from tossing a die many times.
12. salaries of teachers at Roosevelt High School where there are many newly hired teachers.

13. the life expectancy of a sample of microwaves.

14. the heights of a sample of high school seniors, where there are an equal number of males and females.
15. In a distribution that is skewed to the right, which has the greatest value—the mean, median, or mode? Which has the smallest value? Explain.
16. In a distribution skewed to the left, which has the greatest value—the mean, median, or mode? Which has the smallest value? Explain.
17. List three populations other than those given in the text that may be normally distributed.
18. List three populations other than those given in the text that may not be normally distributed.
19. In a normal distribution, what is the relationship between the mean, median, and mode?
20. What does the z, or standard score, measure?
21. When will a z-score be negative?
22. Explain in your own words how to find the z-score.
23. What is the value of the z-score of the mean of a set of data?
24. In a normal distribution, approximately what percent of the data is between
    a) one standard deviation below the mean to one standard deviation above the mean?
    b) two standard deviations below the mean to two standard deviations above the mean?

PRACTICE THE SKILLS

In Exercises 25–36, use Table 13.7 to find the specified area.

25. Above the mean
26. Below the mean
27. Between two standard deviations below the mean and one standard deviation above the mean
28. Between 1.10 and 1.80 standard deviations above the mean
29. To the right of $z = 1.73$
30. To the left of $z = 1.16$
31. To the left of $z = -1.78$
32. To the right of $z = -1.78$
33. To the right of $z = 2.08$
34. To the left of $z = 1.96$
35. To the left of $z = -1.62$
36. To the left of $z = -0.70$

In Exercises 37–46, use Table 13.7 to find the percent of data specified.

37. Between $z = 0$ and $z = 0.78$
38. Between $z = -0.20$ and $z = -0.92$
39. Between $z = -1.34$ and $z = 2.24$
40. Less than $z = -1.90$
41. Greater than $z = -1.90$
42. Greater than $z = 2.66$
43. Less than $z = 1.96$
44. Between $z = 0.72$ and $z = 2.14$
45. Between $z = -1.53$ and $z = -1.82$
46. Between $z = -2.15$ and $z = 3.31$

PROBLEM SOLVING

English Exam Scores *In Exercises 47 and 48, suppose the results on an English exam are normally distributed. The z-scores for some students are shown below.*

Jake	1.3	Marie	0.0	Justin	−1.9	Kevin	0.0
Sarah	1.7	Omar	−2.1	Carol	0.8	Kim	−1.2

47. a) Which of these students scored above the mean?
    b) Which of these students scored at the mean?
    c) Which of these students scored below the mean?
48. a) Which student had the highest score?
    b) Which student had the lowest score?

Heights of 18-Year-Old Males *In Exercises 49–52, assume that the heights of 18-year-old-males are normally distributed with a mean of 69 in. and a standard deviation of 6 in.*

49. What percent of 18-year-old males are between 69 in. and 75 in. tall?
50. What percent of 18-year-old males are less than 75 in. tall?
51. What percent of 18-year-old males are between 63 in. and 75 in. tall?
52. If 1000 18-year-old males are selected at random, how many will be less than 72 in. tall?

Cellular Batteries *In Exercises 53–58, assume that Motorola has tested a new cellular battery. The number of hours that a newly charged battery remains charged, if on standby (no calls made or received), is normally distributed with a mean of 48 hours and a standard deviation of 4 hours. Find the percent of batteries that will remain charged the following number of hours.*

53. less than 50

54. greater than 54

55. between 50 and 54

56. less than 40

57. between 44 and 49

58. greater than 41

Tire Mileage *In Exercises 59–62, the wearout mileage of a certain tire is normally distributed with a mean of 35,000 miles and standard deviation of 2500 miles.*

59. Find the percent of tires that will last between 30,000 miles and 37,500 miles.

60. Find the percent of tires that will last at least 39,000 miles.

61. If the manufacturer guarantees the tires to last at least 30,000 miles, what percent of tires will fail to live up to the guarantee?

62. If 200,000 tires are produced, how many will last at least 39,000 miles?

Placement Exam *In Exercises 63–68, a placement exam is given to all entering students at a certain college. The scores are normally distributed with a mean of 80 and a standard deviation of 8.*

63. What percent of the students scored above 80?

64. What percent of the students scored between 74 and 86?

65. What percent of the students scored below 70?

66. What percent of the students scored above 92?

67. If 200 students are selected at random, how many will score below 70?

68. If 200 students are selected at random, how many will score above 92?

Vending Machine *In Exercises 69–72, a vending machine is designed to dispense a mean of 7.6 oz. of coffee into an 8-oz. cup. If the standard deviation of the amount of coffee dispensed is 0.4 oz. and the amount is normally distributed, find the percent of times the machine will*

69. dispense from 7.4 oz. to 7.7 oz.

70. dispense less than 7.0 oz.

71. dispense less than 7.7 oz.

72. result in the cup overflowing.

Light Bulbs *In Exercises 73–77, the life expectancy of nondefective GE light bulbs is normally distributed, with a mean life of 1500 hr and a standard deviation of 100 hr.*

73. Find the percent of bulbs that will last more than 1450 hr.

74. Find the percent of bulbs that last between 1400 hr and 1550 hr.

75. Find the percent of bulbs that last less than 1480 hr.

76. If 80,000 of these bulbs are produced, how many will last 1500 hr or more?

77. If 80,000 of these bulbs are produced, how many will last between 1400 hr and 1600 hr?

78. *Weight Loss* A weight-loss clinic guarantees that its new customers will lose at least 5 lb by the end of their first month of participation or their money will be refunded. If the loss of weight of customers at the end of their first month is normally distributed, with a mean of 6.7 lb and a standard deviation of 0.81 lb, find the percent of customers that will be able to claim a refund.

79. *Appliance Warranty* The warranty on the motor of a dishwasher is 8 yr. If the breakdown times of this motor are normally distributed, with a mean of 10.2 yr and a standard deviation of 1.8 yr, find the percent of motors that can be expected to require repair or replacement under warranty.

80. *Coffee Machine* A vending machine that dispenses coffee does not appear to be working correctly. The machine rarely gives the proper amount of coffee. Some of the time the cup is underfilled, and some of the time the cup overflows. Does this variation indicate that the mean number of ounces dispensed has to be adjusted or that the standard deviation of the amount of coffee dispensed by the machine is too large? Explain your answer.

81. *Grading on a Normal Curve* Mr. Boccolucci marks his class on a normal curve. Those with z-scores above 1.8 will receive an A, those between 1.8 and 1.1 will receive a B, those between 1.1 and -1.2 will receive a C, those between -1.2 and -1.9 will receive a D, and those under -1.9 will receive an F. Find the percent of grades that will be A, B, C, D, and F.

82. *Grading on a Normal Curve* Professor Malinowski marks his classes on the normal curve. His statistics class has a mean of 72, with a standard deviation of 8. He has decided that 10% of the class will get an A, 20% will get a B, 40% will get a C, 20% will get a D, and 10% will get an F. Find

a) the minimum grade needed to get an A.

b) the minimum grade needed to pass the course (D or better).

c) the range of grades that will result in a C.

CHALLENGE PROBLEMS/GROUP ACTIVITIES

83. *Salesperson Promotion* The owner at Kim's Home Interiors is reviewing the sales records of two managers who are up for promotion, Katie and Stella, who work in different stores. At Katie's store, the mean sales have been \$23,200 per month, with a standard deviation of \$2170. At Stella's store, the mean sales have been \$25,600 per month, with a standard deviation of \$2300. Last month, Katie's store sales were \$28,408 and Stella's store sales were \$29,510. At both stores, the distribution of monthly sales is normal.
   a) Convert last month's sales for Katie's store and for Stella's store to z-scores.
   b) If one of the two was to be promoted based solely on the increase in sales last month, who should be promoted? Explain.

84. *Chebyshev's Theorem* How can you determine whether a distribution is approximately normal? A statistical theorem called **Chebyshev's theorem** states that the minimum percent of data between plus and minus K standard deviations from the mean ($K > 1$) in *any distribution* can be found by the formula

$$\text{Minimum percent} = 1 - \frac{1}{K^2}$$

Thus, for example, between ± 2 standard deviations from the mean, there will always be a minimum of 75% of data. This minimum percent is true for any distribution. For $K = 2$,

$$\text{Minimum percent} = 1 - \frac{1}{2^2}$$
$$= 1 - \frac{1}{4} = \frac{3}{4}, \quad \text{or} \quad 75\%$$

Likewise, between ± 3 standard deviations from the mean, there will always be a minimum of 89% of the data. For $K = 3$,

$$\text{Minimum percent} = 1 - \frac{1}{3^2}$$
$$= 1 - \frac{1}{9} = \frac{8}{9}, \quad \text{or} \quad 89\%$$

The following table lists the minimum percent of data in *any distribution* and the actual percent of data in *the normal distribution* between ± 1.1, ± 1.5, ± 2.0, and ± 2.5 standard deviations from the mean. The minimum percents of data in any distribution were calculated by using Chebyshev's theorem. The actual percents of data for the normal distribution were calculated by using the area given in the standard normal, or z, table.

	$K = 1.1$	$K = 1.5$	$K = 2$	$K = 2.5$
Minimum (for any distribution)	17.4%	55.6%	75%	84%
Normal distribution	72.8%	86.6%	95.4%	99.8%
Given distribution				

The third row of the chart has been left blank for you to fill in the percents when you reach part (e).

Consider the following 30 pieces of data obtained from a quiz.

1, 1, 1, 1, 2, 2, 2, 2, 3, 3, 4, 4, 4, 5, 6,
6, 6, 7, 7, 7, 7, 8, 8, 8, 8, 9, 9, 9, 10, 10

a) Find the mean of the set of scores.
b) Find the standard deviation of the set of scores.
c) Determine the values that correspond to 1.1, 1.5, 2, and 2.5 standard deviations above the mean. (For example, the value that corresponds to 1.5 standard deviations above the mean is $\bar{x} + 1.5s$.)
Then determine the values that correspond to 1.1, 1.5, 2, and 2.5 standard deviations below the mean. (For example, the value that corresponds to 1.5 standard deviations below the mean is $\bar{x} - 1.5s$.)
d) By observing the 30 pieces of data, determine the actual percent of quiz scores between

± 1.1 standard deviations from the mean.
± 1.5 standard deviations from the mean.
± 2 standard deviations from the mean.
± 2.5 standard deviations from the mean.

e) Place the percents found in part (d) in the third row of the chart.
f) Compare the percents in the third row of the chart with the minimum percents in the first row and the normal percents in the second row, then make a judgment as to whether this set of 30 scores is approximately normally distributed. Explain your answer.

85. *Using Data from Your Class* Obtain a set of test scores from your teacher.
   a) Find the mean, median, mode, and midrange of the test scores.
   b) Find the range and standard deviation of the set of scores. (You may round the mean to the nearest tenth when finding the standard deviation.)
   c) Construct a frequency distribution of the set of scores. Select your first class so that there will be between 5 and 12 classes.

d) Construct a histogram and frequency polygon of the frequency distribution in part (c).
e) Does the histogram in part (d) appear to represent a normal distribution? Explain.
f) Use the procedure explained in Exercise 84 to determine whether the set of scores represented is a normal distribution. Explain.

RESEARCH ACTIVITY

86. In this project you actually become the statistician.
   a) Select a project of interest to you in which data must be collected.
   b) Write a proposal and submit it to your instructor for approval. In the proposal, discuss the aims of your project and how you plan to gather the data to make your sample unbiased.
   c) After your proposal has been approved, gather 50 pieces of data by the method you proposed.
   d) Rank the data from smallest to largest.
   e) Compute the mean, median, mode, and midrange.
   f) Determine the range and standard deviation of the data. You may round the mean to the nearest tenth when computing the standard deviation.
   g) Construct a frequency distribution, histogram, frequency polygon, and stem-and-leaf display of your data. Select your first class so that there will be between 5 and 12 classes. Be sure to label your histogram and frequency polygon.
   h) Does your distribution appear to be normal? Explain your answer. Does it appear to be another type of distribution discussed? Explain.
   i) Determine whether your distribution is approximately normal by using the technique discussed in Exercise 84.

13.8 LINEAR CORRELATION AND REGRESSION

In this section, we discuss two important statistical topics: correlation and regression. **Correlation** is used to determine whether there is a relationship between two quantities and, if so, how strong that relationship is. **Regression** is used to determine the equation that relates the two quantities. Although there are other types of correlation and regression, in this section we discuss only linear correlation and linear regression. We begin by discussing linear correlation.

LINEAR CORRELATION

Do you believe that there is a relationship between

a) the time a person studied for an exam and the exam grade received?

b) the age of a car and the value of the car?

c) the height and weight of adult males?

d) a person's IQ and income?

Correlation is used to answer questions of this type. The **linear correlation coefficient**, r, is a unitless measure that describes the strength of the linear relationship between two variables. A positive value of r, or a positive correlation, means that as one variable increases, the other variable also increases. A negative value of r, or a negative correlation, means that as one variable increases, the other variable decreases. The correlation coefficient, r, will always be a value between -1 and 1 inclusive. A value of 1 indicates the strongest possible positive correlation, a value of -1 indicates the strongest negative correlation, and a value of 0 indicates no correlation (Fig. 13.35).

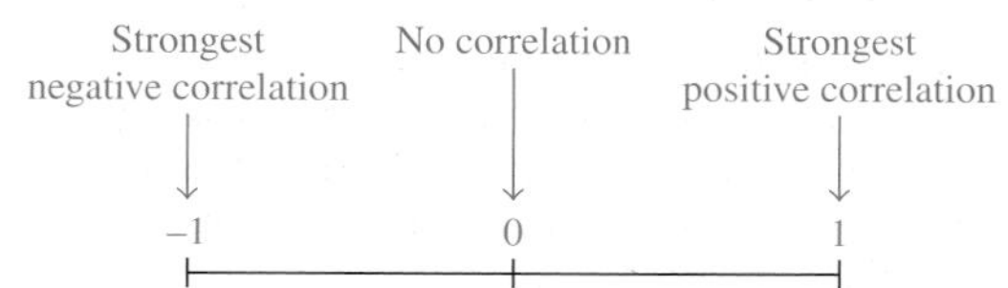

Figure 13.35

A visual aid used with correlation is the **scatter diagram**, a plot of data points. To explain how to construct a scatter diagram, consider the following data from Eddie's Electronics Superstore. During an 8-week period, Eddie's Electronics Superstore has advertised color televisions on TV. The store has kept a weekly record of the number of advertisements per week and the number of televisions sold per week. The information is provided in the following chart.

Week	**1**	**2**	**3**	**4**	**5**	**6**	**7**	**8**
Number of advertisements	8	20	0	14	25	16	15	10
Number of TVs sold	15	28	8	15	27	20	22	12

For each of the 8 weeks, there are two pieces of data, number of advertisements and number of TVs sold. We call the set of data **bivariate data**. When we have a set of bivariate data, we generally denote the quantity that can be controlled, the **independent variable**, x. The other variable, the **dependent variable**, is denoted as y. In this problem, the number of advertisements can be controlled, so we call the number of advertisements x and the number of TVs sold y. If we plot the eight pieces of bivariate data in the Cartesian coordinate system, we get a scatter diagram, as shown in Fig. 13.36.

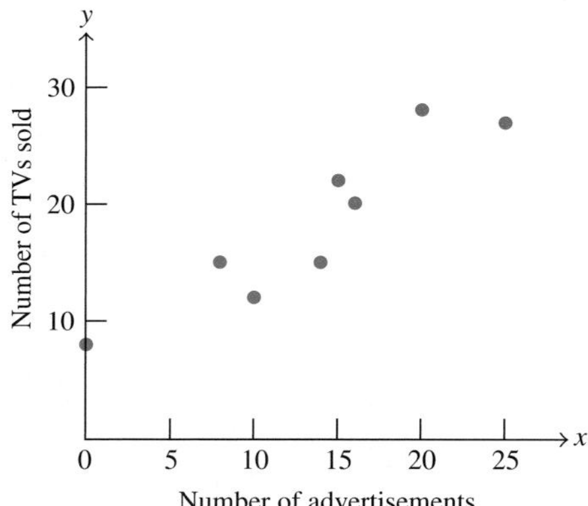

Figure 13.36

Figure 13.36 shows that, generally, the more Eddie's Electronics Superstore advertised, the more TVs they sold.

In Fig. 13.37 on page 715, we show some scatter diagrams and indicate the corresponding strength of correlation between the quantities on the horizontal and vertical axes.

Earlier we mentioned that r will always be a value between -1 and 1 inclusive. A value of $r = 1$ is obtained only when every point of the bivariate data on a scatter diagram lies in a straight line and the line is increasing from left to right (see Fig. 13.37a). In other words, the line has a positive slope, as discussed in Section 6.6.

A value of $r = -1$ will be obtained only when every point of the bivariate data on a scatter diagram lies in a straight line and the line is decreasing from left to right (see Fig. 13.37e). In other words, the line has a negative slope.

The value of r is a measure of how far a set of points varies from a straight line. The greater the spread, the weaker the correlation and the closer the value of r is to 0. Figure 13.37 shows that the more the dots diverge from a straight line, the weaker the correlation becomes.

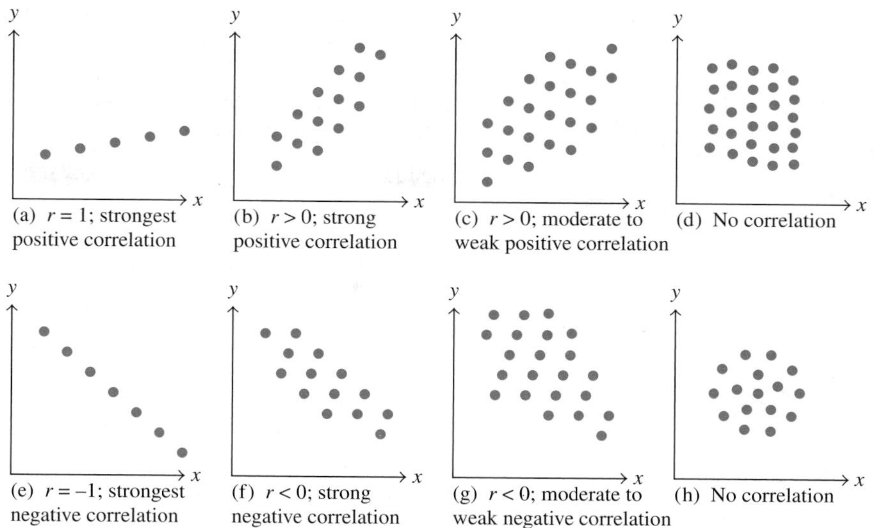

Figure 13.37

The following formula is used to calculate r.

$$r = \frac{n(\Sigma xy) - (\Sigma x)(\Sigma y)}{\sqrt{n(\Sigma x^2) - (\Sigma x)^2}\sqrt{n(\Sigma y^2) - (\Sigma y)^2}}$$

To determine the correlation coefficient, r, and the equation of the line of best fit (to be discussed shortly), a statistical calculator may be used. On the calculator you simply enter the ordered pairs, (x, y), and press the appropriate keys.

In Example 1, we show how to determine r for a set of bivariate data without the use of a statistical calculator. We will use the same set of bivariate data given on page 714.

EXAMPLE 1 *Advertising vs. Sales*

Eddie's Electronics Superstore provided the following data about the number of TV advertisements and the number of TVs sold for an 8-week period. Determine the correlation coefficient between the number of advertisements and the number of TVs sold.

Week	**1**	**2**	**3**	**4**	**5**	**6**	**7**	**8**
Number of advertisements	8	20	0	14	25	16	15	10
Number of TVs sold	15	28	8	15	27	20	22	12

SOLUTION We plotted this set of data on the scatter diagram in Fig. 13.36. Since the number of advertisements can be controlled, we will call it x. We will call the number of TVs sold y. We list the values of x and y and calculate the necessary sums: Σx, Σy, Σxy, Σx^2, and Σy^2. We determine the values in the column labeled x^2 by squaring the x's (multiplying the x's by themselves). We determine the values in the column labeled y^2 by squaring the y's. We determine the values in the column labeled xy by multiplying each x value by its corresponding y value.

Number of advertisements x	Number of TVs sold y	x^2	y^2	xy
8	15	64	225	120
20	28	400	784	560
0	8	0	64	0
14	15	196	225	210
25	27	625	729	675
16	20	256	400	320
15	22	225	484	330
10	12	100	144	120
108	147	1866	3055	2335

Thus, $\Sigma x = 108$, $\Sigma y = 147$, $\Sigma x^2 = 1866$, $\Sigma y^2 = 3055$, and $\Sigma xy = 2335$. In the formula for r, we use both $(\Sigma x)^2$ and Σx^2. Note that $(\Sigma x)^2 = (108)^2 = 11{,}664$ and that $\Sigma x^2 = 1866$. Similarly, $(\Sigma y)^2 = (147)^2 = 21{,}609$ and $\Sigma y^2 = 3055$.

The n in the formula represents the number of pieces of bivariate data. Here, $n = 8$. Now let's determine r.

$$
\begin{aligned}
r &= \frac{n(\Sigma xy) - (\Sigma x)(\Sigma y)}{\sqrt{n(\Sigma x^2) - (\Sigma x)^2}\sqrt{n(\Sigma y^2) - (\Sigma y)^2}} \\
&= \frac{8(2335) - (108)(147)}{\sqrt{8(1866) - (108)^2}\sqrt{8(3055) - (147)^2}} \\
&= \frac{18{,}680 - 15{,}876}{\sqrt{8(1866) - 11{,}664}\sqrt{8(3055) - 21{,}609}} \\
&= \frac{2804}{\sqrt{14{,}928 - 11{,}664}\sqrt{24{,}440 - 21{,}609}} \\
&= \frac{2804}{\sqrt{3264}\sqrt{2831}} \approx 0.92
\end{aligned}
$$

Since the maximum possible value for r is 1.00, a correlation coefficient of 0.92 is a strong positive correlation. This result implies that generally, the more Eddie's Electronics Superstore advertises on TV, the more TVs they sell. ▲

In Example 1, had we found r to be a value greater than 1 or less than -1, it would have indicated that we had made an error. Also, from the scatter diagram, we should realize that r should be a positive value and not negative.

In Example 1, there appears to be a cause–effect relationship. That is, the more advertisements on TV, the more TVs were sold. *However, a correlation does not necessarily indicate a cause–effect relationship.* For example, there is a positive correlation between police officers' salaries and the cost of medical insurance over the past 10 years (both have increased), but that does not mean that the increase in police officers' salaries caused the increase in the cost of medical insurance.

Suppose in Example 1 that r had been 0.53. Would this value have indicated a correlation? What is the minimum value of r needed to assume that a correlation

Table 13.8 Correlation Coefficient, r

n	$\alpha = 0.05$	$\alpha = 0.01$
4	0.950	0.990
5	0.878	0.959
6	0.811	0.917
7	0.754	0.875
8	0.707	0.834
9	0.666	0.798
10	0.632	0.765
11	0.602	0.735
12	0.576	0.708
13	0.553	0.684
14	0.532	0.661
15	0.514	0.641
16	0.497	0.623
17	0.482	0.606
18	0.468	0.590
19	0.456	0.575
20	0.444	0.561
22	0.423	0.537
27	0.381	0.487
32	0.349	0.449
37	0.325	0.418
42	0.304	0.393
47	0.288	0.372
52	0.273	0.354
62	0.250	0.325
72	0.232	0.302
82	0.217	0.283
92	0.205	0.267
102	0.195	0.254

The derivation of this table is beyond the scope of this course. It shows the critical values of the Pearson correlation coefficient.

exists between the variables? To answer this question, we introduce the term *level of significance*. The **level of significance**, denoted α (alpha), is used to identify the cutoff between results attributed to chance and results attributed to an actual relationship between the two variables. Table 13.8 gives **critical values*** (or cutoff values) that are sometimes used for determining whether two variable are related. The table indicates two different levels of significance: $\alpha = 0.05$ and $\alpha = 0.01$. A level of significance of 5%, written $\alpha = 0.05$, means that there is a 5% chance that, when you say the variables are related, they actually are *not* related. Similarly, a level of significance of 1%, or $\alpha = 0.01$, means that there is a 1% chance that, when you say the variables are related, they actually are *not* related. More complete critical value tables are available in statistics books.

To explain the use of the table, we use **absolute value**, symbolized $|\quad|$. The absolute value of a nonzero number is the positive value of the number and the absolute value of 0 is 0. Therefore,

$$|3| = 3, \quad |-3| = 3, \quad |5| = 5, \quad |-5| = 5, \quad \text{and} \quad |0| = 0$$

If the absolute value of r, written $|r|$, is *greater than* the value given in the table under the specified α and appropriate sample size n, we assume that a correlation does exist between the variables. If $|r|$ is less than the table value, we assume that no correlation exists.

Returning to Example 1, if we want to determine whether there is a correlation at a 5% level of significance, we find the critical value (or cutoff value) to the right of $n = 8$ (there are 8 pieces of bivariate data) and under the $\alpha = 0.05$ column. Here, the critical value is 0.707. From the formula we had obtained $r = 0.92$. Since $|0.92| > 0.707$, or $0.92 > 0.707$, we assume that a correlation between the variables exists.

Note in Table 13.8 that the larger the sample size, the smaller is the value of r needed for a significant correlation.

EXAMPLE 2 *Amount of Drug Remaining in the Bloodstream*

To test the length of time that an infection-fighting drug stays in a person's bloodstream, a doctor gives 300 milligrams of the drug to 10 patients. Once each hour, for 8 hours, one of the 10 patients is selected at random and that person's blood is tested to determine the amount of the drug remaining in the bloodstream. The results are as follows.

Patient	**1**	**2**	**3**	**4**	**5**	**6**	**7**	**8**	**9**	**10**
Time (hr)	1	2	3	4	5	6	7	8	9	10
Drug remaining (mg)	250	230	200	210	140	120	210	100	90	85

Determine at a level of significance of 5% whether a correlation exists between the time elapsed and the amount of drug remaining.

SOLUTION Let time be represented by x and the amount of drug remaining by y. We first draw a scatter diagram (Fig. 13.38).

*This table of values may be used only under certain conditions. If you take a statistics course, you will learn more about which critical values to use to determine whether a linear correlation exists.

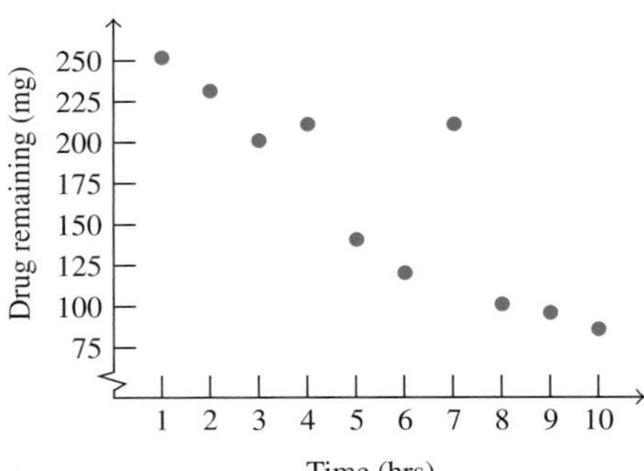

Figure 13.38

The scatter diagram indicates that, if a correlation exists, it will be negative. We now construct a table of values and calculate r.

x	y	x^2	y^2	xy
1	250	1	62,500	280
2	230	4	52,900	460
3	200	9	40,000	600
4	210	16	44,100	840
5	140	25	19,600	700
6	120	36	14,400	720
7	210	49	44,100	1470
8	100	64	10,000	800
9	90	81	8,100	810
10	85	100	7,225	850
55	1635	385	302,925	7500

$$
\begin{aligned}
r &= \frac{n(\Sigma xy) - (\Sigma x)(\Sigma y)}{\sqrt{n(\Sigma x^2) - (\Sigma x)^2}\sqrt{n(\Sigma y^2) - (\Sigma y)^2}} \\
&= \frac{10(7500) - (55)(1635)}{\sqrt{10(385) - (55)^2}\sqrt{10(302{,}925) - (1635)^2}} \\
&= \frac{-14{,}925}{\sqrt{825}\sqrt{355{,}025}} \approx \frac{-14{,}925}{17{,}114.19} \\
&\approx -0.872
\end{aligned}
$$

From Table 13.8, for $n = 10$ and $\alpha = 0.05$, we get 0.632. Since $|-0.872| = 0.872$ and $0.872 > 0.632$, a correlation exists. The correlation is negative, which indicates that the longer the time period, the smaller is the amount of drug remaining. ▲

LINEAR REGRESSION

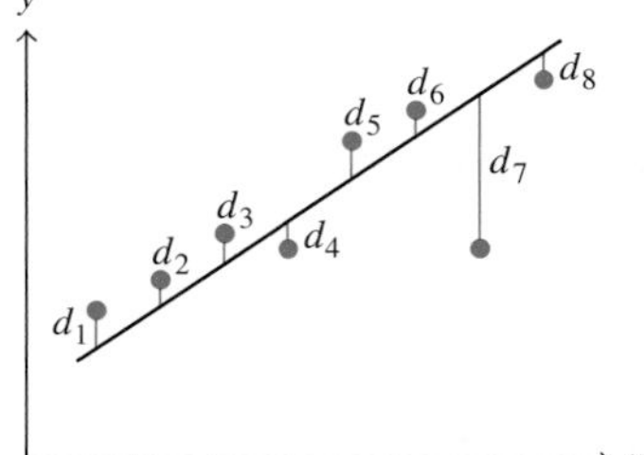

Figure 13.39

Let's now turn to regression. **Linear regression** is the process of determining the linear relationship between two variables. Recall from Section 6.6 that the slope–intercept form of a straight line is $y = mx + b$, where m is the slope and b is the y intercept.

Using the set of bivariate data, we will determine the equation of **the line of best fit**. The line of best fit is also called **the regression line**, or **the least squares line**. The *line of best fit* is the line such that the sum of the vertical distances from the line to the data points (on the scatter diagram) is a minimum, as shown in Fig. 13.39.

In Fig. 13.39, the line of best fit minimizes the sum of d_1 through d_8. To determine the equation of the line of best fit, $y = mx + b$*, we must find m and then b. The formulas for finding m and b are as follows.

The equation of the line of best fit is $y = mx + b$, where

$$m = \frac{n(\Sigma xy) - (\Sigma x)(\Sigma y)}{n(\Sigma x^2) - (\Sigma x)^2}, \quad \text{and} \quad b = \frac{\Sigma y - m(\Sigma x)}{n}$$

Note that the numerator of the fraction used to find m is identical to the numerator used to find r. Therefore, if you have previously found r, you do not need to repeat the calculation. Also, the denominator of the fraction used to find m is identical to the radicand of the first square root in the denominator of the fraction used to find r.

EXAMPLE 3 *The Line of Best Fit*

a) Use the data in Example 1 to find the equation of the line of best fit that relates the number of advertisements and the number of TVs sold.
b) Graph the equation of the line of best fit on a scatter diagram that illustrates the set of bivariate points.

SOLUTION

a) In Example 1, we found $n(\Sigma xy) - (\Sigma x)(\Sigma y) = 2804$ and $n(\Sigma x^2) - (\Sigma x)^2 = 3264$. Thus,

$$m = \frac{n(\Sigma xy) - (\Sigma x)(\Sigma y)}{n(\Sigma x^2) - (\Sigma x)^2} = \frac{2804}{3264} \approx 0.86$$

Now we find the y intercept, b. In Example 1, we found $n = 8$, $\Sigma x = 108$, and $\Sigma y = 147$.

$$\begin{aligned} b &= \frac{\Sigma y - m(\Sigma x)}{n} \\ &= \frac{147 - 0.86(108)}{8} = \frac{54.12}{8} \\ &\approx 6.77 \end{aligned}$$

Therefore, the equation of the line of best fit is

$$\begin{aligned} y &= mx + b \\ y &= 0.86x + 6.77 \end{aligned}$$

where x represents the number of advertisements and y represents the predicted number of TVs sold.

*Some statistics books use $y = ax + b$, $y = b_0 + b_1x$, or something similar for the equation of the line of best fit. In any case, the letter next to the variable x represents the slope of the line of best fit, and the other letter represents the y intercept of the graph.

b) To graph $y = 0.86x + 6.77$, we need to plot at least two points. We will plot three points and then draw the graph.

$$y = 0.86x + 6.77$$

$$x = 10 \qquad y = 0.86(10) + 6.77 = 15.37$$

$$x = 15 \qquad y = 0.86(15) + 6.77 = 19.67$$

$$x = 20 \qquad y = 0.86(20) + 6.77 = 23.97$$

x	y
10	15.37
15	19.67
20	23.97

These three calculations indicate that if Eddie's Electronics Superstore advertises 10 times per week, the predicted number of TVs sold is about 15. If they advertise 15 times per week, the predicted number of TVs sold is about 20; and if they advertise 20 times per week, the predicted number of TVs sold is about 24. Plot the three points (the three black dots) then draw a straight line through the three points. The scatter diagram and graph are plotted in Fig. 13.40.

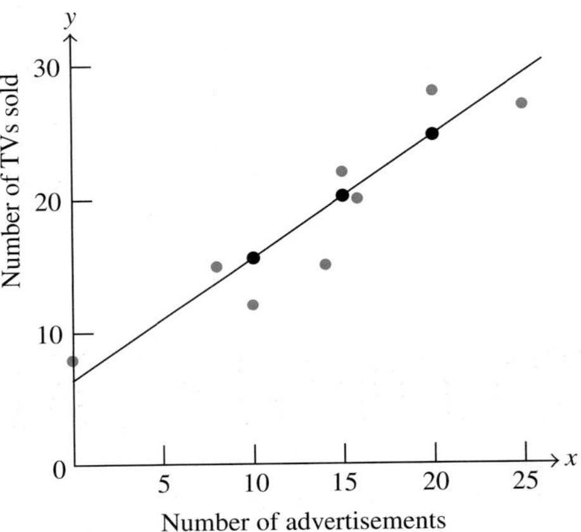

Figure 13.40

In Example 3, the line of best fit intersects the y-axis at 6.77, the value we determined for b in part (a).

EXAMPLE 4 *Line of Best Fit for Example 2*

a) Determine the equation of the line of best fit between the time elapsed and the amount of drug remaining in a person's bloodstream in Example 2.

b) If the average person is given 300 mg of the drug, how much will remain in the person's bloodstream after 5 hr?

SOLUTION

a) From the scatter diagram on page 718 we see that the slope of the line of best fit, m, will be negative. In Example 2, we found $n(\Sigma xy) - (\Sigma x)(\Sigma y) = -14{,}925$ and that $n(\Sigma x^2) - (\Sigma x)^2 = 825$. Thus,

$$\begin{aligned} m &= \frac{n(\Sigma xy) - (\Sigma x)(\Sigma y)}{n(\Sigma x^2) - (\Sigma x)^2} \\ &= \frac{-14{,}925}{825} \\ &\approx -18.1 \end{aligned}$$

From Example 2, $n = 10$, $\Sigma x = 55$, and $\Sigma y = 1635$.

$$\begin{aligned} b &= \frac{\Sigma y - m\Sigma x}{n} \\ &= \frac{1635 - (-18.1)(55)}{10} \\ &\approx 263.1 \end{aligned}$$

Thus, the equation of the line of best fit is

$$\begin{aligned} y &= mx + b \\ y &= -18.1x + 263.1 \end{aligned}$$

where x is the elapsed time and y is the amount of drug remaining.

b) We evaluate $y = -18.1x + 263.1$ at $x = 5$.

$$\begin{aligned} y &= -18.1x + 263.1 \\ y &= -18.1(5) + 263.1 = 172.6 \end{aligned}$$

Thus, after 5 hr, about 173 mg of the drug remains in the average person's bloodstream.

SECTION 13.8 EXERCISES

CONCEPT/WRITING EXERCISES

1. What does the correlation coefficient measure?
2. What is the purpose of regression?
3. What value of r represents the maximum positive correlation?
4. What value of r represents the maximum negative correlation?
5. What value of r represents no correlation between the variables?
6. What does a positive correlation between two variables indicate?
7. What does a negative correlation between two variables indicate?
8. What does the line of best fit represent?
9. What does the level of significance signify?
10. What is a scatter diagram?

In Exercises 11–14, do you believe the correlation between the quantities on the horizontal and vertical axis is a strong positive correlation, a strong negative correlation, a weak positive correlation, a weak negative correlation, or no correlation? Explain your answer.

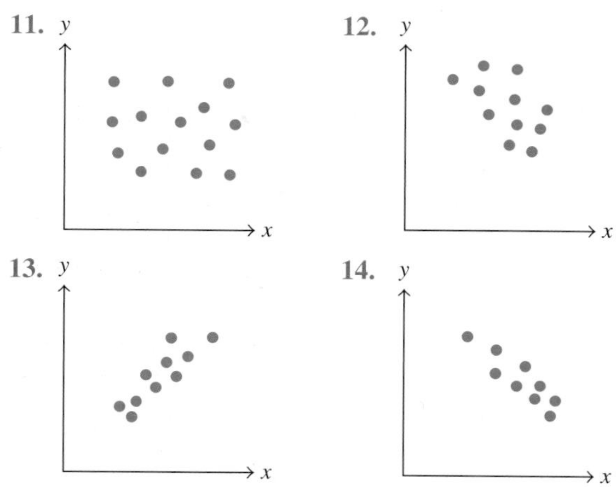

PRACTICE THE SKILLS

In Exercises 15–22, assume that a sample of bivariate data yields the correlation coefficient, r, indicated. Use Table 13.8 for the specified sample size and level of significance to determine whether a linear correlation exists.

15. $r = 0.73$ when $n = 13$ at $\alpha = 0.01$
16. $r = 0.57$ when $n = 22$ at $\alpha = 0.01$

17. $r = -0.63$ when $n = 8$ at $\alpha = 0.05$

18. $r = -0.49$ when $n = 11$ at $\alpha = 0.05$

19. $r = -0.23$ when $n = 102$ at $\alpha = 0.01$

20. $r = -0.49$ when $n = 18$ at $\alpha = 0.01$

21. $r = 0.82$ when $n = 6$ at $\alpha = 0.01$

22. $r = 0.96$ when $n = 5$ at $\alpha = 0.01$

In Exercises 23–30, (a) draw a scatter diagram; (b) find the value of r, rounded to the nearest thousandth; (c) determine whether a correlation exists at $\alpha = 0.05$; and (d) determine whether a correlation exists at $\alpha = 0.01$.

23.

x	y
3	6
4	9
5	11
6	11
9	13

24.

x	y
6	10
8	9
11	7
14	8
17	6

25.

x	y
23	29
35	37
31	26
43	20
49	39

26.

x	y
90	3
80	4
60	6
60	5
40	5
20	7

27.

x	y
5.3	10.3
4.7	9.6
8.4	12.5
12.7	16.2
4.9	9.8

28.

x	y
12	15
16	19
13	45
24	30
100	60
50	28

29.

x	y
100	2
80	3
60	5
60	6
40	6
20	8

30.

x	y
90	90
70	70
65	65
60	60
50	50
40	40
15	15

In Exercises 31–38, find the equation of the line of best fit from the data in the exercise indicated. Round both the slope and the y-intercept to the nearest tenth.

31. Exercise 23
32. Exercise 24
33. Exercise 25
34. Exercise 26
35. Exercise 27
36. Exercise 28
37. Exercise 29
38. Exercise 30

PROBLEM SOLVING

39. *Cost of Long-Distance Calls* The number of long-distance calls made by Ann Kuick monthly, for 6 months, and the corresponding monthly telephone costs, in dollars, are shown below.

Number of calls	10	15	12	20	25	17
Cost (dollars)	37	43	37	49	54	45

a) Determine the correlation coefficient between the number of calls and the cost.
b) Determine whether a correlation exists at $\alpha = 0.05$.
c) Find the equation of the line of best fit for the number of calls and the cost.

40. *Amount of Time Spent Studying* Six students provided the following data about the lengths of time they studied for a psychology exam and the grades they received on the exam.

Time studied (minutes)	20	40	50	60	80	100
Grade received (percent)	40	45	70	76	92	95

a) Determine the correlation coefficient between the length of time studied and the grade received.
b) Determine whether a correlation exists at $\alpha = 0.01$.
c) Find the equation of the line of best fit for the length of time studied and the grade received.

41. *Child Care Expenses* The following table shows the estimated annual expenditures for child care and education based on age in 1997. (*Source:* U.S. Agriculture Department.)

Age (years)	1	4	7	10	13	16
Expense (dollars)	690	780	460	280	200	330

a) Determine the correlation coefficient for age and expense.
b) Determine whether a correlation exists at $\alpha = 0.05$.

c) Find the equation of the line of best fit for age and expense.
d) Use the equation in part (c) to estimate the annual expenditure on child care and education for a 9-year-old child.

42. *Depreciation of a Car* The blue book value of a Ford Explorer, based on its age from 1–8 years, is shown below. (*Source: Kelly's Blue Book.*)

Age (years)	1	2	3	4	5	6	7	8
Value (thousands of dollars)	26.0	23.9	21.8	19.5	15.2	13.8	12.3	11.1

a) Determine the correlation coefficient for age and value.
b) Determine whether a correlation exists at $\alpha = 0.05$.
c) Find the equation of the line of best fit for age and value.
d) Use the equation in part (c) to estimate the blue book value of the Explorer when it is 4.5 years old.

43. *City Muggings* In a certain section of a city muggings have been a problem. The number of police officers patrolling that section of the city has varied. The following chart shows the number of police officers and the number of muggings for 10 successive days.

Police officers	20	12	18	15	22	10	20	12
Muggings	8	10	12	9	6	15	7	18

a) Determine the correlation coefficient for number of police officers and number of muggings.
b) Determine whether a correlation exists at $\alpha = 0.05$.
c) Find the equation of the line of best fit for number of police officers and number of muggings.
d) Use the equation in part (c) to estimate the average number of muggings when 14 police officers are patrolling that section of the city.

44. *Weight Loss* The cumulative weight loss of a customer after several weeks on a diet plan is shown below.

Number of weeks	1	2	3	4	5	6	7
Cumulative weight loss (pounds)	3	5	5	6	5	7	9

a) Determine the correlation coefficient for number of weeks on a diet and cumulative weight loss.
b) Determine whether a correlation exists at $\alpha = 0.05$.
c) Find the equation of the line of best fit for number of weeks and cumulative weight loss.
d) Use the equation in part (c) to estimate the average cumulative weight loss after 5 weeks.

45. *Selling Popcorn at the Movies* The number of tickets sold and the number of units of popcorn sold at Regal Cinema for 8 days is shown below.

Ticket sales	89	110	125	92	100	95	108	97
Units of popcorn	22	28	30	26	22	21	28	25

a) Determine the correlation coefficient between ticket sales and units of popcorn sold.
b) Determine whether a correlation exists at $\alpha = 0.05$.
c) Determine the equation of the line of best fit for tickets sold and units of popcorn sold.
d) Use the equation in part (c) to estimate the units of popcorn sold if 115 tickets are sold.

46. *Movie Ratings* A popular newspaper rates movies from one to four stars (four stars is the highest rating). The following table shows the ratings of 10 movies selected at random and the gross earnings of each movie.

Rating (stars)	4	4	3	2	1	3	4	2	4	1
Earnings (millions of dollars)	100	67	80	120	40	90	60	60	90	100

a) Determine the correlation coefficient between number of stars and the movies' earnings.
b) Determine whether a correlation exists at $\alpha = 0.05$.
c) Determine the equation of the line of best fit for the number of stars and the movies' earnings.

47. *Chlorine in a Swimming Pool* A gallon of chlorine is put into a swimming pool. Each hour later for the following 6 hr the percent of chlorine that remains in the pool is measured. The following information is obtained.

Time	1	2	3	4	5	6
Chlorine remaining (percent)	80.0	76.2	68.7	50.1	30.2	20.8

a) Determine the correlation coefficient for time and percent of chlorine remaining.
b) Determine whether a correlation exists at $\alpha = 0.01$.
c) Determine the equation of the line of best fit for time and amount of chlorine remaining.
d) Use the equation in part (c) to estimate the average amount of chlorine remaining after 4.5 hr.

48. a) Match the first 9 digits in your phone number (including area code) with the 9 digits in your social security number. To do this, match the first digit in your phone number with the first digit in your social security number to get one ordered pair. Match the second digits to get a second ordered pair. Continue this process until you get a total of nine ordered pairs.
b) Do you believe that this set of bivariate data has a positive correlation, a negative correlation, or no correlation? Explain your answer.
c) Construct a scatter diagram for the nine ordered pairs.
d) Calculate the correlation coefficient, r.
e) Is there a correlation at $\alpha = 0.05$? Explain.
f) Calculate the equation of the line of best fit.
g) Use the equation in part (f) to estimate the digit in a social security number that corresponds with a 7 in a telephone number.

49. a) Do you believe there is a positive correlation, a negative correlation, or no correlation between speed of a car and stopping distance when the brakes are applied? Explain.
b) Do you believe there is a stronger correlation between speed of a car and stopping distance on wet or dry roads? Explain.

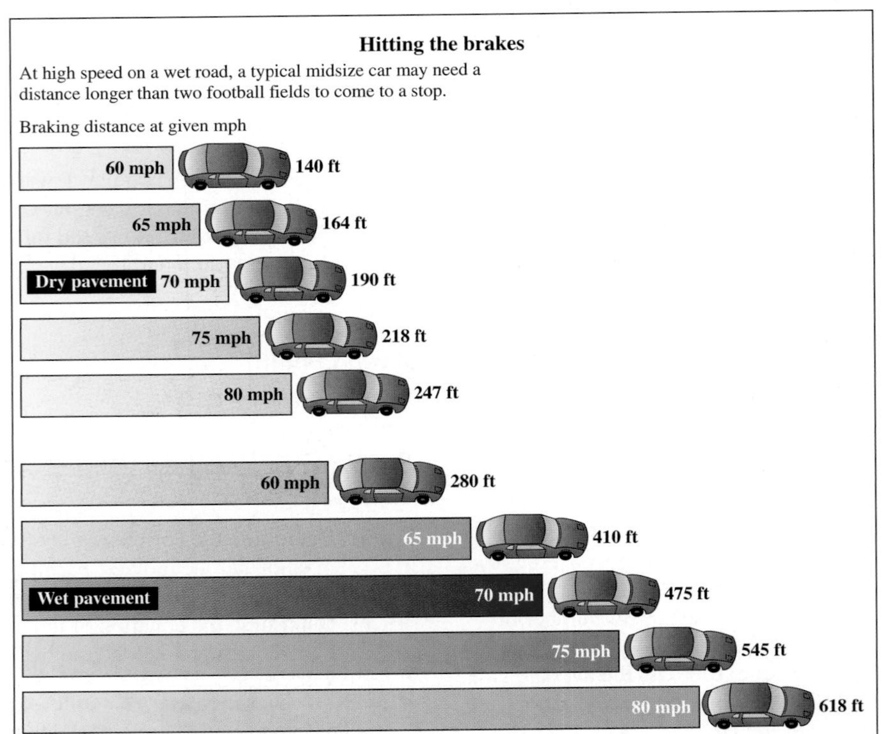

Source: *Car and Driver*, American Automobile Association

c) Use the figure on the preceding page to construct two scatter diagrams, one for dry pavement and the other for wet pavement. Place the speed of the car on the horizontal axis.
d) Compute the correlation coefficient for speed of the car and stopping distance for dry pavement.
e) Repeat part (d) for wet pavement.
f) Were your answers to parts (a) and (b) correct? Explain.
g) Determine the equation of the line of best fit for dry pavement.
h) Repeat part (g) for wet pavement.
i) Use the equations in parts (g) and (h) to estimate the stopping distance of a car going 77 mph on both dry and wet pavements.

CHALLENGE PROBLEMS/GROUP ACTIVITIES

50. a) Assume that a set of bivariate data yields a specific correlation coefficient. If the x and y values are interchanged and the correlation coefficient is recalculated, will the correlation coefficient change? Explain.
b) Make up a table of five pieces of bivariate data and determine r using the data. Then switch the values of the x's and y's and recompute the correlation coefficient. Has the value of r changed?

51. a) Do you believe that a correlation exists between a person's height and the length of a person's forearm? Explain.
b) Select 10 people from your class and measure (in inches) their heights and the lengths of their forearms.
c) Plot the 10 ordered pairs on a scatter diagram.
d) Calculate the correlation coefficient, r.
e) Determine the equation of the line of best fit.
f) Estimate the length of the forearm of a person who is 58 in. tall.

52. a) Have your group select a category of bivariate data that it thinks has a strong positive correlation. Designate the independent variable and the dependent variable. Explain why your group believes the bivariate data have a strong positive correlation.
b) Collect at least 10 pieces of bivariate data that can be used to determine the correlation coefficient. Explain how your group chose these data.
c) Plot a scatter diagram.
d) Calculate the correlation coefficient.
e) Does there appear to be a strong positive correlation? Explain your answer.
f) Calculate the equation of the line of best fit.
g) Explain how the equation in part (f) may be used.

53. Use the following table.

Year	90	91	92	93	94	95
CPI	130	136	140	144	148	153

a) Calculate r.
b) If 90 is subtracted from each year, the table obtained becomes:

Year	0	1	2	3	4	5
CPI	130	136	140	144	148	153

If r is calculated from these values, how will it compare with the r determined in part (a)? Explain.
c) Calculate r from the values in part (b) and compare the results with the value of r found in part (a). Are they the same? If not, explain why.

54. a) There are equivalent formulas that can be used to find the correlation coefficient and the equation of the line of best fit. A formula used in some statistics books to find the correlation coefficient is

$$r = \frac{SS(xy)}{\sqrt{SS(x)SS(y)}}$$

where

$$SS(x) = \Sigma x^2 - \frac{(\Sigma x)^2}{n}$$

$$SS(y) = \Sigma y^2 - \frac{(\Sigma y)^2}{n}$$

and $$SS(xy) = \Sigma xy - \frac{(\Sigma x)(\Sigma y)}{n}$$

Use this formula to find the correlation coefficient of the set of bivariate data given in Example 1 on page 715.
b) Compare your answer with the answer obtained in Example 1.

RESEARCH ACTIVITIES

55. a) Obtain a set of bivariate data from newspaper or magazine.
b) Plot the information on a scatter diagram.
c) Indicate whether you believe that the data show a positive correlation, a negative correlation, or no correlation. Explain your answer.
d) Calculate r and determine whether your answer to part (b) was correct.
e) Determine the equation of the line of best fit for the bivariate data.

56. Find a scatter diagram in a newspaper or magazine and write a paper on what the diagram indicates. Indicate whether you believe that the bivariate data show a positive correlation, a negative correlation, or no correlation and explain why.

CHAPTER 13 SUMMARY

IMPORTANT FACTS

Rules for data grouped by classes

1. The classes should be the same width.
2. The classes should not overlap.
3. Each piece of data should belong to only one class.

Measures of central tendency

The **mean** is the sum of the data divided by the number of pieces of data: $\bar{x} = \dfrac{\Sigma x}{n}$.

The **median** is the value in the middle of a set of ranked data.

The **mode** is the piece of data that occurs most frequently (if there is one).

The **midrange** is the value halfway between the lowest and highest values: midrange $= \dfrac{L + H}{2}$.

Statistical graphs

Circle graph
Histogram
Frequency polygon
Stem-and-leaf display

Measures of dispersion

The **range** is the difference between the highest value and lowest value in a set of data.

The **standard deviation,** s, is a measure of the spread of a set of data about the mean: $s = \sqrt{\dfrac{\Sigma(x - \bar{x})^2}{n - 1}}$.

z-scores

$$z = \frac{x - \bar{x}}{s}$$

Chebyshev's theorem

Minimum percentage $= 1 - \dfrac{1}{k^2}$, $k > 1$

Linear correlation and regression

Linear correlation coefficient, r, is

$$r = \frac{n(\Sigma xy) - (\Sigma x)(\Sigma y)}{\sqrt{n(\Sigma x^2) - (\Sigma x)^2}\sqrt{n(\Sigma y^2) - (\Sigma y)^2}}$$

Equation of the line of the best fit is $y = mx + b$, where

$$m = \frac{n(\Sigma xy) - (\Sigma x)(\Sigma y)}{n(\Sigma x^2) - (\Sigma x)^2}, \text{ and}$$

$$b = \frac{\Sigma y - m(\Sigma x)}{n}$$

CHAPTER 13 REVIEW EXERCISES

13.1

1. **a)** What is a population?
   b) What is a sample?
2. What is a random sample?

13.2

In Exercises 3 and 4, tell what possible misuses or misinterpretations may exist in the statements.

3. The Stay Healthy Candy Bar indicates on its label that it has no cholesterol. Therefore, it is safe to eat as many of these candy bars as you want.
4. More copies of *Time* magazine are sold than are copies of *Money* magazine. Therefore, *Time* is a more profitable magazine than *Money*.
5. The ABC-TV network had a prime time rating of 8.4 in 1997. In 1998, the ABC-TV network had a prime time rating of 8.1. Draw a graph that appears to show **(a)** a small decrease in rating and **(b)** a large decrease in rating.

13.3–13.4

6. Consider the following set of data.

35	37	38	41	43
36	37	38	41	43
36	37	39	41	43
36	37	39	41	44
37	37	39	42	45

   a) Construct a frequency distribution letting each class have a width of 1.
   b) Construct a histogram.
   c) Construct a frequency polygon.

7. *High Temperatures in 40 U.S. Cities* Consider the following high temperatures for 40 cities in the United States on November 4, 1999.

40	51	54	61	66	69	72	76
46	52	54	64	67	70	72	80
46	53	55	65	68	70	75	81
48	53	56	66	68	71	75	81
50	54	57	66	69	72	75	87

a) Construct a frequency distribution. Let the first class be 40–49.
b) Construct a histogram.
c) Construct a frequency polygon.
d) Construct a stem-and-leaf display.
(*Source: Rochester Democrat & Chronicle.*)

13.5–13.6

In Exercises 8–13, for the following test scores 63, 76, 79, 83, 86, 93, find the

8. mean. **9.** median.
10. mode. **11.** midrange.
12. range. **13.** standard deviation.

In Exercises 14–19, for the set of data 4, 5, 12, 14, 19, 7, 12, 23, 7, 17, 15, 21, find the

14. mean. **15.** median.
16. mode. **17.** midrange.
18. range. **19.** standard deviation.

13.7

In Exercises 20–24, assume that anthropologists have determined that a certain type of primitive animal had a mean head circumference of 42 cm with a standard deviation of 5 cm. Given that head sizes were normally distributed, determine the percent of heads

20. between 37 and 47 cm. **21.** between 32 and 52 cm.
22. less than 50 cm. **23.** greater than 50 cm.
24. greater than 39 cm.

In Exercises 25–28, assume that the life of a compact disc (CD) player is normally distributed with a mean of 4.2 years and a standard deviation of 0.5 years. Find the percent of CD players with a life

25. between 4.2 years and 4.7 years.
26. less than 4 years.
27. between 4.4 years and 5.4 years.
28. If the manufacturer warranties its CD players for 3 years, what percent will need to be replaced under warranty?

13.8

In Exercises 29 and 30, use the following chart (and graph) that shows both the U.S. savings rate for the years 1990–1998 and the year-end Dow Jones industrial average for the years 1990–1998.

Year	U.S. savings rate (percent)	Dow Jones Industrial Average (in hundreds)
90	5.1	26
91	5.6	31
92	5.8	33
93	4.4	37
94	3.6	38
95	3.5	51
96	3.0	62
97	2.1	79
98	0.5	91

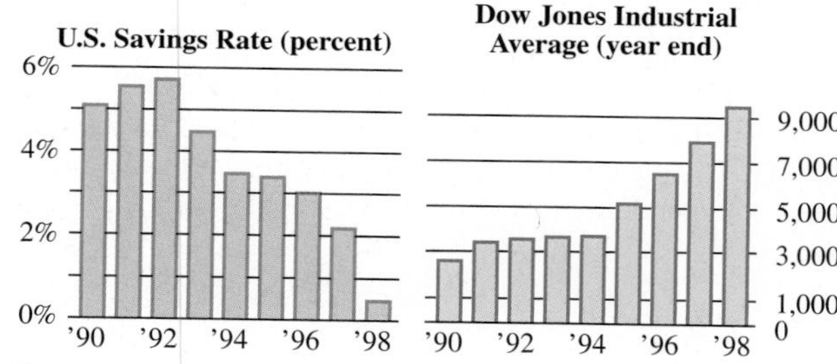

Sources: Bureau of Economic Analysis, Dow Jones

29. a) *Savings Rate* Construct a scatter diagram for the U.S. savings rate with the year on the horizontal axis.
b) Use the scatter diagram in part (a) to determine whether you believe a correlation exists between year and savings rate. If so, is it a positive or a negative correlation? Explain.
c) Calculate the correlation coefficient between the year and the U.S. savings rate.
d) Determine whether a correlation exists at $\alpha = 0.05$. Explain how you arrived at your answer.
e) Determine the equation of the line of best fit between year and savings rate.

30. a) *Dow Jones Industrial Average* Construct a scatter diagram for the Dow Jones industrial average with the year on the horizontal axis.
b) Use the scatter diagram in part (a) to determine whether you believe a correlation exists between year and Dow Jones industrial average. If so, is it a positive or a negative correlation? Explain.

c) Calculate the correlation coefficient between the year and the Dow Jones industrial average.
d) Determine whether a correlation exists at $\alpha = 0.01$. Explain how you arrived at your answer.
e) Determine the equation of the line of best fit between year and Dow Jones industrial average.

31. *Clothing Expenditures* The following table shows the estimated annual expenditures on clothing for a child, by age, in 1997. (*Source:* U.S. Department of Agriculture.)

Age (years)	1	4	7	10	13	16
Expense (dollars)	370	360	410	450	760	670

a) Construct a scatter diagram with age on the horizontal axis.
b) Use the scatter diagram in part (a) to determine whether you believe a correlation exists between age and annual expenditures on clothing. If so, is it a positive or a negative correlation? Explain.
c) Calculate the correlation coefficient between age and annual expenditures.
d) Determine whether a correlation exists at $\alpha = 0.05$. Explain how you arrived at your answer.
e) Determine the equation of the line of best fit between age and annual expenditures.
f) Use the equation in part (d) to estimate the annual expenditure on clothing for a 12-year-old child.

13.5–13.7

Men's Weight *In Exercises 32–39, use the following data obtained from a study of the weights of adult men.*

Mean	187 lb	First quartile	173 lb
Median	180 lb	Third quartile	227 lb
Mode	175 lb	86th percentile	234 lb
Standard deviation	23 lb		

32. What is the most common weight?

33. What weight did half of those surveyed exceed?

34. About what percent of those surveyed weighed more than 227 lb?

35. About what percent of those surveyed weighed less than 173 lb?

36. About what percent of those surveyed weighed more than 234 lb?

37. If 100 men were surveyed, what is the total weight of all men?

38. What weight represents two standard deviations above the mean?

39. What weight represents 1.8 standard deviations below the mean?

13.2–13.7

Presidential Offspring *The following list shows the names of the 41 U.S. presidents and the number of children in their families.*

Washington	0	Cleveland	5
J. Adams	5	B. Harrison	3
Jefferson	6	McKinley	2
Madison	0	T. Roosevelt	6
Monroe	2	Taft	3
J. Q. Adams	4	Wilson	3
Jackson	0	Harding	0
Van Buren	4	Coolidge	2
W. H. Harrison	10	Hoover	2
Tyler	14	F. D. Roosevelt	6
Polk	0	Truman	1
Taylor	6	Eisenhower	2
Fillmore	2	Kennedy	3
Pierce	3	L. B. Johnson	2
Buchanan	0	Nixon	2
Lincoln	4	Ford	4
A. Johnson	5	Carter	4
Grant	4	Reagan	4
Hayes	8	Bush	6
Garfield	7	Clinton	1
Arthur	3		

In Exercises 40–51, use the data above to determine the following.

40. Mean number of children

41. Mode

42. Median

43. Midrange

44. Range

45. Standard deviation (round the mean to the nearest tenth).

46. Construct a frequency distribution; let the first class be 0–1.

47. Construct a histogram.

48. Construct a frequency polygon.

49. Does this distribution appear to be normal? Explain.

50. On the basis of this sample, do you think the number of children per family in the United States would have a normal distribution? Explain.

51. Do you believe that this sample is representative of the population? Explain.

CHAPTER 13 TEST

In Exercises 1–6, for the set of data 15, 31, 31, 33, 40, find the

1. mean.
2. median.
3. mode.
4. midrange.
5. range.
6. standard deviation.

In Exercises 7–9, use the set of data

26	28	35	46	49	56
26	30	36	46	49	58
26	32	40	47	50	58
26	32	44	47	52	62
27	35	46	47	54	66

to construct

7. a frequency distribution; let the first class be 25–30.
8. a histogram of the frequency distribution.
9. a frequency polygon of the frequency distribution.

Statistics on Salaries *In Exercises 10–16, use the following data on weekly salaries at Maxwell Mechanical Contractors.*

Mean	\$600	First quartile	\$550
Median	\$570	Third quartile	\$605
Mode	\$595	79th percentile	\$612
Standard deviation	\$40		

10. What is the most common salary?
11. What salary did half the employees exceed?
12. About what percent of employees' salaries exceeded \$550?
13. About what percent of employees' salaries was less than \$612?
14. If the company has 100 employees, what is the total weekly salary of all employees?
15. What salary represents one standard deviation above the mean?
16. What salary represents 1.5 standard deviations below the mean?

Mileage of 5-Year-Old Cars *In Exercises 17–20, the mileage of 5-year-old cars is normally distributed with a mean of 75,000 and a standard deviation of 12,000 miles.*

17. What percent of 5-year-old cars have mileage between 50,000 and 70,000 miles?
18. What percent of 5-year-old cars have mileage greater than 60,000 miles?
19. What percent of 5-year-old cars have mileage greater than 90,000 miles?
20. If a random sample of 300 five-year-old cars is selected, how many would have mileage between 60,000 and 70,000 miles?
21. *Population Below the Poverty Level* The following chart shows the percent of persons living below the poverty level in the United States for the years 1992–1997.

Persons Living Below Poverty Level

Year	Percent
1992	14.8
1993	15.1
1994	14.5
1995	13.8
1996	13.7
1997	13.3

Source: U.S. Census Bureau.

a) Construct a scatter diagram placing the year on the horizontal axis.
b) Do you believe a correlation exists between the year and the percent of persons living below the poverty level? If so, is it positive or negative? Explain.
c) Determine the correlation coefficient (use only the last two numbers in each year in your calculations; for example, use 92 for 1992).
d) At $\alpha = 0.05$, does a correlation exist?
e) Determine the equation of the line of best fit between year and percent of persons below the poverty level.
f) Use the equation in part (e) to predict the percent of persons below the poverty level in 1998.

GROUP PROJECTS

WATCHING TV

1. Do you think that men or women, aged 17–20, watch more hours of TV weekly, or do you think that they watch the same number of hours?
   a) Write a procedure to use to determine the answer to that question. In your procedure use a sample of 30 men and 30 women. State how you will obtain an unbiased sample.
   b) Collect 30 pieces of data from men aged 17–20 and 30 pieces of data from women aged 17–20. Round answers to the nearest 0.5 hr. Follow the procedure developed in part (a) to obtain your unbiased sample.
   c) Compute the mean for your two groups of data to the nearest tenth.
   d) Using the means obtained in part (c), answer the question asked at the beginning of the problem.
   e) Is it possible that your conclusion in part (d) is wrong? Explain.
   f) Compute the standard deviation of both groups of data. Round each mean to the nearest tenth. How do the standard deviations compare?
   g) Do you believe that the distribution of data from either or both groups resembles a normal distribution? Explain.
   h) Add the two groups of data to get one group of 60 pieces of data. If these 60 pieces of data are added and divided by 60, will you obtain the same mean as when you add the two means from part (c) and divide the sum by 2? Explain.
   i) Compute the mean of the 60 pieces of data by using both methods in part (h). Are they the same? If so, why? If not, why not?
   j) Do you believe that this group of 60 pieces of data represents a normal distribution? Explain.

BINOMIAL PROBABILITY EXPERIMENT

2. **a)** Have your group select a category of bivariate data that it thinks has a strong negative correlation. Indicate the variable that you will designate as the independent variable and the variable that you will designate as the dependent variable. Explain why your group believes the bivariate data have a strong negative correlation.
   b) Collect at least 10 pieces of bivariate data that can be used to determine the correlation coefficient. Explain how your group chose these data.
   c) Plot a scatter diagram.
   d) Calculate the correlation coefficient.
   e) Is there a negative correlation at $\alpha = 0.05$? Explain your answer.
   f) Calculate the equation of the line of best fit.
   g) Explain how the equation in part (f) may be used.

APPENDIX: GRAPH THEORY

> **DID YOU KNOW**
>
> ## Making Connections
>
> 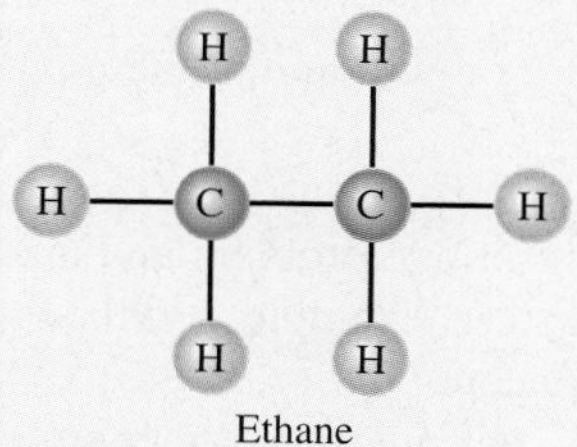
>
> Ethane
>
> A chemist sketches the relationship among atoms in a molecule. An engineer draws a schematic diagram of an electrical network. A genealogist charts a family tree. A sociologist maps the power structure of a large corporation. What do all of these drawings have in common? Each one is a system of points (atoms, electrical components, people, segments) connected by line segments. The German mathematician Denes Konig, who made the first systematic study of such patterns, named them "graphs."

In the eighteenth-century Prussian town of Königsberg (now the city of Kaliningrad, Russia near the Baltic Sea) seven bridges crossed the Prigel River (Fig. A.1a). Individuals in the area tried to determine whether it was possible to walk a path that would cross each of the seven bridges exactly once. They found that they ended up either not crossing one of the bridges, or crossing one of the bridges more than once. The problem was brought to the attention of the Swiss mathematician Leonhard Euler (pronounced "oiler," 1707–1783). His study of this problem, now known as the Königsberg bridge problem, laid the groundwork for a modern branch of mathematics called *graph theory,* a topic in a more general area called **topology**. To solve the problem, Euler drew figures called **graphs** *or* **networks** like that shown in red in Fig. A.1(b). Each dot represented a plot of land and each line a bridge or path. Can you find a path that crosses each bridge exactly once?

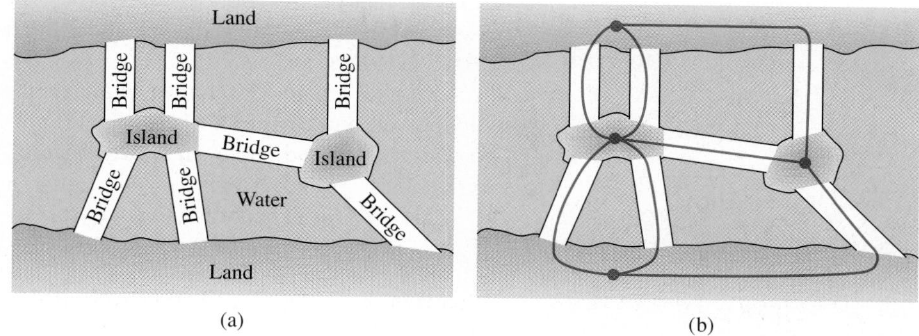

Figure A.1

Before we determine whether there is a path that will cross each bridge exactly once, let's consider Example 1.

EXAMPLE 1

In Fig. A.2 start at any point and try to trace each figure without retracing a line and without removing your pencil from the paper. If you succeed, indicate your starting point and ending point.

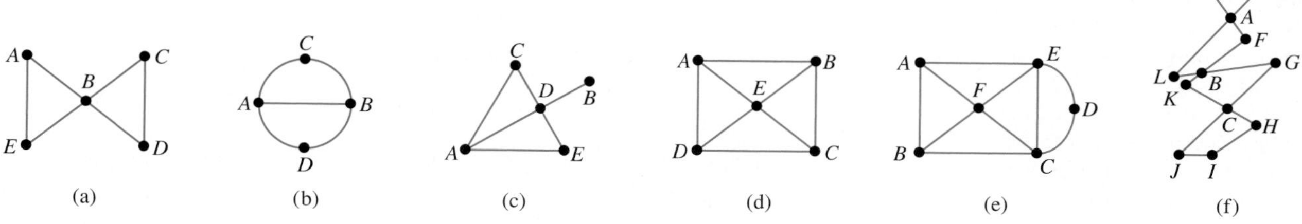

Figure A.2

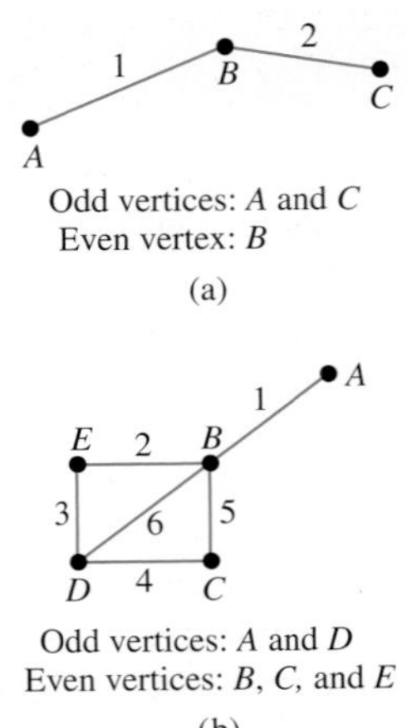

Figure A.3

SOLUTION

a) The figure can be traced if you start at any point. You will end at the point at which you started.

b) The figure can be traced, but only if you start at point A or point B. If you start at point A, you will end at point B, and vice versa.

c) The figure can be traced, but only if you start at point A or point B. If you start at point A, you will end at point B, and vice versa.

d) The figure cannot be traced without retracing a line.

e) The figure can be traced, but only if you start at point A or point B. If you start at point A, you will end at point B, and vice versa.

f) The figure can be traced if you start at any point. You will end at the point at which you started.

In order for you to be able to answer the Königsberg bridge problem and understand why we were able to trace all but one of the figures in Example 1 without retracing a line, we must introduce some new terms and concepts.

A **vertex** is any designated point. An **edge** (or an **arc**) is any line, either straight or curved, that begins and ends at a vertex. Figure A.3(a) has three designated vertices, A, B, and C, and two edges, 1 and 2. Figure A.3(b) has five designated vertices, A, B, C, D, and E, and six edges.

A vertex with an odd number of attached edges is called an **odd vertex**. A vetex with an even number of attached edges is called an **even vertex**. Figure A.3(a) has two odd vertices and one even vertex. Figure A.3(b) has two odd vertices and three even vertices.

A **network** is any continuous (not broken) system of edges and vertices.

A network is said to be **traversable** if it can be traced without removing the pencil from the paper and without tracing an edge more than once.

After completing Example 1, do you have an intuitive feeling as to when a figure is traversable? Think about odd and even vertices. Leonhard Euler discovered an important scientific principle concealed in the Königsberg bridge problem. He presented his simple and ingenious solution of that problem to the Russian Academy at St. Petersburg in 1735. Euler developed the following rules of traversability in solving the problem.

Rules of Traversability

1. A network with no odd (all even) vertices is traversable; you may start from any vertex, and you will end where you began.
2. A network with exactly two odd vertices is traversable; you must start at either of the odd vertices and finish at the other.
3. A network with more than two odd vertices is not traversable.

The network in the Königsberg bridge problem (Fig. A.1b) has four odd vertices, so it cannot be traversed. Therefore crossing each bridge only once is impossible. Note that it is impossible for a network to contain an odd number of odd vertices. (If you don't believe this statement, try to construct such a network.)

DID YOU KNOW

Networks

Today, graph (network) theory is a flourishing field, useful for dealing with a variety of everyday problems. It is used by the airline industry to arrange the complex network of flights connecting cities and by sports officials planning tournaments and playoff schedules. City planners use it to work out new roads and traffic patterns. Tests are underway in traffic-clogged cities such as Los Angeles to link cars, via an on-board navigational computer, to the city's central traffic-control computer. In this way motorists can be rerouted to less congested roadways.

Now go back to Example 1 and determine which figures are traversable, using these rules. Note that Figs. A.2(a) and A.2(f) are traversable from any point because they contain only even vertices. Figures A.2(b), (c), and (e) have exactly two odd vertices and can be traversed but only by starting at one of the odd vertices, either point *A* or point *B*. Figure A.2(d) contains more than two odd vertices and therefore cannot be traversed.

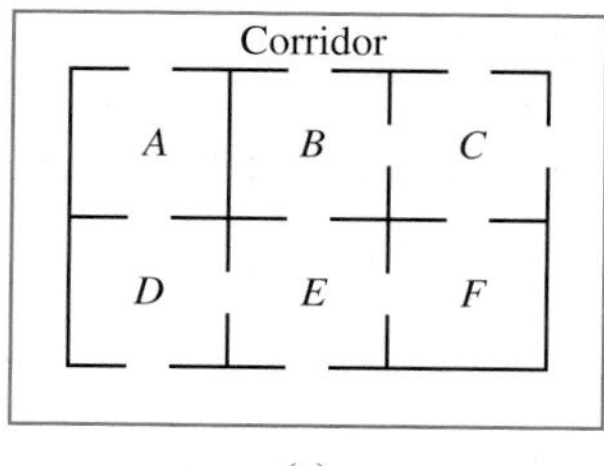

(a)

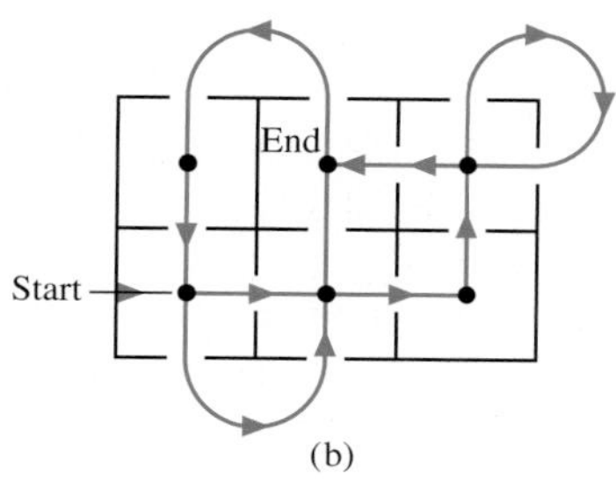

(b)

Figure A.4

EXAMPLE 2

The floor plan of a six-gallery art museum is shown in Fig. A.4(a). The openings represent doors, and the letters represent galleries.

a) Determine the galleries that contain an odd number of doors; an even number of doors.
b) Each gallery can be represented as an odd or even vertex. Use this information to determine whether it is possible to walk through each gallery by using each door only once.
c) Determine a path to walk through each gallery by using each door only once.

SOLUTION

a) Galleries *B* and *D* contain three doors each. Galleries *A* and *F* contain two doors each. Galleries *C* and *E* have four doors.
b) There are only two odd vertices, *B* and *D*, so the figure is traversable, and you can walk through the museum by using each door only once.
c) You must start in either Gallery *B* or *D* (see Fig. A.4b). If you start in *B*, you will end in *D*, and vice versa.. When you leave Gallery *B*, you can leave by any of the three doors.

The floor plan in Example 2 can be reduced to a map, where the rooms are the vertices and the doors are the paths. Construct a map of that floor plan now.

APPENDIX EXERCISES

CONCEPT/WRITING EXERCISES

1. What is a vertex?
2. What is an edge (or an arc)?
3. Explain how to determine whether a vertex is odd or even.
4. In your own words, explain the rules for determining whether a graph is traversable.

PRACTICE THE SKILLS

In Exercises 5–8, determine the number of vertices and the number of edges.

5. 

6. 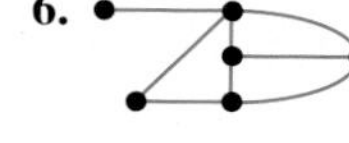

7. 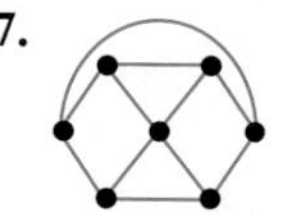

8. 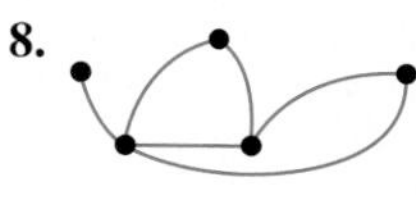

In Exercises 9 and 10, explain why these two figures represent the same graph.

9. 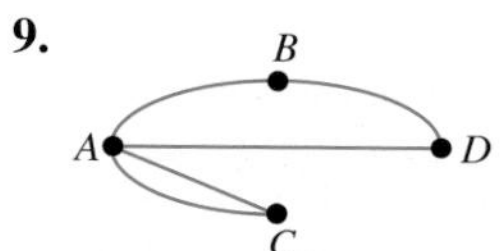

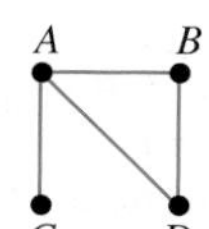

10. 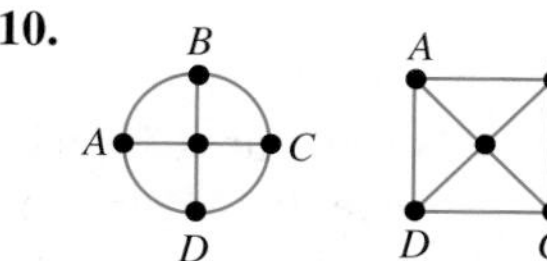

In Exercises 11 and 12, list the vertices that are odd and the vertices that are even.

11.

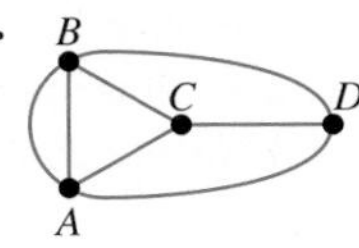

12.

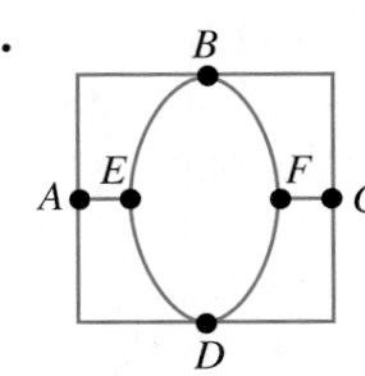

PROBLEM SOLVING

In Exercises 13–20, determine whether the network is traversable. If it is, state the points from which you may start and end.

13.

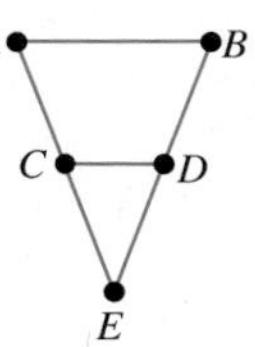

14.

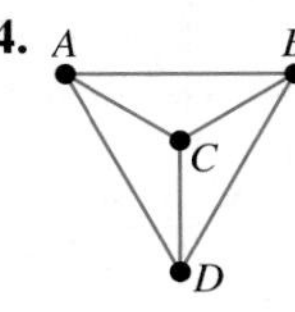

15.

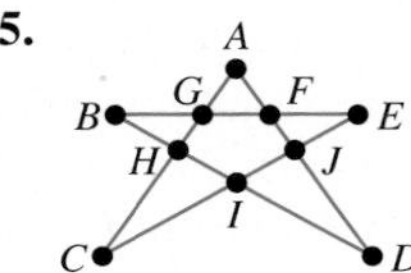

16.

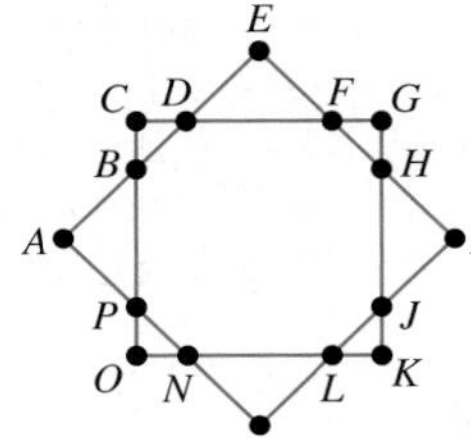

17.

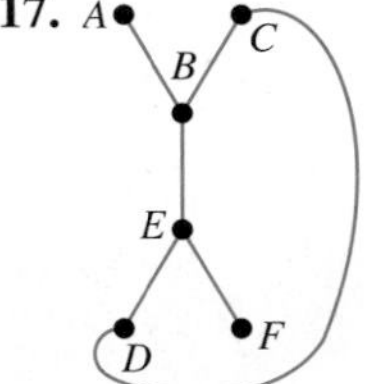

18.

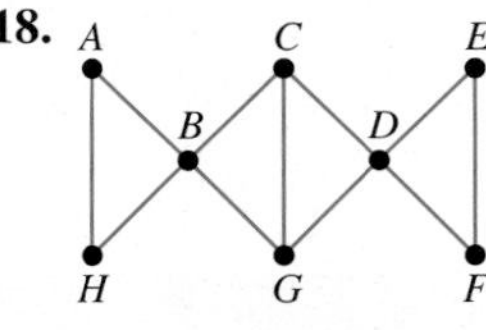

19.

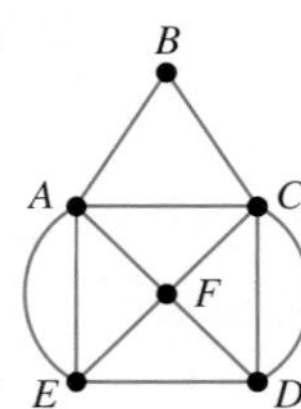

20.

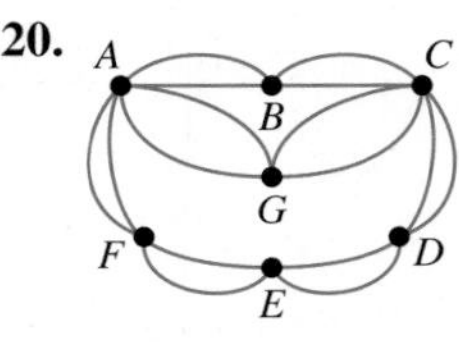

In Exercises 21–28, the floor plan of a building is shown.

a) *Determine the number of rooms that contain an odd number of doors; an even number of doors.*

b) *Use the rules of traversability to determine whether it is possible to walk through the building using each door only once.*

c) *If the answer to part (b) is yes, indicate where you can start and where you will end and describe one such path (for example, A to D to B to . . .etc.).*

21.

22.

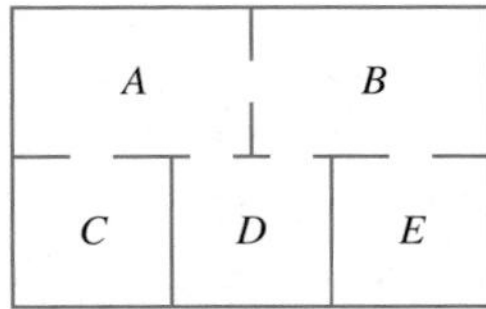

23.

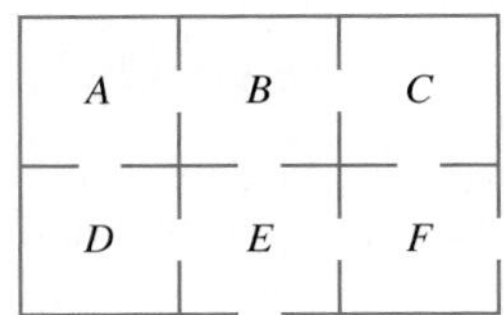

24.

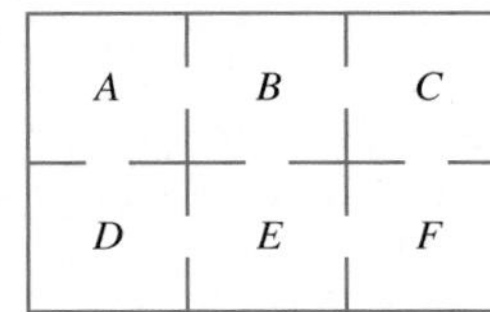

25.

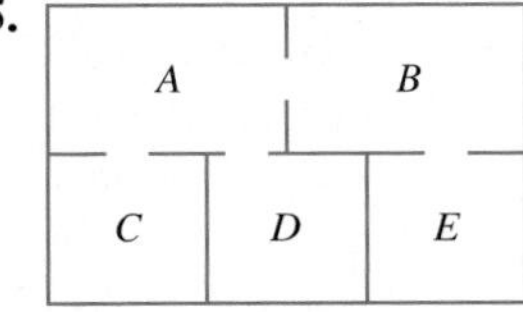

26.

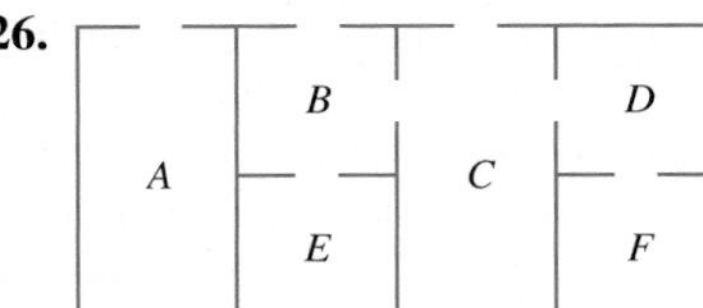

27.

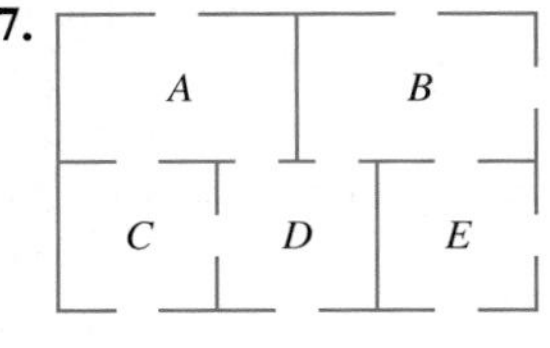

28.

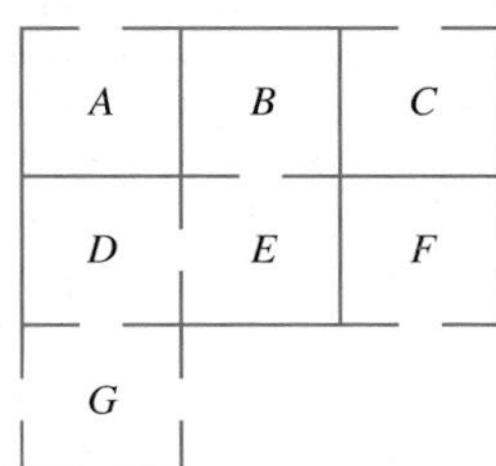

In Exercises 29 and 30, the floor plan for a suite of rooms is shown. Add an exit door in one of the rooms so that the security guard can enter through the door marked enter, pass through each door only once locking it behind him, and then exit by the door you added. Explain why there is only one possible room in which the door may be placed.

29.

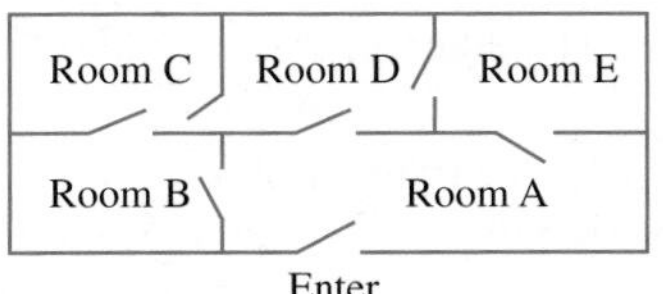

30.

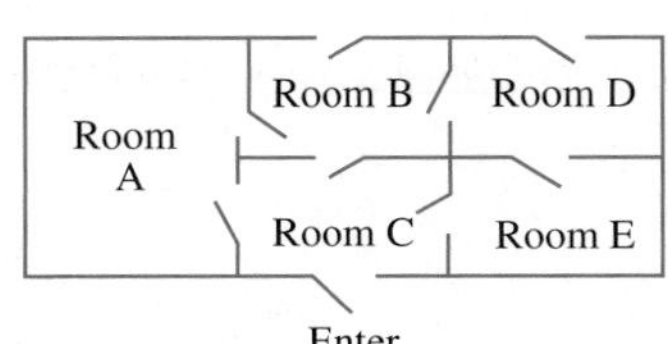

In Exercises 31 and 32, is it possible to cross each bridge exactly once? If it is possible, indicate where the person can start and where the person will finish. Explain your answer.

31.

32.

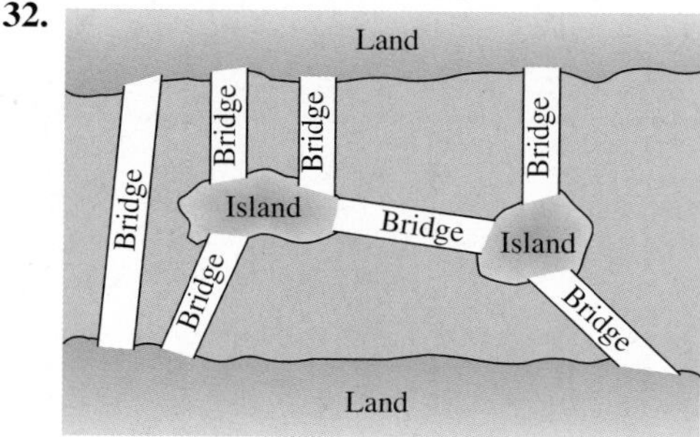

33. Draw a graph that contains four vertices and is traversable from exactly two points.

34. Draw a graph that contains five vertices that is traversable from exactly two points.

35. In the figure, lines connecting two states indicate that the two states share a common border. Which of these states share a common border with (a) Tennessee? (b) Missouri?

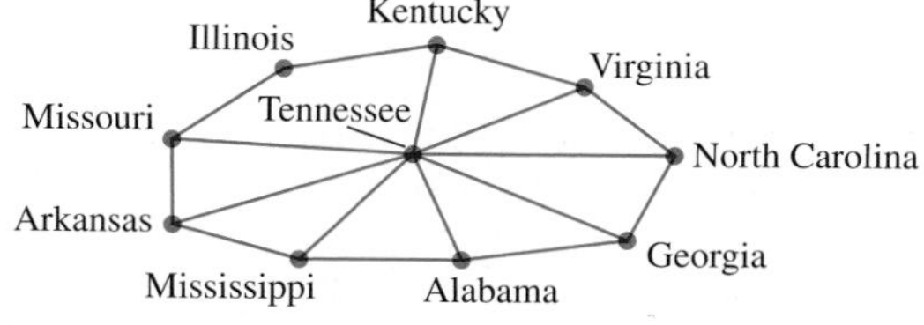

36. In the figure, a line connecting two countries indicates that they share a common border. Which of these countries share a common border with (a) Brazil? (b) Bolivia?

37. The remaining games in a soccer league are illustrated in the graph.

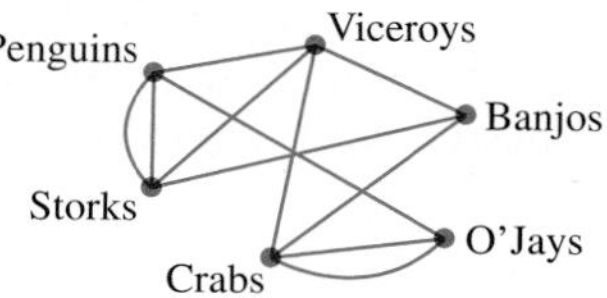

a) How many games do the Viceroys have remaining?
b) How many games do the Penguins have remaining?
c) How many games still need to be played?

38. Dawn, Jessica, Pam, Bill, Ed, and Scott go to a dance. Dawn dances with Bill and Scott. Jessica dances with all the boys and Pam dances only with Scott. Draw a graph that displays this information.

39. France has common borders with Belgium, Germany, Switzerland, Italy, and Spain. Belgium has common borders with the Netherlands, France, and Germany. Germany has common borders with Belgium, Poland, the Czech Republic, Austria, Switzerland, and France. Switzerland has common borders with France, Italy, Austria, and Germany. Draw a graph that displays this information.

40. Can you draw a graph that contains an odd number of odd vertices? If you answer yes draw such a graph.

CHALLENGE PROBLEMS/GROUP ACTIVITIES

41. Gretchen's Delivery service, located in town C, makes deliveries every day to each town shown on the map. The map also shows the highways connecting the towns and the distances between towns. The delivery service's expenses are lowest when the driver's route is the shortest.

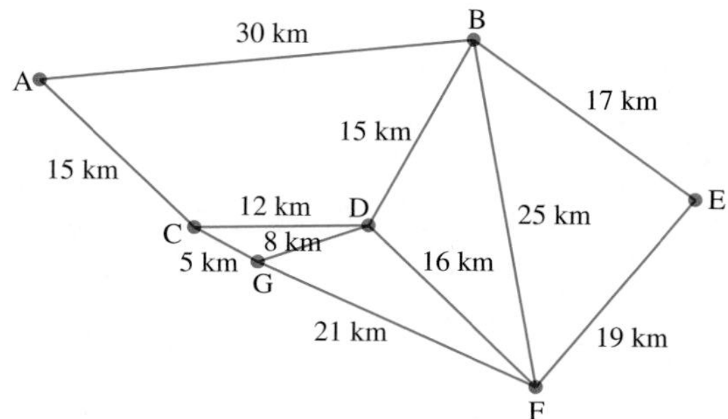

a) Is the map traversable? Explain.
b) The least number of miles that can be driven, starting at C and returning to C, that covers every city is 110 miles. Determine a path that covers all the cities and whose distance is 110 miles.

42. Networks (a) and (b) on page AA-6 illustrate a relationship between the number of edges, regions, and vertices. Network (a) has three vertices (A, B, and C), four edges (1, 2, 3, and 4), and three regions (x, y, and z). Network (b) has five vertices (A, B, C, D, and E), seven edges (1, 2, 3, 4, 5, 6, 7), and four regions (w, x, y, and z). Using these networks and others that you can make up yourself, develop a formula expressing the number of edges in terms of the number of vertices and the number of regions. This formula is known as **Euler's formula for networks.**

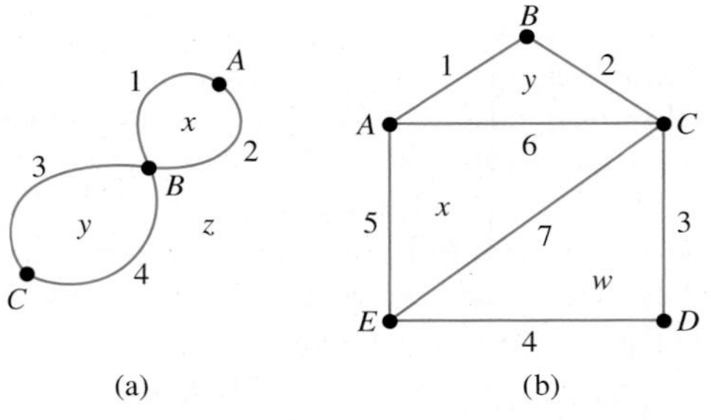

RESEARCH ACTIVITY

43. In 1856 William Rowan Hamilton (1805–1865) introduced a problem similar to the Königsberg bridge problem. Hamilton turned his problem into a game that was marketed in 1856. Write a short paper explaining the game. (References include encyclopedias and history of mathematics books and the Internet.)

ANSWERS

CHAPTER 1

SECTION 1.1, PAGE 5

1. a) 1, 2, 3, 4, 5, . . . **b)** Counting numbers

3. A conjecture is a belief based on specific observations that has not been proven or disproven.

5. Deductive reasoning is the process of reasoning to a specific conclusion from a general statement.

7. Inductive reasoning, because a generalization was made from specific cases

9. 1 5 10 10 5 1

11. $1 + 2 + 3 + 4 + 5 + 6 = 21$

13. △ **15.**

17. 13, 15, 17 **19.** −1, 1, −1 **21.** 1/16, 1/32, 1/64

23. 36, 49, 64 **25.** 34, 55, 89

27. a) Answers will vary.
b) The sum is 9.
c) The sum of the digits in the product when a one or two digit number is multiplied by 9 is 9.

29. a) 36, 49, 64
b) square 6, 7, 8, 9 and 10
c) No, 72 is between 8^2 and 9^2, so it is not a square number.

31. Blue: 1, 5, 7, 10, 12 Purple: 2, 4, 6, 9, 11 Yellow: 3, 8

33. a) 8% **b)** Gained 1.5% in each of 2 previous years

35.

37. a) You should obtain the original number.
b) You should obtain the original number.
c) The result is the original number.
d) $n, 3n, 3n + 6, \dfrac{3n + 6}{3} = n + 2, n + 2 - 2 = n$

39. a) 5
b) You should obtain the number 5.
c) The result is always the number 5.
d) $n, n + 1, \dfrac{n + (n + 1) + 9}{2} = \dfrac{2n + 10}{2} = n + 5,$
$n + 5 - n = 5$

41. $99 \times 99 = 9801$

43. $(3 + 2)/2 = 5/2$, which is not an even number.

45. $3 \times 3 = 9$, and 9 is not even.

47. a) The sum of the measures of the angles should be 180°.
b) Yes, the sum of the measures of the angles should be 180°.
c) The sum of the measures of the angles of a triangle is 180°.

49. 129, the numbers in positions are found as follows:
$a \quad b$
$c \quad a + b + c$

SECTION 1.2, PAGE 13

Answers in this section will vary depending on how you round your numbers. All answers are approximate.

1. 1060 **3.** 1,200,000,000 **5.** 8000 **7.** 100

9. 2,400,000,000 **11.** 1,200,000,000 **13.** 600 mi

15. \$2.33 **17.** \$320 **19.** 510 lb **21.** \$25,500

23. 3000 lb **25.** 85,000 miles **27.** 30 lb

29. 19,200 grubs **31. a)** 2.5 miles **b)** 4 kilometers

33. a) \$9840 **b)** \$15,990

35. a) 4 million **b)** 98 million
c) 59 million **d)** 275 million

37. a) Luxury cars, minivans, large cars
b) \$5460
c) \$12,000

39. 20 **41.** 90 marbles **43.** 150°

45. 10% **47.** 9 square units **49.** 45 ft

SECTION 1.3, PAGE 26

1. 33.75 ft **3.** 19.36 ft **5.** \$7.55

7. a) 38 cents per pound
b) \$3740
c) Loss of \$6160

9. About 770 mph

11. a) \$20
b) Weekly rate, \$10
c) \$16

13. \$12.75 **15.** \$718,750

17. a) 9.2 min
b) 62 min
c) 40 min
d) 150 min

19. a) 30,063,000 **b)** 97,000 **c)** 29,100

21. \$82.08

23. a) \$74.40
b) \$264
c) \$64

25. \$56,091.07

27. a) \$845
b) \$442

29. a) \$705
b) \$755
c) 50

31. \$505.62

33. a) 48 rolls
b) \$198 if she purchases four 10 packs and two 4 packs

35. a) Water/milk: 3 cups; salt: $\frac{3}{8}$ tsp; cream: 9 tbsp (or $\frac{9}{16}$ cup)
b) Water/milk: $2\frac{7}{8}$ cups; salt: $\frac{3}{8}$ tsp; cream: $\frac{5}{8}$ cup (or 10 tbsp)
c) Water/milk: $2\frac{3}{4}$ cups; salt: $\frac{3}{8}$ tsp; cream: $\frac{9}{16}$ cup (or 9 tbsp)
d) Differences exist in water/milk because the amount for 4 servings is not twice that for 2 servings. Differences also exist in Cream of Wheat because $\frac{1}{2}$ cup is not twice 3 tbsp.

37. 144 square inches **39.** The area is 4 times as large.

41. $7 + 7 - (7 \div 7) = 13$ **43.** 12 zebras and 6 cranes

45. a) 30
b) 140

47. a) Place the object, 1g, and 3g on one side and 9g on the other side.
b) Place the object, 9g, and 3g on one side and 27g and 1g on the other side.

49.

15	1	11
5	9	13
7	17	3

51. Multiply the middle number by 3.

53. 6 ways **55.** 20 cubes

57.

	7	
3	1	4
5	8	6
	2	

Other answers are possible, but 1 and 8 must appear in the center.

59.

1	2	3	4	5
2	3	4	5	1
3	4	5	1	2
4	5	1	2	3
5	1	2	3	4

Other answers are possible.

61. Mary is the skier. **63.** 714 sq units

REVIEW EXERCISES, PAGE 32

1. 23, 28, 33 **2.** 25, 36, 49 **3.** 64, −128, 256

4. 25, 32, 40 **5.** 10, 4, −3 **6.** $\frac{3}{8}, \frac{3}{16}, \frac{3}{32}$

7. **8.**

9. a) The original number and the final number are the same.
b) The original number and the final number are the same.
c) The final number is the same as the original number.
d) $n, 2n, 2n + 10, \dfrac{2n + 10}{2} = n + 5, n + 5 - 5 = n$

10. This process will always result in an answer of 3.

11. $6^2 - 4^2 = 36 - 16 = 20$

The answers to Exercises 12-25 will vary, depending upon how you round the numbers. All answers are approximate.

12. 410,000,000 **13.** 38 **14.** 2150 **15.** 200

16. Answers will vary. **17.** \$104 **18.** \$12 **19.** 3 mph

20. \$14.00 **21.** 2 mi **22.** 68 **23.** 5950 **24.** 13 sq units

25. Length = 22 ft; height = 8 ft **26.** \$2.10 **27.** \$1.70

28. \$1.20 **29.** Eurich's is cheaper by \$20.00. **30.** \$11.15

31. \$311 **32.** 7.05 mg **33.** \$665.00 **34.** 6 hr 45 min

35. July 26, 11:00 A.M.

36. a) 6.45 cm^2 **b)** 16.39 cm^3 **c)** 1 cm ≈ 0.39 in.

37. 201

38.

21	7	8	18
10	16	15	13
14	12	11	17
9	19	20	6

39.

23	25	15
13	21	29
27	17	19

40. 59 min 59 sec **41.** 6

42. \$25 Room
\$ 3 Men
\$ 2 Clerk
\$30

43. 140 lb

44. Yes; 3 quarters and 4 dimes, or 1 half dollar, 1 quarter and 4 dimes, or 1 quarter and 9 dimes. Other answers are possible.

45. 216 cm^3

46. Place six coins in each pan with one coin off to the side. If it balances, the heavier coin is the one on the side. If the pan does not balance, take the six coins on the heavier side and split them into two groups of three. Select the three heavier coins and weigh two coins. If the pan balances, it is the third coin. If the pan does not balance, you can identify the heavier coin.

47. 125,250 **48.** 16 blue **49.** 90

50. The fifth figure will be an octagon with sides of equal length. Inside the octagon will be a 7-sided figure with each side of equal length. The figure will have one antenna.

51. 61

52. Some possible answers are shown. Others are possible.

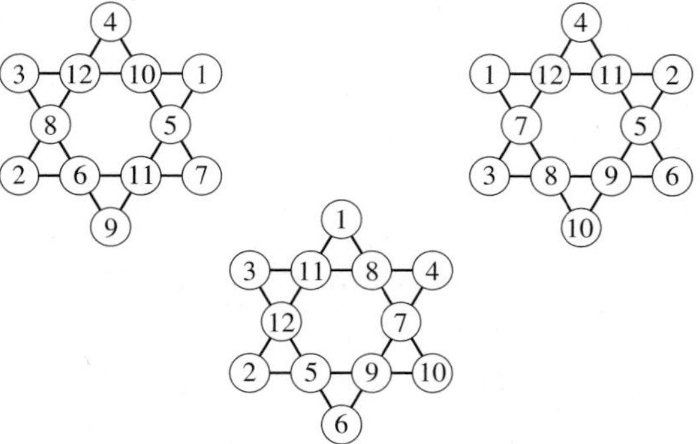

53. a) 2 b) 6 c) 24 d) 120
e) $n(n-1)(n-2)\ldots 1$, (or $n!$), where n = the number of people in line

CHAPTER TEST, PAGE 35

1. 18, 21, 24 2. $\frac{1}{18}, \frac{1}{243}, \frac{1}{729}$
3. a) The result is the original number plus 1.
b) The result is the original number plus 1.
c) The result will always be the original number plus 1.
d) $n, 5n, 5n+10, \dfrac{5n+10}{5} = n+2, n+2-1 = n+1$

The answers for Exercises 4–6 are approximate.

4. 4,000 5. 30,000,000 6. 7 sq units
7. a) $\approx$ 18% b) 16%
8. 153 therms 9. 24 cans 10. $7\frac{1}{2}$ min
11. $\approx$ 30 in. by 22 in. (The actual dimensions are 77 cm by 53 cm.)
12. $49.00
13.

40	15	20
5	25	45
30	35	10

14. Less time if she had driven at 45 mph for the entire trip
15. $2 \cdot 6 \cdot 8 \cdot 9 \cdot 13$; 11 does not divide 11,232.
16. 243 beans
17. a) $11.97 b) $11.81
c) Save 16 cents by using the 25% off coupon.
18. 8

CHAPTER 2

SECTION 2.1, PAGE 43

1. A set is a collection of objects.
3. Description, roster form, and set-builder notation; the set of even counting numbers less than 7, {2, 4, 6}, and $\{x \mid x \in N \text{ and } x < 7\}$
5. A set is finite if it either contains no elements or the number of elements in the set is a natural number.
7. Two sets are equivalent if they contain the same number of elements.
9. $N = \{1, 2, 3, 4, 5, \ldots\}$
11. The universal set is the set that contains all the elements for any specific discussion.
13. Not well defined 15. Well defined 17. Well defined
19. Infinite 21. Finite 23. Infinite
25. {Nebraska, Nevada, New Hampshire, New Jersey, New Mexico, New York, North Carolina, North Dakota}
27. {11, 12, 13, 14, . . . , 177} 29. $C = \{4\}$ 31. $\emptyset$
33. $E = \{6, 7, 8, 9, \ldots, 71\}$
35. $A = \{x \mid x \in N \text{ and } x < 10\}$ or $A = \{x \mid x \in N \text{ and } x \leq 9\}$
37. $C = \{x \mid x \in N \text{ and } x \text{ is a multiple of } 3\}$
39. $E = \{x \mid x \in N \text{ and } x \text{ is odd}\}$
41. $C = \{x \mid x$ is one of the three manufacturers of calculators with the greatest sales in the United States$\}$
43. A is the set of natural numbers less than or equal to 7.
45. L is the set of Great Lakes in the United States.
47. B is the set of the five tallest buildings in the United States.
49. E is the set of natural numbers greater than 5 and less than or equal to 12.
51. False; $\{b\}$ is a set, and not an element of the set.
53. False; h is not an element of the set.
55. False; 3 is an element of the set.
57. True 59. 4 61. 0 63. Both
65. Neither 67. Equivalent
69. a) A is the set of natural numbers greater than 2. B is the set of all numbers greater than 2.
b) Set A contains only natural numbers. Set B contains other types of numbers, including fractions and decimal numbers.
c) $A = \{3, 4, 5, 6, \ldots\}$
d) No; set B cannot be written in roster form since we cannot list all the elements in set B.
71. Cardinal 73. Ordinal
75. Answers will vary. 77. Answers will vary.

SECTION 2.2, PAGE 49

1. Set A is a subset of set B, symbolized $A \subseteq B$, if and only if all the elements of set A are also elements of set B.
3. If $A \subseteq B$, then every element of set A is an element of set B. If $A \subset B$, then every element of set A is an element of set B and set $A \neq$ set B.
5. The number of proper subsets is determined by the formula $2^n - 1$, where n is the number of elements in the set.
7. False; English is an element of the set, not a subset.
9. True 11. True
13. False; the set $\{\emptyset\}$ contains the element $\emptyset$. 15. True
17. False; the set {0} contains the element 0.
19. False; {5} is a subset of the set, not an element of the set.
21. False; no set is a proper subset of itself.
23. True 25. $B \subseteq A, B \subset A$ 27. $A \subseteq B, A \subset B$
29. $B \subseteq A, B \subset A$ 31. $A = B, A \subseteq B, B \subseteq A$ 33. { }
35. { }, {car}, {boat}, {car, boat}
37. a) { }, $\{a\}, \{b\}, \{c\}, \{d\}, \{a, b\}, \{a, c\}, \{a, d\}, \{b, c\}, \{b, d\}, \{c, d\}, \{a, b, c\}, \{a, b, d\}, \{a, c, d\}, \{b, c, d\}, \{a, b, c, d\}$
b) $\{a, b, c, d\}$

39. True **41.** False **43.** True **45.** True **47.** True

49. True **51.** 64 **53.** 128 **55.** $E = F$

57. a) Yes **b)** No **c)** Yes

SECTION 2.3, PAGE 56

1.

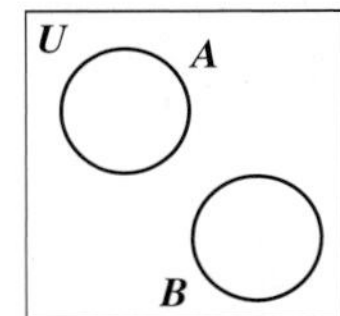

3.

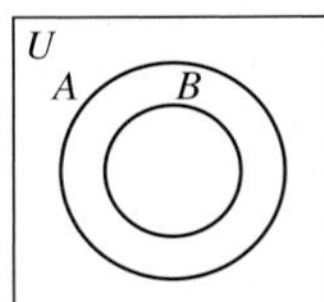

5.

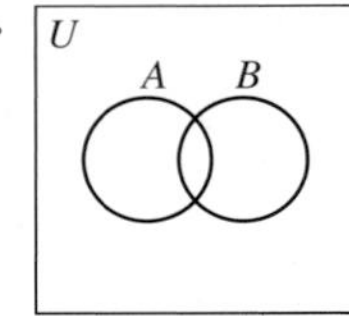

7. Combine the elements from set A and set B into one set. List any element that is contained in both sets only once.

9. a) *or* is generally interpreted to mean *union*.
b) *and* is generally interpreted to mean *intersection*.

11. Region II, the intersection of the two sets.

13.

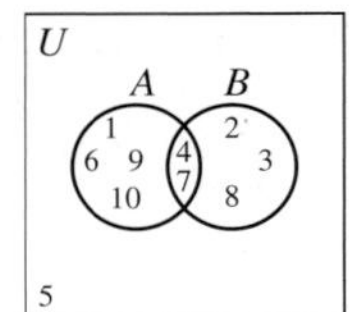

15.

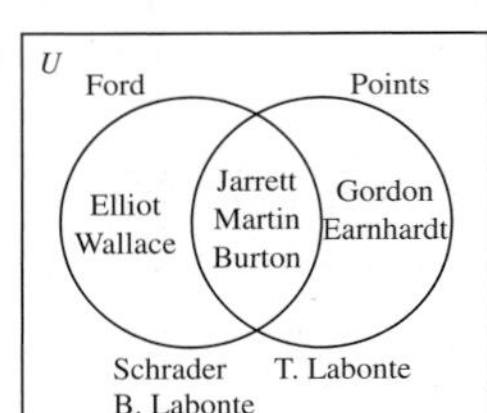

17. A' is the set of U.S. senators who did not vote in favor of the Hartley–Domingo bill.

19. The set of universities in the United States that do not have the word *State* in their name

21. The set of universities in the United States that have the word *State* or the word *South* in their name

23. The set of universities in the United States that have the word *State* in their name and do not have the word *South* in their name

25. The set of U.S. corporations whose headquarters are in New York State and whose chief executive officer is a woman

27. The set of U.S. corporations whose chief executive officer is a woman and who do not employ at least 100 people

29. The set of U.S. corporations whose headquarters are in New York State and whose chief executive officer is a woman and who employs at least 100 people

31. $\{a, f, g, h, r\}$ **33.** $\{c, w, b, t, a, h, f, g, r, p, m, z\}$

35. $\{c, w, b, t, a, h, f, g, r\}$ **37.** $\{c, w, b, t, f, g, r, p, m, z\}$

39. {L, △, @, *, $} **41.** {L, △, @, *, $, R, □, ∞, Σ, Z}

43. {*, $} **45.** {R, □}

47. $\{1, 2, 3, 4, 5, 6, 8\}$ **49.** $\{1, 5, 7, 8\}$ **51.** $\{7\}$

53. { } **55.** $\{7\}$ **57.** {b, e, h, j, k} **59.** {a, f, i}

61. {b, c, d, e, g, h, j, k} **63.** {a, c, d, e, f, g, h, i, j, k}

65. {b} **67.** { } **69.** $\{2, 4, 6, 8\}$, or B **71.** $\{7, 9\}$

73. $\{1, 3, 5, 6, 7, 8, 9\}$ **75.** $\{1, 2, 3, 4, 5, 7, 9\}$

77. $\{2, 4, 6, 8\}$, or B **79.** { }

81. A set and its complement will always be disjoint. For example, if $U = \{1, 2, 3\}$, $A = \{1, 2\}$, and $A' = \{3\}$, then $A \cap A' = \{\ \}$.

83. 44 **85. a)** $8 = 4 + 6 - 2$ **b)** and **c)** Answers will vary.

87. A **89.** B **91.** C

93. $\{2, 6, 10, 14, 18, \ldots\}$ **95.** $\{2, 6, 10, 14, 18, \ldots\}$

97. U **99.** A **101.** $\varnothing$ **103.** A **105.** $B \subseteq A$

107. A and B are disjoint sets. **109.** $A \subseteq B$

111. $\{e, f, h\}$ **113.** $\{d, j, k\}$ **115.** $\{13\}$

117. $\{1, 2, 3, 4, 5, 6, 7, 8, 9, 10, 11, 12, 14, 15\}$

119. $\{2, 3, 4, 5, 7, 9, 10, 11, 12, 13, 14, 15\}$

SECTION 2.4, PAGE 63

1. V, the intersection of all three sets **3.** 4

5. $(A \cup B)' = A' \cap B'$, $(A \cap B)' = A' \cup B'$

7.

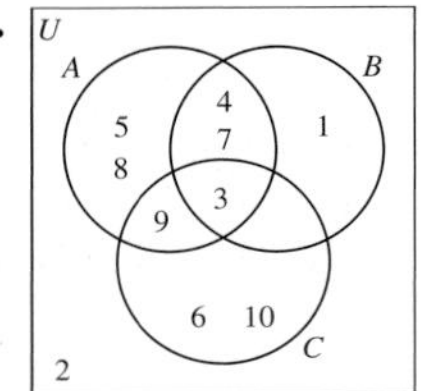

9.

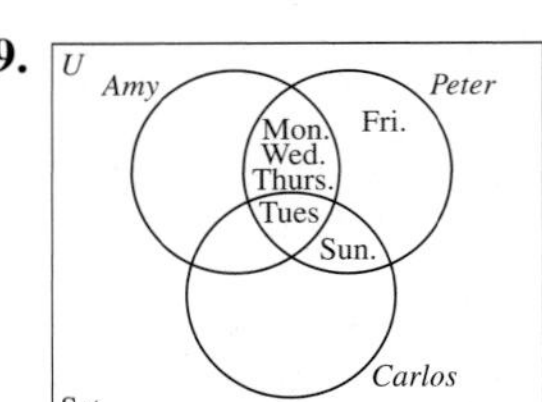

11.

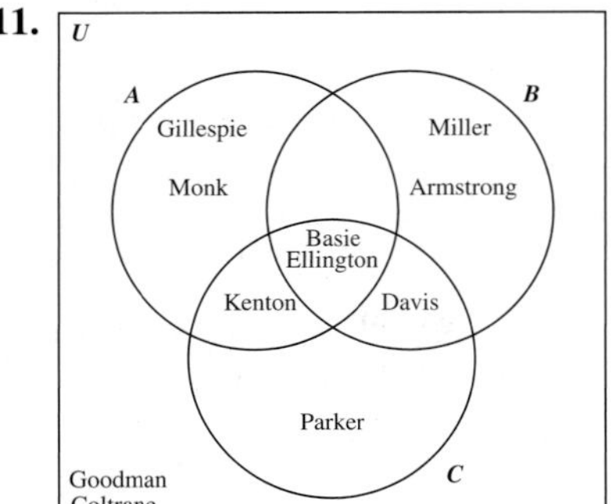

13.

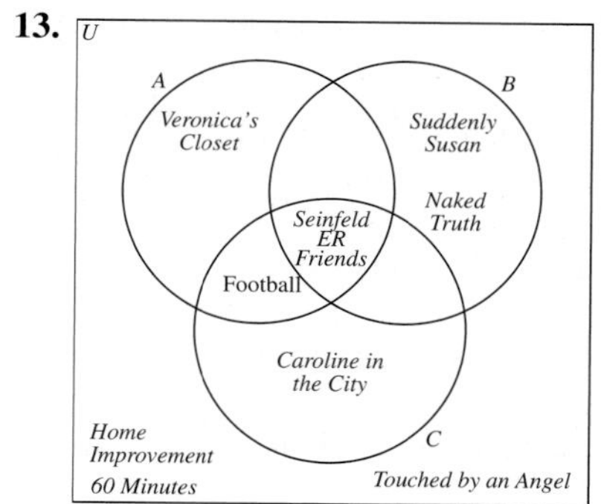

15. V **17.** VI **19.** IV **21.** VII **23.** VI **25.** V **27.** VI **29.** IV **31.** I **33.** V **35.** II **37.** VII **39.** I **41.** VIII **43.** VI **45.** {1, 2, 3, 4, 5, 6} **47.** {4, 5, 6, 7, 8, 10} **49.** {3, 4, 5} **51.** {1, 2, 3, 6, 9, 10, 11, 12} **53.** {1, 2, 3, 4, 5, 6, 7, 8, 9, 12} **55.** {9, 11, 12} **57.** {7, 8, 9, 10, 11, 12} **59.** Yes **61.** No **63.** No **65.** Yes **67.** No **69.** Yes **71.** Yes **73.** Yes **75.** No

77. **a)** Both equal {6, 7}. **b)** Answers will vary. **c)** Both are represented by the regions IV, V, VI.

79.

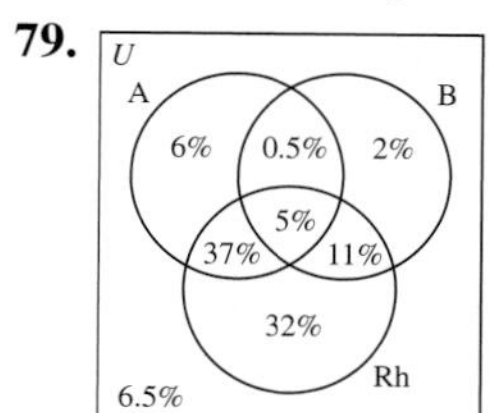

81. a)

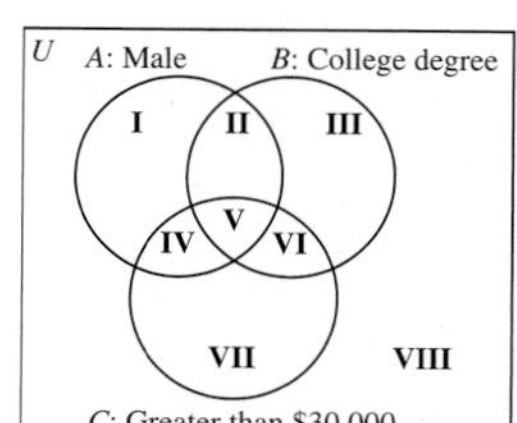

b) V; $A \cap B \cap C$
c) VI; $A' \cap B \cap C$
d) I; $A \cap B' \cap C'$

83. $n(A \cup B \cup C) = n(A) + n(B) + n(C) - 2n(A \cap B \cap C) - n(A \cap B \cap C') - n(A \cap B' \cap C) - n(A' \cap B \cap C)$

SECTION 2.5, PAGE 71

1.

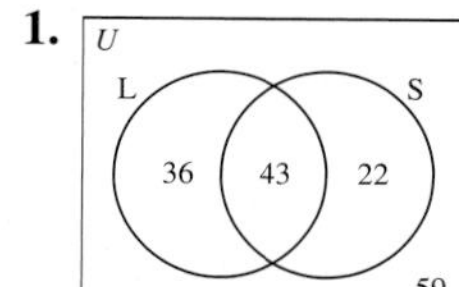

a) 36
b) 22
c) 59

3.

a) 39
b) 27
c) 101

5.

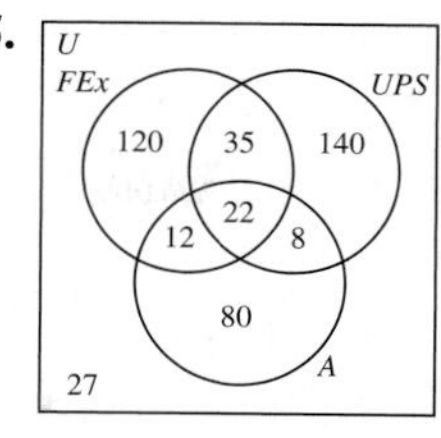

a) 27
b) 80
c) 340
d) 55
e) 337

7.

a) 2
b) 6
c) 22
d) 11
e) 12

9.

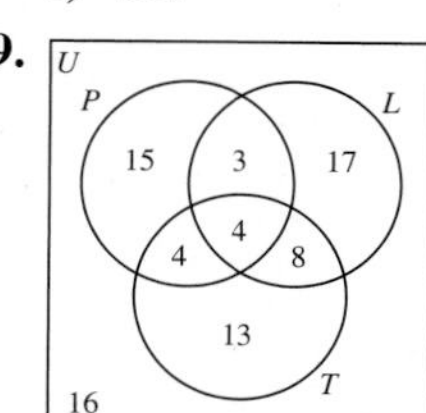

a) 13
b) 45
c) 64
d) 15
e) 16

11.

a) 185
b) 10
c) 25
d) 401

13. In a Venn diagram, regions II, IV, and V contain a total of 37 cars driven by women. This total is greater than the 35 cars driven by women, as given in the exercise.

15.

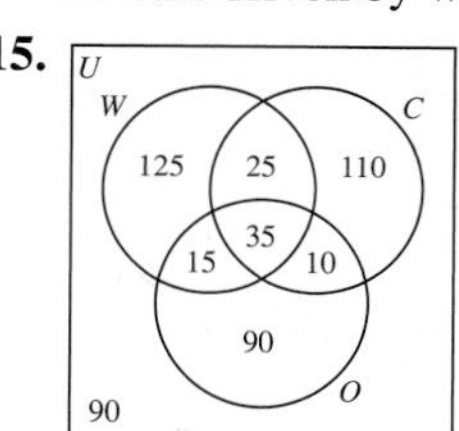

a) 410 **b)** 35 **c)** 90 **d)** 50

SECTION 2.6, PAGE 77

1. An infinite set is a set that can be placed in a one-to-one correspondence with a proper subset of itself.

3. {4, 5, 6, 7, . . . , $n + 3$, . . .}
↓ ↓ ↓ ↓ ↓
{5, 6, 7, 8, . . . , $n + 4$, . . .}

5. {6, 8, 10, 12, . . . , $2n + 4$, . . .}
↓ ↓ ↓ ↓ ↓
{8, 10, 12, 14, . . . , $2n + 6$, . . .}

7. {4, 7, 10, 13, . . . , $3n + 1$, . . .}
↓ ↓ ↓ ↓ ↓
{7, 10, 13, 16, . . . , $3n + 4$, . . .}

9. { 6, 11, 16, 21, 26, . . . , $5n + 1$, . . .}
↓ ↓ ↓ ↓ ↓ ↓
{11, 16, 21, 26, 31, . . . , $5n + 6$, . . .}

11. $\left\{1, \frac{1}{3}, \frac{1}{5}, \frac{1}{7}, \ldots, \frac{1}{2n-1}, \ldots\right\}$
$\downarrow\ \downarrow\ \downarrow\ \downarrow\ \qquad \downarrow$
$\left\{\frac{1}{3}, \frac{1}{5}, \frac{1}{7}, \frac{1}{9}, \ldots, \frac{1}{2n+1}, \ldots\right\}$

13. $\{1, 2, 3, 4, \ldots, n, \ldots\}$
$\downarrow\ \downarrow\ \downarrow\ \downarrow\ \qquad \downarrow$
$\{3, 6, 9, 12, \ldots, 3n, \ldots\}$

15. $\{1, 2, 3, 4, \ldots, n, \ldots\}$
$\downarrow\ \downarrow\ \downarrow\ \downarrow\ \qquad \downarrow$
$\{4, 6, 8, 10, \ldots, 2n + 2, \ldots\}$

17. $\{1, 2, 3, 4, \ldots, n, \ldots\}$
$\downarrow\ \downarrow\ \downarrow\ \downarrow\ \qquad \downarrow$
$\{2, 5, 8, 11, \ldots, 3n - 1, \ldots\}$

19. $\{1, 2, 3, 4, \ldots, n, \ldots\}$
$\downarrow\ \downarrow\ \downarrow\ \downarrow\ \qquad \downarrow$
$\{5, 8, 11, 14, \ldots, 3n + 2, \ldots\}$

21. $\{1, 2, 3, 4, \ldots, n, \ldots\}$
$\downarrow\ \downarrow\ \downarrow\ \downarrow\ \qquad \downarrow$
$\left\{\frac{1}{3}, \frac{1}{4}, \frac{1}{5}, \frac{1}{6}, \ldots, \frac{1}{n+2}, \ldots\right\}$

23. $\{1, 2, 3, 4, \ldots, n, \ldots\}$
$\downarrow\ \downarrow\ \downarrow\ \downarrow\ \qquad \downarrow$
$\{1, 4, 9, 16, \ldots, n^2, \ldots\}$

25. $\{1, 2, 3, 4, \ldots, n, \ldots\}$
$\downarrow\ \downarrow\ \downarrow\ \downarrow\ \qquad \downarrow$
$\{3, 9, 27, 81, \ldots, 3^n, \ldots\}$

REVIEW EXERCISES, PAGE 78

1. True

2. False; the word *best* makes the statement not well defined.

3. True

4. False; no set is a proper subset of itself.

5. False; the elements 6, 12, 18, 24, . . . are members of both sets.

6. True

7. False; both sets do not contain exactly the same elements.

8. True **9.** True **10.** True **11.** True **12.** True

13. True **14.** True **15.** $A = \{7, 9, 11, 13, 15\}$

16. $B = \{$TX, NM, CO, KS, MO, AR$\}$

17. $C = \{1, 2, 3, 4, \ldots, 296\}$

18. $D = \{9, 10, 11, 12, \ldots, 96\}$

19. $A = \{x \mid x \in N \text{ and } 72 < x < 100\}$

20. $B = \{x \mid x \in N \text{ and } x > 85\}$

21. $C = \{x \mid x \in N \text{ and } x < 3\}$

22. $D = \{x \mid x \in N \text{ and } 23 \le x \le 41\}$

23. *A* is the set of capital letters in the English alphabet from E through M, inclusive.

24. *B* is the set of U.S. coins with a value of less than a dollar.

25. *C* is the set of the last three lowercase letters in the English alphabet.

26. *D* is the set of numbers greater than or equal to 3 and less than 9.

27. {5, 7} **28.** {1, 3, 5, 7, 11, 15} **29.** {9, 13}

30. {1, 7, 11, 13, 15} **31.** 16 **32.** 15

33.

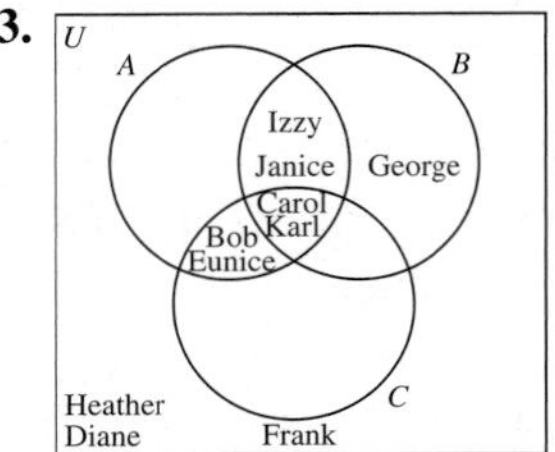

34. {a, c, d, e, f, g, i, k} **35.** {i, d}

36. {a, b, c, d, e, f, g, i, k} **37.** {e} **38.** {a, d, e}

39. {a, b, d, e, f, g} **40.** True **41.** True **42.** \$450

43.

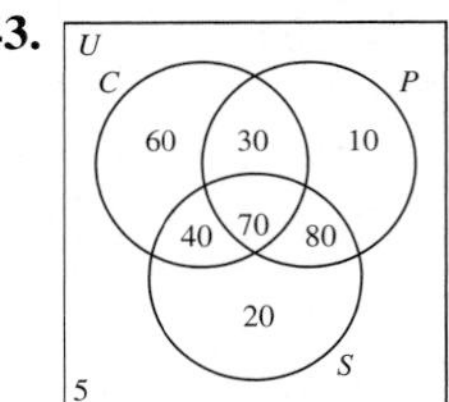

a) 315 **b)** 10 **c)** 30 **d)** 110

44.

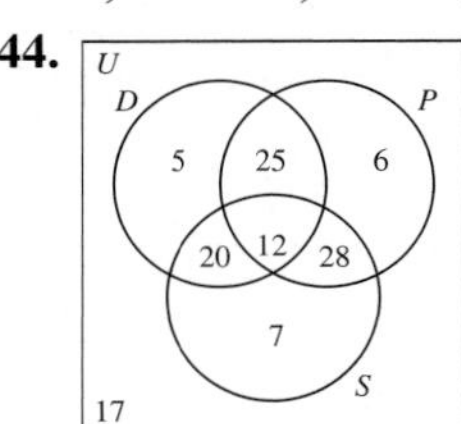

a) 5 **b)** 18 **c)** 73 **d)** 28 **e)** 41

45. $\{2, 4, 6, 8, \ldots, 2n, \ldots\}$
$\downarrow\ \downarrow\ \downarrow\ \downarrow\ \qquad \downarrow$
$\{4, 6, 8, 10, \ldots, 2n + 2, \ldots\}$

46. $\{3, 5, 7, 9, \ldots, 2n + 1, \ldots\}$
$\downarrow\ \downarrow\ \downarrow\ \downarrow\ \qquad \downarrow$
$\{5, 7, 9, 11, \ldots, 2n + 3, \ldots\}$

47. $\{1, 2, 3, 4, \ldots, n, \ldots\}$
$\downarrow\ \downarrow\ \downarrow\ \downarrow\ \qquad \downarrow$
$\{5, 8, 11, 14, \ldots, 3n + 2, \ldots\}$

48. $\{1, 2, 3, 4, \ldots, n, \ldots\}$
$\downarrow\ \downarrow\ \downarrow\ \downarrow\ \qquad \downarrow$
$\{4, 9, 14, 19, \ldots, 5n - 1, \ldots\}$

CHAPTER TEST, PAGE 79

1. True

2. False; the sets do not contain exactly the same elements.

3. True

4. False; the second set has no subset that contains the element 7.

5. False, the empty set is a proper subset of every set except itself.

6. False; the set has 2^3, or 8 subsets. **7.** True

8. False; for any set A, $A \cup A' = U$, not { }.

9. True

10. $A = \{1, 2, 3, 4, 5, 6, 7\}$

11. Set A is the set of natural numbers less than 8.

12. $\{7, 9\}$ **13.** $\{3, 5, 7, 9, 13\}$

14. $\{3, 5, 7, 9\}$, or A **15.** 2

16.

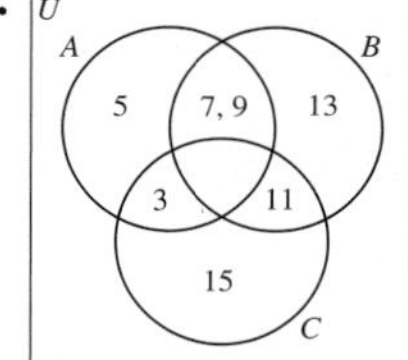

17. Equal

18. a)

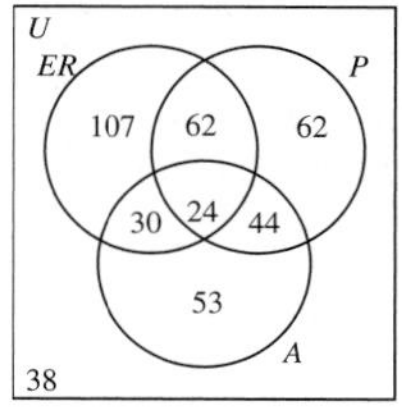

b) 222 **c)** 38 **d)** 136 **e)** 160 **f)** 231 **g)** 62

19. $\{7, 8, 9, 10, \ldots, n + 6, \ldots\}$
$\downarrow \; \downarrow \; \downarrow \; \downarrow \qquad \downarrow$
$\{8, 9, 10, 11, \ldots, n + 7, \ldots\}$

20. $\{1, 2, 3, 4, \ldots, \quad n, \ldots\}$
$\downarrow \; \downarrow \; \downarrow \; \downarrow \qquad \downarrow$
$\{1, 3, 5, 7, \ldots, \quad 2n - 1, \ldots\}$

CHAPTER 3

SECTION 3.1, PAGE 92

1. A statement that conveys only one idea

3. a) Some are **b)** All are **c)** Some are not **d)** None are

5. $\sim p$. (p: The ink *is* purple.)

7. When two simple statements are on the same side of the comma, they are placed together in parentheses when translated into symbolic form.

9. Compound; conjunction, $\wedge$

11. Compound; biconditional, $\leftrightarrow$

13. Compound; disjunction, $\vee$

15. Simple statement

17. Compound; negation, $\sim$

19. Compound; conjunction, $\wedge$

21. Compound; negation, $\sim$

23. No flowers are yellow.

25. Some fish do not swim.

27. All dogs have fleas.

29. Some books are round.

31. All rain forests are being destroyed.

33. No students maintain an A average.

35. $\sim p$ **37.** $\sim q \vee \sim p$ **39.** $\sim p \rightarrow \sim q$

41. $\sim q \leftrightarrow \sim p$ **43.** $\sim p \wedge \sim q$ **45.** $\sim(q \rightarrow \sim p)$

47. Firemen do not wear red suspenders.

49. Firemen wear red suspenders or firemen work hard.

51. If firemen do not work hard, then firemen wear red suspenders.

53. It is false that firemen wear red suspenders or firemen work hard.

55. Firemen do not work hard and firemen do not wear red suspenders.

57. $(p \wedge q) \vee r$ **59.** $(r \leftrightarrow \sim p) \vee \sim q$ **61.** $(r \wedge q) \rightarrow p$

63. $(r \leftrightarrow q) \wedge p$ **65.** $q \rightarrow (p \leftrightarrow r)$

67. The water is 70° and the sun is shining, or we go swimming.

69. If the sun is shining then the water is 70°, or we go swimming.

71. If we do not go swimming, then the sun is shining and the water is 70°.

73. If the sun is shining then we go swimming, and the water is 70°.

75. The sun is shining if and only if the water is 70°, and we go swimming.

77. Not permissible, you cannot have both soup and salad. The *or* used on menus is the exclusive *or.*

79. Not permissible, you cannot have both potatoes and pasta. The *or* used on menus is the exclusive *or.*

81. a) $(\sim p) \rightarrow q$ **b)** Conditional

83. a) $(\sim q) \wedge (\sim r)$ **b)** Conjunction

85. a) $(p \vee q) \rightarrow r$ **b)** Conditional

87. a) $r \rightarrow (p \vee q)$ **b)** Conditional

89. a) $(\sim p) \leftrightarrow (\sim q \rightarrow r)$ **b)** Biconditional

91. a) $(r \wedge \sim q) \rightarrow (q \wedge \sim p)$ **b)** Conditional

93. a) $\sim[(p \wedge q) \leftrightarrow (p \vee r)]$ **b)** Negation

95. a) $c \vee \sim d$ **b)** Disjunction

97. a) $\sim(p \rightarrow \sim a)$ **b)** Negation

99. a) $(f \vee v) \rightarrow e$ **b)** Conditional

101. a) $o \leftrightarrow (\sim s \vee f)$ **b)** Biconditional

103. a) $(e \leftrightarrow s) \vee f$ **b)** Disjunction

105. $[(\sim q) \rightarrow (r \vee p)] \leftrightarrow [(\sim r) \wedge q]$; Biconditional

107. a) The conjunction and disjunction have the same dominance. **b)** and **c)** Answers will vary.

SECTION 3.2, PAGE 103

1. a) 4 **b)**

p	q
T	T
T	F
F	T
F	F

3. a)

p	q	$p \wedge q$
T	T	T
T	F	F
F	T	F
F	F	F

b) Only when both simple statements are true

5.	**7.**	**9.**	**11.**	**13.**	**15.**
F	T	F	T	T	T
F	F	T	F	F	F
	T	T	T	T	T
	T	T	T	T	T
				F	T
				F	F
				T	T
				T	F

17.	**19.**	**21.** $p \wedge q$	**23.** $p \wedge \sim q$
F	T	T	F
F	T	F	T
T	T	F	F
F	T	F	F
F	F		
F	F		
T	T		
T	F		

25. $(p \vee q) \wedge \sim r$	**27.** $(p \wedge q) \wedge \sim r$	**29.** $p \wedge (q \vee \sim q)$
F	F	T
T	T	T
F	F	F
T	F	F
F	F	
T	F	
F	F	
F	F	

31. False **33.** False **35.** True **37.** True **39.** False

41. True **43.** True **45.** True **47.** True **49.** False

51. False **53.** True

55. $\sim p \wedge q$	**57.** $p \vee \sim q$
F	T
F	T
T	F
F	T

55. True in case 3 when p is false and q is true.

57. True in cases 1, 2, and 4, when p is true, or when p and q are both false.

59. $(p \wedge q) \vee r$	**61.** $q \vee (p \wedge \sim r)$
T	T
T	T
T	F
F	T
T	T
F	T
T	F
F	F

59. True in cases 1, 2, 3, 5, and 7. True except when p, q, r have truth values of TFF or FTF or FFF.

61. True in cases 1, 2, 4, 5, and 6. True except when p, q, r have truth values TFT, FFT, or FFF.

63. a) Ms. Duncan and Mrs. Tuttle qualify.
b) The Furmans do not qualify, since their combined income is less than $46,000.

65. a) Michael Bolinder qualifies; the other four do not.
b) Gina Vela is returning on April 2. Laura Griffin Heller is returning on a Monday. Christos Giakoumopoulas is not staying over on a Saturday. Alex Chang is returning on a Monday.

67.
T
T
T
T
F
T
F
T

69. Yes

SECTION 3.3, PAGE 112

1. a)

p	q	$p \rightarrow q$
T	T	T
T	F	F
F	T	T
F	F	T

b) The conditional is false only when the antecedent is true and the consequent is false.

3. a) Substitute the truth values for the simple statements. Then evaluate the compound statement, using the assigned truth values.
b) False

5. A tautology is a compound statement that is always true.

7.	**9.**	**11.**	**13.**	**15.**
F	F	F	T	F
T	T	T	T	T
T	T	T	F	T
T	F	F	T	T

17.	**19.**	**21.**	**23.**	**25.**
T	T	F	T	T
T	F	F	T	T
T	F	T	T	T
F	T	F	T	F
T	F	T	T	T
T	F	T	F	T
T	T	F	F	F
T	T	T	F	T

27. $p \rightarrow (q \wedge r)$	**29.** $(p \leftrightarrow \sim q) \vee r$
T	T
F	F
F	T
F	T
T	T
T	T
T	T
T	F

31. $(\sim p \rightarrow q) \vee r$
T
T
T
T
T
T
T
F

33. Neither **35.** Self-contradiction **37.** Tautology

39. Not an implication **41.** Implication **43.** Implication

45. True **47.** True **49.** False **51.** True **53.** True

55. True **57.** True **59.** False **61.** True **63.** True

65. True **67.** False **69.** False **71.** True **73.** True

75. True **77.** No

79.
T
F
F
T
T
F
F
F

81. A tautology

83.

Tiger	Boots	Sam	Sue
Blue	Yellow	Red	Green
Nine Lives	Whiskas	Friskies	Meow Mix

SECTION 3.4, PAGE 124

1. $\Leftrightarrow$

3. Construct truth tables for each statement. If both have the same truth values in the answer columns of the truth tables, the statements are equivalent.

5. $\sim(p \wedge q) \Leftrightarrow \sim p \vee \sim q$
$\sim(p \vee q) \Leftrightarrow \sim p \wedge \sim q$

7. a and c, b and d

9. Equivalent **11.** Not equivalent **13.** Equivalent

15. Equivalent **17.** Equivalent **19.** Not equivalent

21. Equivalent **23.** Equivalent **25.** Not equivalent

27. Not equivalent

29. Both $(p \rightarrow q) \wedge (q \rightarrow p)$ and $p \leftrightarrow q$ have the same truth values, T, F, F, T. Therefore, they are equivalent.

31. The boat is not at the dock and the boat will not depart.

33. It is false that the house has one phone line and the house has two phone lines.

35. It is false that the novel is written by Dinya Floyd or it is well illustrated.

37. If we go to Cozumel, then it is false that we will not go snorkeling and we will go to Senor Frogs.

39. You do not drink milk or your bones will be strong.

41. If John did not paint the picture, then Ada did not purchase the picture.

43. The noise is too loud and the police will not come.

45. You are 18 years old if and only if you are eligible to vote.

47. If the animal is a mammal then it is warm blooded and if an animal is warm blooded then it is a mammal.

49. Converse: If we go fishing, then the fish are biting.
Inverse: If the fish are not biting, then we will not go fishing.
Contrapositive: If we do not go fishing, then the fish are not biting.

51. Converse: If you have to give up your phone, then the phone bill is large.
Inverse: If the phone bill is not large, then you will not have to give up your phone.
Contrapositive: If you do not have to give up your phone, then the phone bill is not large.

53. Converse: If I will not get out of the car, then the dog is not friendly.
Inverse: If the dog is friendly, then I will get out of the car.
Contrapositive: If I will get out of the car, then the dog is friendly.

55. Converse: If we go down to the marina and take out the sailboat, then the sun is shinning.
Inverse: If the sun is not shinning, then we will not go down to the marina or we will not take out the sailboat.
Contrapositive: If we do not go down to the marina or we will not take out the sailboat, then the sun is not shinning.

57. If two angles of a triangle are equal, then the triangle is isosceles. True

59. If 2 divides the units digit of the counting number, then 2 divides the counting number. True

61. If two lines are not parallel, then the two lines intersect in at least one point. True

63. If the polygon is a quadrilateral, then the sum of the interior angles of the polygon measure 360°. True

65. b) and c) are equivalent.

67. a) and c) are equivalent.

69. a) and b) are equivalent.

71. a) and b) are equivalent.

73. a), b), and c) are equivalent.

75. None are equivalent.

77. a) and c) are equivalent.

79. a) and b) are equivalent.

81. Yes **83.** Yes

SECTION 3.5, PAGE 133

1. The conclusion necessarily follows from the premises.

3. Yes, if the conclusion does not follow from the set of premises.

5. If $(p_1 \wedge p_2) \rightarrow c$ is a tautology, then the argument is valid.

7. $p \rightarrow q$
 $\underline{p}$
 $\therefore q$

9. $p \rightarrow q$
 $\underline{\sim q}$
 $\therefore \sim p$

11. $p \rightarrow q$
 $\underline{q}$
 $\therefore p$

13. Invalid **15.** Valid **17.** Invalid **19.** Invalid

21. Invalid **23.** Valid **25.** Valid **27.** Invalid

29. Invalid **31.** Valid

33. $t \rightarrow b$
 $\underline{t}$
 $\therefore b$
 Valid

35. $w \vee r$
 $\underline{\sim r}$
 $\therefore w$
 Valid

37. $r \vee p$
 $\underline{p \rightarrow \sim r}$
 $\therefore \sim p$
 Invalid

39. $\sim t$
 $\underline{t \rightarrow m}$
 $\therefore m$
 Invalid

41. $g \vee f$
 $\underline{\sim f \rightarrow g}$
 $\therefore f \vee g$
 Valid

43. $c \rightarrow d$
 $\underline{d \leftrightarrow \sim w}$
 $\therefore c \rightarrow \sim w$
 Valid

45. $c \rightarrow m$
 $\underline{\sim m}$
 $\therefore \sim c$
 Valid

47. $b \rightarrow p$
 $\underline{\sim p}$
 $\therefore \sim b$
 Valid

49. $c \wedge \sim h$
 $\underline{h \rightarrow c}$
 $\therefore h$
 Invalid

51. $f \rightarrow d$
 $\underline{d \rightarrow \sim s}$
 $\therefore f \rightarrow s$
 Invalid

53. $l \rightarrow \mathrm{p}$
 $\underline{u \rightarrow p}$
 $\therefore l \rightarrow u$
 Invalid

55. Therefore, you will get an A.

57. Therefore, John fixes the car.

59. Therefore, you did not close the deal.

61. Therefore, if you do not pay off your credit card bills, then the bank makes money.

63. No. The conditional statement will always be true, and therefore it will be a tautology, and a valid argument.

SECTION 3.6, PAGE 139

1. A syllogism

3. The conclusion necessarily follows from the premises.

5. Yes, if the conclusion necessarily follows from the premises, the argument is valid.

7. Valid **9.** Valid **11.** Invalid **13.** Valid

15. Invalid **17.** Invalid **19.** Invalid **21.** Valid

23. Invalid **25.** Invalid **27.** Valid **29.** Invalid

REVIEW EXERCISES, PAGE 142

1. No people drink milk.

2. All dogs have fleas. **3.** Some butterflies bite.

4. No locks are keyless. **5.** Some pens do not use ink.

6. Some rabbits wear glasses.

7. The coffee is Maxwell House or the coffee is hot.

8. The coffee is not hot and the coffee is strong.

9. The coffee is Maxwell House if and only if the coffee is not strong.

10. If the coffee is hot, then the coffee is strong and the coffee is not Maxwell House.

11. The coffee is Maxwell House or the coffee is not hot, and the coffee is not strong.

12. The coffee is not Maxwell House, if and only if the coffee is strong and the coffee is not hot.

13. $p \rightarrow r$

14. $r \wedge q$

15. $(r \rightarrow q) \vee \sim p$

16. $(q \leftrightarrow p) \wedge \sim r$

17. $(r \wedge q) \vee \sim p$

18. $\sim(r \wedge q)$

19.	20.	21.	22.	23.	24.
F	T	T	F	T	F
F	F	F	T	T	T
T	F	T	F	T	T
F	F	T	F	T	T
		F	T	T	T
		F	T	F	T
		F	T	F	T
		F	T	T	T

25. False **26.** True **27.** False **28.** True **29.** True

30. True **31.** True **32.** True **33.** False **34.** False

35. Equivalent **36.** Not equivalent

37. Equivalent **38.** Not equivalent

39. If the stapler is not empty, then the stapler is jammed.

40. It is not true that the boy did not sing bass or the girl did not sing alto.
41. *Newsweek* is not a comic book and *Time* is an almanac.
42. There is water in the vase or the flowers will wilt.
43. It is not true that I went to the party or I finished my special report.
44. If you do not have to stop, then the railroad crossing light is not flashing red.
45. If John's eyes do not have to be checked, then John is not having difficulty seeing.
46. If I am not at work, then today is a holiday.
47. If the carpet stains, then the carpet is not Scotch-Guarded or the carpet is not properly cared for.
48. Converse: If I get a passing grade, then I studied.
Inverse: If I do not study, then I will not get a passing grade.
Contrapositive: If I do not get a passing grade, then I did not study.
49. a), b), and c) are equivalent. **50.** None are equivalent.
51. a) and c) are equivalent. **52.** None are equivalent.
53. Invalid **54.** Valid **55.** Invalid **56.** Valid
57. Invalid **58.** Invalid

CHAPTER TEST, PAGE 144

1. $(p \wedge r) \vee \sim q$
2. $(r \rightarrow q) \vee \sim p$
3. $\sim(r \leftrightarrow \sim q)$
4. It is false that if Celion is the president then Ron is not the secretary.
5. Celion is the president, if and only if Sheldon is the vice president and Ron is the secretary.

6.	7.
F	T
T	T
F	T
F	T
F	F
F	T
F	T
F	F

8. True **9.** True **10.** True **11.** True
12. Equivalent **13.** a) and b) are equivalent.
14. a) and b) are equivalent.
15. $s \rightarrow f$ **16.** Invalid
$$\begin{array}{l} f \rightarrow p \\ \hline \therefore s \rightarrow p \end{array}$$
Valid
17. Some leopards are not spotted.
18. No people are funny.
19. Inverse: If the apple is not red, then it is not a delicious apple.
Converse: If it is a delicious apple, then the apple is red.
Contrapositive: If it is not a delicious apple, then the apple is not red.
20. Yes

CHAPTER 4

SECTION 4.1, PAGE 153

1. A number is a quantity, and it answers the question, "How many?" A numeral is a symbol used to represent the number.
3. 𓍢, C, 百, ρ, 100
5. The Hindu–Arabic numeration system
7. In a multiplicative system, there are numerals for each number less than the base and for powers of the base. Each numeral less than the base is multiplied by a numeral for the power of the base, and these products are added to obtain the number.
9. 232 **11.** 2423 **13.** 334,214 **15.** 𓍢𓍢𓍢𓍢𓎆𓎆𓎆𓏺𓏺𓏺𓏺𓏺𓏺𓏺
17. 𓆼𓆼𓎆𓎆𓎆𓎆𓏺𓏺𓏺𓏺𓏺 **19.** 𓆐𓂭𓂭𓂭𓂭𓆼𓆼𓆼𓍢𓍢𓍢𓍢𓍢𓍢𓍢𓍢𓎆𓎆𓎆𓎆𓏺𓏺𓏺𓏺𓏺
21. 14 **23.** 547 **25.** 1492 **27.** 1945 **29.** 12,666
31. 9464 **33.** XLVII **35.** CLXIV **37.** MM
39. $\overline{\text{IV}}$DCCXCIII **41.** $\overline{\text{IX}}$CMXCIX **43.** $\overline{\text{XX}}$DCXLIV
45. 94 **47.** 4081 **49.** 7650
51. 四十七 **53.** 三百七十八 **55.** 三千五百七十
57. 264 **59.** 22,505 **61.** 9607
63. μ ζ **65.** ψ κ ϛ **67.** λ′ε′ψδ
69. Advantage: Can write some numbers more compactly.
Disadvantage: There are more numerals to memorize.
71. Advantage: Can write some numbers more compactly.
Disadvantage: There are more numerals to memorize.
73. 1936, 𓍢𓍢𓍢𓍢𓍢𓍢𓍢𓍢𓍢𓎆𓎆𓎆𓏺𓏺𓏺𓏺𓏺𓏺, α′πλϛ, 一千九百三十六
75. 422, 𓍢𓍢𓍢𓍢𓎆𓎆𓏺𓏺, CDXXII, 四百二十二
77. π′Q′θ′πQθ

SECTION 4.2, PAGE 159

1. A base 10 place-value system
3. A symbol for zero and for each counting number less than the base are required.
5. Write each digit times its corresponding positional value.
7. 11 and 660
9. 1, 20, 18×20, $18 \times (20)^2$, $18 \times (20)^3$
11. $(5 \times 10) + (7 \times 1)$
13. $(3 \times 100) + (5 \times 10) + (9 \times 1)$
15. $(8 \times 100) + (9 \times 10) + (7 \times 1)$
17. $(5 \times 1000) + (2 \times 100) + (6 \times 10) + (2 \times 1)$
19. $(1 \times 10{,}000) + (0 \times 1000) + (7 \times 100) + (3 \times 10) + (2 \times 1)$
21. $(3 \times 100{,}000) + (4 \times 10{,}000) + (6 \times 1000) + (8 \times 100) + (6 \times 10) + (1 \times 1)$

23. 24 **25.** 784 **27.** 4868 **29.** <<▼▼▼▼▼ **31.** ▼▼ ▼
33. ▼ ▼ <<▼▼▼▼▼ **35.** 51 **37.** 4321 **39.** 4000
41. ••• / ═ **43.** ═ / ⊝ **45.** ••• / ═ / •

47. Advantages: In general, they are more compact; large and small numbers can be written more easily; there are fewer symbols to memorize.
Disadvantages: If many of the symbols in the numeral represent zero, the place value system may be less compact.
49. 33, • / •••

51. (⌽ × ⎔²) + (□ × ⎔) + (△ × 1)

53. a) No largest number
b) ▼▼▼▼ <<<<▼▼▼▼ <<<< ▼▼▼▼▼▼ <<<<▼▼

55. ▼▼ <<<<<▼▼▼▼▼ **57.** ••• / •••• / •••

SECTION 4.3, PAGE 164

3. 7 **5.** 13 **7.** 11 **9.** 100 **11.** 243 **13.** 1367
15. 867 **17.** 83 **19.** 6597 **21.** 1000_2 **23.** 10110_2
25. 1161_7 **27.** 1239_{12} **29.** 1021_8 **31.** $17TE_{12}$
33. 1111110011_2 **35.** 4403_8 **37.** 2086 **39.** 447,415
41. $19C_{16}$ **43.** 1566_{16} **45.** 11111010001_2 **47.** 31001_5
49. $11T9_{12}$
51. The numeral is written incorrectly; there is no 5 in the set of numerals for base 5.
53. Written correctly
55. 13 **57.** 73 **59.** ⊖⦶$_5$ **61.** ⦶⦶⦶$_5$
63. 7 **65.** 36 **67.** ●●$_4$ **69.** ●●●$_4$

71. b) 10213_5 **c)** 1373_8
73. Answers will vary.
75. $b = 6$

SECTION 4.4, PAGE 173

1. a) $1, b, b^2, b^3, b^4$ **b)** $1, 6, 6^2, 6^3, 6^4$
5. 123_5 **7.** 3323_4 **9.** $9E5_{12}$ **11.** 2200_3
13. 24001_7 **15.** 10001_2 **17.** 203_4 **19.** 1134_5
21. 644_{12} **23.** 11_2 **25.** 3616_7 **27.** 1011_3 **29.** 123_5
31. 2403_7 **33.** 21020_6 **35.** 6072_9 **37.** 100011_2
39. 6031_7 **41.** 110_2 **43.** 22_5 **45.** 123_4 **47.** 33_4 R1_4
49. 41_5 R1_5 **51.** 45_7 R2_7 **53.** ⊖○$_5$ **55.** ⊖⊖⦶$_5$
57. ●●$_4$ **59.** ●●●$_4$ **61.** ●●$_4$ **63.** ●●●$_4$
65. 2302_5, 327 **67.** 13_5
69. a) 21252_8 **b)** 306 and 29 **c)** 8874 **d)** 8874 **e)** Yes

SECTION 4.5, PAGE 177

1. Duplation and mediation, the galley method, and Napier rods
3. b) 11,421
5. 493 **7.** 1458 **9.** 8260 **11.** 7225 **13.** 2555
15. 2332 **17.** 900 **19.** 204,728 **21.** 208 **23.** 438
25. 625 **27.** 60,678
29. a) 253×46 **b)** 11,638
31. a) 4×382 **b)** 1528
33. 99∩∩∩∩∩∩∩∩||||||
35. 2222_3

REVIEW EXERCISES, PAGE 178

1. 2102 **2.** 1112 **3.** 1311 **4.** 2114 **5.** 3214 **6.** 2312
7. *bbbbbbaaaaaaa* **8.** *cbbaaaaa* **9.** *ccbbbbbbbbbaaa*
10. *dda* **11.** *dddddcccccccbbbbba* **12.** *ddcccbaaaa*
13. 35 **14.** 27 **15.** 749 **16.** 4068 **17.** 5648 **18.** 4809
19. *gxd* **20.** *byixe* **21.** *hyfxb* **22.** *bzbx* **23.** *fzd*
24. *bza* **25.** 76 **26.** 308 **27.** 568 **28.** 46,883
29. 40,082 **30.** 60,529 **31.** *mb* **32.** *xpe* **33.** *Vrc*
34. *BArg* **35.** *ODvog* **36.** *OFvrf* **37.** 𓆼9999∩∩∩∩∩∩||
38. MCDLXII **39.** 一千四百六十二 **40.** α′υξβ **41.** <<▼▼▼▼ <<▼▼
42. •••• / • / •• **43.** 122,025 **44.** 8254 **45.** 585 **46.** 1991
47. 1277 **48.** 1971 **49.** 49 **50.** 5 **51.** 28
52. 1510 **53.** 1451 **54.** 186 **55.** 13033_4 **56.** 122011_3
57. 111001111_2 **58.** 3323_5 **59.** 327_{12} **60.** 717_8

61. 141_6 **62.** 101111_2 **63.** 176_{12} **64.** 1023_7
65. 12102_5 **66.** 10423_8 **67.** 3411_7 **68.** 100_2
69. $3E4_{12}$ **70.** 3324_5 **71.** 450_8 **72.** 1102_3 **73.** 143_5
74. 1203_4 **75.** 5656_{12} **76.** 21102_3 **77.** 110111_2
78. 13632_8 **79.** 1011_2 **80.** 130_4 **81.** 30_5 **82.** 433_6
83. 411_6 R 1_6 **84.** 664_8 R 2_8 **85.** 3408 **86.** 3408
87. 3408

CHAPTER TEST, PAGE 180

1. A number is a quantity and answers the question "How many?" A numeral is a symbol used to represent the number.

2. 2647 **3.** 1275 **4.** 8090 **5.** 944 **6.** 22,142

7. 9999 **8.** 999∩∩∩∩∩|| **9.** β' υ o ϟ

10. ••••
•••
••
═ **11.** <<▼▼▼▼▼▼▼ <<<▼▼▼▼▼▼▼ **12.** MMCCCLXXVIII

13. In an additive system, the number represented by a particular set of numerals is the sum of the values of the numerals.

14. In a multiplicative system, there are numerals for each number less than the base and for powers of the base. Each numeral less than the base is multiplied by a numeral for the power of the base, and these products are added to obtain the number.

15. In a ciphered system, the number represented by a particular set of numerals is the sum of the values of the numerals. There are numerals for each number up to and including the base and multiples of the base.

16. In a place-value system, each number is multiplied by a power of the base. The position of the numeral indicates the power of base by which it is multiplied.

17. 31 **18.** 103 **19.** 45 **20.** 305 **21.** 100100_2
22. 314_5 **23.** 1444_{12} **24.** 11365_7 **25.** 1122_5
26. 241_7 **27.** 2003_6 **28.** 220_5 **29.** 392 **30.** 8428

CHAPTER 5

SECTION 5.1, PAGE 190

1. Number theory is the study of numbers and their properties.

3. a) *a* divides *b* means that *b* divided by *a* has a remainder of zero.
b) *a* is divisible by *b* means that *a* divided by *b* has a remainder of zero.

5. A composite number is a natural number that is divisible by a number other than itself and 1.

7. a) The GCD of a set of natural numbers is the largest natural number that divides every number in that set.
b) Answers will vary. **c)** 8

9. Mersenne primes are prime numbers of the form $2^n - 1$, where *n* is a prime number.

11. Goldbach's conjecture states that every even number greater than or equal to 4 can be represented as the sum of two (not necessarily distinct) prime numbers.

13. The prime numbers between 1 and 75 are 2, 3, 5, 7, 11, 13, 17, 19, 23, 29, 31, 37, 41, 43, 47, 53, 59, 61, 67, 71, and 73.

15. True **17.** True

19. False; 42 is divisible by 7. **21.** True

23. False; if a number is divisible by 3, then the sum of the digits of the number is divisible by 3.

25. True **27.** 48,324 is divisible by 2, 3, 4, and 6.

29. 2,763,105 is divisible by 3 and 5.

31. 1,882,320 is divisible by 2, 3, 4, 5, 6, 8, and 10.

33. 60 (other answers are possible)

35. $44 = 2^2 \times 11$ **37.** $72 = 2^3 \times 3^2$ **39.** $303 = 3 \times 101$

41. $513 = 3^3 \times 19$ **43.** $1336 = 2^3 \times 167$

45. $2001 = 3 \times 23 \times 29$

47. a) 3 **b)** 90 **49. a)** 14 **b)** 168

51. a) 20 **b)** 1800 **53. a)** 4 **b)** 5088

55. a) 8 **b)** 384 **57.** 17, 19, and 29, 31

59. $4 = 2 + 2$, $6 = 3 + 3$, $8 = 3 + 5$, $10 = 3 + 7$, $12 = 5 + 7$, $14 = 7 + 7$, $16 = 3 + 13$, $18 = 5 + 13$, $20 = 3 + 17$.

61. 5, 17, and 257 are all prime.

63. 120 days **65.** 90 days **67.** 96 cars

69. a) 5, 7, 11, 13, 17, 19, 23, and 29
b) Every prime number greater than 3 differs by 1 from a multiple of 6.
c) This conjecture should appear to be correct.

71. 5 **73.** 2 **75.** 30 **77.** No **79.** Yes

81. a) 12 **b)** 1, 2, 3, 4, 5, 6, 10, 12, 15, 20, 30, 60

83. For any three consecutive natural numbers, one of the numbers is divisible by 2 and another number is divisible by 3. Therefore, the product of the three numbers would be divisible by 6.

85. Yes

87. $8 = 2 + 3 + 3$, $9 = 3 + 3 + 3$, $10 = 2 + 3 + 5$, $11 = 2 + 2 + 7$, $12 = 2 + 5 + 5$, $13 = 3 + 3 + 7$, $14 = 2 + 5 + 7$, $15 = 3 + 5 + 7$, $16 = 2 + 7 + 7$, $17 = 5 + 5 + 7$, $18 = 2 + 5 + 11$, $19 = 3 + 5 + 11$, $20 = 2 + 7 + 11$

SECTION 5.2, PAGE 199

1. Begin at zero. Represent the first addend with an arrow. Draw the arrow to the right if the addend is positive, to the left if negative. From the tip of the first arrow, represent the second addend with a second arrow. The sum of the two integers is at the tip of the second arrow.

3. The product of two numbers with like signs is positive. The product of two numbers with unlike signs is negative.

5. 0 **7.** 3 **9.** 6 **11.** -5 **13.** 2 **15.** -21 **17.** -3

19. -13 **21.** -2 **23.** -6 **25.** -2 **27.** -18 **29.** 49
31. 96 **33.** -60 **35.** -720 **37.** 3 **39.** -1 **41.** -7
43. -15 **45.** -48

In Exercises 47–55, false answers can be modified in a variety of ways. We give one possible answer.

47. False; every integer is not a natural number.

49. False; the difference of two negative integers may be a positive integer, a negative integer, or zero.

51. True **53.** True

55. False; the sum of a positive integer and a negative integer may be a positive integer, a negative integer, or zero.

57. 7 **59.** -17 **61.** -12 **63.** -5 **65.** -6
67. $-9, -5, -3, -1, 0, 7$ **69.** $-6, -5, -4, -3, -2, -1$
71. 26°F **73.** 1769 ft **75.** gained 1 point
77. a) 9 hours **b)** 2 hours **79.** -1
81. $0 + 1 - 2 + 3 + 4 - 5 + 6 - 7 - 8 + 9 = 1$

SECTION 5.3, PAGE 210

1. The set of rational numbers is the set of numbers of the form p/q, where p and q are integers and $q \neq 0$.

3. a) Divide both the numerator and the denominator by their greatest common factor.
b) $\frac{4}{15}$

5. Divide the numerator by the denominator. The quotient is the integer part of the mixed number. The fractional part of the mixed number is the remainder divided by the divisor.

7. a) The reciprocal of a number is 1 divided by the number.
b) $-\frac{1}{5}$

9. a) To add or subtract two fractions with a common denominator, perform the indicated operation on the numerators. Keep the common denominator. Reduce the new fraction to lowest terms, if possible.
b) $\frac{7}{9}$

11. Answers will vary. **13.** $\frac{2}{3}$ **15.** $\frac{9}{14}$ **17.** $\frac{21}{32}$ **19.** $\frac{7}{11}$
21. $\frac{1}{11}$ **23.** $\frac{21}{8}$ **25.** $-\frac{11}{4}$ **27.** $-\frac{79}{16}$ **29.** $\frac{17}{8}$ **31.** $\frac{15}{8}$
33. $1\frac{15}{16}$ **35.** $-42\frac{3}{5}$ **37.** $-58\frac{8}{15}$ **39.** 0.25
41. $0.\overline{571428}$ **43.** 0.375 **45.** $4.\overline{3}$ **47.** $5.\overline{6}$ **49.** $\frac{6}{10} = \frac{3}{5}$
51. $\frac{52}{1000} = \frac{13}{250}$ **53.** $\frac{62}{10} = \frac{31}{5}$ **55.** $\frac{1452}{1000} = \frac{363}{250}$ **57.** $\frac{30{,}001}{10{,}000}$
59. $\frac{1}{3}$ **61.** $\frac{3}{1}$ **63.** $\frac{15}{11}$ **65.** $\frac{46}{45}$ **67.** $\frac{574}{165}$ **69.** $\frac{6}{35}$ **71.** $\frac{2}{5}$
73. $\frac{49}{64}$ **75.** $\frac{36}{35}$ **77.** $\frac{5}{14}$ **79.** $\frac{11}{30}$ **81.** $\frac{23}{110}$ **83.** $\frac{23}{54}$ **85.** $\frac{17}{144}$
87. $-\frac{109}{600}$ **89.** $\frac{17}{12}$ **91.** $\frac{41}{28}$ **93.** $\frac{19}{24}$ **95.** $\frac{23}{60}$ **97.** $\frac{4}{11}$ **99.** $\frac{23}{42}$
101. $\frac{59}{60}$ **103.** $\frac{1}{10}$ **105.** 27 pages **107.** $2\frac{3}{8}$ points
109. $18\frac{3}{4}$ cups **111.** $50\frac{1}{16}$ in. **113.** $12\frac{7}{16}$ in. **115.** $26\frac{5}{32}$ in.
117. a) $37\frac{5}{6}$ ft **b)** 88 ft^2 **c)** $806\frac{2}{3}$ ft^3

In Exercises 119–125 an infinite number of answers are possible. We give one answer.

119. 0.255 **121.** -2.1755 **123.** 3.1234505
125. 4.8725 **127.** $\frac{7}{10}$ **129.** $\frac{3}{40}$ **131.** $\frac{9}{40}$ **133.** $\frac{11}{200}$
135. a) $1\frac{3}{8}$ cup water (or milk) and $\frac{3}{4}$ cup oatmeal
b) $1\frac{1}{2}$ cup water (or milk) and $\frac{3}{4}$ cup oatmeal

SECTION 5.4, PAGE 219

1. A rational number can be written as a ratio of two integers. Real numbers that cannot be written as a ratio of two integers are irrational numbers.

3. A perfect square is any number that is the square of a natural number.

5. a) To add or subtract two or more square roots with the same radicand, add or subtract their coefficients and then multiply the sum or difference by the common radical.
b) $8\sqrt{3}$

7. To rationalize a denominator means to write an equivalent expression that does not contain a radical in the denominator.

9. Rational **11.** Rational **13.** Irrational
15. Rational **17.** Irrational **19.** 9 **21.** 7
23. -13 **25.** -15 **27.** -10
29. Rational number, integer, natural number
31. Rational number, integer, natural number
33. Rational number **35.** Rational number
37. Rational number **39.** $2\sqrt{3}$ **41.** $2\sqrt{13}$
43. $3\sqrt{7}$ **45.** $4\sqrt{5}$ **47.** $9\sqrt{2}$ **49.** $7\sqrt{5}$
51. $-2\sqrt{7}$ **53.** $-13\sqrt{3}$ **55.** $4\sqrt{3}$ **57.** $23\sqrt{2}$
59. $\sqrt{6}$ **61.** $3\sqrt{15}$ **63.** $10\sqrt{2}$ **65.** $\sqrt{2}$ **67.** 3
69. $\dfrac{7\sqrt{2}}{2}$ **71.** $\dfrac{\sqrt{65}}{13}$ **73.** $\dfrac{2\sqrt{15}}{3}$
75. $\dfrac{3\sqrt{2}}{2}$ **77.** $\dfrac{\sqrt{15}}{3}$

79. $\sqrt{15}$ is between 3 and 4 since 15 is between 9 and 16. $\sqrt{15}$ is between 3.5 and 4 since 15 is closer to 16 than to 9.

81. $\sqrt{107}$ is between 10 and 11 since 107 is between 100 and 121. $\sqrt{107}$ is between 10 and 10.5 since 107 is closer to 100 than to 121.

83. $\sqrt{170}$ is between 13 and 14 since 170 is between 169 and 196. $\sqrt{170}$ is between 13 and 13.5 since 170 is closer to 169 than to 196.

In Exercises 85–89, false answers can be modified in a variety of ways. We give one possible answer.

85. True

87. False. The sum of two irrational numbers may be a rational number or an irrational number.

89. False. The product of a rational number and an irrational number may be a rational number or an irrational number.

91. $3\sqrt{2} + 5\sqrt{2} = 8\sqrt{2}$ **93.** $\sqrt{2} \cdot \sqrt{3} = \sqrt{6}$

95. $\sqrt{3} \neq 1.732$ since $\sqrt{3}$ is irrational and 1.732 is rational.

97. $\sqrt{9 + 16} \neq 3 + 4, 5 \neq 7$ **99.** $\frac{\pi\sqrt{7}}{7} \approx 1.2$ sec

101. a) 2.5 sec **b)** 5 sec **c)** 7.5 sec **d)** 10 sec

103. No. The sum of two irrational numbers may not be an irrational number.

SECTION 5.5, PAGE 224

1. The real numbers are the union of the rational numbers and the irrational numbers.

3. If whenever the operation is performed on two elements of a set the result is also an element of the set, then the set is closed under that operation.

5. $a \cdot b = b \cdot a$, the order in which two numbers are multiplied is immaterial. One example is $4(5) = 5(4)$.

7. $(a \cdot b) \cdot c = a \cdot (b \cdot c)$, when multiplying three numbers, you may place parentheses around any two adjacent numbers. One example is $(1 \times 2) \times 3 = 1 \times (2 \times 3)$.

9. No **11.** Yes **13.** Yes **15.** Yes **17.** Yes

19. No **21.** No **23.** No **25.** Yes **27.** Yes

29. Commutative property of addition. The only difference between the expressions on both sides of the equal sign is the order of the 7 and 8 being added.

31. $(-3)(-4) = (-4)(-3) = 12$ **33.** No. $3 \div 4 \neq 4 \div 3$.

35. $[(-2)(-3)](-4) = (-2)[(-3)(-4)] = -24$

37. No. $(16 \div 8) \div 2 \neq 16 \div (8 \div 2)$.

39. No. $(81 \div 9) \div 3 \neq 81 \div (9 \div 3)$.

41. Associative property of addition

43. Commutative property of multiplication

45. Associative property of addition

47. Associative property of addition

49. Commutative property of addition

51. Commutative property of addition

53. Distributive property

55. Commutative property of addition

57. $3y + 12$ **59.** $3\sqrt{2} + 2\sqrt{3}$

61. $x\sqrt{5} + 5$ **63.** $3 - 3\sqrt{2}$

65. Distributive property

67. Distributive property

69. Commutative property of addition

71. Distributive property

73. Associative property of addition

75. No **77.** Yes **79.** No **81.** Yes

83. No **85.** No **87.** No

89. Yes. The man can take off his jacket and then his sweater, or he can take off his sweater and then his jacket. Either way, the end result is the same.

91. No. $0 \div a = 0$ (when $a \neq 0$), but $a \div 0$ is undefined.

SECTION 5.6, PAGE 234

1. The 4 is the base and the 6 is the exponent or power.

3. a) To multiply two exponential expressions with the same base, add the exponents and use this sum as the exponent on the common base.
b) $3^3 \times 3^5 = 3^{3+5} = 3^8$

5. a) A non-zero expression raised to a negative exponent equals 1 divided by that expression raised to that positive exponent.
b) $3^{-5} = \frac{1}{3^5}$

7. a) Any base with an exponent raised to another exponent is equal to the base raised to the product of the exponents.
b) $(4^4)^3 = 4^{4(3)} = 4^{12}$

9. a) Move the decimal point in the original number to the right or left until you obtain a number greater than or equal to 1 and less than 10. Count the number of places the decimal was moved. If it was moved to the left, the count is a positive number; if it was moved to the right, the count is a negative number. Multiply the number obtained in the first step by 10 raised to this count.
b) 4.26×10^{-4}

11. a) The number is greater than or equal to 10.
b) The number is greater than or equal to 1 but less than 10.
c) The number is less than 1.

13. 9 **15.** 25 **17.** -32 **19.** $\frac{16}{25}$ **21.** 16 **23.** 72

25. 25 **27.** $\frac{1}{49}$ **29.** 1 **31.** 81 **33.** $\frac{1}{9}$ **35.** 4096

37. 121 **39.** 16 **41.** -16 **43.** $\frac{1}{64}$ **45.** 1.2×10^5

47. 4.5×10^1 **49.** 5.3×10^{-2} **51.** 1.9×10^4

53. 1.86×10^{-4} **55.** 4.23×10^{-6} **57.** 7.11×10^2

59. 1.53×10^{-1} **61.** 84,000 **63.** 0.012

65. 0.0000213 **67.** 0.312 **69.** 9,000,000 **71.** 231

73. 35,000 **75.** 10,000 **77.** 120,000,000

79. 0.0153 **81.** 320 **83.** 0.0021 **85.** 20

87. 4.2×10^{12} **89.** 4.5×10^{-7} **91.** 2.0×10^3

93. 2.0×10^{-7} **95.** 3.0×10^8

97. 8.3×10^{-4}; 3.2×10^{-1}; 4.6; 5.8×10^5

99. 8.3×10^{-5}; 0.00079; 4.1×10^3; 40,000

101. a) 4,752,000,000 people **b)** 4.752×10^9

103. a) 11.95 hours **b)** 1.195×10^1

105. a) 290,000,000 cells **b)** 2.9×10^8

107. a) 8,640,000,000 ft^3 **b)** 8.64×10^9

109. a) 1.8×10^{10} **b)** 3,332,000 miles

111. a) \$1,360,000,000 **b)** \$1,360,000,000 **c)** \$340,000,000 **d)** \$340,000,000

113. 1,000,000 mg = 1 kg

115. a) 12,000,000,000 people **b)** 469,667 people per day

117. a) 1×10^6, 1×10^9, 1×10^{12} **b)** 1000 days (about 2.74 years) **c)** 1,000,000 days (about 2739.73 years) **d)** 1,000,000,000 days (about 2,739,726.03 years) **e)** 1000 times greater

119. a) 1024 bacteria **b)** about 1448 bacteria

SECTION 5.7, PAGE 242

1. A sequence is a list of numbers that are related to each other by a given rule. One example is 1, 3, 5, 7, 9, ...

3. It is one in which each term differs from the preceding term by a constant amount. One example is 4, 7, 10, 13, 16, ...

5. It is one in which the ratio of any two successive terms is a constant amount. One example is 3, 6, 12, 24, ...

7. 2, 6, 10, 14, 18 **9.** −3, 0, 3, 6, 9 **11.** 5, 3, 1, −1, −3

13. $\frac{1}{2}, 1, \frac{3}{2}, 2, \frac{5}{2}$ **15.** 16 **17.** 11 **19.** $-\frac{91}{5}$ **21.** 9

23. $a_n = 2n$ **25.** $a_n = 10n - 4$ **27.** $a_n = \frac{1}{3}n - 2$

29. $a_n = \frac{3}{2}n - \frac{9}{2}$ **31.** $s_{14} = 105$ **33.** $s_9 = 225$

35. $s_8 = -52$ **37.** $s_8 = 60$ **39.** 2, 8, 32, 128, 512

41. 4, −12, 36, −108, 324 **43.** −3, 3, −3, 3, −3

45. −16, 8, −4, 2, −1 **47.** 3072 **49.** 10,935 **51.** 7290

53. −2187 **55.** $a_n = 3 \cdot 3^{n-1}$ **57.** $a_n = -5 \cdot (-1)^{n-1}$

59. $a_n = \frac{1}{4} \cdot (2)^{n-1}$ **61.** $a_n = 9 \cdot \left(\frac{1}{3}\right)^{n-1}$ **63.** 45

65. 27,305 **67.** −620,011 **69.** −3,188,648 **71.** 1275

73. 2500 **75. a)** \$28,600 **b)** \$195,200 **77.** 52.4288 g

79. 78 times **81.** 32,768 layers **83.** Visitors 36, Home 255

85. a) \$32, \$31 **b)** \$320, \$310 **c)** \$1024, \$1023 **d)** \$10,240; \$10,230 **e)** It is dangerous because it is likely that you won't have enough money to continue doubling your previous bet if you lose several times in a row. (Also, there may be a maximum amount you can bet in a game of chance.)

87. $a_n = 180n - 360, n \geq 3$ **89.** $r = 3, a_1 = 8$

SECTION 5.8, PAGE 249

1. The first and second terms are one. Each term thereafter is the sum of the previous two terms.

3. 0.6180 **5.** Answers will vary.

7. a) 1.618 **b)** 0.618 **c)** 1

9. $\frac{1}{1} = 1, \frac{2}{1} = 2, \frac{3}{2} = 1.5, \frac{5}{3} \approx 1.667, \frac{8}{5} = 1.6, \frac{13}{8} = 1.625, \frac{21}{13} \approx 1.615, \frac{34}{21} \approx 1.619, \frac{55}{34} \approx 1.6176, \frac{89}{55} \approx 1.6182$. The consecutive ratios alternate, increasing and decreasing about the golden ratio.

11. Each number in the Fibonacci sequence is either a prime number or is relatively prime with the number preceding or succeeding it in the sequence. Therefore, the GCF of any two consecutive Fibonacci numbers is 1.

13. Answers will vary. **15.** Answers will vary.

17. Answers will vary. **19.** Answers will vary.

21. Answers will vary. **23.** Yes; 50, 81 **25.** No

27. Yes; 105, 170 **29.** Yes; −1, −1

31. Answers will vary. **33.** Answers will vary.

35. a) 1, 3, 4, 7, 11, 18, 29, 47 **b)** 8 + 21 = 29, 13 + 34 = 47 **c)** It is the Fibonacci sequence.

37. Answers will vary. **39.** Answers will vary.

REVIEW EXERCISES, PAGE 252

1. 2, 3, 4, 5, 6, 8, 10 **2.** 2, 3, 4, 6, 9 **3.** $2^3 \times 41$

4. $2 \times 5^2 \times 7$ **5.** $2^3 \times 3 \times 5 \times 7$ **6.** $2 \times 3^2 \times 7^2$

7. $2^2 \times 3 \times 11^2$ **8.** 12; 36 **9.** 4; 936 **10.** 5; 2250

11. 40; 6720 **12.** 4; 480 **13.** 36; 432 **14.** 45 days

15. −2 **16.** 2 **17.** −4 **18.** −6 **19.** −9 **20.** 3

21. 0 **22.** 4 **23.** 24 **24.** −21 **25.** −15 **26.** 5

27. −2 **28.** 6 **29.** 6 **30.** 3 **31.** 0.8 **32.** $0.\overline{7}$

33. 0.75 **34.** 3.25 **35.** $0.\overline{428571}$ **36.** $0.58\overline{3}$

37. 0.375 **38.** 0.875 **39.** $0.\overline{714285}$ **40.** $\frac{7}{40}$ **41.** $\frac{1}{3}$

42. $\frac{531}{100}$ **43.** $\frac{235}{99}$ **44.** $\frac{12{,}083}{1000}$ **45.** $\frac{21}{5000}$ **46.** $\frac{211}{90}$ **47.** $\frac{11}{2}$

48. $\frac{51}{4}$ **49.** $-\frac{13}{4}$ **50.** $-\frac{283}{8}$ **51.** $6\frac{3}{4}$ **52.** $3\frac{1}{4}$ **53.** $-1\frac{5}{7}$

54. $-27\frac{1}{5}$ **55.** $\frac{10}{21}$ **56.** $\frac{5}{12}$ **57.** $\frac{79}{84}$ **58.** $\frac{2}{11}$ **59.** $\frac{35}{54}$

60. $\frac{53}{28}$ **61.** $\frac{1}{6}$ **62.** $\frac{13}{40}$ **63.** $\frac{8}{15}$ **64.** $2\frac{7}{32}$ tsp **65.** $2\sqrt{5}$

66. $4\sqrt{2}$ **67.** $8\sqrt{5}$ **68.** $-3\sqrt{3}$ **69.** $8\sqrt{2}$

70. $-20\sqrt{3}$ **71.** $8\sqrt{3}$ **72.** $3\sqrt{2}$ **73.** $4\sqrt{3}$ **74.** 3

75. $2\sqrt{7}$ **76.** $\frac{3\sqrt{2}}{2}$ **77.** $\frac{\sqrt{15}}{5}$ **78.** $15 + 5\sqrt{5}$

79. $4\sqrt{3} + 3\sqrt{2}$ **80.** $3\sqrt{2} + 3\sqrt{5}$

81. Commutative property of addition

82. Commutative property of multiplication

83. Associative property of addition

84. Distributive property

85. Commutative property of addition

86. Commutative property of addition

87. Associative property of multiplication

88. Commutative property of multiplication

89. Distributive property

90. Commutative property of multiplication

91. Yes **92.** Yes **93.** No **94.** Yes **95.** No **96.** No

97. 16 **98.** $\frac{1}{8}$ **99.** 7 **100.** 125 **101.** 1 **102.** $\frac{1}{64}$

103. 64 **104.** 81 **105.** 2.3×10^5 **106.** 1.58×10^{-5}

107. 2.75×10^{-3} **108.** 4.95×10^6 **109.** 25,000

110. 0.000139 **111.** 0.000175 **112.** 100,000

113. 8.5×10^2 **114.** 1.0×10^5 **115.** 2.1×10^1

116. 3.0×10^0 **117.** 15,000,000,000 **118.** 0.7

119. 3200 **120.** 5 **121.** 25 times **122.** $5,555.56

123. Arithmetic; 21, 26 **124.** Geometric; −243, 729

125. Arithmetic; −15, −18 **126.** Geometric; $\frac{1}{32}, \frac{1}{64}$

127. Arithmetic; 16, 19 **128.** Geometric; −2, 2

129. 9 **130.** −34 **131.** 25 **132.** 216 **133.** $\frac{1}{4}$

134. −48 **135.** 200 **136.** −25 **137.** 632 **138.** 57.5

139. 28 **140.** 80 **141.** 33 **142.** −21

143. Arithmetic; $a_n = -3n + 10$

144. Arithmetic; $a_n = 5n - 5$

145. Arithmetic; $a_n = -\frac{3}{2}n + \frac{11}{2}$

146. Geometric; $a_n = 3(2)^{n-1}$

147. Geometric; $a_n = 4(-1)^{n-1}$

148. Geometric; $a_n = 5\left(\frac{1}{3}\right)^{n-1}$

149. Yes; 13, 21 **150.** Yes; 17, 28 **151.** No **152.** No

CHAPTER TEST, PAGE 254

1. 2, 3, 4, 6, 8, 9 **2.** $2^2 \times 3 \times 5 \times 7$ **3.** −7 **4.** −20

5. −175 **6.** $\frac{37}{8}$ **7.** $19\frac{5}{9}$ **8.** 0.625 **9.** $\frac{129}{20}$ **10.** $\frac{121}{240}$

11. $\frac{19}{40}$ **12.** $9\sqrt{3}$ **13.** $\frac{\sqrt{30}}{6}$

14. Yes; the product of any two integers is an integer.

15. Associative property of addition

16. Distributive property **17.** 36 **18.** 1024 **19.** $\frac{1}{81}$

20. 8.0×10^6 **21.** $a_n = -4n + 2$ **22.** −187 **23.** 243

24. 1023 **25.** $a_n = 3(2)^{n-1}$ **26.** 1, 1, 2, 3, 5, 8, 13, 21, 34, 55

CHAPTER 6

SECTION 6.1, PAGE 260

1. Letters of the alphabet used to represent numbers are called variables.

3. An algebraic expression is a collection of variables, numbers, parentheses, and operation symbols. An example is $5x^2y - 11$.

5. **a)** The 4 is the base and the 5 is the exponent.
b) Answers will vary.

7. 12 **9.** 16 **11.** −49 **13.** 686 **15.** −3 **17.** 28

19. 15 **21.** $-\frac{10}{9}$ **23.** 7 **25.** 13 **27.** 0 **29.** No

31. Yes **33.** No **35.** Yes **37.** Yes

39. **a)** $52.40 **b)** $131 **41.** $13,125

43. 780 baskets of oranges **45.** 1.71 in.

47. The two expressions are not equal.

SECTION 6.2, PAGE 271

1. The parts that are added or subtracted in an algebraic expression are called terms. In $3x - 2y$, the $3x$ and $-2y$ are terms.

3. The numerical part of a term is called its numerical coefficient. For the term $3x$, 3 is the numerical coefficient.

5. A linear equation is one in which the exponent on the variable is 1. An example is $4x + 6 = 10$.

7. If $a = b$, then $a - c = b - c$ for all real numbers a, b, and c. If $2x + 3 = 5$, then $2x + 3 - 3 = 5 - 3$.

9. If $a = b$, then $a/c = b/c$ for all real numbers a, b, and c where $c \neq 0$. If $4x = 8$, then $\frac{4x}{4} = \frac{8}{4}$.

11. A ratio is a quotient of two quantities. An example is $\frac{7}{9}$.

13. Yes. They have the same variable and the same exponent on the variable.

15. $11x$ **17.** $10x - 11$ **19.** $3x + 11y$ **21.** $-8x + 2$

23. $-5x + 3$ **25.** $6.3x - 5.8$ **27.** $\frac{1}{12}x - 2$

29. $13x - 7y + 3$ **31.** $8s - 17$ **33.** $-1.4x - 3.8$

35. $\frac{7}{20}x + \frac{11}{10}$ **37.** $4.52x - 13.5$ **39.** 5 **41.** 5 **43.** $\frac{24}{7}$

45. $\frac{2}{3}$ **47.** 3 **49.** 3 **51.** 21 **53.** −15 **55.** No solution

57. All real numbers **59.** $\frac{4}{3}$ **61.** −3 **63.** 4 **65.** $68.84

67. $1064.12 **69.** **a)** 14,000 toys **b)** 12 hours

71. **a)** 1.6 kph **b)** 56.25 mph **73.** 0.3 cc

75. **a)** Answers will vary. **b)** −1

77. **a)** An equation that has no solution.
b) You will obtain a false statement.

SECTION 6.3, PAGE 279

1. A formula is an equation that typically has a real-life application.

3. Subscripts are numbers (or letters) placed below and to the right of variables. They are used to help clarify a formula.

5. An exponential equation is of the form $y = a^x$, $a > 0$, $a \neq 1$.

7. 60 **9.** 56 **11.** 25 **13.** 83 **15.** 37.1 **17.** 2

19. 3000 **21.** 8 **23.** 25 **25.** 2 **27.** 200 **29.** 7.2

31. $-\frac{3}{2}$ **33.** 3240 **35.** 0.5 **37.** 14

39. $y = \frac{7x - 15}{6}$ or $y = \frac{7}{6}x - \frac{5}{2}$

41. $y = \dfrac{-4x + 14}{7}$ or $y = -\dfrac{4}{7}x + 2$

43. $y = \dfrac{2x + 6}{3}$ or $y = \dfrac{2}{3}x + 2$

45. $y = \dfrac{3x + 20}{-2}$ or $y = -\dfrac{3}{2}x - 10$

47. $y = \dfrac{9x + 4z - 7}{8}$ or $y = \dfrac{9}{8}x + \dfrac{1}{2}z - \dfrac{7}{8}$

49. $b = \dfrac{A}{h}$ **51.** $a = p - b - c$

53. $w = \dfrac{V}{lh}$ **55.** $r = \dfrac{C}{2\pi}$

57. $b = y - mx$ **59.** $w = \dfrac{P - 2l}{2}$

61. $c = 3A - a - b$ **63.** $T = \dfrac{PV}{K}$

65. $C = \frac{5}{9}(F - 32)$ **67.** $s = \dfrac{S - \pi r^2}{\pi r}$

69. a) \$105 **b)** \$3605 **71.** $\approx$ 18.41 in.^{3}

73. $\approx$ 18.8 mg **75.** $\approx$ \$236,756,624,900,000

77. 1051.47 in.^{3}

SECTION 6.4, PAGE 284

1. A mathematical expression is a collection of variables, numbers, parentheses, and operation symbols. An equation is two algebraic expressions joined by an equal sign.

3. $9 - 6x$ **5.** $6r + 5$ **7.** $15 - 2r$ **9.** $x + 6$

11. $\dfrac{3 + n}{8}$ **13.** $(5y - 6) + 3$ **15.** $x + 7 = 19$; 12

17. $x - 10 = 25$; 35 **19.** $12 + 5x = 47$; 7

21. $8x + 16 = 88$; 9 **23.** $x + 11 = 3x + 1$; 5

25. $x + 3 = 5(x + 7)$; -8 **27.** $x + 2x = 9000$; \$6000

29. $6200 + 500x = 12{,}200$; 12 years

31. $x - 0.10x = 15.72$; \$17.47

33. $x + 3x = 12$; Samantha = 3, Josie = 9

35. $0.20x + 60 = 100$; 200 mi

37. a) $x + x + 3x = 45{,}000$; 9000 ft^2, 9000 ft^2, 27,000 ft^2 **b)** Yes

39. $x + \frac{1}{6}x = 203$; 174 lb

41. $3w + 2(2w) = 140$; Width = 20 ft, length = 40 ft

43. $70x = 760$; $\approx$ 11 months **45.** $\dfrac{r}{2} + 0.07r = 227$; \$426.13

47. Deduct \$720 from Mr. McAdams's income and \$2920 from Mrs. McAdams's income.

49.
$$x + (x + 1) + (x + 2) = 3(x + 2) - 3$$
$$3x + 3 = 3x + 6 - 3$$
$$3x + 3 = 3x + 3$$

SECTION 6.5, PAGE 292

1. Direct variation: As one variable increases, so does the other, and as one variable decreases, so does the other.

3. Joint variation: One quantity varies directly as the product of two or more other quantities.

5. Inverse **7.** Direct **9.** Direct **11.** Inverse

13. Inverse **15.** Inverse **17.** Direct **19.** Direct

21. Answers will vary.

23. a) $r = ks$ **b)** 33 **25. a)** $y = \dfrac{k}{x^2}$ **b)** 5

27. a) $R = \dfrac{k}{W}$ **b)** $\frac{1}{20}$ **29. a)** $F = kDE$ **b)** 210

31. a) $T = \dfrac{kD^2}{F}$ **b)** 51.2 **33. a)** $Z = kWY$ **b)** 100

35. a) $H = kL$ **b)** 3 **37. a)** $A = kB^2$ **b)** 405

39. a) $F = \dfrac{kq_1q_2}{d^2}$ **b)** 672

41. a) $F = km$ **b)** 1600 newtons

43. a) $l = \dfrac{k}{d^2}$ **b)** 80 dB

45. a) $R = \dfrac{kA}{P}$ **b)** 4800 tapes

47. a) $W = kI^2R$ **b)** 40 watts

49. a) $N = \dfrac{kp_1p_2}{d}$ **b)** $\approx$ 121,528 calls

51. a) Inversely **b)** Stays 0.3 **53.** \$124.92

SECTION 6.6, PAGE 299

1. $a < b$ means that a is less than b, $a \le b$ means that a is less than or equal to b, $a > b$ means that a is greater than b, $a \ge b$ means that a is greater than or equal to b.

3. When both sides of an inequality are multiplied or divided by a negative number, the direction of the inequality symbol must be reversed.

5. Yes, the inequality symbol points to the x in both cases.

7. 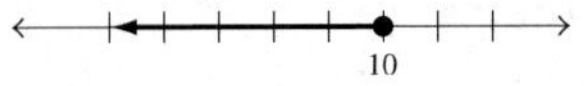 10

9. 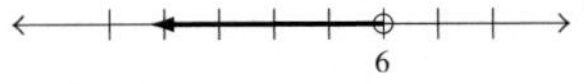 6

11. 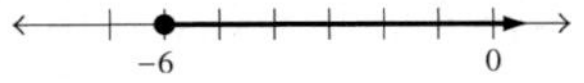 −6 0

13. 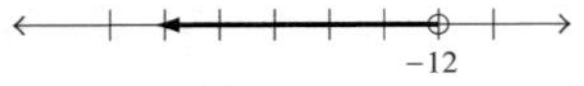 −12

15. 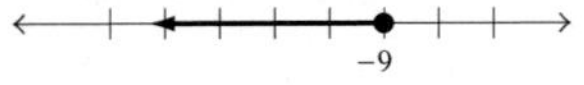 −9

17. 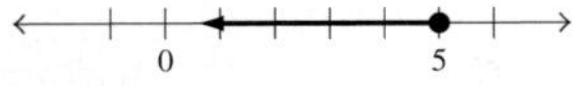 0 5

19.
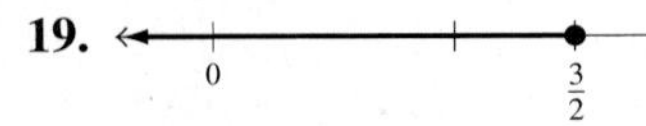

21.
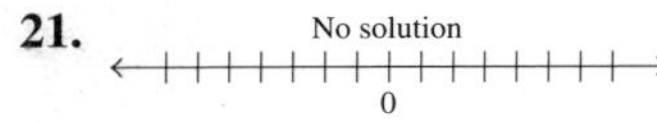

23.

25.

27.

29.

31.

33.

35.

37.

39.

41.

43.

45. Less than 360 mi

47. a) $180 + 60x \leq 1200$
b) 17 boxes

49. more than 20

51. $94 \leq x \leq 100$, assuming 100 is the highest grade possible

53. minimum 58, maximum 91

55. $81.6 \leq x \leq 100$, assuming that 100 is the highest grade possible

SECTION 6.7, PAGE 310

1. A graph is an illustration of all the points whose coordinates satisfy an equation.

3. To find the x-intercept, set $y = 0$ and solve the equation for x.

5. a) Answers will vary. **b)** $-\frac{1}{3}$

For Exercises 7, 9, 11, and 13, see the following figure.

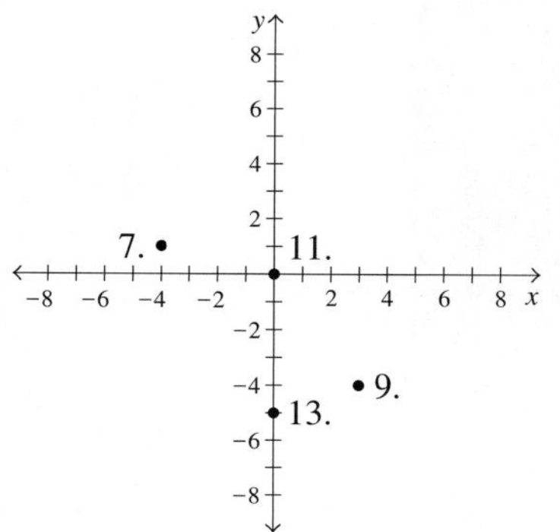

For Exercises 15, 17, 19, and 21 see the following figure.

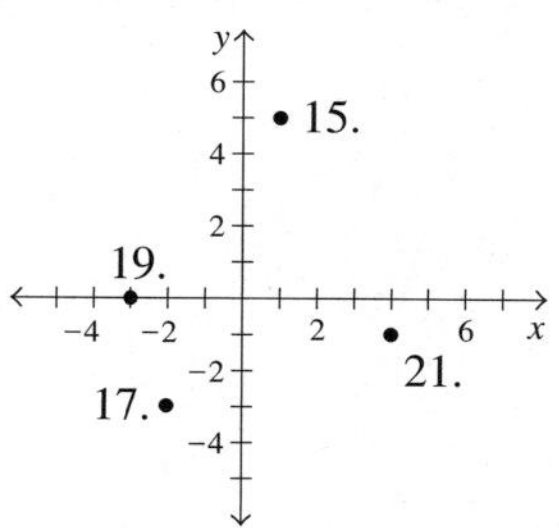

23. (2, 2) **25.** (0, 2) **27.** (−?, 0) **29.** (−5, −3)

31. (2, −3) **33.** (2, −2) **35.** (0, 4), $(1, \frac{5}{2})$

37. (−3, −2) **39.** $(0, \frac{8}{3})$, (4, 0)

41.

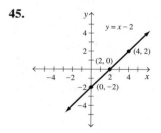

slope: undefined

43.

slope is 0

45.

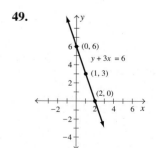

47.

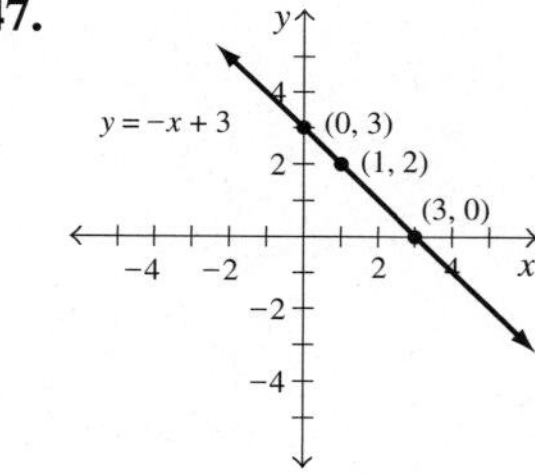

49.

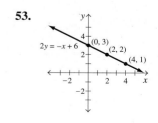

51.

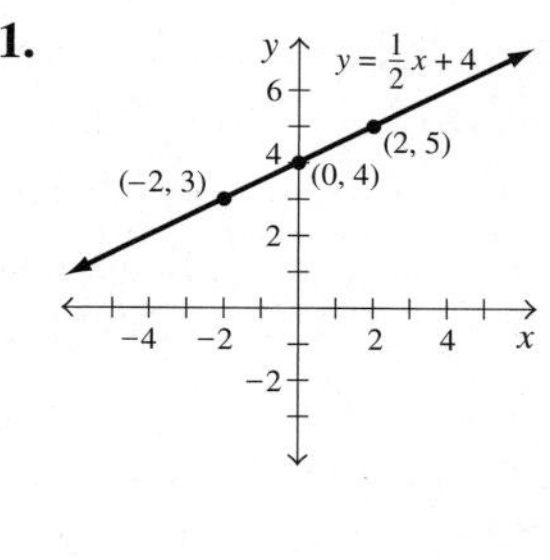

53.

55.

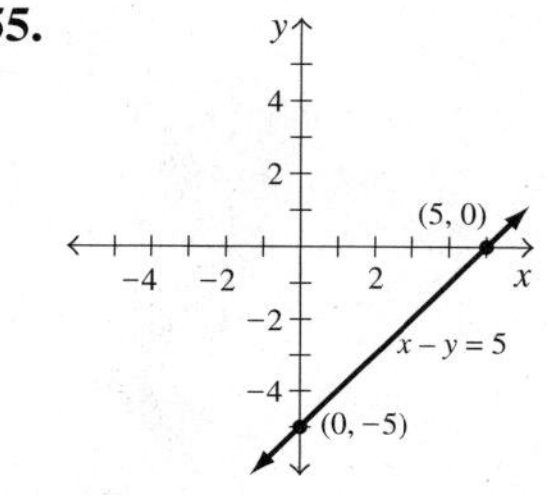

57.

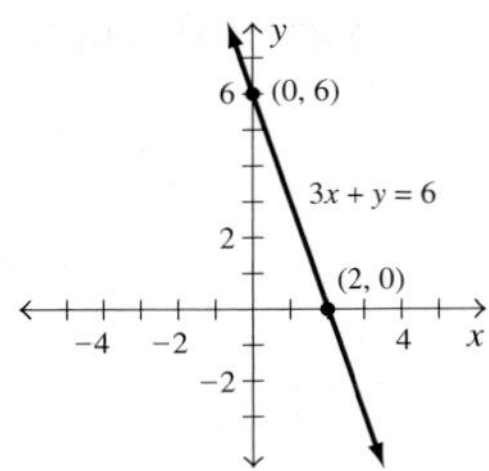

59.

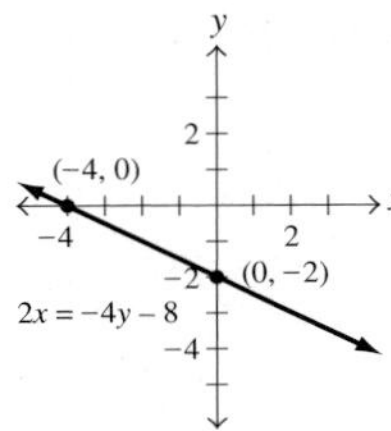

61.

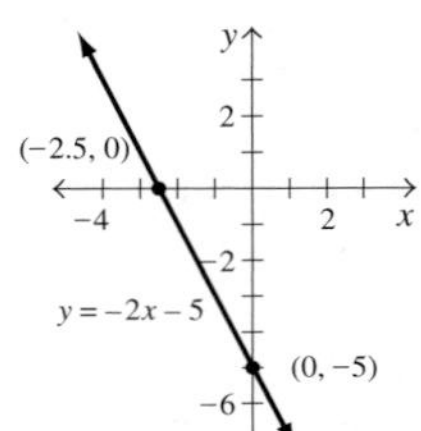

63.

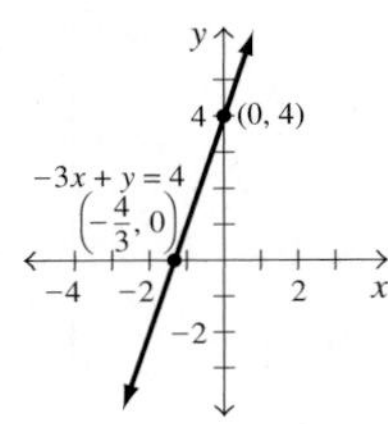

65. 4 **67.** $\frac{11}{4}$ **69.** 0 **71.** Undefined **73.** $-\frac{4}{3}$

75.

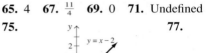

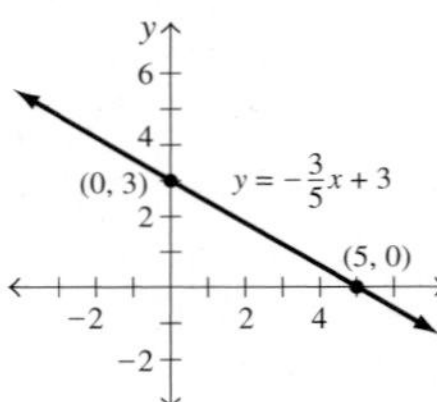

77.

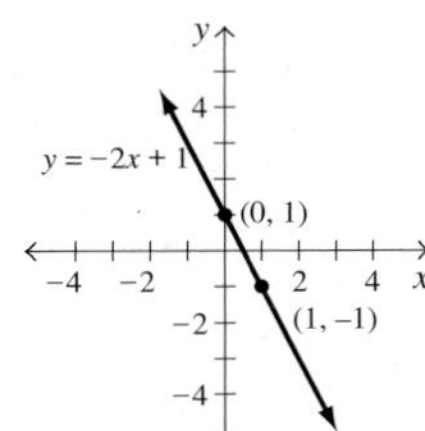

79.

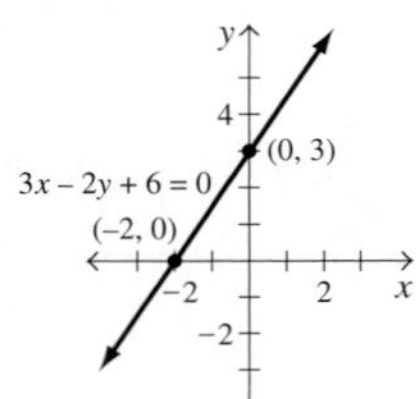

81.

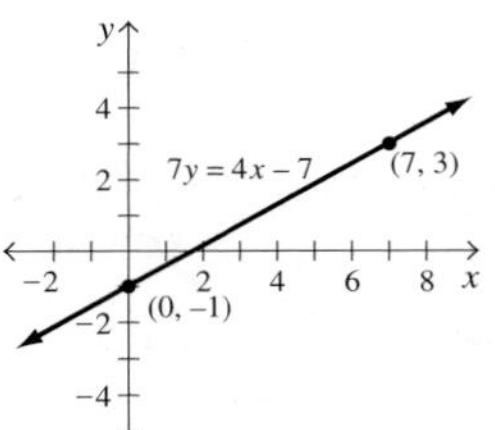

83.

85. $y = -\frac{3}{4}x + 3$

87. $y = \frac{1}{2}x - 1$

89. a) $D(-2, -1)$ **b)** $A = 24$ square units

91. (7, 2) or (−1, 2)

93. 8 **95.** 3

97. a)

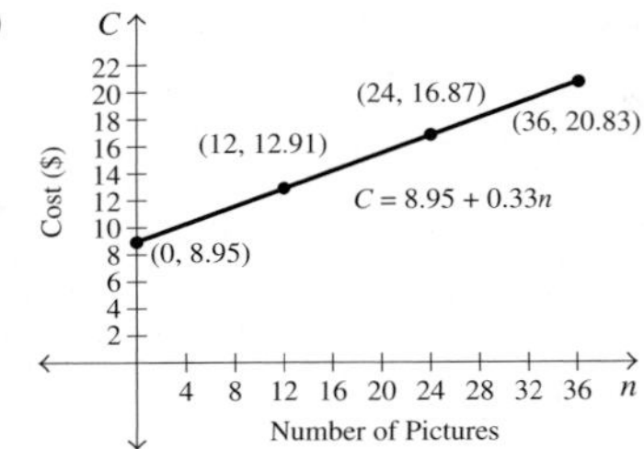

b) \$15.55 **c)** 36

99. a)

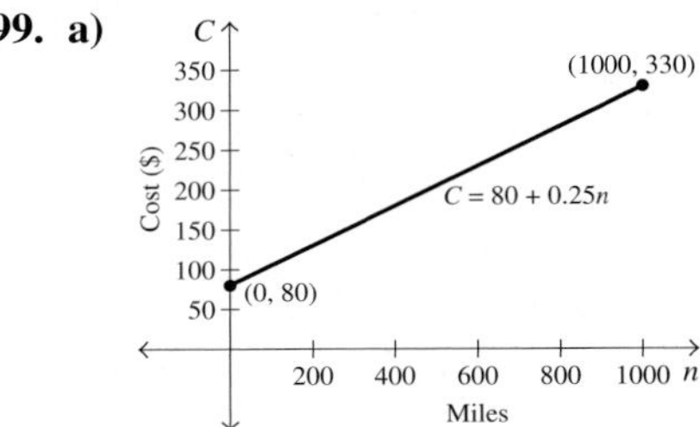

b) \$230.00 **c)** 400 mi

101. a) 10.75 **b)** $y = 10.75x + 53$
c) 85.25 **d)** ≈ 2.5 hours

103. a) ≈ 24.71 **b)** $y = 24.71x + 370$
c) \$518.26 **d)** ≈ 9.31 years after 1980, or in 1989

105. a) Solve the equations for *y* to put them in slope–intercept form. Then compare the slopes and *y* intercepts. If the slopes are equal but the *y* intercepts are different, then the lines are parallel.
b) The lines are parallel.

SECTION 6.8, PAGE 315

1. (1) Mentally substitute the equal sign for the inequality sign and plot points as if you were graphing the equation. (2) If the inequality is $<$ or $>$, draw a dashed line through the points. If the inequality is $\leq$ or $\geq$, draw a solid line through the points. (3) Select a test point not on the line and substitute the *x*- and *y*-coordinates into the inequality. If the substitution results in a true statement, shade in the area on the same side of the line as the test point. If the test point results in a false statement, shade in the area on the opposite side of the line as the test point.

3.

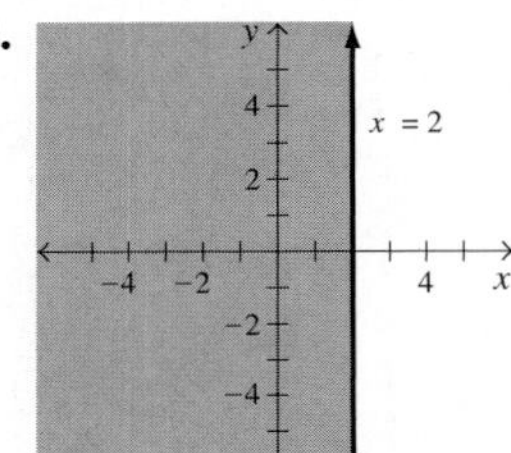

5.

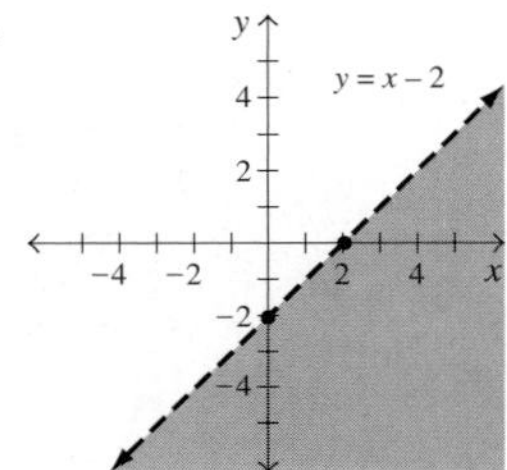

7.
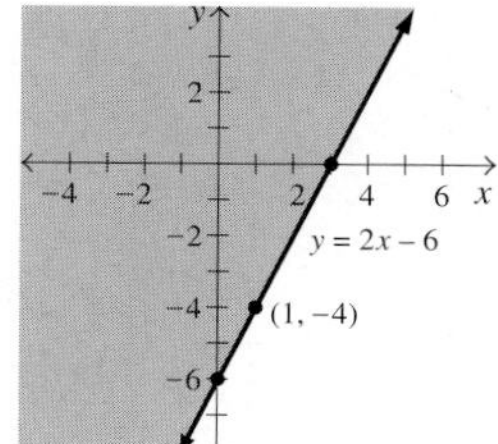

9.
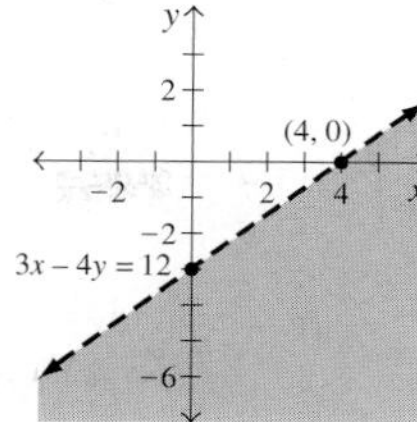

11.
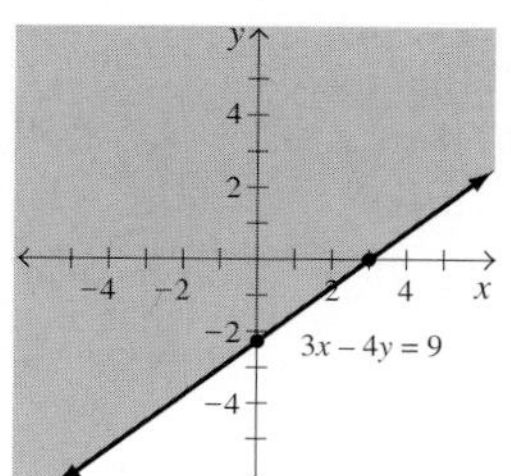

13.
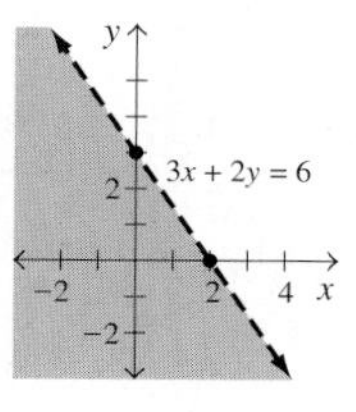

15.
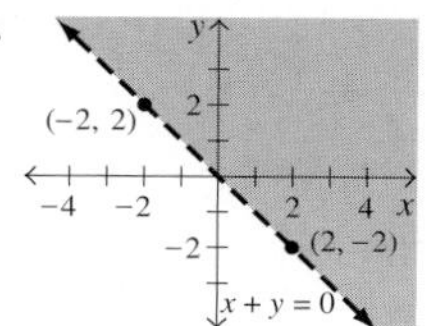

17.
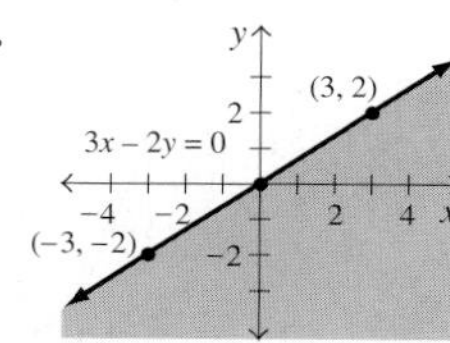

19.
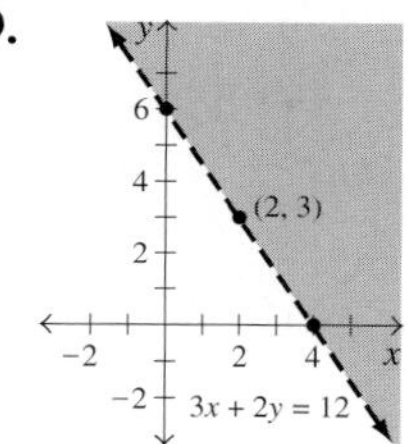

21.
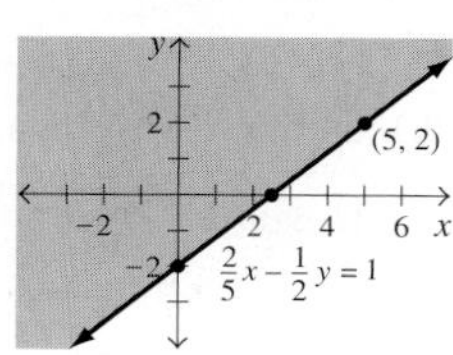

23.
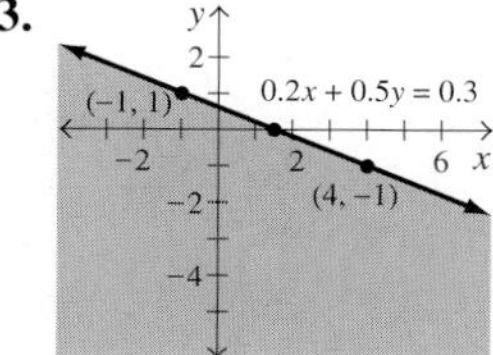

25. a) $2l + 2w \le 40,\ 0 \le l \le 20,\ 0 \le w \le 20$
b)
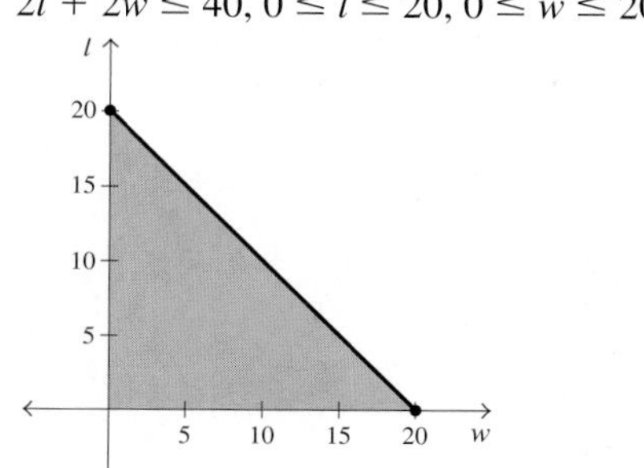

27. a) No, you cannot have a negative number of shirts.
b)
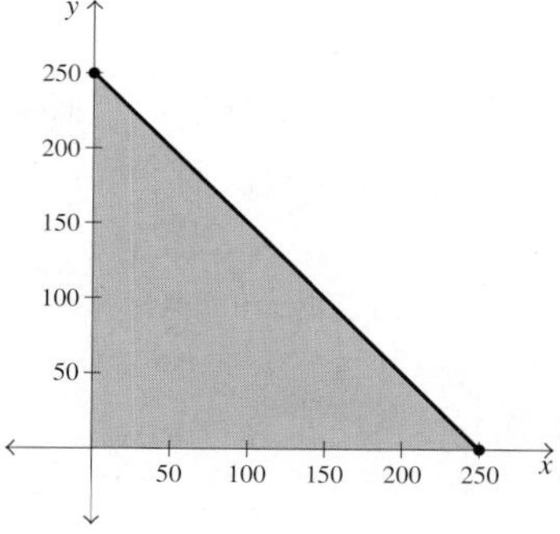

c) Answers will vary.

SECTION 6.9, PAGE 324

1. A binomial is an expression that contains two terms in which each exponent that appears on the variable is a whole number. $2x + 3,\ x - 7$

3. Answers will vary. **5.** Answers will vary.
7. $(x + 3)(x + 7)$ **9.** $(x - 5)(x + 1)$
11. $(x + 6)(x - 4)$ **13.** $(x + 1)(x - 3)$
15. $(x - 7)(x - 3)$ **17.** $(x - 4)(x + 4)$
19. $(x + 7)(x - 4)$ **21.** $(x + 9)(x - 7)$
23. $(2x + 3)(x + 1)$ **25.** $(3x + 1)(x - 5)$
27. $(5x + 2)(x + 2)$ **29.** $(5x + 2)(x - 3)$
31. $(5x - 3)(x - 2)$ **33.** $(3x + 4)(x - 6)$
35. $-2, 5$ **37.** $\frac{3}{2}, -\frac{7}{3}$ **39.** $-3, -2$ **41.** $4, 2$ **43.** $5, -3$
45. $3, 1$ **47.** $9, -9$ **49.** $-9, 4$ **51.** $-2, -\frac{1}{3}$ **53.** $-\frac{1}{5}, -2$
55. $\frac{1}{3}, 1$ **57.** $\frac{1}{4}, 2$ **59.** $6, -5$ **61.** $5, -2$ **63.** $9, -1$
65. No real solution
67. $2 \pm \sqrt{2}$ **69.** $\dfrac{1 \pm \sqrt{33}}{4}$
71. $\dfrac{1 \pm \sqrt{17}}{8}$ **73.** $-1, -\frac{5}{2}$
75. $\frac{7}{3}, 1$ **77.** No real solution **79.** 205 air conditioners
81. a) Two **b)** One **c)** None

SECTION 6.10, PAGE 335

1. A function is a special type of relation where each value of the independent variable corresponds to a unique value of the dependent variable.

3. The domain of a function is the set of values that can be used for the independent variable.

5. If a vertical line touches more than one point on the graph, then for each value of x there is not a unique value for y and the graph does not represent a function.

7. Function, domain: $x = -2, -1, 1, 2, 3$; range: $y = -1, 1, 2, 3$

9. Function, domain: $\mathbb{R}$; range: $\mathbb{R}$

11. Function, domain: $\mathbb{R}$; range: $y = 2$

13. Function, domain: $\mathbb{R}$; range: $y \geq -4$

15. Not a function

17. Function, domain: $0 \leq x < 12$; range: $y = 1, 2, 3$

19. Not a function

21. Function, domain: $\mathbb{R}$; range: $y > 0$

23. Yes **25.** No **27.** Yes **29.** 15 **31.** 1

33. -6 **35.** 52 **37.** 4 **39.** -38 **41.** -48

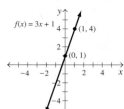

43. $f(x) = 3x + 1$; points $(1, 4)$, $(0, 1)$

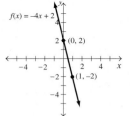

45. $f(x) = -4x + 2$; points $(0, 2)$, $(1, -2)$

47. $f(x) = \frac{3}{2}x - 1$; points $(2, 2)$, $(0, -1)$

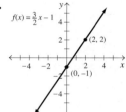

49. a) Upward **b)** $x = 0$ **c)** $(0, -1)$
d) $(0, -1)$ **e)** $(-1, 0)$, $(1, 0)$
f)

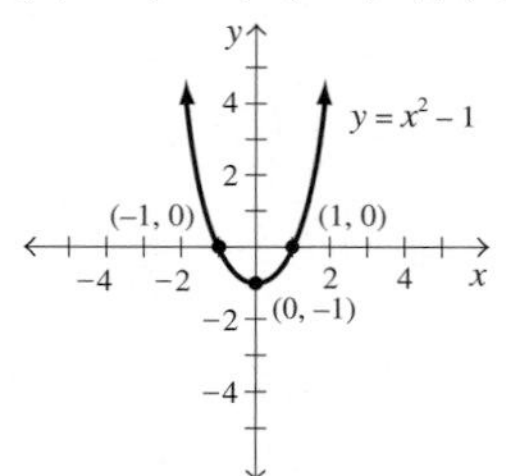

g) Domain: $\mathbb{R}$; range: $y \geq -1$

51. a) Downward **b)** $x = 0$ **c)** $(0, 4)$
d) $(0, 4)$ **e)** $(-2, 0)$, $(2, 0)$
f)

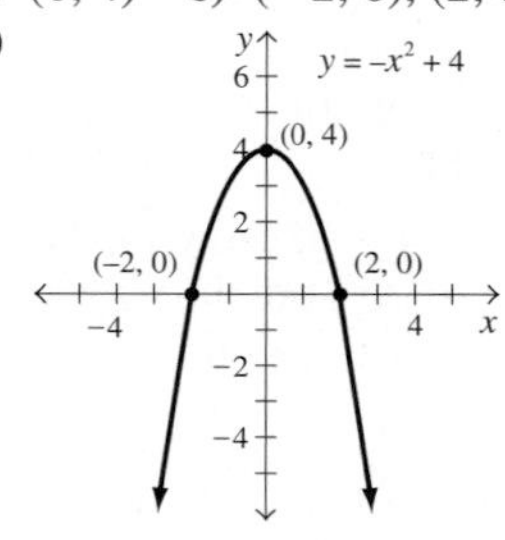

g) Domain: $\mathbb{R}$; range: $y \leq 4$

53. a) Downward **b)** $x = 0$ **c)** $(0, -4)$
d) $(0, -4)$ **e)** No x-intercepts
f)

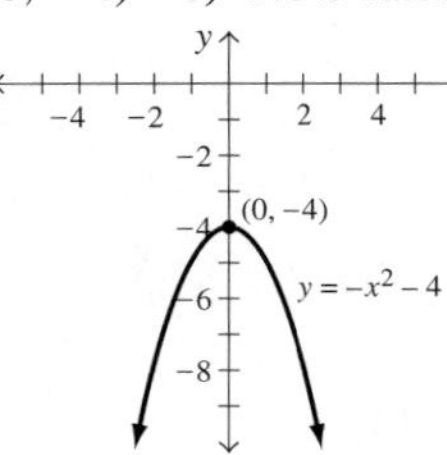

g) Domain: $\mathbb{R}$; range: $y \leq -4$

55. a) Upward **b)** $x = 0$ **c)** $(0, -3)$
d) $(0, -3)$ **e)** $(-1.22, 0)$, $(1.22, 0)$
f)

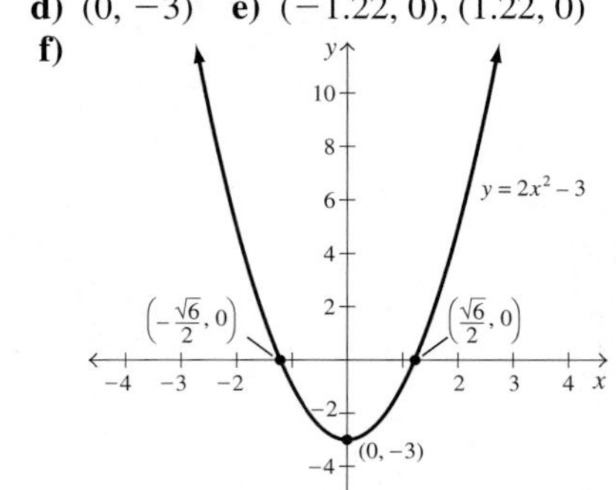

g) Domain: $\mathbb{R}$; range: $y \geq -3$

57. a) Upward **b)** $x = -2$ **c)** $(-2, 6)$
d) $(0, 10)$ **e)** No x-intercepts
f)

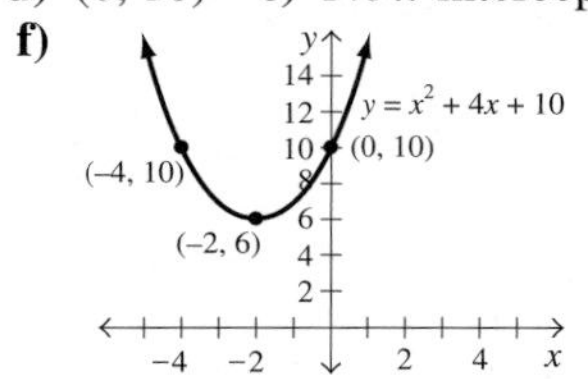

g) Domain: $\mathbb{R}$; range: $y \geq 6$

59. a) Upward **b)** $x = -\frac{5}{2}$ **c)** $(-2.5, -0.25)$
d) $(0, 6)$ **e)** $(-3, 0)$, $(-2, 0)$
f)

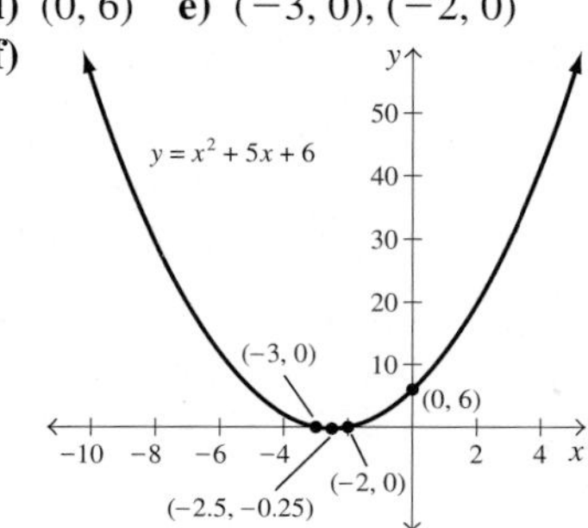

g) Domain: $\mathbb{R}$; range: $y \geq -0.25$

61. **a)** Downward **b)** $x = 2$ **c)** $(2, -2)$
d) $(0, -6)$ **e)** No x-intercepts
f)

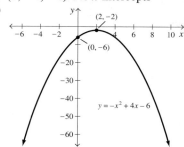

g) Domain: $\mathbb{R}$; range: $y \leq -2$

63. **a)** Downward **b)** $x = \frac{7}{3}$ **c)** $(\frac{7}{3}, \frac{25}{3})$
d) $(0, -8)$ **e)** $(\frac{2}{3}, 0)$, $(4, 0)$
f)

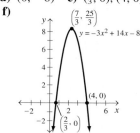

g) Domain: $\mathbb{R}$; range: $y \leq \frac{25}{3}$

65. Domain: $\mathbb{R}$; range: $y > 0$

67. Domain: $\mathbb{R}$; range: $y > 0$

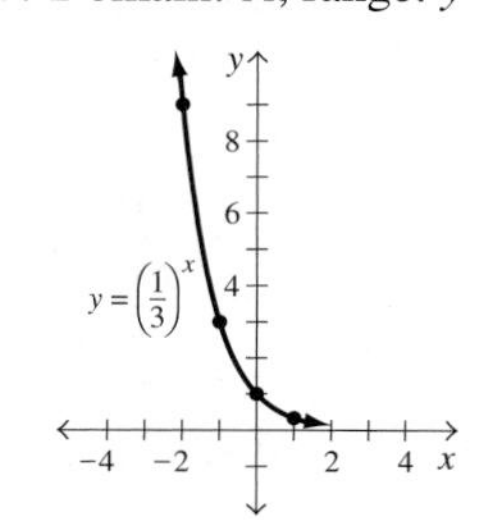

69. Domain: $\mathbb{R}$; range: $y > 1$

71. Domain: $\mathbb{R}$; range: $y > 1$

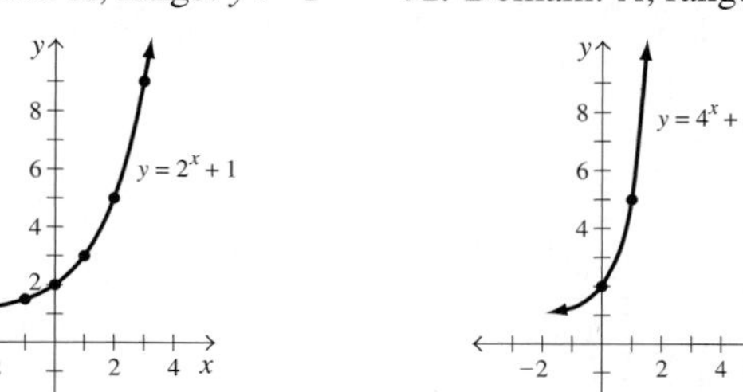

73. Domain: $\mathbb{R}$; range: $y > 0$

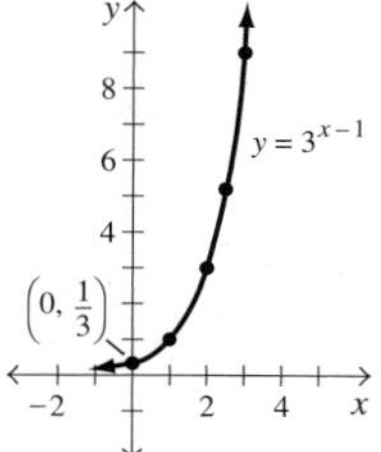

75. Domain: $\mathbb{R}$; range: $y > 0$

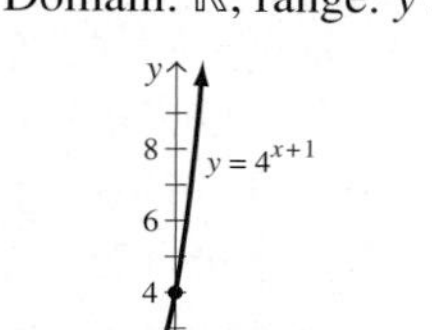

77. **a)** 180 mi **b)** 420 mi **79.** **a)** 77.8°F **b)** 75.32°F
81. **a)** 5200 **b)** $\approx$ 14,852 **83.** **a)** Yes **b)** $\approx$ \$700,000
85. **a)** 23.2 cm **b)** 55.2 cm **c)** 69.5 cm
87. **a)** 170 beats per minute
b) $\approx$ 162 beats per minute
c) $\approx$ 145 beats per minute
d) 136 beats per minute
e) 120 years of age

REVIEW EXERCISES, PAGE 339

1. 23 **2.** -1 **3.** 17 **4.** $\frac{1}{4}$ **5.** -65 **6.** 23
7. $x - 3$ **8.** $13x - 8$ **9.** $3x - \frac{13}{2}$ **10.** -14
11. 7 **12.** 11 **13.** -31 **14.** $\frac{54}{5}$ **15.** $\frac{1}{2}$ cup
16. 375 min, or 6 hr 15 min **17.** 104 **18.** $\approx$ 173.1
19. 101.5 **20.** 10

21. $y = \dfrac{4x - 12}{6}$ or $y = \dfrac{2}{3}x - 2$

22. $y = \dfrac{-5x + 18}{6}$ or $y = -\dfrac{5}{6}x + 3$

23. $y = \dfrac{2x + 22}{3}$ or $y = \dfrac{2}{3}x + \dfrac{22}{3}$

24. $y = \dfrac{-3x + 5z - 4}{4}$ or $y = -\dfrac{3}{4}x + \dfrac{5}{4}z - 1$

25. $l = \dfrac{A}{W}$

26. $w = \dfrac{P - 2l}{2}$

27. $l = \dfrac{L - 2wh}{2h}$ or $l = \dfrac{L}{2h} - w$

28. $d = \dfrac{a_n - a_1}{n - 1}$

29. $7 - 4x$ **30.** $5x - 3$ **31.** $10 + 3r$ **32.** $\dfrac{8}{q} - 11$
33. $12 - 3x = 21$; $x = -3$ **34.** $3x + 8 = x - 6$; $x = -7$
35. $5(x - 4) = 45$; $x = 13$ **36.** $10x + 14 = 8(x + 12)$; $x = 41$
37. $x + \frac{1}{3}x = 48{,}000$; Wesley's income is \$12,000.
38. $9.50x + 15{,}000 = 95{,}000$, 8421 chairs

39. $x + (x + 12{,}000) = 68{,}000$; \$28,000 for B and \$40,000 for A

40. $15x = 300$; 20 hours **41.** 2 **42.** 240

43. 20 **44.** ≈ 426.7 **45.** 4 in. **46.** \$119.88

47. 400 ft **48.** 200.96

49. −1 0

50. −1 0

51. 16

52. −4 6

53. −2 −1 0 1 2 …

54. −8 −7 −6 −5 −4 −3 −2 −1 …

55. −1 0 1 2 3 4 5 6 7 8

56. −10 −8 −6 −4 −2 0 2 4 6

For Exercises 57, 58, 59, and 60, see the following figure.

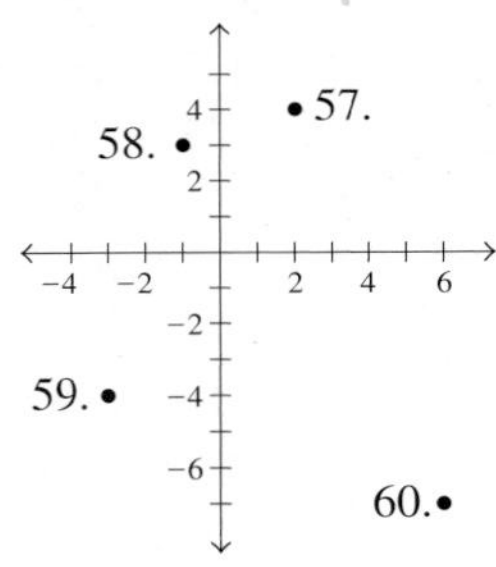

61. $D(-3, -1)$; area = 20 square units

62. $D(4, 1)$; area = 21 square units

63.

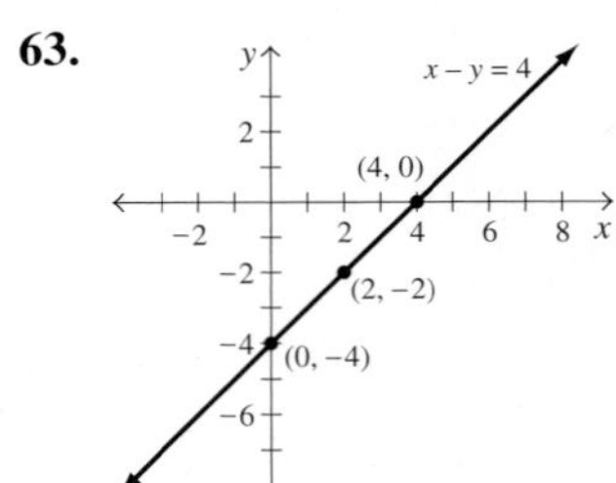

64.

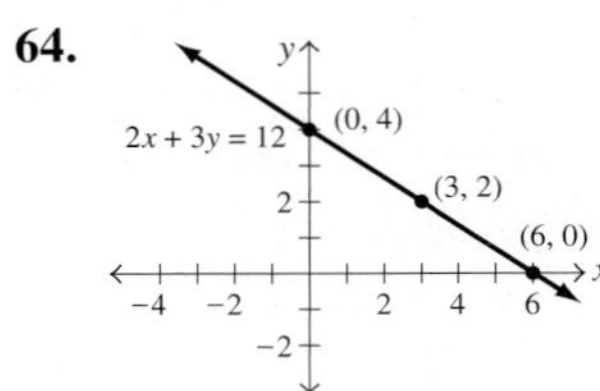

65.

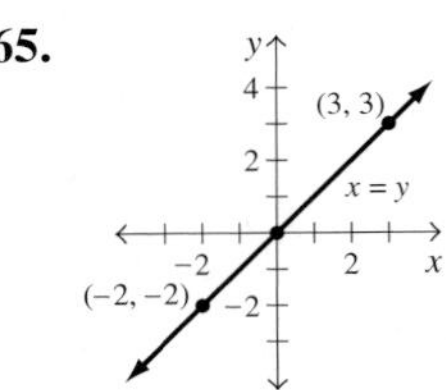

66.

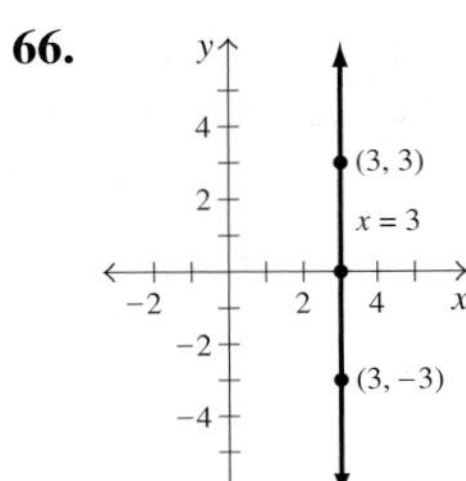

67.

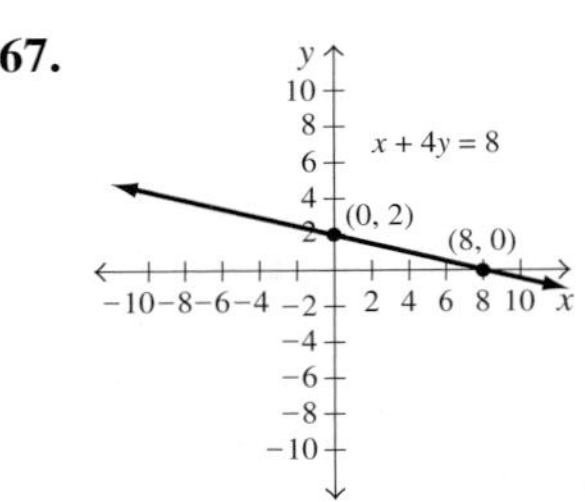

68.

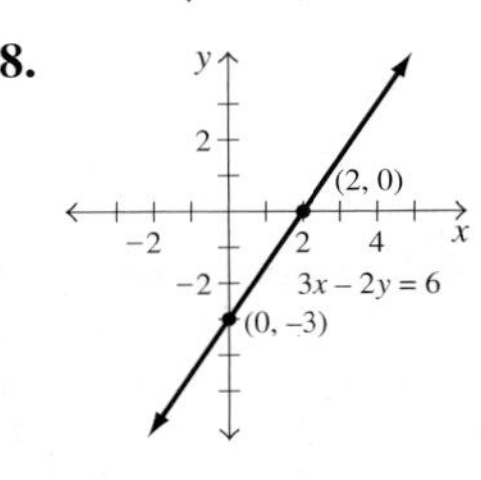

69.

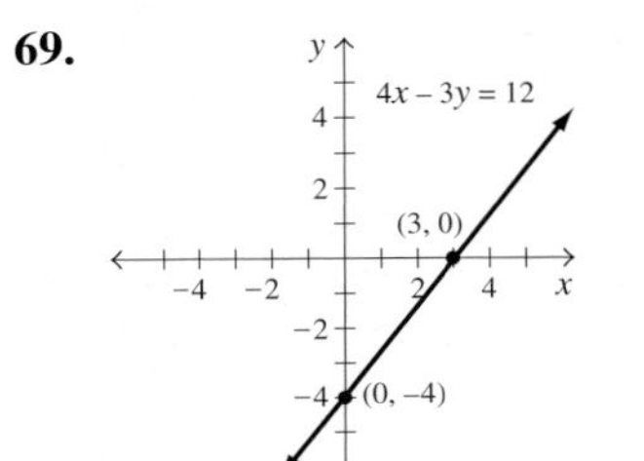

70.

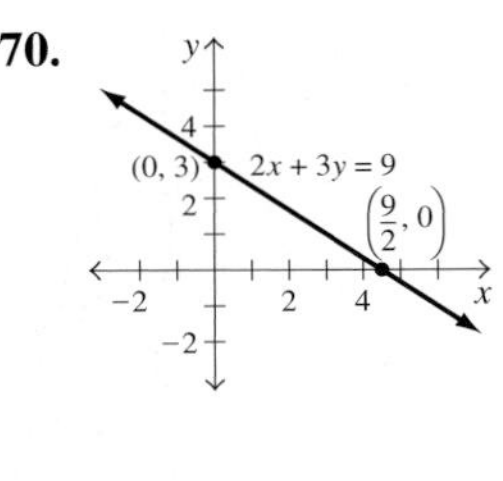

71. $-\frac{1}{5}$ **72.** $-\frac{3}{2}$ **73.** $\frac{7}{6}$ **74.** Undefined

75.

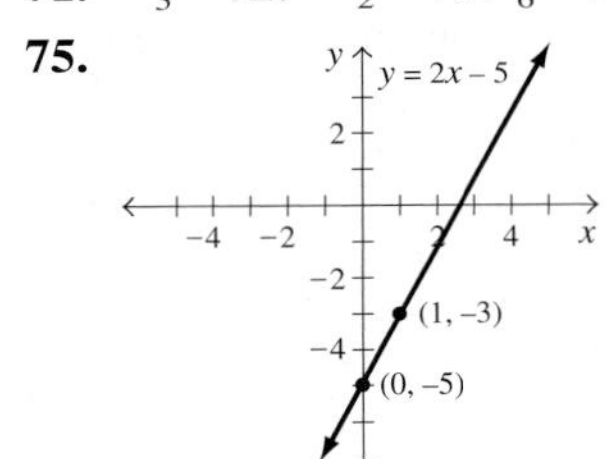

76.

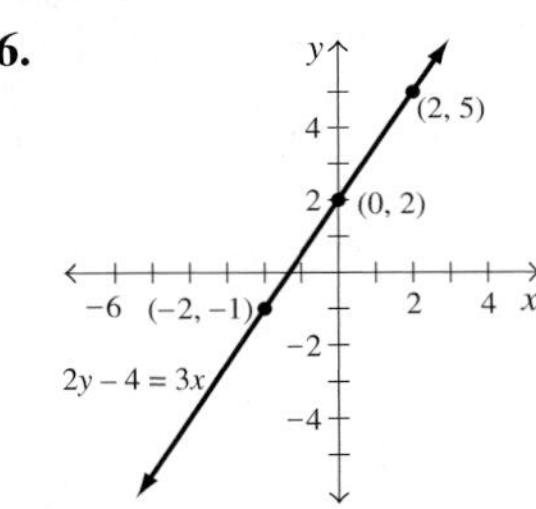

77.

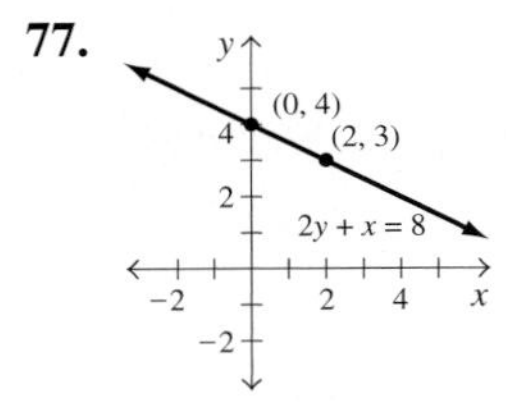

78.

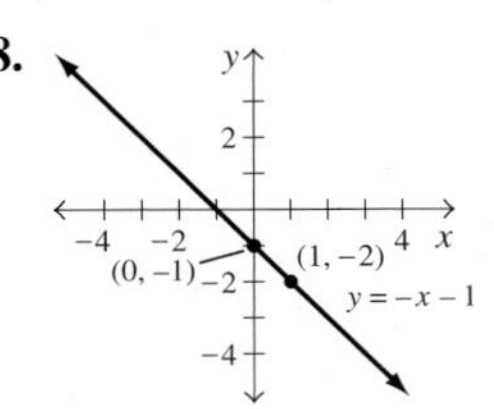

79. $y = 2x + 4$ **80.** $y = -x + 1$

81. a)

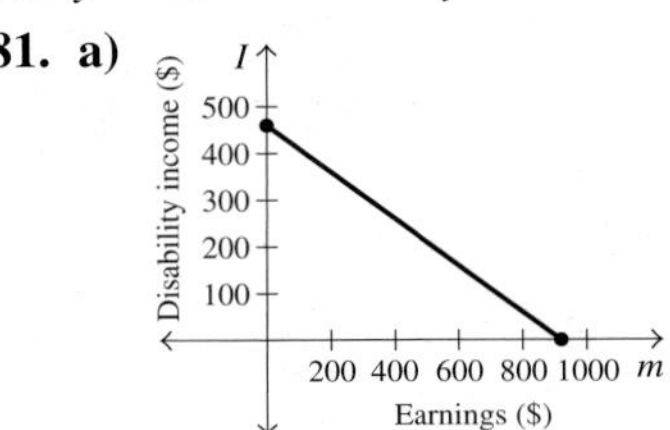

b) About \$160 **c)** About \$160

82. a)

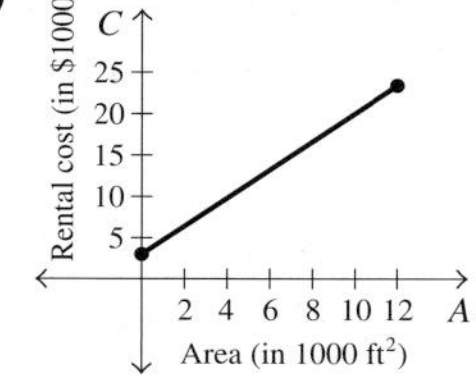

b) About \$6400 **c)** About 4120 ft^2

83.

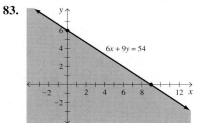

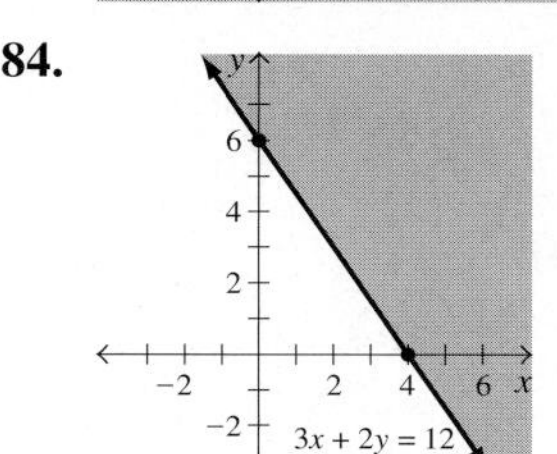

84.

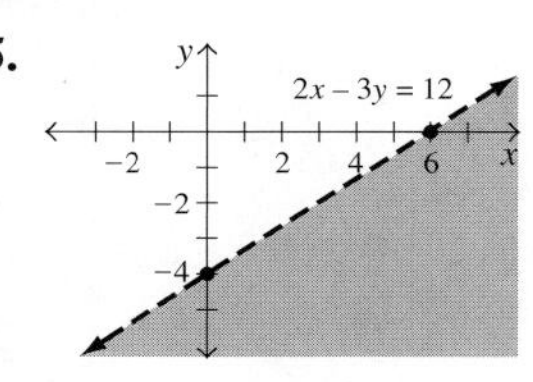

85.

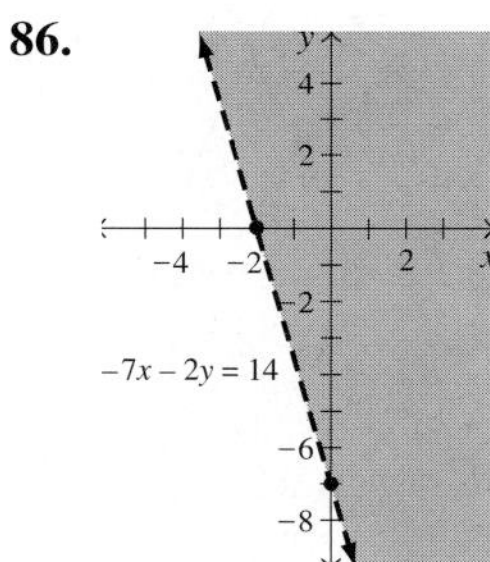

86.

87. $(x + 3)(x + 6)$ **88.** $(x + 5)(x - 4)$

89. $(x - 6)(x - 4)$ **90.** $(x - 5)(x - 4)$

91. $(x - 3)(2x + 7)$ **92.** $(3x - 1)(x + 2)$

93. $-3, -2$ **94.** 1, 5

95. $\frac{2}{3}, 5$ **96.** $-2, -\frac{1}{3}$

97. $\dfrac{3 \pm \sqrt{37}}{2}$ **98.** 1, 2

99. No real solution **100.** $-1, \frac{3}{2}$

101. Function, domain: $x = -2, -1, 2, 3$; range: $y = -1, 0, 2$

102. Not a function **103.** Not a function

104. Function, domain: $\mathbb{R}$; range: $\mathbb{R}$

105. 4 **106.** 14 **107.** 39 **108.** -27

109. a) Downward **b)** $x = -2$ **c)** $(-2, 25)$
d) $(0, 21)$ **e)** $(-7, 0), (3, 0)$
f)

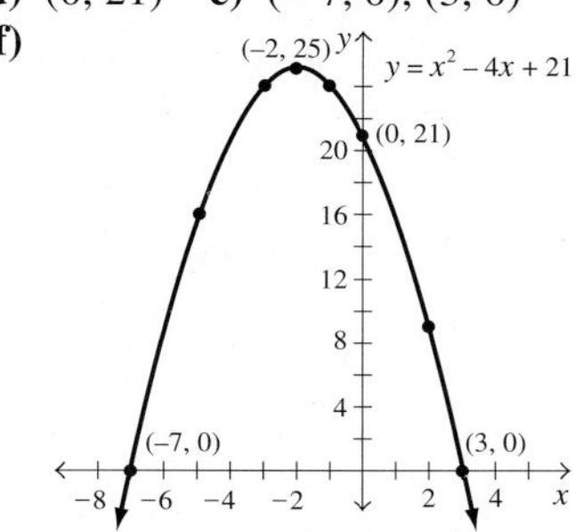

g) Domain: $\mathbb{R}$; range: $y \leq 25$

110. a) Upward **b)** $x = 4$ **c)** $(4, -78)$ **d)** $(0, -30)$
e) $(4 - \sqrt{26}, 0), (4 + \sqrt{26}, 0)$
f)

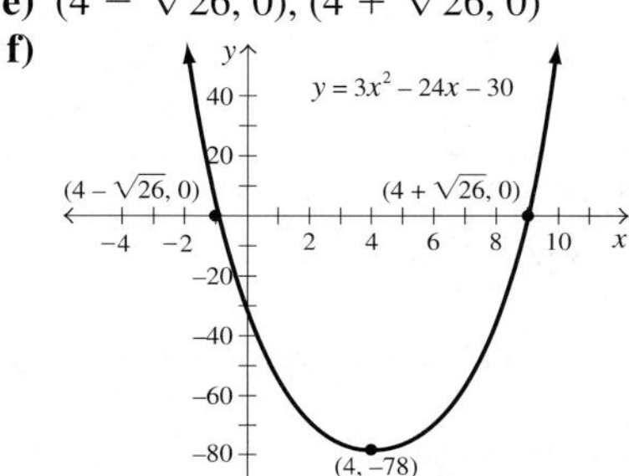

g) Domain: $\mathbb{R}$; range: $y \geq -78$

111. Domain: $\mathbb{R}$; range: $y > 0$

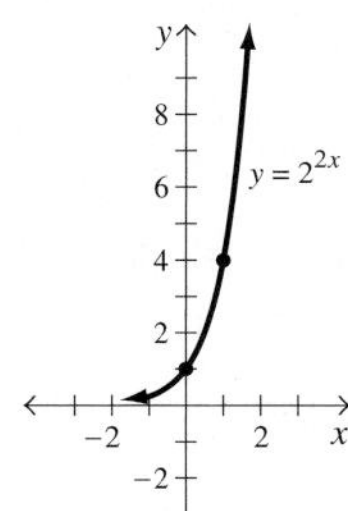

112. Domain: $\mathbb{R}$; range: $y > 0$

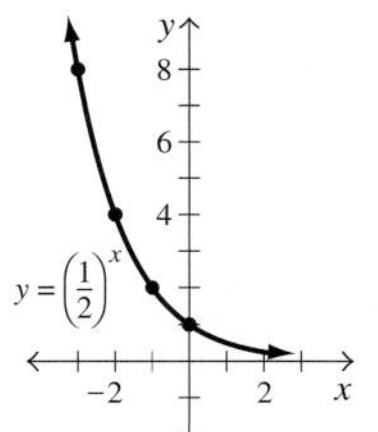

113. 22.8 mpg

114. a) 4208 **b)** 4250

115. 68.7%

CHAPTER TEST, PAGE 342

1. 9 **2.** $\frac{19}{5}$ **3.** 18 **4.** $3x - 10 = 11$; 7
5. $7.75x - 4.35x = 60$, ≈ 18 units **6.** 84
7. $y = \dfrac{5x - 17}{8}$ or $y = \dfrac{5}{8}x - \dfrac{17}{8}$ **8.** $3\frac{1}{3}$ **9.** 6.75 ft

10. 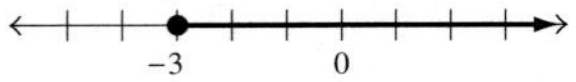

11. $-\dfrac{31}{13}$

12. 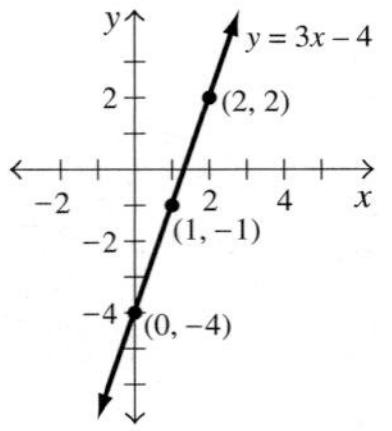

13. 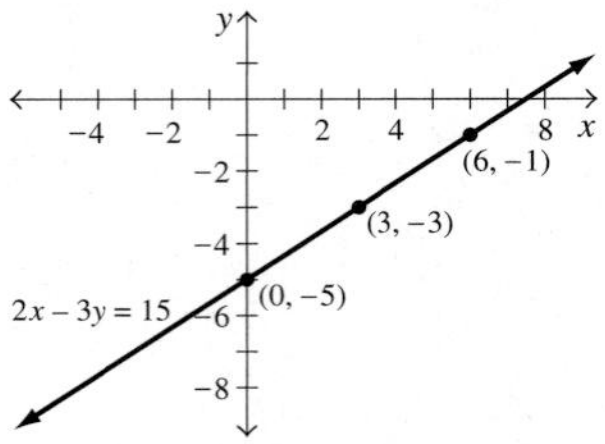

14. 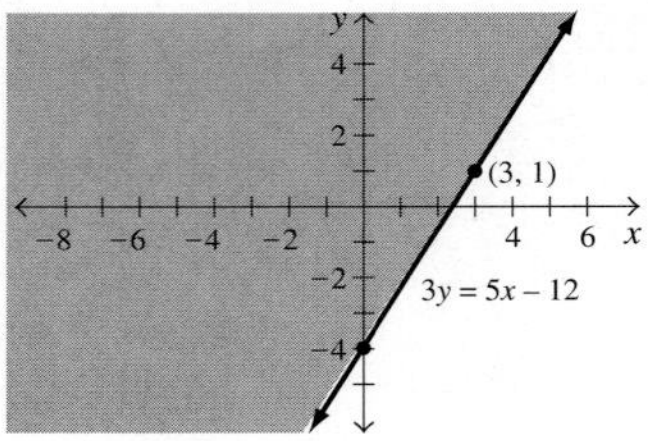

15. 7, −4 **16.** $\frac{4}{3}$, −2 **17.** It is a function **18.** 11
19. a) Upward **b)** $x = 1$ **c)** (1, 3) **d)** (0, 4)
e) No x-intercepts
f) 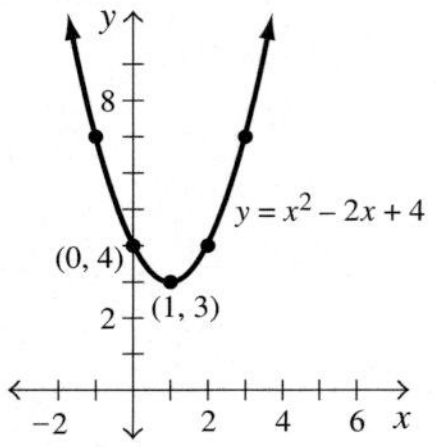

g) Domain: $\mathbb{R}$; range: $y \geq 3$

CHAPTER 7

SECTION 7.1, PAGE 351

1. Two or more linear equations form a system of linear equations.

3. A consistent system of equations is a system that has a solution.

5. An inconsistent system of equations is one that has no solution.

7. No

9. 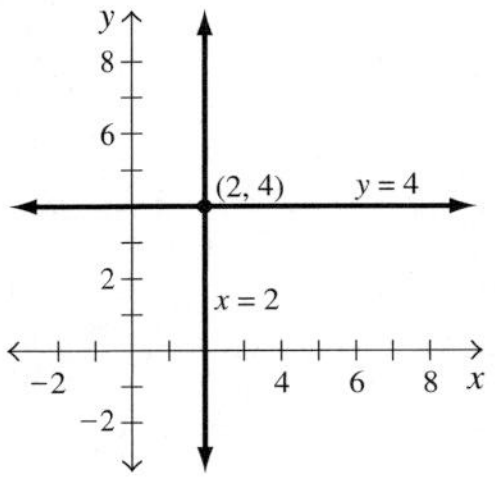

11. 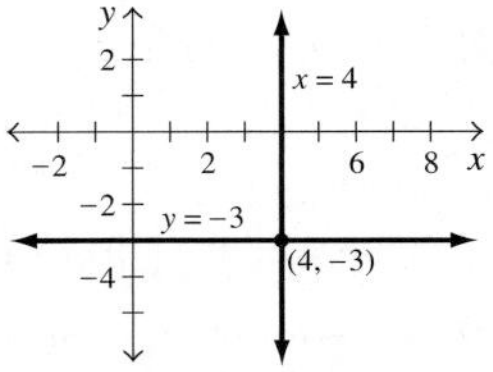

13. 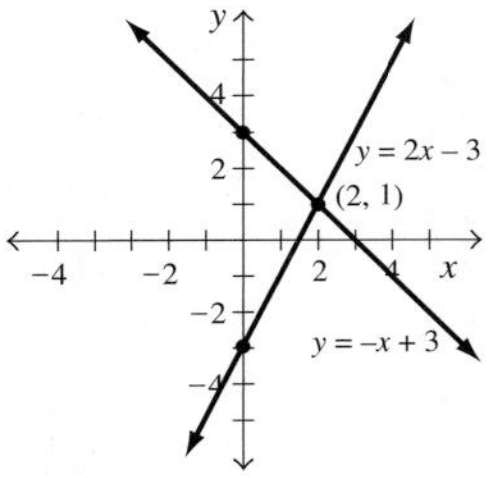

15. 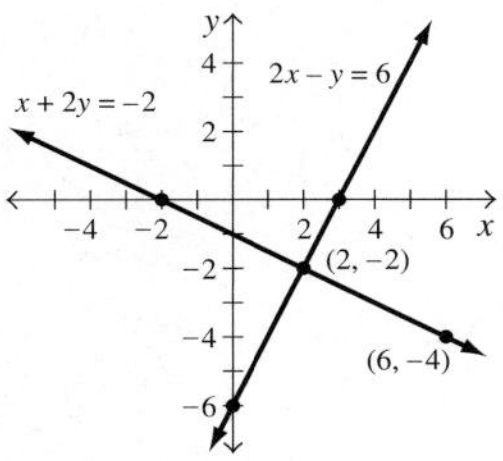

17. 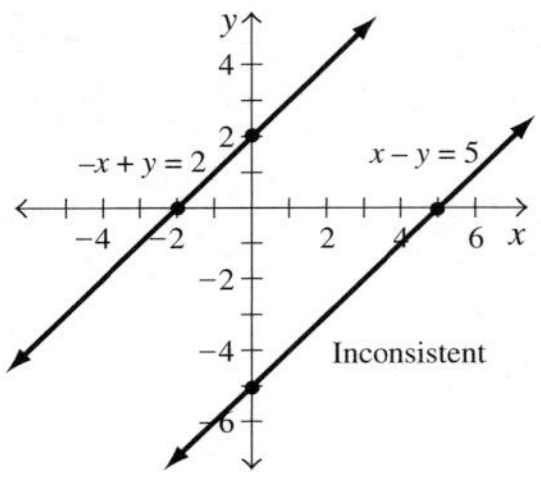

19. 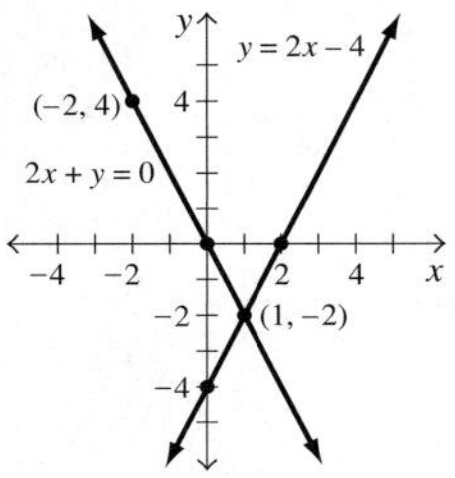

21. 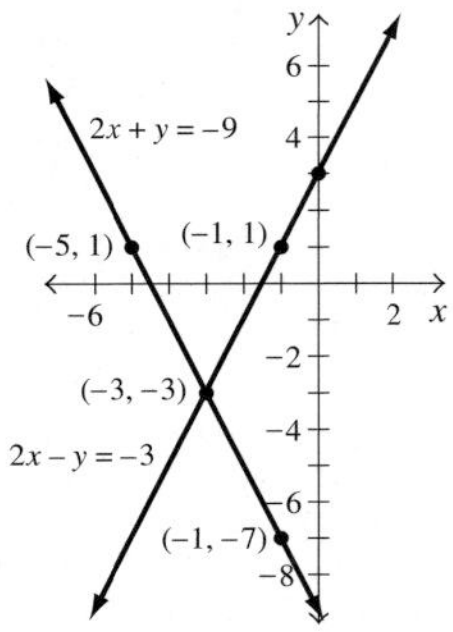

23. 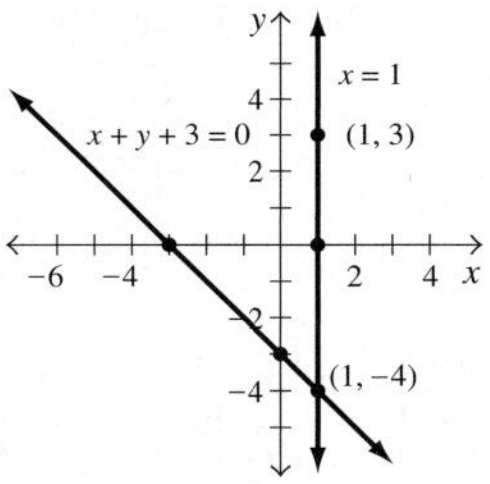

25. 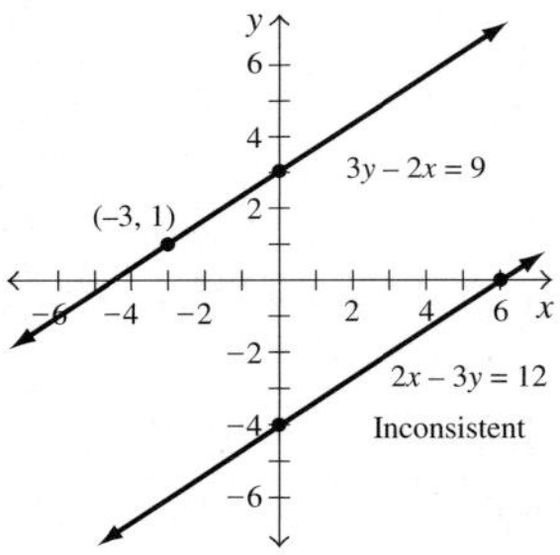

27. 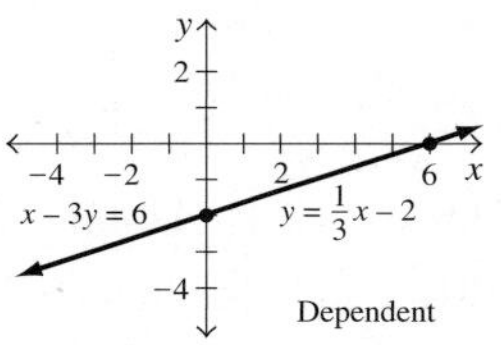

29. a) One unique solution; the lines intersect at one and only one point.
b) No solution; the lines do not intersect.
c) Infinitely many solutions; the lines coincide.

31. An infinite number of solutions **33.** One solution

35. No solution **37.** One solution

39. An infinite number of solutions **41.** One solution

43. Not perpendicular **45.** Perpendicular

47. a) Tom's Tree and Landscape: $C = 60h + 200$
Lawn Perfect: $C = 25h + 305$

b)

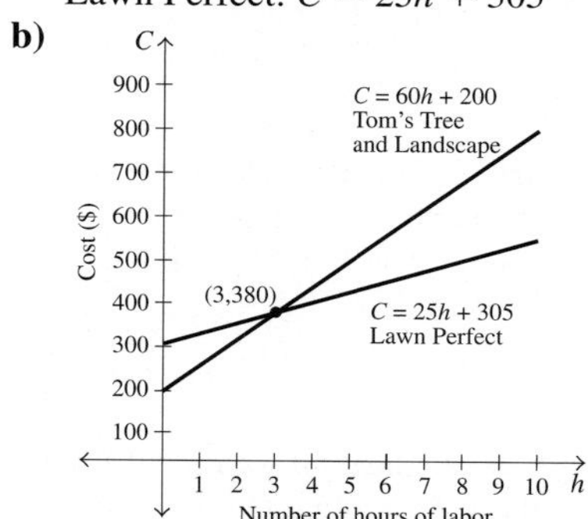

c) 3 hours

49. a) Sivle: $C = 6b + 1600$
Yelserp: $C = 8b + 1200$

b)

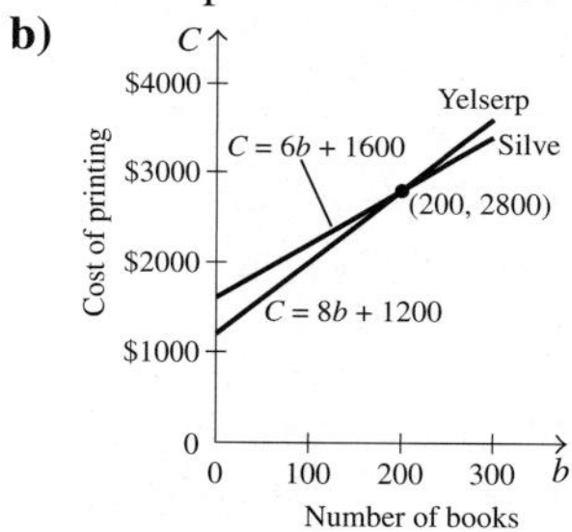

c) 200 books
d) Yelserp is less expensive.

51. a) $C = 95x + 8400$
$R = 165x$

b)

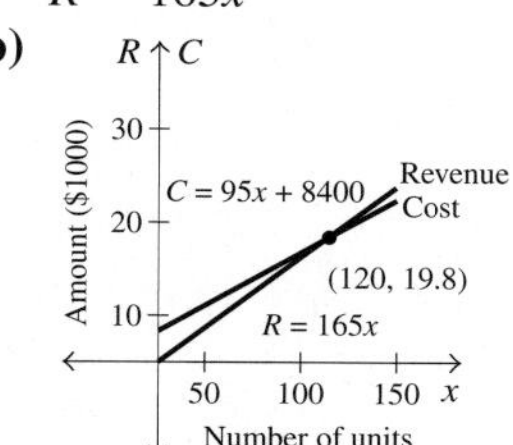

c) 120 units
d) $P = 70x - 8400$
e) Loss of $1400
f) $\approx$ 138 units

53. a) Job 1: $s = 0.15x + 300$
Job 2: $s = 450$

b)

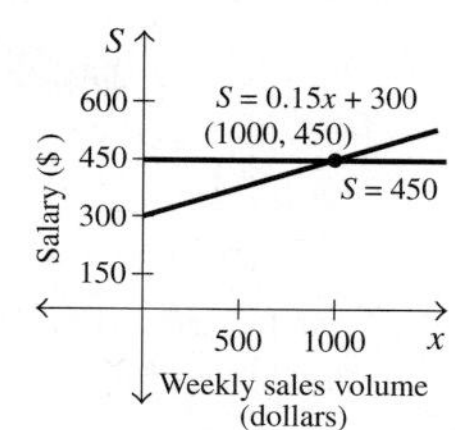

c) $1000 sales volume

55. a) One **b)** Three **c)** Six **d)** Ten
e) To find the number of points of intersection for n lines add $n - 1$ to the number of points of intersection for $n - 1$ lines. For example, for 5 lines there were 10 points of intersection. Therefore, for 6 lines (n = 6) there are $10 + (6 - 1) = 15$ points of intersection.

SECTION 7.2, PAGE 362

1. Answers will vary.

3. The system is inconsistent if you obtain a false statement.

5. $(5, -1)$ **7.** $(0, -3)$

9. No solution; inconsistent system **11.** $(3, 3)$

13. An infinite number of solutions; dependent system

15. $(-5, 2)$ **17.** $(\frac{11}{5}, -\frac{13}{5})$ **19.** $(-\frac{1}{5}, -\frac{8}{5})$

21. No solution; inconsistent system

23. $(2, 1)$ **25.** $(-3, 2)$ **27.** $(-2, 0)$

29. $(-1, 8)$ **31.** $(3, 5)$ **33.** $(1, -2)$

35. No solution; inconsistent system **37.** $(5, -3)$

39. $w = 300 + 0.04s$
$w = 0.16s$
$2500 in weekly sales

41. $2x + y = 58$
$x = y + 23$
Won 27 games and tied 4 games

43. $x + y = 300$
$0.16x + 0.07y = 0.10(300)$
Mix 100 lb of soybean meal with 200 lb of corn meal.

45. $y = 18 + 0.02x$
$y = 24 + 0.015x$
1200 copies

47. $x + y = 20$
$3x + y = 30$
Mix 5 lb of nuts with 15 lb of pretzels.

49. $0.10A + 0.20B = 20$
$0.06A + 0.02B = 6$
80 g of mix A, 60 g of Mix B

51. a) $\approx$ 3.5 years after 1988 or in 1991; $\approx$ 350 million units shipped

53. $(\frac{1}{2}, \frac{1}{3})$ **55.** Answers will vary.

SECTION 7.3, PAGE 371

1. A matrix is a rectangular array of elements.

3. A square matrix contains the same number of rows and columns.

5. a) Answers will vary. **b)** $\begin{bmatrix} 9 & 8 & 5 \\ -1 & 5 & 6 \end{bmatrix}$

7. a) The number of columns of the first matrix must be the same as the number of rows of the second matrix.
b) 2×3

9. a) $I = \begin{bmatrix} 1 & 0 \\ 0 & 1 \end{bmatrix}$ **b)** $I = \begin{bmatrix} 1 & 0 & 0 \\ 0 & 1 & 0 \\ 0 & 0 & 1 \end{bmatrix}$

11. $\begin{bmatrix} -1 & 2 \\ 9 & 7 \end{bmatrix}$ **13.** $\begin{bmatrix} 1 & 3 \\ 5 & 4 \\ 7 & 1 \end{bmatrix}$

15. $\begin{bmatrix} -6 & 7 \\ 12 & -4 \end{bmatrix}$ **17.** $\begin{bmatrix} 1 & 0 & -7 \\ 9 & 8 & -7 \\ 6 & 1 & -9 \end{bmatrix}$

19. $\begin{bmatrix} 6 & 4 \\ 10 & 0 \end{bmatrix}$ **21.** $\begin{bmatrix} 0 & 13 \\ 22 & 0 \end{bmatrix}$

23. $\begin{bmatrix} 13 & 0 \\ 7 & 0 \end{bmatrix}$ **25.** $\begin{bmatrix} 4 & 14 \\ 3 & 12 \end{bmatrix}$

27. $\begin{bmatrix} 15 \\ 22 \end{bmatrix}$ **29.** $\begin{bmatrix} 5 & 1 & 6 \\ -2 & 3 & 1 \\ 4 & 7 & 2 \end{bmatrix}$

31. $A + B = \begin{bmatrix} 6 & 3 & 1 \\ 5 & -2 & 5 \end{bmatrix}$; cannot be multiplied.

33. Cannot be added; $A \times B = \begin{bmatrix} 26 & 38 \\ 24 & 24 \end{bmatrix}$

35. Cannot be added; $\begin{bmatrix} 1 \\ -1 \end{bmatrix}$

37. $A + B = B + A = \begin{bmatrix} 5 & 8 \\ 8 & -1 \end{bmatrix}$

39. $A + B = B + A = \begin{bmatrix} 8 & 0 \\ 6 & -8 \end{bmatrix}$

41. $(A + B) + C = A + (B + C) = \begin{bmatrix} 4 & 11 \\ 4 & 13 \end{bmatrix}$

43. $(A + B) + C = A + (B + C) = \begin{bmatrix} 5 & 5 \\ 7 & -37 \end{bmatrix}$

45. No **47.** No **49.** Yes

51. $(A \times B) \times C = A \times (B \times C) = \begin{bmatrix} 35 & 17 \\ 44 & 20 \end{bmatrix}$

53. $(A \times B) \times C = A \times (B \times C) = \begin{bmatrix} 16 & -10 \\ -24 & 2 \end{bmatrix}$

55. $(A \times B) \times C = A \times (B \times C) = \begin{bmatrix} 17 & 0 \\ -7 & 0 \end{bmatrix}$

57. Small Large **59.** [\$36.04 \$47.52]
$\begin{bmatrix} 38 & 50 \\ 56 & 72 \\ 17 & 26 \\ 10 & 14 \end{bmatrix}$

61. No **63.** Yes **65.** False

67. a) \$28.70 **b)** \$60.10 **c)** $\begin{bmatrix} 28.7 & 24.6 \\ 41.3 & 35.7 \\ 69.3 & 60.1 \end{bmatrix}$

SECTION 7.4, PAGE 379

1. a) An augmented matrix is a matrix formed with the coefficients of the variables and the constants. The coefficients of the variables are separated from the constants by a vertical bar.
b) $\left[\begin{array}{cc|c} 1 & 3 & 7 \\ 2 & -1 & 4 \end{array}\right]$

3. If you obtain an augmented matrix in which a 0 appears across an entire row, the system of equations is dependent.

5. $(4, -1)$ **7.** $(2, 1)$

9. An infinite number of solutions; dependent system

11. $(-2, 1)$ **13.** $(2, 1)$

15. No solution; inconsistent system **17.** $(3, 5)$

19. Poster board: \$1; marker: \$2

21. Cherries: \$4 per pound; mints: \$5 per pound

23. Nonrefillable pencils: 125; refillable pencils: 75

SECTION 7.5, PAGE 382

1. The solution set of a system of linear inequalities is the set of points that satisfy all inequalities in the system.

3.

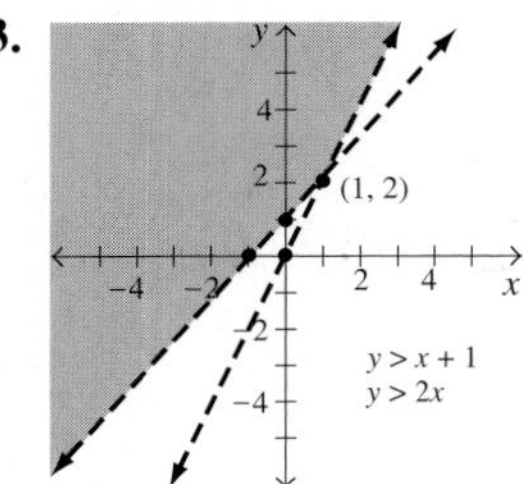

5.

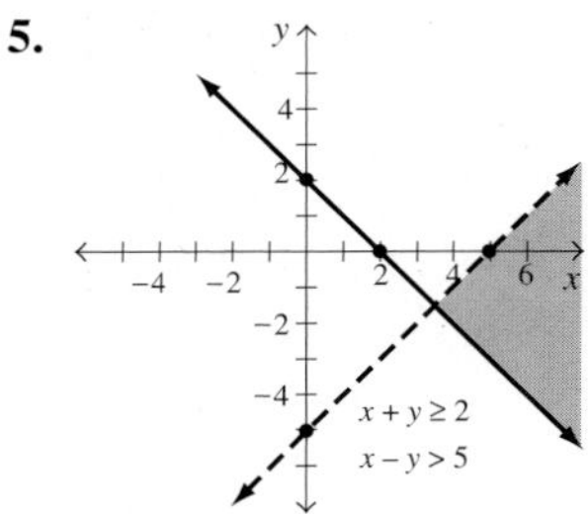

7.

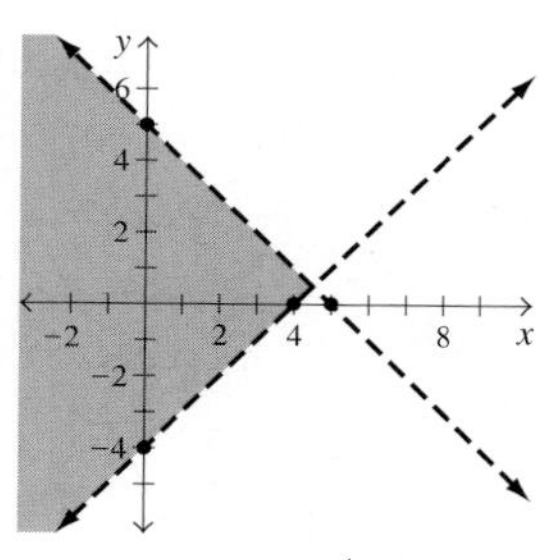

$x - y < 4$
$x + y < 5$

9.

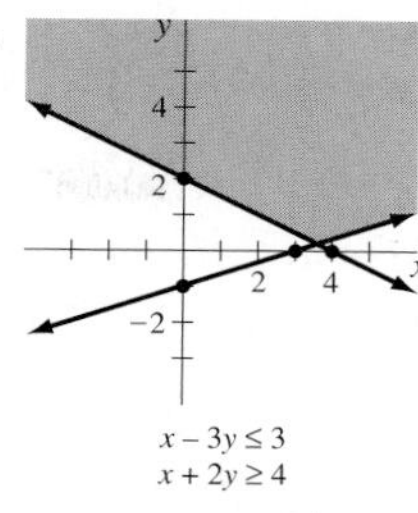

$x - 3y \leq 3$
$x + 2y \geq 4$

11.

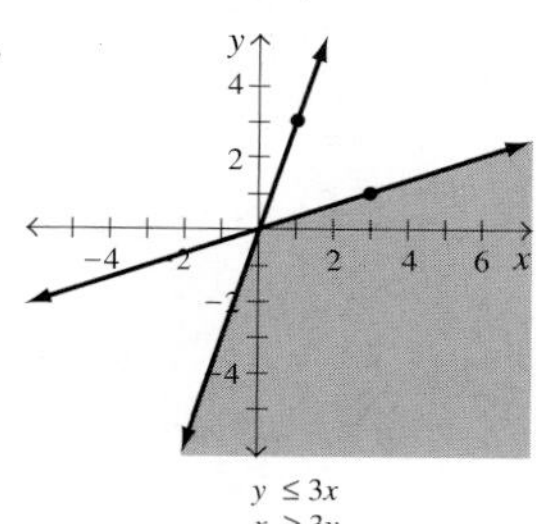

$y \leq 3x$
$x \geq 3y$

13.

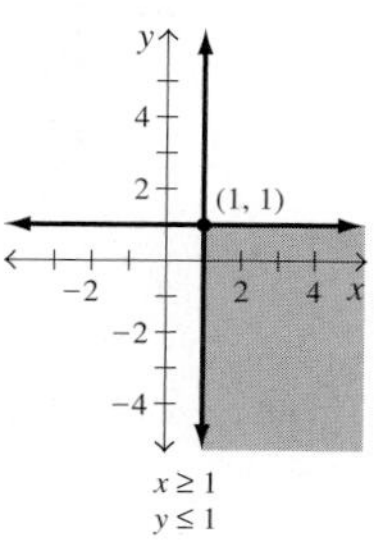

$x \geq 1$
$y \leq 1$

15.

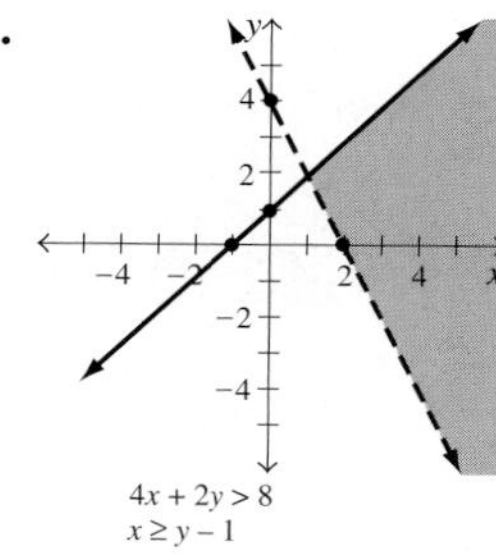

$4x + 2y > 8$
$x \geq y - 1$

17.

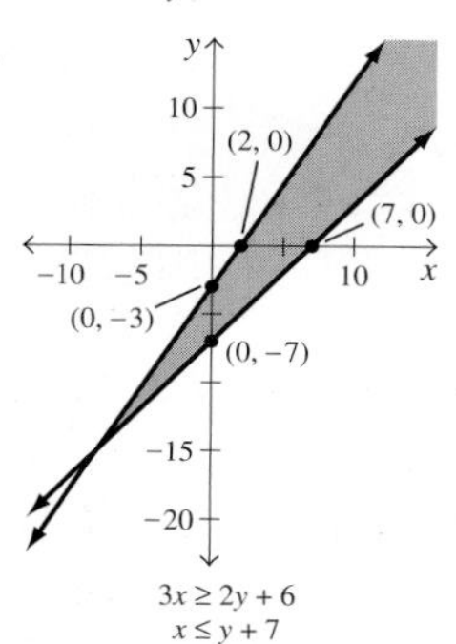

$3x \geq 2y + 6$
$x \leq y + 7$

19. a) $x + y < 500, x \geq 150, y \geq 150$
b)

Rice (calories)
Chicken (calories)

c) One example is (220, 220), approximately 3.7 oz of chicken, 8.8 oz of rice

21. a) No **b)** One example: $x + y > 4$
$x + y < 1$

23. No, every line divides the plane into two half planes, only one of which can be part of the solution.

SECTION 7.6, PAGE 387

1. Constraints are restrictions that are represented as linear inequalities.

3. Vertices **5.** Answers will vary.

7. Maximum is 18 at (2, 3), minimum is 0 at (0, 0).

9. a)

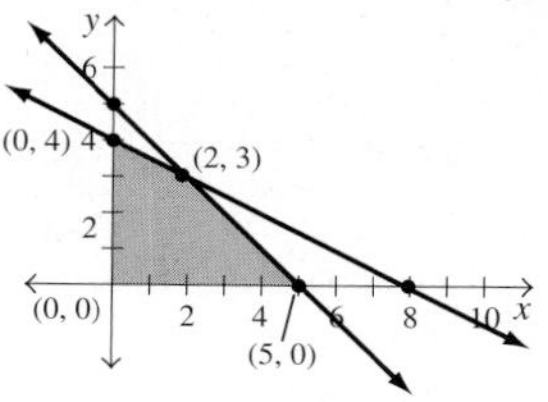

b) Maximum is 21 at (2, 3), minimum is 0 at (0, 0).

11. a)

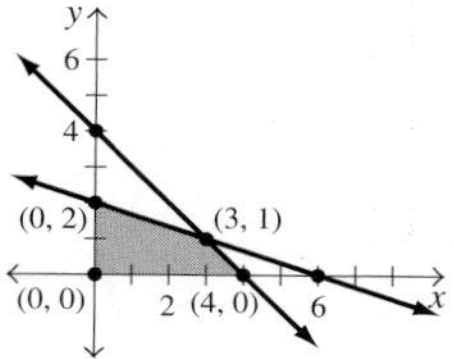

b) Maximum is 28 at (4, 0), minimum is 0 at (0, 0).

13. a)

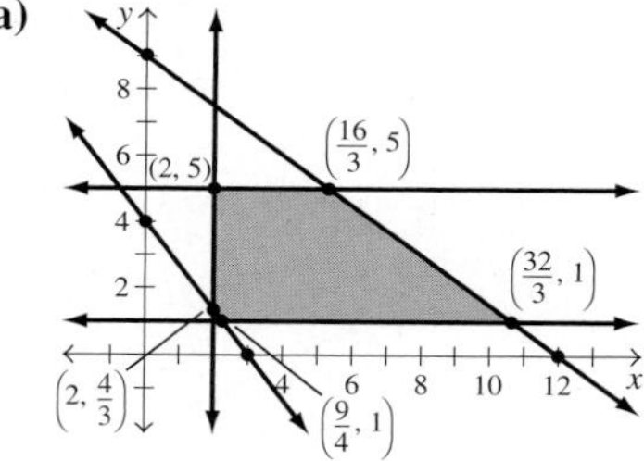

b) Maximum is ≈ 25.12 at $(\frac{32}{3}, 1)$, minimum is 6.6 at $(\frac{9}{4}, 1)$ or $(2, \frac{4}{3})$.

15. a) $x + y \leq 20, x \geq 3, x \leq 6, y \geq 2$
b) $P = 25x + 20y$
c)

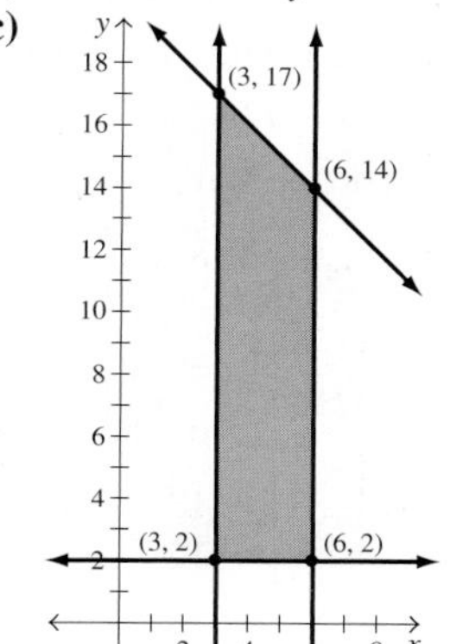

d) (3, 17), (6, 14), (6, 2), (3, 2)
e) six skateboards and 14 pairs of in-line skates
f) \$430

17. a) $60x + 50y \geq 300$, $8x + 20y \geq 80$, $6x + 30y \geq 90$, $x \geq 0$, $y \geq 0$

b) $C = 0.25x + 0.32y$

c)

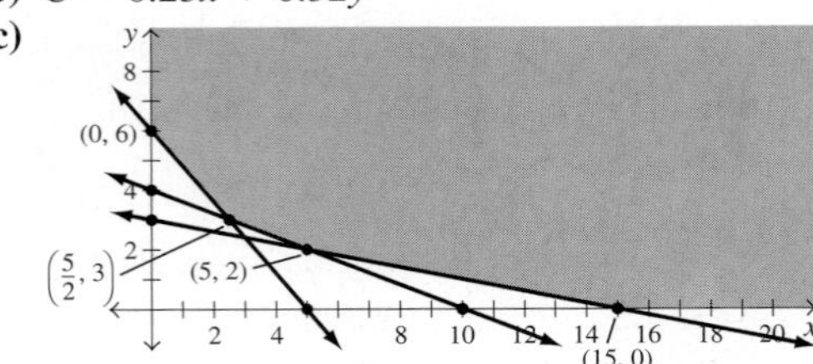

d) $(0, 6)$, $(\frac{5}{2}, 3)$, $(5, 2)$, $(15, 0)$

e) Trimfit: 2.5 cups; Usave: 3 cups

f) \$1.59

19. Two 4-cylinder and seven 6-cylinder, \$2050

REVIEW EXERCISES, PAGE 390

1.

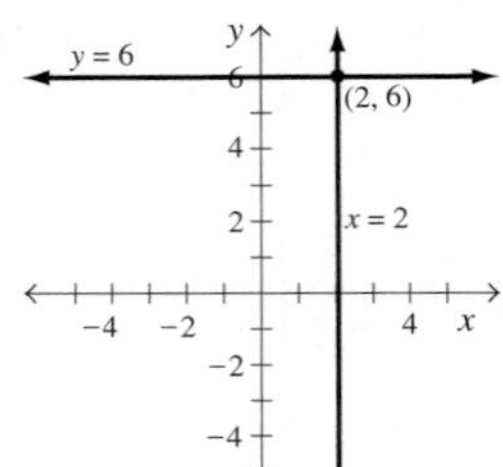

2.

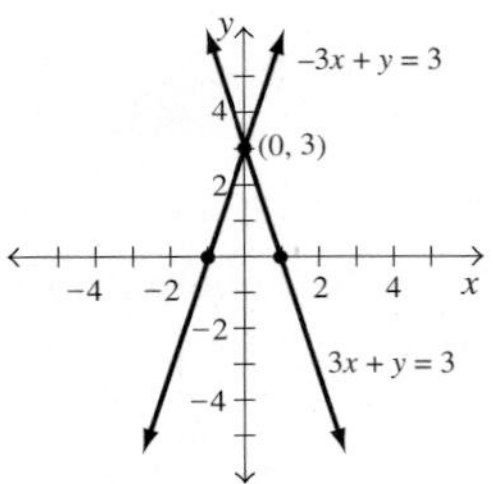

3.

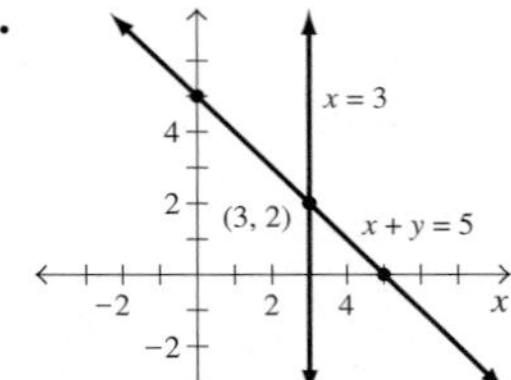

4.

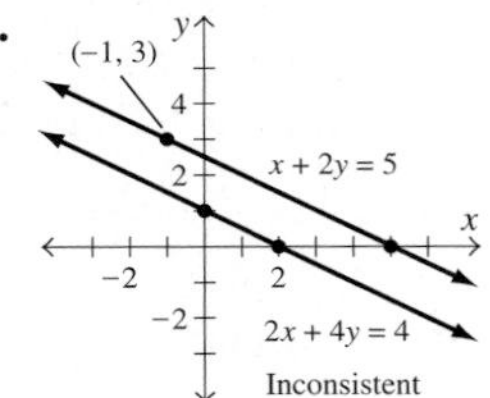

5. An infinite number of solutions **6.** No solution

7. One solution **8.** One solution **9.** $(3, 1)$ **10.** $(-1, -5)$

11. $(-2, -8)$ **12.** No solution; inconsistent **13.** $(-9, 3)$

14. $(-7, 16)$ **15.** $(4, -2)$ **16.** $(2, 0)$ **17.** $(30, -15)$

18. An infinite number of solutions; dependent

19. $\begin{bmatrix} -1 & -8 \\ 8 & 7 \end{bmatrix}$ **20.** $\begin{bmatrix} 3 & 2 \\ -4 & 1 \end{bmatrix}$

21. $\begin{bmatrix} 2 & -6 \\ 4 & 8 \end{bmatrix}$ **22.** $\begin{bmatrix} 7 & 1 \\ -6 & 6 \end{bmatrix}$

23. $\begin{bmatrix} -20 & -14 \\ 20 & 2 \end{bmatrix}$ **24.** $\begin{bmatrix} -12 & -14 \\ 12 & -6 \end{bmatrix}$

25. $(0, 2)$ **26.** $(-2, 2)$ **27.** $(3, -3)$

28. $(1, 0)$ **29.** $(\frac{12}{11}, \frac{7}{11})$ **30.** $(3, 1)$

31. a) 4 hr **b)** All-Day parking lot

32. Mix $83\frac{1}{3}$ ℓ of 80% acid solution with $16\frac{2}{3}$ ℓ of 50% acid solution.

33. \$500 salary, 4% commission rate

34. 30 dimes, 10 quarters

35. a) 32.5 months **b)** Model 6070B

36.

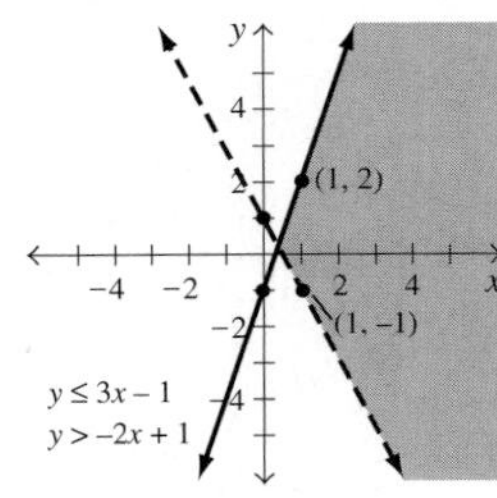

37.

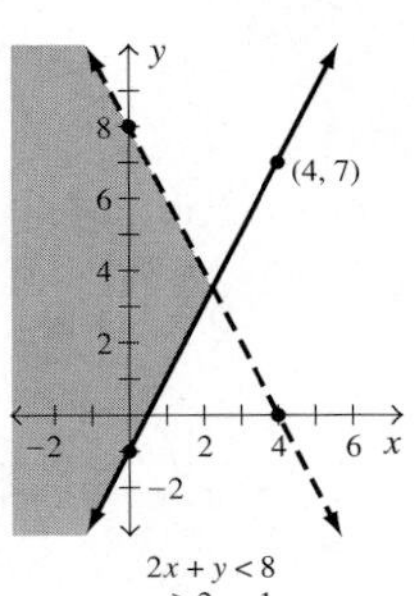

38.

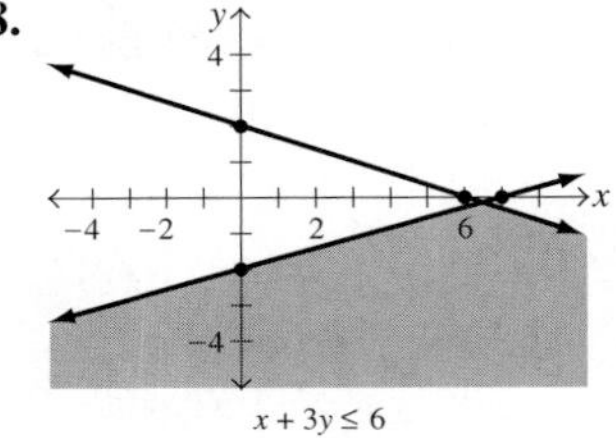

39.

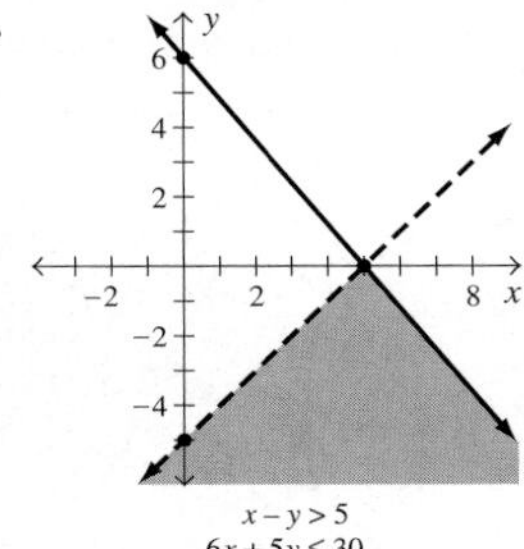

40. The maximum is 54 at $(9, 0)$.

CHAPTER TEST, PAGE 391

1. If the lines do not intersect (are parallel) the system of equations is inconsistent. The system of equations is consistent if the lines intersect. If both equations represent the same line, then the system of equations is dependent.

2.

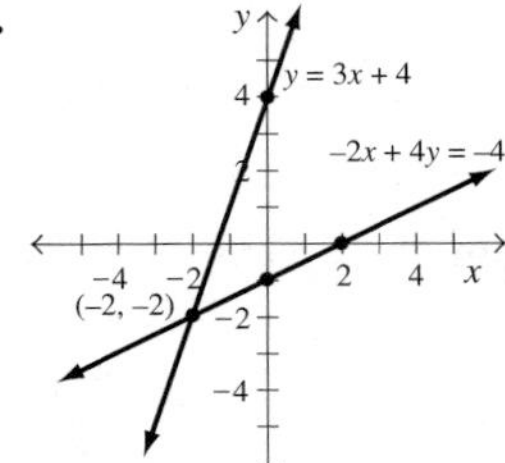

3. One solution **4.** $(2, -3)$ **5.** $(6, 30)$

6. $(3, -1)$ **7.** $(-1, 3)$ **8.** $(\frac{1}{2}, 1)$

9. $(-2, 2)$ **10.** $\begin{bmatrix} 1 & -2 \\ 6 & 11 \end{bmatrix}$

11. $\begin{bmatrix} 7 & -18 \\ 10 & 13 \end{bmatrix}$ **12.** $\begin{bmatrix} -12 & -19 \\ 8 & 42 \end{bmatrix}$

13. 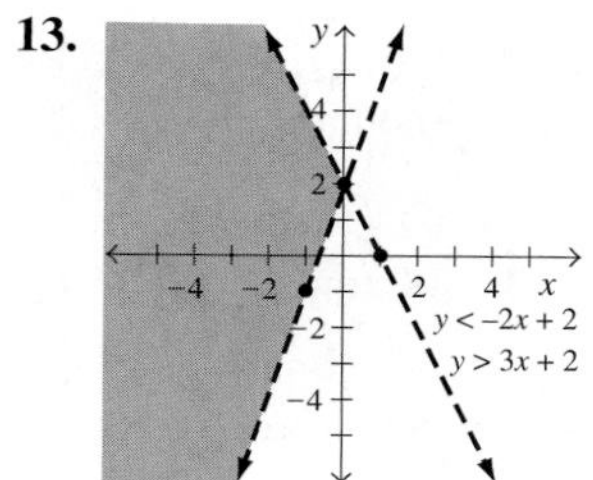

14. 20 lb at \$7.50, 10 lb at \$6.00

15. 12 one-bedroom, 8 two-bedroom

16. a) 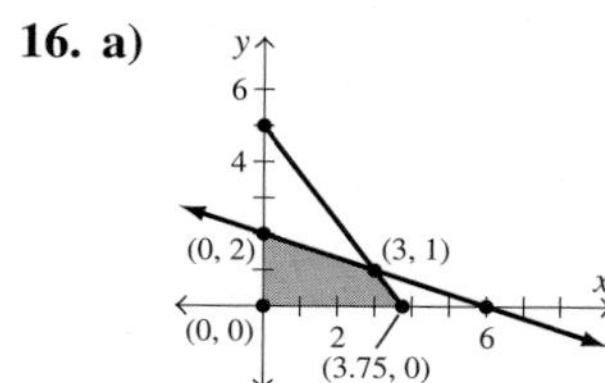

b) Maximum is 9 at (3, 1), minimum is 0 at (0, 0).

CHAPTER 8

SECTION 8.1, PAGE 398

1. The metric system

3. It is the standard of measurement accepted worldwide. There is only one basic unit of measurement for each quantity. It is based on the number 10, which makes many calculations easier than the U.S. customary system.

5. a) Answers will vary. **b)** 0.004 972 km **c)** 30 800 dm

9. a) 100 times greater **b)** 100 dm **c)** 0.01 dam

11. 2 **13.** 5 **15.** (b) **17.** (d) **19.** (e) **21.** decigram

23. gram **25.** dekaliter **27.** Centimeter **29.** Degree Celsius

31. a) 10 **b)** $\frac{1}{100}$ **c)** $\frac{1}{1000}$ **d)** $\frac{1}{10}$ **e)** 1000 **f)** 100

33. mg; $\frac{1}{1000}$ gram **35.** dg; $\frac{1}{10}$ gram **37.** hg; 100 g

39. 320 000 g **41.** 900 **43.** 3570 **45.** 0.0246

47. 9.74 **49.** 13 400 **51.** 7300 mm **53.** 895 000 mℓ

55. 0.001 30 km **57.** 0.8472 daℓ **59.** 680 m, 514 hm, 62 km

61. 420 cℓ, 4.3 ℓ, 0.045 kℓ **63.** 0.032 kℓ, 460 dℓ, 48 000 cℓ

65. The side with the 5 kg would tip down.

67. The pump that removes 1 daℓ per minute

69. a) 346 cm **b)** 3460 cm **71. a)** 3500 mg **b)** 3.5 g

73. a) 6.417 km/ℓ **b)** 6417 m/ℓ

75. 500 mℓ **77.** \$23.33 **79.** 1000

81. 1×10^{24} = 1 000 000 000 000 000 000 000 000

83. ≈ 2.8 cups **85.** 24 500 g **87.** 9 dam

89. 4 dm **91.** 2 dam

SECTION 8.2, PAGE 406

1. Length **3.** Area **5.** Volume **7.** Volume

9. Area **11.** Volume **19.** A cubic decimeter

21. A cubic centimeter **23.** 2.5 acres

25. Meters or centimeters **27.** Kilometers

29. Centimeters **31.** Millimeters **33.** Meters

35. Centimeters or millimeters **39.** (a) **41.** (a)

43. (b) **45.** (c) **53.** Centimeter, kilometer

55. Meter **57.** A centimeter **59.** Square meters

61. Square millimeters or square centimeters

63. Hectares or square meters

65. Square centimeters or square millimeters

67. Square centimeters **69.** (a) **71.** (a) **73.** (c)

75. (a) **83.** Kiloliters **85.** Milliliters **87.** Milliliters

89. Cubic meters **91.** Liters **93.** (c) **95.** (c) **97.** (b)

99. (b) **101. b)** 0.75 m^3 **103. b)** 7.85 cm^3

105. Longer side = 4 cm, shorter side = 1.8 cm, height = 1.5 cm; perimeter = 11.6 cm; area = 6 cm^2

107. 2984 cm^2 **109. a)** 5.25 km^2 **b)** 525 ha

111. a) 450 m^3 **b)** 450 kℓ **113.** \$304

115. 10,000 times larger **117.** 1000 times larger

119. 1 000 000 **121.** 100 **123.** 10 000

125. 1 000 000 **127.** 620 **129.** 76 **131.** 6700

133. a) 4,014,489,600 sq in. **b)** Answers will vary.

SECTION 8.3, PAGE 415

1. Kilogram **3.** 5 **5.** 35°C **9.** Kilograms or grams

11. Grams **13.** Grams **15.** Metric tonnes **17.** Grams

19. (b) **21.** (c) **23.** (b) **25.** Answers will vary.

27. Answers will vary. **29.** (c) **31.** (b) **33.** (b)

35. (c) **37.** 68°F **39.** 33.3°C **41.** 176.7°C **43.** 98.6°F

45. −10.6°C **47.** 113°F **49.** 45°C **51.** 71.6°F

53. 95.18°F **55.** 64.04°F–74.30°F **57.** \$3.15 **59.** 444 g

61. a) 2304 m^3 **b)** 2304 kℓ **c)** 2304 t **63.** 0.0036

65. 42 600 000 **67.** Yes; 78°F is about 25.6°C.

69. a) 5.625 ft^3 **b)** ≈ 351.6 lb **c)** ≈ 42.4 gal

SECTION 8.4, PAGE 423

1. Dimensional analysis is a procedure used to convert from one unit of measurement to a different unit of measurement.

3. $\frac{60 \text{ seconds}}{1 \text{ minute}}$ or $\frac{1 \text{ minute}}{60 \text{ seconds}}$ **5.** $\frac{1 \text{ lb}}{0.45 \text{ kg}}$ **7.** $\frac{0.8 \text{ m}^2}{1 \text{ yd}^2}$

9. 91.875 mi **11.** 1.26 m **13.** 12 m^2 **15.** 62.4 km

17. 1687.5 acres **19.** 2.496 ℓ **21.** 1.52 fl oz **23.** 54 kg

25. 28 grams, 0.45 kilograms

27. 2.54 centimeters, 1.6 kilometers

29. 157.48 centimeters (or $\approx$ 1.57 meters)

31. 9 meters **33.** $\approx$ 561.11 yd **35.** $\approx$ 1146.67 ft

37. 11.25 mi **39.** 43.2 m^2 **41.** 112 kph **43.** 240 $m\ell$

45. 360 m^3 **47.** \$0.495 per pound **49.** $\approx$9078.95 gal

51. a) $\approx$ 50.91 kg **b)** $\approx$ 113.13 lb

53. a) -8460 cm **b)** -84.6 m

55. a) 10.89 ft^2 **b)** 35.937 ft^3 **57.** 25.2 mg

59. 6840 mg, or 6.84 g **61. a)** 25 mg **b)** 900 mg

63. 0.12 ℓ graham crackers, 336 g nuts, 224 g chocolate, 0.32 ℓ coconut, 0.32 ℓ milk; use 22.86 cm $\times$ 33.02 cm baking pan; bake at 176.7°C; makes about 2 doz 3.81 cm $\times$ 7.62 cm bars.

65. 1.0 cc, or b) **67. a)** 3600 cm^3 **b)** $\approx$ 219.7 $in.^3$

REVIEW EXERCISES, PAGE 426

1. $\frac{1}{100}$ of base unit **2.** 1000 $\times$ base unit

3. $\frac{1}{1000}$ of base unit **4.** 100 $\times$ base unit

5. 10 times base unit **6.** $\frac{1}{10}$ of base unit **7.** 0.080 g

8. 320 $c\ell$ **9.** 1.97 mm **10.** 1 kg **11.** 4620 ℓ

12. 19 260 dg **13.** 3000 $m\ell$, 14 630 $c\ell$, 2.67 $k\ell$

14. 0.047 km, 47 000 cm, 4700 m **15.** Centimeters

16. Kilograms or grams **17.** Degrees Celsius

18. Millimeters **19.** Square meters

20. Milliliters or cubic centimeters **21.** Millimeters

22. Kilograms or tonnes **23.** Kilometers **24.** Liters

27. (c) **28.** (b) **29.** (c) **30.** (a) **31.** (a)

32. (b) **33.** 1.64 t **34.** 6 300 000 g **35.** 82.4°F

36. 20°C **37.** ≈ -21.1°C **38.** 102.2°F

39. $l = 4$ cm, $w = 1.6$ cm, $P = 11.2$ cm, $A = 6.4$ cm^2

40. $b = 3.2$ cm, $h = 2.5$ cm, hypotenuse $= 4.1$ cm, $P = 9.8$ cm, $A = 4$ cm^2

41. a) 80 m^3 **b)** 80 000 kg

42. a) 660 m^2 **b)** 0.000 66 km^2

43. a) 96 000 cm^3 **b)** 0.096 m^3 **c)** 96 000 $m\ell$ **d)** 0.096 $k\ell$

44. 10,000 times larger **45.** 68.58 **46.** $\approx$ 233.3

47. 74.7 **48.** $\approx$ 111.1 **49.** 72 **50.** 90 **51.** 57

52. $\approx$ 52.6 **53.** 32.4 **54.** 3.8 **55.** 11.4 **56.** 99.2

57. 0.9 **58.** 82.55 **59. a)** 1050 kg **b)** $\approx$ 2333.3 lb

60. 32.4 m^2 **61. a)** 1154 km **b)** 721.25 mi

62. a) 56 kph **b)** 56 000 meters per hour

63. a) 252 ℓ **b)** 252 kg **64.** \$1.24 per pound

CHAPTER TEST, PAGE 428

1. 67 000 000 mm **2.** 0.0096 hg **3.** 100 times greater

4. 1.8 km **5.** (b) **6.** (a) **7.** (c) **8.** (b) **9.** (b)

10. 10,000 times greater **11.** 1,000,000,000 times greater

12. 1148.08 cm **13.** 5 mph **14.** 10°C **15.** 122°F

16. 360 cm or 365.76 cm, depending on which conversion factor you used

17. a) 3200 m^3 **b)** 3 200 000 ℓ (or 3200 $k\ell$) **c)** 3 200 000 kg

18. \$245

CHAPTER 9

SECTION 9.1, PAGE 438

1. An axiom is a statement that is accepted as being true on the basis of its "obviousness" and its relation to the physical world. A theorem is a statement that has been proven using undefined terms, definitions, and axioms.

3. Two lines in the same plane that do not intersect are parallel lines.

5. Two angles in the same plane are adjacent angles when they have a common vertex and a common side but no common interior points.

7. Two angles the sum of whose measure is 180° are called supplementary angles.

9. An angle whose measure is less than 90° is an acute angle.

11. An angle whose measure is 180° is a straight angle.

13. Ray, $\overrightarrow{BA}$ **15.** Line, $\overleftrightarrow{AB}$ **17.** Ray, $\overrightarrow{AB}$

19. Half open line segment, $\overset{\circ\!-\!\bullet}{AB}$ **21.** $\overleftrightarrow{EG}$ **23.** $\overleftrightarrow{AD}$

25. $\varnothing$ **27.** $\{C\}$ **29.** $\overset{\circ\!-\!\circ}{BC}$ **31.** $\triangle BCF$ **33.** $\overrightarrow{ED}$

35. $\overleftrightarrow{DE}$ **37.** $\measuredangle FBE$ **39.** $\{B\}$ **41.** $\varnothing$ **43.** $\{B\}$

45. Straight **47.** Acute **49.** None of these

51. Right **53.** 39° **55.** $64\frac{1}{2}°$ **57.** 1° **59.** 104°

61. 45° **63.** $80\frac{4}{5}°$ **65.** b **67.** f **69.** a

71. $m\measuredangle 2 = 11.25°$, and $m\measuredangle 1 = 78.75°$ **73.** 127° and 53°

75. Angles 3, 4, and 7 are each 125°; angles 1, 2, 5, and 6 are each 55°.

77. Angles 2, 4, and 5 are each 120°; angles 1, 3, 6, and 7 are each 60°.

79. $m\measuredangle 1 = 76°$, $m\measuredangle 2 = 14°$

81. $m\measuredangle 1 = 31°$, $m\measuredangle 2 = 59°$

83. $m\measuredangle 1 = 132°$, $m\measuredangle 2 = 48°$

85. $m\measuredangle 1 = 33°$, $m\measuredangle 2 = 147°$

87. a) An infinite number **b)** An infinite number

89. An infinite number

For Exercises 91–95, the answers given are one of many possible answers.

91. Plane $AGB \cap$ plane $GBC = \overleftrightarrow{BG}$

93. Plane $HGD \cap$ plane $FGD \cap$ plane $BGD = \overleftrightarrow{GD}$

95. $\overleftrightarrow{AB} \cap$ plane $ABG = \overleftrightarrow{AB}$

97. Always true. If any two lines are parallel to a third line, then they must be parallel to each other.

99. Sometimes true. Vertical angles are only complementary when each is equal to 45°.

101. Sometimes true. Alternate interior angles are only complementary when each is equal to 45°.

103.

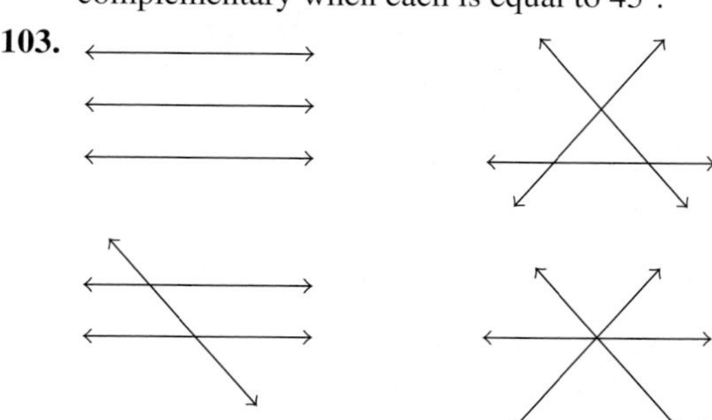

105.

1. Three parallel lines, two on the same plane (l_1 parallel to l_3 parallel to l_9).
2. Three lines on the same plane intersecting in three distinct points (l_9, l_{10}, l_{13}).
3. Three lines in the same plane, two are parallel and the third intersects the other two (l_2 parallel to l_4 and l_1 intersects l_2 and l_4).
4. Three lines intersect at a point, two lines on the same plane (l_1, l_2, l_6).
5. Two lines intersect. A third line is parallel to the first line and skewed to the second. (l_9 and l_{10} intersect, l_2 parallel to l_{10}, l_2 and l_9 are skewed.)
6. Three lines that are skewed to each other (l_3, l_6, l_{13}).
7. Three parallel lines in the same plane (l_2 parallel to l_4 parallel to l_{14}).
8. Three lines in the same plane intersecting in a single point (l_{10}, l_{11}, l_{13}).
9. Two parallel lines and a third line skewed to the other two. (l_6 parallel to l_8, and l_{13} is skewed to l_6 and l_8.)

SECTION 9.2, PAGE 447

1. A polygon is a closed figure in a plane determined by three or more straight line segments.

3. A regular polygon is one whose sides are all the same length and whose interior angles all have the same measure; other polygons may have sides of different length and interior angles with different measures.

5. Similar figures are figures that have the same shape but may be of different sizes.

7. Octagon **9.** Pentagon **11.** Equilateral

13. Scalene **15.** Isosceles **17.** Right **19.** Obtuse

21. Right **23.** Rectangle **25.** Trapezoid

27. Square **29.** 58° **31.** 65°

33. $m\measuredangle 1 = 90°$, $m\measuredangle 2 = 50°$, $m\measuredangle 3 = 130°$, $m\measuredangle 4 = 50°$, $m\measuredangle 5 = 50°$, $m\measuredangle 6 = 40°$, $m\measuredangle 7 = 90°$, $m\measuredangle 8 = 130°$, $m\measuredangle 9 = 140°$, $m\measuredangle 10 = 40°$, $m\measuredangle 11 = 140°$, $m\measuredangle 12 = 40°$

35. 1080° **37.** 900° **39.** 90°, 90° **41.** 108°, 72°

43. $128\frac{4}{7}°$, $51\frac{3}{7}°$ **45.** 55° **47.** 35° **49.** $x = 4$, $y = \frac{16}{5}$

51. $x = \frac{25}{12}$, $y = \frac{12}{5}$ **53.** $\frac{10}{3}$ **55.** 4 **57.** 14

59. 28 **61.** 28° **63.** 8 **65.** 16 **67.** 70°

69. $x = 50°$, $y = 130°$ **71.** 80° **73.** 100°

75. a) $m\measuredangle CED = m\measuredangle ABC$; $m\measuredangle ACB = m\measuredangle DCE$; $m\measuredangle BAC = m\measuredangle CDE$
b) $\approx$2141.49 ft

77. Answers will vary.

SECTION 9.3, PAGE 457

1. a) Answers will vary.
b) Answers will vary.
c)

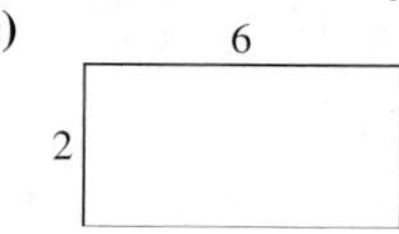

The area of this rectangle is 12 square units. The perimeter of this rectangle is 16 units.

3. 35 in.^{2} **5.** 4.5 ft^2 or 0.5 yd^2

7. Area = 105 ft^2; perimeter = 44 ft

9. Area = 288 in.^{2}; perimeter = 74 in.

11. Area = 132 in.^{2}; perimeter = 48 in.

In Exercises 13–15 and the other exercises involving π, we used the π key on a scientific calculator to determine the answer. If you use 3.14 for π, your answers may vary slightly.

13. Area ≈ 78.54 in.^{2}; circumference ≈ 31.42 in.

15. Area ≈ 38.48 ft^2; circumference ≈ 21.99 ft

17. 13 ft **19.** 36 m **21.** 21.46 m^2 **23.** 8 ft^2

25. 30.54 m^2 **27.** 114.90 ft^2 **29.** ≈1.69 yd^2

31. 164.7 ft^2 **33.** 147,000 cm^2 **35.** 0.0608 m^2

37. $108,550 **39.** $807.30 **41.** 684 ft^2

43. $38.93 **45.** 2.56 yd^2

47. The legs are each 12 cm and hypotenuse is ≈16.97 cm.

49. **a)** $A = s^2$ **b)** $A = 4s^2$ **c)** four times larger

51. **a)** $A = bh$ **b)** $A = 4bh$ **c)** four times larger

53. ≈188.50 in.^{2} **55.** 24 cm^2

SECTION 9.4, PAGE 467

In this section we use the π key on the calculator to determine answers in calculations involving pi. If you use 3.14 for π your answers may vary slightly.

1. Volume is a measure of the capacity of a figure.

3. A polyhedron is a closed surface formed by the union of polygonal regions. A regular polyhedron is one whose faces are all regular polygons of the same size and shape.

5. Answers will vary. **7.** 8 ft^3 **9.** 2714.34 in.^{3}

11. 131.95 cm^3 **13.** 384 in.^{3} **15.** 1150.35 cm^3

17. 524.33 cm^3 **19.** 106.67 in.^{3} **21.** 85.84 yd^3

23. 59.43 m^3 **25.** 791.68 cm^3 **27.** 284.46 cm^3

29. 135 ft^3 **31.** ≈7.85 yd^3 **33.** 2,700,000 cm^3

35. 4 m^3 **37.** Tubs ≈ 141.37 in.^{3}; boxes = 125 in.^{3}

39. The Magic Burger has greater volume by ≈0.98 in.^{3}

41. **a)** The container with the larger diameter holds more **b)** ≈188.50 in.^{3}

43. **a)** 270 m^3 **b)** 270 kℓ **45.** ≈27.83 in.^{3}

47. **a)** ≈ 333.33 ft^3 **b)** ≈ 12.35 yd^3 **49.** 14.14 in.^{3}

51. **a)** 61.58 m **b)** 18,103.11 m^3 **53.** Six faces

55. Six vertices **57.** Fourteen edges

59. The new volume is eight times the original volume.

61. The new volume is eight times the original volume.

63. Yes. The cylinder has three times the volume of the cone, but the customer is only charged twice as much.

65. ≈21.46%

67. **a)** Answers will vary.
b) $V_1 = a^3$; $V_2 = a^2b$; $V_3 = a^2b$; $V_4 = ab^2$; $V_5 = a^2b$; $V_6 = ab^2$; $V_7 = b^3$
c) ab^2

SECTION 9.5, PAGE 475

1. A Möbius strip is a one-sided, one-edged surface.

3. A Klein bottle is a topological object that resembles a bottle but has only one side.

5. **a)** Six **b)** Seven

7. Two figures are topologically equivalent if one figure can be elastically twisted, stretched, bent, or shrunk into the other figure without ripping or puncturing the original figure.

9. One **11.** One

13. **a)** No, it has an inside and an outside. **b)** Two **c)** Two **d)** Two strips, one inside the other

15. No, it does not. **17–23.** Answers will vary.

25. Outside **27.** Outside **29.** Outside **31.** Inside

33. 1 **35.** 5 **37.** 0 **39.** 4 **41.** 3

43. Answers will vary.

SECTION 9.6, PAGE 483

1–5. Answers will vary.

7. **a)** The sum of the measures of the angles of a triangle is 180°.
b) The sum of the measures of the angles of a triangle is less than 180°.
c) The sum of the measures of the angles of a triangle is greater than 180°.

9. A sphere

11. Each type of geometry can be used in its own frame of reference.

13. Spherical-elliptical geometry; flat-Euclidean geometry; saddle shaped-hyperbolic geometry

15.

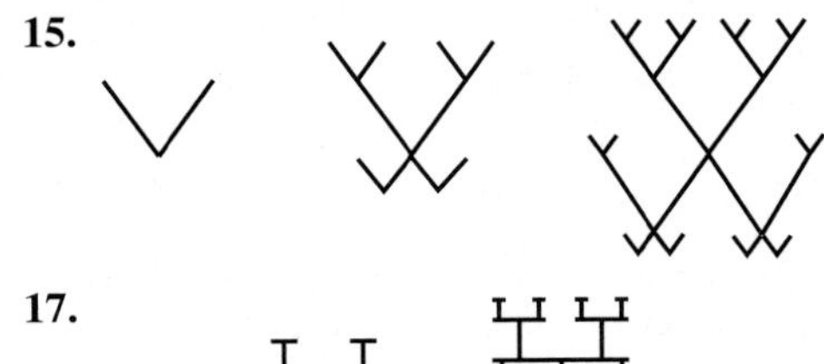

17.

19. **a)**

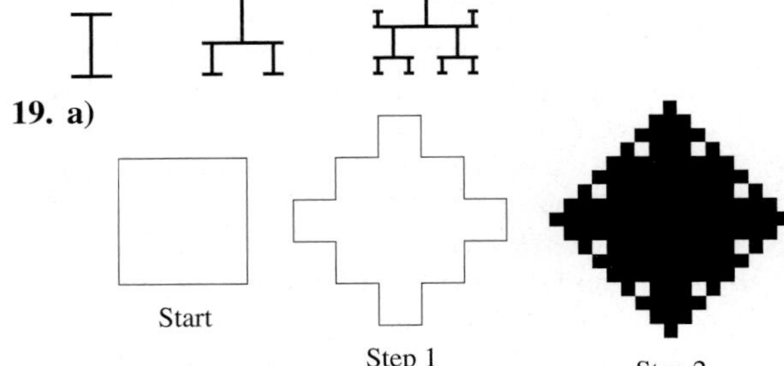

b) Infinite **c)** Finite

REVIEW EXERCISES, PAGE 484

1. $\{F\}$ **2.** $\triangle BFC$ **3.** $\overline{BC}$ **4.** $\overleftrightarrow{BH}$ **5.** $\{F\}$

6. { } **7.** 63.7° **8.** 74.8° **9.** 10.2 in.

10. 2 in. **11.** 58° **12.** 92°

13. $\measuredangle 1 = 75°$, $\measuredangle 2 = 60°$, $\measuredangle 3 = 120°$, $\measuredangle 4 = 75°$, $\measuredangle 5 = 105°$, $\measuredangle 6 = 75°$

14. 720° **15.** 28 cm^2 **16.** 56 $in.^2$ **17.** 13 $in.^2$ **18.** 84 $in.^2$

19. $\approx$ 380.13 cm^2 **20.** $462.50 **21.** 75.40 $in.^3$ **22.** 120 cm^3

23. 28 ft^3 **24.** 540 m^3 **25.** 392.70 mm^3 **26.** 523.60 ft^3

27. a) 67.88 ft^3 **b)** 4617.5 lb; yes **c)** 511.1 gal

28. 4 **29.** Answers will vary. **30.** Outside

31. Euclidean: Given a line and a point not on the line, one and only line can be drawn parallel to the given line through the given point. Elliptical: Given a line and a point not on the line, no line can be drawn through the given point parallel to the given line. Hyperbolic: Given a line and a point not on the line, two or more lines can be drawn through the given point parallel to the given line.

32.

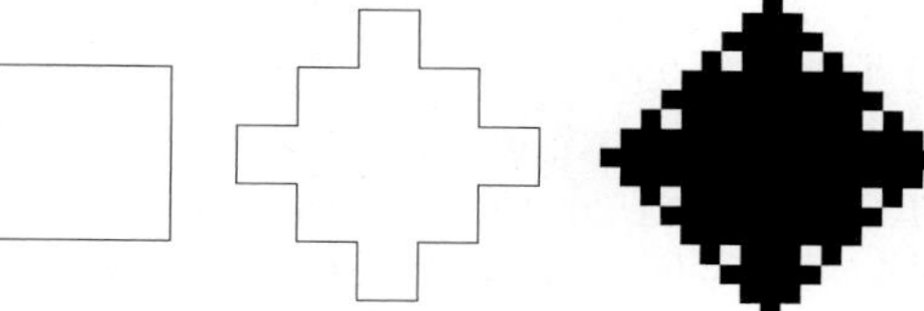

CHAPTER TEST, PAGE 486

1. $\overleftrightarrow{EF}$ **2.** $\triangle BCD$ **3.** $\{D\}$ **4.** $\overleftrightarrow{AC}$ **5.** 72.6°

6. 86.4° **7.** 63° **8.** 1080° **9.** 5.38 cm

10. a) 12 in. **b)** 30 in. **c)** 30 $in.^2$

11. 2144.66 cm^3 **12.** $\approx$ 168.55 yd^3 **13.** 126 ft^3

14. Answers will vary.

15. A surface with one side and one edge

16. Answers will vary.

CHAPTER 10

SECTION 10.1, PAGE 495

1. A binary operation is an operation, or rule, that can be performed on two and only two elements of a set. The result is a single element.

3. a) When we add two numbers, the sum is one number: $4 + 5 = 9$.
b) When we subtract two numbers, the difference is one number: $5 - 4 = 1$.
c) When we multiply two numbers, the product is one number: $5 \times 4 = 20$.
d) When we divide two numbers, the quotient is one number: $20 \div 5 = 4$.

5. A mathematical system is a commutative group if all five conditions hold.
1. The set of elements is closed under the given operation.
2. An identity element exists for the set.
3. Every element in the set has an inverse.
4. The set of elements is associative under the given operation.
5. The set of elements is commutative under the given operation.

7. If a binary operation is performed on any two elements of a set and the result is an element of the set, then that set is *closed* under the given binary operation. For all integers a and b, $a + b$ is an integer. Therefore, the set of integers is closed under the operation of addition.

9. When a binary operation is performed on two elements in a set and the result is the identity element for the binary operation, then each element is said to be the *inverse* of the other. The additive inverse of 2 is -2 since $2 + (-2) = 0$, and the multiplicative inverse of 2 is $\frac{1}{2}$ since $2 \times \frac{1}{2} = 1$.

11. $(a + b) + c = a + (b + c)$, for any elements a, b, and c; $(3 + 4) + 5 = 3 + (4 + 5)$.

13. $a + b = b + a$, for any elements a and b; $3 + 4 = 4 + 3$.

15. $7 - 3 \neq 3 - 7$
$4 \neq -4$

17. $(6 - 4) - 1 \neq 6 - (4 - 1)$
$2 - 1 \neq 6 - 3$
$1 \neq 3$

19. No; no identity element **21.** Yes

23. No; not associative **25.** No; not closed **27.** Yes

SECTION 10.2, PAGE 503

1. Numbers are obtained by starting at the first addend (on a clock face), then moving clockwise the number of hours equal to the second addend.

3. a) Add 6 + 9 to get 3, then add 3 + 5.
b) 8

5. a) Add 12 to 3 to get 15 − 10.
b) 5
c) Since 12 is the identity element, you can add 12 to any number without changing the answer.

7. Yes, the sum of any two numbers in clock 12 arithmetic is a number in clock 12 arithmetic.

9. Yes; 1:11, 2:10, 3:9, 4:8, 5:7, 6:6, 7:5, 8:4, 9:3, 10:2, 11:1, and 12:12

11. Yes, 6 + 9 = 9 + 6 since both equal 3.

13. 12 **15.** 5 **17.** 5 **19.** 8 **21.** 6

23. 6 **25.** 6 **27.** 9 **29.** 7

31.

+	**1**	**2**	**3**	**4**	**5**	**6**
1	2	3	4	5	6	1
2	3	4	5	6	1	2
3	4	5	6	1	2	3
4	5	6	1	2	3	4
5	6	1	2	3	4	5
6	1	2	3	4	5	6

33. 1 **35.** 3 **37.** 2 **39.** 5

41.

+	**1**	**2**	**3**	**4**	**5**	**6**	**7**
1	2	3	4	5	6	7	1
2	3	4	5	6	7	1	2
3	4	5	6	7	1	2	3
4	5	6	7	1	2	3	4
5	6	7	1	2	3	4	5
6	7	1	2	3	4	5	6
7	1	2	3	4	5	6	7

43. 2 **45.** 6 **47.** 4 **49.** 6

51. Yes, it satisfies the five required properties.

53. a) {0, 1, 2, 3}
b) ✈
c) Yes
d) Yes; 0
e) Yes: 0–0, 1–3, 2–2, 3–1
f) (2 ✈ 3) ✈ 1 = 2 ✈ (3 ✈ 1)
g) Yes; 2 ✈ 3 = 3 ✈ 2
h) Yes

55. a) {5, 8, 9, 11}
b) ⌖
c) Yes
d) Yes; 9
e) Yes; 5–5, 8–11, 9–9, 11–8
f) (5 ⌖ 8) ⌖ 11 = 5 ⌖ (8 ⌖ 11)
g) Yes; 5 ⌖ 8 = 8 ⌖ 5
h) Yes

57. a) No; no identity element
b) (1 ⌐W 3) ⌐W 4 ≠ 1 ⌐W (3 ⌐W 4)

59. Not associative:
(△◫□)◫△ ≠ △◫(□◫△),
Not commutative: □◫△ ≠ △◫□

61. No inverse for ~ or for *, not associative

63. No identity element, no inverses, not associative, not commutative

65. a)

+	*E*	*O*
E	*E*	*O*
O	*O*	*E*

b) Yes; it satisfies the five properties.

69. a) Is closed; identity element is 6; inverses: 1–5, 2–2, 3–3, 4–4, 5–1, 6–6; is associative—for example,
$(2 \infty 5) \infty 3 = 2 \infty (5 \infty 3)$
$3 \infty 3 = 2 \infty 2$
$6 = 6$
b) $3 \infty 1 \neq 1 \infty 3$
$2 \neq 4$

71. $5^3 = 125$

SECTION 10.3, PAGE 511

1. A modulo m system consists of m elements, 0 through $m - 1$, and a binary operation.

3. 5;

0	**1**	**2**	**3**	**4**
0	1	2	3	4
5	6	7	8	9
10	11	12	13	14
.	.	.	.	.
.	.	.	.	.

5. 12 **7.** Wednesday **9.** Saturday **11.** Saturday
19. 2 **21.** 3 **23.** 4 **25.** 3 **27.** 1 **29.** 2 **31.** 1
33. 0 **35.** 2 **37.** 6 **39.** 1 **41.** 9 **43.** 2 **45.** 1
47. 3 **49.** 5 **51.** { } **53.** 6 **55.** 10

57. a) 2012, 2016, 2020, 2024, 2028 **b)** 3004
c) 2552, 2556, 2560, 2564, 2568, 2572

59. a) 5 **b)** No **c)** 54 weeks from this week

61. a) Evening **b)** Day **c)** Day

63. a)

+	**0**	**1**	**2**	**3**
0	0	1	2	3
1	1	2	3	0
2	2	3	0	1
3	3	0	1	2

b) Yes **c)** Yes, 0 **d)** Yes, 0–0, 1–3, 2–2, 3–1
e) (1 + 2) + 3 = 1 + (2 + 3)
f) Yes, 2 + 3 = 3 + 2 **g)** Yes **h)** Yes

65. a)

×	**0**	**1**	**2**	**3**
0	0	0	0	0
1	0	1	2	3
2	0	2	0	2
3	0	3	2	1

b) Yes **c)** Yes, 1
d) No; no inverse for 0 or for 2, inverse of 1 is 1, inverse of 3 is 3
e) (1 × 2) × 3 = 1 × (2 × 3)
f) Yes, 2 × 3 = 3 × 2 **g)** No

67. 2 **69.** 1, 2, 3 **71.** 0 **73.** 2 **75.** math is fun.

REVIEW EXERCISES, PAGE 513

1. A mathematical system consists of a set of elements and at least one binary operation.

2. A binary operation is an operation that can be performed on two and only two elements of a set. The result is a single element.

3. Yes, the sum of any two integers is an integer.

4. No, for example $2 - 3 = -1$ and -1 is not a natural number.

5. 3 **6.** 5 **7.** 8 **8.** 10 **9.** 9 **10.** 11

11. Closure, identity element, inverses, and associative property

12. A commutative group **13.** Yes

14. No; no inverse for any integer except 1 **15.** Yes

16. No; no inverse for 0 **17.** No identity element

18. Not associative. For example,
(! ◡ *p*) ◡ ? ≠ ! ◡ (*p* ◡ ?)

19. Not every element has an inverse; not associative. For example, (P ? P) ? 4 ≠ P ? (P ? 4).

20. a) { ≻, ⊙, ?, △} **b)** ⊥ **c)** Yes **d)** Yes, ≻
e) Yes; ≻ – ≻, ⊙ –△, ?–?, △–⊙
f) (⊙ ⊥ ?) ⊥ △ = ⊙ ⊥ (? ⊥ △)
g) Yes; ⊙ ⊥ ? = ? ⊥ ⊙ **h)** Yes

21. 3 **22.** 7 **23.** 6 **24.** 3 **25.** 4 **26.** 2 **27.** 1

28. 12 **29.** 9 **30.** 9 **31.** 7 **32.** 3 **33.** { } **34.** 5

35. 0, 2, 4, 6 **36.** 3 **37.** 5 **38.** 3 **39.** 7 **40.** 8

41.

+	0	1	2	3	4	5
0	0	1	2	3	4	5
1	1	2	3	4	5	0
2	2	3	4	5	0	1
3	3	4	5	0	1	2
4	4	5	0	1	2	3
5	5	0	1	2	3	4

Yes, a commutative group

42.

×	0	1	2	3
0	0	0	0	0
1	0	1	2	3
2	0	2	0	2
3	0	3	2	1

No; no inverse for 0 or 2

43. a) No, she will be off.
b) Yes, she will have the evening off.

CHAPTER TEST, PAGE 514

1. A set of elements and a binary operation

2. Closure, identity element, inverses, associative property, commutative property

3. No, not all elements have inverses.

4.

+	1	2	3	4	5
1	2	3	4	5	1
2	3	4	5	1	2
3	4	5	1	2	3
4	5	1	2	3	4
5	1	2	3	4	5

5. Yes, it is a commutative group. **6.** 4 **7.** 3

8. a) □ **b)** Yes **c)** Yes, *T* **d)** *S* **e)** *S*

9. No, not closed. **10.** Yes, it is a commutative group.

11. Yes, it is a commutative group. **12.** 1 **13.** 3

14. 1 **15.** 3 **16.** 5 **17.** 2 **18.** { } **19.** 5

20. a)

×	0	1	2	3	4
0	0	0	0	0	0
1	0	1	2	3	4
2	0	2	4	1	3
3	0	3	1	4	2
4	0	4	3	2	1

b) No; no inverse for 0

CHAPTER 11

SECTION 11.1, PAGE 524

1. A percent is a ratio of some number to 100.

3. Multiply the decimal number by 100 and add a percent sign.

5. $$\text{Percent change} = \frac{\left(\begin{array}{l}\text{amount in the latest period} - \\ \text{amount in the previous period}\end{array}\right)}{\text{amount in the previous period}} \times 100$$

7. 37.5% **9.** 62.5% **11.** 0.8% **13.** 378% **15.** 0.12

17. 0.0375 **19.** 0.0025 **21.** 0.002 **23.** 0.01 **25.** 14.3%

27. 249,805.8 miles **29.** $17,918,623,800

31. 15,926,629 African-American workers

33. 29.6% **35.** 18.0%

37. a) ≈8.7% increase **b)** About 293.8 million people

39. a) 9.1% increase **b)** 6.8% increase
c) 0.5% increase **d)** 17.2% increase

41. a) 1976 to 1986 **b)** 1976 to 1986

43. 300 **45.** 0.63 **47.** 25%

49. a) $2.61 **b)** $46.11 **c)** $6.92 **d)** $53.03

51. 12 students **53.** $39,055

55. ≈5.3% decrease **57.** ≈9.4% decrease

59. ≈18.6% decrease **61.** $3750

63. He will have a loss of $10. **65.** $21.95

SECTION 11.2, PAGE 534

1. Interest is the money the borrower pays for the use of the lender's money.

3. Security or collateral is anything of value pledged by the borrower that the lender may sell or keep if the borrower does not repay the loan.

5. A personal note is a document that states the terms and conditions of the loan agreement between the borrower and the lender.

7. The difference between ordinary interest and interest calculated using the Banker's rule is the way in which time is used in the simple interest formula. Ordinary interest: A month is 30 days, and a year is 360 days. Banker's rule: Any fractional part of a year is the exact number of days, and a year is 360 days.

9. \$113.40 **11.** \$8.75 **13.** \$15.85 **15.** \$80.06

17. \$1721.52 **19.** 4.0% **21.** \$600 **23.** 2 years

25. \$2091.20 **27. a)** \$131.25 **b)** \$3631.25

29. a) \$182.50 **b)** \$3467.50 **c)** ≈7.9%

31. \$23,793.75 **33.** 190 days **35.** 139 days

37. 264 days **39.** July 17 **41.** March 24 **43.** \$2254.17

45. \$6301.63 **47.** \$5278.99 **49.** \$850.64 **51.** \$2646.24

53. a) November 3, 1999 **b)** \$978.06
c) \$21.94 **d)** ≈4.44%

55. a) ≈409.0% **b)** ≈204.5% **c)** ≈102.3%

57. a) 6.663% **b)** \$6663 **c)** 7.139% **d)** \$6996.15

SECTION 11.3, PAGE 542

1. An investment is the use of money or capital for income or profit.

3. A variable investment is one in which neither the principal nor the interest is guaranteed.

5. The effective annual yield is the simple interest rate that gives the same amount of interest as a compound rate over the same period of time.

7. a) \$4326.40 **b)** \$326.40

9. a) \$3840.25 **b)** \$840.25

11. a) \$1728.28 **b)** \$228.28

13. a) \$2831.95 **b)** \$331.95

15. a) \$7006.98 **b)** \$2006.98

17. \$419,614.45 **19.** \$6734.28

21. a) \$129,210.47 **b)** \$134.88

23. \$117.60 **25.** \$5816.85 **27.** \$662.74 **29.** ≈5.76%

31. a) \$126.97, \$26.97 **b)** \$253.95, \$53.95
c) \$507.89, \$107.89 **d)** Yes, the interest also doubles.

33. ≈7.71%

35. The accounts have the same effective rate when rounded to three decimal places.

37. \$45,250.17 **39.** \$7020.14 **41.** \$1.90

43. a) 24 years **b)** 12 years **c)** 9 years
d) 6 years **e)** 3.27%

45. \$27,550.11 **47. a)** \$55,726.01 **b)** \$55,821.15

SECTION 11.4, PAGE 554

1. With an installment plan, the borrower repays the principal plus the interest with weekly or monthly payments that usually begin shortly after the loan is made. With a personal note, the borrower repays the principal plus the interest as a single payment at the end of the specified time period.

3. The APR is the true rate of interest charged on a loan.

5. The total installment price is the sum of all the monthly payments and the down payment, if any.

7. The unpaid balance method and the average daily balance method

9. a) \$8769.60 **b)** \$626.16

11. a) \$809.20 **b)** \$80.15

13. a) \$628.40 **b)** 9.0%

15. a) \$35.92 **b)** 13%

17. a) 11% **b)** \$1423.28 **c)** \$8230.02

19. a) \$4305.40 **b)** \$730.71 **c)** \$544.43 **d)** \$8954.80

21. a) \$2025.75 **b)** \$195.22 **c)** \$1147.24 **d)** \$6075.90

23. a) \$27.63 **b)** \$1032.10

25. a) \$18 **b)** \$630.86

27. a) \$31 **b)** \$553.02

29. a) \$19.76 **b)** \$743.41

31. a) \$1.56 **b)** \$133.11

33. a) \$512.00 **b)** \$6.66 **c)** \$638.43

35. a) \$121.78 **b)** \$1.52 **c)** \$133.07
d) The interest charged using the average daily balance method is \$0.04 less than the interest charged using the unpaid balance method.

37. a) \$20.70 **b)** \$1520.70

39. a) \$9.25 **b)** \$16.28 **c)** 12.5% **d)** 12.0%

41. a) 6 months **b)** \$74.62
c) Using the credit card they save \$13.38.

43. Since Martina's billing date is June 25th, she can buy the camera from June 26th through June 29th and the purchase will appear on her July 25th bill. Since she has a 20-day grace period, she can pay for the camera on August 5th without paying interest.

SECTION 11.5, PAGE 566

1. A mortgage is a long-term loan in which the property is pledged as security for payment of the difference between the down payment and the sale price.

3. The major difference is that the interest rate for a conventional loan is fixed for the duration of the loan, whereas the interest rate for a variable-rate loan may change every period, as specified in the loan agreement.

5. A buyer's adjusted monthly income is found by subtracting any fixed monthly payments with more than 10 months remaining from the gross monthly income.

7. An add-on rate, or margin, is the percent added to the interest rate on which the adjustable rate mortgage is based.

9. Equity is the difference between the appraised value of your home and the loan balance.

11. a) \$18,000 **b)** \$558

13. a) \$14,800 **b)** \$476.56

15. a) \$14,025 **b)** \$79,475 **c)** \$2384.25

17. a) \$1934.80 **b)** Yes

19. a) \$361,520 **b)** \$201,520 **c)** \$75.90

21. a) \$31,780 **b)** \$2451.60 **c)** \$4330 **d)** \$1212.40 **e)** \$789.42 **f)** \$916.09 **g)** Yes. **h)** \$108.42

23. Bank B

25. a) \$113,095.24 **b)** \$150,793.65

27. a) 805
b)

Payment number	Interest	Principal	Balance of loan
1	\$750.00	\$55.00	\$99,945.00
2	\$749.59	\$55.41	\$99,889.59
3	\$749.17	\$55.83	\$99,833.76

c) 9.38%
d)

Payment number	Interest	Principal	Balance of loan
4	\$780.37	\$24.63	\$99,809.13
5	\$780.17	\$24.83	\$99,784.30
6	\$779.98	\$25.02	\$99,759.28

e) 9.46%

REVIEW EXERCISES, PAGE 570

1. 25% **2.** 66.7% **3.** 62.5% **4.** 3.9%

5. 0.98% ≈ 1.0% **6.** 314.1% **7.** $0.2\overline{6}$ **8.** 0.121

9. 1.23 **10.** 0.004 **11.** $0.008\overline{3}$ **12.** 0.0000045

13. ≈9.6% **14.** ≈11.0% **15.** 31.25% **16.** 275

17. 91.8 **18.** \$6.42 **19.** 40 people **20.** 26.7%

21. \$14 **22.** 9.5% **23.** \$450 **24.** 0.5 years

25. \$4410 **26. a)** \$162 **b)** \$3162

27. a) \$1380 **b)** \$4620 **c)** ≈14.9%

28. a) $7\frac{1}{2}$% **b)** \$830 **c)** \$941.18

29. a) \$2007.34, \$507.34 **b)** \$2020.28, \$520.28 **c)** \$2023.28, \$523.28

30. \$5076.35 **31.** 5.76% **32.** \$13,415.00

33. a) 11% **b)** \$493.02 **c)** \$4350.73

34. a) \$109.18 **b)** \$2014.11

35. a) 11% **b)** \$81.91 **c)** \$1471.20

36. a) \$6.31 **b)** \$847.61

37. a) \$5.36 **b)** \$549.68

38. a) \$3725 **b)** \$11,175 **c)** \$2905.50 **d)** 12%

39. a) \$5.48 **b)** 10%

40. a) \$33,925 **b)** \$4805.33 **c)** \$1345.49 **d)** \$825.40 **e)** \$1142.07 **f)** Yes

41. a) \$13,485 **b)** \$756.51 **c)** \$24.20 **d)** \$285,828.60 **e)** \$195,928.60

42. a) \$550.46 **b)** 8% **c)** 7.75%

CHAPTER TEST, PAGE 572

1. \$80 **2.** 3 years **3.** \$637.50 **4.** \$5637.50

5. \$1997.50 **6.** \$270.50 **7.** 12.5%

8. a) \$105.05 **b)** \$3155.90

9. a) \$6.93 **b)** \$612.70

10. a) \$12.30 **b)** \$1146.57

11. \$2523.20 **12.** \$123.20 **13.** \$11,144.61, \$3644.61

14. \$3036.68, \$536.68 **15.** \$21,675 **16.** \$6603.33

17. \$1848.93 **18.** \$1123.85 **19.** \$1428.02

20. Yes **21. a)** \$426,261 **b)** \$281,761

CHAPTER 12

SECTION 12.1, PAGE 581

1. An experiment is a controlled operation that yields a set of results.

3. Empirical probability is the relative frequency of occurrence of an event. It is determined by actual observation of an experiment.

$$P(E) = \frac{\text{Number of times event has occurred}}{\text{Number of times experiment was performed}}$$

5. Answers will vary.

7. The official definition by the Weather Service is **(d)**. The specific location is the place where the rain gauge is. The Weather Service uses sophisticated mathematical equations to calculate these probabilities.

9. No, it means that the average person with traits similar to Mr. Reebe's will live another 42.94 years.

11. Answers will vary. **13.** Answers will vary.

15. a) $\frac{1}{12}$ **b)** $\frac{7}{15}$ **c)** $\frac{1}{5}$ **17. a)** $\frac{9}{31}$ **b)** $\frac{3}{31}$ **c)** $\frac{15}{31}$

19. a) 0.372 **b)** 0.294 **c)** 112 **21. a)** 1 **b)** Yes

23. a) 0.147 **b)** 0.13 **c)** 0.085, 0.147

25. a) $\frac{6}{20} = \frac{3}{10}$ **b)** $\frac{14}{20} = \frac{7}{10}$ **c)** $\frac{14}{20} = \frac{7}{10}$ **d)** $\frac{2}{20} = \frac{1}{10}$

27. a) 0 **b)** $\frac{50}{250} = 0.2$ **c)** 1 **29. a)** 0.24 **b)** 0.76

SECTION 12.2, PAGE 588

1. If each outcome of an experiment has the same chance of occurring as any other outcome, they are said to be equally likely outcomes.

3. $P(A) + P(\text{not } A) = 1$ **5.** Answers will vary. **7.** 0 and 1

9. a) $\frac{1}{5}$ **b)** $\frac{1}{4}$ **11.** $\frac{1}{48}$ **13.** $\frac{1}{13}$

15. $\frac{12}{13}$ **17.** $\frac{1}{4}$ **19.** 1 **21.** $\frac{3}{13}$

23. a) $\frac{1}{2}$ **b)** $\frac{1}{4}$ **c)** $\frac{1}{4}$ **25. a)** $\frac{1}{3}$ **b)** $\frac{1}{3}$ **c)** $\frac{1}{3}$

27. $\frac{25}{100} = \frac{1}{4}$ **29.** $\frac{85}{100} = \frac{17}{20}$ **31.** $\frac{1}{12}$ **33.** $\frac{1}{3}$

35. $\frac{3}{10}$ **37.** $\frac{7}{10}$ **39.** $\frac{2}{5}$ **41.** $\frac{3}{5}$ **43.** $\frac{3}{10}$ **45.** $\frac{3}{5}$ **47.** $\frac{9}{10}$

49. $\frac{1}{8}$ **51.** $\frac{55}{603}$ **53.** $\frac{1}{26}$ **55.** $\frac{3}{26}$ **57.** $\frac{310}{575} = \frac{62}{115}$

59. $\frac{396}{575}$ **61.** $\frac{73}{575}$ **63.** 159 **65.** $\frac{39}{159} = \frac{13}{53}$

67. $\frac{93}{159} = \frac{31}{53}$ **69.** $\frac{13}{36}$ **71.** $\frac{1}{3}$ **73.** $\frac{23}{36}$

75. a) $\frac{1}{4}$ **b)** $\frac{1}{2}$ **c)** $\frac{1}{4}$ **77. a)** $\frac{1}{4}$ **b)** $\frac{1}{4}$ **c)** $\frac{1}{4}$

SECTION 12.3, PAGE 595

1. Answers will vary. **3.** Odds against

5. 15 to 2 **7. a)** $\frac{1}{2}$ **b)** $\frac{1}{2}$

9. a) $\frac{11}{24}$ **b)** $\frac{13}{24}$ **c)** 13:11 **d)** 11:13

11. 5:1 **13.** 2:1 **15.** 12:1, 1:12 **17.** 10:3, 3:10

19. 1:1 **21.** 5:3 **23. a)** 5:4 **b)** 4:5

25. 8:7 **27.** 14:1 **29.** 8:7 **31.** 1:35

33. a) $\frac{7}{12}$ **b)** $\frac{5}{12}$ **35.** $\frac{9}{14}$ **37.** $\frac{1}{5}$ **39.** 1:4

41. 74:1 **43.** 0.06 **45.** 47:3 **47.** 4:21

49. 3:2 **51.** 1:4 **53. a)** 20:1 **b)** $\frac{20}{21}$

55. Horse 1, $\frac{2}{9}$; Horse 2, $\frac{1}{3}$; Horse 3, $\frac{1}{16}$; Horse 4, $\frac{5}{12}$; Horse 5, $\frac{1}{2}$

SECTION 12.4, PAGE 602

1. The expected value is the expected gain or loss of an experiment over the long run.

3. The fair price is the amount that should be charged for the game to be fair and result in an expectation of 0.

5. No, the fair price is the price to pay to make the expected value 0. The expected value is the expected outcome of an experiment when the experiment is performed many times.

7. \$0.50. Since you would lose \$1.00 on average for each game you played, the price of the game should be \$1.00 less. Then the expectation would be \$0, and the game would be fair.

9. a) −\$2 **b)** Lose \$2 **11.** \$35,000 **13.** 70 points

15. a) $\frac{7}{16}$ in., or 0.4375 in. **b)** $\frac{217}{16}$ in., or 13.5625 in.

17. a) 23% **b)** \$77

19. a) No, it is a disadvantage **b)** Yes, it is an advantage.

21. a) −\$0.50 **b)** \$0.50 **c)** \$500

23. a) \$3 **b)** \$1 **25. a)** \$4.25 **b)** \$2.25

27. \$40.00 gain

29. a) 2.9 points **b)** 2.9 points **c)** 8.7 points

31. \$38,400; yes **33.** 440 **35.** \$22,000 **37.** gain \$1700, yes

39. a) $\frac{9}{16}, \frac{1}{4}, \frac{1}{8}, \frac{1}{16}$ **b)** \$11.81 **c)** \$11.81

41. An amount greater than \$1200 **43.** − \$0.053 or −5.3¢

45. a) \$458.33 **b)** \$308.33 **47.** Answers will vary.

SECTION 12.5, PAGE 609

1. If a first experiment can be performed in M distinct ways and a second experiment can be performed in N distinct ways, then the two experiments in that specific order can be performed in $M \cdot N$ distinct ways.

3. 15

5. The first selection is made. Then the second selection is made before the first selection is returned to the group of items being selected.

7. a) 2500 **b)** 2450 **9. a)** 25 **b)** 20

11. a) 4

b)

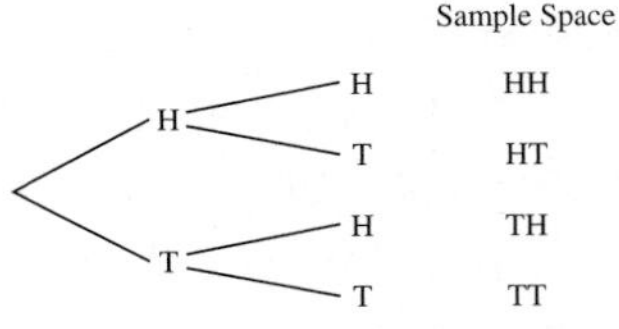

c) $\frac{1}{4}$ **d)** $\frac{1}{2}$ **e)** $\frac{1}{4}$

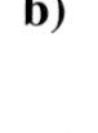

13. a) 6

b)

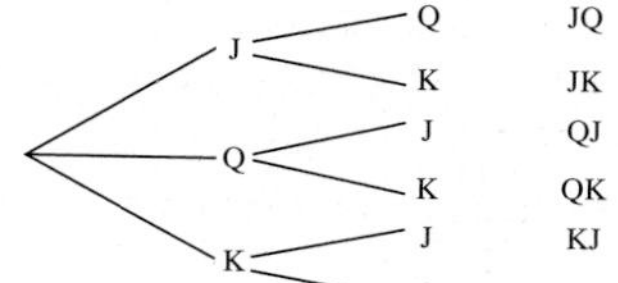

c) 0 **d)** $\frac{1}{6}$ **e)** $\frac{2}{3}$

15. a) 12

b)

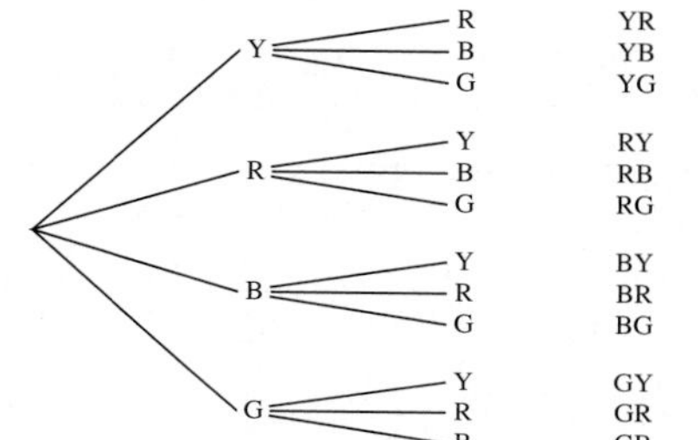

c) $\frac{1}{2}$ **d)** 1 **e)** $\frac{1}{2}$

17. a) 36

b)

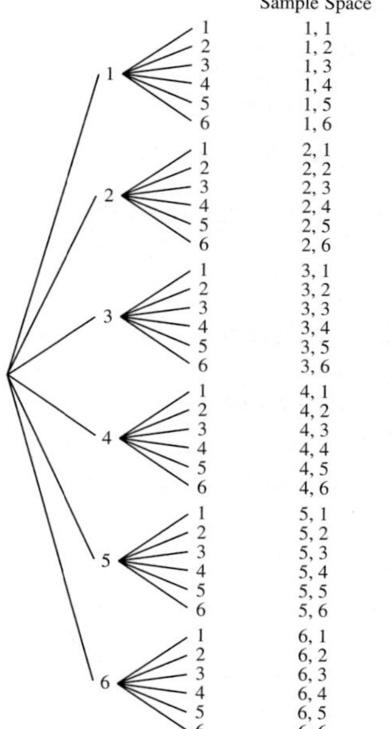

c) $\frac{1}{6}$ **d)** $\frac{1}{6}$ **e)** $\frac{1}{36}$ **f)** No

19. a) 27

b)

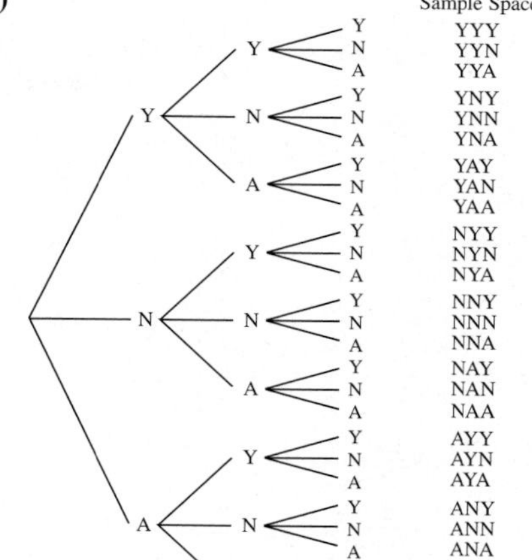

c) $\frac{1}{27}$ **d)** $\frac{2}{9}$ **e)** $\frac{19}{27}$

21. a) 18

b)

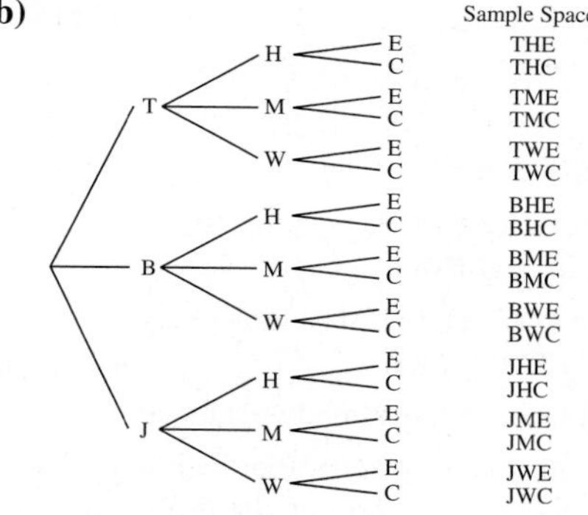

c) $\frac{1}{3}$ **d)** $\frac{2}{3}$ **e)** $\frac{1}{6}$

23. a) 16

b)

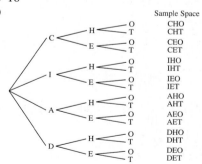

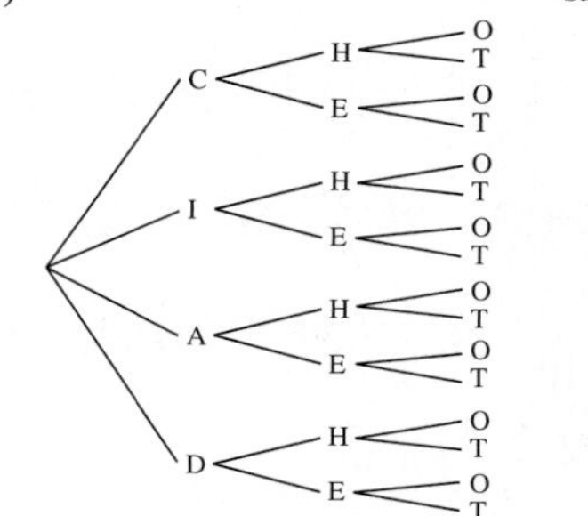

c) $\frac{1}{4}$ **d)** $\frac{1}{2}$ **e)** $\frac{1}{8}$

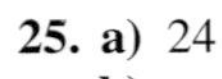

25. a) 24
b)

Sample Space

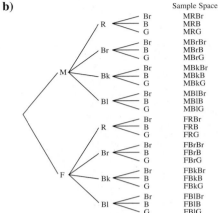

			Sample Space
M	R	Br	MRBr
		B	MRB
		G	MRG
	Br	Br	MBrBr
		B	MBrB
		G	MBrG
	Bk	Br	MBkBr
		B	MBkB
		G	MBkG
	Bl	Br	MBlBr
		B	MBlB
		G	MBlG
F	R	Br	FRBr
		B	FRB
		G	FRG
	Br	Br	FBrBr
		B	FBrB
		G	FBrG
	Bk	Br	FBkBr
		B	FBkB
		G	FBkG
	Bl	Br	FBlBr
		B	FBlB
		G	FBlG

c) $\frac{1}{24}$ **d)** $\frac{1}{8}$

27. a) *m* or *n* **b)** 3 or 4 **c)** *m*3, *m*4, *n*3, *n*4
d) No, not unless we know that all the outcomes are equally likely
e) No, not unless we know that all the outcomes are equally likely
f) Yes

SECTION 12.6, PAGE 619

1. a) At least one event, A *or* B, must occur.
b) Both events, A *and* B, must occur.

3. a) Events that cannot happen simultaneously
b) $P(A \text{ or } B) = P(A) + P(B)$

5. We assume that event *A* has already occurred.

7. Two events are dependent when the probability of one item being selected has an effect on the probability of a second item being selected.

9. a) No, both events can occur at the same time.
b) No, being healthy can affect your happiness.

11. Answers will vary.

13. 0.7 **15.** 0.5 **17.** $\frac{1}{3}$ **19.** $\frac{1}{2}$ **21.** $\frac{2}{13}$ **23.** $\frac{8}{13}$ **25.** $\frac{37}{52}$

27. a) $\frac{1}{25}$ **b)** $\frac{3}{95}$ **29. a)** $\frac{1}{16}$ **b)** $\frac{5}{76}$ **31. a)** $\frac{1}{40}$ **b)** $\frac{1}{38}$

33. a) $\frac{4}{25}$ **b)** $\frac{14}{95}$ **35.** $\frac{7}{10}$ **37.** $\frac{2}{5}$ **39.** $\frac{1}{4}$ **41.** $\frac{1}{8}$ **43.** $\frac{9}{64}$

45. $\frac{1}{8}$ **47.** $\frac{3}{8}$ **49.** $\frac{1}{8}$ **51.** $\frac{1}{8}$ **53. a)** $\frac{1}{32}$ **b)** $\frac{1}{2}$

55. a) $\frac{1}{6}$ **b)** $\frac{1}{5}$ **57. a)** $\frac{3}{4}$ **b)** $\frac{4}{5}$ **59.** 0.0529 **61.** 0.0675

63. 0.008 **65.** 0.00682 **67.** 0.7 **69.** 0.343 **71.** $\frac{1}{4}$

73. $\frac{27}{1024}$ **75.** $\frac{243}{1024}$ **77.** $\frac{5}{22}$ **79.** $\frac{9261}{10{,}648}$ **81.** $\frac{1}{8}$

83. $\frac{5}{12}$ **85.** 0.36 **87.** 0.36

89. a) No **b)** 0.001
ci) 0.00004 **cii)** 0.00096 **ciii)** 0.000999 **civ)** 0.998001

91. $\frac{49}{62{,}500}$ or 0.000784 **93.** $\frac{59{,}049}{62{,}500}$ or 0.944784

95. $\frac{1}{3}$ **97.** $\frac{1}{2}$ **99.** $\frac{14}{45}$ **101.** $\frac{45}{91}$ **103.** $\frac{1}{2}$

SECTION 12.7, PAGE 626

1. The probability of E_2 given that E_1 has occurred

3. $\frac{1}{3}$ **5.** $\frac{1}{3}$ **7.** $\frac{1}{2}$ **9.** $\frac{2}{3}$ **11.** $\frac{1}{5}$ **13.** $\frac{1}{3}$ **15.** $\frac{3}{5}$ **17.** $\frac{3}{7}$

19. $\frac{1}{16}$ **21.** $\frac{1}{7}$ **23.** $\frac{1}{6}$ **25.** $\frac{1}{6}$ **27.** $\frac{2}{3}$ **29.** $\frac{18}{25}$ **31.** $\frac{2}{3}$

33. 0 **35.** $\frac{19}{27}$ **37.** $\frac{5}{14}$ **39.** $\frac{107}{200}$ **41.** $\frac{9}{10}$ **43.** $\frac{10}{11}$ **45.** $\frac{3}{19}$

47. $\frac{44}{47}$ **49.** $\frac{11}{21}$ **51.** $\frac{10}{23}$ **53.** $\frac{18}{23}$ **55.** $\frac{1}{3}$ **57.** $\frac{1}{3}$ **59.** 1.00

61. a) 140 **b)** 120 **c)** $\frac{7}{10}$ **d)** $\frac{3}{5}$ **e)** $\frac{2}{3}$ **f)** $\frac{4}{7}$
g) Because A and B are not independent events

63. a) 0.3 **b)** 0.4 **c)** Yes; $P(A \mid B) = P(A) \cdot P(B)$

SECTION 12.8, PAGE 635

1. Answers will vary. **3.** Answers will vary.

5. The number of permutations of *n* items taken *r* at a time.

7. $\frac{n!}{n_1!\, n_2! \ldots n_r!}$ **9.** 120 **11.** 362,880 **13.** 1

15. 1 **17.** 40,320 **19.** 40,320 **21.** 10,000

23. a) $5^5 = 3125$ **b)** $\frac{1}{3125} = 0.00032$

25. a) 840 **b)** 924 **27.** 720 systems

29. a) 120 **b)** 120 **c)** 24 **d)** 6

31. 336 **33.** 10,000,000,000 **35.** 6,760,000

37. 504,000 **39.** 676,000 **41.** 104,000

43. a) 8,000,000 **b)** 6,400,000,000
c) 64,000,000,000,000

45. 1320 **47.** 5040 **49.** 720 **51.** 5040

53. 2520 **55.** 630 **57.** 6720

59. a) 40,320 **b)** 362,880

61. a) 3125 **b)** ≈128 **c)** 0.00032

63. 12,600 sec, or 3.5 hr **65.** No

SECTION 12.9, PAGE 641

1. Answers will vary. **3.** ${}_nC_r = \frac{n!}{(n-r)!r!}$

5. Answers will vary. **7.** 20

9. a) 35 **b)** 210 **11. a)** 28 **b)** 56 **13. a)** 120 **b)** 720

15. $\frac{1}{6}$ **17.** 2 **19.** 1140 **21.** 5 **23.** 495 **25.** 70 **27.** 6

29. 6 **31.** 28 **33.** 7350 **35.** 560 **37.** 294,000

39. 28,512 **41.** 1200 **43. a)** 45 **b)** 56

45. a) The order is important. Since the numbers may be repeated, it is not a true permutation lock.
b) 64,000 **c)** 59,280

47. a) 9,366,819 **b)** 10,737,573
c) 12,271,512 **d)** 13,983,816
e) No **f)** 15,890,700

49. a) 24 **b)** 24

SECTION 12.10, PAGE 646

1. $\frac{_6C_4}{_{10}C_4}$ **3.** $\frac{_{19}C_{12}}{_{34}C_{12}}$ **5.** $\frac{_{22}C_8}{_{70}C_8}$ **7.** $\frac{_{14}C_9}{_{30}C_9}$ **9.** $\frac{5}{42}$ **11.** $\frac{1}{14}$

13. $\frac{1}{12}$ **15.** $\frac{1}{14}$ **17.** $\frac{1}{9{,}366{,}819}$ **19.** $\frac{1}{10}$ **21.** $\frac{7}{10}$

23. $\frac{6}{115}$ **25.** $\frac{27}{230}$ **27.** $\frac{646{,}646}{3{,}910{,}797{,}436} \approx 0.0001653$

29. $\frac{923{,}410{,}488}{3{,}910{,}797{,}436} \approx 0.236$ **31.** $\frac{10}{91}$ **33.** $\frac{59}{65}$ **35.** $\frac{1}{77}$

37. $\frac{5}{77}$ **39.** $\frac{5}{506} \approx 0.010$ **41. a)** $\frac{1}{123{,}760}$ **b)** $\frac{1}{30{,}940}$

43. $\frac{5}{729}$ **45.** $\frac{821}{1458}$ **47. a)** $\frac{1}{2{,}162{,}160}$ **b)** $\frac{1}{6435}$ **49.** $\frac{1}{24}$

SECTION 12.11, PAGE 654

1. A probability distribution shows the probability associated with each specific outcome of an experiment. In a probability distribution every possible outcome must be listed and the sum of all the probabilities must be 1.

3. $P(x) = (_nC_x)p^xq^{n-x}$ **5.** 0.2916 **7.** 0.3456 **9.** 0.0625

11. a) $P(x) = (_nC_x)(0.15)^x(0.85)^{n-x}$
b) $P(2) = (_{12}C_2)(0.15)^2(0.85)^{10}$

13. 0.4096 **15.** 0.110592 **17.** 0.068773724

19. $\frac{32}{81} \approx 0.395061728$

21. a) 0.01024 **b)** 0.98976

23. a) $\approx$0.1119 **b)** $\approx$0.2966

REVIEW EXERCISES, PAGE 656

1. Answers will vary. **2.** Answers will vary.

3. $\frac{11}{12}$ **4.** Answers will vary. **5.** $\frac{29}{50}$ **6.** $\frac{1}{2}$ **7.** $\frac{4}{5}$

8. 1 **9.** $\frac{1}{5}$ **10.** $\frac{17}{50}$ **11.** $\frac{3}{10}$ **12.** $\frac{9}{25}$ **13.** $\frac{4}{5}$

14. a) 9:1 **b)** 1:9 **15.** 5:3 **16.** $\frac{3}{5}$ **17.** 3:2

18. a) −\$1.20 **b)** −\$3.60 **c)** \$0.80

19. a) −\$0.23 **b)** \$0.23 **c)** Lose \$23.08

20. 660 people

21. a) T: J, G, C; J: T, G, C; G: T, J, C; C: T, J, G

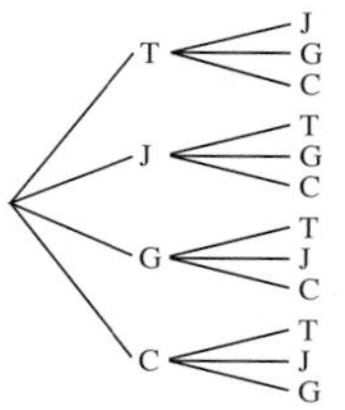

b) Sample Space
TJ
TG
TC
JT
JG
JC
GT
GJ
GC
CT
CJ
CG

c) $\frac{1}{12}$

22. a) H: R, B, G, P; T: R, B, G, P

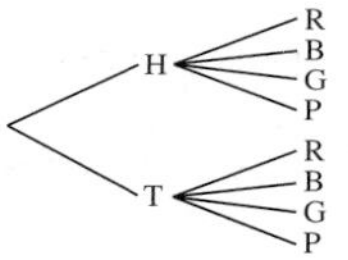

b) Sample Space
HR
HB
HG
HP
TR
TB
TG
TP

c) $\frac{1}{4}$

23. $\frac{1}{4}$ **24.** $\frac{9}{64}$ **25.** $\frac{5}{16}$ **26.** $\frac{7}{8}$ **27.** 1 **28.** $\frac{3}{16}$

29. $\frac{1}{22}$ **30.** $\frac{14}{55}$ **31.** $\frac{41}{55}$ **32.** $\frac{1}{22}$ **33.** $\frac{1}{4}$

34. Against, 3:1; in favor, 1:3 **35.** \$1.25 **36.** $\frac{1}{8}$ **37.** $\frac{5}{8}$

38. In favor, 3:5; against, 5:3 **39.** \$3.75 **40.** $\frac{7}{8}$ **41.** $\frac{89}{106}$

42. $\frac{55}{74}$ **43.** $\frac{19}{74}$ **44.** $\frac{17}{106}$ **45.** $\frac{23}{40}$ **46.** $\frac{3}{17}$ **47.** $\frac{3}{4}$ **48.** $\frac{12}{17}$

49. a) 24 **b)** \$2900

50. 30 **51.** 336 **52.** 504 **53.** 20

54. a) 3003 **b)** 3,628,800

55. a) $\frac{1}{2{,}118{,}760}$ **b)** $\frac{1}{76{,}275{,}360}$ **56.** 120 **57.** 560

58. $\frac{1}{221}$ **59.** $\frac{1}{12}$ **60.** $\frac{1}{18}$ **61.** $\frac{1}{24}$ **62.** $\frac{11}{12}$

63. $\frac{5}{91}$ **64.** $\frac{15}{91}$ **65.** $\frac{3}{13}$ **66.** $\frac{10}{13}$

67. a) $P(x) = (_nC_x)(0.6)^x(0.4)^{n-x}$
b) $P(75) = (_{100}C_{75})(0.6)^{75}(0.4)^{25}$

68. 0.0512

69. a) 0.0256 **b)** 0.9744

CHAPTER TEST, PAGE 660

1. $\frac{7}{10}$ **2.** $\frac{2}{9}$ **3.** $\frac{5}{9}$ **4.** $\frac{7}{9}$ **5.** $\frac{1}{3}$ **6.** $\frac{1}{6}$

7. $\frac{1}{6}$ **8.** $\frac{5}{18}$ **9.** $\frac{5}{12}$ **10.** $\frac{8}{13}$ **11.** 18

12.

Sample Space

1 — a, b, c: 1a, 1b, 1c
2 — a, b, c: 2a, 2b, 2c
3 — a, b, c: 3a, 3b, 3c
4 — a, b, c: 4a, 4b, 4c
5 — a, b, c: 5a, 5b, 5c
6 — a, b, c: 6a, 6b, 6c

13. $\frac{1}{18}$ **14.** $\frac{4}{9}$ **15.** $\frac{2}{3}$ **16.** 608,400 **17.** 3:5

18. $\frac{2}{7}$ **19.** \$0 **20.** $\frac{4}{7}$ **21.** 120 **22.** $\frac{14}{95}$

23. $\frac{81}{95}$ **24.** $\frac{175}{396}$ **25.** $\frac{216}{625} = 0.3456$

CHAPTER 13

SECTION 13.1, PAGE 666

1. Descriptive statistics is concerned with the collection, organization, and analysis of data. Inferential statistics is concerned with making generalizations or predictions from the data collected.

3. Answers will vary. **5.** Answers will vary.

7. a) A population is all items or people of interest.
b) A sample is a subset of the population.

9. a) A systematic sample is a sample obtained by selecting every nth item on a list or production line.
b) Use a random number table to select the first item, then select every nth item after that.

11. a) A stratified sample is one that includes items from each part (or strata) of the population.
b) Select a random sample from each strata

13. An unbiased sample is one that is a small replica of the entire population with regard to income, education, sex, race, religion, political affiliation, age, etc.

15. random sample **17.** cluster sample

19. systematic sample **21.** convenience sample

23. cluster sample **25.** Answers will vary.

SECTION 13.2, PAGE 670

1. Answers will vary.

3. Every employee may not make the average salary. Some may have a salary far below the average or far above the average.

5. The fact that Morgan's is the largest department store does not imply it is inexpensive.

7. Most driving is done close to home. Thus, one might expect more accidents close to home.

9. Averages apply to a set of data. Thus, although the female average score may be greater, some males have higher scores than some females.

11. There may be deep sections in the pond, so it may not be safe to go wading.

13. Half the students in a population are expected to be below average.

15. Just because some prefer it does not mean they buy it. Other factors, such as cost, must be considered.

17. a)

Number of U.S. Hospitals, 1991–1996

Number of hospitals: 1000, 2000, 3000, 4000, 5000, 6000, 7000
Year: '91 '92 '93 '94 '95 '96

b)

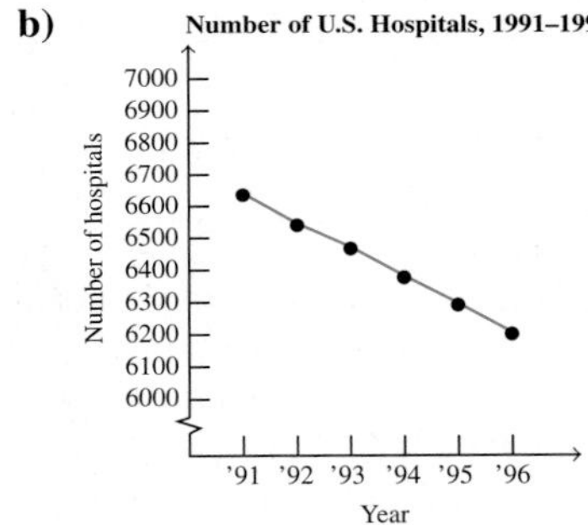

19. a)

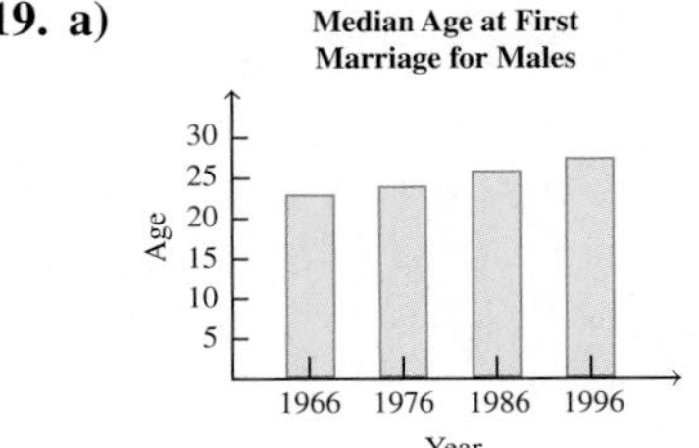

b)

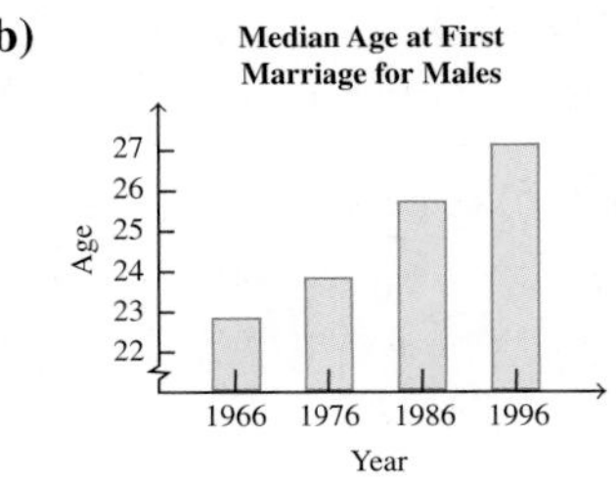

21. a)

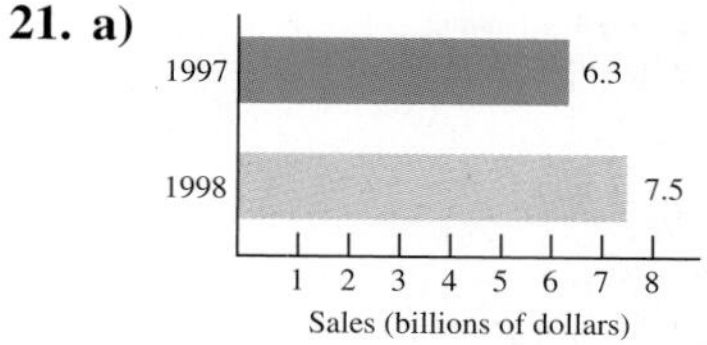

b) Answers will vary.

SECTION 13.3, PAGE 675

1. A frequency distribution is a listing of observed values and the corresponding frequency of occurrence of each value.
3. a) 7 b) 16–22 c) 16 d) 22
5. The modal class is the class with the greatest frequency.
7. a) 17 b) 5 c) 16 d) 14–18 e) 39–43

9.

Number sold	Number of days
0	3
1	8
2	3
3	5
4	2
5	7
6	2
7	3
8	4
9	1
10	2

11.

IQ	Number of students
78–86	2
87–95	15
96–104	18
105–113	7
114–122	6
123–131	1
132–140	1

13.

IQ	Number of students
80–90	8
91–101	22
102–112	11
113–123	7
124–134	1
135–145	1

15.

Age	Number of presidents
40–45	2
46–51	12
52–57	16
58–63	6
64–69	5

17.

Age	Number of presidents
42–46	4
47–51	10
52–56	12
57–61	9
62–66	4
67–71	2

19.

Cost (millions of dollars)	Number of companies
90–199	18
200–309	15
310–419	5
420–529	2
530–639	8
640–749	0
750–859	1
860–969	1

21.

Cost (millions of dollars)	Number of companies
50–149	12
150–249	14
250–349	7
350–449	5
450–549	3
550–649	7
650–749	0
750–849	1
850–949	0
950–1049	1

23.

Population (to nearest 100,000)	Number of cities
5.4–7.4	10
7.5–9.5	5
9.6–11.6	10
11.7–13.7	4
13.8–15.8	1
15.9–17.9	4
18.0–20.0	0
20.1–22.1	0
22.2–24.2	0
24.3–26.3	0
26.4–28.4	1

25.

Population (to nearest 100,000)	Number of cities
5.1–7.6	11
7.7–10.2	8
10.3–12.8	10
12.9–15.4	1
15.5–18.0	4
18.1–20.6	0
20.7–23.2	0
23.3–25.8	0
25.9–28.4	1

27.

Percent with bachelor's degree	Number of states
14.6–18.5	6
18.6–22.5	20
22.6–26.5	11
26.6–30.5	11
30.6–34.5	3

29.

Percent with bachelor's degree	Number of states
14.6–18.0	5
18.1–21.5	14
21.6–25.0	14
25.1–28.5	13
28.6–32.0	2
32.1–35.5	3

Did you know?, page 673: There are 6 F's.

SECTION 13.4, PAGE 682

1. Answers will vary. **3.** Answers will vary.

5. a) Answers will vary.

b)

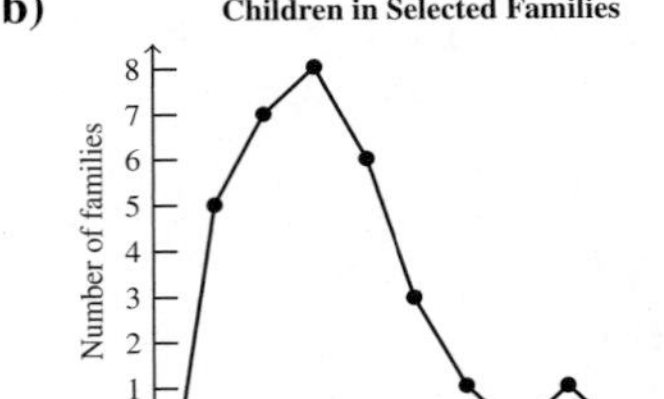

7. a) Answers will vary.

b)

Observed values	Frequency
45	3
46	0
47	1
48	0
49	1
50	1
51	2

9. None: 145; one: 155; two: 70; three: 50; four or more: 80.

11.

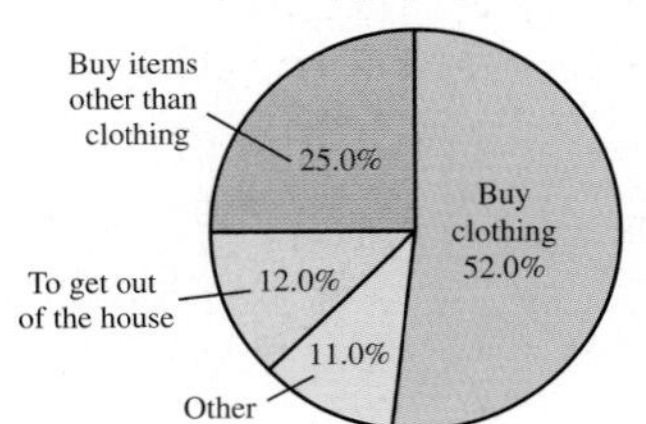

13. a) and **b)**

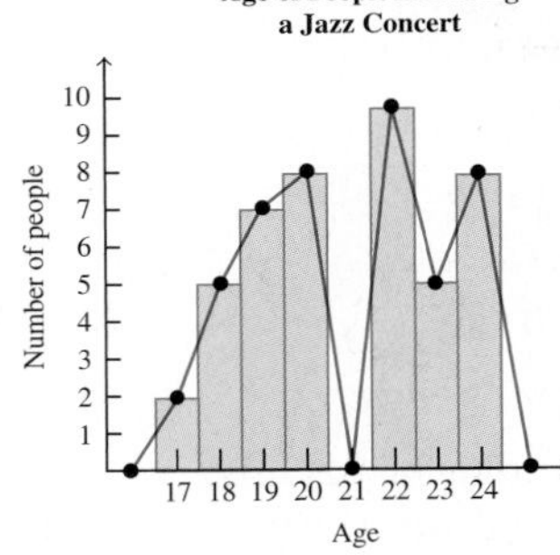

15. a) and **b)**

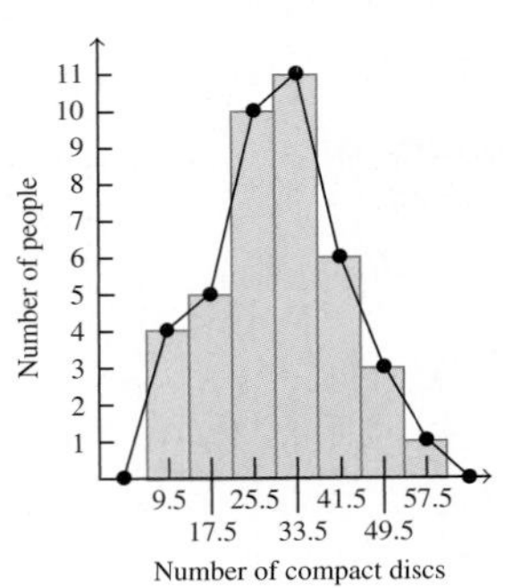

17. a) 30 **b)** 4 **c)** 2 **d)** 79

e)

Number of books	Number of people
0	2
1	5
2	8
3	7
4	4
5	3
6	1

19. a) 7 **b)** 16 **c)** 36

d)

Response time (min)	Number of calls	Response time (min)	Number of calls
3	2	7	3
4	3	8	8
5	7	9	6
6	4	10	3

e)

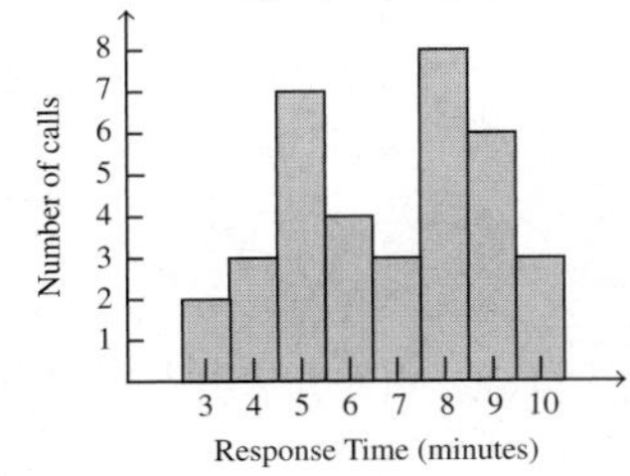

21.

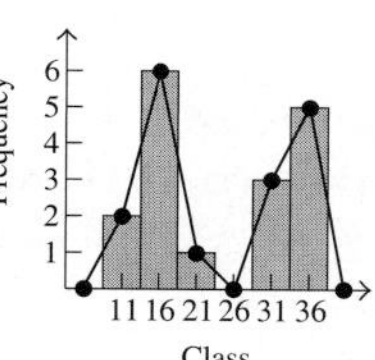

23. 1 | 5 represents 15

Stem	Leaves
1	0 5 7
2	4 4
3	6 0 3
4	8 5 2 5 8
5	3 4
6	0 2 0

25. a)

Salaries (1000's of dollars)	Number of companies
25	1
26	7
27	4
28	3
29	2
30	3
31	3
32	2

b) and **c)**

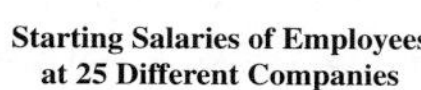

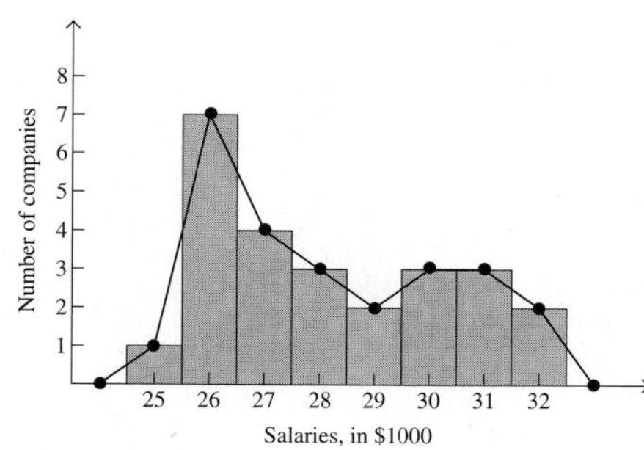

d) 2 | 3 represents 23

Stem	Leaves
2	5 6 6 6 6 6 6 6 7 7 7 7 8 8 8 9 9
3	0 0 0 1 1 1 2 2

27. a)

Sales (millions of dollars)	Number of magazines
63–122	29
123–182	4
183–242	8
243–302	1
303–362	2
363–422	2
423–482	1
483–542	1
543–602	2

b) and **c)**

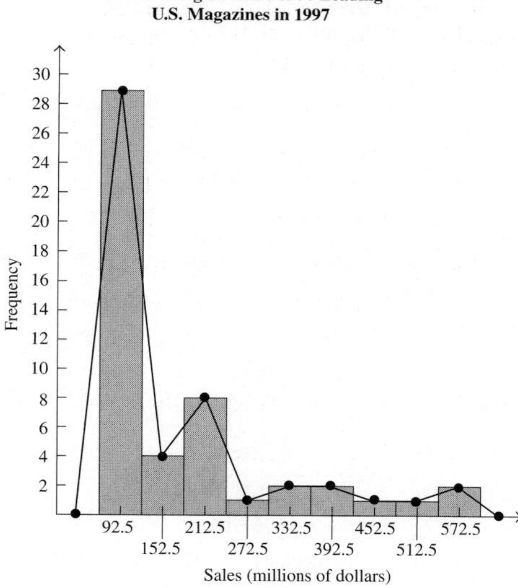

SECTION 13.5, PAGE 691

1. The mean is the balancing point of a set of data. It is the sum of the data divided by the number of pieces of data.

3. The median is the value in the middle of a set of ranked data. To find the median, rank the data and select the value in the middle.

5. The mode is the most common piece of data. The piece of data that occurs most frequently is the mode.

7. The median should be used when there are some values that differ greatly from the rest of the values in the set, for example, salaries.

9. The mean is used when each piece of data is to be considered and "weighed" equally, for example, weights of adult males.

11. 10, 8, 11, 19 **13.** 74.7, 80, none, 70.5

15. 8, 8, none, 8 **17.** 72.9, 60, none, 85

19. 11.9, 12.5, 13, 11.5 **21.** 33.1, 31, none, 34

23. a) 4.9, 5, 5, 6 **b)** 5.3, 5, 5, 6
c) only the mean **d)** the mean and the midrange

25. A 79 on 10 exams gives a total of 790 points. An 80 average on 10 exams requires a total of 800 points. Thus, Jim missed a B by 10 points not 1 point.

27. a) 178.5 ft **b)** 190 ft **c)** None **d)** 189.5 ft
e) Median—when there are a few "extreme" values, the median may be the best measure of central tendency

29. a) 8.2 million **b)** 7.45 million
c) None **d)** 9.65 million

31. 468 **33.** One example is 72, 73, 74, 76, 77, 78.

35. a) yes **b)** no **c)** no **d)** yes
e) mean = 200; midrange = 260

37. a) 33 or greater
b) It is not possible if 100 is the maximum possible grade.
c) 22 or greater **d)** 82 or greater

39. One example: 1, 2, 3, 3, 4, 5, changed to 1, 2, 3, 4, 4, 5.

41. No, by changing only one piece of the 6 pieces of data you cannot alter both the median and the midrange.

43. The data must be ranked.

45. He is taller than approximately 35 percent of all kindergarten children.

47. Second quartile, median

49. a) \$490 **b)** \$500 **c)** 25%
d) 25% **e)** 17% **f)** \$51,000

51. a)

Ruth	Mantle
0.290	0.300
0.359	0.365
0.301	0.304
0.272	0.275
0.315	0.321

b) Mantle's is greater in every case.
c) Ruth: 0.316; Mantle: 0.311; Ruth's is greater.
d) Answers will vary.
e) Ruth: 0.307; Mantle: 0.313; Mantle's is greater.
f) Answers will vary. **g)** Answers will vary.

SECTION 13.6, PAGE 698

1. Range = high value − low value **3.** Answers will vary.

5. Answers will vary. **7.** Answers will vary.

9. The first set will have the greater standard deviation because the scores have a greater spread about the mean.

11. The sum of the values in the (Data − Mean)^2 column will always be greater than or equal to 0.

13. 11, $\sqrt{16.5} \approx 4.06$ **15.** 6, $\sqrt{4.67} \approx 2.16$

17. 5, $\sqrt{7} \approx 2.65$ **19.** 5, $\sqrt{3} \approx 1.73$

21. 7, $\sqrt{6.22} \approx 2.49$ **23.** \$32, $\sqrt{137.78} \approx$ \$11.74

25. a) \$63, $\sqrt{631.6} \approx$ \$25.13 **b)** Answers will vary.
c) Answers remain the same, range: \$63, standard deviation $\approx$ \$25.13.

27. a)–c) Answers will vary.
d) If each number in a distribution is multiplied by n, the mean and standard deviation of the new distribution will be n times that of the original distribution.
e) The mean of the second set is $4 \times 5 = 20$, and the standard deviation of the second set is $2 \times 5 = 10$.

29. a) The standard deviation increases. There is a greater spread from the mean as they get older.
b) $\approx$133 lb **c)** $\approx$21 lb
d) Mean: $\approx$100 lb; normal range: $\approx$60 to 140 lb
e) Mean: $\approx$62 in.; normal range: $\approx$53 to 68 in.
f) 5%

31. a)

East		West	
Number of footware sold	Number of weeks	Number of footware sold	Number of weeks
15–20	2	15–20	0
21–26	2	21–26	0
27–32	5	27–32	6
33–38	4	33–38	9
39–44	7	39–44	4
45–50	1	45–50	6
51–56	1	51–56	0
57–62	2	57–62	0
63–68	1	63–68	0

31. b)

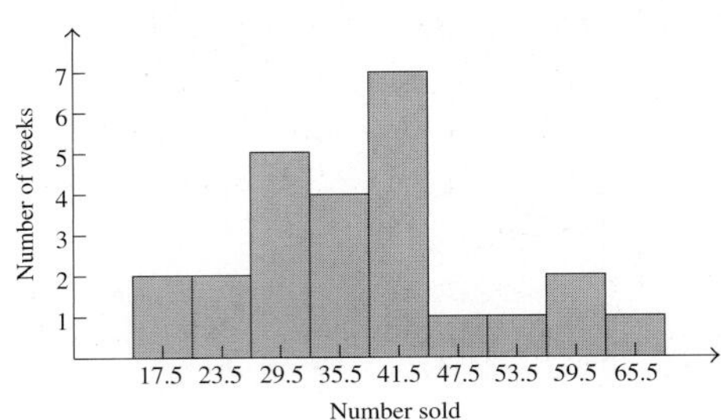

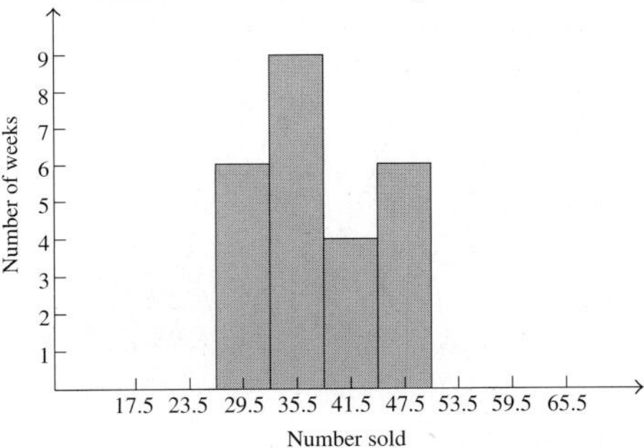

c) They appear to have about the same mean since they are both centered around 38.
d) The distribution for East is more spread out. Therefore, East has a greater standard deviation.
e) East: 38, West: 38 **f)** East: $\approx$12.64, West: $\approx$5.98

SECTION 13.7, PAGE 709

1. A rectangular distribution is one where all the values have the same frequency.

3. A bimodal distribution is one where two nonadjacent values occur more frequently than any other values in a set of data.

5. A distribution skewed to the left is one that has "a tail" on its left.

7. Answers will vary. **9.** Answers will vary.

11. Rectangular **13.** Normal

15. The mean is the greatest value. The median is lower than the mean. The mode is the lowest value.

17. Answers will vary.

19. They all have the same value.

21. A z-score will be negative when the piece of data is less than the mean.

23. 0 25. 0.500 27. 0.818 29. 0.042 31. 0.037

33. 0.019 35. 0.053 37. 28.2% 39. 89.8%

41. 97.1% 43. 97.5% 45. 2.9%

47. a) Jake, Sarah, Carol
 b) Marie, Kevin
 c) Omar, Justin, Kim

49. 34.1% 51. 68.2% 53. 69.2% 55. 24.1%

57. 44.0% 59. 81.8% 61. 2.3% 63. 50.0%

65. 10.6% 67. ≈21 69. 29.1% 71. 59.9%

73. 69.2% 75. 42.1% 77. 54,560 79. 11.1%

81. 3.6%, A; 10.0%, B; 74.9%, C; 8.6%, D; 2.9%, F.

83. a) Katie: $z = 2.4$; Stella: $z = 1.7$
 b) Katie. Her z-score is higher than Stella's z-score. This means her sales are further above the mean than Stella's sales.

85. Answers will vary.

SECTION 13.8, PAGE 721

1. The correlation coefficient measures the strength of the relationship between the quantities.

3. 1 5. 0

7. A negative correlation indicates that as one quantity increases, the other quantity decreases.

9. The level of significance is used to identify the cutoff between results attributed to chance and results attributed to an actual relationship between the two variables.

11. Answers will vary. 13. Answers will vary.

15. Yes 17. No 19. No 21. No

The answers in the remainder of this section may differ slightly from your answers, depending upon how your answers are rounded and which calculator you used. The answers given here were obtained from a Texas Instruments TI-36x solar calculator.

23. a)

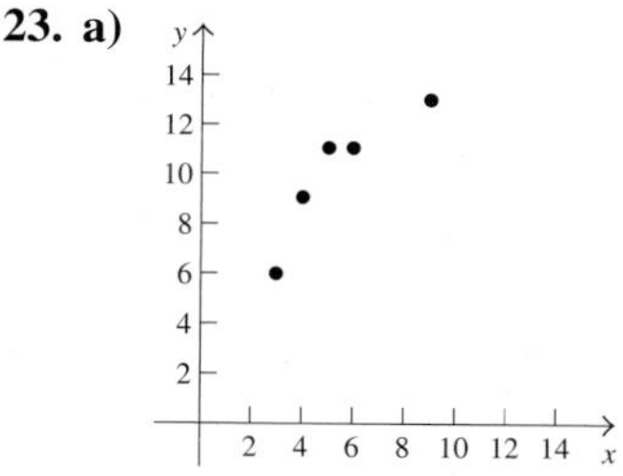

 b) 0.903 c) Yes d) No

25. a)

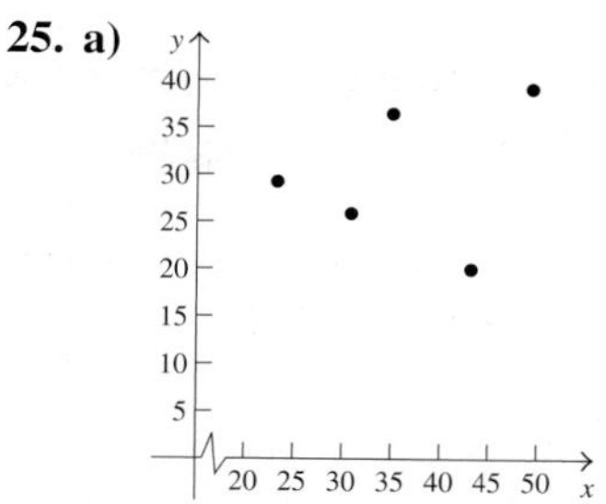

 b) 0.228 c) No d) No

27. a)

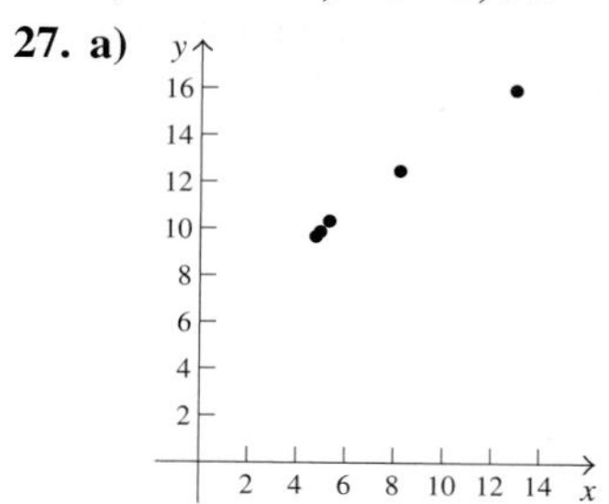

 b) 0.999 c) Yes d) Yes

29. a)

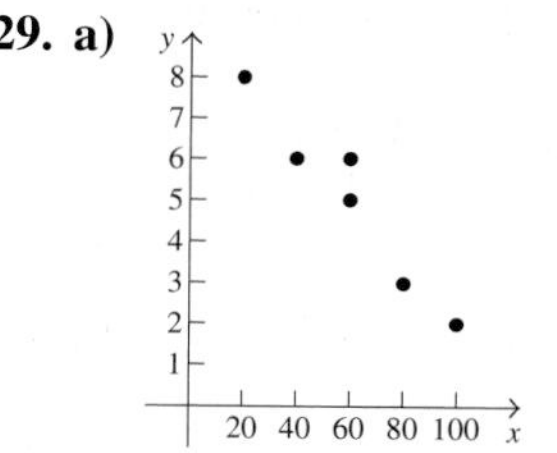

 b) −0.968 c) Yes d) Yes

31. $y = 1.0x + 4.4$ 33. $y = 0.2x + 23.8$

35. $y = 0.8x + 5.8$ 37. $y = -0.1x + 9.5$

39. a) 0.990 b) Yes c) $y = 1.2x + 24.1$

41. a) −0.852 b) Yes
 c) $y = -35.4x + 757.8$ d) ≈ $439.20

43. a) −0.782 b) Yes c) $y = -0.7x + 22.3$ d) 12.5

45. a) 0.800 b) Yes c) $y = 0.2x + 2.3$ d) ≈25 units

47. a) −0.977 b) Yes c) $y = -12.9x + 99.6$ d) 41.6%

49. a) and b) Answers will vary.
 c)

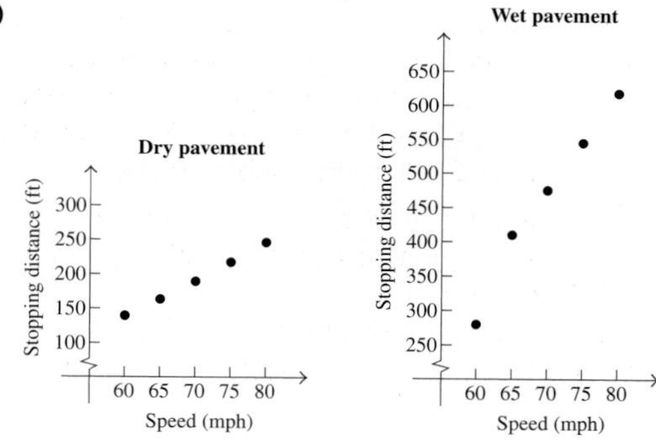

d) 0.999 **e)** 0.990 **g)** $y = 5.4x - 183.4$
h) $y = 16.2x - 669.8$ **i)** Dry, 232.4 ft; wet, 577.6 ft

51. Answers will vary.

53. a) 0.998 **b)** Should be the same. **c)** 0.998

REVIEW EXERCISES, PAGE 726

1. a) A population consists of all items or people of interest.
b) A sample is a subset of the population.

2. A random sample is one where every item in the population has the same chance of being selected.

3. The candy bars may have lots of calories, or fat, or salt. Therefore, it may not be healthy to eat them.

4. Sales may not necessarily be a good indicator of profit. Expenses must also be considered.

5. a)

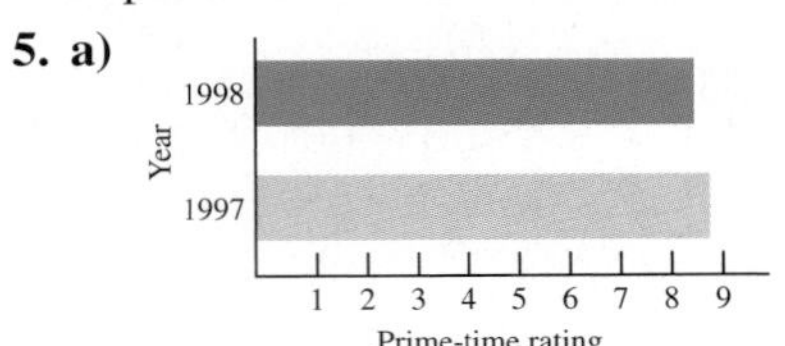

b)

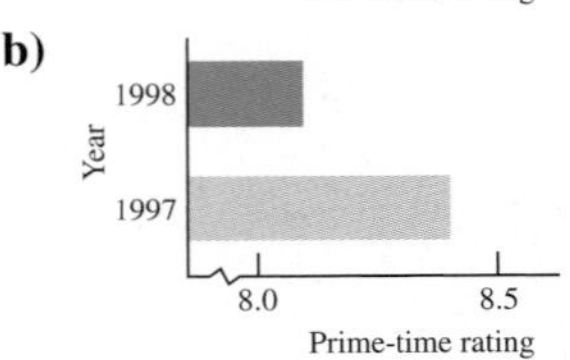

6. a)

Class	Frequency
35	1
36	3
37	6
38	2
39	3
40	0
41	4
42	1
43	3
44	1
45	1

b) and **c)**

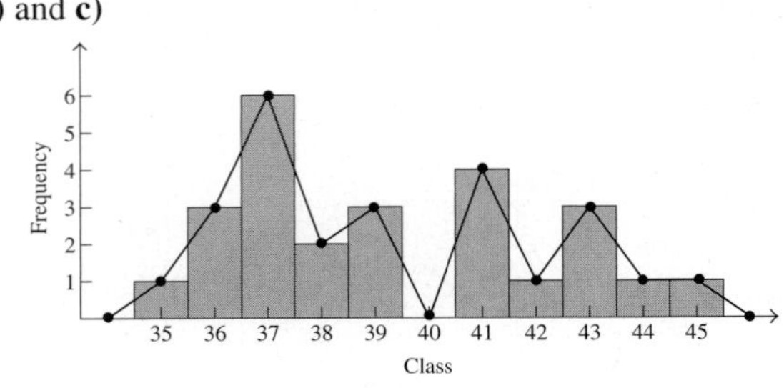

7. a)

High Temperature	Number of Cities
40–49	4
50–59	11
60–69	11
70–79	10
80–89	4

b) and **c)**

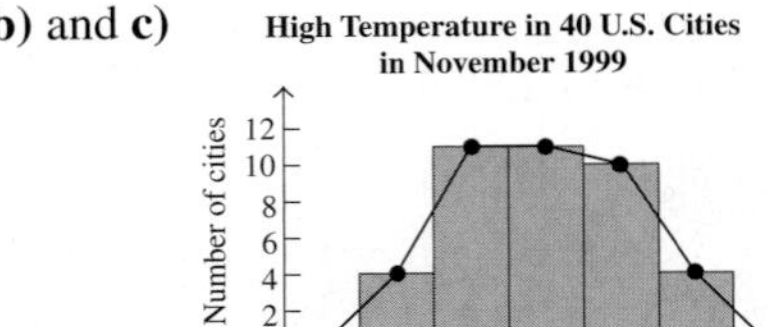

d) 4 | 6 represents 46

4	0 6 6 8
5	0 1 2 3 3 4 4 4 5 6 7
6	1 4 5 6 6 6 7 8 8 9 9
7	0 0 1 2 2 2 5 5 5 6
8	0 1 1 7

8. 80 **9.** 81 **10.** None **11.** 78 **12.** 30
13. $\sqrt{104} \approx 10.20$ **14.** 13 **15.** 13 **16.** None
17. 13.5 **18.** 19 **19.** $\sqrt{40} \approx 6.32$ **20.** 68.2%
21. 95.4% **22.** 94.5% **23.** 5.5% **24.** 72.6%
25. 34.1% **26.** 34.5% **27.** 33.7% **28.** 0.8%

29. a)

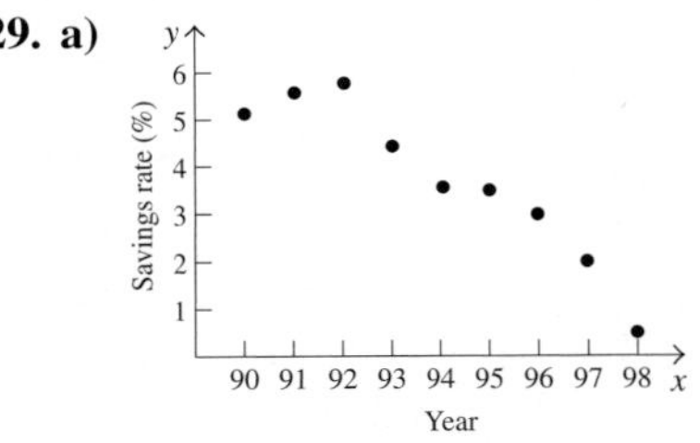

b) Answers will vary.
c) −0.94 **d)** Yes **e)** $y = -0.6x + 59.2$

30. a)

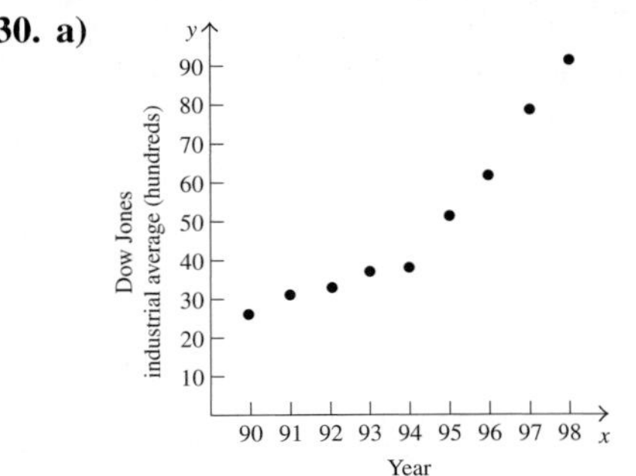

b) Answers will vary.
c) 0.95 **d)** Yes **e)** $y = 7.9x - 696.0$

31. a)

b) Answers will vary.
c) 0.86 **d)** Yes **e)** $y = 26.1x + 281.5$ **f)** \$594.70

32. 175 lb **33.** 180 lb **34.** 25% **35.** 25% **36.** 14%
37. 18,700 lb **38.** 233 lb **39.** 145.6 lb **40.** ≈ 3.610
41. 2 **42.** 3 **43.** 7 **44.** 14 **45.** $\sqrt{8.244} \approx 2.87$

46.

Class	Frequency
0– 1	8
2– 3	14
4– 5	10
6– 7	6
8– 9	1
10–11	1
12–13	0
14–15	1

47. and **48.**

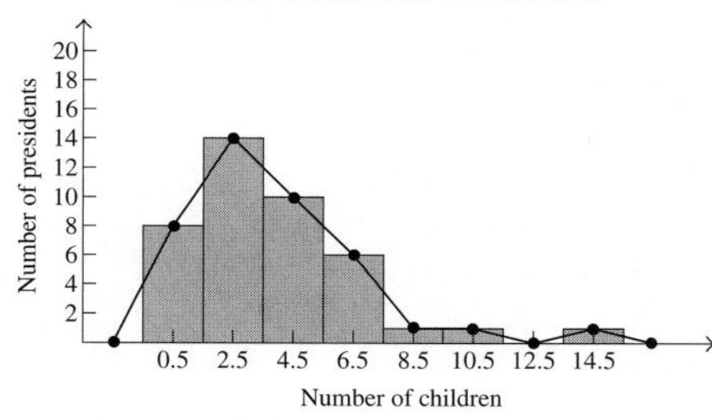

49. No **50.** No **51.** No

CHAPTER TEST, PAGE 729

1. 30 **2.** 31 **3.** 31 **4.** 27.5 **5.** 25 **6.** $\sqrt{84} \approx 9.17$

7.

Class	Frequency
25–30	7
31–36	5
37–42	1
43–48	7
49–54	5
55–60	3
61–66	2

8.

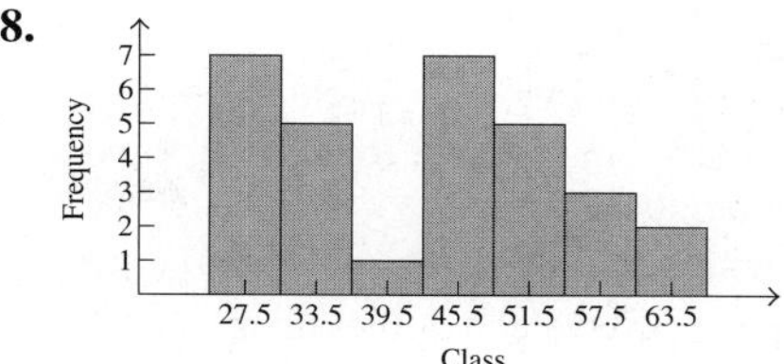

9.

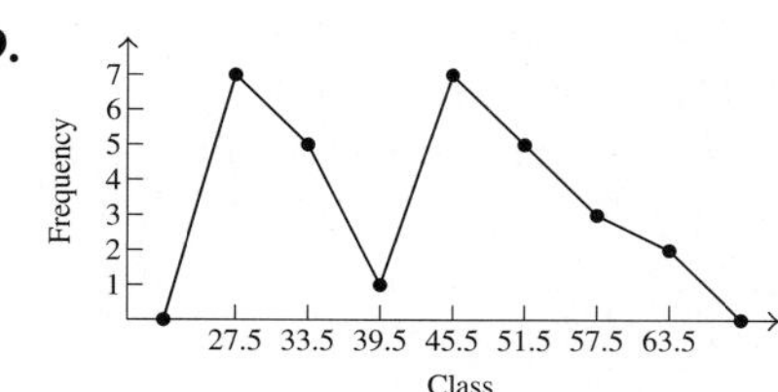

10. \$595 **11.** \$570 **12.** 75% **13.** 79%
14. \$60,000 **15.** \$640 **16.** \$540 **17.** 31.8%
18. 89.4% **19.** 10.6% **20.** $\approx$69 cars

21. a)

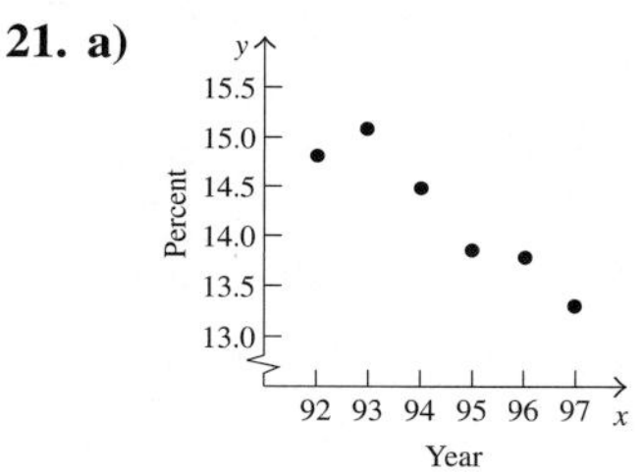

b) Answers will vary. **c)** -0.94 **d)** Yes
e) $y = -0.4x + 47.7$ **f)** 8.5%

APPENDIX, PAGE AA-1

1. A vertex is a designated point.

3. To determine whether a vertex is odd or even, count the number of edges attached to the vertex. If the number of edges is odd, the vertex is odd. If the number of edges is even, the vertex is even.

5. 5 vertices, 7 edges **7.** 7 vertices, 11 edges

9. Each graph has the same number of edges from the corresponding vertices.

11. odd vertices: *C*, *D*; even vertices: *A*, *B*

13. Yes; start at *C* and end at *D*, or start at *D* and end at *C*.

15. Yes; start at any point and end where you started.

17. No.

19. Yes; start at *A* and end at *C*, or start at *C* and end at *A*.

21. a) 0 odd, 5 even **b)** Yes
c) Start in any room and end where you began. One path is *A* to *D* to *B* to *C* to *E* to *A*.

23. a) 2 odd, 4 even **b)** Yes
c) Start at *B* and end at *F*, or start at *F* and end at *B*. One path is *B* to *C* to *F* to *E* to *D* to *A* to *B* to *E* to *F*.

25. **a)** 4 odd, 1 even **b)** Not possible

27. **a)** 3 odd, 2 even **b)** Not possible

29. The door must be placed in room D. Room D is the only room with an odd number of doors.

31. Yes; there are two odd vertices. Begin at either the island on the left or on the right and end at the other island.

33.

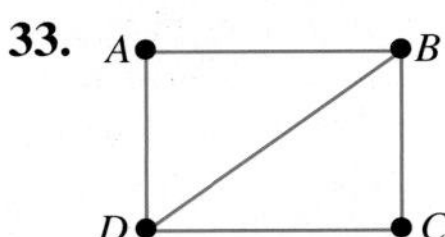

35. **a)** Kentucky, Virginia, North Carolina, Georgia, Alabama, Mississippi, Arkansas, Missouri
b) Illinois, Arkansas, Tennessee

37. **a)** 4 **b)** 4 **c)** 11

39.

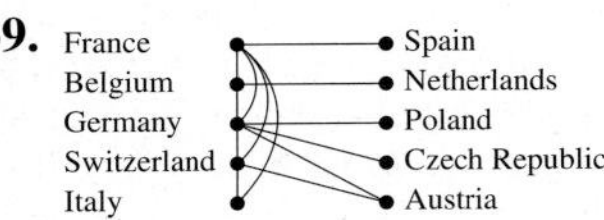

41. **a)** Yes, the graph has exactly 2 odd vertices.
b) One possiblity is C, A, B, E, F, D, G, C

CREDITS

Page 1, photo from *Apollo* 13 movie, © 1995 Universal City Studios, Inc. All rights reserved Kobal Collection; Page 4, photo of Leaning Tower of Pisa © Amanda Merullo/Stock, Boston; Page 6, graph "Health Care Inflation" © 7/12/99 U.S. NEWS & WORLD REPORT; Page 6, graph "Average Family Income vs. Purchase Price for New Cars" from CONSUMER REPORTS, December 1998 © 1998 Consumers Union of U.S., Inc., Yonkers, NY. Reprinted with permission for educational purposes only. No commercial use or photocopying permitted. Page 10, map of Grand Canyon courtesy Grand Canyon National Park, National Park Service; Page 11, photo of woman analyzing blood samples © Susan Van Etten; Page 12, photo of birds © Art Wolfe/Tony Stone Images; Page 13, photo of birds © Art Wolfe/Tony Stone Images; Page 14, map of portion of Boston adapted from the Automobile Association of America; Page 14, map of portion of California adapted from the Automobile Association of America; Page 15, data in graphs "Source of Retirement Income" from Social Security Administration; Page 15, data in graph "What's in a Necklace?" from U.S. NEWS & WORLD REPORT; Page 15, data in graph "Population by Age" from NEWSWEEK; Page 15, data in graph "Getting Older Usually Means Getting Bigger" from Mayo Clinic Newsletter; Page 16, data in graph "Average Vehicle Depreciation" from CONSUMER REPORTS; Page 16, photo of marbles © Mike Anich/Visuals Unlimited; Page 16, photo of ladybugs © Albert J. Copley/Visuals Unlimited; Page 17, photo of giraffe © Allen R. Angel; Page 17, photo of Mount Rushmore © Allen R. Angel; Page 18, data in table on water use from U.S. Environmental Protection Agency; Page 20, photo of George Polya © Palo Alto Weekly; Page 22, photo of women being served in a restaurant © Bruce Ayres/Tony Stone Images; Page 26, data in graph "Farmers Get Less for Their Pigs" from NEWSWEEK, 12/21/98 © 1998 Newsweek, Inc.; Page 26, data in graph "U.S. Cosmetics Retail Sales" from Euromonitor Intl. Inc.; Page 27, data in Logan Airport table courtesy of Massport; Page 27, graph "Demographics and Doctor" © 12/21/98, U.S. NEWS & WORLD REPORT; Page 28, data in shipping table from CONSUMER REPORTS © 1998 Consumers Union of U.S., Inc. © 1998 Consumers Union of U.S., Inc., Yonkers, NY. Reprinted with permission for educational purposes only. No commercial use or photocopying permitted. Page 29, data in table on profit margins from CONSUMER REPORTS; Page 29, "Internet Access" graphs from NEWSWEEK, 11/23/98 © 1998 Newsweek, Inc. All rights reserved. Reprinted by permission. Page 33, data in graph "A Boom in Multiple Births" from NEWSWEEK; Page 35, graph "Percentage with Migraines" from NEWSWEEK, 1/11/99 © 1999 Newsweek, Inc. All rights reserved. Reprinted by permission. Page 35, photo of the Mona Lisa by Leonardo da Vinci © Scala/Art Resource, NY; Page 39, photo of Georg Cantor © The Granger Collection, New York; Page 42, illustration of Creature Card by Bill Hull © 1966, 1968 by Education Development Center, Inc.; Page 44, scene from Dinsey's *Snow White* © Walt Disney, 1937 Kobal Collection; Page 44, photo of Carreras, Domingo, Pavoroti © Bill Pugliano/Liaison Agency; Page 45, photo of Mariah Carey © Roberta Parkin-FSP/Liaison Agency; Page 48, photo of Wendy's hamburger © Young Wolff/PhotoEdit; Page 50, photo of computer circuit board © Bruce Forster/Tony Stone Images; Page 52, photo of New York City © Jake Rajs/Tony Stone Images; Page 56, photo of Winston Cup/NASCAR racing © AP/Wide World Photos; Page 57, data in tables on racing standings and championship golfers from 1999 SPORTS ILLUSTRATED ALMANAC; Page 58, photo of Hollywood Bowl © Michael Newman/PhotoEdit; Page 64, data in table on Olympic medals from 1999 WORLD ALMANAC; Page 65, data in tables on men's and women's top brands from THE WALL STREET JOURNAL ALMANAC 1999, NY: Ballantine Publishing Group, Inc., a division of Random House, Inc.; Page 72, photo of professional football players © SuperStock International; Page 72, photo of Tom Hanks © Evan Agostini/Liaison Agency; Page 73, photo of woman with koala bear © Allen R. Angel; Page 75, photo of Leopold Kronecker © The Granger Collection, New York; Page 79, photo of Sears © Terry Farmer/Tony Stone Images; Page 80, photo of Ally McBeal cast © Greg Gorman/Fox/Kobal Collection; Page 82, lithograph illustration of Ptolemy © The Granger Collection, New York; Page 84, photo of Golden Gate Bridge over San Francisco Bay © Glen Allison/Tony Stone Images; Page 86, photo of physician performing acupuncture © A. Reininger/Contact Stock Images; Page 88, illustration of George Boole © The Granger Collection, New York; Page 93, photo of people grilling food © Gary A. Conner/PhotoEdit; Page 102, "Junk E-mail" graph from NEWSWEEK, 6/21/99 © 1999 Newsweek, Inc. All rights reserved. Reprinted with permission. Page 104, photo of Elvis' birthplace © Allen R. Angel; Page 104, data in table on food consumption from NEWSWEEK, 8/2/99 © 1999 Newsweek, Inc. All rights reserved. Reprinted by permission. Page 104, chart "Hours of Sleep" from NEWSWEEK, 8/2/99 © 1999 Newsweek, Inc. All rights reserved. Reprinted by permission. Page 105, photo of plane taking off © Mark Wagner/Tony Stone Images; Page 109, data in graph "Financing of One-Family Homes" from THE WALL STREET JOURNAL ALMANAC 1999, NY: Ballantine Publishing Group, Inc., a division of Random House, Inc.; Page 112, photo of girl mowing grass © Peter Cade/Tony Stone Images; Page 113, graph of Jupiter and Saturn moons from TIME; Page 113, data in graph "Cosmetic Surgery" from NEWSWEEK; Page 117, illustration from *Alice in Wonderland* © The Granger Collection, New York; Page 117, photo of Charles Dodgson © The Granger Collection, New York; Page 118, photo of Augustus De Morgan © The Granger Collection, New York; Page 122, photo of marina © Kevin Fleming/Corbis; Page 124, photo of people posing for photo © Don Smetzer/Tony Stone Images; Page 125, photo of people snorkeling © Nikolas Konstantinou/Tony Stone Images; Page 125, photo of sailing yacht © Allen R. Angel; Page 126, photo of police officer with motorist © SuperStock International; Page 127, photo of camera © Michael Newman/PhotoEdit; Page 131, photo of blindfolded scales of justice © Reza Estakhrian; Page 132, photo of dalmation in a Purina dog food ad © Ralston Purina—Nature's Course Dog Food; Page 134, photo of Bryce Canyon © Allen R. Angel; Page 134, photo of college football game © Allen R. Angel; Page 136, illustration of Leonhard Euler source unknown; Page 138, photo of UPC on can label © Phil Degginger/Tony Stone Images; Page 140, photo of professional golfer Vijay Singh © Reuters Newsmedia Inc./Corbis; Page 140, photo of country singer Bonnie Raitt © Lee Celano/Liaison Agency; Page 142, photo of Lincoln Memorial © Allen R. Angel; Page 144, photo of leopard © Mitch Reardon/Tony Stone Images; Page 146, photo of logic game © Ontario Science Centre, Toronto; Page 148, photo of Denise Schmandt-Besserat courtesy of Denise Schmandt-Besserat; Page 149, photo of the Rhind Papyrus courtesy the Ancient Art and Architecture Collection Ltd.; Page 150, photo of clock © Allen R. Angel; Page 151, photo of popcorn in a Japanese movie theater © Bill Gallery/Stock, Boston; Page 155, photo of counting boards © Dean Conger/Corbis; Page 156, photo of ship navigational instruments © Jeff Greenberg/PhotoEdit; Page 158, photo of Jan Fleck's painting *Numbers* showing numerals of the world © Allen R. Angel; Page 165, photo of computers in use © AP/Wide World Photos; Page 167, photo from milk carton © Susan Van Etten; Page 171, "Messages to Outer Space" figure courtesy SETI Institute; Page 176, illustration of John Napier © The Granger Collection, New York; Page 182, photo of computer lab © Leon Santow/Tony Stone Images; Page 186, photo of Carissan's factoring machine, housed at the Conservatoire National des Arts et Metiers in Paris. Photo by Jeffrey Shallit, University of Waterloo, Ontario © F. Morain; Page 187, photo of Srinivasa Ramanujan © The Granger Collection, New York; Page 189, photo of bank vault © Andrew Sacks/Tony Stone Images; Page 190, photo of Nayan Hajratwala courtesy Nayan Hajratwala; Page 191, photo of planes landing and taking off © Hideo Kurihara/Tony Stone Images;

Page 193, photo of etching of the poet Virgil © Hulton Getty/Tony Stone Images; Page 196, photo of Mt. Everest © Keren Su/Stock, Boston; Page 209, photos of vibrating strings © Richard Megna/Fundamental Photographs, New York; Page 215, illustration of Pythagoras of Samos © Hulton Getty/Tony Stone Images; Page 220, photo of automobile on country road © Derek Gardner/Tony Stone Images; Page 226, photos of man demonstrating how to remove sweater without removing jacket © Allen R. Angel; Page 230, photo of galaxy © UCO/Lick Observatory Photo/Image; Page 235, photo of crowd © Jim Pickerell/Tony Stone Images; Page 235, photo of Niagara Falls © Cosmo Condina/Tony Stone Images; Page 236, data in graphs on mutual funds from Ibbotson Associates; Page 237, photo of U.S. currency © AP/Wide World Photos; Page 237, photo of Dr. John Allen Paulos of Temple University courtesy Dr. John Allen Paulos; Page 238, photo of Carl Friedrich Gauss © Corbis; Page 244, photo of casino betting table © A. Ramey/PhotoEdit; Page 245, photo of Leonardo (Fibonacci) of Pisa © The Granger Collection, New York; Page 245, photo of sunflower © Renee Lynn/Tony Stone Images; Page 246, photo of the Great Pyramid of Gizeh © Hugh Sitton/Tony Stone Images; Page 247, photo of the Parthenon © George Grigouriou/Tony Stone Images; Page 247, photo of "La Parade de Cirque" by George Seurat (Metropolitan Museum of Art, bequest of Stephen C. Clark, 1960) © The Metropolitan Museum of Art; Page 247, photo of nautilus shell © Susan Van Etten; Page 252, photo of freight train © The Milwaukee Road Historical Association; Page 252, photo of Cajun turkey dinner © John Kelly/Tony Stone Images; Page 253, photo of Galena, Illinois, street scene © Peter Pearson/Tony Stone Images; Page 256, "Math Phobic's Nightmare" cartoon, THE FAR SIDE © 1990 FARWORKS, INC. Used by permission. All rights reserved. Page 259, "The Figure 5 in Gold" by Charles Demuth, 1928, the Metropolitan Museum of Art, Alfred Stieglitz Collection, 1949. Photograph © 1986 the Metropolitan Museum of Art. Page 260, photo of cashier at work in a retail store © David Young Wolff/PhotoEdit; Page 271, photo of sprinkler watering garden © Rick McClain/Liaison Agency; Page 272, photo of Niagara Falls © Allen R. Angel; Page 274, photo of man watching e-trading on computer monitors © Jeff Scheid/Liaison Agency; Page 274, graph "Global E-commerce Revenue" from NEWSWEEK, 9/20/99 © 1999 Newsweek, Inc. All rights reserved. Reprinted by permission. Page 277, photo of Albert Einstein © Lucien Aigner/Corbis; Page 278, photo of statue of Sophie Germain © Stock Montage; Page 280, photo of Campbell's soup can © Rich LaSalle/Tony Stone Images; Page 285, photo of woman in pet supply store © David Young Wolff/PhotoEdit; Page 292, photo of Allen R. Angel © Allen R. Angel; Page 293, photo of stereo speakers © Anthony Meshkinyar/Tony Stone Images; Page 302, photo of illustration of René Descartes courtesy International Business Machines Corporation; Page 304, photo of archaeological dig © Margot Granitsas/Photo Researchers; Page 305, photo of Jani Soininen © Cary M. Prior/AllSport USA; Page 312, photo of compact disc © Tony Freeman/PhotoEdit; Page 313, graph "The Value of Education" from THE WALL STREET JOURNAL ALMANAC 1999, NY: Ballantine Publishing Group, Inc., a division of Random House, Inc. Reprinted with permission. Page 313, graph "Book Sales" from THE WALL STREET JOURNAL ALMANAC 1999, NY: Ballantine Publishing Group, Inc., a division of Random House, Inc. Reprinted with permission. Page 323, time-lapse photo of free fall of a ball © Harold and Esther Edgerton Foundation, 2000, courtesy of Palm Press, Inc. Page 328, photo of computer model that simulates thunderstorms © Information and Publications; Page 334, photo "You Have Mail" © Doug Armand/Tony Stone Images; Page 335, graph "U.S. Mobile Phone Users" from NEWSWEEK, 9/20/99 © 1999 Newsweek, Inc. All rights reserved. Reprinted by permission. Page 337, graph "National Football League Average Salaries" from THE WALL STREET JOURNAL ALMANAC 1999: Ballantine Publishing Group, Inc., a division of Random House, Inc. Reprinted with permission. Page 338, graph "U.S. Average PC Cost" from NEWSWEEK, 9/20/99 © 1999 Newsweek, Inc. All rights reserved. Reprinted by permission. Page 345, photo of telecommunications © Michael Rosenfeld/Tony Stone Images; Page 349, photo of artist sketching © Paddy Eckersley/Stone; Page 352, photo of landscaper © Allen R. Angel; Page 353, photo of girls using payphone © Bob Torrez/Tony Stone Images; Page 362, photo of hockey players © Rick Steward/AllSport USA; Page 363, photo of *Beauty and the Beast* cast in Disney parade © Allen R. Angel; Page 363, graph "Units Shipped by Record Manufacturers" from THE WALL STREET JOURNAL ALMANAC 1999, NY: Ballantine Publishing Group, Inc., a division of Random House, Inc. Reprinted with permission. Page 364, photo of trail sign in Yosemite © Allen R. Angel; Page 367, photo of spreadsheet on computer screen © R. P. Ramirez/PhotoEdit; Page 368, photo of James Sylvester © The Granger Collection; Page 368, illustration of William Rowan Hamilton © Stock Montage; Page 368, illustration of Arthur Cayley © Stock Montage; Page 373, photo of cookie shop display © Allen R. Angel; Page 380, photo of candy shop © Jeff Greenberg/Visuals Unlimited; Page 384, photo of World War II soldiers © Brown Brothers; Page 386, photo of customers using washing machine © David Young Wolff/PhotoEdit; Page 388, photo of people roller blading © Allen R. Angel; Page 390, photo of students in chemistry class © Dana White/PhotoEdit; Page 391, photo of coffee store sales display © Allen R. Angel; Page 393, photo of 2000 Chevrolet Corvette © Allen R. Angel; Page 394, photo of *Mars Climate Orbiter* © NASA Media Services; Page 395, photo of plastic bottles © Allen R. Angel; Page 396, photo of person using computer © Allen R. Angel; Page 396, photo of Canadian metric road sign © Allen R. Angel; Page 397, photo of New Zealand metric road sign © Allen R. Angel; Page 399, photo of Swiss metric sign © Allen R. Angel; Page 400, photo of Picasso painting © Allen R. Angel; Page 401, photo of guitar player © Merritt Vincent/PhotoEdit; Page 404, photo of hot air balloon © Dan Ham/Tony Stone Images; Page 407, photo of football field © Allen R. Angel; Page 407, photo of Great Smoky Mountains National Park © Tom Till/Tony Stone Images; Page 408, photo of view of Niagara Falls © Allen R. Angel; Page 415, photo of water-covered land © Allen R. Angel; Page 416, photo of Fiji market prices © Allen R. Angel; Page 419, photo of Leaning Tower of Pisa © AP/Wide World Photos; Page 420, photo of man entering low walkway © Allen R. Angel; Page 421, photo of land for sale in Fiji © Allen R. Angel; Page 421, photo of fruit for sale © Allen R. Angel; Page 422, photo of speed limit sign © Allen R. Angel; Page 424; photo of km mile marker © Allen R. Angel; Page 424, photo of rice for sale © Allen R. Angel; Page 425, photo of person on scale © Allen R. Angel; Page 425, photo of elevation sign Badwater, Death Valley © Allen R. Angel; Page 426, photo of Space Shuttle courtesy of NASA; Page 428, photo of fruit for sale © Allen R. Angel; Page 428, road sign © Allen R. Angel; Page 430, image "Gaussian Hills that Never Were" by Richard F. Voss from THE FRACTAL GEOMETRY OF NATURE by Benoit B. Madelbrot. Copyright by B. B. Mandelbrot. NY: W. H. Freeman & Co., 1982. Reprinted by permission of B. B. Mandelbrot. Page 431, photo of frontispiece from *The Elements of Euclid* from Edward R. Tufte's ENVISIONING INFORMATION © Edward R. Tufte (Cheshire, CT: Graphics Press, 1990); Page 432, photo of Ryan Morgan courtesy of Ryan Morgan; Page 433, photo of George Eastman on deck of U.S.S. Gallia. Photo by Frederick Church © George Eastman House; Page 444, photo of Alhambra © Robert Frerck/Odyssey, Chicago; Page 445, photo of soccer ball © Mark M. Lawrence/Stock Market; Page 454, photo of Pierre de Fermat source unknown; Page 454, photo of Andrew J. Wiles © AP/Wide World Photos; Page 462, photo of Picasso's *Les Demoiselles d'Avignon* © 2000 The Museum of Modern Art, New York; Page 463, photo of gems © Runk/Schoenberger/Grant Heilman Photography, Inc.; Page 468, photo of aquarium © Carolyn A. McKeone/Photo Researchers, Inc.; Page 469, photo of Pyramid of Cheops © Vanni Archive/Corbis; Page 469, photo of decorated cakes © Great American Stock Index/Stock Imagery; Page 469, photo of the Leaning Tower of Pisa © Don Schimmel/Index Stock Imagery; Page 472, photo of Klein Bottle © Alan Bennett/Hi-Q Glass; Page 478, photo of Ivanovich Lobachevsky © The Granger Collection, New York; Page 478, photo of G. F. Bernhard Riemann

© The Granger Collection, New York; Page 480, photo of Albert Einstein courtesy of the Archives California Institute of Technology; Page 480, photos of fractal images © Robert L. Devaney, Boston University; Page 481, photo of fractal triangle © Andy Ryan/Andy Ryan Photography; Page 481, photo of square fractal antenna in cordless phone from SCIENTIFIC AMERICAN, July 1999 © 1999 Scientific American; Page 489, photo of Native American woman © Tony O'Brien/Stock, Boston; Page 492, photo of computer simulation of protons and neutrons © Fermilab/Peter Arnold, Inc.; Page 493, photo of Niels Abel, source unknown; Page 493, photo of Évariste Galois, source unknown; Page 497, photo of Venice clock © Allen R. Angel; Page 502, illustration of patterns adapted from SYMMETRIES OF CULTURE by Dorothy K. Washburn and Donald W. Crowe (University of Washington Press, 1988); Page 508, photo of Amalie Emmy Noether © The Granger Collection, New York; Page 512, photo of theater lobby © Allen R. Angel; Page 517, photo of new U.S. currency © Marc Schlossman/Tony Stone Images; Page 517, photo of Canadian currency © Andrew Klapatiuk; Page 517, photo of Mexican currency © G. Payne/Liaison Agency; Page 520, photo of The White House © Jon Riley/Index Stock Imagery; Page 520, photo of Oakland Coliseum © Diane Johnson/AllSport USA; Page 521, data in chart "Labor Union Membership" from U.S. Bureau of Labor Statistics; Page 523, photo of boy playing chess © AP/Wide World Photos; Page 524, photo of Shrine of Imam El-Hussein at Karbala © Franciose de Mulder/Corbis; Page 524, data in chart "Inventory Shrinkage" from National Retail Federation and Center for Retailing Education, University of Florida; Page 525, data in chart "Projected Percent of Workers in the Year 2006" from U.S. Bureau of Labor; Page 525, data in chart "Top Five Advertisers" from Competitive Media Reporting and Publishers Information Bureau; Page 525, chart "Five Leading States in Corn Production" from THE WALL STREET JOURNAL ALMANAC 1999, NY: Ballantine Publishing Group, Inc., a division of Random House, Inc. Reprinted with permission. Page 525, data in chart "Median Household Income" from U.S. Census Bureau; Page 526, data in chart "Median Age at First Marriage" from U.S. Census Bureau; Page 526, chart "Growing Array of Mutual Funds" from THE WALL STREET JOURNAL ALMANAC 1999, NY: Ballantine Publishing Group, Inc., a division of Random House, Inc. Reprinted with permission. Page 526, photo of fishing boat © Tony Arruza/Tony Stone Images; Page 528, photo of GTE Federal Credit Union © Allen R. Angel; Page 529, photo of pawn shop © Douglas Barrows/Liaison Agency; Page 535, photo of Republic Bank © Allen R. Angel; Page 535, photo of pawnbroker storefront courtesy of Arrow Building Corp.; Page 535, photo of hot air balloon © Allen R. Angel; Page 538, photo of illustration of Benjamin Franklin © The Granger Collection, New York; Page 539, photo of Manhattan © Frank Wing/Stock, Boston; Page 541, photo of painting of George Washington at Valley Forge © The Granger Collection, New York; Page 542, photo of tractor in farm field © Jeremy Walker/Tony Stone Images; Page 542, photo of water tower © Wayne Eastep/Tony Stone Images; Page 543, photo of waitress taking food order © Index Stock Imagery; Page 544, photo of golfers © Allen R. Angel; Page 547, photo of appliances in store © Frank Siteman/Stone; Page 547, photo of restored Ford Mustang © Ariel Skelley/Stock Market; Page 550, photo of busy shopping mall © Jon Riley/Stone; Page 551, data in chart "Plastic" from CardWeb Inc. web site, CardData; Page 553, photo of ATM machine © Tomas del Amo/Index Stock Imagery; Page 554, photo of canoe © Allen R. Angel; Page 555, photo of sport utility vehicle © AP/Wide World Photos; Page 555, photo of Portland, OR, © Rick Schafer/Index Stock Imagery; Page 557, photo of family watching television © Robert E. Daimmrich/Tony Stone Images; Page 559, photo of house © S. Van Etten/PhotoEdit; Page 560, photo of house © George Hunter/Tony Stone Images; Page 564, photo of sign © Bill Aron/PhotoEdit; Page 566, photo of house © Bob Torrez/Tony Stone Images; Page 567, photo of bank customers © Jon Riley/Stone; page 568, photo of house on waterfront © Gay Bumgarner/Stone; Page 570, photo of cars © Allen R. Angel; Page 570, photo of skateboarder © Joe McBride/Stone; Page 572, photo of skiers © Mark Gibson/Index Stock Imagery; Page 572, photo of house for sale © Ken Fisher/Stone; Page 573, photo of person fishing © Robert Brenner/PhotoEdit; Page 576, photo of painting of Jacob Bernoulli © Bettmann; Page 578, photo of dog © Allen R. Angel; Page 579, photo of Queen Victoria and family © Bettmann; Page 580, photo of Larry Walker © AP/Wide World Photos; Page 581, photo of lightning in sky © AP/Wide World Photos; Page 582, photo of bird at feeder © H. P. Smith, Jr./ANSP/VIREO; Page 583, chart "Makeup of the Population" from NEWSWEEK Special Edition on Women's Health © 1999 Newsweek, Inc. All rights reserved. Reprinted by permission. Page 583, data in chart "Frequency of Repair of Dishwashers" from CONSUMER REPORTS; Page 586, photo of bird © RO-MA Stock/Index Stock Imagery; Page 587, photo of playing cards © Mary Boucher; Page 589, photo of tennis courts © Allen R. Angel; Page 590, data in chart on hazardous waste sites data from U.S. Environmental Protection Agency; Page 593, data in graph on least expensive colleges from The College Board; Page 594, photo of man at Lotto machine, source unknown; Page 596, photo of bingo board © Allen R. Angel; Page 596, chart "Reasons for Unscheduled Absences" from CCH Inc.; Page 597, photo of horse racing © Patrick Donahue/Tony Stone Images; Page 598, photo of police officer writing a parking ticket © Bob Daemmrich; Page 598, photo of jet in flight courtesy Guy G. Caron; Page 602, photo of women's college basketball players © Todd Warshaw/AllSport USA; Page 611, photo of Yogi Bear with children © Timothy O'Keefe/Index Stock Imagery; Page 612, photo of pea plant © David Young Wolff/PhotoEdit; Page 615, photo of family © Bettmann; Page 616, photo of large group of people © Jeff Greenberg/Unicorn Stock Photos; Page 618, data in graph "U.S. Population by Age" from NEWSWEEK; Page 620, data in graph "Expected Population by Age" from NEWSWEEK; Page 621, photo of slot machine © J & M Studios/Liaison Agency; Page 624, photo of lightening © Bill Gillette/Stock, Boston; Page 626, photo of U.S. currency © Myrleen Ferguson/PhotoEdit; Page 627, data in table on NY Yankees payroll from USA TODAY, 10/22/99; Page 628, photo of Peter Jennings © AP/Wide World Photos; Page 630, photo of computer display © AP/Wide World Photos; Page 633, photo of cyclists © Myrleen Ferguson/PhotoEdit; Page 637, photo of Monet painting © Alexander Burkatowski/Corbis; Page 644, photo of Wild Bill Hancock's last poker hand © Allen R. Angel; Page 646, data on Florida lottery sales from Florida State Lottery; Page 647, photo of farm © Rainer Grosskopf/Stone; Page 647, photo of professional baseball game © Allen R. Angel; Page 655, photo of dolphin © Stewart Cohen/Index Stock Imagery; Page 657, photo of baseball field © Allen R. Angel; Page 659, photo of Space Shuttle © NASA/Index Stock Imagery; Page 660, photo of dalmation puppies © Myrleen Ferguson/PhotoEdit; Page 662, photo of puppy and kittens © Tim Davis/Photo Researchers, Inc.; Page 663, photo of family watching television © Nielsen Media Research; Page 667, photo of bingo board © Miguel Gandert/Corbis; Page 670, chart "Percentage of Adults Who Are Overweight" from NEWSWEEK, 7/13/98 © 1998 Newsweek, Inc. All rights reserved. Reprinted by permission. Page 671, data in table on number of hospitals from American Hospital Association; Page 671, data in table on health benefit costs from Foster Higgins; Page 671, data in table on age at first marriage from U.S. Census Bureau; Page 671, data in chart "Gateway Computer Sales" from Gateway; Page 671, chart "America Now and in 2050" from US NEWS & WORLD REPORT, 8/14/95 © 1995 US News & World Report; Page 674, data in table on company costs from Security and Exchange Commission; Page 675, photo of San Francisco © Allen R. Angel; Page 676, photo of car stereo display © Allen R. Angel; Page 677, data in table on population with bachelor's degree from U.S. Census Bureau; Page 677, photo of new graduates © A. Ramey/Unicorn Stock Photos; Page 677, chart "How Americans Describe Their Holiday Spending" from USA TODAY, 12/9/99 © 1999 USA Today. Reprinted with permission. Page 683, data in chart "Online Households" from Jupiter Communications; Pages 685–86, data in table on magazine sales from Audit Bureau of Circulations and Magazine Publishers of America; Page 688, photo of San Francisco

© Allen R. Angel; Page 688, photo of Orlando © Timothy O'Keefe/Index Stock Imagery; Page 689, photo of smiling baby © Susan Van Etten; Page 692, photo of Busch Gardens © Allen R. Angel; Page 693, photo of Babe Ruth © AP/Wide World Photos; Page 694, photo of Mickey Mantle © AP/Wide World Photos; Page 700, data in tables on professional athletes' salaries from Major League Baseball Players Association and National Basketball Association; Page 710, photo of man using microwave oven © David Young Wolff/PhotoEdit; Page 711, photo of coffee vending machine © Allen R. Angel; Page 722, photo of student © Bonnie Kamin/PhotoEdit; Page 723, photo of young children at playground © Myrleen Ferguson/PhotoEdit; Page 724, photo of swimming pool © Allen R. Angel; Page 724, data for figure "Hitting the brakes" from U.S. NEWS & WORLD REPORT © 12/18/95 U.S. News & World Report. Page 727, data for charts on U.S. savings rate and Dow Jones Industrial Average from U.S. NEWS & WORLD REPORT © 3/22/99 U.S. News & World Report.

INDEX

Page numbers followed by *n* refer to footnotes.